HANDBOOK OF COASTAL ENGINEERING

HANDBOOK OF COASTAL ENGINEERING

John B. Herbich, Ph.D., P.E., Editor

W. H. Bauer Professor Emeritus
Ocean Engineering Program
Department of Civil Engineering
Texas A&M University
College Station, TX
and
Vice-President
Consulting and Research Services, Inc.
Wailuku, HI and College Station, TX

McGRAW-HILL
New York San Francisco Washington, D.C. Auckland Bogotá
Caracas Lisbon London Madrid Mexico City Milan
Montreal New Delhi San Juan Singapore
Sydney Tokyo Toronto

Library of Congress Cataloging-in-Publication Data

Handbook of coastal engineering / [edited by] John B. Herbich
   p. cm.
   ISBN 0-07-134402-0
   1. Coastal engineering—Handbooks, manuals, etc. I. Herbich, John B.
~~TA332.H53 1999~~
   627'.58—dc21   99-043805
                                                                                                     CIP

McGraw-Hill

A Division of The **McGraw·Hill** Companies

Copyright © 2000 by The McGraw-Hill Companies, Inc. All rights reserved. Printed in the United States of America. Except as permitted under the United States Copyright Act of 1976, no part of this publication may be reproduced or distributed in any form or by any means, or stored in a data base or retrieval system, without the prior written permission of the publisher.

1 2 3 4 5 7 8 9 0   DOC/DOC   0 6 5 4 3 2 1 0

ISBN 0-07-134402-0

The sponsoring editor for this book was Larry Hager and the production supervisor was Sherri Souffrance. It was set in Times Roman by Ampersand Graphics, Ltd. Printed and bound by R. R. Donnelley & Sons.

McGraw-Hill books are available at special quantity discounts to use as premiums and sales promotions, or for use in corporate training programs. For more information, please write to Director of Special Sales, McGraw-Hill, Two Penn Plaza, New York, NY 10121. Or contact your local bookstore.

This book was printed on recycled, acid-free paper containing a minimum of 50% recycled de-inked fiber.

Information contained in this work has been obtained by The McGraw-Hill Companies, Inc. ("McGraw-Hill") from sources believed to be reliable. However, neither McGraw-Hill nor its authors guarantee the accuracy or completeness of any information published herein and neither McGraw-Hill nor its authors shall be responsible for any errors, omissions, or damages arising out of use of this information. This work is published with the understanding that McGraw-Hill and its authors are supplying information but are not attempting to render engineering or other professional services. If such services are required, the assistance of an appropriate professional should be sought.

*To my wife, Margaret Pauline,
and children,
Ann, Barbara, Gregory, and Patricia,
who have been a source of encouragement
and inspiration during the preparation
of this book.*

CONTENTS

Contributors xi
Preface xiii
Acknowledgments xv
Publisher's Note xvii

Chapter 1 Numerical Solution of Coastal Water Wave Equations 1.1
Xiping Yu and Masahiko Isobe

Chapter 2 Revetment Protection for Coastal and Shoreline Structures Exposed to Wave Attack 2.1
N. W. H. Allsop and K. J. McConnell

Chapter 3 Design of Dikes and Revetments— Dutch Practice 3.1
Krystian W. Pilarczyk

Chapter 4 Wave Forces on Vertical and Composite Walls 4.1
N. W. H. Allsop

Chapter 5 Offshore (Detached) Breakwaters 5.1
John B. Herbich

Chapter 6 Wave Overtopping of Coastal and Shoreline Structures 6.1
P. Besley and N. W. H. Allsop

Chapter 7 Sediment Transport and Beach Profile Change Due to Random Waves 7.1
Shinji Sato

Chapter 8 Coastal Protection Methods 8.1
John Headland, W. Gray Smith, Peter Kotulak, and Santiago Alfageme

Chapter 9	Shoreline Protection Methods—Japanese Experience 9.1
	John R. C. Hsu, Takaaki Uda, and Richard Silvester

Chapter 10	Beach Nourishment Design 10.1
	Billy L. Edge

Chapter 11	Navigation Channels—Design and Operation 11.1
	Joop A. Zwamborn

Chapter 12	Maintenance Dredging in Channels and Harbors 12.1
	John Headland, Peter Kotulak, and Santiago Alfageme

Chapter 13	Subaqueous Capping of Contaminated Sediment 13.1
	Michael R. Palermo

Chapter 14	Modeling the Physical and Chemical Stability of Underwater Caps in Rivers and Harbors 14.1
	Ram K. Mohan

Chapter 15	The USACE Dredging Operations and Environmental Research (DOER) Program 15.1
	Clark McNair and Lyndell Hales

Chapter 16	Numerical Models for Predicting the Fate of Dredged Material Placed in Open Water 16.1
	Hans R. Moritz, Billy H. Johnson, and Norman W. Scheffner

Chapter 17	Removal of Contaminated Sediment by Dredging 17.1
	John B. Herbich

Chapter 18	Marine Aggregate Dredging 18.1
	Henry Bokuniewicz

Chapter 19 Methodology for Delineation of Coastal Hazard Zones and Development Setback for Open Duned Coasts 19.1
Terry R. Healy and Robert G. Dean

Appendix A: Automated Coastal Engineering System (ACES) A.1

Appendix B: Tables B.1

Author Index AI.1

Subject Index SI.1

CONTRIBUTORS

Alfageme, Santiago, *Moffat & Nichol Engineers, New York, NY* (Chapters 8, 12)

Allsop, N. W. H., *University of Sheffield, c/o HR Wallingford, Wallingford, UK* (Chapters 2, 4, 6)

Besley, P. B., *HR Wallingford, Wallingford, UK* (Chapter 6)

Bokuniewicz, Henry, *Marine Science Center, State University of New York, Stony Brook, New York, NY* (Chapter 18)

Dean, Robert G., *Department of Coastal and Oceanographic Engineering, University of Florida, Gainesville, FL* (Chapter 19)

Edge, Billy L., *Head, Ocean Engineering Program, W. H. Bauer Professor of Dredging Engineering, Texas A&M University, College Station, TX* (Chapter 10)

Hales, Lyndell, *Research Hydraulic Engineer, USAE Waterways Experiment Station, Vicksburg, MS* (Chapter 15)

Headland, John R., *Moffat & Nichol Engineers, New York, NY* (Chapters 8 and 12)

Healy, Terry R., *Research Professor of Coastal Environmental Science, The University of Waikato, Hamilton, New Zealand* (Chapter 19)

Herbich, John B., *W. H. Bauer Professor Emeritus, Ocean and Civil Engineering, Texas A&M University, College Station, TX* (editor; Chapters 5 and 17)

Hsu, John R. C., *Senior Lecturer in Coastal Engineering, The University of Western Australia, Nedlands, Western Australia* (Chapter 9)

Isobe, Masahiko, *Department of Civil Engineering, University of Tokyo, Tokyo, Japan* (Chapter 1)

Johnson, Billy H., *USAE Waterways Experiment Station, Vicksburg, MS* (Chapter 16)

Kotulak, Peter, *Moffat & Nichol Engineers, Baltimore, MD* (Chapters 8, 12)

McConnell, Kirsty J., *Engineer, HR Wallingford, Wallingford, UK* (Chapter 2)

McNair, E. Clark, *DOER Operational Program Manager, USAE Waterways Experiment Station, Vicksburg, MS* (Chapter 15)

Mohan, Ram K., *Senior Engineer, Gahagan & Bryan Associates, Inc., Baltimore, MD* (Chapter 14)

Moritz, Hans Rod, *Civil Engineer, US Army Corps of Engineers, Portland District, Portland, OR* (Chapter 16)

Palermo, Michael R., *USAE Waterways Experiment Station, Vicksburg, MS* (Chapter 13)

Pilarczyk, Krystian W., *Head, Hydraulic Engineering Research and Directorate-General for Public Works and Water Management, Delft, The Netherlands* (Chapter 3)

Sato, Shinji, *Senior Researcher, Public Works Research Institute, Ibaraki, Japan* (Chapter 7)

Scheffner, Norman W., *USAE Waterways Experiment Station, Vicksburg, MS* (Chapter 16)

Silvester, R., *Department of Civil Engineering, The University of Western Australia, Nedlands, Western Australia* (Chapter 9)

Smith, W. Gray, *Moffat & Nichol Engineers, Baltimore, MD* (Chapter 8)

Uda, Takaaki, *Public Works Research Institute, Ministry of Construction, Ibaraki, Japan* (Chapter 9)

Yu, Xiping, *Department of Engineering Mechanics, Shanghai Jiao Tong University, Shanghai, PRC* (Chapter 1)

Zwamborn, Joop A., *Coastal and Hydraulic Engineer, Stellenbosch, South Africa* (Chapter 11)

PREFACE

There is an obvious need to develop a handbook of coastal engineering. Major progress has been made in the field of coastal engineering in recent decades through new research and dissemination of information from the biannual International Coastal Engineering Conferences. Other specialty engineering conferences have been held under various auspices in the United States and abroad. Relatively few handbooks have been published covering coastal engineering topics. One of the main purposes for writing this handbook was to collect all available relevant information under one cover.

The topics covered can be classified as follows:

- Wave equations (Chapter 1)
- Design of revetments, dikes, and breakwaters (Chapters 2–6)
- Beach erosion and protection (Chapters 7–10)
- Navigational channels (Chapters 11 and 12)
- Dredging and handling of contaminated sediments (Chapters 13–18)
- Coastal hazard zones and setbacks (Chapter 19)

This volume represents the efforts of twenty-seven experts from around the world. The contributors are from Australia, China (PRC), United Kingdom, Japan, New Zealand, South Africa, The Netherlands, and the United States. All chapters were peer-reviewed, corrected, and finally reviewed by the Editor.

An extensive Appendix includes the ACES computer program; how the U.S. Army Corps of Engineers programs are conceived, authorized, funded, and implemented; the Regulatory Program Applicant Information; government regulations; and miscellaneous useful tables.

The *Handbook of Coastal Engineering* is a compendium of expert information on today's coastal engineering theory and practice. It is intended for:

- Civil engineers who have not taken courses in coastal engineering. This handbook provides extensive coverage of theory and applications in the design of dikes, breakwaters, and other coastal engineering structures; beach erosion and protection; navigational channels and dredging engineering; marine mining; and the removal of contaminated sediment.
- Senior coastal engineering students and graduate students.
- Graduates from noncoastal engineering curricula, who are often asked to design coastal and offshore structures.
- Consulting engineering firms that are called upon to design coastal structures, navigational channels, harbors, and offshore pipelines, or to replenish eroding beaches, etc.

It is our hope that this *Handbook* will serve as an invaluable resource for all those mentioned above.

Finally, the *Handbook* is meant to impress upon its readers the fact that coastal areas throughout the world are subject to forces (waves, currents, tsunamis, and hurricanes) far more severe than the forces on land.

John B. Herbich, Ph.D., P.E

ACKNOWLEDGMENTS

Deepest gratitude is extended to all contributors to the *Handbook*. Great appreciation is also extended to all the reviewers: Dr. Henry Bokuniewicz, State University of New York, Stony Brook, NY; Dr. Zeki Demirbilek, Coastal and Hydraulic Laboratory, U.S. Army Engineer Waterways Experiment Station, Vicksburg, MS; Dr. Wayne Dunlap, Offshore Technology Center, Texas A&M University, College Station, TX; Dr. Billy L. Edge, Ocean Engineering Program, Texas A&M University, College Station, TX; Dr. Jim Kirby, Ocean Engineering Laboratory, Center for Applied Coastal Research, University of Delaware, Newark, DE; Dr. Nickolas Kraus, Coastal and Hydraulic Laboratory, U.S. Army Engineer Waterways Experiment Station, Vicksburg, MS; Dr. Alan W. Niedoroda, Environmental Science and Engineering Inc., Gainesville, FL; Mr. Thomas W. Richardson, Coastal and Hydraulic Laboratory, U.S. Army Engineer Waterways Experiment Station, Vicksburg, MS; Ms. Julie Rosati, Coastal and Hydraulic Laboratory, U.S. Army Engineer Waterways Experiment Station, Vicksburg, MS.

The Editor is most grateful to Ms. Joyce Ann Hyden, Technical Secretary, Ocean Engineering Program, Texas A&M University, College Station, TX, who assembled the manuscripts for publication, processed several chapters of the book, and corresponded with all the contributors. Without her invaluable assistance, the *Handbook* would have taken much longer to produce.

I also wish to thank the many publishers and individuals who have kindly granted permission to reprint copyrighted materials.

John B. Herbich, Ph.D., P.E.

PUBLISHER'S NOTE

The Handbook of Coastal Engineering is a collective effort involving many technical specialists. It brings together a wealth of information from worldwide sources to help scientists, engineers, and technicians solve current and long-range problems.

Great care has been taken in the compilation and production of this volume, but it should be made clear that no warranties, express or implied, are given in connection with the accuracy or completeness of this publication, and no responsibility can be taken for any claims that may arise.

The statements and opinions expressed herein are those of the individual authors and are not necessarily those of the editor or the publisher. Furthermore, citation of trade names and other proprietary marks do not constitute an endorsement or approval of the use of such commercial products or services, nor of the companies that provide them.

CHAPTER 1
NUMERICAL SOLUTION OF COASTAL WATER WAVE EQUATIONS

XIPING YU
Department of Engineering Mechanics,
Shanghai Jiao Tong University,
Shanghai, China

MASAHIKO ISOBE
Department of Civil Engineering,
University of Tokyo,
Tokyo, Japan

1. Introduction	1.1
2. Mathematical Models	1.2
2.1. Wave Equations by Direct Integration	1.5
2.2. Wave Equations by Variational Approach	1.11
2.3. Dissipation Models	1.15
2.4. Boundary Conditions	1.20
3. Numerical Models for Boundary Value Problems	1.23
3.1. High-Order Finite Difference Methods for Reduced MSWE	1.23
3.2. Finite Element Methods for MSWE	1.30
4. Numerical Models for Initial-Boundary Value Problems	1.43
4.1. Time-Dependent Linear Wave Equations	1.43
4.2. Quasilinear Hyperbolic Wave Equations	1.50
4.3. Boussinesq Equations	1.59
5. Summary	1.67
Acknowledgments	1.67
References	1.67

1. INTRODUCTION

One of the most outstanding characteristics of the development of coastal water wave mechanics during the last decades is probably the ever-increasing importance of numerical models. The background of this situation, which has much in common with many other branches of applied physics, is twofold. First, differential or integral representation of the physical reality is generally well established, but not in a form suitable for general analytical treatment, particularly when practical problems with complex geometry of the solution domain are concerned. Second, significant progress in computational methods for differential and integral equations with complicated properties has been made in the recent years, and the accessibility of high-speed computers to practicing engineers has been remarkably improved. The objective in writing this chapter is to provide a state-of-the-art review of the theories and applications of the various numerical models that have been distributed in the various publication media. Owing to the rapid expansion of the literature, a relatively complete review would be voluminous and will not be performed here. Instead, the focus will concentrate on the models that are widely applicable to coastal and ocean engineering practice. Some emphasis will be given to the high-resolution methods.

A *numerical model* in this chapter must be understood as any algebraic representation of a differential equation or an equation system that depicts the whole or some aspects of a surface wave in coastal waters. The relevant physical phenomenon of interest is referred to as the *physical model* and the wave equation or equation system as the *mathematical model*. A *numerical scheme* or *numerical algorithm* is employed when it is intended to describe the intrinsic structure of the numerical model, whereas *numerical method* is reserved for use when it is intended to emphasize the formation strategy of the numerical scheme.

Under given conditions, a numerical model results in a discrete solution, called the *numerical solution*, to the relevant mathematical model, which should reasonably represent the exact solution of the problem that is usually continuous or zoned continuous in both space and time, as far as the numerical model is consistent with the mathematical model and well-conditioned. How well the exact solution of the mathematical model is approximated by the numerical solution is described by the *accuracy of the numerical solution*, which is essentially different from, but closely related to the *accuracy of the numerical model*. The latter implies how precisely the mathematical model is discretized.

The mathematical model of coastal water waves in the most general sense is composed of the continuity equation and the Navier–Stokes equations, in addition to the kinematic and the dynamic free surface conditions, the sea bottom condition, and the lateral boundary condition. Although a significant amount of research has also been devoted to the establishment of an equally general numerical model in the field of computational fluid dynamics (CFD), the predominant tendency in coastal and ocean engineering practices has been the development and application of the less general but more tractable wave equations derived from the general formulation under various assumptions. Among a large variety of such wave equations, the category, called *vertically integrated water wave equations* or, simply, *water wave equations*, is to be explored here. Horizontally two-dimensional problems, which have also been called two-plus dimensional problems in the literature because the vertical structure of the wave-induced flows is considered, will be discussed.

In the following section, a brief derivation of several mathematical models for coastal water waves is provided. Next, the numerical models for steady-state wave problems that are usually formulated as boundary value problems are discussed. Then, a survey of the numerical models for transient wave problems that often appear as initial boundary value problems is made.

2. MATHEMATICAL MODELS

The physical model for surface water waves over a gradually varying sea bottom is illustrated in Figure 1.1. For simplicity, the sea bottom is considered to be impermeable and

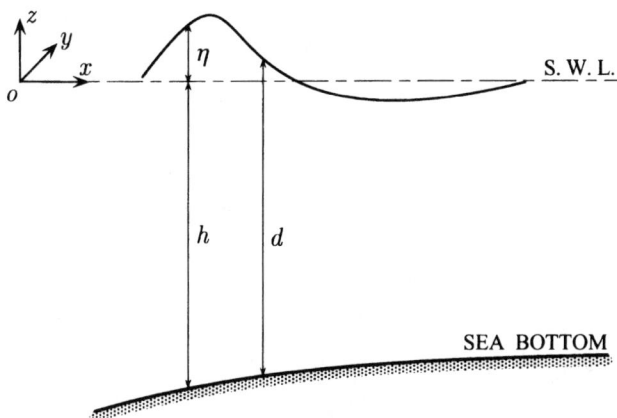

FIGURE 1.1. The general physical model for surface water waves over a gradually varying sea bottom.

undeformable, and the water surface continuous. Following the usual practice in studying surface water wave motions, it is assumed that the water is inviscid and incompressible. A general mathematical representation of the physical reality should then include the following continuity equation and Euler equations:

$$\frac{\partial u_j}{\partial x_j} + \frac{\partial w}{\partial z} = 0 \tag{1}$$

$$\frac{\partial u_i}{\partial t} + \frac{\partial}{\partial x_j}(u_j u_i) + \frac{\partial}{\partial z}(w u_i) + \frac{\partial}{\partial x_i}\left(\frac{\tilde{p}}{\rho}\right) = 0 \tag{2}$$

$$\frac{\partial w}{\partial t} + \frac{\partial}{\partial x_j}(u_j w) + \frac{\partial}{\partial z}(w w) + \frac{\partial}{\partial z}\left(\frac{\tilde{p}}{\rho}\right) = 0 \tag{3}$$

where the summation convention is valid but the indicial suffixes i and j ($i, j = 1, 2$) apply only to the horizontal directions; (x_1, x_2) [$\equiv (x, y)$] and z are the horizontal and vertical coordinates, respectively; (u_1, u_2) [$\equiv (u, v)$] and w are the horizontal and vertical velocity components, respectively; $\tilde{p} = p + \rho g z$ is the wave-induced hydrodynamic pressure, with p being the total pressure; ρ is the density of water; g is the gravitational acceleration; and t is the time. The kinematic conditions at the free surface and at the bottom can be written as

$$\frac{\partial \eta}{\partial t} + u_j^{(s)} \frac{\partial \eta}{\partial x_j} - w^{(s)} = 0 \tag{4}$$

$$u_j^{(b)} \frac{\partial h}{\partial x_j} + w^{(b)} = 0 \tag{5}$$

where η is the free surface elevation measured from the still water level; h is the still water depth; the bracketed superscripts s and b denote the values at the free surface ($z = \eta$) and at the bottom ($z = -h$), respectively. If the surface tension is neglected, the dynamic condition at the free surface requires

$$\tilde{p}^{(s)} = \rho g \eta \tag{6}$$

In many cases, the wave-induced water motion can be assumed irrotational. Then, the velocity potential ϕ can be defined such that

$$\frac{\partial \phi}{\partial x_j} = u_j \quad \text{and} \quad \frac{\partial \phi}{\partial z} = w \tag{7}$$

The continuity equation (1) thus becomes the Laplace equation in terms of ϕ:

$$\frac{\partial^2 \phi}{\partial x_j \partial x_j} + \frac{\partial^2 \phi}{\partial z^2} = 0 \tag{8}$$

and the equations of motion (2) and (3) can be integrated to give the following Bernoulli equation:

$$\frac{\partial \phi}{\partial t} + \frac{1}{2}\left[\frac{\partial \phi}{\partial x_j}\frac{\partial \phi}{\partial x_j} + \left(\frac{\partial \phi}{\partial z}\right)^2\right] + \frac{\tilde{p}}{\rho} = 0 \tag{9}$$

In terms of the velocity potential, the free surface and bottom boundary conditions (4), (5), and (6) can be written as

$$\frac{\partial \eta}{\partial t} + \frac{\partial \phi^{(s)}}{\partial x_j}\frac{\partial \eta}{\partial x_j} - \frac{\partial \phi^{(s)}}{\partial z} = 0 \tag{10}$$

$$\frac{\partial \phi^{(b)}}{\partial x_j}\frac{\partial h}{\partial x_j} + \frac{\partial \phi^{(b)}}{\partial z} = 0 \tag{11}$$

$$\frac{\partial \phi^{(s)}}{\partial t} + \frac{1}{2}\left[\frac{\partial \phi^{(s)}}{\partial x_j}\frac{\partial \phi^{(s)}}{\partial x_j} + \left(\frac{\partial \phi^{(s)}}{\partial z}\right)^2\right] + g\eta = 0 \tag{12}$$

The advantage of the formulation in terms of the velocity potential, as compared to that for the velocity and the pressure, is the decoupling of the kinematic and the dynamic aspects of the wave-induced flow. One need only solve equation (8) for the velocity potential under the boundary conditions (10)–(12). The pressure can be directly derived from the solution of the velocity potential based on the Bernoulli equation (9).

A number of the well-known wave equations have been derived, starting with equations (8)–(12). Readers interested in those derivations may refer to Mei (1989) and Dingemans (1997). In this section, two other techniques, which seem to be somewhat more systematic, are followed to derive the mathematical models of interest. One of these techniques is to directly integrate equations (1) and (2) in the vertical direction; the other is based on a variational principle for equations (8)–(12). These techniques give essentially the same but sometimes different results, as will be evident in the following sections.

2.1. Wave Equations by Direct Integration

2.1.1. Integration of Basic Equations. Equation (1) can be readily integrated in the vertical direction from the bottom to the free surface. As Leibnitz's theorem for the differentiation of an integral is applied and the boundary conditions (4) and (5) are also taken into account, the following well-known form of the continuity equation for unsteady free-surface flows (e.g., Abbott, 1979) is obtained:

$$\frac{\partial \eta}{\partial t} + \frac{\partial q_j}{\partial x_j} = 0 \qquad (13)$$

where $q_j = \int_{-h}^{\eta} u_j dz$ is the volume discharge or volume flux of the fluid in the x_j direction. The vector form of (13) is

$$\eta_t + \nabla \cdot \boldsymbol{q} = 0 \qquad (14)$$

where $\boldsymbol{q} = (q_1, q_2)$ is the volume flux vector; the subscript t denotes differentiation with respect to the time t; and $\nabla = (\partial/\partial x_1, \partial/\partial x_2)$ is the horizontal gradient operator. A similar integral operation to (2) leads to

$$\frac{\partial q_i}{\partial t} + \frac{\partial}{\partial x_j}\int_{-h}^{\eta} u_j u_i dz + \frac{\partial}{\partial x_i}\int_{-h}^{\eta} \frac{\tilde{p}}{\rho} dz - \frac{\tilde{p}^{(s)}}{\rho}\frac{\partial \eta}{\partial x_i} - \frac{\tilde{p}^{(b)}}{\rho}\frac{\partial h}{\partial x_i} = 0 \qquad (15)$$

Since it is intended to have a closed set of differential equations in terms of η and $\boldsymbol{q}$ (or their correspondents), further manipulation of equation (15) is necessary. For such a manipulation, however, a *similarity law* on the dynamic pressure is needed that can replace the equation of motion in the vertical direction (3) and correlate the pressure with the instantaneous water surface elevation and, if necessary, also the volume flux or the horizontal velocity. An assumption on the vertical dependence of the horizontal velocity components may also be necessary if finite amplitude waves are considered. As will be shown, the desired similarity laws do exist for a large number of problems with scientific or engineering background.

2.1.2. Classical Wave Equation. The classical wave equation is based on the following linear relation between the hydrodynamic pressure and the water surface elevation:

$$\tilde{p} = \rho g \eta f(z) \qquad (16)$$

where $f(z)$ is the *pressure–response function* expressed by

$$f(z) = \frac{\cosh k(h+z)}{\cosh kh} \qquad (17)$$

In equation (17), k is the wavenumber governed by the following dispersion equation:

$$\sigma^2 = gk \tanh kh \qquad (18)$$

where σ is the angular frequency of the wave motion. Within the framework of the small amplitude wave theory, η, u_i, and $\tilde{p}$, when nondimensionalized, are all small quantities of the same order, i.e., the order of the wave steepness. If the bottom of the topography is horizontal or otherwise, its slope is vary small as compared to the wave steepness, and interest is restricted to the local phenomena, equation (15) can be reasonably simplified as

$$\boldsymbol{q}_t + \left[\int_{-h}^{0} f(z)\, dz\right] \nabla \eta = 0 \tag{19}$$

Evaluating the integral yields

$$\boldsymbol{q}_t + c^2\, \nabla \eta = 0 \tag{20}$$

where $c = \sigma/k$ is the wave celerity. By eliminating $\boldsymbol{q}$ from equations (14) and (20), the *classical wave equation* (CWE) is obtained for η:

$$\eta_{tt} = c^2 \nabla^2 \eta \tag{21}$$

Note that c can be a very slow function of the horizontal coordinates in cases where the sea bottom is not exactly flat.

Application of the classical wave equation to the description of coastal waves over varying bottoms can only be found in earlier literature (Ito and Tanimoto, 1972). This is because, on the one hand, the equation does not ensure the conservation of energy flux and, thus, cannot reproduce the shoaling effects (Watanabe and Maruyama, 1986), which may account for a few percent of the wave height, and on the other hand, it is of no significant advantage, particularly from the numerical point of view, over the mild-slope wave equation to be discussed in the Section 2.1.3.

For sinusoidal waves, the water surface elevation η can be denoted by

$$\eta = \hat{\eta} e^{-i\sigma t} \tag{22}$$

where $\hat{\eta}$ is the *complex wave amplitude* with its magnitude representing the conventional wave amplitude and its argument expressing the relative phase of the water surface oscillation. Inserting equation (22) into equation (21) results in the following *Helmholtz equation* in terms of $\hat{\eta}$:

$$\nabla^2 \hat{\eta} + k^2 \hat{\eta} = 0 \tag{23}$$

For a steady-state wave problem, equation (23) may have advantages over equation (21).

2.1.3. Mild-Slope Wave Equation. Since the effects of topography change on wave transformation can be cumulative, as in a shoaling process, a critical consideration of the accuracy of the wave equation in terms of the bottom slope has been recognized to be necessary to ensure the validity of the mathematical model in relatively large domains. Without any additional restriction on the bottom steepness to that required by the validity of equation (16), equation (15) should give, instead of equation (19), the following equation of motion:

$$\boldsymbol{q}_t + \nabla(c^2 \eta) - \frac{g\eta}{\cosh kh} \nabla h = 0 \tag{24}$$

Referring to the following relations derived from the dispersion equation (18):

$$\nabla(kh) = \frac{k}{2n} \nabla h \tag{25}$$

$$\nabla(c^2) = \frac{4c^2}{\sinh 2kh} \nabla(kh) \tag{26}$$

where $n = 1/2 + kh/\sinh 2kh$ is a *shallowness parameter*, one can then transform equation (24) into

$$q_t + c^2 \nabla \eta - \frac{\nabla \chi}{\chi} c^2 \eta = 0 \tag{27}$$

where χ is a function of kh determined by

$$\frac{\nabla \chi}{\chi} = \frac{4 \sinh kh - 2 \sinh 2kh - 2 kh}{\sinh 2kh \sinh kh} \nabla(kh) \tag{28}$$

Integration of equation (28) yields

$$\chi(kh) = \exp\left(\int_0^{kh} \frac{4 \sinh \Theta - 2 \sinh 2\Theta - 2\Theta}{\sinh 2\Theta \sinh \Theta} d\Theta \right) \tag{29}$$

The χ-function is shown in Figure 1.2. As the last two terms on the left hand side may be combined, equation (27) may also be written as

$$q_t + \frac{c^2}{\chi} \nabla(\chi \eta) = 0 \tag{30}$$

Equations (14) and (15) form a hyperbolic system for η and q, called the *time-dependent mild-slope wave equations*. These equations are very similar to those proposed by Watanabe and Maruyama (1986) for a hyperbolic representation of the elliptic mild-slope

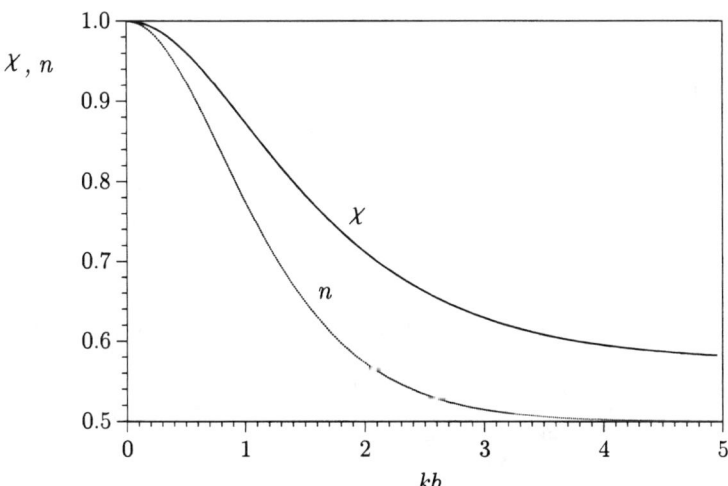

FIGURE 1.2. The χ-function and its comparison with the shallowness parameter.

wave equation (MSWE) originally derived by Berkhoff (1972). Watanabe and Maruyama used the shallowness parameter n in the place of the χ-function and showed their equations are accurate up to the second order of the bottom slope. A comparison of n with χ is presented in Figure 1.2. Similarity between the two functions is evident. Since the evaluation of n is much easier, Watanabe and Maruyama (1986) are followed in the succeeding development, although this may cause inaccuracy in some special cases as reported for other second-order mild-slope wave equations (Chamberlain and Porter, 1995).

As q is eliminated from equations (14) and (30) while χ is replaced by n, the following second-order differential equation for η is obtained:

$$\eta_{tt} - \nabla \cdot \left[\frac{c^2}{n} \nabla(n\eta) \right] = 0 \qquad (31)$$

For further simplification, Radder (1979) may be followed by introducing the *wave function* ψ, which is related to η through

$$\psi = \sqrt{cc_g}\,\eta = \sqrt{nc}\,\eta \qquad (32)$$

where $c_g = nc$ is the group velocity. Since

$$\nabla(n\eta) = \nabla\left(\frac{\sqrt{n}}{c}\psi\right) = \frac{\sqrt{n}}{c} \nabla\psi + \psi \nabla\left(\frac{\sqrt{n}}{c}\right) \qquad (33)$$

one has

$$\nabla \cdot \left[\frac{c^2}{n} \nabla(n\eta) \right] = \nabla \cdot \left[\frac{c}{\sqrt{n}} \nabla\psi - \psi \nabla\left(\frac{c}{\sqrt{n}}\right) \right] = \frac{c}{\sqrt{n}} \nabla^2\psi - \psi \nabla^2\left(\frac{c}{\sqrt{n}}\right) \qquad (34)$$

Noting that $\nabla^2(c/\sqrt{n})$ is a small quantity of the second order in terms of the bottom slope and can thus be neglected without affecting the overall accuracy, one may formally transform equation (31) into the classical wave equation in terms of ψ:

$$\psi_{tt} = c^2 \nabla^2 \psi \qquad (35)$$

For sinusoidal waves with

$$\psi = \hat{\psi} e^{-i\sigma t} \qquad (36)$$

where $\hat{\psi} = \sqrt{nc}\,\hat{\eta}$ is the complex amplitude of the wave function, equation (35) reduces to the Helmholtz equation in terms of $\hat{\psi}$

$$\nabla^2 \hat{\psi} + k^2 \hat{\psi} = 0 \qquad (37)$$

The time-dependent mild-slope wave equations are among the most useful mathematical models for coastal water waves. Different from the classical wave theory, they ensure the conservation of not only mass and momentum but also energy flux, and can thus represent a large category of the wave propagation and transformation processes in coastal waters, including reflection, diffraction, shoaling, refraction, and their interactions. Addition of a breaking model is also easy. Moreover, development of a relevant numerical model is usually straightforward.

It should be mentioned that the time-dependent mild-slope wave equations are essentially different from a hyperbolic representation of the elliptic mild-slope wave equation proposed by Berkhoff (1972). The latter (Copeland, 1985; Madsen and Larsen, 1987) is only physically feasible when a steady-state solution is required.

2.1.4. Nonlinear Shallow-Water Wave Equations. The nonlinear shallow-water wave equations are based on the assumption that the vertical acceleration of a wave-induced flow is negligibly small, which is accurate when the depth-to-wavelength ratio of the problem is infinitesimal. The equation of motion in the vertical direction (3) then implies a hydrostatic pressure distribution:

$$\tilde{p} = \rho g \eta \tag{38}$$

From the irrotationality condition of the flow, a complementary assumption on the vertical uniformity of the horizontal velocity components can also be deduced:

$$u = \bar{u} \tag{39}$$

where the bar denotes vertical average. Substituting equations (38) and (39) into equation (15) one obtains

$$\frac{\partial}{\partial t}(\bar{u}_i d) + \frac{\partial}{\partial x_j}(\bar{u}_j \bar{u}_i d) + gd \frac{\partial \eta}{\partial x_i} = 0 \tag{40}$$

where $d = \eta + h$ is the instantaneous total water depth, $\bar{u}_i = q_i/d$ is the vertically averaged velocity. If the continuity equation (13) is invoked, equation (40) can be further simplified to

$$\frac{\partial \bar{u}_i}{\partial t} + \bar{u}_j \frac{\partial \bar{u}_i}{\partial x_j} + g \frac{\partial d}{\partial x_i} = g \frac{\partial h}{\partial x_i} \tag{41}$$

or, in vector form,

$$\bar{u}_t + (\bar{u} \cdot \nabla)\bar{u} + g \nabla d = g \nabla h \tag{42}$$

Correspondingly, the continuity equation (14) can be written as

$$d_t + \bar{u} \cdot \nabla d + d \nabla \cdot \bar{u} = 0 \tag{43}$$

Equations (42) and (43) form a quasilinear hyperbolic system for d and $\bar{u}$, called the *nonlinear shallow-water wave equations*, or, *nonlinear long-wave equations*. They are valid only when the representative depth-to-wavelength ratio of the problem is very small. In practice, they may be applied to estuarine tidal waves, flood waves, tsunamis, or other infragravity waves.

Waves described by the nonlinear shallow-water wave equations are nondispersive. It is recognized that discontinuities may be developed within their solution under certain conditions, though the initial and boundary conditions may be continuous. Such a discontinuity is generally related to the striking phenomenon of wave breaking in coastal hydrodynamics (Stoker, 1957).

2.1.5. Boussinesq Equations. The Boussinesq equations are useful because they extend the validity range of the nonlinear shallow-water wave equations by including the effects of the vertical acceleration of the wave-induced flows. The equations are based on improved assumptions on the pressure and the horizontal velocity obtained following a perturbative analysis of equation (3). For the classical form of the Boussinesq equations, which is accurate up to the first order of the wave-height to water-depth ratio (i.e., the relative wave height), or equivalently, the second order of the water-depth to wavelength ratio (i.e., the relative water depth), these assumptions are noted to be

$$\tilde{p} = \rho g \eta + \rho z \frac{\partial}{\partial x_j}\left(h \frac{\partial u_j}{\partial t}\right) + \frac{\rho z^2}{2} \frac{\partial^2 u_j}{\partial x_j \partial t} \qquad (44)$$

$$u_i = \bar{u}_i - \frac{h + 2z}{2} \frac{\partial^2 (h\bar{u}_j)}{\partial x_i \partial x_j} + \frac{h^2 - 3z^2}{6} \frac{\partial^2 \bar{u}_j}{\partial x_i \partial x_j} \qquad (45)$$

where $\bar{u}_i = q_i/(h + \eta)$ is the vertically averaged velocity. Substituting equations (44) and (45) into equation (15) and neglecting the terms equivalent to, or higher than the second order of the relative wave height and the fourth order of the relative water depth, one obtains

$$\frac{\partial}{\partial t}[(h + \eta)\bar{u}_i] + \frac{\partial}{\partial x_j}[(h + \eta)\bar{u}_j\bar{u}_i] + g(h + \eta)\frac{\partial \eta}{\partial x_i}$$

$$= \frac{h(h + \eta)}{2} \frac{\partial^2}{\partial x_i \partial x_j}\left(h \frac{\partial \bar{u}_j}{\partial t}\right) - \frac{h^2(h + \eta)}{6} \frac{\partial^3 \bar{u}_j}{\partial x_i \partial x_j \partial t} \qquad (46)$$

If the continuity equation, which is expressed as

$$\frac{\partial \eta}{\partial t} + \frac{\partial}{\partial x_j}[(h + \eta)\bar{u}_j] = 0 \qquad (47)$$

is referred, equation (46) can be simplified to give

$$\frac{\partial \bar{u}_i}{\partial t} + \bar{u}_j \frac{\partial \bar{u}_i}{\partial x_j} + g \frac{\partial \eta}{\partial x_i} = \frac{h}{2} \frac{\partial^2}{\partial x_i \partial x_j}\left(h \frac{\partial \bar{u}_j}{\partial t}\right) - \frac{h^2}{6} \frac{\partial^3 \bar{u}_j}{\partial x_i \partial x_j \partial t} \qquad (48)$$

In vector form, equations (47) and (48) read

$$\eta_t + \nabla \cdot [(h + \eta)\bar{\boldsymbol{u}}] = 0 \qquad (49)$$

$$\bar{\boldsymbol{u}}_t + (\bar{\boldsymbol{u}} \cdot \nabla)\bar{\boldsymbol{u}} + g\nabla \eta = \frac{h}{2} \nabla \cdot [\nabla(h\bar{\boldsymbol{u}}_t)] - \frac{h^2}{6} \nabla(\nabla \cdot \bar{\boldsymbol{u}}_t) \qquad (50)$$

Equations (47) and (48) or (49) and (50) form a system for the water surface elevation η and the vertically averaged velocity $\bar{\boldsymbol{u}}$. They are the *classical form of the Boussinesq equations* (Peregrine, 1967). These equations are different from the shallow-water wave equations only by the two third-order differential terms in the momentum equation (50). These additional terms are called Boussinesq terms. They cause a dependence of the wave celerity on the wavenumber. Hence, the Boussinesq equations depict nonlinear dispersive waves.

The classical form of the Boussinesq equations are a significant improvement on the shallow-water wave equations, but their applicability is still restricted to situations with the relative water depth smaller than about 0.2 for a reasonable accuracy. This limitation is certainly not tolerable for practical purposes. Fortunately, recent research efforts have been directed to further enhance the classical Boussinesq equations and have resulted in some significant progress (Witting, 1984; Madsen et al., 1991; Madsen and Sorensen, 1992; Nwogu, 1993; Schaffer and Madsen, 1995; Wei et al., 1995).

Nwogu (1993) recommended the following expressions for the pressure and horizontal velocity:

$$\tilde{p} = \rho g \eta + \rho z \nabla \cdot (h u_t^{(r)}) + \frac{\rho z^2}{2} \nabla \cdot u_t^{(r)} \qquad (51)$$

$$u = u^{(r)} - (\xi h - z)\nabla[\nabla \cdot (h u^{(r)})] + \frac{\xi^2 h^2 - z^2}{2} \nabla(\nabla \cdot u^{(r)}) \qquad (52)$$

where the bracketed superscript r denotes values at the reference level $z^{(r)} = -\xi h$, and $0 \leq \xi \leq 1$ represents the relative position of the reference level and is a parameter to be determined. With these relations, the vertically integrated continuity equation and the momentum equation can be obtained as

$$\eta_t + \nabla \cdot [(h + \eta)u^{(r)}] = \nabla \cdot \left\{ \frac{(1 - 3\xi^2)h^3}{6} \nabla(\nabla \cdot u^{(r)}) + \frac{(2\xi - 1)h^2}{2} \nabla[\nabla \cdot (h u^{(r)})] \right\} \qquad (53)$$

$$u_t^{(r)} + (u^{(r)} \cdot \nabla)u^{(r)} + g\nabla \eta + \frac{\xi^2 h^2}{2} \nabla(\nabla \cdot u^{(r)}) - \xi h \nabla[\nabla \cdot (h u^{(r)})] = 0 \qquad (54)$$

The *enhanced Boussinesq equations* (53) and (54) are different from the classical Boussinesq equations by the tactical inclusion of the free parameter ξ. The value of ξ can be adjusted to optimize the dispersivity of the equation system. Nwogu (1993) showed that the dispersion relation implied by equations (53) and (54) is the "best" approximation to that governed by the small amplitude wave theory when $\xi = 0.531$.

The enhanced Boussinesq equations are one of the most useful mathematical models for coastal water waves, particularly when the wave nonlinearity should be taken into account. The equations describe weakly nonlinear and weakly dispersive waves, with upper limits on both the wave steepness and the relative water depth. The theoretical limitations are, however, not very critical in a large number of practical problems, because that on the wave steepness can not be very far from its possibly largest value permitted by a breaking condition, whereas that on the relative water depth is not smaller than the deepwater wave criterion beyond which the waves do not "feel" the bottom at all.

2.2. Wave Equations by Variational Approach

2.2.1. Lagrangian Formulation of Water Wave Problems.
A Lagrangian formulation of the water wave problems was first explicitly given by Luke (1967). Luke found that the

formulation for general free-surface flows in terms of the velocity potential, i.e., equations (8)–(12), is equivalent to the following Hamilton's principle

$$\delta \int_{t_1}^{t_2} \int_{\Omega} L \, d\Omega \, dt = 0 \tag{55}$$

where t_1 and t_2 are the starting and ending time of the phenomenon of interest, Ω denotes the solution domain in the horizontal plan, and

$$L = \int_{-h}^{\eta} \left[\frac{\partial \phi}{\partial t} + \frac{1}{2} \frac{\partial \phi}{\partial x_j} \frac{\partial \phi}{\partial x_j} + \frac{1}{2} \left(\frac{\partial \phi}{\partial z} \right)^2 + gz \right] dz \tag{56}$$

is the Lagrangian density.

When the Hamilton's principle (55) is utilized to derive a wave equation, an assumption on the z separated form for the velocity potential ϕ is required:

$$\phi = f_j(z) \zeta_j(x_i, t) \tag{57}$$

where $\zeta_j(x_i, t)$ are water-surface-elevation-like functions and $f_j(z)$, the vertical dependence functions, must be given. Direct integration of the Lagrangian density L can thus be performed to give

$$L = L\left(\eta, \frac{\partial \eta}{\partial t}, \frac{\partial h}{\partial x_j}, \zeta_i, \frac{\partial \zeta_i}{\partial t}, \frac{\partial \zeta_i}{\partial x_j} \right) \tag{58}$$

The Euler–Lagrange equations for the explicit Lagrangian density, i.e.,

$$\frac{\partial L}{\partial \eta} = \frac{\partial}{\partial t} \left[\frac{\partial L}{\partial (\partial \eta/\partial t)} \right] + \frac{\partial}{\partial x_j} \left[\frac{\partial L}{\partial (\partial \eta/\partial x_j)} \right] \tag{59}$$

$$\frac{\partial L}{\partial \zeta_i} = \frac{\partial}{\partial t} \left[\frac{\partial L}{\partial (\partial \zeta_i/\partial t)} \right] + \frac{\partial}{\partial x_j} \left[\frac{\partial L}{\partial (\partial \zeta_i/\partial x_j)} \right] \tag{60}$$

must then lead to a particular wave equation.

2.2.2. Mild-Slope Wave Equation. The mild-slope wave equation is based on the following assumption on the velocity potential:

$$\phi = \zeta(x_i, t) f(z) \tag{61}$$

where $f(z)$ is the pressure response function expressed by equation (17) and ζ is an unknown function independent of the vertical coordinate z. For small amplitude waves, ζ is a small quantity of the same order as the water surface elevation η when both are nondimensionalized. Substituting equation (61) into equation (56) and carrying out the integration (with only the terms up to the third order of the wave steepness retained) gives

$$L = L\left(\eta, \zeta, \frac{\partial \zeta}{\partial t}, \frac{\partial \zeta}{\partial x_j} \right)$$

$$= \left(\frac{c^2}{g} + \eta \right) \frac{\partial \zeta}{\partial t} + \frac{cc_g}{2g} \frac{\partial \zeta}{\partial x_j} \frac{\partial \zeta}{\partial x_j} + \frac{(1-n)\sigma^2}{2g} \zeta^2 + \frac{g}{2} (\eta^2 - h^2) \tag{62}$$

Note that the dispersion equation (18) has been invoked. The Euler–Lagrange equations then yield

$$\frac{\partial \eta}{\partial t} + \frac{\partial}{\partial x_j}\left(\frac{cc_g}{g}\frac{\partial \zeta}{\partial x_j}\right) - \frac{(1-n)\sigma^2}{g}\zeta = 0 \tag{63}$$

$$\frac{\partial \zeta}{\partial t} + g\eta = 0 \tag{64}$$

Eliminating ζ from equations (63) and (64) one obtains

$$\frac{\partial^2 \eta}{\partial t^2} - \nabla\cdot(cc_g\nabla\eta) + (1-n)\,\sigma^2\eta = 0 \tag{65}$$

which is the well-known form of the time-dependent mild-slope wave equation first proposed by Smith and Sprinks (1975) and rederived by Kirby et al. (1992). For a sinusoidal wave, it reduces to

$$\nabla\cdot(cc_g\nabla\hat{\eta}) + k^2 cc_g\hat{\eta} = 0 \tag{66}$$

which is the first successful mathematical model for the combined effects of wave diffraction and refraction, originally obtained by Berkhoff (1972) and rederived by many others (e.g., Mei, 1989). In terms of the wave function $\hat{\psi}$, equation (66) can also be written as

$$\nabla^2\hat{\psi} + k^2\hat{\psi} = 0 \tag{67}$$

which is identical to equation (37).

2.2.3. Nonlinear Shallow-Water Wave Equations. In the nonlinear shallow-water wave theory, the velocity potential is assumed to be independent of the vertical coordinate z. Namely,

$$\phi = \phi(x_i, t) \tag{68}$$

Substituting equation (68) into equation (56) yields

$$\mathcal{L} = \mathcal{L}\left(\eta, \frac{\partial \phi}{\partial t}, \frac{\partial \phi}{\partial x_j}\right)$$

$$= (h+\eta)\frac{\partial \phi}{\partial t} + \frac{h+\eta}{2}\frac{\partial \phi}{\partial x_j}\frac{\partial \phi}{\partial x_j} + \frac{g(\eta^2 - h^2)}{2} \tag{69}$$

From the Euler–Lagrange equations one obtains

$$(\eta + h)_t + \nabla\cdot[(\eta + h)\nabla\phi] = 0 \tag{70}$$

$$\phi_t + g\eta + \frac{1}{2}\nabla\phi\cdot\nabla\phi = 0 \tag{71}$$

Let

$$\bar{u}_i = \nabla \phi \qquad (72)$$

Equation (70) then becomes

$$\eta_t + \nabla \cdot [(h + \eta)\bar{u}] = 0 \qquad (73)$$

whereas equation (71), by taking the gradient on both sides, yields

$$\bar{u}_t + (\bar{u} \cdot \nabla)\bar{u} + g\nabla \eta = 0 \qquad (74)$$

Equations (73) and (74) are again the well-known nonlinear shallow-water wave equations.

2.2.4. Boussinesq Equations. Consider only waves over an essentially horizontal bottom. The velocity potential may then be assumed to have the following simple form:

$$\phi = \zeta(x_i, t) + (h + z)^2 \zeta'(x_i, t) \qquad (75)$$

where ζ and ζ' are unknown functions independent of the vertical coordinate z. No attempt will be made to correlate ζ and ζ' at this time. Substituting equation (75) into equation (56) yields

$$\mathcal{L} = \mathcal{L}\left(\eta, \frac{\partial \zeta}{\partial t}, \frac{\partial \zeta}{\partial x_j}, \zeta', \frac{\partial \zeta'}{\partial t}, \frac{\partial \zeta'}{\partial x_j}\right)$$

$$= (h + \eta)\frac{\partial \zeta}{\partial t} + \frac{(h + \eta)^3}{3}\frac{\partial \zeta'}{\partial t} + \frac{h + \eta}{2}\frac{\partial \zeta}{\partial x_j}\frac{\partial \zeta}{\partial x_j} + \frac{(h + \eta)^3}{3}\frac{\partial \zeta}{\partial x_j}\frac{\partial \zeta'}{\partial x_j}$$

$$+ \frac{(h + \eta)^5}{10}\frac{\partial \zeta'}{\partial x_j}\frac{\partial \zeta'}{\partial x_j} + \frac{2(h + \eta)^3}{3}(\zeta')^2 + \frac{g}{2}(\eta^2 - h^2) \qquad (76)$$

Note that the condition $|\nabla h| \approx 0$ has been taken into account. The Euler–Lagrange equations then lead to

$$\zeta_t + (h + \eta)^2 \zeta'_t + \frac{1}{2}\nabla \zeta \cdot \nabla \zeta + (h + \eta)^2 \nabla \zeta \cdot \nabla \zeta' + \frac{(h + \eta)^4}{2} \nabla \zeta' \cdot \nabla \zeta'$$

$$+ 2(h + \eta)^2 (\zeta')^2 + g\eta = 0 \qquad (77)$$

$$\eta_t + \nabla \cdot [(h + \eta)\nabla \zeta] + \frac{1}{3}\nabla \cdot [(h + \eta)^3 \nabla \zeta'] = 0 \qquad (78)$$

$$(h + \eta)^2 \eta_t + \frac{1}{3}\nabla \cdot [(h + \eta)^3 \nabla \zeta] + \frac{1}{5}\nabla \cdot [(h + \eta)^5 \nabla \zeta'] = \frac{4(h + \eta)^3}{3}\zeta' \qquad (79)$$

It can be readily confirmed that equations (78) and (79) become identical if ζ and ζ' are related to each other through

$$\zeta' + \frac{1}{10}(h+\eta)^2 \nabla^2 \zeta' = -\frac{1}{2}\nabla\zeta \tag{80}$$

Reversing equation (80) one has

$$\zeta' = -\frac{1}{2}\nabla^2\zeta + \frac{(h+\eta)^2}{20}\nabla^4\zeta - \ldots \tag{81}$$

Substituting equation (81) into equations (77) and (78) and retaining only the terms up to the first order of the relative wave height and the second order of the relative water depth gives rise to

$$\zeta_t - \frac{h^2}{2}\nabla^2\zeta_t + \frac{1}{2}\nabla\zeta\cdot\nabla\zeta + g\eta = 0 \tag{82}$$

$$\eta_t + \nabla\cdot[(h+\eta)\nabla\zeta] - \frac{h^3}{6}\nabla^4\zeta = 0 \tag{83}$$

Note that

$$\boldsymbol{u}^{(b)} = \nabla\zeta \tag{84}$$

Equation (83) can thus be written as

$$\eta_t + \nabla\cdot[(h+\eta)\,\boldsymbol{u}^{(b)}] = \frac{h^3}{6}\nabla^2(\nabla\cdot\boldsymbol{u}^{(b)}) \tag{85}$$

while equation (82), after taking the gradient on both sides, yields

$$\boldsymbol{u}_t^{(b)} + (\boldsymbol{u}^{(b)}\cdot\nabla)\boldsymbol{u}^{(b)} + g\nabla\eta = \frac{h^2}{2}\nabla^2\boldsymbol{u}_t^{(b)} \tag{86}$$

Equations (85) and (86) are the original form of the Boussinesq equations (e.g., Dingemans, 1997).

2.3. Dissipation Models

A real fluid is viscous. In addition, a flow in practice is often turbulent. The viscosity of the fluid and the turbulence of the flow together add two important effects to a surface water wave: *diffusion* and *dissipation*.

It is well known that turbulence can be conceptually interpreted as the eddy viscosity. As both the molecular viscosity and the eddy viscosity, or a comprehensive viscosity, is considered, the equations of motion (2) for general free-surface flows should be modified to

$$\frac{\partial u_i}{\partial t} + \frac{\partial}{\partial x_j}(u_j u_i) + \frac{\partial}{\partial z}(w u_i) + \frac{\partial}{\partial x_i}\left(\frac{\tilde{p}}{\rho}\right) = K\frac{\partial^2 u_i}{\partial x_j \partial x_j} + K'\frac{\partial^2 u_i}{\partial z^2} \qquad (87)$$

where the horizontal and the vertical diffusion coefficients K and K' are determined by the comprehensive viscosity that depends on both the fluid properties and the local flow structure. K and K' are usually not constants but functions of the coordinates and probably also of the time.

Study of the effects of the viscous terms on surface water wave motion by the variational approach is known to be difficult. Fortunately, the discussion can proceed by performing the direct integration inclusively. It is simple to show that the continuity equation (14) remains unchanged whether the fluid is inviscid or not. However, the vertically integrated equation of motion (15) must be modified under the generalized conditions. Roughly evaluated, one must add the following terms on the right-hand side of equation (15):

$$\boldsymbol{D} = K(h + \eta)\nabla^2 \boldsymbol{u}^{(r)} - \frac{\boldsymbol{\tau}_0}{\rho} \qquad (88)$$

where $\boldsymbol{D}$ stands for the additional terms, $\boldsymbol{\tau}_0$ is the bottom friction, $\boldsymbol{u}^{(r)}$ is the velocity at a reference level, and K is a vertically averaged horizontal diffusion coefficient. The first term on the right-hand side of equation (88) causes a diffusion of momentum as the consequence of horizontal mixing, while the second term produces dissipation of wave energy to counterbalance the bottom resistance. Expressions that properly describe both the diffusion and the dissipation in a vertically integrated water wave equation have been called *dissipation models*. Equation (88) seems to be the starting equation for the establishment of an effective dissipation model.

In general, the effects of diffusion and those of dissipation on a surface water wave are essentially different, though both of them result in a decay of the wave height. Perceptually, dissipation is more reasonably compared to a damping mechanism, whereas the diffusion is more or less analogous to a smoothing process. A special case is the sinusoidal waves, to which the effects of the diffusion and dissipation reduce to the same. This last fact justifies the practices in many previous studies that give little emphasis to the diffusion process.

2.3.1. Bottom Friction Models. The *bottom friction* is usually expressed by the following quadratic law:

$$\frac{\boldsymbol{\tau}_0}{\rho} = C_f |\boldsymbol{u}^{(b)}| \boldsymbol{u}^{(b)} \qquad (89)$$

where C_f is a dimensionless *friction coefficient*. Equation (89) can be formally derived by resolving the oscillatory boundary layer flow over a flat bottom. It is also consistent with Chezy's law, an empirical formula widely used in open-channel hydraulics for over a century. Since it is closely related to the wave friction coefficient f_w introduced by Jonsson (1966) (also Sleath, 1984), determination of C_f under laboratory conditions is not particularly difficult by following a number of successful researches, including that of Kamphuis (1975).

When linearized, equation (89) can be altered to

$$\frac{\boldsymbol{\tau}_0}{\rho} = C_f' \sigma \boldsymbol{q} \qquad (90)$$

where C_f' is an alternative friction coefficient, or the *linear friction coefficient*. Although C_f' can be determined from C_f by a principle of equivalent work, such an approach is seldom adopted because C_f' has a direct relation with the spatial rate of wave height decay.

After the linear friction term is included, the equation of motion in the classical wave theory, i.e., equation (20), must be modified to

$$q_t + c^2 \nabla \eta + C_f' \sigma q = 0 \tag{91}$$

The resulting wave equation will then become

$$\eta_{tt} + C_f' \sigma \eta_t - c^2 \nabla^2 \eta = 0 \tag{92}$$

or,

$$\nabla^2 \hat{\eta} + (1 + i C_f') k^2 \hat{\eta} = 0 \tag{93}$$

for steady-state sinusoidal waves.

In the time-dependent mild-slope wave equations, one has to replace equation (30) by the following equation to include the effects of bottom friction:

$$q_t + \frac{c^2}{\chi} \nabla(\chi \eta) + C_f' \sigma q = 0 \tag{94}$$

The consequent equation for the wave function ψ, then, should read

$$\psi_{tt} + C_f' \sigma \psi_t - c^2 \nabla^2 \psi = 0$$

or,

$$\nabla^2 \hat{\psi} + (1 + i C_f') k^2 \hat{\psi} = 0$$

for steady-state sinusoidal waves. The steady-state wave equation corresponding to equation (66) is written as

$$\nabla \cdot (c c_g \nabla \hat{\eta}) + (1 + i C_f') k^2 c c_g \hat{\eta} = 0 \tag{97}$$

It is evident that the value of C_f' will have effects on not only the wave height but also the wavelength. However, the physics related to the effects of friction on wavelength has not yet been well understood. To avoid confusion, particularly when a fictitious friction is introduced to simulate the effects of other factors that cause a decay of wave height, it has been suggested to modify equations (96) and (97) to the following form (Yu and Togashi, 1994; 1996; Yu, 1996a):

$$\nabla^2 \hat{\psi} + (1 + i\beta)^2 k^2 \hat{\psi} = 0 \tag{98}$$

$$\nabla \cdot (cc_g \nabla \hat{\eta}) + (1 + i\beta)^2 k^2 cc_g \hat{\eta} = 0 \qquad (99)$$

respectively. It can be readily shown (Yu, 1996a) that β coincides with the *wave decay coefficient* studied by many researchers (e.g., Sleath, 1984). The relation between C_f' and β can be found in Dalrymple et al. (1984):

$$C_f' = 2\beta\sqrt{1 + \beta^2} \qquad (100)$$

When the nonlinear shallow-water wave equations or the Boussinesq equations are concerned, the quadratic law (89) is directly admissible provided the bottom velocity is converted to either the vertically averaged velocity or the velocity at the reference level, depending on the unknowns of the system concerned. In the nonlinear shallow-water wave theory, the equation of motion may then be written as

$$\bar{u}_t + \bar{u} \cdot \nabla \bar{u} + g\nabla d = g\nabla h - C_f \frac{|\bar{u}|\bar{u}}{d} \qquad (101)$$

The equation of motion of the enhanced Boussinesq theory will become

$$u_t^{(r)} + (u^{(r)} \cdot \nabla)u^{(r)} + g\nabla \eta + \frac{\xi^2 h^2}{2} \nabla(\nabla \cdot u^{(r)}) - \xi h \nabla[\nabla \cdot (hu^{(r)})] + C_f \frac{|u^{(r)}|u^{(r)}}{h + \eta} = 0 \qquad (102)$$

2.3.2. Breaking Models. A breaking model is supposed to provide information on whether and where waves break within the solution domain (i.e., the *breaking criterion*) as well as on the excessive horizontal mixing associated with the breaking (i.e., the *surf zone model*), which is often accompanied by a remarkable decay of wave height.

Breaking Criteria. Although recent studies have shown that the behavior of a breaker can be successfully captured in a numerical model as long as comprehensive nonlinear wave equations are adopted and an appropriate solution method is employed (Yu et al., 1992a), an accurate and generally valid criterion governing wave breaking is still necessary in dealing with many coastal water wave problems. A significant number of research efforts of coastal engineers have been devoted to the establishment of such a criterion for nearly a century, though most of the well-known formulas are limited to either permanent waves or unidirectional waves on sloping beaches. For a detailed review of these results the reader is referred to Galvin (1972), Sawaragi (1973) and Horikawa (1989).

One of the most frequently applied breaking criteria was developed by Goda (1970; 1975), which was formulated to fit a large number of reliable experimental data available at that time. With a slight change of parameters the formula can be written as

$$\frac{\sigma^2 H}{g} = A\left\{1 - \exp\left[-\frac{3(1 + 15s^{4/3})}{4} \Phi^2\right]\right\} \qquad (103)$$

where $A = 1.07$ is a constant; H is the wave height, s is the bottom slope, $\Phi = \sqrt{\sigma^2 h/g}$, and all local variables are evaluated at the breaking point. Similarly good formulas authored by other researchers can also be found in literature (e.g., Ostendorf and Madsen, 1979). In equation (103) the breaking index is expressed by $\sigma^2 H/g$. There are reports (Yu, et al., 1995) that recognized the parameter gH/c^2, which reduces to the wave height-to-depth ratio at shallow water conditions and is proportional to the wave steepness in deep water, being an even better breaking index, but a general criterion for that index has not yet been established. It has also been found (Watanabe et al., 1984) that, in a model which requires the velocity to be directly computed, the ratio of the fluid velocity to the wave celerity is a more convenient and less dependent breaking index. In terms of this index, Goda's criterion can be approximately converted to (Isobe, 1986):

$$\frac{\hat{u}^{(m)}}{c} = 0.53 - 0.3 \exp(-1.2\Phi) + 5s^{3/2} \exp[-7.2(\Phi - 2.5)^2] \tag{104}$$

where the bracketed superscript m denotes the value at the mean water level.

It may be necessary to point out that most of the existing breaking criteria are for phase-averaged breaking indices. They do not allow tracing of the instantaneous breaking process. For the phase-resolving wave models, establishment of a breaking criterion based on the instantaneous wave profile, such as that involved in the surface roller model proposed by Schaffer et al. (1992; 1993), may be important.

Surf Zone Models. Waves in a surf zone are characterized by the strong horizontal mixing owing to the breaking-induced turbulence. A *surf zone model* is primarily aimed to semiempirically represent this complex process.

With a linear theory, coastal engineers are often interested in sinusoidal waves, which satisfy the following relation:

$$\nabla^2 \boldsymbol{u} \sim k^2 \boldsymbol{u} \tag{105}$$

The diffusion term can therefore be combined into the bottom friction term and formally hidden. Therefore,

$$\boldsymbol{D} = C_d \sigma \boldsymbol{q} \tag{106}$$

where C_d is called the *fictitious friction coefficient*. It is evident that both the classical water wave equation and the mild-slope wave equations differ solely in the friction coefficient, as only the linearized bottom friction and diffusion are considered. When the diffusion is included, the fictitious friction coefficient C_d replaces the physical friction coefficient C_f'. However, it is necessary to note that C_d and C_f' are conceptually different. C_d represents the combined effects of the bottom friction and the turbulent mixing-induced diffusion. What may be more important is that the formally hidden diffusion usually plays the dominant role in the surf zone. One may wish to formulate C_d through a careful investigation on the detailed characteristics of the turbulence in the surf zone. A similarly reasonable and probably more effective method to determine C_d has been seen as fitting the vastly available experimental data on the rate of decay of wave height or the *energy dissipation rate* for the surf-zone waves. One of the useful expressions so obtained is due to Watanabe and Maruyama (1986):

$$C_d = As\Phi^{-1} \sqrt{\frac{4|q|}{\sqrt{gh^3}} - 1} \tag{107}$$

where $A = 2.5$ is an empirical constant.

When a nonlinear coastal water wave theory is concerned, one expects accurate numerical results on not only the primary wave parameters such as the wave height but also the time-averaged quantities, including the wave-induced currents and the wave setup/setdown. On such an occasion, a direct treatment of the diffusion effects seems to be indispensable (Kabiling and Sato, 1993).

Accurate estimate of the diffusion coefficient requires a careful investigation into the instantaneous diffusion process relative to the wave motion in the surf zone, which does not seem to have been done. One may argue on the usefulness of the phase-averaged diffusion coefficient, on which a significant number of studies have been carried out, and many reliable results have been accumulated because of the need in the study for nearshore currents. It is recommended that the best representation of the rate of the decay of wave height in the surf zone should be given the first priority when a formula is selected.

Implementation of Breaking Models. Implementation of a breaking model requires the surf zone, which can only be determined as a part of the solution, to be prescribed. In a time-dependent numerical model, resolution of this contradiction is not difficult, because the breaking line can be predicted with the most newly computed data in a wave period. Note that a significant amount of computer resources must be allocated to temporarily store the computational results and update them at each computational time step in order to evaluate the phase-averaged breaking index. Otherwise, a rather sophisticated scheme may have to be developed for evaluating the breaking index from the instantaneous spatial wave profile.

In a steady-state wave model, the procedure of trial-and-error or, at least, prediction-and-correction is necessary for the surf zone determination. As a predictor, one may omit the surf zone model. Based on the primary solution so obtained it is possible to calculate the breaking index throughout the solution domain. If the bottom friction coefficient includes the wave height as a parameter, such as that used by Yu and Togashi (1996), repeated corrections of the primary solution are necessary. Since the omission of the surf zone model will not alter the nonbreaking zone, the breaking line can be rigorously determined from the primary solution. Once the breaking line is determined, the surf zone model can be added to obtain the final solution.

2.4. Boundary Conditions

2.4.1. Impermeable Walls. At an impermeable wall, the normal velocity component vanishes. This is expressed by

$$\boldsymbol{n}\cdot\boldsymbol{u}^{(r)} = 0 \qquad (108)$$

where $\boldsymbol{u}^{(r)}$ is the velocity vector at any reference level, and $\boldsymbol{n}$ is the unit vector in the outward normal direction of the boundary. In terms of the volume flux $\boldsymbol{q}$, the condition can be written as

$$\boldsymbol{n}\cdot\boldsymbol{q} = 0 \qquad (109)$$

The condition is also equivalent to

$$\frac{\partial \eta}{\partial n} = 0 \qquad (110)$$

which implies

$$\frac{\partial \hat{\eta}}{\partial n} = 0 \qquad (111)$$

2.4.2. Porous Breakwaters and Partial Reflective Walls. At a porous breakwater whose physical thickness is very small when compared to the local wavelength, as shown in Figure 1.3, the boundary condition to be satisfied by steady-state waves is well established. By assuming that the flow within the porous structure is governed by the porous-medium wave theory of Sollitt and Cross (1972), Yu and Chwang (1994) (also Yu, 1995) obtained

$$\left(\frac{\partial \hat{\eta}}{\partial n}\right)_+ = \left(\frac{\partial \hat{\eta}}{\partial n}\right)_- = -ikZ(\hat{\eta}_+ - \hat{\eta}_-) \tag{112}$$

where n is one of the normal directions of the breakwater and the subscripts + and − denote the positive and the negative face of the breakwater, respectively; $Z = \gamma/[kb(C_R - iC_I)]$ is the porous impedance, with b, γ, C_R, and C_I being the physical thickness, the porosity, the resistance coefficient, and the inertia coefficient of the porous structure, respectively; $C_I = [1 + C_m(1 - \gamma)/\gamma]$, and C_m is the added-mass coefficient. When the inertial force acting on the fluid by the porous structure is insignificant as compared to the resistance, equation (112) reduces to the Darcy's law, which states that the fluid velocity in a porous medium is proportional to the pressure gradient.

At a partially reflective wall with known reflection coefficient, which may correspond to the physical situation where wave energy absorption devices are placed in front of a sea wall, the following relation is appropriate:

$$\frac{\partial \hat{\eta}}{\partial n} - i\mu k \hat{\eta} = 0 \tag{113}$$

where $\mu = (1 - K)/(1 + K)$ and K is the reflection coefficient.

2.4.3. Nonreflective Boundaries. The complete mathematical statement of a coastal water wave problem often includes a *nonreflective boundary condition*, which admits outgoing waves only. If a nonreflective boundary is located at infinity where the water depth is either infinity or constant, the classical *Sommerfeld condition* is applicable:

$$\lim_{r \to \infty} \sqrt{r}\left(\frac{\partial \hat{\eta}}{\partial r} - ik\hat{\eta}\right) = 0 \tag{114}$$

where r is the radial direction with the origin at the center of the subdomain where waves undergo significant transformation. The Sommerfeld condition can be directly incorporated into a boundary integral equation method (Lee, 1971; Hwang and Tuck 1970), an infinite element method (Chen, 1990), or a hybrid method (Mei and Chen, 1975).

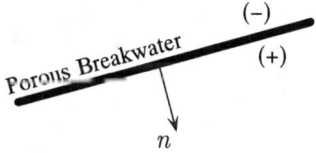

FIGURE 1.3. A thin porous breakwater.

For time-dependent wave problems, it is usually impractical to deal with a boundary at infinity. If the following relation, which is equivalent to the Sommerfeld condition

$$\frac{\partial \eta}{\partial t} + c\frac{\partial \eta}{\partial r} = 0 \tag{115}$$

is utilized to approximate a nonreflective condition, the relevant boundary should then be placed at least several wavelengths away from the subdomain where waves undergo significant transformation.

A successful condition that is practically applicable to a nonreflective boundary not far away from the region where waves are deformed is the so called *one-way wave equation* (Engquist and Majda, 1977; Trefethen and Halpern, 1986; Yu, 1998b; also Blaschak and Kriegsmann, 1988; Givoli, 1991), expressed as

$$\frac{\partial^2 \eta}{\partial t^2} + Mc\frac{\partial^2 \eta}{\partial t \partial n} + Nc^2\frac{\partial^2 \eta}{\partial s^2} = 0 \tag{116}$$

where M and N are constants and s is the tangential direction of the boundary. For a sinusoidal wave, equation (116) reduces to

$$M\frac{\partial \hat{\eta}}{\partial n} - ik\hat{\eta} = -i\frac{N}{k}\frac{\partial^2 \hat{\eta}}{\partial s^2} \tag{117}$$

There are also reports on effective radiation of outgoing waves by specially designed numerical schemes (Orlanski, 1976). A nonreflective boundary can also be handled in the same way as in a laboratory basin. That is, one may specify an arbitrary condition at the position that is a certain distance behind a fictitious nonreflective boundary and install a sponge layer (a special subregion of the solution domain with artificial resistance) to dissipate the outgoing wave energy. It has been demonstrated that the combined use of an effective sponge layer with the Sommerfeld condition can suppress the boundary reflection to a practically acceptable level (Sato et al., 1990; Oyama and Nadaoka, 1991; Eric et al., 1997).

2.4.4. Numerical Wavemakers. The simplest numerical wavemaker is the forced water surface elevation or the forced volume flux at a segment of the boundary of the problem concerned. Because any reflected wave from the domain will be rereflected to disturb the solution, the applicability of such a wavemaking boundary condition is limited to diffraction problems or problems of which the numerical computation can be terminated before the front of the reflected waves from the interior reaches the position of the wavemaker.

Most of the active wavemaking theories developed for physical wavemakers in laboratory wave basins are directly applicable to numerical wave models. Because of the need for an advanced feedback scheme, however, such a practice does not seem to have been reported.

The most popular formulation of nonreflective wavemaking boundaries in computational practice has been based on the linearized nonreflective boundary condition for a possible reflection. Denoting the nonreflective condition by

$$\mathcal{D}(\eta_r) = 0 \tag{118}$$

where $\mathcal{D}$ is an appropriate linear differential operator and η_r represents the reflected wave, the wavemaking boundary condition should then be written as

$$\mathcal{D}(\eta) = \mathcal{D}(\eta_0) \qquad (119)$$

where η_0 denotes the incident wave. If the nonreflective condition happens to be the one-way wave equation, equation (119) reads

$$\frac{\partial^2 \eta}{\partial t^2} + Mc\frac{\partial^2 \eta}{\partial t \partial n} + Nc^2\frac{\partial^2 \eta}{\partial s^2} = (1 + M\cos\theta_0 + N\sin^2\theta_0)\frac{\partial^2 \eta_0}{\partial t^2} \qquad (120)$$

where θ_0 is the approaching angle of the incident wave.

3. NUMERICAL MODELS FOR BOUNDARY VALUE PROBLEMS

3.1. High-Order Finite Difference Methods for Reduced MSWE

In this section the high-order finite difference methods are examined for the reduced mild-slope wave equation in terms of the complex amplitude of the wave function $\hat{\psi}$:

$$\hat{\psi}_{xx} + \hat{\psi}_{yy} + \kappa^2 \hat{\psi} = 0 \qquad (121)$$

where $\kappa = (1 + i\beta)k$ is, in general, a slow function of the coordinates. All the difference schemes are assumed to be effective only over symmetric rectangular grid elements, as shown in Figure 1.4. Since the dimension of the grid elements must be much smaller than the length scale of the spatial variation of $\hat{\psi}$ (which has been assumed to be much smaller than the length scale of the spatial variation of κ), there is enough reason to treat κ as a constant locally in a grid element. Then, by the symmetry of the problem, the finite difference scheme to be developed can be prescribed to take the following form:

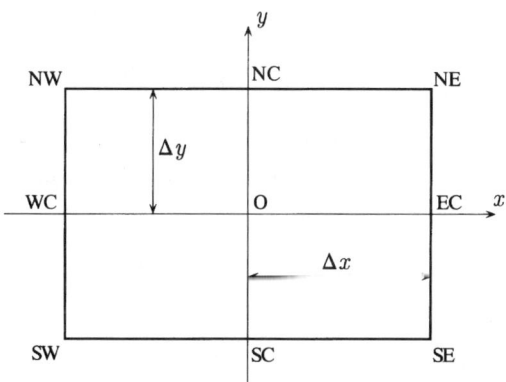

FIGURE 1.4. Grid element for high-order finite difference schemes.

$$\hat{\psi}_0 = \alpha_1(\hat{\psi}_{EC} + \hat{\psi}_{WC}) + \alpha_2(\hat{\psi}_{NC} + \hat{\psi}_{SC}) + \alpha_3(\hat{\psi}_{NE} + \hat{\psi}_{SE} + \hat{\psi}_{NW} + \hat{\psi}_{SW}) \tag{122}$$

where α_1, α_2, and α_3 are constants to be resolved.

It may be necessary to mention that the high-order finite difference schemes illustrated in the following subsections, as the high-resolution methods usually are, are accurate but less flexible if a varying mesh system is necessary either to fit the boundary configuration or to emphasize particular subdomains. Further development is evidently necessary or one should combine these schemes with other lower-order methods when dealing with various practical problems.

3.3.1. Finite Analytic Scheme.

The finite analytic method for deriving finite difference schemes of partial differential equations was proposed by Chen and his co-workers (Chen and Li, 1979; Chen and Chen, 1984). The basic stance of this method is to take advantage of the local analytic solution to the governing partial differential equation within the infinitesimal grid element that is geometrically regular. When the standard procedure is followed, the wave function $\hat{\psi}$ is decomposed as $\hat{\psi} = \hat{\psi}_1 + \hat{\psi}_2 + \hat{\psi}_3 + \hat{\psi}_4$, with each of the four unknown functions $\hat{\psi}_1$, $\hat{\psi}_2$, $\hat{\psi}_3$, and $\hat{\psi}_4$ satisfying one of the following nonhomogeneous boundary conditions along one side, and the homogenous condition along the other three sides, of the grid element:

$$\hat{\psi} = \hat{\psi}_N(x) \quad \text{at} \quad y = \Delta y \tag{123}$$

$$\hat{\psi} = \hat{\psi}_S(x) \quad \text{at} \quad y = -\Delta y \tag{124}$$

$$\hat{\psi} = \hat{\psi}_E(y) \quad \text{at} \quad x = \Delta x \tag{125}$$

$$\hat{\psi} = \hat{\psi}_W(y) \quad \text{at} \quad x = -\Delta x \tag{126}$$

Should $\hat{\psi}$ along each side of the grid element be interpolated in terms of its nodal values in the following manner (Chwang and Chen, 1987):

$$\hat{\psi}(x) = \frac{\hat{\psi}_E + \hat{\psi}_W}{2} + \frac{\hat{\psi}_E - \hat{\psi}_W}{2} \sin \frac{\pi x}{2\Delta x} + \frac{2\hat{\psi}_C - \hat{\psi}_E - \hat{\psi}_W}{2} \cos \frac{\pi x}{2\Delta x} \tag{127}$$

$$\hat{\psi}(y) = \frac{\hat{\psi}_N + \hat{\psi}_S}{2} + \frac{\hat{\psi}_N - \hat{\psi}_S}{2} \sin \frac{\pi x}{2\Delta y} + \frac{2\hat{\psi}_C - \hat{\psi}_N - \hat{\psi}_S}{2} \cos \frac{\pi y}{2\Delta y} \tag{128}$$

the finite analytic solution for $\hat{\psi}$ can be derived. As $\hat{\psi}_0$ is evaluated, one must then reach equation (122) with

$$\alpha_1 = \frac{1}{2} \operatorname{sech} \sqrt{\left(\frac{\pi}{2\epsilon}\right)^2 - \left(\frac{\delta}{\epsilon}\right)^2} \tag{129}$$

$$\alpha_2 = \frac{1}{2} \operatorname{sech} \sqrt{\left(\frac{\pi\epsilon}{2}\right)^2 - \delta^2} \tag{130}$$

$$\alpha_3 = \sum_{l=1}^{\infty} \frac{\sin l\pi}{l\pi} \left[\text{sech} \sqrt{\left(\frac{l\pi}{2\epsilon}\right)^2 - \left(\frac{\delta}{\epsilon}\right)^2} + \text{sech} \sqrt{\left(\frac{l\pi\epsilon}{2}\right)^2 - \delta^2} \right] - \frac{1}{2}(\alpha_1 + \alpha_2) \tag{131}$$

where $\epsilon = \Delta y/\Delta x$ is the aspect ratio and $\delta = \kappa \Delta y$ is the relative size of the grid element. It can be shown that equation (131) is reasonably approximated by

$$\alpha_3 = \frac{1}{2}(\alpha_1 + \alpha_2)\left[\left(\text{sech}\,\frac{\pi}{2\epsilon} + \text{sech}\,\frac{\pi\epsilon}{2}\right)^{-1} - 1\right] \tag{132}$$

which is numerically much easier.

The coefficients expressed by equations (129), (130), and (132) defined the *finite analytic scheme* for equation (121). As $|\kappa|\Delta x$ and $|\kappa|\Delta y$ are small, these coefficients may also be written as

$$\alpha_1 = \bar{\alpha}_1 + \tilde{\alpha}_1 \kappa^2 \Delta x \Delta y \tag{133}$$

$$\alpha_2 = \bar{\alpha}_2 + \tilde{\alpha}_2 \kappa^2 \Delta x \Delta y \tag{134}$$

$$\alpha_3 = \bar{\alpha}_3 + \tilde{\alpha}_3 \kappa^2 \Delta x \Delta y \tag{135}$$

where

$$\bar{\alpha}_1 = \frac{1}{2} \text{sech}\,\frac{\pi}{2\epsilon} \tag{136}$$

$$\bar{\alpha}_2 = \frac{1}{2} \text{sech}\,\frac{\pi\epsilon}{2} \tag{137}$$

$$\bar{\alpha}_3 = \frac{1}{4} - \frac{1}{2}(\bar{\alpha}_1 + \bar{\alpha}_2) \tag{138}$$

$$\tilde{\alpha}_1 = \frac{1}{2\pi} \tanh\,\frac{\pi}{2\epsilon}\, \text{sech}\,\frac{\pi}{2\epsilon} \tag{139}$$

$$\tilde{\alpha}_2 = \frac{1}{2\pi} \tanh\,\frac{\pi\epsilon}{2}\, \text{sech}\,\frac{\pi\epsilon}{2} \tag{140}$$

$$\tilde{\alpha}_3 = \frac{\tilde{\alpha}_1 + \tilde{\alpha}_2}{4(\bar{\alpha}_1 + \bar{\alpha}_2)} - \frac{1}{2}(\tilde{\alpha}_1 + \tilde{\alpha}_2) \tag{141}$$

Since $\bar{\alpha}_1, \bar{\alpha}_2, \bar{\alpha}_3, \tilde{\alpha}_1, \tilde{\alpha}_2$, and $\tilde{\alpha}_3$ are only functions of ϵ, the computational efforts for evaluating the coefficients $\alpha_1, \alpha_2,$ and α_3 can be significantly reduced if equations (133)–(135) are used instead of equations (129)–(132).

Optimal Scheme. The optimal scheme is derived by expanding all the nodal values of $\hat{\psi}$ involved in equation (122) in terms of $\hat{\psi}$ and its derivatives at the central node of the grid element and then allowing the expanded equation to be as accurate as possible.

When the governing equation (121) is invoked to convert the differentiation with respect to y to that with respect to x, the expansion of equation (122) takes the following form:

$$\hat{\psi}_0 = (2\alpha_1 + 2A_0\alpha_2 + 4B_0\alpha_3)\hat{\psi}_0$$

$$+ (\alpha_1 + 2A_2\alpha_2 + 4B_2\alpha_3)(\hat{\psi}_{xx})_0 \Delta x^2$$

$$+ \left(\frac{1}{12}\alpha_1 + 2A_4\alpha_2 + 4B_4\alpha_3\right)(\hat{\psi}_{xxxx})_0 \Delta x^4 + \cdots \qquad (142)$$

where

$$A_0 = 1 - \frac{1}{2}\delta^2 + \frac{1}{24}\delta^4 + \cdots \qquad (143)$$

$$A_2 = \left(-\frac{1}{2} + \frac{1}{12}\delta^2\right)\epsilon^2 + \cdots \qquad (144)$$

$$A_4 = \frac{1}{24}\epsilon^4 + \cdots \qquad (145)$$

$$B_0 = A_0 = 1 - \frac{1}{2}\delta^2 + \frac{1}{24}\delta^4 + \cdots \qquad (146)$$

$$B_2 = \frac{1}{2}A_0 + A_2 = \frac{1}{2}(1 - \epsilon^2) - \frac{1}{12}(3 - \epsilon^2)\delta^2 + \cdots \qquad (147)$$

$$B_4 = \frac{1}{24}A_0 + \frac{1}{2}A_2 + A_4 = \frac{1}{24}(1 - 6\epsilon^2 + \epsilon^4) + \cdots \qquad (148)$$

The validity of equation (142) then requires

$$2\alpha_1 + 2A_0\alpha_2 + 4B_0\alpha_3 = 1 \qquad (149)$$

$$\alpha_1 + 2A_2\alpha_2 + 4B_2\alpha_3 = 0 \qquad (150)$$

$$\frac{1}{12}\alpha_1 + 2A_4\alpha_2 + 4B_4\alpha_3 = 0 \qquad (151)$$

Solving these linear equations leads to

… NUMERICAL SOLUTION OF COASTAL WATER WAVE EQUATIONS …

$$\alpha_1 = \frac{A_2B_4 - A_4B_2}{\frac{1}{12}A_0B_2 + 2A_2B_4 + A_4B_0 - A_0B_4 - \frac{1}{12}A_2B_0 - 2A_4B_2} \tag{152}$$

$$\alpha_2 = \frac{\frac{1}{24}B_2 - \frac{1}{2}B_4}{\frac{1}{12}A_0B_2 + 2A_2B_4 + A_4B_0 - A_0B_4 - \frac{1}{12}A_2B_0 - 2A_4B_2} \tag{153}$$

$$\alpha_3 = \frac{\frac{1}{4}A_4 - \frac{1}{48}A_2}{\frac{1}{12}A_0B_2 + 2A_2B_4 + A_4B_0 - A_0B_4 - \frac{1}{12}A_2B_0 - 2A_4B_2} \tag{154}$$

As equations (143)–(148) are substituted and the unnecessarily high-order terms are neglected, equations (152)–(154) become

$$\alpha_1 = \frac{-6\epsilon^2 + 30\epsilon^4 + \epsilon^2\delta^2 - 3\epsilon^4\delta^2}{60\epsilon^2 + 60\epsilon^4 - 6\delta^2 - 34\epsilon^2\delta^2 - 6\epsilon^4\delta^2 + 3\delta^4 + 3\epsilon^2\delta^4} \tag{155}$$

$$\alpha_2 = \frac{30\epsilon^2 - 6\epsilon^4 - 3\delta^2 + \epsilon^2\delta^2}{60\epsilon^2 + 60\epsilon^4 - 6\delta^2 - 34\epsilon^2\delta^2 - 6\epsilon^4\delta^2 + 3\delta^4 + 3\epsilon^2\delta^4} \tag{156}$$

$$\alpha_3 = \frac{6\epsilon^2 + 6\epsilon^4 - \epsilon^2\delta^2}{120\epsilon^2 + 120\epsilon^4 - 12\delta^2 - 68\epsilon^2\delta^2 - 12\epsilon^4\delta^2 + 6\delta^4 + 6\epsilon^2\delta^4} \tag{157}$$

These coefficients define the *optimal finite difference scheme* for equation (121). It is obviously of the sixth order in general.

It is known that the accuracy of the optimal finite difference scheme can be further enhanced in case Δx and Δy are of different order in magnitude (i.e., $\epsilon \gg 1$ or $\epsilon \ll 1$). When $\epsilon \gg 1$, the higher-order terms of Δy can be taken into account to yield

$$A_0 = 1 - \frac{1}{2!}\delta^2 + \frac{1}{4!}\delta^4 - \cdots = \cos\delta \tag{158}$$

$$A_2 = \epsilon^2\left(-\frac{1}{2!} + \frac{2}{4!}\delta^2 - \frac{3}{6!}\delta^4 + \cdots\right) = -\frac{\epsilon^2}{2\delta}\sin\delta \tag{159}$$

$$A_4 = \epsilon^4\left(\frac{1}{4!} + \frac{3}{6!}\delta^2 + \frac{6}{8!}\delta^4 - \cdots\right) = -\frac{\epsilon^4}{8\delta^2}\left(\cos\delta - \frac{1}{\delta}\sin\delta\right) \tag{160}$$

instead of equations (143), (144), and (145). Since $B_0, B_2, B_4, \cdots$ can be derived from A_0, $A_2, A_4, \cdots$, the coefficients of the scheme become

$$\alpha_1 = \frac{1}{2}\frac{9\epsilon^2\delta - 3\epsilon^2\delta\cos 2\delta \quad 3\epsilon^2 \sin 2\delta \quad \delta^2 \sin 2\delta}{9\epsilon^2\delta \quad 3\epsilon^2\delta\cos 2\delta - 3\epsilon^2\sin 2\delta + 5\delta^2\sin 2\delta} \tag{161}$$

$$\alpha_2 = \frac{3\epsilon^2\delta\cos\delta - 3\epsilon^2\sin\delta + 5\delta^2\sin\delta}{9\epsilon^2\delta - 3\epsilon^2\delta\cos 2\delta - 3\epsilon^2\sin 2\delta + 5\delta^2\sin 2\delta} \tag{162}$$

$$\alpha_3 = -\frac{1}{2} \frac{3\epsilon^2\delta\cos\delta - 3\epsilon^2\sin\delta - \delta^2\sin\delta}{9\epsilon^2\delta - 3\epsilon^2\delta\cos 2\delta - 3\epsilon^2\sin 2\delta + 5\delta^2\sin 2\delta} \quad (163)$$

As the truncation error is differently ordered in terms of Δx and of Δy, the scheme defined by equations (161)–(163) is no longer symmetric with respect to x and y. This can be readily confirmed since $\alpha_1 \neq \alpha_2$ at $\epsilon = 1$.

Comparison of Various Schemes. The classical finite difference scheme for the reduced mild-slope wave equation is the *five-point scheme* (Lapidus and Pinder, 1982). When recast into (122), the five-point scheme requires

$$\alpha_1 = \frac{\epsilon^2}{2 + 2\epsilon^2 - \delta^2} \quad (164)$$

$$\alpha_2 = \frac{1}{2 + 2\epsilon^2 - \delta^2} \quad (165)$$

$$\alpha_3 = 0 \quad (166)$$

Since α_3 vanishes, the scheme involves only five nodes of the grid element.

Nine-point schemes with the fourth-order accuracy were recommended by Rosser (1975) and recently by Arad et al. (1996). These schemes can also be represented by equation (122). For *Rosser's scheme*, the coefficients are expressed by

$$\alpha_1 = \frac{10\epsilon^2 - 2 + \delta^2}{20 + 20\epsilon^2 - 8\delta^2} \quad (167)$$

$$\alpha_2 = \frac{10 - 2\epsilon^2 + \delta^2}{20 + 20\epsilon^2 - 8\delta^2} \quad (168)$$

$$\alpha_3 = \frac{1 + \epsilon^2}{20 + 20\epsilon^2 - 8\delta^2} \quad (169)$$

The coefficients of various finite difference schemes over square grid elements, i.e., at $\epsilon = 1$, are comparatively plotted in Figures 1.5 and 1.6. It is noted from these figures that when δ is less than about 0.5, the nine-point schemes (the optimal scheme, the finite analytic scheme, and Rosser's scheme) are fairly close to each other. The five-point scheme is, however, intrinsically different from the others. As δ becomes larger than about 1, all the schemes show rapid change of the coefficients with δ. A numerical solution of the reduced mild-slope wave equation may thus depend on the grid size to a large extent. This implies that a grid that is so coarse that the value of δ becomes larger than about 1 should never be employed in practical computations. In theory, this means that local approximations involved in the finite difference schemes are accurate only within a neighborhood whose radius is about one-tenth of the local wavelength.

When visualizing the variation of α_1, α_2, and α_3 at different values of ϵ, one may also find that the optimal scheme and the fourth-order schemes give negative α_2 when ϵ is

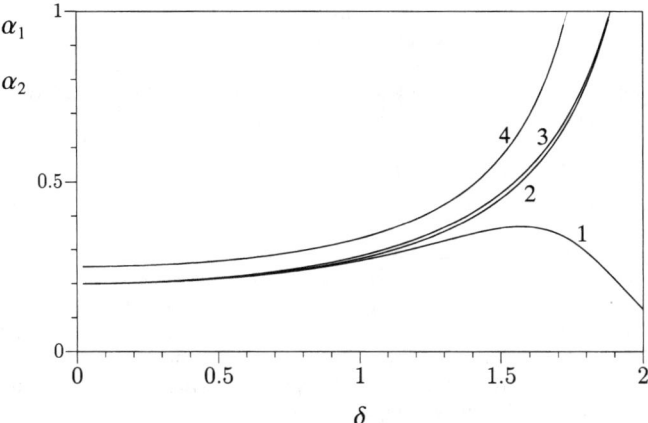

FIGURE 1.5. Coefficients α_1 and α_2 of (1) the optimal scheme, (2) the finite analytic scheme, (3) Rosser's scheme, and (4) the five-point scheme, for the reduced mild-slope wave equation.

large, which is not preferred from the numerical point of view (Chwang and Chen, 1987). On the other hand, the coefficients of the finite analytic scheme are always positive.

It has also been shown that the finite difference equation system is ill-conditioned if a fine grid is adopted. The finer the grid, the worse the condition of the equation system. This inherent contradiction in the finite difference method for boundary value problems should be taken into account when one tries to determine the grid size. On the one hand, the grid should be fine enough so that the truncation error of the scheme can be controlled. On the other hand, the grid should not be allowed to be so fine so that system of the difference equation becomes ill-conditioned.

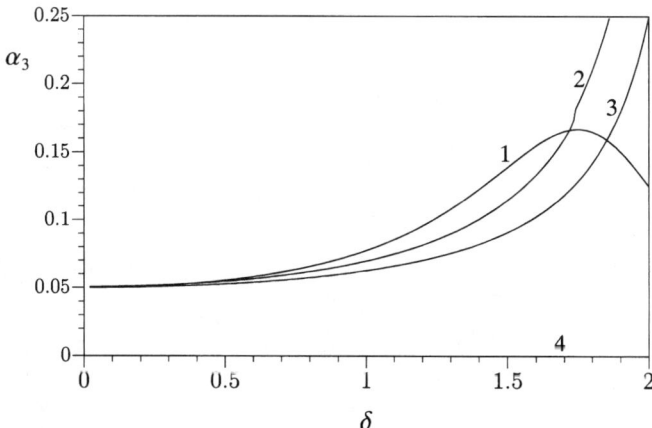

FIGURE 1.6. Coefficient α_3 of (1) the optimal scheme, (2) the finite analytic scheme, (3) Rosser's scheme, and (4) the five-point scheme, for the reduced mild-slope wave equation.

3.1.4. Applications. There are quite a number of reports on successful applications of the five-point scheme to the computation of coastal water wave problems. Williams et al. (1980) solved the original form of the mild-slope wave equation to study the diffraction of waves by a submerged shoal. They found their numerical results agreed very well with experimental data and also with the solution by the ray method, but outside the caustic. Panchang et al. (1991) studied, in addition to wave diffraction by a shoal, also wave-induced oscillation in a rectangular harbor and found that the numerical solution on the resonance in the harbor compares well with the classical solution by the boundary integral equation method.

Yu (1996a; 1998a) applied the optimal scheme and the finite analytic scheme to the computation of wave diffraction by the spherical shoal in a wave channel, as shown in Figure 1.7, which has been extensively studied by Ito and Tanimoto (1972) with different methods. The shoal covers a circular region bounded by $x^2 + y^2 = 0.8^2$ (all in meters). The water depth is 0.15 m over the flat bottom and 0.05 m over the top of the shoal. The incident wave propagates in the positive x direction with a wavelength $L = 0.40$ m before being diffracted. The incident wave height H_0 varies from 0.6 cm to 1.4 cm. Figure 1.8 shows the wave height distribution computed with the optimal scheme. The numerical result was obtained over a square grid mesh with $\Delta x = \Delta y = 0.025$ m. The incidence boundary was given by a forced oscillation, and a sponge layer was employed in the downwave margin. Comparison of the numerical results with the experimental data obtained by Ito and Tanimoto (1972) at different cross-sections is shown in Figures 1.9 and 1.10 with satisfactory agreement. For reference, the numerical solution by the five-point scheme is also displayed.

3.2. Finite Element Methods for MSWE. In this section the finite element methods (FEM) applied to the mild-slope wave equation are reviewed:

$$(cc_g \hat{\eta}_x)_x + (cc_g \hat{\eta}_y)_y + \kappa^2 cc_g \hat{\eta} = 0 \qquad (170)$$

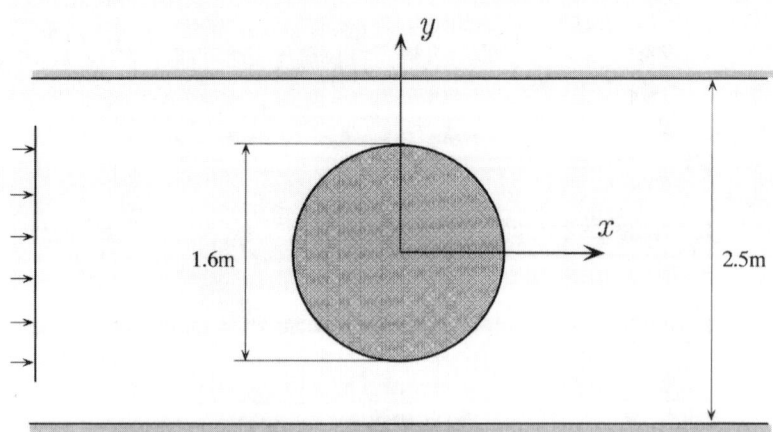

FIGURE 1.7. Submerged shoal in a wave channel.

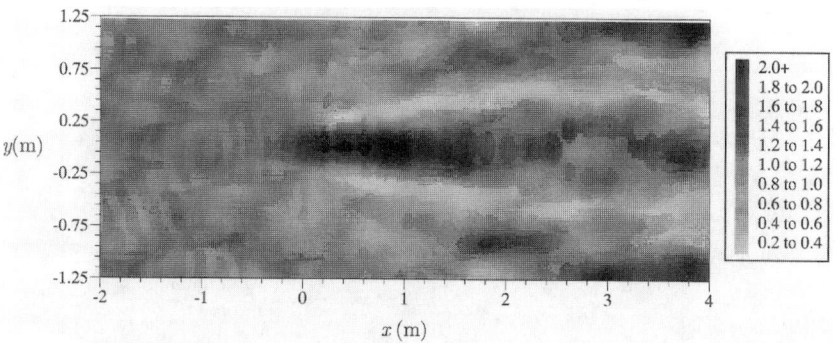

FIGURE 1.8. Wave height over the submerged shoal.

The domain of definition of $\hat{\eta}$ is denoted by Ω and the boundary of Ω by Γ. It is known that boundary conditions that can be readily incorporated into a finite element formulation include

$$\hat{\eta} = \Psi \qquad (\text{on } \Gamma_1) \tag{171}$$

$$\frac{\partial \hat{\eta}}{\partial n} + \lambda \hat{\eta} = J \qquad (\text{on } \Gamma_2) \tag{172}$$

where $\Gamma_1 + \Gamma_2 = \Gamma$; λ and J are, in general, complex-valued constants related to both the physical boundary properties and the local wave parameters. It has been recognized that,

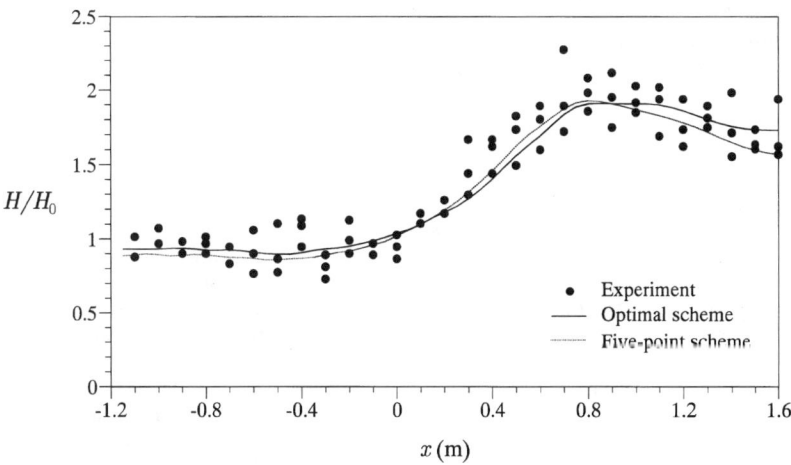

FIGURE 1.9. Wave height over the submerged shoal at $y = 0$.

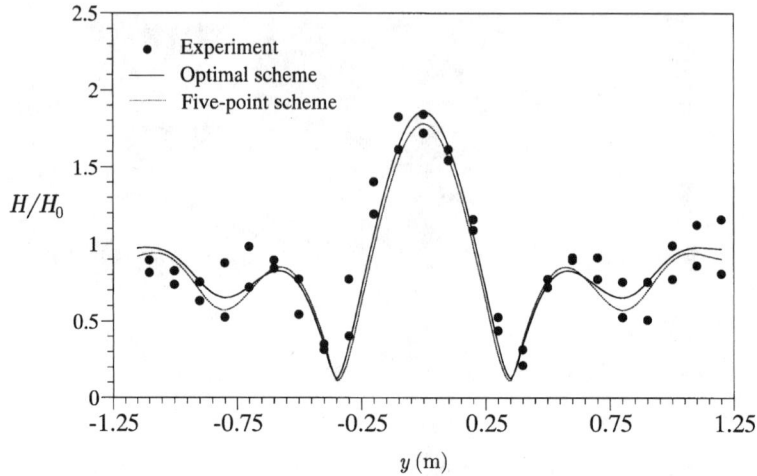

FIGURE 1.10. Wave height behind the submerged shoal at $x = 1.2$ m.

by carefully choosing the expressions for Ψ, λ, and J, many boundaries encountered in coastal and ocean engineering practice, including most of the those mentioned in the previous section on mathematical models, can be formulated as either equation (171) or equation (172).

3.2.1. Integral Representations and FE Equations. A finite element method is necessarily based on an *integral representation* of the physical problem or an equivalent global principle. It has become well known that two methods, namely, the *variational method* and the *weighted residual method*, are useful in converting the mild-slope wave equation (170) into such an integral formulation.

Variational Method. It can be readily shown that an admissible solution of equation (170), being subject to the boundary condition (172), stagnates the following functional:

$$\Pi = \frac{1}{2} \int_{\Omega} [cc_g(\hat{\eta}_x^2 + \hat{\eta}_y^2) - \kappa^2 cc_g \hat{\eta}^2]\, d\Omega + \int_{\Gamma_2} cc_g \left(\frac{1}{2} \lambda \hat{\eta}^2 - J\hat{\eta} \right) d\Gamma \qquad (173)$$

With the finite element method one begins with seeking for an approximant of Π which is expressed as a function of the values of $\hat{\eta}$ at prescribed nodes that possess a certain distribution in Ω. A relevant procedure of extremalization is then expected to yield an algebraic equation system, called the *finite element equations*, which result in a numerical solution of the problem concerned. Specifically, the domain Ω is discretized into finite and, if necessary, infinite elements with regular shapes. In each element, *interpolation functions* are introduced to express $\hat{\eta}$ as a weighted average of these functions, with the nodal values of $\hat{\eta}$ being the weight. The interpolation functions are also called *element shape functions* because, in many cases, the relative nodal coordinates or the shape of the element uniquely determines them. By virtue of the element interpolation functions, it is possible to construct *global shape functions* associated to each node (Hughes, 1987) so that

$$\hat{\eta} = \mathbf{L}^T \hat{\boldsymbol{\eta}} = \hat{\boldsymbol{\eta}}^T \mathbf{L} \tag{174}$$

where $\hat{\boldsymbol{\eta}}$ is the vector formed by all the nodal values of $\hat{\eta}$ and $\mathbf{L}$ is the vector formed by the global shape functions; the superscript T denotes the transpose of matrices. Substituting equation (174) into equation (173) gives

$$\Pi = \frac{1}{2} \int_\Omega \hat{\boldsymbol{\eta}}^T [cc_g(\mathbf{L}_x^T \mathbf{L}_x + \mathbf{L}_y^T \mathbf{L}_y) - \kappa^2 cc_g \mathbf{L}^T \mathbf{L}] \hat{\boldsymbol{\eta}} \, d\Omega$$

$$+ \int_{\Gamma_2} cc_g \left(\frac{1}{2} \lambda \hat{\boldsymbol{\eta}}^T \mathbf{L}^T \mathbf{L} \hat{\boldsymbol{\eta}} - J\mathbf{L}^T \hat{\boldsymbol{\eta}} \right) d\Gamma \tag{175}$$

Equation (175) can also be written in the following bilinear form:

$$\Pi = \frac{1}{2} \hat{\boldsymbol{\eta}}^T \mathbf{K} \hat{\boldsymbol{\eta}} - \mathbf{F}^T \hat{\boldsymbol{\eta}} \tag{176}$$

where

$$\mathbf{K} = \int_\Omega [cc_g(\mathbf{L}_x^T \mathbf{L}_x + \mathbf{L}_y^T \mathbf{L}_y) - \kappa^2 cc_g \mathbf{L}^T \mathbf{L}] \, d\Omega + \int_{\Gamma_2} \lambda cc_g \mathbf{L}^T \mathbf{L} \, d\Gamma \tag{177}$$

$$\mathbf{F} = \int_{\Gamma_2} Jcc_g \mathbf{L} \, d\Gamma \tag{178}$$

Numerical evaluation of $\mathbf{K}$ and $\mathbf{F}$ is often a routine assembling of element integrals, to which analytic expressions may be derived or an appropriate numerical quadrature algorithm applied. For Π expressed by equation (176), the stationary condition of functional is equivalent to

$$\frac{\delta \Pi}{\delta \hat{\boldsymbol{\eta}}} = 0 \tag{179}$$

Substituting equation (176) into equation (179) and noting that $\mathbf{K}$ is symmetric, one finally obtains

$$\mathbf{K} \hat{\boldsymbol{\eta}} = \mathbf{F} \tag{180}$$

Note that the equations resulting from the variation with respect to any nodal value of $\hat{\eta}$ on boundary Γ_1 should be replaced by the forced boundary condition equation (171). After this modification, an appropriate linear equation solver, which must take into account the spareness of the coefficient matrix $\mathbf{K}$ for the purpose of efficiency, will finally lead to the numerical solution of interest.

Weighted Residual Method. The weight residual method is the most generally applicable method for converting differential equations into a FEM-oriented integral representation of the relevant problem. It requires the difference between the numerical solution to be sought and the exact solution, i.e., the *residual*, to vanish within the solution domain in an overall sense. For the present problem this is written as

$$\int_\Omega w[(cc_g\hat{\eta}_x)_x + (cc_g\hat{\eta}_y)_y + \kappa^2 cc_g\hat{\eta}]\, d\Omega - \int_{\Gamma_2} wcc_g\left(\frac{\partial \eta}{\partial n} + \lambda\hat{\eta} - J\right) d\Gamma = 0 \qquad (181)$$

where the *weighting function* w is necessarily an arbitrary function of the coordinates. It is not difficult to show that, if w is arbitrary, equation (181) is equivalent to the mild-slope wave equation (170), subject to the boundary condition (172). Upon partial integration, equation (181) can be transformed into

$$\int_\Omega [cc_g(w_x\hat{\eta}_x + w_y\hat{\eta}_y) - \kappa^2 cc_g w\hat{\eta}]\, d\Omega + \int_{\Gamma_2} wcc_g(\lambda\hat{\eta} - J)\, d\Gamma = 0 \qquad (182)$$

which is called a *weak formulation* because of the eased requirement on the continuity of the numerical solution of $\hat{\eta}$ to ensure the integrability of the basic formulation. Again, the numerical solution of $\hat{\eta}$ is expressed by equation (174). Note that the components of the vector **L** form a base set by which all possible numerical solutions of $\hat{\eta}$ can be expressed. It is then enough to let equation (181) be satisfied by each component of **L** instead of an arbitrary function w to ensure the validity of the integral formulation. This is known as the *Galerkin method*. When the Galerkin method is employed, equation (182) leads to

$$\int_\Omega [cc_g(\mathbf{L}_x^T\mathbf{L}_x + \mathbf{L}_y^T\mathbf{L}_y) - \kappa^2 cc_g\mathbf{L}^T\mathbf{L}]\hat{\eta}\, d\Omega + \int_{\Gamma_2} cc_g(\lambda\mathbf{L}^T\mathbf{L}\hat{\eta} - J\mathbf{L}^T)\, d\Gamma = 0 \qquad (183)$$

which is identical to equation (180).

3.2.2. Finite and Infinite Elements. The most flexible and tactical step in the application of a finite element method to the mild-slope wave equation is probably the selection of the element type and the related interpolation functions. Since, for short-crested wave problems, the dimension of an ordinary element should usually be much smaller than the local wavelength to ensure the accuracy of the numerical solution. This implies that a practical problem with relatively large computational domain may be impossible to deal with without a particular consideration to the element representation. In addition, coastal water wave problems are usually formulated as exterior problems, the solution domain of which extends to infinity. Careful selection of elements is also necessary to properly account for the wave behavior at infinity.

Conventional Finite Elements. With a brief inspection of the previous work, it becomes evident that the conventional triangular elements with linear interpolation functions and the quadratic elements with bilinear interpolation functions have been mostly preferred when finite element methods were applied to coastal water wave problems. An indisputable reason for this situation is the easiness and robustness of the less complicated elements. However, high-order elements are certainly necessary when practical problems with huge computational domains are confronted. One may employ the conventional high-order elements (Zienkiewicz and Morgan, 1983). But a noted factor that reduces the effectiveness of the conventional high-order techniques is that most of them do not take into account the periodicity of the wave phenomena in space, which is fatal when the element dimension is not necessarily smaller than a few fractions of the local wavelength. Although there are reports on periodic interpolation functions defined on rectangular elements (Yu and Togashi, 1995) that are obviously superior to the bilinear functions, further development of effective high-order elements is still necessary.

Infinite Elements. Infinite elements have been developed to deal with exterior water wave problems with radiation at infinity. A number of researchers have contributed to this topic and a few practically useful algorithms have been reported.

Chen (1990) proposed a two-node infinite element as shown in Figure 1.11. The interpolation functions associated with each node are defined by

$$N_i = L_i(\theta)R(r) \tag{184}$$

where $i = 1, 2$, and

$$L_1(\theta) = \frac{\theta_2 - \theta}{\theta_2 - \theta_1} \tag{185}$$

$$L_2(\theta) = 1 - L_1(\theta) \tag{186}$$

$$R(r) = \sqrt{\frac{r_0}{r}}\, e^{i\kappa(r-r_0)} \tag{187}$$

Chen's two-node infinite element has a few advantages. It is simple in form. In addition, the element integrals that appear during the course of forming the finite element equations can often be analytically evaluated. However, the element is applicable only in the region where water depth is constant or almost infinite. The two-node element can be easily generalized to the case with additional nodes in between nodes 1 and 2. The generalized interpolation functions include the high-order correspondents of the linear Lagrangian interpolation functions L_1 and L_2 as multipliers.

Bettess and Zienkiewicz (1977) recommended a nine-node infinite element shown in Figure 1.12, where the three far nodes [(3,1), (3,2), and (3,3)] should be located far enough from the others. The interpolation function associated with each node is assumed to be

$$N_{ij} = L_i(s)R_j(r) \tag{188}$$

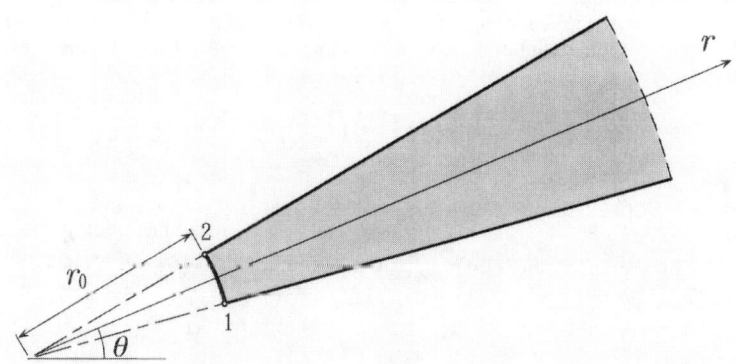

FIGURE 1.11. A two-node infinite element.

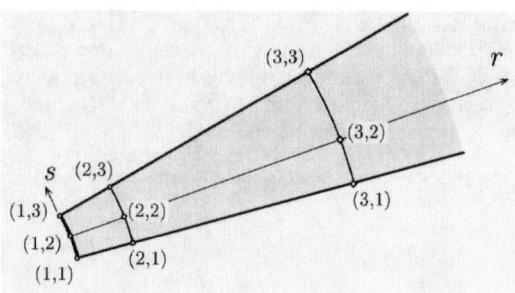

FIGURE 1.12. A nine-node infinite element.

where $i, j = 1, 2, 3$; $L_i(s)$ represent the quadratic Lagrangian interpolation functions; and

$$R_j(r) = L_j(r)e^{r_j - r}e^{ikr} \qquad (j = 1, 2) \tag{189}$$

$$R_3(r) = 1 - R_1 - R_2 \tag{190}$$

In equation (189) the summation convention does not apply. When evaluating element integrals one may need the following identity:

$$\int_0^\infty r^n e^{(-\alpha + i\beta)r}\, dr = \left(\frac{\alpha + i\beta}{\alpha^2 + \beta^2}\right)^n (n-1)! \tag{191}$$

where α and β are constants.

It may be worthwhile to point out that Bettess and Zienkiewicz's nine-node infinite element is effective only in terms of the implementation of radiation conditions. The numerical solutions within the infinite elements are of no physical meaning because the decay rate of the physical variables in the r direction of the element is artificial.

Zienkiewicz et al. (1983) proposed another useful nine-node infinite element based on a local mapping in the radial direction of the element, as illustrated in Figure 1.13. The mapping can also be expressed by

$$r = \Xi(\xi) \equiv (2r_1 - r_2)N_0 + r_2 N_2 \tag{192}$$

where ξ is a new coordinate and

$$N_0 = -\frac{\xi}{1 - \xi} \tag{193}$$

$$N_2 = 1 - N_0 \tag{194}$$

The same mapping should also be applied to the physical variables defined over the infinite element. It can be readily shown that a polynomial in terms of ξ:

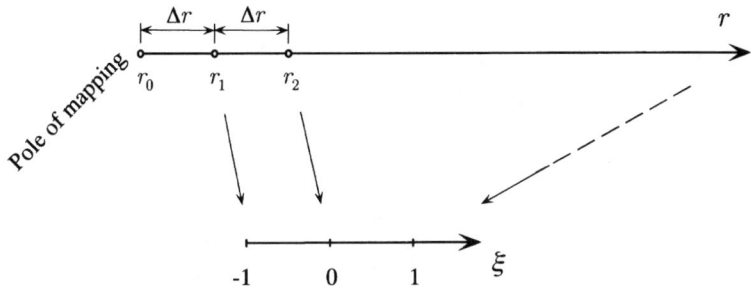

FIGURE 1.13. Ξ-mapping.

$$P(\xi) = \alpha_0 + \alpha_1 \xi + \alpha_2 \xi^2 + \cdots \qquad (195)$$

with $\alpha_0 + \alpha_1 + \alpha_2 + \cdots = 0$, will be mapped onto a reciprocal function in the physical coordinates:

$$P(r) = \frac{\beta_1}{r} + \frac{\beta_2}{r^2} + \cdots \qquad (196)$$

If the following mapping for the s coordinate is employed:

$$s = s_j L_j(\zeta) \qquad (197)$$

the interpolation function associated with each node of the infinite element can then be reasonably expressed by

$$N_{ij} = L_i(\zeta) \Xi_j(\xi) M(\xi) \qquad (198)$$

where $M(\xi)$ is a factor designated to capture the behavior of the wave motion in the infinite element:

$$M(\xi) = \sqrt{r} e^{i\kappa r} = \sqrt{\frac{2\Delta r}{1-\xi}} \exp\left[\frac{2i\kappa\Delta r}{1-\xi}\right] \qquad (199)$$

The Super Element. The super element has also been developed to deal with exterior problems with a radiation condition at infinity. The concept is essentially straightforward. By a circular interface Γ_0 the solution domain is divided into the inner subregion Ω_0 and outer subregion Ω_s, as illustrated in Figure 1.14. The inner subregion is to be discretized into conventional finite elements. In the outer subregion, which can be of a constant or infinite water depth, one intends to find a parametric analytic solution or a boundary integral equation on the complex amplitude of the water surface elevation involved in the formulation. The outer subregion in which such a procedure succeeds is usually called a *super element*. When a super element is involved in the analysis, a finite element method has been called the *hybrid method*. A detailed review of the hybrid methods may be found in Mei (1978).

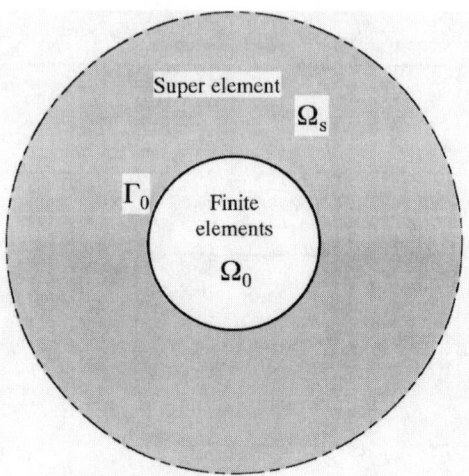

FIGURE 1.14. A super element.

Following the conventional practice (Berkhoff, 1972; Chen and Mei, 1974) one may enforce the following formal condition on Γ_0 to establish the finite element equation at each node on Γ_0 and in the inner subregion:

$$\frac{\partial \hat{\eta}}{\partial n} = J \tag{200}$$

The consequent finite element equation system should then include the nodal values of J on Γ_0 as unknowns and possess of the following form:

$$\mathbf{K}\hat{\eta} = \mathbf{G}\mathbf{J} + \mathbf{F} \tag{201}$$

where $\mathbf{J}$ is a vector formed by the nodal values of J on Γ_0 and $\mathbf{G}$ is a coefficient matrix.

In constant water depth, the mild-slope wave equation reduces to the Helmholtz equation. The water surface in the super element can thus be represented by

$$\hat{\eta} = \sum_{n=0}^{\infty} (A_n \cos n\theta + B_n \sin n\theta) H_n^{(1)}(\kappa r) \tag{202}$$

where $H_n^{(1)}$ is the Hankel function of the first kind and nth order; r is measured from the center of Ω_0; A_n and B_n are constants to be determined. By equation (202) it is possible to derive each nodal value of J as a linear function of A_n and B_n. Let the nodal values of $\hat{\eta}$ expressed by equation (202) be identical to those involved in equation (201) at the same nodes on Γ_0. Another set of linear equations that relate the nodal values of $\hat{\eta}$ on Γ_0 to A_n and B_n is obtained. A simultaneous solution of equation (201) with these two sets of linear equations involving A_n and B_n will finally result in the numerical solution of $\hat{\eta}$ on Γ_0 and in Ω_0, and also the values of A_n and B_n. Once A_n and B_n are solved, $\hat{\eta}$ in the super element may also be determined upon demand.

An alternative to the parametric analytic solution in the super element is known to be the boundary integral equation for the exterior:

$$\omega_p \hat{\eta} = \frac{i}{4} \int_{\Gamma_0} \left[\hat{\eta} \frac{\partial H_0^{(1)}(\kappa r)}{\partial n} - H_0^{(1)}(\kappa r) \frac{\partial \hat{\eta}}{\partial n} \right] d\Gamma \qquad (203)$$

where $\hat{\eta}$ on the left hand side is evaluated at the singular point of $H_0^{(1)}$, i.e., the source; r is the distance between the source and the relevant boundary point; ω_p is a coefficient that equals to 1 if the source is within Ω_0 and equals ½ if it is on Γ_0; the positive direction of Γ_0 is the anticlockwise direction. As equation (203) is applied to each node on Γ_0 and the boundary integral is evaluated by a boundary element method (e.g., Becker, 1992), a linear equation system correlating the nodal values of $\hat{\eta}$ and J on Γ_0 can be obtained. A simultaneous solution of this equation system with equation (201) also yields the numerical solution of $\hat{\eta}$ on Γ_0 and in Ω_0, along with the nodal values of J.

3.2.3. Applications. The finite element methods have been the most popular choice in solving the mild-slope wave equation for engineering purposes. Berkhoff (1972) employed the hybrid method to solve various problems with combined diffraction and refraction in his original paper on the development of the mild-slope wave equation. Chen and Mei (1974) applied the hybrid element method to the computation of harbor oscillations. Their numerical results were in very good agreement with experimental data and with the well-known results by the boundary integral equation methods as well. Bettess and Zienkiewicz (1977), Bettess et al. (1984), and Behrendt and Jonsson (1984) applied the infinite element method. Sakai and Tsukioka (1975) investigated a harbor oscillation, Pos and Kilner (1987) studied the diffraction by a breakwater gap, and Houston (1978) simulated the interaction of tsunamis with the Hawaiian Islands, all by the finite element method.

Yu and Togashi (1996) computed the infragravity wave-induced oscillations in Nagasaki Bay on the western cost of Japan. The bay is famous for a resonant oscillation called *abiki* by the local people. It is open to the sea through a narrow entrance and connected to a mid-sized rive at the far end. The still-water depth varies from about 40 m at the entrance to about 10 m at the river mouth. Figure 1.15 demonstrates the triangular mesh used in the computation. Figure 1.16 is the computed amplification factor at Ohato. It is noted that the most severe resonances occur at about 33.2 min, 17.9 min, 10.1 min, etc., which correspond to the natural modes of the bay at 30.60 min, 17.28 min, 10.08 min, etc., respectively. The resonant wave periods were in good agreement with observations. The slight difference between the natural periods and the resonant wave periods (the smaller the natural period, the larger the difference) is known to be an effect of the radiation damping through the bay entrance (Miles and Munk, 1961; Yu, 1996b). The amplification factor at the most-severe-resonance condition of the 30.60-minute bay mode is about 4.5, which also compares well with field data.

Yu et al. (1992b) applied a finite element method to the computation of waves with breaking. The physical model studied was the wave transformation around a detached breakwater on a sloping beach, as shown in Figure 1.17. Incident waves of 2 cm in height and 1.2 s in period were generated normal to the still-water shoreline. The computation was carried out for a half domain, owing to the symmetry of the problem in the alongshore direction. A sponge layer was placed along the lateral boundary to handle the non-reflecting condition. The bottom friction simulated the effect of a laminar boundary layer. Triangular elements with a dimension of about 1/15 of the deepwater wavelength were used. Figure 1.18 is a comparison of the computed breaking line with measured data. The

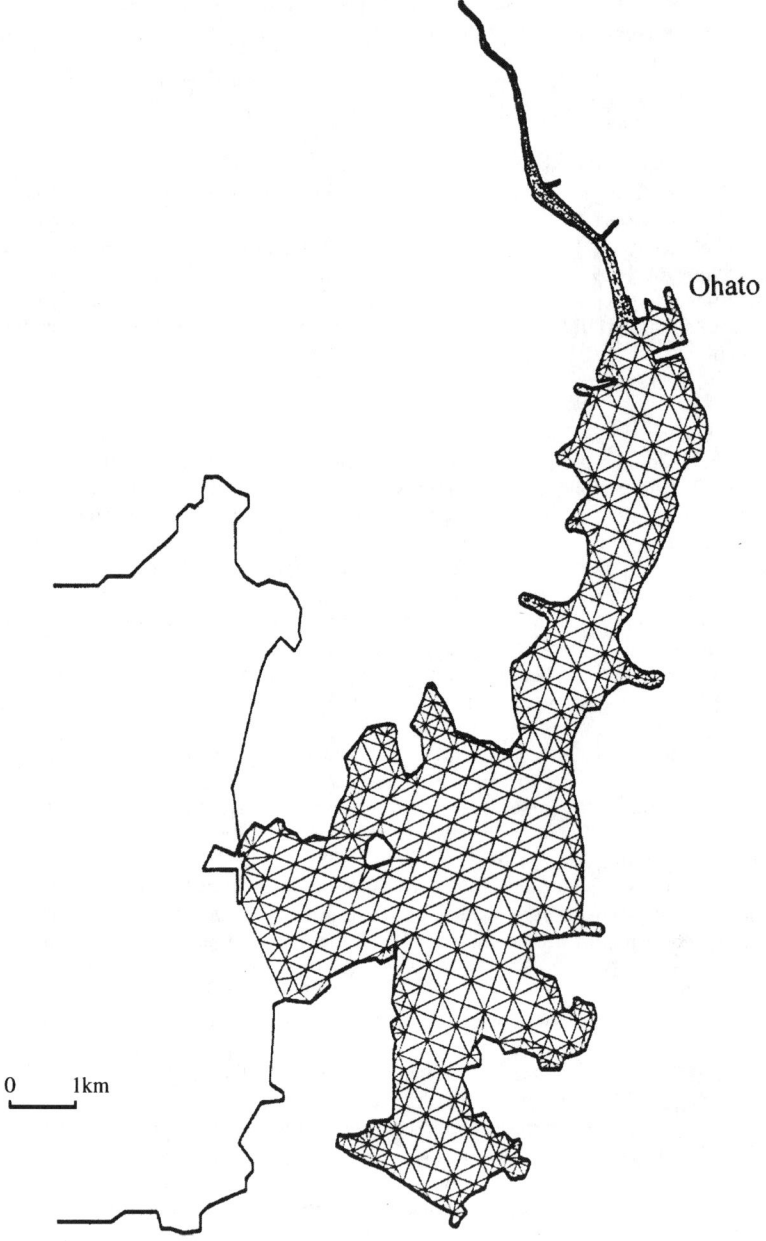

FIGURE 1.15. Computational mesh for wave-induced oscillation in Nagasaki Bay.

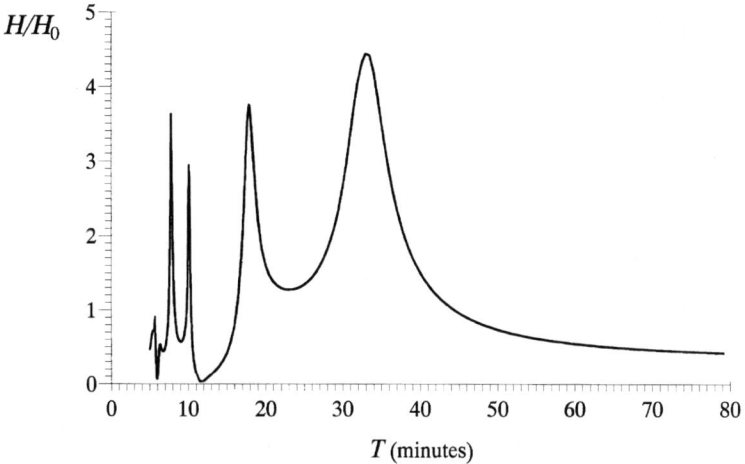

FIGURE 1.16. Amplification factor of Nagasaki Bay at Ohato.

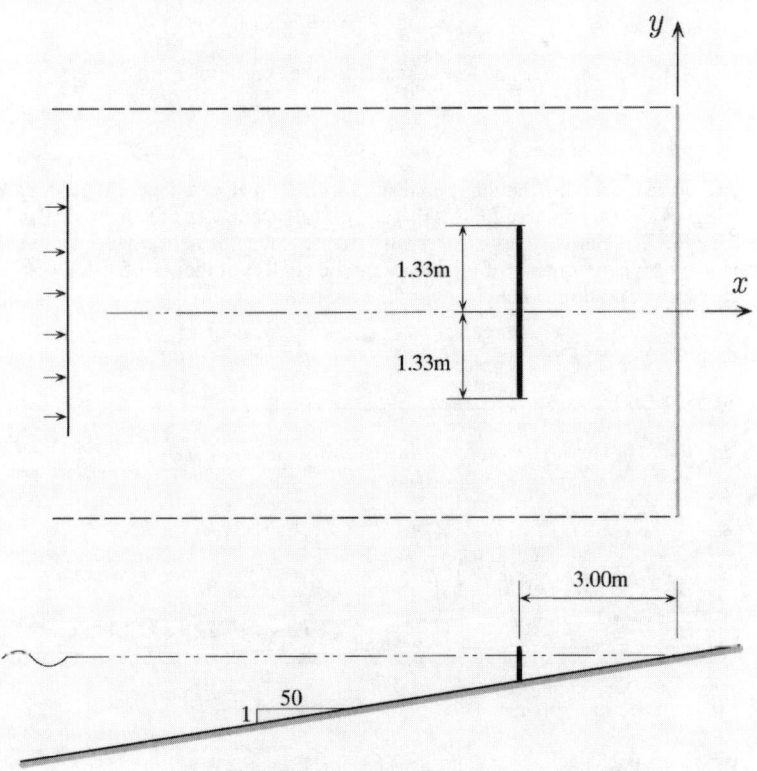

FIGURE 1.17. Physical model of a detached breakwater.

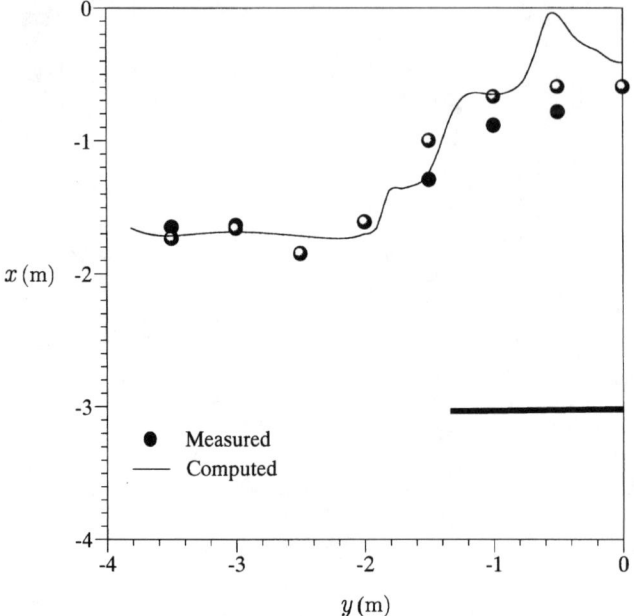

FIGURE 1.18. Breaking line behind a detached breakwater.

computed wave height distribution is compared with the measurement in Figures 1.19 and 1.1.20. It is evident from these figures that the overall accuracy of the numerical solution is acceptable. The relatively large discrepancy in the wave height in front of the breakwater seems to be a consequence of the inappropriate neglect of the wall friction that resists the vertical water motion in the vicinity of the breakwater.

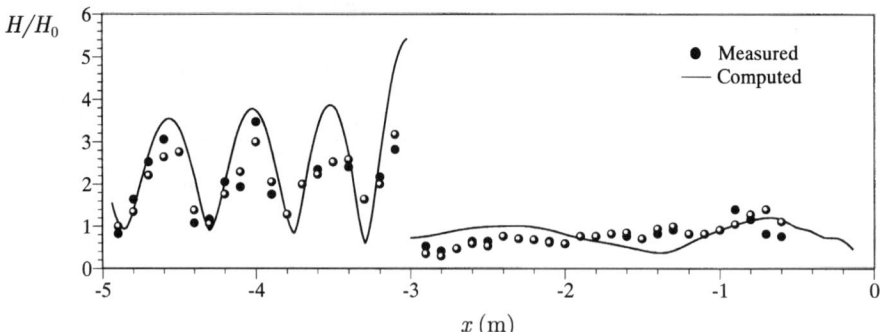

FIGURE 1.19. Wave height around the detached breakwater at $y = -1$ m.

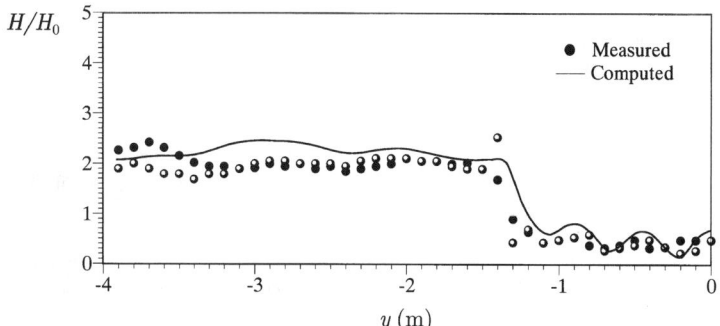

FIGURE 1.20. Wave height behind the detached breakwater at $x = -3$ m.

4. NUMERICAL MODELS FOR INITIAL-BOUNDARY VALUE PROBLEMS

4.1. Time-Dependent Linear Wave Equations

Numerical methods for linear wave equations are usually simple and well established. Most of the stable second-order schemes discussed in standard textbooks, explicit or implicit, should work well. It must be emphasized that, in general, the first-order schemes cannot be used without verification, since the numerical viscosity may result in significant nonphysical wave dissipation. Important factors affecting the selection of a scheme include the available computer resources, the easiness of implementation of the scheme, and its flexibility with various kinds of boundary conditions.

It is well recognized that there is no general superiority of either the explicit or the implicit schemes in terms of the cost performance. The explicit schemes need less computational time at each temporal step but they are usually subject to a stability condition that restricts the amount of time stepping. On the other hand, the implicit schemes are often unconditionally stable but they require much more computer resources at each temporal step. For the coastal water wave problems, particularly short waves, however, there has been a tendency for the implicit schemes to be less emphasized. This situation is somehow related to the fact that the resolution of the numerical results to a wave phenomenon is governed by the larger part of $\sigma \Delta t$ and $k \Delta x$. Therefore, the temporal step in a computation can not be much larger than $\Delta x / c$ to retain the resolution. This implies that the time step for the Courant number to depart significantly from unity should not be encouraged. In other words, the implementation of a scheme with balanced resolution may satisfy the stability condition of an explicit scheme automatically.

It is recommended that highly accurate schemes that do not necessitate a significant increase of the computational efforts should always be given the first priority. With such a high-order scheme it is possible to employ relatively large grid size to save computer resources, or the numerical error may be reduced. One may argue on the necessity of applying a high-order scheme to the wave equation that does not possess of a very high order. However, since the factors that affect the accuracy of the mathematical model and those that determine the accuracy of the numerical model are generally independent of each other, it is not appropriate to reject an accurate numerical method because of the limit of the mathematical model.

4.1.1. A Hybrid Method for MSWE. There is an evident upsurge in the development and the application of the *finite volume methods* (FVM) in the recent years. The momentum behind this is obviously the advantages of the methods. They can be used on an unstructured mesh, similar to the finite element methods, but still retain the flexibility of the finite difference methods. In addition, local conservation of mass and momentum can be assured. In this subsection, a hybrid method is presented for the time-dependent mild-slope wave equations, based on a finite volume scheme for the continuity equation and a finite difference scheme for the equation of motion. The method is essentially based on unstructured meshes.

The time-dependent mild-slope wave equations (Watanabe and Maruyama, 1986) are written as

$$\eta_t + \nabla \cdot \boldsymbol{q} = 0 \tag{204}$$

$$\boldsymbol{q}_t + \frac{c^2}{n} \nabla(n\eta) + C_d \sigma \boldsymbol{q} = 0 \tag{205}$$

Suppose the solution domain Ω has been discretized into irregular elements, with each being an arbitrary polygon. A representative element $\Delta\Omega$ is shown in Figure 1.21. The center and the vertexes of the element polygon are defined as the *nodes*, where η is to be computed. From the perpendicular bisectors of the line segments connecting each pair of the nodes, an auxiliary discretization of the solution domain is obtained. Each element resulting from the auxiliary discretization is obviously also a polygon and will be called a *control volume*. A representative control volume $\Delta\Omega^c$, with the boundary being denoted

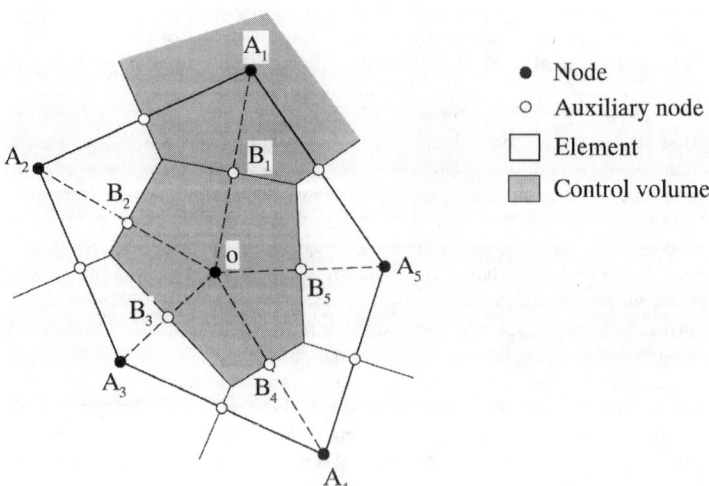

FIGURE 1.21. Unstructured mesh for the hybrid method of the time-dependent mild-slope wave equation.

by $\Delta\Gamma^c$, is also indicated in Figure 1.21. The midpoints of the line segments that connect each pair of the nodes are defined as the *auxiliary nodes*, where the velocity component is locally defined within each control volume, with the outward direction being positive. The continuity equation (204) may now be integrated over the control volume $\Delta\Omega^c$ and over the time interval $[m\Delta t, (m + 1)\Delta t]$. Referring to Gauss' theorem, this gives

$$\frac{\partial}{\partial t} \int_{\Delta\Omega^c} \eta \, d\Omega + \int_{\Delta\Gamma^c} q_n \, d\Gamma = 0 \qquad (206)$$

Using the value of η at the central node to represent the mean values of η within the control volume and the values of q_n at the auxiliary nodes to represent the mean value of q_n at the relevant side of the control volume, equation (206) yields

$$\eta_o^{m+1} = \eta_o^m - \frac{\Delta t}{\Delta S} \sum_B [q_n]_B^{m+1/2} \Delta l_B \qquad (207)$$

where the summation is carried out with respect to the number of the auxiliary nodes associated to the control volume; ΔS is the area and Δl_B is the length of the relevant side of $\Delta\Omega^c$. $[q_n]_B^{m+1/2}$ in equation (207) should be computed from the following leap-frog finite difference approximation to equation (205), projected at the normal direction of $\Delta\Gamma^c$:

$$[q_n]_B^{m+1/2} = [q_n]_B^{m-1/2} - \frac{\Delta t}{\Delta r}\left[\frac{c^2}{n}\right]_B (n_A \eta_A^m - n_o \eta_o^m) - \frac{\sigma \Delta t}{2} [C_d]_B ([q_n]_B^{m+1/2} + [q_n]_B^{m-1/2}) \qquad (208)$$

where $\Delta r = \sqrt{(x_A - x_o)^2 + (y_A - y_o)^2}$.

When the solution domain is primarily discretized into rectangular elements with nine nodes, as shown in Figure 1.22, the scheme described above reduces to the conventional staggered-grid finite difference scheme used by many researchers to solve the mild-slope wave equations (Ito and Tanimoto, 1972; Watanabe and Maruyama, 1986; Copeland, 1985). The reduced form of the scheme, written with the notations for a structured mesh, becomes

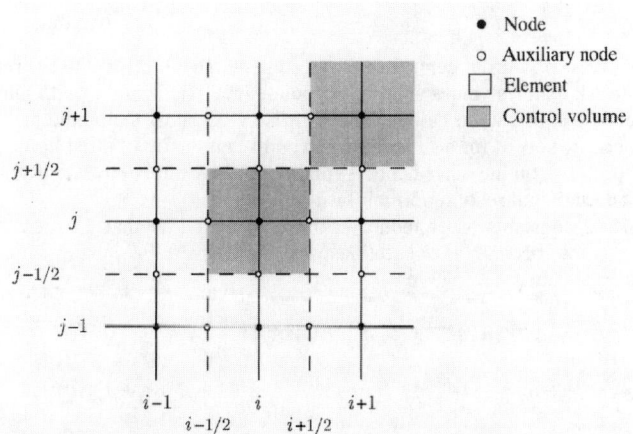

FIGURE 1.22. Staggered grid as a special case of the unstructured mesh.

$$[q_x]_{i+1/2,j}^{m+1/2} = [q_x]_{i+1/2,j}^{m-1/2} - (c_{i+1/2,j})^2 \frac{\Delta t}{\Delta x}(\eta_{i+1,j}^m - \eta_{i,j}^m) - \sigma\Delta t[C_d]_{i+1/2,j}(q_{i+1/2,j}^{m+1/2} + q_{i+1/2,j}^{m-1/2}) \quad (209)$$

$$[q_y]_{y,i+1/2}^{m+1/2} = [q_y]_{y,i+1/2}^{m-1/2} - (c_{i,j+1/2})^2 \frac{\Delta t}{\Delta y}(\eta_{i,j+1}^m - \eta_{i,j}^m) - \sigma\Delta t[C_d]_{y,j+1/2}(q_{i,i+1/2}^{m+1/2} + q_{i,j+1/2}^{m-1/2}) \quad (210)$$

$$\eta_{i,j}^{m+1} = \eta_{i,j}^m - \frac{1}{n_{i,j}} \frac{\Delta t}{\Delta x}(n_{i+1/2,j}[q_x]_{i+1/2,j}^{m+1/2} - n_{i-1/2,j}[q_x]_{i-1/2,j}^{m+1/2})$$

$$- \frac{1}{n_{i,j}} \frac{\Delta t}{\Delta y}(n_{i,j+1/2}[q_y]_{i,j+1/2}^{m+1/2} - n_{i,j-1/2}[q_y]_{i,j-1/2}^{m+1/2}) \quad (211)$$

4.1.2. High-Order ADI Method for CWE. In this subsection, Fairweather and Mitchell's (1965) method is followed to establish a high-order *alternating direction implicit* (ADI) scheme for the classical wave equation with a slowly varying celerity:

$$\eta_{tt} = c^2(\eta_{xx} + \eta_{yy}) \quad (212)$$

The general form of the scheme to be developed is assumed to be composed of a predictor and a corrector as follows:

$$\eta_{i,j}^* - 2\eta_{i,j}^m + \eta_{i,j}^{m-1} = \delta_x^2[A_1\eta_{i,j}^* + \left(B_1 - \frac{B_3}{A_1}\right)\eta_{i,j}^m + \left(C_1 - \frac{C_3}{A_1}\right)\eta_{i,j}^{m-1}]$$

$$+ \delta_y^2\left[\left(B_2 - \frac{B_4}{A_1}\right)\eta_{i,j}^m + \left(C_2 - \frac{C_4}{A_1}\right)\eta_{i,j}^{m-1}\right] \quad (213)$$

$$\eta_{i,j}^{m+1} - \eta_{i,j}^* = \delta_y^2\left(A_2\eta_{i,j}^{m+1} + \frac{B_4}{A_1}\eta_{i,j}^m + \frac{C_4}{A_1}\eta_{i,j}^{m-1}\right) + \delta_x^2\left(\frac{B_3}{A_1}\eta_{i,j}^m + \frac{C_3}{A_1}\eta_{i,j}^{m-1}\right) \quad (214)$$

where δ^2 is the second-order central-difference operator with respect to the relevant variable; the asterisk denotes values at an intermediate level; A, B, and C with numerical suffixes are scheme constants to be determined. It is evident that the predictor (213) should be simultaneously solved for all i with j fixed and is said to be implicit in the x direction. The corrector (214), on the other hand, is implicit in the y direction. The whole scheme is thus in the alternating direction implicit fashion.

The scheme constants are intended to be determnied so that the two steps of the scheme as a whole become as accurate as possible. By eliminating η^*, equations (213) and (214) can be combined to give

$$\eta_{i,j}^{m+1} - 2\eta_{i,j}^m + \eta_{i,j}^{m-1} = \delta_x^2(A_1\eta_{i,j}^{m+1} + B_1\eta_{i,j}^m + C_1\eta_{i,j}^{m-1})$$

$$+ \delta_y^2(A_2\eta_{i,j}^{m+1} + B_2\eta_{i,j}^m + C_2\eta_{i,j}^{m-1}) - \delta_x^2\delta_y^2(A_1A_2\eta_{i,j}^{m+1} + B_4\eta_{i,j}^m + C_4\eta_{i,j}^{m-1})$$

$$- \delta_x^2\delta_x^2(B_3\eta_{i,j}^m + C_3\eta_{i,j}^{m-1}) \quad (215)$$

Let all terms in equation (215) be expanded into Taylor series centered at $x = i\Delta x$, $y = j\Delta y$, and $t = m\Delta t$. In the expansions, the even derivatives of η with respect to t are converted to the derivatives of η with respect to x and y through invoking the governing equation (212). Then all terms on both sides of the equation corresponding to different derivatives become independent. By equating the coefficients of the terms with the same derivatives, one obtains two relations at the second order:

$$A_1 + B_1 + C_1 = \left(c \, \frac{\Delta t}{\Delta x} \right)^2 \equiv \mu_x^2 \tag{216}$$

$$A_2 + B_2 + C_2 = \left(c \, \frac{\Delta t}{\Delta y} \right)^2 \equiv \mu_y^2 \tag{217}$$

three relations at the third order:

$$A_1 - C_1 = 0 \tag{218}$$

$$A_2 - C_2 = 0 \tag{219}$$

three relations at the fourth order:

$$\mu_x^2 (A_1 + C_1) - 2(B_3 + C_3) = \frac{1}{6} \mu_x^2 (\mu_x^2 - 1) \tag{220}$$

$$\mu_y^2 (A_1 + C_1) + \mu_x^2 (B_3 + C_3) - 2(A_1 A_2 + B_4 + C_4) = \frac{1}{3} \mu_x^2 \mu_y^2 \tag{221}$$

$$A_2 + C_2 = \frac{1}{6} (\mu_y^2 - 1) \tag{222}$$

two relations at the fifth order:

$$C_3 = 0 \tag{223}$$

$$A_1 A_2 = C_4 \tag{224}$$

and three relations at the sixth order, which will not be written because not all of them can be satisfied in general. When the scheme constants are determined from equations (216)–(224), equations (213) and (214) give at least a fifth-order scheme for equation (212). Solving equations (216)–(224) one obtains

$$A_1 = C_1 = \frac{1}{12} (\mu_x^2 - \gamma) \tag{225}$$

$$B_1 = \frac{1}{6} (5\mu_x^2 + \gamma) \tag{226}$$

$$A_2 = C_2 = \frac{1}{12}(\mu_y^2 - 1) \tag{227}$$

$$B_2 = \frac{1}{6}(5\mu_y^2 + 1) \tag{228}$$

$$B_3 = \frac{1-\gamma}{12}\mu_x^2 \tag{229}$$

$$C_3 = 0 \tag{230}$$

$$B_4 = -\frac{1}{72}[\mu_x^2 \mu_y^2 + 5\mu_x^2 + 5\gamma\mu_y^2 + \gamma] \tag{231}$$

$$C_4 = \frac{1}{144}[\mu_x^2\mu_y^2 - \mu_x^2 - \gamma\mu_y^2 + \gamma] \tag{232}$$

where γ is a free parameter. A different value of γ leads to a different form of the fifth-order scheme. The value of γ can really be arbitrary only as far as the accuracy of the scheme is concerned. However, one must be aware of that the value of γ has a direct relation with the scheme stability condition. Figure 1.23 shows the difference in the stability range of the scheme at $\gamma = 1$ and $\gamma = 1 + \mu_x^2$. In fact, the value of γ must be chosen to ensure a stable scheme for μ_x and μ_y in a reasonable range.

At $\gamma = 1$, the scheme illustrated, if applied to a square grid, coincides with that of Fairweather and Mitchell (1965), which is of the sixth-order accuracy and conditionally stable

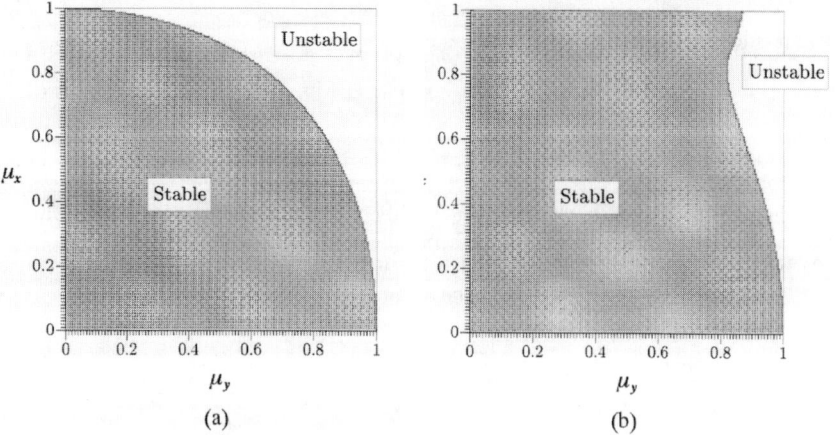

FIGURE 1.23. Stability condition of the high-order ADI scheme for the classical wave equation at (a) $\gamma = 1$ and (b) $\gamma = 1 + \mu_x^2$.

at

$$\mu_x = \mu_y \leq \sqrt{3} - 1 \qquad (233)$$

4.1.3. Applications. The special case of the hybrid method, i.e., the staggered-grid finite difference method, has been utilized to solve the time-dependent mild-slope wave equations or the hyperbolic representation of the elliptic mild-slope wave equation by a number of researchers. Ito and Tanimoto (1972) are probably the first to recommend this method for wave diffraction by a submerged shoal. Copeland (1985) and Madsen and Larsen (1987) followed Ito and Tanimoto. Watanabe and Maruyama (1986) extended this method to problems with breaking.

Shown in Figure 1.24 is the physical model for the transformation of an obliquely incident wave train by a jetty over a sloping beach, numerically studied by Watanabe and Maruyama (1986). The deep-water wave height is 2 cm and the deep-water wave incidence angle is $\pi/3$. The incident wave period is 1.2 s. In the computation, $\Delta x = \Delta y = 5$ cm. The computed breaking line and wave height are demonstrated in Figures 1.25 and 1.26 with an evidently good agreement with the experimental data. The relatively large discrepancy near the jetty was regarded as an effect of wave-induced currents.

Application of the high-order ADI scheme to wave diffraction and refraction was reported by Yu (1998b). Figure 1.27 demonstrates the computed wave height distribution behind the submerged circular shoal shown in Figure 1.7. The computational grid size is the same as that used in the optimal finite-difference scheme for the elliptic mild-slope wave equation. As seen in the figure, the steady-state solution agrees very well with the result from the optimal scheme presented before.

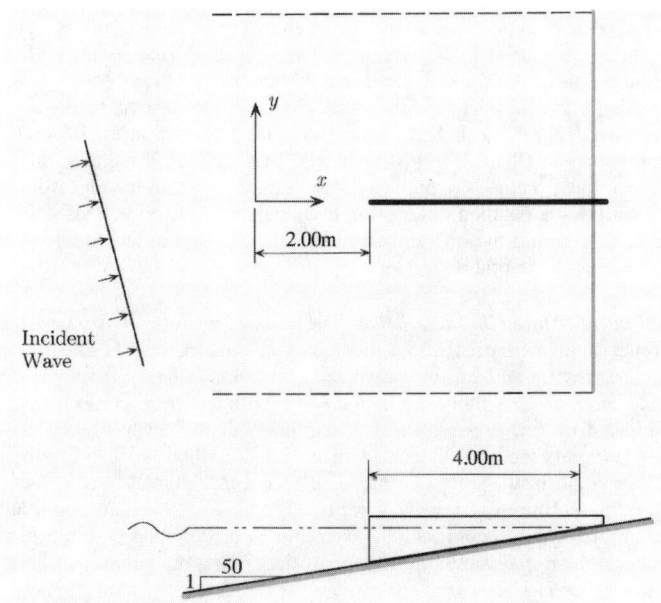

FIGURE 1.24. Physical model of the jetty on a sloping beach.

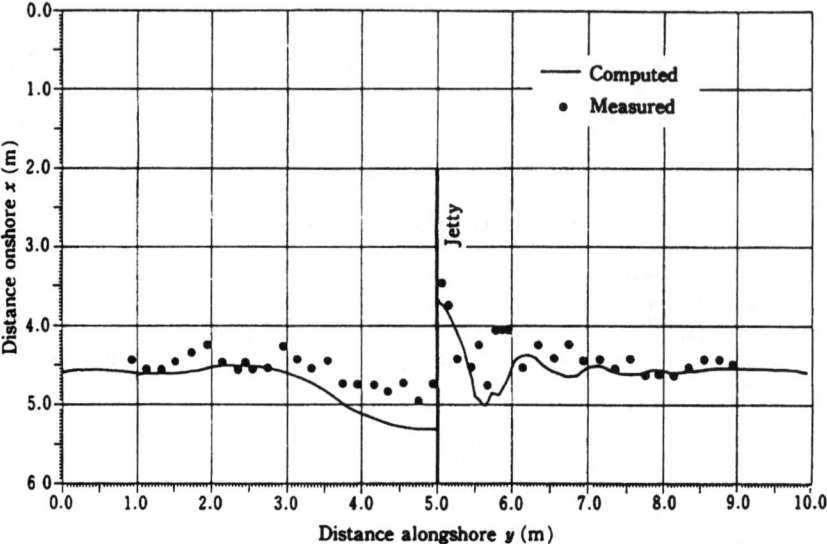

FIGURE 1.25. Breaking line around the jetty.

4.2. Quasilinear Hyperbolic Wave Equations

Numerical solution of the shallow-water wave equations, which are categorized as the quasilinear hyperbolic type in the study of partial differential equations, is a classical subject of coastal and estuarine hydrodynamics. The well-established numerical methods include various forms of the finite difference methods (Liu and Leedertse, 1978; Abbott et al., 1981; Mahmood and Yevjevich, 1975; Tan, 1992), the finite element methods (Zienkiewicz and Taylor, 1991; Kawahara, 1985), the finite volume methods (Zhao et al., 1994; Anastasiou and Chan, 1997; Mingham and Causon, 1998), etc. A relatively complete review of these methods, which can be voluminous, is not the objective of this subsection. Instead, some detailed description will be given to two methods—the method of finite characteristics and the operator-splitting method, which have obvious advantages when compared with the others.

4.2.1. Method of Finite Characteristics. The *method of finite characteristics* is the finite difference implementation of the method of characteristics. It is also referred to as the *characteristic-oriented difference method*, the *method of characteristics at prescribed time interval*, or simply the *method of characteristics*. The concept was originally established by Lin (Rouse, 1950; Lin, 1952) although it has been sometimes referred to as Hartree's method. The excellence of this method is its ability to capture the flow of information in the physical phenomena. It can instantaneously trace a moving boundary or the motion of water over a dry bed, which has been an important issue affecting nearshore hydrodynamics. It is also able to represent a possible discontinuity, i.e., a breaker, which may show up in spite of the continuous initial and boundary conditions (Stoker, 1957).

For the derivation of a characteristic oriented scheme it is usually preferred to write the basic equations in a matrix form as

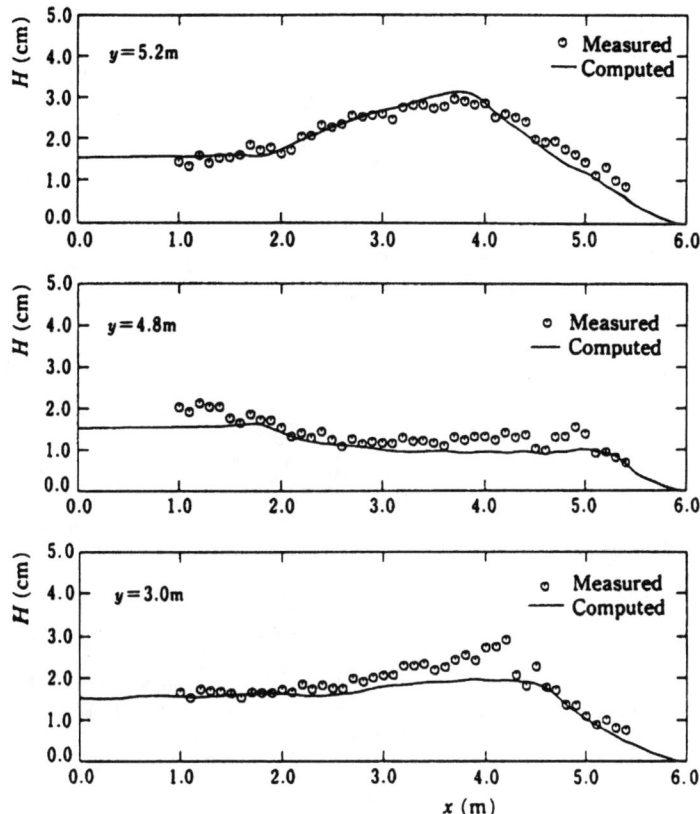

FIGURE 1.26. Wave height distribution around the jetty.

$$\mathbf{I}\frac{\partial \mathbf{Z}}{\partial t} + \mathbf{A}\frac{\partial \mathbf{Z}}{\partial x} + \mathbf{B}\frac{\partial \mathbf{Z}}{\partial y} = \mathbf{R} \qquad (234)$$

where

$$\mathbf{I} = \begin{bmatrix} 1 & 0 & 0 \\ 0 & 1 & 0 \\ 0 & 0 & 1 \end{bmatrix}; \quad \mathbf{A} = \begin{bmatrix} \bar{u} & d & 0 \\ g & \bar{u} & 0 \\ 0 & 0 & \bar{u} \end{bmatrix}; \quad \mathbf{B} = \begin{bmatrix} \bar{v} & 0 & d \\ 0 & \bar{v} & 0 \\ g & 0 & \bar{v} \end{bmatrix}$$

$$\mathbf{Z} = \begin{Bmatrix} d \\ \bar{u} \\ \bar{v} \end{Bmatrix}; \quad \mathbf{R} = \begin{Bmatrix} 0 \\ R_1 \\ R_2 \end{Bmatrix}$$

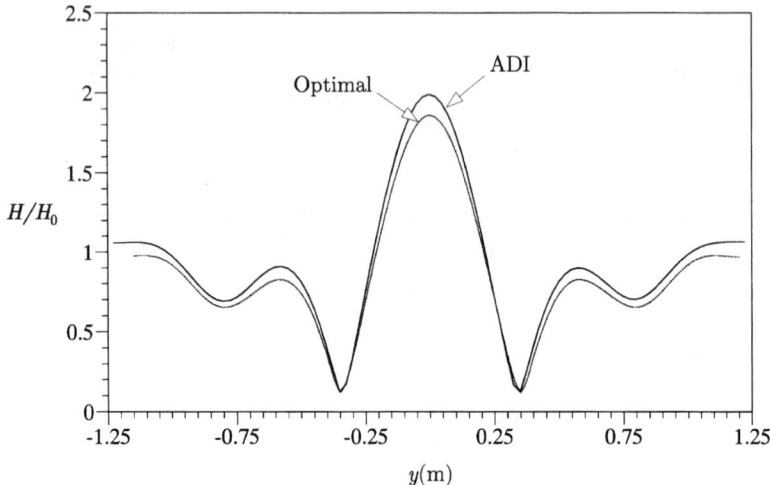

FIGURE 1.27. Comparison of the computed wave height behind the submerged shoal by the high-order ADI scheme and by the optimal scheme.

with $R_1 = gh_x - C_f \bar{u}\sqrt{\bar{u}^2 + \bar{v}^2}/d$, $R_2 = gh_y - C_f \bar{v}\sqrt{\bar{u}^2 + \bar{v}^2}/d$. Equation (234) is known to possess of two families of characteristics in the (x, y, t) space (e.g., Lin, 1980). The first family, i.e., the characteristic conoid, is represented by

$$\left(\frac{dx}{dt} - \bar{u}\right)^2 + \left(\frac{dy}{dt} - \bar{v}\right)^2 = c^2 \tag{235}$$

where $c = \sqrt{gd}$ is the shallow-water wave celerity. The characteristic equation satisfied on the first family of characteristics takes the following form:

$$\frac{g}{c}\frac{Dd}{Dt} + \cos\theta \frac{D\bar{u}}{Dt} + \sin\theta \frac{D\bar{v}}{Dt} - \sin\theta \frac{D\bar{u}}{D\theta} + \cos\theta \frac{D\bar{v}}{D\theta} = R_1 \cos\theta + R_2 \sin\theta \tag{236}$$

where θ is the polar angle of the characteristic conoid and

$$\frac{D}{Dt} = \frac{\partial}{\partial t} + (\bar{u} + c\cos\theta)\frac{\partial}{\partial x} + (\bar{v} + c\sin\theta)\frac{\partial}{\partial y} \tag{237}$$

$$\frac{D}{D\theta} = -c\sin\theta \frac{\partial}{\partial x} + c\cos\theta \frac{\partial}{\partial y} \tag{238}$$

The second family of characteristics, which reduces to the streamline, is determined by

$$\frac{dx}{dt} = \bar{u} \tag{239}$$

$$\frac{dy}{dt} = \bar{v} \tag{240}$$

The characteristic equation satisfied on the second family of characteristics should be written as

$$-\sin\theta \, \frac{D\bar{u}}{Dt} + \cos\theta \, \frac{D\bar{v}}{Dt} + \frac{g}{c}\frac{Dd}{D\theta} = R_1 \sin\theta + R_2 \cos\theta \tag{241}$$

where

$$\frac{D}{Dt} = \frac{\partial}{\partial t} + \bar{u}\frac{\partial}{\partial x} + \bar{v}\frac{\partial}{\partial y} \tag{242}$$

$$\frac{D}{D\theta} = -c\sin\theta\,\frac{\partial}{\partial x} + c\cos\theta\,\frac{\partial}{\partial y} \tag{243}$$

The method of finite characteristics is usually to take advantage of four special bi-characteristics among the first family of characteristics, in addition to the streamline that represents the second family of characteristics, as depicted in Figure 1.28. The four bi-characteristics are expressed by

$$\frac{dx}{dt} = \bar{u} + c\cos\theta_l \tag{244}$$

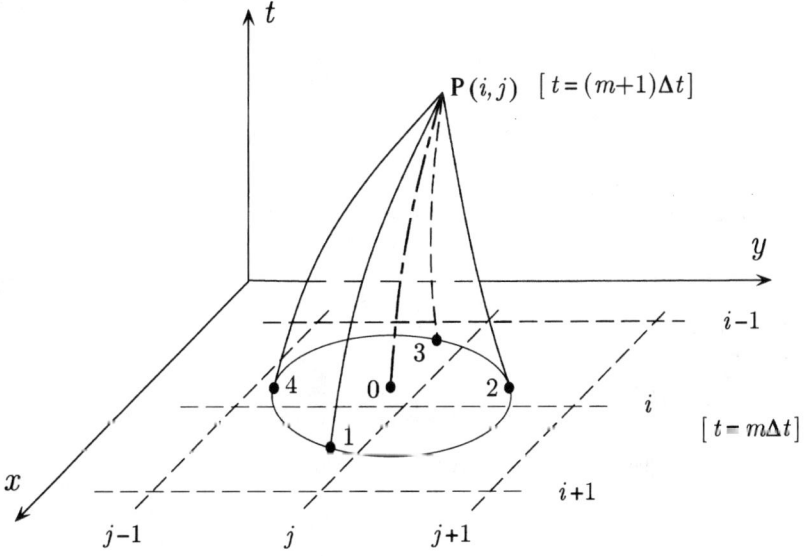

FIGURE 1.28. Bi-characteristics of the shallow-water wave equations.

$$\frac{dy}{dt} = \bar{v} + c \sin \theta_l \qquad (245)$$

where $\theta_l = (l-1)\pi/2$, and $l = 1, 2, 3, 4$.

The numerical scheme includes two steps. The first step is to determine the intersections of the bi-characteristics and the streamline with the xy-plane at the known time level $t = m\Delta t$. The second step is then to properly discretize the characteristic equations.

A first-order scheme is straightforward. It can be described following Townson (1974). Different implementation may be found in Lin et al. (1980) and Katopodes and Strelkoff (1979). Townson's method is to discretize equations (244), (245), (239), and (240) in the following manner:

$$x_l = x_i + (\bar{u}_p + c_p \cos \theta_l) \Delta t \qquad (246)$$

$$y_l = y_j + (\bar{v}_p + c_p \cos \theta_l) \Delta t \qquad (247)$$

$$x_0 = x_i + \bar{u}_p \Delta t \qquad (247)$$

$$y_0 = y_j + \bar{v}_p \Delta t \qquad (248)$$

where $l = 1, 2, 3, 4$. The characteristic equations along the bi-characteristics are written as

$$g(d_p - d_1) + c_p(\bar{u}_p - \bar{u}_1) = -c_p^2(\bar{v}_y)_p \Delta t + c_p(R_1)_p \Delta t \qquad (250)$$

$$g(d_p - d_2) + c_p(\bar{v}_p - \bar{v}_2) = -c_p^2(\bar{u}_x)_p \Delta t + c_p(R_2)_p \Delta t \qquad (251)$$

$$g(d_p - d_3) - c_p(\bar{u}_p - \bar{u}_3) = -c_p^2(\bar{v}_y)_p \Delta t - c_p(R_1)_p \Delta t \qquad (252)$$

$$g(d_p - d_4) - c_p(\bar{v}_p - \bar{v}_4) = -c_p^2(\bar{u}_x)_p \Delta t - c_p(R_2)_p \Delta t \qquad (253)$$

Along the streamline, it is suggested to adopt an alternative relation to equation (241):

$$\frac{g}{c} \frac{Dd}{Dt} + c(\bar{u}_x + \bar{v}_y) = 0 \qquad (254)$$

which can be discretized as

$$g(d_p - d_0) = -c_p^2 [(\bar{u}_x)_p + (\bar{v}_y)_p] \Delta t \qquad (255)$$

An operation to equations (250)–(253) and (255) yields

$$d_p = \frac{1}{2} [d_1 + d_2 + d_3 + d_4 - 2d_0 + (\bar{u}_1 - \bar{u}_3 + \bar{v}_2 - \bar{v}_4) c_p/g] \qquad (256)$$

$$\bar{u}_p = \frac{1}{2}[\bar{u}_1 + \bar{u}_3 - g(d_1 - d_3)/c_p] + (R_1)_p \, \Delta t \tag{257}$$

$$\bar{v}_p = \frac{1}{2}[\bar{v}_2 + \bar{v}_4 - g(d_2 - d_4)/c_p] + (R_2)_p \, \Delta t \tag{258}$$

A trial-and-error procedure will thus allow d_p, $\bar{u}_p$, and $\bar{v}_p$ to be determined. One should also have an interpolation method to evaluate d_l, $\bar{u}_l$, and $\bar{v}_l$ from their values at the grid nodes. Both the bilinear interpolation and the quadratic bivariational interpolation seem to be applicable, but the latter has been strongly recommended.

A second-order finite characteristic scheme is a little more involved. Different implementations can be found in Katopodes and Strelkoff (1978; 1979) and Lai (e.g., 1986). According to Katopodes and Strelkoff (1978), the difference equations for the finite characteristics at this order must be written as

$$x_l = x_i + \frac{1}{2}(\bar{u}_p + c_p \cos \theta_l + \bar{u}_l + c_l \cos \theta_l') \, \Delta t \tag{259}$$

$$y_l = y_j + \frac{1}{2}(\bar{v}_p + c_p \cos \theta_l + v_1 + c_l \sin \theta_l') \, \Delta t \tag{260}$$

$$x_0 = x_i + \frac{1}{2}(\bar{u}_p + \bar{u}_l) \, \Delta t \tag{261}$$

$$y_0 = y_j + \frac{1}{2}(\bar{v}_p + \bar{v}_l) \, \Delta t \tag{262}$$

where $l = 1, 2, 3, 4$. $\theta_l' \neq \theta_l$ because of the possible torsion of the characteristic conoid. The relation between θ_l' and θ_l can be found in Butler (1960) as

$$\theta_l' = \theta_l + \Delta \theta_l \tag{263}$$

$$\Delta \theta_l = \Delta t[\cos^2 \theta_l (\bar{u}_y)_l - \sin^2 \theta_l (\bar{v}_x)_l + \cos \theta_l \sin \theta_l (\bar{v}_y - \bar{u}_x)_l]$$
$$+ \Delta t[\cos \theta_l (c_y)_l - \sin \theta_l (c_x)_l] \tag{264}$$

Substitution of equation (263) into equations (259)–(262) will thus allow the bi-characteristics to be determined. The discretized characteristic equations at the second-order accuracy must be written as

$$g(d_p - d_1) + \frac{1}{2}(c_p + c_1)(\bar{u}_p - \bar{u}_1) + \frac{1}{2} c_1 \Delta \theta_1 (\bar{v}_p - \bar{v}_1)$$
$$= \Delta t \left[\frac{1}{2}(c_p + c_1)(R_1)_p + \frac{1}{2} c_1 (R_2)_1 - (c_p)^2 (\bar{v}_y)_p \right] \tag{265}$$

$$g(d_p - d_2) + \frac{1}{2}(c_p + c_2)(\bar{v}_p - \bar{v}_2) - \frac{1}{2}c_2\Delta\theta_2(\bar{u}_p - \bar{u}_2)$$

$$= \Delta t\left[\frac{1}{2}(c_p + c_2)(R_2)_p - \frac{1}{2}c_2(R_1)_2 - (c_p)^2(\bar{u}_x)_p\right] \quad (266)$$

$$g(d_p - d_3) - \frac{1}{2}(c_p + c_3)(\bar{u}_p - \bar{u}_3) - \frac{1}{2}c_3\Delta\theta_3(\bar{v}_p - \bar{v}_3)$$

$$= \Delta t\left[-\frac{1}{2}(c_p + c_3)(R_1)_p - \frac{1}{2}c_3(R_2)_3 - (c_p)^2(\bar{v}_y)_p\right] \quad (267)$$

$$g(d_p - d_4) - \frac{1}{2}(c_p + c_4)(\bar{v}_p - \bar{v}_4) + \frac{1}{2}c_4\Delta\theta_4(\bar{u}_p - \bar{u}_4)$$

$$= \Delta t\left[-\frac{1}{2}(c_p + c_4)(R_2)_p + \frac{1}{2}c_4(R_1)_4 - (c_p)^2(\bar{u}_x)_p\right] \quad (268)$$

$$g(d_p - d_0) = -\frac{\Delta t}{2}\{(c_p)^2[(\bar{u}_x)_p + (\bar{v}_y)_p] + (c_0)^2[(\bar{u}_x)_0 + (\bar{v}_y)_0]\} \quad (269)$$

From these five equations one can solve for d_p, $\bar{u}_p$, $\bar{v}_p$, $(\bar{u}_x)_p$, and $(\bar{v}_y)_p$ by trial and error. The values of d_l, $\bar{u}_l$, $\bar{v}_l$, $(\bar{u}_x)_l$, and $(\bar{v}_y)_l$ may be determined by the quadratic bivariational interpolation from their values at the grid nodes.

4.2.2. Operator-Splitting Method. The operator-splitting method has been widely preferred because of its flexibility. For a theoretical background the reader is referred to the book by Marchuk (1982). A brief illustration of the concept can be given by considering the model equation:

$$\frac{\partial Y}{\partial t} = \mathcal{M}(Y) = \mathcal{M}_1(Y) + \mathcal{M}_2(Y) \quad (270)$$

where $\mathcal{M}$, $\mathcal{M}_1$, and $\mathcal{M}_2$ are differential operators with respect to the spatial coordinates. Equation (270) can be formally discretized by the following two-step finite difference scheme:

$$\frac{Y^{m+\tau} - Y^m}{\Delta t} = \mathcal{N}_1(Y^{m+\tau_1}) \quad (271)$$

$$\frac{Y^{m+1} - Y^{m+\tau}}{\Delta t} = \mathcal{N}_2(Y^{m+\tau_2}) \quad (272)$$

where τ, τ_1, and τ_2 are fractional numbers; $\mathcal{N}_1$ and $\mathcal{N}_2$ are spatial difference operators corresponding to $\mathcal{M}_1$ and $\mathcal{M}_2$, respectively. Obviously, $Y^{m+\tau}$ and Y^{m+1} in equations (271) and (272) can also be regarded as the discrete solutions of the following differential equations:

$$\frac{\partial Y}{\partial t} = \mathcal{M}_1(Y) \tag{273}$$

$$\frac{\partial Y}{\partial t} = \mathcal{M}_2(Y) \tag{274}$$

within the time interval $[m\Delta t, (m + 1)\Delta t]$, with the solution of equation (273) as the initial condition of equation (274) and the solution of equation (274) as the initial condition of equation (273) at the next time step. A graphic intepretation of this successive solution process is depicted in Figure 1.29.

The flexibility of the operator-splitting method follows from the fact that equations (273) and (274) can be independently and, consequently, most effectively discretized, considering the mathematical and physical properties of $\mathcal{M}_1$ and $\mathcal{M}_2$.

Evidently, the operator-splitting method is not a particular technique for the shallow-water wave equations. It is actually applicable to the solution of various differential equations describing evolution phenomena. Successful application to other types of wave equations has also been reported (Yu et al., 1992c). In this subsection, two distinguished schemes established for the shallow-water wave equations are presented.

Locally One-Dimensional Scheme. This scheme was recommended by He and Lin (1986). It is to split the shallow-water wave equations into the following successive set of equations, with each being equivalent to a reduced one-dimensional problem:

$$\begin{cases} d_t + \bar{u}d_x + d\bar{u}_x = 0 \\ u_t + \bar{u}\bar{u}_x + gd_x = -gh_x - C_f\sqrt{\overline{u^2} + \overline{v^2}}\,\bar{u}/d \\ v_t + \bar{u}\bar{v}_x = 0 \end{cases} \tag{275}$$

$$\begin{cases} d_t + \bar{v}d_y + d\bar{v}_y = 0 \\ \bar{v}_t + \bar{v}\bar{v}_y + gd_y = -gh_y - C_f\sqrt{\overline{u^2} + \overline{v^2}}\,\bar{v}/d \\ \bar{u}_t + \bar{v}\bar{u}_y = 0 \end{cases} \tag{276}$$

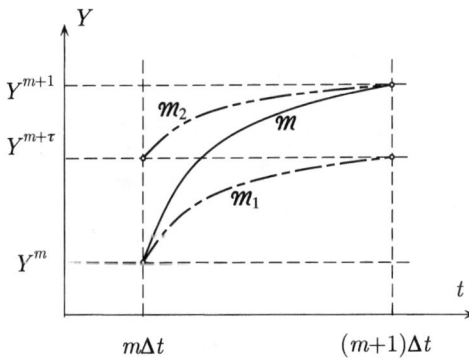

FIGURE 1.29. Illustration of the operator-splitting method.

Effective methods for the simultaneous solution of the first two equations in both (275) and (276) are well documented. The one-dimensional version of the method of finite characteristics (e.g., Mahmood and Yevjevich, 1975; Lai, 1986) should be a good choice if the phenomenon concerned is highly nonlinear. Various successful TVD (total variation diminishing) schemes (Louaked and Hanich, 1998) for one-dimensional flows are also recommended. Accurate schemes for the last equation in both equation (275) and equation (276) can be found in a standard textbook on numerical analysis of partial differential equations.

Physically Decomposed Scheme. This scheme was proposed by Benque et al. (1982). It is to split the shallow-water wave equations by their contributory physics. When the horizontal diffusion is also included in the original formulation, the variation of the instantaneous water depth and the vertically averaged velocity components can be regarded as a result of the propagation, the advection, and the diffusion. The successive set of equations then become

$$\begin{cases} d_t + \bar{u}d_x + d\bar{u}_x + vd_y + d\bar{v}_y = 0 \\ \bar{u}_t + gd_x = -gh_x - C_f \sqrt{\bar{u}^2 + \bar{v}^2}\, \bar{u}/d \\ \bar{v}_t + gd_y = -gh_y - C_f \sqrt{\bar{u}^2 + \bar{v}^2}\, \bar{v}/d \end{cases} \quad (277)$$

$$\begin{cases} \bar{u}_t + \bar{u}\bar{u}_x + \bar{v}\bar{u}_y = 0 \\ \bar{v}_t + \bar{u}\bar{v}_x + \bar{v}\bar{v}_y = 0 \end{cases} \quad (278)$$

$$\begin{cases} \bar{u}_t = (K\bar{u}_x)_x + (K\bar{u}_y)_y \\ \bar{v}_t = (K\bar{v}_x)_x + (K\bar{v}_y)_y \end{cases} \quad (279)$$

For the advection step, Benque et al. (1982) recommended the method of characteristics, which is now extremely simple. One needs to issue only a single characteristic line according to

$$\frac{dx}{dt} = \bar{u} \quad (280)$$

$$\frac{dy}{dt} = \bar{v} \quad (281)$$

An interpolation scheme then immediately gives the required values of $\bar{u}$ and $\bar{v}$ at this step, because they are invariant along the characteristic line. An accurate scheme for the diffusion step can be borrowed from a textbook on numerical heat transfer. There is also no particular difficulty to find an effective scheme for the wave propagation. Since the equations governing the wave propagation are essentially linear, a stable second-order difference scheme may work well.

4.2.3. Applications. Applications of the method of finite characteristics can be found in Townson (1974), Lai (e.g., 1986), Katopodes and Strelkoff (1978; 1979), and Lin et al. (1980) (also, Lin et al., 1982). Katopodes and Strelkoff (1978; 1979) studied the dam break problem. Townson (1974) computed the tidal flow in the River Tay estuary, United

Kingdom, and obtained fairly accurate solutions when compared with laboratory measurements. Lin et al. (1980) studied the tidal flow in Hang Zhou Bay, China, which is famous for forcing tidal bores in the Tsien Tang River (Stoker, 1957). With a rather coarse grid ($\Delta x = \Delta y = 5$ km) as shown in Figure 1.30, the numerical solutions of Lin et al. on both the water surface and the flow velocity were still found to be in very good agreement with field data, as demonstrated in Figure 1.31.

The operator splitting method was used by Benque et al. (1982) to simulate the tidal flow in the Bay of Saint Brieuc and in the River Canche estuary, both in France. It was also used by He and Lin (1986) to study the tidal flow in Hang Zhou Bay, China. He and Lin's results were found to be in good agreement with the field data and also with the solution by the method of finite characteristics, as shown in Figure 1.32.

4.3. Boussinesq Equations

Development of delicate numerical methods for the Boussinesq equations is known to be rather involved. For a numerical solution that is consistent with the high accuracy of the mathematical model, one must control not only the *numerical viscosity*, which originates from the truncated terms with the second-order derivatives, but also the *numerical dispersivity*, which is related to the truncated terms with the third-order derivatives. This implies that a relatively accurate scheme may have to be at least third-ordered for the terms involving lower-order differentiations but can allow a low-order approximation for the high-order differentiations, so that the truncation error will include neither the second-order nor the third-order derivatives.

In this subsection, two major techniques used in the development of high-order finite difference schemes for the Boussinesq equations are described: the *truncation-error elimination method* and the *direct high-order difference method*. Both are reported to be highly effective.

4.3.1. Truncation-Error Elimination Method. This technique was recommended by Abbott and his co-workers (Abbott et al., 1978; Abbott et al., 1985) for the classical

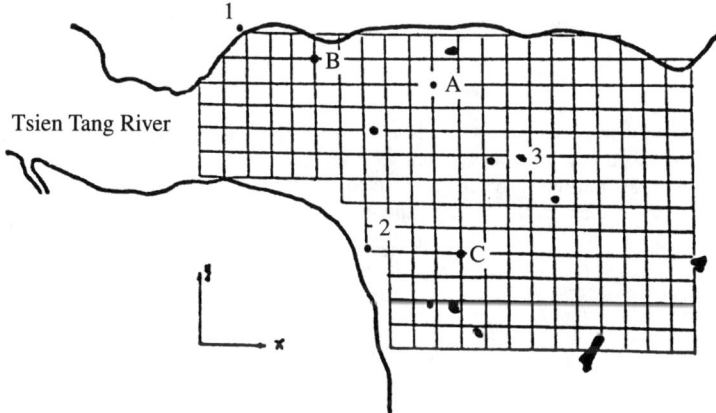

FIGURE 1.30. Computational grid of Hang Zhou Bay.

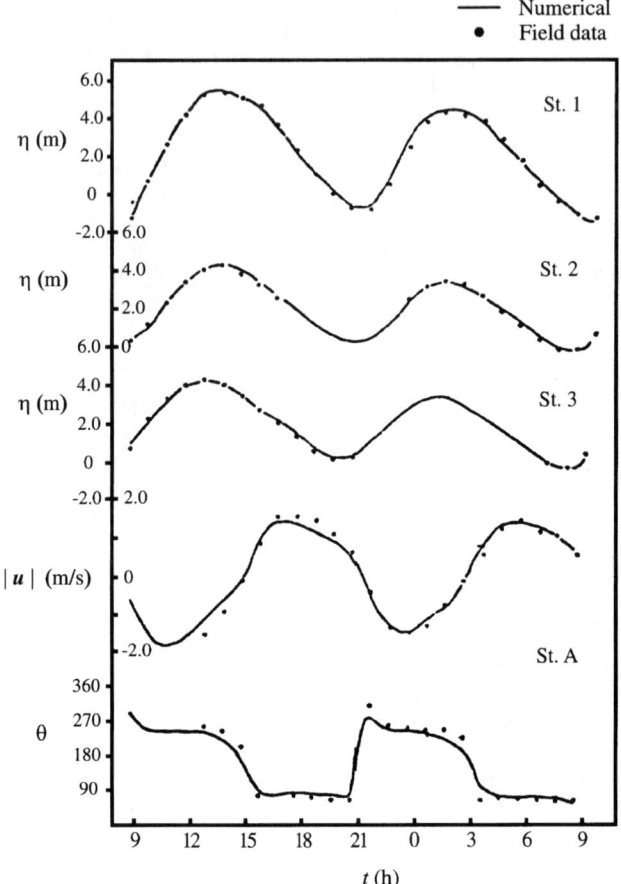

FIGURE 1.31. Water surface elevation and the tidal flow velocity in Hang Zhou Bay.

Boussinesq equations. The basic idea is as follows. First of all, a primary scheme with the second-order accuracy is selected, such as that of Leendertse (Liu and Leedertse, 1978). By Taylor expansion one can then identify the leading-order terms, i.e., the third-order terms in the truncation errors. If the following linear relations

$$\eta_t + h\nabla \cdot \boldsymbol{u}^{(r)} = 0 \tag{282}$$

$$\boldsymbol{u}_t^{(r)} + g\nabla \eta = 0 \tag{283}$$

are utilized to convert the cross derivatives in the third-ordered error terms without any loss of accuracy, it may be possible to eliminate the leading-order truncation errors simply by a modification to the coefficients of the Boussinesq terms.

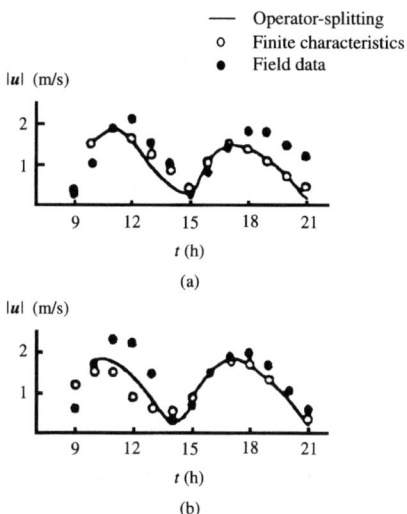

FIGURE 1.32. Flow velocity in Hang Zhou Bay at (a) Station B and (b) Station C.

Construction of a high-order scheme for the Boussinesq equations by the truncation-error elimination technique necessitates some mathematical manipulation but involves no difficulties. Identification of the leading-order errors must be done for each difference representation of derivatives, which is, however, routine. As an example, Abbott et al. (1978) demonstrated the procedure for the following derivative, which is interpreted as a term in the x-implicit step of the ADI-type Preismann scheme applied to the horizontally two-dimensional Boussinesq equations:

$$\left[\frac{\partial u}{\partial t}\right]_{i+1/2,j}^m = \frac{1}{4\Delta t}(u_{i+1,j}^{m+1} - u_{i+1,j}^{m-1} + u_{i,j}^{m+1} - u_{i,j}^{m-1}) + E_l + E_h \tag{284}$$

where E_l and E_h denote the leading-order and high-order error terms, respectively; E_h is to be neglected but E_l, which is expressed as

$$E_l = -\frac{\Delta x^2}{8}\left[\frac{\partial^3 u}{\partial x^2 \partial t}\right]_{i+1/2,j}^m - \frac{\Delta t^2}{6}\left[\frac{\partial^3 u}{\partial t^3}\right]_{i+1/2,j}^m$$

$$= -\left(\frac{\Delta x^2}{8} + \frac{gh\Delta t^2}{6}\right)\left[\frac{\partial^3 u}{\partial x^2 \partial t}\right]_{i+1/2,j}^m \tag{285}$$

will be combined into the Boussinesq terms and then discretized by a second order scheme.

4.3.2. Direct High-Order Difference Method. The direct high-order difference method was first proposed by Wei and Kirby (1995). Zheng et al. (1998) modified Wei and Kirby's scheme to enable the widely preferred spatially staggered grid. Both work considered the enhanced Boussinesq equations derived by Nwogu (1993). For the convenience of notation, the basic equations are written in the following form:

$$\eta_t = E(\eta, u, v) \tag{286}$$

$$[U(u)]_t = F(\eta, u, v) + [P(v)]_t \tag{287}$$

$$[V(v)]_t = G(\eta, u, v) + [Q(u)]_t \tag{288}$$

where $u \equiv u^{(r)}$, $v \equiv v^{(r)}$, and

$$\begin{aligned} E(\eta, u, v) = & -[(h + \eta) u]_x - [(h + \eta) v]_y \\ & - \{A_1 h^3 (u_{xx} + v_{xy}) + A_2 h^2 [(hu)_{xx} + (hv)_{xy}]\}_x \\ & - \{A_1 h^3 (u_{xy} + v_{yy}) + A_2 h^2 [(hu)_{xy} + (hv)_{yy}]\}_y \end{aligned} \tag{289}$$

$$U(u) = u + B_1 h^2 u_{xx} + B_2 h (hu)_{xx} \tag{290}$$

$$F(\eta, u, v) = -g\eta_x - uu_x - vu_y \tag{291}$$

$$P(v) = -B_1 h^2 v_{xy} - B_2 h (hv)_{xy} \tag{292}$$

$$V(v) = v + B_1 h^2 v_{yy} + B_2 h (hv)_{yy} \tag{293}$$

$$G(\eta, u, v) = -g\eta_y - uv_x - vv_y \tag{294}$$

$$Q(u) = -B_1 h^2 u_{xy} - B_2 h (hu)_{xy} \tag{295}$$

with $A_1 = \xi^2/2 - 1/6$, $A_2 = -\xi + 1/2$, $B_1 = \xi^2/2$, $B_2 = -\xi$; ξ represents the relative position of the reference level. The time derivatives in equations (286), (287), and (288) can be discretized by Adams–Bashforth–Mouton's predictor-corrector method. The predictor uses Adams and Bashforth's third-order scheme:

$$\eta^*_{i,j} = \eta^m_{i,j} + \frac{\Delta t}{12} (23 E^m_{i,j} - 16 E^{m-1}_{i,j} + 5 E^{m-2}_{i,j}) \tag{296}$$

$$\begin{aligned} U^*_{i+1/2,j} = & U^m_{i+1/2,j} + \frac{\Delta t}{12} (23 F^m_{i+1/2,j} - 16 F^{m-1}_{i+1/2,j} + 5 F^{m-2}_{i+1/2,j}) \\ & + 2 P^m_{i+1/2,j} - 3 P^{m-1}_{i+1/2,j} + P^{m-2}_{i+1/2,j} \end{aligned} \tag{297}$$

$$V^*_{i,j+1/2} = V^m_{i,j+1/2} + \frac{\Delta t}{12} (23 G^m_{i,j+1/2} - 16 G^{m-1}_{i,j+1/2} + 5 G^{m-2}_{i,j+1/2})$$

$$+ 2Q^m_{i,j+1/2} - 3Q^{m-1}_{i,j+1/2} + Q^{m-2}_{i,j+1/2} \qquad (298)$$

where the asterisk denotes the predicted values. The corrector adopts Adams–Mouton's fourth-order scheme:

$$\eta^{m+1}_{i,j} = \eta^m_{i,j} + \frac{\Delta t}{24}(9E^*_{i,j} + 19E^m_{i,j} - 5E^{m-1}_{i,j} + E^{m-2}_{i,j}) \qquad (299)$$

$$U^{m+1}_{i+1/2,j} = U^m_{i+1/2,j} + \frac{\Delta t}{24}(9F^*_{i+1/2,j} + 19F^m_{i+1/2,j} - 5F^{m-1}_{i+1/2,j} + F^{m-2}_{i+1/2,j})$$

$$+ P^*_{i+1/2,j} - P^m_{i+1/2,j} \qquad (300)$$

$$V^{m+1}_{i,j+1/2} = V^m_{i,j+1/2} + \frac{\Delta t}{24}(9G^*_{i,j+1/2} + 19G^m_{i,j+1/2} - 5G^{m-1}_{i,j+1/2} + G^{m-2}_{i,j+1/2})$$

$$+ Q^*_{i,j+1/2} - Q^m_{i,j+1/2} \qquad (301)$$

The correction step must be repeated until the relative error of the computed η, U, and V at $t = (m + 1)\Delta t$ becomes small enough.

The spatial discretization can be based on a spatially staggered mesh, as shown in Figure 1.33. The continuity equation is then discretized at (i, j) and the momentum equations in the x and y directions are discretized at $(i + \frac{1}{2}, j)$ and $(i, j + \frac{1}{2})$, respectively. Central difference schemes are employed for all derivatives except for those in the advective terms, that is, the fourth-order central difference scheme is used for the first-order derivatives and the second-order central difference schemes are used for the second- and third-order derivatives. For a variable F centered at (i, j) these are expressed as

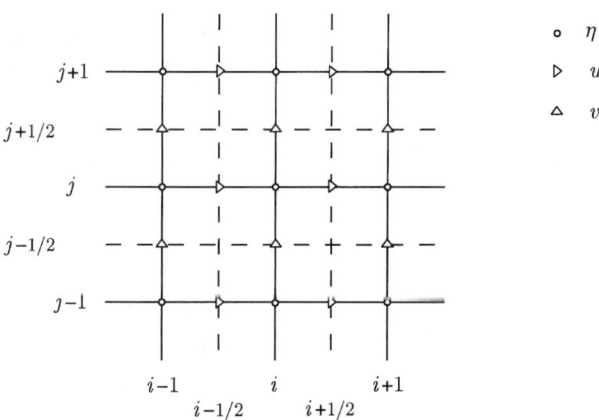

FIGURE 1.33. Spatially staggered grid.

$$\delta_x F_{i,j} = \frac{1}{24\Delta x} (F_{i-3/2,j} - 27F_{i-1/2,j} + 27F_{i+1/2,j} - F_{i+3/2,j}) \qquad (302)$$

$$\delta_y F_{i,j} = \frac{1}{24\Delta y} (F_{i,j-3/2} - 27F_{i,j-1/2} + 27F_{i,j+1/2} - F_{i,j+3/2}) \qquad (303)$$

$$\delta_{xx} F_{i,j} = \frac{1}{\Delta x^2} (F_{i+1,j} - 2F_{i,j} + F_{i-1,j}) \qquad (304)$$

$$\delta_{yy} F_{i,j} = \frac{1}{\Delta y^2} (F_{i,j+1} - 2F_{i,j} + F_{i,j-1}) \qquad (305)$$

$$\delta_{xy} F_{i,j} = \frac{1}{\Delta x \Delta y} (F_{i+1/2,j+1/2} - F_{i+1/2,j-1/2} - F_{i-1/2,j+1/2} + F_{i-1/2,j-1/2}) \qquad (306)$$

$$\delta_{xxx} F_{i,j} = \frac{1}{\Delta x^3} (F_{i+3/2,j} - 3F_{i+1/2,j} + 3F_{i-1/2,j} - F_{i-3/2,j}) \qquad (307)$$

$$\delta_{yyy} F_{i,j} = \frac{1}{\Delta y^3} (F_{i,j+3/2} - 3F_{i,j+1/2} + 3F_{i,j-1/2} - F_{i,j-3/2}) \qquad (308)$$

$$\delta_{xxy} F_{i,j} = \frac{1}{\Delta x^2 \Delta y} (F_{i+1,j+1/2} - F_{i+1,j-1/2} - 2F_{i,j+1/2} + 2F_{i,j-1/2} + F_{i-1,j+1/2} - F_{i-1,j-1/2}) \qquad (309)$$

$$\delta_{xyy} F_{i,j} = \frac{1}{\Delta x \Delta y^2} (F_{i+1/2,j+1} - F_{i-1/2,j+1} - 2F_{i+1/2,j} + 2F_{i-1/2,j} + F_{i+1/2,j-1} - F_{i-1/2,j-1}) \qquad (310)$$

where δ denotes the difference operators. For the advective terms, the second-order upwind scheme can be utilized. The advection of F by the horizontal velocity component W in the x-direction may then be written as

$$(W\delta_x F)_{i+1/2,j} = \frac{1}{4\Delta x} (|\hat{W}| - \hat{W}) (3F_{i+1/2,j} - 4F_{i+3/2,j} + F_{i+5/2,j})$$

$$+ \frac{1}{4\Delta x} (|\hat{W}| + \hat{W}) (3F_{i+1/2,j} - 4F_{i-1/2,j} + F_{i-3/2,j}) \qquad (311)$$

$$(W\delta_y F)_{i+1/2,j} = \frac{1}{4\Delta y} (|\overline{W}| - \overline{W}) (3F_{i+1/2,j} - 4F_{i+1/2,j+1} + F_{i+1/2,j+2})$$

$$+ \frac{1}{4\Delta y} (|\overline{W}| + \overline{W}) (3F_{i+1/2,j} - 4F_{i+1/2,j-1} + F_{i+1/2,j-2}) \qquad (312)$$

where $\hat{W} = W_{i+1/2,j}$ and $\overline{W} = (W_{i,j-1/2} + W_{i,j+1/2} + W_{i+1,j-1/2} + W_{i+1,j+1/2})/4$.

incident waves

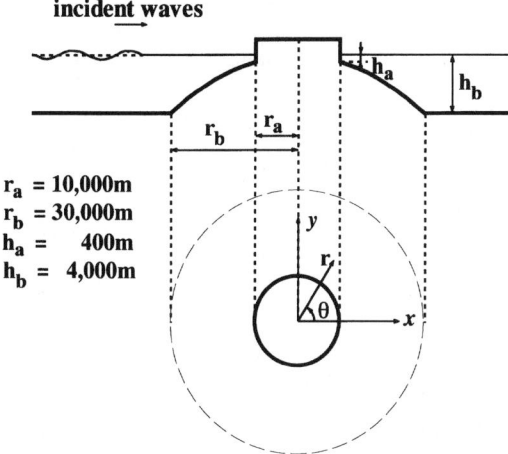

$r_a = 10{,}000\text{m}$
$r_b = 30{,}000\text{m}$
$h_a = 400\text{m}$
$h_b = 4{,}000\text{m}$

FIGURE 1.34. Physical model for waves around the cylindrical island over a parabolic mound.

With the above principle for discretization, equations (286), (287), and (288) can all be converted into linear equations involving only tridiagonal matrices, and can thus be effectively solved.

4.3.3. Applications. Applications of the direct high-order difference schemes to the computation of nonlinear wave transformation under various situations can be found in Wei and Kirby (1995) and Zheng et al. (1998). Shown in Figure 1.34 is the physical model for the combined wave diffraction and refraction by a cylindrical island over a parabolic mound studied by Zheng et al. (1998). The grid size in the computation was $\Delta x = \Delta y = 1000$ m, which is rather coarse as compared to the diameter of the island. The time increment was set to $\Delta t = 0.25$ s. The numerical solution on the wave height distribution around the island was found to be in good agreement with the analytic solution by Vastano and Reid (1976) as demonstrated in Figure 1.35.

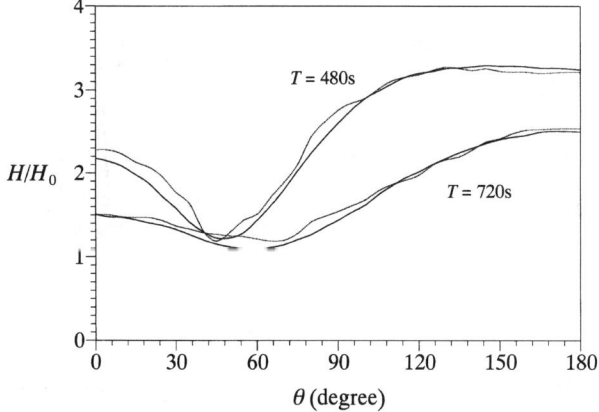

FIGURE 1.35. Wave height distribution around the cylindrical island.

Figures 1.36 and 1.37 show the computational results for the propagation of a initial rise of sea level, which has a background in the simulation of earthquake-induced Tsunami wave propagation in coastal waters. The initial wave profile was given by

$$\eta_0 = 0.045 \exp\{-2[(x - 6.75)^2 + (y - 6.75)^2]\} \quad \text{(m)} \tag{313}$$

The grid size in the computation was $\Delta x = \Delta y = 0.15$ m. The time step was $\Delta t = 0.025$ s. The front of the computed transient wave was noted to be well explained by Mei's (1989) linear analytic solution.

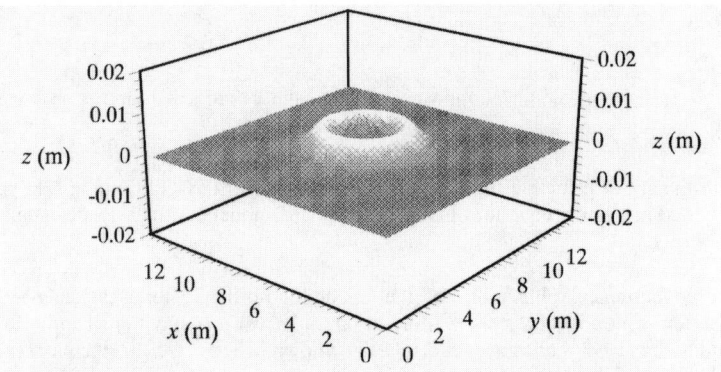

FIGURE 1.36. Cylindrical wave induced by the initial elevation of water surface at $t = 1$ s.

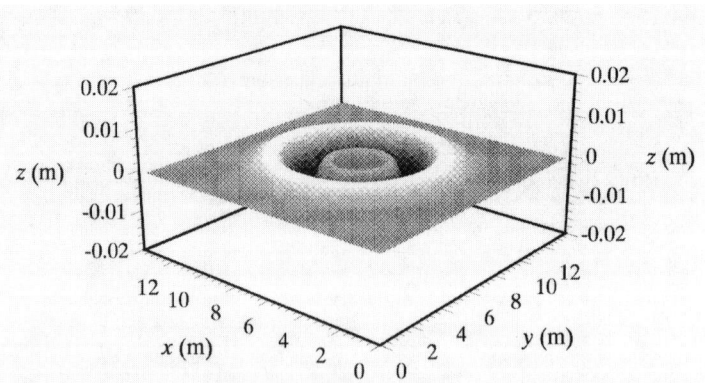

FIGURE 1.37. Cylindrical wave induced by the initial elevation of water surface at $t = 2$ s.

5. SUMMARY

A detailed review has been provided of the numerical models for coastal water waves, particularly those described by the mild-slope wave equations (both in the elliptic form and in the hyperbolic form), the nonlinear shallow-water wave equations, and the Boussinesq equations.

The wave equations of much concern to the development were derived by vertically integrating the basic equations for general free-surface flows and by the variational approach as well. Empirical dissipation models, with the effects of both the bottom friction and the turbulent mixing related diffusion incorporated, were discussed in detail. A brief description of the typical boundary conditions appearing in practical computations was also given.

For steady-state waves over slowly varying bottoms, high-resolution finite difference schemes, including the optimal scheme and the finite analytic scheme, were presented. Finite element methods for the mild-slope wave equation were also examined with an emphasis on the infinite elements and the super element, which are of particular importance in computational water wave mechanics.

For the numerical models for transient waves, an unstructured mesh oriented hybrid method was illustrated for the time-dependent mild-slope wave equation and a high-resolution alternating direction implicit method for the classical wave equation. For the nonlinear shallow-water wave equations, some details were given on the method of finite characteristics and the operator-splitting method. For the Boussinesq equations the high-resolution methods were stressed.

ACKNOWLEDGMENTS

This chapter was prepared when the first author was a faculty member of the Coastal Engineering Laboratory at Department of Civil Engineering, University of Tokyo. The authors would like to thank Professor Akira Watanabe, Department of Civil Engineering, University of Tokyo, for the daily communication in the laboratory, which contributed to the formation of the many ideas presented here. Valuable comments by Professor Pin-nam Lin, Institute for Water Conservancy and Hydroelectric Power Research (IWHR), are also appreciated.

REFERENCES

Abbott, M. B. (1979). *Computational Hydraulics: Elements of the Theory of Free-Surface Flows.* Pitman.

Abbott, M. B., McCowan, A. D., and Warren, I. R. (1981). Numerical modeling of free-surface flows that are two-dimensional in plan. In *Transport Models for Inland and Coastal Waters* (ed. Fischer, H. B.). Academic Press, 222–283.

Abbott, M. B., McCowan, A. D., and Warren, I. R. (1985). Accuracy of short-wave numerical models. *J. Hydraulic Engineering, ASCE, 110*(10), 1287–1301.

Abbott, M. B., Petersen, H. M., and Skovgaard, O. (1978). On the numerical modeling of short waves in shallow water. *J. Hydraulic Research, 16,* 173–203.

Anastasiou, K. and Chan, C. T. (1997). Solution of 2D shallow water equations using the finite volume method on unstructured triangular meshes. *International J. Numerical Methods in Fluids, 24,* 1225–1245.

Arad, M., Yakhot, A., and Ben-Dor G. (1996). High-order accurate discretization stencil for an elliptic equation. *International J. Numerical Methods in Fluids, 23,* 367–377.
Becker, A. A. (1992). *The Boundary Element Method in Engineering.* McGraw-Hill.
Behrendt, L. and Jonsson, I. G. (1984). The physical basis of the mild slope equation. *Proceedings of the 19th Coastal Engineering Conference,* ASCE, 941–954.
Benque, J. P., Cunge, J. A., Feuillet, J., Hauguel, A., and Holly, F. M. (1982). New method for tidal current computation. *J. Waterway, Port, Coastal and Ocean Engineering Division, ASCE, 108*(3), 396-417.
Berkhoff, J. C. W. (1972). Computation of combined refraction and diffraction. *Proceedings of the 13th Coastal Engineering Conference,* ASCE, 471–490.
Bettess, P., Liang, S. C., and Bettess, J. A. (1984). Diffraction of waves by semi-infinite breakwater using finite and infinite elements. *International J. Numerical Methods in Fluids, 11,* 813–832.
Bettess, P. and Zienkiewicz, O. C. (1977). Diffraction and refraction of surface waves using finite and infinite elements. *International J. Numerical Methods in Engineering, 11,* 1271–1290.
Blaschak, J. G. and Kriegsmann, G. A. (1988). A comparative study of absorbing boundary conditions. *J. Computational Physics, 77,* 109–139.
Butler, D. S. (1960). The numerical solution of hyperbolic systems of partial differential equations in three independent variables. *Proceedings of the Royal Society, London, A225,* 232–252.
Chamberlain, P. G. and Porter, D. (1995). The modified mild-slope equation. *J. Fluid Mechanics, 291,* 393–407.
Chen, C. J. and Chen, H. C. (1984). Finite analytic method for unsteady two dimensional Navier–Stokes equations. *J. Computational Physics, 53,* 209–226.
Chen, C. J. and Li, P. (1979). Finite differential method in heat conduction: application of analytic solution technique. ASME paper No. 79-WA/HT-50.
Chen, H. S. (1990). Infinite elements for water wave radiation and scattering. *International J. Numerical Methods in Fluids, 11,* 555–569.
Chen, H. S. and Mei, C. C. (1974). Oscillations and wave forces in a man-made harbor in the open sea. *Proceedings 10th Symposium on Naval Hydrodynamics,* 573–596.
Chwang, A. T. and Chen, H. C. (1987). Optimal finite difference method for potential flows. *J. Engineering Mechanics, ASCE, 113*(11), 1759–1773.
Copeland, G. J. M. (1985). A practical alternative to the mild-slope wave equation. *Coastal Engineering, 9,* 125–149.
Dalrymple, R. A., Kirby, J. T., and Hwang, P. A. (1984). Wave diffraction due to areas of energy dissipation. *J. Waterway, Port, Coastal and Ocean Engineering, ASCE, 110*(1), 67–79.
Dingemans, M. W. (1997). *Water wave Propagation over Uneven Bottoms,* Parts 1 & 2. World Scientific.
Engquist, B. and Majda, A. (1977). Absorbing boundary conditions for the numerical simulation of waves. *Mathematics of Computation, 31,* 629–651.
Eric C. C., Ishikura, M., and Aono, T. (1997). Open boundaries in nonlinear dispersive wave models, *Proceedings of Coastal Engineering, JSCE, 44,* 46–50 (in Japanese).
Fairweather, G. and Mitchell, A. R. (1965). A high accuracy alternating direction method for the wave equation. *J. Institute of Mathematics and its Applications, 1,* 309–316.
Galvin, C. J. (1972). Wave breaking in shallow waters. In *Waves on Beaches* (ed. Meyer, R. E.). Academic Press, 413–456.
Givoli, D. (1991). Non-reflecting boundary conditions. *J. Computational Physics, 94,* 1–29.
Gobbi, M. F. and Kirby, J. T. (1996). A forth order Boussinesq-type wave model. *Proceedings of the 25th Coastal Engineering Conference,* ASCE, 1116–1129.
Goda, Y. (1970). A synthesis of breaker indices. *Proceedings Japan Society of Civil Engineers,* No. 180, 39–49 (in Japanese).
Goda, Y. (1975). New wave pressure formulae for composite breakwaters. *Proceedings of the 14th Coastal Engineering Conference, ASCE,* 1702–1720.
He, S. and Lin, B. (1986). 2-D tidal flow by operator-splitting method. *Acta Oceanologica Sinica, 5*(4), 508–516.
Horikawa, K. (1989). *Nearshore Dynamics and Coastal Processes: Theory, Measurement and Predictive Model.* University of Tokyo Press.

Houston, J. R. (1978). Interaction of tsunamis with the Hawaiian Islands calculated by a finite element numerical model. *J. Physical Oceanography, 8*, 93–102.
Hughes, T. J. R. (1987). *The Finite Element Method*. Prentice-Hall.
Hwang, L. S. and Tuck, E. O., (1970). On the oscillation of harbors of arbitrary shape. *J. Fluid Mechanics, 42*, 447–464.
Isobe, M. (1986). A parabolic refraction-diffraction equation in the ray-front coordinate system. *Proceedings of the 20th Coastal Engineering Conference, ASCE, 1*, 306–317.
Ito, Y. and Tanimoto, K. (1972). A method of numerical analysis of wave propagation. *Proceedings of the 13th Coastal Engineering Conference, ASCE*, 503–522.
Jonsson, I. G. (1966). Wave boundary layers and friction factors. *Proceedings of the 10th Coastal Engineering Conference, ASCE*, 127–148.
Kabiling, M. B. and Sato, S. (1993). Two-dimensional nonlinear dispersive wave-current and three dimensional beach deformation model. *Coastal Engineering in Japan, 36*, 196–212.
Kamphuis, J. W. (1975). Friction factors under oscillatory waves. *Proceedings Waterway, Harbor, and Coastal Engineering Division, ASCE, 101*(2), 135–144.
Katopodes, N. and Strelkoff, T. (1978). Computing two-dimensional dam-break flood waves. *J. Hydraulic Engineering Division, ASCE, 104*(9), 1269–1288.
Katopodes, N. and Strelkoff, T. (1979). Two-dimensional shallow water-wave models. *J. Engineering Mechanics Division, ASCE, 105*(2), 317–334.
Kawahara, M. (1985). *Finite Element Methods for Fluid Flows*. Gihodo Publishing Co. (in Japanese).
Kirby, J. T., Lee, C., and Rasmussen, C. (1992). Time-dependent solutions of the mild-slope wave equation. *Proceedings of the 23th Coastal Engineering Conference, ASCE*, 391–404.
Lai, C. (1986). Numerical modeling of unsteady open-channel flow. In *Advances in Hydroscience*, Volume 14 (ed. Yen, B. C.). Academic Press, 161–333.
Lapidus, L. and Pinder, G. F. (1982). *Numerical Solution of Partial Differential Equations in Science and Engineering*. Wiley.
Lee, J. J. (1971). Wave induced oscillations in harbors of arbitrary geometry. *J. Fluid Mechanics, 45*, 375–394.
Lin, P. (1952). Numerical analysis of continuous unsteady flow in open channels. *Trans. American Geophysical Union, 33*(2), 227–234.
Lin, P. (1980). The state-of-the-art of the research on open channel flows. In *Progress in Hydro-science and Hydropower Development*. Shuili Publishing Co. (in Chinese).
Lin, P., Dai, Z., and Li, K. (1982). Unsteady flow studies in China. *J. Waterway, Port, Coastal and Ocean Engineering Division, ASCE, 108*(3), 343–360.
Lin, P., Zhao, X. H., and Shi, L. B. (1980). Effects of closing off an estuary on the tides in an adjoining bay. *Shuili Xuebao*, No. 3, 16-26 (in Chinese).
Liu, S. K. and Leendertse, J. J. (1978). Multidimensional numerical modeling of estuaries and coastal seas. In *Advances in Hydroscience*, Volume 11 (ed. Chow, V. T.). Academic Press, 95–164.
Louaked, M. and Hanich, L. (1998). TVD scheme for the shallow water equations. *J. Hydraulic Research, 36*(3), 363–378.
Luke, J. C. (1967). A variational principle for a fluid with a free surface. *J. Fluid Mechanics, 27*, 359–397.
Madsen, P. A. and Larsen, J. (1987). An effective finite difference approach to the mild slope equation. *Coastal Engineering, 11*, 329–351.
Madsen, P. A., Murray, R., and Sorensen, O. R. (1991). A new form of Boussinesq equations with improved linear dispersion characteristics. *Coastal Engineering, 15*, 371–388.
Madsen, P. A. and Sorensen, O. R. (1992). A new form of Boussinesq equations with improved linear dispersion characteristics; Part 2, A slowly-varying bathymetry. *Coastal Engineering, 18*, 183–204.
Mahmood, K. and Yevjevich, V. (1975). *Unsteady Flow in Open Channels*, Volumes I & II. Water Resources Publications.
Marchuk, G. I. (1982). *Methods of Numerical Mathematics*. Springer-Verlag.
Mei, C. C. (1978). Numerical methods in water-wave diffraction and radiation. *Annual Review of Fluid Mechanics, 10*, 393–416.

Mei, C. C. (1989). *The Applied Dynamics of Ocean Surface Waves.* World Scientific.
Mei, C. C. and Chen, H. S. (1975). Hybrid element method for water waves. *Proceedings of the 2nd Annual Symposium of Waterway, Harbor and Coastal Engineering Division, ASCE,* 63–81.
Miles, J. and Munk, W. (1961). Harbor paradox. *J. Waterway and Harbor Division, ASCE, 87*(3), 111–130.
Minghan, C. G. and Causon, D. M. (1996). High-resolution finite-volume method for shallow water flows. *J. Hydraulic Engineering, ASCE, 124*(6), 605–614.
Nwogu, O. (1993). Alternative form of Boussinesq equations for nearshore wave propagation. *J. Waterway, Port, Coastal and Ocean Engineering, 119*(6), 618–638.
Orlanski, I. (1976). A simple boundary condition for unbounded hyperbolic flows. *J. Computational Physics, 21,* 251–269.
Ostendorf, D. W. and Madsen, O. S. (1979). An analysis of longshore currents and associated sediment transport in the surf zone. *MIT Report,* Sea Grant 79–13.
Oyama, T. and Nadaoka, K. (1991). Development of a numerical wave tank for analysis of nonlinear and irregular wave field. *Fluid Dynamics Research, 8,* 231–251.
Panchang, V. G., Pearce, B. R., Wei, G., and Benoit, C. R. (1991). Solution of the mild-slope wave problem by iteration. *Applied Ocean Research, 13*(4), 187–199.
Peregrine, D. H. (1967). Long waves on a beach. *J. Fluid Mechanics, 27,* 815–827.
Pos, J. D. and Kilner, F. A. (1987). Breakwater gap wave diffraction: an experimental and numerical study. *J. Waterway, Port, Coastal and Ocean Engineering, ASCE, 113*(1), 1–21.
Radder, A. C. (1979). On the parabolic equation method for water wave propagation. *J. Fluid Mechanics, 95,* 159–176.
Rosser, J. B. (1975). Nine-point difference solutions for Poisson equation. *Computers and Mathematics with Applications, 1,* 351–360.
Rouse, H. (1950). *Engineering Hydraulics.* Wiley.
Sakai, F. and K. Tsukioka (1975). Application of the finite element method to surface wave analysis. *Coastal Engineering in Japan, 18,* 45–52.
Sato, N., Isobe, M., and Izumiya, T. (1990). A numerical model for calculating wave height distribution in a harbor of arbitrary shape. *Coastal Engineering in Japan, 33*(2), 119–131.
Sawaragi, T. (1973). Wave breaking. *Proceedings of the Annual Summer Seminar on Hydroscience, 73-B-2,* 1–38.
Schaffer, H. A., Digaard, R., and Madsen, P. A. (1992). A two-dimensional surf zone model based on the Boussinesq equations. *Proceedings of the 24th Coastal Engineering Conference, ASCE,* 576–589.
Schaffer, H. A. and Madsen, P. A. (1995). Further enhancements of Boussinesq-type equations. *Coastal Engineering, 26,* 1–14.
Schaffer, H. A., Madsen, P. A., and Digaard, R. (1993). A Boussinesq model for wave breaking in shallow water. *Coastal Engineering, 20,* 185–202.
Sleath, J. F. A. (1984). *Sea Bed Mechanics.* Wiley.
Smith, R. and Sprinks, T. (1975). Scattering of surface waves by a conical island. *J. Fluid Mechanics, 72,* 373–384.
Sollitt, C. K. and Cross, R. H. (1972). Wave transmission through porous breakwaters. *Proceedings of the 13th Coastal Engineering Conference, ASCE,* 1827–1846.
Stoker, J. J. (1957). *Water Waves.* Interscience.
Tan, W. (1992). *Shallow Water Hydrodynamics.* Elsevier.
Townson, J. M. (1974). An application of the method of characteristics to tidal calculations in (x-y-t) space. *J. Hydraulic Research,* 12(4), 499–523.
Trefethen L. N. and Halpern, L. (1986). Well-posedness of one-way wave equations and absorbing boundary conditions. *Mathematics of Computation, 47,* 421–435.
Vastano, A. C. and Reid, R. O. (1967). Tsunami response for islands: verification of a numerical procedure. *J. Marine Research, 25,* 129–139.
Watanabe, A., Hara, T., and Horikawa, K. (1984). Study on breaking condition for compound wave train. *Coastal Engineering in Japan, 27,* 71–82.
Watanabe, A. and Maruyama, K. (1986). Numerical modeling of nearshore wave field under combined refraction, diffraction and breaking. *Coastal Engineering in Japan, 29,* 19–39.
Wei, G. and Kirby, J. T. (1995). A time-dependent numerical code for extended Boussinesq equations. *J. Waterway, Port, Coastal and Ocean Engineering, ASCE, 121*(5), 251–261.

Wei, G., Kirby, J. T., Grilli, S. T., and Subramanya, R. (1995). A fully nonlinear Boussinesq model for surface waves; Part 1, Highly nonlinear unsteady waves. *J. Fluid Mechanics, 294,* 71–92.
Williams, R. G., Darbyshire, J., and Holmes, P. (1980). Wave refraction and diffraction in a caustic region: a numerical solution and experimental validation. *Proceedings of the Institute of Civil Engineers, 69,* 635–649.
Witting, J. M. (1984). A unified model for the evolution of nonlinear water waves. *J. Computational Physics, 56,* 203–236.
Yu, X. (1995). Diffraction of water waves by porous breakwaters. *J. Waterway, Port, Coastal and Ocean Engineering, ASCE, 121*(6), 275–282.
Yu, X. (1996a). Finite analytic method for the mild slope wave equation. *J. Engineering Mechanics, ASCE, 122*(2), 109–115.
Yu, X., (1996b). Oscillation in a coupled bay-river system; 1, analytic solution. *Coastal Engineering, 28,* 147–264.
Yu, X. (1998a). Finite difference methods for the reduced water wave equation. *Computer Methods in Applied Mechanics and Engineering, 154,* 265–280.
Yu, X. (1998b). One-way wave equations as nonreflecting boundary conditions. *Proceedings of Coastal Engineering, JSCE, 45,* 31–35 (in Japanese).
Yu, X. and Chwang, A. T. (1994). Wave induced oscillation in harbors with porous breakwaters. *J. Waterway, Port, Coastal and Ocean Engineering, ASCE, 120*(2), 125–144.
Yu, X., Isobe, M. and Watanabe, A. (1992a). Numerical computation of wave transformation on beaches. *Coastal Engineering in Japan, 35*(1), 1–19.
Yu, X., Isobe, M. and Watanabe, A. (1992b). Finite element solution of wave field around structures in nearshore zone. *Coastal Engineering in Japan, 35*(1), 21–33.
Yu, X., Isobe, M. and Watanabe, A. (1992c). A numerical model for nonlinear wave transformation in nearshore zone by multi-step finite characteristic methods. *Coastal Engineering in Japan, 35*(1), 35–49.
Yu, X., Isobe, M., and Watanabe, A. (1995). Wave breaking on submerged plate. *J. Waterway, Port, Coastal and Ocean Engineering, ASCE, 121*(2), 105–113.
Yu, X. and Togashi, H. (1994). Irregular waves over an elliptic shoal. *Proceedings of the 24th Coastal Engineering Conference, ASCE,* 746–760.
Yu, X. and Togashi, H. (1995). A sinusoidal shape function for finite element analysis of wave motion. *Proceedings of the 2nd International Conference on Hydro-Science and Engineering,* 1460–1467.
Yu, X. and Togashi, H., (1996). Oscillation in a coupled bay-river system; 2, numerical method. *Coastal Engineering, 28,* 165–182.
Zhao, D. H., Shen, H. W., Tabios, G. Q., Lai, J. S., and Tan, W. Y. (1994). Finite-volume two-dimensional unsteady flow model for river basins. *J. Hydraulic Engineering, ASCE, 120*(7), 863–883.
Zheng, P., Yu, X., and Isobe, M. (1998). High-order finite difference method for Boussinesq equations. *Proceedings of Coastal Engineering, JSCE, 45,* 21–25 (in Japanese).
Zienkiewicz, O. C., Emson, C., and Bettess, P. (1983). A novel boundary infinite element. *International J. Numerical Methods in Engineering, 19,* 393–404.
Zienkiewicz, O. C. and Morgan, K. (1983). *Finite Elements and Approximation Method.* Wiley.
Zienkiewicz, O. C. and Taylor, R. L. (1991). *The Finite Element Method,* Volumes I & II. McGraw-Hill.

CHAPTER 2
REVETMENT PROTECTION FOR COASTAL AND SHORELINE STRUCTURES EXPOSED TO WAVE ATTACK

N. W. H. ALLSOP
*University of Sheffield
c/o HR Wallingford
Wallingford, UK*

K. J. McCONNELL
*HR Wallingford
Wallingford, UK*

Notation	2.1
1. Introduction	2.3
2. Revetment Systems and Components	2.4
2.1. Types and Elements of Revetment Systems	2.4
2.2. Design Life and Performance	2.6
3. Failure Modes of Revetments	2.6
4. Revetment Design	2.9
4.1. Design Process and Main Parameters	2.9
4.2. Design Methods—Protection Size and Thickness	2.10
4.3. Discussion and Comparison of Design Methods	2.14
4.4. Design Methods—Hydraulic Performance	2.16
Conclusions	2.22
Acknowledgments	2.22
References	2.22

NOTATION

A Slope coefficient for overtopping
A_e Erosion area on cross-section
A_s Area of slab/block
B Slope coefficient for overtopping
b Block width

C	Coefficient for cover layer thickness
C_r	Reflection coefficient
C_w	Mat coefficient for concrete mattresses
D_{f15}	15% sieve value for filter material
D_n	Nominal particle diameter, defined as $(M/\rho_r)^{1/3}$ for rock and $(M/\rho_c)^{1/3}$ for concrete armor
D_{n50}	Nominal particle diameter calculated from the median particle mass M_{50}
D_{85}, D_{50}, D_{15}	Particle diameters at 85%, 50%, and 15% nonexceedance levels
d_s	Scour depth
F_S	Factor of safety
g	Gravitational acceleration
H	Wave height, from trough to crest
H_D	Design wave height
H_{max}	Maximum wave height
H_s	Significant wave height, average of highest one-third of wave heights
H_{sb}	Breaking (significant) wave height
H_{si}	Inshore significant wave height
$H_{2\%}$	Wave height at 2% exceedance level
h	Water depth
h_s	Water depth at toe of structure
L	Wavelength
l	Block length
K_D	Empirical stability coefficient used in Hudson equation
K_{RR}	Stability coefficient used in Hudson's equation for riprap armor
k	Wave number = $2\pi/L$
k_Δ	Layer thickness coefficient used in *Shore Protection Manual* (1984)
k_f	Filter layer permeability
k_g	Permeability of geotextile
k_s	Permeability of underlying material
k'	Cover/armor layer permeability
M	Mass of armor unit
M_{50}	Median mass of armor unit derived from the mass distribution curve
N_z	Number of (zero-crossing) waves
N_s	Stability number, defined as $H_s/\Delta D_{n50}$
$N_{d\%}$	Number of units displaced, expressed as percent of units in area of armor considered
n_v	Volumetric porosity
O_{90}	Pore size of geotextile
P	Notional permeability factor, used in calculation of armor stability
p	Encounter probability
Q^*	Dimensionless overtopping discharge
q	Mean overtopping discharge per unit length of revetment
R_c	Crest freeboard, level of crest less static water level
R^*	Dimensionless run-up
r	Roughness coefficient for wave run-up and overtopping
S_b	Coefficient in Klein Breteler and Bezuijen's (1991) method for blockwork stability
S_c	Coefficient in Yarde et al.'s (1996) method for slabbing/blockwork stability
S_d	Damage number for (rock) armored slopes = A_e/D_{n50}^2
s_m	Steepness of mean wave period = $2\pi H_s/gT_m^2$

s_{mo}	Offshore mean wave steepness
s_p	Steepness of peak wave period = $2\pi H_s/gT_p^2$
T	Wave period of regular wave
T_m	Mean wave period
T_p	Peak wave period
T_R	Return period
t_a	Armor/cover layer thickness
t_f	Filter layer thickness
U	Coefficient of uniformity = D_{60}/D_{10}
w	Width of gap between slabs/blocks
α	Structure front slope angle to horizontal
β	Angle of wave attack to structure/bed contour alignment
β_0	Wave direction in deep water
Δ	Reduced relative density, e.g. $(\rho_r/\rho_w) - 1$
γ_{br}	Breaker index
μ	Coefficient of friction
ρ	Mass density, usually of fresh water
ρ_w	Mass density of sea water
ρ_r, ρ_c	Mass density of rock or concrete armor units
ξ_m	Iribarren number, $\tan\alpha/s_m^{0.5}$, calculated in terms of mean wave steepness
ξ_{mcr}	Critical Iribarren number, distinguishing between plunging and surging waves for van der Meer's rock armor formulae
ξ_p	Iribarren number, $\tan\alpha/s_p^{0.5}$, calculated in terms of peak wave steepness

A wide range of different revetment systems have been devised to protect coastal or shoreline embankments from erosion by waves. The revetment system must resist direct wave attack, reduce hydraulic forces on underlying materials, and contribute to dissipation of wave energy. This chapter identifies the main types of revetment systems, describes the main prediction methods used to design revetment armoring, and for the first time offers a simple method to compare design thicknesses of different systems.

The hydraulic performance of a revetment slope also significantly influences the overall structure cost. This chapter therefore also identifies simple prediction methods to calculate crest levels for given overtopping performance and to estimate wave reflections.

1. INTRODUCTION

Many coastal slopes, embankment dams, and related structures are protected against wave attack by revetment armoring systems. The revetment system must resist direct wave attack, reduce hydraulic forces on underlying materials, and contribute to dissipation of wave energy.

Historically, design methods have often been material-specific, have required different design parameters, and varied considerably in reliability. As a result, engineers have experienced particular difficulties when comparing alternative options for new structures, and risk analysts have been restricted in calculations of residual life and failure risk.

A research group in the United Kingdom, supported by industrial partners, considered this problem and investigated empirical and theoretical methods for revetment design. The design manual edited by McConnell (1998) covers general aspects of revetment systems, gives methods for the determination of hydraulic boundary conditions, and brings together design methods for a number of different construction materials in a more unified framework.

This chapter summarizes results of this research to draw together guidance on design and optimization of revetments to resist direct and indirect effects of wave attack. A simple unified method to compare thicknesses of different systems is then suggested. The overtopping performance of a flood embankment, usually given by the mean overtopping discharge rate, is often the most important factor in setting the structure crest level, hence overall dimensions and cost. Good overtopping performance often requires increased roughness of the revetment armor and/or shallower slope angles. The need to reduce wave reflections, often to reduce the risk of local scour, also requires energy dissipating shallow slopes and/or increased armor roughness. Simple methods to determine hydraulic performance are therefore also described, with particular emphasis on prediction of wave overtopping, and wave reflections.

2. REVETMENT SYSTEMS AND COMPONENTS

2.1. Types and Elements of Revetment Systems

Revetment systems may be constructed on a coastline or along a shoreline to serve a number of different functions:

- protection to exposed faces of coastal flood embankments
- reduction of overtopping discharges, and/or limiting damage to crest/rear faces
- erosion protection at cliff bases and around new reclamations
- protection to windward faces of reservoir dams and embankments

A revetment system will normally include a number of key components (Figure 2.1):

- cover or armor layer
- filter layer
- toe protection
- crest protection

Of these components, the armor layer is generally the main element, dimensioned to ensure stability. A stable and safe construction must ensure that the other elements of the revetment are adequately dimensioned. Those dimensions or material specifications are determined in turn from the thickness or unit dimension of the outer armor layer.

The main types of materials likely to be used in revetment protection are:

- rock armor (narrow graded) or riprap (wider graded)
- stone or concrete blocks—loose or grouted
- concrete blocks—loose, cable-tied, solid, cellular, interlocking, articulated
- concrete slabs—usually cast in situ
- fabric-formed concrete mattresses
- asphaltic materials—open stone asphalt, bitumen grouting, asphaltic concrete

The required revetment performance and definition of structural integrity will depend on the type of structure. For a rigid revetment, minimal distortion or damage will be

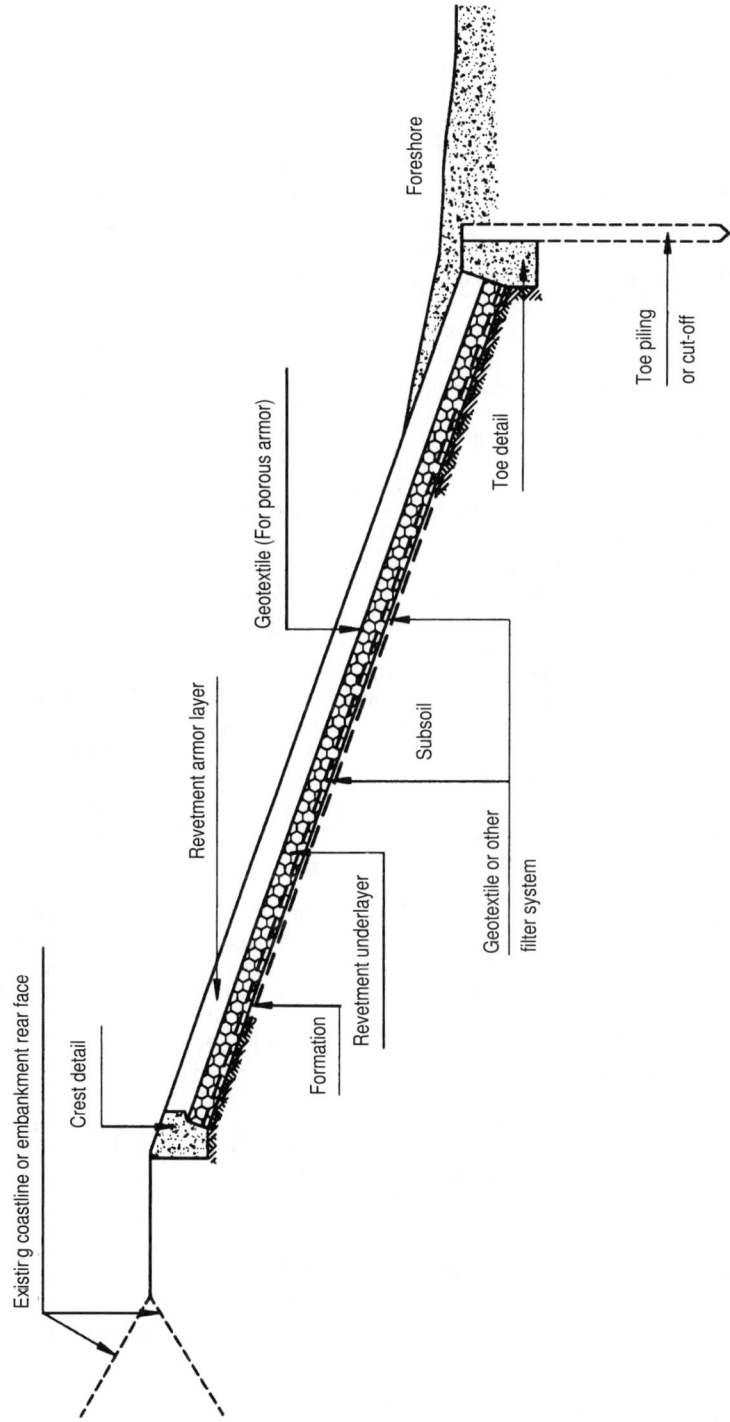

FIGURE 2.1. Revetment system and components.

permissible. For a flexible revetment, a greater degree of movement may be allowed, with structural integrity being maintained. For rock armor structures, some movement of rocks is allowed. This is termed damage, denoted by the parameter S_d, and can be up to 5% without the structure being considered as having failed. Rock armored structures can be "self-healing" with displaced armor elements settling into voids in the armor layer, and helping to maintain stability. This effect is reduced for structures with steeper slopes.

If the structure is required to limit overtopping, then an acceptable degree of overtopping will be determined, based on guidelines given in the CIRIA/CUR *Rock Manual* edited by Simm (1991). Limits chosen depend on land use behind the revetment and the consequences of overtopping.

2.2. Design Life and Performance

Most revetment systems are designed and constructed for a specific design life, which will depend on the function of the revetment. Temporary works will have a relatively short design life and may be removed or demolished when no longer required. A more permanent structure such as a revetment for a flood embankment is likely to have a longer design life, typically 50 to 100 years.

Such structures will be designed to withstand a design event (or events), probably defined by a particular combination of wave condition and water level. Each design event will have a selected return period, T_R, which indicates the annual likelihood of the event being exceeded. For example, a wave height with a return period of 50 years has an annual likelihood of being equalled or exceeded of 0.02.

This likelihood of exceedance of the design event during the life of the structure is the encounter probability. As the return period of the design event increases, the encounter probability decreases. Guidance is given in BS6349 Part 1, British Standards Institution (BSI) (1984), on determining the encounter probability of an event of duration T_R. The encounter probability, p, of an event of a return period, T_R, during the design life, N, can be calculated from a function plotted in BS6349 Part 7, BSI (1991):

$$p = 1 - (1 - 1/T_R)^N \qquad (1)$$

The design return period should normally be significantly longer than the design life. Due to the stochastic nature of wave conditions and water levels, there is always a finite probability that the design event will be exceeded during the design life, so the designer should identify a suitable level of risk of exceedance and design the structure for an event with the corresponding return period. For example, for a design life of 50 years, the 1000 year event has a 5% probability of exceedance, but for a structure designed to a return period equal to the design life, there is a 63% probability that the design event will be exceeded.

Definition of design conditions becomes more complicated when two or more variables (e.g., wave height and water level) are considered. The design return period should then represent the likelihood that both (or all) variables are exceeded at the same time. Specialist studies may be required to establish this joint probability.

3. FAILURE MODES OF REVETMENTS

A number of important failure modes must be considered in revetment design (Figure 2.2):
- scour
- wave uplift pressures

REVETMENT PROTECTION FOR COASTAL AND SHORELINE STRUCTURES

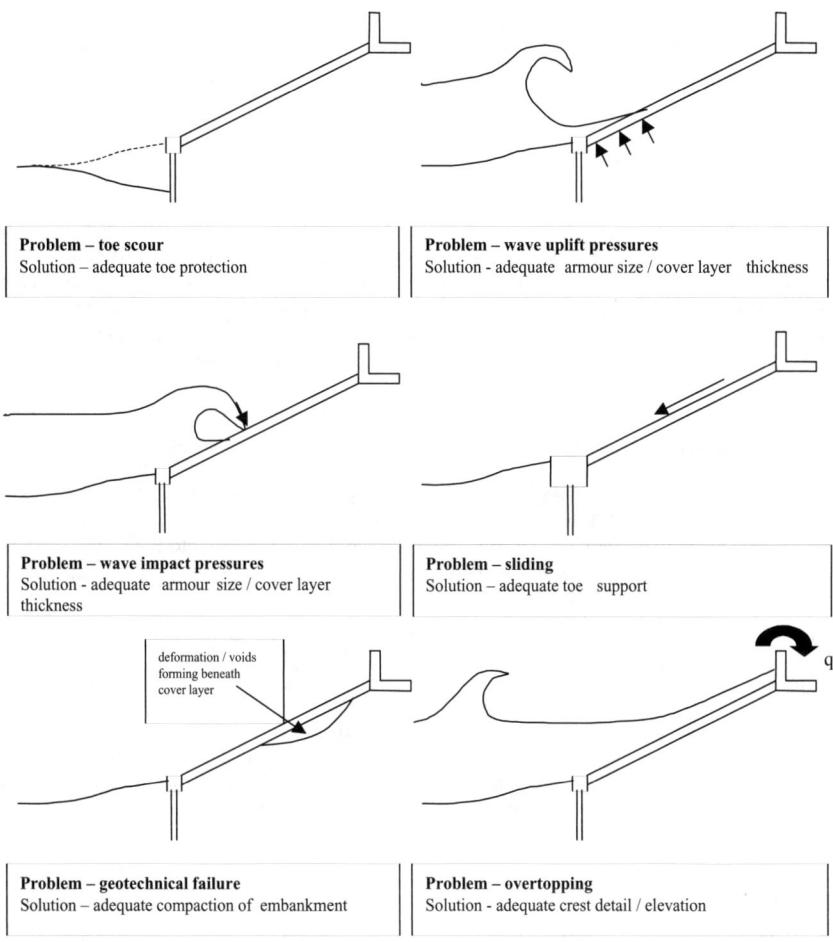

FIGURE 2.2. Revetment failure modes.

- wave impact pressures
- sliding
- geotechnical failure
- overtopping

Scour

Local scour at the toe of the revetment may occur. This may cause structural instability if the toe of the revetment is undermined. In order to avoid this, an adequate toe detail should be constructed. This may comprise a toe beam with sheet piles extending down to below the maximum predicted scour depth (as shown in Figure 2.2) or an extension of the cover layer along the bed, allowing this to drop into the scour hole that is created. This last approach is particularly effective in the case of fabric mattress or cable-tied concrete block revetments.

Wave Uplift Pressures

Generally, most design methods assess the stability of the armor layer of the revetment. One of the main causes of failure of the armor, particularly if it has a low permeability, will be due to uplift pressures acting on the underside of the revetment, particularly during wave draw-down. The majority of stability design methods do not explicitly determine these uplift pressures but they are based on empirical design methods derived from model tests that identify the onset of instability. Indirectly, they therefore determine the armor size or cover layer thickness required to resist these uplift pressures.

Wave Impact Pressures

Localized breaking wave impacts may cause damage to the structure, such as local deformations or loss of individual cover layer elements. The armor or cover layer should be of an adequate size to prevent this. Wave impacts generally occur close to the water level, so there may only be a need to provide increased protection in the region of the structure exposed to the still water level.

Sliding

Sliding of cover layer or armor may occur if the structure does not have adequate resistance at the toe to resist the down-slope component of the armor weight. In particular, excessive uplift pressures may significantly reduce the restraining friction force between the armor and underlying layers, increasing the likelihood of sliding.

Geotechnical Failure

A number of types of geotechnical failure may occur. Slumping of poorly compacted subsoil may cause significant deformation of the armor layer and depending on the type of revetment construction, may lead to brittle failure of the structure.

Settlement of subsoil may also occur. Construction may allow for some settlement, but significant differential settlement along a section of the structure is likely to lead to local failure. Revetment systems should generally be placed on well-compacted subsoil or fill and be configured to resist the external (wave-driven) loadings. Revetment systems should not be expected to contribute to geotechnical stability of the slope or embankment.

Overtopping

If wave overtopping exceeds an acceptable threshold, then this may cause damage to the crest of the structure or dangerous conditions behind the crest. Design should ensure an adequate crest detail to prevent structural failure due to washing out of material from the structure crest or damage to the rear slope. If overtopping is heavy, then the crest detail may comprise a wave wall to reduce overtopping, or the crest level may have to be increased until an acceptable overtopping discharge is achieved. This is discussed further in Section 4.4.

4. REVETMENT DESIGN

4.1. Design Process and Main Parameters

A simple summary of the overall design procedure is shown in Figure 3.3. The reader should be aware that the design process is, however, circular not linear, with each of the main design parameters being revisited two or three times as different factors are refined.

At the first stage, the designer will gather together as much information as possible on the conditions at the site of interest, the desired performance characteristics, and any constraints that may apply. If information on the primary hydraulic conditions is not yet available, then initial estimates of wave and/or current conditions and on water levels will need to be derived. Methods are given in McConnell (1998) and other references such as Simm (1991) and Simm et al. (1995) for the initial derivation of wave conditions and water levels at coastal locations. McConnell (1998) also give guidelines for the derivation of wave conditions on inland reservoirs, based on guidance by Yarde et al. (1996).

The full range of possible construction options should be considered at an early stage. The initial (external) dimensions of structure cross-sections should be determined,

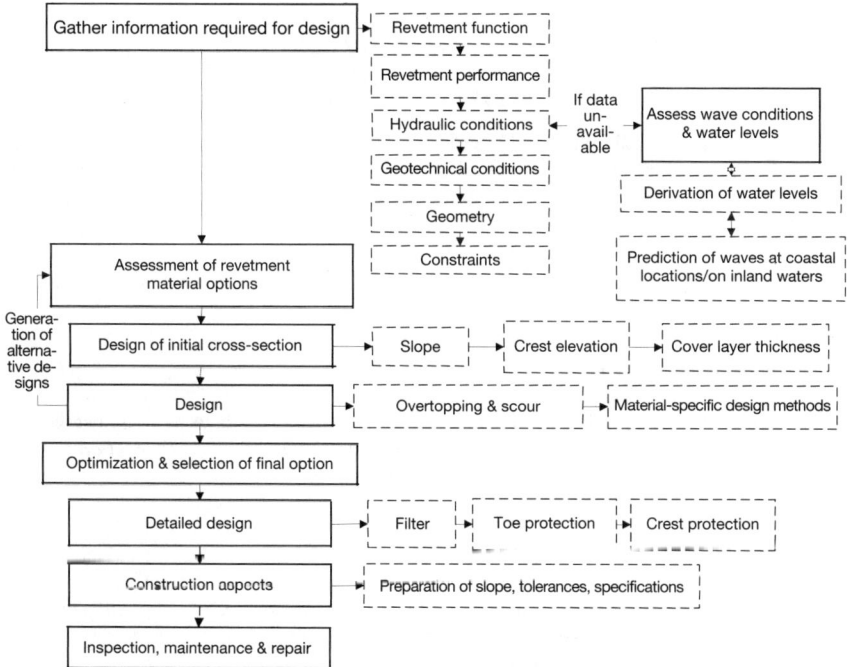

FIGURE 2.3. Flowchart of design process.

particularly the structure slope and crest elevation. Each of the options can then be developed in more detail using available design guidance, before the final option is selected, detailed design completed, and other issues such as construction methodology, maintenance, and repair are addressed.

During the review of design methodologies for revetments, it became apparent that there are many sources of design guidance available, but that this guidance is usually very specific to a particular construction material, as discussed in Section 2. Historically, this diversity of design methods has hindered consideration of different construction options during the design phase, and it has not assisted appraisal of alternative designs at tender stage. The manual edited by McConnell (1998) therefore sought to present design methods for different revetment materials in ways that allow the easiest possible comparisons of results, and that overall approach is used in this chapter.

The main parameters required for design of revetment protection against wave attack are:

- wave height, usually H_s
- wave period, T_m and/or T_p
- structure slope angle (seaward face), α
- armor size/armor layer thickness, M_{50}, D_{n50}, t_a
- structure permeability, k
- local bed and water levels

Information on geotechnical properties of the material on which the revetment will be constructed is also desirable, assisting in identification and prevention of likely failure modes.

4.2. Design Methods—Protection Size and Thickness

The design methods available in the literature for a range of material types were reviewed with a steering committee of revetment designers and suppliers in order to determine the practicalities of applying the design methods. The main methods reviewed were as follows:

- rock armor—Hudson; van der Meer
- concrete blocks—Klein Breteler and Bezuijen; Yarde et al.
- fabric-formed concrete mattresses—Sprague and Koutsourais; Pilarczyk
- asphalt—Rijkwaterstaat

Rock Armor. Hudson (1958) gives a method for the design of rock armor against (originally) regular waves developed in the later 1950s and used in the *Shore Protection Manual,* CERC (1984). Hudson's equation can be expressed as

$$H_s/\Delta D_{n50} = (K_D \cot \alpha)^{1/3} \qquad (2)$$

where Δ is the relative density of the armor material and K_D is a stability coefficient, dependent on the type of waves at the structure—either breaking or nonbreaking—and whether the armor stone is rough or smooth. Although Hudson did not explicitly include the wave period in this equation, consideration of whether waves are breaking or nonbreaking will take some account of this.

Van der Meer (1988) developed further methods for the sizing of rock armor for stability. He considered that two different types of waves may attack the structure—either plunging or surging. The equations given are

$$H_s/\Delta D_{n50} = 6.2\, P^{0.18}\, (S_d/\sqrt{N_z})^{0.2}\, \xi_m^{-0.5} \quad \text{for plunging waves} \qquad (3a)$$

$$H_s/\Delta D_{n50} = 1.0\, P^{-0.13}\, (S_d/\sqrt{N_z})^{0.2}\, \sqrt{\cot \alpha}\, \xi_m^P \quad \text{for surging waves} \qquad (3b)$$

where P is a notional permeability factor; S_d is the damage number $= A_e/D_{n50}^2$, A_e is the erosion area; N_z is the number of zero-crossing waves; ξ_m is the Iribarren number, $\tan \alpha/s_m^{0.5}$ (where α is the slope of the structure and s_m is the mean sea steepness, $s_m = 2\pi H_s/gT_m^2$).

The permeability factor, P, varies from $P = 0.1$ for rock on an impermeable slope to $P = 0.6$ for a mound of rock armor (Figure 2.4).

The transition between plunging and surging waves is given by

$$\xi_{cr} = (6.2\, P^{0.31}\, (\tan \alpha)^{0.5})^{1/(P+0.5)} \qquad (4)$$

where $\xi_m < \xi_{cr}$ indicates plunging waves and $\xi_m > \xi_{cr}$ indicates surging waves. Example curves for a 1 in 2 structure and permeabilities $P = 0.1$ and 0.6 are shown in Figure 2.5.

Hudson's and van der Meer's methods were derived from model tests and have been widely used in coastal engineering for the design of rock armored breakwaters and revetments. Modifications to van der Meer's equations have been developed to allow for con-

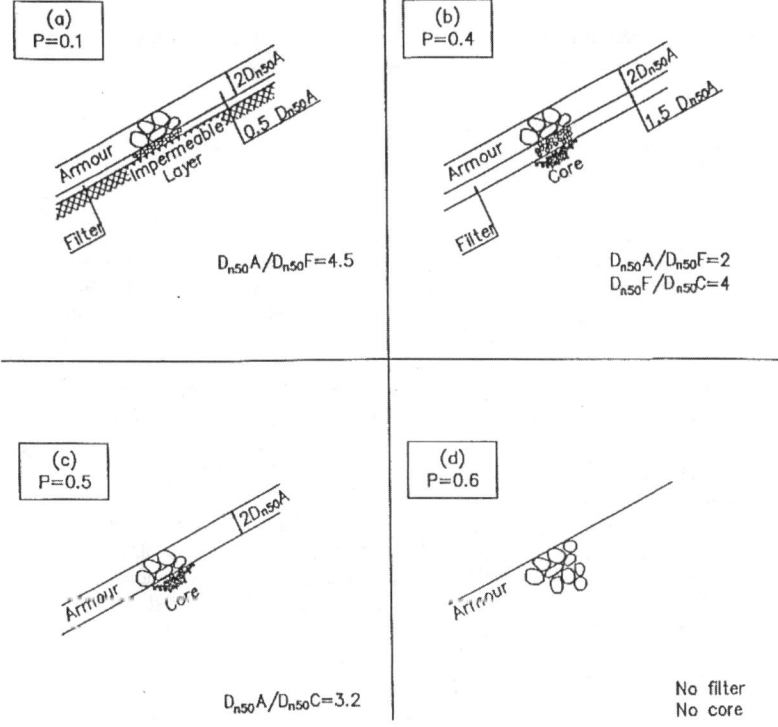

FIGURE 2.4. Values of Van der Meer's permeability factor.

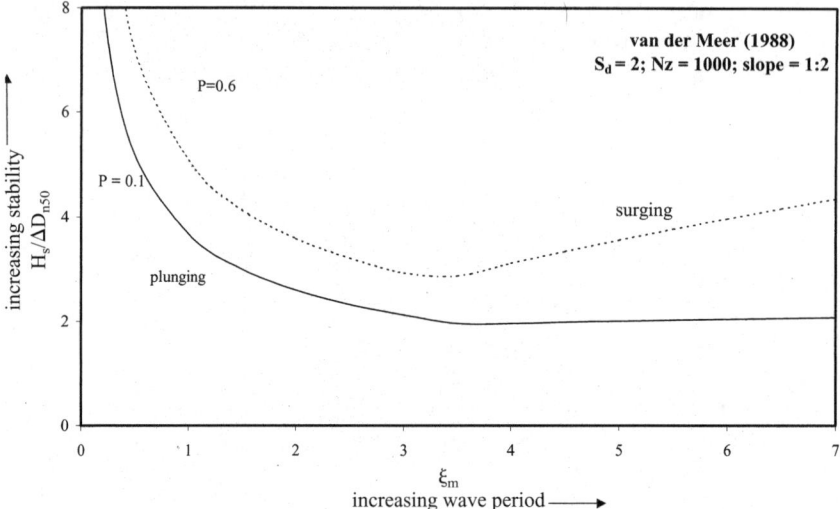

FIGURE 2.5. van der Meer's method for rock armor—influence of permeability.

sideration of different armor layer thicknesses and the influence of rock armor grading; see Allsop (1990), Bradbury et al. (1990), and Simm (1991).

Concrete Blocks and Slabs. Design methods for determining the required concrete blockwork thickness were devised by Klein Breteler and Bezuijen (1991) from model experiments. They give the equation

$$H_s/\Delta t_a = S_b \, \xi_p^{-0.67} \tag{5}$$

where S_b is a stability coefficient that indicates the stability of the revetment as a function of the relative permeabilities of armor and underlayers. Klein Breteler and Bezuijen give upper and lower limits of S_b based on findings of experimental work. For example, for a blockwork revetment of normal stability, with a granular filter beneath, the range of S_b is given as 3.7 to 8 (Figure 2.6).

Using the highest value of S_b will give the slab thickness beyond which the structure will be unstable. Using the lowest value of S_b will give the slab thickness that will be stable under the design conditions. In practice, as little guidance is available on performance of the structure between these two limits, the designer will be likely to use the more conservative value of the two.

Yarde et al. (1996) gave particular consideration to the case of reservoir dams, and to wave conditions generated over limited fetch lengths, such as occur on inland bodies of water, where wave periods are short and steepnesses large. They extended Klein Breteler and Bezuijen's general method for short wave periods and for larger slabs, and suggested a modified equation. This is illustrated in Figure 2.7 for 1 m × 1 m and 3 m × 3 m slabs:

$$H_s/\Delta t_a = S_c/\xi_p \tag{6}$$

Yarde et al. quantified the stability coefficient, S_c, as a function of the dimensions and permeabilities of the cover layer and underlayer:

$$S_c = 3.3 \, \ln[(\sqrt{A_s}/t_f) \, (w/D_{f15})^{0.1}] + 4.0 \tag{7}$$

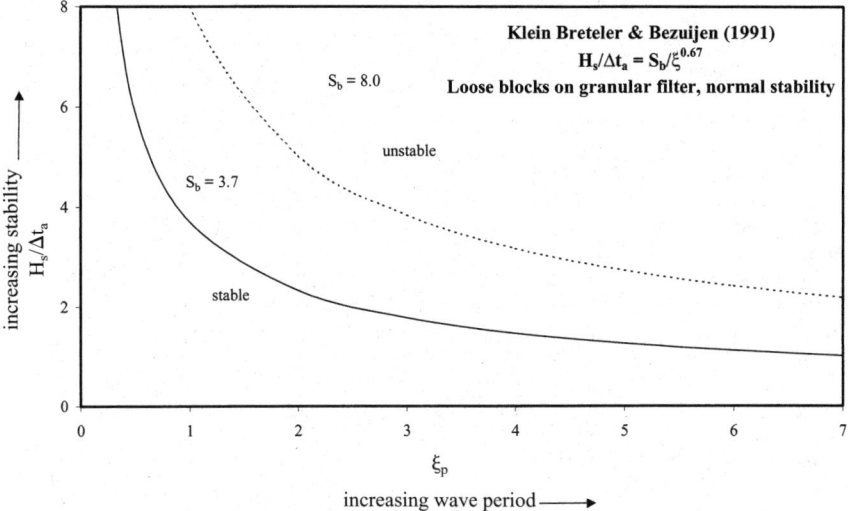

FIGURE 2.6. Klein Breteler and Bezuijen's method for concrete blockwork.

where A_s is the slab area; t_f is the thickness of the filter layer; w is the gap between slabs representing drainage area or cover layer permeability; D_{f15} is the 15% nonexceedance diameter of the filter layer material (obtained from the grading curve), taken to indicate the permeability of the filter layer.

When considering the design methodologies available for blockwork, it is apparent that most methodologies are based on loose or interlocking blocks of low permeability.

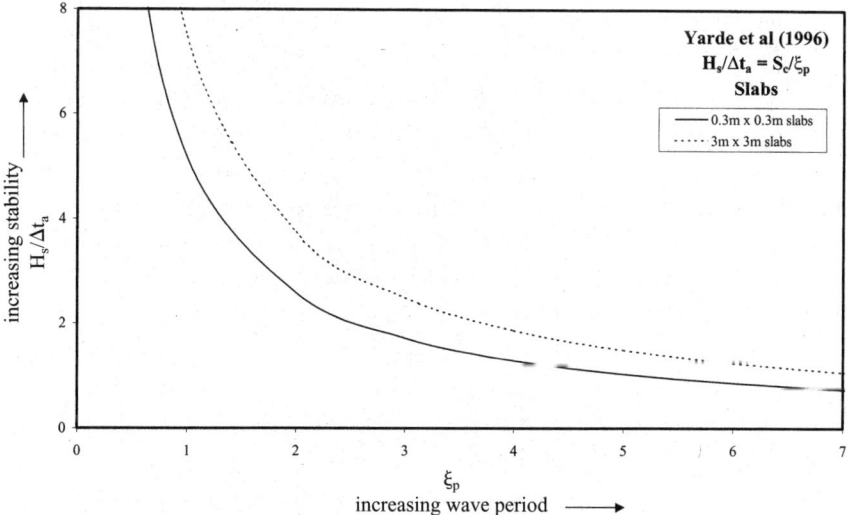

FIGURE 2.7. Yarde et al.'s method for concrete slabs.

Many proprietary cellular block systems are available that have much higher permeabilities, but do not necessarily fit the standard prediction methods. Discussions with the research steering committee and analysis of model test data from Lindenberg (1983) suggested that the method of Klein Breteler and Bezuijen could be applied with careful choice of stability coefficient S_b.

Often, concrete blocks may be tied with nylon or steel cables and used to create mats of blockwork and thus facilitate placement. While it is generally agreed that the cables should not be considered to provide additional strength in the structure, allowing thinner blocks to be used, they may help to provide a restraining force in the event that sliding failure of the revetment occurred.

Model studies by Lindenberg (1983) and practical experience suggest that gravel blinding of blockwork may help provide an increase in the stability of concrete blocks. This enhancement would, however, only work if both the concrete blocks and the blinding material were sufficiently robust/durable to resist crushing over the life of the revetment. There is much debate as to whether this stability increase can be relied upon and it is recommended that this improvement be ignored in ultimate stability calculations.

Fabric-Formed Concrete Mattresses. For fabric-formed concrete mattresses, guidance was given by Sprague and Koutsourais (1992), although this method did not include the influence of wave period. As an alternative, Pilarczyk (1998) discusses the extension of the method provided by Klein Breteler and Bezuijen (1991) for concrete blockwork to the design of fabric-formed concrete mattresses, using modified values of the stability coefficient S_b. Comparison of both methods with field data of existing structures suggested that the methods may be conservative.

Asphaltic Materials. Comprehensive guidance for design of asphaltic structures is given by Rijkwaterstaat (1985). Methods are given for designing open stone asphalt and impermeable asphaltic concrete against wave impact loading and for the design of impermeable asphaltic structures to prevent failure due to excess uplift pressures, and for the design of bitumen-grouted rock armor.

One distinguishing feature of the methods for asphaltic construction in comparison with other methods is that they generally follow the procedure of explicitly calculating the loading on the structure and using the load to determine the structure dimensions. Other methods rely on empirical relationships derived from model data and will not actually calculate the force on the structure. The one exception to this is bituminous grouting of rock armor, where Rijkwaterstaat suggest a modification to the Hudson formula for rock armor due to the increase in strength and stability given by the grout.

4.3. Discussion and Comparison of Design Methods

On examination, it becomes clear that although the design methods are material-specific, many of them, with the exception of methods for asphaltic construction, have a similar form:

$$H_s/\Delta D = f(k, \xi) \qquad (8)$$

The function $H_s/\Delta D$ (or $H_s/\Delta t_a$) is referred to as the stability number and is conveniently used in a number of design methods for different materials to describe the relationship between the armor size or cover layer thickness and the wave height.

Generally, the stability of a revetment system will decrease as wave steepness decreases, when wave period and the Iribarren number both increase (see Figures 2.5 to 2.7). Both Klein Breteler and Bezuijen and Yarde et al.'s methods show this effect, as for a given wave height and armor thickness, an increase in wave period moves a design point from the stable zone into the unstable (or less stable) zone. The exception is that van der Meer's equations show that stability increases following the transition from plunging to surging waves (Figure 2.5). This increase is small for low-permeability systems, $P = 0.1$, but is more marked for higher permeabilities ($P = 0.4$ to 0.6). This may be illustrated by considering the effects of down-rushing flows. For a given wave height, long period (low steepness) waves will give the greatest differential head between the external water level and the phreatic surface within the structure. It is this head difference that is most likely to destabilize the armor. For higher overall permeabilities, this head difference is reduced, so water flows out more freely from the structure, and the armor layer resists a bigger wave height for a given thickness, t_a.

This particular effect is not apparent in the methods for blockwork by Klein Breteler and Bezuijen (Figure 2.6) and for slabs by Yarde et al. (Figure 2.7) as these systems do not mobilize the same effective permeability (to wave action) as rock armor or riprap. Blockwork and slab systems therefore have much lower permeabilities than even the lower permeability rock structures, and do not resist the effects of long-period wave action.

So, these methods may be compared, but only with care, as some of the key parameters differ. First, van der Meer's method uses ξ_m, but the blockwork methods use ξ_p. Relationships between T_m and T_p are not simple, but where such a simplification is needed for first estimates, the manual suggests the relationship $T_m = 0.82\ T_p$.

The main output parameter also differs—the blockwork methods calculate $H_s/\Delta t_a$ but van der Meer's method calculates $H_s/\Delta D_{n50}$. For van der Meer's tests, the layer thickness, t_a, was approximately 2.2 stone diameters, giving $t_a = 2.2\ D_{n50}$. Using these two conversions, and considering only a rock armored revetment of cot $\alpha = 2$, subject to a storm of $N_z = 1000$ waves, and allowable damage $S_d = 2$, van der Meer's equations for $P = 0.1$ and $P = 0.6$ have been superimposed on Klein Breteler and Bezuijen's and Yarde et al.'s methods for blocks and slabs in Figure 2.8 to give a general set of prediction curves for layer thickness.

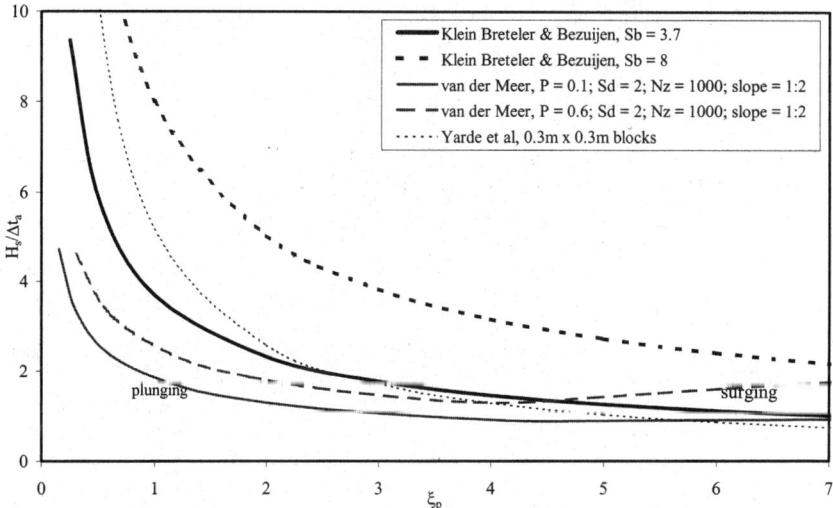

FIGURE 2.8. Comparison of methods.

The greatest layer thickness is required for rock armor on an impermeable foundation. Less, and smaller, rock armor is needed if the overall construction is made more permeable ($P = 0.6$). A particular benefit arises for surging wave conditions, $\xi_p > 4$, where the required layer thickness now reduces. In this region, the thickness of rock armor would now be less than the more conservative estimate of blockwork thickness using Klein Breteler and Bezuijen, $S_b = 3.7$. Over plunging wave conditions, the use of blockwork will require thinner layers, and slabbing thinner still. The wide bands of uncertainty for blockwork implicit by the suggested range of S_b from 3.7 to 8 result in a wide range of possible solutions using this method. The (slightly) more complex method of Yarde et al. suggests that Klein Breteler and Bezuijen's approach may be conservative for plunging waves, but is less so for surging conditions. This may have to be viewed with some caution, noting that Yarde et al. restricted their method to a limited range of wave steepness and slope angles.

These methods illustrate why it is very important not to underestimate the wave period to be used in design. The prediction methods illustrate why lower steepness waves are of more concern for stability and indeed for other hydraulic responses, such as wave run-up and overtopping. The occurrence of swell waves should therefore be carefully considered when identifying the conditions to be used in design discussed by Hawkes et al. (1997a,b).

4.4. Design Methods—Hydraulic Performance

Overtopping. The primary duty of a revetment system is to protect the underlying materials, as discussed in Sections 4.1–4.3. Once that requirement is satisfied, most revetment systems must also contribute to the dissipation of wave energy in order to limit wave run-up and overtopping, and perhaps reflections. Considerations of the limiting overtopping discharges that may be permitted usually set the crest level of the revetment, or require additional elements, such as a wave wall. These factors may significantly influence the overall cost of the revetment, particularly on an embankment, so accurate predictions of overtopping performance are essential to confirm that the design crest level delivers overtopping below safe discharge limits. Such overtopping limits are often set by one or more of the following:

a) the total volume that can be accepted over the defense during a design event

b) the peak discharge rate to limit damage to the crest or rear face

c) the peak discharge rate to maintain protection required for people using the defence and/or to limit damage to the property defended

Overtopping does not itself constitute structural failure, but heavy overtopping may constitute a functional failure of the defense, and may in turn endanger the stability of the crest or other elements of the structure. Very small average discharges can be permitted for rural or urban sea walls where property can be damaged by heavy overtopping, or personnel put at risk.

A number of limiting discharges were suggested by Owen (1980), and were later presented in the form of Figure 2.9 for sea walls and related structures by Simm (1991) and Besley (1999). These suggested discharges include limits for damage to the crest or rear face of an embankment sea wall, and other limits suggested for safety of pedestrians or vehicle drivers in the area immediately behind the defence. This diagram illustrates the exponential nature of wave overtopping where a small change to overtopping is measured by a factor of 2 or 3.

Many rural sea walls are of simple section with seaward slopes of 1:2–1:6. In the Unit-

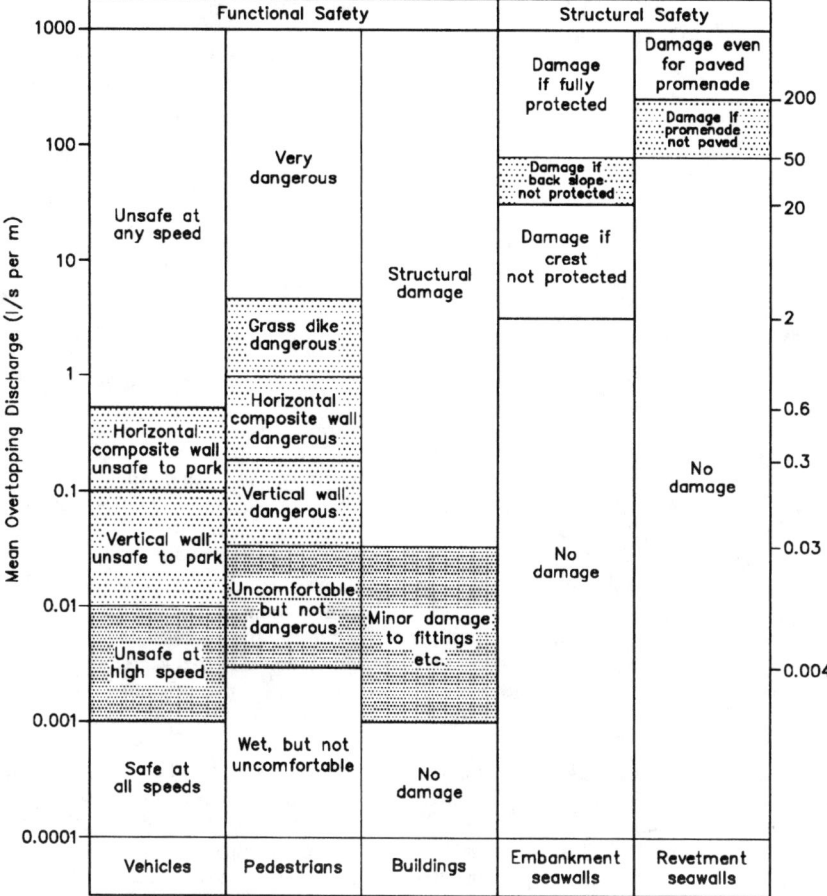

FIGURE 2.9. Suggested limits to overtopping discharges.

ed Kingdom, rural flood embankments often use clay fill and are constructed to slopes between 1:2 and 1:4. Along the North Sea coast of the Netherlands, Germany, and Denmark, the dikes may have been built over sand dunes and seaward slopes are often shallower, 1:4–1:8. Wave walls are relatively seldom included for rural flood defences, but are more common on mass-concrete seawalls defending urban connurbations.

The overtopping performance of embankment sea walls under random waves was studied intensively in the late 1970s, and the mean wave overtopping discharge (q) under random waves was related to the crest freeboard relative to water level (R_c) and wave parameters H_s and T_m. The prediction method developed by Owen (1980) relates dimensionless parameters Q^* and R^* by an exponential equation with a roughness coefficient, r, and coefficients A and B for each slope angle:

$$Q^* = A \exp(-BR^*/r) \qquad (9)$$

TABLE 2.1. Values of coefficients A and B for Owen's formula

Slope	A	B
1 : 1.0	0.0079	20.1
1 : 1.5	0.0102	20.1
1 : 2.0	0.0125	22.1
1 : 3.0	0.0163	31.9
1 : 4.0	0.0192	47.0

where the dimensionless discharge Q^* and freeboard R^* parameters are defined:

$$Q^* = q/(gT_m H_s) \tag{10}$$

$$R^* = R_c/T_m(gH_s)^{0.5} \quad \text{or} \quad (R_c/H_s)(s_m/2\pi)^{1/2} \tag{11}$$

and r is the relative run-up or roughness coefficient, with values given in Table 2.2.

Values of coefficients A and B have been derived for slopes from 1:1.0 to 1:4.0 and are summarised in Table 2.1.

The form of equation (9) is illustrated in Figure 2.10, where Q^* is plotted against R^* using values of A and B from Table 2.1. Over the normal range of freeboards, the discharge characteristics for slopes 1:1, 1:1.15, and 1:2 are similar, but overtopping reduces significantly for slopes shallower than 1:2. For structures with particularly small relative freeboards and/or large wave heights, and very low values of R^*, the prediction lines come together at one point, indicating that slope angle and relative roughness no longer have much influence in controlling overtopping. At this point, the slope is said to be "drowned out."

Owen's method was developed initially for smooth slopes only, but the use of the roughness factor, r, allowed its use for rough and even armored slopes. Since 1980, alternative prediction methods for armored slopes have been explored, but no new method has yet proved more reliable. The chief advantages of Owen's method are its simplicity and availability of data to support particular coefficients. The disadvantages are that the method was not explicitly developed for armored slopes, the coefficient r is not always constant; and values of r have not been measured for some types of armor.

TABLE 2.2. Roughness or relative run-up coefficients

Armor type	r
Smooth, impermeable	1.0
Rough concrete	0.85
Pitched stone in mortar	0.75–0.8
Rock armor, 2 layers	0.5–0.6
Hollow cube armor units, one layer	0.5
Dolos armor units	0.4
Stabit armor units	0.35–0.4
Tetrapods, two layers	0.3

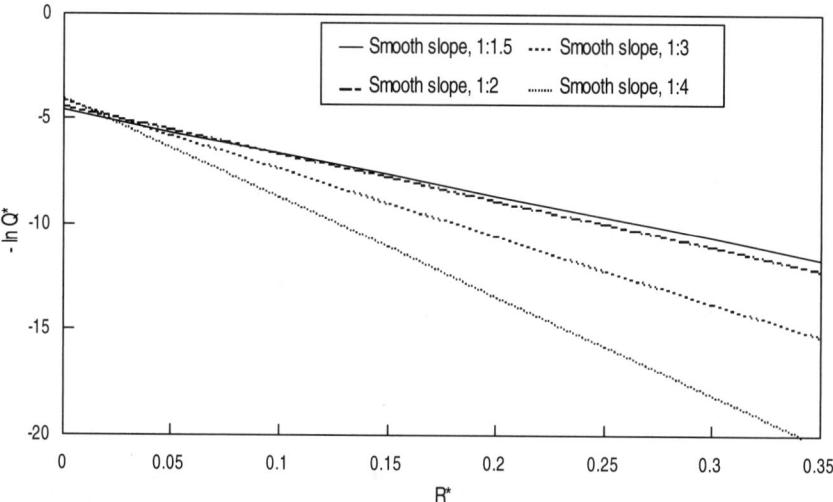

FIGURE 2.10. Wave overtopping, use of Owen's empirical method for smooth slopes.

Wave Reflections. All coastal structures reflect wave energy, often described by a reflection coefficient, C_r, defined in terms of incident and reflected wave heights or energies: $C_r = H_r/H_i$ or $C_r = (E_r/E_i)^{1/2}$. A simple formula derived for regular waves by Seelig (1983) and extended for random waves by Allsop (1990, 1995) gives C_r in terms of the mean surf similarity parameter or Iribarren number ξ_m:

$$C_r = a\xi_m^2/(b + \xi_m^2) \qquad (12)$$

Tests by Allsop and Channell (1989) measured wave reflections for wave conditions in the range $0.004 < s_m < 0.052$ and smooth slopes and for rock armor in the range $0.6 < H_s/\Delta D_{n50} < 1.9$ on a rock underlayer on an impermeable slope. Measurements of C_r for smooth and rock armored slopes are shown in Figure 2.11, and values for coefficients a and b in equation (12) are listed in Table 2.3.

Toe Scour. Analysis of performance of seawalls in the United Kingdom by Stickland and Haken (1986) demonstrated that local scour at the toe of a revetment is one of the most frequent causes of undermining and failure of revetments. The structure should be designed to prevent this, perhaps by including toe protection to a depth greater than the predicted scour depth.

For sand beaches, the *Shore Protection Manual* (CERC, 1984) suggests that the depth of scour might equal the maximum unbroken wave height:

$$d_s = H_{\max b} \qquad (13)$$

where $H_{\max b} = 1.8 H_{sb}$.

Powell (1987) noted that orbital velocities in a scour hole of this depth can still be significant, and this could underpredict scour depths. It has been found that for $0.02 < s_{mo} < 0.04$ the scour depth for steeply sloping or near vertical structures is approximately equal to the incident unbroken wave height. Maximum scour occurs when the structure is at the plunge point of breaking waves.

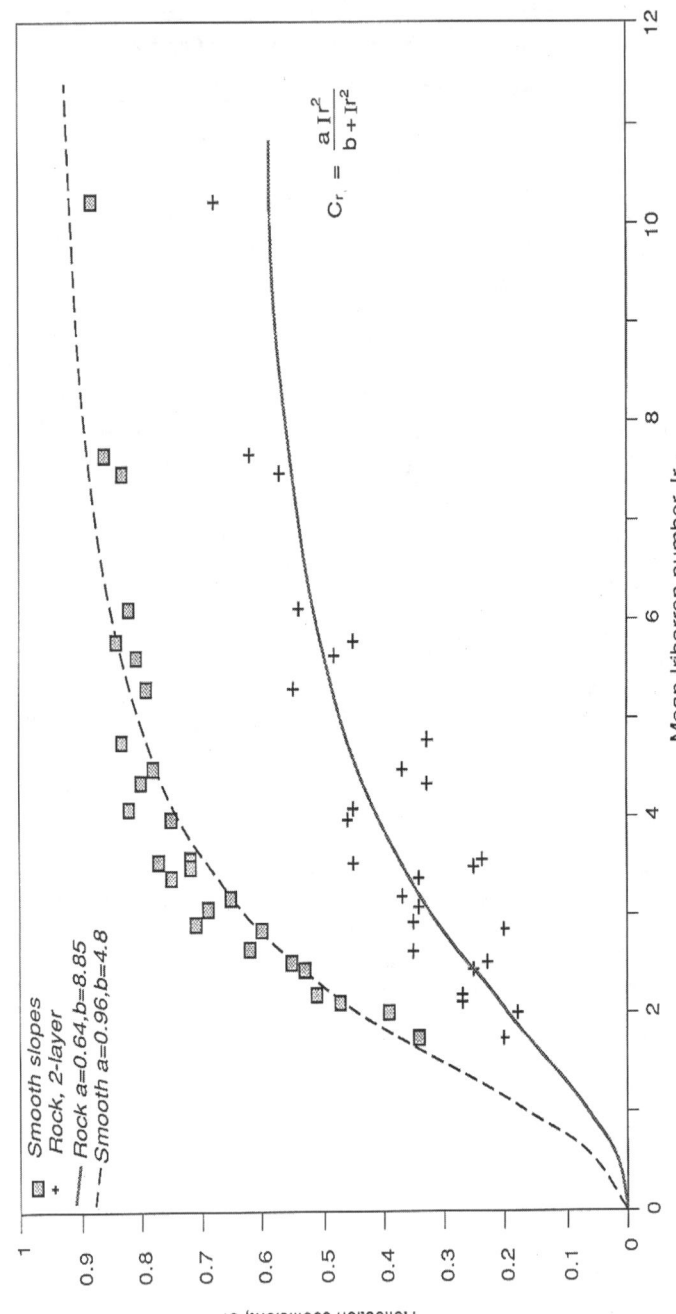

FIGURE 2.11. Reflection performance of smooth and armored slopes.

TABLE 2.3. Coefficients for Allsop's empirical formula for reflections

Armor type	a	b
Smooth slope	0.96	4.80
Rock, two layer	0.64	8.85
Rock, one layer	0.64	7.22

The depth of scour is proportional to the reflection coefficient of the structure. For smooth impermeable structures, reflections can be minimised by adopting shallow slopes, typically flatter than 1 in 3. If the structure is protected by permeable rock armor with two or more layers of rock, steeper slopes can be adopted.

For shingle beaches, Powell (1989) gives dimensionless design graphs for calculation of scour depth. These relate the dimensionless scour depth d_s/H_s to the mean wave steepness s_{mo} and the local water depth h_s/H_s. An example is given in Figure 2.12 for a vertical wall and a storm duration of 3000 waves.

Some further guidance for the calculation of scour at the toe of sloping structures is given by Powell (1989) based on results of model tests.

- For impermeable sloping structures of 1:1.5 to 1:2 there is no significant reduction in scour depth compared to that at a vertical wall.

- Reducing the slope of an impermeable structure to 1:3 can reduce local scour typically by 25% but up to a maximum of 50% compared to a vertical wall.

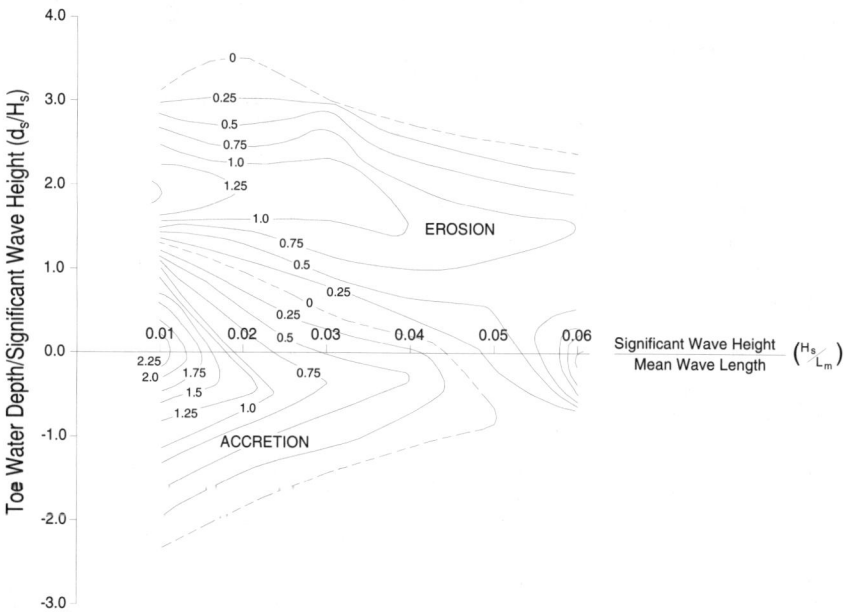

FIGURE 2.12. Prediction graphs for scour depths, after Powell (1989).

- Rock armored revetments generally show no susceptibility to local scour and may even cause accretion.

Toe details should ensure that protection is provided to a depth below that of the predicted scour to prevent undermining of the structure.

If scour appears to be a significant problem, then modifications to the structure to reduce scour may be necessary, such as adopting a shallower slope angle.

CONCLUSIONS

- Generalized design guidance has been compiled, bringing together design methods and practical advice for a wide range of revetment systems.
- Comparison of design methods for rock and blockwork structures has indicated that the influence of wave period is more significant than has been commonly assumed. Careful consideration should be given to uncertainties in the derivation of design wave periods, particulary the likelihood of swell wave.
- At design stage, hydraulic performance criteria can be as important, or more important, than the details of the armor system.

ACKNOWLEDGMENTS

The research studies that produced the "Revetment Manual" edited by McConnell (1998), on which this chapter is based, were completed as part of a project funded by the Department of the Environment, Transport and the Regions Construction Sponsorship Directorate under contract number CI 39/5/98.

The authors gratefully acknowledge guidance given by the project steering group comprising representatives from revetment suppliers: M. Hawkswood (Proserve Ltd.), C. Booth and G. Baker (MMG Civil Engineering Systems Ltd.), B. Howden and D. Ballinger (Grass Concrete Ltd.), C. Cridge (Ruthin Precast Concrete Ltd.), N. Leguit and R. Smith (Hesselberg Hydro).

The authors also acknowledge useful discussions with D. Hay-Smith (R.H. Cuthbertson and Partners), J. Ackers (Binnie Black and Veatch) and P. Besley (HR Wallingford) in the development of this paper and the unified analysis method suggested here for armor thickness.

The first author also wishes to acknowledge the support of the University of Sheffield.

REFERENCES

Allsop, N. W. H. (1990). Rock armoring for coastline and shoreline structures: hydraulic model studies on the effects of armor grading. Report EX 1989, HR Wallingford, UK.

Allsop, N. W. H. (1994). Wave overtopping of sea walls, breakwaters and shoreline structures. Technical Note 633 in Proc. ICE, Water, Maritime and Energy, December 1994, publn. Thomas Telford, London.

Allsop, N. W. H. (1995). Vertical walls and breakwaters: optimization to improve vessel safety and wave disturbance by reducing wave reflections. In *Wave Forces on Inclined and Vertical Wall Structures*, pp. 232–258, Ed. N. Kobayashi & Z. Demirbilek. ASCE, New York.

Allsop, N. W. H. (1998). Hydraulic performance and stability of coastal structures. In *Concrete in Coast Protection,* Ed. R. T. L. Allen. Thomas Telford, London.

Allsop, N. W. H., and Channell, A. R. (1989). Wave reflections in harbours: Reflection performance of rock armoured slopes in radom waves. Report OD 102, Hydraulics Research, Wallingford, UK.

Besley, P. B. (1999). Overtopping of seawalls—design and assessment manual. R & D Technical Report W 178, ISBN 1 85705 069 X, Environment Agency, Bristol.

Bezuijen, A., and Klein Breteler, M. (1996). Design formulae for block revetments. *Journal of Waterway, Port, Coastal and Ocean Engineering, 12,* 6, ASCE, USA.

Bradbury, A. P., Latham, J-P., and Allsop, N. W. H. (1990). Rock armor stability formulae: influence of stone shape and layer thickness. 22nd ICCE, Delft, ASCE, New York.

British Standards Institution (1984). *British Standard Code of Practice for Maritime Structures. Part 1. General Criteria.* BS 6349: Part 1: 1984 (and Amendments 5488 and 5942). British Standards Institution, London.

British Standards Institution (1991). *British Standard Code of Practice for Maritime Structures. Part 7. Guide to the Design and Construction of Breakwaters.* BS 6349: Part 7: 1991. British Standards Institution, London.

CERC (1984). *Shore Protection Manual.* 2 Vols. US Army Corps of Engineers, US Government Printing Office, Washington.

CIRA/CUR (19xx). *Rock Manual.*

Hawkes, P. J., Coates, T. T., and Allsop, N. W. H. (1997a). What happens if your design conditions ignore swell? Proceedings Ocean Wave Measurement and Analysis, publn. ASCE, New York.

Hawkes, P. J., Coates, T. T., and Jones, R. J. (1997b). Impact of bi-modal sea on beaches and control structures. Report SR 507, HR Wallingford.

Herbert, D. M., Lovenbury, H. T., Allsop, N. W. H., and Reader, R. A. (1995). Performance of blockwork and slabbing protection for dam faces. Strategic Research Report SR 345, HR Wallingford with CIRIA, Wallingford.

Husdon, R. Y. (1958). Design of quarry-stone cover layers of rubble mound breakwaters, hydraulic laboratory investigation. Research Report 2-2, Waterways Experiment Station, Vicksburg, MS.

Klein Breteler, M., and Bezuijen, A. (1991). Simplified design method for block revetments. In *Proceedings ICE Conference on Coastal Structures and Breakwaters.* Thomas Telford, London.

Lindenberg, J. (1983). Stability of Armorflex block slope protection mats under wave attack. Report M1910, Delft Hydraulics, The Netherlands.

McConnell, K. J. (1998). *Revetments Against Wave Attack—a Design Manual.* Thomas Telford, London. ISBN 0-7277-2706-0.

McConnell, K. J., and Allsop, N. W. H. (1999). Revetment protection for embankments exposed to wave attack. *Proceedings MAFF Keele Conference of River and Coastal Engineers.* MAFF, London.

Owen, M. W. (1980). Design of sea walls allowing for wave overtopping. Report EX 924, Hydraulics Research, Wallingford.

Pilarczyk, K. W. (1998). Stability criteria for geosystems—an overview. 6th International Geosynthetics Conference, Atlanta.

Powell, K. A., Allsop, N. W. H., and Owen, M. W. (1985). Design of Concrete Block Revetments Subject to Wave Action: Literature Review. Report SR 54, Hydraulics Research, Wallingford.

Powell K. A. (1987). Toe scour at sea wall subject to wave action—a literature review. Report SR 119, HR Wallingford, UK.

Powell, K. A. (1989). The scouring of coarse sediments at the toe of seawalls. Proceedings Seminar on Seawall Design, HR Wallingford, UK.

Rijkswaterstaat (1985). The Use of Asphalt in Hydraulic Engineering. Technical Advisory Committee on Water Defences, Communication no. 37, The Hague, The Netherlands.

Seelig, W. N. (1983). Wave reflection from coastal structures. *Proceedings of Conference on Coastal Structures '83,* pp. 961–973, Arlington, March 1983. ASCE, New York.

Simm, J. D. (Ed.) (1991). *Manual on the Use of Rock in Coastal and Shoreline Engineering.* CIRIA/CUR, Special Publication 83, ISBN: 0 86017 326 7, CIRIA, London.

Simm, J. D., Brampton, A. H., Beech, N. W., and Brooke, J. S. (1996). *Beach Management Manual.* Report 153, ISBN: 0-86017 438 7, CIRIA, London.

Sprague, C. J., and Koutsourais, M. M. (1992) Fabric formed concrete revetment systems. In

Geosynthetics in Filtration, Drainage and Erosion Control. Ed. R. M. Koerner. Elsevier Advanced Technology.
Stickland, I. W., and Haken, I. (1986). Seawalls, Survey of Performance and Design Practice. Tech Note No. 125, Construction Industry Research and Information Association (CIRIA) London.
Thomas, R. S., and Hall, B. (1991). *Sea Wall Design Guidelines.* CIRIA/Butterworths, London.
van der Meer, J. W. (1988). Rock Slopes and Gravel Beaches Under Wave Attack. PhD thesis, Delft Hydraulics, the Netherlands.
Yarde, A. J., Banyard, L. S., and Allsop, N. W. H. (1996). Reservoir dams: wave conditions, wave overtopping and slab protection. Report SR 459, HR Wallingford, Wallingford.

CHAPTER 3
DESIGN OF DIKES AND REVETMENTS – DUTCH PRACTICE

KRYSTIAN W. PILARCZYK
Hydraulic Engineering Division
Rijkswaterstaat, Public Works Department
Delft, the Netherlands

Notation	3.2
1. Introduction	3.4
2. History of Dike Protection	3.4
2.1. History	3.5
2.2. Polders	3.7
2.3. Past Experience	3.8
3. Design Philosophy and Methodology	3.9
3.1. Design Philosophy	3.9
3.2. Design Methodology	3.12
4. Boundary Conditions and Interactions	3.14
4.1. Boundary Conditions	3.14
4.2. Processes and Interactions	3.17
4.3. Effect of Dikes and Sea Walls on the Beach	3.21
5. Design of Dikes and Sloping Sea Walls	3.23
5.1. Structural Aspects and Design Checklist	3.23
5.2. Geometrical Design of Dikes and Sea Walls	3.25
5.3. Wave Run-up	3.26
5.4. Wave Overtopping	3.35
5.5. Slope Protection (Revetments)	3.42
5.6. Optimization of Slope Stability	3.49
5.7. Scour and Toe Protection	3.50
5.8. Protection Against Overtopping	3.51
5.9. Joints and Transitions	3.53
5.10. Geotechnical Aspects	3.54
6. Structural Response of Placed-Block Revetments	3.55
6.1. Applications and Design Aspects	3.55
6.2. Cover Layer Stability	3.59
6.3. Migration of Subsoil	3.70
6.4. Geotechnical Stability	3.72
7. Review of Design Considerations for Dikes and Revetments	3.81
8. Construction Aspects	3.86
8.1. The Dike and Its Components	3.86
8.2. Block Mattresses	3.88

9. Maintenance and Monitoring of Safety — 3.91
   9.1. Goal — 3.91
   9.2. Principles of Management Scheme — 3.91
   9.3. Management Scheme for Flood Control in the Netherlands — 3.94
   9.4. Maintenance and Monitoring System for Dikes — 3.95
10. Conclusions — 3.99
References and Selected Readings — 3.101

NOTATION

B	berm width (m)
B	width of the block (m)
B_p	seaward extent of the toe protection (m)
b	thickness of filter layer (m)
b	exponent in equation (23)
c_v	consolidation coefficient = $k/(\rho g \cdot n_s w')$ (m²/s)
d	thickness of gabion or stone mattress (m)
d_h	berm depth relative to still water level (SWL) (m)
d_x	grain size of subsoil corresponding to x% by weight of finer particles (m)
D	thickness of the cover layer (m)
D_x	grain size of filter corresponding to x% by weight of finer particles (m)
f	coefficient, dependent on Δ, $\tan\alpha$, friction, etc. (–)*
F	total stability factor (–)
F_a	the driving force for sliding (N)
F_f	the maximum friction force (N)
F_n	the normal force between cover layer and subsoil (N)
g	acceleration due to gravity (m/s²)
h	water depth (m)
h_m	water depth at the position of the toe of the structure (m)
h_s	shallow water depth (m)
H	incoming wave height of regular waves (m)
H_b	breaker wave height for regular waves (m)
H_s	incoming significant wave height of irregular waves (m)
H_{scr}	significant wave height at which blocks are lifted (m)
i	maximum gradient in the filter (–)
k	permeability subsoil or filter (m/s)
k'	permeability of the cover layer (m/s)
L	length of the block (m)
L_{II}	the length of area II (m)
L_{III}	the length of area III (m)
L_{es}	consolidation length = $\sqrt{(T \cdot c_v)}$ (m)
L_o	wave length of regular waves at deep water = $gT^2/2\pi$ (m)
L_{op}	wave length of irregular waves at deep water = $gT_p^2/2\pi$ (m)
L_s	length of protection in the splash area (–)
n_f	the porosity of the filter layer (–)
n_s	the porosity of the subsoil (–)
N_W	number of incoming waves (–)

*(–) means dimensionless.

DESIGN OF DIKES AND REVETMENTS—DUTCH PRACTICE 3.3

N_{ow} number of overtopping waves (–)
O_{90} average diameter of the standardized sand fraction, of which 90% remains on the geotextile after a sieve test under defined conditions (m)
p_a atmospheric pressure (1.10^5 N/m²)
Q_b dimensionless overtopping discharge with breaking waves $\xi_{op} < 2$ (–)
Q_n dimensionless overtopping discharge with nonbreaking waves $\xi_{op} > 2$ (–)
q average overtopping discharge per unit crest length (m³/s per m)
R_c crest height relative to still water level (SWL) (m)
$R_{d,2\%}$ wave run-down level (absolute value is exceeded by 2% of incoming waves) (m)
$R_{u,2\%}$ wave run-up level of irregular waves, exceeded by 2% of incoming waves (m)
R_b dimensionless crest height with breaking waves $\xi_{op} < 2$ (–)
R_n dimensionless crest height with nonbreaking waves $\xi_{op} > 2$ (–)
r_B reduction factor for the berm width (–)
r_{dh} reduction factor for the berm elevation (–)
s_{op} wave steepness with L_o based on T_p ($s_{op} = H_s/L_{op}$) (–)
s air content (normally between 1 and 10%)
T average wave period of regular waves (s)
T_m mean period (s)
T_p wave period at peak of spectrum (s)
T_s significant period, average of the highest 1/3 part (s)
V volume of overtopping wave per unit crest width (m³ per m)
w the compressibility of pure water (5.10^{-9} m²/N)
w' compressibility of the pore water with air (m²/N)
y length coordinate along the slope (m)
z_1 phreatic level in filter relative to the point where the wave front meets the revetment (m) (usually $z_1 \approx \phi_b$).
z_o depth (m)
α the slope angle (°).
α_{eq} equivalent slope gradient for a slope with a berm (°)
β angle between potential front of breaking wave and the verticle (°)
ß   angle of wave attack (°)
γ_b reduction factor for a berm (–)
γ_f reduction factor for the roughness (–)
γ_h reduction factor for a shallow foreshore (–)
γ_v reduction factor for a vertical structure (–)
γ_β reduction factor for the angle of wave attack (–)
ξ_{op} surf similarity parameter based on T_p ($\xi_{op} = \tan\alpha/\sqrt{s_{op}}$) (–)
ξ_{eq} equivalent surf similarity parameter ($\xi_{eq} = \gamma_b \xi_{op}$) (–)
Δ $(\rho_c - \rho)/\rho$ = relative density of blocks (–)
$\Delta\phi$ the difference in piezometric head between the location at $z = z_0$ and just above the subsoil (m)
Γ coefficient representing friction, inertia, etc. (–)
τ the shear stress on a plane in the subsoil (kN/m²)
Λ leakage length (m)
ν kinematic viscosity ($1.2 \cdot 10^{-6}$ m²/s)
ζ_o breaker parameter for regular waves = $\tan\alpha/\sqrt{(H/L_o)}$ (–)
ξ_{op} breaker parameter for irregular waves = $\tan\alpha/\sqrt{(H/L_{op})}$ (–)
ρ density of water (kg/m³)
ρ_c density of concrete (kg/m³)
ρ_f density of filter grains (kg/m³)
ρ_s the density of the grains of the subsoil (kg/m³)
σ_b the normal stress by the filter layer and/or cover layer (N/m²)

σ the normal stress on the same plane (kN/m²)
θ steepness angle of potential front of breaking wave (°)
ϕ stability factor or stability function for incipient of motion, defined at $\xi_p = 1$ (–)
ϕ piezometric head in the filter layer (m)
ϕ_T piezometric head on the cover layer (m)
ϕ_w maximum piezometric head over cover layer (m)
ϕ_p stone arrangement packing factor (–)
Φ the friction angle of the material (°)
Φ' the friction angle between the blocks and the filter layer (–)
ψ_u system-determined (empirical) stability upgrading factor ($\psi_u = 1.0$ for riprap as a reference and $\psi_u \geq 1$ for other revetment systems) (–)
Ω relative open area of the revetment (joints between the blocks and gaps in the blocks) (–)

1. INTRODUCTION

Low-lying countries like the Netherlands are strongly dependent on safe coastal defenses (sea dikes and/or dunes). In the past, the design of dikes and revetments was based more on experience instead of generally accepted and valid calculation methods. The increased demand for reliable design methods for protective structures in the Netherlands has resulted in increased research in this field and, as a result, in preparation of a set of design guidelines for design and maintenance of dikes, sea walls, revetments, and other coastal structures.

Sea dikes and sea walls are only two options for coastal defense and must be considered in conjunction with or as an alternative to beach management and other options. These structures are built to protect upland areas (including land reclamation) when resources become endangered by erosion or inundation due to storm surges, waves, and wave overtopping. The proper design of these structures is of major importance. All coastal protection systems have advantages and disadvantages that should be recognized before the choice is made.

The Dutch general design philosophy and the principles of functional and technical design of these structures are discussed in this chapter. Design methods concerning the geometrical design (shape/height of dikes), stability criteria for slope protection (various types of revetments), and protection against overtopping are included. The Dutch experience can be of value for solving similar problems elsewhere.

For a treatment of these matters in greater depth the reader is referred to the original reports and publications (e.g., CUR/RWS, 1995; Pilarczyk, 1998).

2. HISTORY OF DIKE PROTECTION

The coast of the Netherlands consists of about 300 km of dunes and about 100 km of dikes and dams. The original length of sea dikes (until 1932) was about 700 km. This length was drastically shortened by dams closing the Zuyder Sea (1932) and tidal estuaries of the Delta Project (1957–1986). The primary function of the dunes and dikes is to protect the hinterland from flooding during periods of storm surges and heavy wave attack. To understand the historical development of protection by dikes in the Netherlands, it is essential to understand the gradual rise of the sea level with respect to the land and

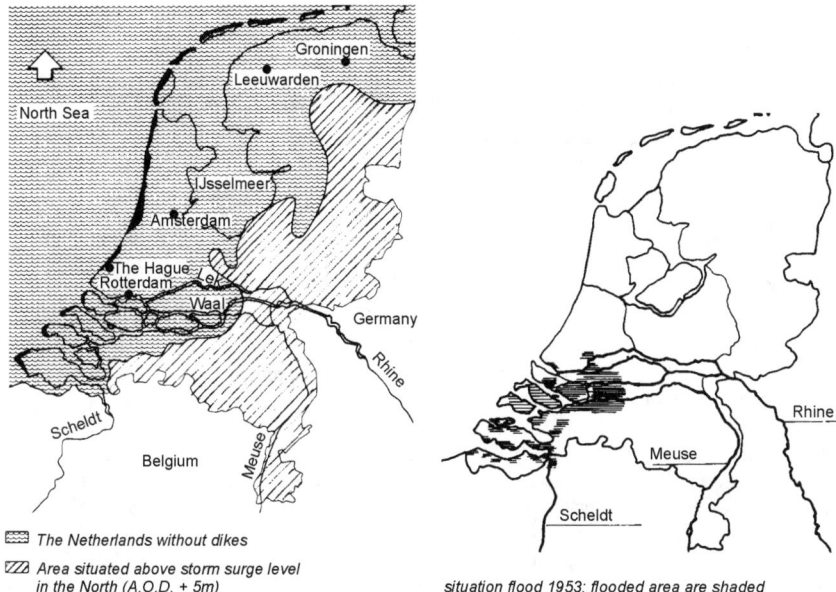

FIGURE 3.1. Dikes protect a major part of the Netherlands.

also the deposits of soil by the North Sea and the rivers (Agema, 1982). About half of the country is below mean sea level and without dikes more than half of the country would be underwater (Figure 3.1). About 60% of the total population lives in low-lying areas. Under present climate conditions (over the last 100 years) the relative rise in sea level (due to sea level rise and subsidence) is about 20 cm per century, and the coast is strongly attacked by tidal currents and storm surges. Increased sea level rise will have serious implications for the safety of the land protected by dikes, dunes, and other defense structures along the coast and the lower parts of the main rivers. Sea level rise increases the risks of overtopping and the ultimate collapse of these structures during storms.

2.1. History

During the last geological periods, the first coastal barrier (dunes) along the North Sea were formed. Between this barrier and the higher pleistocene area in the eastern part of the country, sand and silt were deposited. The rivers Rhine, Meuse, and Scheldt added river silt, building and shaping the land by means of regular inundations. As a result, the geological profile of the western Netherlands shows sand, silt, and peat layers (Figure 3.2). In consequence of that, most dikes and dams are founded on soft soils and subjected to high settlement.

Long before the Christian era, people must have lived in the unprotected lower parts of the Netherlands. In the first centuries, people learned to protect themselves and their stock against storm surges by inhabiting naturally higher parts of the land and erecting artificial clay mounds ("terps") on which their homes and barns were built. Many such early settlements became the nuclei of present villages and towns.

3.6 HANDBOOK OF COASTAL ENGINEERING

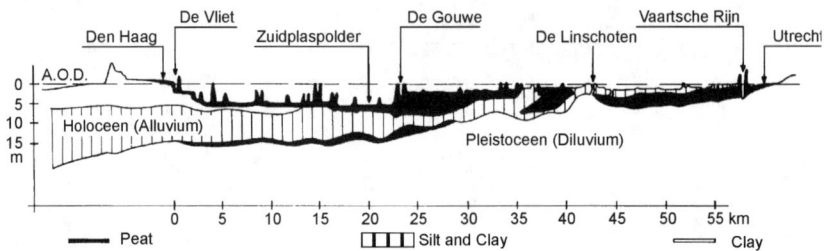

FIGURE 3.2. Geological profile of the western Netherlands.

It can be noted that creation of artificial mounds (escape areas) is still an actual alternative for a number of countries (e.g., Bangladesh).

Considering the simple tools available at that time for digging and transportation of soil and the construction of the terps, one is impressed by the tremendous task these early settlers carried out. An important step was made towards improving living conditions and safety by building dikes. In the ninth century, the first dikes were built. This developed in such a way that by the thirteenth century one can speak of a more organized method of dike construction. An area protected against high water levels by surrounding dikes is called a "polder." Dike construction at that time was undertaken by people who directly faced the elements of nature. At first, the aim of dike construction was only defensive; people protected the land where they lived. In a later phase, the construction of dikes was used in an more offensive way; for example, by reclaiming land from the sea. In this way, from the middle of the thirteenth century until today, in total about 550,000 ha of land was reclaimed. However, during the centuries, much of the previously reclaimed land was lost by attack of the sea, mainly due to storm surges, which many times caused destruction of the dikes. Another phenomenon was the occurrence of landslides along tidal channels (most dikes were situated on loose soils), thus causing disappearance of dikes. During high storm surges, the sea also eroded this land. Nevertheless, every time there was the spirit of the people to push back the sea. Most of the lost land was reclaimed again, despite the ever-occuring storm surges.

The history of the Netherlands is marked by storm surge disasters. The most recent disaster took place on the first day of February 1953, when a northwesterly storm struck the southwestern part of the Netherlands (Delta area). The storm surge level reached 3 to 3.5 m above normal high water and exceeded design storm surge levels by about 0.5 m at some places. Some dikes could not withstand these levels, so that at several hundreds of places the dikes were damaged and/or broken, over a total length of 190 km. Through nearly 90 breaches, 150,000 ha of polder land was inundated. This caused the death of 1850 people and 100,000 persons had to be evacuated; moreover, a lot of livestock drowned and thousands of buildings were damaged or destroyed. This disaster gave a new impulse to improve the whole sea defense system in the Netherlands. The resulting Delta Plan included the strengthening of existing dikes and shortening the length of protection in the Delta area by closing off estuaries and a tidal river. The Rotterdam Waterway and the Western Scheldt, both important fairways for the harbors of Rotterdam and Antwerp, respectively, had to remain open. The original planned barrier dam of the Eastern Scheldt has been changed into an open storm surge barrier because of environmental demands (RWS, 1994). The other estuaries and the tidal river Hollandse IJssel are separated from the North Sea mainly by barrier dams. Some of them have discharge sluices and navigation locks. The basic strengthening of the existing sea defenses in the Netherlands is now completed. Together with the Delta works, the low-lying "polderland" of the Netherlands has reached a relative high degree of safety against storm surges.

2.2. Polders

To understand the system of dikes in relation to polders, the two ways of creating polders are described. Polders are reclaimed lowlands protected by dikes against high water levels and waves. Two types of polders can be defined (Figure 3.3):

a. impoldering of marshlands in tidal waters, with a height of mean high water;
b. reclaiming (parts of) the bottom of lakes, with a depth of several meters below mean sea level.

Impoldering of marshlands (Figure 3a). In areas where hydraulic conditions of tidal waters are appropriate (small currents, quiet wave climate, etc.) sand and silt will deposit. This process accelerates in the phase during which the level of deposits is rising and grasses are starting to grow. Finally, the level of deposits equals the mean high water mark and marshlands are formed. These marshlands are located along the coast, at the ends of estuaries, and along bays etc., and sometimes appear as islands. After surrounding an area of marshland by a dike and so protecting the land against high water and storm

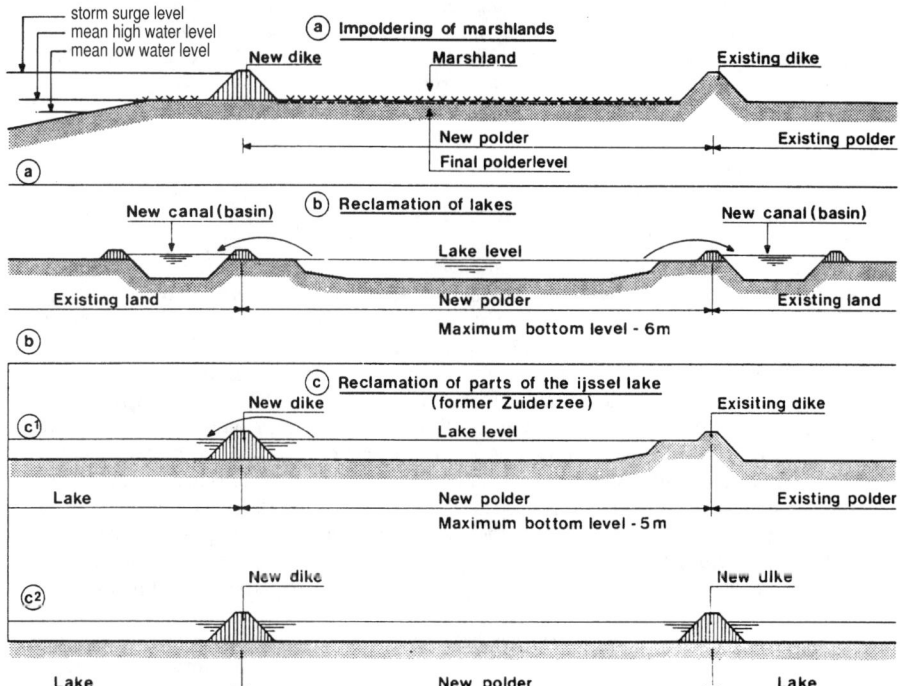

FIGURE 3.3. Impoldering of marshlands and reclamation of lakes.

surges, a polder is created. Draining water is discharged by gravity through sluices in the dike. In course of time, the soil of the polder settles due to lowering of the watertable as a consequence of draining. This means that the level of the polderland and dikes may settle between 1 and 1.5 m. A process of sequential impoldering developed in such a way that large areas of tidal waters and in many cases whole estuaries or bays were gained from the sea. As a consequence dikes of polders behind newly reclaimed land (polders) became "dry" dikes (inner dikes), creating additional safety.

Several lakes in the west of the Netherlands have become polders in the following way. First, on the land around the lake a canal with a dike on each side was constructed and "pumping" stations were installed. Then, the water in the lake was pumped into this canal, acting as a basin, and finally discharged via a system of canals to the sea. These dikes are also acting as inner dikes. This method of reclamation started in 1542, when "pumping" became possible with paddle wheels driven by windmills. Later (1787) steam engines were also applied. These "pumping" stations were also used to drain the polder. An example of this type of reclamation is the Haarlemmermeer (18,000 ha), with a bottom depth of 4.5 m below A.O.D. (Amsterdam Ordnance Datum = Mean Sea Level), where the international airport, Amsterdam–Schiphol, is located. Another one is the Prins Alexanderpolder (depth: A.O.D. –6 m), with new residential areas of the city of Rotterdam.

In the IJssel Lake, formed after closing off the Zuyder Sea by a barrier-dam (1932), four polders (depth: to 4.5 m below A.O.D.) have been reclaimed. In these cases, the polders were made by constructing of a dike with some pumping stations on the IJssel Lake bottom around the parts of the lake to be reclaimed (Figure 3c). The enclosed area of water is simply pumped into the remaining part of the lake and discharged through sluices in the barrier dam. This dam is protecting the polders against storm surges of the North Sea. The polder dikes resist the water of the IJssel Lake; they can also act as a secondary protection against storm surges in case the barrier dam fails.

2.3. Past Experience

The design of dikes and hydraulic structures as drainage sluices, etc. in the past (until 1953) was mainly based on experience in practice. Dikes were generally constructed with clay and showed a trapezoidal cross-section with the steep slope at the inner side. In more recent times, a dike with a wave energy absorbing outer berm was developed and sand was applied as core material. The height and strength of dikes and structures had to be related to a certain storm surge level and wave action, especially the magnitude of the wave run-up; the run-up elevation increases with steepness of the outer slope. Previous storm surge levels were sometimes marked in stones of buildings. For more than a century, water levels have been continuously recorded. Wave characteristics and wave run-up were observed by eye so that reliable data were not available. Only the swag marks on the slope of dikes after storms gave an indication of the level of wave run-up. Later, these swag marks were obtained by leveling. Taking into account the (still) water level, a rough value of run-up could be derived. With these data, it was possible to improve the design to a certain extent. In general, it can be said that the design was based on the highest observed storm surge level and an estimate of wave characteristics (height and period) and wave run-up.

The experience during centuries was that higher storm surge levels and more extreme wave action occurred than those which were predicted. Overflow and/or overtopping of dikes, caused, in most cases, instability of steep inner slopes, which in many cases induced a breach and consequently led to inundation. Depending on the storage capacity of the polder(s) (depth, area, and the composition of the soil under and in the vicinity of the dike), deep and wide tidal channels can develop in these breaches. In about 1920, soil mechanical investigations were started in order to improve dike safety. Also due to other modes of failure—for example, erosion of outer slope, collapse of drainage sluices, cutoffs in dikes etc.—breaches occurred, followed by inundations of the polders. The increase of storm surge levels was not only due to nature but also due to impoldering of relatively large tidal areas, which decreased the storage capacity needed for higher water levels. As a consequence and moreover due to erosion of the wave-reducing foreland of dikes, caused by tidal currents and wave action, wave attack and run-up became more severe. The more severe wave attack necessitated an improvement of the grass protection of the outer slope of dikes. Seaweed, rows of wooden piles, and in more recent years, boulders, pitched stone, concrete blocks, etc. were applied.

The first calculations of the probability of storm surge levels, design waves, and related wave run-up were conducted in the framework of the Zuyder Sea enclosure (1927–1932). However, a new design philosophy was introduced by the Dutch hydraulic engineer Wemelsfelder (1939), who published in 1949 an article about probability of exceedance of storm surge levels along the coast of the Netherlands (Wemelsfelder, 1939, 1960). From this time it was possible, in principle, to establish a design water level that had a predicted probability of exceedance.

3. DESIGN PHILOSOPHY AND METHODOLOGY

3.1. Design Philosophy

Natural or artificial dunes and dikes are functioning in the Netherlands to protect upland (population and economical values) against erosion or inundation due to storm surges. The main purpose of a dike or sea wall is to fix the land and sea boundary; it is not intended to protect either the beach fronting it or adjoining, unprotected beaches. Thus, sea walls neither promote accretion nor reduce the regional trend of the coast to erode, but are constructed for protection of upland under extreme conditions. Dikes are one of various forms of coastal protection that may be used singularly or in combination with other methods.

There is still much misunderstanding on the use of dikes and sea walls and their possible disadvantages related to the disturbance of the natural coastal processes and even acceleration of beach erosion. However, it should be said that in many cases when the upland becomes endangered by inundation (as in The Netherlands) or by high-rate erosion (possible increase of sea level rise) leading to high economical or ecological losses, whether one likes it or not, the dike or sea wall can even be a "must" for survival. The proper coastal strategy to be followed should always be based on the total balance of the possible effects of the countermeasures for the coast considered, including the economical effects or possibilities. It is an "engineering art" to minimize the negative effects of the solution chosen (Kraus and Pilkey, 1988).

Absolute safety against storm surges is nearly impossible to realize. Therefore, it is much better to speak about the probability of failure of a certain defense system. To apply this method, all possible causes of failure have to be analyzed and consequences determined. This method is actually under development in the Netherlands for dike and dune design. The "fault tree" is a good tool for this aim (Figure 3.4). In Figure 3.4 all possible modes of failure of elements can eventually lead to the failure of a dike section

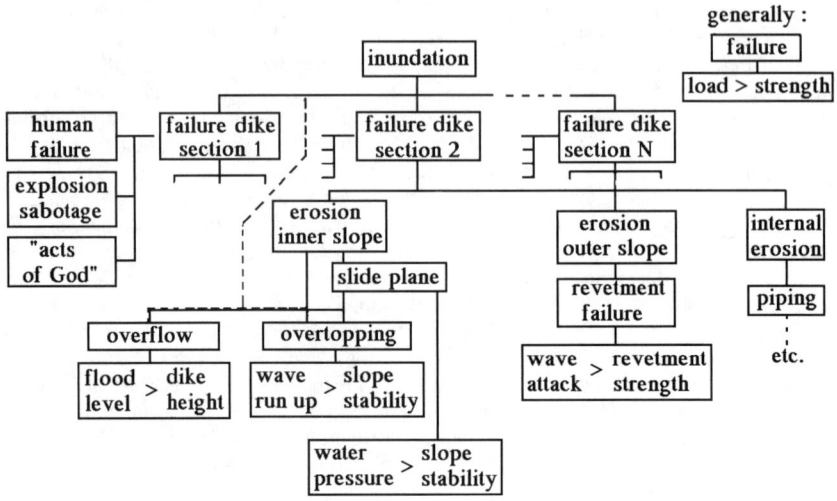

FIGURE 3.4. Simplified fault tree for a dike.

and to inundation. They can also influence the behavior of the revetment if properly designed.

Although all categories of events that may cause the inundation of a polder are equally important for overall safety, the engineer's responsibility is mainly limited to the technical and structural aspects. In the case of the sea dike, the following main events can be distinguished (Figure 3.5):

- overflow or overtopping of the dike
- erosion of the outer slope or loss of stability of the revetment
- instability of the inner slope leading to progressive failure
- instability of the foundation and internal erosion (i.e., piping)
- instability of the whole dike

For all these modes of failure, the situation wherein the forces acting are just balanced by the strength of the construction is considered (the ultimate limit state). In the adapted concept of the ultimate limit state, the probability density function of the "potential threat" (loads) and the "resistance" (dike strength) are combined. The category "potential threat" contains basic variables that can be defined as threatening boundary conditions for the construction, e.g., extreme wind velocity (or wave height and period), water levels, and a ship's impact (collision). The resistance of the construction is derived from the basic variables by means of theoretical or physical models. The relations that are used to derive the potential threat from boundary conditions are called transfer functions (e.g., to transform waves into forces on grains or other structural elements). The probability of occurence of this situation (balance) for each technical failure mechanism can be found by applying mathematical and statistical techniques.

The safety margin between "potential threat" and "resistance" must guarantee a sufficiently low probability of failure. Three different philosophies are currently available in construction practice (TAW, 1990; CUR/CIRIA, 1991; CUR/RWS, 1995a,b):

DESIGN OF DIKES AND REVETMENTS—DUTCH PRACTICE 3.11

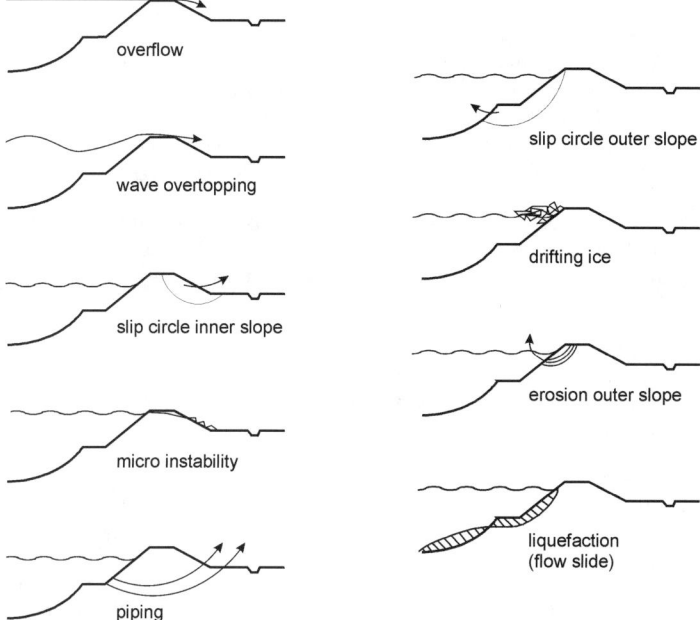

FIGURE 3.5. Failure modes/mechanisms (limit states) for a dike or dam.

1. deterministic
2. quasiprobabilistic
3. probabilistic

For a fully probabilistic approach, more knowledge must still be acquired concerning the complete problems associated with the use of theoretical models relating loads and strength. Studies on all these topics are still underway in the Netherlands. The present Dutch guidelines for dike and dune design follow a philosophy that lies between the deterministic and the quasiprobabilistic approaches.

The ultimate potential threat for the Dutch dikes is derived from extreme storm surge levels with a very low probability of exceedance (1% per century for sea dikes and 10% for river dikes) and equated with the average resistance of the dike without any apparent safety margin. Besides the ultimate limit state, there are situations wherein the ever continuing presence of a (frequent) load causes a deterioration of constructional resistance in time, without any imminence of failure (e.g., fatigue of concrete and steel, creep of erosion of clay under the revetment, clogging or ultraviolet (U.V.) deterioration of geotextiles, corrosion of cabling, unequal settlements of deformations, etc.). However, this deterioration of constructional resistance can cause an unexpected failure in extreme conditions. These are so-called serviceability and fatigue limit states that can also be considered as inspection and maintenance criteria.

As already mentioned, the fully probabilistic approach for dikes based on the limit state concept is rather cumbersome because a theoretical description for various failure modes is not yet available. To overcome this problem, a scheme to simulate nearly all possible combinations of natural boundary conditions in a scale model of the construction and to correlate the damage done to the boundary conditions can be developed (black box approach).

The ultimate potential threat for the Dutch sea defenses is from extreme storm surge levels with a very low probability of exceedance (1% per century for sea dikes and dunes) and equated with the average resistance of the dike (or dune). Under these ultimate load conditions, probability of failure of the dike (sea wall) should not exceed 10%. The probabilistic approach to sea defenses as developed in the Netherlands is treated extensively in TAW report (1990).

Acquiring knowledge of these recent developments can be rather profitable for all countries, especially for estimation of possible risks involved in the realized projects and for finding the optimum between the risks and the investment.

3.2. Design Methodology

When designing coastal structures, the following aspects have to be considered:

- the function of the structure
- the physical environment
- the construction method
- operation and maintenance

The main stages that can be identified during the design process are shown in Figure 3.6. The designer should be aware of the possible constructional and maintenance constrains (Pilarczyk, 1990).

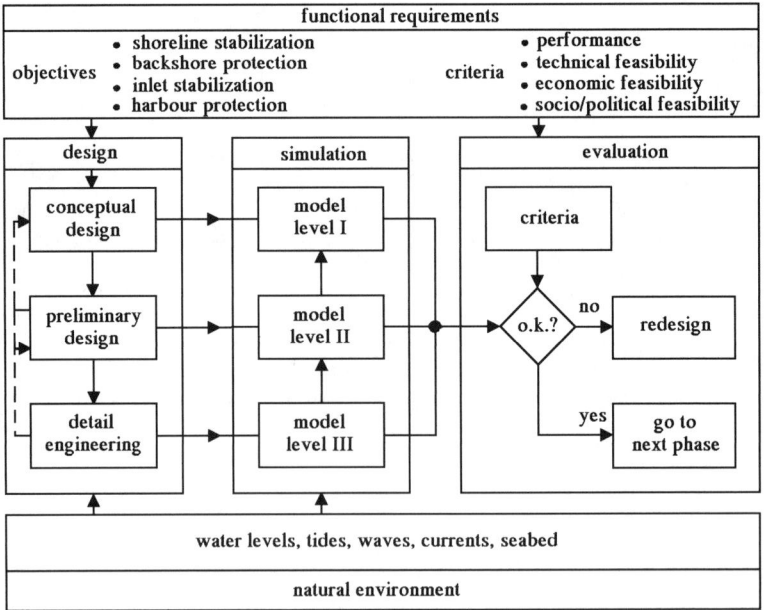

FIGURE 3.6. Design methodology.

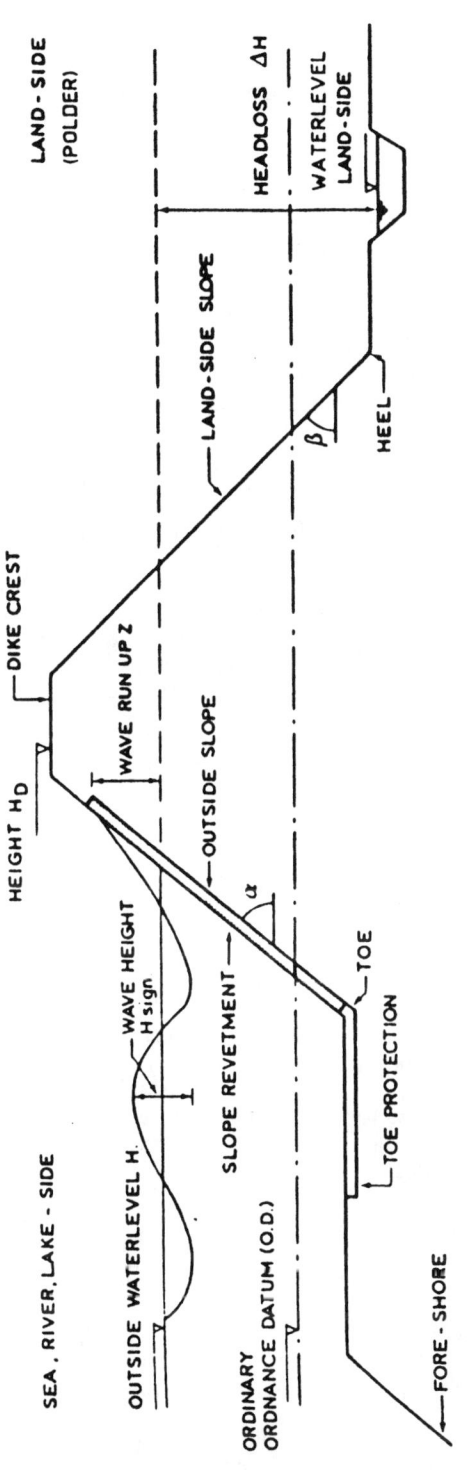

FIGURE 3.7. Schematization of a dike.

Based on the main functional objectives of the coastal structure, a set of technical requirements has to be assessed. When designing a dike/sea wall, the following requirements to be met can be formulated:

1. the structure should offer the required extent of protection against flooding at an acceptable risk
2. events at the dike/sea wall should be interpreted with a regional perspective of the coast
3. it must be possible to manage and maintain the structure
4. requirements resulting from landscape, recreational and ecological viewpoints should also be met when possible
5. the construction cost should be minimized to an acceptable/responsible level
6. legal restrictions

Elaboration of these points depends on specific local circumstances, such as type of upland (lowland or not) and its development (economical value), availability of equipment, manpower, and materials, etc. High dikes/sea walls are needed for protection of lowlands against inundation, whereas lower sea walls are often sufficient in other cases. The cost of construction and maintenance is generally a controlling factor in determining the type of structure to be used. The starting points for the design should be carefully examined in cooperation with the client or future manager of the project.

In this chapter, it is further assumed that the decision based on the conceptual and functional design is in favor of a sloping dike. The basic elements of such a structure are shown in Figure 3.7.

4. BOUNDARY CONDITIONS AND INTERACTIONS

4.1. Boundary Conditions

4.1.1. Assessment of the Existing Situation. A lot of relevant information for a sea wall/dike design can be drawn from files and existing maps. In addition to this, a field reconnaissance and land survey are indispensable, as well as photographic recording of the characteristic points in the area. Special attention should be paid to the position of the beach and/or onshore profiles, and the morphology of the area considered (eroding/accreting coast). The composition of the existing dike body and the geologic structure of the subsoil are also very important. When these data are not available, soil investigation should be considered (soundings, borings, etc.). More detailed information on these subjects can be found in CUR/CIRIA (1991).

4.1.2. Hydraulic Boundary Conditions. In view of the function of (coastal) water defenses, the loads will be mostly due to the actions of long and/or short waves. In broad outline, the following wave phenomena can be distinguished: (a) low-frequency water level changes, such as flood waves, tidal waves, wind set-up gradients, and seiches (Figure 3.8); (b) wind waves and swell; (c) ship's wakes in navigable waterways.

These water level variations strongly influence the area that needs to be protected with hard revetment. The flow diagram on determination of hydraulic boundary conditions is given in Figure 3.8.

Water level variations in canals and water storage channels are comparatively small and are probably only caused by lock water intake, seepage, drainage, and wind effects.

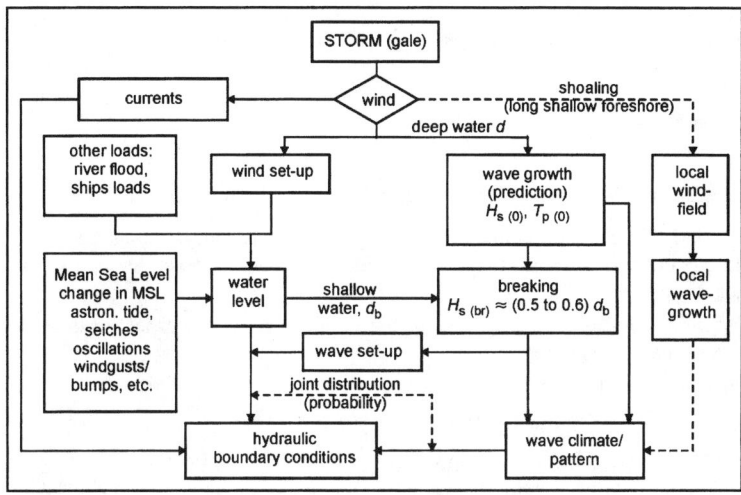

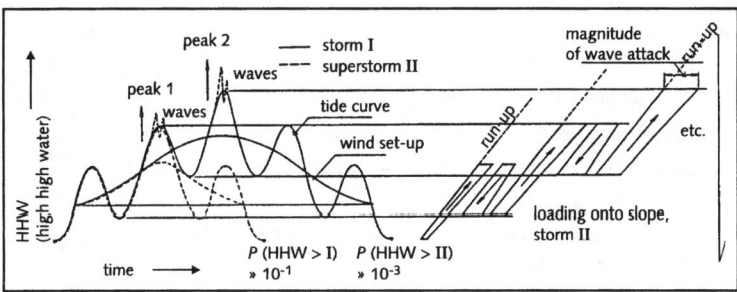

FIGURE 3.8. Flow diagram of hydraulic boundary conditions.

Water levels in lakes can vary as a result of wind set-up, inflow or outlfow of water, and evaporation. Water levels in a reservoir can change markedly due to filling or emptying, but rainfall and wind set-up can also play a role.

Water levels in a river are determined by the river's discharge regime and, for the lower reaches (estuaries) by tides and also wind-set-up. For a coastal defense embankment, water levels are governed by tides and winds. The most complex situation occurs at coastal shores, where water level fluctuations can assume many forms.

A lot of relevant information for sea wall design can be drawn from files and existing maps. In addition to this, a field reconnaissance and land survey are indispensable, as well as photographic recording of the characteristic points in the area. Special attention should be paid to the position of the beach and/or onshore profiles, and the morphology of the area considered (eroding/accreting coast?). The specification of environmental parameters are given in CIRIA (1986). The prediction methods related to hydraulic boundary conditions can be found in SPM (1984).

4.1.3. Geotechnical Conditions. For major structures, a good geological analysis, based on the overall geological structures of the country, is of the utmost importance for an understanding of the geophysical and geohydrological conditions (Quelerij, 1989; CUR/CIRIA, 1991). The most important geological aspects are:

- geological stratification, formation, and history
- groundwater regime
- seismicity

The main questions that a geotechnical investigation has to answer are:

- what kind of soil is found and at what depth; i.e., soft soils, such as sand, clay, and peat, or hard soils, such as limestone and calcareous sandstone, or very hard soils, such as quartzite and basalt
- what are the mechanical properties of the various soils with respect to their strength and deformation characteristics
- is the soil fissured or weathered
- will the soil degrade in (short) time

The first step is to organize and design site investigations. The field program forming part of the site investigation is complemented by laboratory testing and geotechnical calculations. The last and perhaps most difficult step is the integration of the result of the investigations and structural design, resulting in the final foundation design.

At the set-up and organization of the soil investigation program the geotechnical engineer is confronted with the following questions:

- which soil data have to be collected
- at what locations (number and depths)
- which site investigation techniques and laboratory tests should be performed
- when is the program to be carried out and
- who will take care of the contracting work in the field, the laboratory tests, and the interpretation of the results

The answer of these questions will depend, among others, on:

- the boundary conditions stipulated by the client (time and money schedule)
- the knowledge, judgment, and experience of the geotechnical engineer
- the availability of existing data—for example, topographical, geological, and geotechnical maps
- the phase of the design: for a preliminary design only global information over a wide area is needed to recognize the main geotechnical problems; in the final design phase or during the construction period, detailed information on engineering soil parameters is needed
- the type of geotechnical failure mechanisms involved
- the availability and restrictions (including the terrain accessibility) of the investigation tools and the quality of the personnel to handle these tools

A high-quality investigation must be economically efficient in the sense that the cost of the investigation must be money well spent. The investigator must be able to justify each and every item in the site investigation in terms of the value of that item in building up the geotechnical model. The investigator should be able to show good and sufficient reasons for undertaking each part of the investigation.

It is emphasized that there is no standard form of site investigation for a particular engineering work. Each site investigation should be regarded as a completely new venture. Several standardized investigation techniques have, however, been developed, which the geological and geotechnical engineer can use for obtaining the relevant data for basic calculations and design criteria. Four types of site investigation methods can be listed:

1. geophysical measurements from the soil surface
2. penetration tests, such as cone penetration and standard penetration tests
3. borings, including sampling and installation of observation wells
4. specific measurements, such as plate loading tests and nuclear density measurements

4.1.4. Construction Materials. A large number of materials may be used in various forms in the construction of sea walls and dikes, including sand, gravel, quarry rock, industrial waste products (slags, minestone, silex from the cement industry), clay, timber, concrete, asphalt and geotextile. All these materials have to fulfill some structural and environmental specifications that are mostly regulated by the national standards. Useful information can be found in various handbooks and guideline reports (CUR/TAW, 1989; TAW, 1991, 1995; CIRIA, 1986; PIANC, 1987a; CUR/CIRIA, 1989; CUR/TAW, 1995; CUR/RWS, 1995a,b; Pilarczyk 1998).

4.2. Processes and Interactions

4.2.1. Loading Zones. The degree of wave attack on a dike or other defense structure during a storm surge depends on the angle of attack of the storm, the duration and strength of the wind, the extent of the water surface fronting the sea wall, and the bottom topography of the area involved. For coastal areas there is a correlation between the water level (tide plus wind set-up) and the height of the waves, because wind set-up and waves are both caused by wind. Therefore, the joined frequency distribution of water levels and waves seems to be the most appropriate for the design purposes (also from the economical point of view). For sea walls in the tidal region, fronting deep water, the following approximate zones can be distinguished (Figure 9):

I. the zone permanently submerged (not present in the case of a high-level "foreshore")
II. the zone between MLW and MHW; the ever-present wave-loading of low intensity is of importance for the long-term behavior of structure
III. the zone between MHW and the design level; this zone can be heavily attacked by waves but the frequency of such attack reduces as one goes higher up the slope
IV. the zone above design level, where there should only be wave run-up.

In principle, a bank slope revetment functions no differently under normal circumstances than under extreme conditions. The accent is, however, more on the persistent character of the wave attack rather than on its size. The quality of the seaward slope can, prior to the occurrence of the extreme situation, already be damaged during relatively normal conditions to such a degree that its strength is no longer sufficient to provide protection during the extreme storm. The division of the slope into loading zones has not only direct connection with the safety against failure of the revetment and the dike as a whole, but also with different application of materials and execution and maintenance methods for each zone (Figure 3.10).

Alternatives have to be generated during the conceptual, preliminary, and detailed design phases in order to select a most suitable design. It is emphasized that for each design phase, these alternatives should be evaluated at a comparable level of detail. The same ap-

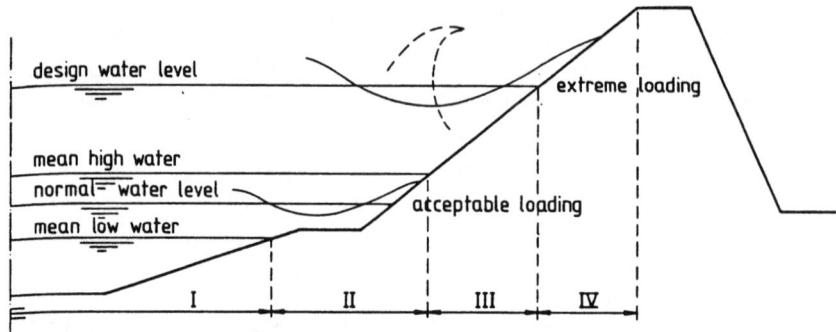

FIGURE 3.9(a). Loading zones on a dike.

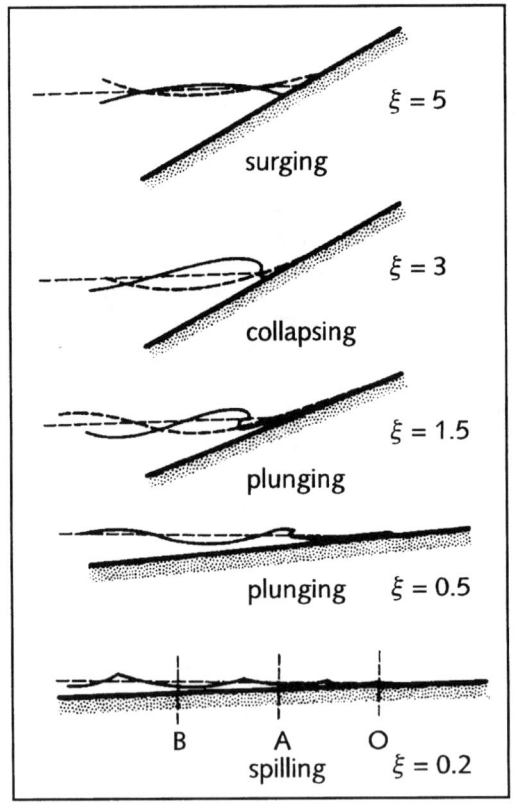

FIGURE 3.9(b). Breaker types on a slope.

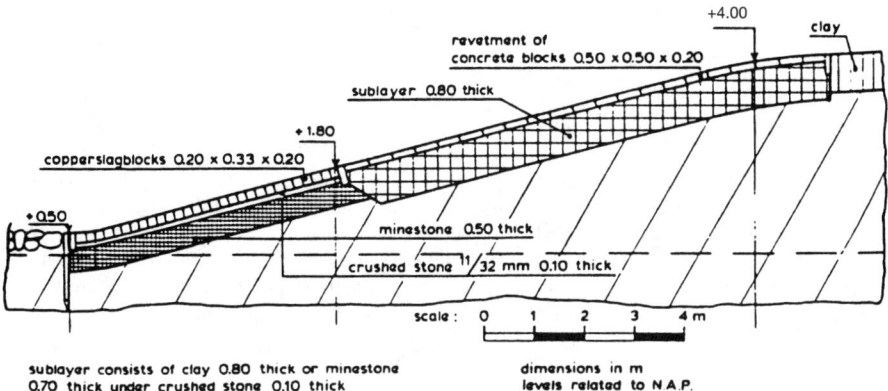

FIGURE 3.10. Example of dike protection (Oester dam, the Netherlands)

plies to the construction alternatives, which may have a great influence on the total structure costs.

4.2.2. Wave–Structure Interaction. The interaction between waves and slopes is dependent on the local wave height and period, the external structure geometry (water depth at the toe), slope with/without berm, the crest elevation, and the internal structural geometry (types, size and grading of revetments and secondary layers). The type of structure–wave interaction is defined by the surf similarity parameter (or breaker parameter), which is defined as (see also Figure 9(b))

$$\xi_{op} = \tan \alpha / \sqrt{s_{op}}$$

with:

ξ_{op} = breaker parameter
α = slope angle
$s_{op} = 2\pi H_s/gT_p^2$ = wave steepness
T_p = wave period, peak period of the wave spectrum
H_s = significant wave height, being the average value of the highest ⅓ part of the wave heights. H_s is the significant wave height at the toe of the structure.

The wave steepness is a fictitious or computation quantity meant to describe the influence of a wave period. This quantity is fictitious as the wave height at the location of the toe is related to the wave length in deep water ($gT_p^2/2\pi$).

Several wave periods can be taken from a spectrum, among them the peak period T_p, the mean period T_m (computed from the spectrum or the time signal), and the significant period $T_{1/3}$. Here the ratio T_p/T_m mostly lies between 1.1 and 1.25 and T_p and $T_{1/3}$ are virtually equal.

With $\xi_{op} < 2$–2.5, the waves break on the slope. This is mostly the case with slopes of 1:3 or milder. For larger values of ξ_{op}, the waves do not break on the slope any longer. In that case, the slopes are often steeper than 1:3 and/or the waves are characterized by a smaller wave steepness (for example, swell).

For large values of the wave length or for large values of α (steep slopes), the wave behaves like a long wave, which reflects against the structure without breaking—a so-called surging wave. For shorter waves and medium slopes, waves will fall short and break, causing plunging breakers ξ_{op} values in the range of 1 to 2.5. This figure is common along the Dutch coast with slope angles of 1 to 3 to 1 to 5, wave periods 6 to 8 s and wave heights of 3 to 5 m. For mild slopes, wave breaking becomes a more continuous process, resulting in a more gradual dissipation of wave energy. This type of breaking is called "spilling." For the design of structures, surging and plunging breakers are of main importance. The area that suffers from wave-loading is bounded by the higher up-rush and the lowest down-rush point. Obviously, this zone varies with the tide.

No reliable formulas are available to predict the maximum velocities during up-rush and down-rush. For surging and spilling breakers, numerical solutions have been obtained; they are, however, not yet operational. As a first approximation, the maximum velocity, U_{max}, on a smooth slope can be computed by the following formula:

$$U_{max} = a\sqrt{(gH_s)}\xi_{op}^b \qquad (2)$$

where: H_s = significant wave height, g = gravity, a = coefficient equal to about 1 for irregular waves, and b = exponent equal roughly to 0.5.

4.2.3. Load Strength Concept. Once the hydraulic design conditions have been established, actual design loads have to be formulated. For a given structure, many different modes of failure may be distinguished, each with a different critical loading condition. Schematically, this is shown in Figure 3.10. For the dike as a whole, instability may occur due to failure of subsoil, front or rear slope. Each of these failure modes may be induced by geotechnical or hydrodynamical phenomena. The present section is restricted to the stability of the front slope. Moreover, only instability as a result of hydrodynamical processes is taken into account.

Starting with the hydraulic input (waves, water levels) and the description of the structure, external pressures on the seaward slope are determined. Together with the internal characteristics of the structure (porosity of revetment and secondary layers), these pressures result in an internal flow field with corresponding internal pressures. The resultant load on the revetment has to be compared with the structural strength, which can be mobilized to resist these loads. If this strength is inadequate, the revetment will deform and may ultimately fail (Figure 3.11).

The phenomena that may be relevant can be divided roughly according to the three components of the system: water, soil, and structure. The interaction between these components can be described using three transfer functions (see Figure 3.11):

I. The transfer function from the overall hydraulic conditions, e.g., wave height H, mean current velocity U, to the hydraulic conditions along the external surface, i.e., the boundary between free water and the protection or soil, e.g., external pressure P.

II. The transfer function from the hydraulic conditions along the external surface to those along the internal surface, i.e., the boundary between protection and soil. The hydraulic conditions along the internal surface can be described as the internal pressure P.

III. The structural response of the protection to the loads along both surfaces.

Information about these functions can be obtained by means of measurements in nature and scale model tests. If quantitative knowledge of the physical phenomena involved is available, or if there is sufficient data, then mathematical models or empirical formulas containing information are formulated and referred to as "models." All three transfer functions can be described in one model, or individually in three separate models, depending on the type of structure and the loading. The distinction between the three func-

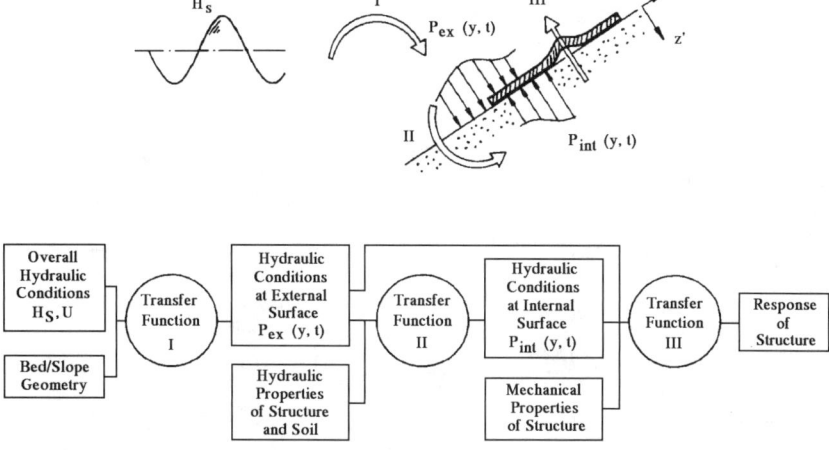

FIGURE 3.11. System approach. Transfer functions.

tions here mainly serves as a framework to describe the different phenomena that are important for the modeling. In many cases, the various processes cannot be described as yet. Therefore a "black box" approach is followed in which the relation between critical strength parameters, structural characteristics, and hydraulic parameters are obtained empirically.

4.3. Effect of Dikes and Sea Walls on the Beach

A sea wall forms a peculiar protection solution. It prevents further erosion of the coastline (it fixes the land–water line) but does not stop the physical processes that cause the erosion (a gradient in the longshore transport and the offshore transport). Therefore, it does not stop the erosion of the inshore zone, and probably even increases it through reflection of the incoming waves. Very often, the local solution by a sea wall leads to the displacement and expansion of the problem down-coast (Figure 3.12a). On an originally eroding coast, the erosion in front of a sea wall can even lead to undermining the sea wall or dikes, and often additional measures are needed (i.e., heavy toe protection and/or groins in front of a sea wall) to prevent this (Figure 3.12b). The maintenance cost of this protection can be, therefore, sometimes very high. This stresses the need for proper studies before the decision to construct a sea wall is taken. In general, it can be stated that a coast protected by a sea wall will only have a beach in front of this wall if sufficient sand supply is available. This happens where no erosion takes place under normal circumstances. The sea wall or dike is constructed then, for protection under extreme conditions. However, it is possible that even under these circumstances wave reflection causes so much erosion in front of the sea wall that the beach will not be restored. Therefore, sea walls are generally not a very feasible solution; but under special circumstances, such as along strongly curved and heavily eroding coasts, they may be the only possibility. By locating the defense line sufficiently landward from the eroding coast, the possibility can be created for observing the natural developments and preparing additional measures if necessary. To

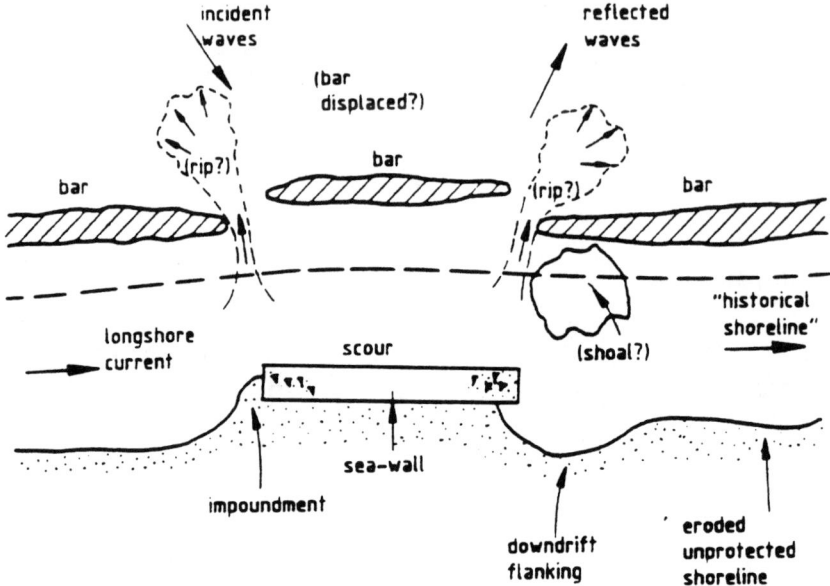

FIGURE 3.12(a). Plan view of potential impacts of sea walls (Kraus and Pilkey, 1988).

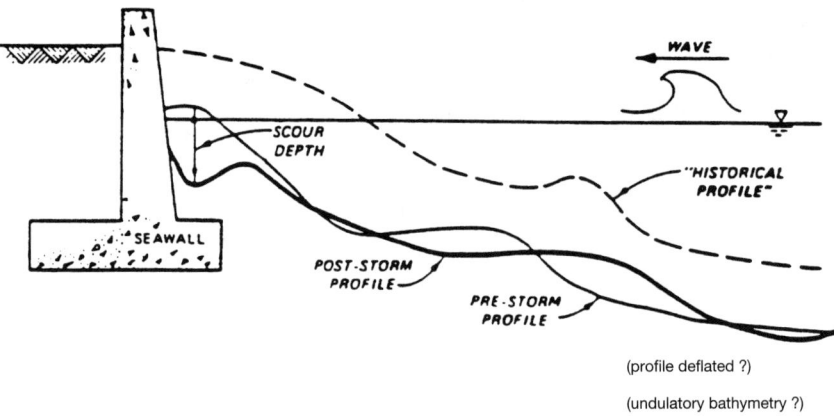

FIGURE 3.12(b). Profile view of potential impacts of sea walls (Kraus and Pilkey, 1988).

minimize the effect of extra scouring due to reflection of waves, the slope of sea walls or dikes should be not more than 1 in 3. That is one of the reasons that Dutch dikes mostly have slopes of 1 in 4 or even milder. Also, wave energy absorbing revetments, for example, rubble structures, will diminish the reflection and thus the scouring intensity.

As a result of controversy over the use of sea walls for coastal protection, the Coastal Sediments '87 conference had as its theme "The Effects of Sea walls on the Coast." The results of this conference have been evaluated and presented in a special issue of *Journal of Coastal Research* (Kraus et al., 1988). The final conclusions are summarized below.

The majority of quantitative-type laboratory and field studies indicate that sea walls neither accelerate nor enhance long-term erosion of beaches, and no systematic difference in post-storm recovery of beaches with and without sea walls was found if there was ample sediment or a wide surf zone existed. If beaches are deficient in sediment or if sea (or lake) level is rising at the site, erosion is more likely to occur at armored beaches as compared to unarmored beaches. The magnitude of the change does not appear to be extremely greater than the variability of change in the natural, unarmored beach system. However, the armoring can cause localized additional storm scour, both in front of and at the ends of the armoring. More study is needed to refine and quantify these conclusions, as well as determine their limit of applicability. In order to better utilize sea walls as an important instrument for shore stabilization, it is recommended that an effort be made to review historical data on shoreline position, beach profile, and littoral processes in the vicinity of existing sea walls, and to initiate comprehensive monitoring programs. Finally, the researcher should attempt to purge all preconceptions and biases in dealing with this controversial subject with the goal of defining and interpreting processes and responses at the study site within the context of that site.

It can be stated that the results of the evaluation of this conference agree firmly with the conclusions drawn by Dean (1986). Dean's final conclusions are cited below:

> Uncertainties resulting from a lack of definitive information has led to considerable speculation and claims regarding the adverse effects of coastal armoring on the adjacent shorelines. Employing sound principles and laboratory and field data, an attempt is made to evaluate the potential adverse effects of armoring. It is concluded that:
>
> 1. There are no factual data to support claims that armoring causes: profile steepening, increased longshore transport, transport of sand to a substantial distance offshore, or delayed post-storm recovery.
>
> 2. The interaction of an armored segment of shoreline with the littoral system is more of a "geometric" or "kinematic" interaction as contrasted to a "dynamic" interaction. The interaction depends on the amount of sand in the system vis-à-vis the equilibrium beach profile for the prevailing tide and wave conditions.
>
> 3. Armoring can cause localized additional storm scour, both in front of and at the ends of the armoring. A simple sediment supply–demand argument is proposed to explain the scour.

A methodology is presented to quantify the potential adverse effects of an armoring installation and appropriate periodic sand additions proposed as a means of mitigation to elevate the installation to one of neutral impact on the adjacent shoreline.

5. DESIGN OF DIKES AND SLOPING SEA WALLS

5.1. Structural Aspects and Design Checklist

The selection of the structural concept depends on the function, the local environmental conditions, and the constructional constraints. The governing criteria are the technical and economic feasibility for the project under consideration.

The function of the sea wall is mainly to protect the hinterland against the adverse effects of high water and waves. If high-water protection is required, the structure should have a height well above the maximum level of wave uprush during storm surge. This normally calls for high crest elevation (TAW 1990). If, however, some overtopping is allowed in view of the character of the hinterland, the design requirement is formulated in terms of the allowable amount of overtopping. Obviously, crest elevation can be reduced considerably in this case. A basic input to the effective planning and designing of any protection scheme is a reliable set of statistics that describe the physical environment against which such protection is necessary. The principal items under consideration are: bathymetry, climate, water levels, wind and wave climate, coastal processes (sediment transport), geotechnical data, construction constraints, etc. The geologic structure of the subsoil is also very important (settlement!). When these data are not available, soil mechanical investigation should be considered (soundings, borings etc.). Primarily, the requirement is for long-term data (water levels, winds, waves, etc.), much of which is readily obtainable from various national data banks, special international organizations (i.e., World Meteorological Organization), publications, and local authorities.

In the past, only local usage and experience have determined the selection of type and dimensions of coastal protection. Often, designs were conservative and too costly or inadequate. The technical feasibility and dimensioning of coastal structures can now be determined more accurately and supported by better experience than in the past. Often, however, the solution being considered should still to be tested in with a scale model, since no generally accepted design rules exist for all possible solutions and circumstances.

The most critical structural design elements are a stability of cover layer (including splash-area), secure foundation to minimize settlement, and toe protection to prevent undermining. All of these have a potential of failure of coastal structures. The checklist of the usual steps needed to develop an adequate structure design is given below:

a. Formulate functional requirements
b. Prepare alternative solutions
c. Select suitable solution
d. Determine the water level range for the site
e. Determine the wave heights and (eventual) currents
f. Detect suitable structure configurations (geometry)
g. Review the possible failure mechanisms
h. Select a suitable armor alternative and armor units size
i. Design the filter and underlayers
j. Determine the potential run-up to set the crest elevation
k. Determine the amount of overtopping expected for low structures
l. Design the toe protection, transitions, and crest protection
m. Design under-drainage features if they are required
n. Provide for local surface run-off and overtopping run-off, and make any required provisions for other drainage facilities, such as culverts and ditches
o. Consider end conditions to avoid failure due to flanking
p. Provide for firm compaction of all fill and backfill materials. This requirement should be included on the plans and in the specifications, and due allowance for compaction must be made in the cost estimate
q. Make final check of your design
r. Develop cost estimate for each alternative
s. Select the final design
t. Prepare specifications for materials and execution including quality control

The existing design rules for some (selected) structural elements (shape, height, cover layer, etc.) are briefly reviewed in the subsequent sections. For detail engineering (dimensioning) the following publications can be of use: Pilarczyk (1990, 1998), CUR/CIRIA (1991), SPM (1984), and CUR/RWS (1995a,b).

DESIGN OF DIKES AND REVETMENTS—DUTCH PRACTICE

The design of coastal protection is not a simple matter. In all cases, experience and sound engineering judgment play an important role in applying these design rules, or else mathematical or physical testing can provide an optimum solution.

5.2. Geometrical Design of Dikes and Sea Walls

Selection of the structural concept depends on the function, the local environmental conditions, and the constructional constraints. The governing criteria are the technical and economic feasibility for the project under consideration. The function of the dike/sea wall is mainly to protect the hinterland against the adverse effect of high water and waves.

If high-water protection is required, the structure should have a height well above the maximum level of wave-uprush during storm surges. This normally calls for high crest elevation. If, however, some overtopping is allowed in view of the character of the hinterland, the design requirement is formulated in terms of the allowable amount of overtopping. Obviously crest elevation can be reduced considerably in this case.

The shape of the cross-sectional profile of a dike/sea wall influences the distribution of wave forces, and thus also influences the choice of material (type of protective units and their dimensions) suitable for slope protection (revetment) and the height of structure. The gradient of the slope must not be so steep that the whole slope of the revetment can lose stability (through sliding). This criterion gives, therefore, the maximum slope angle. More gentle (flatter) slopes lead to a reduced wave force on the revetment and less wave run-up; wave energy is dissipated over a greater length. A similar effect can be obtained by applying a berm (trapezoidal profile).

By using the wave run-up approach for calculation of the crest height of a trapezoidal profile of a dike/sea wall for different slope gradients, the minimum volume of the embankment can be obtained. However, this does not necessarily imply that minimum earth volume coincides with minimum cost. An expensive part of the embankment is the revetment, and the slope area increases as the slope angle decreases. Careful attention is needed however, because the revetment costs are not always independent of the slope angle; e.g., for steep slopes, heavy protection is needed whereas for mild slopes the (cheaper) alternative solutions (i.e., sand mattresses, grass mats, etc.) can often provide sufficient protection.

Another point of economic optimization can be the available space for dike/sea wall construction or improvement. All these factors should be taken into account when optimizing the structure slope. The common Dutch practice is to apply a slope 1 in 3 on the inner slope and between 1 in 3 and 1 in 5 on the outer (seaward) slope. These mild slopes also minimize the scour at the toe of a dike (less reflection). The minimum crest width is 2 m.

Present practice for obtaining a substantial reduction in wave run-up is to place the outer berm at (or close to) design water level. If the berm lies too much below that level, the highest storm flood waves will not break beneath the berm, and the run-up will be inadequately affected, thus providing relatively heavy wave loading on the upper slope. However, the optimum distribution of wave forces is obtained when the (relatively small) berm is about 0.5 to 1 wave height below the design water level. In such a case, the whole slope can be protected with the same (relatively small) units. When the berm is equal or somewhat higher than the design water level, the waves will always break on a lower slope. In this case, the lower slope needs much heavier protection (but along a shorter length) than the upper slope, where very often a grass mat is sufficient. An important function of the berm can also be its use as an access road for maintenance or even for permanent use (promenade, etc.).

The above discussion indicates that the proper choice of the shape is a very important step in the design of a dike/sea wall. Mostly, it is an iterative procedure.

According to the existing safety law in the Netherlands, the design of all sea dikes is based fundamentally on a water level with a probability of exceedance of 10^{-4} per annum. In the Netherlands, the wind set-up is usually incorporated in the estimated storm-surge level. If this is not the case, the wind set-up should be calculated separately and added to the design water level. Besides the design flood level, several other elements also play a role in determining the design crest level of a dike, namely (Figure 3.13):

- Wave run-up (2% of exceedance is applied in the Netherlands) depending on wave height and period, angle of approach, roughness and permeability of the slope, and profile shape
- An extra margin to the dike height to take into account seiches (oscillations) and gust bumps (single waves resulting from a sudden violent rush of wind); in the Netherlands this margin varies (depending on location) from 0 to 3 m for the seiches and 0 to 0.5 m for the gust bumps
- A change in chart datum (NAP) or a rise in the mean sea level in the Netherlands, which until now is assumed to be roughly 0.25 m per century, and in the future ~0.6 m per century (see RWS 1990)
- Settlement of the subsoil and the dike body during its lifetime (Figure 3.14).

The combination of all the factors mentioned above defines the freeboard on the dike (called wake height in Dutch). The recommended minimum freeboard is 0.5 m.

5.3. Wave Run-up

5.3.1. Definitions. Run-up and overtopping are decisive factors influencing the crest height of a dike, whereas run-up and overtopping are strongly related to the shape of a dike (slope angles, berm, roughness, etc.).

Wave run-up is often indicated by $R_{u,2\%}$ (see Figure 3.15). This is the run-up level, vertically measured with respect to the still water level (SWL), which is exceeded by 2% of the incoming waves. Note that the number of exceedance is related here to the number of incoming waves and not to the number of run-up levels.

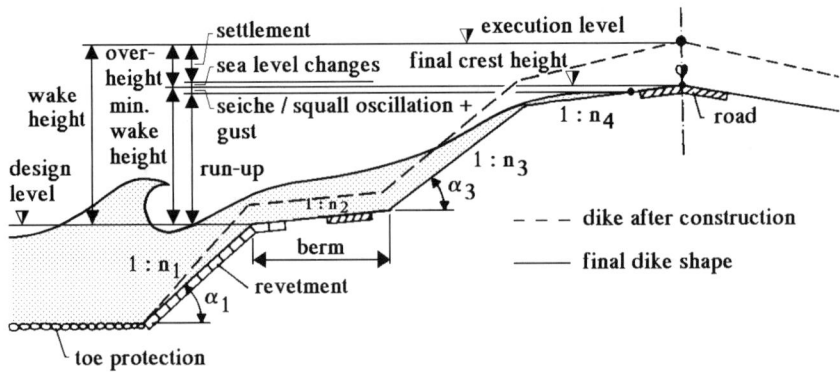

FIGURE 3.13. Determination of dike height.

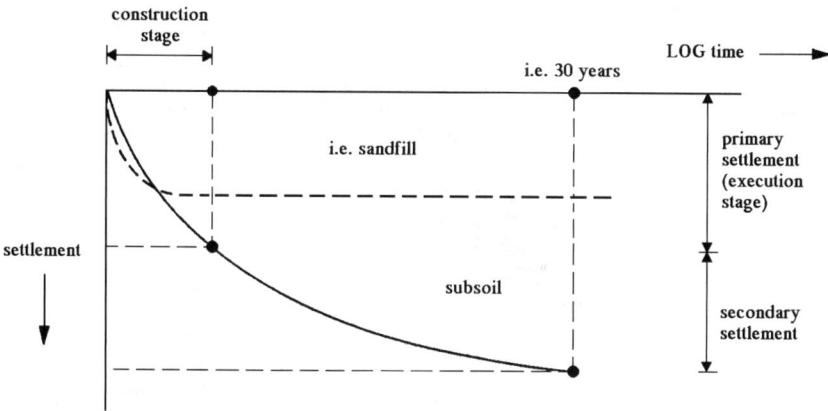

FIGURE 3.14. Settlement as function of time.

The relative run-up is given by $R_{u,2\%}/H_s$, where H_s is the significant wave height, being the average value of the highest ⅓ of the wave heights. This H_s is the significant wave height at the toe of the structure.

The relative run-up is usually given as a function of the surf similarity parameter or breaker parameter, which is defined by Equation 1 (Section 4.2.2).

5.3.2. Model Results. Figure 3.16 shows all available data pertaining to smooth and straight slopes as well to slopes with berms or roughness and obliquely incoming, short-crested waves (Van der Meer and Janssen, 1994). Both, small- and large-scale tests are included. When all the influences are incorporated into one figure, the scatter becomes larger than for only smooth, straight slopes.

The value that can be expected for the average value of the wave run-up for smooth and straight slopes can be described by

$$\frac{R_{u,2\%}}{H_s} = 1.5\gamma_h\gamma_f\gamma_\beta\xi_{eq} \quad \text{with a maximum of } 3.0\gamma_h\gamma_f\gamma_\beta \tag{3}$$

where

$R_{u,2\%}$ = 2% run-up level above the still water line
H_s = significant wave height near the toe of the structure

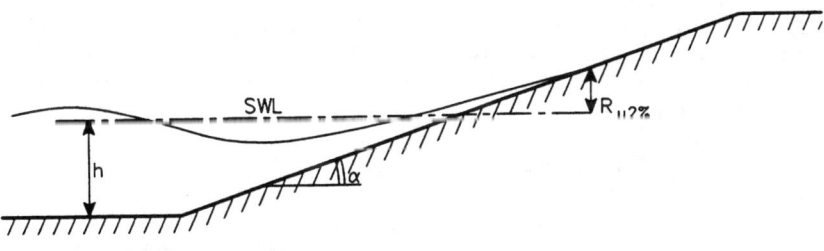

FIGURE 3.15. Wave run-up level $R_{u,2\%}$.

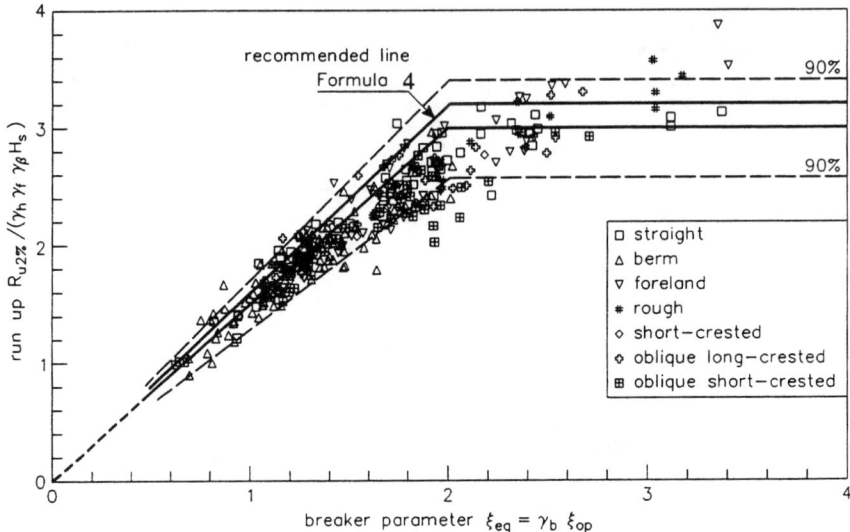

FIGURE 3.16. Wave run-up data, including possible influences.

ξ_{op} = breaker parameter, $\xi_{op} = \tan \alpha / \sqrt{s_{op}} = \tan \alpha / \sqrt{2\pi H_s/(gT_p^2)}$
ξ_{eq} = equivalent breaker parameter for a slope with a berm, $\xi_{eq} = \gamma_b \xi_{op}$
g = acceleration due to gravity
T_p = peak period of the wave spectrum
γ_b = reduction factor for a berm
γ_h = reduction factor for a shallow foreshore
γ_f = reduction factor for slope roughness
γ_β = reduction factor for oblique wave attack

The scatter around Equation 3 can be described by interpreting the coefficient 1.5 as a normally distributed stochastic variable with a mean value of 1.5 and a variaation coefficient (standard deviation divided by the mean value) of $V = \sigma/\mu = 0.06$. If we take the coefficient of 1.5 as a stochastic variable (which is an over-estimation around $\xi_{eq} = 2$) for all results in Figure 3.16, then the variation coefficient can be taken as $V = \sigma/\mu = 0.085$. The 90% confidence bands are presented along with Equation 3.

5.3.3. General Equations. In manuals or guidelines, it is advised that one should not deal with a general trend. Instead, it is recommended that a more conservative approach be used. In many international standards a safety margin of one standard, deviation is applied. Therefore, the proposed general design formula that can be applied for wave run-up on dikes is given by

$$\frac{R_{u,2\%}}{H_s} = 1.6 \gamma_h \gamma_f \gamma_\beta \xi_{eq} \quad \text{with a maximum of } 3.2 \gamma_h \gamma_f \gamma_\beta \quad (4)$$

The formula is valid for the range $0.5 < \xi_{eq} < 4$ or 5. The relative wave run-up $R_{u,2\%}/H_s$ depends on the breaker parameter ξ_{op} and on four reduction factors, namely: for the influence of a shallow foreshore (breaking waves in shallow water), roughness of the slope, oblique wave attack, and for a berm. The influence of a berm is accounted for by an equivalent slope gradient expressed in ξ_{eq}. If there is no berm, then $\xi_{eq} = \xi_{op}$ applies. We shall return later to the question of how the reduction factors should be computed.

Equation 4 is shown in Figure 3.16, where the relative run-up $R_{u,2\%}/(\gamma_h \gamma_f \gamma_\beta H_s)$ is set out against the breaker parameter ξ_{eq}. The relative run-up increases until $\xi_{eq} = 2$ and remains constant for larger values. The latter is the case for relatively steep slopes and/or low wave steepnesses. The theoretical limit for a vertical structure ($\xi_{op} = \infty$) is $R_{u,2\%}/H_s = 1.4$ in Equation 4, but this is far outside the application range considered here.

However, Equation 3 should not be used for the wave run-up when deterministically designing dikes; for that purpose one should use Equation 4. For probabilistic designs, Equation 3 should be taken with the above variation coefficients.

The reduction factors in Equation 4 have all been determined separately in model investigations. It is possible that a combination of these reduction factors produces a very low total reduction. Since combinations of wave run-up reduction factors have not been investigated, a minimum total reduction factor should be established. A minimum reduction factor is proposed of $\gamma_b \gamma_h \gamma_f \gamma_\beta = 0.5$. For the application of Equation 2 this implies that if $\gamma_b \gamma_h \gamma_f \gamma_\beta < 0.5$, the total reduction factor should be set equal to 0.5. If a berm is present, both the γ_b in ξ_{eq} and the combination $\gamma_h \gamma_f \gamma_\beta$ should be multiplied by the same factor in order to come to a value of 0.5. This factor is actually $\sqrt{0.5/(\gamma_b \gamma_h \gamma_f \gamma_\beta)}$. In specific cases, it could be demonstrated by means of investigations that a lower reduction factor is acceptable.

In TAW (1974) an "Old Dutch Formula" is given for mild (milder than 1 in 2.5), smooth, straight slopes and storm waves:

$$R_{u,2\%} = 8 H_s \tan \alpha \quad \text{(in metric system)} \tag{5a}$$

This formula only corresponds to Equation 4 for an average wave steepness of $s_{op} = 0.040$ and a value of 1.0 for all the reduction factors. However, after rearrangement, this formula has the form

$$\frac{R_{u,2\%}}{H_s} = (1.60 \text{ to } 1.75) \xi_{op} \quad (\text{for } \xi_{op} < 2.5) \tag{5b}$$

Note: the numerical value of 1.75, which describes the upper envelope of spreading in test results, was often used in the past designs for safety reasons.

This formula is virtually identical to Equation 4, except for the reduction factors and the limit for steeper slopes, which has here the value of 3.2. In other words, the run-up formula used for twenty years will be maintained and complemented on specific points.

5.3.4. Influence of a Stepped Slope and/or a Berm on Wave Run-up. Figure 3.17 shows an example of a dike with a berm. The front edge of the berm is located at a

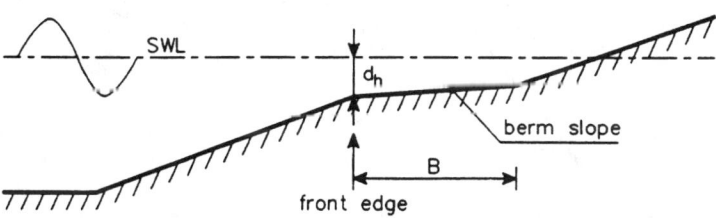

FIGURE 3.17. Front edge, slope, and width of the berm.

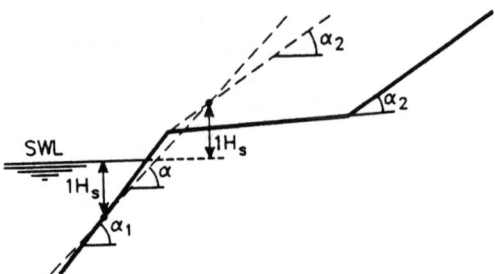

FIGURE 3.18. Definition sketch of the representative slope gradient α with a stepped slope, exclusive of berm influence.

depth d_h under the still water line. Typical berm slopes are about 1:15. The width of the berm is B.

With a stepped slope (with or without berm), the slope gradient is not constant. This makes it necessary to define a representative slope gradient tan α, as shown in Figure 3.18, which leaves the berm out of consideration. The slope gradient that is most influential for run-up is the average slope in the zone between the (SWL − H_s) and the (SWL + H_s) levels. A berm, if present, should not be accounted for when determining the average slope gradient. The procedure is as follows:

a. draw a line parallel to the upper slope, which intersects the front edge of the berm,
b. set a point on the lower slope at H_s below the SWL,
c. set a point on the relocated upper slope at H_s above the SWL,
d. connect the two points with a line. The angle of this line is the averaged slope gradient tan α pertaining to run-up.

The influence of a berm on the run-up can be determined by defining an equivalent slope gradient tan α_{eq}, as shown in Figure 3.19. By doing so, it is also possible to account for the influence of the berm location. A method with an equivalent slope means, in fact,

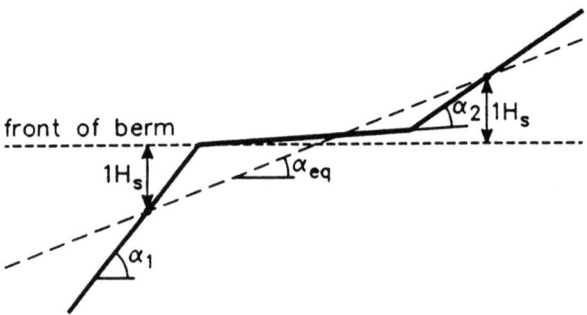

FIGURE 3.19. Determination of the equivalent slope gradient tan α_{eq} with a berm.

that a reduction factor is applied to the breaker parameter itself instead of to the term $R_{u,2\%}/H_s$.

The equation then becomes

$$\xi_{eq} = \gamma_b \xi_{op} \qquad (6)$$

The reduction factor γ_b consists of two other factors—one for the influence of the berm width r_B and one for the location of the berm with respect to the SWL, r_{dh}. The following applies:

$$\gamma_b = 1 - r_B(1 - r_{dh}) \quad \text{and} \quad 0.6 \leq \gamma_b \leq 1.0 \qquad (7)$$

If the berm is located at the water line (SWL) (the most favorable case) then $r_{dh} = 0$ and r_B causes γ_b to be less than 1 (influence of the berm width only). If the berm is not situated at the water line (SWL), the reduction factor r_B will be multiplied by a number less than 1, after which the reduction factor γ_b becomes larger than in the case where the berm is situated at water line (SWL) (that means less reduction in run-up).

In Figure 3.19 the determination of an equivalent slope tan α_{eq} is graphically depicted. This is accomplished by simply drawing a line through two points, namely, one lying at 1 H_s above and below the front of the berm (i.e., not with respect to SWL).

The influence of the berm width can be given by considering the change in slope gradient:

$$r_B = 1 - \frac{\tan \alpha_{eq}}{\tan \alpha} \qquad (8)$$

where tan α is the slope gradient as defined in Figure 3.18. When, for the sake of simplicity, a horizontal berm is considered the following formula can be established:

$$r_B = 1 - \frac{\tan \alpha_{eq}}{\tan \alpha} = \frac{B/H_s}{2 \cot \alpha + B/H_s} \qquad (9)$$

The influence of the berm elevation with respect to SWL can be represented by

$$r_{dh} = 0.5 \left(\frac{d_h}{H_s}\right)^2 \quad \text{with} \quad 0 \leq r_{dh} \leq 1 \qquad (10)$$

where d_h is defined in Figure 3.17. This means that the influence of the berm will become practically nil if the berm lies more than $\sqrt{2}H_s$ below or above SWL. In that case, run-up is computed with the slope situated around SWL, without the berm (Figure 3.8). When the berm is situated higher than $\sqrt{2}H_s$ above SWL, the run-up levels may be overpredicted (higher than the berm location). This will be even more for the case with a substantial berm width. If d_h lies higher than $\sqrt{2}H_s$ above SWL (so $\gamma_b = 1.0$), then it can be stated that $R_{u,2\%} = d_h$ if $B/H_s > 2$ and that for values of B/H_s between 0 and 2 one can interpolate between d_h (with $B/H_s = 2$) and the run-up value with $\gamma_b = 1.0$ (with $B/H_s = 0$).

The berm is most effective when lying at SWL ($r_{dh} = 0$). The influence of the berm can be neglected when lying at more than $\sqrt{2}H_s$ below SWL. An optimum berm width will be obtained if the reduction factor reaches the value of 0.6. In principle, this optimum berm width can be determined with the equations for every berm geometry (with one berm). For a horizontal berm at SWL, Equations 7 and 9 yield the optimum berm width:

$$B = \tfrac{4}{3} H_s \cot \alpha \qquad (11)$$

5.3.5. Influence of a Shallow Foreshore on Run-up. With a fairly substantial water depth at the toe of the slope ($h_m/H_s > 3$–4), the probability distribution of the wave heights corresponds with a so-called Rayleigh distribution. Here, h_m is the water depth at the toe of the structure. With a shallow foreshore ($h_m/H_s < 3$–4), the waves start breaking on this foreshore and the distribution will deviate from that with deep water. This holds particularly for the highest waves. This is schematically shown in Figure 3.20.

For a Rayleigh distribution, the relation $H_{2\%}/H_s = 1.40$, with $H_{2\%}$ the wave height that is exceeded by 2% of the waves. But this ratio will become smaller if waves break on the foreshore and varies roughly from 1.1 to 1.4. The reduction factor for wave run-up with shallow water on a foreshore is given by γ_h. However, since the 2% run-up level is described, it is practical to seek a relation between γ_h and $H_{2\%}/H_s$.

The general conclusion about wave run-up with shallow water on a foreshore is that by using $H_{2\%}$ instead of H_s, the general trend will be best represented. A reduction factor γ_h, which accounts for the relation $H_{2\%}/H_s$ and the water depth at the location of the toe of the structure (h_m), can be used for representing the influence of a shallow foreshore.

Still, there is a need for further investigations into the wave height distribution on a shallow foreshore and research on this matter is currently underway. In the following, a tentative and only empirically based formulation of the reduction factor γ_h or the relation $H_{2\%}/H_s$ is given. In principle, this relation is only valid for a foreshore slope of 1:100:

$$\gamma_h = \frac{H_{2\%}}{1.4\, H_s} = 1 - 0.03\left(4 - \frac{h_m}{H_s}\right)^2 \quad \text{for} \quad 1 < \frac{h_m}{H_s} < 4$$

$$\gamma_h = 1 \quad \text{for} \quad \frac{h_m}{H_s} \geq 4$$

(12)

Equation 12 is shown in Figure 3.21. The reduction factor γ_h has been set out against the relative water depth h_m/H_s. If the foreshore considerably deviates from 1:100, a reduction factor of $\gamma_h = 1.0$ should be taken to be conservative. Also, Equation 12 is based on the wave height at the toe of the structure and not on the deep-water wave height.

5.3.6. Influence of Roughness on Run-up. The influence of roughness on run-up is given by the reduction factor γ_f. Reduction factors for various types of revetments have been published earlier. The origin of these factors dates back to the Russian investigations per-

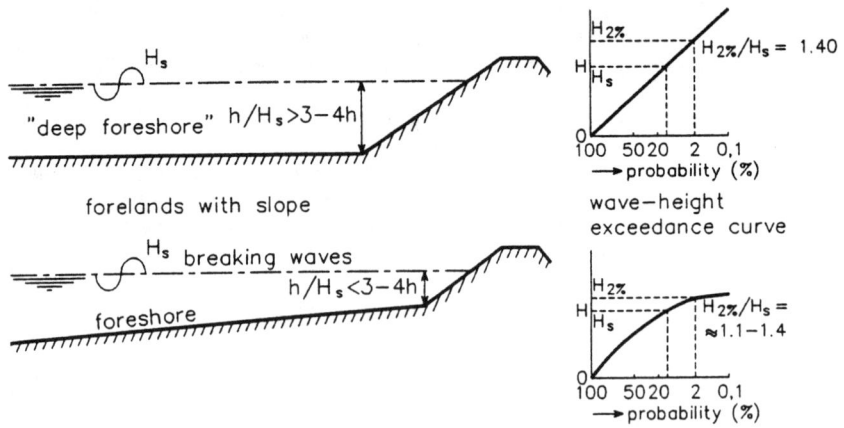

FIGURE 3.20. Effect of a shallow foreshore on the wave height distribution.

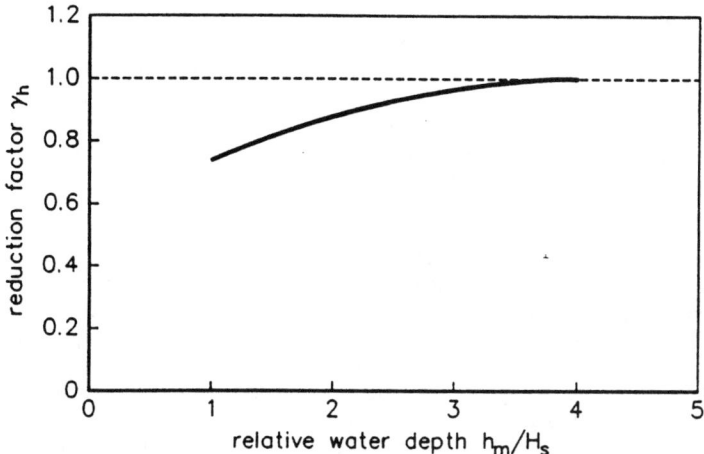

FIGURE 3.21. Reduction factor γ_h for a shallow foreshore (1:100).

formed in the 1950's with regular waves (e.g., Drogosz-Wawrzyniak, 1965). A table of these factors was further developed in TAW (1974) and published in several international manuals. New studies, often large-scale and conducted with random waves, have led to a new table (Table 3.1) of reduction factors for rough slopes. In Table 3.1, smooth slopes are also described on which roughness elements such as blocks and ribs were installed. The width of a block or rib is given by f_b and the rib length by f_L. Installation of the blocks

TABLE 3.1. Reduction factor γ_f for a rough slope.

Type of slope				Reduction factor γ_f	Old reduction factors
Smooth, concrete, asphalt				1.0	1.0
Closed, smooth, block revetment				1.0	0.9
Grass (3 cm)				0.90–1.0	0.85–0.90
1 rubble layer ($H_s/D = 1.5$–3)				0.55–0.60	0.80
2 or more rubble layers ($H_s/D = 1.5$–6)				0.50–0.55	0.50–0.55
Roughness elements on a smooth slope. Height = f_h, width = f_b					
Label	f_h/f_b	f_b/H_s	Surface covered		
1/25 block	0.88	0.12–0.24	1/25	0.75–0.85	
1/9 block	0.88	0.12–0.19	1/9	0.70–0.75	
Half block height	0.44	0.12–0.24	1/25	0.85–0.95	
Only above water line	0.88	0.12–0.18	1/25	0.85–0.95	
Wide block	0.18	0.55–1.10	1/4	0.75–0.85	
Ribs, $f_b/H_s = 0.12$–0.19 and $f_L/f_h = 7$ (optimal), where f_L = distance between the ribs				0.60–0.70	

is determined by the part of the total slope surface covered by these blocks. A rubble mound slope (rock) is characterized by the diameter D. The reduction factors in Table 3.1 apply for $\xi_{op} < 3$–4. For larger values of ξ_{op}, the reduction factors become 1.

5.3.7. Influence of the Angle of Wave Attack on Run-up. The angle of the wave attack ß is defined as the angle of the propagation direction with respect to the normal of the alignment axis of the dike (see Figure 3.22). Perpendicular wave attack is therefore given by ß = 0°.

The reduction factor for the angle of wave attack is given by γ_β. Until recently, few investigations were carried out with obliquely incoming waves but these investigations had been performed with long-crested waves. "Long-crested" means that the length of the wave crest is assumed to be infinite. In investigations with long-crested waves, the wave crest is as long as the wave-generating board and the wave crests propagate parallel to one another.

In nature, waves are mostly short-crested. This implies that the wave crests have a certain length and the waves a certain main direction. The individual waves have a direction around this main direction. The extent to which they vary around the main direction (directional spreading) can be described by a spreading value. Only a long swell, for example, one coming from the ocean, has crests long enough to be called "long-crested." A wave field with strong wind is short-crested.

An investigation by Van der Meer and de Waal (1993) describes wave run-up and overtopping as a function of obliquely incoming waves and directional spreading. Figure 3.23 summarizes the results of the investigation of Van der Meer and de Waal (1993). The reduction factor γ_β has been set out against the angle of wave attack ß.

Long-crested waves with 0° < ß < 30° cause virtually the same wave run-up as a perpendicular attack. Outside of this range, the reduction factor decreases fairly quickly to about 0.6 at ß = 60°. With short-crested waves, the angle of wave attack has apparently less influence. This is mainly due to the fact that within the wave field the individual waves deviate from the main direction ß. For both run-up and overtopping with short-crested waves, the reduction factor decreases linearly to a certain value at ß = 90°. This is $\gamma_\beta = 0.8$ for the 2% run-up and 0.7 for overtopping. So, for wind waves, the reduction factor has a minimum of 0.7–0.8 and not 0.6, as was found for long-crested waves. Since a wave field under storm conditions can be considered to be short-crested, it is recommended that the lines in Figure 3.23 be used for short-crested waves.

For the 2% run-up and overtopping, different reduction factors apply for obliquely incident waves. The reason for this is that here the incoming wave energy per unit length of structure is less than that for perpendicular wave attack. The wave overtopping is defined

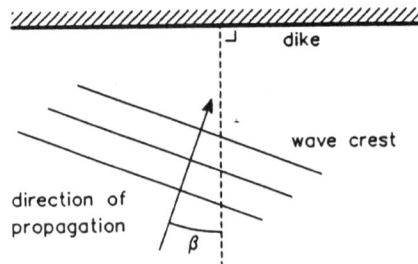

FIGURE 3.22. Definition of the angle of wave attack ß.

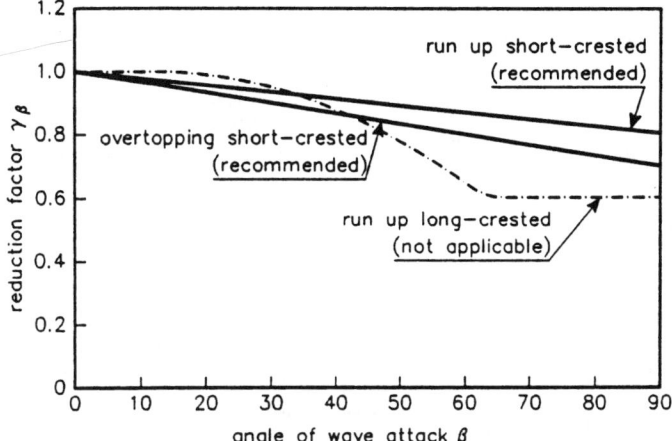

FIGURE 3.23. Reduction factor γ_β for the angle of wave attack.

as a volume per unit length, whereas the run-up does not depend on the structure length. The use of the lines given in Figure 3.13 for short-crested waves is recommended and can be described by the following equations.

For the 2% wave run-up with short-crested waves:

$$\gamma_\beta = 1 - 0.0022\beta \quad (\beta \text{ in degrees}) \qquad (13)$$

For wave overtopping with short-crested waves:

$$\gamma_\beta = 1 - 0.0033\beta \quad (\beta \text{ in degrees}) \qquad (14)$$

As was already mentioned, the height of a dike/sea wall is affected by functional requirements. In the case of a high-crested structure (protection against inundation), the run-up and the necessary crest height can be calculated according to the methods mentioned above. An example of possible variation of a crest height as a function of a dike shape for nonovertopping conditions is given in Figure 3.24.

In the Dutch practice, the upper level of protection is defined by $\leq$ SWL + 0.5 $R_{2\%}$ and $\geq H_s$ (sometimes > 0.5 H_s). Above this elevation, grass usually provides sufficient protection.

5.4. Wave Overtopping

5.4.1. Average Overtopping Discharge. In the case of low-crest elevation (overtopping allowed), the design requirement is formulated in terms of the allowable amount of overtopping and the necessary protection of the splash area (dimensions of protection units, length of protection, etc.). The most recent approach for wave overtopping at dikes (research commissioned by the Rijkswaterstaat) is recently presented by Van der Meer and de Waal (1993) (see also Van der Meer and Janssen, 1994).

With wave overtopping, the crest height is lower than the run-up levels of the highest waves. The parameter to be considered here is the crest freeboard R_c (see Figure 3.25). This is the difference between SWL and the crest elevation. The crest height itself can be

results :	ctg α = 3 ; γ_b = 1	ctg α = 4 ; γ_b = 1	ctg α = 4 ; γ_b ≅ 0.7
storm surge MSL +	5.00	5.00	5.00
run-up	13.30	10.00	7.00
sea-level rise	0.25	0.25	0.25
seiches / oscillations	0.25	0.25	0.25
settlement	0.50	0.50	0.50
dike crest MSL +	19.30	16.00	13.00

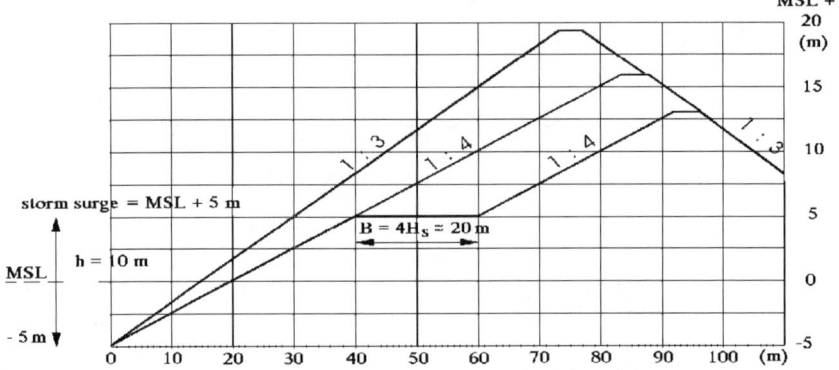

FIGURE 3.24. Example of dike calculation (alternatives); H_s = 4.7 m (depth limited), T_p = 8.5 sec, berm reduction γ_b = 0.7 for ctg α = 4 and $B = 4 H_s$.

given as an absolute crest height h_d, determined with respect to OD (Ordnance Datum). The crest height reduced by the water level (also with respect to OD) then yields the crest freeboard R_c, relative to SWL.

Wave overtopping is usually given as an average discharge q per unit width, in m³/s per m or in L/s per m. The Dutch guideline on river dikes (TAW, 1991) indicates that for relatively heavy seas and with wave heights of up to a few meters the 2% wave run-up criterion yields an overtopping discharge of the order of 1 L/s per m. It becomes 0.1 L/s per m with lower waves such as those occurring in rivers. An acceptable overtopping of 1 L/s per m in the river area, instead of the 2% wave run-up, can then lead to a reduction of the freeboard of the dike. The same guideline further quotes "Which criterion applies depends, of course, also on the design of the dike and the possible presence of buildings. In certain cases, such as a covered crest and inner slopes, sometimes 10 L/s per m can be tol-

FIGURE 3.25. Crest freeboard above SWL with wave overtopping.

erated." In Dutch guidelines it is assumed that the following average overtopping rates are allowable for the inner slope:

a. 0.1 L/s per m for sandy soil with poor turf
b. 1.0 L/s per m for clay soil with relatively good grass
c. 10 L/s per m with a clay protective layer and grass according to the standards for an outer slope or with revetment construction

At present, studies are being carried out to evaluate the relation among 0.1, 1.0, and 10 L/s per m overtopping as well as the condition of the inner slope.

Wave overtopping can be expressed in two formulas: one for breaking waves ($\xi_{op} < 2$) and one for nonbreaking waves ($\xi_{op} > 2$). Figure 3.26 gives an overview of information on breaking waves. In this figure, the important parameters are given along the axes; all existing data points are given with a mean and 95% confidence bands and typical applications are indicated along the vertical axis. In addition to Delft hydraulics data, data for smooth slopes from Owen (1980) and Führböter et al. (1989) are included.

The dimensionless overtopping discharge Q_b (b for breaking waves) is given on the ordinate:

$$Q_b = \frac{q}{\sqrt{gH_s^3}} \sqrt{\frac{s_{op}}{\tan \alpha}} \tag{15}$$

and the dimensionless crest height R_b (application range $0.3 < R_b < 2$) with

$$R_b = \frac{R_c}{H_s} \frac{\sqrt{s_{op}}}{\tan \alpha} \frac{1}{\gamma_b \gamma_h \gamma_f \gamma_\beta} \tag{16}$$

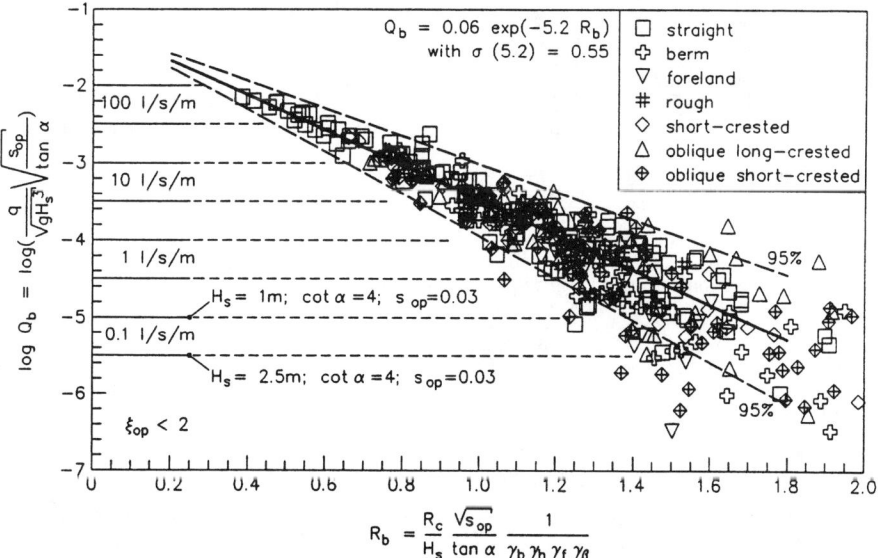

FIGURE 3.26. Wave overtopping data for breaking waves with the mean and confidence bands and with an indication of typical applications ($\xi_{op} < 2$).

where

Q_b = dimensionless overtopping discharge for breaking waves ($\xi_{op} < 2$)
q = overtopping discharge (in m³/s per m width)
g = acceleration due to gravity
H_s = significant wave height (average of highest ⅓ part)
s_{op} = wave steepness = $2\pi H_s/(gT_p^2)$
T_p = peak period
R_b = dimensionless crest height with breaking waves ($\xi_{op} < 2$)
R_c = crest freeboard above SWL
$\gamma_b, \gamma_h, \gamma_f, \gamma_\beta$ = reduction factors for influence of a berm, shallow foreshore, roughness, and angle of wave attack (see previous section). The minimum value using a combination of factors is 0.5.

Both the dimensionless overtopping discharge and the dimensionless crest height are related to the significant wave height, the wave steepness, and the slope gradient. To account for the varying conditions, the dimensionless crest height is increased by the reduction factors $\gamma_b, \gamma_h, \gamma_f,$ and γ_β, which were described earlier: Equation 7 for γ_b, Equation 12 for γ_h, Table 1 for γ_f, and Equation 14 for γ_β. The average of all the observations in Figure 3.26 can be described by

$$Q_b = 0.06 \exp(-5.2 R_b) \quad \text{for } \xi_{op} < 2 \quad (17)$$

The reliability of the equation is given by taking the coefficient 5.2 as a normally distributed stochastic variable with an average of 5.2 and a standard deviation = 0.55. By means of this standard deviation, confidence bands ($\mu \pm x\sigma$) can be drawn for x times the standard deviation (1.64 for the 90% and 1.96 for the 95% confidence limit). The coefficient 0.06 gives the intersection with $R_b = 0$. Also, in Figure 3.26 several overtopping discharges are illustrated, namely, 0.1, 1, 10, and 100 L/s per m. The discharges apply for a 1:4 slope and a wave steepness of $s_{op} = 0.03$. The upper line of the interval applies to a significant wave height of 1.0 m (e.g., river dikes) and the lower one for a wave height of 2.5 m (e.g., sea dikes).

The available data points for nonbreaking waves ($\xi_{op} > 2$) have been set out in Figure 3.27. The dimensionless overtopping discharge is now given along the vertical ordinate by

$$Q_n = \frac{q}{\sqrt{gH_s^3}} \quad (18)$$

and the dimensionless crest height R_n along the abscissa by

$$R_n = \frac{R_c}{H_s} \frac{1}{\gamma_b \gamma_h \gamma_f \gamma_\beta} \quad (19)$$

with

Q_n = dimensionless overtopping discharge for nonbreaking waves ($\xi_{op} > 2$)
R_n = dimensionless crest height for nonbreaking waves ($\xi_{op} > 2$)

As with wave run-up (see Equation 4), the wave steepness and slope gradient have little influence on the overtopping in the case of nonbreaking waves. For γ_b in Equation 19, the influence of a berm vanishes if $\xi_{op} \geq 4$ (since $\gamma_b = 1$). In the interval $2/\gamma_b < \xi_{op} < 4$ a larger value for γ_b must be taken than computed with Equation 7, namely, a value from linear interpolation between γ_b (for $\xi_{op} = 2/\gamma_b$) and 1.0 (for $\xi_{op} = 4$).

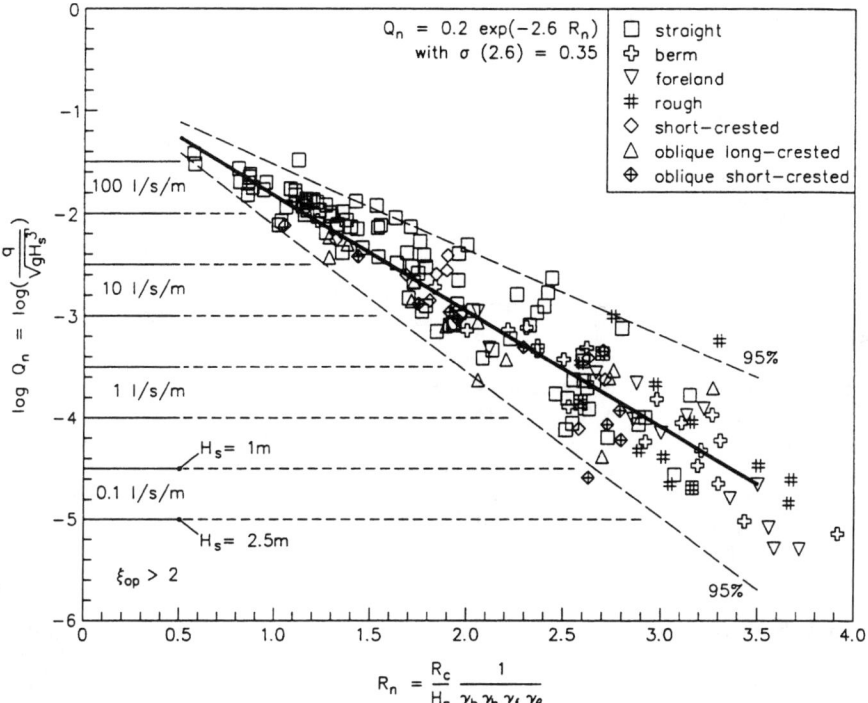

FIGURE 3.27. Wave overtopping data for nonbreaking waves with the mean and confidence bands and with an indication of typical applications (ξ_{op} 2).

The average of all observations in Figure 3.27 is given by:

$$Q_n = 0.2 \exp(-2.6 R_n) \qquad (20)$$

The reliability of the formula can be given by taking the coefficient 2.6 as a normally distributed stochastic variable having a standard deviation of $\sigma = 0.35$. With this standard deviation, the 95% confidence bands have been drawn in Figure 3.27. Now the coefficient 0.2 gives the intersection with $R_n = 0$.

The dimensionless overtopping and crest height as defined by Equations 18 and 19 can also be used for vertical structures, but now with γ_v instead of $\gamma_b\gamma_h\gamma_f\gamma_\beta$. Equation 20 may then be valid for vertical structures and it is possible to plot data in a graph similar to Figure 3.27. With a reduction factor γ_v for vertical structures of about 0.6, the data may be described by Equation 20. Also, in Figure 3.27 the intervals are given along the ordinate that express the overtopping discharges of 0.1, 1, 10, and 100 L/s per m, respectively. These intervals hold for a wave height of $H_s = 1$ m (upper line) and 2.5 m (lower line) and are independent of the slope gradient and wave steepness.

As is the case with wave run-up, a somewhat more conservative formula should be applied for design purposes than the average value. The two recommended formulas for overtopping are: for breaking waves with $\xi_{op} < 2$:

$$Q_b = 0.06 \exp(-4.7 R_b) \qquad (21)$$

and for nonbreaking waves with $\xi_{op} > 2$:

$$Q_n = 0.2 \exp(-2.3 R_n) \qquad (22)$$

with Q_n and R_n as defined in Equations 18 and 19.

Both design formulas are graphically shown in Figures 3.28 and 3.29. In these figures, the recommended lines, the mean, and the 95% conidence limits are given. Also, in Figure 3.28 the formula from TAW (1974) is drawn and is practically the same as the newly recommended line. When assessing a crest height of a dike in a deterministic way, Equations 21 and 22 are recommended. With probabilistic computations, both the given estimates of the mean value (Equations 17 and 20) and the given standard deviations can be applied.

In general, it can be said that the results for run-up and overtopping on smooth, straight slopes differ very little from TAW (1974). With the new formulas, steep slopes are accounted for by considering nonbreaking waves separately. The improvement is mainly the description of the reliability of the formulas and a better description of the influence of berms, a shallow foreshore, roughness, and the angle of wave attack.

In De Waal and Van der Meer (1992), overtopping was related to a "shortage in crest height" $(R_{u,2\%} - R_c)/H_s$ instead of R_c/H_s. The method described there gives similar results to those described here. The approach, however, has a limited application if the overtopping discharge becomes fairly large or if the crest height is much lower than the 2% run-up height, especially when reduction factors for influences of berms, etc. are applied. In those cases, the method presented in this Chapter is preferred.

5.4.2. Wave Overtopping per Wave. The average overtopping discharge does not provide information on the amount of water of a given overtopping wave passing the crest. The overtopping volumes of individual waves deviate considerably from the average dis-

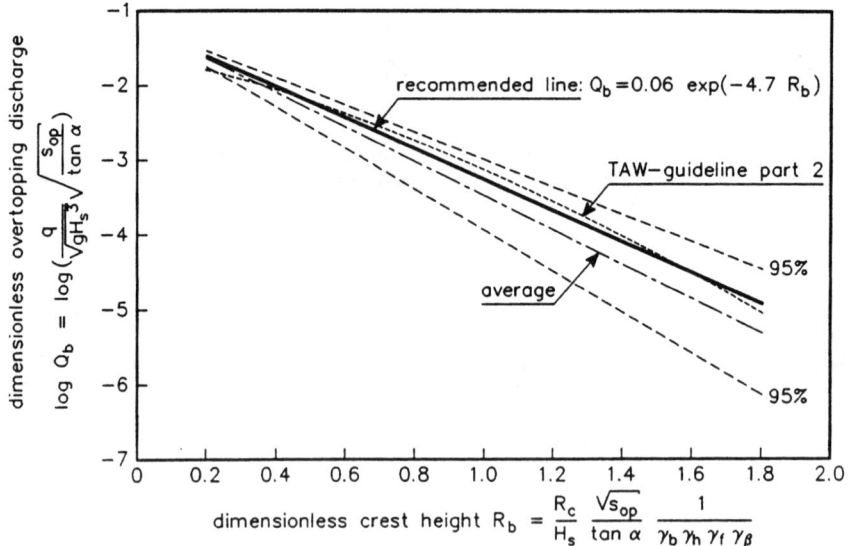

FIGURE 3.28. Wave overtopping with breaking waves, ($\xi_{op} < 2$).

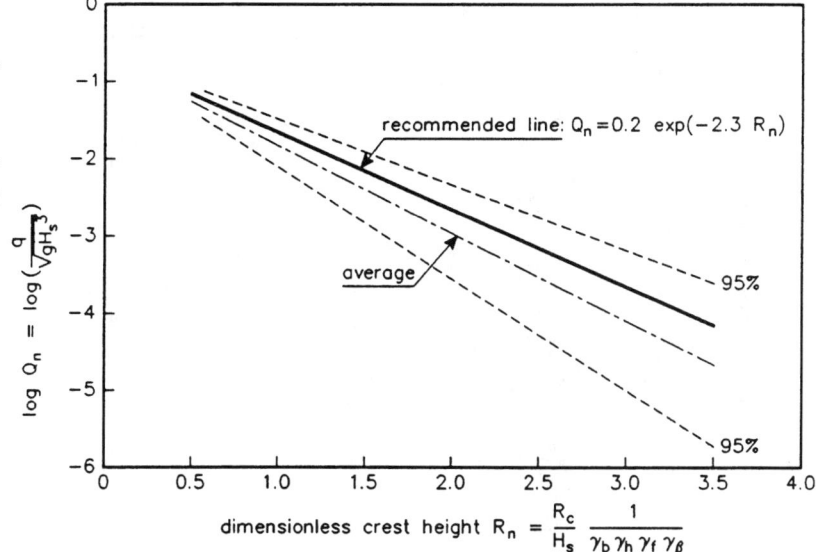

FIGURE 3.29. Wave overtopping with nonbreaking waves, ($\xi_{op} > 2$).

charge. By means of the average overtopping discharge, the probability distribution function of the overtopping discharges can be computed. To give an impression of the relation between the average overtopping discharge q and the expected value of the maximum volume in the largest overtopping wave V_{max}, this relation is given for two situations in Figure 3.30. Conditions are for a storm duration of 1 hour, a slope gradient of 1:4, and a wave steepness of $s_{op} = 0.04$ with a T_p/T_m ratio of 1.15. Relations have been drawn for wave heights of $H_s = 1$ m and 2.5 m. For small average overtopping discharges, the ratio V_{max}/q is of the order of 1000 and for large average overtopping discharges it is of the order of 100, although not dimensionless but with the dimension of seconds.

To get an indication of the instantaneous discharge during the passage of one wave, the maximum volume in overtopping wave should be divided by an effective fraction of wave period. This can be roughly approximated by (0.3 to 0.4) T. It provides an average value of a maximum discharge. This figure can be applied as an input into the stability criteria for protection of splash area and inner slope (i.e., applying criterion of Knauss, 1979).

More information on wave overtopping volumes per wave can be found in Pilarczyk (1990, 1998) and Van der Meer and Janssen (1994).

There are no generally valid recommendations for acceptable levels of overtopping for sea walls and/or dikes. In the standard Dutch practice, a safe value of about 0.002 m³/s for grassed crest and rear slope is recommended. Recent experience indicates that this value can be increased to 0.005 or even to 0.01 m³/s for "good" quality grass mat on clay sublayer. Information on a proper clay specification for a grass mat can be found in the guidelines (TAW, 1991, 1995).

Fukuda et al. (1974), based on field observations, suggest the following figures on allowable overtopping related to inconvenience for persons or vehicles located 3 m behind the breakwater:

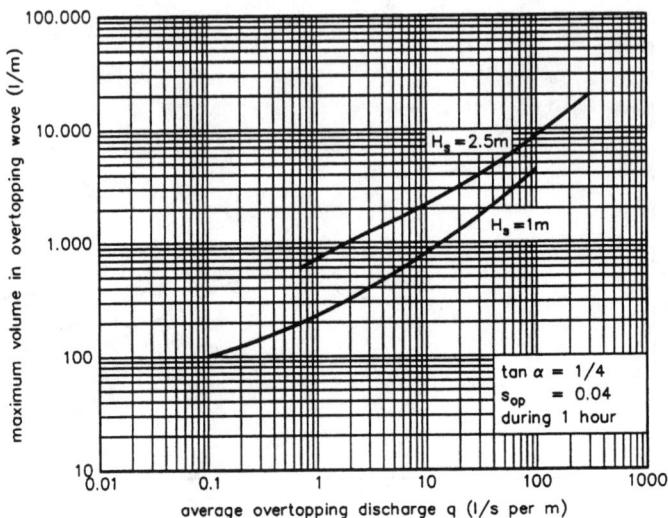

FIGURE 3.30. Relation between average overtopping discharge and maximum volume of the highest overtopping wave.

$q = 4.10^{-6}$ m^3/m/s: inconvenience for walking people

$q = 3.10^{-6}$ m^3/m/s: danger for people and traffic

More information can be found in CUR/CIRIA (1991). The design aspects of protection against overtopping are discussed in the following sections.

5.5. Slope Protection (Revetments)

5.5.1. General Considerations. In the past, only local usage and experience have determined the selection of the type and dimensions of coastal protection. Often, designs were conservative and too costly or were inadequate. The technical feasibility and dimensioning of coastal structures can be actually determined on a more sound basis and supported by better experience than in the past. However, the solution being considered should still be tested with a scale model, since no generally accepted design rules exist for all possible solutions and circumstances.

The types of revetments that are presently being studied are shown in Table 3.2. In this table, the critical modes of failure, the corresponding determinant loads, and the required strength are summarized.

Classical slope revetments may be divided into different categories (in increasing order of protection):

a. natural material (sand, clay, and grass)

b. protected by loose units (gravel, riprap)

c. protected by interlocking units (concrete blocks and mats/mattresses)

d. protected by concrete and asphalt slabs

TABLE 3.2. Review of revetments with critical modes of failure.

Type of cover layer	Critical failure mode	Determinant wave loading	Strength
Sand/gravel	Initiation of motion Transport of material Profile formation	Velocity field in waves	Weight, friction Dynamic "stability"
Clay/grass	Erosion Deformation	Maximum velocity Wave impact	Cohesion Grass roots Quality of clay
Riprap	Initiation of motion Deformation	Maximum velocity Seepage	Weight Friction Permeability of sublayer/core
Gabions and sand, stone, or cement mattresses, including geotextiles	Initiation of motion Deformation Rocking Abrasion/corosion of wires U.V. light	Maximum velocity Wave impact Climate Vandalism	Weight Blocking Wires Large unit Permeability including sublayer
Placed blocks, including block mats	Lifting Bending Deformation Sliding	Overpressure Wave impact	Thickness, friction, interlocking Permeability, including sublayer/geotextile Cabling/pins
Asphalt	Erosion Deformation Lifting	Maximum velocity Wave impact Overpressure	Mechanical strength Weight

The resistance of the protection listed above is derived from friction, cohesion, weight of the units, friction between the units, interlocking, and mechanical strength. As a result of the difference of strength properties, critical loading conditions are also different. Maximum velocities will be determined for clay/grass dikes and gravel/riprap, as they cause displacement of the material. Uplift pressures and impacts, however, are of more importance for paved revetments and slabs, as they tend to lift the revetment. As these phenomena vary both in space and in time, critical loading conditions vary both with respect to the position along the slope and the time during the passage of a wave. Instability for grass/clay and gravel/riprap will occur around the still water level, where velocities are highest during up- and down-rush. Moreover, wave impacts are more intense in the area just below the still water level.

Instability of paved revetments without sufficient interlock occurs at the peak of maximum down-rush, where uplift forces are higher, just before the arrival of the next wave front. If the protection is pervious, uplift forces are strongly reduced. Instability will have occurred due to the combined effect of uplift and impact forces, just after wave breaking. Concrete slabs and asphalt will mainly respond to uplift forces, as maximum loads are distributed more evenly over a layer area, thus causing a higher resistance against uplift compared with loose block pavement.

Stability (or threshold conditions) for loose materials, from sand to rock, has been investigated rather extensively, and the proper design criteria are available. However, the stability of randomly dumped quarried rock can often be substantially improved by using special measures (composite systems—grouting, pitched stone, mattresses, etc.). There are also a number of protective systems related to artificial materials such as concrete (i.e., block mats) and asphalt. Grass mats can also serve as slope protection. For most of these systems, it is possible to give some rough, indicative stability criteria that allow the engineer to make a comparison with a randomly placed rock, and thus to make a proper choice of protection. The following systems will be considered: riprap, placed and/or pitched blocks/stones, grouted (bound) stones, bituminous systems, gabions/stone mattresses, fabric containers (bags, mats), and clay/grass mats.

Because of the great variety of the possible composition of protective systems it is not possible to present a generally valid stability formulation for all these systems. Therefore, only some principles and examples will be given in a general subsection on stability criteria for wave attack where, for comparison of all these systems, the stability of dumped rock will serve as a reference.

Stability of placed block revetments (including geotechnical aspects) is treated more extensively in a separate subsection due to the fact that these systems were investigated in depth in the Netherlands. Moreover, this information also applies to other revetment systems.

5.5.2. General Overview on Stability Criteria for Wave Attack.
The general empirical (approximate) formula derived by Pilarczyk (1990) and supported by large-scale tests is

$$\frac{H_s}{\Delta_m D} \leq \psi_u \phi \frac{\cos \alpha}{\xi_p^b} ; (\text{ctg } \alpha \geq 2) \tag{23}$$

or

$$\Delta_m D = \psi_u^{-1} \phi^{-1} \cos \alpha^{-1} H_s \xi_p^b; (\text{strength}) = (\text{load}) \tag{23}$$

with

ξ_p = breaker similarity index on a slope,

$\xi_p = \tan \alpha \, (H_s/L_o)^{-0.5} = 1.25 \, T_p H_s^{-0.5} \tan \alpha$

in which

ψ_u = system-determined (empirical) stability upgrading factor ($\psi_u = 1.0$ for riprap as a reference and $\psi_u \geq 1$ for other revetment systems) [–]
ϕ = stability factor or stability function for incipient of motion, defined at $\xi_p = 1$ [–]
H_s = significant wave height [m]
T_p = peak wave period [s]
L_o = wave length [m]; $L_o = gT_p^2/2\pi$
D = specific size or thickness of protection unit [m]
α = slope angle [°]
Δ_m = relative density of a system-unit [–]
b = exponent related to the interaction process between waves and revetment type (roughness, porosity/permeability, etc.), $0.5 \leq b \leq 1$. For rough and permeable revetments such as riprap, $b = 0.5$. For smooth and less-permeable placed-block revetments, $b = 1$. The value $b = 2/3$ can be treated as a common representative value for other systems (i.e., more open blocks and block mats, mattresses of special design etc.).

D and Δ_m are defined for specific systems such as:

- rock: $D = D_n = (M_{50}/\rho_s)^{1/3}$ and $\Delta_m = \Delta = (\rho_s - \rho_w)/\rho_w$
- blocks: D = thickness of block and $\Delta_m = \Delta$
- mattresses: $D = d$ = average thickness of mattress and $\Delta_m = (1-n)\Delta$, where n = bulk porosity of fill material and Δ = relative density of fill material. For common quarry stone $(1-n)\Delta \approx 1$.

The formula is applicable up to $\xi_p = 3$ (breaking waves); for $\xi_p > 3$, the sizes calculated at $\xi_p = 3$ can still be applied.

The wave attack on a slope can be roughly transformed into the maximum velocity component on a slope during run-up and run-down, U_{max}, by using the following formula:

$$U_{max} = p\sqrt{(gH_s\xi_p)} \quad \text{(for irregular waves and smooth slopes } p = 1)$$

The stability factor ϕ for loose aggregates can be more generally defined using the van der Meer's formula (1984), namely

$$\phi = 6.2\, P_b^{0.18}\, (S_b^2/N)^{0.1} \quad \text{(for } \xi < 3\text{, breaking waves)}$$

where: P_b = permeability of the core material (i.e., $P_b = 0.5$ for breakwaters and 0.1 for revetments), S_b = damage number, and N = number of waves.

In the case of relatively impermeable core (i.e., sand or clay), $P_b = 0.1$, and limited number of waves ($N \approx 3000$) the following indicative ϕ-values for rock can be determined:

$\phi = 2.0$ for incipient motion of stones (lower stability limit)

$\phi = 2.25$ average value for incipient motion (motion 1 to 3 stones over the width of slope equal to D_n)

$\phi = 3.0$ as a first approximation for maximum tolerable damage for two-layer system on granular filter (damage depth less than or equal to $2D_n$); $\phi = 3$ can also be applied for incipient motion of rock placed on a permeable core (rockfill core or thick granular filter)

The ϕ-value equal to 2.25 will be used as a reference value for the stability comparison with other alternative systems. The difference in stability of rock due to the improving measures will be expressed by the upgrading of the factor ψ_u.

The important difference between the loose rock and the alternative systems concerns the behavior of the systems after the initiated movement (damage). Due to the self-healing effect of the loose rock, a certain displacement of rock units can often be accepted (up to $\phi \leq 3$). In the case of alternative systems, i.e., block revetment, the initial damage (i.e., removal of one block) can easily lead to progressive damage; there is no reserve stability.

The comparison of stability of various systems (parameter ψ_u) and the necessary parameters for calculation purposes are contained in Table 3.3. Other structural requirements and design rules concerning revetments are given by Pilarczyk (1990, 1998).

Comment for Users. (a) Block Revetments and Block Mats.
The use of ψ_u values higher than 2.5 is not advised except when supported by mathematical models and/or large scale tests incorporating geotechnical stability. For older revetments, some increase of stability is often observed due to the increase of natural friction and/or interlocking. However, permeability may decrease, which would act adversely.

The edges of the adjacent block mats, if not properly connected to each other, should be treated as free blocks ($\psi_u = 1.33$ to 1.50). For slopes steeper than 1 in 3, the geotechni-

TABLE 3.3. Indicative categories for protective systems

Criterion			Limits
$\dfrac{H_s}{\Delta_m D} = \Psi_u \phi \dfrac{\cos \alpha}{\sqrt{\xi_p}} = \psi_u\, 2.25 \dfrac{\cos \alpha}{\sqrt{\xi_p}}$			$\phi_{(rock)} = 2.25$ $\operatorname{ctg} \alpha \geq 2$ $\xi_p < 3$
System	ψ_u	Description	Sublayer
Rock	1.0	Riprap (2 layers)	Granular
	1.33	Riprap (tolerable damage)	Granular
Pitched	1.00	Poor quality irregular stone	Granular
stone	1.33	Good quality regular stone	Granular
	1.50	Natural basalt	Granular
Blocks/	1.50	Loose closed blocks; $H_s < 1.5$ m	Geotextile on sand
block mats	1.50	Loose closed blocks	Granular
	1.50	Blocks connected to geotextile	Granular
	2.00	Loose closed blocks	Geotextile on clay
	2.00	Cabled blocks/open blocks (>10%)	Granular
	>2.50	Grouted cabled blocks/Interlocked blocks, adequately designed	Granular
Grout	1.05	Surface grouting (30% of voids)	Granular
	1.50	Pattern grouting (60% of voids)	Granular
Open	2.00	Open stone asphalt; $U_p \leq 6$ m/s	Geotexile on clay
stone–asphalt	2.50	Open stone asphalt; $H_s < 4$ m	Sandasphalt
Gabions	2+3.0	Gabion/mattress as a unit, $H_s < 1.5$	Geotextile on sand
or			
	2+2.5	Stone fill in a basket; $d_{min} = 1.8\, D_n$	Geotextile on clay
Fabric	1.00	$P_m < 1$ less permeable mattress	Sand or
containers	1.50	$P_m \approx 1$ (P_m = ratio permeability of top/sublayer)	clay and
	2.00	$P_m \gg 1$ permeable mattress of special design	geotextile
Grass	—	Grass mat on poor clay; $U_p < 2$ m/s Grass mat on proper clay; $U_p < 3$ m/s/s	Clay (U_p = permissible velocity)

cal (in-)stability (i.e., sliding) can be a decisive factor and it should be examined properly. In the case of top layer placed directly on (compacted-)sandy subsoil and geotextile, the impinging wave height should not be more than 1.5 m because of danger of local profile deformation and/or liquefaction. It should be noted that for practical reasons the minimum thickness of loose blocks is about 0.10 m and for blocks grouted with granular material and block mats it is 0.08 m. More sophisticated approaches to stability aspects of these systems can be found in the guidelines on dimensioning of block revetments (CUR/RWS, 1995a).

b) Grouted Stone. Surface grouting is not advised in the case of highly permeable sublayers. Creation of the completely impermeable surface should be avoided because it may introduce extra lift forces. In the case of pattern grouting, about 50 to 70% of the total surface is filled. The upgrading factor is very dependent on the construction methods and care must be taken to ensure that the grout does not remain on the surface of the stone layer only, or sags completely through the layers. In the area of high wave impact, the grouted lumps themselves can be split by dynamic action, therefore this type of construction should be applied up to $H = 3$ m (frequent loading) and $H < 4$ m (less-frequent loading). In the later case, for safety reasons, it is recommended to use of three layers of broken stone. If a lump of grouted stones is split and washed away, the third layer will still protect the core, since it is held in place by the overlying grouted lumps.

c) Bituminous Systems. In the case of open stone–asphalt placed on a sand asphalt filter, the thickness of the system may be defined as the total thickness of both layers. For the edges of all-bituminous systems, the $\psi_u = 2$ should be applied. Because of possibility of liquefaction, open stone–asphalt on geotextile and sand is recommended only up to $H = 2$ m. For $H > 2$ m, the sand–bitumen filter under the top layer of open stone–asphalt is recommended. Due to the limited resistance of open stone–asphalt against surface erosion (max. velocity $u = 7$ m/s) this system can be applied up to $H = 3$ m, and, for a less-frequent wave loading, up to $H = 4$ m. For practical reasons, the minimum thickness of open stone–asphalt is 0.08 m if prefabricated and 0.10 m if placed in situ. However, the more common thickness are 0.10 and 0.15 m, respectively. Bituminous plate systems (especially impermeable ones) should also be examined to determine the allowable stresses and strains (bending moments) and the uplift criterion. The calculation methods can be found in TAW (1985) and Pilarczyk (1998). The example of thickness of various asphalt revetments related to the allowable stresses is given below for compacted clay with slope 1 in 3:

H_s (m)	Asphalt concrete	Open stone–asphalt	Sand–asphalt
2	0.10 m	0.20 m	0.40 m
3	0.20 m	0.40 m	0.80 m
4	0.30 m	0.65 m	
5	0.40 m		

Note: for compacted sand bed, these figures are on conservative side.

In general, the resistance of the sand–asphalt is limited to the velocity of 3 m/s and the wave height of 1.5 m (or $H < 2$ for less-frequent loading). Currents are not usually a determining factor in the design of asphalt concrete.

d) Gabion Baskets and Mattresses. The primary requirement is that the gabion or mattress of thickness d will be stable as a unit. The thickness of the mattress can be related to the stone size D. In most cases it is sufficient to use two layers of stoone in a mattress ($d = 1.8\ D$) and the upgrading factor can be recommended in the range $2 < \psi_u < 3$ (max). The secondary requirement is that the movement of stones in the basket should not be too high because of the possible deformation of baskets and the loading on the mesh wires. To avoid the situation in which the basket of a required thickness d will be filled by too-fine material, the second criterion, related to D, has been formulated. The choice of $\psi_u = 2$ to 2.5 related to D means that the level of loading of the individual stones in the basket will be limited to roughly twice the loading at the incipient motion conditions. Thus, using $\psi_u \leq 2.5$ and $d \geq 1.8\ D$, the gabion unit is stable, and the individual stone movement is limited.

In systems with more than two layers it is preferable to use a finer stone below the top layers (i.e., up to $D/5$) to create a better filter function and to diminish the hydraulic gradients at the surface of subsoil. The formulations for gabions and mattresses are only valid for waves with a height up to $H = 1.5$ m, or for less-frequent waves up to $H = 2.0$ m. In either case it is important that both the subsoil and the stone infill are adequately compacted. When the current exceeds 3 m/s or the wave height exceeds 1 m, a fine granular sublayer (about 0.2 m thick) should be incorporated. In other cases it is satisfactory to place the mattress directly onto the geotextile and compacted subsoil. For practical reasons, the minimum thickness of mattresses is 0.15 m.

e) Fabric and Other Containers. The stability criterion for fabric mattresses of thickness d filled with sand, sand–cement, or other materials attacked by waves is derived from some (limited) tests and recent knowledge of revetment principles. The value of the upgrading factor (ψ_u) depends on the ratio of the permeabilities of the mattress and the subsoil P:

- for $P \ll 1$: $\psi_u = 1.0$ (less-permeable mattresses)
- for $P \approx 1$: $\psi_u = 1.5$
- for $P \gg 1$: $\psi_u = 2.0$ (permeable mattresses of special design)

The permeability of the mattress should be treated as an integrated permeability of all the components, i.e., geotextile container and fill material together. For wave heights 1 m < H < 2 m and a sandy subsoil, special measures against sliding and/or liquefaction should be taken: extra compaction, extra thickness (50 to 100%), and, eventually, a fine granular sublayer—0.2 m thick (broad-graded), etc. In the case of permeable mattresses (i.e., gravel fill) on sandy subsoil, the underneath part of the container should preferably be made of the sand–tight geotextile (filter function). For slopes steeper than 1 in 3, to avoid sliding, special attention should be paid to anchoring at the top of the mattress and to adequate toe support. Special measures should be taken concerning the transitions (avoid exposed edges), scour protection at the toe, and protection from overtopping in the splash area. Sand mattresses, even properly compacted, are very susceptible to deformation. Therefore, their permanent use should be limited to relatively mild wave attack ($H < 1.5$ m). In general, the use of fabric containers of various (specific) designs exposed to wave heights higher than 2 m and application of ψ_u values different from those mentioned above is only responsible when supported by large-scale or prototype tests, including geotechnical stability. For practical reasons, the minimum thickness of fabric containers/mattresses is 0.15 m. For promoting vegetation through the geotextile mattress sand and/or fine gravel, mixed with cohesive additives and seed, are very suitable as fill material. More information on design of various geotextile systems can be found in Pilarczyk (1995, 1998).

f) Grass Mats. Some of the existing dikes along the Wadden Sea (northern part of the Netherlands) still need reinforcement as these do not yet meet the specific safety requirements. One of the options for reinforcement is a slope protection of grass on a bed of clay, rather than stone, concrete, or asphaltic protection. This option is feasible because vast mud-flats (high foreshore) and grasslands stretch away on the seaside of the existing dikes and are inundated only during storm surges. Moreover, the wave action in the Wadden Sea is much reduced by a row of barrier islands. Due to these factors, the design wave height does not exceed 2 m. Delft Hydraulics was commissioned to investigate the stability of such a grass dike by means of a full-scale model study, which was an absolute requirement as grass cannot be scaled down (Seijffert and Philipse, 1990). The investigations have been completed in the Delta Flume. A five meter wide section of the grass dike was reproduced to full scale. The model consisted of a sand core covered with a clay layer on a slope 1 in 8. Sods of grass with depth of roots of approximately 40 cm were laid on top of the clay layer (the grass was taken from an existing dike that was reinforced 10 years ago). During the tests, the wave heights and periods and water levels (tidal cycles) were varied continuously according to predetermined boundary conditions during the design storm surge. The maximum wave height (H_s) was equal to 1.85 m with T = 5.6 sec (plunging breaker falling on a water cushion). The measured maximum velocity on the slope (1:8) was about 2 m/s. After 30 hours of continuous random wave attack, the condition of the grass dike was still exceptionally good. The surface erosion rate of clay protected by grass was not more than 1 mm per hour. In a number of additional tests, the durability of the grass and the enlargement of holes, previously dug in the grass, were studied. Although wave action considerably enlarged some of these holes, the residual strength of the dike was such that its collapse was not imminent.

The second investigation was carried out in a large (site) flume on slope 1 in 4. Special discharge equipment was used to simulate the run-up and run-down velocities on this slope. Two qualitatively different grass mats on clay were used. The grass mats were tested with the average velocity of 2 m/s (average over 40 hours of test) and the thickness of a water layer of about 0.6 m. The maximum velocity was about 4 m/s. Erosion speed of the clay surface was 1 to 2 mm per hour up to 20 hours, depending on quality of grass mat. After 20 hours of loading, the erosion speed started to grow progressively for a poor-quality grass mat. A similar process took place for a good-quality grass mat, but after 40 hours of loading.

Some additional information on resistance of unprotected clay surface (slope 1 in 3.5) were obtained during the investigation carried out for the Eastern Scheldt dikes. In this case, two qualitatively different clays were used (fat and lean clay). The surging breaker conditions were applied to eliminate the effect of wave impact (H = 1.05 m, T = 12 s, max velocity 3 m/s). The erosion on the upper part of slope was the same for both clay types and equal to about 2–3 cm after about 5 hours of loading. After the same time, the erosion below SWL. was about 7 cm for a good clay, whereas for a lean clay a cavity of about 0.4 m depth was created (probably because of the local nonhomogeneity of clay). Also during this investigation, a number of additional tests on the erosion of different sublayers (including clay) at locally damaged top layers (some protective units were removed) were performed. All the tests mentioned above indicated that the strength of the grass slopes is strongly affected by the quality of clay and the condition of grass and its rooting.

Recently, new large-scale tests on overtopping and erosion of grass dikes with slope 1 in 4 were performed in the Delta Flume of Delft Hydraulics (Meijer and Verheij, 1994; Smith et al., 1994). This investigation has resulted in preparion of a new set of stability criteria for design of grass mats on dikes. The stability due to direct impact of breaking waves was different from the erosion rate due to the run-up and overtopping. The final formulation of both criteria is presented in Pilarczyk (1998). Some additional information on this subject can be found in CIRIA (1976, 1987) and TAW (1991).

5.5.3. Stability Criteria for Current Attack. Coastal morphology, scour process, and, in some cases, stability of protective units (slope and/or bottom protection) are or can be influenced not only by waves but also by currents and a combination of waves and currents. The currents can be of various origins, namely wave-induced currents, tidal currents, ship-induced currents, natural currents (i.e., in the mouths of the rivers). Additional information can be found in Pilarczyk (1990, 1998) and CUR/RWS (1995b)

5.6. Optimization of Slope Stability

The wave forces on a plane continuous slope are distributed rather unequally (the high wave-impact area near the water level, the intermediate uprush area, and the low-attacked area beneath the point of breaking). The wave action on relatively fine materials indicates that nature tries to distribute the forces equally to provide equilibrium S-slopes. The same principle can be applied in designing the shape of sea walls and dikes leading to application of smaller protective units than in the case of a plane slope.

For practical reasons, the "optional" shape will be schematized to a trapezoidal profile. By selection of a proper position of a berm below the design water level and a proper width of a berm, the wave forces will be distributed in such a uniform way that the same material can be used along the whole profile. In this case the increase of stability (50% or more) can be obtained by a berm with a width equal to 0.15 times wave length and situated 0.5–1.0 times wave height below design water level. Based on the results of various studies, indicative design guidelines have been prepared for riprap-bermed slopes and toe protection (Pilarczyk, 1990 and CUR/CIRIA, 1991).

This concept of stability of protection should also be verified at water levels lower than the design level.

5.7. Scour and Toe Protection

Toe protection consists of armoring of the beach or bottom surface in front of a structure, which prevents waves from scouring and undercutting it. Factors that affect the severity of toe scour include wave breaking (when near the toe), wave run-up and backwash, wave reflection, and grain size distribution of the beach or bottom materials.

Toe stability is essential because failure of the toe will generally lead to failure throughout the entire structure. Toe scour is a complex process. Specific (generally valid) guidance for scour prediction and toe design based on either prototype or model results have not been developed as yet, but some general (indicative) guidelines for designing toe protection are given in SPM (1984), CUR/RWS (1995b), and Hoffmans and Verheij (1997).

The maximum scour force occurs where wave downrush on the structure face extends to the toe and/or the wave is breaking near the toe (i.e., shallow-water structure). These conditions may take place when the water depth at the toe is less than twice the height of the maximum expected unbroken wave that can exist at that water depth. The width of the apron for shallow-water structures with a high reflection coefficient, which is generally true for slopes steeper than about 1 in 3, can be planned from the structure slope and the expected scour depth. The maximum depth of a scour trough due to wave action below the natural bed is about equal to the maximum expected unbroken wave at the site. To protect the stability of the face, the toe soil must be kept in place beneath a surface defined by an extension of the face surface into the bottom to the maximum depth of scour. This can be accomplished by burying the toe, when construction conditions permit, thereby extending the face into an excavated trench the depth of the expected scour. Where an apron must be placed on the existing bottom or can only be partially buried, its width should not be less than twice the wave height.

Some solutions for toe protection can be found in the *Shore Protection Manual* (SPM, 1984), CUR/CIRIA (1991), and PIANC (1992).

If the reflection coefficient is lower than the limit (slopes milder than 1 in 3), and/or the water depth is higher than twice the wave height, much of the wave force will be dissipated on the structure face and a smaller apron width may be adequate, but must be at least equal to the wave height (minimum requirement). Since scour aprons generally are placed on very flat slopes, quarry stone of the size (diameter) equal to one-half or even one-third of the primary cover layer probably will be the heaviest required, unless the apron is exposed above the water surface during wave action. Quarry stone of primary cover-layer size may be extended over the toe apron if the stone will be exposed in the troughs of waves, especially breaking waves. The minimum thickness of cover layer over the toe apron should be two quarry stones. Quarry stone is the most favorable material for toe protection because of its flexibility. If geotextile is used as a secondary layer, it should not be extended over the whole width of the apron to provide the flexible edges (at least 1 m) against undermining, or it should be folded back and then buried in cover stone and sand to form a Dutch toe. The size of toe protection against waves can also be roughly estimated by using the common formulas on slope protection and introducing mild slopes (i.e., 1 in 8 to 1 in 10) and local wave height. Some alternative designs of toe protection are shown in Figures 3.32, 3.58, and 3.59.

Hales & Houston (1983) considered the stability of a rock blanket extending seaward from a permeable rubble slope on a 1:25 slope foreshore. They tested with regular waves to determine the conditions at which the rock in the scour blanket was just stable. To these conditions they fitted a mean trend given by

$$\frac{H_b}{\Delta D_n} = (17.5 \div 28.5)\left(\frac{B_p}{L_s}\right)^{2/3} \approx 20 \, (B_p/L_s)^{2/3} = 20\left(\frac{B_p}{T\sqrt{gh_s}}\right)^{2/3} \qquad (24)$$

(coefficient 17.5 represents a more conservative line), where

H_b = breaker wave height ($\approx 0.78 \, h_s$ for regular waves)
L_s = wave length in shallow water, given by $T(gh_s)^{1/2}$ in this instance
B_p = seaward extent of the toe protection
h_s = shallow-water depth

This formulation can be used as a first indication of decreasing stone size (D_n) with distance (B_p).

Toe protection against currents may require smaller protective stone but wider aprons. The necessary design data can be estimated from site hydrography and/or model studies. Special attention must be given to sections of the structure where scour is intensified; i.e., to the head, areas of a section change in alignment, the channel sides of jetties, and the downdrift sides of groins. Where waves and reasonable currents ($u > 1$ m/s) occur together, it is recommended increasing the cover size at least by a factor of 1.3.

Note that the conservatism of the apron design (width and size of cover units) depends on the accuracy of the methods used to predict the waves, current action, and maximum depth of scour. For specific projects, a detailed study of scour of the natural bottom and nearby similar existing structures should be conducted at a planned site, and/or model studies should be considered before determining a final design. In all cases, experience and sound engineering judgment play an important role in applying these design rules.

5.8. Protection Against Overtopping

If a structure (revetment) is overtopped, even by minor splash, the stability can be affected. Overtopping can: (a) erode the area above or behind the revetment, negating the structure's purpose; (b) remove soil supporting the top of the revetment, leading to the unraveling of the structure from the top down, and (c) increase the volume of water in the soil beneath the structure, contributing to drainage problems. The effects of overtopping can be limited by choosing a higher crest level or by armoring the bank above or behind the revetment with a splash apron. For a small amount of overtopping, a grass mat on clay can be adequate. The splash apron can be a filter blanket covered by a bedding layer and, if necessary to prevent scour due to splash, by riprap, concrete units, or asphalt.

No definite method for designing against overtopping is known, due to the lack of the proper method on estimating the hydraulic loading. Pilarczyk (1990) proposed the following, indicative way of design of the splash area (Figure 3.31):

$$\frac{H_s}{\Delta D_n} = \frac{2 \cos \alpha_i}{\Phi_T \xi^{2b} \left(1 - \frac{R_c}{R_u}\right)} \qquad (25)$$

where

H_s = significant wave height
ξ = breaker index; $\xi = \tan \alpha (H_s/L_o)^{-0.5}$
α = slope angle
α_i = angle of crest or inner slope
L_o = wave length
b = coefficient equal to 0.5 for smooth slopes and 0.25 for riprap
R_c = crest height above still water level
R_u = wave run-up on plane slope
D = thickness of protective unit ($D = D_n$ for rock)
ϕ_T = total stability factor equal to 1.0 for rock, 0.5 for placed blocks, and 0.4 for block mats

The length of protection in the splash area, which is related to the energy decay, can be roughly assumed as equal to

$$L_s = \frac{\psi}{5} T \sqrt{g(R_n - R_c)} \geq L_{min} \qquad (26)$$

with a practical minimum (L_{min}) equal at least to the total thickness of revetment (including sublayers) as used on the slope. ψ is an engineering-judgement factor related to the local conditions (importance of structure), $\psi \geq 1$.

Stability of rockfill protection of crest and rear slope of an overtopped or overflowed dam or dike can also be approached with the Knauss formula (Knauss, 1979). The advantage of this approach is that the overtopping discharge, q, can be used directly as input pa-

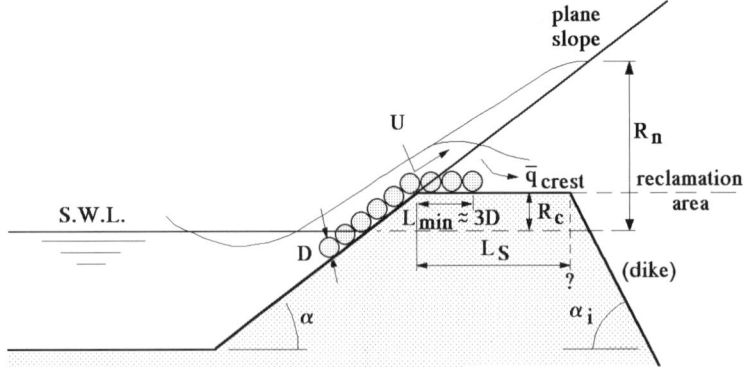

FIGURE 3.31. Definition of splash area.

rameter for calculation. Knauss analyzed steep-shute flow hydraulics (highly aerated/turbulent) for the assessment of stone stability in overflow rockfill dams (impervious barrages with a rockfill spillway arrangement). This kind of flow seems to be rather similar to that occurring during high overtopping. His (simplified) stability relationship can be rewritten to the following form

$$q = 0.625 \sqrt{g} \, (\Delta D_n)^{1.5} \, (1.9 + 0.8\phi_p - 3 \sin \alpha_i) \tag{27}$$

in which

q = maximum admissible discharge (m³/s/m)
g = gravitational acceleration (9.81 m/s²)
D_n = equivalent stone diameter, $D_n = (M_{50}/\rho_s)^{1/3}$
Δ = relative density; $\Delta = (\rho_s - \rho_w)/\rho_w$
α_i = inner slope angle
ϕ_p = stone arrangement packing factor, ranging from 0.6 for natural dumped rockfill to 1.1 for optimal manually placed rock; it seems to be reasonable to assume $\phi_p = 1.25$ for placed blocks

Note: when using Knauss formula, the calculated critical (admissible) discharge should be identified with a momentary overtopping discharge per overtopping fraction of characteristic wave, i.e., volume of water per characteristic wave divided by overtopping time per wave, roughly (0.3 to 0.4) T, and not with the time-averaged discharge (q).

5.9. Joints and Transitions

Despite a well-designed protective system, the construction is only as strong as the weakest section. Therefore, special care is required when designing transitions. In general, slope protection of a dike or sea wall consists of a number of structural parts such as: toe protection, main protection in the area of heavy wave and current attack, upper-slope protection (very often grass mat), berm for run-up reduction or as maintenance road. Different materials and different execution principles are applied for these specific parts. Very often, a new slope protection has to be connected to an already existing protective construction that involves another protective system. To obtain a homogeneous, strong protection, all parts of protective structures have to be taken under consideration.

Experience shows that erosion or damage often starts at joints and transitions. Therefore, an important aspect of revetment construction, which requires special attention, are the joints and the transitions; joints between the same material and between other revetment materials, and transitions to other structures or revetment parts. A general design guideline is that transitions should be avoided as much as possible. If they are inevitable, the discontinuities introduced should be minimized. This holds for differences in elastic and plastic behavior and in the permeability or the sand tightness. Proper execution is essential in order to obtain satisfactory joints and transitions.

When these guidelines are not followed, the joints or transitions may influence loads in terms of forces due to differences in stiffness or settlement, migration of subsoil from one part to another (erosion), or strong pressure gradients due to a concentrated ground water flow. However, it is difficult to formulate more detailed principles and/or solutions for joints and transitions. The best way is to combine the lessons from practice with some physical understanding of systems involved. Examples to illustrate the problem of transitions are given in Figure 3.32.

As a general principle, the transition should be of a strength equal to or greater than the adjoining systems. Very often, it needs a reinforcement in one of the following ways:

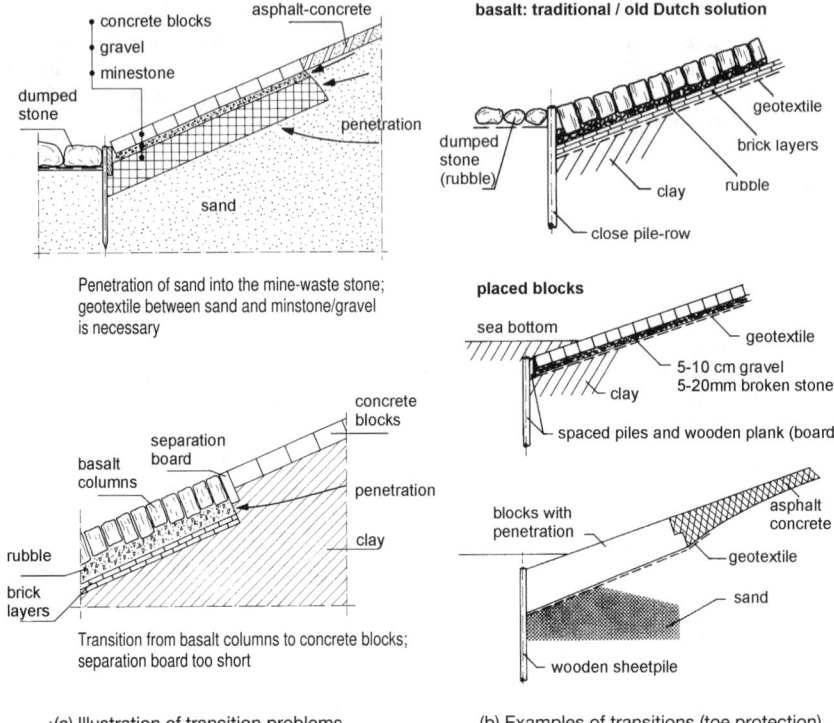

FIGURE 3.32. Transitions in revetments.

(a) increase the thickness of the coverlayer at the transition, (b) grout riprap or block coverlayers with bitumen, and (c) use concrete edge strips or boards to prevent damage from progressing along the structure.

Top edge and flank protection are needed to limit vulnerability of the revetment to erosion continuing around its ends. Extension of the revetment beyond the point of active erosion should be considered but is often not feasible. Care should therefore be taken that the discontinuity between the protected and unprotected areas is as small as possible (use a transition roughness) so as to prevent undermining. In some cases, open cell blocks or open block mats (eventually vegetated) can be used as transition (i.e., from hard protection into grass mat). The flank protection between the protected and unprotected areas needs mostly a thickened or grouted cover layer, or a concrete edge strip with some flexible transition, i.e., riprap.

5.10. Geotechnical Aspects

5.10.1. General. Depending on the functional requirements, coastal structures have to withstand a combination of actions induced by waves, currents, differences in water levels, seismicity, and other specific loadings (such as ship collisions or surcharges). These actions, including the self-weight of the structure, have to be transferred to the subsoil in such a way that (de Quelerij, 1989) (a) the deformations of the structure are acceptable, and (b) the probability of instability is sufficiently low.

The transfer of actions through the structure to the subsoil involves changes in soil stresses (pore water pressures and effective stresses) in the soil layers and, in case of sloping structures, also in the structure itself. Particularly in soft soils, the stress changes will gradually develop during a long period of time. Due to these changes in soil stresses, the underlying and adjacent soil layers will deform vertically and horizontally and the shear strength of the soil may be reduced. As a consequence, any structure built on top of the soil layers will deform too or may even lose its stability. This applies to both the design conditions (i.e., under extreme loadings) and the loadings during the construction period of the structure.

The changes in soil stresses and the associated deformations not only depend on the (hydraulic) loadings, but on the geometry (e.g., slope steepness), the structure weight, and the permeability, stiffness, and shear strength of the subsequent structure and soil layers as well. For this reason the design of coastal structures has to be based on an integral approach of the interaction between the structure and the subsoil. A good knowledge of the main geotechnical properties of the soil layers and the construction materials is therefore required. The main principles of soil mechanics can be found in Lambe and Whitman (1979), RWS (1987), and CUR/TAW (1989).

5.10.2. Geotechnical Limits. A general design approach, which should also be adopted for the geotechnical limits, is given in previous sections. In order to get a proper understanding of the geotechnical failure mechanisms, a review is presented of the main features related to the coastal structures (Figure 3.33).

The main geotechnical limits that should be evaluated in the design of the sloping structures are:

a) macroinstability of slopes due to failure along circular or straight sliding surfaces

b) settlements (and horizontal deformations) due to the self-weight of the structure

c) microinstability of slopes caused by seeping out of groundwater

d) piping or internal erosion due to seepage flow underneath the structure

e) liquefaction caused by erosion (flow sides) or by cyclic loading (wave actions or earthquakes)

f) erosion of revetments at the outer slopes (or underwater slopes) due to unstable filters or local failure of top layer elements

In the TAW guides (1990, 1991) and CUR/RWS (1995b), the main geotechnical limits are discussed, with emphasis on simulation modeling. Since in the design of dikes most of the presented failure modes are relevant, this type of structure will serve as a reference for other coastal structures.

6. STRUCTURAL RESPONSE OF PLACED-BLOCK REVETMENTS

6.1. Applications and Design Aspects

Revetments of placed blocks or block mats (also called mattresses) are often used as a protection of slopes of various coastal structures against wave attack. Extensive studies

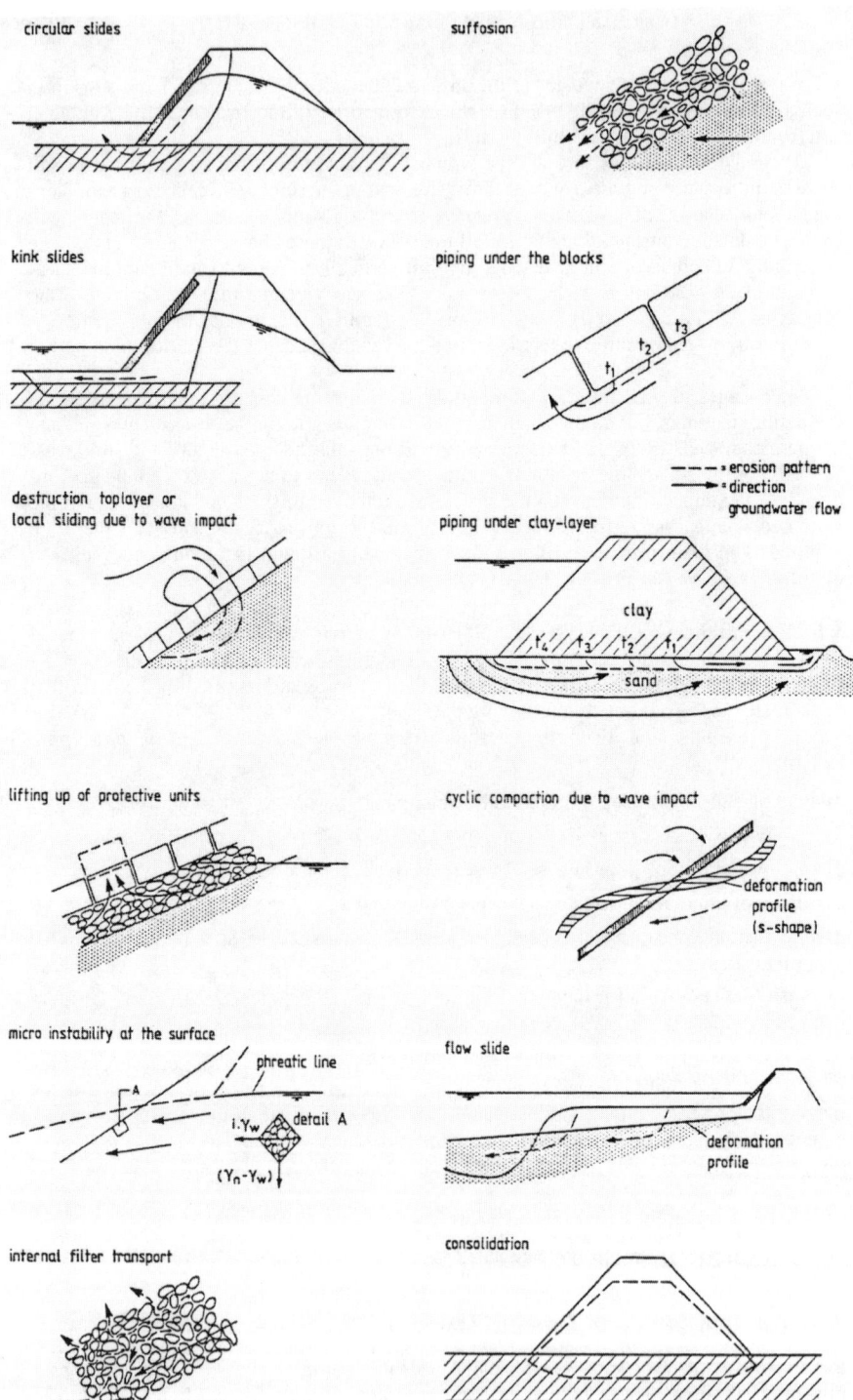

FIGURE 3.33. Review of geotechnical failure modes of a dike.

have been performed during the last decade on the stability of placed-block revetments, mainly in the Netherlands, but also abroad. The studies in the Netherlands have been carried out by Delft Hydraulics and Delft Geotechnics in cooperation with and commissioned by the Netherlands Ministry of Transport and Public Works (Rijkswaterstaat). It has led to the establishment of a large database with results of model and prototype studies, and to the development of a set of design formulas. In addition, it has also resulted in a practical design handbook (CUR/RWS, 1995a). Some of these results are also incorporated in the PIANC report (1992). This section is based on the recent publications by Pilarczyk et al (1995) and Pilarczyk (1998), wherein actual developments on this subject are summarized.

Examples of the structures and blocks considered here are shown in Figures 3.34 and 3.35. Such structures are found on banks of rivers and lakes, sloping sea walls and sea dikes, jetties, and in harbors. The main cause of damage to structures in this application is wave attack. Damage by current is only found when applied in spillways, steep canals, etc., but these are not considered here.

Basalt columns are an example of natural stone that can be applied as placed blocks to form the cover layer of a revetment. Prefabrication of concrete blocks, however, is often a very common alternative for natural blocks.

The advantage of blocks precast on or close to the site is the independence of stone supply. This holds in particular when stones have to come from distant sources. Other ad-

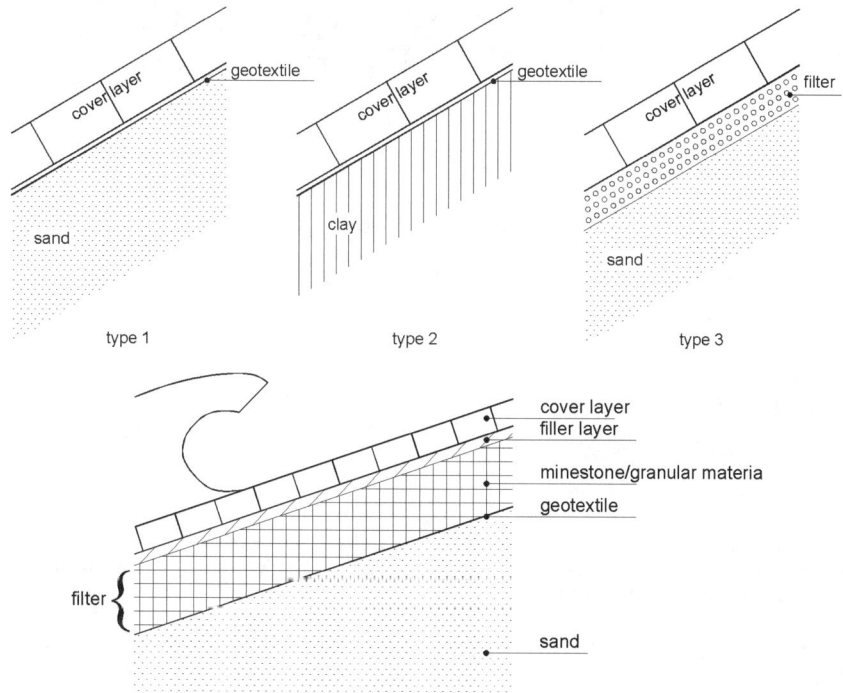

FIGURE 3.34. Examples of block revetment structures (cross-sections).

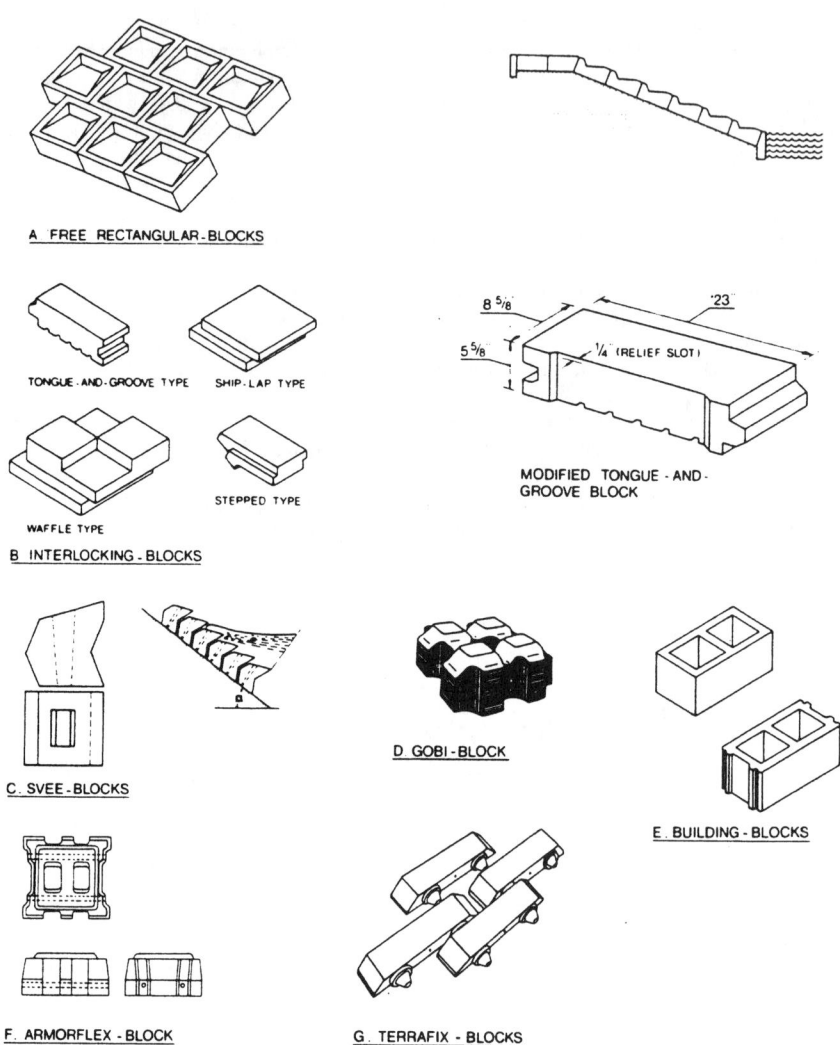

FIGURE 3.35. Examples of revetment blocks.

vantages are that specifications on material size, shape, and density can be met easily through the fabrication process. By choosing the right aggregates, the material characteristics can also be influenced (density, porosity, surface roughness).

Artificial blocks are often designed to provide for additional strength. This additional strength allows for reduction of the weight or size of the stones. The fabrication of specified elements permits the application of cables to connect individual elements, which improves the stability. These cables can be galvanized steel wires or polymer fibers. Even without special measures a mutual friction between closely placed blocks will provide for an extra stability against uplifting. In the design, often no account is taken of frictional forces, which leads to a conservative design. However, in recently developed methods, some friction can be taken into account.

Placed blocks can be applied together with appropriate sublayers. In case of repair, a certain area usually has to be replaced. The placement of preconnected blocks or block mats usually requires special, though rather simple, equipment. Working from floating equipment is limited due to wave-induced movements.

In general, a revetment system will consist of a number of layers, the principal of which are the cover layer, filter layer(s) and, as far as necessary, complementary sublayer(s). In the following, it will be shown that a revetment system must be designed as an integrated system of cover layer, sublayers, and subsoil (Figure 3.34) (PIANC, 1992 and CUR/RWS, 1995a).

The failure mechanisms, which should be considered in designing block revetments, are among others:

a) uplifting of blocks

b) migration of subsoil particles through the granular filter and/or cover layer

c) geotechnical instability of sublayers/subsoil

d) sliding of the revetment

The cover or armor layer is the major protection of the structure and should resist external and internal loadings. The strength against external loadings can primarily be provided for by a sufficient weight of the armor elements.

The internal loadings depend to a large extent on the permeability ratio of cover and filter layers. Further on, the permeability of the core may affect the stability of the cover layer as far as the phreatic level inside the structure is concerned. Additional stability of the cover layer can be obtained by friction, interlocking, or tensile forces. These forces may act between the elements of the armor layer and between the armor elements and the underlayers. Most of the artificial systems have been designed deliberately to mobilize these additional forces. The strength and the capacity of load reduction are often used interchangeably.

To determine the stability of the cover layer, the $H_s/\Delta D$ ratio versus the breaker index ξ_{op} (surf-similarity parameter) can be used, with $\xi_{op} = \tan \alpha/\sqrt{H_s/L_{op}}$ as defined before.

A theory concerning the uplift of individual blocks will be presented briefly in the following sections, resulting in an equation including all significant parameters that influence the stability. This theory will be simplified for practical application and then fitted to the results of large-scale model studies.

6.2. Cover Layer Stability

6.2.1. External Wave Loading. Upon breaking on a slope, regular waves exert cyclic hydraulic loads. On the basis of physical model tests in wave tanks, good knowledge has been obtained of the relevant load phenomena within a wave cycle. For different types of revetments, different moments or periods from the wave cycle are decisive for the stability of the cover layer.

The external loads can be quantified by way of physical model tests and with numerical methods (Petit et al., 1994; Van Gent et al., 1994). The practical use of the numerical simulations is still very limited, especially regarding the wave-induced pressure distribution on a slope (in space and time).

A much simpler approach towards a computation of the relevant wave loads is to abandon a full description of time- and place-dependent wave pressures on the slope (ϕ_s), and to concentrate only on the instant of critical wave loads. For placed-block revetments the most critical load situation occurs at the moment of maximum wave rundown. This proved to have general validity for the structures considered here. The critical pressure front is schematized sufficiently by the parameters θ (maximum gradient), ϕ_b (maximum piezometric head, $\phi_b = \max \phi_s$), and d_s, as shown in Figure 3.36. Empirical equations for ϕ_b and β, based on wave-pressure measurements in a small-scale model with slopes between $\frac{1}{2} < \tan \alpha < \frac{1}{4}$ and wave steepness between $0.01 < H/L_0 < 0.07$ (regular waves), are given in Burger et al. (1990):

$$\frac{\phi_b}{H} = \min\left(\frac{\xi_o}{\sqrt{\tan \alpha}}; 2.2\right) \qquad (28)$$

$$\tan \theta = \cot \beta = \frac{5.9 \tan \alpha}{\xi_o} \qquad (29)$$

$$\frac{d_s}{H} = \min\left(\frac{0.11 \tan \alpha}{(H/L_o)^{0.8}}; 1.5\right) \qquad (30)$$

Model tests have shown that the same equations can be applied with reasonable accuracy for oblique wave attack up to 45° as for perpendicular wave attack.

6.2.2. Internal Loading. The internal hydraulic loading on the revetment can be split up into two items:

a) The pressure under the cover layer, relative to the pressure on top of the cover layer, causing uplift of the blocks

b) The hydraulic gradients under the cover layer (mainly parallel along the slope), which can cause the migration of subsoil particles

We consider two main types of structures: (1) with granular filter between the cover layer and subsoil, and (2) without granular filter. Quite often, a placed-block revetment has a configuration as shown in Figure 3.34. A cover layer is placed on a filter layer of limited thickness lying on the base material. Usually, the permeability of the filter layer is

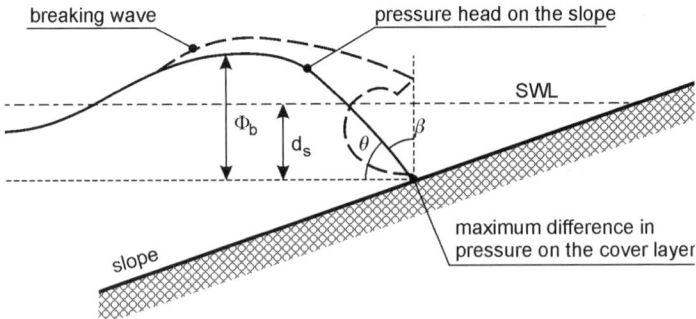

FIGURE 3.36. Schematized damage mechanism caused by pressure potential (piezometric head) over the cover layer.

much larger than the permeability of both the base and the cover layer. The situation of a cover layer that is less permeable than the underlying filter layer can also be present in revetment structures with a block mattress, gabions, sand bags, or gravel-filled geotextile cover layers. The hydraulic loading on this type of structure can be quantified analytically.

The internal loading of revetments without a granular filter has to be described with more sophisticated solution methods. For example, the STEENZET/2 program of Delft Geotechnics, a finite-element program specially developed to calculate the pore-pressure response in the filter layer(s) and subsoil below a placed-block revetment. This program can use measured wave pressures as a boundary condition and it can handle laminar as well as turbulent flow (Hjortnæs-Pedersen et al., 1987).

In the following, we will first consider revetments on a granular filter and then extend the derived formula to revetments without a granular filter.

Revetments with Granular Filter. During the wave rundown there is a large piezometric head gradient on top of the revetment (see Figure 3.36), caused by the simultaneous occurrence of rundown of the preceding wave and the arrival of the present wave. The piezometric head underneath the cover layer is a damped representation of the potential on top of the revetment, causing an uplift pressure at the location of maximum wave rundown. The extent of the damping is influenced by the permeability ratio of the cover layer and the filter layer and also by the compressibility of the air/water mixture in the filter. The latter is important for very fine granular filters ($D_{50} < 3$ mm) and will not be considered here (see, for example, Bezuijen et al., 1986).

The piezometric head over the cover layer during wave rundown can be quantified by considering the mass balance of the water in the filter and the Darcy flow equation (Figure 3.37).

The flow in the filter layer is quasistatic. In the filter layer, a mean potential ϕ can be derived in a plane perpendicular to the slope, assuming that the flow in the filter layer is

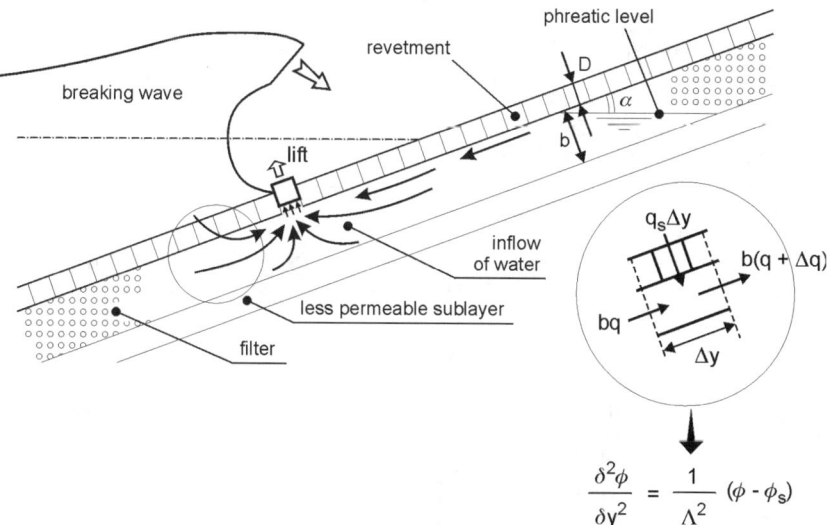

FIGURE 3.37. Mass balance in filter.

parallel to the slope. Further, the flow in the cover layer is assumed to be perpendicular to the slope.

The differential equation can then be written as

$$\frac{d^2\phi}{dy^2} = \frac{\phi - \phi_s}{\Lambda^2} \qquad (31)$$

The leakage length Λ is defined as the length of a piece of protection, in which the flow resistance through cover layer and filter layer are the same. This parameter is a measure of the pressure-head difference (ϕ_w) on the cover layer for given wave forces:

$$\Lambda = \sqrt{\frac{bDk}{k'}} \quad \text{or} \quad \frac{\Lambda}{D} = \sqrt{\frac{b}{D}\frac{k}{k'}} \qquad (32)$$

where D is the thickness of the cover layer, b is the thickness of the filter layer, k is the permeability of the filter layer, and k' is the permeability of the cover layer.

A solution of equation (31) for schematized boundary conditions (Figures 3.36 and 3.37) was presented by Wolsink (see Burger et al., 1990). The pressure head difference ϕ_w on the cover layer is given by

$$\phi_w = \left(\frac{1}{2}\Lambda \cos \alpha \tan \theta\left(1 - \exp\left[-\frac{\phi_b}{\Lambda \cos \alpha \tan \theta}\right]\right) + \frac{1}{2}\Lambda \sin \alpha\right)\left(1 - \exp\left[\frac{-2z_1}{\Lambda \sin \alpha}\right]\right) \qquad (33)$$

The resulting formula for the maximum gradient in the filter layer is:

1. Maximum downward gradient:

$$i = \sin \alpha \qquad (34)$$

2. Maximum upward gradient:

$$i = \cos \alpha \tan \theta\left(1 - \exp\left[\frac{-\phi_b}{2\Lambda \cos^2 \alpha \tan \theta}\right]\right) - \frac{\sin \alpha}{2} - \exp\left[\frac{-\phi_b}{2\Lambda \cos^2 \alpha \tan \theta}\right] \qquad (35)$$

Equations (33) and (35) are presented in Figures 3.38 and 3.39. It is clear that the uplift pressure over the cover layer increases as the leakage length (Λ) increases and the steepness of the wave front (θ) increases. But the larger Λ, the smaller is the maximum upward gradient in the filter (i).

The above equations for the loads are derived for regular wave attack. Unfortunately, little is known about the load ratio of regular and irregular waves. Experiments show that especially the large waves cause instability and that the number of waves during a storm plays a minor role.

On comparing the piezometric head on the slope under regular and irregular wave attack, the following is concluded for the wave height at threshold of damage:

$$\left(\frac{H}{H_s}\right)_{\text{damage}} = 1.4 \qquad (36)$$

The displacement of a block occurs if the uplift pressure exceeds the weight of the block plus the additional forces, such as friction and inertia. The limit state is

$$\phi_w = \Gamma \Delta D \cos \alpha \qquad (37)$$

Introducing the coefficient for irregular wave attack, $H/H_s = 1.4$, into formula (33) yields a complicated stability formula that can be approximated by (Klein Breteler and Bezuijen, 1991):

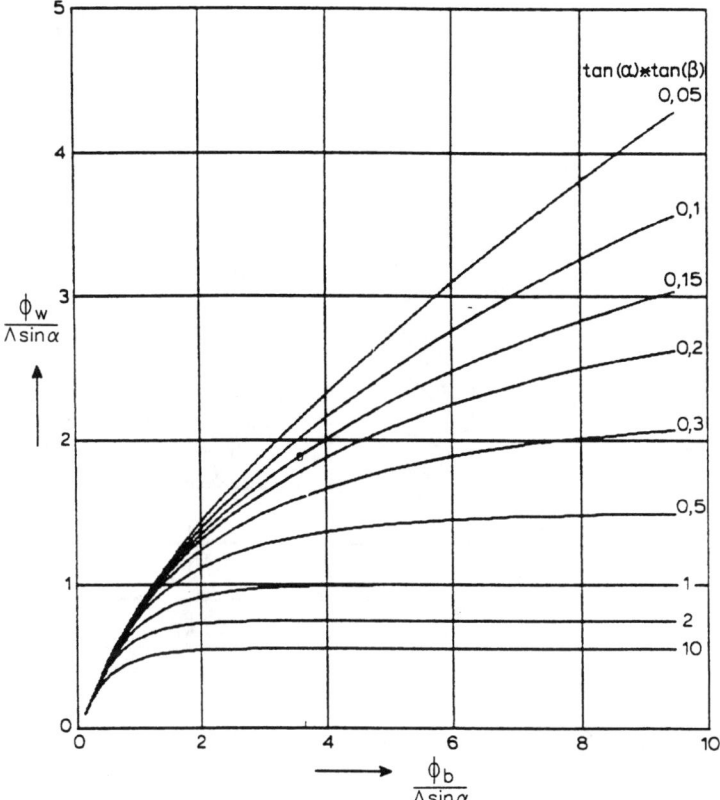

FIGURE 3.38. Uplift.

$$\frac{H_{scr}}{\Delta D} = f\left(\frac{D}{b}\frac{k'}{k}\right)^{0.33}\xi_{op}^{-0.67} \tag{38}$$

The equations work properly for placed/pitched block revetments and block mats within the following range: $0..01 < k'/k < 1$ and $0.1 < D/b < 10$. Moreover, when $D/\Lambda > 1$ use $D/\Lambda = 1$, and when $D/\Lambda < 0.01$ use $D/\Lambda = 0.01$. The range of stability coefficient is $5 < f < 15$; the higher values refer to presence of high friction and/or interlocking of a system.

From these equations, assuming by approximation that f is constant, it appears that:

a) An increase in the volumetric mass, Δ, produces a proportional increase in the critical wave height. If ρ_s is increased from 2300 to 2600 kg/m^3, H_{scr} is increased by about 23%.

b) The breaker parameter, ξ_{op}, comprises the slope angle (tan α) and the wave steepness (H_s/L_{op}). If the slope angle is reduced from 1:3 to 1:4, H_{scr} is increased by about 20%.

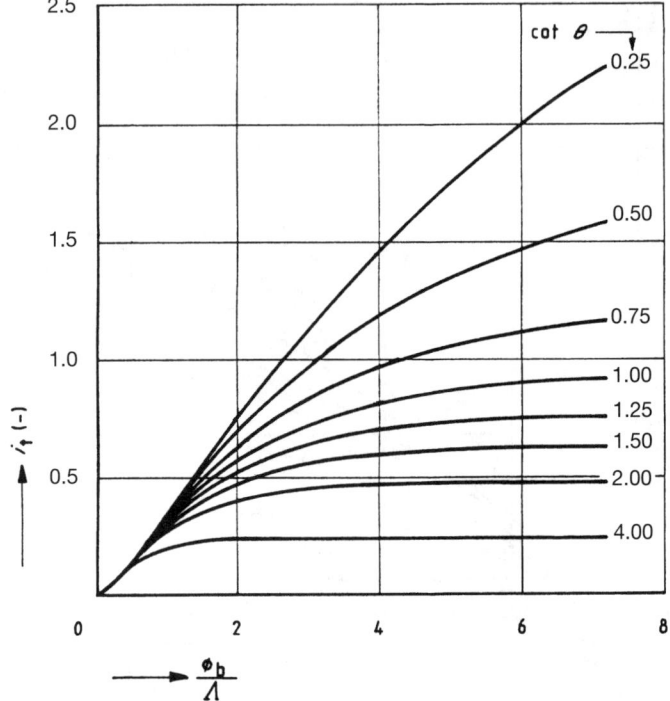

FIGURE 3.39. Maximum upward gradient.

c) An increase of 20% in the thickness of the cover layer, D, increases H_{scr} by ~27%.

d) A 30% reduction in the leakage length, Λ, increases H_{scr} by about 20%. This can generally be achieved by halving the thickness of the filter layer or by doubling the k'/k value. The latter can be achieved by approximation, by:
- reducing the grain size of the filter by about 50%
- doubling the number of holes between the blocks
- making hole sizes 1.5 times larger
- doubling joint width between blocks.

Changing the structural parameters changes the coefficient f slightly; the effect of these parameters can only be evaluated by approximation. It should be noted that changing the structural geometry can mean that failure mechanisms other than blocks being lifted out may govern the stability.

This stability equation contains all our theoretical understanding of the physical phenomena involved. However, the practical applicability is limited:
- It is only useful for block revetments with a granular filter layer underneath the blocks.
- The permeability of the cover layer can be calculated with the formulas in Klein Breteler and Bezuigen (1988), but they are very complicated.

- A lower boundary for Γ has been given by Burger et al. (1990), but a good quantification is still not possible.
- The theoretical and empirical bases for the assumption $H/H_s = 1.4$ is poor.

Most of these problems can be avoided by the application of only the most essential parts of the stability formula and completing it with empirical data from large-scale model studies. Partly based on the general trends in the results of model tests, a selection is made of Equation (38):

$$\frac{H_{scr}}{\Delta D} = F\xi_{op}^{-0.67} \qquad (39)$$

The value of the total stability factor F depends on the type of structure, as follows:

a) Low stability: $(k'/k)(D/b) < 0.05 - 0.1$
b) Normal stability: $0.5 - 1 > (k'/k)(D/b) > 0.05 - 0.1$
c) High stability: $(k'/k)(D/b) > 0.5 - 1$

The conditions for high stability are very difficult to meet and it turns out that there are no model studies performed with this type of structure. Therefore, it has been left out of the present discussion, leaving only two types of structures. With the equations from Klein Breteler and Bezuigen (1988) they can be defined as follows [D_{15} = grain size of filter exceeded by 85% of weight, (m)]:

Low stability, if:
- thick filter layer: $b/D > 0.5$
- course filter material: $D_{15} > 4$ mm
- closed cover layer:

  solid blocks with small joints: open area $\Omega < 2\%$

  blocks with holes with a spacing less than 0.3 m: $\Omega < 5\%$

  blocks with holes with a spacing wider than 0.3 m: $\Omega < 10\%$

  normal stability (structures other than defined as "low stability")

Extension to Other Types of Revetments with Model Tests. Up to now we have dealt with block revetments on a granular filter layer only. However, there are also structures with a cover layer directly placed on clay, or with a geotextile on sand. Furthermore, there are block mats and interlocking cover layers. For these structures there is no such theory as for the blocks on a granular filter. Therefore, we can merely assume that Equation (39) also is valid for these structures. In the next chapter this assumption proves not to be contradictory to the available test results.

We can conclude that the theory has led to a simple stability formula (Equation 39) and a subdivision into eight types of structures:

a) cover layer with loose blocks (without linkage or interlocking)
  a1) cover layer on granular filter, low stability
  a2) cover layer on granular filter, normal stability
  a3) cover layer on geotextile on sand
  a4) cover layer on clay

b) cover layer with linked blocks (i.e., with cables or interlocking, such as block mats)
   b1) cover layer on granular filter, low stability
   b2) cover layer on granular filter, normal stability
   b3) cover layer on geotextile on sand
   b4) cover layer on clay

The theory presented in the previous section is fitted to the results of a large collection of results of model studies from all over the world. Only large-scale studies are used because both the waves and the wave-induced flow in the filter should be well represented in the model. All available tests are summarised in Figures 3.40 to 3.46 and for each type of structure a lower and upper boundary for the value of F is given. The lower boundary gives with Equation (39) a stability curve, below which stability is guaranteed. Between the upper and lower boundary the stability is uncertain. Various unpredictable factors influence the stability of the structure. The upper boundary gives a curve above which instability is (almost) certain.

Comment to the Figures and Discussion

1. Most of the tests have been performed without a berm in the slope, except the studies reported in Oesterdam (1982), Tekmarine (1982), and most of Tekmarine (1985).
2. The studies concerning structure types b2 and b3 were very scarce. Therefore, tests with a rather small scale are used.
3. There are no test data available on structures with linked blocks on clay (b4).
4. The results for structure types a3 and b3 (blocks on geotextile on sand) may only be applied if $H_s < 1 - 1.5$ m or to structures with coarse sand ($d_{50} > 0.3$ mm) and gentle

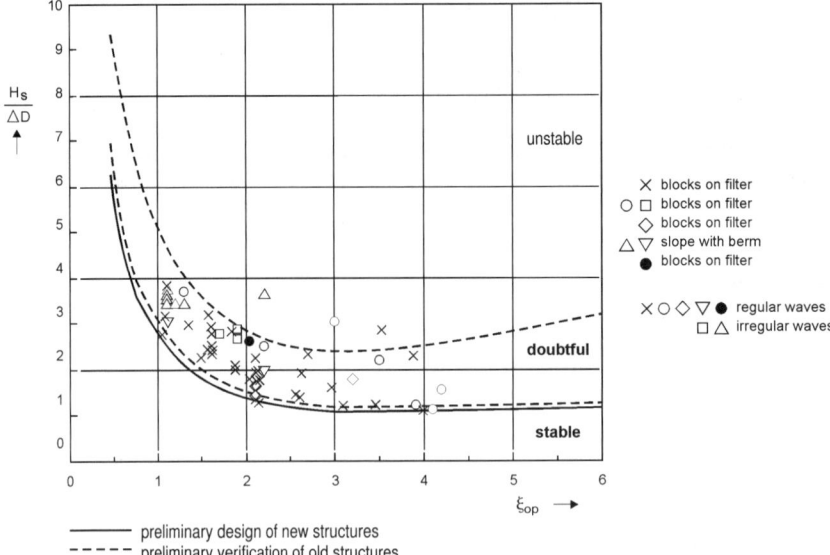

FIGURE 3.40. Test results for type a1 (loose blocks on granular filter, low stability).

DESIGN OF DIKES AND REVETMENTS—DUTCH PRACTICE   3.67

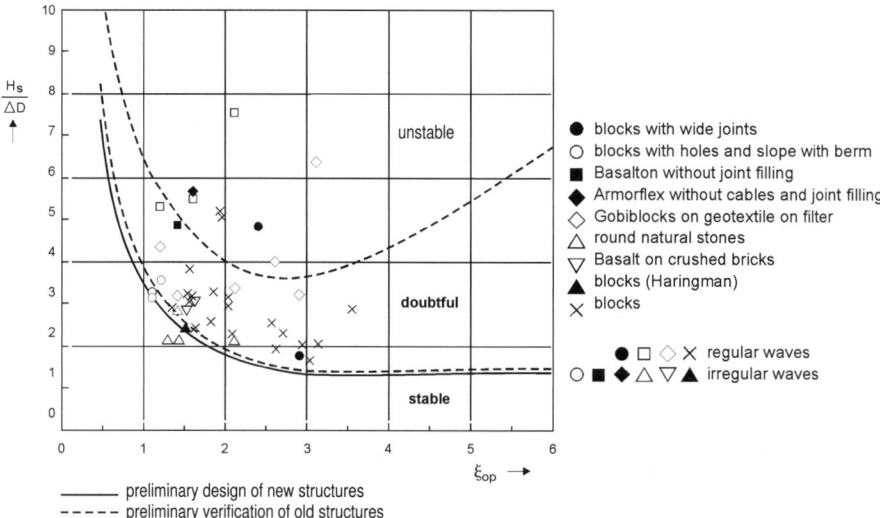

FIGURE 3.41. Test results for type a2 (loose blocks on granular filter, normal stability).

slope (tan α < 0.25), because the uplift of blocks is assumed to be the dominant damage mechanism (instead of soil mechanical failure).
5. The results for structure type a4 can be applied on the condition that clay of high quality is used (erosion resistant). If there is no such clay present, then a geotextile is recommended to prevent erosion during (lengthy) wave loading. The stability is then equal to that of structure type a3.

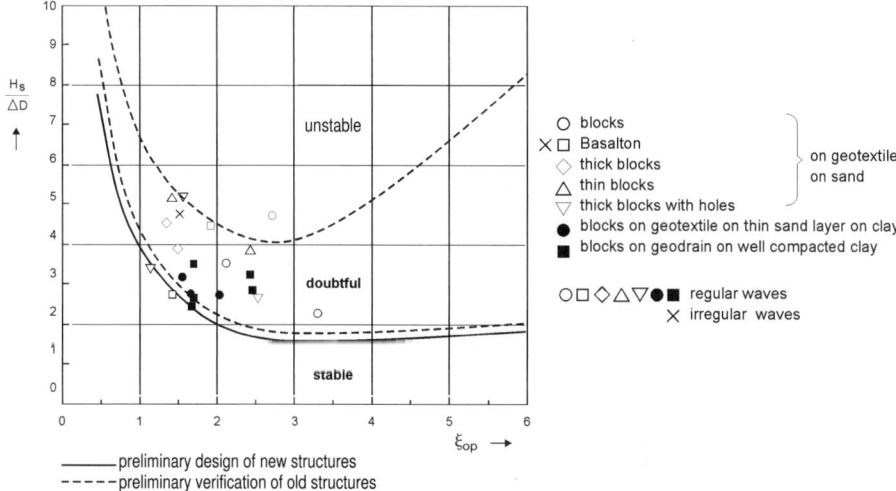

FIGURE 3.42. Test results for type a3 (loose blocks on geotextile on sand).

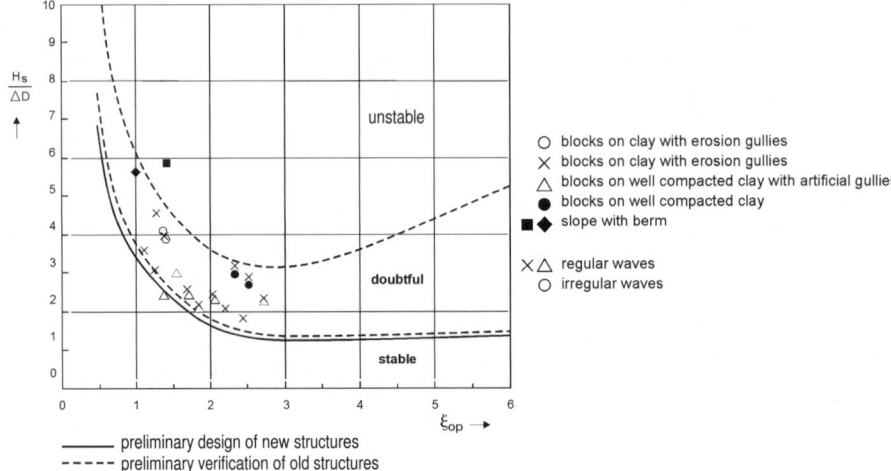

FIGURE 3.43. Test results for type a4 (loose blocks on clay).

The combination of the stability formula, which was derived from theory, together with the results of many large-scale model studies from all over the world has produced a reliable design tool for the preliminary design of placed-block revetments. Its reliability is only influenced by the fact that most tests have been performed with regular waves. An inaccuracy is introduced by the transformation of the regular wave load to an equivalent irregular wave load. Further studies are undertaken to improve the transformation method.

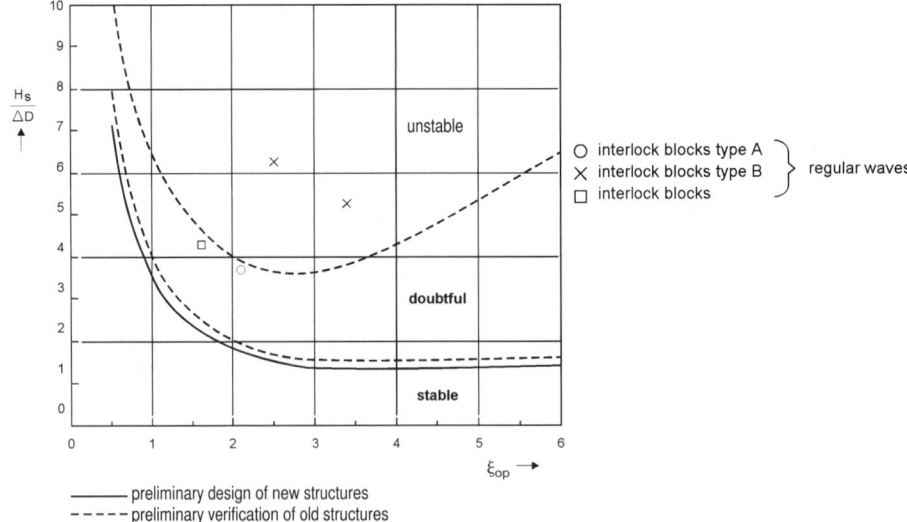

FIGURE 3.44. Test results for type b1 (linked blocks on granular filter, low stability).

DESIGN OF DIKES AND REVETMENTS—DUTCH PRACTICE

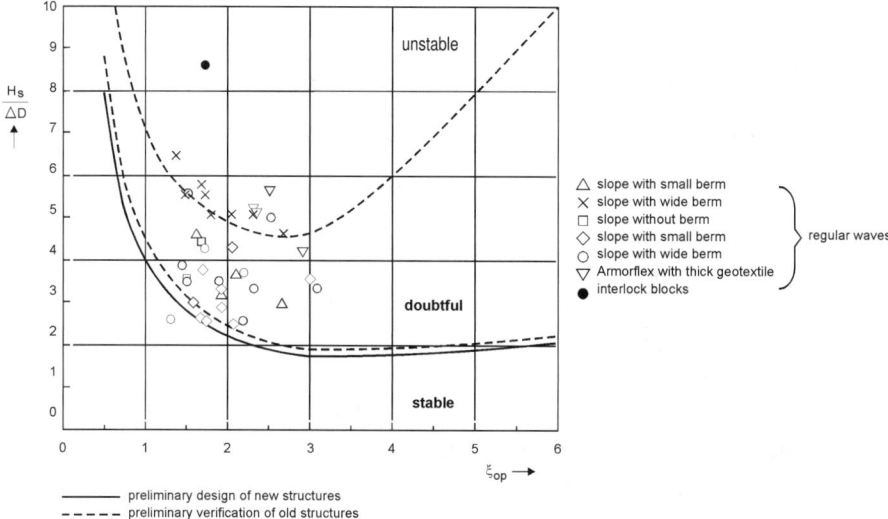

FIGURE 3.45. Test results for type b2 (linked blocks on granular filter, normal stability).

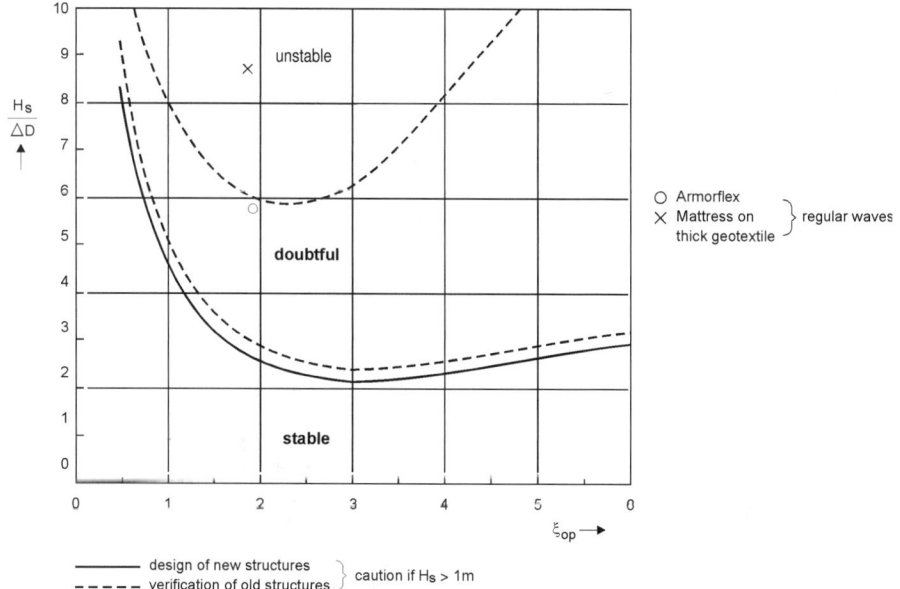

FIGURE 3.46. Test results for type b3 (linked blocks on geotextile on sand).

6.3. Migration of Subsoil

The migration of the subsoil particles through the filter layer or through the cover layer leads to local erosion of the subsoil near the water level and will result in a local settlement of the filter and cover layer. This damage mechanism is manifested by some stones that are sunk further than adjacent stones, or a gradually increasing S-profile. Some minor settlement will hardly affect the stability against wave action, but it will get worse in every serious wave attack (storm). Loss of coherence of the cover layer is the final stage and then failure is at hand.

No problems will arise if the granular filter or geotextile on the subsoil is geometrically sand-tight (D_x is grain size of filter and d_x that of subsoil) and conforms with the following conditions:

a) granular filters on sand: $D_{15}/d_{50} < 5$ (40)

b) geotextiles on sand: $O_{90}/d_{90} < 1$ (41)

c) geotextiles on clay or silt: $O_{90}/d_{90} < 1$ and $O_{90} < 100$ μm (42)

(O_{90} is the opening size of a geotextile corresponding to the average sand diameter of the fraction of which 90% of the weight remains on or in the geotextile). Unfortunately these criteria are often difficult to meet.

A more advanced requirement is based on hydrodynamic sand-tightness, viz., the internal flow must not be capable of washing out the subsoil material (even though the openings of the geotextile are much larger than the subsoil grains). This arises from (a) the hydrodynamic forces on the subsoil are greatly reduced by the geotextile, (b) the cohesion forces of the particles do not allow small particles to be washed away.

The hydrodynamical sand-tightness criteria can be applied to the majority of structures because hydraulic loads are usually low in the proximity of the subsoil (see Figure 3.48). Only in some cases in which the geotextile or subsoil–filter interface is very close to the surface of the structure and the hydraulic loads are heavy (for example, breaking waves), the geotextile or filter should be geometrically sand-tight (Figure 3.47). The critical hydraulic gradient for granular filters on a sand subsoil can be read from Figure 3.49.

The following criteria for geotextiles with O_{90} between 100 and 300 μm on clay or sand are applicable (Klein Breteler et al., 1994).

Good clay (colloid content = 39%; $d_{50} = 9$ μm; $d_{90} = 80$ μm):

$$i_{cr} = \frac{0.03}{n^2 D_{15}} \quad (43)$$

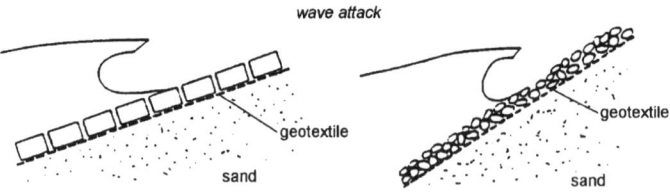

FIGURE 3.47. Examples of structures in which geometrically sand-tight geotextiles are necessary.

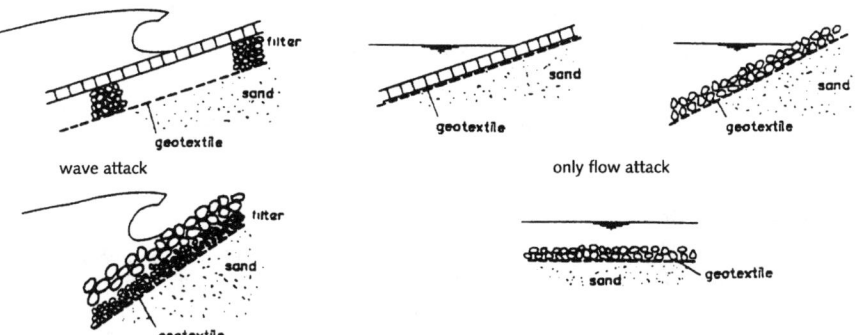

FIGURE 3.48. Examples of structures in which hydrodynamically sand-tight geotextiles can be applied.

Medium and poor clay (colloid content = 20%; 42 μm < d_{50} < 130 μm; 100 μm < d_{90} < 400 μm):

$$i_{cr} = \frac{0.01}{n^2 D_{15}} \qquad (44)$$

Fine sand (d_{50} = 90 μm; d_{90} = 130 μm):

$$i_{cr} = \frac{0.001}{n^2 D_{15}} \qquad (45)$$

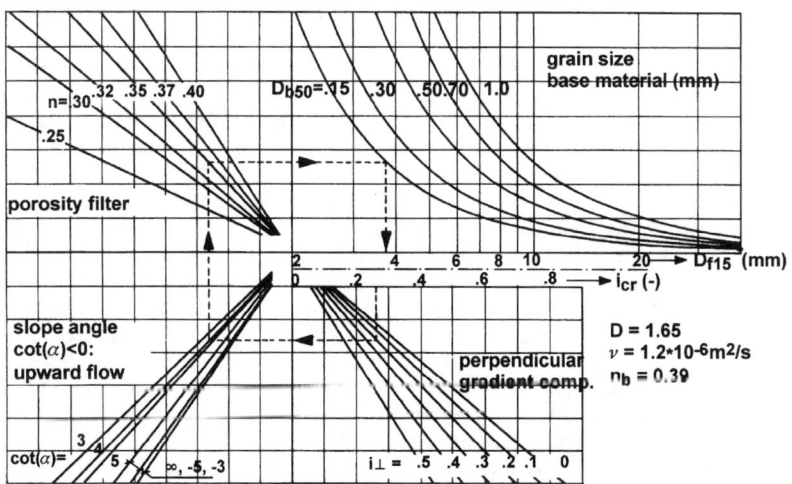

FIGURE 3.49. Diagram for critical gradient for granular filter on sand.

where n is the porosity of the filter layer (usually $0.3 < n < 0.4$) and D_{15} is the grain size of the granular material on the geotextile (m). The value of i can be calculated from Equations (34) and (35).

If gradients larger than i_{cr} can be expected (structures similar to those in Figure 3.47), then a geometrically sand-tight geotextile or filter is recommended.

6.4. Geotechnical Stability

6.4.1. Introduction. Apart from instability of the cover layer and migration of fines, instability of the subsoil (Kohler and Schulz, 1986; Bezuijen et al., 1990) is also a failure mechanism. In such a case, the cover layer does not fail, but the deformation of the subsoil leads to damage to the revetment.

This failure mechanism can be the dominant failure mechanism in the case of a permeable revetment (riprap revetment, block mattresses with holes in the blocks) on a granular subsoil. In the case of a relatively impermeable revetment (asphalt revetments or placed blocks on a highly permeable filter layer), lifting of the cover layer will precede geotechnical instability. As a result, a design method based on lifting of the blocks has to be used for impermeable revetments.

First, a qualitative description of the wave-induced pore pressures is given and the influence on the stability of the revetment is discussed. Thereafter, stability equations are derived and examples are presented. Comparison with the results of large-scale model tests have shown that there is reasonable agreement; the calculated results are on the safe side.

6.4.2. Pore Pressures in the Subsoil. The general structure of a revetment is shown schematically in Figure 3.34. A permeable revetment means that going from the subsoil to the cover layer the permeability of each layer increases. The failure mechanism of such a revetment is governed by the water movement caused by wave attack.

The overall loading on the revetment by the wave action and the pore pressures in both the revetment layers as well as the subsoil are shown in Figure 3.50. At the outside of the revetment there is a fluctuating water level caused by the wave attack. Deep down in the subsoil the phreatic surface is hardly influenced by the fluctuations in wave action. This surface is located somewhat higher than the mean water level (SWL). During wave rundown, the water pressure on the revetment is reduced. This reduction in water pressure on the revetment would not induce any loading on the revetment if the pore pressures were equal to the hydrostatic pressures corresponding with the reduced water level (line I in Figure 3.50). However, somewhere in the subsoil (e.g., at point A and deeper inside the subsoil) the fluctuations of the wave attack are damped and the pore pressure remains nearly constant. At that location, the pore pressures will be equal to a hydrostatic pressure distribution corresponding to the phreatic surface (line II in Figure 3.50). From point A upwards to the top of the revetment, the real pore pressure will be somewhere in between these two lines. This leads to a pressure difference over the revetment (cover layer and sublayer), as is shown in Figure 3.50 by the hatched area. The total loading on the revetment is the pressure difference $(p_0 - p_1)$. This pressure difference can cause instability of the revetment, resulting in lifting or sliding of parts of the revetment. As can be seen in Figure 3.50 this pressure difference is proportional to the difference between the nearly constant phreatic surface and the maximum run-down.

At the location where a revetment has to be built, the design wave is known. The only way to influence the total load on the revetment is to minimize wave rundown. This can be done by changing the slope of the revetment. When the wave rundown is fixed, the to-

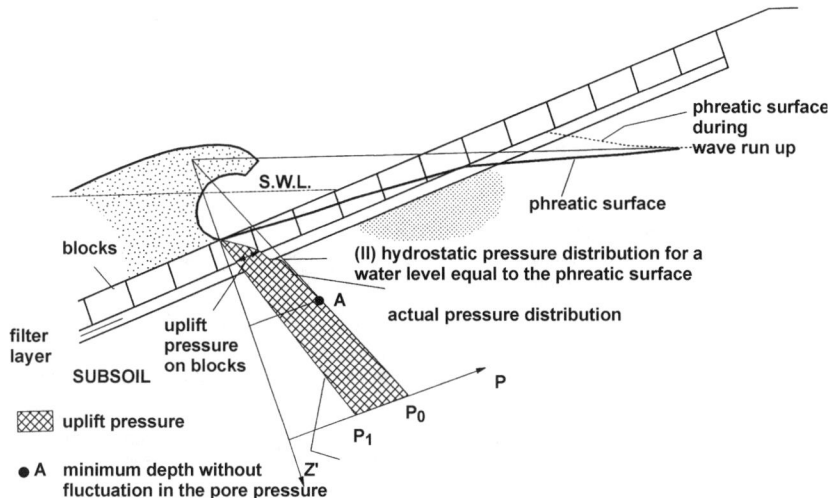

$P_0 - P_1$ represents the local loading on the revetment

FIGURE 3.50. Pore pressures during wave run-down.

tal load on the revetment is also fixed, because this is simply the pressure difference caused by the difference in water head inside and outside. Therefore, the total loading ($p_0 - p_1$) is independent of the design of the structure. However, the distribution of the load can be influenced by the designer. A small permeability of the cover layer will lead to a large pressure difference over the cover layer and thus a considerable loading over that layer. On the other hand, for a permeable revetment the loading will be concentrated on the subsoil.

6.4.3. Strength of Subsoil. To analyze the strength of the subsoil it is assumed that the subsoil consists of granular material that can be described as a friction material. Stability is guaranteed as long as the ratio between shear stress and normal stress is smaller than the tangent of the friction angle, Φ:

$$\frac{\tau}{\sigma'} < \tan \Phi \tag{46}$$

Without any water movement, the calculation of the normal and shear stress in a plane parallel to a slope is straightforward, leading to the well-known relation that the slope angle cannot exceed the friction angle. In the case of water movement in the subsoil and thus a nonhydrostatic pressure distribution, the influence of the pore pressure on the normal and shear stress has to be included in the calculation. This can lead to a failure surface that is different from the plane parallel to the slope. Therefore, generally, the stability has to be evaluated by a slip-circle analysis or finite-element calculation.

Generally, the pore-pressure distribution in the subsoil underneath a revetment under wave attack has to be calculated by numerical methods. For example Hjortnæs-Pedersen et al. (1987), have described a finite element method to simulate the pore-pressure distribution underneath a placed-block revetment. In this method it is possible to use measured wave pressures as a boundary condition. The advantage of a finite-element method is that the influence of layers with different permeabilities and different geometries can be evaluated. A disadvantage is that such methods are generally rather complicated.

Bezuijen (1991) has developed a simplified procedure for permeable revetments that leads to a minimum revetment weight per square meter to prevent subsoil instability, including the influence of the pore-pressure distribution. This method will be described in the next section.

6.4.4. Stability Calculations. In the stability calculation, a slip surface parallel to the slope is assumed at a depth z in the subsoil (see Figure 3.50). The value of z will be determined later. Two different situations have been taken into account:

a) The cover layer has a good toe structure or anchoring (in the case of a block mattress), giving only normal forces to the slope. If the filter layer (if present) is also locked up in a way that will cause no shearing forces, then the only shear stress that can exist is caused by the subsoil layer with thickness z_0,

b) There is no adequate toe structure or anchoring. In this case the revetment (cover layer and the filter layer, if present) will also contribute to the shear stress at $z = z_0$.

From the calculations presented by Bezuijen (1991), it appears that the revetment thickness will be impractically high in the latter case (up to 1 m thickness for cover layer and filter layer for a 1:3 slope loaded with a 1 m significant wave height) and therefore this chapter concentrates on the case with an adequate toe structure, or anchoring.

The normal stress on a slip surface in the subsoil is composed of the weight of the revetment and the weight of the subsoil with a thickness z_0 above the slip surface (the weight under water). This stress has to be reduced by the difference in piezometric head at depth z_0 and just below the revetment (Figure 3.51).

In this calculation, the water flow in the subsoil is supposed to be perpendicular to the slope of the revetment. Finite-element calculations have shown that this is a reasonable assumption in the case of a permeable revetment placed on a relatively impermeable subsoil. Such a flow does not influence the shear stress. The revetment does not contribute to the shear stress because the component of the weight parallel to the slope is counterbalanced by the toe structure.

The shear stress is difficult to determine exactly. The weight component parallel to the slope leads to a shear stress in the plane of the slip surface, but this component can also be partly balanced by the friction between the cover layer and the subsoil. Since the friction

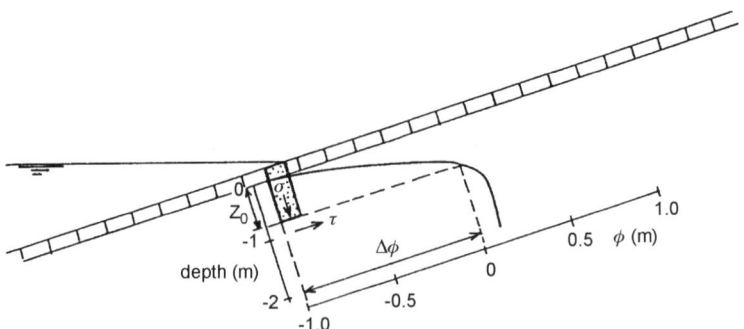

FIGURE 3.51. Definition sketch for stability calculation.

coefficient between the revetment and the subsoil is unknown and since it depends on the material just above the subsoil, it was decided to neglect this influence. This means that in Figure 3.51 the shear stress along the slip surface has to balance the weight component of the subsoil parallel to the slope. With these assumptions the following relations for the normal and vertical stress at the slip surface z_0 are found:

$$\sigma = [\sigma_b + (1 - n_s)(\rho_s - \rho)gz_0] \cos \alpha - \rho g \Delta \phi \quad (47)$$

$$\tau = [(1 - n_s)(\rho_s - \rho)gz_0] \sin \alpha \quad (48)$$

$$\sigma_b = \{(\rho_c - \rho)D + [1 - n](\rho_f - \rho)b\} \quad (49)$$

6.4.5. Stability Criterion. It is now possible with Equations (46), (47), and (48) to derive a minimum value of the weight of the revetment and filter layer (σ_b) necessary to achieve a stable revetment. With a toe protection or anchoring this is

$$\sigma_b \geq \frac{\rho g \Delta \phi}{\cos \alpha} - (1 - n_s)(\rho_s - \rho)gz_0\left(1 - \frac{\tan \alpha}{\tan \Phi}\right) \quad (50)$$

To use this relation it is necessary to define $\Delta \phi$ and the critical depth (z_0) at which the slip surface occurs.

Here $\Delta \phi$ is assumed to be equal to the run down value. According to CUR/CIRIA (1991), the following relation can be used for the rundown ($R_{d,2\%}$) for irregular waves:

$$\frac{R_{d,2\%}}{H_s} = -0.33\, \xi_{op} \quad \text{for } \xi_{op} < 4.5$$

$$\frac{R_{d,2\%}}{H_s} = -1.50 \quad \text{for } \xi_{op} \geq 4.5 \quad (51)$$

z_0 can be determined with consolidation theory (see Bezuijen, 1991):

$$z_0 = \frac{1}{2} L_{es} \sqrt{(\pi)} \quad (52)$$

With

$$L_{es} = \sqrt{Tc_v} \qquad c_v = \frac{k}{\rho_w g n w'} \quad (53)$$

In the equation of c_v it is assumed that the soil skeleton is very stiff compared with the stiffness of the water–air mixture. Normally this assumption is valid, since a few percent air in the pore water decreases the compressibility considerably to values lower than the compressibility of densified sand.

The permeability of the subsoil is most accurately determined from permeability tests. A first approximation can be obtained, based on the grain size and porosity of the soil (Adel, 1989):

$$k = \frac{g}{160\nu} \frac{n_s^3 d_{15}^2}{(1 - n_s)^2} \quad (54)$$

As relation for w' can be used (Verruijt, 1969):

$$w' = w + \frac{s}{p_a} \quad (55)$$

where

w' = compressibility of the pore water with air (m²/N)
w = the compressibility of pure water (5.10^{-9} m²/N)
s = air content in water, normally between 1% and 10%, as a decimal fraction (i.e., 10% = 0.1)
p_a = atmospheric pressure (1.10^5 N/m²)

The air content in the water, s, is seldom known. From simulations of large-scale experiments values between 5 and 10% were found. Lindenberg (1986) has measured this value in fine densified sand (135 μm) and found a value of 6%; in very loose sand, a value of 12% was found. In a normal situation with some compaction, 10% is assumed to be a safe value (the larger the air content, the lower the stability).

The results of this calculation method have been compared with the results of large-scale model tests. Measurements on the pore-pressure distribution in the sand has shown that this is comparable to the assumed pore-pressure distribution for a block revetment placed on sand (Bezuijen, 1991). Since the method neglects friction and clamping forces between the blocks, the resulting block thickness appeared to be conservative.

For wind waves a bit less conservative approach is possible (Klein Breteler and Bezuijen, 1998). When a revetment is loaded with a rather steep wind wave ($\xi < 2.5$), during a certain time, there will always be a wave front during the maximum wave run-down, as is schematized in Figure 3.50. Such a wave front will stabilize the soil directly underneath the wave front. In the one-dimensional calculation as presented up to now, this is not taken into account. In a two-dimensional calculation method this is possible. Performing calculations with this two-dimensional procedure, it appeared that fluidization, not sliding, of the sand underneath the revetment is the critical failure mechanism. If at some location in the revetment the excess pore pressure exceeds the weight of both revetment and the subsoil that is above that location, then fluidization will occur. This fluidization will lead to deformations of the subsoil and thus to deformations of the revetment.

The calculation method to estimate the possibility of fluidization is comparable to the method just described for sliding except that the stability criterion changes from that in Equation (46) to $\sigma' > 0$. This changes Equation (50) to

$$\sigma_b \geq \frac{\rho g \Delta \phi - (1 - n_s)(\rho_s - \rho)g z_0}{\cos \alpha} \qquad (56)$$

Figures 3.52 and 3.53 present the results of calculations with both stability criteria. Equations (50) until (56) are used to prepare the plots shown in these figures. The following parameters are used in the calculations for these plots:

$\rho_s = 2650$ kg/m³ $\qquad \rho = 1000$ kg/m³
$n_s = 0.45$ $\qquad \Phi = 35°$
$s = 0.1$ (10%) $\qquad \nu = 10^{-6}$ m²/s

Figures 3.52 and 3.53 show the calculated weight as a function of the wave height for different slope angles, grain sizes, and wave steepnesses. The plots show clearly the difference between both criteria. The lines marked "ww" are the lines plotted using the stability condition for wind waves ($\xi < 2.5$, wave front present). The other lines present the more conservative approach (i.e., sudden drop in a water level or run-down without a wave front). Because both situations appear during the run-down of wind waves, it is recommended to use at least the average value from both criteria or, to be completely certain, the more conservative one.

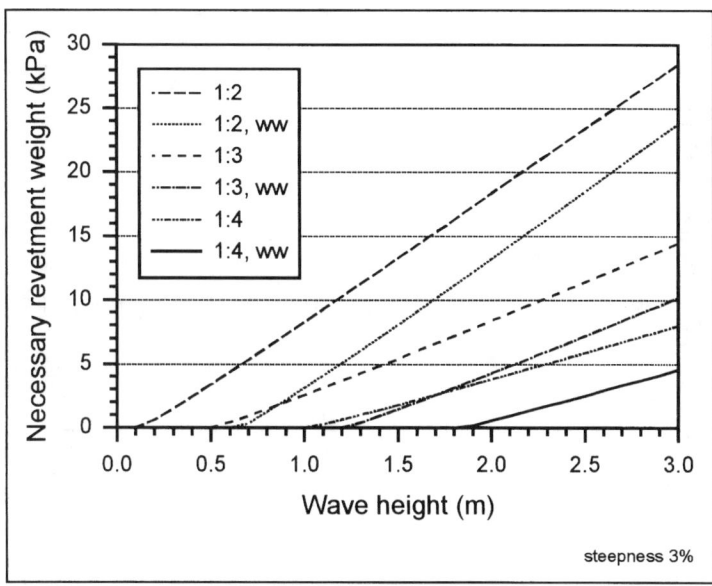

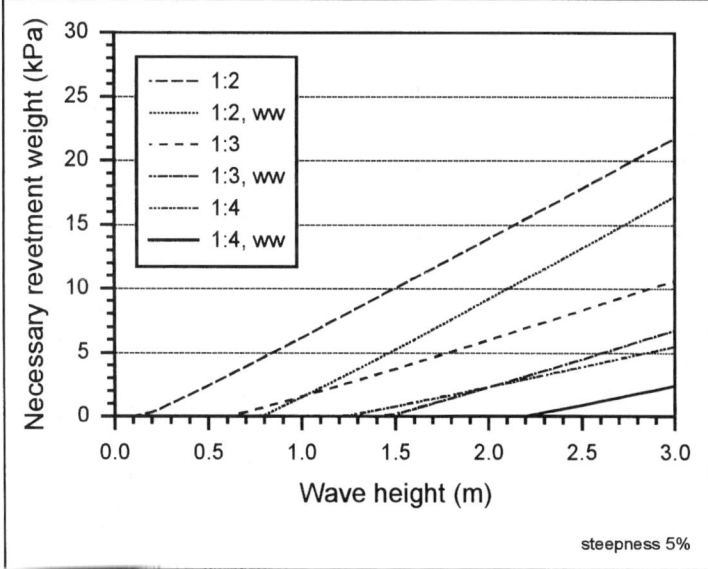

FIGURE 3.52. Result of stability calculation. Weight necessary to prevent instability of the subsoil for different slopes as a function of the wave height; wave steepness 3% and 5%, grain size of subsoil 0.2 mm.

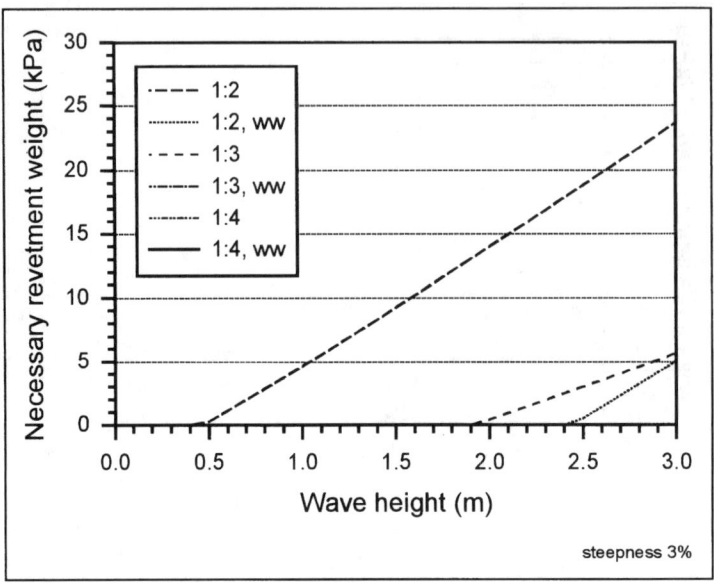

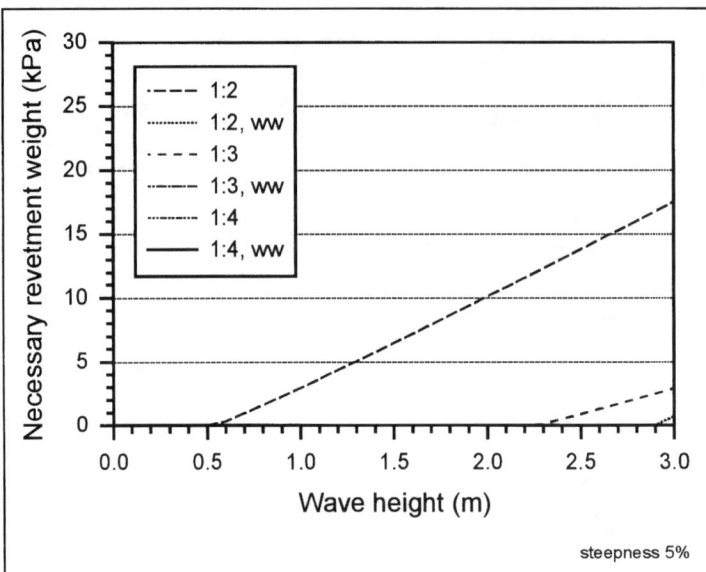

FIGURE 3.53. Result of stability calculation. Weight necessary to prevent instability of the subsoil for different slopes as a function of the wave height; wave steepness 3% and 5%, grain size of subsoil 0.4 mm.

DESIGN OF DIKES AND REVETMENTS—DUTCH PRACTICE   **3.79**

It is further shown that geotechnical instability is a dangerous failure mechanism in the case of fine-graded subsoils and steep slopes. When no lines are drawn, this indicates that geotechnical instability will not be a likely failure mechanism for waves up to 3 m for the situation described. It should be noted that this is only true if permeability is as large as assumed and the blocks are stable against wave attack.

6.4.6. Sliding of a Revetment. Apart from lifting of parts of the cover layer and geotechnical instability, sliding of the cover layer can also be a dangerous failure mechanism, especially on steep slopes and for cover layers that give only little horizontal support (for example, riprap revetments or revetments with bags), and for cover layers with a poor toe structure or anchoring. As for lifting of the blocks, sliding is most likely in the case of an impermeable cover layer placed on a permeable filter layer. A situation that leads to a flow in the filter layer is shown in Figure 3.37.

For the revetments with little horizontal support, only the stability of a single element from the cover layer against sliding has to be evaluated. For revetments with horizontal support, the sliding forces and stabilizing forces of a larger part of the cover layer have to be evaluated. This section presents a simple procedure for determining the stability of a part of a block revetment against sliding.

The loading on the revetment by wave attack can cause sliding of the cover layer. Assume a revetment without any toe protection or anchoring of the blocks. Stability of such a revetment has to be guaranteed by the friction force between the revetment and the subsoil. The maximum friction force is assumed to be related to the normal force between the soil and the subsoil. A linear relation without cohesion is assumed:

$$F_f = F_n \tan \Phi' \qquad (57)$$

Without wave attack F_n is determined by the underwater weight of a block multiplied by the cosine of the slope angle. The driving force for sliding is the sine component of the block weight. The driving force will be smaller than the maximum friction force; otherwise it would be impossible to build the revetment without toe structure or anchoring.

During wave attack, the driving force remains the same. The normal stress changes due to the uplift pressures. The situation for a placed-block revetment is shown in Figure 3.54. The distribution of the difference in piezometric head over the revetment is sketched during wave run-down. At the upper end of the revetment, the difference is negative, meaning that there are no hydrostatic uplift pressures. The weight of the block is still nearly the weight above water and the normal stress between blocks and subsoil is hardly influenced by the wave attack. Lower on the revetment, there is a positive difference in piezometric head that reduces the grain stress and can lead to sliding of the revetment. Below the wave front, there is again a negative difference in piezometric head. The blocks are pushed on the subsoil and have an apparent weight larger than the weight below the water line, but the driving force is reduced (compared with the situation without wave attack), because the blocks are situated below the water line. These blocks will even have more stability against sliding than in the situation without wave attack.

Suppose sliding of the blocks just before the wave front occurs. Then this will not influence the blocks on the upper part of the revetment unless there is some connection by cables or interlocking, but it will influence the lower blocks. Sliding will only occur when the total driving force of all the blocks, below the highest block for which the driving force exceeds the friction force, is higher than the total maximum possible friction force.

In the case of a block mattress connected with cables, the situation will be the other way around. When there is a large joint between the blocks, the blocks in the most heavily loaded area cannot be stabilized by the blocks lower on the revetment and should be stabilized by the cables between the blocks. The cable forces are transmitted

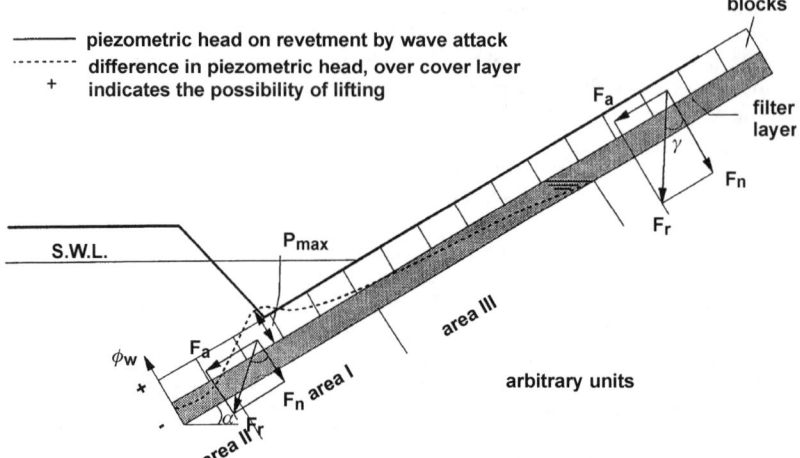

FIGURE 3.54. Possible distribution of piezometric head just before a breaking wave and its influence on the direction of the resultant force of block on the subsoil.

to the blocks higher on the revetment. This means that for the lower part or the upper part of the revetment (depending on the type of revetment—placed blocks or block mattresses) the maximum possible friction force has to be summed and compared with the maximum driving force. If the driving force is higher than the maximum possible friction force, stability has to be obtained by a sufficiently strong toe structure or anchorage of a block mattress.

The method described here can be used qualitatively with different methods to calculate the uplift pressure. Bakker and Meijers (1988) have elaborated this method for the calculation method described in the section on internal loading. They suggested a value for the friction angle: $\Phi' = \tfrac{2}{3} \Phi_f$, where Φ_f is the the friction angle of the filter material. Further, they concluded on the basis of their results that toe structures and/or anchorages will often be indispensable.

Bezuijen et al. (1990) presented a simple method to get a first indication of how large the forces on toe structures can be. In this method, it was assumed that over the height equal to one wave height there is no friction between the cover layer and the subsoil (see Figure 3.54). This is a pessimistic assumption for a revetment with a short leakage length. On the other parts of the revetment, the maximum possible friction force is comparable with the situation without wave attack. Additional strength due to a negative pressure gradient below the water line is neglected. Above the area of wave attack, it is assumed that the influence of the water can be neglected.

Each block in the area of wave attack (area I in Figure 3.54) contributed to the total driving force for sliding:

$$F_a = LBD\rho_c g \sin \alpha \qquad (58)$$

A block in the area without wave attack will be stable, even if it is loaded to some extent with the driving force from the area with wave attack. With F_a the driving force for sliding and F_f the friction force, the maximum possible loading from areas I and III (F_t) on the area without wave attack (area II) before sliding occurs can be written as

$$F_t = F_f - F_a = L_{II}BD(\rho_c - \rho)g[\cos \alpha \tan \Phi' - \sin \alpha] \quad (59)$$

and above the water line (area III in Figure 3.54):

$$F_t = F_f - F_a = L_{III}BD\rho_c g[\cos \alpha \tan \Phi' - \sin \alpha] \quad (60)$$

With these relations the force on a toe structure or anchorage can be calculated, as will be done in the following example. A placed-block revetment on a slope 1:3 is loaded with a wave height of 1 m. The toe structure is placed 1.5 m below the area of wave attack. The revetment consists of blocks with dimensions of 0.5 × 0.5 × 0.25 m. The friction angle between the blocks and the filter layer is 25°. In the loaded area each block contributes with a loading force of 445 N (using a value of 2300 kg/m³ for the density of concrete). The stabilizing area is below the water line, therefore Equation (57) has to be used, leading to a stabilizing force of 100 N for each block. There are approximately 6 blocks in the loaded area and 9 blocks in the area without wave attack. The total force that will load the toe structure is therefore 1770 N for each row of blocks. Since the length of the blocks is 0.5 m, this result leads to a loading of the toe structure of 3540 N/m.

7. REVIEW OF DESIGN CONSIDERATIONS FOR DIKES AND REVETMENTS

The review of the key elements that must be considered in the design (dimensioning) of dikes and revetment structures is illustrated in Figure 3.55. More detailed design methods for these structures are discussed in (PIANC, 1992 and CUR/RWS, 1995a). The most critical structural design elements are :

a) the stability of the cover layer

b) the security of the foundation

c) the minimization of settlement and sliding

d) the toe protection to prevent undermining

All of these are potential causes of failure of coastal structures.

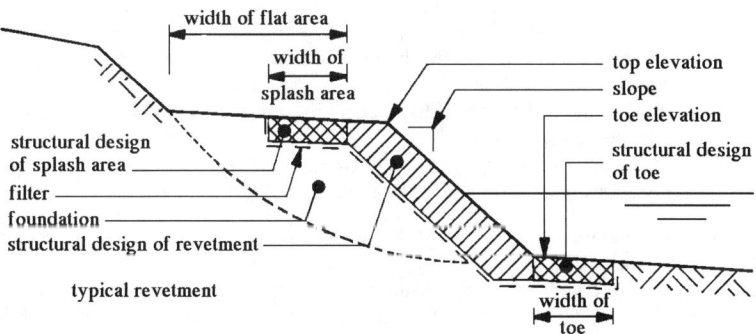

FIGURE 3.55. Design components of typical revetment structure.

The usual steps needed to develop an adequate structure design are:

1. General aspects
   1.1. Formulate purpose of the project
   1.2. Formulate functional requirements
   1.3. Formulate project restraints (materials, labor, equipment, financing, time, risk, etc.)
   1.4. Formulate global description of boundary conditions
2. Alternative general solutions (items 2 and 3 are reversible)
3. Boundary conditions
   3.1. Determine hydraulic boundary conditions and loads (water levels, wave climate, currents, morphology, beach variations, etc.)
   3.2. Determine geotechnical boundary conditions (soil types and relevant parameters)
   3.3. Determine other relevant conditions and loads (ice, earthquake, vegetation, etc).
4. Feasibility studies for generated alternatives
   4.1. Determine suitable structure configurations (geometry), crest, toe and/or berm elevation related to the allowed run-up and overtopping
   4.2. Review the possible failure mechanisms
   4.3. Select a suitable armor alternative and size of armor units
   4.4. Make a preliminary feasibility analysis of alternatives
      develop cost estimate for each alternative
      evaluate possible construction requirements and limitations for the alternative solutions
   4.5. Select the final solution
5. Final design
   5.1. Consider the use of models (improving boundary conditions)
   5.2. Consider the probabilistic approach
   5.3. Make final estimation and evaluation of the structure geometry (slope, berm, crest, toe) and check the allowable figures on run-up and/or overtopping
   5.4. Design the final dimensions of the revetment, cover layer, filter and/or underlayers
   5.5. Make prediction of scour and design toe protection
   5.6. Design transitions and crest protection (splash area)
   5.7. Design other structure-related elements
      drainage features (if required)
      surface run-off and overtopping run-off facilities
      flanking of the structure
   5.8. Check for possible failure mechanisms of the final design and corresponding damage and risk level
   5.9. Go through the overall project and checklist and finally decide whether alternative geometries should be considered
   5.10. Prepare specifications for materials, equipment, execution, and cost, including quality control and motivation of choices

The primary function of any slope protection, flexible or not, is to protect the edge of land and water against hydraulic loads by waves, tides, and currents. The determination of the hydraulic design conditions is the result of a quantification of the local conditions in combination with a certain level of safety. In this way the design conditions are defined and presented in the form of a water level, a wave height, and a wave period, usually completed with some expectation for the form of the energy density spectrum and some-

DESIGN OF DIKES AND REVETMENTS—DUTCH PRACTICE   **3.83**

times even completed with an estimate for the duration of the selected design condition. However, the fact that the design conditions are fixed does not at all mean that the loads on the slope revetment structure are also fixed. Within certain limits, of course, it is possible for the designer to influence and consequently to choose the size, the sort, and the place of attack of the hydraulic loads, by a proper selection of the geometry, lay-out, and materials for the structure. In previous sections the mathematical equations for the calculation of external and internal hydraulic loads were presented. The parameters in the equations can be manipulated by the designer to control the performance and effectiveness of structures.

With reference to the design of block revetments as one of various other alternatives and in view of design optimization, the following design considerations can be used during the design process:

1. Steeper or milder slope gradient

    Steeper slope makes the protective length (revetment) shorter; as a first approximation, the slope length (L) is related to the height of slope to be protected (h) by $L = h/\sin \alpha$.

    For breaking waves ($\xi_{op} < 2.5$) the run-up ($R_{u,2\%}$) on the steeper slope will increase proportionally to $\tan \alpha$, namely: $R_{u,2\%} \leq 8H_s \tan \alpha$; this yields a higher crest position and eventually, a larger volume of the dike.

    The run-down on the steeper slopes also increases, possibly leading to higher overpressures and, thus, thicker protective elements.

    For steeper slopes of loosely placed blocks, the friction between the blocks increases with $\sin \alpha$. However, it is difficult to quantify the consequences of this effect exactly.

    For steeper slopes, the internal gradients increase, leading to more severe requirements concerning the sublayers.

    A steeper slope imposes more severe requirements for the support by a toe protection.

    The damage progress after an initial damage is more rapid for the steep slopes, thus providing more dangers of scouring.

    Steep slopes are more easily damaged by ice, especially when using slopes steeper than 1 in 3; the above considerations should be taken into account for a proper design.

    For steeper slopes, the risk of geotechnical instability increases.

2. Berm or no berm

    Application of a berm reduces the run-up, making possible a lower crest elevation.

    A berm can serve as a maintenance road.

    A berm creates a discontinuity in protection (weak point).

    A berm reduces the phreatic level in a dike with a positive effect in the case of low-permeable or impermeable revetments.

    A berm reduces ice ride-up.

3. High or low permeability of the cover layer

    High permeability, in combination with a proper sublayer, reduces the uplift pressure and leads to thinner units. It is, however, important that the permeability does not decrease during the lifetime (aging).

    When the high permeability is created by large openings in or between blocks, washing out of the sublayers can take place; to avoid this, the following measures can be taken:

    (a) coarser filter; however, this sometimes leads to increase of the hydraulic gradients across the cover layer and thus to thicker units

(b) geotextile underneath the cover layer elements. Attention should be paid to a sufficiently low hydraulic resistance normal to the slope, which should not increase the uplift pressure.
(c) another solution can be the use of bounded filters (sand–bitumen, sand–cement, etc.)

Note: To reduce these disadvantages the permeability should be distributed over the units instead of being concentrated (e.g., in one big hole).

High permeability of the cover layer may increase the hydraulic gradients at the sublayer–subsoil interface or in the subsoil; proper care should be exercised in adequate sublayer design.

High permeability of cover layer reduces the run-up somewhat.

In the case of a very high permeability of block revetments created by large holes, the drag forces along the slope may increase considerably, leading to large forces on the units and thus larger dimensions.

4. Rough or smooth surface

A rough surface (which can also be obtained by using blocks of various heights) reduces the run-up and thus reduces the crest elevation and eventually the volume of a dike. This effect is evident mainly when the whole run-up zone is equipped with roughness elements. When the upper slope is protected by a grass mat, the application of the roughness elements on the lower part of a slope will have a limited effect.

High roughness elements introduce high drag forces, which should be incorporated in the stability calculations.

Rough surface is unfavorable under ice conditions.

5. High or low permeability of sublayers (filter)

Decreasing sublayer permeability reduces the uplift forces on the cover layer. In the case of a cover layer of low permeability, this may lead to reduction of the thickness of the cover layer. However, it should be checked whether this lower permeability of the sublayer and the corresponding reduction of weight is acceptable with respect to the stability of the sublayers.

For noncohesive (granular) materials a decrease of the permeability can be obtained by using:

finer granular material (however, washing out through the cover layer should be avoided and the geotechnical (in-) stability should be checked)
wide-graded material (the internal stability should be examined).

Applying clay as a cohesive sublayer requires formulation of proper specifications on clay properties to avoid erosion, piping, or shrinkage. However, it should be checked whether an impermeable sublayer might cause other problems, e.g., malfunctioning of the toe.

Lower permeability of sublayer/filter increases the hydraulic gradients at the interface with the subsoil or inside it. This can be counteracted by increasing the thickness of the sublayer/filter or by applying a geotextile on top of the subsoil. In addition, the geotechnical stability should be evaluated.

6. Shape of sublayer/filter material

Rounded material is often cheaper than broken material; however, in the case of insufficiently compacted grains, a slightly lower angle of internal friction may lead to geotechnical instability, more settlement, and forces on the toe structure.

7. Thick or thin sublayer/filter

In the case of block revetments of low permeability, reduction of the thickness of sublayer/filter leads to reduction of the uplift forces but simultaneously leads to increase of the hydraulic gradients along the interface with the subsoil or inside it.

8. Shape of blocks

Rectangular blocks	Columns of irregular shape
good alignment/joining	mostly nicer appearance
low permeability	higher permeability
easy mechanical placing	less-easy mechanical placing
problems in bends	easier with bends
difficult to repair	easier repair
washing-in/grouting quite difficult	washing-in/grouting possible
often cheaper	often more expensive
more rapid progress of damage	slower progress of damage

Note: In the case of blocks, the self-healing tendency, as for riprap, is absent; therefore the stability of blocks should be guaranteed under all design conditions.

9. Concrete (or other artificial material) or natural stone
    Natural stone, if available in respect to the required quality and quantity, can often be a favorable solution.
    Concrete blocks (or asphaltic revetments) can often be a good alternative (especially when the natural stone is not locally available) because of:
    often lower cost
    good/constant quality
    uniform size
    mechanical placement
    more choices regarding composition, size, etc.
    Note: The economical optimization including the availability of materials, equipment, and skills is mostly decisive for the choice.

10. Effect of aging and/or wear/fatigue
    During the lifetime of revetment structures their original specifications can change due to climatological effects (wind, rain, frost, abrasion, sedimentation due to waves, marine growth, etc). As far as possible, the course of time should be taken into account in the design process. However, it is not easy to quantify these effects. Some qualitative description is given below:
    Aging of the cover layer
        Due to the wave attack at various water levels, the permeability and the interlocking may change with time. For small interspaces between the blocks, the permeability can decrease due to siltation of sediment and the friction between the blocks may increase.
        Vegetation in the interspaces may also increase the friction/interlocking; however, it is possible that in the case of a heavy wave attack, the silted and/or vegetated interspaces will be cleaned up again, thus providing no additional strength at the moment of design loading on the protective units.
    Aging of the sublayers
        In the case of alternative materials used as sublayers (mine stone, slags, silex, etc.) special attention should be paid to the changes of the physical properties of these materials under influence of air, wave shocks, varying humidity, frost, etc.
        In the case of geotextiles, special attention should be paid to the possibility of clogging and/or blocking (leading to drastic change of permeabilities and, thus, increase of uplift pressures).
        The siltation of the sublayers/filter has, in general, a positive effect; due to the decrease of permeability, the uplift forces decrease.

11. Residual strength of revetments
   Revetments should be designed in such a way that the chance of failure is acceptably low. The quantification of a risk is related to the type of revetment, especially regarding the progress of damage, for example:
   A very rough surface is more sensitive to damage than a smooth surface.
   Application of a strong geotextile retards the extension of damage to the subsoil.
   Cohesive (clay) or bounded sublayers are primary measures used to increase the secondary strength of revetment structures if the permeability of the materials is not a disadvantage for the total stability, and if the cohesive material is of sufficient strength.
12. Cost optimization
   The total costs of a revetment are related to:
   capital costs (execution)
   yearly maintenance
   large/periodic maintenance
   repair of damage
   demolition (after a lifetime)
   A total capitalization of the cost gives the most optimal result.
   In general, revetments with lower capital costs will be damaged more frequently and will need more maintenance. Local subsidy regulations may influence the choice. However, in case of sea defenses, especially along low shores, higher capital costs (stronger protection) should be preferred. In the case of land reclamation or bank protection, the results of the capitalization of the costs can be applied directly.

8. CONSTRUCTION ASPECTS

8.1. The Dike and Its Components

Dike construction works can be subdivided into:

- new works: a completely new dike, for example, when cutting off a bend in a river
- reinforcement of the outer slope (water side)
- reinforcement or the inner slope (land side)

The whole outer slope, including berms, must be considered when designing a new dike or reinforcing an existing dike. When reinforcing the inner slope, generally only the section up to an existing berm or the crest section needs to be considered (see Figure 3.56).

Dike reinforcing begins with placing the essential clay. This can involve:

- excavation of clay from sections of the old dike
- when reinforcing the outer slope, excavation and ripening of clay from the mud flats in front of the old dike and from the foundations of the new dike
- any excavation required in the area behind the dike
- supply of clay to site, if necessary

Retaining dikes are then constructed with the clay (see Figure 3.57). If there is deep water in front of the dike, the retaining dikes can be constructed using mine stone (broadly graded, granular waste product of the mine industry). If the foreshore is high, for example at mean tide level, sand can be pumped into place and bulldozed into retaining dikes. Usually, sand for hydraulic fill should be obtained locally. The sand can be dredged

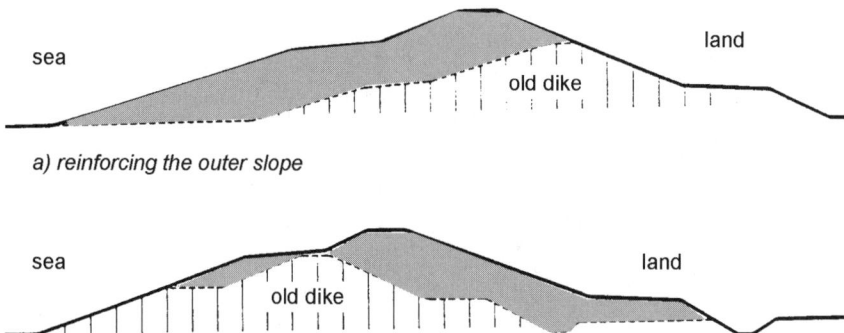

a) reinforcing the outer slope

FIGURE 3.56. Reinforcing the outer and inner slopes of a dike.

offshore and pumped directly into barges by suction dredges. The barges then transport the sand to the coast from where it is pumped to the site using pressure lines. If barges cannot reach the site, long pressure lines have to be used. For direct sand excavation, without barges, pumping has to be carefully controlled because of the danger of silting (if the suction/pumping period is too long, the silt fraction may separate out). Sometimes the sand is first pumped into large storage areas prior to use. Depending on the stability of the sand and the available water pressure, the sand is laid in one, two, or sometimes three layers. The rate of settling is checked by using settlement plates; the water pressure is checked by using water-pressure meters.

The second layer of material can be deposited between retaining dikes constructed from the sand of the first layer, etc. These temporary dikes are protected with plastic sheets, which must be removed after the layer has been pumped into place, since they could form slip surfaces within the dike. After being deposited, the hydraulic fill can be profiled and compacted using bulldozers. The revetment clay is then laid in two layers and also compacted using bulldozers. The total thickness of clay must be at least 80 cm on the slopes and 60 cm on inner slopes (TAW, 1991, 1995; CUR/RWS, 1995a). The best quality clay should be used for the outer slope. Because clay on the outer slopes, is liable to be flushed out, it should only be used directly under blocks above HWST (High Water Spring Tide). Clay can be used in the tidal zone if it is applied under a filter layer, with a geotextile between the clay and the filter (see Figure 3.58).

If blocks have to be placed directly on clay, the subsurface must first be prepared and compacted. A thin layer (1 to 2 cm) of moist clay can then be laid to provide a smooth subsurface for the blocks. This clay should be laid along the dike and can be sprayed if necessary. Blocks are usually placed mechanically using a block claw. If necessary, the

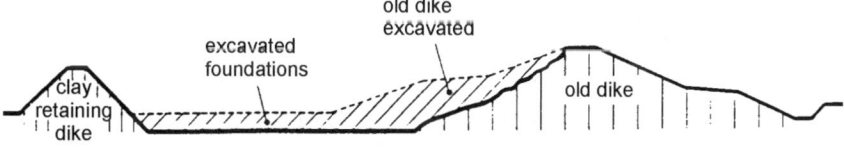

FIGURE 3.57. Retaining dike constructed from clay excavated out of the old dike.

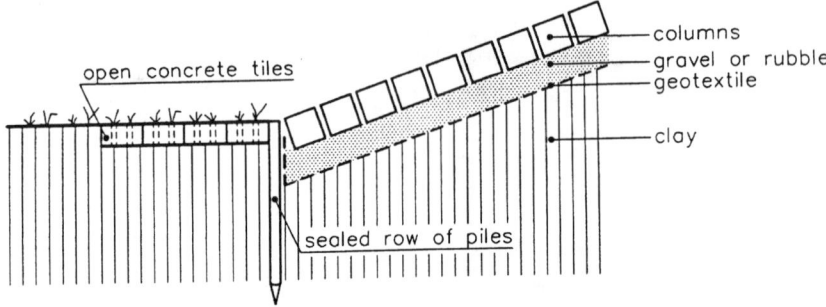

a. columns

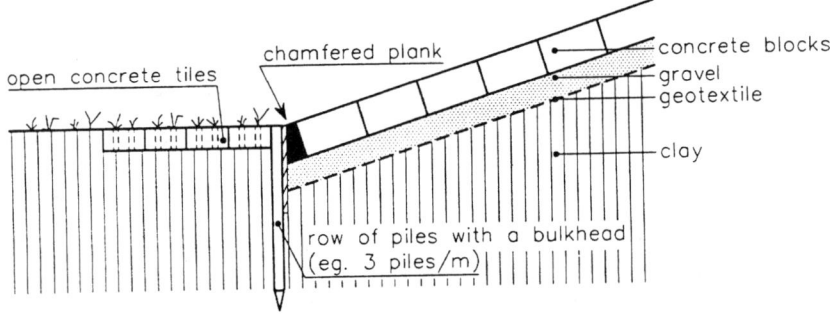

b. blocks

FIGURE 3.58. Example of toe structures with a high foreshore.

blocks can be tamped down. If mine stone is used for the retaining dike, it can be reused, after the sand has been pumped into place, as a filter layer, 0.5 to 1 m thick.

In order to obtain a smooth surface for the pitching, a thin filler layer of gravel or rubble is laid on the mine stone. This layer should be as thin as possible (say 5 to 10 cm), because the thickness can have a detrimental effect on the stability of the revetment.

When strengthening the inner face of a dike, the old revetment, assuming it is still satisfactory, can remain in place unless its profile differs from that required. In order to obtain a good connection, it is generally necessary to take up a 2 m wide strip of the old revetment and re-lay it under the new profile (see Figure 3.60).

8.2. Block Mattresses

Block mattress revetments (block mats) are constructed in three phases:
- construction of the dike body and any revetment sublayers (clay and/or filter layers)
- laying the mattress
- anchoring the mattress and applying the joint filler

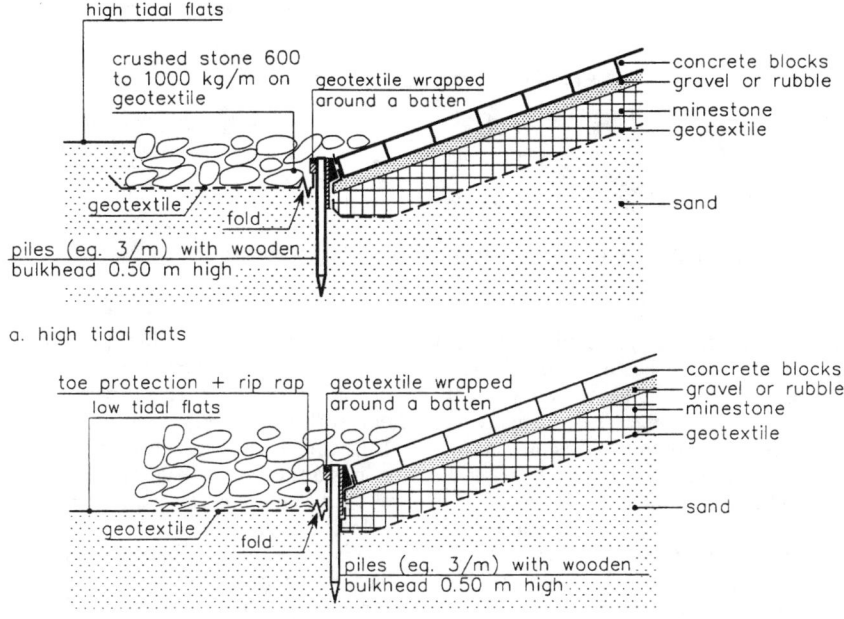

FIGURE 3.59. Examples of toe structures in the tidal zone.

The construction of the dike body is described above. A well-compacted slope is important in order to produce a smooth surface and thus ensure that there is a good connection between the mattress and the subsurface. When laying mattresses on banks, it is strongly recommended that they are laid on undisturbed ground and that areas excavated too deeply are carefully refilled. Before using a geotextile, the slope must be carefully inspected for any projections that could puncture the material. Care must be taken when

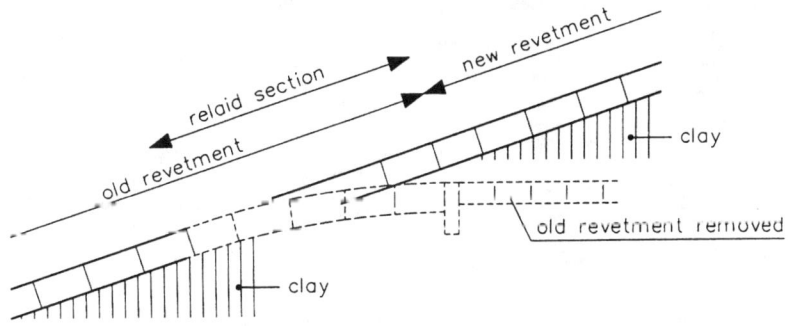

FIGURE 3.60. Transition between old and new revetments when reinforcing a dike.

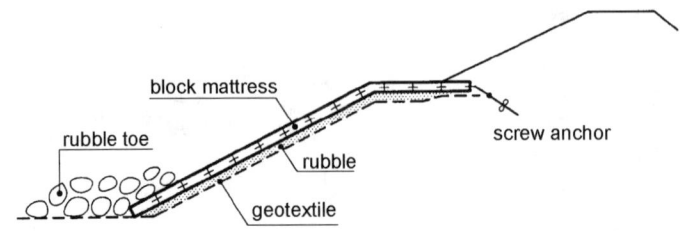

FIGURE 3.61. Block mattress toe construction and anchorage.

laying a mattress on a geotextile to ensure that extra pressures are not applied and that the geotextile is not pushed out of place. Geotextile sheets must be stitched together with an overlap of at least 0.5 to 1.0 m to prevent subsoil being flushed out. This is particularly important if the mattress is laid directly on sand or clay.

Block mattresses are laid using a crane and a balancing beam. The mattress must be in the correct position before it is uncoupled because it is difficult to pick up again and also time-consuming. Provided that part of the mattress can be laid above the water line, it can generally be laid very precisely and joints between adjacent mattreses can be limited to 1 to 2 cm. Laying a mattress completely under water is much more difficult. The spacing between the blocks of adjacent mattresses nonetheless should never be more than 3 cm. Once in place, mattresses should be joined so that the edges and corners cannot bang together under the action of waves. Loose corners are particularly vulnerable. In addition the top and bottom edges of the revetment should be anchored, as shown in Figures 3.61 and 3.62. A toe structure is not needed to stop mattresses from sliding.

More information on execution aspects of dikes and revetments can be found in TAW (1991), CUR (1992, 1993) and CUR/RWS (1995a,b).

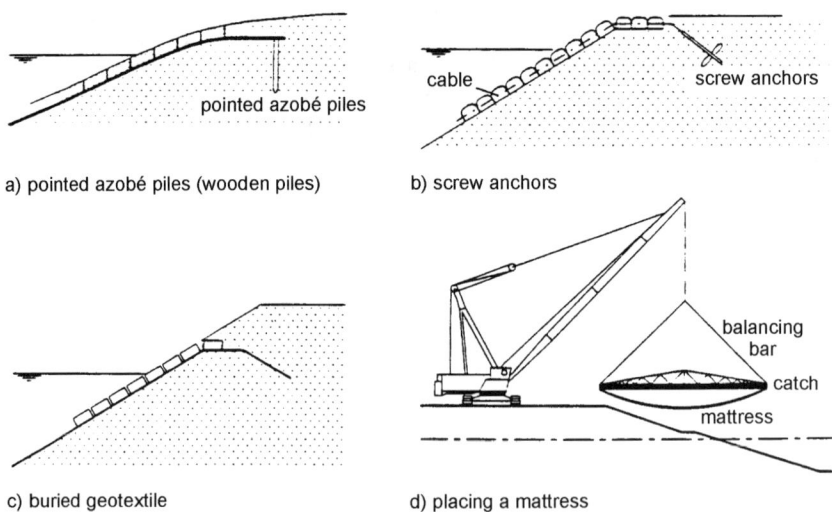

FIGURE 3.62. Placing a block mat (mattress) and some methods of anchoring.

9. MAINTENANCE AND MONITORING OF SAFETY

9.1. Goal

Sea protection measures have to be designed, constructed, and maintained in such a way that for a stated period of time (reference period) they will perform all agreed functions with satisfactory reliability and without excessive maintenance. The reference period may vary between 1 and 50 or 100 years and has to be regarded as a help parameter for design and assay decisions. The true lifetime of the structure may be larger than the reference period (Quelerij, et al., 1990, 1991; van Hijum, 1998).

During the lifetime of the structure, maintenance is a continuous task. The responsible maintenance engineer must devise a maintenance program in which the necessary use and employment of personnel, materials, and equipment (including inspection techniques) are scheduled and accounted for.

The management of coastal defenses is one of the few areas of civil engineering in which there are no standardized solutions and methods. Approaches still rely strongly on empirical methods, experience, and local habits and customs. This was not a problem in a time with less interest in the coastal environment and steady state budgets. However, today various disciplines have become aware of the variety of (potential) functions of the coastal environment. Important functions may include environmental and ecological purposes, economic activities (human settlement, land use, sea-borne transport, and mining), and recreational potentials.

On the other hand the struggle for budgets has become more intensive: more functions, constraints, standards, and criteria and less money to parcel out, after weighing of the various interests. So the managing coastal engineer has to look for a rational approach to this problem and has to design the project in such a way that it will be equally successful in the competition for resources with the demands of other types of public service.

9.2. Principles of Management Scheme

9.2.1. Objectives and Elements. The management scheme has to satisfy the regional council (or other responsible public authorities) that it is deriving maximum value for money spent (social output), and that the project is technically sound. It appears to be almost impossible to identify and evaluate the social output of coastal defense works in an explicit way. Therefore, it is proposed to do this in an implicit way in a REC assessment model. REC stands for "risk, economics, and conservation."

Together with a database (or register) and a management policy plan, this model provides the basic information for the maintenance plan, the annual budget estimates, the annual account, and justification reports. All these elements together form the management scheme.

9.2.2. Database Register. This register contains all physical qualities of the administered objects and related aspects inside the influence zone:

- the boundaries of the influence zone. This zone includes those areas of adjoining seabed and land where processes are linked in some way to the behavior of the defense works.
- a description of the as-built situation and the actual situation, including longitudinal and cross-sections, geotechnical profiles, etc.

- a list of issued licenses
- an ownership and farming-out register
- a map with all cables and pipes owned by public utilities, oil companies, etc.
- a damage record
- a record of executed maintenance
- boundary conditions (hydraulic, geotechnical, traffic, etc.)

9.2.3. Management Policy Plan. Control of the coastline and coastal activities may lay with different authorities and regulatory bodies. In this case it may be difficult to fix responsibility, control, and accountability throughout the considered coastal area, or to decide among proposed uses when questions of permitting, engineering, maintenance, and financial responsibilities arise (see Barret et al., 1989 and Fischer 1990).

A management policy plan is needed to link the objectives of the political arena and users of the coastal zone with those of the defense works management. The plan should contain the following elements:

- a statement on the scope of the plan with the links to all relevant laws and regulations
- the procedure to be followed at the adoption and amending of the plan by the responsible administration
- a statement on the management policy outlines. Policy options are: maintain existing coastline, setback defense line, retreat, or advance. Management and engineering options are: create, modify, reinstate, or do nothing. In some situations it may indeed be desirable to do nothing; future decades may see an increasing use of this "zero option."
- a review of all relevant functions and uses, dealing with environmental aspects, flood defense, industry, recreation, and land use (transport, agriculuture etc). The political weighing of the different competing functions have to be described in the management policy plan. The weighing reveals criteria and boundary conditions for the management (input and assay), functional criteria, hydraulic and geotechnical boundary conditions, defense standards, and acceptable risk levels. Note that original functions may have changed and still may change. In addition some af the boundary conditions may change. For instance, a changing climate will influence the threat to coastal areas, both directly in the hydrodynamic sphere and indirectly via changes in sediment transport and biotic coastal dynamics.
- The management and maintenance strategy and tactics to be followed for the short and long term, including scenarios in case of calamities have to be included

9.2.4. REC-Assessment Model. Management and maintenance will be difficult to justify in narrow benefit–cost terms. Therefore it is proposed to make an evaluation in an implicit way. Figure 3.63 shows the flow plan for a model on risk, economic, and conservation assessment, that is suited to visualize the change in social output value when certain measures are taken, for example, the increase of potential flood risk when the maintenance budget is cut down (van Hijum, 1998).

Essentially, the method contains the following steps:

- definition of objectives and specifications of the structures
- definition of criteria for performance, structural integrity, boundary conditions, accepted risk level, finances, and conservation
- dssessment of input data: structural (resistance) parameters and environmental (surcharge) parameters

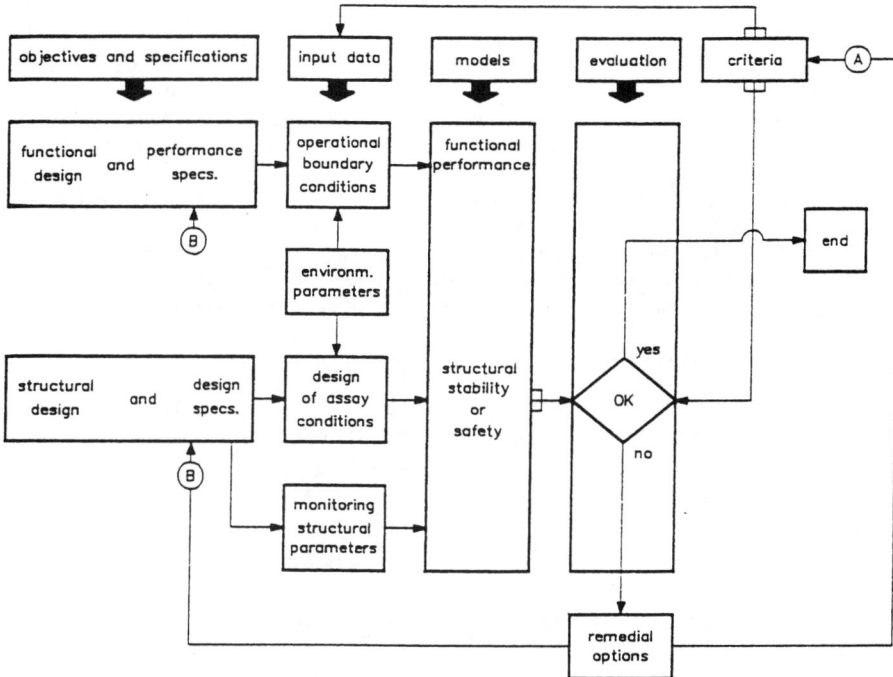

FIGURE 3.63. Flow scheme for a REC assessment model.

- models to describe the behavior of the structures
- model to evaluate the results and to define remedial options

The objectives are based upon the functional and structural requirements and the corresponding performance and design specifications. All this information can be found in the database and the management policy plan. The structural design and/or assay specifications may be expressed in terms of a probability of failure or as a damage level. The model used for the evaluation of the results can be a multicriteria analysis, supplemented by a more sophisticated model to optimize the effect of changes in structural or environmental parameters.

Remedial options, in case the evaluation reveals no satisfactory solution, are (see Figure 3.63):

- adjust criteria, which usually means: accept a higher risk level, accept a lower leisure quality, accept a lower level of naturalness, etc., or increase management budgets
- modify the functional and/or structural design

Decisions on these options have to be made by the responsible authorities; resulting new policy has to be incorporated into the management policy plan.

9.2.5. Annual Budget Estimate. The total annual costs not only consist of the expected maintenance costs (Ei) following from the maintenance plan. The annual costs should also incorporate the capital costs (Ec) that are the annual payments to cover interest and writing-off. In addition, the total annual cost should include an allowance for potential costs consequent upon failure or unserviceability (Ep). These costs may be treated as an equivalent annual insurance premium payable throughout the reference period (p) and just sufficient to compensate for the loss, valued at present day costs and equal to $Ef \cdot Pf/p$, where Ef is the estimated consequential cost and Pf is the total risk of failure in the reference period.

It is possible to determine for each structure an optimum in the relation between the annual cost and the annual probability of failure. The short-term character of the budget estimate does not match the long-term character of maintenance works. Therefore, the annual estimate should always be completed with a detailed review of expected adaptions for the coming five years and a less detailed estimate for even longer periods.

9.2.6. Maintenance Plan. The maintenance plan will generally comprise:

- type of maintenance strategy:

    corrective: maintenance that is performed after malfunctioning or failure of parts of the structure

    preventive: maintenance that is carried out before malfunctioning or failure of the structure

    use-based: preventive maintenance that is carried out after a prescribed period of time or intensity of use

    condition-based: preventive maintenance that is performed when a condition parameter reaches a prescibed limit

- monitoring and inspection program, aiming at a periodical assessment of the actual condition of the structure

- functional requirements and standards that prescribe the norm values (criteria or limits) that the actual conditions have to meet

- prediction models to estimate the alteration of the condition of the structure in the period until the next inspection

- planning of repair measures consisting of reconditioning or restoration of the structure by means of remedial measures

- cost and time schedule of personnel, materials, and equipment

Based on additional information during service life of the structure, it may be necessary to change the originally (during the design stage) developed maintenance plan. It is emphasized that the asssessment of such an altered plan is restricted by a set of actual boundary conditions that give the managing coastal engineer has much less flexibility than during the design stage of the structure.

9.3. Management Scheme for Flood Control in the Netherlands

In the Netherlands, a great deal of the management scheme has been regulated by law. A reassessment of the safety of the low-lying regions behind the sea and river dikes, after the disaster of 1953, has induced an extensive reinforcement of the sea and river defense system (Delta Works). The sea defense works were completed in 1990, the river works will be completed within the next ten to fifteen years. The central government wants to consolidate the safety level as achieved at the time of completion of the Delta Works.

For that purpose, the low-lying regions are divided into so-called dike-rings, each with an acceptable level of risk. These risk levels are laid down in a law and have to be guarded by the managing local administrations (in most cases the Water Boards). This Law on Flood Control (Government of the Netherlands, 1989) also gives rules for:

- the supervision by the regional authority (province)
- the procedure to be followed by the local administration and the province for the justification reports
- the contents of a data bank register, to be set up by the local administration
- the extreme hydraulic boundary conditions, to be provided by the central government
- the responsibilities regarding maintaining of the coastline
- the provision of technical guidelines
- the training of personnel, volunteers (dike "army"), and material for operation under extreme near-failure conditions
- the grant-in-aid for management and maintenance and for new defense works
- the installation of regional boards of flood control

9.4. Maintenance and Monitoring System for Dikes

9.4.1. General Conceptions. According to the REC model, the maintenance plan forms a major instrument for the management of the coastal structure. The elements of the maintenance plan with emphasis on the monitoring system will be described briefly. Although the setup of the plan will be worked out for a dike system and shall be focussed on the main function of flood protection, the set-up is also applicable to other coastal structures and other functions.

A dike system forms part of a total water-retaining system, consisting of different types of structures. If one part of the system fails, flooding of the inland area, the so-called dike-ring, may occur. In order to guarantee an acceptable safety level under extreme loading conditions, the contribution of the total length of the dike sections to the overall probability of failure of the potential flooding area may not exceed a specified value. Each of the dike sections is composed of a number of structure components, termed elements. These elements can be related to the detailed water-retaining functions as derived from the main function. Examples are:

dike element	function
slope revetment	hydraulic erosion control
dike crest	control overflow and overtopping
subsoil layer	ensure geotechnical stability

The elements can be described by a number of characteristic condition parameters, mainly geometrical and material-strength parameters. For example, the dike crest can be described just by the geometrical parameters of height and width of the crest, whereas the description of a slope revetment necessitates specifications of the type, geometry, weight, and strength of the successive layers of the revetment. During the service life of the dike, a number of deterioration mechanisms may occur that will affect the strength of the structural elements. In other words, these deterioration processes may change the values of the condition parameters. Examples of these processes are gradual settlement of the dike crest due to consolidation of the underlaying soft soil layers, washing out of finer particles of a stone revetment during daily tidal conditions, weathering of asphaltic revetments due to ice, wind, and rain effects, and damage effects to slopes due to biological activities such as vegetation and animal burrowing. These alterations caused by deterioration mechanisms, which will be recognized as characteristic damage patterns, can be of such significance that the dike safety will no longer be sufficient. This means that the actual quality of the condition parameters may decrease to a level at which the extreme loading conditions cannot be sustained.

9.4.2. Management Strategy. By means of an adequate maintenance and control system of the dike the managing authority aims to secure that the actual quality of the relevant condition parameters do not decrease to a level lower than the acceptable failure limit. For this purpose the so-called preventive condition-based maintenance strategy has been adopted. This strategy distinguishes three limits of quality levels of the condition parameters (see Figure 3.64). These limits are:

- warning limit: quality level at which a more intensive control of the condition parameter is needed (higher inspection frequency)
- action limit: quality level at which repair measures should be prepared and carried out before the failure limit has been reached
- failure limit: quality level that is just acceptable for the safety requirement. If the condition decreases below this level, the dike system will not provide sufficient safety

The margin between the action limit and the failure limit will depend on the inspection frequency and the mobilization time for the construction of the repair measures. An optimum maintenance and control strategy will be obtained by considering the minimum costs of repair and inspection, on the condition that the probability of exceeding the failure limit is sufficiently low.

Although most of the above-mentioned aspects are familiar in structural engineering (steel and concrete structures) the applications of a systematic condition-based maintenance approach for hydraulic and geotechnical engineering are very limited. This may be caused particularly by the considerable length of the dike sections, the heterogeneous

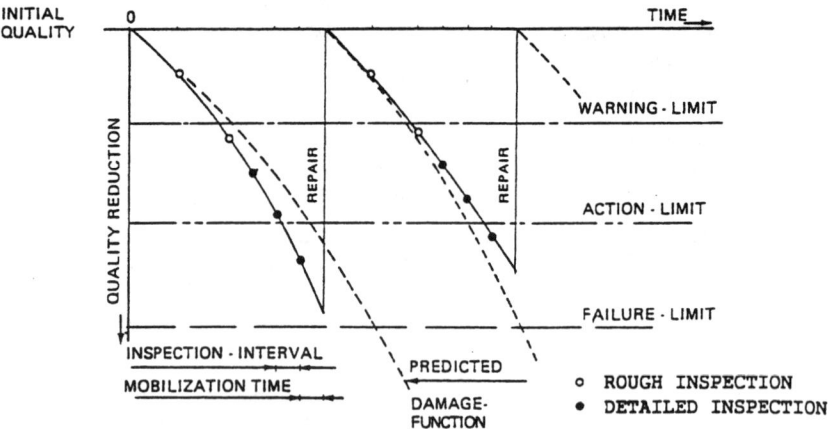

FIGURE 3.64. Preventive condition-based maintenance.

composition of the dike elements, including the variability of the soil layers, and the variability of the failure mechanisms and associated damage patterns.

To obtain a practical approach for the periodic safety control of the Dutch dikes along the coastline, the main rivers, and the lakes, a research program, commissioned by Rijkswaterstaat, has been carried out by the Dutch Civil Centre for Research and Codes (CUR, 1990). For this purpose, a systematic control method has been developed whereby the observable damage patterns, related to both the deterioration and ultimate failure mechanisms, play a major role. Much attention has been paid to a stepped approach of data collecting. In an early stage, only rough information is needed for the total dike length, whereas later on detailed data have to be collected for only a small part of the total dike length. In this way the total amount of data collecting and in-situ monitoring can be reduced significantly.

9.4.3. Setup of Safety Control System. The introduction of the safety control system for dikes can be divided into three stages:

1. The first setup of the control system for the total length of dikes to be maintained. This includes the collection of the basic information on the dike, the assessment of relevant damage patterns, and the inspection strategy. (Activity to be performed once)
2. The periodic inspection and monitoring of the dike and comparison of the actual behavior with the predicted behavior. (Activity to be performed continuously, several times a year)
3. Update of the control system to incorporate new developments with respect to functional design requirements and new knowledge of loading conditions, modeling of physical behavior of the dikes, or monitoring technics. (Activity to be performed every five or ten years, depending on the technological developments)

The first setup of the safety control method for dikes consists of five successive steps. The first two steps deal with raw information and apply to all concerned dike sections. As a result of step two, critical dike sections will be selected. The remaining steps, for which more detailed data are needed, are applicable to a limited number of representative critical dike sections.

The five successive steps are:

1. Rough description of dike elements and boundary conditions
2. Selection of decisive dike sections and dike clusters on the basis of relative strength with respect to ultimate failure mechanisms on one hand and deterioration mechanisms on the other
3. Determination of relevant damage patterns and damage failure limits
4. Assessment of inspection and monitoring strategy
5. Determination of scheme for repair measures.

The practical meaning of these steps has been demonstrated by means of the case study "Hondsbossche Sea Dike" by de Quelerij (1990). Step 4 will be discussed briefly in general terms below.

9.4.4. Inspection and Monitoring Strategy. The inspection strategy should provide information on what (which damage pattern characteristics), where (locations and depth), when (how often), how (which measuring method), and by whom (which organization) data should be collected and evaluated. Monitoring is assumed to form a part of the total inspection and will not be discussed separately.

It is not possible to give a detailed description of the inspection strategy because it depends very much on the actual situation. Alternatively, some general remarks will be given as a guideline for the setup of an inspection system. The principle of the inspection strategy should be based on a phased inspection with respect to the amount of detail. Starting from a rough inspection, a more detailed inspection may be required, depending on the results of the preceeding inspection. A distinction should be made between the periodic inspection of the damage patterns (deterioration mechanisms) according to the maintenance plan on one hand and the incidental inspection after the occurrence of special loading conditions (extreme mechanisms) on the other.

The last type of inspection should be performed after each severe loading condition (such as an extreme storm surge) and will generally start with a rough visual inspection of the total dike length.

The answer to the question "which data on which location have to be periodically inspected" will primarilary follow from the type of damage patterns to be controlled. Each damage pattern has its own observable characteristics. For example the quality of a grass-on-clay revetment can be characterized by the presence of uncovered spots (that is, the area where the grass is absent). The quality with respect to consolidation of the soft soil layers can be characterized by the settlement of the dike crest height. Some damage patterns have to be controlled only at a specific location that is representative of the total dike cluster, whereas other damage types should be controlled along the total dike length.

To decide when the data on the damage patterns should be collected—in other words, which inspection frequency is needed—the actual condition relative to the warning, action and, failure limits, and the velocity and predictability of the expected damage in the next period are important.

To illustrate the effects of these aspects, two extreme combinations of damage patterns and associated inspection strategies are outlined in Table 3.4.

The decision how the data should be collected—that is to say, which measure techniques should be applied—again depends strongly on the type of damage patterns to be controlled. In this context, the availability and terrain accessibility of equipment for geotechnical site investigation may play a major role. For each situation, a cost-effective technique has to be selected.

A summary of the most applicable measurement techniques related to the damage patterns clustered per extreme mechanism are listed below.

Geometry of dike

- conventional geodetical land surveys (including crest height)
- advanced techniques with high resolution (for local settlements)

TABLE 3.4. Inspection strategy by extreme combinations of damage patterns.

Aspects of inspection strategy	Aspects of damage increase	
	Progressive, great uncertainty	Gradual increase, small uncertainty
Inspection method	Detailed	Rough
Inspection frequency	High	Low
Action–failure limit	Great margin	Small margin

Slope stability, seepage, and piping

- inspection of seepage flow of inner slope during high water levels by visual observation or by infrared air photogrammetry
- installation of observation wells and piezometers for the measurements of pore-water pressures and the phreatic level in the dike during both normal and high water levels
- visual inspection of inland sand-boiling wells

Erosion of grass revetment

- visual inspection of uncovered area
- shallow borings and sampling of top layers

Erosion of stone or block revetment

- infiltration test to determine permeability of top layers
- in-situ tensile test to determine local up-lift resistance of the individual stone or block
- geophysical measurement to detect local cavities underneath the top layer (by ground radar or acoustic methods)
- visual observation of the gaps between the stones and blocks or the detection of complete loss of these elements

Erosion of asphaltic revetment

- assessment of layer thickness from sampling or geophysical measurements
- laboratory testing on samples to determine durability effects
- in-situ deflection measurements to determine the stiffness
- nuclear or electrical density measurements in bore holes to determine the quality of the foundation layer underneath the asphaltic top layer
- global visual inspection of superficial cracks

As shown in this list, visual observation plays a major role in the inspection. However special attention should be given to those damage patterns that are not easy detectable from the outside geometry of the dike. An example for this type of damage is the presence of erosion channels underneath the top layer of a revetment. These channels may substantially reduce the strength of the slope revetment under extreme conditions without showing any superficial damage under daily conditions. There is still a need to develop special measuring techniques to control this type of "hidden" damage. Geophysical methods may contribute to the solution of this type of detection problem.

10. CONCLUSIONS

1. The limitation in the length of this chapter does not allow to preparation of a fully detailed evaluation of the available Dutch experience on design of dikes and revetments. The background information can be found in the reports mentioned in the references.
2. Dikes and/or sea walls are only one option for coastal defense and must be considered in conjunction with, or as an alternative to, beach management and other options.

3. The main purpose of a sea wall is to fix the land–sea boundary; it is not intended to protect either the beach fronting it or adjoining, unprotected beaches.
4. There are no factual data to support claims concerning the adverse effects of sea walls, such as disturbance of natural coastal processes and/or accelration of beach erosion.
5. Sea walls can cause localized additional storm scour, both in front of and at the ends of the armoring; special measures should be taken to neutralize these effects.
6. Equations on wave run-up and overtopping can be of use for conceptual and/or final design of dikes and sloping structures. Influences of structure geometry and wave-boundary conditions have been included in a practical manner. Equations can be applied to deterministic design, including some design safety, but also to probabilistic design, giving the average trend and the reliability by employing a variation coefficient.
7. The shape and slope gradient of a dike/sea wall strongly affects the scouring process, run-up (height of a dike), and overtopping, and dimensions of slope protection units.
8. The proper design of toe protection (against scouring), slope protection (against wave forces), and protection against overtopping can be decisive to maintain the total stability of a dike/sea wall. Although definitive design methods are still lacking, the proposed design rules can be of use in particular cases.
9. The combination of the stability equation, which was derived from theory, together with the results of many large-scale model studies from all over the world has produced a reliable design tool for the preliminary design of placed-block revetments. Its reliability is only influenced by the fact that most tests have been performed with regular waves. An inaccuracy is introduced by the transformation of the regular wave load to an equivalent irregular wave load. Further studies are undertaken to improve the transformation method.
10. Considering the geotechnical stability of the subsoil of granular material underneath a revetment under wave attack, the following conclusions can be made:
    a) The calculation method developed in this paper can be used to investigate whether geotechnical instability is a dangerous failure mechanism.
    b) If according to this method geotechnical instability appears to be critical, it can be useful to apply more sophisticated (numerical) methods to analyze the stability.
    c) A revetment needs to have a certain weight to prevent geotechnical instability of the subsoil. A very permeable cover layer (with respect to the subsoil) can prevent lifting of parts of the cover layer, but then geotechnical instability becomes a critical failure mechanism. This phenomenon should be considered when in a design a granular filter (which always has a certain weight) is replaced by a "weightless" geotextile.
    d) In a revetment with a steep slope (1:3 or more) it is absolutely necessary to have an adequate toe structure or anchoring of the blocks (the latter in the case of a block mattress).
    e) Geotechnical stability is most critical in the case of a finely graded subsoil with a large air content in the pore water. Comparison of the calculated results with the results of large-scale model tests showed that the outcome of the method is on the safe side.
11. A good tuning of the permeabilities of the cover layer and sublayers (including geot-

extile) is an essential condition for balanced design. Geotextiles are only one of the components of structural design and must be considered in conjunction with, or as an alternative to granular filters or other options. The replacement of a granular filter by a geotextile results in a much smaller weight on the subsoil and consequently a smaller stability against geotechnical failure.

12. Further prototype verification of developed dimensioning criteria is still needed. Careful evaluation of prototype failure cases may provide useful information and data for verification purposes.

13. In all cases, experience and sound engineering judgement play an important role in applying these design rules; in some cases, mathematical or physical testing can provide an optimum solution.

REFERENCES AND SELECTED READINGS

Adel, H. den, 1989. *Re-analyzed Permeability Measurements with the Forchheimer Relation* (in Dutch), Delft Geotechnics report CO272553/56.
Agema, J., 1982. 30 Years of development of the design criteria for flood protection and water-control works, in *Proceedings of Delta Barrier Symposium,* Rotterdam.
Bakker, K. J. and Meijers, P., 1988. Stability against sliding of flexible revetments, in *International Symposium on Modelling Soil–Water–Structure Interactions,* Delft, A. Balkema.
Barrett, M. G. et al., 1989. Coastal Management, *Proceedings Conference on Coastal Management,* 9–11 May 1989, Bournemouth, U.K.
Battjes, J. A., 1974. Computation of set-up, longshore currents, run-up and overtopping due to wind-generated waves, in *Communications on Hydraulics,* Delft University of Technology, Report no. 74-2. The Netherlands.
Bezuijen A., M. Klein Breteler, and K. W. Pilarczyk, 1986. Large scale tests on a block-revetment placed on sand with a geotextile separation layer, in *Third International Conference on Geotextiles,* Vienna 1986.
Bezuijen, A., M. Klein Breteler, and A. M. Burger, 1990. Placed block revetments, in *Coastal Protection* K. W. Pilarczyk (ed.), Balkema, Rotterdam. ISBN 906191 1273.
Bezuijen, A., 1991. Geotechnical Failure of Revetments, in *Coastal Zone '91,* Los Angeles.
Burger, A. M., M. Klein Breteler, L. Banach, A. Bezuyen, and K. W. Pilarczyk, 1990. Analytical design formulas for relatively closed block revetments, *Journal of Waterway, Port, Coastal and Ocean Eng. ASCE,* vol. 116 no. 5 Sept/Oct. 1990
CIRIA, 1976. *A Guide to the Use of Grass in Hydraulic Engineering Practice,* Construction Industry Research and Information Association (CIRIA), Report 71, London.
CIRIA, 1986a. *Sea Walls; A Literature Review,* Construction Industry Research and Information Association (CIRIA) and Hydraulics Research, Wallingford, London.
CIRIA, 1986b. *Sea Walls; Survey of Performance and Design Practice,* CIRIA, Technical Note 125, London, England.
CIRIA, 1987a. *Design of Reinforced Grass Waterways,* Construction Industry Research and Information Association (CIRIA), Report 116, London.
CIRIA, 1987b. *Reinforcement of Steep Grassed Waterways,* CIRIA, technical note 120.
CUR, 1990. *Periodical Judgement on the Strength of Dikes* (in Dutch), part 1 "methodology" and parts 2 to 5 "case studies," CUR, Gouda, The Netherlands.
CUR, 1992. *Artificial Sand Fills in Water,* CUR, Report 152, Gouda, Netherlands.
CUR, 1993. *Filters in Hydraulic Engineering,* Report no. 161, Civil Engineering Research and Codes (CUR), Gouda, The Netherlands.
CUR/CIRIA, 1991. *Manual on Use of rock in Coastal Engineering,* Centre for Civil Engineering Research and Codes (CUR), Gouda, The Netherlands.
CUR/RWS, 1987. *Manual on Artificial Beach Nourishment,* Centre for Civil Engineering Research and Codes (CUR)/Rijkswaterstaat (RWS), Report 130, P.O.Box 420, 2800 AK Gouda, The Netherlands.

CUR/RWS, 1995a. *Design Manual for Pitched Slope Protection*, Report no. 155, Centre for Civil Engineering Research and Codes (CUR), P.O. Box 420, 2800 AK Gouda, The Netherlands.
CUR/RWS, 1995b. *Manual on Use of Rock in Hydraulic Engineering*, CUR, Gouda.
CUR/TAW, 1989. *Guide to the Assesment of the Safety of Dunes as a Sea Defence*, Centre for Civil Engineering Research and Codes/Technical Advisory Committee on Water Defenses (TAW), Report 140, Gouda, Netherlands.
CUR/TAW, 1989. *Guide to Concrete Dike Revetments*, CUR, Report 119, Gouda.
Dean, R., 1986. Coastal Armoring: effects, principles and mitigations, in 20th Coastal Engineering Conference, Taipei, Taiwan.
De Waal, J. P. and J. W. Van der Meer, 1992. Wave runup and overtopping at coastal structures, in ASCE, Proceedings 23rd ICCE, Venice, Italy.
Drogosz-Wawrzyniak, L., 1965, Calculation of height of wave rushing up the slopes and the range of slope revetments, *Archivium Hydrotechniki, 4,* Polish Academy of Science, Gdansk, Poland.
Fakuda, N., T. Uno, and J. Irie, 1974. Field observations of wave overtopping of wave absorbing revetment, *Coastal Engineering in Japan, 17;* 117–128.
Fischer, D. W., 1990. Governing the US Coastal Zone: an unresolved issue, *Ocean and Shoreline Management, 13,* (1990) 21–34, USA.
Franco, L., M. De Gerloni, and J. W. Van der Meer, 1994. Wave overtopping at vertical and composite breakwaters, in ASCE, *Proceedings. 24th ICCE,* Kobe, Japan.
Führböter, A., U. Sparboom, and H. H. Witte, 1989. Großer Wellenkanal Hannover: Versuchsergebnisse über den Wellenauflauf auf glatten und rauhen Deichböschungen met der Neigung 1:6, Die Küßte, Archive for Research and Technology on the North Sea and Baltic Coast.
Government of the Netherlands, 1989. Law on Flood Control (in Dutch), Draft 1989, Dutch Ministry of Transport, Public Works and Water Management, The Hague, The Netherlands.
Hales, L. Z. and J. R. Houston, 1983. Erosion control of scour during construction, Technical Report HL-80-3, Rep. 4, U.S. Army, Vicksburg, MS.
Hoekstra, A. and K. W. Pilarczyk, 1992. Coastal Engineering Design Codes in The Netherlands, Coastal Engineering Practice '92, ASCE Conference, Long Beach, CA.
Hjortnæs-Pedersen, A. G. I., A. Bezuijen, and H. Best, 1987. Non-stationary flow under revetments using the finite element method, in *Proceedings 9th European Conference Soil Mechanics and Foundation Engineering,* Dublin, Aug./Sept.
Hoffmans, G. I. C. M. and H. J. Verheij, 1997. *Scour Manual,* A. A. Balkema, Rotterdam.
Klein Breteler, M. and A. Bezuijen, 1988. The permeability of closely placed blocks on gravel, in *Proceedings of International Symposium on Modelling Soil–Water–Structure interactions,* SOWAS, Delft, 1988.
Klein Breteler, M. and A. Bezuijen, 1991. Simplified design method for block revetments, in *Proceedings of Coastal Structures and Breakwaters Conference,* London 1991, Thomas Telford, London.
Klein Breteler, M., G. Smith, and K. W. Pilarczyk, 1994. Performance of geotextiles on clay, silt and fine sand in bed and bank protection structures, in *5th International Conference on Geotextiles,* Singapore.
Klein Breteler, M. and A. Bezuijen, 1998. Design criteria for placed block revetments, in *Dikes and Revetments* (K. W. Pilarczyk, ed.), A. A. Balkema, Rotterdam.
Knauss, J., 1979. Computation of maximum discharge at overflow rock-fill dams, in *13th Congress des Grands Barrages,* New Delhi, Q50, R.9.
Kohler, H. J., and H. Schulz, 1986. Use of geotextiles in hydraulic constructions in the design of revetments, in *Proceedings 3rd International Conference on Geotectiles,* Vienna.
Kraus, N.C. and O.H. Pilkey (eds.), 1988. The effect of seawalls on the beach, *Journal of Coastal Research,* Special issue no. 4.
Lambe, T. W. and Whitman, R. V., 1979. *Soil Mechanics.* Wiley, New York.
Lindenberg, J., 1986. *Liquefaction of Sand Below a Block Revetment Slope 1:3 under wave attack* (in Dutch), Delft Geotechnics report CO-416751/16.
Meer, J. W. van der, 1990. Static and dynamic stability of loose materials, in *Coastal Protection* (K. W. Pilarczyk, ed.), Balkema, Rotterdam.
Meer, J. W. van der and C. J. M. Stam, 1991. *Wave Runup on Smooth and Rock Slopes,* Delft Hydraulics, Publication no. 454.
Meer, J. W. van der and J. P. de Waal, 1993. *Water Movement on Slopes* (in Dutch), Delft Hydraulics, Report *H* 1256 (commissioned by Rijkswaterstaat).

Meer, J. W. van der and J. P. F. M. Janssen, 1994. *Wave Run-up and Wave Overtopping at Dikes and Revetments,* Delft Hydraulics Publications, no. 485, (commissioned by Technical Advisory Committee on Water Defenses), Delft.
Meijer, D. G. and H. Verheij, 1994. *Grass Dikes—Analysis of Measurements from Large-Scale Tests,* Delft Hydraulics, draft report Q1584 (in Dutch).
Oesterdam, 1982. *Large-Scale Tests on Oesterdam* (in Dutch); report M1795/M1881, Vol. VI, Delft Hydraulics and Delft Geotechnics.
Owen, M. W., 1980. *Design of Seawalls Allowing for Wave Overtopping,* Report No. EX 924, Hydraulics Research, Wallingford, UK.
Petit, H. A. H., P. van den Bosch, and M. R. A. van Gent, 1994. *SKYLLA: Wave Motion in and on Coastal Structures; Implementation of Impermeable Slopes and Overtopping Boundary Conditions,* Delft Hydraulics, report H1780, October 1994
PIANC, 1986. *Consideration of Risk in Determining Bank Protection,* PIANC PTC I-WG 3, Brussels, Belgium.
PIANC, 1987a. *Guidelines for the Design and Construction of Flexible Revetments Incorporating Geotextiles for Inland Waterways,* PIANC, Supplement to Bulletin no. 57, Secretariat: Boulevard S. Bolivar 30, B-1210 Brussels, Belgium.
PIANC, 1987b, Risk consideration when determining bank protection requirements, in *Guidelines for the Design and Construction of Flexible Revetments Incorporating Geotextiles in Marine Environment,* PIANC, Supplement to Bulletin no. 78/79, Brussels, Belgium.
PIANC, 1992. *Guidelines for the Design and Construction of Flexible Revetments Incorporating Geotextiles in Marine Environment,* PIANC, Supplement to Bulletin 78/79, Brussels, Belgium.
Pilarczyk, K. W. et al, 1987. Application of some waste materials in hydraulic engineering, in *2nd European Conference on Environmental Technology,* Amsterdam, The Netherlands.
Pilarczyk, K. W. (ed.), 1990. *Coastal Protection,* Balkema, Rotterdam.
Pilarczyk, K. W. and M. Klein Breteler, 1994. Designing of revetments incorporating geotextiles, in *5th International Conference on Geotextiles,* Singapore.
Pilarczyk, K. W., 1995a. *Novel systems in coastal engineering: Geotextile systems and other methods,* HYDROpil, Zoetermeer, the Netherlands.
Pilarczyk, K. W., 1995b. Geotextile systems for coastal protection: An overview, in *4th Internernational Conference on Coastal and Port Engineering in Developing Countries (COPEDEC),* Rio de Janeiro, Brasil.
Pilarczyk, K. W., M. Klein Breteler, and A. Bezuijen, 1995. Wave forces and structural response of placed block revetments on inclined structures, in *Wave Forces on Inclined and Vertical Wall Structures* (Demirbilek and Kobayashi, eds.), ASCE, New York.
Pilarczyk, K. W., (ed.), 1998. *Dikes and Revetments,* A. A. Balkema, Rotterdam.
Quelerij, L. de, 1989, Geotechnical aspects, in *Short Course on Design of Coastal Structures,* AIT, Bangkok.
Quelerij, L. de and E. van Hijum, 1990. Maintenance and monitoring of water retaining structures, in *Coastal Protection* (K. W. Pilarczyk, ed.), pp. 369–401, A. A. Balkema, Rotterdam.
Quelerij, L. de, E. van Hijum, and K. W. Pilarczyk, 1991. Performance assessement and maintenance of coastal defenses, in *Proceedings of Coastal Structures and Breakwaters '91,* Thomas Telford, London.
RWS/TAW, 1985. *Guide on the Use of Asphalt in Hydraulic Engineering,* Rijkswaterstaat Communications, no. 37, The Hague.
RWS (Rijkswaterstaat), 1987. *The Closure of Tidal Basins; Closing of Estuaries, Tidal Inlets and Dike Breaches.* Delft University Press, Delft, The Netherlands.
RWS (Rijkswaterstaat), 1990. *A New Coastal Defense Policy for The Netherlands,* Rijkswaterstaat, Tidal Water Division, The Hague.
RWS, 1994, *Design Plan Oosterschelde Storm-Surge Barrier: Overall Design and Design Philosophy,* Ministry of Transport, Public Works and Water Management, Rijkswaterstaat (RWS) (A. A. Balkema, Publisher), the Netherlands.
Seijffert, J. W. and L. Philipse, 1990. Resistance of Grassmat to Wave Attack, in *Proceedings of International Conference on Coastal Engineering,* Delft.
Smith, G. M., J. W. W. Seijffert, and J. W. van der Meer, 1994. Erosion and Overtopping of a Grass Dike: Large Scale Model Tests," in *Proceedings 24th International Conference on Coastal Engineering,* Kobe, Japan.
SPM, 1984. *Shore Protection Manual,* U.S. Army Corps of Engineers, Vicksburg, MS.

TAW, 1974. *Wave Run-up and Overtopping,* Technical Advisory Committee on Water Defenses in The Netherlands, Government Publishing Office, The Hague, The Netherlands.
TAW, 1985. *The Use of Asphalt in Hydraulic Engineering,* Government Publishing Office, The Hague.
TAW, 1990. *Probabilistic Design of Flood Defenses,* Technical Advisory Committee on Water Defenses (TAW), Published by the Centre for Civil Engineering Research and Codes (CUR), Report 141, Gouda, Netherlands.
TAW, 1991. *Guidelines on Design of River Dikes,* Technical Advisory Committee on Water Defenses (TAW), Published by the Centre for Civil Engineering Research and Codes (CUR), Gouda, The Netherlands.
TAW, 1995. *Guidelines on Clay Specifications* (in Dutch), Technical Advisory Committee on Water Defenses (TAW)/Rijkswaterstaat, Delft, the Netherlands.
Tekmarine, 1983. *Large-scale Model Investigation of Compound Slope Profiles,* Tekmarine Inc., Project TCN-024; Sierra Madre, CA.
Tekmarine, 1985. *Two-Dimensional Model Study of Slope Protection Systems for the Sohio Endicott Project,* Tekmarine Inc., January 1985; Sierra Madre, CA.
Van Gent, M. R. A., P. Tonjes, H. A. H. Petit, and P. v.d. Bosch, 1994. Wave action in and on permeable structures, in *Proceedings of the Coastal Engineering Conference,* ASCE, Kobe, Japan.
van Hijum, E, 1998. Aspects of execution and management, in *Dikes and Revetments* (K. W. Pilarczyk, ed.), A. A. Balkema, Rotterdam
Van Santvoort, G., 1994. *Geotextiles and Geomembranes in Civil Engineering,* revised edition, A. A. Balkema, Rotterdam.
Verruijt, A., 1969. Elastic storage of aquifers, in *Flow Through Porous Media* (R. J. M. de Wiest, ed.), Academic Press, the Netherlands.
Wemelsfelder, P. J. 1939. Wetmatigheden in het optreden van stormvloeden (Physical patterns in the development of storm surges), in Dutch, *De Ingenieur, No. 9.*
Wemelsfelder, P. J. 1960. The frequency curves of high water in the tidal area of the Netherlands, (in Dutch), in *Report of the Delta Committee,* The Hague, the Netherlands.

CHAPTER 4
WAVE FORCES ON VERTICAL AND COMPOSITE WALLS

N. W. H. ALLSOP
University of Sheffield
c/o HR Wallingford
Wallingford, UK

Notation	4.1
Overview	4.4
4.1. Introduction	4.4
4.1.1. Vertical and Composite Wall Structures	4.5
4.1.2. Wave Forces on Walls	4.5
4.1.3. Recent Research on Wave Forces on Vertical Walls	4.8
4.2. Types of Wave Loading	4.10
4.2.1. Pulsating Wave Loads	4.10
4.2.2. Wave Impact Loads	4.10
4.2.3. Broken Wave Loads	4.10
4.2.4. Identification of Type of Wave Load	4.11
4.3. Pulsating Wave Loads	4.15
4.3.1. Horizontal Forces/Pressures (Normal Attack)	4.15
4.3.2. Wave Forces under Oblique and Short-Crested Wave Attack	4.18
4.4. Wave Impact Loads	4.22
4.4.1. Horizontal Forces/Pressures (Normal Attack)	4.22
4.4.2. Wave Impact Forces under Oblique and Short-Crested Wave Attack	4.29
4.4.3. Scale Effect Corrections and Practical Application	4.32
4.5. Broken Wave Forces	4.35
4.6. Seaward or Negative Forces	4.35
4.7. Recommendations for Analysis and Design	4.39
4.7.1 Summary of Design Procedures	4.39
4.7.2. Uncertainties, Scale Effects, Safety Factors, and Models	4.41
Acknowledgments	4.42
References and Readings	4.43

NOTATION

A, a	Empirical coefficient
A_c	Armor crest level relative to water level
$A(n)$	Normalization factor used to describe directional spreading

a_e	Air content used by Partenscky in estimation of wave impact pressures
B, b	Empirical coefficients
B_b	Crest width of rubble mound berm
B_c	Width of caisson
B_{cw}	Width of crown wall
B_{eq}	Equivalent width of rubble mound in front of wall, averaged over height of mound
B_{wl}	Structure width at static water level
B_t	Width of rubble mound at toe level
B_{rel}	Parameter giving effect of toe berm on wave breaking
B^*	Relative berm width, $= B_{eq}/L_p$
C_{Fh}	Wave impact force reduction factor
C_r	Coefficient of wave reflection
C_t	Coefficient of wave transmission
$C_{\beta 2}$	Battjes load reduction factor for wave obliquity
d	Water depth over toe mound in front of wall
d_{eff}	Effective water depth
D	Directional distribution function used by Battjes
E_i	Incident wave energy
E_r	Reflected wave energy
E_t	Transmitted wave energy
F_B	Buoyant up-thrust on a caisson or related element
F_S	Factor of safety
F_h	Horizontal force on caisson or crown wall element, often taken as pulsating
$F_{h,A\&V}$	Wave impact force evaluated by Allsop & Vincinanza's (1996) method
$F_{h,\mathrm{impact}}$	Wave impact force, horizontal
$F_{h,1/250}$	Mean of highest 1/250 horizontal wave forces
F_u	Uplift force on caisson or crown wall element
f	Wave frequency
f_m	Frequency of peak of wave energy, $= 1/T_p$
g	Gravitational acceleration
H	Wave height, crest to trough
H_b	Breaking wave height, usually taken as maximum breaking height $H_{\mathrm{max}b}$
H_{bc}	Critical wave height for transition between pulsating and impact by Calabrese's method
H_d	Design wave height, may be set to H_s or H_{max} depending on the design method
H_{max}	Maximum individual wave height in design case
H_{m0}	Significant wave height from spectral analysis, defined $4.0\sqrt{m_0}$
H_{so}	Representative significant wave height at an offshore or nearshore position where it is substantially unaffected by shallow water processes
H_s	Significant wave height, average of highest one-third of wave heights
H_{si}	Incident significant wave height at toe of structure, taking account of refraction, shoaling and depth-limited breaking
H_s^*	Relative wave height to toe berm depth, $= H_{si}/d$
H_{sb}	Breaking significant wave height
h	Water depth
h^*	Relative approach depth to wave height, $= h_s/H_{so}$
h_b	Height of berm above seabed
h_{br}	Water depth at point of breaking
h_c	Height of rubble mound/core beneath caisson/wall

h_f	Exposed height of caisson or crown wall over which wave pressures act
h_s	Water depth at toe of structure
k	Wave number $= 2\pi/L$
k_b	Empirical factor for influence of relative berm length, B_{eq}/d, on breaking wave height.
I	Pressure impulse, taken here as $p \cdot \Delta t$
I_{impact}	Pressure impulse under impact event, taken here as $p_{impact} \cdot \Delta t$
K_L	Wave impact coefficient used by Partenscky $= 5.4[(1/a_e) - 1]$
L	Wavelength, in the direction of propagation
L_c	Length of individual caisson
L_{mo}	Offshore wavelength of mean (T_m) period, usually calculated using deep water assumption
L_o	Deep water or offshore wavelength $= gT^2/2\pi$
L_{pi}	Inshore wavelength of peak (T_p) period
L_{po}	Offshore wavelength of peak (T_p) period, usually calculated using deep water assumption
L_{ps}	Wavelength of peak period in water depth at structure toe
m	Bed slope, 1:m
m_{rel}	Effective bed slope
m_0	Zeroth moment of the wave energy density spectrum
m_2	Second moment of the wave energy density spectrum
N_{wo}	Number of waves overtopping, as proportion or % of total incident
N_z	Number of zero-crossing waves in a record $= T_R/T_m$
$P_b, P_{b\%}$	Proportion (or percentage) of breaking waves
$P_i, P_{i\%}$	Proportion (or percentage) of impact events
p	Wave pressure
p_{av}	Average wave pressure, usually averaged over vertical wall height h_f
p_{dyn}	Dynamic or impact pressure, used by Partenscky
p_{impact}, p_{imp}	Wave impact pressure
p_1, p_2, p_3, p_u	Wave pressures acting at points on wall calculated by Goda's method
q	Mean overtopping discharge, per unit length of structure
Q^*	Owen's dimensionless overtopping parameter
R_c	Crest freeboard, height of crest above static water level
r	Roughness or run-up reduction coefficient, usually relative to smooth slopes
s_m or s_{mo}	Steepness of mean wave periods, $s_{mo} = 2\pi H/gT_{mo}^2$
s_p or s_{op}	Steepness of peak wave periods, $s_{op} = 2\pi H/gT_p^2$
T	Wave period
T_m, T_{mo}	Mean wave period, mean period offshore
T_p	Wave period of spectral peak, inverse of peak frequency
T_R	Length of wave record, duration of sea state
t_d	Duration of wave pressure/load
t_r	Rise time, usually of wave impact
u, v, w	Components of velocity along x, y, z axes
v_c	Wave velocity, used by Blackmore and Hewson
x, y, z	Orthogonal axes, distance along each axis
z	Level in water, usually above seabed
α (alpha)	Structure front slope angle to horizontal
α_1, α_2	Coefficients in Goda's method to predict wave forces on caissons
β (beta)	Angle of wave attack to breakwater alignment
δ_0	Additional run-up height on vertical wall, used in Sainflou's method
η_0, η^*	Extreme (notional) run-up level, used in Goda's and related methods

4.4   HANDBOOK OF COASTAL ENGINEERING

ρ (rho)	Mass density, often of (fresh) water
ρ_w	Mass density of sea water
$\rho_r, \rho_c, \rho_\alpha$	Mass density of rock, concrete, armor units
Δ (delta)	Reduced relative density, e.g. $(\rho_r/\rho_w) - 1$
Δt	Time increment, often used as rise time
λ (lambda)	Model/prototype scale ratio (Froude)
λ	Aeration factor used by Blackmore and Hewson
ξ (xi)	Iribarren number or surf similarity parameter, = $\tan \alpha / s^{1/2}$
ξ_m, ξ_p	Iribarren number calculated in terms of s_m or s_p
ξ_{br}	Breaking parameter used by Calabrese
θ (theta)	Wave direction relative to principal wave direction θ_0
ω (omega)	Wave frequency in radians, used in Battjes method for short-crested seas

OVERVIEW

This chapter identifies prediction methods to calculate wave forces acting on vertical and composite breakwaters and sea walls. It identifies the main types of vertical and composite walls, categorizes wave loads into pulsating, impact, and broken, and describes methods to predict wave forces for each type of loading.

Most wave forces on walls are pulsating, where horizontal wave momentum is imparted to the wall over a duration of a quarter to half of a wave period. Methods to calculate pulsating wave forces for simple structures and wave conditions are relatively well established. New methods are described here to extend or modify well-established prediction methods for a wider range of conditions. These particularly include an improved and validated method to predict wave forces acting outwards or seawards.

Some wave forces are substantially more intense than pulsating wave loads, especially those forces caused by waves breaking directly onto the wall. These wave impact forces occur significantly less frequently, but are often of very high intensity and short duration, giving local peak pressures 10 or more times greater than pulsating wave pressures caused by the same wave height. A new method to identify when wave impact loads may occur on vertical and composite walls is presented, and methods to estimate the magnitude of wave impact loads are described for normal and for oblique or short-crested waves.

Most design methods for vertical/steep walls have been developed for caisson breakwaters, so much of this chapter discusses this type of structure. Most of the design methods can also be applied to sea walls or related shoreline structures where the relative wave height/depth conditions are in the appropriate range.

Many structures in shallow water will be reached by broken waves only under larger storms. This chapter therefore also includes guidance on likely wave forces under broken wave conditions.

4.1. INTRODUCTION

Technical interest on breakwaters and other coastal structures in Europe and America over the last part of the 20th century focused mainly on rubble mound structures. There were significant improvements in technical guidance on the design of rubble mound breakwaters; see particularly the comprehensive review by Burcharth (1994), design methods in the manual edited by Simm (1991), and descriptions of design methods by van der Meer (1995), and Allsop (1995a). Further details of design and construction practice for breakwaters are given in Clifford (1996), and Allsop (1998).

The main concerns in the design of rubble mound breakwaters are associated with the stability limits of the main and subsidiary armor, the (geotechnical) stability of the mound and its foundations, and (to a lesser extent) stability against wave forces of any crown wall elements. Design methods for armoring, be it narrow-graded rock armor, wide-graded rip-rap, or specialized concrete armoring, therefore focus primarily on setting a limiting armor unit mass for (adequate) resistance against movement. There is therefore very little information available on wave forces exerted upon armoring, so this chapter does not deal with rubble mound structures, other than wave forces on crown walls.

In shallow water, embankments or mounds of clay, sand, or rockfill may be protected against wave attack by revetment systems; see especially Pilarczyk et al. (1995) and McConnell (1998). These types of structures are discussed in depth in Chapters 2, 3, 5, and 8 of this handbook, and are not covered further here. The direct wave forces acting on sloping sea walls seldom cause concern in design, and the effects of indirect loadings are covered in Chapters 2 and 3. A useful review of wave forces on embankment structures is given by Fuhrboter (1994), so forces on those types of structures are not discussed here.

4.1.1. Vertical and Composite Wall Structures

Many existing breakwaters and sea walls around the United Kingdom and Italy were formed from stone or concrete blocks with vertical, near-vertical, or steeply sloping faces; see particularly Bray and Tatham (1992) for port structures, Thomas and Hall (1992) for U.K. breakwaters and sea walls (also see Figures 4.1 and 4.2), and Franco (1994, 1996) for breakwaters and sea walls around the Mediterranean.

In Italy and Japan, many recent harbor breakwaters have been formed by concrete caissons with vertical faces on shallow rubble mounds; see especially Franco (1994), Oumeraci (1994), Tanimoto and Goda (1991) and Tanimoto and Takahashi (1994). For some breakwaters in considerable depth (an example in Japan is in up to 60 m depth), the mound may be a relatively substantial proportion of the overall height. Even so, most design interest for this type of structure focuses on the caisson.

Wave loads are resisted primarily by gravity and friction forces, and by bearing capacity of the rubble foundation. In Japan, design methods for caisson breakwaters have become well established; see Goda (1985, 1995) and Tanimoto and Takahashi (1994). Even so, the prediction of wave loads acting on wall elements is still relatively uncertain. Site experience of vertical wall breakwaters has been mixed, with some structures lasting more than 100 years with few signs of distress; but at other sites vertical or composite walls have been destroyed shortly after construction, or even during construction. Even in the last part of the 20th century, vertical breakwaters in the United Kingdom, Italy, and Japan have suffered damage caused directly by large wave forces, see Allsop and Bray (1994), Franco (1994), Hitachi (1994), Oumeraci (1994), and Tanimoto and Takahashi (1994).

4.1.2. Wave Forces on Walls

It has been appreciated for many years that apparently similar wave conditions may give rise to dramatically different wave pressures or forces depending on the form of wave breaking at or close to the wall. In a fully random storm sea, there will inevitably be a wide range of different types of wave breaking, but it has been found convenient to use

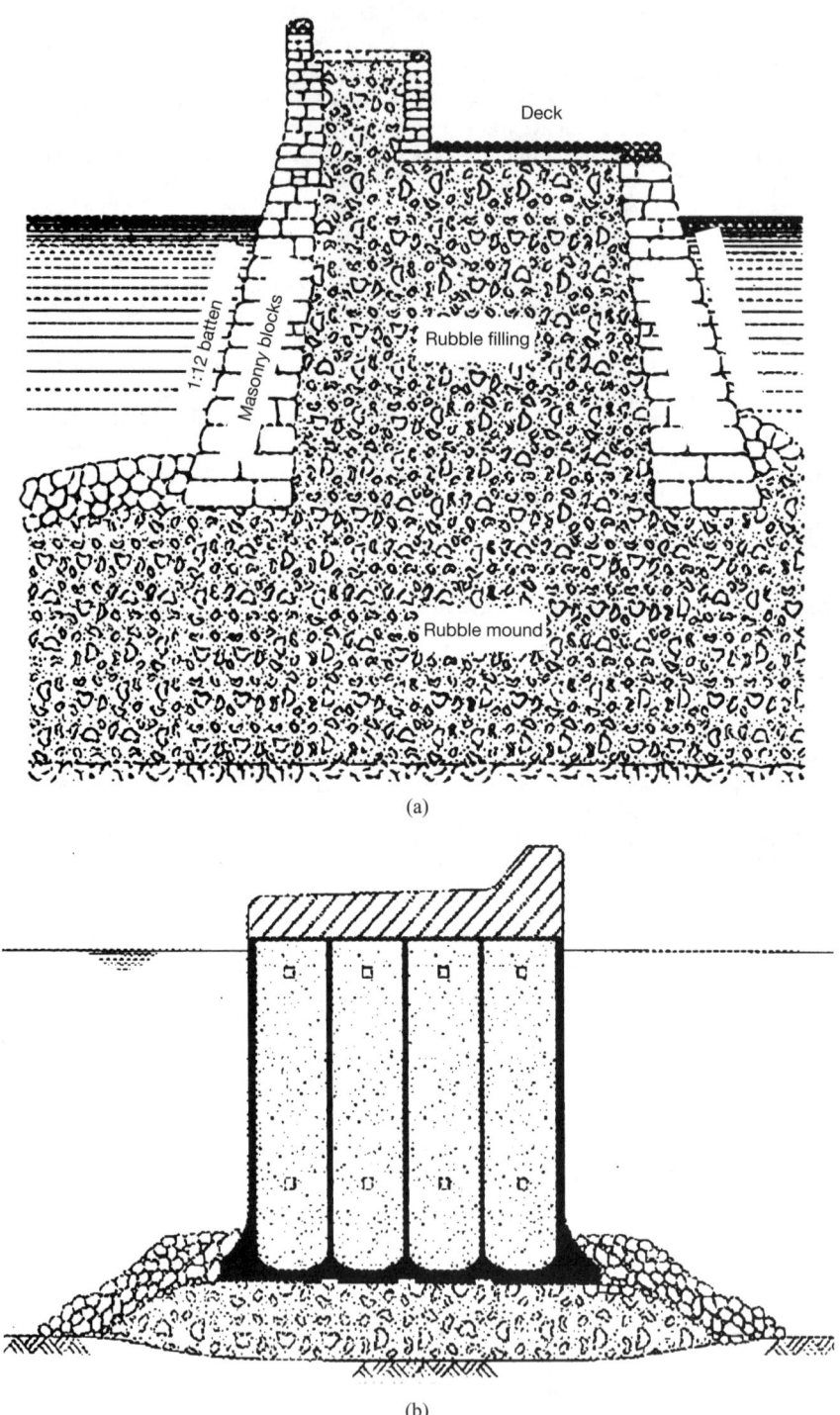

FIGURE 4.1. Blockwork (a) and caisson (b) breakwaters.

WAVE FORCES ON VERTICAL AND COMPOSITE WALLS 4.7

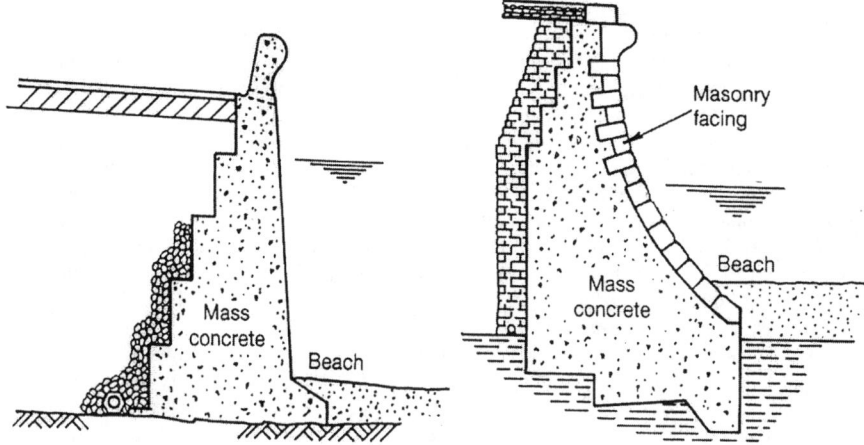

FIGURE 4.2. Example urban sea walls (United Kingdom).

three categories of wave load/breaking condition, the first two of which are illustrated by example measurements of wave pressures in Figure 4.3:

a) Nonbreaking or pulsating
b) Impulsive breaking or impact
c) Broken waves

The simplest case is generally when the wave is nonbreaking, also termed reflecting or pulsating. For this condition, the wave motion is relatively smooth, and some processes can be predicted by simple wave theories. Much greater wave forces arise if the wave can break directly against the wall, termed plunging, breaking, impulsive, or impact. These conditions are more difficult to predict, and exhibit significant variability/uncertainty. Rather lower forces arise if waves have already broken before reaching the wall. The wave motion is turbulent, but often highly aerated. More detailed definitions of each load

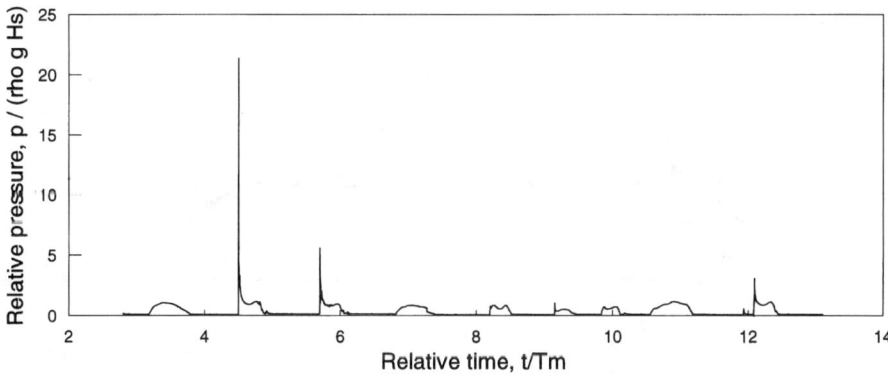

FIGURE 4.3. Example wave pressure traces, model tests.

type, and descriptions of methods to predict occurrence of each load type, are described below. The main geometry parameters are indicated in Figure 4.4.

Prediction of simple pulsating wave loads is relatively easy, but prediction of the occurrence and magnitude of more intense wave impact loads on such structures is substantially more complicated, and calculations of such forces/pressures are often uncertain.

Simple prediction methods for pulsating wave loads by Goda (1974), Ito (1971), and Hiroi (1919) generally predict average pressures up to about $p_{av} = 2\ \rho g H_s$, where H_s is the incident significant wave height. Research studies in Europe, however, have measured local wave impact pressures up to or greater than $p_{\text{impact}} = 40\ \rho g H_s$, much higher than would be predicted by simple design methods; see especially Allsop et al. (1996b, 1996c) and McKenna (1997). Indeed, research by Kirkgoz (1990, 1995) has suggested pressures up to $p_{\text{impact}} = 100\ \rho g H$.

Problems due to wave impacts may appear comparatively small. Vertical breakwaters are massive structures, are unlikely to respond to very short duration impacts, and the incidence of failure is seen as relatively low. Quays or sea walls are usually backed by fill, and do not fail in the direction of the principal wave loading. Fortunately, design wave conditions occur quite rarely, so failures are relatively rare. Even so, recent failures of vertical breakwaters in the United Kingdom, Japan, and Italy have demonstrated that present design methods are insufficient to predict extreme forces, and research in the United Kingdom and Germany has demonstrated that wave impacts may have considerable influence on design loadings. Use of dynamic design methods have been described at length by Oumeraci et al. (2000), and descriptions of simplified dynamic methods have been given by Voortman and Vrijling (1999) and Torrini et al. (1999).

4.1.3. Recent Research on Wave Forces on Vertical Walls

Within Europe, two major international research projects have addressed aspects of the design and performance of vertical and composite breakwaters, supported in turn by

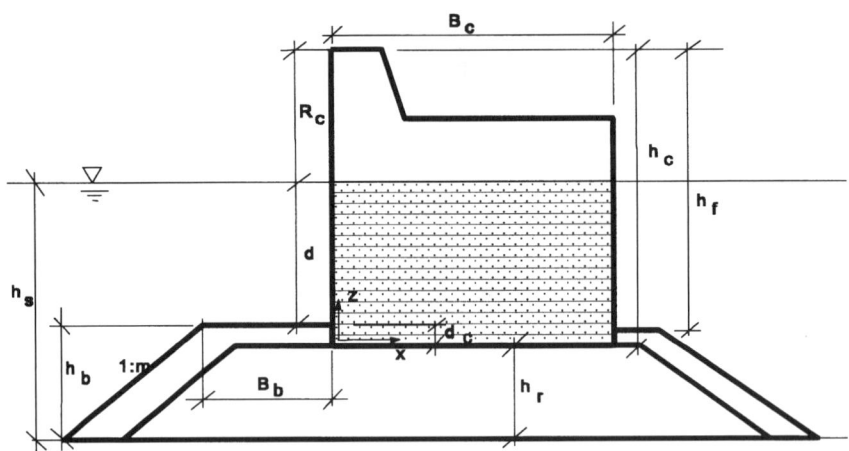

FIGURE 4.4. Definition of main structure parameters.

many more national or local research projects. The MCS-project led by Franzius Institute of the University of Hannover contributed substantially to assembling practical experience and new research information on these classes of coastal/harbor structures under four main headings:

Task 1: Wave impact forces and structure/foundation interaction
Task 2: Scaling problems and air entrainment
Task 3: Local morphological changes
Task 4: Wave overtopping, constructional measures and three-dimensional effects

Studies within the MCS-project covered a number of disparate elements. Considerable advances were made in identifying and explaining movements/failures of some vertical breakwaters; initial development of dynamic design load criteria and methods; and new design methods to dimension foundation mound protection and to predict the onset of scour. New data were derived on wave overtopping and reflections, and especially on the effect of design modifications intended to reduce reflections, overtopping, and/or wave forces.

A further European Union research project on Probabilistic Design Tools for Vertical Breakwaters (PROVERBS) as described by Oumeraci et al. (1999) and Oumeraci and Kortenhaus (1999) developed probabilistic design/analysis methods for vertical breakwaters and related structures. Again, the project was divided into four task areas, but with more focus on integration:

Task 1: Hydrodynamic aspects, including wave impact forces
Task 2: Foundation aspects, including dynamic modeling of foundation response
Task 3: Structural aspects, primarily analysis/design of concrete caissons
Task 4: Probabilistic design tools, and initial testing for example structures.

Within PROVERBS, Task 1 on hydrodynamics was intended to produce new guidance on types and magnitudes of wave loads, global and local, short or long. The primary objective of Task 1 was to supply essential data and prediction methods on hydrodynamic loadings on vertical and composite structures to the other areas of the PROVERBS project. The second objective was to develop new prediction tools to be used more widely by consulting engineers, contractors, and owners in analysis of safety and performance of such structures. Some of those new methods are described here.

Other results of the PROVERBS project not covered here, including results of field data measurements, and standardized response parameter coverage, new data on performance of perforated caissons, identification of effects of steep bed slopes, and new data on wave-induced pressures in rubble foundations and uplift forces beneath caissons are described by Oumeraci et al. (2000), Oumeraci and Kortenhaus (1999), Bélorgy et al. (1999), and Bullock et al. (1999).

A number of different prediction methods for wave forces on vertical/composite walls have been developed, but it has not yet been possible to demonstrate that one particular method is necessarily more complete or more reliable than another. It is incumbent upon any designer/analyst to apply all possible alternative methods, and use engineering judgement and experience to decide which gives the most realistic result for the particular application considered.

4.2. TYPES OF WAVE LOADING

4.2.1. Pulsating Wave Loads

Most wave forces on walls are pulsating; a substantial proportion of the horizontal momentum of the wave is imparted to the wall over a duration of approximately a quarter to half of a wave period. Wave motions at a structure are relatively smooth, with some processes reasonably well predicted by simple wave theories. Methods to calculate wave forces for simple structures and pulsating wave conditions are relatively well established, and are described by Goda (1974, 1985, 1995) and Tanimoto and Goda (1991). Design methods using those prediction methods are used in British Standard BS 6349 (BSI, 1984), and in the CIRIA/CUR manual edited by Simm (1991).

Techniques are described here to predict pulsating forces and to extend or modify well-established prediction methods for a wider range of pulsating wave conditions, particularly including improved and validated methods to predict pulsating wave forces acting outward or seaward.

4.2.2. Wave Impact Loads

Some wave forces are substantially more intense than pulsating wave loads, especially forces caused by waves breaking directly onto the wall or an element. On most structures, wave impact forces occur infrequently, and do not constitute a design condition. For some structures however, wave impact loads may occur sufficiently frequently and with significant intensity that they may need to be considered in the design process. In such instances, the wave may shoal up over a submerged mound or steep approach slope, then form a plunging breaker just seaward of the wall. A major part of the wave momentum is then conveyed to the wall over a very small fraction of the wave period.

Wave impact loads may therefore be very intense, giving local peak pressures 10 or more times greater than pulsating wave pressures, but of significantly shorter durations. Wave impact conditions are therefore of more significance for smaller/lighter elements than for massive caissons, and may be more important for structures on more flexible foundations than for those on very stiff foundations. A method to identify when wave impact loads may occur on vertical and composite walls is presented in Section 4.2.4 below, and methods to estimate the magnitude of wave impact loads are described for normal and for oblique or short-crested waves in Section 4.4.

4.2.3. Broken Wave Loads

For most coastal sea walls and for many breakwaters in shallow water, waves under larger storms may be significantly reduced by depth-limited breaking before they reach the sea wall or breakwater. These broken waves may reform, but are generally well aerated and much less likely to be transformed into the well-shaped breakers that lead to wave impacts. Wave forces under broken waves are therefore much lower than impact loads, and may indeed be lower than pulsating loads. This chapter therefore also includes simple guidance on likely wave forces under broken wave conditions.

4.2.4. Identification of Type of Wave Load

The main focus of research at the end of the 20th century was derivation of probabilistic design methods. These required that loadings and responses be identified over the full potential range of each of the main design parameters. It therefore became more essential that parameter regions where responses change rapidly be well identified.

Problems in identifying types of wave loadings are compounded by any uncertainties in defining conditions that lead to wave impacts. Schmidt et al. (1992) and later Oumeraci (1994) define seven different breaker classifications in terms of H_b/d.

Unfortunately, the breaker height H_b is often difficult to predict with certainty, so these classifications may be of restricted practical use. Early activity in the PROVERBS research described by Oumeraci et al. (1999, 2000) and by Allsop et al. (1999) was therefore to use results of previous research to identify regions in the overall range of structure geometry and wave/water level parameters that lead to wave impacts. An initial decision tree was developed by Allsop et al. (1996) based on rules by Goda (1985) to identify whether particular structures or sea states will cause a risk of impulsive wave conditions; see Figure 4.5. This method gives useful first (although pessimistic) predictions, but gives no explicit guidance on broken waves.

Measurements of wave forces or pressures on simple vertical or composite breakwaters in two-dimensional wave flumes were analyzed to identify different loading cases, primarily for bed slopes of 1:50 or shallower. First, the relative berm height, $h_b^* = h_b/h_s$ determines the type of structure—whether:

Simple vertical wall ($h_b^* < 0.3$)
Composite structure with a low mound ($0.3 \leq h_b^* < 0.9$)
Composite structure with high mound ($0.9 < h_b^* < 1.0$)
Rubble mound with a crown wall ($h_b^* \geq 1.0$)

Then the main wave/structure parameters that influence the occurrence and severity of wave impacts are given by:

a) Relative approach depth to (offshore) wave height, $h^* = h_s/H_{so}$. This parameter indicates whether waves are near breaking, or are likely to have broken before reaching the structure.

b) Relative wave height to toe berm depth, $H_s^* = H_{si}/d$. This parameter significantly affects the type of wave breaking onto the wall, giving particular emphasis to the influence of the toe mound/berm.

c) Relative berm length to wavelength, $B^* = B_{eq}/L_p$. This parameter (using berm width defined halfway up the berm, $B_{eq} = B_b + h_b/2 \tan \alpha$) describes whether the berm is long enough in direction of wave travel to cause breaking onto the wall, or before the wall.

The influences of these different parameter groups were initially discussed by Allsop et al. (1996a,b), and an alternative version was discussed by McKenna and Allsop (1998). The version shown in Figure 4.6 was derived by Allsop et al. (1999) for bed slopes no steeper than 1:50, and was used for the main simulations in PROVERBS. The parameter map indicates that wave impacts are most likely to occur for three categories of conditions:

- Vertical walls with large waves ($H_{si}/d > 0.35$)
- Low mound composite breakwaters with large waves ($0.65\ H_{si}/d < 1.3$)
- High mound composite breakwaters with moderate berm widths ($0.14 < B_{eq}/L_{pi} < 0.4$) and large waves ($0.65 < H_{si}/d < 1.3$).

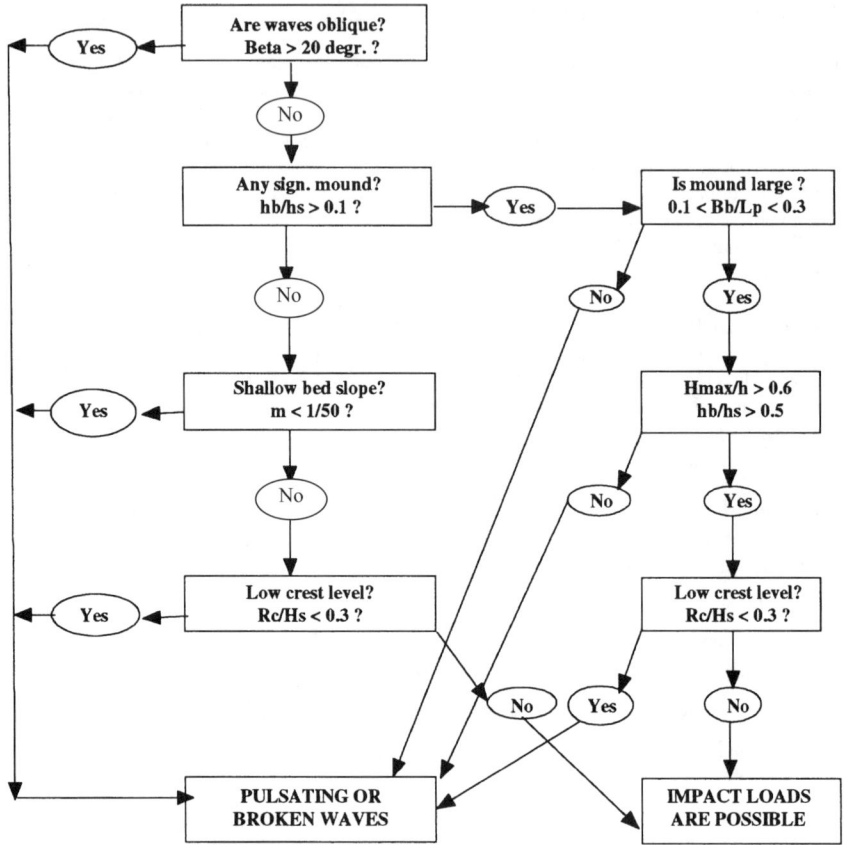

FIGURE 4.5. Decision tree for impulsive wave loads, after Goda (1985) and Allsop et al. (1996c).

Broken waves occur when the local water depth is insufficient to support unbroken waves. For simple vertical walls with no significant mound, waves may start to break when the local wave height to depth exceeds $H_{si}/d > 0.35$. Close to this limit, only a few waves will break, so the risk of impacts is high. As the local wave condition more closely approaches a breaking limit (approximately $H_{si}/d \approx 0.55$ on shallow bed slopes), the proportion of broken waves increases, and the probability of a large but unbroken wave reduces.

A simple model of wave breaking was developed within PROVERBS by Calabrese (1998, 2000) to give estimates of the likely proportion or percentage of wave impacts on a vertical or composite wall. Wave breaking is assumed to occur when, at the location of the structure, the incident wave height with exceedance probability of 0.4%, $(H_{99.6})$ is higher than a critical breaker height H_{bc}, defined as the transition wave height between breaking and nonbreaking waves in front of the structure. The procedure again uses the equivalent berm width, B_{eq}:

$$B_{eq} = B_b + (h_b/2 \tan \alpha) \tag{1}$$

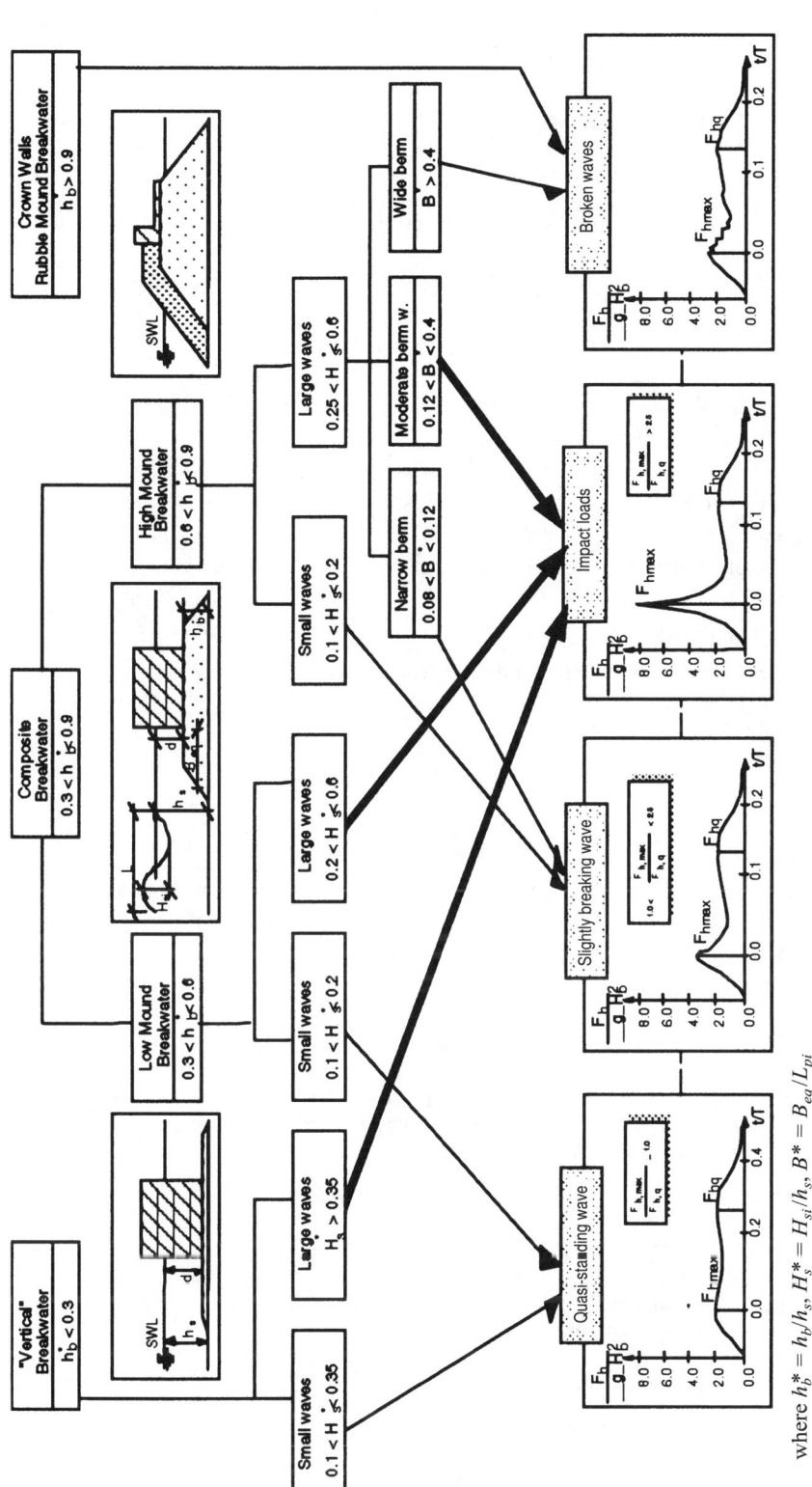

where $h_b^* = h_b/h_s$, $H_s^* = H_{si}/h_s$, $B^* = B_{eq}/L_{pi}$

FIGURE 4.6. Parameter map to predict occurrence of wave impact loads.

Design wave conditions are given by H_{si} and T_p, taking account of effects on local wave heights caused by refraction and shoaling. The peak period wavelength L_{pi} in the local water depth h_s is derived by solving the following:

$$L_{pi} = (gT_p^2/2\pi) \tanh(2\pi h_s/L_{pi}) \tag{2}$$

The critical wave height at breaking, H_{bc}, is defined for the range of relative depths where wave breaking occurred in tests by Allsop et al. (1996c), $0.07 < h_s/L_{pi} < 0.225$:

$$H_{bc} = (0.1025 + 0.0217\ C^*)\ L_{pi} \tanh(2\pi k_b h_s/L_{pi}) \tag{3}$$

$$k_b = 0.0076(B_{eq}/d)^2 - 0.1402(B_{eq}/d) + 1 \quad \text{for } 0 \le B_{eq}/d < 10 \tag{4}$$

$$C^* = (1 - C_r)/1 + C_r) \tag{5}$$

Values of the reflection coefficient C_r may be estimated for the particular structure combination, or from guidance by Allsop (1995b):

For simple vertical walls and small mounds $C_r = 0.95$
For low-crest walls ($0.5 < R_c/H_{si} < 1.0$) $C_r = 0.8 + 0.1\ R_c/H_{si}$
For composite walls, large mounds, heavy breaking $C_r = 0.5$ to 0.7

There are still many uncertainties in the prediction of onset of breaking in front of vertical/composite breakwaters, so a conservative approach will be to assume $C_r = 1$, in which case equation (3) will reduce to

$$H_{bc} = 0.1025\ L_{pi} \tanh(2\pi k_b h_s/L_{pi}) \tag{6}$$

The incident wave height, H_{si}, may then be compared with the breaking height, H_{bc}, calculated above to give categories of breaking:

$H_{si}/H_{bc} \le 0.6$ No evident breaking occurs and wave load is pulsating
$0.6 < H_{si}/H_{bc} < 1.2$ Wave breaking occurs and waves may give impacts
$H_{si}/H_{bc} \ge 1.2$ Heavy breaking or waves may give broken loads

The "surf zone" boundary may be identified:

$$(H_{si}/H_{bc})_{bro} = 0.74 + 0.040(B_{eq}/d) + 0.013(B_{eq}/d)^2 \tag{7}$$

Comparing the ratio H_{si}/H_{bc} with $(H_{si}/H_{bc})_{bro}$ two cases can occur. For $H_{si}/H_{bc} < (H_{si}/H_{bc})_{bro}$ impacting waves are likely to reach the structure). The percentage of breaking waves $P_{b\%}$ may be estimated from the following:

$$P_{b\%} = \exp[-2(H_{bc}/H_{si})^2] \cdot 100\% \tag{8}$$

For $H_{si}/H_{bc} \ge (H_{si}/H_{bc})_{bro}$ a number of waves will arrive at the structure already broken, so these broken waves must be subtracted from the total number of breaking waves to give a true estimate of potential impacts on the structure. The percentage of impact waves $P_{i\%}$ may then be estimated from the following:

$$P_{i\%} = \{\exp[-2(H_{bc}/H_{si})^2] - 0.58 \exp[-1.93(H_{bc}/H_{si})^2]\} \cdot 100\% \tag{9}$$

Values of $P_{i\%}$ may then be used to reappraise the likely loading case:

$P_{i\%} < 2\% \Rightarrow$ Little breaking, wave loads are primarily pulsating

$2 < P_{i\%} < 10 \Rightarrow$ Breaking waves give impacts

$P_{i\%} > 10\% \Rightarrow$ Heavy breaking may give impacts or broken loads

In all of these calculations, it should be noted that H_{bc} is a fictional rather than measured parameter, and may differ significantly from breaking significant wave heights determined by other methods; see particularly Weggel (1972), Owen (1980), and Allsop and Durand (1998). The simple methods described here do not include effects of shoaling, bed friction, or refraction. It should also be noted that no equivalent method has yet been derived for steep bed slopes where greater shoaling and later breaking may allow much larger unbroken wave heights to persist close to a structure. The methods suggested here should therefore be used with caution, and should not be applied out of range.

4.3. PULSATING WAVE LOADS

The main problem to be addressed in initial design is to dimension the components of the breakwater, its elements, and its foundation to resist wave action and its effects, and to deliver the required level of hydraulic performance, chiefly reduction of wave action. For the main structure, the problem is then to dimension the caisson large enough to resist sliding or overturning forces, yet small enough to ensure optimum cost for performance. In the first instance, this has historically been achieved by deriving an equivalent sliding load, then configuring the caisson wide enough to generate sufficient resistance to sliding. The main methods used in design manuals to estimate pulsating wave forces on upright walls, breakwaters or sea walls, have been derived by Hiroi (1919), Ito (1971), and Goda (1974, 1985) for simple and composite breakwaters, Sainflou (1928) for simple walls in deep water, and Jensen (1983) and Bradbury et al. (1988) for crown walls.

4.3.1. Horizontal Forces/Pressures (Normal Attack)

The most widely used prediction method for wave forces on vertical walls was developed by Goda (1974, 1985). In Europe, Goda's method is cited by British Standard BS6349 Pt 1, BSI (1984), and by the CIRIA/CUR rock manual edited by Simm (1991). It should be noted that this method was developed to calculate the horizontal sliding force for concrete caissons on rubble mound foundations, and was calibrated against laboratory tests and back-analysis of historic failures. It does not therefore purport to give wave pressures, even though equivalent or idealized wave pressures are calculated.

Before considering Goda's method in detail, it is useful to review briefly previous methods, particularly those by Ito, Hiroi, and Sainflou; see Ito (1971). Hiroi's formula gives a uniform wave pressure on the front face up to 1.25 H above still water level:

$$p_{av} = 1.5 \, \rho_w g H \tag{10}$$

where p_{av} = the average wave pressure and H is a design wave height, usually assumed to be H_{max}, but see discussion below.

Ito discusses the use of Hiroi's formula where the relative water depth over the mound $d/H_s < 2$, and Sainflou's methods when $d/H_s > 2$. It is interesting to note that Sainflou's method generally gives pressures of about $p_{av} = 0.8 - 1.0\ \rho_w gH$, rather smaller than Hiroi's.

In use in Japan, there was some uncertainty whether Hiroi's method gave safe results, whether the design wave height should be taken as $H = H_s$ or $H = H_{max}$, and over the effects of waves breaking over the mound. A simple method by Ito, discussed by Goda (1985), gave a rectangular distribution of horizontal pressures acting on the front face of the caisson, calculated in terms of H_{max}. The value of H_{max} was then taken as $H_{max} = 2H_s$, or H_b if waves are depth-limited. The average pressure, p_{av}, is then determined for two different regions of relative water depth, H/h_s. Ito assumed a triangular uplift pressure distribution, but uniform pressures on the vertical face, later approximated by

$$p_{av} = 0.7\ \rho_w gH_{max} \qquad \text{for } H_{max} < d \qquad (11a)$$

$$p_{av} = \rho_w gH_{max}(0.15 + 0.55\ H/d) \quad \text{for } H_{max} > d \qquad (11b)$$

Sainflou's method derives a pressure distribution with maximum, p_1 at static water level, tapering off to zero at a maximum notional run-up height, termed a clapotis height, above static water level (s.w.l.) of $H + \delta_0$, and reducing linearly with depth from p_1 to p_2 at the rubble base:

$$p_1 = (p_2 + \rho_w gh)(H + \delta_0)/(h + H + \delta_0) \qquad (12a)$$

$$p_2 = \rho_w gH/[\cosh(2\pi h/L)] \qquad (12b)$$

$$\delta_0 = (\pi H^2/L)\coth(2\pi h/L) \qquad (12c)$$

The *Shore Protection Manual* (CERC, 1984) suggests that Sainflou's method may overestimate wave forces for shorter nonbreaking waves, and uses the Miche–Rundgren formulae to derive the height of the clapotis from which an (assumed) linear hydrostatic pressure is calculated. The accompanying uplift pressure is assumed to be triangular from the front face, with the pressure at the seaward corner consistent for front face or underside. For long waves of low steepness, *SPM* recommends Sainflou's method, showing design curves varying with H_i/gT^2.

The most robust (and most widely accepted) prediction method for wave loads on vertical and composite breakwaters/sea walls is that developed by Goda (1974, 1985). This method assumes that wave pressures on the front face can be represented by a trapezoidal distribution, reducing from p_1 at s.w.l. to p_3 at the caisson base; see Figure 4.7. At points above s.w.l., pressures reduce to zero at the notional run-up point given by a height η^*. Underneath the caisson, uplift pressures at the seaward edge are determined by a separate expression, and may be less than pressures calculated for the toe of the seaward face. In Goda's method, uplift pressures are distributed triangularly from the seaward edge to zero at the rear heel.

The method was developed from hydraulic model tests in which wave pressures were measured, and from a larger set of tests of sliding of model caissons. The resulting prediction formulae were then calibrated by comparisons with field experience. The main response parameters determined in Goda's method are

$$\eta^* = 0.75(1 + \cos\beta)H_{max} \qquad (13a)$$

$$p_1 = 0.5(1 + \cos\beta)(\alpha_1 + \alpha_2 \cos^2\beta)\rho_w gH_{max} \qquad (13b)$$

$$p_2 = p_1/[\cosh(2\pi h/L)] \qquad (13c)$$

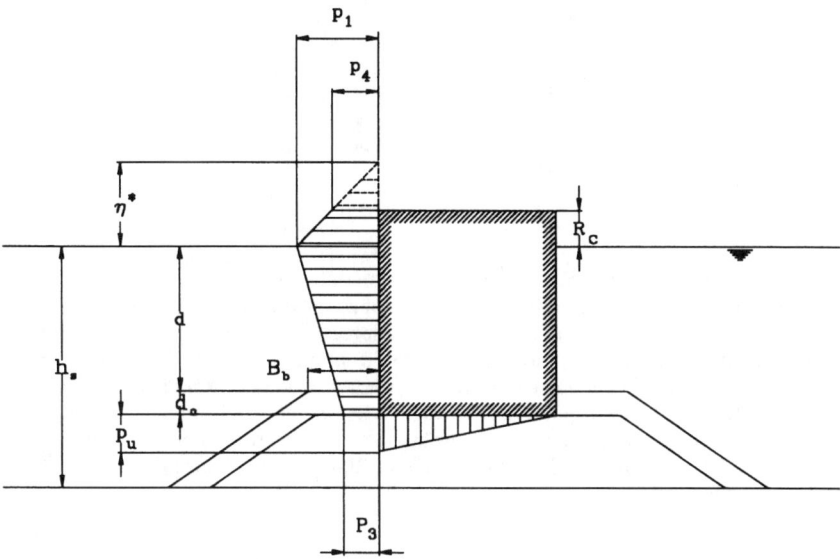

FIGURE 4.7. Pressure distributions in Goda's method.

$$p_3 = \alpha_3 p_1 \tag{13d}$$

$$p_u = 0.5(1 + \cos\beta)(\alpha_1 \alpha_3)\rho_w g H_{max} \tag{13e}$$

The coefficients α_1, α_2, and α_3 are determined from

$$\alpha_1 = 0.6 + 0.5\,[(4\pi h/L)/\sinh(4\pi h/L)]^2 \tag{14a}$$

$$\alpha_2 = \min\{[(h_b - d)/3h_b](H_{max}/d)^2,\, 2d/H_{max}\} \tag{14b}$$

$$\alpha_3 = 1 - (h'/h)[1 - 1/\cosh(2\pi h/L)] \tag{14c}$$

where η^* is the maximum elevation above s.w.l. to which pressure could be exerted (taken by Goda as $\eta^* = 1.5\,H_{max}$ for normal wave incidence) and β is the angle of wave obliquity in plan; see Section 4.3.2. The design wave height, H_{max}, is taken as $1.8\,H_s$ for all positions seaward of the surf zone. In conditions of broken waves, H_{max} should be taken as $H_{max b}$. The water depth h is taken at the toe of the mound, and d over the mound at the front face of the caisson, but h_b is taken 5 H_s seaward of the structure.

The total horizontal force, F_h, is calculated by integrating pressures p_1, p_2, and p_3 over the height h_f of the front face. Similarly, the total uplift force is calculated by integrating from $p = p_u$ at the front edge to $p = 0$ at the rearward edge, giving a total uplift force F_u of $F_u = 0.5\,p_u B_c$

All force and pressures calculated by Goda's method represent a 1/250 exceedance level, $F_{1/250}$. Forces at other exceedance levels $F_{i\%}$ may be estimated from the following ratios of $F_{i\%}/F_{1/250}$ using the 1/250 value, and assuming a Rayleigh distribution:

Exceedance level	$F_{i\%}/F_{1/250}$
50%	0.33
90%	0.59
98%	0.77
99%	0.84
99.5%	0.90
99.8%	0.97
99.9%	1.03

The methods described above generally assume normal wave attack relative to the wall, $\beta = 0°$, and that the point (or vertical line) of application has no finite width. These give the most pessimistic estimate of likely wave forces. In some instances, waves will arrive obliquely to the wall, or may be short-crested. In both instances the effective wave load will be reduced. Reduction factors for oblique waves and/or short-crestedness are described in Section 4.3.2 for pulsating waves, and in Section 4.4.2 for wave impacts. For convenience, and because the mechanisms are very similar, reduction factors that apply when wave loads are averaged over structures of finite length are also described in Sections 4.3.2 and 4.4.2.

4.3.2. Wave Forces under Oblique and Short-Crested Wave Attack

Effective wave loadings on a straight vertical or composite wall will reduce if waves arrive at the wall at an oblique angle rather than at $\beta = 0°$, or if waves are short-crested rather than long-crested. Two processes govern these reductions. At a single position along a wall, oblique (long-crested) waves simply convey less momentum normal to the wall, so effective wave pressures acting normally to the wall are reduced.

For short-crested waves, the same argument applies to those waves arriving obliquely, but some waves will still arrive normal to the wall. If all waves were of the same height and period (simply spread in direction), then some waves would indeed arrive normally to the wall at the selected position, and the peak wave pressure/force would not be reduced. In realistic seas however, wave energy is spread over a range of wave periods, and appears as a range of wave heights. In a short-crested sea, the probabilities of waves that give the largest wave pressures arriving normal to the wall is reduced, so at any given probability level corresponding to design storm length, extreme wave pressures/forces are themselves reduced.

There is a second cause of reduction of effective forces. Most sea walls or vertical breakwaters are composed of large elements, usually caissons, or of elements joined together to behave monolithically over a certain length. For instance, caissons may be cast in lengths of 10 m, 20–30 m, or even as long as 120 m. For the longer caissons, waves of finite length will only apply maximum wave forces over a proportion of the caisson length, so the effective load averaged over the caisson is again reduced. For pulsating waves, discussed in this section, these reductions apply for oblique or short-crested waves.

Methods to determine these reductions for pulsating wave loads on vertical and composite walls have been developed by Goda (1985) and by Battjes (1982).

Oblique Waves. In Goda's method, discussed in Section 4.3.1, the effects of wave obliquity appear in expressions for η^*, and for p_1 and p_u in which a general reduction factor of $0.5(1 + \cos \beta)$ is used. This gives a very simple way to estimate the effect of wave obliquity on the reduction of effective momentum for pulsating waves. The coefficient α_2 is modified by the reduction factor $\cos^2 \beta$ to cover the influence of obliquity on impulsive pressures. On this point, Goda (1985) argues that the effect of impulsive components decrease rapidly at larger obliquities due to the decrease in the normal component of the momentum of the wave, proportional to $\cos^2 \beta$. Impulsive pressures further reduce with obliquity as the effective duration of the impact peak pressure on the caisson increases.

The effects of obliquity β on the force relative to normal wave attack using Goda's method are illustrated in Figure 4.8 for wave steepness of $s_{mo} = 0.06$ and a simple vertical wall. Wave forces reduce steadily with obliquity. At $\beta = 45°$ wave forces predicted by Goda's method reduce to 80–85% of those due to normal waves.

Wave steepness has a small effect on the force decay, becoming greater with obliquity, but the maximum variation in force decay for waves of realistic steepness for structure design is of a few percentage points only. The influence of wave steepness on force decay with obliquity may therefore be ignored in all practical instances.

The effect of small values of obliquity are also relatively mild. In Japanese practice, it is argued that it is difficult to estimate angles of obliquity with precision, so it is recommended that the threshold angle should be $\beta = 15°$. For incident angles of $\beta \leq 15°$, wave attack is assumed to be normal (as if $\beta = 0°$), and no reduction is applied. For obliquities $\beta > 15°$, Japanese practice as described by Goda (1995a) recommends that the true angle of obliquity be reduced by 15°, taking account of the relatively wide directional sectors in which wave directions may be determined from wind information. In practice in Japan, Goda's full prediction method discussed above is therefore only used for values of

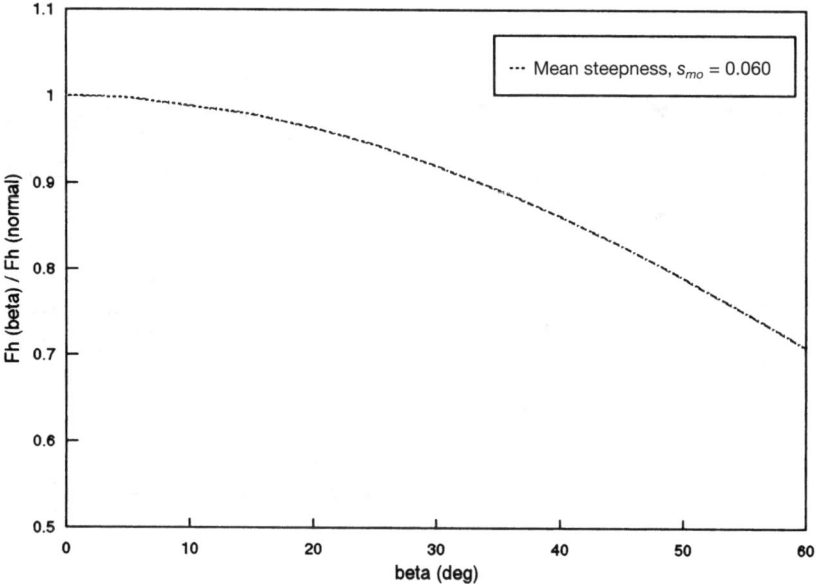

FIGURE 4.8. Effect of wave obliquity on pulsating wave forces, after Goda (1985).

$\beta \geq 15°$. This adjustment of incident wave angle is intended to cover both uncertainties in estimation of wave direction and directional spreading.

Battjes (1982) developed a theoretical model to calculate the reduction in effective load as a function of caisson length relative to incident wavelength and wave obliquity. The reduction in effective load shown in Figure 4.9 is calculated by

$$C_\beta = \{\sin[\pi \sin \beta (L_c/L)]/[\pi \sin \beta (L_c/L)]\} \qquad (14)$$

where β is the wave obliquity and L_c/L is the relative caisson length to wavelength.

The form of this relationship is presented for values of $\beta = 0, 15, 30,$ and $45°$ in Figure 4.9. Battjes' analysis is based on the use of linear wave theory, and therefore assumes nonbreaking waves. This method also predicts that the wave obliquity has no effect on an infinitesimally short segment of the wall, thus $C_\beta = 1$ at a relative caisson length of zero, $L_c/L = 0$. It may be noted that Goda's method does give reduced pressures under oblique attack irrespective of caisson length, but that method was developed from model tests and field data analysis in which the caissons always had finite lengths, typically 15–30 m.

Short-Crested Waves. In many situations, waves at a breakwater in deep water will be short-crested, with a distribution of energy over a range of angles θ around the mean wave direction θ_0. Even if the mean wave direction is itself normal to the structure, much of the incident energy over the other directions will be oblique to the wall. Furthermore, any wave crest will be relatively short, so it will act over only a part of the length of a caisson. It might, therefore, be expected that short-crested waves will give rise to smaller loads than long-crested waves.

Battjes (1982) developed a theoretical model to estimate the load reduction over relative caisson lengths L_c/L for short-crested waves. Wave spreading of a short-crested sea is expressed by a directional distribution function, D, where

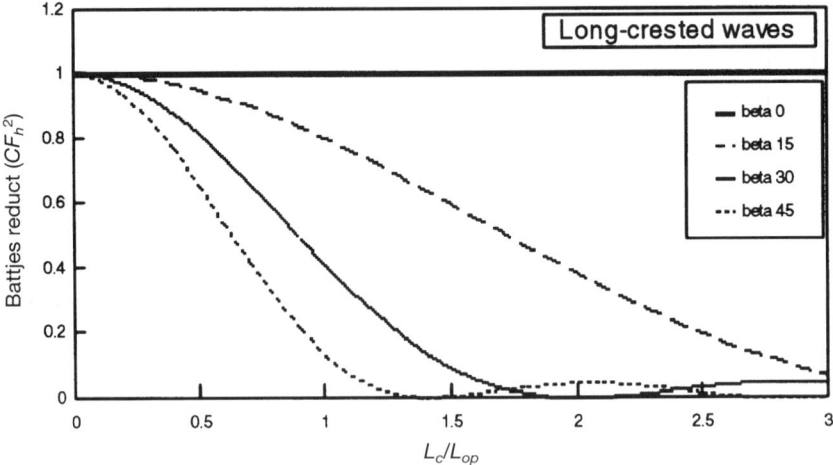

FIGURE 4.9. Effects of caisson length and wave obliquity on pulsating wave forces, after Battjes (1982).

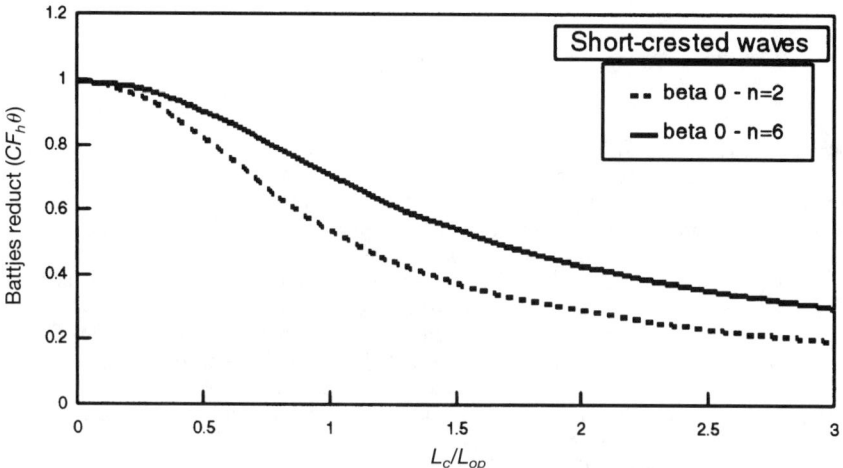

FIGURE 4.10. Effect of short-crested waves on pulsating wave forces, after Battjes (1982).

$$D = D_n(\theta, \theta_0) = A(n) \cos^n (\theta - \theta_0) \quad |\theta - \theta_0| \leq \pi/2 \quad (16a)$$

$$D = 0 \quad |\theta - \theta_0| > \pi/2 \quad (16b)$$

The directional spreading index, n, is assumed to be constant, i.e., not variable with wave frequency. The value of $A(n)$, a normalization factor, given by

$$A(n) = \tfrac{1}{2} \quad \text{where } n = 1 \quad (17a)$$

$$A(2) = 2/\pi \quad \text{where } n = 2 \quad (17b)$$

$$A(n) = [n/(n-1)] A(n-2) \quad \text{where } n = 3, 4, 5 \quad (17c)$$

such that

$$\int_{-\pi}^{+\pi} D(\theta; \omega) \, d\theta = 1 \quad (18)$$

for all ω where ω is the angular frequency.

The effect of the short-crestedness on the load is expressed by

$$C_\theta = \int_{-\pi}^{+\pi} C_\beta^2 D(\theta; \omega) \, d\theta \quad (19)$$

This reduction factor is applied to the load per unit length, given values of the directional spreading index and mean wave direction. The reduction of C_θ with increasing L_c/L_p is shown in Figure 4.10. The reduction of wave forces increases with wave spreading, as shown by the curve for $n = 2$ compared to $n = 6$. This method predicts that the wave spreading has no effect on an infinitely short segment of the wall, thus $C_\beta = 1$ at a relative caisson length of zero, $L_c/L = 0$.

The maximum horizontal force F_h on a structural element of length L_c is given by

$$F_h(L_c) = C_\theta^{0.5} F_h(L_c = 0) \tag{20}$$

Recent Experimental Studies. Experimental studies by Franco et al. (1996) investigated some of these issues in hydraulic model tests in a large wave basin, measuring wave forces and pressures for a number of adaptations of a simple caisson on a small mound. Wave forces were measured at one point along the caisson with a force plate, and pressures were measured with transducers at other locations along the series of caissons.

A single water level and a single wave height were used by Franco et al., equivalent to $H_{si}/h_s = 0.23$ at wave steepnesses of $s_{op} = 0.02$ and 0.04. The tests used three degrees of wave spreading, with normal attack and with six angles of obliquity. An example of load decay with caisson length is shown by Franco et al., and is discussed by Allsop and Calabrese (1998, 1999), who note that the results show reductions in the load with increasing caisson length, even under long-crested normal waves. Under oblique short-crested waves, the reduction is greater, that is, the value of C_θ is smaller.

Franco found that Goda's method gave reasonable estimates of the (pulsating) horizontal force under three-dimensional waves, but that there was considerable scatter in the results. For the long-crested waves, the reduction of horizontal load with wave obliquity was fairly well described by Goda. Under short-crested waves, however, the experimental results showed no reduction in the wave forces with wave obliquity. This effect is not described by Goda. Franco suggests a constant reduction factor for short-crested waves at all obliquities of 10%, so $C_\beta = 0.9$.

The reduction of load with increasing caisson length under long-crested waves for large values of L_c/L_p was found to be greater than Battjes' predictions based on linear wave theory. Under short-crested waves, however, agreement with Battjes' predictions was fairly good.

4.4. WAVE IMPACT LOADS

4.4.1. Horizontal Forces/Pressures (Normal Attack)

In Europe, engineers noted very large forces on some sea walls; see particularly Rouville et al. (1938) who recorded forces in excess of 600 kN/m² on the sea wall at Dieppe. Some conditions led to very large pressures, perhaps larger than could be accommodated by engineering of that era. Many researchers since have tried to describe wave impact forces, but studies in the laboratory have often led to results that appear to be too large for practical use.

It is generally accepted that dynamic loads can be very important, but it is often argued that most structures are substantially unaffected by short-duration, high-intensity loads. Schmidt et al. (1992) remind us that despite more than 80 years of research on impact loading on vertical structures, there are still two negative attitudes to the role of wave impact loadings in the design of such structures. The first attitude simply assumes that impact pressures are not important and thus should not be adopted in the design. The second attitude is to skip the problem of evaluating the design impact load by assuming that the structure can be designed in such a way that impact pressure will not occur.

The main problems are in identifying the magnitudes and durations (both are needed) of wave impact loads, and then applying those loads using dynamic response characteristics of the structure to derive effective loads. Applying impact loads using static load analysis will always give over-pessimistic results. These types of structures are of considerable inertia, and therefore of relatively long response period. Very short period excitations will therefore be strongly damped. It is now, however, clear that simply ignoring wave impacts is not a safe approach.

Research by Oumeraci and co-workers has shown that repeated impacts may give incremental sliding. Other researchers have shown how some foundation materials may respond to repeated impact pressures, especially if pore pressures can be driven up. Lastly, the use of smaller and/or lighter elements necessarily requires greater rigor in analysing dynamic effects. The safe approach is therefore to conduct a dynamic analysis of the structure, and its foundation, and of the applied loads, an approach strongly argued by Oumeraci (1994), Oumeraci and Kortenhaus (1994), and Oumeraci et al. (1995). Problems arise in the high level of data required, both in the time series of loadings, but also in the dynamic response characteristics of the mound and its foundation. Development of overall stability models is still therefore at a relatively early stage, but advances in simpler methods have been described by Torrini et al. (1999) and Besley et al. (1999).

Prediction Methods by Minikin and Shore Protection Manual. Bagnold (1939) postulated an analytical model in which momentum from an incoming wave compresses a trapped air pocket. The wave slows and stops as pressure in the trapped air rises. At maximum pressure, the wave momentum has been converted to pressure over the impact rise time. Bagnold's approach required the identification of the thickness of the air pocket, and of the virtual length of the water piston. Neither of these could be measured with reliability at the time.

Minikin's (1950, 1963) method was developed in the early 1950s to estimate local wave impact pressures caused by waves breaking directly onto a vertical breakwater or sea wall, and therefore addressed the problems of impact pressures. Minikin used Bagnold's piston model and calibrated a version of this model with Rouville et al.'s (1938) pressure measurements on a sea wall at Dieppe to give maximum peak pressures for typical wave impact events. The resulting expression for p_{max} may be written

$$p_{max} = 0.5\ C_{mk} \pi \rho_w g H_{max}(1 + d/h)(d/L) \tag{21a}$$

where C_{mk} is a coefficient defined to allow fitting to Rouville's data, and accounting for the typical size of an air pocket. Minikin suggests a value of $C_{mk} = 2$, which is then cancelled within equation 21a to give the simpler version used by BS6349 Pt1 (1984):

$$p_{max} = \pi \rho_w g H_{max}(1 + d/h)(d/L) \tag{21b}$$

This expression was then rewritten by Minikin with $\pi \rho_w g$ replaced by 2.9, and this was further compounded the *Shore Protection Manual*, CERC (1984), which rewrote Minikin's formula with πg replaced by 101. Added confusion arose over the use of units.

This apparently minor omission became more serious when later authors implied that the version using 101 could be used in other units than f.p.s, and thus propagated an erroneous version of Minikin's method ever since! The effect is that calculations using Minikin's method as in the *Shore Protection Manual* give substantially different results from those intended. Another serious confusion is introduced in the use of the (quasi) hydrostatic element in the total horizontal force, compounded by errors by Minikin himself in example calculations. Errors in applying Minikin's method are discussed at length by McKenna (1997) and by Allsop et al. (1996c) who illustrate the differences introduced by (mis-)interpreting the *SPM* version. The comparative effect is shown in Figure 4.11 as a

4.24 HANDBOOK OF COASTAL ENGINEERING

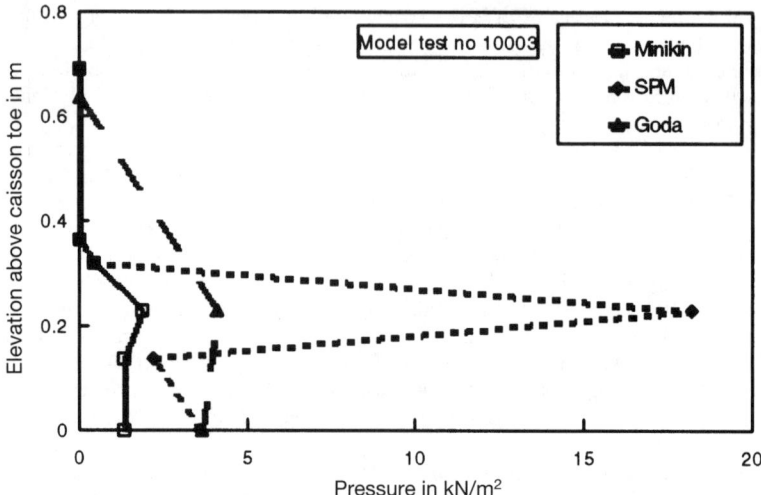

FIGURE 4.11. Vertical distribution of pressures using Goda, Minikin, and *SPM* methods compared with measurements in a hydraulic model test (Allsop, 1996c).

vertical pressure distribution (for a model test example), as predicted by Goda, Minikin as above, and the *SPM* version of Minikin.
Reverting to dimensionally correct expressions, the integration of Minikin's formulae for wave pressures may give an expression for total horizontal force:

$$F_{hmax} = 0.5 \, C_{mk} \pi \rho_w g H_{max} d \{(1+d/h) \, H/3L + \tfrac{1}{2}\pi + H/(8\pi d)\} \qquad (22)$$

In practice, it has been found by other reviewers, and perhaps by practicing engineers, that the *SPM* version of Minikin's method gave substantially greater pressures than other formulae, and its use for calculations of wave forces for practical design has been very limited. Misconceptions on the validity/use of Minikin's method were highlighted by discussions by Ergin and Abdalla (1995), responded to by Allsop and Muller (1995) and Goda (1995b). The most realistic view on Minikin's formulae is probably epitomized by Goda (1995a), who summarized the informed view on this method as "can be considered to belong to a group of pressure formulae of historical interest."

Hannover/Braunschweig Methods for Predicting Impact Forces. Much attention has been devoted to pursuing the goal of quantifying impact pressures. At small scale, very large (comparatively) pressures may be measured if small fast-responding transducers are sampled very rapidly. There has, however, been much doubt that this would be found at large scale. Partensky (1988) quoting Oumeraci uses results from the large wave channel at Hannover/Braunschweig (known as the GWK) to suggest that impact pressures of very short durations (0.01 to 0.03 s) may be calculated from

$$p_{dyn} = K_L \rho_w g H_b \qquad (23a)$$

where H_b is the breaking wave height and the coefficient K_L is given in terms of the air content a_e of the breaking wave:

$$K_L = 5.4\,[(1/a_e) - 1] \tag{23b}$$

For smooth slopes on embankments of slope 1:2 to 4, Partenscky (1988) recommends air contents up to $a_e = 0.6$ to 0.8, but for vertical sea walls and breakwaters, Partenscky recommends $a_e = 0.1$ to 0.35, giving values of the coefficient $K_L = 10$ to 45. Partenscky also derives formulae for the vertical distribution of wave impact pressures, but these take no account of air content.

A problem that arises in applying Partenscky's method, and indeed a problem that arises in interpreting all of the prediction methods/data from the GWK, is the derivation of the single-valued breaking wave height, H_b. The uncertainties in obtaining this value were later partially solved for shallow bed slopes by the development of Calabrese's method to predict H_b (see Section 4.2.4), but there remain significant uncertainties in the definition of H_b for other slopes, and in the allocation of an appropriate air content a_e.

Within PROVERBS, it was recommended that wave impact forces also be estimated using the procedure developed by Kortenhaus et al. (1999) if $P_{i\%} > 1\%$. This uses a very simple fundamental equation for the maximum horizontal force calculated by

$$F_{h,\text{impact}} = F^*_{h,\text{impact}}\,\rho g H_b^2 \tag{24a}$$

where H_b is the wave height at breaking, taken as equal to H_{bc} from Section 4.2.4. The relative maximum wave force $F^*_{h,\text{impact}}$ act on the wall can then be calculated using a generalized extreme value (GEV) distribution as follows:

$$F^*_{h,\text{impact}} = \frac{\alpha}{\gamma}\{-[\ln P(\hat{F}^*_{h,\text{impact}})]^\gamma\} + \beta \tag{24b}$$

where $P(F^*_{h,\text{impact}})$ is the probability of nonexceedance of impact forces, which generally may be taken as 90%; α, β, and γ are the statistical parameters for the GEV distribution, which can be taken from Table 4.1.

The rise time t_r of a triangular force–time curve may be assumed as follows:

$$t_r = k'\left(\frac{\sqrt{d_{\text{eff}}/g}}{F^*_{h,\text{impact}}}\right) \tag{25a}$$

The factor k' can then be described by a log-normal distribution with a mean value of 0.086 and a standard deviation of 0.084. The effective water depth in front of the structure d_{eff} may be calculated from

$$d_{\text{eff}} = d + B_{rel}\,m_{rel}\,(h_s - d) \tag{25b}$$

TABLE 4.1. Values of α, β, and γ for GEV distribution of relative horizontal force

Bed slope	Number of waves	α	β	γ
1:7	116	2.90	6.98	−0.53
1:10	159	10.21	12.76	−0.063
1:20	538	3.74	7.60	−0.295
1:50	3321	1.91	3.27	−0.232

where B_{rel} is the part of the berm width which influences the effective water depth (B_{rel} = 1 for no berm):

$$B_{rel} = \begin{cases} 1 & \text{for } B_b/L \leq 1 \\ 1 - 0.05 \left(\dfrac{B_b}{L}\right) & \text{for } B_b/L > 1 \end{cases} \quad (25c)$$

and m_{rel} is the part of the berm slope influencing the effective water depth (m_{rel} = 0 for simple vertical walls):

$$m_{rel} = \begin{cases} 1 & \text{for } m < 1 \\ \dfrac{1}{\sqrt{m}} & \text{for } m \geq 1 \end{cases} \quad (25d)$$

Prediction Methods for Impact Pressures from Wallingford. The methods developed at Hannover/Braunschweig suffer from relying on a breaking wave height H_b. This cannot always be defined unambiguously from the incident wave height H_{si}, and calculations of H_b using methods by Calabrese and others often differ from measurements of H_{max} in hydraulic model tests, and from predictions by other methods, see discussion in Section 4.2.4.

A simple and robust method to predict wave impact pressures was derived by Allsop and Vicinanza (1996) based on testing by Allsop et al. (1996b,c) and McKenna (1997). They noted that for waves close to breaking given by $0.35 < H_{si}/d < 0.6$, other prediction methods underestimate measured forces. Differences are greatest where the incident wave conditions approach the breaking limit, approximated for shallow bed slopes by $H_{si}/h_s \approx 0.55$. A simple prediction curve was fitted to test results for $0.35 < H_{si}/d < 0.6$:

$$F_{h,1/250} = F_{h,A\&V} = 15 \, \rho_w g d^2 \, (H_{si}/d)^{3.134} \quad (26)$$

Fortuitously, this equation seems also to give a good description of wave impact forces for walls on low mounds given by $0.3 < h_b/h_s < 0.6$, and higher relative wave heights given by $0.6 < H_{si}/d \leq 1.3$.

Measurements of wave forces from a later set of model tests reported by McConnell (2000), termed HR97, for a simple vertical wall with four different bed approach slopes were processed to determine $F_{h,1/250}$. Forces were plotted against H_{si}/h_s (equivalent to H_{si}/d used by Allsop et al. 1996), termed HR94. The results were compared with Allsop and Vicinanza's (1996) prediction method (Figures 4.12 and 4.13). These new data suggested that Allsop and Vicinanza's method can underpredict forces if wave heights can exceed $H_{si}/d = 0.6$. That method therefore also underpredicts forces for steeper bed slopes. A modified equation for 1:50 bed slopes has therefore been developed by McConnell (2000):

$$F_{h,1/250}/\rho g d^2 = 20.23 \, (H_{si}/d)^{3.504} \quad (27a)$$

A new prediction method is proposed for bed slopes steeper than 1:50:

$$F_{h,1/250}/\rho g d^2 = 22.947 (H_{si}/d)^{2.798} \quad (27b)$$

The data for individual wave impacts for all tests for the four bed slopes were plotted as $F_h/\rho g H_{si}^2$ against t_r/T_m; see Figure 4.14. An approximate upper-bound curve was fitted to

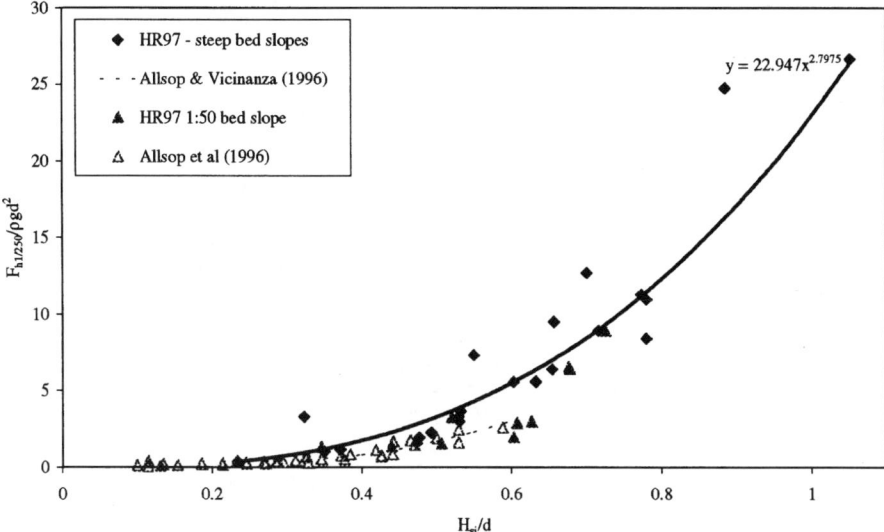

FIGURE 4.12. Wave impact forces, steep and 1:50 bed slopes.

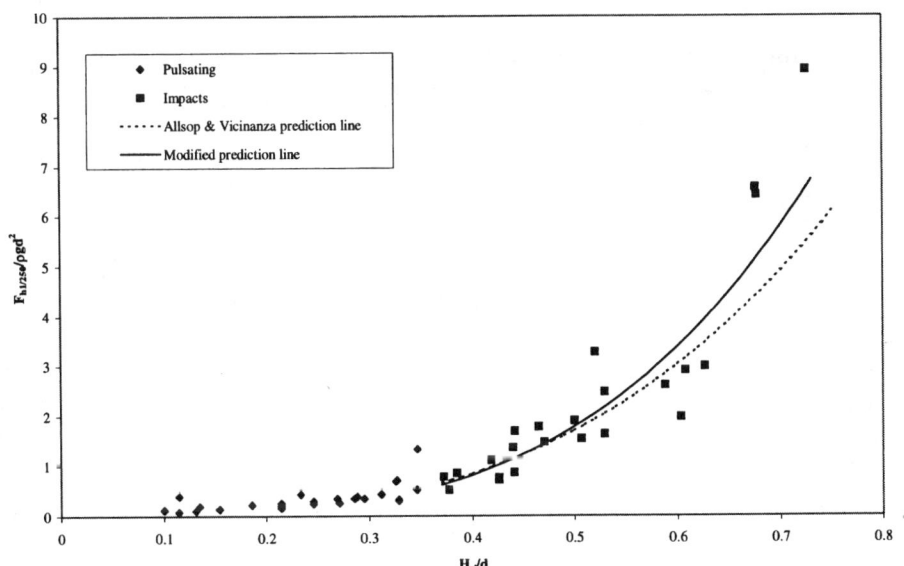

FIGURE 4.13. Wave impact forces, 1:50 bed slope.

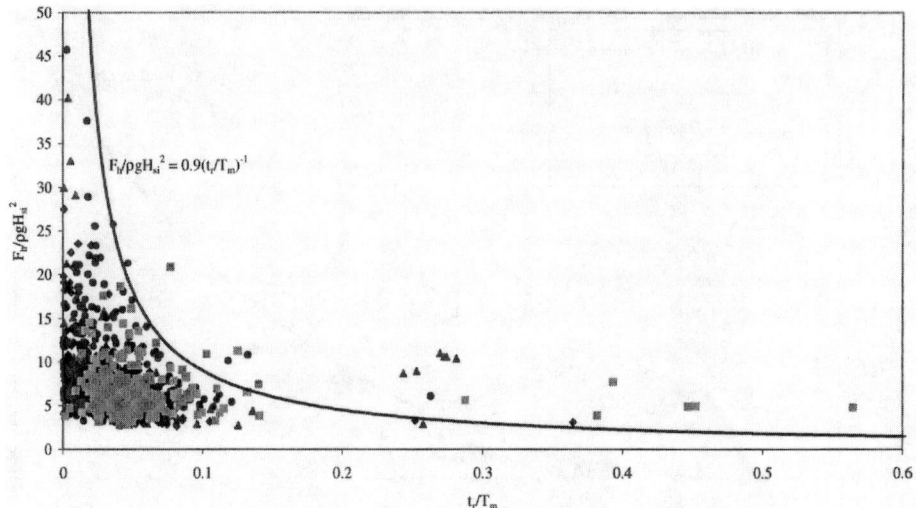

FIGURE 4.14. Wave impact rise times and forces.

the data, giving a method to relate wave impact rise time to impact force, valid in the region of wave impacts up to $t_r/T_m \approx 0.15$:

$$F_h/\rho g H_{si}^2 = 0.9(t_r/T_m)^{-1} \qquad (27c)$$

Shorter rise times, and hence impact durations, can occur on the steeper slopes, but the adoption of the upper limit given above from the complete data set will offer some conservatism by increasing the predicted duration of the impact force. This is likely to bring it closer to the (larger) natural frequency of massive caisson structures. Care should be taken if the method is being applied to much smaller structural elements, where actual impact durations may be closer to the shorter natural periods of those elements of concern.

Pressure Gradients. The analysis by Allsop et al. (1996c) demonstrated that pulsating wave conditions give relatively low absolute values of the wave pressure, so pressure gradients are relatively mild, seldom exceeding values of $dp/dz > 1$. The situation is, however, dramatically different for impact conditions. For wave impacts, values of the peak local pressure gradients (as measured between adjoining pressure transducers) on simple vertical walls varied over the range $dp/dz = 2$ to 70. These values increased slightly for walls on low mounds to $dp/dz = 5$ to 90, and for high mounds to $dp/dz = 2$ to 80. The mean value of these results and coefficients of variation are summarized below:

Structure	Range of dp/dz	Mean (dp/dz)	Coefficient of variation (dp/dz)
Vertical	2–70	13	120%
Low mound	5–90	30	90%
High mound	2–80	22	80%

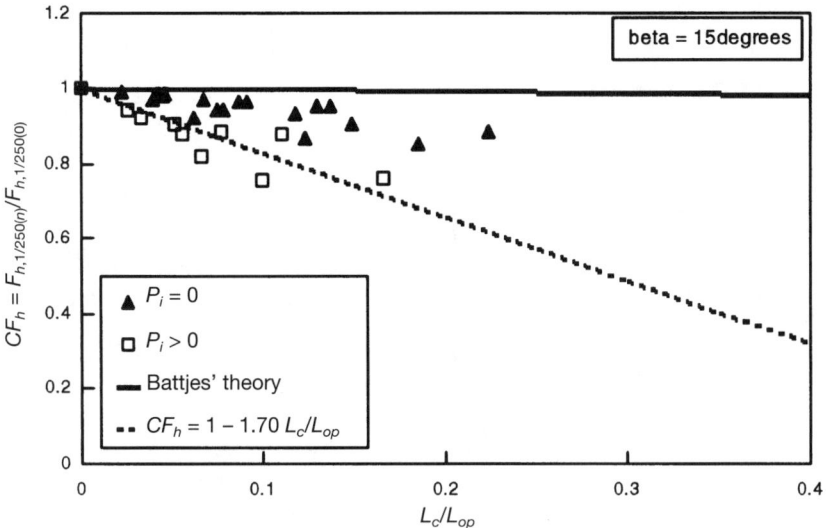

FIGURE 4.15. Effect of obliquity on wave impact forces.

For impact waves, the greatest relative local pressure, and the steepest pressure gradient, measured by Allsop et al. (1996c) in laboratory tests are given by

$$p_{max}/(\rho_w g H_{si}) \leq 50 \quad (28a)$$

$$(dp/dz)_{max} \leq 90 \quad (28b)$$

4.4.2. Wave Impact Forces under Oblique and Short-Crested Wave Attack

The effects of wave obliquity and/or short-crestedness on pulsating waves has been predicted by Goda and Battjes, and tested against laboratory measurements, as discussed in Section 4.3.2. No equivalent theoretical analysis is available for wave impact conditions, except an expectation that effective wave impact forces will reduce with obliquity, and caisson length, and possibly with short-crestedness.

Studies by Allsop and Calabrese (1999) in the large U.K. Coastal Research Facility wave basin (CRF) measured impulsive loads under oblique and short-crested waves and evaluated the effects of caisson length. For *simple vertical walls*, impacts under *long-crested normal* waves compared with results from two-dimensional tests with little variation. The onset of impacts occurred at $H_{si}/h_s = 0.35$ for long-crested waves at $\beta = 0°$.

For short-crested waves at $\beta = 0°$, the onset of breaking again showed no changes in comparison with long-crested waves. Impacts begin at $H_{si}/h_s \geq 0.35$, although there were some indications that impacts do not increase as rapidly with increasing H_{si}/h_s in short-crested waves as for long-crested waves.

For *oblique long-crested* waves, $\beta = 15°$, $30°$, and $45°$, there were much fewer impacts

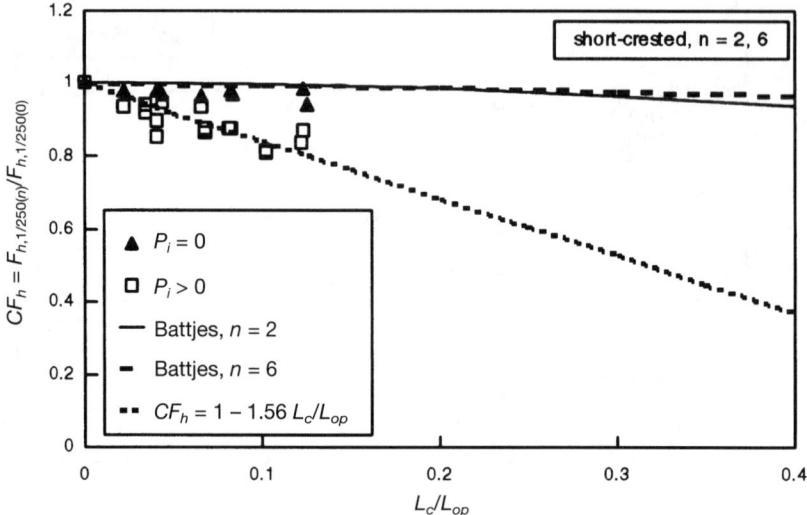

FIGURE 4.16. Effect of caisson length and short-crestedness on wave impact forces.

than for normal long-crested waves (see example in Figure 4.15). For larger waves, $H_{si}/h_s > 0.35$, impacts were less frequent with oblique waves than for normal attack, $\beta = 0°$.

A berm or slope in front of a simple wall may increase impacts substantially in number and severity. The start of impact events on walls on low mounds, $0.3 < h_b/h_s < 0.6$, occurs at or above $H_{si}/h_s \geq 0.35$, with P_i increasing rapidly at higher values of H_{si}/h_s. Long-crested waves at $\beta = 0°$ show impacts for low mounds at $H_{si}/h_s \leq 0.30$. Tests on high mound composite walls at $\beta = 0°$ confirmed that impacts increase further with high mounds, $0.6 < h_b/h_s < 0.9$. Impacts start at smaller values of H_{si}/h_s, as low as 0.25.

Pulsating wave forces for *long-crested normal* waves agree well with results from three-dimensional tests by Franco et al. (1996). Comparisons with Goda's predictions show relatively good agreement over the pulsating zone. An upper limit to wave impact forces under normal or long-crested oblique waves is given by the simple prediction by Allsop and Vicinanza (1996); see Section 4.4.1 above.

For *short-crested waves*, forces show no significant change in local force for the range of conditions tested, dispersion index ($n = 2$ or 6), compared with loads generated by long-crested waves of the same height (see example in Figure 4.15), but averaged forces are reduced.

The influences of *oblique long-crested* waves on forces on any narrow strip of the caisson are more significant. Over the pulsating zone, $H_{si}/h_s \leq 0.35$, forces are very similar to those for normal approach, even though the component of force perpendicular to the caisson might have been expected to reduce. In the impact region however, wave loadings diminish considerably under oblique attack.

These comparisons show consistent increases in $F_{h,1/250}$, with reduced averaging. Allsop and Vicinanza's simple formula gives a reasonable representation of forces averaged over typical caisson widths of 10–20 m, but underestimates the "local" force over a single narrow strip, even for normal and long-crested wave attack. Values of $F_{h(\text{peak})}/F_{h(\text{av})}$ reached 1.2–1.3 for normal long-crested attack. Under long-crested oblique attack, most results were lower, not exceeding $F_{h(\text{peak})}/F_{h(\text{av})} = 1.15$, but with a single test giving 1.4.

Under short-crested waves the ratio $F_{h(\text{peak})}/F_{h(\text{av})}$ never exceeded 1.15, suggesting that peak forces are unlikely to exceed those analyzed in this research by any substantial margin, except under conditions of normal attack.

Effect of Caisson Length. Battjes (1982) argued that oblique or short-crested wave attack on caisson of length L_c will give further reductions in effective force relative to normal and/or long-crested attack, and relative to loads on a narrow strip (modeled in Allsop & Calabrese's (1999) tests as a single column of transducers).

Results from the UKCRF tests were combined with results from Franco et al. (1996), which show little decay over caisson lengths L_c/L_{op} up to 0.4. Measurements from the CRF, however, show up to 10% decay, ie C_{Fh} down to 0.9 for nonimpact conditions for relative caisson lengths up to $L_c/L_{op} = 0.15$. Wave impact conditions ($H_{si}/h_s > 0.35$), however, gave substantially greater reductions in the effective force, even over short caisson lengths, $0.005 < L_c/L_{op} < 0.2$. A simple regression line gives the reduction factor C_{Fh} in terms of relative caisson length with a coefficient $B = 1.35$ for long-crested waves and $\beta = 0°$:

$$C_{Fh} = 1 - B(L_c/L_{op}) \qquad (29)$$

Under slightly oblique attack, $\beta = 15°$, forces for nonimpacting conditions show more significant reductions than for $\beta = 0°$, but there is only slightly greater change for impact conditions. The same simple form of regression line gives C_{Fh} in terms of L_c/L_{op}: for $\beta = 15°$, yielding $B = 1.70$. At greater obliquities, the force reduction is more marked for pulsating conditions. Measurements at $\beta = 30°$ show slightly greater reduction for impacts.

Effects of short-crested waves show no significant effect of spreading between $n = 2$ and $n = 6$. The regression for $\beta = 0°$ gives $B = 1.56$, steeper than for long-crested waves at $\beta = 0°$, but less severe than for long-crested waves and $\beta = 15°$.

These results suggest that Battjes' model may be used to give conservative predictions in the pulsating zone, but that force reductions under impacts are much more significant than predicted by linear methods. Calculations of the mean decay function on F_h for impacting conditions can be summarized by the simple equation relating decay to relative caisson width, L_c/L_{op}, given in equation (29), where coefficient B is defined for each test case in Table 4.2.

Under oblique or short-crested waves, the variation of peak forces relative to those averaged over a short length equivalent to a single caisson of about 20 m are relatively small, not exceeding a ratio of 1.2. The variation of peak force on a single narrow strip under normal wave attack is more substantial, with peak forces up to 1.3 times greater than the average.

Battjes' method for estimating the decay of average force with longer caissons gives

TABLE 4.2. Impact force reduction coefficients

Wave condition	Coefficient B	Coefficient of Variation (%)
Long-crested, $\beta = 0°$	1.35	6.6
Long-crested, $\beta = 15°$	1.69	9.2
Long-crested, $\beta = 30°$	1.69	10.4
Short-crested, $n = 2$	1.55	10.6
Short-crested, $n = 6$	1.58	9.3
Short-crested, $n = 2, 6$	1.56	6.8

very small reductions for most practical caisson lengths. The tests with pulsating conditions show that Battjes' predictions are generally conservative. However, for impact conditions, average forces reduce significantly with caisson length, giving reductions of 25% or so over relative caisson lengths of only 0.2. A simple reduction factor for F_h under impacting conditions as a function of L_c/L_{op} has been developed. Values of a coefficient B are presented here in Table 4.2 for long-crested waves at different obliquities, and for short-crested waves.

These and other studies also suggest the following initial conclusions on spatial correlation of impact forces under oblique/short-crested waves:

- For heavy impacts ($F_{\text{impact}}/F_{\text{Goda}} \gg 2.5$), and small obliquity or spreading, assume a typical coherence length $\leq L/16$
- For light impacts ($F_{\text{impact}}/F_{\text{Goda}} < 2$), normal wave attack ($\beta = 0°$), and little spreading, assume a typical coherence length $\leq L/4$

4.4.3. Scale Effect Corrections and Practical Application

The main restrictions to the application of these test results are given by the limits of data covered, distortions of the responses arising from scale effects or uncertainties in scaling from model tests to prototype, and differences in the response characteristics of different structures or elements relative to those used in making or analyzing the measurements. The ranges of parameters, and of the main dimensionless parameter groups covered by these studies have been summarized in earlier sections of this chapter. Application of the wave impact predictions depend critically on the reliability with which laboratory measurements of pressures may be applied to full scale in sea water.

The principal results of the research studies were wave forces/pressures. For pulsating wave pressures, the relationships between wave momentum, pressure impulse, and total horizontal force are relatively simple. The assumption of Froude scaling is realistic, so no scale corrections are required for pulsating load conditions.

For wave impact pressure, scaling is less simple. It is well accepted that wave impacts in small scale hydraulic model tests if scaled by Froude will be greater in magnitude but shorter in duration than their equivalents at full scale in (invariably aerated) sea water. It is probable therefore that high impact pressures measured in model tests should be scaled to lower values, but that the impulse durations must be scaled to longer values.

Many of these uncertainties have been studied at large scale in large wave flumes like the GWK at Hannover/Braunschweig Universities, or using salt and fresh water in experiments at Plymouth University. It has been argued that the addition of only small fractions of air may dramatically change pressure transmission characteristics of the water, thus substantially modifying pressures that might be experienced by the structure. Two studies, however, suggest that the effect is very much smaller. There is some indication from studies at Plymouth described by Walkden et al. (1995) that even quite high levels of aeration in the model only reduced wave pressures measured in the model by about 20%. Numerical modeling studies by Peregrine (1995), can also be used to suggest that scale errors due to air effects might be limited to about 50%.

In contrast, analysis by Allsop et al. (1996b,c) suggests that even small changes of relative mound level will change wave impact pressures by factors of 3–5, suggesting that influences of small changes in geometry may be of greater effect than uncertainties introduced by scale effects. It is, however, still necessary to assess the likely contribution to overall uncertainties arising from any scale errors.

WAVE FORCES ON VERTICAL AND COMPOSITE WALLS   4.33

Scale Corrections to Impact Pressures and Rise Times. Fieldwork measurements of wave impacts described by Allsop et al. (1995c) recorded 3270 impact events, and of those, Howarth et al. (1996) analyzed 632 impacts. A 1:32 scale model was subjected to wave conditions chosen to represent waves/water levels measured in the field. Impact pressures were measured on a model armor unit in the same position as the field unit. Comparisons of these data of impact pressures and rise times for full scale in sea water and small scale (1:32) in fresh water were used by Allsop et al. (1996b,c) to calculate pressure impulse, estimated by $p \cdot \Delta t$. Values of pressure impulse showed close agreement over nonexceedance levels of 92–99.9%, supporting the thesis that pressure impulse can be scaled by Froude, even where pressures or rise times cannot.

Comparison of impact pressures from field and model suggested that there is a relatively constant relationship between field and laboratory pressures over these exceedance levels, and that impact pressures measured in hydraulic laboratories in fresh water need to be corrected by factors between about 0.40–0.45 over nonexceedance levels of 92–99%.

A similar approach was taken on pressure rise times, taken here to indicate also the effect on impact durations. The variability of rise times were wider than for impact pressures, and more care will be needed in interpreting these results to take account of limitations in the data. The correction factors derived above may be summarized:

Nonexceedance level	Correction factors	
	Impact pressure	Rise time/duration
95%	0.45	7
99%	0.40	4

Use of these measurements also requires information on the response time of the structure. It has been shown by Muraki (1966) that typical periods of oscillation for (Japanese) caissons is 0.2–0.4 seconds. It is, however, probable that small elements such as stone or concrete blocks in older walls will respond to much shorter periods. This is illustrated by application of these methods to a simple design case. It should be noted that an alternative approach is taken in this example where the effective load is estimated using force impulses, so no scale correction is used.

Application of Impact Prediction Methods Wave effects in inland reservoirs in the United Kingdom have required measures to limit wave overtopping on some embankment dams without raising the overall dam crest, perhaps by adding a wave wall at the crest of slope. Such walls attract wave forces, and some wall configurations may give impulsive wave breaking onto the wall. The methods developed above were applied to determine wave loadings on a section of wave wall.

For the example shown in Figure 4.17, wave loads on the wave parapet walls were caused by broken waves only for most water levels. At high water levels, the revetment slope causes waves to break directly against the upper parapet wall giving wave impact forces estimated as approaching $F_{h,\text{impact}} = 200$ kN/m, approximately 5–20 times greater than pulsating wave loads. As wave impact forces are of short duration and of limited spatial extent, effective forces on a length of wall will be rather lower. Structural analysis of the wall taken as a freestanding cantilever of 5 m width, gave natural response periods $t_{re-sp} = 0.01$–0.04 s.

Wave conditions at the dam give $H_{si}/h_s = 0.9$, which substantially exceeds the limit $H_{si}/h_s = 0.35$ in the parameter map (see Figure 4.6 in Section 4.2.4), so local wave impacts are probable. First, however, the Goda force was calculated using the simple pressure dis-

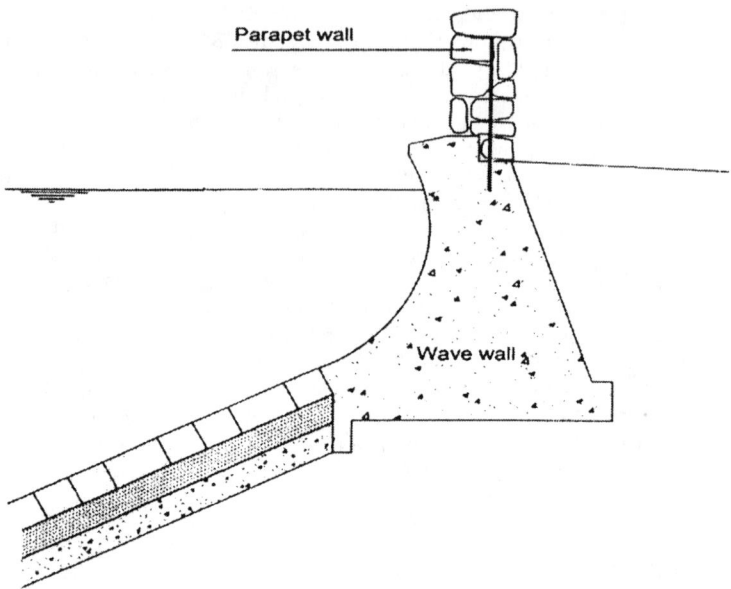

FIGURE 4.17. Wave wall used in example calculation.

tribution shown previously in Figure 4.7. The total force over the wave wall and parapet wall for the design wave condition was $F_{h,1/250}$ = 66 kN/m, and the component on the parapet wall alone gave $F_{h,\text{Goda}}$ = 20 kN/m.

Wave impact forces on the parapet wall were then estimated using Allsop and Vicinanza's method (equation 26), giving wave impact forces of $F_{h,\text{impact}}$ = 180 kN/m. Wave impact events from which the prediction method was derived had rise times typically of 0.001 T_m. Using these durations and values of $F_{h,\text{impact}}$ gave force impulses of I_{impact} = 0.18 kNs/m. Applying the calculated impulses over durations t_{resp} = 0.01–0.04 s gave effective wave loadings of $F_{h,\text{equiv}} \leq$ 40 kN/m.

Whereas pulsating forces are usually taken to act simultaneously over significant lengths of the parapet wall, wave impact loadings are limited spatially. The simple guidance from Allsop and Calabrese (1998, 1999) summarized above suggests that strong wave impact forces will not spread over widths greater than about 1/16 of a wavelength. The test suggested for "heavy impacts" is where the impact force is greater than 2.5 times the Goda force ($F_{\text{impact}} > 2.5\ F_{h,\text{Goda}}$). For the conditions considered here, this occurs if $F_{\text{impact}} >$ 45 kN/m, significantly exceeded by the forces near to 200 kN/m calculated above.

Application of this rule suggested that instantaneous impact forces might be limited to a width of wall of approximately 1 m. Allsop and Calabrese (1998, 1999) also suggest that light impacts will not spread over more than $L/4$, about 3m for this case. Considering a 5 m stretch of wall, loading on the rest of the wall can be determined using Goda's method. The wave impact load might pessimistically be considered as a rectangular distribution spread over the central 2 m. The effective load averaged over a 5 m section of wall at the center of the dam was then F_h = 28 kN/m, approximately 40% greater than the Goda load, but only 14% of the unfactored wave impact load.

TABLE 4.3. Aeration coefficients for broken wave loads (Blackmore and Hewson, 1984)

Bed slope	1:5 to 1:10	1:30 to 1:50	1:100
Foreshore conditions			
Smooth bed, sand	1.5	0.9	0.7
Rough, rocky	0.5	0.3	0.24
Very rough, emergent rocks	0.13	0.18	0.14

4.5. BROKEN WAVE FORCES

For many coastal sea walls, and for some breakwaters, the design wave condition may be limited by depth in front of the structure. In these cases, the larger waves at the structure will be broken (see parameter map previously in Figure 4.6), and it is most unlikely that wave impact loads will be caused. A method to estimate an average wave pressure from broken wave loads was developed by Blackmore and Hewson (1984).

$$p_{i,\max} = \lambda \rho T_p C_b^2 \quad (30a)$$

where λ is an aeration for which values are suggested in Table 4.3, ρ is the water density, T_p is the wave spectral peak period, and C_b is the velocity of the breaker at the wall. The simplest formula for breaker celerity may be given by shallow water wave theory:

$$C_b = (gd)^{1/2} \quad (30b)$$

These methods may be used to make an initial estimate of the horizontal wave force under broken waves, $F_{h,\text{broken}}$, to be applied only if $F_{h,\text{broken}} < F_{h,\text{Goda}}$:

$$F_{h,\text{broken}} = h_f \cdot p_{i,\max} = h_f \lambda \rho T_p C_b^2 \quad (30c)$$

4.6. SEAWARD OR NEGATIVE FORCES

Most design methods for caisson and other vertical breakwaters concentrate on forces that act landward, usually termed positive forces. It has, however, been shown that some breakwaters have failed by sliding or rotation seaward, indicating that net seaward or negative forces may indeed be greater than positive forces.

Previous prediction methods by Sainflou and Goda have been discussed and extended by McConnell et al. (1999). Both of the early methods were based on (relatively) deep water, and nonbreaking or pulsating waves. Sainflou's method used trochoidal theory to provide pressure distributions at wave crest and trough (Figure 4.18):

$$p_1' = \rho g(H - h_0) \quad (31a)$$

$$p_2' = \rho g H / \cosh(2\pi h/L) \quad (31b)$$

$$h_0 = (\pi H^2/L) \coth(2\pi h/L) \quad (31c)$$

4.36 HANDBOOK OF COASTAL ENGINEERING

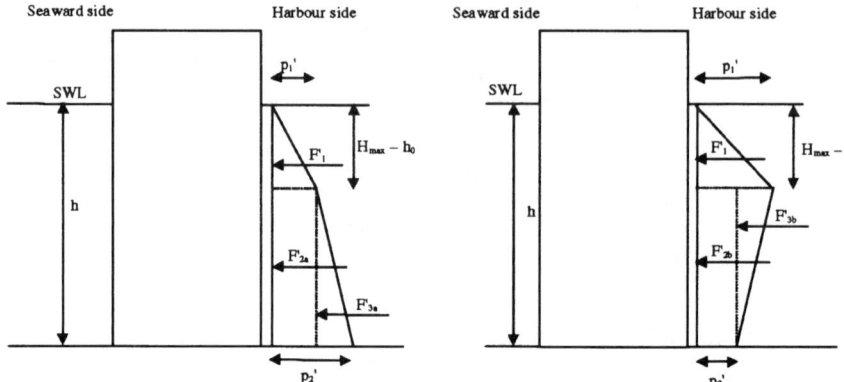

FIGURE 4.18. Net negative pressure distribution according to Sainflou.

The parameter h_0 takes into consideration the asymmetry of waves in front of a structure. The wave height used in Sainflou's formulae is assumed to be H_{max}. From these formulae may be calculated the net negative horizontal force using Sainflou's method, F_{hS}.

$$F_{hS} = (H_{max} - h_0) p_1'/2 + (p_1' + p_2')(h - H_{max} + h_0)/2 \tag{32}$$

Goda (1967) developed a simple diagram of positive and negative wave forces on caissons, shown in Figure 4.19. This gives a rapid assessment of likely wave forces, but there are some problems with this diagram, such as the concept of waves of zero steepness. The diagram does, however, indicate one important point: for values of the relative depth, h/L < 0.25, it indicates higher net negative (–ve) forces than positive (+ve) forces for most wave steepnesses. This is initially somewhat surprising, as it implies that for these conditions seaward forces under wave troughs govern the primary design response. This has been tested by examining data from tests within PROVERBS. Values of $F_{h,\min(1/250)}/F_{h,\max(1/250)}$ showed that there is some risk that –ve forces exceed +ve forces for small relative wave heights, H_{si}/h_s < 0.3. For deeper water conditions, H_{si}/h_s < 0.2, most test data give measured –ve forces that are greater than the conventional landward forces ($F_{h,\min(1/250)}/F_{h,\max(1/250)}$ > 1).

Measurements derived from the two data sets previously termed HR94 and HR97 (see Section 4.4.1) are compared with the prediction curves in Figure 4.20. Only the curves for wave steepnesses, s_{mo} = 0.02, 0.04, and 0.06, are shown as these encompass the steepnesses used in the two test series. Sainflou's curves suggest an underprediction for H_{si}/d < 0.3 and H_{si}/d > 0.5, whereas between these two regions the test measurements and Sainflou's prediction show reasonably good agreement.

Ratios of $F_{h,\min(1/250)}$ to $F_{h,\min,\text{Sainflou}}$ were calculated for each measurement set, and plotted against H_{si}/d in Figure 4.21. A simple regression analysis gives an almost horizontal line through the data points, suggesting that the correlation between test measurements and predictions is relatively constant.

The modified prediction method therefore requires the calculation of $F_{h,\min}$, derived from Sainflou's formula, which is then multiplied by a factor, itself dependent on whether a probabilistic or a deterministic approach is adopted.

Values of $F_{h,\min,1/250}$ have been calculated using mean and standard deviations from the test data. These gave a mean value of the ratio of measured to Sainflou predictions of

WAVE FORCES ON VERTICAL AND COMPOSITE WALLS   4.37

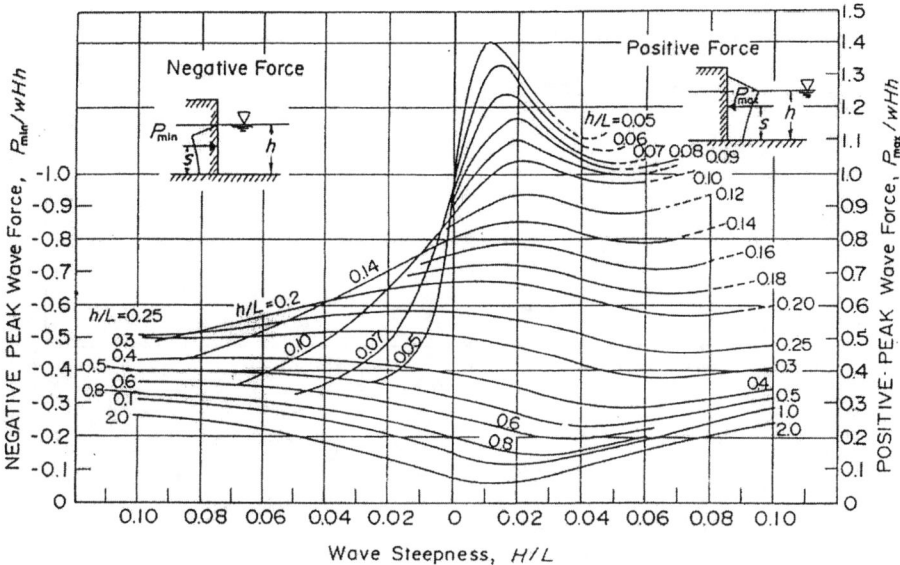

FIGURE 4.19. Design diagram for onshore and offshore wave forces by Goda (1967).

1.126 and a coefficient of variation of 13%. The resulting probabilistic formula for negative forces on simple vertical walls with a 1:50 approach slope, and for $H_{si}/h_s < 0.6$. is

$$F_{h,\min} = 1.126\, F_{hS} \pm 13\% \tag{33a}$$

Deterministic calculations of $F_{h,\min,1/250}$ use the following for simple vertical walls with $H_{si}/h_s < 0.6$:

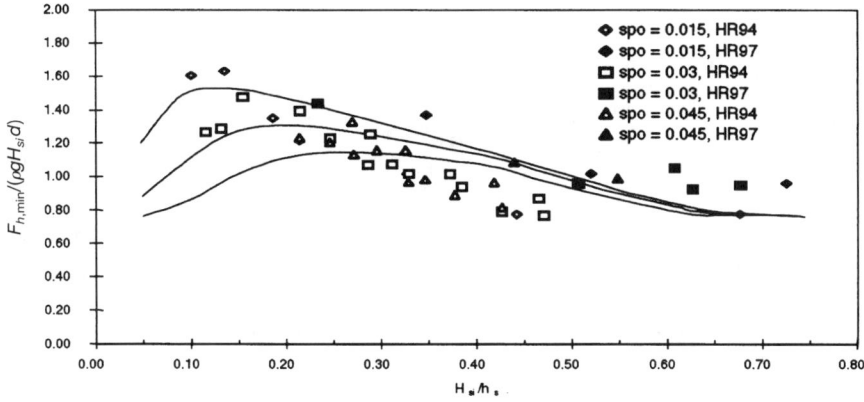

FIGURE 4.20, Measured –ve forces compared with Sainflou prediction lines.

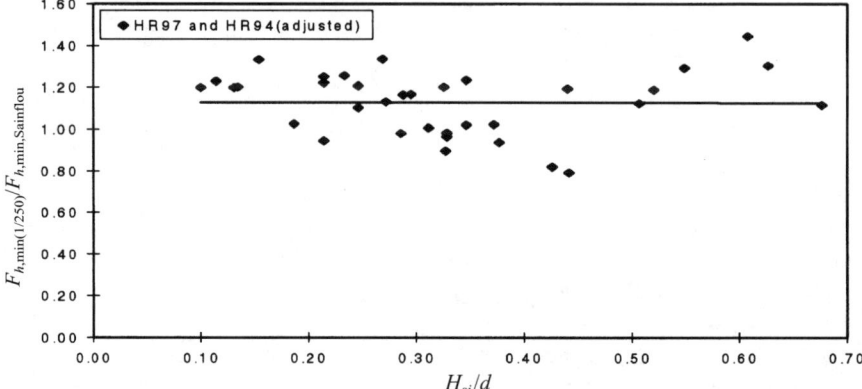

FIGURE 4.21. Regression line for $F_{h,\min(1/250)}/F_{h,\min,\text{Sainflou}(1/250)}$ against H_{si}/d.

$$F_{h,\min} = 1.27 \cdot F_{hS} \tag{33b}$$

In practice, although many old walls were originally of relatively simple construction, most newly built caissons are of composite construction, and many old structures have had added toe berms or mounds. The revised design methods were therefore compared with seaward force measurements for composite walls. Data from the HR94 tests were used and additional data were provided by ENEL-CRIS, and from tests reported by Franco et al. (1996). These data sets have been nondimensionalized as $F_{h,\min}/(\rho g H_{si} d)$ against H_{si}/d, where d is the depth of water at the vertical face, i.e., on top of the rubble mound. These are compared with the data for vertical walls in Figure 4.22.

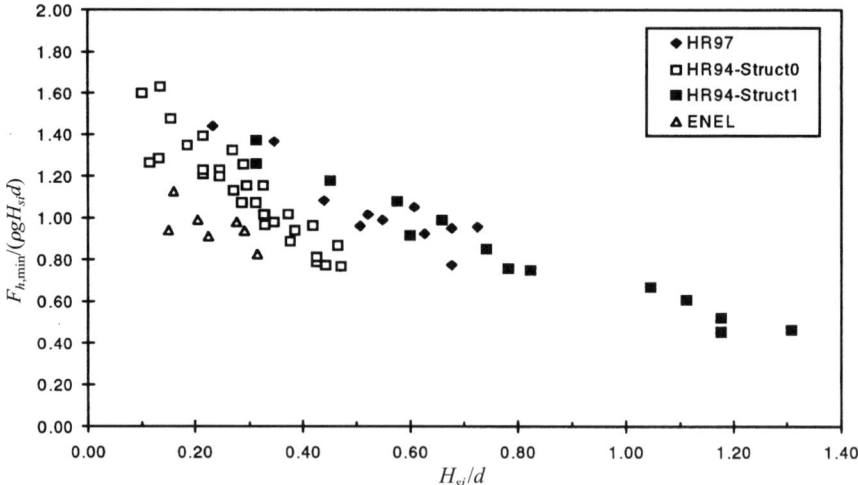

FIGURE 4.22. Comparison of results for simple vertical and composite walls.

This shows fairly good agreement between the results for composite structures and those for vertical walls, suggesting that the new methods in equations 34a and 34b may be safely adopted for desk studies and initial design.
Using a probabilistic approach:

$$F_{h,\min} = 1.126 \left\{ (H_{\max} - h_0) \frac{p'_1}{2} + (p'_1 + p'_2) \frac{(d - H_{\max} + h_0)}{2} \right\} \pm 13\% \quad (34a)$$

Using a deterministic approach:

$$F_{h,\min} = 1.27 \left\{ (H_{\max} - h_0) \frac{p'_1}{2} + (p'_1 + p'_2) \frac{(d - H_{\max} + h_0)}{2} \right\} \quad (34b)$$

It is important to note that in all calculations of net negative force described here, it is assumed that the same static water level acts on both sides of the breakwater, and that there are no additional wave-induced forces acting on the harbor side of the structure. Wave action within the harbor, or overtopping impacts behind the breakwater, may generate forces additional to those discussed here.

4.7. RECOMMENDATIONS FOR ANALYSIS AND DESIGN

4.7.1. Summary of Design Procedures

The first steps in determining wave loads on a breakwater or sea wall are to identify the key geometric and hydraulic parameters, and to make initial estimates of the force magnitude. The estimation of wave forces on vertical and composite walls may be determined in a step-by-step procedure.

Step 1: Main geometric and wave parameters
Width, height, and slope of front of berm in front of wall B_b, h_b, and α
Crest freeboard above water level, height of caisson face R_c and h_f
Equivalent berm width $B_{eq} = B_b + (h_b/2 \tan \alpha)$
Depth of water over the berm for design water level d
Obliquity of structure to (design) wave direction β

It should be noted that some parameters take different values for different water levels, so those parameters may need to be applied at more than one water level.

Wave condition(s) given by H_{si} and T_p should include effects of wave shoaling and refraction, and of depth-limited breaking. The peak period wavelength L_{pi} should be derived in the water depth of the structure, h_s. Goda's simple breaking method may be used to calculate $H_{\max} = 1.8 H_s$ or $H_{\max,b}$, where the breaking wave depth h_{break} is taken 5 H_s seaward of the structure.

Step 2: First estimate wave force/mean pressure
Use Hiroi's formula (equation 10) to estimate an equivalent uniform wave pressure p_{av} on the front face over a wall height h_f up to 1.25 H_s above still water level, and hence the total horizontal force F_{Hiroi}.

Use F_{Hiroi} to give first estimate of breakwater width B_c to resist sliding, assuming no dynamic uplift pressures, but including buoyant up-thrust F_B, and friction $\mu = 0.5$.

Step 3: Improve calculation of horizontal and uplift forces

Use Goda's method to predict horizontal and uplift forces at 1/250 level, $F_{h,\text{Goda}}$ and $F_{u,\text{Goda}}$, and related pressure distribution. Wave pressures on the front are distributed trapezoidally, reducing from p_1 at the water level to p_2 at the caisson base. Above the water level, $p = 0$ at the notional run-up point at height η^*. The total horizontal force F_h (per m length of breakwater) is calculated by integrating pressures p_1, p_2, and p_3 over the front face.

Uplift pressures are distributed triangularly from the seaward edge to zero at the rear heel. The total uplift force F_u (per m length of breakwater) is given by $F_u = 0.5\, p_u B_c$.

Step 4: Revise estimates of caisson size

Use both horizontal and uplift forces $F_{h,\text{Goda}}$, $F_{u,\text{Goda}}$, to revise estimate of caisson width., assuming friction $\mu = 0.5$ or other given value.

Use simple methods to check crest elevation against required wave transmission or overtopping limits, and confirm or revise crest freeboard, R_c.

Step 5: Identify loading case using parameter map

The main parameters used in the parameter map for determination of the wave loading on the structure are the relative berm height, h_b/h_s, the relative wave height, H_{si}/d, and the relative berm width, B_{eq}/L_{pi}. The wave parameters H_{si} and L_{pi} are determined in the water depth h_s, and L_{pi} is determined by linear wave theory.

The relative berm height, h_b/h_s, determines the type of structure—whether a simple vertical wall, a composite structure with a low mound, a composite structure with a high mound or a rubble mound with a crown wall.

Calculate key decision parameters:

Relative berm height $(h_b{}^* = h_b/h_s)$

Relative wave height $(H_s{}^* = H_{si}/h_s)$

Relative berm width, $(B^* = B_{eq}/L_p)$

Use parameter map to determine loading case type: pulsating, transition, impact, or broken.

Step 6: Estimate $P_{i\%}$

Estimate $P_{i\%}$ using method in Section 4.2.4 by Calabrese (1998, 2000). Calculate maximum breaking wave height, $H_{99.6\%bC}$, and significant (breaking) wave height H_{sibC}, and derive estimate of $P_{i\%}$. Note that H_{sibC} is a fictional rather than measured parameter, and may differ significantly from breaking significant wave heights determined by other methods, see particularly Weggel (1972), Owen (1980), and Allsop and Durand (1998).

Use $P_{i\%}$ to decide loading case

$P_{i\%} < 2\% \Rightarrow$ Little breaking, wave loads are primarily pulsating

$2 < P_{i\%} < 10 \Rightarrow$ Breaking waves give impacts

$P_{i\%} > 10\% \Rightarrow$ Heavy breaking may give impacts or broken loads

Step 7: Initial calculation of impact force

If the parameter map indicates "transition" or "impact" conditions, then use equations (27a–c) to calculate impact force, $F_{h,A\&V}$, at 1/250 level, and an estimate of force rise time. Use this impact force if $F_{h,A\&V}/F_{h,\text{Goda}} > 1.2$.

If wave attack is oblique, methods described in Section 4.4.2 based on results by Allsop and Calabrese (1998, 1999) and Franco et al. (1996), show that wave impact conditions ($H_{si}/h_s > 0.35$) give substantial reductions in effective force, even over short caisson lengths, $0.005 < L_c/L_{op} < 0.2$. A simple regression line gives the reduction factor C_{Fh} to be applied to wave impact loads in terms of relative caisson length.

Step 8: Scale corrections

If the condition in step 6 and/or 7 is pulsating, wave forces $F_{h,\text{Goda}}$ and $F_{u,\text{Goda}}$ are unaffected by scale effects, so they may be scaled directly by Froude, i.e., scale correction factor of unity is applicable.

If forces are Impact in steps 5 and 6, then a scale correction factor should be applied to impact pressures computed using the simple prediction methods described earlier. A correction factor of 0.4–0.5 on $F_{h,\text{impact}}$ has been derived by Allsop et al. (1996c) based on field measurements of impact pressures. A more complete, but more complicated general approach in four stages is suggested by Kortenhaus et al. (1999):

a) estimate aeration from level of $P_{i\%}$
b) estimate attenuation of $F_{h,\text{impact}}$ from level of aeration
c) apply scale correction to $F_{h,\text{impact}}$ based on aeration-induced attenuation
d) scale rise time and impact duration, t_r and T_D, by duration correction

Step 9: Pressure distributions

If conditions in step 5 and/or 6 are pulsating, plot pressures calculated by Goda's method. If forces are impact, then use new general method; see Kortenhaus et al. (1999), or examples described by Allsop et al. (1996).

4.7.2. Uncertainties, Scale Effects Safety Factors, and Models

The prediction methods described above have been developed to give initial estimates of wave pressures/forces, or of effective wave loads. These methods are primarily based on results from small-scale hydraulic model tests, quantifying movement of idealized caissons, or measurements of wave force or wave pressures. In each instance, the tests considered a limited range of simplified structures under a limited range of wave loadings. Some studies have used oblique waves, and a few used short-crested waves, but the majority of design methods have only been developed for normal wave attack.

Even with these simplifications, there remain considerable differences between different prediction methods, and there are also significant uncertainties in their application to nonidealized structures under realistic wave conditions.

The uncertainties are lowest for pulsating wave loads and simple caisson breakwaters, using Goda's method where appropriate to yield predictions of horizontal sliding force and uplift forces. These predictions do not need to be corrected for scale effects, but in any practical application should be multiplied by an appropriate factor of safety, F_S. For most conditions, it is generally recommended that values be chosen for $F_S = 1.5$–2.0. Where the design case differs noticeably from the idealization used for the design method, safety factors greater than $F_S = 1.5$–2.0 should be considered.

For wave impacts, it has been shown that wave impact pressures/forces may substantially exceed pulsating loads. Wave impact forces are, however, inherently more variable, depending significantly on the form of wave breaking, the shape of the individual wave, degree of aeration, etc. It is therefore important that any calculations of wave impact pressures/forces should explore the likely uncertainties by repeating the calculations using different methods where available, or by sensitivity calculations over the likely range of input parameters. Individual factors of safety greater than $F_S = 1.5 - 2.0$ may need to be considered.

For large structures, where the main structural elements behave monolithically, the effect of wave impact forces may be assessed using simple dynamic design methods. A number of dynamic models have been developed within the PROVERBS project. A simple methods using pressure impulse calculations has been discussed here. Alternative approaches have been described elsewhere; see especially Torrini et al. (1999).

In recent years, a number of numerical models have been developed to describe wave action at structures, see for instance the Volume of Fluid model NEWMOTICS described by Waller and Allsop (1997). There are three main types of numerical models of wave action:

a) *Depth-averaged.* These models are efficient and quick to run. They only operate for waves in shallow water when it can be assumed that wave velocities do not differ much over the water depth. These models do not reproduce the processes of wave overturning and breaking so will not give wave forces correctly. Most of these models have been produced for a single section (known as one-dimensional models or numerical wave flumes).

b) *Overturning wave models.* These models generally use boundary integral or boundary element methods (BIM or BEM). They calculate pressures and velocities throughout depth, and will handle wave overturning. They will not, however, handle any form of breaking re-entry, collision or separation, so cannot be used beyond the first moment of breaking wave contact. Most of these models have been produced for a single section (known as 2-dV models or numerical wave flumes), but some will handle three-dimensional simulations.

c) *Full wave breaking models.* These use volume of fluid or similar methods, and will reproduce full wave breaking including fluid separation. Most of these models are two-dimensional, but three-dimensional versions will follow in time.

There is a significant problem with all numerical models of wave forces, particularly those involving a collision between a fast-moving air–water interface and a structural element. In the numerical model, the fluid is represented as incompressible, but the structure is described as completely fixed/rigid. At the moment of collision between incompressible water and rigid structure, local pressures will be infinitely high. It is only by averaging processes over spatial cells, and over a time step, that apparently realistic values for pressures may be computed. The problem arises in justifying the correct time and space steps to give realistic pressure/force results. This may be overcome by determining pressure or force impulses, using these as input to dynamic calculations of structure response.

For all the work on empirical design methods discussed above, there remains no simpler or more reliable method to assess wave loads than hydraulic modeling using (usually Froudian) scale models. Such models, typically at scales between 1:10 and 1:50, will use fully random waves, short-crested, or long-crested waves as appropriate. Wave pressures should be recorded at sample rates of 500 Hz or faster, and measurements should usually be made over at least 1000 waves. The results will usually be analyzed statistically, giving deterministic and/or probabilistic descriptions of the measured parameters

ACKNOWLEDGMENTS

Information and research results that have been used to develop this chapter have arisen from projects funded in the United Kingdom by the Construction Sponsorship Directorate of Department of the Environment, Transport and the Regions (DETR) under contract numbers PECD 7/6/263 and 312, and CI 39/5/96, and part by the European Union MAST programme under the MCS-Project, contract MAS2-CT92-0047, and later the PROVERBS project, contract MAS3-CT95-0041. Additional support was given by the Department of Hydraulics of University of Naples Frederico II, and by the Department of Civil and Structural Engineering of University of Sheffield. Further funding for visiting researchers at Wallingford was awarded from the National Council for Research in Italy, CNR. Support for extended analysis was provided by the Ministry of Agriculture, Fisheries and Food under Flood and Coastal Defence Commission FD0201.

The author wishes to acknowledge the support of the University of Sheffield in preparation of many of the papers cited here, and in drafting this chapter, and to thank colleagues at Wallingford, within the PROVERBS projects, and visiting researchers for assistance in testing and in processing data, especially Phillip Besley, Alberto Bracci-Laudiero, Mario Calabrese, Adrian Channel, Marina Fillipone, Hartmut Flohr, Rob Jones, Janice McKenna, Paul McConkey, Kirsty McConnell, Lucia Torrini, and Diego Vicinanza.

Suggestions on spatial limitations and coherence were distilled from discussions with Professor Alberto Lamberti from Bologna in Italy, with additional comments from Professor Hans Burcharth and Dr Peter Frigaard of Aalborg University in Denmark.

The author wishes to record his particular thanks to Andreas Kortenhaus, without whose good humor, tolerance and tireless work the PROVERBS project would have been much more difficult, and to Kirsty McConnell who helped maintain considerable amounts of measurement data, complete further analysis, and keep track of frequently shifting opinions.

REFERENCES AND READINGS

Allsop N. W. H. (1995a) "Stability of rock armor and rip-rap on coastal structures" Chapter 14 in *River, Coastal and Shoreline Protection: erosion control using rip-rap and armorstone*, ed. Thorne, Abt, Barends, Maynord, Pilarczyk, pp. 213–229, ISBN 0-471-94235-9, John Wiley & Sons, Chichester, U.K..

Allsop N. W. H. (1995b) "Vertical walls and breakwaters: optimisation to improve vessel safety and wave disturbance by reducing wave reflections." Chapter 10 in *Wave Forces on Inclined and Vertical Wall Structures* pp. 232–258, ed. Kobayashi N. & Demirbilek Z., ISBN 0-784400080-6, ASCE, New York.

Allsop N. W. H., Vann A. M., Howarth M., Jones R. J. & Davis J. P. (1995c) "Measurements of wave impacts at full scale: results of fieldwork on concrete armour units." *Conference on Coastal Structures and Breakwaters '95*, pp. 287–302, Institution of Civil Engineers/Thomas Telford, London.

Allsop N. W. H. (Ed.) (1998) "Coastlines, structures and breakwaters" *Proceedings of Conference at ICE, 19–20 March 1998*, ISBN 0 7277 2668 4, Thomas Telford, London.

Allsop N. W. H. & Bray R. N. (1994) "Vertical breakwaters in the United Kingdom: historical and recent experience" *Proceedings of Workshop on Wave Barriers in Deep Waters*, pp. 76–100, Port and Harbour Research Institute, Yokosuka, Japan.

Allsop N. W. H. & Calabrese M. (1998) "Impact loadings on vertical walls in directional seas " Proceedings 26th ICCE, Copenhagen, ASCE, New York.

Allsop N. W. H. & Calabrese M. (1999) "Forces on vertical and composite breakwaters—effects of oblique or short-crested waves." Research Report SR 465, HR Wallingford, Wallingford, UK.

Allsop N. W. H., Calabrese M., Vicinanza D. & Jones R. J. (1996a) "Wave impact loads on vertical and composite breakwaters." Proceedings 10th Congress of the Asia and Pacific Division of IAHR, 26–29 August 1996, Langkawi Island, Malaysia

Allsop N. W. H. & Durand N. (1998) "Prediction of depth-limited wave breaking over steep

bed slopes." Strategic Research Report SR 516, HR Wallingford (spring 1998), Wallingford, UK.
Allsop N. W. H. & McConnell K. J. (2000) "Wave action at seawalls: wave forces, overtopping and armouring design." Strategic Research Report SR 509, HR Wallingford (spring 2000), Wallingford, UK.
Allsop N. W. H., McKenna J. E., Vicinanza D. & Whittaker T. J. T. (1996b) "New design formulae for wave loadings on vertical breakwaters and seawalls." *25th International Conference on Coastal Engineering, September 1996, Orlando*, ASCE, New York.
Allsop N. W. H. & Müller G. (1995) Discussion to Technical Note 2575: "Comparative study on breaking wave forces on vertical walls" by Ergin A. & Abdalla S., *Proc. ASCE, J. Waterway, Port, Coastal and Ocean Engineering Division*, September/October 1995, ASCE, New York
Allsop N. W. H., Kortenhaus, A., McConnell K. J. & Oumeraci H. (1999) "New design methods for wave loadings on vertical breakwaters under pulsating and impact conditions." *Proceedings Coastal Structures '99, Santander, Spain*, ASCE, New York.
Allsop N. W. H., Vicinanza D., & McKenna J. E. (1996c) "Wave forces on vertical and composite breakwaters." Research Report SR 443, HR Wallingford, Wallingford, UK.
Allsop N. W. H. & Vicinanza D. (1996) "Wave impact loadings on vertical breakwaters: development of new prediction formulae." *Proceedings 11th International Harbour Congress*, pp. 275–284, Royal Flemish Society of Engineers, Antwerp, Belgium.
Bagnold R.A. (1939) "Interim report on wave pressure research" *J. Institution of Civil Engineers, 12*, 202–226, ICE, London
Battjes J. A. (1982) "Effects of short-crestedness on wave loads on long structures." *Applied Ocean Research, 4*, 3, 165–172.
Bélorgey M., Bergmann H., De Gerloni M., Franco L., Passoni, G. & Tabet-Aoul E. H. (1999): Perforated caisson breakwaters: hydraulic performance and wave forces. *Proceedings Coastal Structures '99, Santander, Spain*, ASCE, New York.
Besley P. B. (1999) *Overtopping of Seawalls—Design and Assessment Manual.* R & D Technical Report W 178, ISBN 1 85705 069 X, Environment Agency, Bristol, UK.
Besley P. B., Allsop N. W. H., Ackers J. A., Hay-Smith D. & McKenna J. E. (1999) "Waves on reservoirs, and their effects on dam protection." *Proceedings of Annual General Meeting presentation to British Dam Society*, ICE, London.
Blackmore P. A. & Hewson P. (1984) "Experiments on full scale impact pressures." *Coastal Engineering, 8*, 331–346, Elsevier, Amsterdam.
Bradbury A. P. & Allsop N. W. H. (1988) "Hydraulic effects of breakwater crown walls." *Proceedings of Conference on Design of Breakwaters*, pp. 385–396, Institution of Civil Engineers, Thomas Telford, London.
Bradbury A. P., Allsop N. W. H. & Stephens R. V. (1988) "Hydraulic performance of breakwater crown walls." HR Report SR 146, Hydraulics Research, Wallingford, UK.
Bray R. N. & Tatham P. F. B. (1992) *Old Waterfront Walls: Management, Maintenance and Rehabilitation.* ISBN: 0 419 17640 3, CIRIA/E & FN Spon, London.
British Standards Institution (1984) *British Standard Code of Practice for Maritime Structures, Part 1. General Criteria.* BS 6349: Part 1: 1984 (and Amendments 5488 and 5942), British Standards Institution, London.
British Standards Institution (1991) *British Standard Code of Practice for Maritime Structures, Part 7. Guide to the Design and Construction of Breakwaters.* BS 6349: Part 7: 1991, British Standards Institution, London.
Bullock, G. N., Hewson, P. J., Crawford, A. R., Bird, P. A. D. (1999) "Field and laboratory measurements of wave loads on vertical breakwaters." *Proceedings Coastal Structures '99, Santander, Spain*, ASCE, New York.
Burcharth H. F. (1994) "The design of breakwaters." Chapter 29 in *Coastal, Estuarial and Harbour Engineers' Reference Book*, pp. 381–424, Ed. M. B. Abbott & W. A. Price, ISBN 0-419-15430-2, E & FN Spon, London.
Calabrese M. (1998) "Onset of breaking in front of vertical and composite breakwaters." *8th International Conference ISOPE*, Montreal.
Calabrese M. (2000) "A breaking criterion for waves in front of simple vertical and composite breakwaters." Paper submitted to *Coastal Engineering*, in press Spring 2000, Elsevier, Amsterdam.
CERC (1984) *Shore Protection Manual.* 2 Vols., U.S. Army Corps of Engineers, U.S. Govt. Printing Office, Washington.

Clifford J. E. (Ed.) (1996) *Advances in Coastal Structures and Breakwaters.* Proceedings of conference at ICE, 27–29 April 1995, ISBN 0 7277 2509 2, Thomas Telford, London.
Ergin A. & Abdalla S. (1995) "Comparative study on breaking wave forces on vertical walls." *Proc. ASCE, J. Waterway, Port, Coastal and Ocean Engineering Division,* September/October 1995, ASCE, New York
Franco L. (1994) "Vertical breakwaters: the Italian experience." *Coastal Engineering, Special Issue on Vertical Breakwaters, 22,* 31–55, Elsevier Science BV, Amsterdam.
Franco L (1996) "Ancient Mediterranean harbours: a heritage to preserve." *Ocean & Coastal Management, 30,* 115–152, Elsevier Science, Oxford.
Franco C., Meer J. W. van der, & Franco L (1996) "Multi-directional wave loads on vertical breakwaters." Proceedings 25th International Conference on Coastal Engineering, Orlando, ASCE, New York.
Fuhrboter I. A. (1994) "Wave loads on sea dikes and sea-walls." Chapter 27 in *Coastal, Estuarial and Harbour Engineers' Reference Book,* pp. 351–367, Ed. M. B. Abbott & W. A. Price, ISBN 0-419-15430-2, E & FN Spon, London.
Goda Y. (1967) "The fourth order approximation to the pressure of standing waves." *Coastal Engineering in Japan,* Vol. 10, pp. 1–11, JSCE, Tokyo.
Goda Y. (1974), "New wave pressure formulae for composite breakwaters." *Proceedings 14th International Conference on Coastal Engineering,* pp. 1702–1720, ASCE, New York.
Goda Y. (1985) *Random Seas and Maritime Structures,* University of Tokyo Press, Tokyo.
Goda Y. (1994), "Dynamic response of upright breakwaters to inpulsive breaking wave forces," *Coastal Engineering 22,* 135–158.
Goda Y. (1995a) "Japan's design practice in assessing wave forces on vertical breakwaters." Chapter 6 in *Wave Forces on Inclined and Vertical Wall Structures,* pp. 140–155, ed. Kobayashi N. & Demirbilek Z., ISBN 0-784400080-6, ASCE, New York.
Goda Y (1995b) Discussion to Technical Note 2575: "Comparative study on breaking wave forces on vertical walls" by Ergin A. & Abdalla S., *Proc. ASCE, J. Waterway, Port, Coastal and Ocean Engineering Division,* September/October 1995, ASCE, New York.
Hattori M., Arami A. & Yui T. (1994), "Wave impact pressure on vertical walls under breaking waves of various types." *Coastal Engineering, 22,* 79–114.
Hayashi T. & Hattori M. (1958), "Pressure of the breaker against a vertical wall." *Coastal Engineering in Japan,* pp. 25–37, JSCE, Tokyo.
Hiroi I. (1919) "On the method of estimating the force of waves." *Memoirs of the Engineering Faculty, Imperial University of Tokyo,* Vol. X, No. 1, p. 19.
Hitachi S. (1994) "Case study of breakwater damages—Mutsu-Ogawara Port." *Proceedings of Workshop on Wave Barriers in Deep Waters,* pp. 308–331, Port and Harbour Research Institute, Yokosuka.
Howarth M. W., Vann A. M., Davis J. P., Allsop N. W. H. & Jones R.J . (1996) "Comparison of wave impact pressures on armour units at propototype and model scale." *25th International Conference on Coastal Engineering, Orlando,* ASCE, New York.
Ito Y. (1971) "Stability of mixed type breakwaters—A method of probable sliding distance." *Coastal Engineering in Japan,* Vol. 14, pp. 53–61, JSCE, Tokyo.
Jensen O. J. (1983) "Breakwater superstructures." *Proceedings of Conference on Coastal Structures '83,* Arlington, pp. 272–285, ASCE, New York.
Jenson O. J. (1984) *A Monograph on Rubble Mound Breakwaters,* Danish Hydraulic Institute, Horsholm.
Kirkgoz M. S. (1990), "An experimental investigation of a vertical wall response to a breaking wave impact." *Ocean Engineering, 17*(4), 379–391, Elsevier Science, Oxford.
Kirkgoz M. S. (1995) "Breaking wave impact on vertical and sloping coastal structures." *Ocean Engineering, 22,* 1, 35–48, Elsevier Science, Oxford.
Kortenhaus A., Oumeraci H., Allsop N. W. H., McConnell K. J., Van Gelder P. H. A. J. M. & Hewson P. J. (1999) "Wave impact loads—pressures and forces." In *Final Proceedings, MAST III, PROVERBS-Project: Vol. IIa: Hydrodynamic Aspects,* Leichtweiss Institute, University of Braunschweig, Germany.
Meer, J. W. van der (1995) "A review of stability formulae for rock and rip-rap slopes under wave attack." Chapter 13 in *River, Coastal and Shoreline Protection: Erosion Control Using Rip-rap and Armourstone,* ed. Thorne, Abt, Barends, Maynord, Pilarczyk, pp. 191–212, ISBN 0-471-94235-9, John Wiley & Sons, Chichester, UK.

Meer, J. W. van der & Janssen J. P. F. M. (1995) "Wave run-up and wave overtopping at dikes." " Chapter 1 in *Wave Forces on Inclined and Vertical Wall Structures*, ed. Kobayashi N. & Demirbilek Z., ISBN 0-7844-0080-6, ASCE, New York.
McConnell K. J. (2000) "Wave forces on walls: the influence of wind/swell seas and steep approach slopes." Technical Research Report, HR Wallingford (spring 2000), Wallingford, UK.
McConnell K. J. (1998) *Revetments Against Wave Attack: A Design Manual*. ISBN 0-7277-2706-0, Thomas Telford, London.
McConnell K. J., Allsop N. W. H. & Flohr H. (1999) "Seaward wave loading on vertical coastal structures." *Proceedings Coastal Structures '99, Santander, Spain*, ASCE, New York.
McKenna J. E. (1997) *Wave Forces on Caissons and Breakwater Crown Walls*. Ph.D thesis, Queen's University of Belfast, September 1997, Belfast, UK.
McKenna J. E. & Allsop N. W. H. (1998) "Statistical distribution of horizontal wave forces on vertical breakwaters." *Proceedings of 26th International Conference on Coastal Engineering, June 1998, Copenhagen*, ASCE, New York, pp. 2082–2095.
Miche R. (1944) "Mouvement ondulatoire de la mer." *Annales des Ponts et Chaussèes, 144*, 25–61, Paris.
Minikin R. R. (1950) *Winds, Waves and Maritime Structures: Studies in Harbour Making and the Protection of Coasts*. Charles Griffin & Co., London.
Minikin R. R. (1963) *Winds, Waves and Maritime Structures*, 2nd ed. Charles Griffin & Co., London.
Muraki Y. (1966) "Field investigations on the oscillations of breakwaters caused by wave action." *Coastal Engineering In Japan, 9*, JSCE, Tokyo.
Oumeraci H. (1994) "Review and analysis of vertical breakwater failures—lessons learned." *Coastal Engineering, Special Issue on Vertical Breakwaters, 22*, 3–29, Elsevier Science BV, Amsterdam.
Oumeraci H, Allsop N. W. H., de Groot M. B., Crouch R. S & Vrijling J. K (1999) "Probabilistic Design Methods for Vertical Breakwaters (PROVERBS)." *Proceedings Coastal Structures '99, Santander, Spain*, ASCE, New York.
Oumeraci H. & Kortenhaus A., (1994) "Analysis of the dynamic response of caisson breakwaters." *Coastal Engineering, 22*, 159–183.
Oumeraci H., Kortenhaus A. & Klammer P. (1995) "Displacement of caisson breakwaters induced by breaking wave impacts." *Proceedings of International Conference on Advances in Coastal Structures and Breakwaters*, pp. 50–63, Institution of Civil Engineers, Thomas Telford, London.
Oumeraci, H. & Kortenhaus, A. (eds.) (1999) *Final Report: Probabilistic Design Tools for Vertical Breakwaters. Final Proceedings, MAST III, PROVERBS-Project: Probabilistic Design Tools for Vertical Breakwaters*, Vol. II. University of Braunschweig, Germany.
Oumeraci, H.; Kortenhaus, A.; Allsop, N. W. H.; De Groot, M. B.; Crouch, R. S.; Vrijling, J. K.; Voortman, H. G. (2000) *Probabilistic Design Tools for Vertical Breakwaters*.Balkema, Rotterdam.
Owen M. W. (1980) "Design of sea walls for wave overtopping." Report EX 924, Hydraulics Research, Wallingford, UK.
Owen M. W. (1982) "The hydraulic design of sea-wall profiles." In *Proceedings ICE Conference on Shoreline Protection, September 1982*, pp. 185–192, Thomas Telford, London.
Owen M. W. (1982) "Overtopping of sea defences." In *Proceedings of Conference on Hydraulic Modelling of Civil Engineering Structures, September 1982*, Coventry, BHRA, Bedford.
Partenscky H-W. (1988) "Dynamic forces due to waves breaking at vertical coastal structures." *Proceedings 2nd International Symposium on Wave Research and Coastal Engineering*, pp. 261–275, University of Hannover, Hannover.
Peregrine D. H. (1995) "Water wave impact on walls and the associated hydro-dynamic pressure field." Chapter 11 in *Wave Forces on Inclined and Vertical Wall Structures*, pp. 259–281, ed. Kobayashi N. & Demirbilek Z., ASCE, New York.
Pilarczyk K. W., Klein Breteler M. & Bezuijen A.(1995) "Wave forces and structural response of placed block revetments on inclined structures." Chapter 3 in *Wave Forces on Inclined and Vertical Wall Structures*, ed. Kobayashi N. & Demirbilek Z., ISBN 0-7844-0080-6, ASCE, New York.
Rouville M. A., Besson P. & Petry P. (1938) "Etudes internationales sur les efforts dus aux lames." *Annales des Ponts et Chaussees, 108*, 5–113, Paris.

Sainflou, G. (1928) "Essai sur les digues maritimes verticales." *Annales des Ponts et Chausees,* Vol. 98, Paris.
Schmidt R., Oumeraci H. & Partenscky H-W. (1992) "Impact loads induced by plunging breakers on vertical structures." *Proceedings of 23rd ICCE, Venice,* ASCE, New York.
Simm J. D. (Ed.) (1991) *Manual on the Use of Rock in Coastal and Shoreline Engineering."* CIRIA/CUR, Special Publication 83, ISBN: 0 86017 326 7, CIRIA, London.
Takahashi, S. (1996) *Design of Vertical Breakwaters.* Port and Harbour Research Institute, Yokosuka, Japan; also notes for Short Course, 25th ICCE, Orlando.
Tanimota K. & Goda Y. (1991) "Historical development of breakwater structures in the world." *Proceedings Conference Coastal Structures and Breakwaters, November 1991,* ICE, pp. 193–206, Thomas Telford, London.
Tanimoto K. & Takahashi S. (1994) "Design and construction of caisson breakwaters—the Japanese experience." *Coastal Engineering,* 22, 57–78, Elsevier Science BV, Amsterdam.
Thomas R. S. & Hall B. (1992) *Seawall Design.* ISBN 0-7506-1053-0, CIRIA/Butterworth–Heinemann, Oxford.
Torrini L., McConnell K. J. & Allsop N. W. H. (1999) "Simplified dynamic analysis of wave impact loadings on vertical/composite breakwaters." *Proceedings Coastal Structures '99, Santander, Spain,* ASCE, New York.
Voortman H. G. & Vrijling J. K. (1999) "Vertical breakwaters: The foundation problem." In *Proceedings of Coastal Structures '99, Santandar, Spain,* Balkema, Rotterdam.
Vrijling J. K., Burcharth H. F., Voortman H. G., & Sørenssen J. D. (1999) "The design philosophy for a vertical breakwater." In *Proceedings of Coastal Structures '99, Santandar, Spain,* Balkema, Rotterdam.
Walkden M. J. A., Crawford A. R., Hewson, P. J. Bullock G. N., & Bird P. A. D. (1995), "Wave impact loading on vertical structures," *Proceedings Coastal Structures and Breakwaters '95,* Thomas Telford, London.
Walkden M. A., Hewson P., & Bullock G. N. (1996) "Wave impulse prediction for caisson design." *Proceedings 25th International Conference on Coastal Engineering, Orlando,* ASCE, New York.
Waller M. N. H. & Allsop N. W. H. (1997) "Wave dynamics at coastal structures: Development of numerical modelling methods." Research report SR 496, July 1997, HR Wallingford, Wallingford, UK.
Weggel J. R., Hall W. & Maxwell C. (1970), "Numerical model for wave pressure distributions." *J. Waterways, Harbours, and Coastal Engineering Division,* 96, WW3, 623–642, ASCE.
Weggel J. R. (1972) "Maximum breakwater height." *Journal Waterway, Harbor and Coastal Engineering Division, ASCE,* Vol. 98, No. WW4, 529–548.

CHAPTER 5
OFFSHORE (DETACHED) BREAKWATERS

JOHN B. HERBICH, PH.D., P.E.
W. H. Bauer Professor Emeritus
Ocean Engineering Program
Civil Engineering Department
Texas A&M University
College Station, TX 77843-3136, USA

Notation	5.1
Introduction	5.2
Literature Review	5.9
Emergent Breakwaters	5.9
Submerged Breakwaters	5.12
Reef-type Breakwaters	5.15
Model Studies	5.16
Emergent Breakwaters	5.16
Mathematical Models	5.18
Comparison of Model and Field Studies	5.23
Description of Selected Projects	5.34
United States Breakwaters	5.34
Foreign Breakwaters	5.47
Analysis	5.50
Additional Comments	5.51
Summary and Statistical Analysis	5.51
Design Considerations	5.57
Empirical Relationships	5.57
Summary and Conclusions	5.75
Appendix A	5.79
Appendix B	5.83
Bibliography	5.90

NOTATION

d, D Design water depth at structure
d' Approximation to d
d_{b5} Breaking water depth at site based on H_{05} and T_5
g Acceleration of gravity (9.81 m/sec^2)

H_{05} Deepwater wave height calculated by averaging the five largest nonstorm waves that occur in a year (m)
H_s Significant wave height
I Beach slope
L_5 Wavelength at structure corresponding to deepwater conditions H_{05} and T_5
L_c Alongshore salient width, measured at original shoreline
L_g Gap distance between adjacent breakwater segments
L_{05} Deepwater wavelength corresponding to T_5
L_p Alongshore project length (length of shoreline to be protected)
L_s Breakwater segment length
SAR   Salient area ratio
T_5 Wave period corresponding to H_{05}
X Breakwater segment distance from original shoreline
X' Approximation to X
X_g Shoreline change at gap
$\underline{X_s}$ Desired shoreline advancement, salient length
$\overline{X_s}'$ Approximation to X_s
X_s Average salient length for project

INTRODUCTION

Beach erosion and accretion has proceeded for many thousands of years, ever since the first landmass was exposed to waves, currents, and winds. Man's interest in preserving or stabilizing the shoreline began with his first development in the coastal region, such as ports, harbors, residential communities, etc. Although these coastal developments have been around for hundreds, if not thousands, of years, coastal erosion problems have reached severe proportions only within the past thirty to forty years. The reasons for this significant increase in shoreline erosion can be attributed to a variety of occurrences, ranging from increased development along the coastline since the 1920s, the continuous sea level rise (due to the greenhouse effect), and the damming of rivers, which, at one time, supplied coastal sediment.

The first attempts to mitigate shoreline erosion problems were by the use of sea walls, revetments, and groins. Groins are built approximately perpendicular to the shoreline and serve to trap a portion of the longshore sediment-laden current or the littoral drift. Long stretches of coastline may be protected by constructing a series of groins, known as groin fields. The objective in groin field construction is an extended beach within successive groins; however, a zone of erosion often occurs in the immediate areas downdrift of the groin field. Well-designed groin fields may be sufficient in areas sheltered from storm waves, otherwise the groins will act as reflection basins for high-energy waves, resulting in sand being funneled offshore where it may be beyond recovery.

Revetments and sea walls are constructed along eroding shorelines, frequently resulting in the disappearance of beaches. Additional studies on whether the sea walls and revetments cause beach erosion have been underway, sometimes with conflicting results.

More recent developments in shore protection structures are offshore, or detached, breakwaters. Prototype experience indicates that offshore breakwaters can be an important alternative for shore stabilization and accretion.

The primary objectives of an offshore breakwater system are to increase the longevity of a renourished beach, provide a wider beach for recreation, and provide protection to upland areas from waves and flooding (EM 1110-2-1617).[151] Offshore breakwaters can also be deployed to create and stabilize wetland areas. The combination of low-crested breakwaters and planted marsh grasses may be used to establish wetlands and control erosion along estuarine shorelines. The offshore breakwater design should seek to minimize negative impacts of the structure on the downdrift shoreline. Groins may impound sediment, whereas properly designed offshore breakwaters can allow continued movement of longshore transport through the project area.

The primary consideration in offshore breakwater design is the resulting shoreline configuration due to the breakwater(s). An offshore breakwater will either form a tombolo (connecting shoreline to the breakwater), a sand salient, or will have a limited effect. A limited effect occurs if there is an inadequate sediment supply, or if the breakwater is constructed too far offshore to influence a shoreline change. Offshore breakwaters are generally designed either for a resulting salient formation or a tombolo.

The resulting shoreline configuration depends on many factors, including the longshore sediment transport rate, sediment supply, sediment size, underwater beach slope, wave climate, currents, tidal range, and the breakwater system parameters, including structure length, gap distance, distance from shore, depth at structure, and the structure's crest height.

When properly designed, the formation of a sand salient (salient) begins soon after construction of the breakwater. Approximately 25 to 50% of the sand volume may be deposited in the first year, with a steady state usually being reached after three to four years. After that time, the system will remain relatively stable, undergoing minor or moderate seasonal changes.[45] The process of beach accretion may be accelerated by depositing sand on the shoreline between the breakwater and the shoreline, employing either pumping (dredging) or trucking methods.

In the last twenty years the number of offshore breakwaters in Japan has increased approximately 1000%, while the increase in groin-type structures has increased only 100%. Between 1962 and 1981, 2100 offshore breakwaters were built, or about 105 breakwaters per year. This shows a strong trend in that country toward the use of offshore breakwaters over groins as a means of beach stabilization and protection. Offshore breakwaters have been constructed in Australia, Brazil, Canada, Cyprus, Denmark, France, India, Israel, Italy, Japan, Mexico, Monaco, Morocco, Nicaragua, Portugal, Singapore, South Africa, Spain, Sri Lanka, and Ukraine. In the United States, offshore breakwaters have been constructed in California, Florida, Hawaii, Illinois, Louisiana, Maryland, Massachusetts, New Jersey, Ohio, Pennsylvania, and Virginia. Some 225 detached breakwater segments existed along the continental United States and Hawaiian coasts, as compared with 4000 segments along Japan's (9400-km) coastline.[117] There are many variations of the offshore breakwater concept: reef breakwaters (designed with low crest elevation and homogeneous rock or stone size), headland breakwaters (to promote formation of a headland), submerged sill, or a perched beach.[117] A summary of U.S. offshore breakwater projects is shown in Table 5.1.[18]

An offshore breakwater is built as a barrier parallel to the shoreline and therefore it is not attached to the coast. This "emergent" offshore breakwater is designed so that waves will not overtop the structure, i.e., the crest height of the breakwater is significantly above the highwater level. The theory behind emergent offshore breakwaters is that they serve to diminish incoming wave energies that tend to lift and carry away beach materials while allowing the littoral drift to continue along undisturbed.[79] Wave energy is then refracted and diffracted around each end of the emergent breakwater and transmitted into the breakwater's shadow zone. Diffracted waves converge behind the breakwater, forming a calm water area, resulting in sand being deposited in the sheltered region. This ideal state has been proven difficult to establish because of reduced wave energy in the lee of the structure and the littoral drift's dependency upon this energy for sediment resuspension

Table 5.1. Summary of U.S. offshore breakwaters[18]

Coast	Project	Location	Date of construction	Number of segments	Project length	Segment length	Gap length	Distance offshore preproject	Water depth	Fill placed	Beach response*	Constructed by	Maintained by
Atlantic	Winthrop Beech (low tide)	Massachusetts	1935	5	825m	91m	30m	Unknown	3.0 (mlw)	No	1	State of Massachusetts	State of Massachusetts
Atlantic	Winthrop Beech (high tide)	Massachusetts	1935	1		100	30	305	3.0 (mhw)	No	3	State of Massachusetts	
Atlantic (Potomac River)	Colonial Beach (Central Beach)	Virginia	1982	4	427	81	46	64	1.2	Yes	2	USACE	USACE
Atlantic (Potcmac River)	Coloniel Beach (Castlewood Park)	Virginia	1982	3	335	61,93	26,40	46	1.2	Yes	1	USACE	USACE
Chesapeake Bay	Elms Beach (wetland)	Maryland	1985	3	335	47	53	44	0.6-0.9	Yes	1	State of Maryland	State of Maryland
Chesapeake Bay	Elk Neck State Park (wetland)	Maryland	1989	4	107	15	15		0.6-0.9	No	2-4	USACE	USACE
Chesapeake Bay	Terrapin Beach (wetland)	Maryland	1989	4		23	15, 31, 23	38.1	0.8-0.9	Yes	5	USACE	USACE
Chesapeake Bay	Eastern Neck (wetland)	Maryland	1992–1993	26	1676	31	23		0.3-0.6	Yes		US Fish and Wildlife Service, USACE	US Fish and Wildlife Service
Chesapeake Bay	Bay Ridge	Maryland	1990–1991	11	686	31	31	42.7		Yes	4	Private	Private
Gulf of Mexico	Redington Shores	Florida	1985–1986	1	100	100	0	104		Yes	1	USACE	USACE
Gulf of Mexico	Holly Beach	Louisiana	1985	6	555	46, 51, 50	93, 89	78,61	2.5	No	4	State of Louisiana	State of Louisiana
Gulf of Mexico	Holly Beach	Louisiana	1991-1993	76		46, 53	91, 84	122,183	1.4, 1.6	Yes	3	State of Louisiana	State of Louisiana
Gulf of Mexico	Grand Isle	Louisiana		4	84	70	21	107	2	No	3	City of Grand Isle	City of Grand Isle
Lake Erie	Lakeview Park	Ohio	1977	3	403	76	49	152	3.7	Yes	4	USACE	City of Lorain
Lake Erie	Presque Isle	Pennsylvania	1978	3	440	38	61, 91	60	0.9-1.2	Yes	2	USACE	USACE
Lake Erie	Presque Isle	Pennsylvania	1989–1992	55	8300	46	107	76-107	1.5-2.4 (lwd)	Yes	3-4	USACE	USACE
Lake Erie	Lakeshore Park	Ohio	1982	3	244	38	61	120	2.1	Yes	5	USACE	City of Ashtabula
Lake Erie	Eest Herbor	Ohio	1983	4	550	46	90, 105, 120	170	2.3	No	5	State of Ohio	State of Ohio
Lake Erie	Maumee Bay (headland)	Ohio	1990	5	823	61	76		1.3	Yes	1	USACE	State of Ohio
Lake Erie	Sims Park (headland)	Ohio	1992	3	975	38	49		2.5	Yes	1	USACE	City of Euclid
Pacific	Venice	California	1905	1	180	180	0	370		No	5	Private	Private
Pacific	Haleiwea Beach	Hawaii	1965	1	49	49	0	90	2.1(msl)	Yes	3	USACE/State of Hawaii	USACE
Pacific	Sand Island	Hawaii	1991	3	110	21	23					USACE	USACE

*Beach response is coded as follows: 1 = permanent tombolos, 2 = periodic tombolos, 3 = well developed selients, 4 = subdued salients. 5 = no sinuosity.

5.4

and transport. The result is sediment deposition in the lee of the offshore breakwater. The quantity of sand deposited and the final, or equilibrium, shoreline configuration depend upon several factors, including the wave environment before and after breakwater construction, breakwater design and placement, initial littoral drift concentration and energy level, together with sedimentary properties. A cuspate salient (Figure 5.1) is usually formed in the lee of the breakwater. This shoreline configuration will allow a small portion of the littoral drift to pass between the salient and the offshore breakwater. However, as a general "rule of thumb," if the breakwater is designed so that the distance from shore is less than 80% of the breakwater's length, the salient will likely grow until contact is made with the breakwater. The shoreline configuration is known as a tombolo (Figure 5.1). If tombolo formation occurs, a significant proportion of littoral drift materials is deposited behind and just upcurrent of the offshore breakwater; erosion occurs downstream of the offshore breakwater system until a new state of equilibrium is reached. The time required to reach equilibrium depends upon the breakwater design and placement as well as environmental conditions; however, it is not uncommon for equilibrium to be reached within three to five years. During this equilibrium-establishing period, neighboring downdrift beaches may erode to meet the sediment supply demands of shorelines further down the coast. The offshore, or detached breakwater, in its emergent configuration, has proven itself effective in coastal erosion control throughout the world; however, due to its potential tombolo formation and relatively high cost of construction and maintenance, a submerged breakwater may be desirable. At present, there is much interest in the construction of detached breakwaters having the structure's crest height at or below the water's surface. Submerged structures and their shoreline responses have been examined in field studies conducted in Australia,[70] Japan,[77] and South Africa.[125] Submerged breakwaters are beginning to receive worldwide acceptance for beach erosion control. An example of a submerged offshore breakwater in the United States is the one constructed at Redington Shores in Florida.[18]

The submerged breakwater is designed as a wave filter, e.g., waves of small steepness pass over unhindered, whereas waves of large steepness break on the structure, thus

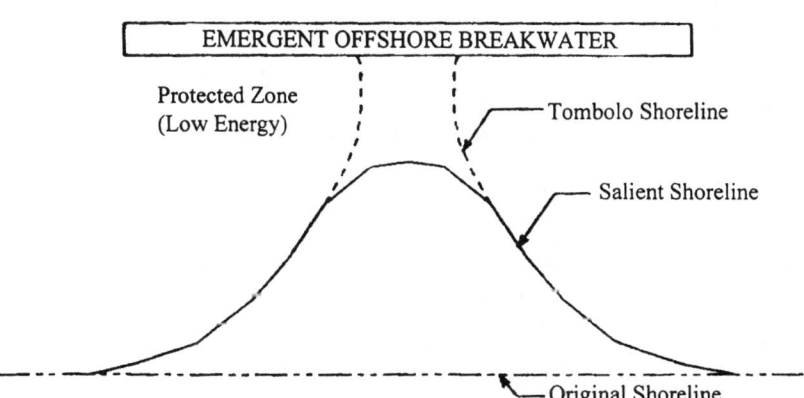

FIGURE 5.1. Definition sketch of salient and tombolo.

losing much of their energy. Laboratory measurements indicated that wave steepness is a determining factor in beach erosion and accretion. Storm waves are steeper and tend to erode the shoreface, while summer waves are of a lesser steepness, thereby causing sand movement onshore and restoring the beach to a fuller condition. In the case of submerged breakwaters, the basic mechanism forming tombolos or salients is dependent on the breakwater's degree of submergence and geometry.

Despite the offshore breakwater's proven effectiveness in the control of beach erosion, such breakwaters are not as popular in the United States as they are in other countries. The reasons for this can be attributed to the somewhat higher cost of construction (as compared with groin construction), difficulties in offshore placement, and the required unit armor weight necessary to withstand breaking wave forces. Model studies, conducted at Texas A&M University[20,56,57,58,81,82,142] and other research institutions, have identified many of the significant emergent offshore breakwater design parameters and their effects on shoreline configuration. These and other studies, once formalized, are expected to enhance the popularity of offshore breakwaters in the U.S. as shore protection methods by providing design curves for efficient offshore breakwater design and placement.

Problems associated with emergent offshore breakwater construction are significantly reduced in the case of a submerged breakwater. For example, fewer materials are required for construction, because the submerged breakwater does not extend above the water's surface. The required unit armor weight is also reduced, because the wave is allowed to overtop the structure, rather than break on its seaward face. However, in areas where the tidal range is significant, the unit armor weight must be sufficient to withstand breaking wave forces.

Presently, submerged breakwaters are being considered as an alternative erosion control method in many countries. Design criteria, however, for the breakwater's degree of submergence, spacing, gap between successive breakwaters, and offshore placement distance are not well defined. At present, there is also interest in the design of offshore breakwaters with crest heights slightly above or below the mean water level, thereby reducing construction costs. During major storms, breakwaters are submerged by storm surges, thus reducing wave forces on the breakwaters.

Design procedures have been largely empirical and have varied from project to project. This has resulted in highly successful projects in most cases; however, some projects have experienced beach erosion on the downstream side of the breakwaters. Additional research both in the laboratory and in the field is needed to develop more comprehensive design criteria.

An offshore breakwater can be constructed as a single structure for localized shore protection or as multiple breakwaters to protect a longer section of beach. Figure 5.2 shows the general characteristics of single and multiple offshore breakwaters.

Spacing between breakwater segments (gap width) should be carefully evaluated by increasing the gap width according to the amount of wave energy entering the area behind breakwaters.

The ratio of gap width to wavelength of incoming waves is also important. Increasing the gap/wavelength ratio also increases the penetration of wave energy behind the breakwaters.[22]

A numerical model (GENESIS) computes wave diffraction at a gap and calculates both diffraction and refraction for random waves and accounts for wave shoaling and breaking.[51]

The "exposure ratio" is a ratio of gap width to the sum of breakwater length and gap width. Exposure ratios for various prototype projects are listed in Table 5.2 and range from 0.25 to 0.66.

Houston[69] discussed the economic value of beaches and indicated that travel and

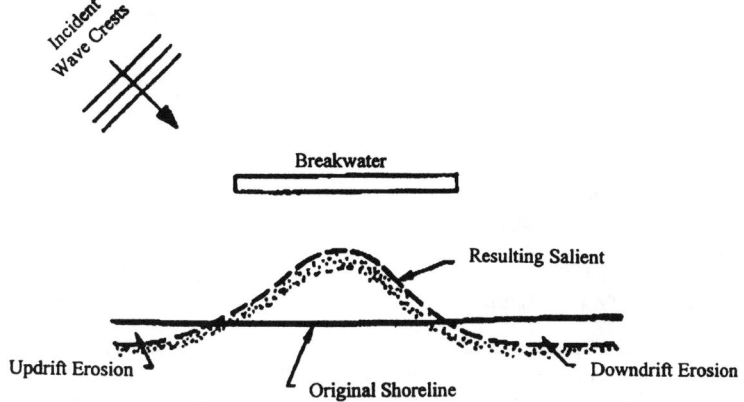

a. Single offshore breakwater

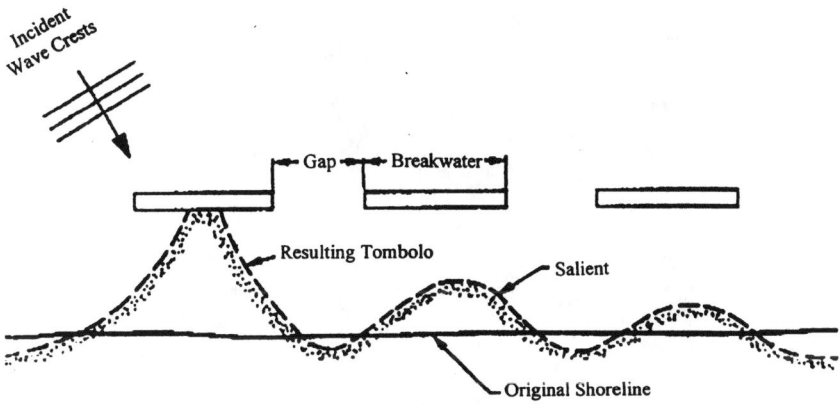

b. Multiple offshore breakwaters

FIGURE 5.2. Definition of offshore breakwaters.[118]

tourism is the United States's leading industry, employer, producer of new jobs and earner of foreign exchange. Recreational beaches are a leading factor in travel and tourism. Travel and tourism contribute $746 billion to the United States's Gross Domestic Product according to the *Wall Street Journal* (1995).

Beaches are the leading tourist destination in the United States; coastal states receive about 85% of tourist-related revenues.

Figure 5.3 shows the multiple breakwater system providing beach protection at Kaike, Tottori, Japan. Figure 5.4 shows a tombolo developed behind the offshore breakwater in Denmark.

5.8 HANDBOOK OF COASTAL ENGINEERING

TABLE 5.2. "Exposure ratios" for various prototype multiple breakwater projects*
(modified from EM 1110-2-1617)

Project	Exposure ratio	Shoreline response
Winthrop Beach, MA	0.25	Permanent tombolos (low tide); well-developed salients (high tide)
Lakeview Park, Lorain, OH	0.36	Subdued salients
Castlewood Park, Colonial Beach, VA	0.31 to 0.38	Permanent tombolos
Central Beach, Colonial Beach, VA	0.39 to 0.45	Periodic tombolos
East Harbor, State Park, OH	0.56	Limited
Presque Isle, Erie, PA (experimental prototype)		Permanent tombolos
(Hydraulic model)	0.60	

*The "exposure ratio" is defined as the ratio of gap width to the sum of the breakwater length and gap width. It is the fraction of shoreline directly exposed to waves and is equal to the fraction of incident wave energy reaching the shoreline through the gaps. A "sheltering ratio" that is the fraction of incident wave energy intercepted by the breakwaters and kept from the shoreline can also be defined. It is equal to 1 minus the "exposure ratio."

FIGURE 5.3. Photo of multiple breakwater system at Kaike, Tottori, Japan.

FIGURE 5.4. Photo of tombolo behind offshore breakwater in Denmark.

LITERATURE REVIEW

Emergent Breakwaters

Existing literature can be subdivided into three sections:
1. Hydraulic model studies
2. Numerical simulation models
3. Field evaluation of existing breakwaters

The first offshore breakwater in the United States was constructed in Venice, California in 1905, followed by breakwaters in Santa Barbara, California (1929) and Santa Monica, California (1934). An extensive annotative bibliography on detached breakwaters was compiled by Lesnick[89] in 1979. The bibliography was presented to assist in the development of reliable design procedures for offshore breakwaters. Two hundred and seventeen entries were annotated in the bibliography.

Shoreline changes have been monitored in the field after the construction of offshore breakwaters. Walker et al.[170] examined a series of three offshore breakwaters built in the fall of 1977 at Lakeview Park, Ohio on Lake Erie. The purpose of these breakwaters was to protect a replenished beach used for recreation and shore protection. The study showed that the project functioned well with very little loss of sand from the system and without adverse effects on the downdrift side.

Nir[104] investigated fifteen offshore breakwaters built in the late sixties and early seventies along the Israeli coast for recreational purposes. Most of the lengths of the Israeli breakwaters were just less than their distances offshore. This caused the formation of tombolos. Almost no sand accumulated when the distance offshore of the breakwater was greater than twice the length of the breakwater. From his field data, Nir[104] developed a "least squares" line equation for the average sand thickness in the tombolo as a function of the breakwater's distance from the original shoreline divided by the length of the breakwater. Most tombolos reached the breakwater in one or two years and matured to an equilibrium configuration in about five years. The sand of these tombolos was shown to come from nearby beaches and shallow water.

The most comprehensive field study of offshore breakwaters was conducted in Japan by Toyoshima.[145,146] Toyoshima states that before the 1940's in Japan, littoral drift caused most of the beach erosion. Since that time, waves coming straight at the beach have been causing severe beach erosion by carrying large amounts of sand offshore from the beach during storms. Toyoshima's study showed that offshore breakwaters tend to stabilize the variability of shoreline changes over the years. The Kaike Coast has gained a stable coastline by virtue of a detached breakwater system (Figure 5.3).

Toyoshima's statistical study of offshore breakwaters constructed in Japan produced the following design guidelines:

1. An offshore breakwater should be placed in the surf zone, which means shoreward from the breaking waves.

2. If a tombolo is desired, the distance of the breakwater from the original shoreline divided by the breakwater length should be less than 0.74.

3. The length of the breakwater should be two to six times the wavelength or 200 (61 m) to 650 feet (198 m) long while the gap distance between breakwaters should be one wavelength or 65 (20 m) to 165 feet (50 m).

Toyoshima found from surveys that much of the sand was transported from the offshore zone to the nearby area of the detached breakwater. An accretion in the volume of sand was discovered not only on the onshore side but also on the offshore side of the breakwater, which was an important finding.

Perlin[110] used a numerical model to predict the beach plan in the lee of an offshore breakwater. His one-line finite difference scheme model handles the case where the tombolo connects to the breakwater and takes into account both diffraction and refraction. The model allows the designer to predict the optimum length and orientation of the structure.

Mimura, et al.[95] combined both a hydraulic model and a numerical model to study the influence of a detached breakwater on the neighboring beach. Relevant quantities related to the wave and current fields, sand transport, and bottom topography were measured as accurately as possible in a hydraulic model. The shoreline change was also simulated by using a one-line numerical model. Agreement between the numerical model and the hydraulic model was found to be good if the predictive expression for the longshore sand transport rate included the longshore variation in wave height.

Comparisons between field and hydraulic model studies have also been conducted by researchers.

Fried[38] examined offshore breakwaters along the coast near Tel Aviv, Israel and a hydraulic model of the same coastline. The study showed that the accelerated sedimentological process in the model reproduces quite well the natural process in the prototype. The study also corroborates its preliminary assumption that the newly developed beach will not be widened on account of the neighboring shorelines, but as a result of changes in the longshore transport. This reduces only the quantity of sand that would otherwise be carried offshore due to local hydrographical and sedimentological circumstances. Fried tested fourteen alternatives of a series of six offshore breakwaters in the model. It was shown that the building schedule of the offshore breakwaters had a direct influence on the rate of sand accretion but not on the final equilibrium state. Rosen and Vajda[120] hypothesized that a morphological and sedimentological equilibrium is reached behind a detached breakwater when the shape of the contour lines become such that along the sheltered beach the diffracted waves have components of momentum opposed to the gradients of the mean sea level (induced by radiation stress due to nonuniform wave heights along the wave fronts approaching from both sides). The study indicated that the significant parameters characterizing the dimensions of the salient or tombolo in the equilibrium state are shown to be the relative length of the breakwater (as compared to its distance to the original shoreline), the relative distance to the original shoreline (as compared to the position of the breaker line) and the relative height of the breakwater crest (above mean water level compared to the incident wave height). Data from the field and the hydraulic model were compiled and relationships from the above variables were represented graphically.

Other researchers have relied on only hydraulic model tests to study the effect of offshore breakwaters. The advantage of a model study is that it is relatively cheap, adaptable, and of fairly short duration.

Silvester[134] conducted one of the earliest model tests on offshore breakwaters in 1957. His model tests were designed for wave patterns with two different lengths of breakwaters and waves of two different steepnesses.

In South Africa, Schoones and Moller[125] designed a model test for eight offshore breakwaters for developing a recreational beach at False Bay, South Africa. The preliminary study showed that if breakwaters are built closer to the shoreline at the end of the offshore breakwater series, they act as groins by forming tombolos. This will decrease the current velocity through the gaps between the end breakwaters and will improve the circulation.

Noda[106] attempted to clarify the physical parameters controlling the development of tombolos behind offshore breakwaters on the basis of the results obtained by laboratory experiments. The parameters are wave characteristics, sediment size, water depth, placement of the structure and its dimensions. Noda showed that when an offshore breakwater is placed so that the distance from the structure to the original shoreline divided by the distance from the wave breaking line to the original shoreline is 0.56, the maximum deposition occurs in the tombolo.

Similar model tests were conducted by Shinohara and Tsubaki[129] and Harris.[56] The model tests were conducted to clarify the changes of the shoreline from the construction of an offshore breakwater or a series of two offshore breakwaters. The amount of sand deposited within the sheltered region and movement of the sandy beach were monitored for both storm and swell waves. Shinohara and Tsubaki's model tests consisted of a single offshore breakwater while Harris' study consisted of two breakwaters separated by a gap. Both studies proved again that the morphological equilibrium state is influenced by wave steepness and the relative distance from the initial shoreline. Criteria were determined in these studies for the best positioning of the breakwaters.

Shinohara and Tsubaki[129] showed a close relation between diffracted crestline and final shoreline and they concluded that the main cause of the changes of the shoreline and the sand movement of the beach was the diffraction of the intruding waves. Harris showed the importance of the gap width in a series of offshore breakwaters. As the gap narrows, more sand is accumulated behind the offshore breakwater.

Cords[20] observed shoreline change behind a series of three offshore breakwaters in a three-dimensional wave basin with a sloped, movable-bed beach. The study indicated that the area and volume gained behind a breakwater will be greater as the breakwater is moved closer to shore and the gap distance between adjoining breakwaters is decreased.

Most of the sand movement and salient formation occurs during short periods of storm wave action.

Submerged Breakwaters

Natural parameters involved in establishing the performance of a submerged breakwater at a particular location include sediment availability, sediment grain size, wave characteristics, local bathymetry, design and placement of the breakwater(s), and relative submergence. Because it is difficult to accurately model all of the parameters that influence shoreline response to a submerged breakwater, prototype analysis of existing breakwater projects is extremely valuable. Although the number of existing installations of submerged breakwaters is significantly less than existing emergent breakwaters, the submerged type has been employed worldwide.

The breakwater's relative submergence plays a key role in regulating wave energy transmission over the breakwater. In an experiment conducted by Johnson, Fuchs, and Morison,[74] a series of wave steepnesses were conducted for particular values of relative submergence. It was found that the breakwater's wave damping effectiveness increased along with increasing wave steepness. This phenomenon can be attributed to the fact that waves near their critical steepness are easily tripped on the structure, whereas waves of low steepness can pass over the submerged structure relatively undisturbed.

A second problem associated with the breakwater's degree of relative submergence is the "pile-up" phenomenon. The pile-up phenomenon is an expression of the quasiequilibrium reached between the main rate of water flowing into the protected zone by waves breaking over the low or submerged breakwater, and that of water flowing out of the protected zone as a result of the difference in mean water level inside and outside. The pile-up phenomenon has been observed in both field and laboratory[67] investigations, and is found to increase in severity, i.e., there are greater differences in the breakwater's seaward and landward water level as the breakwater's relative submergence and length are increased. This pile-up phenomenon has been shown, experimentally,[67] to rapidly increase as the breakwater's relative submergence exceeds the value of 0.7, i.e., breakwater height is 70% of the water's depth.

Lyzlov[90] conducted a submerged breakwater model study that examined three values of relative submergence, 0.7, 0.85, and 1.0. Lyzlov found that the average dissipation of wave height (72 tests) was registered progressively at 33, 39, and 46% at corresponding relative submergence values of 0.7, 0.85, and 1.0. Lyzlov also discovered that for values of wave steepness greater than 0.05, the dissipative effect of the submerged breakwater decreases for all values of H_s/D.

Dattatri[25] performed wave height dissipation studies for a series of relative submergence, he concluded that as H_s/D approached a value of 0.6, or the breakwater height is 60% of the water's depth, the wave energy transmission approaches 75285%. Hence, Dattatri suggested that the submersion of the structure should never exceed 40% of the water's depth ($H_s/D = 0.6$) in order to be effective in shoreline protection. The relationships between breakwater crest width and incident wave length (W/L), significant wave height and water depth H_s/D, and the energy transmission coefficient (KT) for a permeable breakwater of rectangular cross section are shown in Figure 5.5.

Krafft and Herbich[81] conducted a performance evaluation and literature review of thirteen U.S. breakwater projects and four foreign breakwater projects. The projects evaluated included both emergent and submerged breakwaters. Several of these breakwater projects and additional applications are described herein.

Kabelac[75] describes the implementation of an Akhun unit, an underwater breakwater

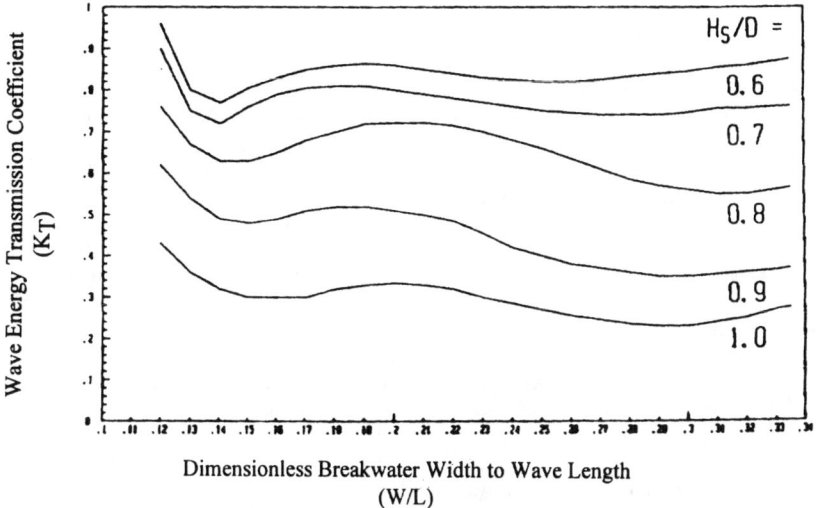

Dimensionless Breakwater Width to Wave Length
(W/L)

FIGURE 5.5. Transmission coefficient versus relative crest width and relative submergence for a permeable submerged breakwater of rectangular cross-section.[142]

unit of trapezoidal shape with a vertical shoreward face, installed at a location in the Black Sea to protect an eroding shoreline that supports nearby railroad tracks. Within a three-year period an estimated 479,911 ft^3 (13,600 m^3) of sediments had been accumulated and shoreline stability was maintained. The design was based on a hydraulic model. In fact, frequent use has been made of submerged breakwaters along the Black Sea coast, and significant research concerning submerged breakwaters has been completed in the former USSR.

Success with artificially located underwater mounds of sand is documented by Zwamborn et al.[172] A design based on moveable bed models and prototype tests resulted in construction of a 14,760 ft- (4.5 km) long mound completed in 1970 along the coast of Durban, South Africa. The underwater mound acts as a submerged breakwater and was constructed to prevent siltation of a nearby harbor, and to enlarge and protect Durban's beaches. The breakwater was constructed using dredged material from the harbor itself.

Due to undesirable wave conditions at the port of Leixoes, Portugal, a submerged breakwater comprised of a rubble mound core topped with concrete blocks was erected in 1936. More favorable wave conditions resulted in the harbor, allowing for improved ship maneuverability.

Japan has used offshore detached breakwaters for shore protection more widely than any other country. Toyoshima[143] presented a study of breakwaters in Japan wherein he documents the existence of eight submerged breakwaters. The performance of these specific breakwaters is not detailed. By 1981 Japan had a total of 2,305 breakwaters, both emergent and submerged.

The predominant use of submerged breakwaters has been for shore protection. Thus, sediment entrapment capabilities are an important design factor. Protective capabilities of the beach increase as the shoreline extends seaward. Model tests have shown that submerged breakwaters constructed with sloping seaward surfaces and vertical leeward surfaces (Akhun type) maintained dissipation effectiveness and were geometrically suitable for sediment accumulation. The sloping front face allows sediments to overtop, whereas the vertical backside does not promote sediment exchange. Moreover, it was found that breakwaters having a sloped seaward face attenuate waves as efficiently as rectangular breakwaters and reduce seafloor scour by reducing wave reflection.

The relative distance offshore, which is the ratio of breakwater distance from shoreline to the effective length of the breakwater, X/B, has been found to be a significant factor in sediment movement behavior. Some guidelines have been developed recommending that the ratio be larger than one to prevent tombolo formation for emergent breakwaters. Tallent[142] found that although tombolo formation did not occur for submerged breakwaters, sediment entrapment effectively increased until a value of relative distance offshore of about 0.8 was reached.

Extensive research has been conducted concerning interaction between wave action and submerged offshore breakwaters. Characteristics of the behavior of overtopping waves have been well documented. The variables that control the amount of wave energy overtopping are breakwater height, h; freeboard, F; water depth, d; deepwater wave height, H_0, or incident wave height, H_i; deepwater wave length, L_0, or incident wave length, L; wave period, T; and breakwater crest width, C. One notable characteristic is the relative wave energy loss with increasing wave steepness, H/L. This phenomenon suggests the submerged breakwater behaves as a filter, attenuating steeper waves with higher erosional capability.

Breakwater crest width has also been found to have an important effect on wave attenuating characteristics. Maximum wave attenuation occurring at a ratio of wavelength to breakwater crest width of 3 to 4 has been documented.

The relative level of submergence, which is the ratio of breakwater height to water depth, h/d, has been identified as having a minimum critical value of 0.7 to 0.8 to succeed in providing significant wave energy attenuation. Since there exists a close relationship between energy attenuation and sediment deposition, this minimum value of submergence would then be required for accretive abilities as found in laboratory experiments.

Adams and Sonu[3] have found agreement between model tests of the Santa Monica breakwater and Tanaka's[143] empirical results. Tanaka describes wave transmission coefficients as a function of both the ratio of breakwater freeboard, F, to deepwater wave height, H_0, and of relative breakwater crest width. His results show that negligible wave attenuation is attained until F/H_0 is approximately -1.0, and that the transmitted wave height continues to be diminished as the ratio F/H_0 approaches a value of approximately 1.0. At this upper limit of unity, F/H_0 ceases to be a governing parameter.

The wave field seaward of a structure is comprised of the incident and the reflected wave from the structure. Wave energy reflection characteristics of breakwaters have been measured and analyzed using both the wave height envelope method and spectral technique, among others. Additionally, overtopping of a submerged breakwater causes an area of scattered wave heights, which has been shown to shift part of the energy to higher frequencies and produce secondary waves. Spectral analysis methods may be employed for evaluation of these phenomena.

Spectral analysis relates the time domain wave elevation time history into a frequency domain distribution of wave amplitude and phase. The analysis may be performed with the use of the fast Fourier transform. The discrete Fourier Transform can be described as:

$$S(\omega) = 1/(2/\pi N) \left| \sum_{t=1}^{N} X_t e^{(-i\omega t)} \right|^2 \quad (1)$$

where

$S(\omega)$ = modified periodogram (asymptotically unbiased estimate of the spectral density function)
X_t = the centered and padded data
t = time
ω = wave frequency

The relationship between the incident wave spectrum, $S_i(\omega)$, and the transmitted wave spectrum, $S_t(\omega)$, can be described by the transmitted frequency response function:

$$H_t(\omega) = S_t(\omega)/S_i(\omega) \quad (2)$$

The coefficient of transmission is related to the transmitted frequency response function by the relationship:

$$K_t(\omega) = \{H_t(\omega)\}^{1/2} \quad (3)$$

Similarly, the reflection frequency response function and reflection coefficient may be described by:

$$H_r(\omega) = S_r(\omega)/S_i(\omega) \quad (4)$$

$$K_r(\omega) = \{H_r(\omega)\}^{1/2} \quad (5)$$

A least-squares spectral analysis method of evaluating the incident and reflected wave elevations in a random wave field using a three-wave gauge method is described by Mansard and Funke.[92] A version of this method has been used with success for periodic waves by Mansard.

Reef-type Breakwaters

A reef breakwater is a low-crested rubble-mound unit *without the* traditional multilayer section. This type of breakwater is constructed of homogeneous stones with individual stone weights similar to those used in armor and the first underlayer of conventional breakwaters.[5]

Because of their high porosity and low crest, reef breakwaters are stable and, if they are high enough, can dissipate wave energy. Since they have no core, they are allowed to adjust and deform to an approximate equilibrium condition (Figure 5.6).

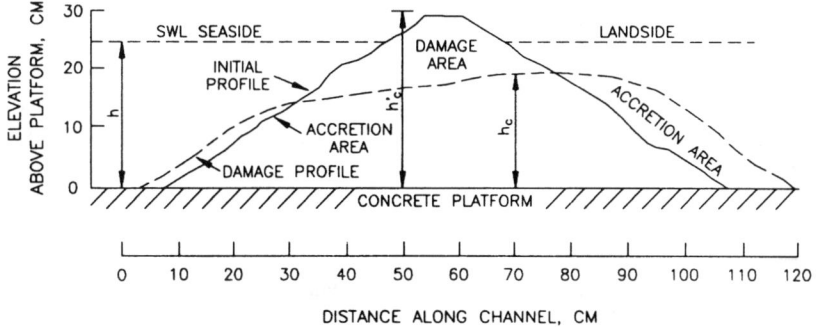

FIGURE 5.6. Typical reef profile, as built and after adjustment to severe wave conditions (Ahrens, 1987).[4]

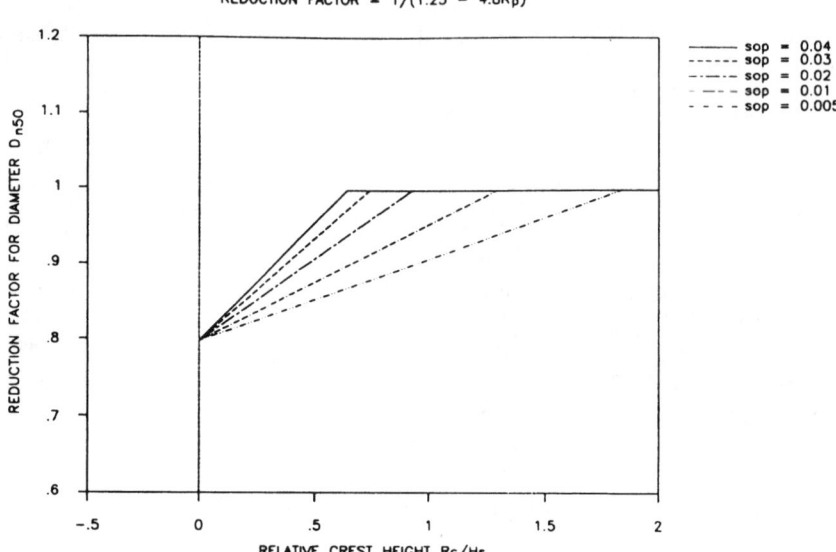

FIGURE 5.7. Design graph with reduction factor for the stone diameter of a low-crested structure as a function of relative crest height and wave steepness (Van der Meer, 1991).[169]

A reduction factor for the size of stone diameter of a low-crested structure as a function of relative crest height and wave steepness is predicted in Figure 5.7.

MODEL STUDIES

A number of laboratory studies were performed in various countries; some of these are summarized in this section.

Emergent Breakwaters

The performance of a detached breakwater depends on the geometry of breakwater placement, wave properties, and sediment characteristics. Dimensional analysis of relevant parameters yields the following relationship:

$$Q_b/XBD = fn(X/B, X_{br}/B, H_0/L_0, G/B, \alpha, \ldots) \qquad (6)$$

where

Q_b = volume of sand moved into the sheltered volume of the breakwater defined by $(X)(B)(D)$
X = distance from original shoreline to seaward edge of breakwater
B = length of breakwater

D = depth of water at seaward edge of breakwater
X_{br} = distance from shoreline to breaker line
G = gap distance between successive breakwaters
α = angle of incident wave crests

The functional expression indicates a dependence on the geometric parameters of the breakwater in determining the amount of sand deposited in the sheltered volume of the breakwater.

The effects of G/B and X/B on Q_b were examined by Harris and Herbich.[57,58] Generally, sand volume increases as X/B decreases. For all field data and most model data examined, tombolos formed for values of X/B less than one. No tombolos formed for X/B equal to or greater than one. The effect of G/B is significant but not as pronounced as that of X/B. Sand volume trapped in the sheltered area generally increases as G/B decreases.

The effect of the location of the breaker line in relation to the location of the breakwater was reported by Rosen and Vajda[120] who showed the importance of having the breaker line seaward of the breakwater to get a maximum salient length. Shinohara and Tsubaki[129] showed the importance of deep-water wave steepness (H_0/L_0) in determining the amount of sand deposited in the sheltered area of a breakwater. Their data clearly showed that as wave steepness increases the amount of sand deposited in the sheltered area increases.

Figure 5.8 shows a relationship between Q_b/XBD and X/B for different values of G/B.[56,57] Figure 5.9 shows the effect of gap spacing accreted behind each breakwater and the gap Q_{b+g} divided by the control volume $XD(G + B)$ for different values of X/B and G/B. The figure indicates that as the gap gets smaller, more sand is accumulated. This plot is dimensionless and may be used to determine expected sand volumes in the sheltered area of one breakwater and one gap. Of particular note is that there can be a net loss of sand in the sheltered area for large gap spacings. Figure 5.10 is a dimensionless plot of

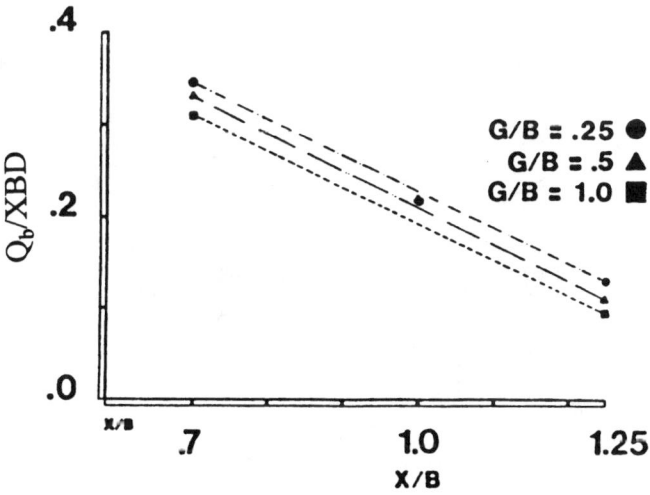

FIGURE 5.8. Effect of gap spacing on sand accreted behind each breakwater.[56,57]

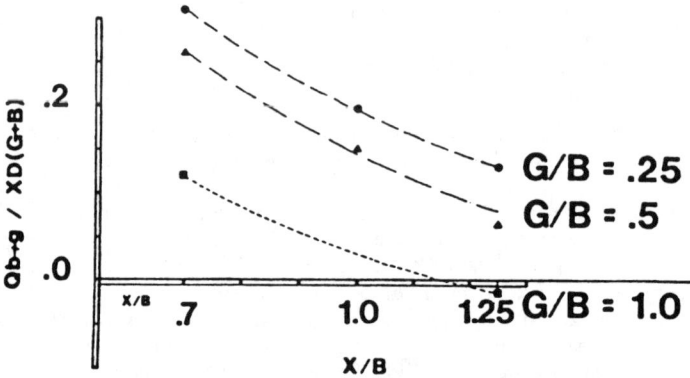

FIGURE 5.9. Effect of gap spacing on sand accreted behind each breakwater and gap.[56,57]

sand accumulated or lost in the gap (Q_b/XDG) as a function of X/B and G/B. As seen in the figure, for the range of X/B shown, the sand will erode for large gap spacings and accrete for small gap spacings.

Mathematical Models

Herbich et al. 1996[64] reported on a simulation study employing the GENESIS mathematical simulation model to evaluate the effect of different types of coastal protection struc-

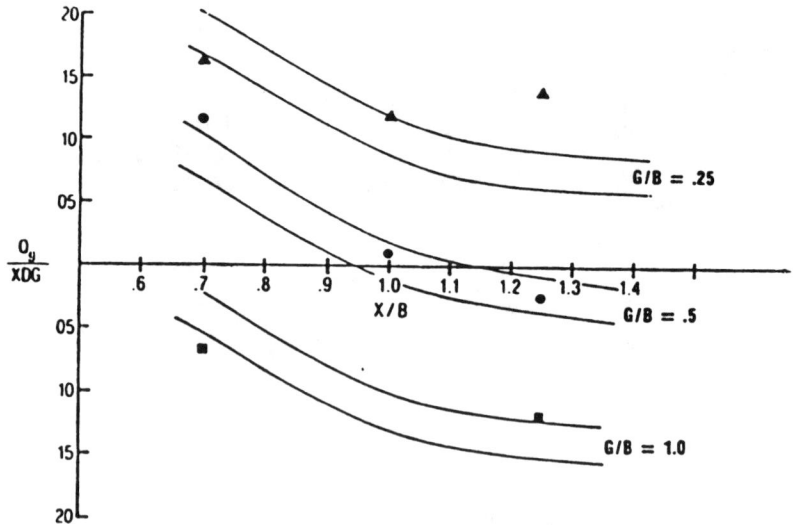

FIGURE 5.10. Effect of gap spacing on sand accreted behind each gap.[56,57]

tures, including offshore breakwaters. The model outline consists of six calculation modules that function semiindependently, allowing modification of any module without affecting or requiring detailed knowledge of any other modules. The principal quantities calculated were:

a) Wave height, wave angle, mean water level, and runup
b) Longshore current
c) Cross-shore sediment transport rate
d) Longshore sediment transport rate
e) Bottom contour orientation and topography change

The procedure of simulation had the following steps:

1. The shoreline data were digitized using Autocad.
2. The wave data were prepared using the Coastal Research Institute (CRI) and the Suez Canal Authority (SCA) data, which were recorded over various periods of time.
3. A computational program (GENESIS) was used to conduct the required simulation. A simulation of 13 years was selected for each run to establish a case of relatively stable state, undergoing only minor or moderate changes. The results were calculated and plotted once each year.

The following numerical model arrangement was used for all cases:

a) The total length of simulated coast was 3 km. This length was divided into 60 equal cells, each 50 m in width.
b) Simulation starting date was November 30, 1986, ending on November 30, 1999.
c) Time step used was 6 hours with a total of 18,992 time steps per run.
d) Shoreline position was calculated and presented once each year.
e) Longshore sediment transport coefficients K_1 and K_2 were 0.5 and 0.25 respectively.
f) The depth of the offshore wave input was 10.0 m.
g) The average effective sand diameter was 0.25 mm.
h) The average height of land above mean water level was 1.5 m, and the limiting depth of profile movement seaward was 6.0 m.
i) Average beach slope was 1 in 20.
j) It was assumed that the breakwaters and groins were impermeable, and there was no wave reflection from the structures.

The series of simulations conducted were very illustrative in showing the long-term effect of different structures on the shoreline position. Figure 5.11 illustrates the effect of constructing several offshore breakwaters on the shoreline (note the distorted scale).

Prediction of beach evolution by numerical modeling has proven to be a powerful technique that can be used to evaluate different designs and to assist in the final project design.

Hanson and Kraus[53] performed a generalized calibration of the shoreline change employing the GENESIS model against observed field response to a single offshore breakwater.

Effect of varying significant wave height is depicted in Figure 5.12; effect of varying wave period on salient's change is shown in Figure 5.13.

Figure 5.14 presents results for single breakwater at different X/L (X = length of breakwater, L = wave length) and H_0/D (H_0 = deep water wave height, D = water depth) values.

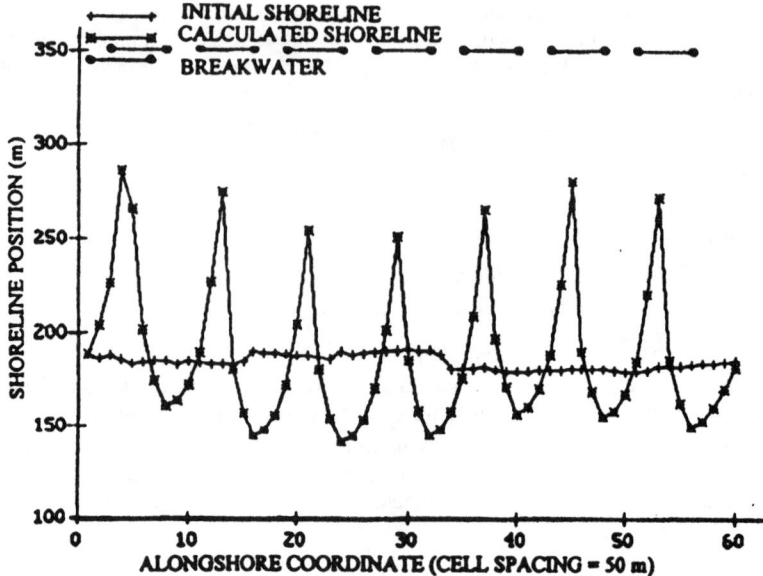

FIGURE 5.11. Protection of Ras-Elbar area shoreline.[64]

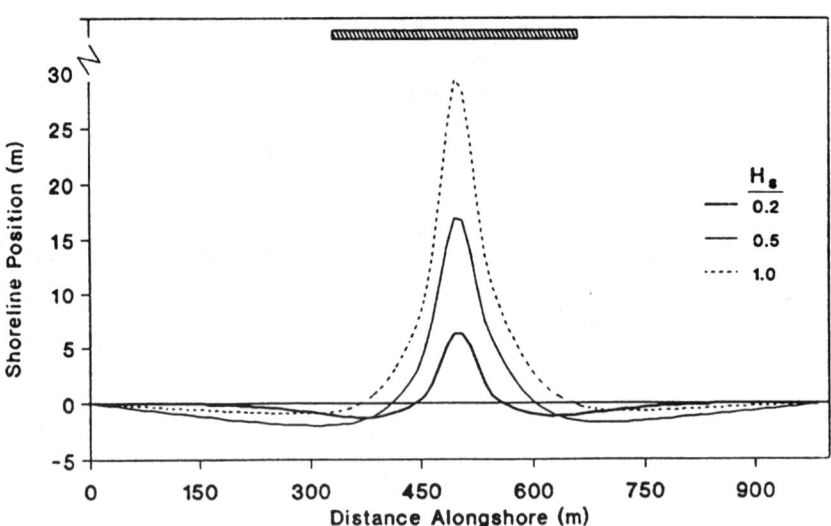

FIGURE 5.12. Influence of varying significant wave height on shoreline change behind a single detached breakwater.[53]

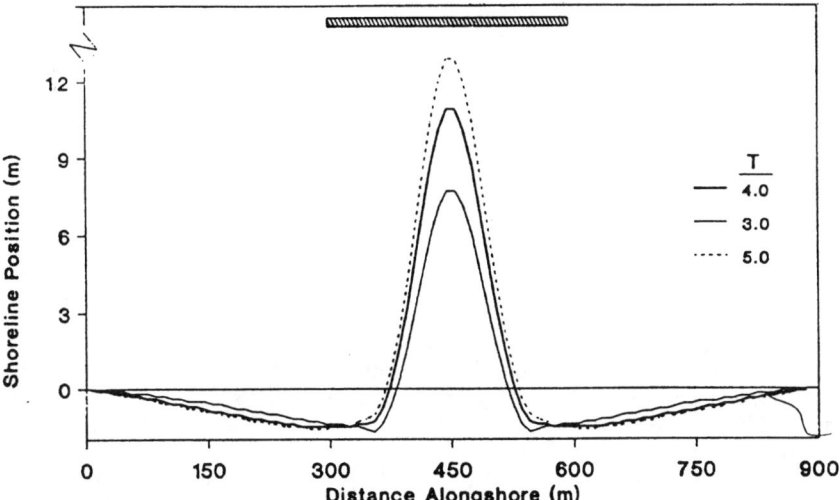

FIGURE 5.13. Influence of varying wave period on shoreline change behind a detached breakwater.[53]

Hanson and Kraus[54] separated the different response types as a function of wave transmissivity. As breakwater transmissivity increases, shoreline response decreases. Based on these results, the criterion for a salient to form was found to be

$$\frac{X}{L} \leq 48(1 - K_T)\frac{H_0}{D} \tag{7}$$

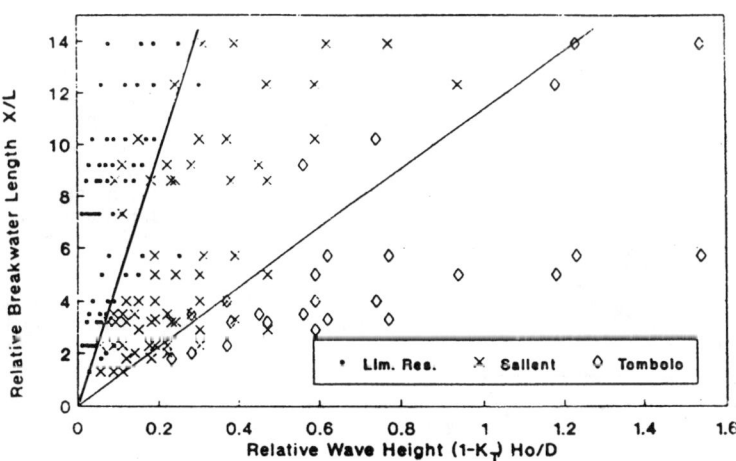

FIGURE 5.14. Calculated shoreline response to a single detached breakwater.[53]

which separates regions of limited shoreline response and clear salient development. Similarly, the criteria for a tombolo to form was found to be

$$\frac{X}{L} \leq 11(1-K_T)\frac{H_0}{D} \qquad (8)$$

where X = length of breakwater
L = wavelength
K_T = structure segment transmissivity
H_0 = incident wave height, and
D = depth at the structure segment

which separates regions of salient and tombolo formation.

As evidence for the validity of the proposed relationships, the prototype measurements are plotted in Figure 5.15. The prototype measurements fit well within the domains of the proposed criteria and provide at least limited validation of the calculated results.

Similar scaling effects are shown in Figure 5.15, where the effect of gap width to wave length is portrayed. The diffraction effects resulting from smaller value of G/L are much like those of larger B/L values. For a given wave height, energy is reduced in the lee of the structure as the gap to wavelength ratio decreases. Cords[20] investigated the change of salient length and volumetric deposition when maintaining a constant X/B, G/B, H_0, and D and decreasing G/L from 0.45 to 0.39. The salient length decreased 56% and volumetric deposition decreased 75% when G/L was reduced.

The GENESIS shoreline change model was also employed to numerically simulate shoreline changes along the Gulfward shoreline of Grand Isle.[45]

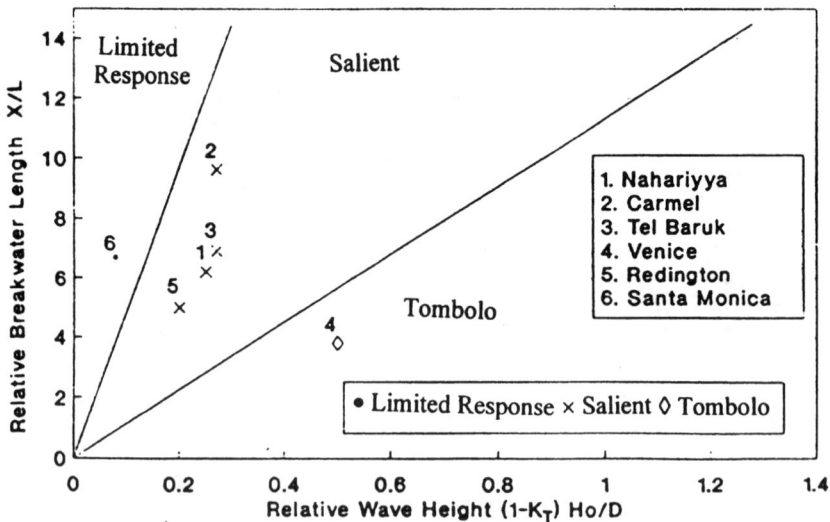

FIGURE 5.15. Field measurements and proposed criteria.[53]

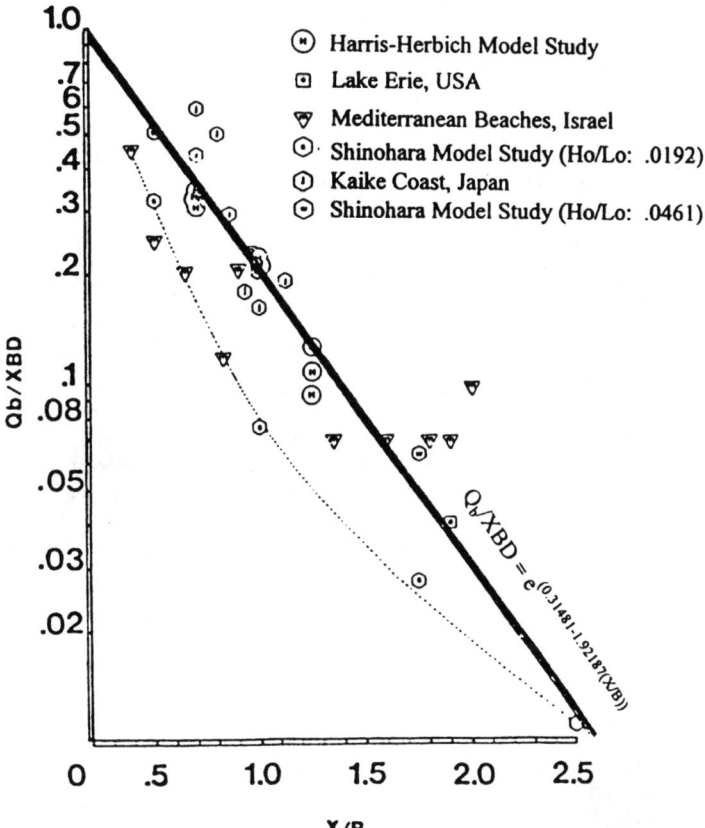

FIGURE 5.16. Comparison of field data and model studies, Q_b/SBD versus X/B.

COMPARISON OF MODEL AND FIELD STUDIES

Harris and Herbich[57,58] compared field and laboratory data plotted in Figure 5.16. The agreement appears to be reasonably good and the equation for the line is

$$Q_b/XBD = e^{[0.31481-1.92187(X/B)]} \qquad (9)$$

This formula is applicable for X/B between 0.5, and 2.5. For X/B less than 0.5 very little information is available but the curve should not go above a theoretical maximum of $0.5 = Q_b/XBD$ for subaqueous volume only and the slope should reverse so that Q_b/XBD would equal zero at $X/B = 0$. Toyoshima[147] reported that for continuous breakwaters with small X/B values tombolos did not form.

Krafft and Herbich[81] compared the laboratory and field data for several existing offshore breakwaters. In spite of data point scatter, the relationship between the volume of sediment deposition (or the volume of sediment accreted since construction) Q has a reasonable relationship with sheltered volume defined as

$$XBD \qquad (10)$$

where

X = distance from original shoreline to breakwater
B = breakwater length
D = water depth at the structure

Figure 5.17 summarizes all data from the model studies and the field measurements. The relationship between the two volumes is

$$Q = 0.42XBD \qquad (11)$$

Data for prototype studies alone are shown in Figure 5.18.

Native and beach fill sediments play an important role in shoreline response. Shores with coarser-grained sediments are associated with higher energy environments than those with fine sediments. These two beach types will respond at different rates under a new wave climate. Regional littoral drift determines sediment supply to the project, thus the accretion rates.

Volumetric deposition, Q, is defined as the volume of sediments accreted since con-

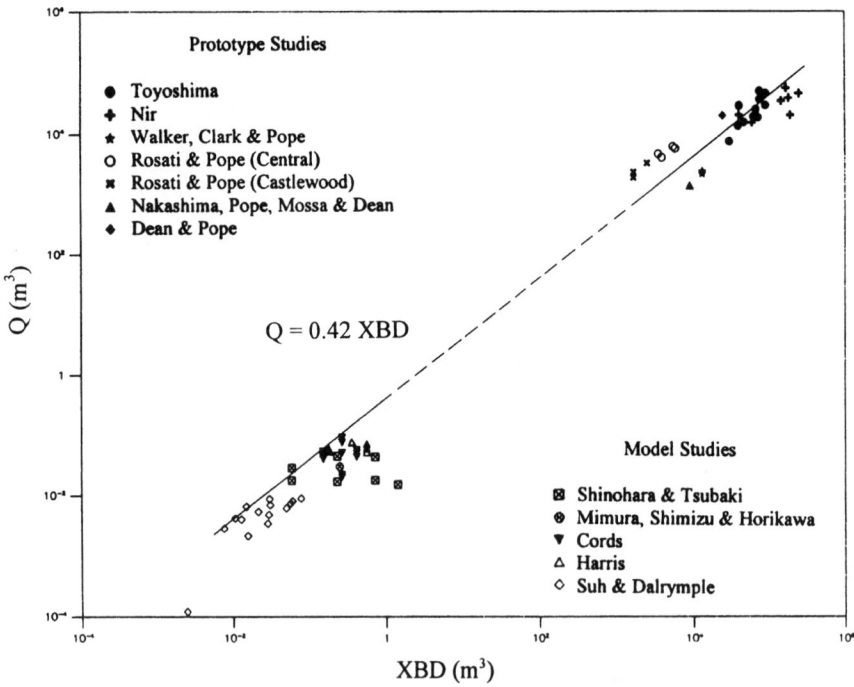

FIGURE 5.17. Sheltered volume relative to sediment deposition for all studies.[81]

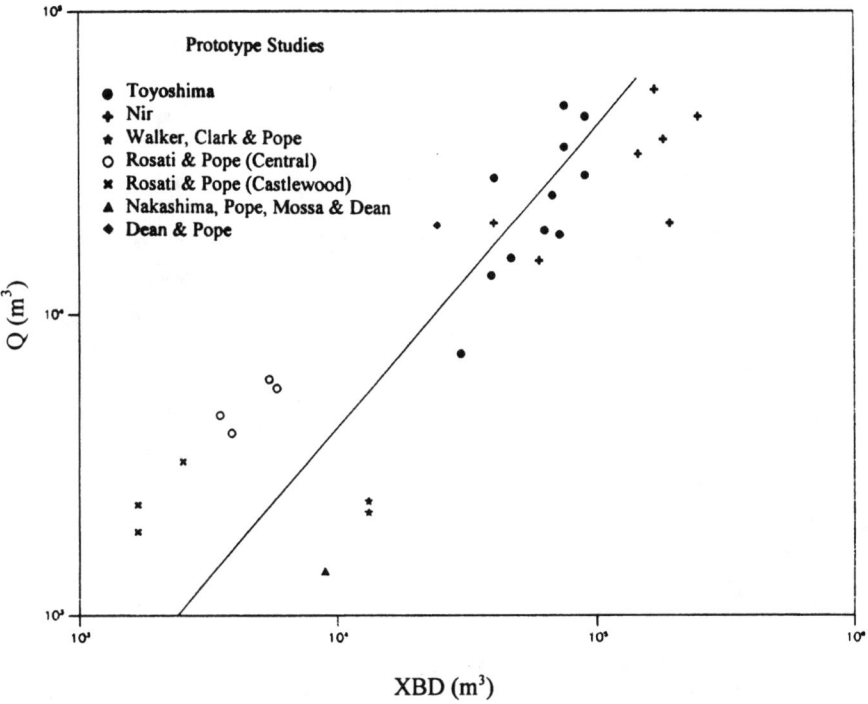

FIGURE 5.18. Sheltered volume relative to sediment deposition for prototype studies.[81]

struction; however, this volume usually includes artificial fill. Volumetric deposition, Q, is defined as the sediments that are deposited within the region parallel to the shoreline a distance equal to breakwater length, B; leeward of the structure for the preconstruction distance, X; and subsequently to a depth, D, as shown in Figure 5.19.

Shoreline response is generally measured by planform response and/or total deposition of sediment. This analysis is a comparison of total deposition of sediment in the sheltered region shoreward of the breakwater. In order to estimate and compare planform response for a given volumetric deposition, it was assumed the salient took the form of Figure 5.19. Thus, volumetric deposition can be described by

$$Q = \tfrac{1}{2} BDX_s \quad (12)$$

where X_s = distance from original shoreline to peak of salient.

Rearranging and solving for salient length yields

$$X_s = \frac{2Q}{DB} \quad (13)$$

Several researchers have documented salient length and volumetric deposition and some information has been drawn from Suh and Dalrymple.[141] This information allowed calculation of salient length as shown in Figures 5.20 and 5.21.

Sediment deposition for prototype data was evaluated first with respect to X, B, and D,

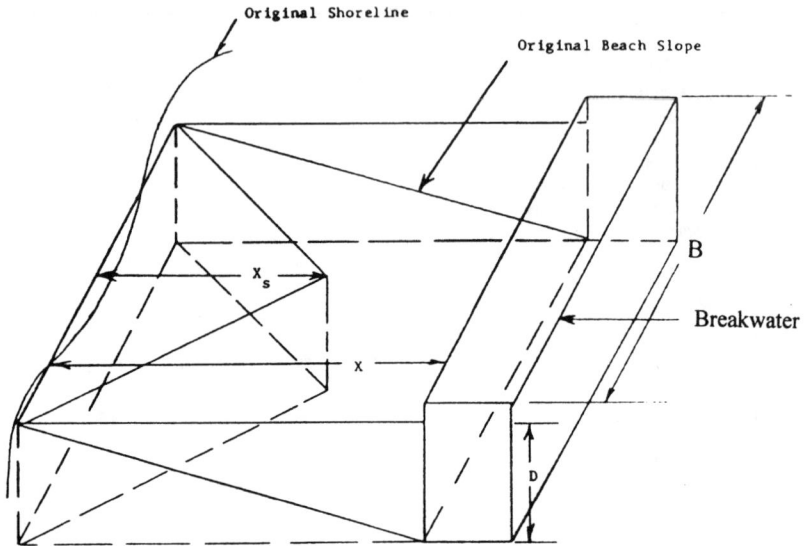

FIGURE 5.19. Breakwater definition sketch.

the three dimensions that define the sheltered region. The effect of breakwater length on sediment deposition for seven field studies is shown in Figure 5.22. Deposition is shown to increase with increasing breakwater length. The same relationship is seen for the effect of water depth and distance offshore (Figures 5.23 and 5.24), although the scatter is significant. Due to the fact that there is generally a reduction in the sheltering effectiveness

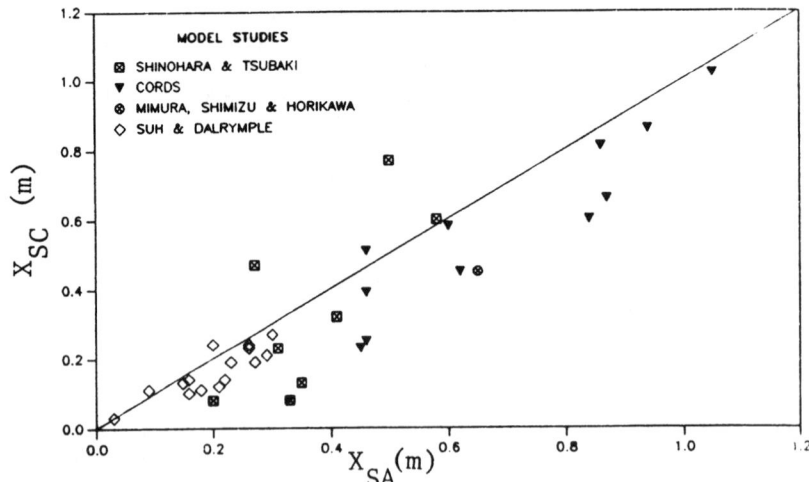

FIGURE 5.20. Actual versus calculated salient length.

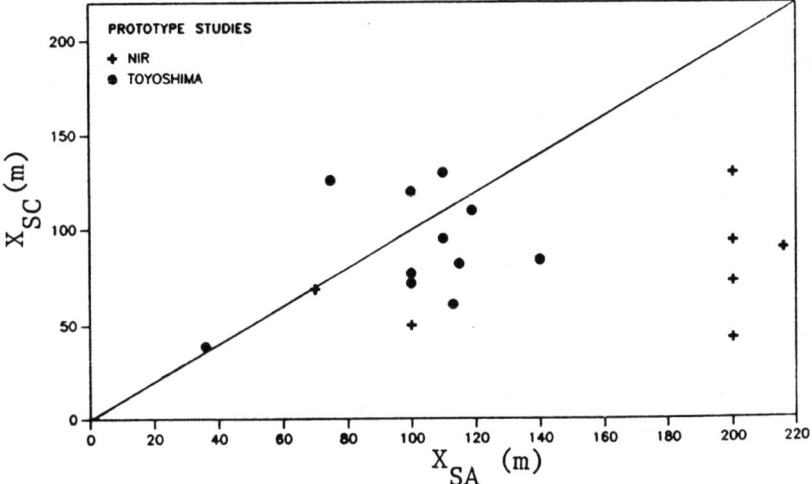

FIGURE 5.21. Actual versus calculated salient length.

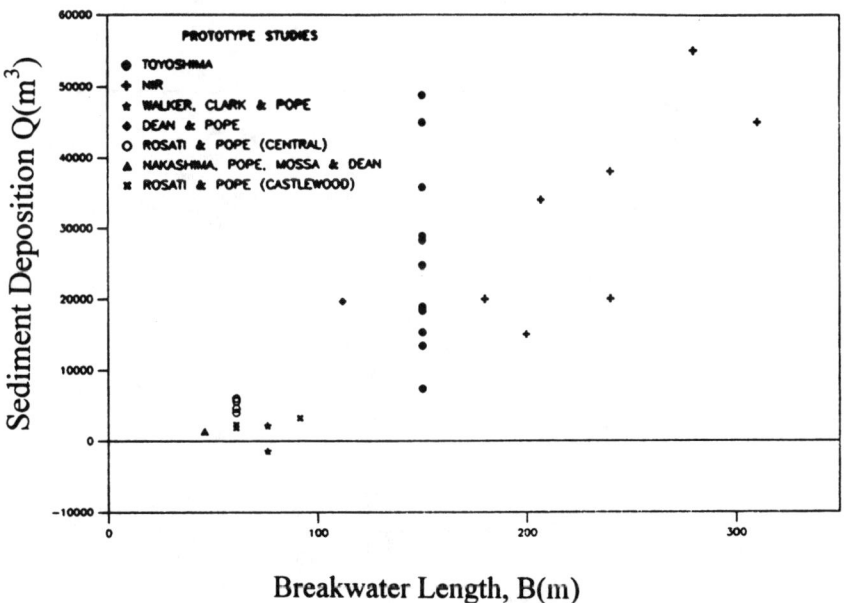

FIGURE 5.22. Effect of breakwater length and sediment deposition.

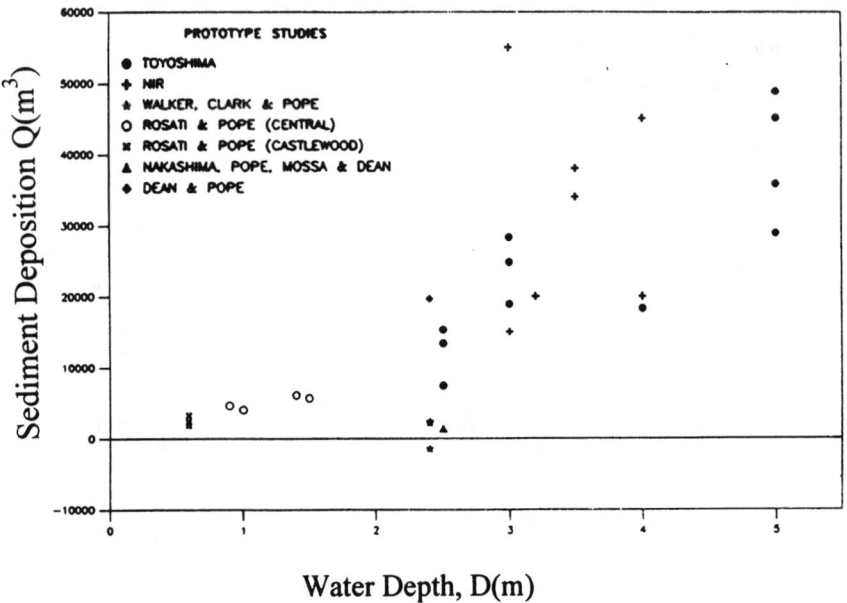

Water Depth, D(m)

FIGURE 5.23. Effect of water depth on sediment deposition.

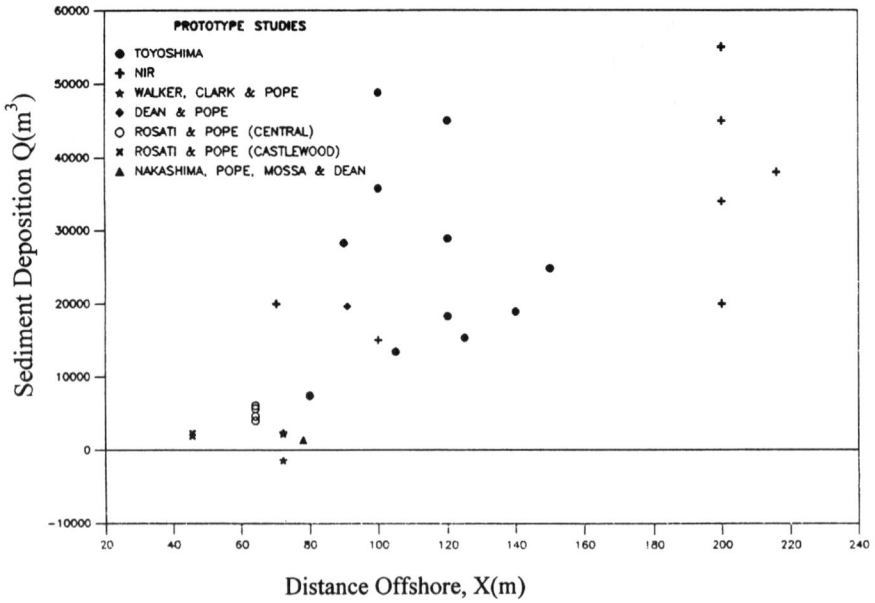

Distance Offshore, X(m)

FIGURE 5.24. Effect of distance offshore on sediment deposition.

of a detached breakwater as the distance offshore (X) is increased for a given breakwater length (B), the combination of these particular projects indicate a bias towards a limited range of X/B values, as will be shown later. The complete sheltered volume relative to total accretion is shown in Figure 5.25. The high degree of scatter seen in the larger values of Nir's data[104] can be attributed to shore-connected groins attached to several of the breakwaters, which subsequently caused narrow tombolos and relatively low deposition. These points are plotted for comparison only. The scatter throughout the figure is a result of numerous inherent errors associated with measuring large quantities of sediment deposition, the random nature of coastal dynamics and different breakwater designs.

The effect of the ratio of breakwater length to wavelength on sediment accretion in the lee of the breakwater is shown in Figure 5.26. The graph depicts a trend of increasing deposition as the B/L ratio increases. Higher depositional rates are a result of reduced diffraction coefficients in the sheltered area in conjunction with lower incident wave height. This phenomenon creates an environment favorable for deposition.

Breakwater relative distance offshore, X/B is a critical design parameter, and several researchers have developed design criteria on the basis of this ratio (*Shore Protection Manual*[132], Dally and Pope[22], and Harris and Herbich[57]). These researchers have generally found that a relatively small value of X/B, less than or equal to one, indicates diffraction patterns for a normally incident wave are such that the diffracting waves around either end of the breakwater superimpose to a low degree before reaching shore. This results in a low-energy region conducive to sediment deposition. For larger values of X/B, greater than two or so, the shoreline is exposed to a much greater extent, to both the diffracted waves that have superimposed in the lee of the breakwater and to the directionali-

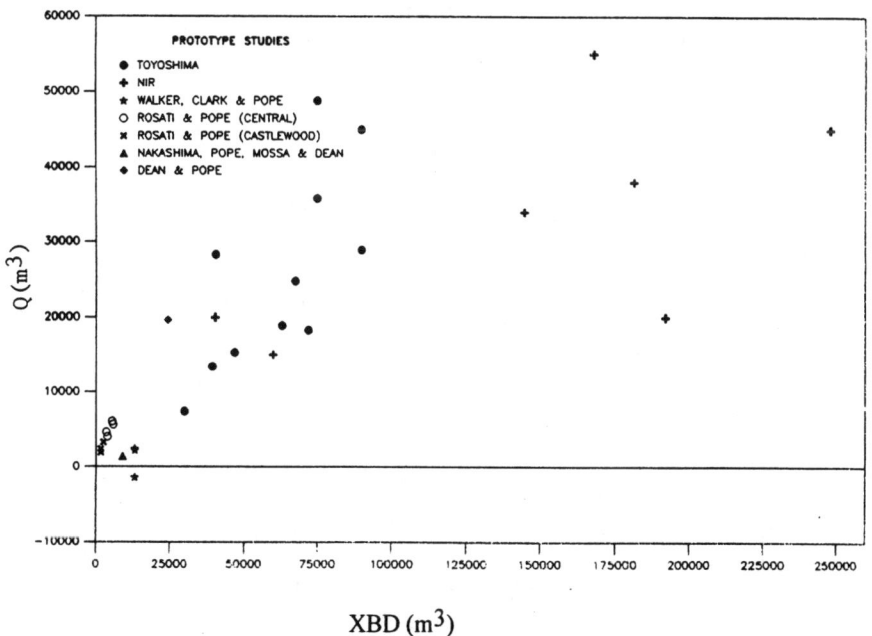

FIGURE 5.25. Sheltered volume relative to sediment deposition.

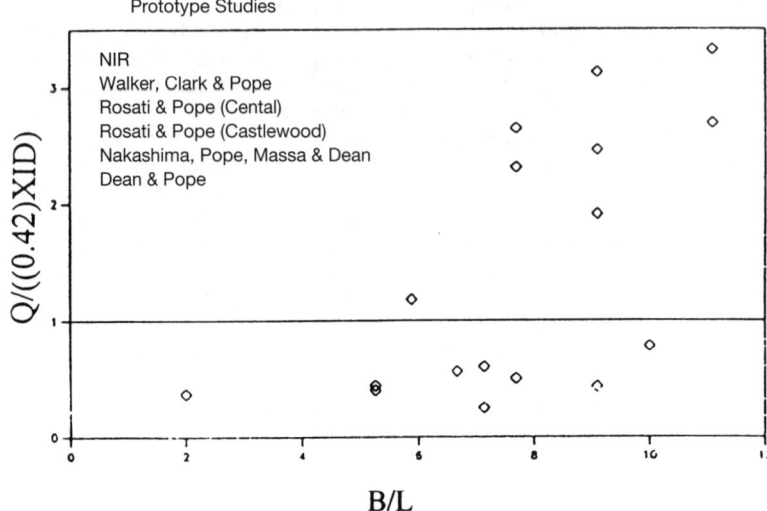

FIGURE 5.26. Effect of breakwater length to wave length ratio.

ty of the local sea. A diffraction analysis is necessary to determine the energy in the lee of a particular breakwater design at each project site.

A relative distance offshore versus relative shoreline response is plotted in Figure 5.27. Most of the studies analyzed here are shown to fall within a limited range of X/B and exhibit a high degree of scatter, thus, no conclusive relationship is apparent. However,

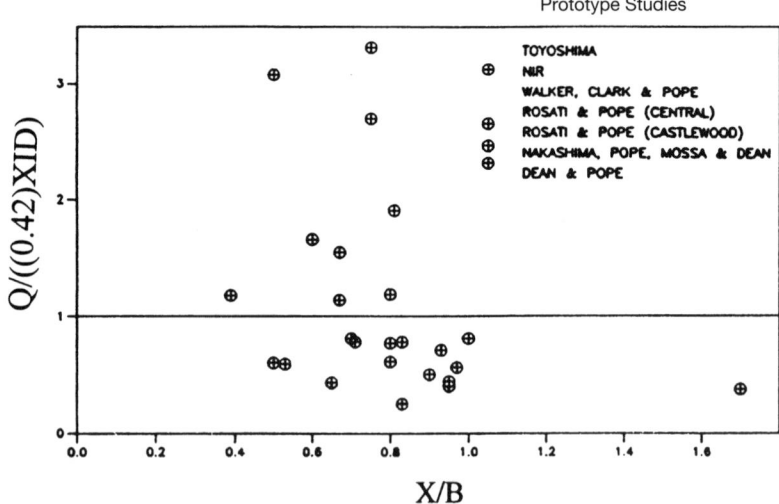

FIGURE 5.27. Effect of distance offshore to breakwater length ratio.

two observations can be made from the result of this. First, for the most part, the other influencing parameters can be analyzed with some confidence that the majority of the field data were not highly prejudiced by the critical design parameter, X/B. Second, the equation determined as representative for an average for these breakwaters was not necessarily representative of a cross-section of breakwater designs.

Model data were incorporated and compared to prototype data nondimensionally. Due to the method of approximating the line describing the data (which favored prototype data), the model studies mainly fell below the line.

A significant factor influencing lower sediment deposition can be seen in Figure 5.28. Model studies have generally included an investigation of a range X/B values and those studies with larger values reflect lower deposition, as expected. Also, several model studies with small values of X/B correspond to the highest deposition. The larger range of X/B ratios investigated in laboratory experiments contributes in part to the smaller volumetric deposition seen in model studies.

The effect of wave steepness is shown in Figure 5.29. Model studies typically result in an increase in deposition with an increase in wave steepness, as shown by Cords[20] and Shinohara and Tsubaki.[129] However, a model study performed by Rosen and Vajda[120] showed a decrease in salient length as wave steepness increased for the same X/B values. In the same model study, the steeper waves resulted in the breaker line being moved further shoreward of the breakwater, which has also been shown to decrease deposition.

A comparison of laboratory experiments and project performance is shown as a function of breakwater length to wave length ratio (Figure 5.30). Clearly, the model studies are not fully representative of these prototype projects. Smaller values of B/L increase the diffraction effects in the lee of the breakwater, and, for a given wave height, increase the wave energy in the lee of the breakwater. All of the model studies evaluated were unscaled laboratory experiments performed for pure research. Site-specific modeling of

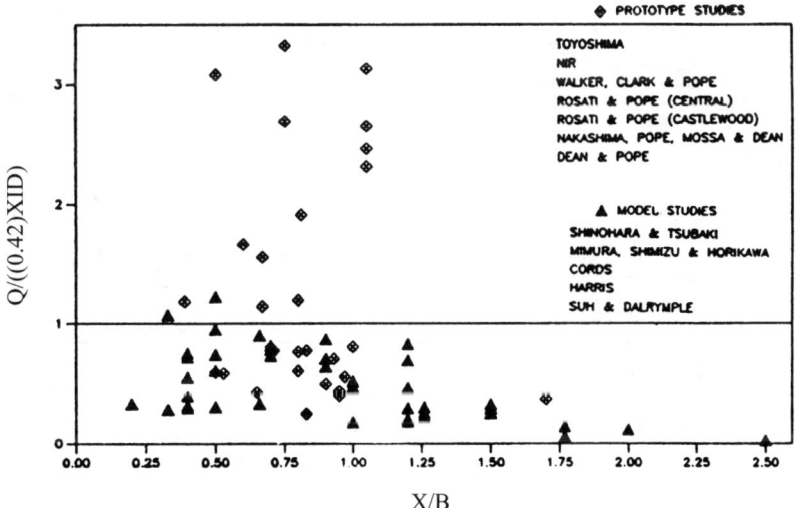

FIGURE 5.28. Effect of distance offshore to breakwater length ratio.

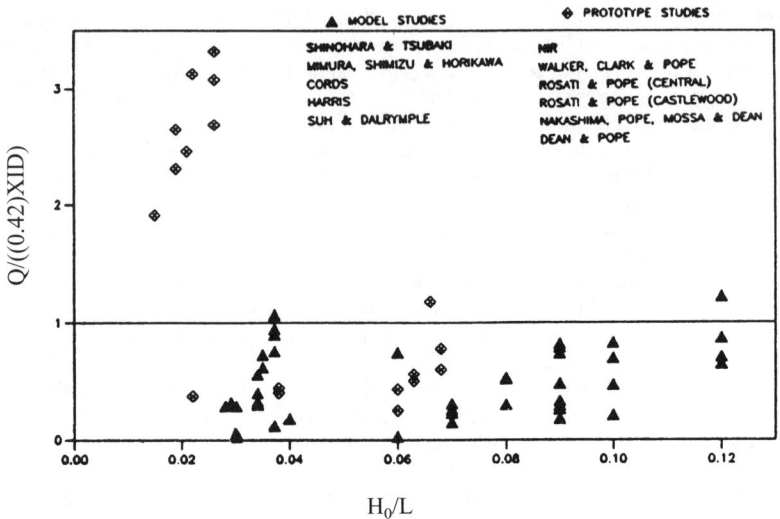

FIGURE 5.29. Effect of wave steepness.

project performance has been shown to accurately portray (Seabergh[127]) and even overpredict (Nir[104]) shoreline changes.

Similar scaling effects are shown in Figure 5.31, where the effect of gap width to wavelength is portrayed. The diffraction effects resulting from smaller values of G/L are much like those of larger B/L values. For a given wave height, energy is reduced in the

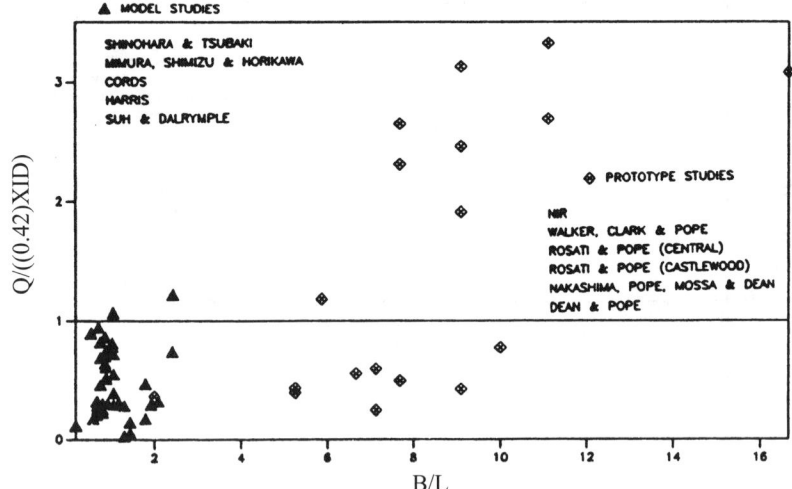

FIGURE 5.30. Effect of breakwater length to wavelength ratio.

OFFSHORE (DETACHED) BREAKWATERS   5.33

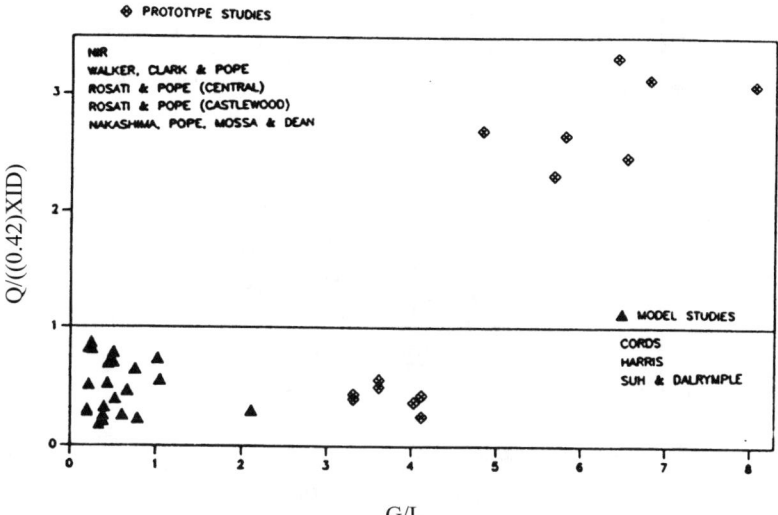

FIGURE 5.31. Effect of gap width to wavelength ratio.

lee of the structure as the gap to wavelength ratio decreases. Cords[20] appears to be the only researcher to investigate the change of salient length and volumetric deposition when maintaining a constant X/B, G/B, H_0, and D, and decreasing G/L from 0.45 to 0.39. The salient length decreased 56% and volumetric deposition decreased 75% when G/L was reduced.

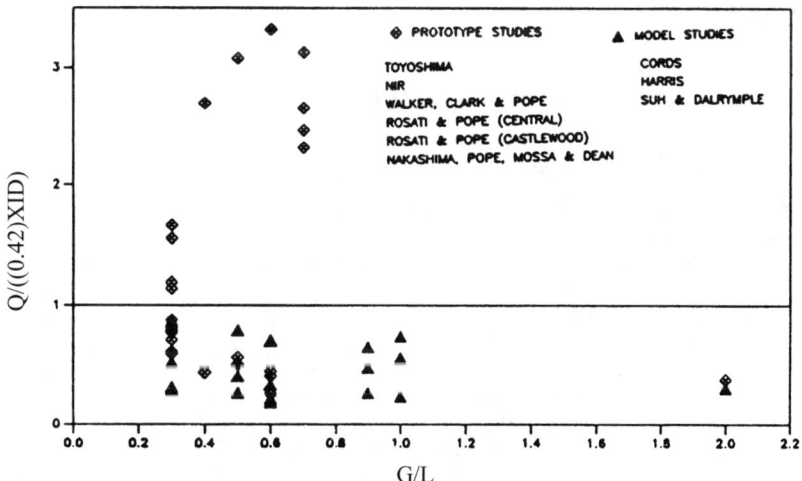

FIGURE 5.32. Effect of gap width to breakwater length ratio.

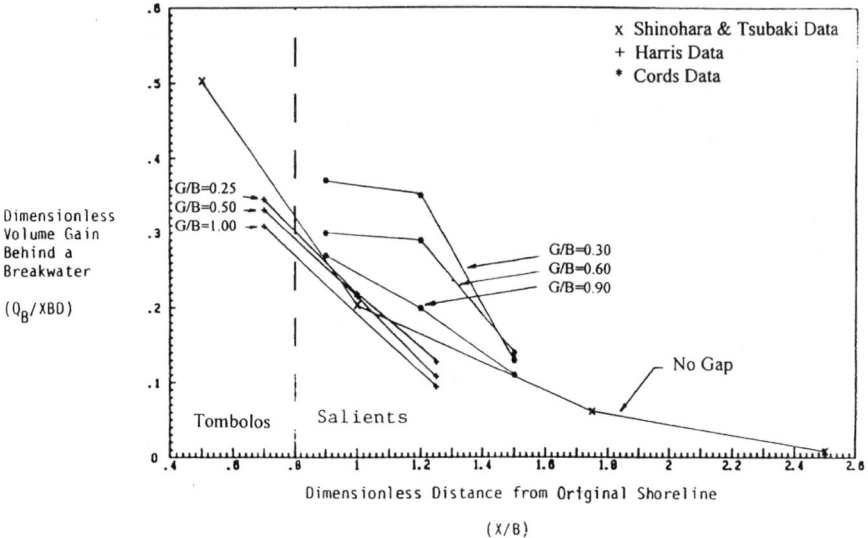

FIGURE 5.33. Comparison of model studies of Shinohara and Tsubaki, Harris, and Cords for one, two, and three breakwaters, respectively.

The model studies have represented a typical range of G/B ratios for the field studies evaluated (Figure 5.32).

A comparison has been made by Cords[20] between model tests for a single breakwater (Shinohara and Tsubaki[129]), two breakwaters (Harris and Herbich[58]), and three breakwaters (Cords[20]). The average volume gain behind a breakwater in a series of three breakwaters is substantially greater than the cases of a single breakwater or a series of two breakwaters (Figure 5.33).

DESCRIPTION OF SELECTED PROJECTS[81]

United States Breakwaters

a) Location: Lakeview Park, Lorain, Ohio, on Lake Erie[111,170]
   Date of Construction: 1977
   Purpose: The objective of the project is to protect placed beach fill and provide and maintain a recreational park.
   Design and Construction: The rubble mound, three-segment breakwater project is designed to protect an average of 61 m of beach fill. The design procedure included diffraction diagrams of predominant wave directions to: (1) ensure maintenance of natural transport to downdrift beaches; (2) predict shoreline response based on the assumption that equilibrium is attained when the wave rays approach the shoreline normally, thus creating no longshore current; and (3) establish that the intersection of diffraction coefficients equal to 0.3 occurred between the breakwater and the salient to prevent tombolo formation.

An existing groin was heightened and extended to 91 m to maintain fill but not augmented sufficiently to trap the littoral drift. Three offshore detached breakwater units were constructed 76.2 m in length, at an average distance offshore of 146.2 m before sediment fill. Gap widths between sections were 49 m each. The structures were constructed in a water depth of about 2.4 m LWD. Breakwater crest height and width was 2.4 m LWD and 4.3 m, respectively. The design allowed for a transmitted wave height of only 0.3 m to insure that wave setup and trip currents in the vicinity of the structures are minimal.

Littoral Transport: Wave conditions create a potential littoral transport of 16,450 m^3/yr but only approximately 6,500 m^3/yr is realized, due to low sediment availability. Net littoral transport is to the east but potential littoral drift from easterly waves is nearly two-thirds of that from westerly waves.

Wave Climate: Wave heights and periods up to 2.4 m and 7.0 seconds occur annually. Average nearshore breaking wave height was found to be 0.8 m and average period is 3.5 seconds. Prevalent wave direction is to the east.

Net wave energy flux was determined from waves originating from a direction 60 degrees west of north and predominant wave direction used for breakwater design was from 40 degrees west of north.

Beach Fill: 84,200 m^3 of beach fill were initially placed and periodic nourishment of 3,800 m^3 per year was allowed for. However, beach replenishment was only required twice during the first five years of project life, once in July 1980 and again in September 1981.

Shoreline Response: Five years after completion, two breakwaters maintained the placement of fill in addition to gaining about 7500 m^3 each. However, the beach has been aligned in a more westerly direction than originally designed and the west breakwater maintains only a narrow width of beach. Atypical westerly storms of near design intensity are responsible for removing a significant portion of the fill, but normal wave activity tends to display accretive characteristics. Although the project does not have widely spaced segments, the asymmetry of the waves has formed a crenelate-like shoreline upon which cuspate salients have formed in the lee of the center and eastern segments. The shoreline has adjusted to a very stable configuration that varies with incident wave conditions and accretes about 2,294 m^3 of sediment per year (Figure 5.34).

Performance: The project is performing successfully as a recreational facility. The downdrift shoreline has experienced slight erosion.

b) Location: Haleiwa, Oahu, Hawaii
Date of Construction: 1965
Purpose: The project, composed of a rubble mound, offshore breakwater and nearby groin, was designed to develop and protect a recreational beach.
Design and Construction: The detached, single breakwater is 49 m in length and was constructed 91 m offshore of the original shoreline. Crest elevation and water depth at the structure is 1.2 m and 2.1 m MSL, respectively.
Littoral Transport: Undocumented
Wave Climate: This region of the Island of Oahu is exposed to large, high-energy winter waves.
Beach Fill: A total of 69,000 cubic meters of beach fill was placed soon after construction of the breakwater. This positioned the breakwater about 49 m from the shoreline. Beach replenishment was again performed in 1970 after severe storms damaged both breakwater and groin.
Shoreline Response: The shoreline has maintained a cuspate salient in the lee of the breakwater while the breakwater has been intact. However, damage to the structure during storm attack has caused erosion and the need for additional fill.

FIGURE 5.34. Breakwaters at Lakeview Park.

Performance: Haleiwa Beach is a popular recreational beach. The breakwater has provided a sheltered swimming area. Wave damage to the structure occurred both in 1969 and 1974. The breakwater was repaired after these incidents.

c) Location: Santa Monica, California
Date of Construction: 1933
Purpose: The original intention of the breakwater was to harbor small boats; however, sediment accretion in the lee of the breakwater drastically reduced the number of small boat anchorages.
Design and Construction: The rubble mound offshore breakwater was constructed 610 m offshore with a length of 640 m. Breakwater crest elevation was 3 m above MLLW with a crest width of 3 m. Water depth at the structure is about 8 m. Readjustment of the breakwater began soon after construction and wave damage was documented in 1934. By 1983, crest elevation at some locations was 1.6 m below MLLW and crest width was approximately 13.4 m; however, at other locations the breakwater pierced the water surface.
Littoral Drift: An average littoral drift of 206,000 m^3/yr is characteristic of the region.[131]
Wave Climate: Venice Beach, located several miles south, has a mean significant wave height and period of 0.37 m and 10.5 sec, respectively. Significant wave heights used for hydraulic model testing of wave overtopping ranged from 2 m to 4 m.
Beach Fill: None
Shoreline Response: Sediment deposition behind the breakwater extended the shoreline approximately 240 m^{22} and virtually ceased the passing of littoral material to downdrift beaches. Downdrift regions suffered heavy sediment losses. The area was dredged on several occasions; however, accretion continued until equilibrium was attained. Struc-

tural degradation led to higher wave transmission and eventually littoral transport resumed to downdrift beaches. A broad salient remains in the lee of the breakwater and accretion along the shoreline north of the breakwater resulted.
Performance: The breakwater provides a wide protected recreational beach and protected anchorage for small boats. During early stages of deposition, enormous losses were incurred by land- and homeowners downdrift of the breakwater.[3]

d) Location: Channel Islands Harbor, California
Date of Construction: 1960
Purpose: The objective of the project is to provide protection for a small craft harbor and entrap sediment for bypassing.
Design and Construction: Channel Island Harbor's jetties were designed and constructed in conjunction with the breakwater. Design requirements dictated that the breakwaters dimensions and placement must provide for: (1) sufficient volume for sediment entrapment, (2) maneuvering space between jetties and breakwater, and (3) year-round protection from waves. A diffraction analysis was performed on the dominant wave governing littoral transport in order to predict the shape of the salient as a function of breakwater length. Breakwater design crest height was 4.3 m. The structure is about 700 m long and is located in a water depth of 9 m. It was constructed 550 m offshore.
Littoral Transport: The region experiences a very high transport rate of 895,000 m^3/yr.
Wave Climate: Winds are predominantly from the north and northwest, creating breaker heights averaging from 1 to 3 m. The design significant wave height for the breakwaters was 4.8 m. The region is characterized by long wave periods ranging from 11 to 12 sec.[16]
Beach Fill: None
Shoreline Response: Soon after construction, the shoreline experienced rapid accretion. During the first seven years, 10,712,000 m^3 of sediment was bypassed.
Performance: The breakwater satisfies the stated objectives very well. Channel Islands Harbor is considered to be a calm harbor even under stormy conditions and shoaling is prevented until the entrapment area fills to such a capacity that sediments spill over the north jetty.

e) Location: Winthrop Beach, Massachusetts[109]
Date of Construction: 1931–1933
Purpose: The breakwater project was constructed to stabilize an eroding coastline and preserve a sea wall that was built to protect the City of Winthrop.
Design and Construction: The five-segment, rubble mound breakwater is composed of granite armor units. The segments are 100 m in length, the gap width between units is 30 m during MHW, and total breakwater length is about 625 m. Crest width and elevation are 3.7 m and 4.1 m respectively. The structure was originally located 300 m offshore. Water depth at MLW is 3 m and the region experiences a mean normal tide range of 2.7 m.
Littoral Drift: The region receives a low quantity of littoral material. Dominant direction of drift is north to south.
Wave Climate: Winthrop Beach is exposed to northeasterly to southeasterly wind waves. A significant wave height of 7 m is exceeded 0.1% of time during the winter months at Boston.
Beach Fill: Fill had been placed both north and south of the breakwater in conjunction with groins within the first 20 years of breakwater construction. In 1971, beach fill was again placed in the region behind the breakwater. The fill continues to be lost during storms or to the littoral current.

Shoreline Response: A cuspate salient approximately the length of the breakwater is formed during high tide. At low tide, the salients are shown to connect to a one-time island, Winthrop Head drumlin,[91] thus creating an inundated tombolo. Some erosion has occurred behind the south end of the breakwater, south of the breakwater, and north of the structure. Erosion along the southern vicinity is due to refraction and diffraction patterns but the southern segment does serve to protect the northern segments form the oblique incident waves. The region north of the breakwater does not receive littoral material during times when the littoral current shifts.

Performance: The breakwater has succeeded in widening the beach and protecting the sea wall. The beach now also serves as a recreational area. Although some subsidence of the breakwater has occurred, no maintenance has been necessary.

f) Location: Ala Moana Park, Honolulu, Hawaii[101]
Date of Construction: 1964
Purpose: The rubble mound breakwater was initially part of a larger public park recreational development that was never fully completed.
Design and Construction: A model study of the breakwater project was completed by the U.S. Army Engineer Waterways Experiment Station.[15] The design incorporates two shore-attached breakwater units at either end of four detached offshore breakwaters placed in an overlapping fashion. A diffraction analysis was performed to determine the expected incident wave height within the enclosed beach area. Breakwaters vary in length from approximately 15 to 20 m.
Wave Climate: The project is within a sheltered region and waves are generally mild. However, storm waves in the area can be severe and the final design wave had a deep water wave height and period of 2.4 m and 20 sec respectively. Waves are predominantly from the SE and SW.
Littoral Transport: Not documented
Beach Fill: An artificial beach was created with beach fill.
Shoreline Response: A high percentage of the beach fill has remained intact. Some erosion along the eastern shore has occurred and a subdued salient has developed behind the innermost breakwater.
Performance: The project functions very well as a recreational beach.

g) Location: Colonial Beach, Virginia[27,116]
Date of Construction: 1982
Purpose: The objective of the project was to: (1) protect a cohesive bluff and road, and (2) create a recreational beach while avoiding increased shoaling of a downdrift navigation channel.
Design and Construction: Two segmented breakwater projects were constructed. The Central Beach project consists of a four-segment detached, rubble mound breakwater and the Castlewood Park Beach project is composed of a three-segment, detached rubble mound breakwater. Design procedures for determining breakwater length and gap spacings included the development of diffraction diagrams for purposes similar to those described in the Lakeview Park project, as well as information obtained from Toyoshima's[145] shallow-water depth system. The breakwaters are constructed perpendicular to design wave direction at a crest elevation of 0.91 m, allowing for 0.3 m of settlement. To prevent loss of fill material, a 100-ft terminal groin was constructed at Castlewood Park Beach. Both projects have a design crest width of 1.8 m. The Central Beach section has four segments 61 m long with 46 m gaps. This breakwater was located 64 m from the original, unfilled shoreline. The Castlewood Park section has three segments of which two are 61 m and the third is 91 m in length. Gap distances between these breakwaters are

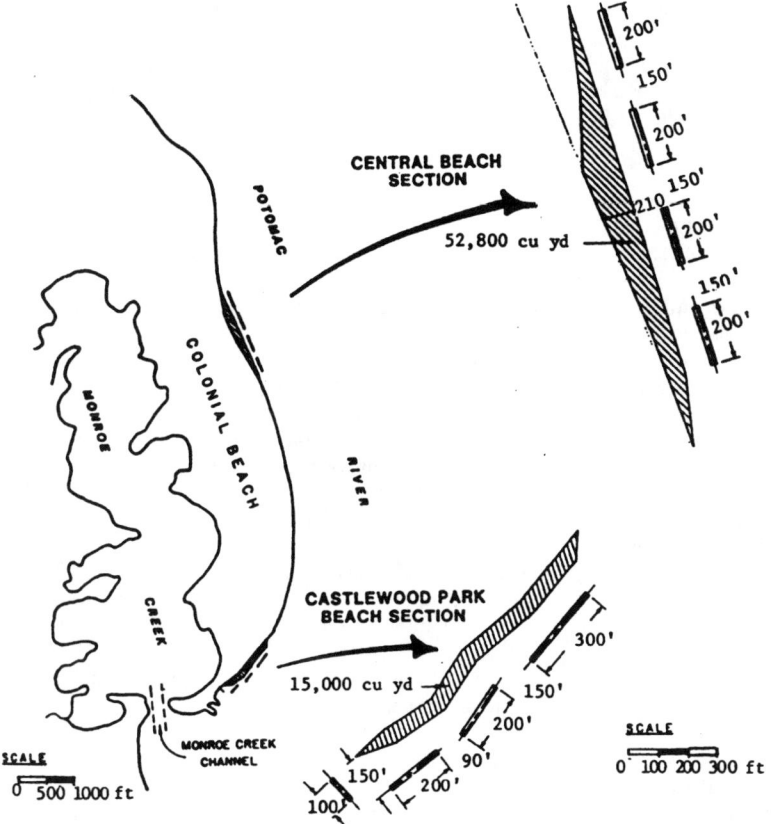

FIGURE 5.35. Colonial Beach Peninsula, location of project areas and site parameters.[27]

27 m and 46 m. This project was located 46 m from the original unfilled shoreline (Figure 5.35).

Littoral Transport: Low quantities of littoral material are available in this area.

Wave Climate: The mild wave climate in this area is a result of the limited fetch. Average wave heights vary from 0–0.3 m, with periods from 2–3 sec. Local storms generate wave heights from 0.5–1.4 m, with periods of 2.3–4.6 sec.

Beach Fill: 40,400 m^3 of beach fill were placed at Central Beach and 11,500 m^3 of fill were placed on Castlewood Park Beach. The fill had a median grain diameter of 0.5 mm, whereas the median grain diameters for Central and Castlewood Park Beaches are 0.8 mm and 0.3 mm, respectively. During storm conditions the back beach region of Central Beach became inundated with water; therefore, 2,300 m^3 of sand were used for renourishment.

Shoreline Response: Tombolos or near tombolos formed behind all seven of the breakwaters. This is due to the combination of fill and the small ratio of distance from shoreline to length of breakwater.

Performance: Full tombolos have formed behind several of the breakwaters.

Castlewood Park Beach lacks significant exchange of water through the breakwater gap area, and a marshy tidal flat has developed in the embayments.
The protection region of the Central Beach Project experiences water level setup that causes overwash depressions in the tombolos and salients. Both projects have successfully protected the bluff and road and Central Beach serves as an attractive recreational beach.

h) Location: Venice, California[109]
Date of Construction: 1905
Purpose: The offshore detached breakwater was constructed for protection of an amusement pier, which has since been removed.
Design and Construction: The single, rubble mound breakwater is 183 m in length and was constructed 370 m from the shoreline at MHW. During the mid 1960's a timber groin was erected immediately behind the breakwater, creating a T-groin. Breakwater crest elevation and water depth are approximately 3.1 m MSL and 1.8 m MLW, respectively.[132]
Littoral Transport: Regional littoral transport is approximately 206,000 m³/yr. Littoral transport was reduced in 1933 due to construction of the Santa Monica breakwater.
Wave Climate: Nearshore mean significant wave height is 0.37 m. Annual mean period is 10.5 sec. The largest storm waves range from 3.7 m to 4.6 m.[132]
Beach Fill: 11 million cubic meters of dredged sand was deposited in the vicinity during 1948.
Shoreline Response: The shoreline advanced considerably after the addition of fill material and a tombolo formed. The tombolo eventually eroded to form a salient. The construction of a T-groin encouraged the formation of another tombolo, which remains to date.[22]
Performance: The breakwater performed well in protecting the Venice amusement pier and now shelters a wide recreational beach (Figure 5.36).

i) Location: Redington Shores, Florida[26]
Date of Construction: 1985
Purpose: The objective of the project was to: (1) protect a sea wall, county park, and beach fill, and (2) create a recreational beach.
Design and Construction: A numerical shoreline response model was employed in the design of a single rubble mound breakwater. The model was used to evaluate: (1) breakwater length, (2) distance offshore, (3) transmissibility, and (4) the effectiveness of an additional groin. Specific goals were to avoid erosion to downdrift regions due to breakwater placement and stabilization of the protected shoreline. The final design resulted in a low-crested, 0.5 m MSL, highly permeable, 50% transmissibility, breakwater. The structure has a section of length 46 m at a forty-five degree angle to the shoreline (termed "dog-leg"), constructed to reduce local downdrift erosion, and a 79 m section parallel to the shoreline.
Owing to erosion problems downdrift of the breakwater, the crest elevation was reduced to increase overtopping of the structure.
Littoral Transport: The area receives a yearly average of 38,000 m³ of littoral drift that is transported in a southerly direction.
Wave Climate: The coastal, yearly mean wave height varies from 6 cm to 30 cm within a period of 2 to 4 sec. Infrequent hurricanes occur throughout the region. Generally, a low wave energy environment is characteristic of the region.
Beach Fill: Final design requirements for beach fill called for a minimum and maximum of 20,300 m³ and 60,800 m³ of sand, respectively. 23,000 m³ of fill were available for placement on the beach. The before-fill shoreline was 3 m from the seawall.

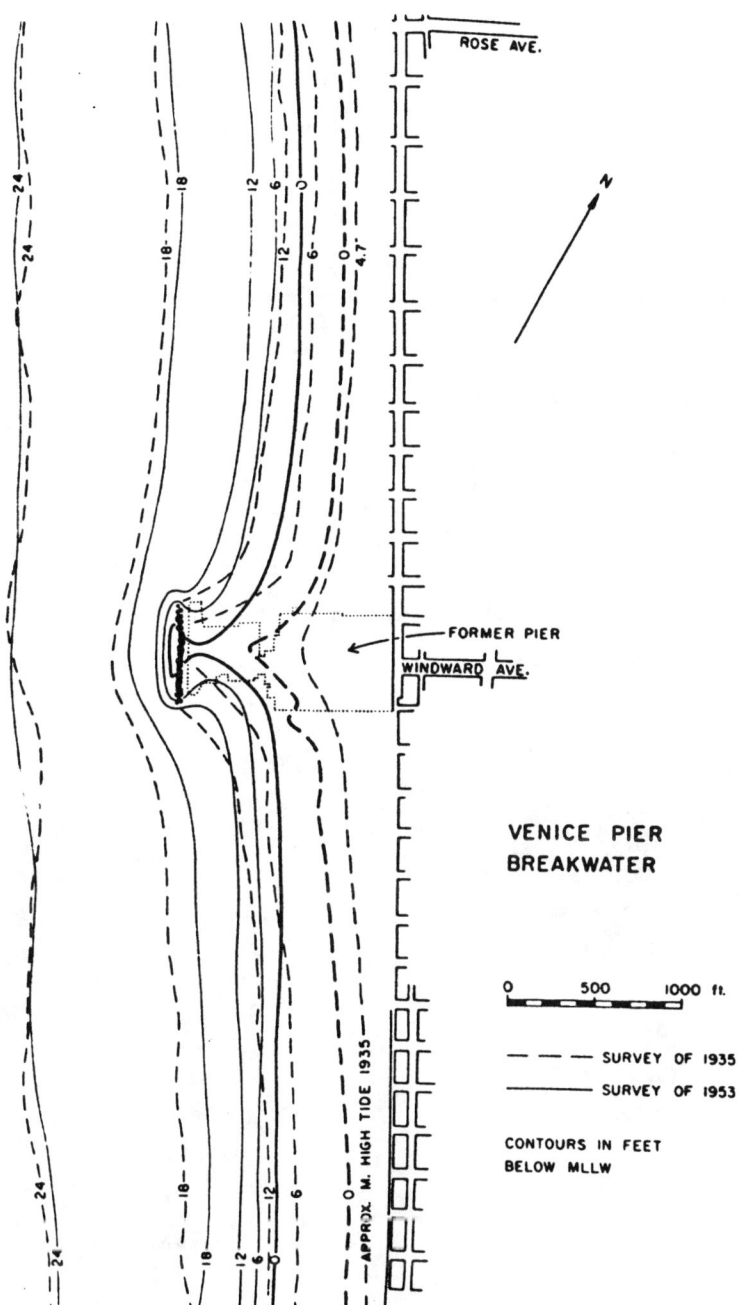

FIGURE 5.36. Venice breakwater as of 1953.[70]

Shoreline Response: A well-developed salient formed in the lee of the offshore breakwater, except at low tides when a tombolo can be seen.

Performance: The resulting shoreline formation functions well for both shore protection and recreational purposes. At a distance of about 200 m downdrift of the structure, the shoreline has completely retreated to the seawall. The erosion problem instigated the removal of some stone armor units.[115]

j) Location: Presque Isle Peninsula, Pennsylvania[22,113,127]
Date of Construction: 1978
Purpose: Presque Isle Peninsula protects Erie Harbor from the predominant westerly waves approaching the harbor. Shore protection has been necessary since the early 1800's to prevent the formation of a gap and wave energy intrusion. Beach fill and groins with periodic renourishment have been implemented since the 1950's. Fifty-eight breakwater segments with beach fill were constructed to insure the stability of the peninsula and to provide recreational beaches for the public.

Design and Construction: Originally, three rubble mound offshore breakwater segments had been constructed as a field test. Evaluation of the performance of these segments has assisted in the design of the complete project. The units were constructed in an average water depth of 1.0 m long-term average lake level and have a crest height of 1.2 m. The units are 40 m in length and have gaps of 60 m and 90 m. A movable-bed hydraulic model was constructed and calibrated to replicate shoreline response at the test site.[127] This permits more accurate modeling and prediction of shoreline response to the construction of the remainder of the project. Periodic renourishment is planned (Figure 5.37).

Littoral Drift: The longshore current travels in an easterly direction.

FIGURE 5.37. Presque Isle Peninsula project.

Wave Climate: During the months of April through November, waves of height less than 0.6 m occur 88% of the time. The most significant waves come from the western sector.

Beach Fill: Fill was placed along the beach, which extended the shoreline to a distance of 45 m to 60 m from the structures. Median grain diameter of the fill was 1.8 mm, whereas natural lake sediments in the vicinity have a median grain diameter that ranges from 0.11 to 0.25 mm.

Shoreline Response: Shoreline response varies seasonally. At times a tombolo forms that tends to partially obstruct the longshore current from passing to downdrift structures. Low crest elevation allows for sufficient overtopping to disconnect the tombolo. Sinuous salients are the norm, with the updrift structure maintaining the largest cuspate formation.

Performance: The segmented breakwater system has adequately sheltered the placed beach fill and aids in maintaining a relatively stable shoreline. The beach area is used for recreational purposes.

k) Location: Holly Beach, Louisiana[45,99]
Date of Construction: December 1985
Purpose: The segmented breakwater project was constructed to protect a highway and threatened revetment.

Design and Construction: The segmented breakwater system is an experimental project that combines various configurations of rock, riprap (Figure 5.38), timber piles, and used tires. The table below outlines the description of each of the six segments. Gap widths between units ranged from 89 to 93 m.

Littoral Drift: Littoral transport of 47,000–76,000 m^3/yr is documented for an area

FIGURE 5.38. Holly Beach, Louisiana experimental project.

west of the project, which may give an indication of local longshore transport. The direction of transport is east to west.

Wave Climate: Holly Beach is exposed to a relatively moderate wave climate, with an average breaking wave height and period of 0.5 m and 5 sec, respectively. The waves are generally wind waves received from the south and southeast direction at an average of 20% of the time. Diurnal and spring tides are approximately 0.6 m and 0.75 m, respectively.

Beach Fill: None

Shoreline Response: Shoreline response in the sheltered area of the breakwater is seen to be related to wave energy transmission through the units. Based on profile surveys, Figure 5.39, the riprap structure with the least wave transmission accreted the maximum volume of sediments, whereas the unit with the highest wave transmissibility offered little or no shoreline protection. The shoreline behind the riprap structure has adopted a well-developed salient at MSL and the submerged tombolo appears at low tide. The riprap structure is located at the extreme downdrift end of the project, and during directional shifts of the longshore current, deposition occurs in the shadow zone of this low-permeable structure. Thus, in the early stages of salient development, littoral material is not being transported east to the more permeable units. Breakwater segments having low to moderate permeability have developed cuspate salients on their leeside. The number two segment allows a relatively high transmission of wave energy and the shoreline in its shadow zone has had a mixed response.

Performance: Five of the six breakwater segments have caused deposition of sediments in front of the revetment and highway. Based on an initial project performance evaluation, the system has proven to be a low-cost temporary alternative for shore protection. Deterioration of the tires from exposure to the elements and damage to the timber pile–tire units was apparent and the project was in need of maintenance only eight months after completion.

Gravens and Rosati[45] compared the various functional parameters for several offshore breakwaters at Grand Isle, Louisiana.

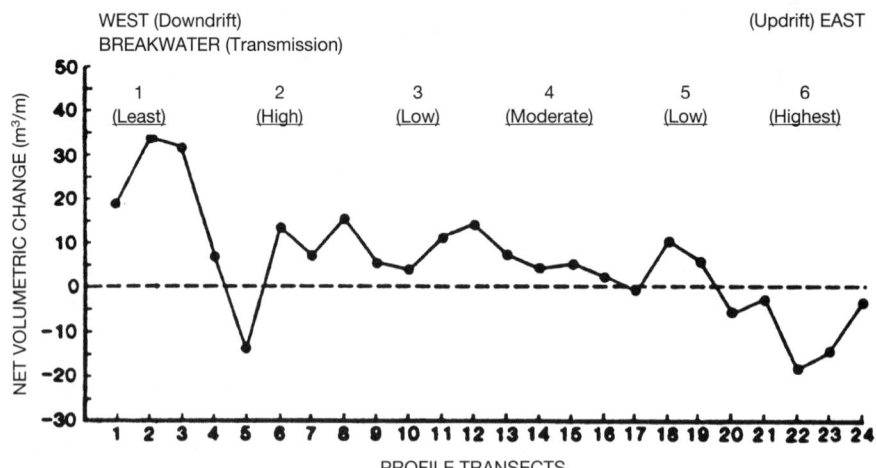

FIGURE 5.39. Qualitative wave transmission classification and net volumetric changes in sediment between 1/86 and 8/86 relative position of the breakwater system.[99]

Table 5.3 compares the offshore breakwater parameters and observed morphologic response. The structure length (L_s) to gap length (L_g) for the proposed Grand Isle design is lower than for both the East and West Cameron Parish projects, thus providing a greater wave energy blocking. Consequently, there should be a greater influence of shoreline change than is occurring at the Cameron Parish projects. The comparison of numerically (GENESIS Model) predicted shoreline response was quite comparable with the observed shoreline responses at the Cameron Parish project (Table 5.3).[45]

l) Location: Gloucester Point, Virginia, on the York River[55]
Date of Construction: 1983
Purpose: The project was proposed to mitigate erosion caused by the placement of a pair of groins.
Design and Construction: A trisectional gabion and riprap breakwater was constructed within an embayment created by two groins. The southern and central sections were placed 12.2 m and 10.7 m offshore of the MTL, respectively, and are 11.0 m in length. The spacing between these sections is 14.6 m. The northern unit is 7.3 m in length and is placed 15.2 m offshore of the MTL. This unit is 18.3 m from the central unit (Figure 5.40).
The breakwater crest height is approximately at MHW, with the northern section being slightly higher.
Littoral Transport: Not documented.
Wave Climate: Wind waves are predominantly from the north and northwest. Wind velocity from this direction averages 10 to 12 knots in the winter months.
Beach Fill: None
Shoreline Response: The southern and central units have developed tombolos during MLW. During higher water levels, the tombolos become submerged and cuspate salients remain. The northern section, which is shorter and further offshore, has only developed a cuspate salient during MLW. Sediment loss was experienced north of the north unit, hence, the addition of a section or an extension of the northern section is being considered.
Performance: The project has functioned well by stabilizing at least 75% of the shoreline within the embayment.

Table 5.3. Comparison of breakwater parameters and observed morphologic response[45]

Project/Design	L_s (ft)	L_g (ft)	X (ft)	Depth (ft)	L_s/L_g	L_s/X	Observed response
West Cameron Parish Project	150	290	560	6	1.9	0.27	Minor ($X_s \sim -20$ to 20 ft)
East Cameron Parish Project	150	300	360	6	2	0.42	Salient ($X_s \sim 10$ to 120 ft)
Old Holly Beach Project, rubble mound structure	150	300	240	6	2	0.63	Salient/low tide tombolo
Mayors' Project, Grand Isle	70	70	350	3.5	1	0.20	Tombolo
Design D1, Grand Isle	200	317	600	6	1.6	0.33	Predicted salient ($X_s \sim 60$ to 80 ft)

5.46   HANDBOOK OF COASTAL ENGINEERING

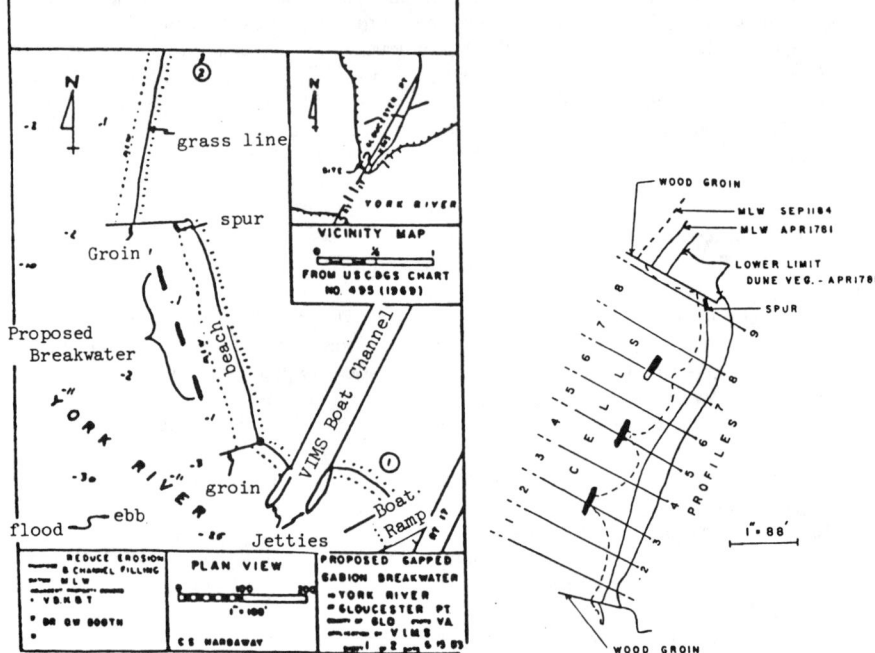

a) VIMS West Gapped Breakwater System as Depicted on Army Corps Permit Application.

b) VIMS West Gapped Breakwater System. Base map showing profile and cell locations.

FIGURE 5.40. Gloucester Point Project.[55]

m) Location: Lake Forest, Illinois, located on the west shore of Lake Michigan[7]
Date of Construction: 1987
Purpose: The goal of the project is the protection of eroding bluffs and preservation of beach fill placed at a public beach.
Design and Construction: The project design was based on an undistorted, three-dimensional movable-bed model at a scale of 1:20. Optimum design of breakwater length and position was contingent upon obtaining a minimum loss of fill material and maintaining a low degree of agitation in a boat launch basin. Due to the high degree of skewness of the predominant wave direction, the breakwaters could not be placed perpendicular to the prevailing incident wave and adequately protect the designed length of shoreline and fill. The final design resulted in two breakwaters at either end of the project being shore connected and three offshore, gapped breakwaters being placed such that they are intercepted by shore-connected groins at approximately mid-length, termed T-groins. Thus, the beach is divided into cells that reduce southerly transfer of fill due to the high degree of skewness of the waves. Design breakwater crest height is 2.4 m. This height allows for considerable overtopping during storms. Two offshore breakwaters are 76 m in length and have shore-connected groins that are 46 m in length. The third breakwater is 61 m

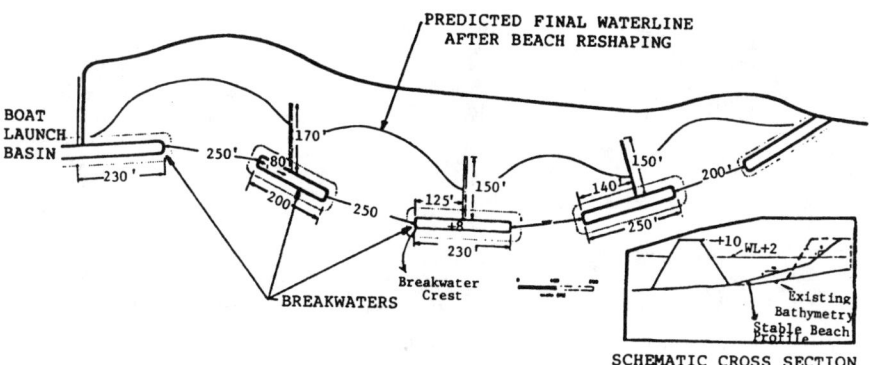

FIGURE 5.41. Lake Forest Project.[7]

long and is connected to a groin that is 52 m in length. Distance between breakwaters is 61 and 76 m (Figure 5.41).

Littoral Transport: The southerly littoral drift near Lake Forest is virtually nonexistent due to shoreline construction north of the vicinity.

Wave Climate: A combined 10 year wind-wave hindcast and refraction-shoaling analysis resulted in a maximum, deepwater, significant wave height of 3 m and period of 6.8 sec, from an azimuth of NE 60 degrees.

Beach Fill: Sufficient beach fill was placed on the shoreline to satisfy the approximate shoreline response. The fill has a median and maximum diameter of 2.8 and 5 mm, respectively.

Shoreline Response: The shoreline maintains well-developed sinuous salients.

Performance: Very little, if any, beach fill has been lost and downdrift erosion due to the project has not occurred. Performance as a public recreational beach is very good. During storm conditions with high water levels, the coarse fill material forms a 3 m berm that tends to protect the shoreline under intense wave action. The berm is periodically graded.

Foreign Breakwaters

a) Location: Kaike Coast, Japan[146]
Date of Construction: 1971–1981
Purpose: Eleven segmented offshore breakwaters were constructed to protect a long stretch of eroding shoreline after both a groin system and sea wall had failed to do so.

Design and Construction: The tetrapod breakwater project was designed from both empirical knowledge gained from similarly constructed projects and from utilizing the first several structures as experimental designs. The first six breakwaters were alternately constructed upstream and downstream of the littoral drift, over a six-year period. This scheme was adopted in order to reduce localized downdrift erosion. The remaining five breakwaters were constructed sequentially. The breakwaters are 150 m in length and have a gap of 50 m. Their original distance offshore varies from about 80 to 150 m and they are built in an average depth of approximately 4 m (Figure 5.42).

Littoral Drift: The westerly littoral drift stems largely from a nearby river and has been drastically reduced in part by construction of dams and agricultural weirs. A marked

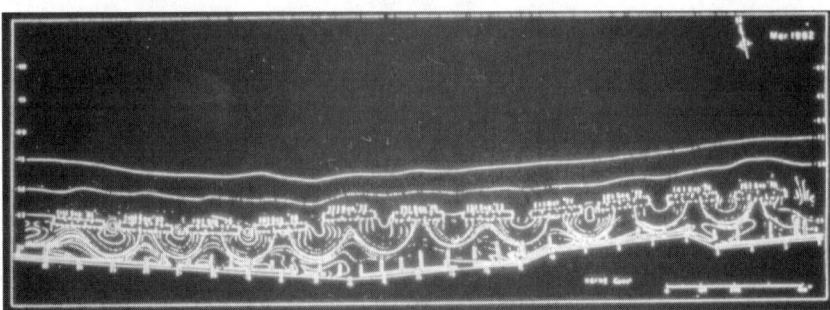

FIGURE 5.42. Kaike Coast offshore breakwaters.

reduction in littoral supply has also occurred to the breakwaters located in the downdrift region of the project. The detached breakwater system was completed in 1981.

Wave Climate: Kaike is located within the Miho Bay, which is oriented such that large waves are refracted into the region.

Beach Fill: None

Shoreline Response: Complete tombolo formation has occurred in the lee of 73% of the breakwaters and well-developed salients have developed shoreward of all the remaining breakwaters except for the extreme downdrift structure. The eleventh breakwater had only accreted a cuspate salient.

Performance: The Kaike coast has been stabilized and a wide beach formed in the lee of each structure. Offshore of the updrift breakwaters, large sediment accumulation has occurred. Offshore of the downdrift breakwaters, sediment losses were created. These conditions may be a result of some sediment being forced offshore and around the tombolo formations. There has been some undesirable subsidence of the breakwaters.

b) Location: Mediterranean Coast, Israel[104]

Date of Construction: 1971–1975

Purpose: Seven offshore detached breakwaters were constructed in order to develop wider beaches for shore protection and recreation.

Design and Construction: The rubble mound segmented breakwater system, Tel Aviv breakwaters numbers three through nine, were designed based on the results of a series of distorted, movable-bed hydraulic model experiments. The horizontal and vertical scales were 1:50 and 1:75 respectively. The scale for the modeling sediment was 1:1. The sediment had a specific gravity of 1.8. The final design included seven breakwaters 130 m in length with gap spacings of 120 m. Five of the breakwaters are 250 m offshore and two are 200 m offshore. A short, unconnected groin is constructed shoreward of breakwater number six and a terminal groin was placed at the extreme downdrift end of the project.

Littoral Drift: The rate of longshore transport in the Tel Aviv region was estimated at 320,000 m^3/yr. However, due to groin construction and tombolo formation, the littoral drift has been reduced.

Wave Climate: The predominant wave directions are from the NW and SW sectors. The region experiences an average wave height and a mean significant wave height of 0.98 m and 1.9 m, respectively.

Beach Fill: None

Shoreline Response: The two breakwater segments that were constructed 200 m offshore have developed sinuous salients. Shoreward of one of these segments is a short groin that aids in entrapping sediments. The remaining five units have generally aided in accreting the shoreline but have only encouraged the formation of subdued salients.

c) Location: Mediterranean Coast, Israel[38,104]
Date of Construction: 1965–1968
Purpose: The goal of the projects was to reduce erosion of recreational beaches and prevent scour of adjoining retaining walls.
Design and Construction: a) The single rubble mound Tel-Baruch breakwater is 100 m in length and is shore-connected by a 100 m long groin. The breakwater is constructed in water depth of 3 m MSL and has a crest height of 1.0 m MSL. b) Tel Aviv breakwaters numbers one and two are also shore-attached by groins of length equal to 200 m. They were constructed at a water depth of 4 m and have a gap width of 135 m. Tel Aviv breakwater number one is 310 m in length and breakwater number two is 240 m in length. These breakwaters have a maximum crest height of 1.25 m.
Shoreline Response: a) Due to the small ratio of distance from shoreline to breakwater length, 0.5, and the presence of the shore-attached groin, the Tel-Baruch breakwater created a tombolo on its leeward side. b) For similar reasons, tombolos formed in the lee of Tel Aviv breakwaters numbers one and two.
Performance: The newly created shorelines perform as intended, permitting protected swimming and providing wide, sandy recreational beaches. However, the groins and tombolos have severely interrupted the littoral current by both trapping and moving sediments offshore that would have otherwise fed downdrift beaches. Damage has been sustained by downdrift sediment-starved shorelines.

d) Location: Southern Coastline of Singapore[136]
Date of Construction: 1971
Purpose: The project was designed to protect a region of reclaimed land.
Design and Construction: A total of 36 gabion-type offshore breakwaters and eight riprap offshore breakwaters were constructed seaward of a length of reclaimed land that supports housing complexes. The failure of a rock revetment led to implementation of artificial headlands. The gabion breakwaters are constructed of layered units of stone encased in steel cages that have a crest height of 0.3 m above HWL and are approximately 30 m in length. The riprap structures are constructed of a base mound of fill covered by a "vinylon" sheet upon which an inverse filter of stone is placed. The overlaying stone is 0.6 m in diameter. The riprap structures have a crest elevation of 0.9 m above HWL and are about 37 m in length.
Both types of breakwaters were constructed at a depth of 0.6 m below MWL and are oriented parallel to incident swell at high tide. The artificial headlands are spaced an average of 244 m apart. Project design was based on the Silvester and Ho[136] stable-bay principle.
Littoral Drift: The direction of predominant wave incidence in combination with a tracer test performed with fluorescent sand indicates that sediment transport is to the west. The littoral drift in this region is generally strong.
Wave Climate: The southern shore of Singapore is protected from the prevailing northeastern winds. Thus, swell waves from the South China Sea are the dominant wave source. An average wave height and period in the month of March was found to be 0.3 m and 6 sec respectively. The maximum wave height never exceeded 0.9 m.
Beach Fill: The vicinity protected is all reclaimed land. Construction dimensions are after fill conditions.
Shoreline Response: The protected shoreline has adopted tombolo formations during low and medium tides. Cuspate salients are visible during high tide.

Analysis

The angle of the incident wave to the shoreline forms the downdrift component of the total energy flux. In a region where sediment is readily available, this component creates a longshore current that carries sediment alongside the shoreline. A key facet of breakwater design is the optimum planform orientation to the shoreline and to the angle of wave approach. The breakwater should be oriented to best reduce the wave energy reaching the shoreline and to alter the wave angle such that wave diffraction effects result in moderate or negligible longshore transport depending upon the desired effect on the local and adjacent beaches. Locating the structure parallel to the shoreline maximizes the effective length of the breakwater, whereas orientation normal to the predominant angle of wave approach is generally most efficient in reducing the longshore component of wave energy flux. If the predominant or most significant waves were approaching approximately in a shore-normal direction, the natural and most effective design would be a shore-parallel breakwater. However, if the wave field is highly seasonal or the design wave approach is not nearly at right angles to the shoreline, then an alternative design may be necessary. Also, the wave field leeward of the breakwater is undergoing refraction and diffraction effects. This will mean that the near symmetry of wave pattern normal incidence will not occur under highly oblique wave conditions. Seven examples illustrating these effects and viable design solutions are presented.

Owing to the high skewness of wave approach relative to the segmented project at Lake Forest, Illinois,[7] groins were constructed to maintain beach fill. Construction of groins allowed more flexibility in designing the orientation of the breakwaters, which are nearly shore-parallel. This project is located in a region where there is negligible littoral drift; thus, no adverse effects downdrift have occurred. Also, currents within the embayments have not been found to pose a threat to swimmers. The project was designed on the basis of the results of three-dimensional model experiments.

The segmented breakwater project at Lorain, Ohio[111] was designed based on an incident wave originating from 40 degrees west of north, or about 5 degrees west of normal incidence from the project site. The three units were constructed parallel to the shoreline. The project has repeatedly been exposed to westerly storms that have eroded a portion of the fill placed behind the westernmost structure. The shoreline protected by the central and eastern units has experienced a net gain of sediment. The performances of the central and eastern structures have potentially been improved by a) the western unit behaving as a sacrificial unit in that the shoreline in its lee has eroded but energy reaching the downdrift units from westerly storm waves has been considerably reduced, and b) a terminal groin constructed at the eastern-most end of the project aids in maintaining fill but does not behave as a complete littoral barrier. Although the Lakeview, Ohio segmented breakwater project does not have widely spaced breakwaters, the asymmetry of the waves has formed a crenelate-like shoreline.

Channel Islands Harbor breakwater was constructed to shelter the harbor and reduce shoaling by trapping the littoral drift north of the north jetty, forming a sediment trap. The breakwater is oriented approximately 30 degrees west of north (parallel to the shoreline), and predominant waves are from the northwest and west. In terms of sediment accretion, the jetties serve in a dual capacity. The north jetty traps littoral drift in the low-energy region within the shadow of the breakwater and the southern jetty serves to protect the downdrift region, where some deposition has also occurred. Equilibrium of the salient is not attained, as sediment bypassing procedures are performed to prevent shoaling of the channel. Owing to the small ratio of distance offshore to breakwater length, 0.77, and the adequate supply of littoral material, the formation of a salient would be anticipated without the construction of the jetties.

Refraction–diffraction patterns of an oblique wave around a shore-parallel, offshore breakwater are depicted from the aerial photograph of Channel Islands Harbor breakwater. The waves diffracting around the updrift end of the breakwater become more shore-parallel, whereas those passing the downdrift edge of the breakwater become more shore-normal.

Silvester[134,136] has discussed the use of crenelate-shaped bays to stabilize a shoreline exposed to a skewed wave approach. Application of headland breakwaters oriented at oblique angles to the shoreline and to the direction of predominate wave approach was successful along the southern Singapore coastline. This method utilizes widely spaced breakwaters that are located relatively close to shore such that tombolo formation occurs. The crenelate-shaped bays that form between the headlands are a result of wave diffraction patterns creating a tombolo and shaping the shoreline until the waves approach the newly formed shoreline at right angles. This is both an accretionary and an erosional process.

Additional Comments

The orientation of a breakwater to both the incident wave angle and the shoreline contributes significantly to its success. Several design alternatives and lessons learned from existing projects have been outlined. Design alternatives for offshore breakwater projects that are exposed to an asymmetric wave field may combine a shore-parallel breakwater(s), for optimum length of shoreline protection, in addition to one or more of the following designs:

1. Breakwater-connected groins can be constructed for protection of fill. If adjacent beach erosion is a concern, connecting groins are only viable in regions where littoral drift is virtually nonexistent.

2. Short, nonconnecting groins may also be considered for protection of beach fill. The length of the groin should be sufficiently short to allow much of the natural littoral drift to pass. Terminal groins such as the one located at Lakeview Park, Ohio or groins located immediately behind the structure, within the sheltered region, may provide satisfactory protection from a strong longshore current.

3. The extension of a single breakwater or the addition of a unit to a segmented project can be placed updrift of the project and offshore of a region that is not intended for shoreline protection. The offshore breakwater will provide only a small amount of protection to the shoreline in its lee, but will reduce the wave energy reaching the sheltered region of the downdrift breakwaters. The planiform shoreline response for this configuration would be crenelate such as for artificial headlands, with the exception of tombolo formation.

Alternatively, orientation of the breakwater normal or near normal to the oblique incident wave will assist in reducing the longshore component of energy flux as was done along the southern coast of Singapore. However, in order to avoid deleterious erosional problems downdrift of the project, salient formation should be considered in lieu of tombolo formation. In all cases a diffraction–refraction analysis should be performed.

Summary and Statistical Analysis

A summary of United States offshore breakwaters is given in Table 5.4.

Seiji, et al.[128] conducted a statistical study on the effect and stability of detached breakwaters in Japan. The coastal zone is highly utilized in Japan and offshore breakwaters play a leading role in diminishing or preventing beach erosion.

Table 5.4. Summary of characteristics of United States detached breakwaters

Project	Type of breakwater*	When built	Total length of breakwater project (B)	No. of segments	Length of segments (B_s)	Gap width (G)	Distance offshore (X)
Venice, CA	Si	1905	183m	N/A	N/A	N/A	366 m 213 m
Santa Barbara, CA	Si	1929	434 m	N/A	N/A	N/A	305 m
Santa Monica, CA	Si	1934	610 m	N/A	N/A	N/A	610 m
Haleiwa Beach, HI	Si	1965	49 m	N/A	N/A	N/A	91 m original 49 m w/fill
Winthrop Beach, MA	Se	1935	625 m	5	91m	30m	305 m
Lakeview Park, OH	Se	1977	403 m	3	62 m	49 m	137 m original 76 m w/fill
Presque Isle, PA	Se	1978	440 m	3	38 m	60,91 m	46 m
Colonial Beach, VA (Central)	Se	1982	427 m	4	61 m	45 m	64 m
Colonial Beach, VA (Castlewood)	Se	1982	335 m	3	61,93 m	26, 40 m	46 m
Lakeshore Park, OH	Se	1982	244 m	3	38 m	60 m	120 m original 75 m w/fill
East Harbor, OH	Se	1982	244 m	3	46 m	90, 105, 120 m	180 m
Lincoln Park, IL	Si	1939	457 m	N/A	N/A	N/A	183 m
Channel Islands, CA	Si	1960	700 m	N/A	N/A	N/A	550 m
Waikiki Beach, HI	Si	1938	213 m	N/A	N/A	N/A	76 m
Holly Beach, LA	Se	1985	555 m	6	46, 50 m	93, 89 m	78, 61 m
Redington Shores, FL	Si	1986	112 m	N/A	N/A	N/A	91.5 m
York River, Gloucester Point, VA	Se	1983	62.2 m	3	11 m 11 m 7.3 m	14.6 18.3 m	12.2 m 107 m 15.2 m

*Si = single. Se = segmented
†Datum used is local MLW (unless otherwise stated) or low-water datum (LWD) for the Great Lakes.

Table 5.4. *Continued*

Water† depth (D)	Crest† elevation	Tombolo (T) or salient (S)	Comments	Fill placed	B_s/X	B/X
1.8m	3.7m	S	Pre 1940s	No	N/A	0.5
		T	1940s–1960s Groin installed early 1960s			0.9
7.6 m	3.7 m	T	Project dredged and shoreward connection added to prevent tombolo formation	No	N/A	1.4
7.6 m	3.0 m	S	Periodic dredging prevents tombolo formation	No	N/A	1.0
2.4 m	1.5 m	S	159 m groin nearby	No	N/A	0.5
		Near T		Yes		1.0
3 m MLW	5.5 m MLW	T	Two beach planforms as a result of 2.7 m tidal range	No	0.3	2.0
5.7 m MHW	2.8 m MHW	S			0.3	
3.0 m	2.4 m	S	Terminal groins at both ends	No	0.5	2.9
				Yes	0.8	5.3
0.3 m	1.8 m	T/S	Tombolos form during low wave energy condition, removed by storms	Yes	0.8	9.6
1.2 m	0.4 m	T/S	Tombolos behind breakwaters, salients behind others	Yes	0.9	6.7
0.61m	0.4m	T/S			1.3,.2.0	6.7
1.5 m	20 m	S	Still adjusting, fine-size fill being lost longshore	Yes	0.3	2.0
					0.5	3.3
1.5 m	24 m	S	Still adjusting, low sand supply	No	0.3	3.1
3.7 m	−1.2 m	F	Fill placed and held satisfactory	Yes	N/A	2.5
4.3 m						
9.1 m MLLW	+43 m	T	Tombolo is periodically bypassed	No	N/A	13
—	0 m	S	Fill placed, which eroded slowly over 8-year period	Yes	N/A	2.8
2.5 m	1.2,. 2.4 m	S	1 riprap, 5 pile w/used tire	No	0.8	9.1
2.4 m	0.46 m	S	Accumulated 24 m of beach in less than 20 days	Yes	N/A	3.8
0.41 m	0.3 m	T	Estuary unattached groins nearby	No	0.9	4.90
	0.3 m	S				
	0.5 m					

5.53

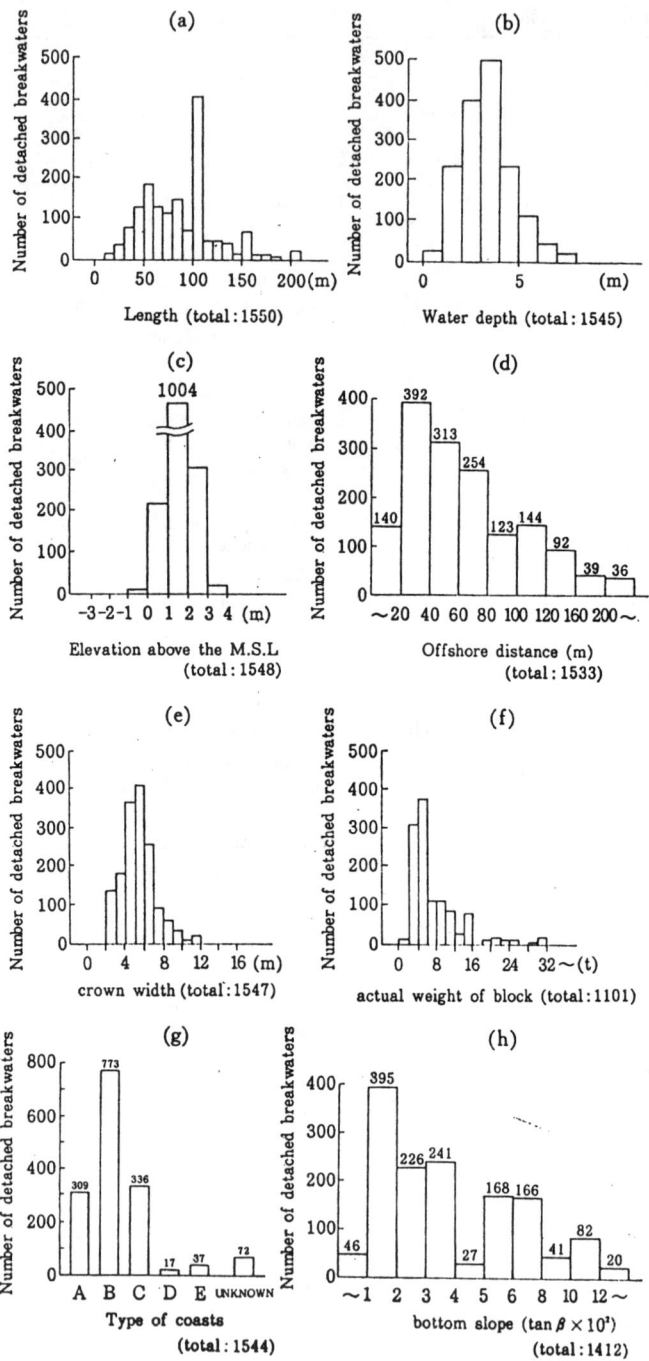

FIGURE 5.43. Frequency distribution of various dimensions of the offshore breakwaters; (a) length, (b) water depth, (c) elevation above the MSL, (d) offshore distance, (e) crown height, (f) weight of block, (g) type of coasts, and (h) bottom slope.[128]

OFFSHORE (DETACHED) BREAKWATERS

A survey was conducted seeking information on some 1,552 offshore breakwaters constructed in Japan prior to 1982. The results of the survey are shown in Figure 5.43(a) to (h):

a) The maximum frequency of breakwater length is between 100 and 110 m
b) The maximum water depth is between 3 and 4 m (note that 90% of breakwaters are in water depth below 5 m)
c) The most frequent elevation of breakwaters above MSL is 1–2 m,
d) The most frequent distance between the breakwater and original shore is between 20 and 40 m
e) The most frequent crown width is between 5 and 7 m
f) The most frequent weight of concrete blocks is between 4 and 6 metric tons
g) The most breakwaters were built on the coasts with a fairly well-developed bar-trough bathymetry
h) The most common slopes were between 0.01 and 0.02

Figure 5.44 shows the tombolo formation in 60% of the cases.
Figure 5.45 shows the frequency distribution of maximum shoreline advance. The maximum frequency is between 10 and 20 m.
Figures 5.46–5.50 are photographs showing deployment of offshore breakwaters alone or in combination with groins, designed to provide recreational beaches and prevent beach erosion.

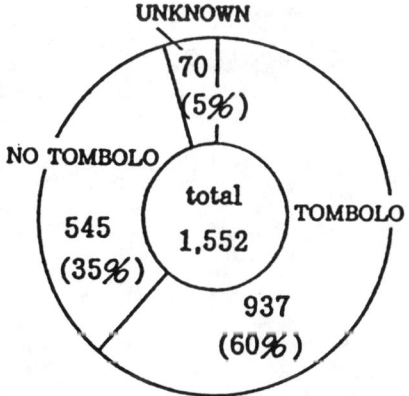

FIGURE 5.44. Percentage of tombolo formation.[128]

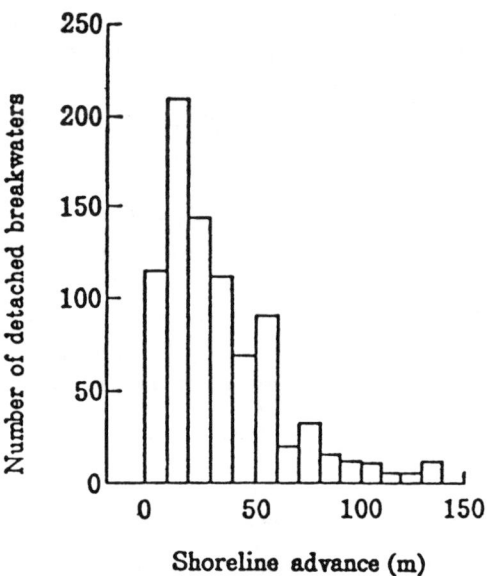

FIGURE 5.45. Frequency distribution of the maximum shoreline advance due to tombolo formation.[128]

a) before

b) after

FIGURE 5.46. Effect of offshore breakwaters (courtesy of *Danish Hydraulics*, No. 7, November 1986).

FIGURE 5.47. Combination of offshore breakwaters and groins designed to form recreation beaches and prevent beach erosion (Spain).

DESIGN CONSIDERATIONS

Rosati[115] summarized the techniques for detached breakwater design in a U.S. Army Corps of Engineers' report. Several sections of the report are reproduced below:

Empirical Relationships

A breakwater project can be designed in segments to protect a long length of shoreline, for salient or tombolo formation, or both. In the case of a site with a significant longshore transport rate adjacent to beaches that cannot tolerate much project-induced beach change, the design beach response should be a salient. A salient allows littoral movement of material to continue shoreward of the structures, minimizing project impacts on adja-

a) before

b) after

FIGURE 5.48. Offshore breakwaters designed to form tombolos (Spain).

cent shores. If impacts to adjacent beaches are not of concern, the breakwater system could be designed for tombolo formation, maximizing project beach accretion. A common nomenclature for all relationships will be used throughout this chapter and is presented in Figure 5.51. Simple relationships between structure length and distance offshore have been proposed by various researchers.

1. Inman and Frautschy.[72] This is based on observations of beach response at the 434 m long single-detached breakwater at Venice in Santa Monica, CA. The authors observed that pronounced accretion did not occur for the following condition:

$$\frac{L_s}{X} \leq 0.17 - 0.33 \tag{14}$$

where L_s is the breakwater segment length, and X is the breakwater segment distance from the original shoreline.

FIGURE 5.49. Combination of groins and offshore breakwaters, Benalmadener Coast (Spain).

FIGURE 5.50. Offshore breakwaters constructed to provide a recreational beach, Pedregalejo Beach, Malaga, Spain.

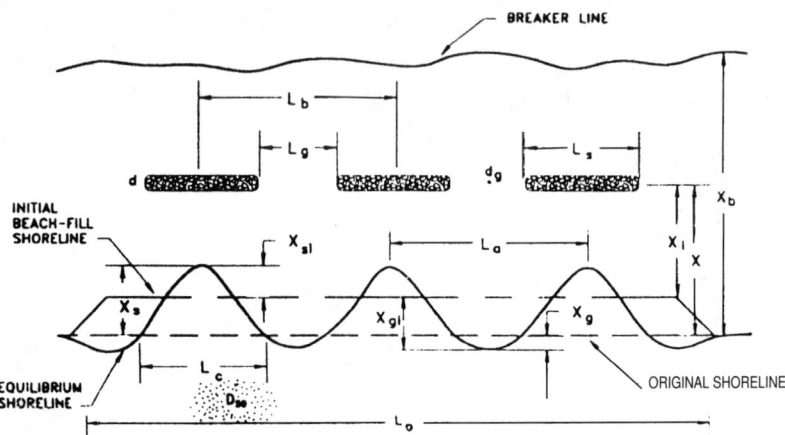

FIGURE 5.51. Definition of terms used in U.S. Army Corps of Engineers report.[115]

2. Nir.[104] This research is based on project performance of 12 breakwaters on the Israeli coast, Nir[102] concluded that accretion is very low or does not occur at all if the segment length to offshore distance ratio has a value of 0.5 or less:

$$\frac{L_s}{X} \leq 0.5 \tag{15}$$

Assuming that the planiform of the tombolo/salient formed a trapezoid, Nir calculated the area and volume of each salient out to the depth of structure. Dividing the volume by the area, Nir obtained the average sand thickness in the tombolo/salient z_s and related it to the structure–distance ratio as follows (note that the equation is dimensionally inconsistent):

$$z_s(\text{m}) = 1.786 - 0.809\left(\frac{X}{L_s}\right) \tag{16}$$

The relationship was developed from prototype data with X/L_s values ranging from 0.38 to 2.0, resulting in sand thicknesses ranging from approximately 0.3 to 1.7 m. Seven out of the twelve structures developed tombolos. Beach response for the Israeli breakwaters was observed to reach a mature age about 5 years after construction was completed; adjacent beaches suffered severe erosion during the first 3–4 years (or longer) after construction.

3. Gourlay.[43] Gourlay presents similar relationships based on physical-model and field observations. Based on the laboratory studies of Adachi, Sawaragi, and Ogo[2], Shinohara and Tsubaki,[129] and others, Gourlay reaches the following conclusions for laboratory response to detached breakwaters:

a) Tombolos can form only if the structure is located in the surf zone.

b) Noninterfering diffraction patterns, resulting in two independent current systems (and possibly double tombolo formation), exist for the following structure–distance ratio:

OFFSHORE (DETACHED) BREAKWATERS

$$\frac{L_s}{X} > 2.0 \tag{17}$$

c) Diffraction patterns interfered and reduced longshore currents for a lower range of the ratio. Beach response was observed to be either a salient if bed load was the dominant transport mechanism (fall velocity parameter, $H_0/VT < 1.5$, where H_0 is the deepwater wave height, V is the sediment fall velocity, and T is the wave period), or a more complex feature if suspended load transport dominated ($H_0/VT > 1.5$):

$$\frac{L_s}{X} < 0.4 \text{ to } 0.5 \tag{18}$$

d) Noninterfering currents strong enough to transport material resulted in tombolo formation for an intermediate ratio value:

$$\frac{L_s}{X} = 0.67 \text{ to } 1.0 \tag{19}$$

Gourlay makes the following observations about beach response to prototype breakwaters based on performance of structures in California, Japan, and Israel:

a) Tombolo formation can only occur if the structure is inside the surf zone and

$$\frac{L_s}{X} > 0.67 \tag{20}$$

b) For cases in which the structure is outside the surf zone, the seaward extent of the beach response (salient) is determined by the breaking point location.

4. Rosen and Vajda.[120] Rosen and Vajda graphically present relationships to predict the equilibrium salient and tombolo size based on structure and site parameters. Relationships to predict salient formation are based on a physical model study of beach response (0.64 mm sand) to an impermeable breakwater under normal wave incidence conducted by Rosen and Vajda, including laboratory data measured by Shinohara and Tsubaki[129] and prototype data from Israeli, U.S., and Japanese structures. Given a desired salient length in the on-offshore direction (measured from original shoreline) X_s, and a dimensionless breaker distance X_b/S, where X_b is the distance from the original shoreline to the mean breaker line, the recommended ratio L_s/X can be graphically determined (Figure 5.52).

Several relationships are also presented for determining tombolo size in the equilibrium state, based on the field data of Inman and Frautschy,[72] Toyoshima,[146] and Nir[104] (Figure 5.53). Given project variables L_s, X, and beach slope I, the following parameters can be estimated: the volume of sand trapped in the tombolo Q, length of tombolo in alongshore direction (measured at structure) L_T, and area of tombolo A_T.

5. Shore Protection Manual (SPM).[132] Based on the pattern of diffracting wave crests in the lee of a breakwater, the U.S. Army Corps of Engineers' Shore Protection Manual recommends that the structure length be less than the distance offshore to inhibit tombolo formation:

$$\frac{L_s}{X} < 1.0 \quad \text{(tombolo formation prevented)} \tag{21}$$

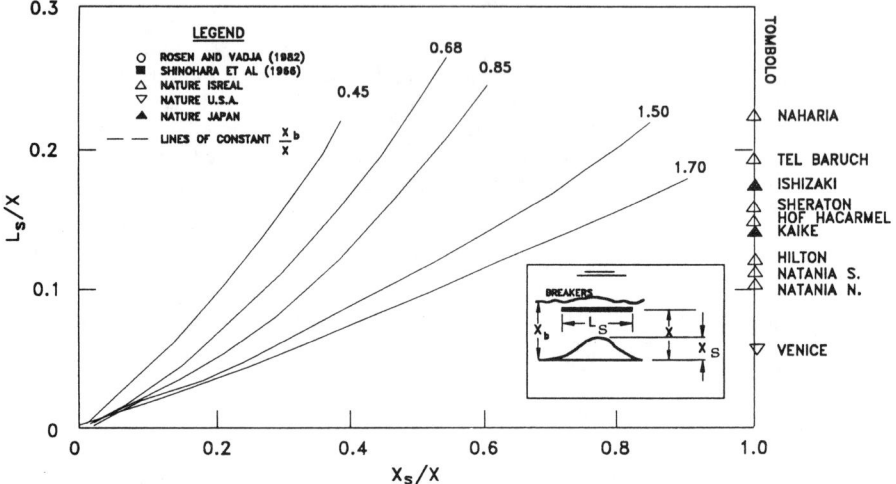

FIGURE 5.52. Equililbrium salient formation (after Rosen and Vajda[120]).

This structure–distance ratio usually allows diffracting wave crests to intersect in the shadow zone of the structure prior to undistorted wave crests reaching the adjacent beach. For normal wave approach, the approximate location of the salient apex is the intersection of the diffracting wave crests as they reach the shoreline.

The SPM recommends the following limit to ensure tombolo formation:

$$\frac{L_s}{X} > 2.0 \quad \text{(tombolo formation prevented)} \tag{22}$$

6. Dally and Pope.[22] Dally and Pope recommend limits of the structure–distance ratio based on the type of shoreline advance desired (either nonuniform, where salients or tombolos occur, or uniform, in which an equal advance of the beach occurs), and the length of the beach to be protected. For nonuniform protection over relatively short project distances, a single impermeable detached breakwater is recommended, with a structure length as long as the project beach. If the depth of construction implied by the calculated distance offshore (see relationships below) is too great, Dally and Pope recommend moving the structure shoreward to a feasible depth and increasing structure transmissibility. Longer project shorelines should be protected with a relatively impermeable segmented system for nonuniform shoreline advance.

a) The following structure–distance ratios (and gap distance, for segmented systems) are recommended for tombolo formation:

$$\frac{B}{X} = 1.5 \text{ to } 2 \quad \text{(for single breakwater)} \tag{23}$$

$$\frac{B}{X} = 1.5, \quad L \leq B_g \leq B \quad \text{(for segmented breakwater)} \tag{24}$$

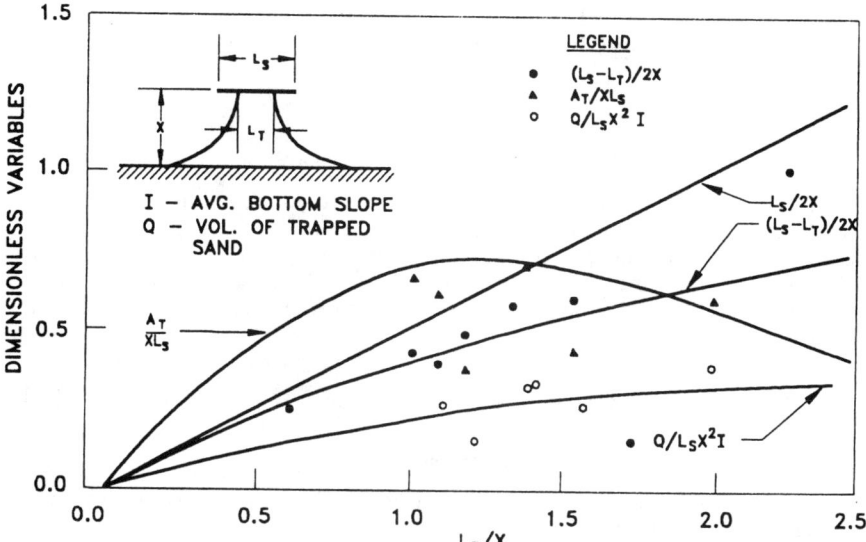

FIGURE 5.53. Equilibrium tombolo formation (after Rosen and Vajda[120]).

where L is the wave length at the structure and B_g is the gap distance between adjacent breakwater segments.

b) The structure–distance ratio is reduced for salient formation:

$$\frac{B}{X} < 0.5 \quad \text{for single breakwater, or segmented breakwaters} \quad (25)$$

c) For uniform protection of a site, a structure located well outside the normal surf zone is recommended. Either a highly permeable, partially submerged structure (60% transmissibility is recommended) or an impermeable, frequently segmented structure will allow the required degree of wave energy to enter the shadow zone of the structure for uniform shoreline advance.

7. Toyoshima.[145,146] Toyoshima recommends breakwater design guidance based on experience gained through observation of prototype performance of 86 single and segmented breakwater systems (217 segments total) along the Japanese coast. The relationships are categorized based on the proposed depth at the structure: the shoreline system, where the structure is placed at or near the waterline; the shallow-water system, with a structure depth less than or equal to 1 m; and the median-depth system, for structure in a depth of 2 to 4 m. Relationships are not given for a deepwater system, which is described as being located at the outside of the surf zone for the purpose of dissipating wave energy rather than sediment deposition.

a) The shoreline system is recommended for sites with either a steep beach slope, where only small-scale sand deposition is required, to protect an existing structure from wave attack (i.e., sea wall), or to maintain the existing beach width. The "group island type" (segmented) breakwater is recommended, constructed of two-layer pile armor blocks. A problem noted with shoreline system structures is large-scale scour on the seaward face of the structure, especially if the structure is exposed to significant wave attack.

$$\frac{L_s}{L} = 2 \text{ to } 3 \quad \text{(segmented breakwater (recommended } L_s = 40 \text{ to } 60 \text{ m))} \quad (26)$$

$$L_g = L \quad \text{(recommended } L_g = 20 \text{ m)} \quad (27)$$

b) The shallow-water depth system (constructed in depths less than 1 m) is recommended for gently sloping beaches or at sites where the structures are expected to be constructed from the beach using a track and crane method. A "group island type" (segmented) system is recommended, constructed of armor blocks with a crest elevation approximately one-half the wave height above high-water level (1 to 1.5 m above high-water level).

$$\frac{L_s}{L} = 3 \text{ to } 5 \quad \text{(segmented breakwater (recommended } L_s = 60 \text{ to } 100 \text{ m))} \quad (28)$$

$$L_g = L \quad \text{(recommended } L_g = 20 \text{ m)} \quad (29)$$

If an island breakwater is used (single breakwater), Toyoshima recommends the following segment length ratio:

$$\frac{L_s}{L} \leq 10 \quad \text{(single breakwater)} \quad (30)$$

c) The median-depth system (structures constructed in the surf zone, 2- to 4-m depth) was the most common type of system observed in Toyoshima's survey of Japanese breakwaters. For reduction of wave energy, a continuous structure (long single breakwater) is recommended, whereas either an island (single) or group island type (segmented) breakwater is recommended for the primary purpose of sediment deposition. Construction is through the use of a floating plant; a rubble mound or composite type rather than armor block construction is recommended to reduce costs. As was recommended for the shallow-water depth system, the structure crown height should be one-half wave height above the high-water level (1 to 1.5 m above high water). For single breakwaters, the following structure ratio is recommended:

$$\frac{L_s}{L} = 3 \text{ to } 10 \quad \text{(recommended } L_s = 100 \text{ to } 300 \text{ m)} \quad (31)$$

For a segmented system, the structure lengths are decreased:

$$\frac{L_s}{L} = 2 \text{ to } 6 \quad \text{(recommended } L_s = 60 \text{ to } 200 \text{ m)} \quad (32)$$

$$L_g = L \quad \text{(recommended } L_g = 20 \text{ to } 50 \text{ m)} \quad (33)$$

For both single and segmented systems, the following structure–distance ratio is recommended as a standard:

$$\frac{L_s}{X} = 1 \text{ to } 3.3 \quad (34)$$

For a continuous breakwater,

$$\frac{X}{L} = 1 \text{ to } 3 \quad (\text{recommended } X = 30 \text{ to } 100 \text{ m}) \tag{35}$$

Toyoshima used the described guidance to design more than 20 systems along the Kaike Coast, Japan, and concludes that the structures were successful in accomplishing their intended purpose. He recommends, however, that a countermeasure for beach erosion be considered for the beach opposite the breakwater gaps, as this area is often eroded after breakwater installation. In later discussions of the Kaike Coast project (Toyoshima[145,146]), performance of the breakwaters is still considered successful, although subsidence of the structures is reported to be a significant contributor in reducing their efficiency.

8. Walker, Clark, and Pope.[171] Walker, Clark, and Pope discuss the method used to design the Lakeview Park, Lorain, OH, segmented detached breakwater project for salient formation. This technique, hereafter referred to as the diffraction energy method, involves construction of diffraction coefficient K_D isolines for representative waves from predominant directions. Detached breakwaters along the California coast, particularly the Venice breakwater, were observed to have shorelines approximated by the intersection of K_D isolines equal to 0.3. Walker, Clark, and Pope theorize that storm waves, usually an order of a magnitude greater than the average wave conditions, are able to transport material at least as far shoreward as the average wave K_D equal to 0.3 isolines. Thus, these isolines are a good indicator of the likely shoreline position. The Lakeview Park project was designed such that the $K_D = 0.3$ isolines intersected lakeward of the project beach fill but shoreward of the three breakwater segments, ensuring that tombolo formation would not occur. Actual shoreline response to the structures was slightly different than had been expected; i.e., the west end of the project beach was continually eroded, despite beach-fill replenishment. The most likely reason for the discrepancy between design and actual shoreline response, however, may not be the design method itself but the wave climate chosen for design (USAED, Buffalo[160]; Pope and Dean[113]), as waves from the northwest were inappropriately weighted.

9. Pope and Dean.[113] Pope and Dean present bounds of observed beach response based on prototype performance of seven U.S. detached breakwater projects. The occurrence of tombolo, salient, or nonsinuous beach response is given as a function of two dimensionless parameters: the segment length-to-gap ratio and an effective distance-offshore-to-depth-at-structure ratio, where d is the depth at the structure. High- and low-water conditions are differentiated in Figure 5.54 for those projects with a large water level range (Presque Isle and Winthrop Beach) by the symbols (H) and (L), respectively.

10. Ahrens and Cox.[5] Ahrens and Cox used the beach response index classification scheme presented by Pope and Dean[113] to develop a relationship for expected morphological response as a function of the segment-to-gap ratio:

$$I_s = e^{[1.72 - 0.41(Ls/X)]} \tag{36}$$

where I_s is the beach response index, coded as follows:

$I_s = 1$ (permanent tombolo formation)
$I_s = 2$ (periodic tombolos)
$I_s = 3$ (well-developed salients)
$I_s = 4$ (subdued salient)
$I_s = 5$ (no sinuosity)

11. Berenguer and Enriquez.[13] Berenguer and Enriquez developed design guidance for pocket beaches, projects where sediment transport is laterally and frontally limited, based on performance of 24 projects along the Spanish Mediterranean coast. The authors noted that the longest alongshore distance between adjacent salient/tombolos, L_a, was approximately twice the on-offshore distance between the gap and shoreline $(X + X_g)$, where X_g is the erosion/accretion opposite the gap, measured from the original shoreline:

$$L_a = 2(X + X_g) \tag{37}$$

Relating this gap erosion distance to the length of gap, the authors found that

$$(X + X_g) = 256 + 0.85 L_g \tag{38}$$

and

$$(XL_b = 2.5 (X + X_g)^2 \tag{39}$$

where L_b is the alongshore center-to-center distance between adjacent segments. Assuming a semicircular beach response in the area defined by L_b and X, the maximum surface area (i.e., beach fill) that could be stable, S_p, can be calculated:

$$S_p = 0.37 \, XL_b \tag{40}$$

For values of $L_g d_g$ between 150 and 500 m², where d_g is the depth at gap, Berenguer and

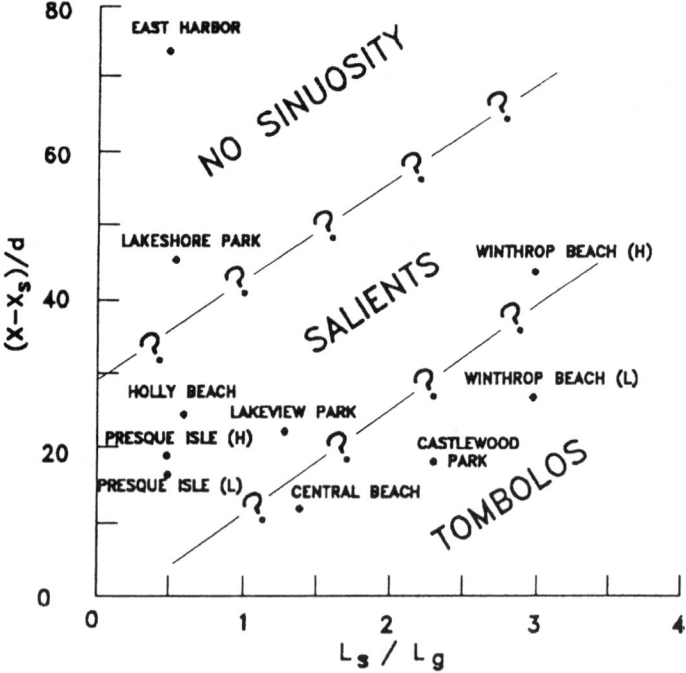

FIGURE 5.54. Morphological response as a function of structural parameters (after Pope and Dean[113]).

Enriquez developed a relationship to predict the gap erosion $(X + X_g)$, given the average sediment size at the site D_{50} (Figure 5.55).

12. Ahrens.[5] Ahrens extended the results of Berenguer and Enriquez[13] by performing a regression analysis of the Mediterranean pocket beach data and making some simplifying assumptions about wave height at the gap, H_g. The wave height was assumed equal to 0.78 times the depth at the gap, d_g, with a maximum value of 3.0 m. The variables H_g, d_g, and L_g should be in meters, and D_{50} in millimeters (the relationship is not dimensionless). The gap erosion $(X + X_g)$ is predicted well, with a squared correlation coefficient of 0.78:

$$(X + X_g) = 6.67 \frac{d_g^{0.8} L_g^{0.4} H_g^{0.2}}{D_{50}} \quad (41)$$

13. Seiji, Uda, and Tanaka.[128] Based on a survey of over 1,500 breakwaters in Japan, Seiji, Uda, and Tanaka predict the following gap erosion relationships, where gap erosion is defined as retreat of shoreline to the lee of the gap from the initial (preproject) shoreline position:

$$\frac{L_g}{X} < 0.8 \quad \text{(no erosion opposite gap)} \quad (42)$$

$$0.8 \le \frac{L_g}{X} \le 1.3 \quad \text{(possible erosion opposite gap)} \quad (43)$$

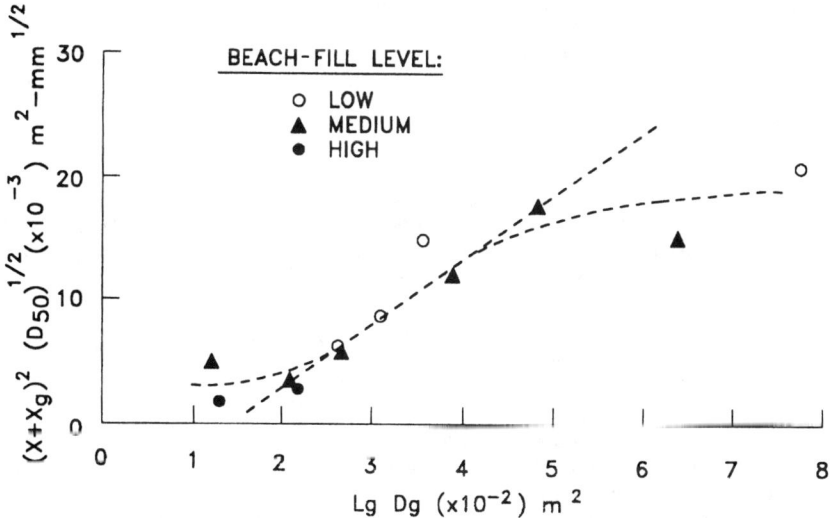

FIGURE 5.55. Relationship to predict gap erosion for pocket beaches (after Berenguer and Enriquez[13]).

$$\frac{L_g}{X} \geq 1.3 \quad \text{(certain erosion opposite gap)} \tag{44}$$

14. Noble.[105] Noble evaluated the effects of detached structures along the California coast and those discussed in the literature and concluded that "detached breakwaters produce only minimal impact when the offshore distance of the structure is greater than six times the breakwater length." That is

$$\frac{L_s}{X} \leq \frac{1}{6} \quad \text{(minimal impact)} \tag{45}$$

15. Noda.[106] Noda conducted a series of movable-bed (0.2 mm sand) laboratory experiments evaluating physical parameters controlling the development of tombolos and salients, especially due to on-offshore transport. Noda used both storm and swell-type waves and concluded that maximum deposition occurs when

$$0.5 \leq \frac{X}{X_b} \leq 1.0 \tag{46}$$

Well-developed tombolos were observed to form when

$$\frac{X}{X_b} = 0.56 \tag{47}$$

and a bimodal tombolo developed for

$$\frac{X}{X_b} = 0.39 \tag{48}$$

The total amount of sand deposition per unit width of gap was observed to be about three times the quantity of material transported shoreward through a permeable breakwater, when dimensions of the structures are "chosen appropriately." When structures were placed seaward of the breaker line, no sand was observed to move seaward through the structures; when structures were placed near the shoreline, Noda observed that sand may move offshore.

16. Hallermeier.[46] In a review of field and laboratory sediment depositional patterns in the vicinity of coastal structures, Hallermeier developed a relationship for the depth limit of sediment transport. Hallermeier recommends the following depth as a guide when positioning detached breakwaters when tombolo formation is deemed undesirable:

$$d_{sa} = \frac{2.9\,H_\theta}{\sqrt{(S-1)}} - \frac{110\,H_\theta^2}{(S-1)gT_\theta^2} \quad \text{(depth for salient formation)} \tag{49}$$

where

d_{sa} = annual seaward limit of the littoral zone
H_e = deepwater wave height exceeded 12 hr per year
S = ratio of sediment to fluid density
g = acceleration of gravity
T_e = wave period corresponding to H_e

For headland structures (tombolo development), structures should be sighted near

$$d = \frac{d_{sa}}{3} \quad \text{(headland structures)} \tag{50}$$

17. Suh and Dalrymple.[141] Based on unscaled monochromatic movable-bed laboratory tests, Suh and Dalrymple give relationships for salient length, X_s, given structure length L_s and location in surf zone for single offshore breakwaters:

$$X_s = 0.156 \, L_s \quad \frac{X_b}{X} < 0.5 \quad R = 0.98 \tag{51}$$

$$X_s = 0.317 \, L_s \quad 0.5 < \frac{X_b}{X} < 1.0 \quad R = 0.85 \tag{52}$$

$$X_s = 0.377 \, L_s \quad \frac{X_b}{X} \geq 1.0 \quad R = 0.75 \tag{53}$$

where R is the correlation coefficient.

Combining the laboratory results with available prototype data, the following relationship for salient length is obtained:

$$X_s = X(14.8)\frac{L_g X}{L_s^2} e^{[-2.83(L_g X/L_s^2)]} \tag{54}$$

Tombolos usually formed for single prototype breakwaters when

$$\frac{L_s}{X} \geq 1.0 \tag{55}$$

For multiple offshore breakwaters, tombolos formed when

$$\frac{L_g X}{L_s^2} \approx 0.5 \tag{56}$$

18. Harris and Herbich.[58] Harris and Herbich conducted a series of movable bed (D_{50} = 0.63 mm) monochromatic waves physical model tests to evaluate the effect of detached breakwater gap spacing on sand entrapment. Relationships for the average quantity of sand deposited to the lee of each breakwater, Q_b, both to the lee of the breakwater and in the gap area, ($Q_b + Q_g$), and in the gap area alone are presented in Figures 5.56 through 5.58. Combining model results with prototype data, Harris and Herbich relate the dimensionless accreted volume to structure parameters as follows:

$$\frac{Q_b}{XL_s d} = e^{[0.31-1.92(X/L_s)]} \tag{57}$$

The relationships is based on data with values of X/L_s ranging from 0.5 to 2.5.

19. Sonu and Warwar.[138] Sonu and Warwar empirically relate the growth of the Santa Monica, CA tombolo through time as follows:

$$Q = Q_0(1 - e^{-At}) \quad (58)$$

where Q_0 is the volume at final equilibrium state, equal to 2,100,000 yd³ for the Santa Monica breakwater and A is an empirical coefficient, determined as 0.104 year⁻¹. The authors conclude that, without dredging events, the Santa Monica breakwater would have reached 90% growth in 22 years.

20. Silvester and Ho.[136] For beaches that have equilibrated with headland structures, Silvester and Ho present a relationship to predict the maximum gap indentation $(X + X_g)$, given a gap distance L_g and predominant wave direction β (degrees) (Figure 5.59).

21. Japanese Ministry of Construction.[73] The empirical method of breakwater design described by the JMC in their "Handbook of Offshore Breakwater Design" has several apparent advantages over the empirical relationships presented previously. The JMC method is a step-by-step iterative procedure, allowing the designer to follow specific guidelines towards the final design. Rosati and Truitt[117] discuss the JMC method of breakwater design and present several example problems.

However, the JMC method has its disadvantages for use in design of U.S. detached breakwater systems. Approximately 60% of the projects on which the method is based resulted in tombolo formation; therefore, it is more appropriate for headland or pocket-beach-type systems, rather than detached breakwater or reef systems. Unlike U.S. breakwater projects, beach fill is not placed as a part of the JMC's projects; therefore, there is no provision for beach fill in the iterative method. All structures considered in the JMC study were permeable; however, the effect of increased or decreased structure transmissibility is not included in the JMC method. For construction of a highly transmissible struc-

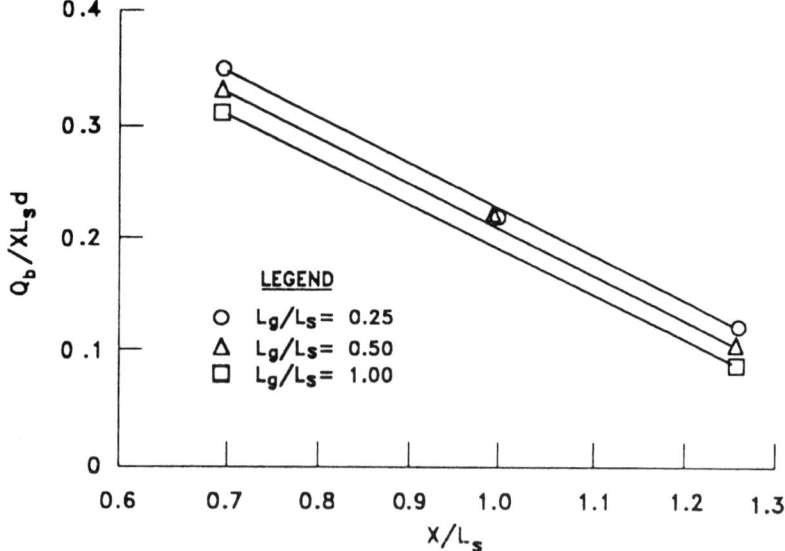

FIGURE 5.56. Effect of gap spacing on sand deposition to lee of breakwater (after Harris and Herbich[58]).

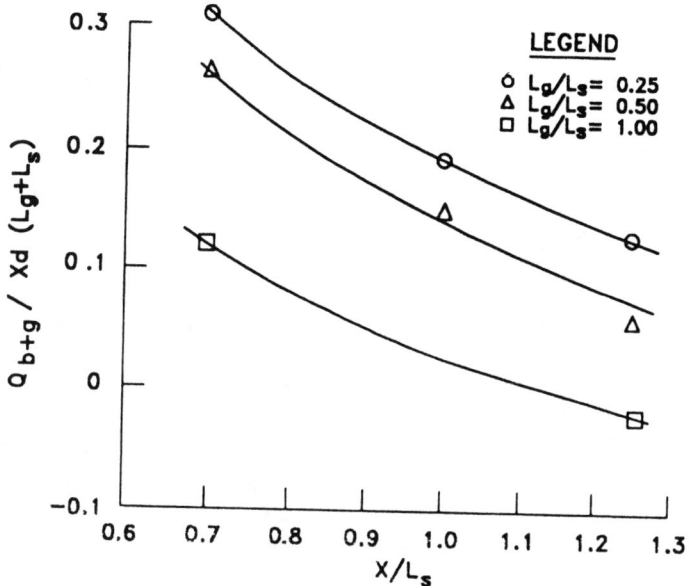

FIGURE 5.57. Effect of gap spacing on sand deposition for each breakwater gap pair (after Harris and Herbich[58]).

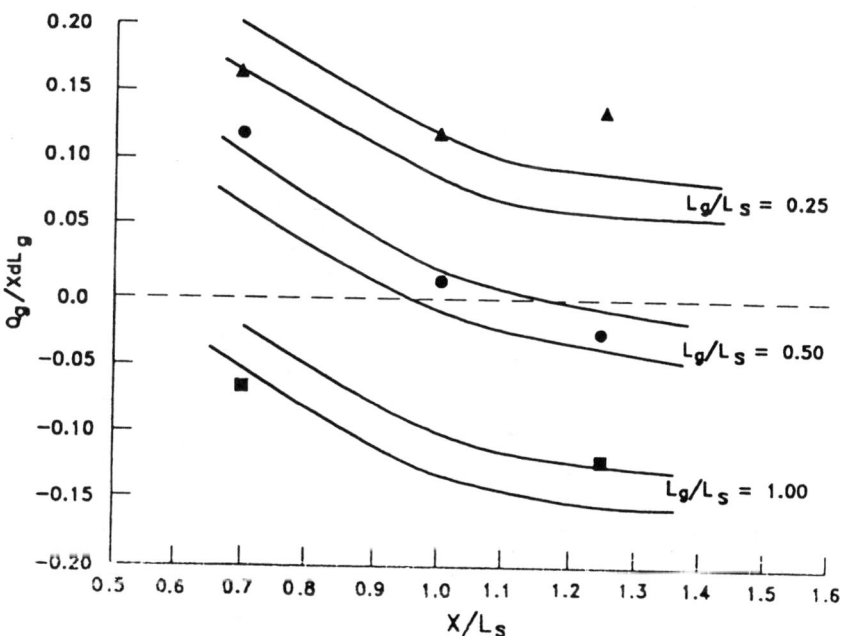

FIGURE 5.58. Effect of gap spacing on sand deposition for gap region (after Harris and Herbich[58]).

ture such as a reef breakwater or perched beach, use of the JMC method for design is most likely inappropriate. The wave height required for input is the average of the highest five nonstorm waves that occur in a typical year and corresponding wave period, often difficult parameters to extract from typically available hindcast wave data sets. The effects of a variable water level on design are not explicitly accounted for in the procedure.

22. Evaluation of Various Methods

Morphological Response. (a) A trend is apparent in the prototype data for deposition to increase as the structure length to distance offshore ratio increases. However, the ability of the relationships to accurately predict observed response is at best fair. If other types of design tools are available, it is recommended that these simple empirical relationships be used only as a general guide when designing a project.

(b) Suh and Dalrymple's[141] relationship for salient length was applied to all segmented projects. The relationship tends to overpredict the seaward excursion of the salient for the majority of prototype data evaluated (Figure 5.60), but appears to predict very well for the pocket-beach-type structures with periodic tombolo formation (Central Beach and Castlewood Park Beach). The correlation coefficient R is equal to 0.67.

(c) Because data defining the equilibrium volume and salient thickness to the lee of each breakwater segment were not readily available for U.S. projects, a direct comparison between Harris and Herbich's volumetric and Nir's thickness relationships, respectively, could not be performed However, the measured salient area was directly compared with a predicted area value using Harris and Herbich's volumetric relationship divided by Nir's salient thickness parameter. It should be recognized that this combined evaluation is not a true assessment of either method; however, it does give an indication of whether the combined relationships are accurate in predicting salient area. The measured salient area is

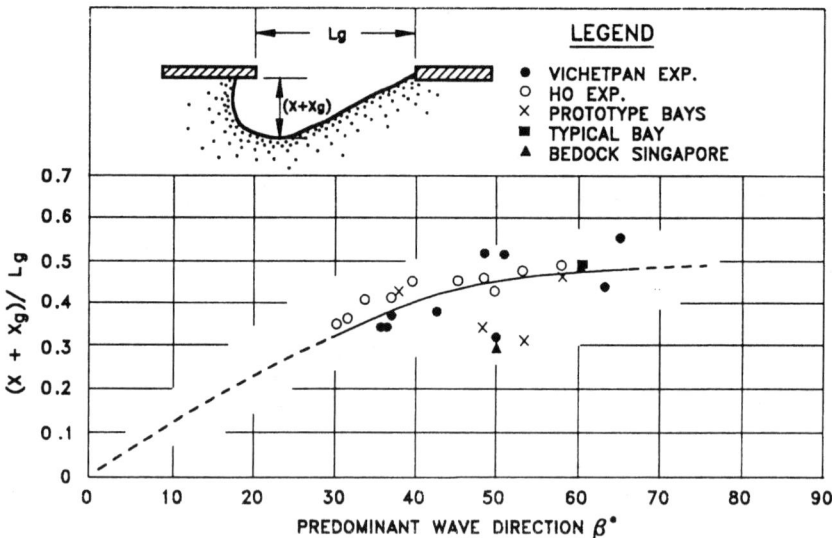

FIGURE 5.59. Relationship to predict gap erosion for headland structures (after Silvester and Ho[136]).

calculated by assuming a triangular shape of the salient/tombolo. Much scatter is apparent in the comparison (Figure 5.61), with a correlation coefficient equal to 0.18.

(d) Seiji, Uda, and Tanaka's ranges to predict gap erosion were evaluated (Figure 5.62). The lower boundary for "no erosion" ($L_g/X \leq 0.8$) was a reasonably good predictor of either accretion or very little erosion. Gap erosion occurred for ratios of L_g/X greater than 0.8.

(e). Berenguer and Enriquez'[13] and Ahrens' formulae to predict gap erosion for pocket beaches were evaluated using data from the two sites at Colonial Beach (headland structures). Both relationships are inversely proportional to the data (Berenguer and Enriquez: $R = -0.99$; Ahrens: $R = -0.24$) (Figure 5.63).

(f) Hallermeier's[47] relationship for recommended structure depth was evaluated using the recommended depth for salient formation for all sites except Colonial Beach, where the recommended depth for tombolo formation was used (Figure 5.64). An excellent correlation between depth at the structure and Hallermeier's recommended depth exists for all but the Lakeshore Park data, resulting in a correlation coefficient $R = 0.55$.

JMC Method.[145] (g) The JMC method of breakwater design has limitations for use with typical U.S. projects: beach fill is not included as a part of the design method; guidance is based on data from sites at which tombolo formation occurred for the majority of cases; wave conditions required for design are the wave height and period corresponding to the "average of the largest five nonstorm waves occurring in a typical year," and the effects of structural transmissibility and water-level variations are not parameters in the method. A comparison of the JMC and typical U.S. design using data from the Lakeview Park, Lorain, OH, detached breakwater project was presented by

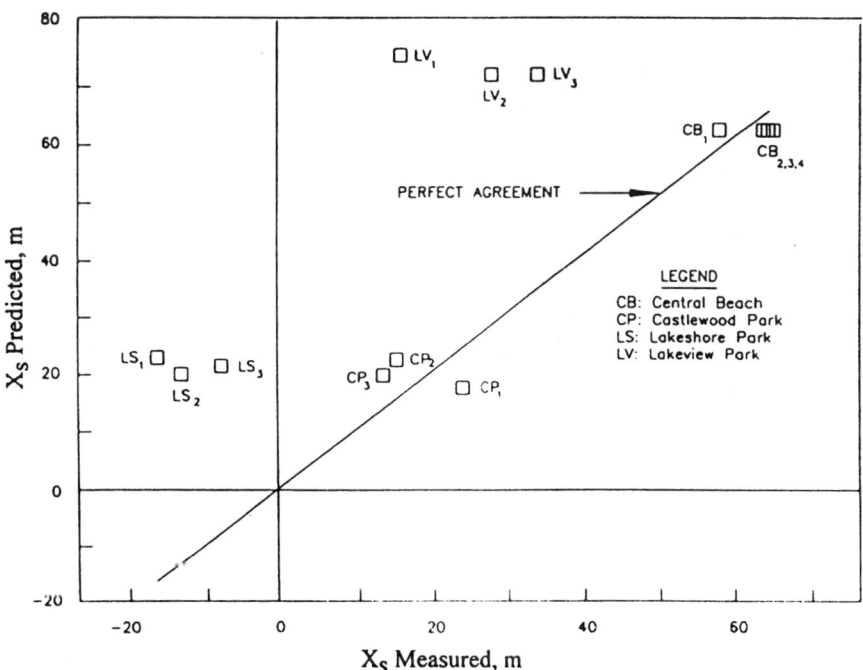

FIGURE 5.60. Evaluation of Suh and Dalrymple's[141] relationship for salient length.

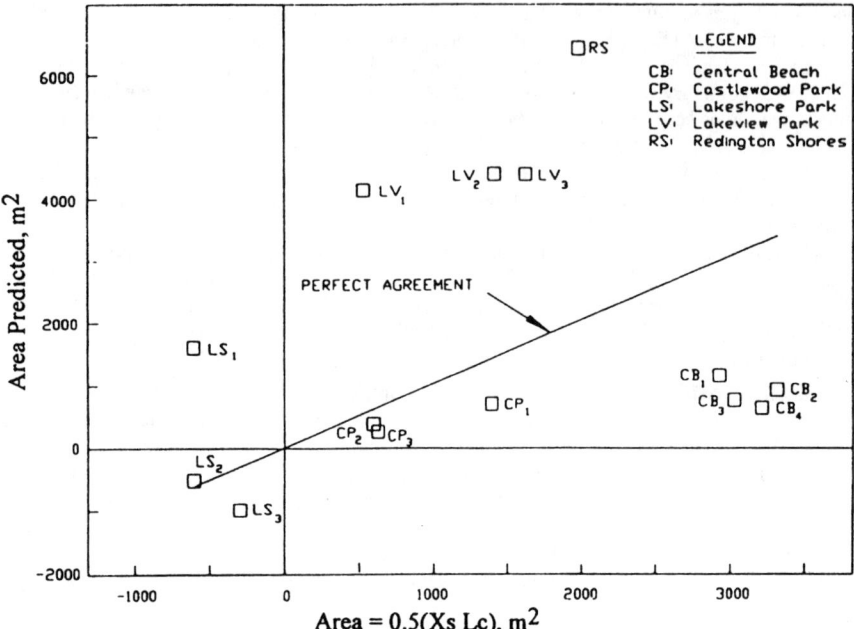

FIGURE 5.61. Combined evaluation of Harris and Herbich's[58] and Nir's[104] relationships.

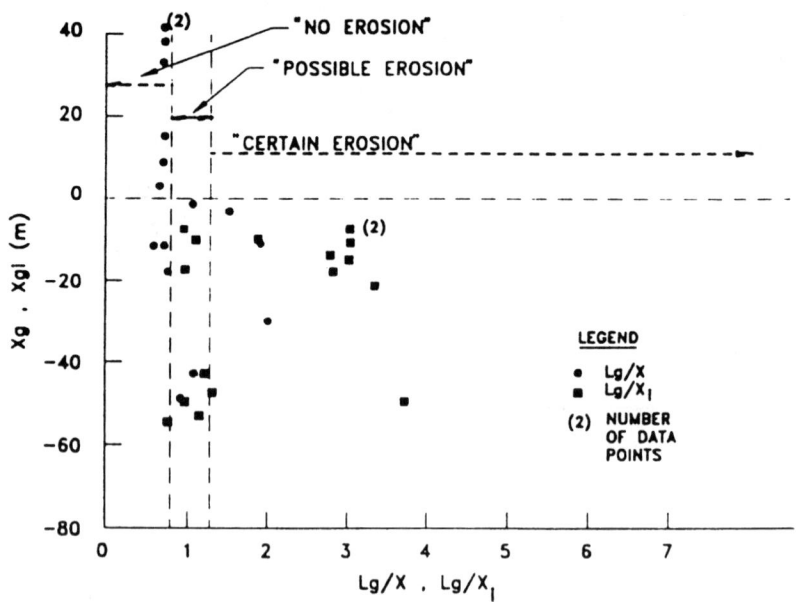

FIGURE 5.62. Evaluation of Seiji, Uda, and Tanaka's[128] limits for gap erosion.

Rosati and Truitt.[117] Figure 5.65 presents project parameters resulting from their JMC method, which can be compared with the as-constructed parameters resulting from the diffraction energy method (Figure 5.66). The four example problems conducted by Rosati and Truitt indicated that, for the site parameters evaluated, use of the JMC design tended to result in "more numerous, shorter length segments with a decreased gap width . . . structures are placed closer to shore than observed in U.S. projects."

The empirical relationships for offshore breakwater design are summarized in Table 5.4. EM 1110-2-1617[151] summarizes the conditions for the three types of beach response as predicted by the various relationships described in Table 5.5. Tables 5.6, through 5.8 present conditions for the type of beach response. Table 5.6 presents L_s/y values for the formation of tombolos, Table 5.7 presents the L_s/y values for the formation of salients, and Table 5.8 summarizes the L_s/y values for minimal shoreline response (L_s is the offshore breakwater length and y is the distance of the breakwater from the shoreline).

SUMMARY AND CONCLUSIONS

1. An offshore breakwater is a structure built parallel to the shoreline and detached from it. The breakwater is designed to provide protection from wave action, thus causing deposition of sediment on the leeward side of the breakwater.

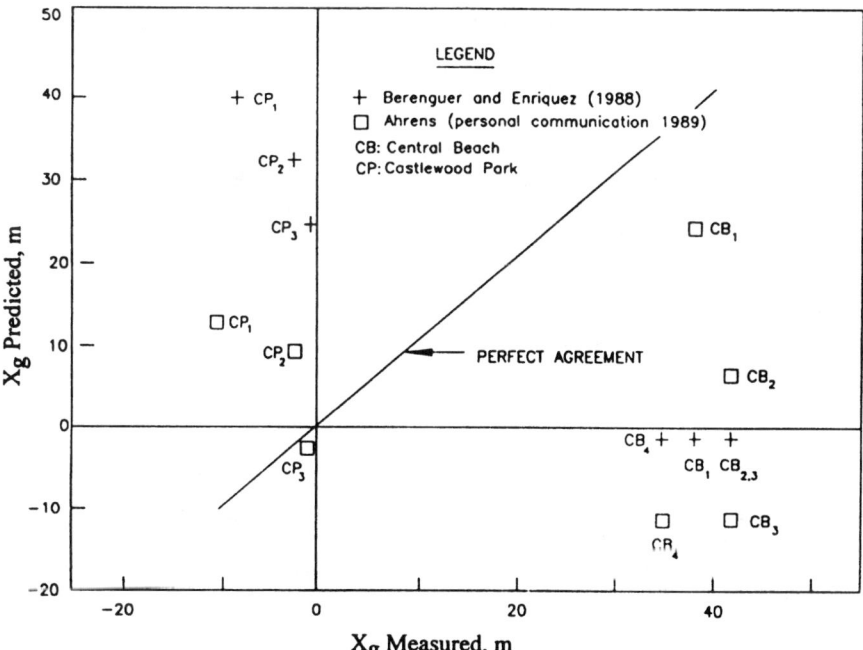

FIGURE 5.63. Evaluation of Berenguer and Enriquez's[13] and Ahrens'[4] relationships for pocket beach gap erosion.

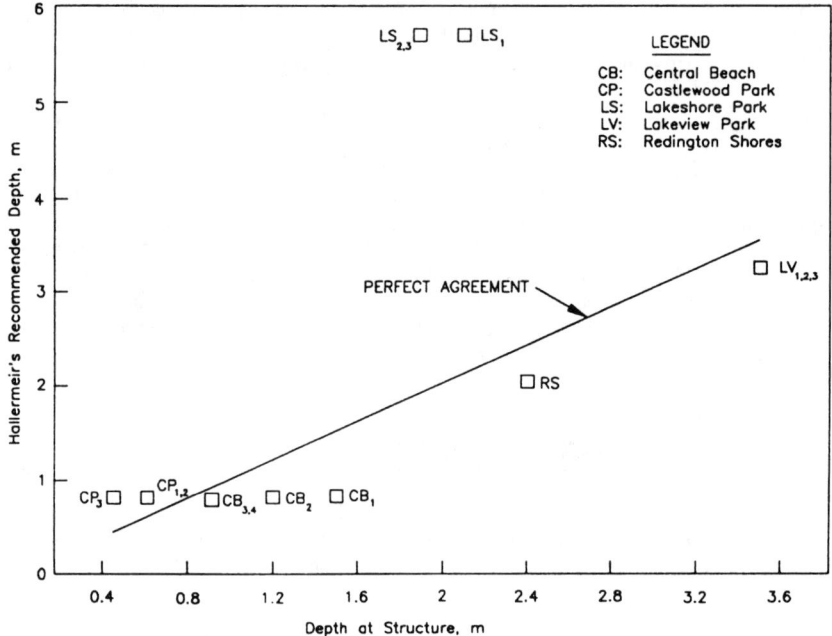

FIGURE 5.64. Evaluation of Hallermeier's[47] relationship for structure design depth.

2. Offshore breakwaters are built to protect the shoreline and replenish beach sand by interrupting longshore and wave-generated currents. When properly designed, an offshore breakwater will either form a tombolo or a sand salient. Formation of the salient begins soon after construction of the breakwater. Approximately 50% of the sand volume may be deposited in the first year and a steady "semiequilibrium" state usually is reached after 3 to 4 years. The process of beach accretion may be accelerated by depositing sand on the shoreline between the breakwater and the shoreline.

3. There have been thousands of offshore breakwaters constructed all over the world. Most offshore breakwater installations have been successful.

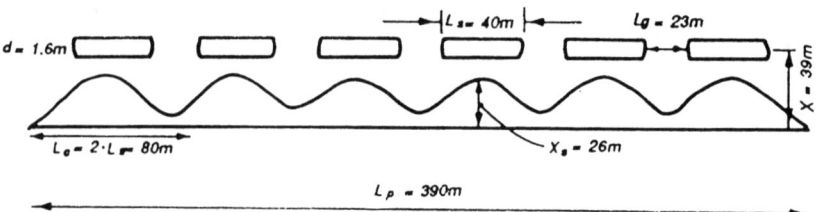

FIGURE 5.65. Lakeview Park project parameters as designed with JMC method (Rosati and Truitt[117]).

TABLE 5.5. Empirical relationships for offshore breakwater design

Inman and Fautschy (1966)	Predicts accretion condition; based on beach response at Venice in Santa Monica, CA
Toyoshima (1972, 1974)	Recommends design guidance based on prototype performance of 86 breakwater systems along the Japanese coast
Noble (1978)	Predicts coastal impact of structures in terms of offshore distance and length; based on California prototype breakwaters
Walker, Clark, and Pope (1980)	Discusses method used to design the Lakeview Park, Lorain, OH segmented system for salient formation; develops the diffraction energy method based on diffraction coefficient isolines for representative waves from predominant directions
Gourlay (1981)	Predicts beach response; based on physical model and field observations
Nir (1982)	Predicts accretion condition; based on performance of 12 Israeli breakwaters
Rosen and Vadja (1982)	Graphically presents relationships to predict equilibrium salient and tombolo size; based on physical model/prototype data
Hallermeier (1983)	Develops relationships for depth limit of sediment transport and prevention of tombolo formation; based on field/laboratory data
Noda (1984)	Evaluates physical parameters controlling development of tombolos/salients; especially due to on-offshore transport; based on laboratory experiments
Shore Protection Manual (1984)	Presents limits of tombolo formation from structure length and distance offshore; based on the pattern of diffracting wave crests in the lee of a breakwater
Dally and Pope (1986)	Recommends limits of structure–distance ratio based on type of shoreline advance desired and length of beach to be protected
Harris and Herbich (1986)	Presents relationship for average quantity of sand deposited in lee and gap areas; based on laboratory tests
Japanese Ministry of Construction (1986); also Rosati and Truitt (1990)	Develops step-by-step iterative procedure, providing specific guidelines for final design; tends to result in tombolo formation; based on Japanese breakwaters
Pope and Dean (1986)	Presents bounds of beach response based on prototype performance; response given as a function of segment length-to-gap ratio and effective distance offshore-to-depth at structure ratio; provides beach response index classification
Seiji, Uda, and Tanaka (1987)	Predicts gap erosion; based on performance of 1500 Japanese breakwaters
Sonu and Warwar (1987)	Presents relationship for tombolo growth at the Santa Monica, CA breakwater
Suh and Dalrymple (1987)	Gives relationship for salient length given structure length and surf zone location; based on lab tests and prototype data
Berenguer and Enriquez (1988)	Presents various relationships for pocket beaches, including gap erosion and maximum stable surface area (i.e., beach fill); based on projects along the Spanish coast
Ahrens and Cox (1990)	Uses Pope and Dean (1986) to develop a relationship for expected morphological response as function of segment-to-gap ratio

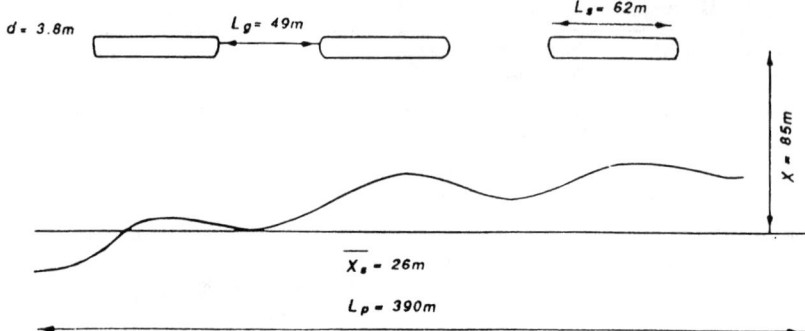

FIGURE 5.66. Lakeview Park project as constructed (diffraction energy method) (Rosati and Truitt[154]).

TABLE 5.6. Conditions for the formation of tombolos

Condition	Comments	Reference
$L_s/y > 2.0$		SPM (1984)
$L_s/y > 2.0$	Double tombolo	Gourlay (1981)
$L_s/y > 0.67$ to 1.0	Tombolo (shallow water)	Gourlay (1981)
$L_s/y > 2.5$	Periodic tombolo	Ahrens and Cox (1990)
$L_s/y > 1.5$ to 2.0	Tombolo	Dally and Pope (1986)
$L_s/y > 1.5$	Tombolo (multiple breakwaters)	Dally and Pope (1986)
$L_s/y > 1.0$	Tombolo (single breakwater)	Suh and Dalrymple (1987)
$L_s/y > 2\ b/L_s$	Tombolo (multiple breakwaters)	Suh and Dalrymple (1987)

TABLE 5.7. Conditions for the formation of salients

Condition	Comment.	Reference
$L_s/y < 1.0$	No tombolo	SPM (1984)
$L_s/y < 0.4$ to 0.5	Salient	Gourlay (1981)
$L_s/y = 0.5$ to 0.67	Salient	Dally and Pope (1986)
$L_s/y < 1.0$	No tombolo (single breakwater)	Suh and Dalrymple (1987)
$L_s/y < 2\ b/L_s$	No tombolo (multiple breakwaters)	Suh and Dalrymple (1987)
$L_s/y < 1.5$	Well-developed salient	Ahrens and Cox (1990)
$L_s/y < 0.8$ to 1.5	Subdued salient	Ahrens and Cox (1990)

OFFSHORE (DETACHED) BREAKWATERS 5.79

TABLE 5.8. Conditions for minimal shoreline response

Condition	Comments	Reference
$L_s/y \leq 0.17$ to 0.33	No response	Inman and Fautschy (1966)
$L_s/y \leq 0.27$	No sinuosity	Ahrens and Cox (1990)
$L_s/y \leq 0.5$	No deposition	Nir (1982)
$L_s/y \leq 0.125$	Uniform protection	Dally and Pope (1986)
$L_s/y \leq 0.17$	Minimal impact	Noble (1978)

APPENDIX A

I. Offshore Breakwater Design (Harris, Cords, Herbich Method)

Given:

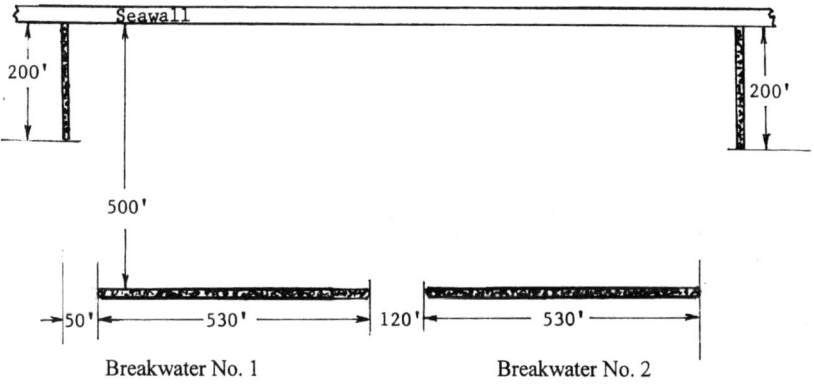

FIGURE 5.A-1. Breakwater geometry.

Design Parameters

$X = 500$ ft

$B = 530$ ft

$G = 120$ ft

$H_s = 9.0$ ft

$T = 7.7$ sec

1. Find where wave is breaking

$$\frac{H_0'}{5.12T^2} = \frac{9.0 \text{ ft}}{5.12(7.7)^2} = 0.0296$$

$$m = \frac{d}{x} \quad \text{(From } Shore\ Protection\ Manual\ 2\text{-}65\text{)}$$

Find depth in front of Breakwater No. 1

$$y_1 = 315 \text{ ft} \quad \frac{0.2 \text{ ft}}{500} = \frac{z}{185} \quad z = 0.1$$

$$d_1 = 7.0 \text{ ft}$$

Find depth in front of Breakwater No. 2

$$y = 965 \text{ ft} \quad \left(\frac{0.2}{500} + \frac{0.4}{500}\right)\frac{1}{2} = \frac{z}{35}$$

$$d_2 = 6.7 \text{ ft}$$

Find depth in gap

$$y = 640 \text{ ft} \quad \frac{0.2 \text{ ft}}{500} = \frac{z}{140} \quad z = 0.1$$

$$d_g = 6.8 \text{ ft}$$

Find m's

$$m_1 = 7.0 \text{ ft}/500 \text{ ft} = 0.0140$$

$$m_g = 6.8 \text{ ft}/500 \text{ ft} = 0.0136$$

$$m_2 = 6.7 \text{ ft}/500 \text{ ft} = 0.0134$$

$$\frac{H_b}{H_0'} = 1.08 \quad \text{(From } Shore\ Protection\ Manual\ 2\text{-}65\text{)}$$

$$H_b = 9.7 \text{ ft}$$

$$\frac{H_b}{gT^2} = \frac{9.7}{32.2(7.7)^2} = 0.005$$

$$\frac{d_b}{H_b} = 1.20 \quad \text{for Breakwaters Nos. 1 and 2 and gap} \quad \text{(From } Shore\ Protection\ Manual\ 2\text{-}66\text{)}$$

$$d_b = (1.20)(9.7 \text{ ft}) = 11.6 \text{ ft}$$

$$X_B = d_b/m = 11.6/0.0140 = 830 \text{ ft} \quad \text{for Breakwater No. 1}$$

$$X_B = d_b/m = 11.6/0.0136 = 860 \text{ ft} \quad \text{for gap}$$

$$X_B = d_b/m = 11.6/0.0134 = 870 \text{ ft} \quad \text{for Breakwater No. 2}$$

Breakwaters in breaking zone.

OFFSHORE (DETACHED) BREAKWATERS   5.81

2. Calculate volume of sand trapped Breakwaters Nos. 1 and 2

$$X = 500 \qquad X/B = 0.943$$
$$B = 530 \qquad G/B = 0.227$$
$$G = 120$$
$$Q/XBD = e^{[0.31481 - 1.92187(X/B)]}$$
$$Q/XBD = e^{(0.31481 - 1.81232)}$$
$$Q = 0.2237 \, (XBD) = 0.2237(500)(530)D$$
$$Q = 59276.7D$$

Breakwater No. 1 $Q = 59276.7(7.0)$
$$= 414{,}937 \text{ ft}^3$$
$$= 15{,}368 \text{ yd}^3$$
$$\frac{0.2237}{0.500} = \% \text{ gain} = 45\%$$

Breakwater No. 2 $Q = 59{,}276.7(6.7)$
$$= 397{,}153.9 \text{ ft}^3$$
$$= 14{,}710 \text{ yd}^3$$
$$\% \text{ gain} = 45\%$$

3. Calculate volume of sand gained/lost through the gap

$$G/B = 0.23 \qquad X/B = 0.943$$
$$Q_{b+g}/XD(G + B) = 0.225 \qquad \text{(See Figure 5.9)}$$
$$Q_{b+g} = 0.225(500)6.8(530 + 120)$$
$$Q_{b+g} = 497{,}250 \text{ ft}^3$$
$$= 18{,}416 \text{ yd}^3$$
$$Q_G = Q_{b+g} - Q_b$$
$$Q_G = 18{,}416 - 15{,}368 \text{ yd}^3 \qquad \text{(For Breakwater No. 1)}$$
$$= 3{,}048 \text{ yd}^3$$
$$\text{Average } Q_G = 3{,}380 \text{ yd}^3$$
$$Q_G = 18{,}416 - 14{,}710 \text{ yd}^3 \qquad \text{(For Breakwater No. 2)}$$
$$= 3{,}706 \text{ yd}^3$$
$$Q_G/XDG = 0.125$$
$$Q_G = (500)6.8(120)0.125$$
$$= 51{,}000 \text{ ft}^3 = 1{,}890 \text{ yd}^3$$

Volume in gap = 7,555 yd^3

% gain = 45%

4. Calculate tombolo shape

use Rosen diagrams[120]

$Y_B/X_B = B/X = 500/500 = 1.06$

$(Y_B - Y_T)/2X_B = 0.42$ (tombolo length)

$Y_B - Y_T = 0.42(2)500$

$Y_B - Y_T = 420$

$Y_T = 530 - 420 = 110$ ft

$\dfrac{A_T}{X_B Y_B} = 0.71$

$A_T = 0.71(530)(500)$

$= 188,150$ ft^2

$= 20,900$ yd^2

70% coverage

Determine increase in shoreline length

$265 - 55 = 210$

$210^2 + 500^2 = C^2$

$C = 540$ ft

→ Shoreline increases from 530 ft to approximately 1080 ft

II. Conclusion

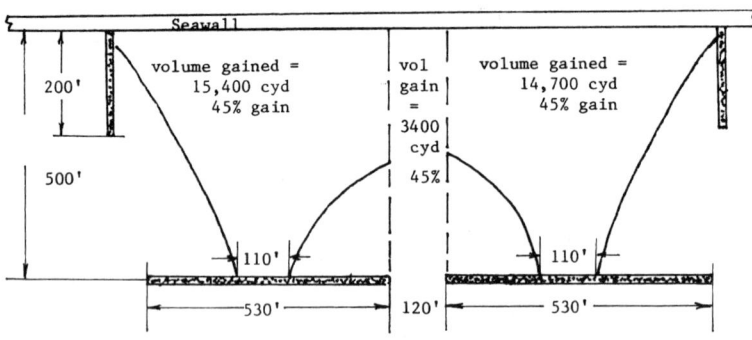

FIGURE 5.A-2. Tombolo formation.

OFFSHORE (DETACHED) BREAKWATERS 5.83

Note: Two breakwaters increase shoreline from 530 ft behind each breakwater to approximately 1,080 ft.

APPENDIX B

B-1. JMC Design Method

The Japanese Ministry of Construction (JMC) published a handbook for the design of offshore breakwaters based on a survey of 1522 projects. The JMC presents a series of steps through which offshore breakwaters can be designed (Figure 5B-1). Figure 5B-2 depicts the variables used in the JMC method. Figure 5B-3 shows the deepwater wave steepness as a function of nearshore steepness for various beach slopes, Figure 5B-4 provides the salient area ratio as a function of site parameters for beach type B, and Figure 5B-5 provides the salient area ratio for beach type C. Figure 5B-6 provides the relationship between dimensionless parameters and shoreline change at the gap between breakwaters.

Table 5B-1 defines the beach type for use in the JMC method.

B-2. Offshore Breakwater Design (JMC Method)

Example Problem. Given: Average of five highest deepwater wave heights occurring in a year H_{05} = 2.5 m, corresponding wave period T_5 = 12.0 sec, desired salient length X_s = 15 m, length of project shoreline L_p = 380 m, beach slope I = 1/30. The beach has a well-developed offshore bar, with sand-sized material.

a) Because the beach is mildly sloped with well-developed bar and sand-sized beach material, classify it as a Beach Type B.
b) Wave parameters and length of shoreline are given.
c) Desired salient length is given.
d) Calculate the deepwater wavelength L_{05} and deepwater steepness H_{05}/L_{05}:

$$L_{05} = \frac{gT_5^2}{2\pi} = \frac{(9.81)(12)^2}{2(3.14)} = 224.8 \text{ m}$$

$$\frac{H_{05}}{L_{05}} = \frac{2.5}{224.8} = 0.011$$

Using Figure 5B-3 with I = 1/30 and H_{05}/L_{05} = 0.011, estimate

$$\frac{d_{b5}}{H_{05}} = 1.8$$

Therefore,

$$d_{b5} = 1.8(2.5) = 4.5 \text{ m}$$

e) Make initial guess of design water depth at structure, d':

$$d' = \frac{d_{b5} + X_s I}{2} = \frac{4.5 + 15\left(\frac{1}{30}\right)}{2} = 2.5 \text{ m}$$

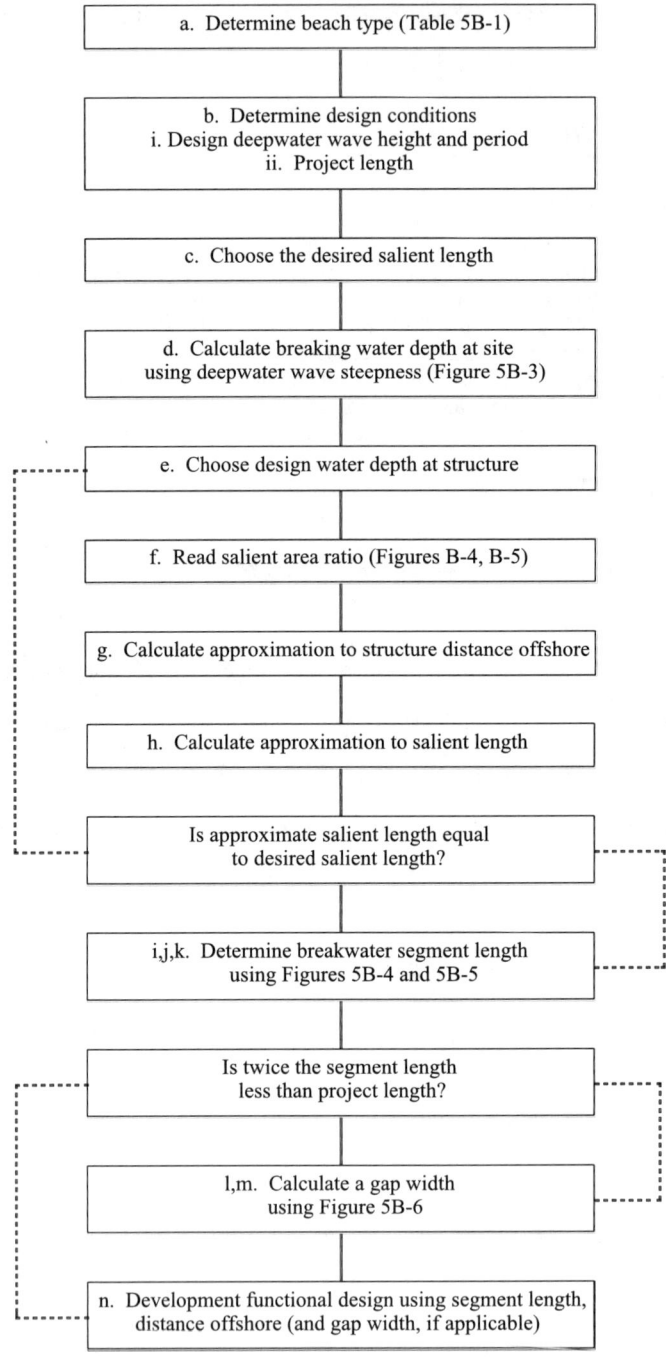

FIGURE 5B-1. JMC design method.

OFFSHORE (DETACHED) BREAKWATERS

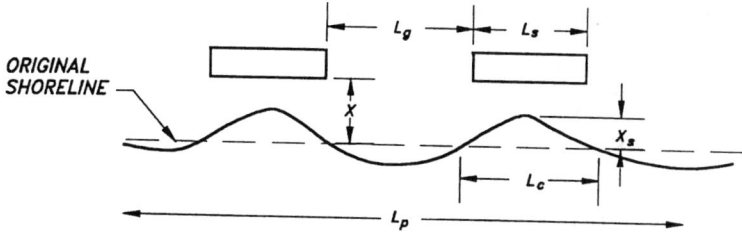

a. Plan view

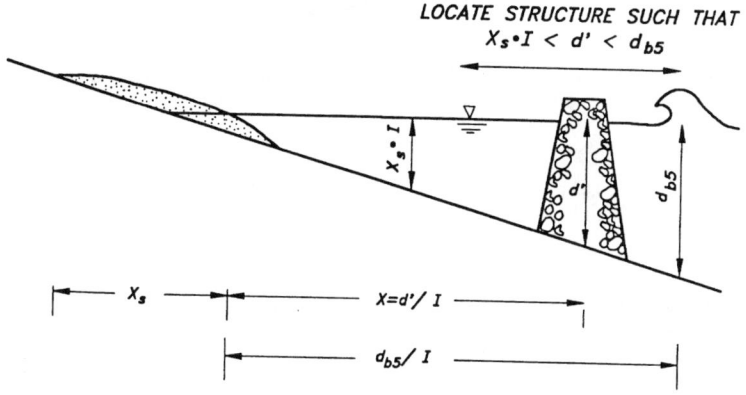

b. Cross-sectional view

FIGURE 5B-2. Variables used in JMC design method.

f) Use Figure 5B-4 to estimate SAR:

$$\frac{d'}{d_{b5}} = \frac{2.5}{4.5} = 0.56$$

From Figure 5B-4:

$$SAR = 0.6$$

g) Calculate the first approximation to structure distance offshore X':

$$X' = \frac{d'}{I} = \frac{2.5}{\left(\frac{1}{30}\right)} = 75 \text{ m}$$

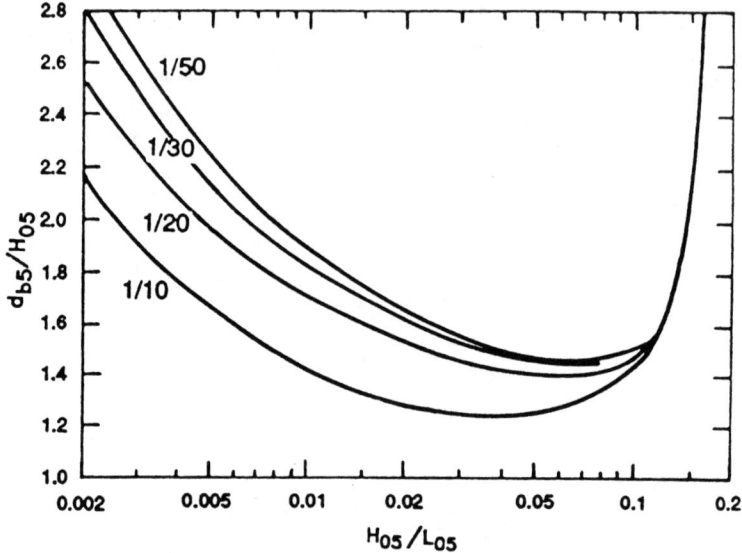

FIGURE 5B-3. Deepwater wave steepness versus nearshore steepness for various beach slopes (Goda[39]).

h) Calculate the first approximation to salient length X_s:

$$X_s' = \text{SAR } X' = 0.6(75) = 45 \text{ m}$$

Since the first approximation to salient length ($X_s' = 45$ m) was not equal to the desired salient length ($X_s = 15$ m), repeat steps (e) through (h) with a second estimate of structure depth, d'. Let water depth at structure $d' = 1.5$ m; then, using $d'/d_{b5} = 0.33$, estimate SAR = 0.35. The structure distance offshore is then $X' = 1.5/(1/30) = 45$ m, and the estimated salient length $X_s' = 0.35(45) = 15.8$ m, approximately equal to desired salient length (15 m). Therefore, $X_s = X_s' = 15.8$ m, $X = X' = 45$ m, and $d = d' = 1.5$ m.

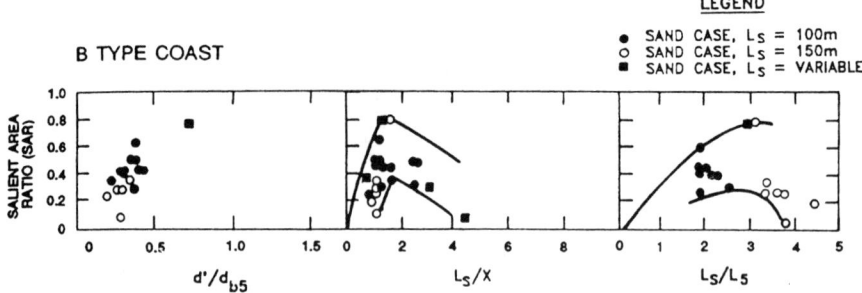

FIGURE 5B-4. Salient area ratio versus site parameters for Beach Type B (modified from JMC).

OFFSHORE (DETACHED) BREAKWATERS 5.87

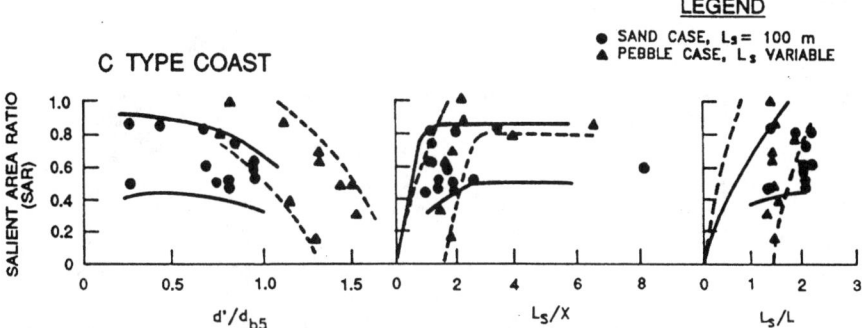

FIGURE 5B-5. Salient area ratio versus site parameters for Beach Type C (modified from JMC).

i) Calculate the local wavelength at the structure:

$$L_5 = T_5\sqrt{gd} = 12.0\sqrt{9.81(1.5)} = 46.0 \text{ m}$$

Calculate the ranges of structure length:

$$1.8\, L_5 < L_s < 3.0\, L_5$$

$$1.8\,(46.0) < L_s < 2.5\,(45.0)$$

$$36.0 \text{ m} < L_s < 112.5 \text{ m}$$

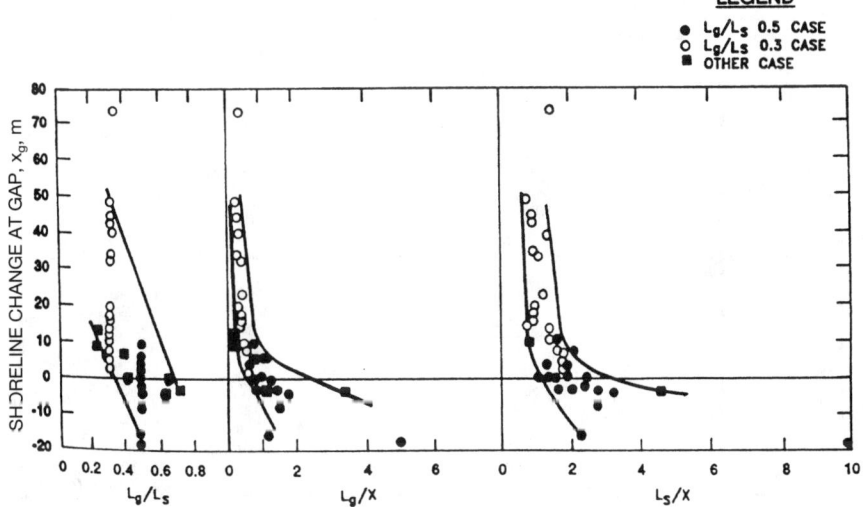

FIGURE 5B-6. Relationship between nondimensional parameters and shoreline change at gap (JMC).

TABLE 5B-1. Definition of beach type for use in JMC design method (modified from JMC)

TYPE OF BEACH	PROFILE	CHARACTERISTICS
A	AOMORI COAST (profile diagram, distance offshore 0–400 m, elevation 5 to −10 m)	• WATER DEPTH IS SHALLOW AT THE HORIZONTAL PORTION OF THE BOTTOM. • WAVE HEIGHT IS SMALL, AND DEPTH OF THRESHOLD FOR SEDIMENT MOVEMENT SMALL. $H_{05} < 0.5$ m $I = 1/30$ FINE SAND
B	ISHIKAWA COAST (profile diagrams)	• BAR IS WELL DEVELOPED. • BEACH SLOPE IS GENTLE AT DEPTH FOR THRESHOLD OF SEDIMENT MOTION. • COASTLINE IS PERPENDICULAR TO AVERAGE WAVE DIRECTION. $H_{05} \geq 0.5$ m $I = 1/30$ SAND
C	SHIMOSHINKAWA COAST / SURUGA COAST (profile diagrams)	• BOTTOM SLOPE IS RELATIVELY STEEP AND THERE IS NO BAR. $H_{05} \geq 0.5$ m $I = 1/15$ SAND AND PEBBLES
D	FUJI COAST (profile diagram, elevation 10 to −30 m)	• BOTTOM SLOPE IS STEEP $H_{05} \geq 0.5$ m $I = 1/3$ TO $1/10$ PEBBLES
E	KOOCHI COAST (profile diagram)	• SIMILAR TO TYPE-C, BUT IN SOME CASES THERE IS A BAR OFFSHORE. $H_{05} \geq 0.5$ m $I = 1/15$ PEBBLES

LEGEND

——— AVERAGE BEACH PROFILE
- - - - - AVERAGE DEVIATION

j) Ranges for structure length are:

$$0.8\,X < L_s < 3.0\,X$$

$$0.8\,(45.0) < L_s < 2.5\,(45.0)$$

$$36.0 \text{ m} < L_s < 112.5 \text{ m}$$

k) Obtain ranges for structure length, L_s:

$$82.8 \text{ m} < L_s < 112.5 \text{ m}$$

Structure length is calculated as the average of the extremes:

$$L_s = \frac{82.8 + 112.5}{2} = 98 \text{ m}$$

l) Calculate the gap width:

$$0.7\,X < L_g < 1.8\,X$$

$$0.7\,(45) < L_g < 1.8\,(45)$$

$$31.5 \text{ m} < L_g < 81.0 \text{ m}$$

$$0.5\,L_5 < L_g < 1.8\,L_5$$

$$0.5\,(46.0) < L_g < 1.0\,(46.0)$$

$$23.0 \text{ m} < L_g < 46.0 \text{ m}$$

m) Obtain ranges for gap width

$$31.5 \text{ m} < L_g < 46.0 \text{ m}$$

The gap width is calculated as the average of the two values:

$$L_g = \frac{31.5 + 46.0}{2} = 39 \text{ m}$$

LEGEND
H_{05} = 2.5m
T_5 = 12.0 sec
X_s = 224.8m
L_{05} = 15m

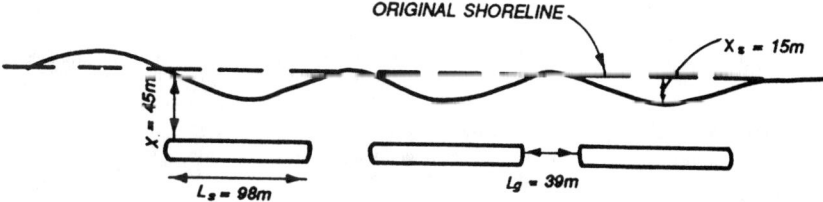

FIGURE 5B-7. Design example 1 using the JMC method.

n) To protect the length of the project shoreline L_p = 380 m, three breakwater segments with length L_s = 98 m are required, with a corresponding gap width L_g = 39 m (Figure 5B-7).

BIBLIOGRAPHY

1. Abecasis, B. 1964. The Structural Behavior and the Sheltering Efficiency of the Submerged Breakwater at the Entrance to the Port of Leixoes, Maintenance Charges and Effects, *Proc. 9th Coastal Engineering Conference*, ASCE, pp. 596–599.
2. Adachi, S., Sawaragi, T., and Ogo, A. 1959. The Effects of Coastal Structures on the Littoral Sand Drifts, *Coastal Engineering in Japan*, Vol. 2, Japanese Society of Civil Engineers, pp. 85–98.
3. Adams, C. and Sonu, C. 1986. Wave Transmission Across Near-Surface Breakwaters, *Proc. 20th Coastal Engineering Conference*, ASCE, pp. 1729–1738.
4. Ahrens, J. P. 1987. Characteristics of Reef Breakwaters, Technical Report CERC-87-17, U.S. Army Engineer Waterways Experiment Station, Coastal Engineering Research Center, Vicksburg, MS.
5. Ahrens, J. P. and Cox, J. 1990. Design and Performance of Reef Breakwaters, *Journal of Coastal Research*, Special issue no. 7, Spring, 61–75.
6. Allsop, N. W. H. and Kalmus, D. C. 1985. Plymouth Marine Events Base: Performance of Wave Screens, Report No. EX 1327, Hydraulics Research, Wallingford, UK.
7. Anglin, C. D., MacIntosh, K. J., Baird, W. F., and Werren, D. J. 1987. Artificial Beach Design, Lake Forest, Illinois, *Proc. Coastal Zone '87*, ASCE, pp. 1121–1129.
8. Bade, P. and Kaldenhoff, H. 1980. Energy Transmission Over Breakwater—A Design Criterion?, *Proc. 17th Coastal Engineering Conference*, ASCE, pp. 1885–1897.
9. Beach Erosion Control, Colonial Beach, Virginia, 1980. Detailed Project Report, U.S. Army Corps of Engineers, Baltimore District, May.
10. Begley, S., Carey, J., and Bailey, E. 1984. The Vanishing Coasts, *Newsweek*, September 24.
11. Bender, T. 1992. An Overview of Segmented Offshore/Headland Breakwater Projects Constructed by the Buffalo District, *Coastal Engineering Practice '92*, ASCE, Long Beach, CA, March.
12. Bender, T. A. 1985. Lakeshore Park, Ashtabula, Ohio, Segmented Offshore Breakwater Project, *Proc. 42nd Meeting of the Coastal Engineering Research Board*, R. Whalin, Editor, U.S. Army Engineer Waterways Experiment Station, pp. 135–141.
13. Berenguer, J. and Enriquez, J. 1988. Design of Pocket Beaches: The Spanish Case, *Proc. 21st International Coastal Engineering Conference*, ASCE, pp. 1411–1425.
14. Bottin, R. R., Jr. 1982. Lakeview Park Beach Erosion Study, Ohio, Letter Report, U.S. Army Engineer Waterways Experiment Station, Coastal Engineering Research Center, Vicksburg, MS.
15. Brasfield, C. W. and Chatham, C. E., Jr. 1967. Magic Island Complex Including Kewala Basin and Ala Wai Harbor, Honolulu, Oahu, Hawaii, TR 20-767, U.S. Army Engineer Waterways Experiment Station, Vicksburg, MS.
16. Bruno, R. O., Watts, G. M., and Gable, C. G. 1977. Sediments Impounded by an Offshore Breakwater, *Proc. Coastal Sediments '77*, ASCE, pp. 1006–1025.
17. Chandler, P. and Sorensen, R. 1972. Transformation of Waves Passing a Submerged Bar, *Proc. 13th Coastal Engineering Conference*, ASCE, pp. 385–403.
18. Chasten, M. A., Rosati, J. D., McCormick, J. W., and Randall, R. E. 1993. Engineering Design Guidance for Detached Breakwaters as Shoreline Stabilization Structures, T. R. CERC-93-19.
19. Cooperative Beach Erosion Control Project for Lakeview Park, Lorain, Ohio. 1975. General Design Memo, Phase II, Project Design, U.S. Army Corps of Engineers District, Buffalo, NY.
20. Cords, D. A. 1986. Model Study of Shoreline Changes Due to a Series of Offshore Breakwaters, M. S. Thesis (unpublished), Ocean Engineering, Texas A&M University, College Station, TX, May.
21. Curren, C. R. and Chatham, C. E., Jr. 1977. Imperial Beach, California, Design of Structures

for Beach Erosion Control; Hydraulic Model Investigation, Technical Report CERC-77-15, U.S. Army Engineer Waterways Experiment Station, Vicksburg, MS.
22. Dally, W. R. and Pope, J. 1986. Detached Breakwaters for Shore Protection, Technical Report CERC-86-1, U.S. Army Engineer Waterways Experiment Station, Vicksburg, MS, Jan., 88 pp.
23. Dalrymple, R. A., Driscoll, A. M., and Ramsey, J. S. 1991. Laboratory Testing of the Beachsaver Breakwater System, Center for Applied Coastal Research, University of Delaware, Newark, February, 38 pp.
24. Danish Hydraulic Institute. 1988. Offshore Breakwaters, Negombo: Report on Physical Model Tests, Report prepared for Coast Conservation Department of Sri Lanka, Chapters 1–4.
25. Dattatri, J. 1978. Analysis of Regular and Irregular Waves and Performance Characteristics of Submerged Breakwaters, Ph. D. Dissertation presented to the Indian Institute of Technology at Madras, India.
26. Dean, J. and Pope, J. 1987. The Redington Shores Breakwater Project: Initial Response, *Proc. Coastal Sediments '87*, ASCE, pp. 1369–1384.
27. Dean, J. L., Pope, J. and Fulford, E. 1986. The Use of Segmented Detached Breakwaters to Protect Cohesive Shores: Colonial Beach, VA, *Proc. Symposium on Cohesive Shores*, May 5–7, Burlington, Ontario, National Research Council, Ottawa, Canada, pp. 292–308.
28. Dean, R. G. 1973. Heuristic Models of Sand Transport in the Surfzone, *Conference on Engineering Dynamics in the Surf Zone*, Sydney, Australia, pp. 208–214.
29. Dean, R. G. 1974. Evaluation and Development of Water Wave Theories for Engineering Application, SR-1, Vol. II, Coastal Engineering Research Center, U.S. Army Engineer Waterways Experiment Station, Vicksburg, MS.
30. Dean, R. G. 1978. Diffraction Calculation of Shoreline Planforms, *Proc. 16th International Conference on Coastal Engineering*, ASCE, Hamburg, Germany.
31. Dement, L. 1977. Two New Methods of Erosion Protection for Louisiana, *Shore and Beach*, 45:31–38.
32. Dick, T. and Brebner, A. 1968. Solid and Permeable Submerged Breakwaters, *Proc. 11th Coastal Engineering Conference*, ASCE, pp. 1141–1158.
33. Diskin, M. H., Vajda, M. L., and Amir, I. 1970. Piling-Up Behind Low and Submerged Permeable Breakwaters, *Journal of the Waterways and Harbors Division*, ASCE, Vol. 96, No. WW2, May, pp. 359–372.
34. Dyer, K. 1986. *Coastal and Estuarine Sediment Dynamics*, Wiley-Interscience, New York, NY, p. 191.
35. Eighty-First Congress, First Session, Document No. 333. 1949. Colonial Beach, VA, Beach Erosion Control Study, Letter from the Secretary of the Army, Committee on Public Works, 34 pp.
36. Environmental Geologic Atlas of the Texas Coastal Zone—Galveston-Houston Area, Bureau of Economic Geology. 1972. The University of Texas at Austin, Austin, TX.
37. Environmental Geologic Atlas of the Texas Coastal Zone—Port Lavaca, Area, Bureau of Economic Geology. 1976. The University of Texas at Austin, Austin, TX.
38. Fried, I. 1976. Protection by Means of Offshore Breakwater, *Proc. 15th Coastal Engineering Conference*, ASCE, pp. 1407–1424.
39. Fulford, E. T. 1992. Bay Ridge, Anne Arundel County, Maryland Offshore Breakwater and Beach Fill Design, *Coastal Engineering Practice '92*, ASCE, Long Beach, CA, March.
40. Goda, Y. 1985. *Random Seas and Design of Maritime Structures*, University of Tokyo Press, Japan.
41. Goda, Y. and Abe, Y. 1968. Apparent Coefficient of Partial Reflection of Finite Amplitude Waves, Report 7, No. 3, Port and Harbor Research Institute, Yokosuka, Japan.
42. Gorecki, R. 1985. Evaluation of Presque Isle Offshore Breakwaters for Beach Stabilization, *Proc. 42nd Meeting of the Coastal Engineering Research Board*, R. Whalin, Editor, U.S. Army Engineer Waterways Experiment Station, Vicksburg, MS, pp. 83–122.
43. Gourlay, M. R. 1980. Beach Processes in the Vicinity of Offshore Breakwaters, *Proc. 17th Coastal Engineering Conference*, ASCE, pp. 1320–1339.
44. Grace, P. J. 1989. Investigation of Breakwater Stability at Presque Isle Peninsula, Erie, PA, Technical Report CERC-89-3, U.S. Army Engineer Waterways Experiment Station, Vicksburg, MS, May.
45. Gravens, M. B. and Rosati, J. D. 1994. Numerical Model Study of Breakwaters at Grand Isle, Louisiana, U.S. Army Engineer Waterways Experiment Station, M. P. CERC-94-16, 75 pp.

46. Hallermeier, R. 1984. Added Evidence on New Scale Law for Coastal Models, *Proc. 19th Coastal Engineering Conference*, ASCE, pp. 1227–1243.
47. Hallermeier, R. J. 1983. Sand Transport Limits in Coastal Structure Designs, *Proc. Coastal Structures '83*, ASCE, March 9–11, Arlington, VA, pp. 703–716.
48. *Handbook of Detached Breakwater Design*. 1986. (Rigantci sekkei no tebiki.) River Bureau, Ministry of Construction, Government of Japan (in Japanese).
49. Hanson, H. 1989. GENESIS—A Generalized Shoreline Change Numerical Model, *Journal of Coastal Research*, Vol. 5(1), pp. 1–27.
50. Hanson, H. and Kraus, N. C. 1986. Forecast of Shoreline Change Behind Multiple Coastal Structures, *Coastal Engineering in Japan*, 29, pp. 195–213.
51. Hanson, H. and Kraus, N. C. 1989. GENESIS: Generalized Model for Simulating Shoreline Change. Report 1: Technical Reference, Technical Report CERC-89-19, U.S. Army Engineer Waterways Experiment Station, Coastal Engineering Research Center, Vicksburg, MS.
52. Hanson, H. and Kraus, N. C. 1990. Shoreline Response to a Single Transmissive Detached Breakwater, *Proc. 22nd Coastal Engineering Conference*, ASCE.
53. Hanson, H. and Kraus, N. C. 1991. Numerical Simulation of Shoreline Change at Lorain, Ohio, *Journal of Waterways, Port, Coastal and Ocean Engineering*.
54. Hanson, H., Kraus, N. C., and Nakashima, L. D. 1989. Shoreline Change Behind Transmissive Detached Breakwaters, *Proc. Coastal Zone '89*, ASCE, pp. 568–582.
55. Hardaway, C. S. 1985. Estuarine Shore Erosion Control: Gapped Breakwaters, *Proc. Coastal Zone '85*, ASCE.
56. Harris, M. M. 1984. Offshore Breakwater Placement Criteria, Master of Engineering Report (unpublished), Ocean Engineering Program, Texas A&M University, August.
57. Harris, M. M. and Herbich, J. B. 1986. Effects of Breakwater Spacing on Sand Entrapment, Symposium on Scale Effects in Modeling Sediment Transport Phenomena, *Symposium '86*, International Association for Hydraulic Research, Vol. 24, No. 5, Toronto, Canada, August 25–28, pp. 347–357.
58. Harris, M. M. and Herbich, J. B. 1986. Effects of Breakwater Spacing on Sand Entrapment, *Journal of Hydraulic Research*, Vol. 24, No. 5.
59. Hayashi, T., et al. 1966. Hydraulic Research on Closely Spaced Pile Breakwaters, *Proc. 10th Coastal Engineering Conference*, Vol. II, Chapter 50, pp. 873–884.
60. Herbich, J. B. 1983. Remarks: A Public Meeting Regarding the Shore Erosion Study for Galveston County and Surfside Beach in Brazoria County, October.
61. Herbich, J. B. 1989. Shoreline Changes Due to Offshore Breakwaters, *Proc. XXIII Congress*, Vol. C, International Association of Hydraulic Research, C-317-C-327.
62. Herbich, J. B. 1990. Pile and Offshore Breakwaters, *Handbook of Coastal and Ocean Engineering*, Vol. 1, J. B. Herbich, Editor, Gulf Publishing Company, Houston, TX, pp. 895–920.
63. Herbich, J. B. and Douglas, B. 1989. Wave Transmission Through a Double-Row Pile Breakwater, *Proc. 21st International Conference on Coastal Engineering*, ASCE, Torremolinos, Spain, Chapter 165, pp. 2229–2241.
64. Herbich, J. B., Elfiky, A-E., Elmongy, A-E., and Elsaeed, G. 1996. Shore Protection Studies Using Mathematical Models for the Ras-Elbar Area, Egypt, *Proc. 25th International Coastal Engineering Conference*, Vol. 4, Chapter 307, pp. 3976–3985, ASCE, Orlando, FL.
65. Herbich, J. B. and Trivedi, D. 1992. Effect of Detached Breakwaters on Shoreline and Bathymetry When Subjected to Oblique Wave Attack, COE Report No. 320, Ocean Engineering Program, Texas A&M University, College Station, TX, April.
66. Herron, W. J. and Harris, R. L. 1966. Littoral Bypassing and Beach Restoration in the Vicinity of Port Hueneme, California, *Proc. Coastal Engineering Conference*, ASCE, pp. 651–675.
67. Hom-ma, M. and Horikawa, K. 1961. A Study on Submerged Breakwaters, *Coastal Engineering in Japan*, Tokyo, Japan, Vol. IV, pp. 85–102.
68. Horikawa, K. and Sonu, C. 1987. Experimental Study of a Submerged Breakwater, Japanese Society of Civil Engineers, 12th Annual Convention, June.
69. Houston, J. R. 1995. The Economic Value of Beaches, The CERCular, CERC-95-4, Coastal Engineering Research Center, U.S. Army Engineer Waterways Experiment Station, Vicksburg, MS, December.
70. Hsu, J. R. C. and Sylvester, R. 1990. Accretion Behind Single Offshore Breakwater, *Journal of Waterway, Port, Coastal and Ocean Engineering*, Vol. 116, No. 3, pp. 362–380.

71. Imms, K. J. 1972. Submerged Platform Breakwater, Master of Science Thesis, University of Western Australia.
72. Inman, L. D. and Frautschy, J. D. 1965. Littoral Processes and the Development of Shorelines, *Proc. Coastal Engineering*, Santa Barbara, pp. 511–536.
73. Japanese Ministry of Construction. 1986. *Handbook of Offshore Breakwater Design,* River Bureau of the Ministry of Construction, Japanese Government, translated from Japanese for the Coastal Engineering Research Center, U.S. Army Engineer Waterways Experiment Station, Vicksburg, MS.
74. Johnson, J. W., Fuchs, R. A. and Morison, J. R. 1951. The Damping Action of Submerged Breakwaters, *Transactions of the American Geophysical Union,* Vol. 32, No. 5, pp. 704–717.
75. Kabelac, O. 1963. Model Texts of Coastal Protective Structures in USSR, *Journal of Waterways and Harbors Division*, ASCE, pp. 21–34.
76. Kamphuis, J. W. and Nairn, R. B. 1984. Scale Effects in Large Coastal Mobile Bed Models, *Proc. 19th International Conference on Coastal Engineering*, ASCE, pp. 2322–2338.
77. Kawakami, T., Irie, I., and Katayama, T. 1974. Performance of Offshore Breakwaters of the Niigata Coast, *Coastal Engineering in Japan*, Vol. 17, pp. 129–139.
78. Kilpatrick, W. S. 1984. Wave Transmission Through a Row of Rigid, Vertical Piles, (unpublished report) Ocean Engineering Program, Texas A&M University.
79. Kohno, H., Uda, T. and Yabusaki, Y. 1986. On the Scattering of Concrete Armor Units of Detached Breakwaters Due to Waves, *Proc. 20th International Conference on Coastal Engineering*, ASCE.
80. Komar, P. 1976. *Beach Processes and Sedimentation,* Prentice-Hall, Englewood Cliffs, NJ, pp. 429.
81. Krafft, K. M. and Herbich, J. B. 1988. Literature Review and Evaluation of Offshore Detached Breakwaters, for U.S. Army Engineer Waterways Experiment Station, CERC, Vicksburg, MS, Texas A&M University, Report No. COE-297, September, 98 pp.
82. Krafft, K. M. 1993. Wave Attenuation and Shoreline Alteration Due to Submerged Offshore Breakwaters, M. S. Thesis (unpublished), Ocean Engineering, Texas A&M University, College Station, TX.
83. Kraus, N. C. 1983. Applications of a Shoreline Prediction Model, *Proc. Specialty Conference on the Design, Maintenance and Performance of Coastal Structures,* ASCE, Arlington, VA, March 9–11.
84. Kraus, N. C. 1984. Estimate of Breaking Wave Height Behind Structures, *Journal Waterway, Port, Coastal and Ocean Engineering,* ASCE, 110(2), pp. 276–282.
85. Kraus, N. C. and Harikai, S. 1983. Numerical Model of Shoreline Change at Oarai Beach, *Coastal Engineering,* 7(1), pp. 1–28.
86. Kraus, N. C., Hanson, H., and Harikai, S. 1984. Shoreline Change at Oarai Beach—Past, Present and Future, *Proc. 19th Coastal Engineering Conference,* ASCE, pp. 2107–2123.
87. Kriebel, D., Dally, W., and Dean, R. 1986. Undistorted Froude Model for Surf Zone Sediment Transport, *Proc. 20th Coastal Engineering Conference,* ASCE, pp. 1296–1310.
88. LeMéhauté, B. 1969. An Introduction to Hydrodynamics and Water Waves, Water Wave Theories, Vol. II, Technical Report ERL-118-POL-3-2, U.S. Department of Commerce, ESSA, Washington, DC.
89. Lesnick, J. R. 1979. An Annotated Bibliography on Detached Breakwaters and Artificial Headlands, CERC Miscellaneous Report 79-1, U.S. Army Engineer Waterways Experiment Station, Vicksburg, MS.
90. Lyzlov, A. I. 1963. Model Test of Coastal Protective Structures in USSR, *Journal of the Waterways and Harbor Division,* Vol. 89, No. WWI, February, pp. 21–34.
91. Magoon, O. T. 1976. Offshore Breakwaters at Winthrop Beach, Massachusetts, *Shore and Beach,* Vol. 44, No. 3, October, pp. 34.
92. Mansard, E. P. D. and Funke, E. R. 1980. The Measurement of Incident and Reflected Spectra Using a Least Squares Method, *Proc. 17th International Conference on Coastal Engineering,* ASCE, pp. 154–172.
93. Mansard, E. P. D., Sand, S. E., and Funke, E. R. 1985. Reflection Analysis of Non-Linear Regular Waves, Technical Report TR-HY-001, NRC No. 25144, National Research Council Canada.
94. Massel, S. R. 1983. Harmonic Generated by Waves Propagating Over a Submerged Step, *Coastal Engineering,* Elsevier Science Publishers B. V., Amsterdam, Holland, pp. 357–380.

95. Mimura, N., Shimizu, T., and Horikawa, K. 1983. Laboratory Study on the Influence of Detached Breakwater on Coastal Change, *Proc. Coastal Structures '83*, ASCE, pp. 740–752.
96. Mohr, M. C. and Ippolito, M. 1991. Initial Shoreline Response at the Presque Isle Erosion Control Project, *Coastal Sediments*, Vol. II.
97. Murray, M. and Sayao, O. J. 1990. Offshore Breakwater for the Sergipe Marine Terminal, Brazil, *Proc. 22nd International Conference on Coastal Engineering*, ASCE, Delft, the Netherlands, pp. 3207–3222.
98. Nakamura, M., Shiraishi, H. and Sasaki, Y. 1966. Wave Damping Effect of Submerged Dike, *Proc. 10th Conference on Coastal Engineering*, ASCE, Vol. 1, pp. 254–267.
99. Nakashima, L. D., Pope, J., Mossa, J., and Dean J. L. 1987. Performance Evaluation of a Segmented Breakwater System, Holly Beach, Louisiana, *Proc. Coastal Sediments '87*, ASCE.
100. Nakashima, L. D., Pope, J., Mossa, J., and Dean, J. L. 1987. Initial Response of a Segmented Breakwater System, Holly Beach, Louisiana, *Proc. Coastal Sediments '87*, ASCE, pp. 1399–1414.
101. Nance, T. F. and Hirota, P. M. 1974. Magic Island . . . Ten Years After, *Shore and Beach*, Vol. 42, No. 2, October, pp. 19–22.
102. Natal'chishin, G. D. 1974. Determination of Wave Forces on Submerged Shore-Protection Breakwaters in Shallow Waters, *Gidrotekhnicheskoe Stroitel'stro*, No. 4, April, pp. 42–44.
103. National Shoreline Study, Regional Inventory Reports, South Atlantic—Gulf Region, Puerto Rico and the Virgin Islands. 1971. U.S. Army Engineer South Atlantic Division, Atlanta, Georgia, August.
104. Nir, Y. 1982. Offshore Artificial Structures and Their Influence on the Israel and Sinai Mediterranean Beaches, *Proc. 18th International Coastal Engineering Conference*, ASCE, pp. 1837–1856.
105. Noble, R. M. 1978. Coastal Structures' Effects on Shoreline, *Proc. 17th International Coastal Engineering Conference*, ASCE, pp. 2069–2085.
106. Noda, E. K. 1982. Equilibrium Beach Profile Scale-Model Relationship, *Journal of Waterways, Harbors and Coastal Engineering Division*, ASCE, November.
107. Noda, H. 1978. Scale Relations for Equilibrium Beach Profiles, *Proc. 16th Coastal Engineering Conference*, ASCE, pp. 1531–1541.
108. Numata, J. 1975. Experimental Study on Wave Dissipating Effect of the Breakwater Composed by Concrete Armor Units, *Proc. 22nd Japanese Conference on Coastal Engineering*, JSCE, pp. 501–505 (in Japanese).
109. Peraino, J., Chase, B. L., Plodowski, T., and Amy, L. 1975. Features of Various Offshore Structures, Miscellaneous Paper No. 3-75, U.S. Army Corps of Engineers, Coastal Engineering Research Center, Fort Belvoir, VA, April.
110. Perlin, M. and Dean, R. G. 1983. A Numerical Model to Simulate Sediment Transport in the Vicinity of Coastal Structures, Miscellaneous Report 83-10, U.S. Army Engineer Waterways Experiment Station, Vicksburg, MS.
111. Pope, J. and Rowen, D. D. 1983. Breakwaters for Beach Protection at Lorain, OH, *Proc. Coastal Structures '83*, ASCE, pp. 753–768.
112. Pope, J. 1985. Segmented Offshore Breakwaters: An Alternative for Beach Erosion Control, *Proc. of the Ninth Annual Conference of the Coastal Society*, October 14–17, 1984, The Coastal Society, Bethesda, MD (also in *Shore and Beach*, Vol. 54, No. 4, October 1986, pp. 3–6).
113. Pope, J. and Dean, J. L. 1986. Development of Design Criteria for Segmented Breakwaters, *Proc. 20th International Coastal Engineering Conference*, ASCE, November 9–14, Taipei, Taiwan, pp. 2144–2158.
114. Pope, J. and Gorecki, R. G. 1982. Geologic and Engineering History of Presque Isle Peninsula, PA, Field Trip Guidebook for New York State Geological Association 54th Annual Meeting, pp. 183–216.
115. Rosati, J. D. 1990. Functional Design of Breakwaters for Shore Protection: Empirical Methods, T. R. CERC-90-15, U.S. Army Engineer Waterways Experiment Station, Vicksburg, MS, September.
116. Rosati, J. D. and Pope, J. 1989. The Colonial Beach, VA, Detached Breakwater Project, M. P. CERC-89-2, U.S. Army Engineer Waterways Experiment Station, Vicksburg, MS, January.
117. Rosati, J. D. and Truitt, C. L. 1990. An Alternative Design Approach for Detached Breakwater

Projects, Tech. Report CERC-90-7, U.S. Army Engineer Waterways Experiment Station, Vicksburg, MS, September, 22 pp.
118. Rosati, J. D., Gravens, M. B., and Chasten, M. A. 1992. Development of Detached breakwater Design Criteria Using a Shoreline Response Model, *Coastal Engineering Practice '92*, ASCE, Long Beach, CA, March.
119. Rosati, J. D., Pope, J., Bender, T., and Truitt, C. L. 1989. The Lakeshore Park, Ashtabula, Ohio, Breakwater Project, Misc. Paper CERC-89-5, U.S. Army Engineer Waterways Experiment Station, Vicksburg, MS.
120. Rosen, D. S. and Vajda, M. 1982. Sedimentological Influences of Detached Breakwaters, *Proc. 18th Coastal Engineering Conference*, ASCE, pp. 1930–1949.
121. Sawaragi, T. and Deguchi, I. 1978. Distribution of Sand Transport Rate Across a Surf, *Proc. 16th International Conference on Coastal Engineering*, ASCE, Hamburg, Germany, pp. 1596–1613.
122. Sayao, O. J. and Chow, K. C. A. 1992. Application of a Beach Plan Evolution Model in Sergipe, Brazil, *Coastal Engineering Practice '92*, ASCE, Long Beach, CA, March 1992.
123. Scheffner, N. W. and Dean, J. L. A User's Guide to the N-Line Model: A Numerical Model to Simulate Sediment Transport in the Vicinity of Coastal Structures, U.S. Army Engineer Waterways Experiment Station, 55 pp.
124. Schneider, C. 1981. The Littoral Environment Observation Program (LEO) Data Collection Program, CETA-81-5, U.S. Army Coastal Engineering Research Center, Ft. Belvoir, VA, March.
125. Schoones, J. S. and Moller, J. P. 1982. Design and Calibration of False Bay Sediment Model, *Proc. 18th Coastal Engineering Conference*, ASCE, pp. 1161–1180.
126. Sea Coast Division, River Bureau and Coastal Engineering Division, Public Works Research Institute, Ministry of Construction. 1985. Study on Effect and Stability of the Detached Breakwaters, *Proc. 39th Annual Technical Meeting of Ministry of Construction*, pp. 1–10 (in Japanese).
127. Seabergh, W. C. 1983. Design for Beach Erosion at Presque Isle Beaches, Erie, Pennsylvania, CERC Technical Report HL-83-15, U.S. Army Engineer Waterways Experiment Station, Vicksburg, MS.
128. Seiji, M., Uda, T., and Tanaka, S. 1987. Statistical Study on the Effect and Stability of Detached Breakwaters, *Coastal Engineering in Japan*, Vol. 30, No. 1, pp. 131–141.
129. Shinohara, K. and Tsubaki, T. 1966. Model Study on the Change of Shoreline of Sandy Beach by the Offshore Breakwater, *Proc. 10th Coastal Engineering Conference*, ASCE, pp. 550–563.
130. Shiraishi, N., Numata, A., and Hase, N. 1960. The Effect and Damage of Submerged Breakwaters in Niigata Coast, *Coastal Engineering in Japan*, Vol. 3, pp. 89–99.
131. *Shore Protection Manual*, 3rd Edition. 1977. U.S. Army Waterways Experiment Station, CERC, Government Printing Office, Washington, DC.
132. *Shore Protection Manual*, 4th Edition. 1984. U.S. Army Waterways Experiment Station, CERC, U.S. Government Printing Office, Washington, DC, pp. 4–94.
133. Silvester, R. 1957. Offshore Breakwaters, *Journal of the Waterways and Harbor Division*, ASCE, Vol. 83, No. WW3, September, pp. 1368-1-1368-15.
134. Silvester, R. 1973. Submerged Platform Breakwaters, *Proc. First Australian Conference on Coastal Engineering*, pp. 182–189.
135. Silvester, R. 1976. Headland Defense of Coasts, *Proc. 15th Coastal Engineering Conference*, ASCE, pp. 1394–1406.
136. Silvester, R. and Ho, S. 1972. Use of Crenelate Shaped Bays to Stabilize Coasts, *Proc. 13th Coastal Engineering Conference*, ASCE, pp. 1347–1365.
137. Sivard, F. L. 1971. Building a Beach with an Offshore Fill, *Shore and Beach*, Vol. 39, April.
138. Sonu, C. J. and Warwar, J. F. 1987. Evolution of Sediment Budgets in the Lee of a Detached Breakwater, *Proc. Coastal Sediments '87*, ASCE, pp. 1361–1368.
139. Sorensen, R. M. 1991. Monitoring and Evaluation of the "Beachsaver" Breakwater System, Sea Isle City, New Jersey, Report IHL-130-91, Lehigh University, Bethlehem, PA, June, 43 pp.
140. Sorenson, R. M. 1992. Field Monitoring of a Modular Detached Breakwater System, *Coastal Engineering Practice '92*, ASCE, Long Beach, CA.
141. Suh, K. and Dalrymple, R. A. 1987. Offshore Breakwaters in Laboratory and Field, *Journal of Waterway, Port, Coastal and Ocean Engineering*, ASCE, Vol. 113, No. 2, pp. 105–121.

142. Tallent, J. 1986. The Submerged Offshore Breakwater and its Effect on Seafloor Topography, Masters Thesis (unpublished), Civil Engineering, Texas A&M University, College Station, TX.
143. Tanaka, N. 1976. Effects of Submerged Rubble-Mound Breakwater on Wave Attenuation and Shoreline Stabilization, *Proc. 23rd Japanese Conference on Coastal Engineering*, pp. 152–157.
144. Terry, J. B. and Howard, E. 1986. Redington Shores Beach Access Breakwater, *Shore and Beach*, Vol. 54, No. 4, pp. 7–9.
145. Toyoshima, O. 1972. *Coastal Engineering for Practicing Engineers—Beach Erosion*, Morikita Publishing Company, Tokyo, Japan, 320 pp. (English translation available through the Coastal Engineering Research Center for Chapter 8 on Offshore Breakwaters, pp. 227–317), U.S. Army Engineer Waterways Experiment Station, Vicksburg, MS.
146. Toyoshima, O. 1974. Design of a Detached Breakwater System, *Proc. 14th Coastal Engineering Conference*, ASCE, pp. 1419–1431.
147. Toyoshima, O. 1982. Variation of Foreshore Due to Detached Breakwaters, *Proc. 18th Coastal Engineering Conference*, ASCE, pp. 1872–1892.
148. U.S. Army Corps of Engineers. 1950. Accretion of Beach Sand Behind a Detached Breakwater, Technical Memorandum No. 16, Beach Erosion Board, May.
149. U.S. Army Corps of Engineers. 1976. National Shoreline Study: Shore Management Guidelines, Department of the Army, Corps of Engineers, Washington, DC, pp. 56.
150. U.S. Army Corps of Engineers. 1984. Use of Segmented Offshore Breakwaters for Beach Erosion Control, Coastal Engineering Technical Note CETN-III-22.
151. U.S. Army Corps of Engineers. 1992. Coastal Groins and Nearshore Breakwaters, Engineer Manual 1110-2-1617, Washington, DC.
152. U.S. Army Engineer District, Baltimore. 1980. Colonial Beach, VA, Detailed Project Report, Baltimore, MD.
153. U.S. Army Engineer District, Buffalo. 1975. Lakeview Park, Lorain, Ohio General Design Memorandum, Buffalo, NY.
154. U.S. Army Engineer District, Buffalo. 1975. Cooperative Beach Erosion Control Project at Lakeview Park, Lorain, Ohio, General Design Memorandum, Phase II—Project Design, Buffalo, NY, June.
155. U.S. Army Engineer District, Buffalo. 1975. Lakeview Park, Lorain, Ohio, Phase II General Design Memorandum, Buffalo, NY.
156. U.S. Army Engineer District, Buffalo. 1980. Presque Isle Peninsula, Eries, Pennsylvania, Final Phase I General Design Memorandum, Buffalo, NY.
157. U.S. Army Engineer District, Buffalo. 1982. Lakeshore Park, Ashtabula, Ohio—Beach Erosion Control and Shoreline Protection Study, Detailed Project Report, Buffalo, NY.
158. U.S. Army Engineer District, Buffalo. 1983. Supplemental Report to Phase I General Design Memorandum—Cooperative Beach Erosion Control Project at Presque Isle Peninsula, Erie, Pennsylvania, Buffalo, NY.
159. U.S. Army Engineer District, Buffalo. 1984. Presque Isle Pennsylvania—Appraisal Report, Buffalo, NY.
160. U.S. Army Engineer District, Buffalo. 1985. Cooperative Beach Erosion Control Project Reformulation Report and Draft Supplement to the Final Environmental Impact Statement—Presque Isle Peninsula, Erie, Pennsylvania, Buffalo, NY.
161. U.S. Army Engineer District, Buffalo. 1986. Presque Isle Peninsula General Design Memorandum, Phase II, Detailed Project Design, Buffalo, NY.
162. U.S. Army Engineer District, Buffalo. 1986. Sims Park, Euclid, Ohio—Detailed Project Report on Shoreline Erosion/Beach Restoration on Lake Erie, Buffalo, NY.
163. U.S. Army Engineer District, Buffalo. 1988. Maumee Bay State Park, Ohio General Design Memorandum—Detailed Project Design, Buffalo, NY.
164. U.S. Army Engineer District, Jacksonville. 1984. Pinella County, Florida Beach Erosion Control Project, Sand Key Segment, Jacksonville, FL, 54 pp.
165. U.S. Army Engineer District, Los Angeles. 1978. Imperial Beach, Erosion Control Project, San Diego County, California, General Design Memorandum 4, Los Angeles, CA, April.
166. U.S. Army Engineer District, New Orleans. 1971. Survey of Holly Beach and Vicinity, Louisiana, Series No. 95, 19 pp.
167. U.S. Army Engineer Division, Pacific Coast. 1979. Water Resources Development, Hawaii.

168. Uda, T. 1989. Statistical Analysis of Detached Breakwaters in Japan, *Proc. 21st International Coastal Engineering Conference*, ASCE, pp. 2028–2042.
169. van der Meer, J. W. 1991. Stability and Transmission at Low-crested and High-crested Structures. Delft Hydraulics Publication No. 453. Delft, The Netherlands.
170. Vellinga, P. 1978. Movable Bed Model Test on Dune Erosion, *Proc. 16th Coastal Engineering Conference*, ASCE, pp. 2020–2039.
171. Walker, J., Clark, D. and Pope, J. 1980. A Detached Breakwater System for Beach Protection, *Proc. 17th Coastal Engineering Conference*, ASCE, pp. 1968–1987.
172. Zwamborn, J. A., Fromme, G. A. W., and Fitzpatrick, J. B. 1970. Underwater Mound for the Protection of Durban's Harbor, *Proc. of the 12th Conference on Coastal Engineering,* ASCE, Vol. 2, pp. 975–994.

CHAPTER 6
WAVE OVERTOPPING OF COASTAL AND SHORELINE STRUCTURES

P. B. BESLEY
HR Wallingford
Wallingford, U.K.

N. W. H. ALLSOP
University of Sheffield
c/o HR Wallingford,
Wallingford, U.K.

Notation	6.1
Overview	6.2
1.1 Wave Run-up and Mean Overtopping of Structures	6.2
1.1.1. Simply Sloping Smooth Impermeable Slopes	6.3
1.1.2. Armored Rubble Mounds	6.7
1.1.3. The Influence of Crest Walls	6.10
1.1.4. Vertical Sea Walls and Caissons	6.13
1.2. Number of Waves Overtopping	6.16
1.2.1. Vertical Walls	6.16
1.2.2. Sloped Structures	6.16
1.3. Peak Overtopping Discharge	6.17
1.4. Tolerable Overtopping Discharge	6.17
1.4.1. Tolerable Mean Overtopping Discharge	6.18
1.4.2. Tolerable Peak Overtopping Discharge	6.18
1.5. Wave Overtopping	6.19
1.6. Acknowledgments	6.20
1.7. References and Suggested Readings	6.20

NOTATION

A Slope coefficient for overtopping
B Slope coefficient for overtopping
H_s Significant wave height, average of highest one-third of wave heights
H_{sb} Breaking (significant) wave height
H_{si} Inshore significant wave height
h Water depth
h_s Water depth at toe of structure

L Wavelength
$Q^\#$ Dimensionless overtopping discharge
q Mean overtopping discharge per unit length of revetment
R_c Crest freeboard, level of crest less static water level
$R^\#$ Dimensionless run-up
r Roughness coefficient for wave run-up and overtopping
s_m Steepness of mean wave period = $2\pi H_s/gT_m^2$
s_{mo} Offshore mean wave steepness
s_p Steepness of peak wave period = $2\pi H_s/gT_p^2$
T_m Mean wave period
T_p Peak wave period
α Structure front slope angle to horizontal
β Angle of wave attack to structure/bed contour alignment
ξ_m Iribarren number tan $\alpha/s_m^{0.5}$, calculated in terms of mean wave steepness
ξ_{mcr} Critical Iribarren number, distinguishing between plunging and surging waves for van der Meer's rock armor formulae
ξ_p Iribarren number tan $\alpha/s_p^{0.5}$, calculated in terms of peak wave steepness

OVERVIEW

Reliable prediction of the hydraulic performance of any coastal or shoreline structure is one of the most important design activities. The overtopping performance significantly influences the crest level and/or seaward slope angle, and therefore the overall cost of the structure. This chapter identifies prediction methods to calculate crest levels for given overtopping performance, and discusses limiting overtopping discharges that may be tolerated for different structures or uses.

1.1. WAVE RUN-UP AND MEAN OVERTOPPING OF STRUCTURES

Many coastal slopes, embankment dams and related structures are constructed to protect low-lying areas against wave overtopping and hence flooding by high tides and/or storm surges. In the analysis or design of these structures, the main hydraulic responses of interest are the wave run-up levels, the number of overtopping waves and the mean overtopping discharge.

Historically, crest levels of many flood defence structures were set with reference to "maximum" wave run-up levels. In reality, waves heights are distributed randomly and storm water levels are influenced by tides and surges, so most practical structures will experience some overtopping under extreme conditions. In the design of these structures, calculations of run-up levels are therefore less useful than calculations of crest levels to give set overtopping discharges for given return periods. Recent research in the United Kingdom and Europe has therefore concentrated on wave overtopping.

Prediction of overtopping discharges is now regarded as one of the most important processes affecting the design of a sea wall or breakwater and is principally influenced by:

- Crest freeboard relative to water level, A_c or R_c
- Slope angle of the armor, cot α

- Roughness/porosity of the armor system, given by relative roughness, r, often a function of the type and size (D_{n50}) of the armor units
- Height and shape of the crown wall above the armor
- Width of the armor crest berm, G
- Incident wave height and wave period, H_s and T_m
- Plan angle of wave attack, β

The two most common methods of describing overtopping at present are by the number of waves passing over the crest, N_{ow}, or by the mean overtopping discharge per unit length of structure, q, usually given in m³/s per meter run or l/s per meter run. Of these, the mean overtopping discharge q is by far the most reliable. Over the design range of performance for a typical sea wall or breakwater, values of q may be expected to vary over orders of magnitude. Yet values of q can be determined with reasonable reliability and accuracy from relatively inexpensive and rapid hydraulic model tests.

1.1.1. Simply Sloping Smooth Impermeable Slopes

Wave Run-up. Measurements of wave run-up on smooth slopes have been analyzed to give very simple prediction lines. Relative run-up levels vary with incident wave height and with the surf similarity parameter, defined as:

$$\xi = \tan \alpha / s^{0.5} \tag{1}$$

where α is the angle of the sea wall slope to the horizontal
s_0 is the nominal sea steepness ($= 2\pi H_s/gT^2$)
H_s is the nearshore significant wave height

It may be noted that the definition above is not complete, as the wave period to be used has not been defined. Different prediction methods use either the mean wave period, T_m, or the peak period, T_p, from which may be calculated s_{om} or s_{op}, and in turn ξ_{om} or ξ_{op}.

The sea steepness is described as nominal because it relates inshore wave height to offshore wavelength. The following equations may be used to estimate significant run-up levels, R_{us}, on smooth slopes for various ranges of the surf similarity parameter:

$$0 < \xi_{op} < 2.0 \quad R_{us}/H_s = 1.35 \, \xi_{op} \tag{2}$$

$$\xi_{op} \geq 2.0 \quad R_{us}/H_s = 3.0 - 0.25 \, \xi_{op} \tag{3}$$

For simple configurations with slopes of between 1:1.133 and 1:2.5, a Raleigh distribution may be assumed to relate 2% run-up and significant run-up levels.

$$R_{u2\%} = 1.4 \, R_{us} \tag{4}$$

Wave Overtopping of Smooth Slopes. A considerable number of studies have been undertaken into the overtopping performance of sea walls. Many of the early studies in the United States used regular wave physical model studies, but studies in Europe from the 1970s used random wave tests to develop more realistic prediction methods. The most widely applied results are those described by Owen (1980), who developed a simple empirical method to predict the overtopping performance of smooth impermeable simply sloping structures (see Figure 6.1).

Owen's method is based on model test measurements of overtopping discharges for a range of simple and bermed embankments with seaward slope angles of 1:1, 1:2, and 1:4. The bermed structures used in this study all had berms sited at or below still water level and had the same slope angle above and below the berm. Overtopping volumes

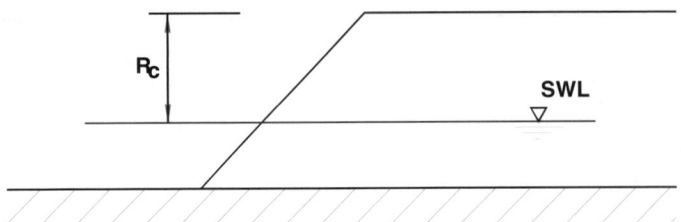

FIGURE 6.1. Smooth, impermeable, simply sloped sea wall.

were collected over 500 waves, and mean discharges are described per unit length of sea wall.

Owen proposed a design method, now widely used in coastal engineering, to calculate the mean discharge overtopping a simply sloping sea wall. In this method the discharge and freeboard are nondimensionalised to give the two main parameters R^* and Q^*:

$$R^* = R_c/[T_m (gH_s)^{0.5}] \quad (5)$$

$$Q^* = q/(T_m gH_s) \quad (6)$$

where q is the mean overtopping discharge rate per meter run of sea wall
T_m is the wave period at the toe of the wall
g is acceleration due to gravity
H_s is the significant wave height at the toe of the wall
R_c is the freeboard of the wall (crest height above still water level)

The two dimensionless parameters are combined in the equation:

$$Q^* = A \exp(-BR^*) \quad (7)$$

where A and B are empirical coefficients that depend on the profile of the sea wall.

Owen derived, or interpolated, values of A and B for simple sea walls ranging in slope angle from 1:1 to 1:5. The original coefficients proposed by Owen were later revised as further research results became available; see Table 6.1.

Owen also found that the equations used for simply sloping sea walls could equally be applied to the bermed structures such as that shown in Figure 6.2. Owen proposed that the modified empirical coefficients shown in Table 6.2 should be used.

TABLE 6.1. Empirical coefficients—smooth impermeable simply sloping sea walls

Sea wall slope	A	B
1.1	7.94×10^{-3}	20.1
1:1.5	8.84×10^{-3}	19.9
1:2	9.39×10^{-3}	21.6
1:2.5	1.03×10^{-2}	24.5
1:3	1.09×10^{-2}	28.7
1:3.5	1.12×10^{-2}	34.1
1:4	1.16×10^{-2}	41.0
1:4.5	1.20×10^{-2}	47.7
1:5	1.31×10^{-2}	55.6

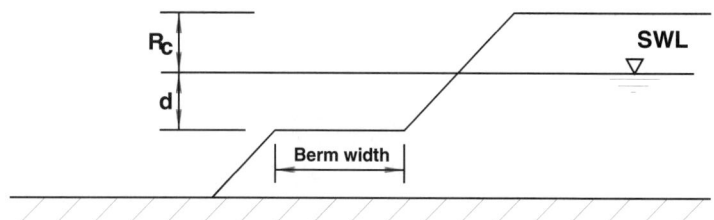

FIGURE 6.2. Bermed sea wall.

TABLE 6.2. Empirical coefficients—bermed sea walls—berm at or below SWL

Seawall slope	Berm elevation	Berm width	A	B
1:1			6.40E-3	19.50
1:2	−4.0	10	9.11E-3	21.50
1:4			1.45E-2	41.10
1:1			3.40E-3	16.52
1:2	−2.0	5	9.80E-3	23.98
1:4			1.59E-2	46.63
1:1			1.63E-3	14.85
1:2	−2.0	10	2.14E-3	18.03
1:4			3.93E-3	41.92
1:1			8.80E-4	14.76
1:2	−2.0	20	2.00E-3	24.81
1:4			8.50E-3	50.40
1:1			3.80E-4	22.65
1:2	−2.0	40	5.00E-4	25.93
1:4			4.70E-3	51.23
1:1			2.40E-4	25.90
1:2	−2.0	80	3.80E-4	25.76
1:4			8.80E-4	58.24
1:1			1.55E-2	32.68
1:2	−1.0	5	1.90E-2	37.27
1:4			5.00E-2	70.32
1:1			9.25E-3	38.90
1:2	−1.0	10	3.39E-2	53.30
1:4			3.03E-2	79.60
1:1			7.50E-3	45.61
1:2	−1.0	20	3.40E-3	49.97
1:4			3.90E-3	61.57
1:1			1.20E-3	49.30
1:2	−1.0	40	2.35E-3	56.18
1:4			1.45E-4	63.43
1:1			4.10E-5	51.41
1:2	−1.0	80	6.60E-5	66.54
1:4			5.40E-5	71.59
1:1			8.25E-3	40.94
1:2	0.0	10	1.78E-2	52.80
1:4			1.13E-2	68.66

Recently van der Meer and de Waal (1992) have proposed an alternative series of equations to estimate overtopping for simply sloping and bermed sea walls. Wave overtopping is expressed by one of two formulae, depending on whether the structure is subject to plunging (breaking) or surging (nonbreaking) waves:

Plunging waves are defined where $\xi_{op} < 2$

Surging waves are defined where $\xi_{op} \geq 2$

For plunging waves:

$$R^{\#} = R_c[s_{op}/(H_s \tan \alpha)]^{0.5} \qquad (8)$$

$$Q^{\#} = q[s_{op}/(gH_s^3 \tan \alpha)]^{0.5} \qquad (9)$$

The main prediction equation then becomes

$$Q^{\#} = 0.06 \exp(-5.2 R^{\#'}/\gamma) \qquad (10)$$

where γ is the total reduction factor ($= \gamma_b \gamma_f \gamma_\beta \gamma_h$)
γ_b is the reduction factor for berms
γ_f is the reduction factor for roughness
γ_β is the reduction factor for oblique short-crested wave attack
γ_h is the reduction factor for depth-limited wave attack

For surging waves the parameters are modified as follows:

$$R^{\#'} = R_c/H_s \qquad (11)$$

$$Q^{\#'} = q/(gH_s^3)^{0.5} \qquad (12)$$

and are related by

$$Q^{\#'} = 0.2 \exp(-2.6 R^{\#'}/\gamma) \qquad (13)$$

The methods of Owen and van der Meer have different advantages and disadvantages for the designer. The Owen equation is simple to use but requires different empirical coefficients for each different wall profile. Van der Meer's overtopping equations are more complicated to use, but do not require different coefficients for alternative sea wall profiles. The disadvantage of reducing the number of empirical coefficients in the van der Meer equations is, however, that the sensitivity of the response to different parameters has been much reduced by averaging in the derivation of the equations, resulting in less definition in calculated overtopping discharges for any particular sea wall profile.

A further difference is in the input wave height. Owen's method requires that input wave conditions be defined at the toe of the sea wall, so any depth-limited breaking must be determined before entering the wave height into Equations 5–7. Van der Meer's method in Equations 8–13 uses nearshore wave conditions and then allows for (local) wave breaking through the use of γ_h. The method of calculating γ_h proposed by van der Meer is, however, only applicable to foreshore slopes of 1:100 or shallower, so this may significantly underestimate conditions on steeper beaches of coarse sand, shingle/gravel, or rock.

After much deliberation, the authors consider that the method proposed by Owen is still the most appropriate means of estimating overtopping discharges at smooth simply sloping and bermed sea walls for the wider range of conditions that pertain around the U.K. coastline. The use of a large number of empirical coefficients is a minor drawback in comparison to the increased accuracy of the results.

Owen (1980) and de Waal and van der Meer (1992) investigated the performance of simply sloping and bermed sea walls under angled wave attack Figure 6.3. Under long-

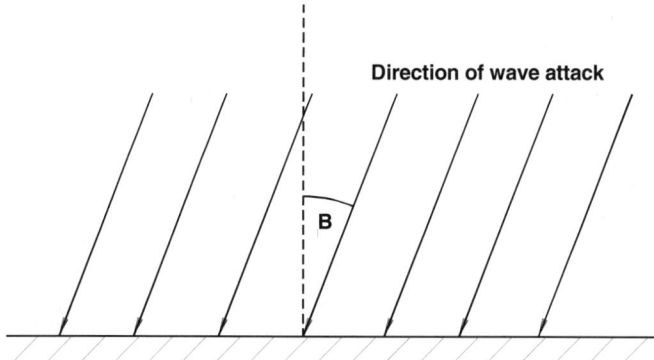

FIGURE 6.3. Angled wave attack.

crested oblique waves, Owen noted that overtopping could increase for angles up to $\beta = 30°$ with the maximum increase occurring at $\beta = 15°$. Owen derived correction factors for A and B for β from 0° up to 60°.

De Waal and van der Meer derived equations to describe the reduction factor for overtopping under oblique wave attack, γ_β. For long-crested waves and $\gamma_\beta \geq 0.60$ and $\gamma_\beta = 1.0$ for $0° \leq \beta \leq 60°$:

$$\gamma_\beta = \cos^2(\beta - 10°) \qquad (14)$$

and for short-crested waves:

$$\gamma_\beta = 1 - 0.0033\,\beta \qquad (15)$$

Banyard and Herbert (1995) reported different equations to describe the behavior of simply sloping and bermed sea walls in short-crested seas. These equations enable the overtopping ratio, O_r, to be calculated where O_r is defined as the ratio of overtopping at a given wave angle to that predicted under normal wave attack. For simply sloping sea walls:

$$O_r = -0.000152\,\beta^2 + 1 \qquad (16)$$

and for bermed sea walls:

$$O_r = 1.99 - 1.93\{1.0 - [(\beta - 60.0)/69.8]^2\}^{0.5} \qquad (17)$$

The behavior of the two types of sea wall differed considerably, with the bermed structure exhibiting a greater reduction in overtopping for a given wave angle than the simple slope embankments. The difference in performance was particularly noticeable at small angles of wave attack, with the bermed sea wall equation predicting a significant reduction in overtopping compared to normal wave attack.

For simply sloping sea walls in short-crested seas, the equations of Banyard and de Waal gave similar results for $\beta < 30°$, but at more oblique angles the equations diverged significantly, with de Waal predicting larger overtopping discharges.

1.1.2. Armored Rubble Mounds

Wave Run-up of Armored Mounds. Armored mounds generally dissipate significantly more wave energy than the equivalent smooth slope, thus giving lower run-up levels and overtopping discharges. In instances where methods are not available to predict overtopping discharges, extreme run-up levels at various exceedance levels may estimated using equations from the CIRIA/CUR rock manual edited by Simm (1991) based on tests by van der Meer on rock armored slopes:

$$R_{ux}/H_s = a\xi_m \quad \text{for } \xi_m < 1.5 \quad (18)$$

$$R_{ux}/H_s = b\xi_m^c \quad \text{for } \xi_m > 1.5 \quad (19)$$

where ξ_m is the mean surf similarity parameter ($= \tan \alpha/s_m^{0.5}$)
α is the angle of the sea wall slope to the horizontal
s_m is the (fictional) mean sea steepness ($= 2\pi H_s/gT_m^2$)
H_s is the incident significant wave height and T_m is the mean wave period
a, b, and c are coefficients (Table 6.3)
For permeable structures, $P > 0.4$, the upper limit on R_{ux}/H_s is given by

$$R_{ux}/H_s = d \quad (20)$$

Research studies on the distribution of run-up levels by Allsop et al. (1985) demonstrate that run-up levels fit the Raleigh distribution, giving the distribution of $R_{ui\%}$ relative to R_{us}, where $R_{ui\%}$ is the run up level exceeded by $i\%$ of waves:

$$R_{ui\%}/R_{us} = [-\ln(i\%/100)/2]^{0.5} \quad (21)$$

Wave Overtopping of Rough/Armored Slopes. Owen (1980) extended his work on simply sloping and bermed sea walls to cover rough impermeable and armored (rough permeable) structures. Owen related his previous dimensionless parameters, Q^* and R^*, as given in Equations 5 and 6, by the following equation:

$$Q^* = A \exp(-BR^*/r) \quad (22)$$

Where A and B are the empirical coefficients applicable to a smooth slope and r is the roughness coefficient for the surface of the wall. Values of roughness vary from $r = 1.0$ for smooth slopes to $r = 0.55$ for a rock armored slope. Smaller values of r may be justified for a few specialized rubble mound armor systems of high roughness. Owen produced typical values of the roughness coefficient based upon the relative run-up performance of alternative types of construction (Table 6.4).

A similar approach to that of Owen regarding the use of a roughness coefficient was adopted by de Waal and van der Meer (1992) in order to estimate overtopping on rough and armored structures (Equation 10). Although other authors have attempted to derive a more suitable method than that of Owen, none have so far succeeded and hence the Owen method is suggested for use on rough and armored slopes.

Studies of the overtopping discharge performance of rough and armored sea walls under angled wave attack are mainly limited to site-specific studies. Juhl and Sloth (1994)

TABLE 6.3. Coefficients for wave run-up predictions for different exceedence levels

Exceedence level, x	a	b	c	d
0.1%	1.12	1.34	0.55	2.58
2%	0.96	1.17	0.46	1.97
Significance (13.6%)	0.72	0.88	0.41	1.35

TABLE 6.4. Typical roughness coefficients

Type of sea wall	Roughness coefficient, r
Smooth concrete or asphalt	1.0
Smooth concrete blocks with little or no drainage	1.0
Stone blocks, pitched or mortared	0.95
Concrete stepped structure	0.95
Turf	0.9–1.0
One layer of rock armor on impermeable base	0.8
One layer of rock armor on permeable base	0.55–0.60
Two layers of rock armor	0.50–0.55

used long-crested seas to investigate the effect of wave angle on the overtopping performance of breakwaters. They noted that for small angles of wave attack a few tests exhibited overtopping ratios, O_r, greater than unity, although on average a reduction in overtopping was found. This was similar to the behavior of smooth slopes noted by Banyard and Herbert (1995). Juhl and Sloth (1994) concluded that the overtopping ratio was dependent upon the freeboard, R_c, but derived no empirical equations to describe the overtopping performance.

In summary, the recommended method for predicting the overtopping discharge over an armored slope is

$$R^* = R_c/[T_m(gH_s)^{0.5}] \qquad (23)$$

$$Q^* = A \exp(-BR^*/r) \qquad (24)$$

$$Q = Q^* T_m g H_s \qquad (25)$$

where R_c is the freeboard (the height of the crest of the wall above still water level)
H_s is the significant wave height at the toe of the wall
T_m is the mean wave period
r is the roughness coefficient (see Table 6.4)
g is acceleration due to gravity
A and B are empirical coefficients dependent upon the cross-section of the sea wall (see Table 6.1)
Q is the mean overtopping discharge rate per meter run of sea wall

The geometry of the structure and the wave and water level conditions enable R^* and Q^* to be calculated using Equations 23 and 24 along with the coefficients given in Tables 6.1 and 6.4. Use of Equation 25 then enables the mean overtopping discharge, Q, to be calculated.

Armored sea walls often include a crest berm that will dissipate significant wave energy and thus reduce overtopping. The application of Equations 23–25 does not take into account crest berms and hence discharges are overpredicted. In order to conservatively incorporate the performance of a crest berm, it is suggested that an imaginary slope be constructed between the still water level/sea wall intersection point and the rear of the crest berm. Equations 23–25 may then be applied to this imaginary slope.

Due to the limited information on the performance of rough and armored sea walls under angled wave attack, it is suggested that the methods developed for smooth sea walls, described in Section 1.2, be used.

Recent analysis by Allsop and Franco (1992) under the G6-S MAST research project has derived alternative values of A and B for the Owen equation (assuming $r = 1$) for a number of armored structures. For *rock-armored* slopes of 1:1.5 and 1:2 with a crest width of approximately 3 stones, a crown wall elevation equal to the armor crest ($F_c = 0$), and subject to normal wave attack, tests results suggest new values of the coefficients in Equation 22:

$$Q^* = 0.0102 \exp(-62.4R^*) \quad \text{for 1:1.5 slopes with } G = 3D_{n50} \quad (26)$$

$$Q^* = 0.00478 \exp(-57.7R^*) \quad \text{for 1:2.0 slopes with } G = 3D_{n50} \quad (27)$$

Test results for wave overtopping on *tetrapod-armored* structures have also been analyzed. Results for 1:15 slopes with $G = 3.4D_{n50}$ cover the cases where the crown wall reaches the same level as the armor, $F_c = 0$:

$$Q^* = 0.0217 \exp(-72.6R^*) \quad \text{for } G = 3.4D_{n50} \text{ and } F_c = 0 \quad (28)$$

1.1.3. The Influence of Crest Walls

Crest Walls on Impermeable Sea walls. Limited research work has been completed into the performance of wave-return walls sited at the crest of impermeable sea walls. Owen and Steele (1991) undertook the most comprehensive study, investigating the performance of recurved wave-return walls on top of 1:2 and 1:4 simply sloping sea walls (see Figure 6.4).

Owen found that the ratio of the discharge overtopping the recurved wall to the discharge that would have occurred if the recurved wall had been absent was dependent upon the structural configuration of the sea wall and the dimensionless wall height, W^*. The dimensionless wall height was defined as

$$W^* = W_h/A_c \quad (29)$$

where W_h is the height of the recurved wall illustrated in Figure 6.4
A_c is the height of the top of the seaward slope (or base of return wall) above still water level

The following design method is applicable to recurved return-wall profiles as proposed by Berkeley-Thorn and Roberts (1981) and illustrated in Figure 6.5. This is a very efficient type of recurved wall and alternative profiles may be significantly less efficient.

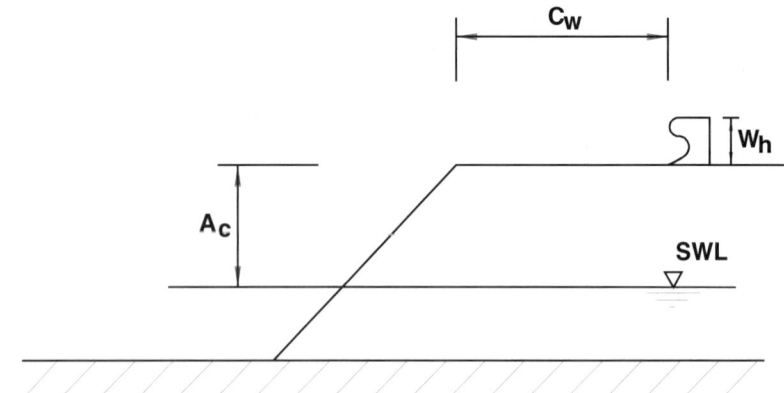

FIGURE 6.4. Wave-return wall at the crest of an impermeable sea wall.

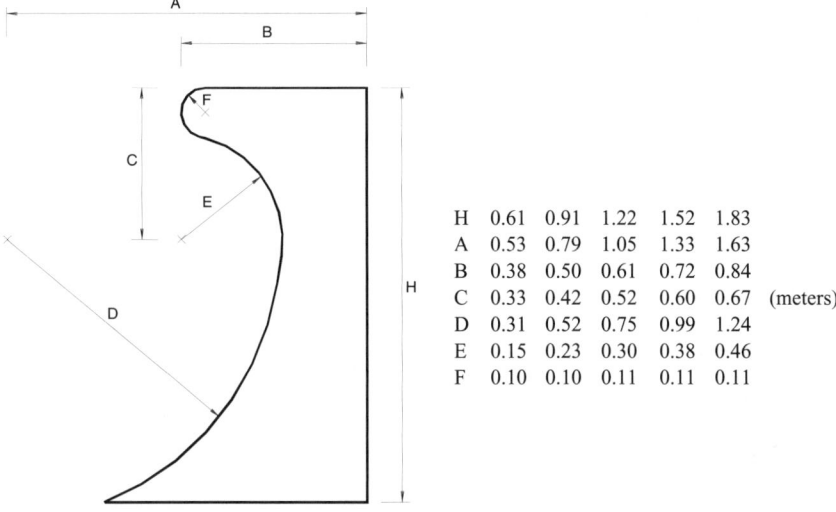

FIGURE 6.5. Typical form of a recurved wall profile.

H	0.61	0.91	1.22	1.52	1.83
A	0.53	0.79	1.05	1.33	1.63
B	0.38	0.50	0.61	0.72	0.84
C	0.33	0.42	0.52	0.60	0.67 (meters)
D	0.31	0.52	0.75	0.99	1.24
E	0.15	0.23	0.30	0.38	0.46
F	0.10	0.10	0.11	0.11	0.11

The discharge, which would occur if the crest wall were absent, is first calculated using Owen's method:

$$A_c^* = A_c/[T_m(gH_s)^{0.5}] \qquad (30)$$

$$Q_c^* = A \exp(-BA_c^*) \qquad (31)$$

$$Q_c = Q_c^* \, T_m g H_s \qquad (32)$$

where A_c is the height of the top of the seaward slope (or base of return wall) above still water level
  H_s is the significant wave height at the toe of the sea wall, and T_m is the mean wave period
  g is acceleration due to gravity
  A and B are empirical coefficients dependent upon the cross-section of the sea wall (see Table 6.1)
  Q_c is the mean discharge per meter run of sea wall which would have overtopped the structure if the wave-return wall had been absent

Also, $W^* = W_h/A_c$, where W_h is the height of the wave return wall. (33)

Knowledge of the dimensionless wall height, W^*, the seaward slope of the sea wall and the distance of the return wall behind the top of the seaward slope, C_w, allows an adjustment factor, A_f, to be obtained from Table 6.5. The adjusted dimensionless freeboard, X^*, is then given by

$$X^* = A_f A_c^* \qquad (34)$$

The values of X^* and W^* allow the use of Figure 6.6 in order to obtain a discharge factor, D_f, which is defined as

$$D_f = Q_w/Q_c \qquad (35)$$

TABLE 6.5. Adjustment factors—wave-return walls

$W_h/R_c \geq 0.6$		
Sea wall slope	Crest berm width, C_w (m)	A_f
1:2	0	1.00
1:2	4	1.07
1:2	8	1.10
1:4	0	1.27
1:4	4	1.22
1:4	8	1.33
$W_h/R_c < 0.6$		
Sea wall slope	Crest berm width, C_w (m)	A_f
1:2	0	1.00
1:2	4	1.34
1:2	8	1.38
1:4	0	1.27
1:4	4	1.53
1:4	8	1.67

where Q_w is the mean discharge per meter run of sea wall overtopping the wave-return wall.

Figure 6.5 shows that a crest wall is more efficient (i.e., D_f is low) when the volume of water arriving at its base is small (i.e., X^* is high).

Adjustment factors are only presently available for slopes of 1:2 and 1:4 and therefore some interpolation will be required for alternative slopes. It is recommended that for slopes between 1:1 and 1:2.5 adjustment factors applicable to the 1:2 slope be employed in the analysis. For slope angles between 1:2.5 and 1:4, it is suggested that linear interpo-

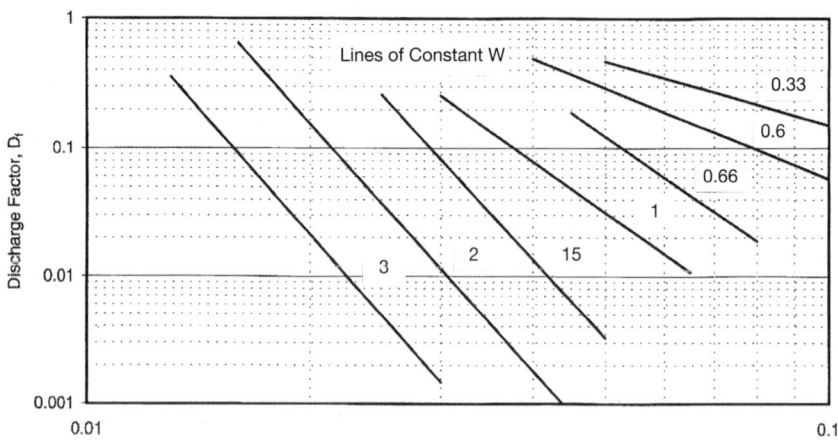

FIGURE 6.6. Discharge factors, walls on impermeable slopes.

lation be completed between the available adjustments factors based upon the cotangent of the slope of the structure. For slopes shallower than 1:4, a conservative solution will be ensured if adjustment factors applicable to a 1:4 slope are employed. Although the adjustment factors may be obtained by interpolation, the value of $Q_c{}^*$ in this analysis should always be calculated using coefficients of A and B applicable to the precise sea wall profile.

For smooth-bermed and rough impermeable sea walls, the relevant slope should be converted to an equivalent smooth, simply sloping structure which, for the same wave conditions, water level, and crest level, would give the same overtopping discharge. The method of determining the overtopping performance of smooth-bermed and rough impermeable sea walls is respectively outlined in Sections 1.1.1 and 1.1.2. Once an equivalent sea wall slope has been obtained, the performance of the return wall can then be assessed using the method outlined above for simply sloping structures.

Banyard and Herbert (1995) studied the performance of recurvee walls under angled wave attack. The recurve walls exhibited similar behavior in both short- and long-crested seas. For angles of wave attack up to and including 30°, increases in overtopping were noted. The largest increases in overtopping were over six times that predicted for normal wave attack.

In order to determine overtopping due to angled wave attack in short- or long-crested seas, the discharge, assuming normal attack, should first be calculated. For angles of wave attack from 0°–45° the overtopping ratio, O_r, may then be determined from the following equation:

$$O_r = -1.18 \ln(D_f) - 0.40 \qquad (36)$$

where O_r is the ratio of the discharge under angled wave attack to that under normal attack.

A minimum value of $O_r = 0.1$ should be assumed if a value of $O_r < 0.1$ is calculated from Equation 36. For angles beyond 45°, an overtopping ratio of 0.1 should be assumed.

Crest Walls Located on Permeable Sea Walls. A similar design method can be used for crest walls located on permeable slopes. Bradbury and Allsop (1988) measured discharges overtopping nonrecurved crown walls mounted on the top of rock breakwaters. In all cases, the breakwater had a slope of 1:2 and a crest berm width of at least two rocks.

Bradbury and Allsop's data was recently reanalyzed in terms of discharge factors, D_f, (Besley, 1999). The results are plotted on Figure 6.7. In order to use Figure 6.7, the discharge without the crest wall in place, Q_c, should first be calculated. The discharge factors are plotted against Q_c^* rather than against the crest freeboard. This reflects the fact that the volume of water arriving at the base of the crest wall (and hence its efficiency) is dependent on the roughness of the slope. The discharge factors, D_f, obtained for the walls located on permeable slopes were considerably better than those obtained for the equivalent walls located on smooth slopes, despite the less efficient return-wall profile. This is due to the ability of the crest berm on the permeable structure to drain water. The results may also be applied conservatively to recurved crest walls.

No data is available on angled wave attack. An estimate of the discharge may be obtained by using the method outlined above for angled wave attack on impermeable slopes with recurved walls.

1.1.4. Vertical Sea Walls and Caissons

Goda (1985) investigated the overtopping performance of plain vertical sea walls (see Figure 6.8) for approach slopes of 1:10 and 1:30 and offshore sea steepnesses (s_{om} =

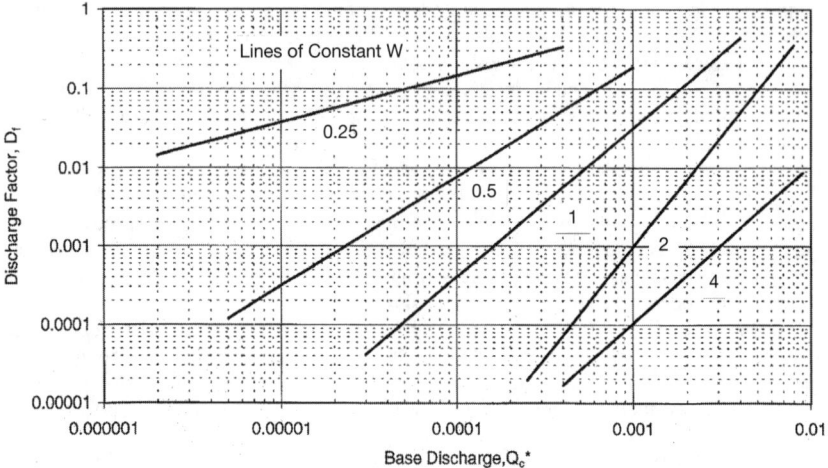

FIGURE 6.7. Discharge factor walls on permeable slopes.

$2\pi H_{so}/gT_{mo}^2$, where H_{so} and T_{mo} are, respectively, the offshore significant wave height and period) ranging from 0.012 to 0.036. Goda proposed a design method based on a graphical format. For a given approach bathymetry and offshore sea steepness, a dimensionless discharge, $Q^{\#}$, was plotted on the y-axis against h/H_{so} on the x-axis, where:

$$Q^{\#} = Q/(2gH_{so}^3)^{0.5} \qquad (37)$$

where h is the water depth at the toe of the structure.

Lines of constant relative freeboard, R_c/H_{so}, were illustrated on each graph, where R_c is the height of the crest of the wall above still water level.

Herbert (1993) has confirmed and extended the work of Goda covering approach bathymetries of 1:10, 1:30, and 1:100 and offshore sea steepnesses of $0.017 < s_{om} < 0.060$. The weakness of the work of Goda and Herbert is that significant interpolation is required. The influence of the offshore bathymetry must be assessed by selecting the nearest simple seabed slope and graphs are only available for a limited number of offshore sea steepnesses.

Recently, Franco et al. (1994) investigated the overtopping on vertical breakwaters in

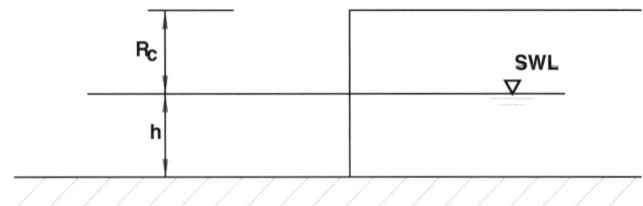

FIGURE 6.8. Typical plain vertical wall.

deep water. The following equation, which does not include any effect of wave period, was derived for relative freeboards ranging from $0.9 < R_c/H_s < 2.2$:

$$Q/(gH_s^3)^{0.5} = 0.2 \exp(-4.3 R_c/H_s) \tag{38}$$

where H_s is the significant wave height at the toe of the structure.

Allsop et al. (1995) have derived an empirical equation of the same form as that of Franco et al. as follows:

$$Q/(gH_s^3)^{0.5} = 0.03 \exp(-2.05 R_c/H_s) \tag{39}$$

This equation, which covers a range of relative freeboards of $0.03 < R_c/H_s < 3.2$, is applicable to vertical walls in both deep and shallow water and is therefore more appropriate to coastal structures than the work of Franco et al.

Further analysis of the combined data sets of Allsop et al (1995) and de Waal (1994) determined that the overtopping performance of vertical walls is dependent upon the type of incident wave conditions. In deep water, waves hit the structure and are generally reflected back seaward (so-called reflected waves). However, as the waves become limited by the available water depth, they are prone to break over the sea wall (so-called impact waves), causing a change in the overtopping performance.

A wave breaking parameter, h^*, was defined that dictates whether waves at the structure are dominated by impact waves or by reflecting waves. h^* is defined by

$$h^* = (h/H_s)(2\pi h/gT_m^2) \tag{40}$$

Reflecting waves dominate when $h^* > 0.3$. Impacting waves dominate when $h^* \leq 0.3$. The formulation of h^* reflects the fact that waves are more likely to break if the wavelength or the wave height is large compared to the water depth, h.

Separating their data according to h^*, Allsop et al. (1995) determined that, for $h^* > 0.3$, the mean overtopping discharge was accurately described by an equation of the form of Equation 38, but with new empirical parameters:

$$Q/(gH_s^3)^{0.5} = 0.05 \exp(-2.78 R_c/H_s) \tag{41}$$

For $h^* \leq 0.3$, however, a different relationship was determined. To reflect the importance of wave breaking, new dimensionless discharge and freeboard parameters, both incorporating h^*, were defined:

$$Q_h = Q/(gh^3)^{0.5}/h^{*2} \tag{42}$$

$$R_h = (R_c/H_s)h^* \tag{43}$$

Examination of all the results from both Allsop et al's (1995) and de Waal's (1994) data sets for which $h^* \leq 0.3$ produced the following relationship:

$$Q_h = 1.37 \times 10^{-4} R_h^{-3.24} \tag{44}$$

Overtopping equations were subsequently derived for both types of wave action. The goodness of fit of the data was superior to all previous equations and, for this reason, the method described immediately above is recommended for use with vertical walls.

The influence of wave angle on the overtopping performance of a vertical wall in short- and long-crested seas has been studied by Banyard and Herbert (1995). Overtopping under oblique waves was found to be up to eight times greater than the value for normal wave attack. The value of the overtopping ratio, O_r, was found to be dependent upon on relative freeboard, R_c/H_s. The overtopping ratio increased as the relative freeboard increased, i.e., the greatest increases in overtopping were for conditions where the normal wave attack discharge is small.

1.2. NUMBER OF WAVES OVERTOPPING

1.2.1. Vertical Walls

From the results of model tests, Franco et al. (1994) developed the following equation to predict the number of waves overtopping a vertical wall:

$$N_{ow}/N_w = \exp[-(R_c/H_s/0.91)^2] \qquad (45)$$

where N_{ow} is the number of waves overtopping
N_w is the number of incident waves
R_c is the crest freeboard
H_s is the significant wave height at the toe of the structure

The tests were conducted in relatively deep water so that nonbreaking wave conditions predominated.

The number of waves overtopping the structure equates to the number of run-up events exceeding the crest level. As run-up level is closely dependent on wave height, and wave heights are Rayleigh-distributed, Equation 45 is in the form of a Raleigh distribution.

A similar series of experiments was conducted at HR Wallingford (Allsop et al., 1995). The main difference between HR Wallingford's tests and those conducted by Franco et al. (1994) were that HR Wallingford's incorporated both shallow and deep-water conditions. The deep water results produced a very similar result to Equation 45.

When the water is shallow and breaking waves predominate, however, Equation 45 underpredicts N_{wo}. There are two principle reasons for this. First, where waves break in significant numbers, the distribution of individual wave heights diverges from the Raleigh form. Second, waves that break onto the face of the structure generate a different water motion than those which reflect and run up it. The results for breaking conditions must be distinguished from those for nonbreaking conditions, and require a different method of analysis. The wave breaking parameter, $h*$, given by Equation 40, can be used to separate the two cases.

The data of Allsop et al. (1995), for which $h* \leq 0.3$, produced the following equation:

$$N_{ow}/N_w = 0.031 \, R_h^{-0.99} \qquad (46)$$

Where R_h is defined by Equation 43. Equation 46 is therefore recommended for cases where breaking waves predominate (i.e., when $h* \leq 0.3$).

1.2.2. Sloped Structures

Owen (1982) showed that for smooth slopes the proportion of waves overtopping could be described by the following equation:

$$N_{ow}/N_w = \exp(-CR*^2) \qquad (47)$$

where C is a parameter that depends on the sea wall slope (Table 6.6)
$R*$ is the dimensionless freeboard $\{= R_c/[T_m(gH_s)^{0.5}]\}$
Since the run-up level of waves is a function of slope roughness, Equation 47 can be modified to take account of roughness, thus:

$$N_{ow}/N_w = \exp[-C(R*/r)^2] \qquad (48)$$

Equation 48 was validated by the results of model tests conducted on sea walls armored with various types of concrete units and with slopes of 1:1.5 (Besley, 1995). Unfortunately, the amount of data available from studies with slopes other than 1:1.5 is limited.

TABLE 6.6. Parameter for Equation 47

Seawall slope	C
1:1	63.8
1:2	37.8
1:4	110.5

1.3. PEAK OVERTOPPING DISCHARGE

Overtopping limits have traditionally been specified in terms of mean discharge rates. When attempting to assess safety levels, however, this approach is questionable, as the maximum individual event is expected to be of greater significance. For a given level of mean discharge, the volume of the largest overtopping event will vary with wave conditions and structural type. It is thus inconsistent to specify safety levels with sole reference to mean discharge levels.

An estimate of individual wave overtopping volumes can be made using the expressions given earlier for the number of waves overtopping. The average volume, V_{bar}, per unit length of the structure in each overtopping wave is given by

$$V_{bar} = (QT_m N_w)/N_{ow} \quad (49)$$

The relationship between the average volume per overtopping wave, V_{bar}, and the maximum volume in the sequence, V_{max}, can be determined by examining the distribution of individual overtopping volumes. By assuming a Weibull distribution, Franco et al. (1994) derived the following equation for a vertical wall:

$$V_{max} = 0.84 \, V_{bar}[\ln(N_{ow})]^{1/0.75} \quad (50)$$

This formula is valid for reflecting waves. The wave breaking parameter h^* from equation (40) can be used to determine whether waves are reflecting from or impacting onto a vertical sea wall. For waves impacting onto a vertical wall, Besley (1995) derived the following equation:

$$V_{max} = 0.92 \, V_{bar}[\ln(N_{ow})]^{1/0.85} \quad (51)$$

For typical sloped structures, Besley (1999) derived a similar relationship and showed that V_{max} was affected by the wave steepness:

$$V_{max} = 0.85 \, V_{bar}[\ln(N_{ow})]^{1/0.76} \quad \text{for sp} = 0.02 \quad (52)$$

$$V_{max} = 0.96 \, V_{bar}[\ln(N_{ow})]^{1/0.92} \quad \text{for sp} = 0.04 \quad (53)$$

These equations can be used to determine the maximum individual overtopping volume expected, and hence to assess the hazard level which it represents.

1.4. TOLERABLE OVERTOPPING DISCHARGE

Allowable run-up and run-down criteria may be defined indicating the maximum allowable water levels for extreme and design criteria. A more useful measurable parameter is the overtopping discharge. Overtopping acceptance conditions will depend on the location and the use of the structure.

1.4.1. Tolerable Mean Overtopping Discharge

Damage to sea walls, buildings, or infrastructure has previously been defined as a function of the mean discharge. The guidelines given by Simm (1991) are reproduced in Table 6.7. Different limits are given for embankments (with back slopes) and revetments (without back slopes).

Recently, however, it has been acknowledged that a consideration of individual wave overtopping events, rather than simply mean discharge rates, is important when designing a structure for safe usage. For example, a very low mean discharge may be the result of a small number of large (and therefore dangerous) waves overtopping the structure.

1.4.2. Tolerable Peak Overtopping Discharge

Unfortunately, data correlating individual overtopping events with hazard levels are rare. One of the few studies was carried out by Franco et al. (1994), who conducted experiments that investigated safe overtopping limits for pedestrians and vehicles. It was demonstrated, by means of model tests and experiments on volunteers, that the danger level that an individual overtopping event represents can be directly related to its volume.

Franco et al. (1994) discovered that the safe limit varied with structural type. The "safe" limit was defined as that which created a less than 10% chance of the individual subjected to it falling over. A given volume overtopping a vertical structure was found to be more dangerous than the same volume overtopping a sloped, armored structure. The safe limit for a vertical wall was found to be 0.1 m^3/m, whereas for a sloped, armored structure it was 0.75 m^3/m. However, Franco et al (1994) also noted that a volume as low as 0.05 m^3/m could unbalance an individual when striking their upper body without warning. The latter figure was determined from experiments on volunteers rather than from model tests and can thus be considered more realistic.

Smith et al. (1994) reported on full-scale tests conducted on grass dykes. An observer stood on the crest of the dyke as it was being tested. The experiment was intended to determine safe overtopping limits for personnel carrying out inspection and repair work. Smith et al. (1994) concluded that work was unsafe when the mean discharge exceeded 10 l/s/m. From examination of Smith et al.'s data this corresponded to a V_{max} of approximately 1.6 m^3/m. This is considerably higher than the limits determined by Franco et al. (1994), and is in accord with Franco et al.'s observation that safe limit of V_{max} varies with

TABLE 6.7. Tolerable mean discharges (m^3/s per meter run)

Buildings:			
		$Q < 1 \times 10^{-6}$	No damage
1×10^{-6}	<	$Q < 3 \times 10^{-5}$	Minor damage to fittings, etc.
		$Q > 3 \times 10^{-5}$	Structural damage
Embankment Sea walls:			
		$Q < 0.002$	No damage
0.002	<	$Q < 0.02$	Damage if crest not protected
0.02	<	$Q < 0.05$	Damage if back slope not protected
		$Q > 0.05$	Damage even if fully protected
Revetment Sea walls:			
		$Q < 0.05$	No damage
0.05	<	$Q < 0.2$	Damage if promenade not paved
		$Q > 0.2$	Damage even if promenade paved

structural type. One reason for this variation may be, as suggested above, the different way in which the water strikes the individual. Smith et al. (1994) reported that the vast majority of the overtopping discharge acted on the individual's legs only. It must also be borne in mind that the safety limits for trained personnel working on a structure and anticipating overtopping are higher than those for other users.

Information on prototype safety is available from Herbert (1996) who monitored overtopping at a vertical sea wall. During the installation and operation of the monitoring apparatus it was noted that personnel could work safely on the crest of the wall during mean discharges of up to 0.1 l/s/m. Individual overtopping volumes were not measured. However, the methods described in the preceding sections of this report can be used to provide an estimate of V_{max}, given that the mean discharge and the incident wave conditions were measured. This results in an estimated V_{max} of approximately 0.04 m^3/m for the sea state that caused the 0.1 l/s/m discharge. This is in close agreement with Franco et al.'s estimate of the volume (0.05 m^3/m), which could cause someone to lose their balance.

Herbert (1996) also noted that overtopping became a danger to vehicles when the mean discharge exceeded 0.2 l/s/m. Using the process described above it was determined that this corresponds to a V_{max} of approximately 0.06 m^3/m. It is thus recommended that this be adopted as the upper safe limit for vehicles driven at any speed.

To conclude, an individual tolerable volume of 0.04 m^3/m should be applied to all structures for pedestrians, despite the fact that tests have suggested that the limit varies with structural type. The authors consider that it is possible for the most dangerous mode of impact to occur on all types of structures. Whenever green-water overtopping occurs, then it should not exceed these limits.

1.5. WAVE OVERTOPPING

Any situation in which the limits given above are exceeded is likely to be potentially hazardous to pedestrians and vehicles. These limits imply that many situations in which even a single green-water overtopping event occurs are potentially hazardous. In these cases, safety can only be assured when no overtopping events take place. Because of the random nature of waves, it is difficult to specify a situation in which overtopping events are completely eliminated. A probabilistic approach is therefore required. The probability of an individual wave overtopping is given by N_{ow}/N_w. The probability that there will be at least one overtopping event during the sequence is given by:

$$P(\text{overtopping}) = 1 - (1 - N_{ow}/N_w)^{N_w} \tag{54}$$

When analysis of individual overtopping volumes indicates that very small numbers of overtopping events create unsafe conditions, the structure should be optimized by limiting the probability of an overtopping event taking place to an acceptable level. The acceptable risk of an overtopping event occurring may depend on the use of the structure in question. In situations where there is a high risk of green-water overtopping events occurring, then they must be limited to the limits given above.

1.6. ACKNOWLEDGMENTS

The research studies that produced this chapter were completed as part of projects funded by the Environment Agency and Ministry of Agriculture, Fisheries and Food.

The second author also wishes to acknowledge the support of the University of Sheffield in preparation of this chapter.

1.7. REFERENCES AND SUGGESTED READINGS

Ahrens J. (1981). Irregular wave run-up on smooth slopes. Technical paper No. 81-17 US Army Corps of Engineers, Coastal Eng. Res. Center, Fort Belvoir, Virginia.
Allsop N. W. H., Hawkes P. J., Jackson F. A., and Franco L. (1985). Wave run-up on steep slopes—model tests under random waves. Report SR 2, HR Wallingford Ltd.
Allsop N. W. H. (1990). Reflection performance of rock armored slopes in random waves. Proc. 22nd ICCE, Delft, ASCE, New York.
Allsop N. W. H. (1994). Wave overtopping of sea walls, breakwaters and shoreline structures. Technical Note 633 in Proc. ICE, Water, Maritime and Energy, December 1994, Thomas Telford, London.
Allsop N. W. H. (1998). Hydraulic performance and stability of coastal structures. In Concrete in Coast Protection, Ed. R. T. L. Allen, Thomas Telford, London.
Allsop N. W. H., Besley P., and Madurini, L. (1995). Overtopping performance of vertical and composite breakwaters, seawalls and low reflection alternatives. Paper 4.6 in *Final Proceedings of MCS Project,* University of Hanover, Hanover, Germany.
Allsop N. W. H., Durand N., and Hurdle D. P. (1998). Influence of steep seabed slopes on breaking waves for structure design. Proc. 26th ICCE, Copenhagen, ASCE, New York.
Allsop N. W. H. and Franco C. (1992). MAST G6-S Coastal Structures Topic R3: Performance of rubble mound breakwaters. Paper 3.12 to G6-S Final Workshop, Lisbon, November 1992.
Allsop N. W. H., Hawkes P. J., Jackson F. A., and Franco L. (1985). Wave run-up on steep slopes, model tests under random waves. Report SR 2, Hydraulics Research, Wallingford, U.K.
Allsop N. W. H., McKenna J. E., Vicinanza D., and Whittaker T. J. T. (1996). New design formulae for wave loadings on vertical breakwaters and seawalls. 25th International Conference on Coastal Engineering, September 1996, Orlando, ASCE, New York.
Aminti P. L. and Franco L (1988). Wave overtopping on rubble mound breakwaters.. Proc. 21st ICCE, Malaga, June 1988, ASCE, New York.
Banyard L. and Herbert D. M. (1995). The effect of wave angle on the overtopping of sea walls. Report SR396, HR Wallingford, Wallingford, U.K.
Berkeley-Thorn R. and Roberts A. C. (1981). *Sea Defence and Coast Protection Works,* Thomas Telford Ltd, London.
Besley P. B. (1999). Overtopping of seawalls—design and assessment manual. R & D Technical Report W 178, ISBN 1 85705 069 X, Environment Agency, Bristol.
Besley P. B., Stewart T., and Allsop N. W. H. (1998). Overtopping of vertical structures: new methods to account for shallow water conditions. Proc. Int. Conf. on Coastlines, Structures & Breakwaters '98, March 1998, Institution of Civil Engineers, Thomas Telford, London.
Bradbury A. P. and Allsop N. W. H. (1988). Hydraulic effects of breakwater crown walls. Proceedings Conference on Design of Breakwaters, pp. 385–396, Institution of Civil Engineers, Thomas Telford, London.
Bradbury A. P., Allsop N. W. H., and Stephens R. V. (1988). Hydraulic performance of breakwater crown walls. HR Report SR 146, Hydraulics Research, Wallingford, U.K.
Coastal Engineering Research Center (1984). *Shore Protection Manual,* Vols. I–II, 4th Edition, U.S. Government Printing Office, Washington.
Durand N. and Allsop N. W. H. (1998). Effects of steep bed slopes on depth-limited wave breaking. Proceedings of Waves '97 Conference, November 1997, Virginia Beach, ASCE, New York.
Franco, L, de Gerloni, M., and van der Meer, J. W. (1994). Wave overtopping at vertical and composite breakwaters. Proc. 24th ICCE, Kobe, ASCE, New York.
Goda Y. (1985). *Random Seas and Design of Maritime Structures.* University of Tokyo Press, Tokyo.

Herbert D. M. (1993). Wave overtopping of vertical walls. Report SR 316, HR Wallingford, Wallingford, U.K.
Herbert D. M. (1996). The overtopping of seawalls, a comparison between prototype and physical model data. Report TR22, HR Wallingford, Wallingford, U.K.
Herbert D. M., Owen M. W., and Allsop N. W. H. (1994). Overtopping of seawalls under random waves. Proc. 24th Int. Conf. Coastal Eng., Kobe, October 1994, ASCE, New York.
Juhl J. and Sloth P. (1994). Wave overtopping of breakwaters under oblique waves. Proc. 24th ICCE, Kobe, ASCE, New York.
Meer J. W. van der, Tonjes P., and de Waal J. P. (1998). A code for dike height design and examination. Paper 1 in Coastlines, Structures and Breakwaters '98, Int. Conf. at Institution of Civil Engineers, March 1998, Thomas Telford, London.
Owen M. W. (1980). Design of sea walls allowing for wave overtopping. Report EX 924, Hydraulics Research, Wallingford, U.K.
Owen M. W. (1982). Overtopping of sea defences. Proc. Conf. Hydraulic Modelling of Civil Engineering Structures, BHRA, Coventry, U.K.
Owen M. W. and Steele A. A. J. (1991). Effectiveness of recurved wave return walls. Report SR 261, HR Wallingford, Wallingford, U.K.
Simm J. D. (Ed.) (1991). Manual on the use of rock in coastal and shoreline engineering. CIRIA/CUR, Special Publication 83, ISBN: 0 86017 326 7, CIRIA, London.
Smith G. M., Seijffert J. W. W., and van der Meer J. W. (1994). Erosion and overtopping of a grass dike, large scale model tests. International Conference on Coastal Engineering, Kobe, Japan.
Yarde A. J., Banyard L. S., and Allsop N. W. H. (1996). Reservoir dams: wave conditions, wave overtopping and slab protection. Report SR 459, HR Wallingford, Wallingford, U.K.
Waal, J. P. de (1994). Wave overtopping of vertical coastal structure: Influence of wave breaking and wind. Paper presented to 2nd MCS Workshop, Milan, 1994.
Waal J. P. de and Meer J. W. van der (1992). Wave runup and overtopping on coastal structures. Proc. 23rd ICCE, Venice, ASCE, New York.

CHAPTER 7
SEDIMENT TRANSPORT AND BEACH PROFILE CHANGE DUE TO RANDOM WAVES

SHINJI SATO
University of Tokyo
Tokyo, Japan

1. Introduction — 7.1
2. Classification of Beach Profiles — 7.2
3. Near-Bottom Velocities under Random Waves — 7.5
   3.1. Asymmetry in Velocity Variation — 7.5
   3.2. Short Waves — 7.6
   3.3. Undertow — 7.7
   3.4. Long Waves — 7.7
4. Effects of Long Waves on Cross-Shore Sediment Transport — 7.9
   4.1. Suspended Sediment Flux — 7.9
   4.2. Suspended Sediment Flux Offshore and Seaward of the Surf Zone — 7.11
   4.3. Suspended Sediment Flux in the Surf Zone — 7.11
   4.4. Effect of Long Waves on Beach Profile Change — 7.18
5. Numerical Modeling of Beach Profile Change — 7.19
6. Summary — 7.22
References — 7.22

1. INTRODUCTION

Beach erosion has been a serious problem in the past several decades. It has been especially significant in recent years as a consequence of reduction of sand supply from rivers following reservoir construction and dredging of submarine sand and minerals. Increasing needs for utilization of coastal zones and the construction of port facilities also block the littoral drift, resulting in severe erosion on the down-drift side and excessive accretion on the up-drift side. According to Bird [6], more than 70% of sandy coasts around the world are categorized as eroding beaches. The recent trend of sea level rise due to global warming may accelerate the erosion. In order to find effective countermeasures against beach erosion, we have to understand the mechanism of sediment transport in detail. However, our knowledge of sand transport mechanisms on natural beaches is still qualitative, which sometimes forces us to rely on experience.

In order to predict beach evolution quantitatively, we have to understand first the dynamics of waves and currents in nearshore environments. Waves in nature are, however, random and sediment transport mechanisms under random waves are complex. Laboratory experiments in wave flumes have been used as powerful techniques to investigate the mechanisms under well-controlled environments. A number of valuable data on the mechanics of the wave boundary layer, incipient motion of sand particles, and suspended sand concentration above sand ripples have been accumulated by laboratory measurements using monochromatic waves. However, they may not be used directly to explain sediment transport in nature owing to the presence of scale effects and wave randomness. Large wave flumes equipped with random wave generators and oscillatory flow tanks have been constructed during the last decade, making possible the study of sediment transport mechanisms with quasiprototype scales. Field investigations and numerical modeling have also been performed extensively with new measuring instruments and computer facilities.

In this chapter, recent progress in the field of coastal dynamics under random waves is reviewed. The role of long-wave components in the coastal dynamics and numerical modeling under random waves is described in detail. Sand transport in nearshore areas is usually treated as longshore transport and cross-shore transport. Longshore transport is considered to be significant in long-term beach evolution, whereas cross-shore transport is considered to be responsible for short-term or seasonal variation. Both types of transport are affected by wave randomness; frequency spectra are more associated with cross-shore sediment movement and longshore transport is more associated with directional spectra. Long waves of the trapped and leaky modes may also affect both cross-shore and longshore transport, playing a significant role in the formation of three-dimensional rhythmic topographies. As longshore transport was reviewed in detail by Komar [17], this chapter will concentrate on cross-shore transport mechanisms and resulting beach profile changes under random waves.

2. CLASSIFICATION OF BEACH PROFILES

It is widely accepted that beach profiles are classified as summer (accretive) profile and winter (erosional) profile (e.g., Komar [16], Horikawa [11]). The conditions determining which type will be developed have been studied by Johnson [13], Saville [35], and Iwagaki and Noda [12]. Several parameters, including offshore wave steepness, sediment grain size, and bed slope, have been identified as essential.

Dean [9] introduced the following parameters to determine the resultant beach profile types:

$$C_1 = \left(\frac{H_0}{L_0}\right)\left(\frac{\pi w_s}{gT}\right)^{-1} \quad (1)$$

$$C_2 = \frac{H_0}{w_s T} \quad (2)$$

where H_0 is the deep-water wave height, L_0 the deep-water wavelength, w_s the settling velocity of sand particle, T the wave period, and g the gravity acceleration. Dean [9] showed that the erosional beach profile developed for $C_1 > 1.7$ or $C_2 > 0.85$ and the accretive profile developed otherwise.

Kriebel et al. [19] carried out small-scale laboratory experiments to verify whether small-scale experiments are capable of reproducing prototype scale experiments. The sediment size in the small-scale model was determined so that the undistorted Froude model had the same C_2 value as the prototype scale experiments. They concluded that the beach profiles under the undistorted model were similar to those under prototype scale experiments and that the beach profiles in small-scale experiments and prototype experiments can be classified by the same C_1 or C_2 values, although the critical values they found were different from those originally proposed by Dean [9]. They suggested the difference is due to the scale effects incorporated in Dean's original criteria.

Kraus et al. [18] evaluated the applicability of various criteria on the basis of laboratory data in large wave tanks using monochromatic waves and field data. They examined the classification of beach profiles by the combination of various nondimensional parameters, including C_1 and C_2, and concluded that simple criteria successfully predict the beach profiles for both large wave tank data using monochromatic waves and field data using random waves if the mean wave height was used in the field application.

Sunamura and Horikawa [41] classified the beach profiles into three types—erosional, accretive, and the intermediate—and showed that the development of each type was dependent on the following C_3 value:

$$C_3 = \left(\frac{H_0}{L_0}\right)\left(\frac{D}{L_0}\right)^{-0.67} (\tan \beta)^{0.27} \qquad (3)$$

where D is the sediment grain size and $\tan \beta$ the initial beach slope. They showed that beach profiles change from accretive to intermediate and intermediate to erosional as C_3 increases, and found the boundary C_3 values to be 4 and 7, respectively, for the condition of laboratory monochromatic waves. The boundaries were 9 and 18 for field data, where significant wave height was used to calculate C_3.

Mimura et al. [22] performed a series of small-scale laboratory experiments to study beach profile change due to random waves. Figure 7.1 is the classification of beach

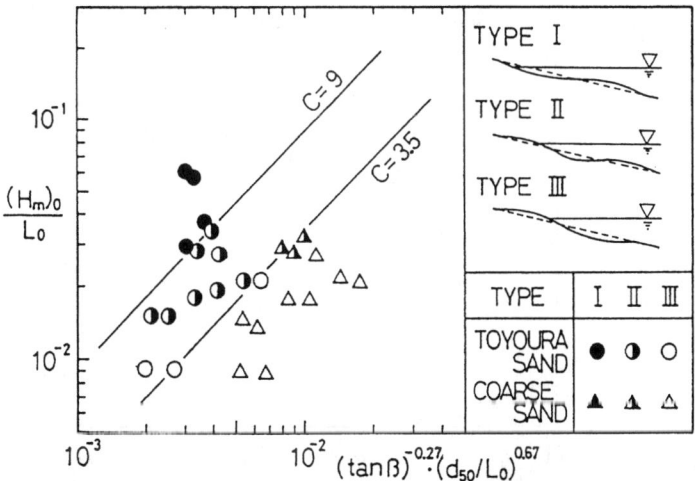

FIGURE 7.1. Classification of beach profiles for random waves in a laboratory flume [22].

profiles. Circular symbols are for 0.18 mm fine sand (Toyoura sand in the figure) and triangles are for 0.75 mm coarse sand. Mimura et al. [22] found that the boundary between accretive and intermediate profiles was expressed by $C_3 = 3.5$ and that between intermediate and erosional profiles by $C_3 = 9$ when the C_3 value was estimated on the basis of mean wave height. When the significant wave was used instead of the mean wave, the critical values of C_3 changed from 3.5 to 5 and 9 to 13, respectively. The correspondence between the boundary C_3 laboratory data values obtained with random waves and those obtained with monochromatic waves is better when the mean wave height is used, which agrees with the conclusion of Kraus et al. [18]. The difference between the boundary C_3 laboratory data values ($C_3 = 5$ and 13) obtained by random waves and those obtained by field data ($C_3 = 9$ and 18) is considered to be due to the scale effect and three-dimensionality in the field.

The use of simple criteria to determine beach profile types is thus found to be promising even for the condition of random waves to understand the macroscopic trend of beach profile change. However, in order to predict the dynamic response of beach profiles to variable sea states, more sophisticated models based on the physical processes of local sand transport are required.

Mimura et al. [22] also observed that beach transformation under random waves is different from that under monochromatic waves as follows:

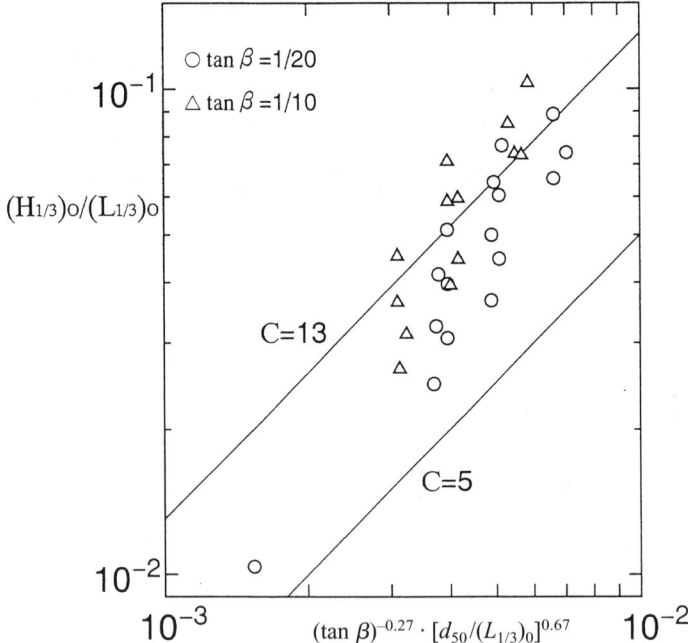

FIGURE 7.2. Classification of beach profiles for bichromatic grouping waves in a laboratory flume.

1. Intermediate beach profiles are more frequently developed
2. The speed of beach deformation is slower

These differences are considered to be closely associated with characteristics of randomly breaking waves. The variation in breaking points will directly affect the locations of sand suspension and the process of bar formation in the outer surf zone, and consequently affect the sediment transport in the inner surf zone since it influences the development of long waves and undertow.

Effects of the variation in breaking points and the development of long waves on cross-shore sediment movement can also be investigated in laboratory experiments by using bichromatic grouping waves. Figure 7.2 shows the classification of beach profiles due to bichromatic grouping waves in laboratory experiments, where the second-order Stokes wave theory was used to generate the incident wave. In all runs of experiments shown in Figure 7.2, the resultant beach profiles were of the intermediate type, even for large C_3 values, indicating that the intermediate beach profiles are more dominant under grouping waves than random waves. This suggests that the long wave, which is more pronounced in grouping waves, tends to contribute to the berm formation near the shoreline.

3. NEAR-BOTTOM VELOCITIES UNDER RANDOM WAVES

3.1. Asymmetry in Velocity Variation

The estimation of near-bottom velocity is important, since it provides fundamental information on sand transport mechanisms. The near-bottom velocity u of random waves involves fluctuation over a wide range of frequencies and is expressed as follows:

$$u = U + u_w + u_l + u_t \qquad (4)$$

where U is steady current component or undertow, u_w the short-wave components originally dominated in the incident wave spectrum, u_l the long-wave components incident as free waves and/or induced through the nonlinear interaction of short-wave components, and u_t turbulence generated by wave breaking as well as that generated in the bottom boundary layer. The onshore velocity is defined to be positive in this chapter. In order to predict the velocity variation with good accuracy, proper modeling for each component is required.

In oscillatory flow tank tests using asymmetric oscillations, Sato and Horikawa [32] and Ribberink and Al-Salem [28] confirmed that the net sand transport rate was closely correlated with the velocity amplitude as well as the asymmetry in velocity variations. The asymmetry in velocity variation under shallow-water waves is realized in various ways, as shown in Figure 7.3. Undertow itself is considered to be an asymmetry superimposing a bias in near-bottom velocities. Wave nonlinearity will produce a skewed variation with sharp crest and flat trough at short-wave velocities. Wave nonlinearity will also tend to tilt the wave in the surf zone, producing asymmetry in the acceleration. Phase-dependency of long waves on the short-wave envelope will produce other type of asymmetry, which will influence the slowly varying transport of sediments. Turbulence may induce another asymmetry when it has a specific phase-dependency on wave-orbital velocities. These asymmetries are mixed together on natural beaches and influence sand transport in various manners. It is considered, for example, that the asymmetry in short-

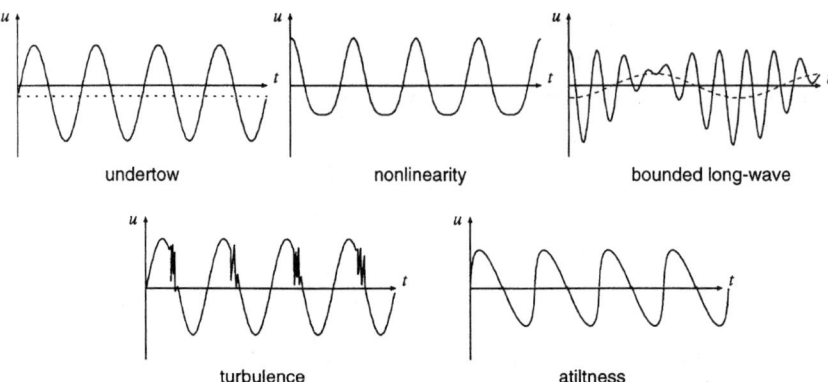

FIGURE 7.3. Various asymmetry in velocity variation.

wave components will primarily influence sediment transport near the bed, whereas the asymmetry due to bound long waves and turbulence will influence the transport of suspended sediment.

Many models have been presented to estimate near-bottom velocities under random waves. The first model to be established was the model that simulates the transformation of short waves due to random wave breaking. Undertow and long-wave components are consequently modeled on the basis of the short-wave computation. The estimates of u_w, U, and u_l are briefly described herein. Turbulence will not be discussed in this chapter, since the definition of turbulence in random wave fields is still uncertain, although it plays an important role in sand suspension. It should be noted here that nonlinear wave models were also used recently in beach evolution modeling, in which long wave and steady components were solved simultaneously with short-wave computation [34].

3.2. Short Waves

Many models have been proposed to estimate velocity variation due to short-wave component both in the frequency domain and in the physical domain [3]. In the frequency domain, the amplitude $\hat{u}_w(f)$ of near-bottom velocity variation under random waves is related to the amplitude $\hat{\eta}(f)$ of surface elevation as

$$\hat{u}_w(f) = \frac{2\pi f}{\sinh kh} \hat{\eta}(f) \qquad (5)$$

where k is the wave number at frequency f and h is the water depth. Since Equation 5 is based on the linear wave theory, no asymmetry is involved in the simulated velocity variation as long as the surface elevation is simulated as a sum of sinusoidal waves. In order to simulate asymmetric velocities, the interaction of each wave component should be incorporated. Sato and Horikawa [32] applied the second-order Stokes wave theory to the estimation of asymmetric near-bottom velocity under random waves. However, the application is limited to nonbreaking waves in relatively deep water.

Models in the physical domain have been more frequently used in beach evolution modeling since they can more easily incorporate the wave-height decay due to wave breaking in surf zone. In the physical domain models, the cross-shore distribution of short-wave heights in surf zone was computed by introducing heuristic wave decay models [31, 39] or was computed numerically on the basis of depth-integrated mass and momentum equations [45, 33]. The asymmetry in near-bottom velocity was then introduced with various nonlinear wave theories, such as stream-function theory [31], cnoidal wave theory [33], and vocoidal theory [39]. In a recent model based on the Boussinesq equation, asymmetric velocity variations were simulated directly [34].

The asymmetry in velocity due to short waves is likely to produce net onshore transport, since more sand will be transported shoreward by larger onshore velocity under the wave crest. However, the direction will be reversed for the transport of fine sand above rippled beds [32].

3.3. Undertow

Undertow is breaking-induced, offshore-directed flow below trough level that compensates onshore mass flux near the surface due to the propagation of breaking bores. The estimation of undertow is important in modeling offshore transport [7, 40]. Although sophisticated models predict the vertical profile of undertow velocity on the basis of eddy viscosity assumptions [43, 25], the undertow velocity near the bottom is the most important in sand transport modeling, since suspended sand concentration becomes exponentially large near the bottom.

The steady component U of near-bottom velocity can be computed as a compensating return flow of onshore mass flux due to breaking bores. When we assume that the cross-sectional area of a breaking bore is proportional to the square of wave height and that the compensating flow is uniform over the water depth, we can roughly estimate the steady velocity U by

$$U = -A\frac{H^2}{dT} \qquad (6)$$

where H is the local wave height, d the mean water depth, and T the wave period. The coefficient A was determined to be *4* on the basis of laboratory data for monochromatic waves [33]. The steady velocity under random waves can be estimated from the sum of mass fluxes due to individual breaking waves. For random waves, H^2 in Equation 6 is replaced by the sum of H^2 of breaking waves, while T is replaced by the sum of all wave periods. Figure 7.4 shows a comparison of undertow velocity under random waves. In spite of rough assumptions introduced, the model reproduces the cross-shore distribution of undertow quite accurately. It should be noted here that wave randomness tends to smooth out various discontinuities at the breaking point of monochromatic waves. The estimation of undertow by Equation 6 is a typical example; it predicts a large gap at the breaking point for monochromatic waves but with rather smooth distribution, as in Figure 7.4 for random waves.

3.4. Long Waves

Long waves or infragravity waves, with periods several times as long as short waves, become pronounced as they increase their heights near the shoreline, where short waves decrease their heights owing to the energy dissipation due to wave breaking. Mechanisms of

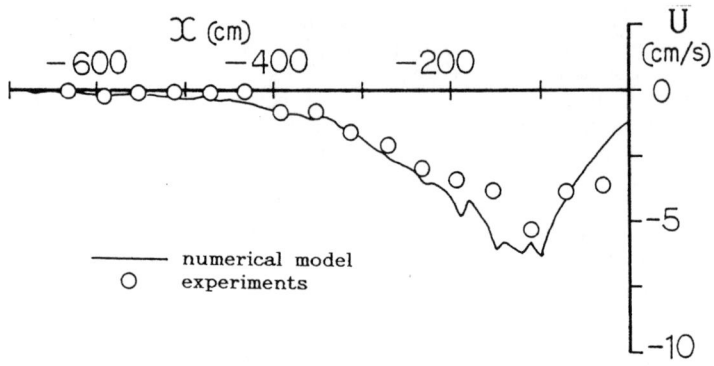

FIGURE 7.4. Distribution of undertow velocity [33].

the development of long waves have been investigated by many researchers [44, 21, 23]. Three major mechanisms are suggested:

1. Incidence of free long waves
2. Temporal variation of set-up due to time-varying breaking point
3. Transition of the bounded long wave to free wave as a result of breaking of large waves in wave groups

The second and third mechanisms are associated with the grouped nature of the incident wave and can be modeled by the long-wave equation with a radiation stress term as a forcing term [33, 29]. Governing equations for simulating the development of the long-wave component are

$$\frac{\partial \eta_l}{\partial t} + \frac{\partial}{\partial x}(u_l d) = 0 \tag{7}$$

$$\frac{\partial u_l}{\partial t} + u_l \frac{\partial u_l}{\partial x} + \frac{1}{\rho d}\frac{\partial S_{xxl}}{\partial x} = -g\frac{\partial \eta_l}{\partial x} - \frac{\tau_{xl}}{\rho d} \tag{8}$$

where ρ is the density of water, S_{xxl} the radiation stress, and τ_{xl} the bottom shear stress. The subscript l indicates that the term corresponds to the long-wave component. The radiation stress is estimated from the results of short-wave computation as

$$S_{xxl} = \rho g (\eta_w^2)_l \left(\frac{1}{2} + \frac{2kd}{\sinh 2kd} \right) \tag{9}$$

where η_w is the surface elevation due to the short-wave component and $(\eta_w^2)_l$ denotes the low-pass-filtered variation of η_w^2. The symbol k is the wave number for the significant wave computed by the linear theory. The bottom shear stress τ_{xl} can be computed by the following model proposed by Nishimura [24] for the wave-current field:

$$\tau_{xl} = \rho C_f (w + \hat{u}_w^2/w) u_l \tag{10}$$

$$w = (u_l + 2\hat{u}_w/\pi + |u_l - 2\hat{u}_w/\pi|)/2 \tag{11}$$

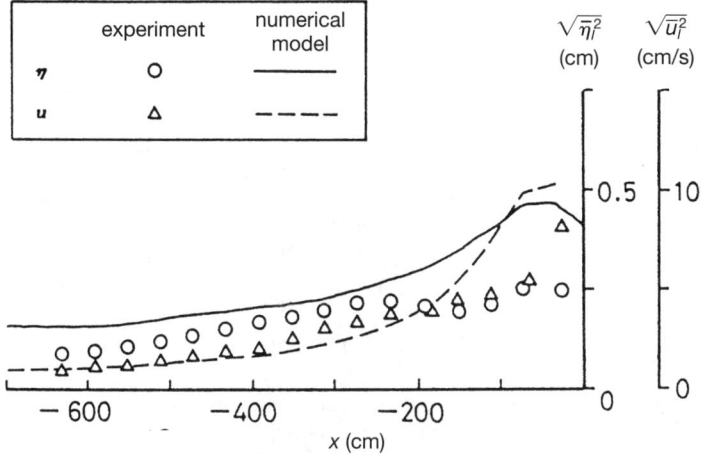

FIGURE 7.5. Development of long-wave component [33].

where C_f is the friction coefficient and $\hat{u}_w$ the amplitude of the near-bottom velocity due to the short-wave component.

The incident long wave at the offshore boundary can be estimated by the results of short-wave computation by

$$\eta_l = -\frac{S_{xxl}}{\rho(gd - C_g^2)} + \text{constant} \tag{12}$$

where the value of the constant is determined so that the mean water level of the incident wave train becomes zero.

Based on the results of short-wave computation, the development of the long-wave component can be estimated by the numerical integration of Equations 7 and 8. Figure 7.5 shows a comparison of cross-shore distributions of the root-mean-square amplitudes of η_l and u_l estimated on the basis of the model described in above. The development of the long-wave component towards the shoreline is simulated, although the model tends to overestimate the long wave in the vicinity of the shoreline. The phase-dependency of the simulated long wave on the short-wave envelope, which is essential in sediment transport, will be described in the next section.

4. EFFECTS OF LONG WAVES ON CROSS-SHORE SEDIMENT TRANSPORT

4.1. Suspended Sediment Flux

Effects of long waves on cross-shore sand transport have been studied. The oscillatory flow induced by long-wave components exert a greater influence on suspended loads than on bed loads. The contribution of long waves to suspended sand transport is evaluated by the suspended sediment flux, expressed as the average of the product of instantaneous velocity and sediment concentration as follows:

$$\overline{uc} = \overline{UC} + \overline{u'c'} \qquad (13)$$

where u and c are cross-shore velocity and suspended sand concentration, respectively, U and C are average time, and u' and c' are fluctuating components, respectively. The contribution $\overline{u'c'}$ of fluctuating components can be expressed as the sum of contributions from various frequencies. The relative contribution at different frequencies can be estimated by the cospectrum, cosp_{uc} between u and c, since it gives the cross-product between them as a function of frequency. The suspended sediment flux is therefore expressed by

$$\overline{uc} = \overline{UC} + \int_0^\infty \text{cosp}_{uc} df \qquad (14)$$

Figure 7.6 shows a schematic diagram of suspended sand transport under bounded long waves. It was suggested by Shi and Larsen [36] that long waves bounded by wave groups would result in seaward transport. The mechanism is based on the reasonable assumption that c' is large under a series of large waves when u' is directed offshore. The sediment flux under large waves thus contributes significantly to seaward transport. As the velocity variation u' due to bounded long waves is directed onshore under small waves when c' is small, the net sand transport will be directed offshore. It should be noted, however, that the seaward transport is only realized by bounded long waves. As discussed in the previous section, long waves of various types are coexistent in nature, such as free incident waves, reflected waves from shoreline, and edge waves, which tend to obscure the seaward transport contributed by bounded long waves.

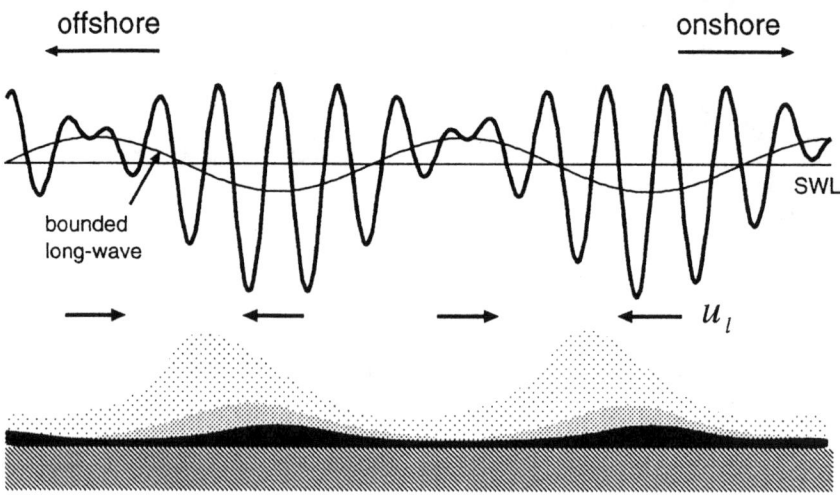

FIGURE 7.6. Schematic diagram of suspended sand transport under a bounded long wave.

4.2. Suspended Sediment Flux Offshore and Seaward of the Surf Zone

Sato and Horikawa [32] found on the basis of experiments with an oscillatory flow tunnel that the direction of sediment transport was determined by the asymmetry in velocity history. They simulated near-bottom velocity variations due to long-wave components by using the second-order Stokes wave theory. The simulated velocities showed the typical phase-dependency of long waves bounded to short-wave envelopes. Their data indicated that the net sand transport tended to be directed offshore as the relative magnitude of bounded long waves increased.

Figure 7.7 shows an example of field data obtained at the Ajigaura Coast facing the Pacific Ocean, Japan. The figure shows low-pass-filtered variations of near-bottom cross-shore velocity u_l (top), velocity squared $(u^2)_l$ (middle), and suspended sand concentration c_l (bottom) measured at 10 cm above the bottom near the breaking point. As the significant wave period was 8s, the long-wave component was extracted by using a numerical low-pass filter with cut-off frequency f_c of 0.033 Hz. It is apparent that the temporal variation of c_l is in phase with $(u^2)_l$, indicating that sand suspension was more pronounced under a series of large waves.

Figure 7.8 illustrates the joint probability density between u_l and $(u^2)_l$ and that between c_l and $(u^2)_l$ for the data shown in Figure 7.7. The bracket < > implies that the variable was respectively normalized by the standard deviation. A strong positive correlation is noticed between suspended sand concentration c_l and wave energy level $(u^2)_l$ but the negative correlation between the cross-shore velocity u_l and wave energy level $(u^2)_l$ is less obvious. This indicates that only part of the long waves are bounded to the wave group and long waves generated by other mechanisms are also superimposed. Figure 7.9 shows the cospectrum between u and c. Suspended sediment transport due to the short-wave component is directed shoreward, while that due to long-wave component in a range 0.01 Hz $< f <$ 0.05 Hz is directed seaward, which suggests that bounded long waves are dominant in that frequency band.

The shoreward transport by short waves and the seaward transport by long waves were also reported by Osborne and Greenwood [26]. Figure 7.10 shows a cospectrum of the near-bed cross-shore velocity and suspended sand concentration observed seaward of the surf zone of a nonbarred shore face. The figure shows that short-wave components produce strong onshore transport at wind-wave frequencies (0.1 $< f <$ 0.5 Hz), whereas long-wave components produce offshore transport at frequencies (0 $< f <$ 0.1 Hz) lower than the wind-wave band. It should be noted however that the contribution of long waves is weak compared to the short-wave contribution. It is also noted that the contribution of long waves may be dependent on wave conditions. It may temporarily change its direction in swell-dominated waves and at various stages in a storm [46, 47].

4.3. Suspended Sediment Flux in the Surf Zone

Figure 7.11 shows the cospectra between u and c estimated in a random-wave experiment. In the experiment, cross-shore velocities and suspended sand concentration at 0.5 cm above bed were measured for the condition of random waves breaking on a 1/20 uniformly sloping bed filled with 0.2 mm sand. The dimension of the incident wave was $(H_0)_{1/3} = 11.7$ cm and $T_{1/3} = 1.05$ s. The value of x in Figure 7.11 denotes the distance from the still-water shoreline. The cospectrum at $x = -30$ cm, located in the vicinity of the shoreline, shows onshore-directed transport for the whole range of frequencies. The

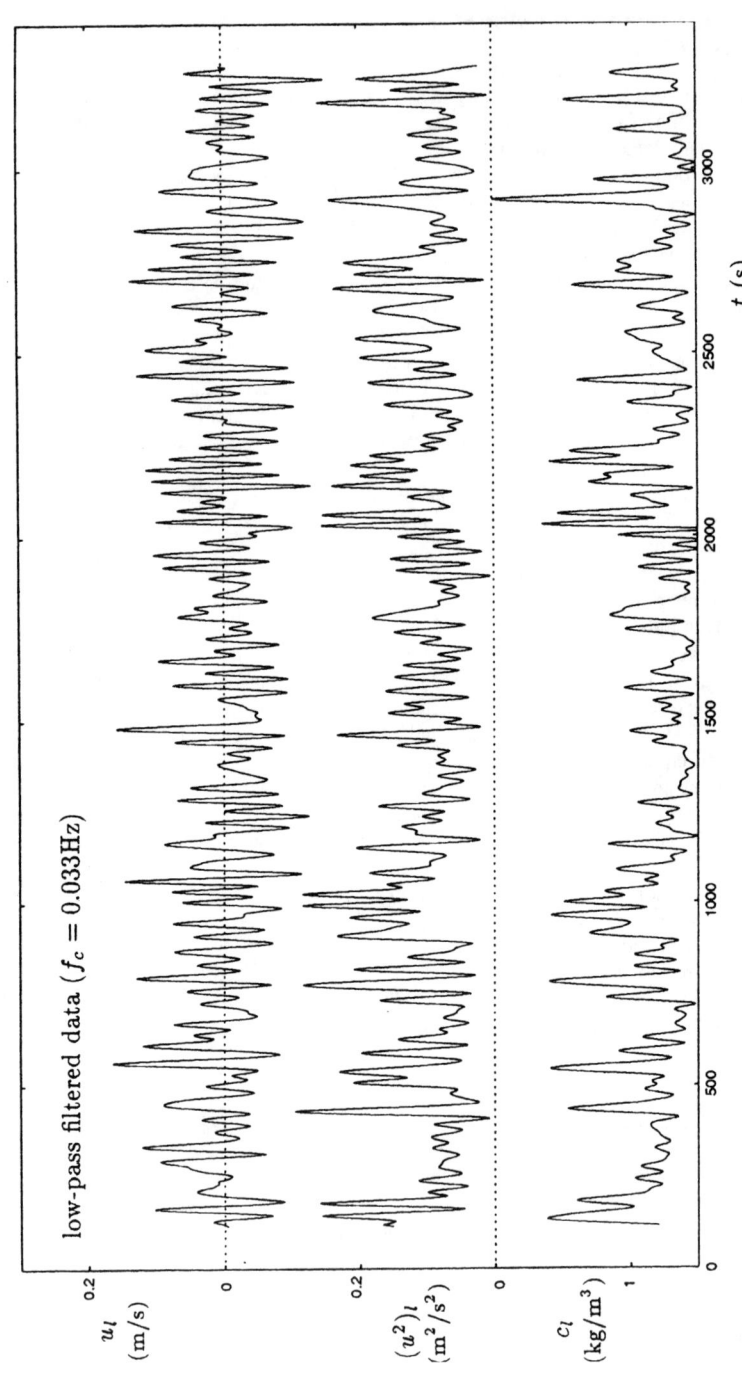

FIGURE 7.7. Low-pass-filtered variations of u, u^2, and c.

SEDIMENT TRANSPORT AND BEACH PROFILE CHANGE

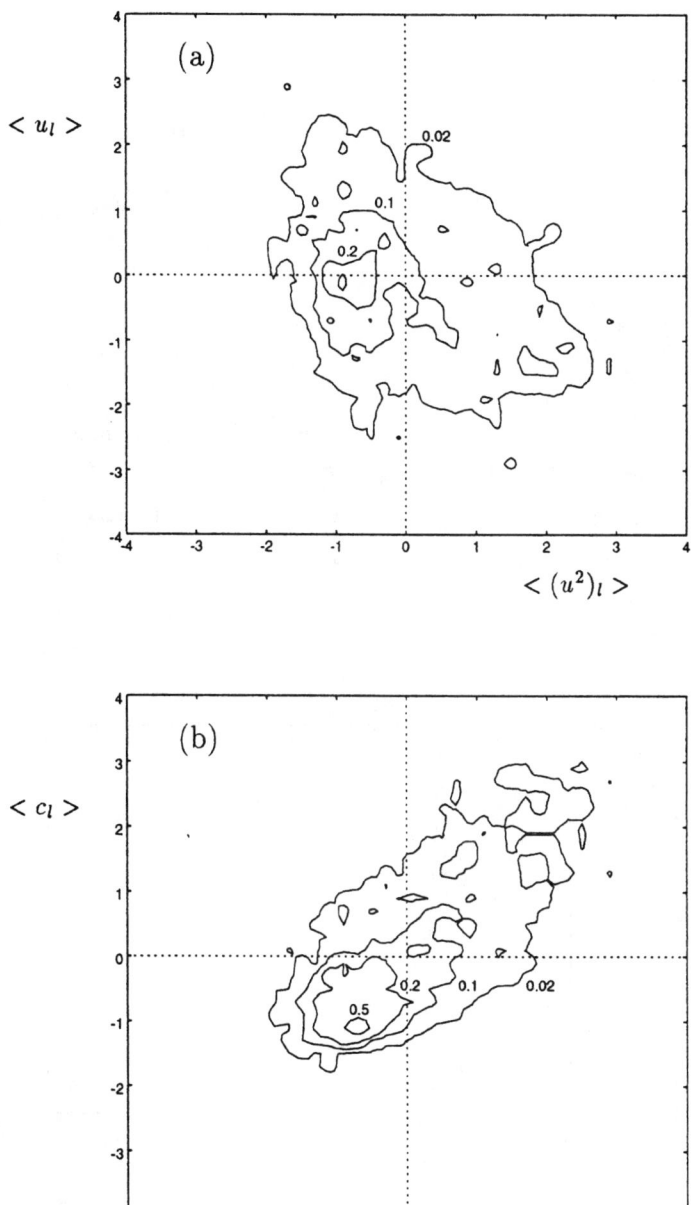

FIGURE 7.8. Probability density of u, u^2, and c.

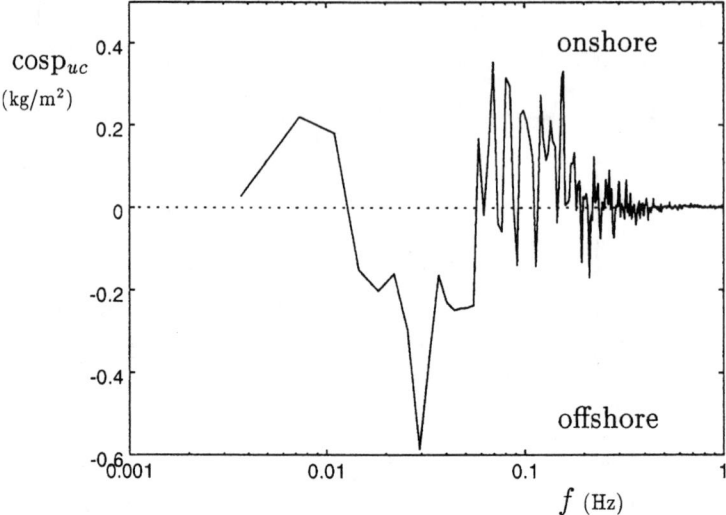

FIGURE 7.9. Cospectrum between u and c.

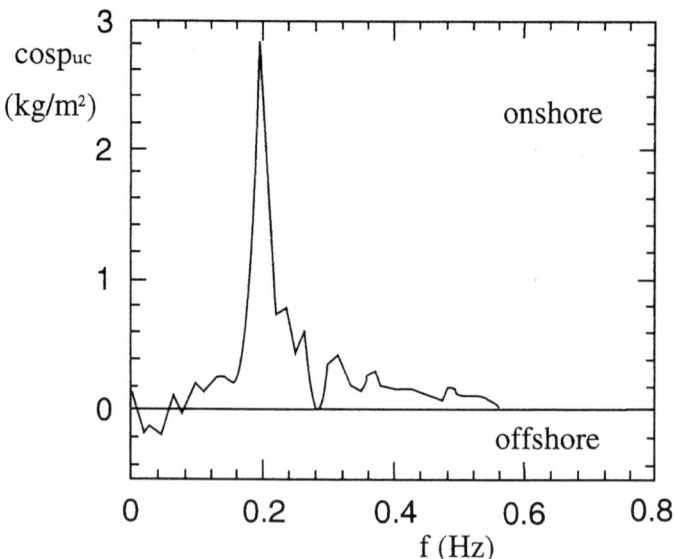

FIGURE 7.10. Co-spectrum between u and c [26].

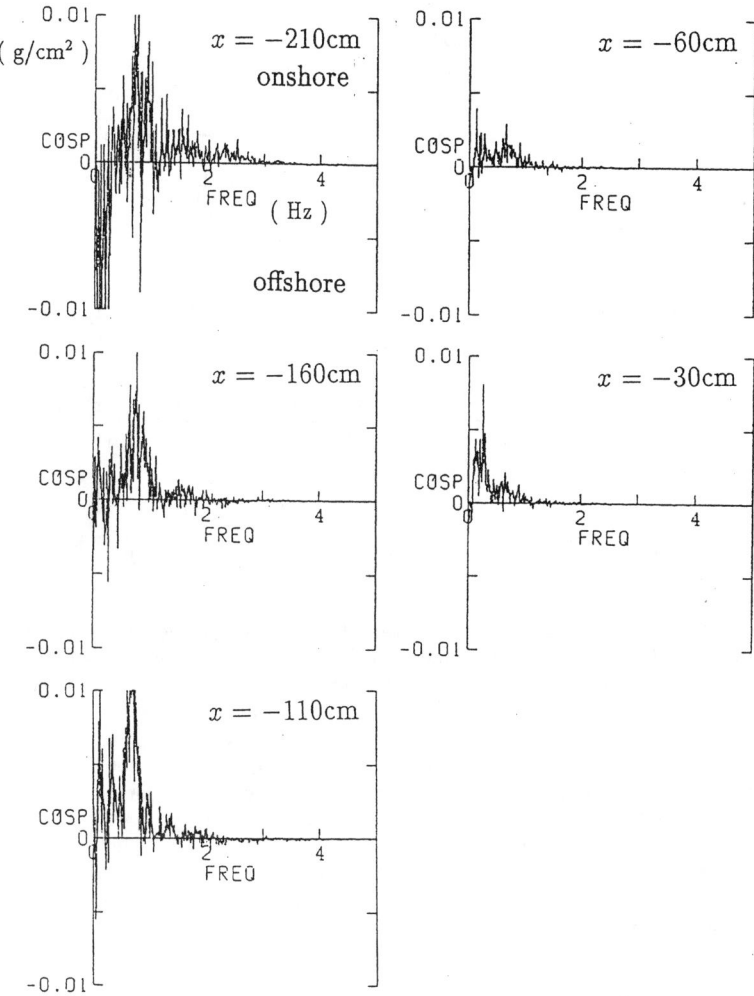

FIGURE 7.11. Cospectrum between u and c in a laboratory wave flume.

contribution by the long-wave component ($0 < f < 0.25$ Hz) is directed onshore near the shoreline but gradually tends to be directed offshore with increasing distance from the shoreline. The contribution of each frequency band to the total sediment flux is illustrated in Figure 7.12, in which the steady component is estimated by UC and the contributions of short waves $(uc)_w$ and long-waves $(uc)_l$ are respectively estimated by

$$(uc)_w = \int_{f_c}^{\infty} \text{cosp}_{uc} df \tag{15}$$

$$(uc)_l = \int_{-\infty}^{f_c} \text{cosp}_{uc} df \tag{16}$$

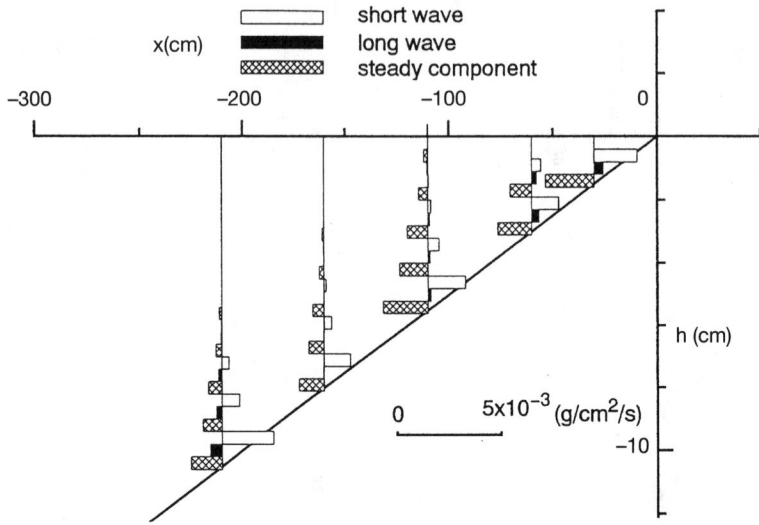

FIGURE 7.12. Contributions of various components to suspended sediment flux.

where f_c denotes the critical frequency between the long- and short-wave components, determined to be 0.25 Hz. Figure 7.12 indicates that the steady component, always directed offshore, demonstrates the largest contribution at all the points. The contribution of the short-wave components is always directed onshore and is almost as large as the contribution of the steady component. The contribution of the long-wave component is relatively small but directed offshore in the outer surf zone and onshore near the shoreline.

Beach and Steinberg [5] estimated the cospectra between u and c on the basis of field data obtained at the east and west coasts of the United States. Figure 7.13 shows cospectra for data obtained near the bottom of the outer surf zone (Figure 7.13a) and the inner surf zone (Figure 7.13b). The contribution of the long-wave component is directed offshore in the outer surf zone and onshore in the inner surf zone, which is consistent with laboratory measurements shown in Figure 7.11.

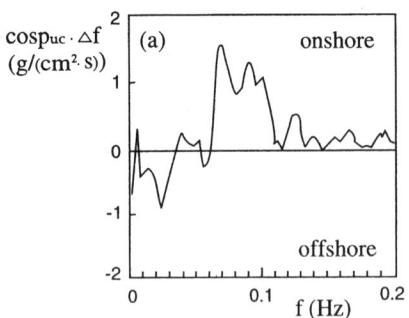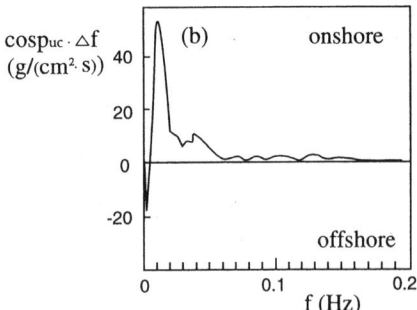

FIGURE 7.13. Cospectrum of u and c [5].

It is interesting that the phase-dependency of long waves on wave groupedness varies in the cross-shore direction, since the onshore transport due to long waves observed near the shoreline in Figures 7.11, 7.12, and 7.13b is considered to be associated with the phase-dependency. Sato and Mitsunobu [33] calculated the correlation coefficient C_{Hu} between the wave height H of short-wave component and the cross-shore velocity u_l of long-wave component through laboratory experiments as well as a numerical model. Figure 7.14 shows the cross-shore distribution of C_{Hu} for random waves breaking on a 1/20 slope. The circles denote the estimation based on laboratory measurements and the solid line denotes the correlation calculated by the numerical models described in the previous section. The correlation is negative in the offshore region, indicating that u_l is directed offshore under large waves, but changes to positive near the shoreline. The change in the sign of the correlation factor between the long-wave and the short-wave envelope in the surf zone was also described in Abdelrahman and Thornton [2] and Roelvink and Stive [31]. The reversal of the phase-dependency may be explained by the interaction between short waves and the currents induced by the long-wave component. Since the magnitude of long wave increases toward the shoreline and the celerity of short waves decreases toward the shoreline, the transformation of short waves near the shoreline will be more strongly influenced by the currents induced by long-wave components. Near the shoreline, it is thought that large waves are likely to ride on the following currents rather than on the opposing currents, thus resulting in the positive correlation between H and u_l. Owing to the positive correlation between H and u_l, onshore transport of sediment is considered to occur near the shoreline, where sediments are suspended under large waves when the velocity induced by long waves is shoreward. The onshore transport near the shoreline realized by the reversed phase-dependency is considered to exert strong influences on the berm formation frequently observed under random waves. It should be noted, however, that the phase-dependency and the contribution of long waves to sediment transport are highly variable in space on natural beaches, especially on beaches with multiple sand bars [1, 27].

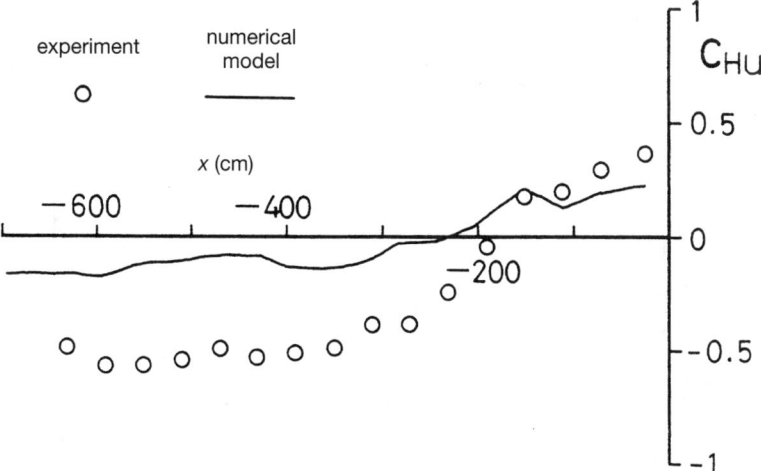

FIGURE 7.14. Correlation between short-wave height and velocity due to long-wave component [33].

4.4. Effect of Long Waves on Beach Profile Change

Effects of long waves on beach profile change have been studied through field investigations and laboratory experiments. In contrast to the accretive process suggested by the onshore transport near the shoreline described in the previous section, Katoh and Yanagishima [15] observed that berms were eroded by infragravity waves incident in the early stages of storms, as shown in Figure 7.15. Figure 7.15a shows the variation of offshore

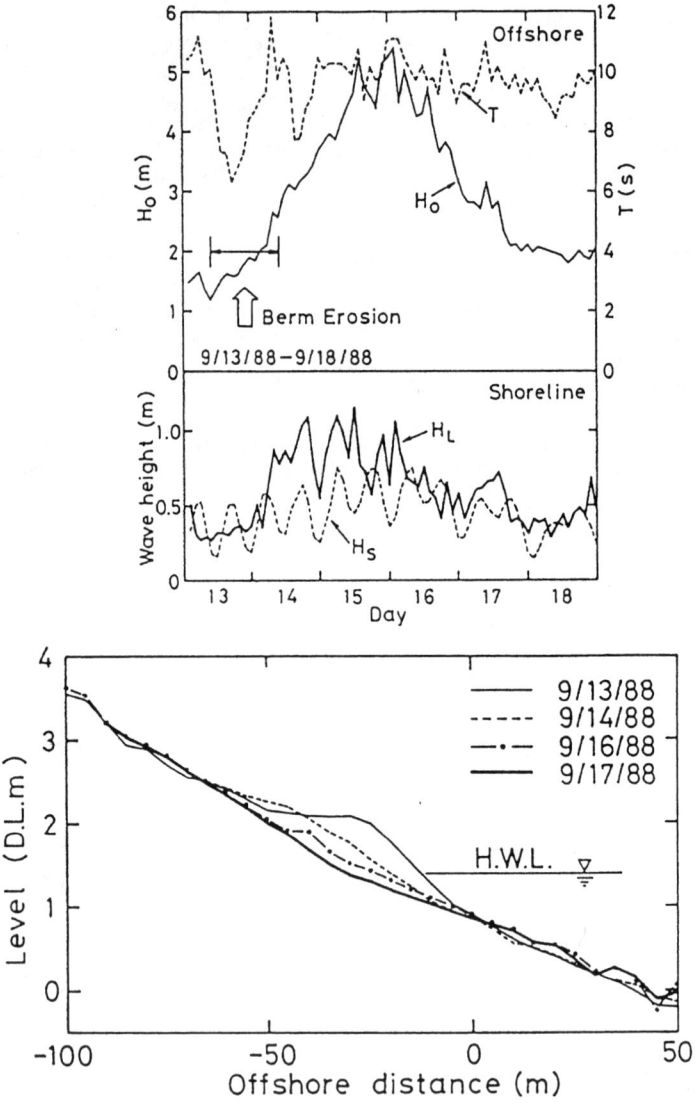

FIGURE 7.15. Berm erosion due to infragravity waves [15].

wave height H_0 and period T, as well as wave heights H_L of the long-wave component and H_S of the short-wave component observed near the shoreline. It is noticed that berm erosion occurred at the early stage of the storm, when the offshore waves were not well developed but significant long waves were incident on the beach. The difference between these conflicting results is considered to be due to the magnitude of the long-wave component. In Katoh and Yanagishima's [15] observation, the velocity due to long waves was very large near the shoreline, since observations were made when free long waves with large heights were incident prior to the storm, which flushed the berm away by large shear stress induced by long wave itself. In the mechanism described in the previous section, on the other hand, the role of long waves is secondary, transporting sand suspended by short waves.

Long waves are also considered to be associated with the formation of sand bars. Roelvink and Stive [31] showed by a numerical model and Dally [8] showed by laboratory experiments that the location of a sand bar was shifted when the development of long waves was suppressed. The relationship between standing long waves and the development of offshore bars was discussed in Short [38], Katoh [14], and Aagaard and Greenwood [1].

5. NUMERICAL MODELING OF BEACH PROFILE CHANGE

The numerical modeling of nearshore waves and the transformation of beach profiles is one of the most important subjects in the field of coastal engineering. Among various models presented so far, deterministic models, which are based on the modeling of physical processes in the surf zone, are promising since they have a potential ability to simulate dynamic beach evolution for a wide range of conditions. The deterministic models are mostly composed of the following submodels:

1. Wave transformation, including shoaling and breaking
2. Velocity field, especially near the bottom
3. Estimation of local sand transport rate
4. Bottom topography change based on mass conservation of sediments.

The change in beach profile will influence wave transformation, hence closed-loop iterations are required. In this section, the application of sand transport models under random waves will be described. The intercomparison of beach profiles predicted by various deterministic models was reviewed elsewhere [30].

Sato and Mitsunobu [33] tested the applicability of sediment transport formulas to random-wave conditions for four models—those proposed by Bailard [4], Sunamura [42], Shibayama and Horikawa [37], and Dibajnia and Watanabe [10]. The sediment transport formula proposed by Sunamura [42] was derived on the basis of laboratory measurements with monochromatic waves, and was expressed in terms of the Ursell parameter U_r and nondimensional velocity amplitude Ψ'' near the bottom, as follows:

$$U_r = \frac{gHT^2}{d^2}, \qquad \Psi'' = \frac{H^2}{sdD} \qquad (17)$$

where H is the wave height, T the wave period, d the water depth, D the sediment diameter, and s the specific gravity of sand particles in water. The net sand transport rate was expressed by

$$\frac{Q}{w_s D} = \begin{cases} 0 & : \Psi'' \leq 17 \\ -3.00 \times 10^{-6} U_r (\Psi'' - 17)^2 & : \Psi'' > 17, U_r \leq 50 \\ -3.75 \times 10^{-1} U_r^{-2} (\Psi'' - 17)(\Psi'' - 0.048 U_r^{1.5}) & : \Psi'' > 17, 50 < U_r \leq 230 \\ -6.9 \times 10^{-8} U_r^{-0.2} \Psi''(\Psi'' - 0.13 U_r) & : \Psi'' > 17, U_r \geq 230 \end{cases} \quad (18)$$

where Q is the volumetric net sand transport rate and w_s the settling velocity of sand particles.

Dibajnia and Watanabe's [10] model is based on the assumption that the net sand transport rate is proportional to the product of excessive bottom shear stress and velocity amplitude. The net sand transport rate is expressed by a sum of wave-induced transport and turbulence-induced transport due to wave breaking, as follows:

$$\hat{u} = \frac{\pi H}{T \sinh kd'} \qquad \tau = \frac{1}{2} \rho f_w \hat{u}^2 \quad (19)$$

$$Q = \left(A_w \frac{\tau - \tau_c}{\rho g} \hat{u} + A_{wb} \frac{f_D E}{\rho g} \right) F_D \quad (20)$$

$$F_D = \tanh \left[\kappa_d \frac{(f_w \Pi)_c - f_w \Pi}{(f_w \Pi)_c} \right], \qquad \Pi = \frac{\hat{u}^2}{sgD} \frac{d}{L_0} \quad (21)$$

where τ is the bottom shear stress, f_w, the wave friction factor, τ_c the critical shear stress for the movement of sand particles, $\hat{u}$ the velocity amplitude, f_D the energy dissipation factor due to wave breaking, E the wave energy density, and F_D is a function that determines the direction of net sand transport. Both the Sunamura [42] and Dibajnia and Watanabe [10] models express net sand transport due to individual waves in terms of parameters of individual waves. In order to apply them to random waves, the total net sand transport rate is estimated by a sum of contributions separately calculated for individual waves. The estimation of the net sand transport rate under random waves by the superposition of individual wave contributions was also adopted by Larson [20] by using a different sand transport model.

Bailard's [4] model assumes that sand transport rates are expressed in terms of moments of near-bottom velocity. The net sand transport rates are expressed as a sum of bed load and suspended load, as follows:

$$Q = \frac{\overline{q(t)}}{(\rho_s - \rho) g} \quad (22)$$

$$q(t) = \rho C_f \frac{\epsilon_B}{\tan\phi} \left[u(t)|u(t)|^2 - \frac{\tan\beta}{\tan\phi} |u(t)|^3 \right] + \rho C_f \frac{\epsilon_S}{w_s} \left[u(t)|u(t)|^3 - \frac{\epsilon_S}{w_s} \tan\beta |u(t)|^5 \right] \quad (23)$$

where $q(t)$ is the instantaneous sand transport rate in immersed weight, C_f the drag coefficient, ϵ_B and ϵ_S coefficients for bed load, and suspended load respectively, $\tan\phi$ the angle of repose of sand particles.

Shibayama and Horikawa's [37] model is based on the Shields parameter Ψ and the ratio $\hat{u}/w_s$ of velocity amplitude to settling velocity for a half wave cycle. The sand transport rate during the half wave cycle is expressed by the following formula, depending on sand transport types also categorized by Ψ and $\hat{u}/w_s$.

$$\frac{Q}{w_s D} = \begin{cases} 19\Psi^3 & : \text{bed load} \\ (1 - \alpha - N\alpha) 19\Psi^3 & : \text{bed load–suspended load transition} \\ -19N\Psi^3 & : \text{suspended load} \\ 19\Psi^3 & : \text{sheetflow} \end{cases} \quad (24)$$

SEDIMENT TRANSPORT AND BEACH PROFILE CHANGE   7.21

where $N(=d_0/\lambda)$ is the ratio of orbital diameter d_0 of wave-induced particle motion near the bottom to the wavelength λ of sand ripples, and α is a parameter determined by Ψ and $\hat{u}/w_s$. Sato and Mitsunobu [33] modified the Shibayama and Horikawa [37] model, since their model predicted unrealistic change of Q when the transport type changed from suspended load to sheet flow. The modified sand transport rate in the transition region is expressed by

$$\frac{Q}{w_s D} = (1 - \Lambda)19\,\Psi^3 + \Lambda(-19N\Psi^3) \qquad (25)$$

$$\Lambda = [1 - (\Psi_{rms}/0.6)^2] \times \min(1, 2/N) \qquad (26)$$

where Ψ_{rms} is the Shields parameter based on the root-mean-square amplitude of velocity variation near the bottom.

In the Bailard [4] and Shibayama and Horikawa [37] models, net sand transport rates are estimated on the basis of velocity variation near the bottom. The direction and the amount of the net sand transport rates are therefore sensitive to the asymmetry in velocities discussed in the previous sections. A numerical model proposed by Sato and Mitsunobu [33], which simulated the wave height decay in the surf zone, the development of undertow, and the interaction of u_w and u_l under random waves, as described in the previous section, was applied to the estimation of cross-shore distribution of net sand transport rate under a random-wave condition on a uniform 1/20 slope, shown in Figure 7.16.

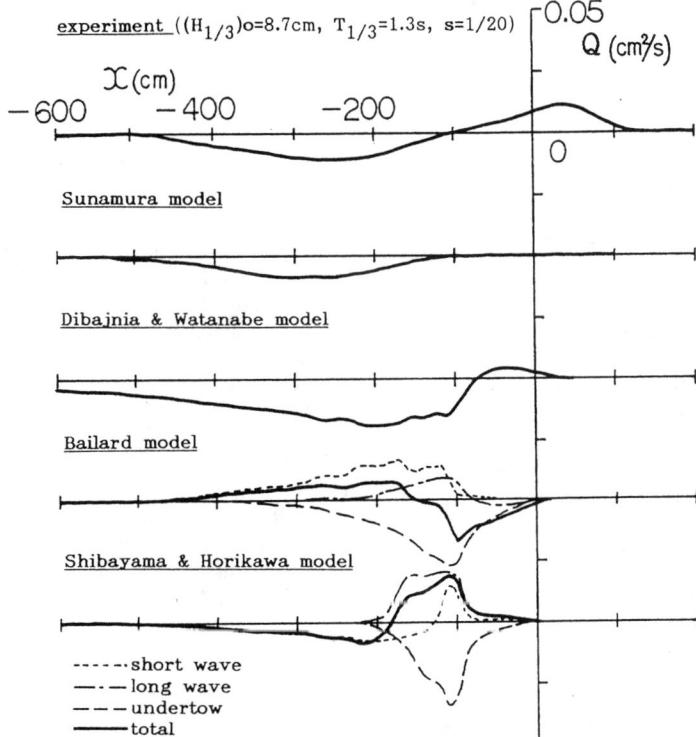

FIGURE 7.16. Cross-shore distributions of net sand transport rate [33].

On the top of Figure 7.16 is shown the net sand transport rate calculated from laboratory measurements [22]. It is noticed that seaward transport is dominant around the breaking point and shoreward transport is dominant near the swash zone. The resultant beach profile in this case is that of the intermediate type. In the offshore region, the Sunamura [42], Dibajnia and Watanabe [10], and Shibayama and Horikawa [37] models give similar seaward transport with slight differences in their magnitudes. The Bailard [4] model, however, predicts shoreward transport in the offshore zone. This is considered to be due to the energetic approach, which neglects the reverse transport of fine sand caused by vortices over sand ripples. Regarding surf zone transport, the Sunamura [42] model fails to estimate shoreward transport, but both the Dibajnia and Watanabe [10] and Shibayama and Horikawa [37] models predict shoreward transport, although the location of the maximum onshore transport is shifted to seaward.

Since the Bailard [4] and Shibayama and Horikawa [37] models are based on temporal variations of near-bottom velocities, it is interesting to identify contributions of various components. As the sand transport process is nonlinear, the contribution Q_l of the long-wave component is calculated, for example, by the following equation,

$$Q_l = Q(U + u_l + u_w) - Q_{l0}(U + u_w) \tag{27}$$

where Q_{l0} is the net sand transport rate computed by using velocity variations without long-wave components. It is noticed in Figure 7.16 that the steady component contributes to seaward transport and the long-wave component mainly to shoreward transport. The difference between these two models is primarily due to the difference in the short-wave contribution. It is interesting that the contribution by the long-wave component demonstrates similar distributions in both models. In order to determine the contribution of each component on natural beaches, further study is needed on the application of other sand transport models to random-wave conditions and on the comparison with large-scale random-wave experiments.

6. SUMMARY

Various models of beach transformation due to random waves were reviewed. They included simple parameter methods and sophisticated deterministic models composed of numerical submodels for waves, currents, and sediment movement. It was shown that deterministic models can simulate the beach profile change due to random waves provided that proper models are used in the estimation of the asymmetry in nonlinear wave orbital motion and undertow. The role of long waves in cross-shore sediment transport was also described.

Although deterministic models are still in the developing stage, they are promising, since the validity of the submodels can be checked and improved for each physical process. Among the physical processes that crucially affect sediment movement, sediment suspension due to wave breaking needs to be studied intensively in the near future in order to improve the estimation of the sand transport rate. It is stressed as a concluding remark that no numerical models can be developed without careful observations of the phenomena and informative data obtained in the experiments. Much attention should therefore be paid to measurements in laboratories and in the field to establish a rational beach evolution model.

REFERENCES

1. Aagaard, T. and B. Greenwood. 1994. Suspended sediment transport and the role of infragravity waves in a barred surf zone, *Marine Geology, 118,* 23–48.

SEDIMENT TRANSPORT AND BEACH PROFILE CHANGE   7.23

2. Abdelrahman, S. M. and E. B. Thornton. 1987. Changes in the short wave amplitude and wavenumber due to the presence of infragravity waves, *Proceedings Specialty Conference on Coastal Hydrodynamics,* pp. 458–478.
3. Abbott, M. B. 1991, Numerical Modeling, in *Handbook of Coastal and Ocean Engineering,* ed. J. B. Herbich, Vol. 2, Gulf Publishing Company.
4. Bailard, J. A. 1982, Modeling on-offshore sediment transport in the surf zone, *Proceedings 18th Conference on Coastal Engineering,* ASCE, pp. 1419–1438.
5. Beach, R. A. and R. W. Sternberg. 1991. Infragravity driven suspended sediment transport in the swash, inner and outer-surf zone, *Coastal Sediments '91,* ASCE, pp. 114–128.
6. Bird, E. C. F. 1985. *Coastline Changes, A Global Review,* New York: Wiley, 219 pp.
7. Dally, W. R. and R. G. Dean. 1984., Suspended sediment transport and beach profile evolution, *J. Waterways, Port, Coastal and Ocean Engineering,* ASCE, Vol. 110, pp. 15–33.
8. Dally, W. R. 1991. Long wave effects in laboratory studies of cross-shore transport, *Coastal Sediments '91,* ASCE, pp. 85–99.
9. Dean, R. G. 1973. Heuristic models of sand transport in the surf zone, *Proceedings Conference on Engineering Dynamics in the Surf Zone,* ASCE, pp. 208–214.
10. Dibajnia, M. and A. Watanabe 1987. A numerical model of wave deformation and beach transformation in surf zone, *Proceedings 34th Japanese Conference on Coastal Engineering,* JSCE, pp. 291–295. (in Japanese)
11. Horikawa, K. 1978. *Coastal Engineering—An Introduction to Ocean Engineering,* University of Tokyo Press, 402 pp.
12. Iwagaki, Y. and H. Noda 1963. Laboratory study of scale effects in two-dimensional beach processes, *Proceedings 8th Conferenceon Coastal Engineering,* ASCE, pp. 194–210.
13. Johnson, J. W. 1949. Scale effects in hydraulic models involving wave motion, *Trans. A.G. U., 30*(4), 517–525.
14. Katoh, K. 1984. Multiple longshore bars formed by long period standing waves, *Rep. of Port and Harbour Res. Inst., 23*(3), 3–46.
15. Katoh, K. and S. Yanagishima. 1990. Berm erosion due to long period waves, *Proceedings 22nd Conference on Coastal Engineering,* ASCE, pp. 2073–2086.
16. Komar, P. D. 1976. *Beach Processes and Sedimentation,* Englewood Cliffs: Prentice-Hall, 429 pp.
17. Komar, P. D. 1991, Littoral sediment transport, in *Handbook of Coastal and Ocean Engineering,* ed. J. B. Herbich, Gulf Publishing Company, Vol.2, pp. 681–714.
18. Kraus, N. C., M. Larson, and D. L. Kriebel. 1991, Evaluation of beach erosion and accretion predictors, *Coastal Sediments '91,* ASCE, pp. 572–587.
19. Kriebel, D. L., W. R. Dally, and R. G. Dean. 1986. Undistorted Froude model for surf zone sediment transport, *Proceedings 20th Conference on Coastal Engineering,* ASCE, pp. 1296–1310.
20. Larson, M. 1994. Prediction of beach profile change at mesoscale under random waves, *Proceedings 24th Conference on Coastal Engineering,* ASCE, pp. 2252–2266.
21. Longuet-Higgins, M. S. and R. W. Stewart. 1962. Radiation stress and mass transport in gravity waves, with application to "surf beat," *J. Fluid Mech., 13,* 481–504.
22. Mimura N., Y. Ohtsuka, and A. Watanabe. 1986. Laboratory study on two-dimensional beach transformation due to irregular waves, *Proceedings 20th Conference on Coastal Engineering,* ASCE, pp. 1393–1406.
23. Mizuguchi, M. 1982. A field observation of wave kinematics in the surf zone, *Coastal Engineering in Japan,* Vol. 25, JSCE, pp. 91–107.
24. Nishimura, H. 1981. A model for rocky coast (I), NERC Report, No. 13. (in Japanese)
25. Okayasu, A., A. Watanabe, and M. Isobe. 1990. Modeling of energy transfer and undertow in the surf zone, *Proceedings 22nd Conference on Coastal Engineering,* ASCE, pp. 123–135.
26. Osborne, P. D. and B. Greenwood. 1992a. Frequency dependent cross-shore suspended sediment transport, 1. A non-barred shoreface, *Marine Geology, 106,* 1–24.
27. Osborne, P. D. and B. Greenwood 1992b, Frequency dependent cross-shore suspended sediment transport, 2. A barred shoreface, *Marine Geology, 106,* 25–51.
28. Ribberink, J. A. and A. A. Al-Salem. 1994. Sediment transport in oscillatory boundary layers in cases of rippled beds and sheetflow, *J. Geophys. Res., 99*(C6), 12707–12727.
29. Roelvink, J. A. 1993. Surf beat and its effect on cross-shore profiles, Ph.D Thesis, Delft University of Techology, 150 pp.

30. Roelvink, J. A. and I. Brøker. 1993. Cross-shore profile models, *Coastal Engineering*, Elsevier, Vol. 21, pp. 163–191.
31. Roelvink, J. A. and M. J. F. Stive. 1989. Bar-generating cross-shore flow mechanisms on a beach, *J. Geophys. Res., 94*(C4), 4785–4800.
32. Sato, S. and K. Horikawa. 1986. Laboratory study on sand transport due to asymmetric oscillatory flows, *Proceedings 20th Conference on Coastal Engineering*, ASCE, pp. 1481–1495.
33. Sato, S. and N. Mitsunobu. 1991. A numerical model of beach profile change due to random waves, *Coastal Sediments '91*, ASCE, pp. 674–687.
34. Sato, S. and M. Kabiling. 1994. A numerical simulation of beach evolution based on a nonlinear dispersive wave-current model, *Proceedings 24th Conference on Coastal Engineering*, ASCE, pp. 2557–2570.
35. Saville, T., Jr. 1957. Scale effects in two-dimensional beach studies, *Proceedings 7th General Meeting*, IAHR, pp. 1–8.
36. Shi, N. C. and L. H. Larsen. 1984. Reverse sediment transport by amplitude-modulated waves, *Marine Geology, 54*, 181–200.
37. Shibayama, T. and K. Horikawa. 1985. A numerical model for two-dimensional beach transformation, *Proceedings JSCE*, Vol. 357/11-3, pp. 167–176.
38. Short, A. D. 1975. Multiple offshore bars and standing waves, *J. Geophys. Res., 80*(27), 3838–3840.
39. Southgate, H. N. and R. B. Nairn 1993. Deterministic profile modeling of nearshore processes. Part 1. Waves and Currents, *Coastal Engineering*, Elsevier, Vol. 19, pp. 27–56.
40. Stive, M. J. F. 1986. A model for cross-shore sediment transport, *Proceedings 20th Conference on Coastal Engineering*, ASCE, pp. 1550–1564.
41. Sunamura, T. and K. Horikawa. 1974. Two-dimensional beach transformation due to waves, *Proceedings 14th Conference on Coastal Engineering*, ASCE, pp. 920–938.
42. Sunamura, T. 1984. Prediction of on/offshore sediment transport rate in the surf zone including swash zone, *Proceedings 31st Japanese Conference on Coastal Engineering*, JSCE, pp. 316–320. (in Japanese)
43. Svendsen, I. A. 1984. Mass flux and undertow in a surf zone, *Coastal Engineering*, Elsevier, Vol. 8, pp. 347–365.
44. Symonds, G., D. A. Huntley, and A. J. Bowen 1982. Two-dimensional surf beat: Long wave generation by a time-varying breakpoint, *J. Geophys. Res., 87*(C1), 492–498.
45. Watanabe, A. and M. Dibajnia. 1988. Numerical modeling of nearshore waves, cross-shore sediment transport and beach profile change, *Proceedings IAHR Symp. on Mathematical Modeling of Sediment Transport in the Coastal Zone*, pp. 166–174.
46. Wright, L. D., J. D. Boon, S. C. Kim, and J. H. List. 1991. Modes of cross-shore sediment transport on the shoreface of the Middle Atlantic Bight, *Marine Geology, 96*, 19–51.
47. Wright, L. D., J. P. Xu, and O. S. Madsen. 1994. Across-shelf benthic transports on the inner shelf of the Middle Atlantic Bight during the "Halloween storm" of 1991, *Marine Geology, 118*, 61–77.

CHAPTER 8
COASTAL PROTECTION METHODS

JOHN HEADLAND
Moffatt & Nichol Engineers
New York, NY

W. GRAY SMITH
Moffatt & Nichol Engineers
Baltimore, MD

PETER KOTULAK
Moffatt & Nichol Engineers
Baltimore, MD

SANTIAGO ALFAGEME
Moffatt & Nichol Engineers
New York, NY

8.1.	Introduction	8.2
8.2.	Physical Environmental Conditions	8.2
	8.2.1. Water Levels	8.3
	8.2.2. Wind	8.6
	8.2.3. Waves	8.6
	8.2.4. Currents	8.8
	8.2.5. Bathymetry and Topography	8.9
	8.2.6. Littoral Regime	8.10
	8.2.7. Soils	8.18
8.3.	Shore Protection Methods	8.19
	8.3.1. Introduction	8.19
	8.3.2. Beach Nourishment	8.19
	8.3.3. Structures to Reduce Cross-Shore Erosion and Inundation	8.20
	8.3.4. Structures to Reduce Longshore Sediment Transport	8.25
8.4.	Functional Design	8.28
	8.4.1. Structures to Reduce Cross-Shore Erosion and Inundation	8.28
	8.4.2. Structures to Reduce Longshore Sediment Transport	8.30
8.5.	Structural Design (Deterministic Design, Design Optimization, and Probabilistic Design	8.45
	8.5.1. Design Conditions	8.46
	8.5.2. Wave Runup and Overtopping	8.47
	8.5.3. Rubble Mound Armor Design	8.50
	8.5.4. Scour Protection	8.54
	8.5.5. Underlayers and Filters	8.54

8.5.6. Optimal and Probabilistic Design 8.54
8.5.7. Probabilistic Design 8.60
8.5.8. Construction Materials 8.62
8.5.9. Construction Methods 8.63
References 8.64

8.1. INTRODUCTION

Coastal protection systems are engineering works designed to prevent or minimize the damages to near-shore areas stemming from, wave attack, erosion, and coastal flooding. Ideally, there would be little need for shore protection. A good understanding of long-term behavior would, along with prudent planning, minimize the risk of damages to infrastructure in coastal areas. For example, coastal engineering knowledge could be used to avoid siting of infrastructure in: (1) low-lying areas such as barrier islands, or (2) in areas characterized by high rates of long-term erosion. Shore protection is necessary in the real world where decisions have been made without or despite the benefit of the coastal engineering knowledge and with a view towards maximizing the economic value of near-shore property.

Much of the infrastructure built along the coast in developed countries predates the field of coastal engineering. In the absence of expertise, many near-shore communities developed during the 19th and through the mid 20th centuries were located in areas vulnerable to coastal flooding. Notwithstanding the rise of coastal engineering knowledge, one can find many modern examples of shore developments that needed shore protection soon after they were constructed. This stems from the real estate values along the coast. The pressure to develop these valuable areas is unending and has led to the compromises that render near-shore infrastructure vulnerable to the sea.

Inasmuch as shore protection cannot be avoided in many cases, this chapter provides an overview of planning/design techniques for various coastal protection methods. Two fundamental shore protection methods are addressed, namely: (1) beach nourishment and (2) beach nourishment augmented by coastal structures. Beach nourishment is discussed in Chapter 10 of this book. Accordingly, this chapter presents shore protection schemes that feature coastal structures. It is emphasized, however, that coastal structures by themselves are seldom a successful means for shore protection without some form of beach nourishment.

As mentioned above, this chapter provides a summary of design methods for coastal protection works. A basic distinction is made between: (1) structures that reduce cross-shore erosion and inundation, including sea walls, revetments, dikes, dunes, and perched beaches and (2) structures that reduce longshore sediment transport, including offshore breakwaters, groins, T-groins, and artificial headlands. There has been no attempt to include all of the possible design techniques and references available. Instead, the authors focus on methods that they have used in practice, which have proven to be reasonably reliable in field applications.

8.2. PHYSICAL ENVIRONMENTAL CONDITIONS

Shore protection works cannot be designed without a good understanding of local environmental site conditions. Factors governing design are discussed below and include:

COASTAL PROTECTION METHODS   **8.3**

- Water levels (astronomical and storm tides)
- Winds and waves
- Currents
- Bathymetry and topography
- Littoral regime
- Soils

8.2.1. Water Levels

Shore protection design requires a good understanding of water levels. The static water level refers to the mean water surface in the absence of waves.

Figure 8.1 shows a beach profile subject to storm waves. Water levels associated with a normal high tide and the design storm tide are also shown in Figure 8.1. The importance of water level is apparent in Figure 8.1 and shows that the beach berm and dune would not be inundated during a normal high tide. In contrast, each of these features would be inundated and subject to waves during a storm. Accordingly, it is imperative that water levels be known to a relatively high degree of accuracy.

Design water levels are influenced by the following factors:

- Astronomical tides
- Storm tide (or surge)
- Wave setup
- Tsunamis
- Seiches
- Climatological variations
- Secular variations

Normal water level variations are dominated by astronomical tides. Extreme water levels, on the other hand, tend to be dictated by storm tides and wave setup, although there are many locations where storm tides can be ignored (e.g., portions of the Pacific coast of the United States).

Astronomical tides are generated by the gravitational attractions between the earth, moon, and sun. These tides are well understood and can be accurately predicted for most locations throughout the world. Semidiurnal tides (two tides per day) are characterized by a nearly sinusoidal oscillation of the water surface with a period of roughly 12 hours and 25 minutes. A tide that occurs once per day is known as a diurnal tide. Some areas experience a combination of diurnal and semidiurnal tides (mixed tides). Significant differences between consecutive high tides may also be experienced (diurnal inequality). Large tides develop during the new and full moons; these tides are known as spring tides. The lowest tides occur during the last quarters of the moon's phase and are known as neap tides.

Water levels (i.e., datums) are normally based on 19 years of tidal recordings. Tides are referenced to a variety of tidal datums including:

Mean High Water Springs (MHWS)—Average height of spring high water

Mean Higher High Water (MHHW)—Average height of higher high water

Mean High Water (MHW)—Average height of the high water

Mean Sea Level (MSL)—Average height of the water surface for all tide stages

Mean Tide Level (MTL)—Tide level midway between MHW and MLW

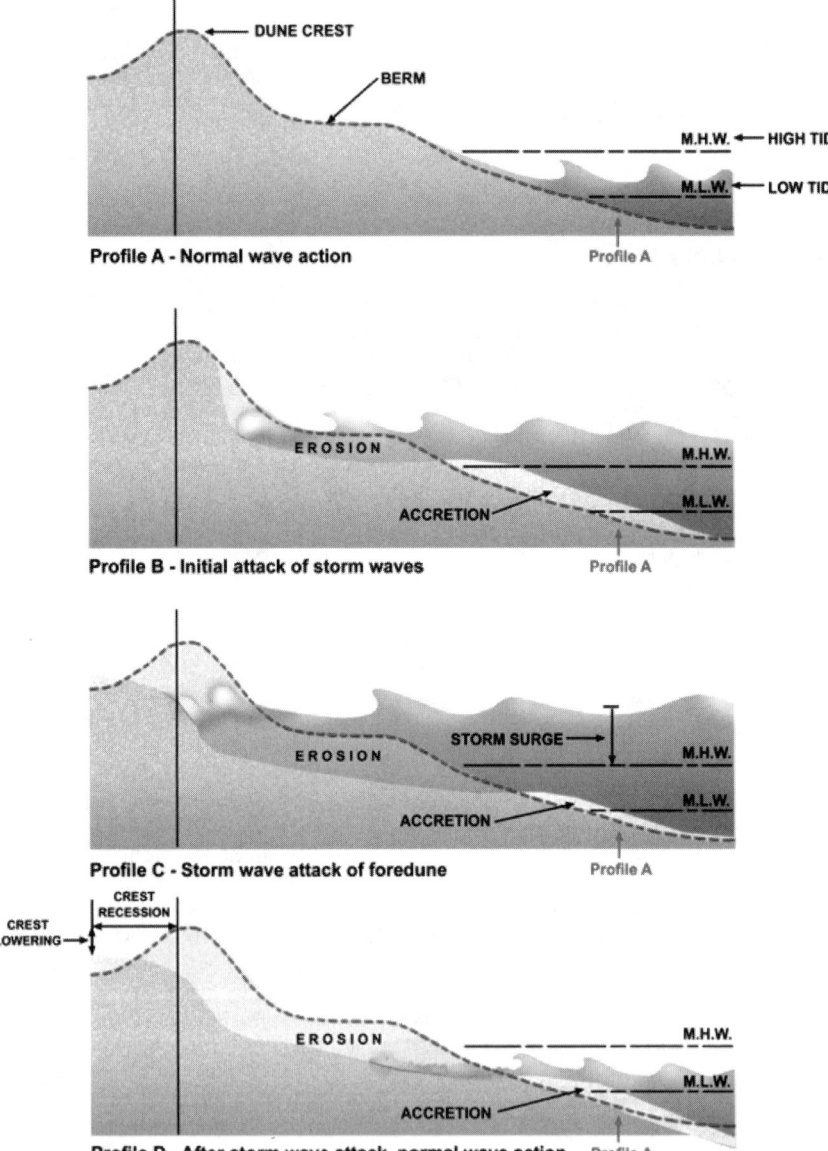

FIGURE 8.1. Storm wave attack on a beach and dune (after *SPM*, 1984).

Mean Low Water (MLW)—Average height of low water
Mean Lower Low Water (MLLW)—Average height of lower low water
Mean Low Water Springs (MLWS)—Average height of spring low water

Storm tide refers to the super-elevation of the water surface that results from: (1) barometric pressure differentials and (2) storm wind stress. Wave setup is a term used to describe the rise in water level that attends wave breaking. Specifically, the change in momentum associated with the breaking of waves results in a surf zone force that raises water levels at the shoreline.

Design storm tides can be estimated from statistical analyses of historic water level records. Unfortunately, tide gages may not be reliable sources due to operational difficulties during storms. Additionally, tide gage records may not cover a sufficient number of years to accurately estimate extreme events. Nonetheless, measured water levels are often used in combination with numerical models to estimate design storm tides. Typically, storm tides are described in terms of stage–frequency curves that express the annual probability (or return) period of a given water surface elevation. An example is provided in Figure 8.2.

Tsunamis are unusually large waves generated by submarine earthquakes, landslides, or volcanoes. Tsunamis can travel great distances and achieve heights exceeding 30 meters in coastal regions. While tsunamis can have devastating impacts, these events are somewhat rare and are seldom considered in shore protection schemes. Exceptions are areas frequently subjected to tsunamis and critical facilities that must be designed for remote events (e.g. bridges, power plants, etc.) Seiche is a type of standing wave in enclosed basins, and can be important in design of shore protection for harbor facilities. Climatological and secular water level variations stem from seasonal or long-term (i.e., hundreds of years) climate changes. Long-term sea level rise is an example of such variations. Estimates of long-term sea level rise are available for many locations and should be considered in shore protection design.

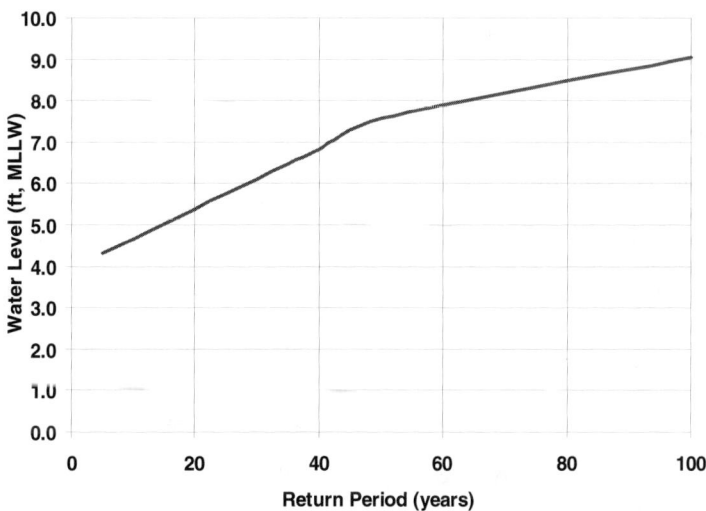

FIGURE 8.2. Plot of water level versus return period.

8.2.2. Wind

Wind has a strong influence on the storm tides and waves that characterize a location. Information on wind can normally be obtained from local weather stations. Wind is used to hindcast waves; however, this subject is not treated in the present work. The effects of wind are implicit in the design water levels and waves at a site. Wind itself is seldom used in calculations for storm protection design. Cases where wind is used directly include wave overtopping analyses and wind loading on monolithic structures.

8.2.3. Waves

Waves, along with water levels, dictate the design of storm protection schemes. Wave theory (including wave mechanics and wave hindcasting) is the subject of numerous other texts (e.g., *Shore Protection Manual,* Dean and Dalyrmple, 1984). The following paragraphs provide a modest amount of information useful in developing a typical shore protection design. The approach taken here is predicated on a fundamental characteristic of most locations requiring shore protection. Specifically, most shore protection structures are located in relatively shallow water (i.e., 20 feet or less) and are subject to depth-limited waves that break with heights on the order of the water depth. The water depth that fronts the structure is the sum of the local bottom elevation and the design water level (usually the sum of storm tide and wave setup). For example, a 100-year hurricane may produce offshore waves with maximum heights on the order of 50 feet or more. However, if the total water depth fronting a coastal structure is limited to 10 feet, then the waves that actually reach the structure will only be on the order of 10 feet, no matter how large the offshore waves. A prudent designer will take advantage of this fact and locate the toe of a shore protection structure in as shallow water as practicable to minimize the waves that can reach the structure. This phenomenon results in the need to accurately predict the water levels and bottom elevations that will exist at the toe of a coastal structure during a design event. The phenomenon also places an emphasis on the prediction of the breaking wave conditions at the toe of the structure. Wave transformation mechanics such as shoaling, diffraction, and refraction are not covered in this chapter, but are treated in Dean and Dalrymple (1984).

A sea state is normally composed of a spectrum of waves, with varying heights and periods, that may range from relatively long waves to short ripples. Wave height, H, is the difference in surface elevation from wave crest to wave trough. Wave period, T, is the time between successive wave crests. In order to summarize the spectral characteristics of a sea state, it is customary to represent a wave spectrum in terms of a distribution of wave energy over a range of wave periods. Having made this distribution, it is convenient to represent the wave spectrum by a single representative wave height and period. The representative wave height is the significant wave height, H_s, while the representative wave period is the peak spectral wave period, T_p. The significant wave height, H_s, is defined as the average of the highest one-third of the waves in the spectrum. Depending on the duration of the storm condition represented by the wave spectrum, maximum wave heights may be as high as 1.8 to 2 times the significant wave height. The peak spectral period, T_p, is the wave period that corresponds to the maximum wave energy level in the wave spectrum. Other important wave characteristics include the wavelength, L, which is the distance between successive wave crests, and the wave celerity, C, which is the speed at which the wave advances. Wave celerity C is equal to the wavelength L divided by the wave period T.

As mentioned above, waves in the deepwater wave spectrum may be as much as twice the significant wave height. Given the relatively shallow depths fronting most coastal protection structures, these structures are normally exposed to some breaking waves. The random wave model of Goda (1985) can be used to examine the maximum and significant waves that can reach a structure founded in a given water depth. The Goda model accounts for wave breaking and wave setup and provides wave heights from deepwater to the shoreline.

The first step in examining depth-limited wave conditions is to compute the total water depth. The total depth, h_b, is the sum of the selected water elevation above a given tidal datum and the bottom elevation below the same tidal datum. The maximum breaker height that can be supported in this depth is computed using the following formulae published in the *Shore Protection Manual* (1984):

$$h_b = \frac{H_b}{B_w - A_w \dfrac{H_b}{gT^2}}$$

$$A_w = 43.75(1 - e^{-19m})$$

$$B_w = \frac{1.56}{(1 + 19.5\,m)}$$

where H_b = breaking wave height at the outer edge of the surf zone
m = tangent of beach slope
h_b = local water or breaker depth
g = acceleration due to gravity
T = wave period

Goda's analyses require an estimate of the equivalent offshore significant wave height, which is computed from the above maximum breaking wave height and the linear shoaling coefficient in accordance with the equations below. Shoaling refers to the increase in wave height that results when waves move from deep to shallow water (see Dean and Dalrymple, 1984).

$$H_s \approx \frac{H_b}{1.8}$$

$$H_0' = \frac{H_s}{K_s}$$

$$K_s = \sqrt{\frac{1}{\tanh\left(2\pi \dfrac{h_b}{L}\right)\left(1 + \dfrac{4\pi \dfrac{h_b}{L}}{\sinh\left(4\pi \dfrac{h_b}{L}\right)}\right)}}$$

$$L = \frac{gT^2}{2\pi} \tanh\left(2\pi \frac{h_b}{L}\right)$$

where H_s = approximate significant wave height at breaking
H_0' = equivalent unrefracted deepwater significant wave height
K_s = shoaling coefficient
L = local wave length

The maximum wave in the surf zone is denoted by H_{max} and can be computed using the following equations published by Goda (1985):

$$H_{max} = 1.8 K_s H_0' \qquad \frac{h}{L_0} \geq 0.2$$

and

$$H_{max} = \min[(\beta_0^* H_0' + \beta_1^* h),\ \beta_{max}^* H_0',\ 1.8\, K_s H_0'] \qquad \text{for } \frac{h}{L_0} < 0.2$$

where

$$\beta_0^* = 0.052 \left(\frac{H_0'}{L_0}\right)^{-0.38} e^{20.0 m^{1.5}}$$

$$\beta_{max}^* = \max\left(1.65,\ 0.53 \left(\frac{H_0'}{L_0}\right)^{-0.29} e^{2.4m}\right)$$

Similar equations are available for computing H_s:

$$H_s = K_s H_0' \qquad \text{for } \frac{h}{L_0} \geq 0.2$$

and

$$H_s = \min[(\beta_0^* H_0' + \beta_1^* h),\ \beta_{max}^* H_0',\ 1.8\, K_s H_0'] \qquad \text{for } \frac{h}{L_0} < 0.2$$

where

$$\beta_0 = 0.028 \left(\frac{H_0'}{L_0}\right)^{-0.38} e^{20.0 m^{1.5}}$$

$$\beta_1 = 0.52 e^{3.8m}$$

$$\beta_{max} = \max\left[1.92,\ 0.32 \left(\frac{H_0'}{L_0}\right)^{-0.29} e^{2.4m}\right]$$

An example wave analysis is provided in Figure 8.3.

8.2.4. Currents

Currents relevant to coastal protection design include tidal currents and wave driven currents (i.e., longshore and cross-shore currents). Tidal currents can be very important near tidal inlets and can have a significant bearing on functional and structural design of a shore protection scheme. Tidal currents are best estimated using a combination of field measurements and numerical modeling.

Longshore currents are generated by breaking waves and are important on the open coast. Longshore currents work together with waves to transport sediment parallel to

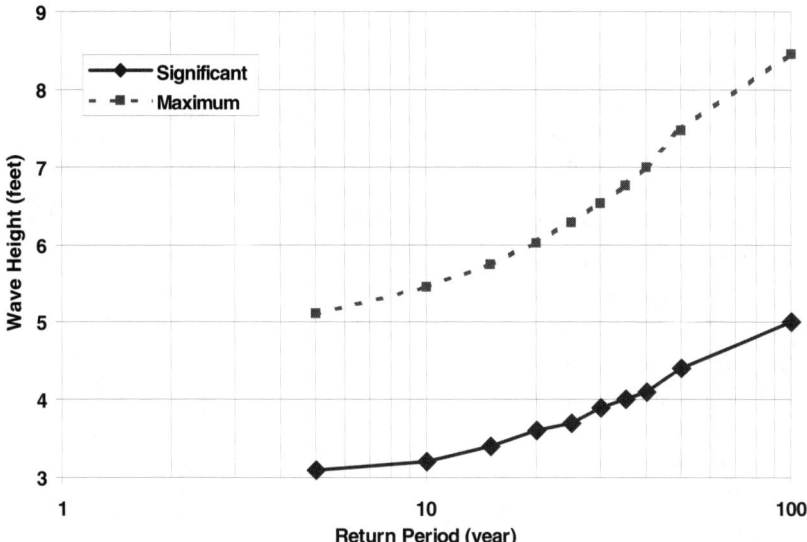

FIGURE 8.3. Goda wave analysis results.

shore (i.e., longshore transport). Longshore currents seldom dictate structural design, but longshore sediment transport is a very important factor in coastal structure design. While longshore currents are best computed using a numerical model (e.g., NMLONG or MIKE 21) a reasonable approximation can be made using the following formula (Pilarczyk and Zeidler, 1996):

$$V_{ls} = K_l \sqrt{gH_b} \sin(2\theta_b)$$

where V_{ls} = longshore current
K_l = coefficient ranging from 0.3 to 0.6
θ_b = breaking wave angle

Cross-shore currents influence the geometry of a beach profile during a storm event. Cross-shore currents are incorporated implicitly in several of the available numerical models for beach profile evolution (e.g. SBEACH and EDUNE.)

8.2.5. Bathymetry and Topography

As emphasized above, the design of a coastal protection scheme is often dictated by water depth. Accordingly, the bathymetry fronting a protected area can have a large bearing on the functional and structural design of coastal protection works. Lower costs generally result when the fronting bathymetry is shallow insofar as the shallow near-shore bathymetry acts to break waves before they reach the structure. Subaerial (i.e., above water) topography can also influence the design of a coastal protection scheme. For example, a scheme may be less expensive if the area has a protective dune system or is fronted by a wide beach.

8.2.6. Littoral Regime

A discussion of the coastal processes relevant to shore protection design could be the subject of an entire chapter in this handbook. Since the focus of the present chapter is coastal protection methods, the discussion presented in this section only addresses that necessary to provide a context for coastal protection planning and design. Further discussions of littoral process can be found in various available handbooks and texts (*SPM*, 1984, Pilarczyk and Zeidler, 1996, Van Rijn, 1993).

Factors important to the design and planning of coastal protection works include:

- Regional geomorphology
- Long-term shoreline changes
- Sediment budget analyses and shoreline modeling
- Sediment characteristics
- Storm-induced shoreline changes

8.2.6.1. Regional Geomorphology. Coastal protection works should not be designed for a particular site without knowledge of regional geology and geomorphology. Typical coastal engineering problems are located on barrier islands, in the vicinity of a coastal inlet, within an embayment between two headlands, or adjacent to some form of natural or unnatural littoral barrier (e.g., harbor). It is critical for the designer to have a basic geologic/geomorphologic understanding of the area in order to judge whether it is prudent to attempt to address a particular problem through the use of coastal protection. Normally, one can deduce a basic cause and effect relationship between a particular coastal engineering problem and the regional littoral system. The most common problems include:

- Erosion near man-made littoral barriers such as improved inlets, harbors, dammed rivers, or coastal protection works. Sometimes, the littoral barriers are natural (e.g., submarine canyon, inlet).
- Nearshore development that is vulnerable to wave attack and coastal flooding because it is located too close the shoreline. In many cases, the nearshore development was unwittingly located in an area vulnerable to long-term or short-term erosion in ignorance.

Both of the above problems are often characterized by chronic long-term erosion. There are examples of the latter problem, however, where the site is only vulnerable to short term, storm-induced, erosion.

8.2.6.2. Long-term Shoreline Changes. Once a general understanding of the regional geomorphology has been gained, the coastal engineer should quantify long-term shoreline movements. Such information is critical to design inasmuch as it will establish long-term shoreline erosion and accretion rates. This information should be obtained over as long a period as practicable, preferably several decades, in order to gain a solid understanding of long-term changes. Detailed shoreline position surveys may be available in developed areas, although it is generally necessary to resort to shoreline positions from historical coastal charts or aerial photography. It may be necessary to set up a monitoring system in remote areas prior to the development of coastal protection works.

Typically, historical information (surveys or photographs) will provide some insight into long-term shoreline erosion rates and, in many cases, shoreline analyses will clearly identify problems. One should be aware, however, that information from historical shoreline movements does not always present a clear understanding of a particular problem. For example, some sites are characterized by alternative periods of erosion and accretions. Additionally, historical information may only be available for storm events and the large erosion associated with storms may make it impossible to extract reliable information on long-term erosion rates.

Notwithstanding the paucity of data available at some sites, the coastal engineer must have a good understanding of long-term shoreline change rates. Design of coastal protection works without a reasonable knowledge of chronic erosion rates can be irresponsible and coastal engineers must be prepared to advise owners to invest in studies and field data collection necessary to gain a reasonable understanding of long-term shoreline movements. Figure 8.4 summarizes the type of information that can be obtained from a shoreline change analysis.

8.2.6.3. Sediment Budget Analyses and Shoreline Evolution Modeling. A sediment budget is a summary of losses and gains of littoral sediments for a region. A summary of a method for developing sediment budgets is provided below along with an example.

Sediment budgets are normally prepared through analyses of: (1) historical shoreline movements and attendant estimates of volumetric changes, (2) offshore losses due to sea level rise, and (3) estimates of longshore sediment transport rates.

Sediment budget analysis commences with a separation of the study area into shoreline segments or littoral cells that can be identified as distinct regions by virtue of characteristic shoreline changes, areas with critical infrastructure or coastal structures, or areas with characteristic wave conditions. Volumetric change rates for each sediment budget cell are often computed on the basis of historical shoreline change rates using equilibrium profile concepts. Specifically, any shoreline movement can be assumed to result in an equivalent translation of the entire active beach profile extending from the offshore seaward limit of significant sediment transport (see Section 8.4.2.3) to the landward limit of sediment transport on the subaerial beach. The depth of the active beach for the study area

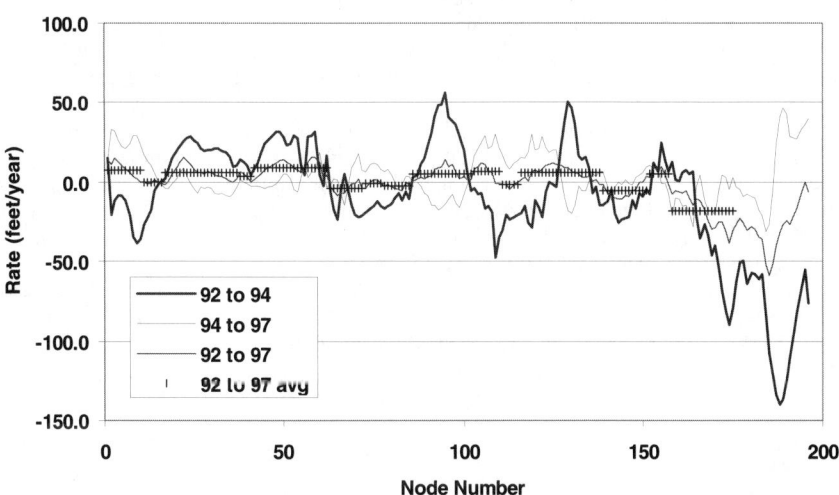

FIGURE 8.4. Example shoreline change analysis (Moffatt & Nichol Engineers, 1999a)

can be computed using an estimate of the seaward limit of sediment transport and the landward limit at the upper limit of normal wave runup. The total depth of active beach can be used to estimate the volume associated with shoreline movement. A standard rule of thumb is that 1 foot of shoreline change corresponds to 1 cubic yard of volumetric change. This rule of thumb is equivalent to a total active beach depth of 27 feet (i.e., 1 foot of shoreline movement × 27 feet of active depth = 27 cubic feet = 1 cubic yard). The rule of thumb can be more accurately defined for a specific site using several years of beach profile and shoreline surveys.

The resulting volume changes are usually presented as an annual change (i.e., m^3/year). These changes include not only the affects of longshore sediment transport but also permanent additions or losses associated with material moving normal to the beach that might occur as a result of barrier island overwash, sea level rise, and/or the presence of coastal structures. Long-term increases in sea level give rise to permanent offshore losses of sediment. Brunn (1962) developed the following formula for evaluating losses due to sea level rise:

$$\Delta x = \Delta \zeta \frac{L}{(h+d)}$$

where Δx = shoreline erosion rate due to sea level rise
$\Delta \zeta$ = rate of sea level rise
L = distance from 0 NGVD to closure depth
d = closure depth
h = height of landward limit of active profile

Volumetric losses due to sea level rise are often small in comparison to the historical volumetric changes.

Longshore sediment transport rates are often estimated on the basis of longshore energy flux computations. Such computations require a reasonably detailed wave climatology. Directional wave gage measurements constitute the best means for establishing the wave climatology for an area. Unfortunately, such data are often unavailable. Wave hindcasts can be used in the absence of measurements.

It is necessary to transform wave hindcast or gage data to the surf zone. This can be accomplished using the Goda model described above. The steps involved in the wave analysis are described below. Results provide an estimate of the average annual longshore sediment transport rate.

Step 1. Start with the deepwater significant wave height, H_s, peak spectral wave period, T_p, and wave direction and transfer deepwater conditions to an appropriate point near the surf zone using a wave transformation model, for example, REFDIF (Kirby and Dalrymple, 1993) or MIKE 21 (Danish Hydraulic Institute, 1998).

Step 2. Compute the maximum offshore wave height in the spectrum, H_{max-o}, by multiplying the significant wave height, H_s by 1.8.

Step 3. Estimate the breaking depth, h_b, associated with H_{max-o}, accounting for the effects of wave refraction assuming straight and parallel bottom contours. This step, which establishes the outer edge of the surf zone for a given offshore wave spectrum, requires iterative solution of the following equations:

$$H'_0 = H_0 K_r$$

$$h_b = \frac{H_b}{B_w - A_w \frac{H_b}{gT^2}}$$

$$A_w = 43.75(1 - e^{-19m})$$

$$B_w = \frac{1.56}{(1 \pm 19.5m)}$$

$$H_b = 0.76 m^{1/7} \left(\frac{H'_0}{L_0}\right)^{-1/4} \cos(\alpha_0)^{3/8} H'_0$$

$$L = \frac{gT^2}{2\pi} \tanh\left(2\pi \frac{h_b}{L}\right)$$

$$\frac{\sin(\alpha_0)T}{L_0} = \frac{\sin(\alpha_b)T}{L_b}$$

$$K_r = \sqrt{\frac{\cos(\alpha_0)}{\cos(\alpha_b)}}$$

where H_b = breaking wave height at the outer edge of the surf zone
m = tangent of beach slope
H_0 = deepwater wave height
H'_0 = equivalent unrefracted wave height
H_0 = deep water wave length = $gT^2/2\pi$
α_0 = deepwater wave angle
h_b = breaker depth at the edge of the surf zone
α_b = breaking wave angle at the edge of the surf zone
T = wave period

Although the above equations are used in this step to evaluate the maximum wave height at the edge of the surf zone, these equations are also applicable to the significant wave height.

Step 4. Having established the seaward edge of the surf zone, select the location within the surf zone where longshore energy fluxes are to be computed. Compute the equivalent deepwater wave height corresponding to the significant wave height, H'_0, at the selected surf zone depth using the above equations (substituting local surf zone values for water depth, wave angle, and wave length to compute K_r) using the original offshore significant wave height as H_0.

Step 5. Having computed the H'_0 corresponding to the selected surf zone depth, compute the local significant wave height in the surf zone using the above Goda equations.

Step 6. Compute the longshore energy flux at the selected surf zone location:

$$P_{ls} = \frac{\rho g}{16} H_s^2 C_g \sin(2\alpha_b)$$

where P_{ls} = longshore energy flux
C_g = wave group velocity

$$C_g = \sqrt{gh}$$

Step 7. Compute the longshore sediment transport using the following empirical equation:

$$Q = \beta P_{ls}$$

where Q = longshore sediment transport in cubic yards

The above steps are repeated for each of the wave conditions in the hindcast or gage records and weighted by the probability of occurrence of each wave condition to give the average annual longshore wave energy fluxes and attendant longshore sediment transport values. Published values for the SPM constant, β, can be used, although it is preferable to use wave energy flux at the borders of each littoral cell to solve for the constant. The latter approach provides a means for "calibrating" the sediment budget.

As stated previously, the overall intent of a sediment budget is to develop a balance of losses and gains along a segment of the coast. An example sediment budget is shown in Figure 8.5. Output from a sediment budget will include: (1) average annual longshore sediment transport rates in both directions, (2) net longshore sediment transport, which is the difference in the directional components of transport, and (3) the gross longshore sediment transport, which is the sum of the absolute values of the two directional components of transport.

Each of the above sediment transport values is important to engineering design. The directional components provide an indication of the directional variation in transport conditions at a particular site. The net sediment transport represents the predominant direction and magnitude of longshore sediment transport.

Net longshore sediment transport is important to the design of shore protection systems for reducing longshore sediment transport. For example, a coastline with a north–south alignment might have periods of longshore sediment transport to the north or south but may have a net or predominant longshore sediment transport rate to the south. For this particular example, the term updrift (analogous to upstream) refers to being north of a particular point, whereas downdrift (analogous to downstream) refers being south of the same point. References to updrift and downdrift are common in discussions regarding coastal protection systems. The net longshore sediment transport rate provides an immediate understanding as to which side of a littoral barrier (e.g., groin) is likely to accrete (updrift side) and which side is likely to erode (downdrift side).

The gross longshore sediment transport rate represents the total amount of sediment that passes a particular point on the shoreline. Inlets and other sinks (e.g., a new breach in a barrier island, an unfilled groin field) may trap longshore sediment transport in both directions (i.e., the gross sediment transport rate). One of the more challenging conditions that may face a designer is the case where the net transport rate is very small compared to the gross transport rate. Such a condition often results when the directional components of transport are large but the absolute value of each directional component is comparable. Design of coastal protection systems under such conditions must give careful consideration to the ramifications associated with large reversals in longshore sediment transport.

A good sediment budget will provide the designer with a reasonable understanding of littoral processes in the area. Furthermore, the sediment budget can be used to establish long-term erosion and evaluate various corrective actions. Such information will help guide the design of coastal protection works.

As a final note, numerical models of shoreline evolution (e.g., GENESIS) and cross-shore profile dynamics such as SBEACH (Hanson and Kraus, 1991) can be helpful tools in the examination of littoral processes in a region. Example GENESIS results are presented in Figures 8.6 and 8.7. Additional information regarding shoreline evolution modeling is presented in Section 8.4.

8.2.6.4. Sediment Characteristics. Most beaches are characterized by fine to coarse sand. There are a wide variety of natural beach conditions, however, and these conditions can range from soft muds to shingle or cobble beaches. The equilibrium beach profile is

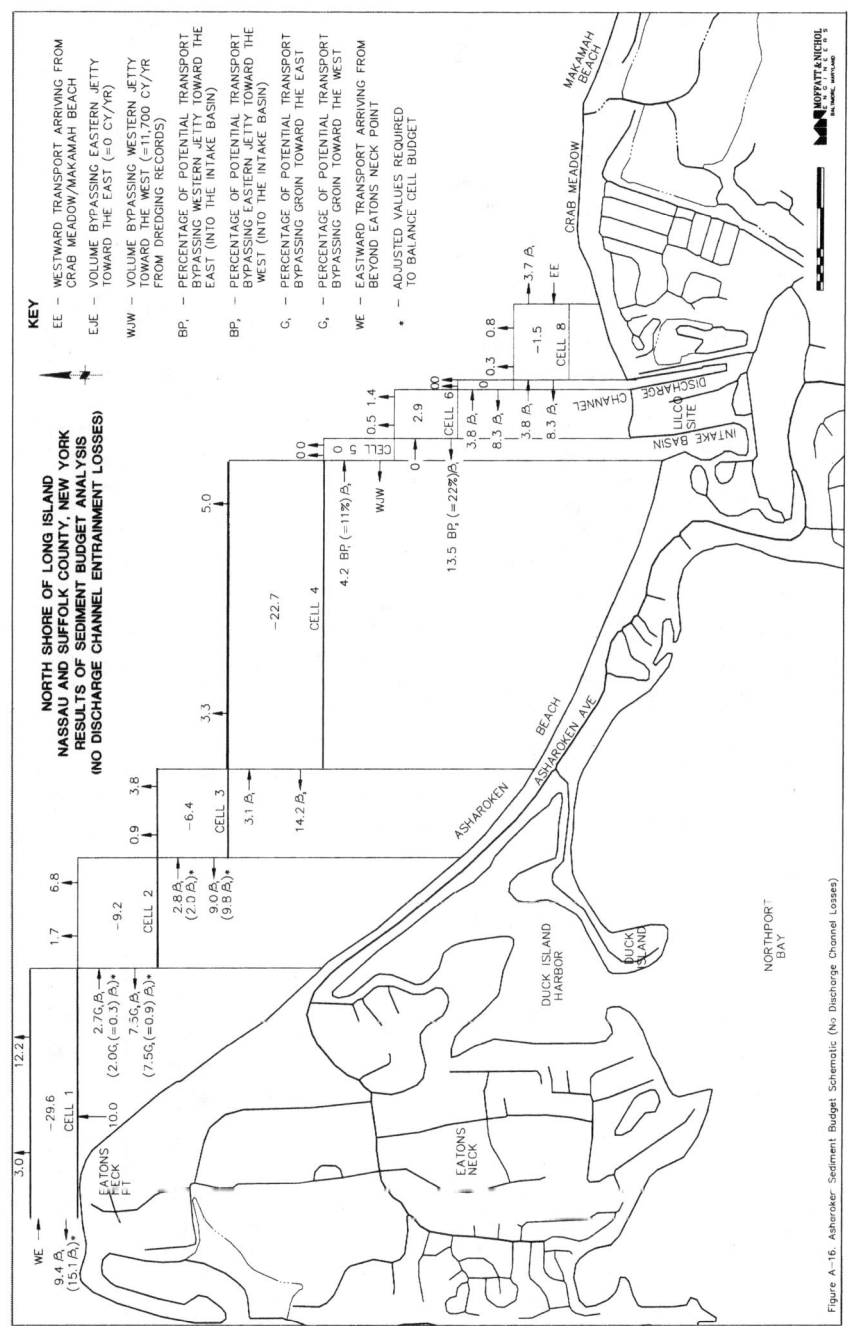

FIGURE 8.5. Sediment budget for North Shore of Long Island (Moffatt & Nichol Engineers, 1995).

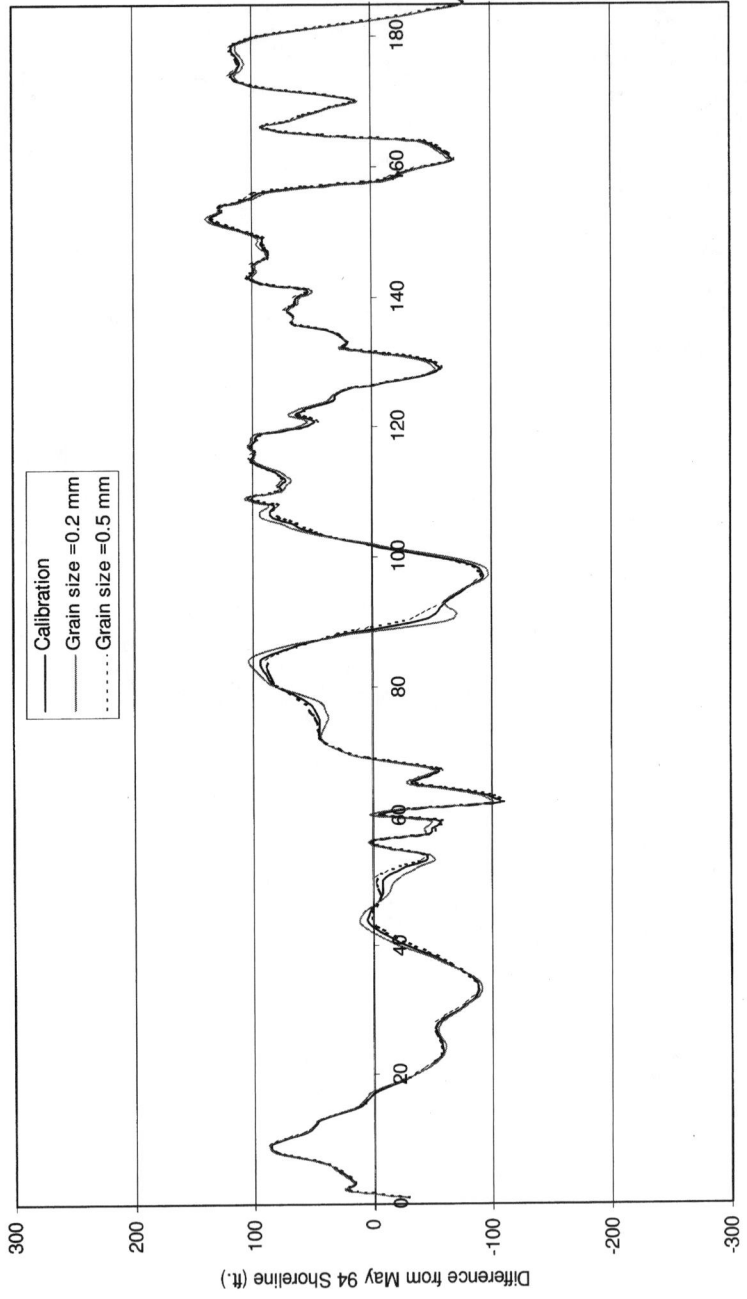

FIGURE 8.6. Genesis results—sensitivity study, shoreline differences (Moffatt & Nichol Engineers, 1999a).

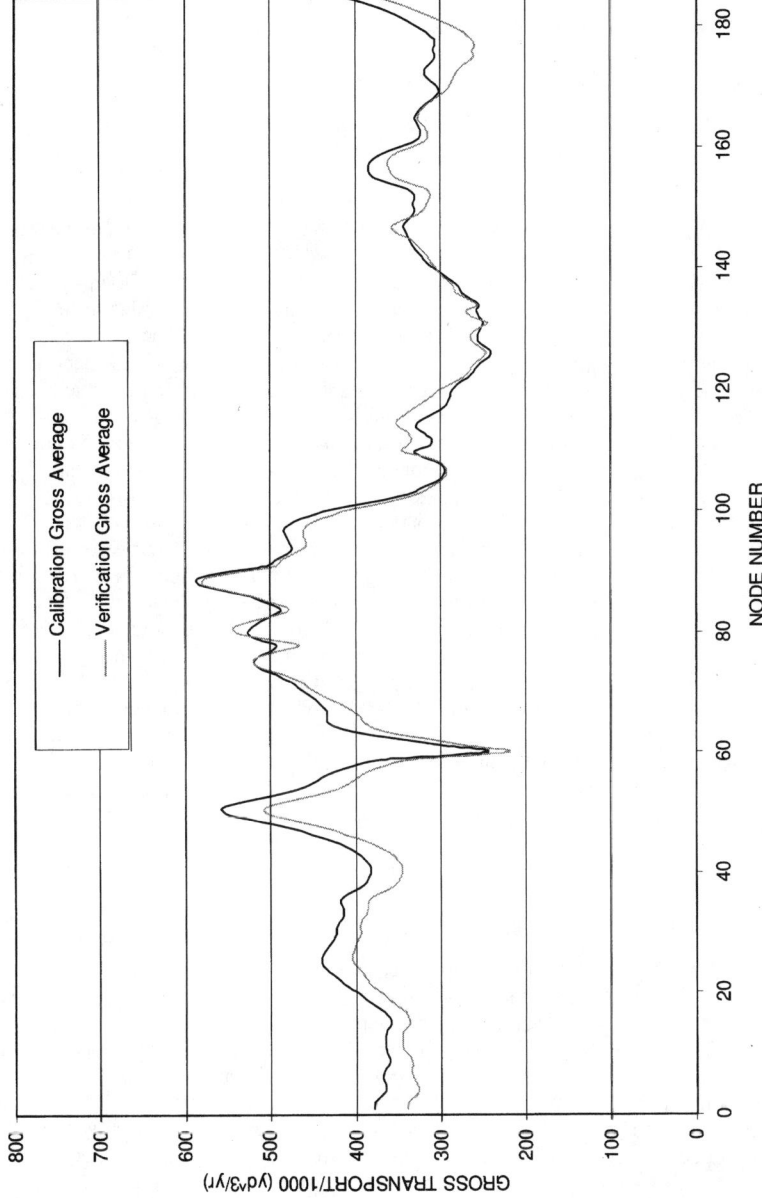

FIGURE 8.7. Bolsa Chica preliminary engineering studies. Annual gross transport by node (Moffatt & Nichol Engineers, 1999a).

generally a function of the sediment size (Dean and Dalrymple, 1993) and can be estimated using the following formula:

$$h = Ay^{2/3}$$

where h = vertical position of the profile
  y = horizontal position of the profile
  A = coefficient that is a function of grain size

In general, flatter beaches are associated with finer grain sizes, steeper beaches, with coarser grain sizes.

8.2.6.5. Storm-Induced Shoreline Changes. Shoreline areas are vulnerable to storm flooding and two types of shoreline erosion discussed above, namely: (1) long-term or gradual shoreline erosion and (2) short-term, storm-induced erosion. Flooding occurs when storm tides and waves overtop coastal dunes and structures as shown in Figure 8.1. Long-term erosion, which is measured in terms of decades, has been discussed above and is generally associated with the wave-driven movement of sediment alongshore and the offshore movement of sediment resulting from long-term sea level rise. Intense tropical or extratropical storms can cause a massive amount of cross-shore erosion in a relatively short time (< 12 hours). Such erosion, which can cause the dune line to retreat as much as 75 feet or more, occurs as the beach responds to the elevated water levels and accompanying wave heights of the storm system. Sand is removed from the dune face and is deposited offshore as shown schematically in Figure 8.1. The beach will rebuild after passage of the storm and in many cases will resume its prestorm geometry. In this sense, storms produce a temporary erosion of the beach. However, such erosion presents a very real threat to structures located along the shore for two reasons. First, beach erosion can undermine the foundation of a nearshore building. Second, an eroded beach allows more storm tide and wave overtopping than an uneroded beach, thus increasing the risk of damage associated with flooding and wave impacts. This is particularly true when the dune has been breached or completely removed by erosion.

Cross-shore numerical erosion models such as EDUNE (Kreibel and Dean, 1985) or SBEACH (Larson and Kraus, 1989) provide a means for estimating storm-induced shoreline erosion. Some applications of these models will be presented in subsequent sections.

8.2.7. Soils

Soils underlying many sites along the open coast consist of relatively firm sands. As a result, there is often a good chance that coastal protection schemes can be developed without undue concern for geotechnical problems such as slope stability or excessive settlement. This is not always the case and there many examples of muddy coasts where the geotechnical issues are of paramount importance. In either case, it is imperative to conduct an appropriate geotechnical investigation and factor the results of that investigation into design of the protection works. From a practical point of view, such analyses will produce recommendations for stable structure slopes and requirements for ameliorating any short- or long-term settlement of the structure. Details on geotechnical aspects of coastal structure design can be found in a number of texts (e.g., Bowles, 1979.)

8.3. SHORE PROTECTION METHODS

8.3.1. Introduction

An adequately sized protective beach is the fundamental element of any coastal protection plan. The shore protection offered by a beach can be augmented with either shore parallel structures or shore-perpendicular structures. The overall intent is to use coastal structures to gain either a higher level of protection for the beach nourishment project (cross-shore and longshore structures) or increase the longevity of beach nourishment projects (longshore structures). The use of coastal structures with beach nourishment is normally only justified when the total annualized costs of the combined actions is less than the cost of beach nourishment alone. Typically, this is the case when long-term erosion rates in an area are relatively high (longshore structures) or when the beach is too narrow to provide an adequate level of protection (cross-shore structures).

Shore-parallel structures are generally used to prevent: (1) the cross-shore movement of sand that occurs during storms, and (2) the inundation of hinterland areas. Examples include: sea walls, revetments, dikes, and the perched beach concept. Shore-parallel structures can also be used to reduce longshore sediment transport (e.g., offshore breakwaters). Shore-perpendicular structures, which include groins and T-groins, offshore breakwaters, and artificial headlands, are generally used to reduce longshore sediment transport. Structures that reduce longshore sediment transport can be very effective in reducing maintenance requirements for beach nourishment. For example, if it were necessary to replenish a beach once every three years without structures, properly designed structures might extend the period between replenishment to 5 or more years. An economic analysis must be conducted to establish the optimal beach nourishment/structure plan.

Beach nourishment is generally needed wherever the shoreline segment to be protected is characterized by long-term erosion. Protective structures should not be used on an eroding coast in the absence of beach nourishment. The structural integrity of a shore-parallel structure will eventually be threatened as the beach retreats towards that structure. A shore-perpendicular structure may remain intact as the beach retreats but will likely cause accelerated erosion downdrift of the structure.

An area characterized by a stable or accreting shoreline will not often require beach nourishment. Where this is the case, shore-parallel structures can be used to provide an additional level of protection against shoreline retreat during storms and storm tides. More often than not, however, most stable or accreting beaches do not require shore protection.

8.3.2. Beach Nourishment

A previously stated, a complete discussion of beach nourishment can be found in Chapter 10 of this book. This chapter will only cover the basic elements of beach fill design in order to provide a context for design of shore protection structures.

Figure 8.1 presented the basic geometrical features of a natural or man-made protective beach. The beach includes an offshore slope, a foreshore area that extends from the waterline to a flat area known as the beach berm, and a dune. Figure 8.1 summarized the response of a beach to storms and showed how the prestorm profile adjusts to waves and elevated water levels. Specifically, sand is eroded from the berm and dune and deposited in an offshore bar. The eroded profile shape is more resistant to wave action and elevated water levels by virtue of the offshore bar (which acts to break waves before they reach the beach) as well as the flattened upper profile geometry. Once the storm has subsided, the profile will rebuild and eventually return to its prestorm geometry.

Beaches offer storm protection through a natural dynamic response to varying waves and water levels. Accordingly, it is difficult to provide better shore protection than that offered by a beach. Problems arise when a natural beach cannot provide a sufficient level of protection because the beach is either too narrow or eroding too fast. In such cases it may be necessary to create an artificial beach (i.e., beach nourishment), construct coastal structures, or both.

Beach nourishment consists of adding sand to an existing beach in order to provide a desired level of storm protection. The overall geometry of the constructed beach is designed to prevent waves and water levels from reaching hinterland areas landward of the beach. As a result, design of the beach nourishment geometry follows from the anticipated beach response during severe storms. Predictive numerical models, such as SBEACH or EDUNE, can be used to estimate beach response during storms.

The principle geometrical features of the beach include the beach berm and the dune volume (i.e., height and width). The dune geometry is of particular importance, inasmuch as the dune acts to prevent wave overtopping and the landward migration of flood waters. The berm height and width also provide an important protective function. In many cases, beach nourishment design focuses on establishing the berm and dune geometry necessary to provide a given level of storm protection. There are many situations in which the beach may provide coastal protection but is also serving as a recreational beach. A relatively wide beach berm is desirable in such cases.

Normally, the dune and berm features of a beach fill are constructed according to the desired geometry above water, where construction equipment can easily shape the above-water fill. On the other hand, it is difficult to shape the offshore portion of the beach nourishment profile. Accordingly, material is normally pumped onto the offshore portion of the profile and allowed to equilibrate to natural conditions. Except in cases where there is a large difference in the size of the native and fill sand, the offshore portion of the beach will eventually parallel the prefill geometry. A detailed discussion of this matter can be found in Dean and Dalrymple (1993).

The above paragraphs focus on the cross-shore behavior of an artificially nourished beach; however, artificially nourished beaches are also subject to longshore sediment transport. In most cases, beach fills have been designed with a maintenance program in mind. Long-term erosional losses are normally associated with longshore sediment transport although major cross-shore losses can occur during storms under some circumstances. Accordingly, it is common to design beach fills with an additional buffer of sand that will be lost during the period between renourishment cycles. This added volume of sand is sometimes called advanced nourishment and is designed to assure the minimum beach cross-section necessary to provide a design level of protection to hinterland areas.

8.3.3. Structures to Reduce Cross-Shore Erosion and Inundation

Structures designed to prevent cross-shore erosion and hinterland flooding are normally used when a natural beach cannot provide adequate protection alone. Example applications include areas where infrastructure is located too close to the shoreline or in areas where long-term erosion is too severe or highly variable to rely on beach nourishment alone. Typical structures include:

- Sea walls (conventional and buried structures)
- Revetments (conventional and buried structures)

COASTAL PROTECTION METHODS   8.21

- Dikes
- Dunes
- Perched Beach

8.3.3.1. Sea Walls. Sea walls are used to protect upland structures from wave impact and erosion damage. These structures are relatively massive and designed to resist the full force of storm waves. Sea walls can be of rubble mound (i.e., rock) or vertical/monolithic (steel or concrete sheetpile, cast-in-place concrete) construction. The front face of a rubble mound sea wall is normally armored with large rock or concrete armor units. It is also common to armor a portion or the backslope of a sea wall to provide protection against overtopping. Sea walls normally require extensive toe protection to prevent scour. Because wave runup is high on such structures, vertical sea walls have to be high in order to provide protection and may be judged to be aesthetically unacceptable. A recurved superstructure is often used to reduce wave overtopping with an attendant reduction in crest elevation. Vertical and monolithic sea walls are vulnerable to catastrophic failure when overloaded or undermined by waves, whereas rubble mound structures tend to fail in a progressive manner. Nonetheless, suitable sea walls can be designed using either rubble mound or vertical/monolithic construction.

Sea walls only protect the land area leeward of the structure and offer no protection to the fronting beach. When located on an eroding beach, sea walls will eventually be threatened by shoreline retreat. Accordingly, it best to employ sea walls in combination with beach restoration.

A buried rubble mound sea wall placed landward of the shoreline is an alternative to a conventional rubble mound or vertical sea wall. Example application of a buried sea wall is described in Headland (1992). The buried sea wall is hidden within the dune and is only exposed during severe events. When used in concert with beachfill, the sea wall provides the last-line-of-defense storm protection, whereas beach restoration combats long-term shoreline erosion. Example sea walls are presented in Figures 8.8 and 8.9.

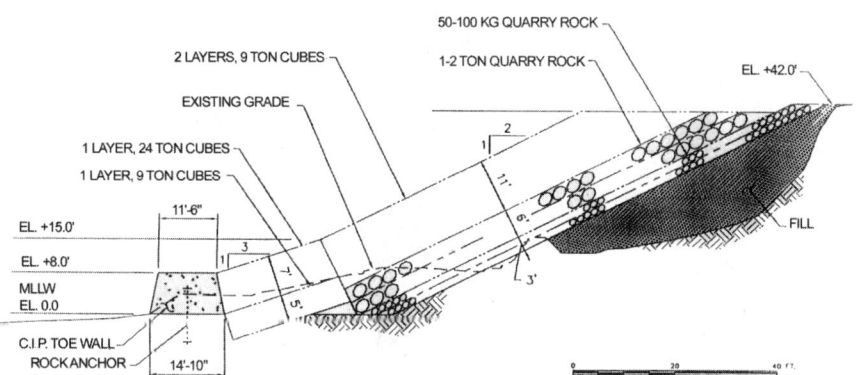

FIGURE 8.8. Rubble mound sea wall with concrete armor and toe wall (Moffatt & Nichol Engineers, 1999b).

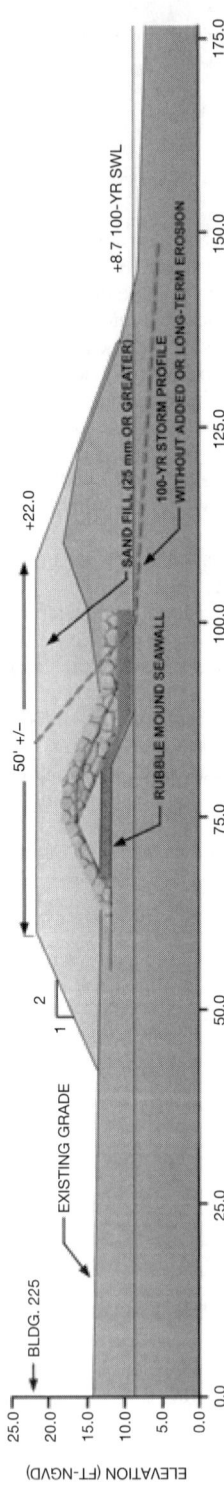

FIGURE 8.9. Dune/sea wall design cross-section (Headland, 1992).

COASTAL PROTECTION METHODS   **8.23**

8.3.3.2. Revetments.   A coastal revetment is similar to a sea wall and there are cases where it is difficult to make a clear distinction between the two structures. A revetment is generally exposed to a lower level of wave attack than a sea wall, does not feature back slope armor, and often fronts an earthen embankment. An example revetment is shown in Figure 8.10.

8.3.3.3. Dikes.   Dikes are similar to revetments and sea walls. They generally feature a large and sometimes relatively impermeable core. Dikes are often used to prevent flooding during storms. Dikes are also used to contain hydraulically placed fill. An example is presented in Figure 8.11.

8.3.3.4. Dunes.   Dunes are found in nature at the landward extent of an active beach, as shown in Figure 8.1. Artificial dunes are often constructed in cases where a natural dune does not exist or in cases where the natural dune does not provide sufficient protection. Dunes can be a highly effective and economical means of providing shore protection. An example dune design is shown in Figure 8.9.

8.3.3.5. Perched Beach.   A perched beach is characterized by a submerged sill structure that acts to retain sand and prevent the offshore movement of sand during a storm event. Perched beaches act to restrict the short-term erosion that occurs during storms. Perched beaches also provide a means for reducing the amount of fill required for beach nourishment. The sill acts to reduce the seaward limit of the equilibrium beach profile. An example is shown in Figure 8.12.

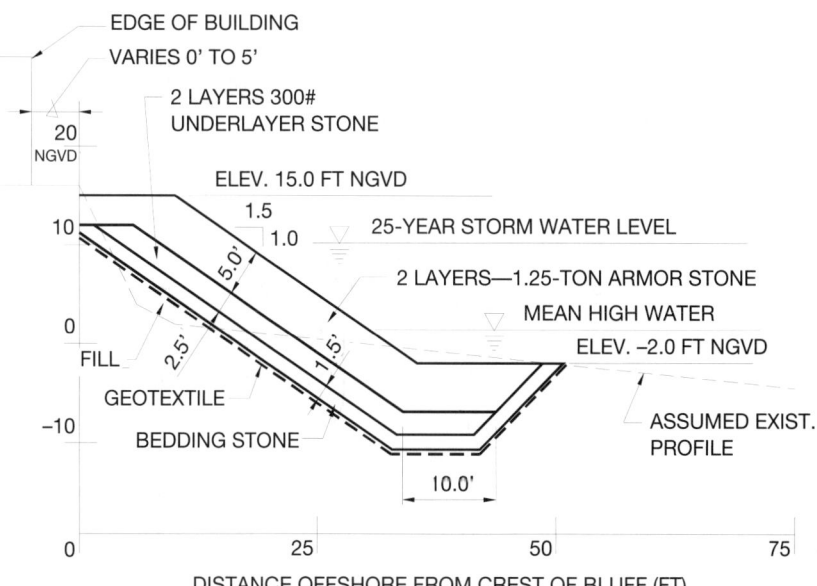

FIGURE 8.10. Typical revetment section (Moffatt & Nichol Engineers, 1994a).

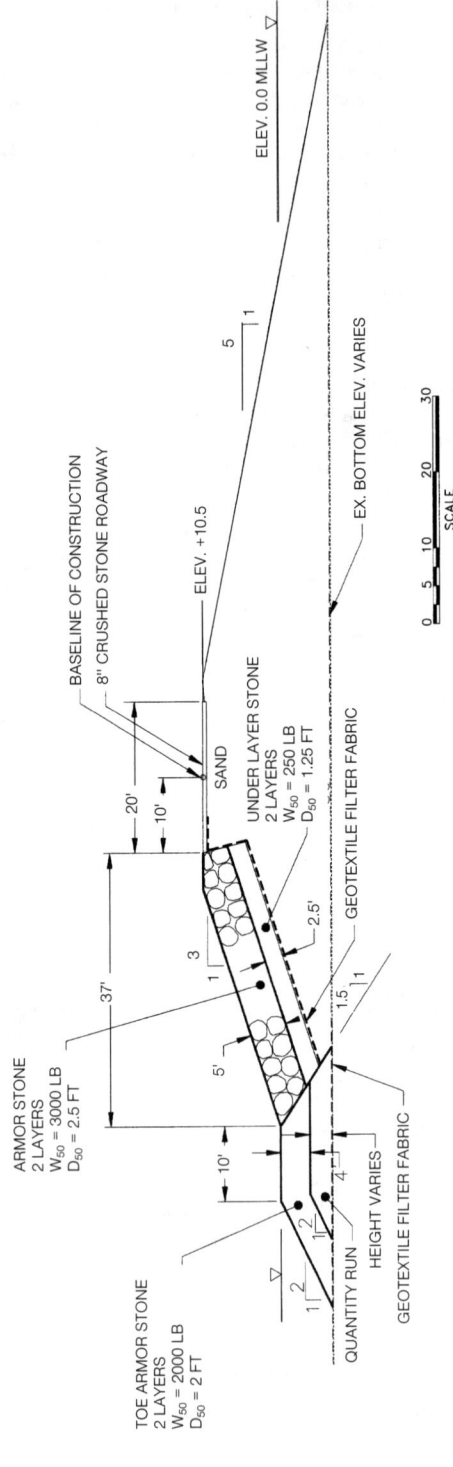

FIGURE 8.11. Typical dike section (after Moffatt & Nichol Engineers, 1996).

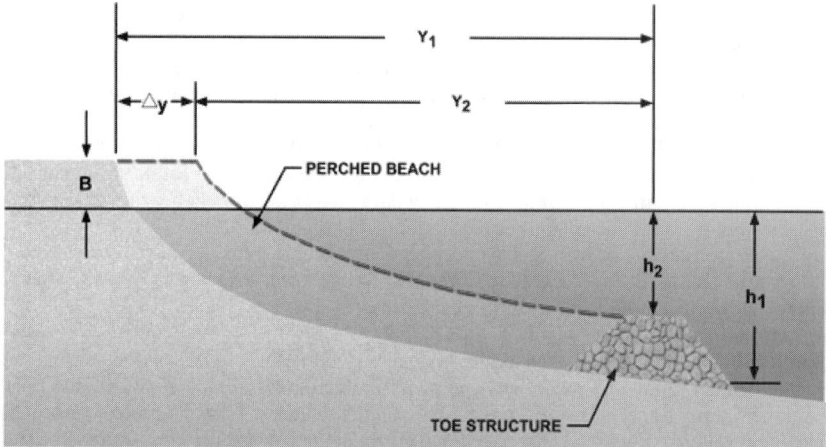

FIGURE 8.12. Perched beach (after Dean and Dalrymple, 1993).

8.3.4. Structures to Reduce Longshore Sediment Transport

8.3.4.1. Offshore Breakwaters and Artificial Headlands. Offshore breakwaters and artificial headlands reduce the amount of wave energy that reaches a beach. The former structure type is the subject of Chapter 5 of this book and will be treated in an abbreviated version here. Both structures are typically of rubble mound construction and are located seaward of and parallel to the shoreline (see Figure 8.13). The overall effect of these structures is to produce gradients in wave energy that promote sediment deposition behind the breakwaters. Offshore breakwaters often consist of a series of short, segmented elements. Depending on the distance of the offshore breakwater from the shoreline, the structure may or may not interrupt longshore sediment transport. Offshore breakwaters are often combined with beach restoration. In such cases, the protective beach serves to reduce storm-induced damages, whereas the offshore breakwater system serves to reduce

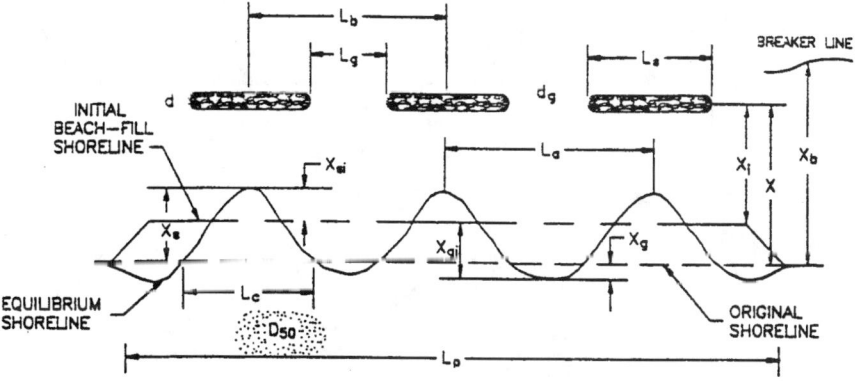

FIGURE 8.13. Offshore breakwaters (after Dean-Rosati, 1990).

long-term erosion of the beach. Breakwaters also reduce the offshore loss of sediments during storms.

Breakwaters may be connected to or detached from the shoreline. Shore-connected breakwaters are constructed close to shore and promote the formation of a tombolo (i.e., a permanent attachment of the shoreline to the breakwater). Tombolos can form naturally or can be artificially constructed. The latter approach will reduce the potential for losses of sediment at adjacent beaches, which could occur if the tombolo is left to form naturally, (i.e., the sand impounded in the tombolo would be lost to adjacent beaches). Detached breakwaters are constructed further offshore than shore-connected breakwaters or headlands and promote the growth of a shoreline bulge or salient in the lee of the structure.

An artificial headland is a variant of the shore-connected breakwater. A series of artificial headlands can be used to form pocket beaches that are characterized by a stable, equilibrium shoreline, see Figure 8.14. Offshore breakwaters or headlands can be used combined with traditional groins to form T-groins, see Figure 8.15 (discussed below).

A distinction must be made between shore protection and beach stabilization structures. The former provides storm protection while the latter stabilizes a protective beach. The breakwater/headland systems discussed above are beach stabilization structures. The beach that is held in place by the structures rather than the structures themselves normally provides storm damage reduction in a breakwater/headland system. This distinction is not always understood and has led to the misuse of offshore breakwaters as a shore protection measure.

8.3.4.2. Groins and T-Groins. Groins are normally constructed perpendicular to the shoreline and act to interrupt or reduce longshore sediment transport. This interruption will produce accretion updrift of a groin(s) and produce concomitant erosion downdrift of the groin(s). When constructed in concert with beach restoration, groins act to reduce advanced maintenance and renourishment requirements. Groins generally extend from the dune/beach interface to water depths on the order of 10 to 12 feet below MSL. At a single

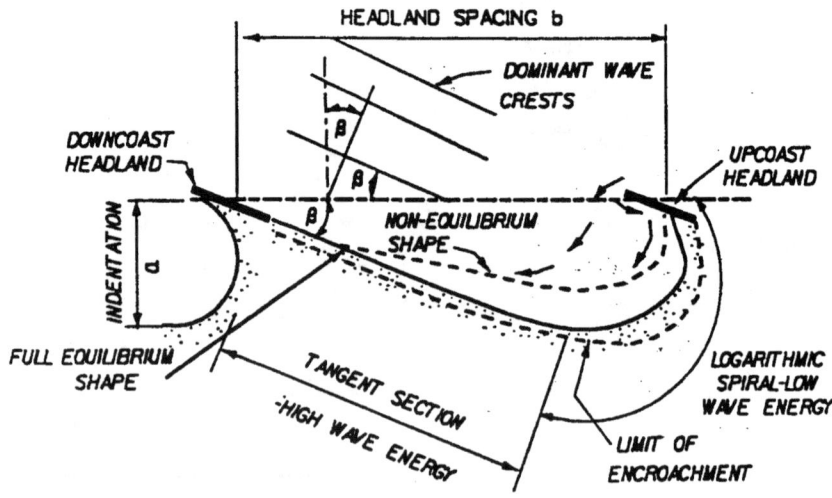

FIGURE 8.14. Artificial headlands (after Dean-Rosati, 1990).

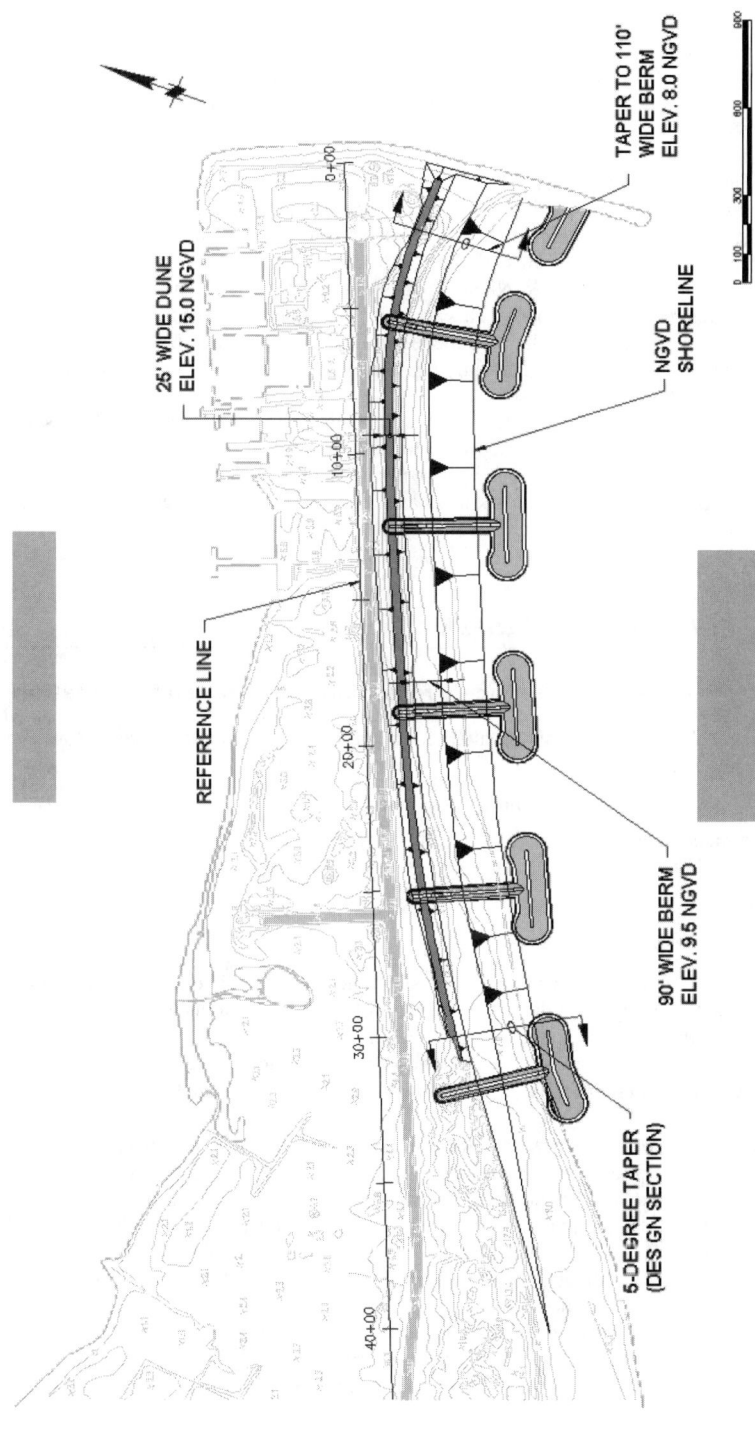

FIGURE 8.15. Beach restoration with T-groins.

groin, the updrift impoundment of sand is generally offset by an equivalent amount of erosion downdrift of the structure. Groins are often constructed in series (i.e., fields) to provide protection for continuous shoreline segments. In this arrangement, erosion is displaced to the most downdrift groin, rendering the downdrift area susceptible to accelerated erosion. Erosion downdrift of a groin field can be mitigated through the use of tapered groin transitions, "low-profile" groins and/or beach nourishment. Filling the groin compartments to capacity is a particularly effective way to mitigate the erosion downdrift of a groin field. Filling will promote bypassing of longshore sediment transport through the groin field. "Low-profile" groins can be constructed to match the template of the design beachfill geometry (Moffatt & Nichol Engineers, 1994). The intent is to maintain the minimum desired beach geometry. This approach differs from conventional groins, which often protrude above the beach and impede sand bypassing. Groins fields can be particularly effective in areas characterized by significant longshore sediment transport rates or high erosion rates.

Unfortunately, groins are vulnerable to storm-induced or offshore erosion losses in the immediate vicinity of the a groin, where longshore currents are trained seaward by the groin. These losses may be reduced by providing a shore-parallel stem at the sea wall end of the groins. The result is what is known as a T-groin. T-groins are effective in areas of severe erosion, (e.g. in the vicinity of tidal inlet.)

8.4. FUNCTIONAL DESIGN

8.4.1. Structures to Reduce Cross-Shore Erosion and Inundation

Section 8.3.3 listed the following alternative structures designed to prevent cross-shore erosion and inundation:

- Sea walls (conventional and buried structures)
- Revetments (conventional and buried structures)
- Dikes
- Dunes
- Perched beach

There are two fundamental aspects of functional design for the above structures, namely: (1) locating the structure on the design beach profile and (2) establishing the toe and crest elevation of the structure so that the desired level of protection is provided by the structure. Lack of a sufficiently low toe elevation or a sufficiently high crest elevation has probably resulted in more failures of coastal protection measures than any other cause.

4.4.1.1. Cross-Shore Structure Location. In many respects, the above alternatives can be classified in terms of their relative location on the active beach profile. A dune is located at the landward boundary of the beach. Moving seaward, a dike or revetment might be located within the dune or just seaward of the dune area. Sea walls are normally located somewhat seaward of a typical revetment location and would be exposed to waves on a near-continuous basis. Finally, the containment structure for a perched beach is located below the waterline.

Where a coastal protection device is located on the active beach profile has a large bearing on its structural requirements. In most applications, each of the above structures would be subject to depth-limited waves. As previously stated, the size of the wave follows directly from the local bottom elevation and the storm tide. For example, a buried revetment, dike, or dune located near the landward boundary of the active beach profile will be exposed to relatively small waves and may only be reached by waves during severe storms. A conventional sea wall may be directly exposed to large waves during normal and severe storms. The containment structure for a perched beach, on the other hand, is exposed to large waves, but is subject to relatively small loading, inasmuch as large waves simply pass over the submerged structure.

4.4.1.2. Design Beach Profile and Structure Toe Elevation. Design of a coastal protection system cannot proceed without a surveyed beach profile and some estimate of the design profile during severe storms. In areas subject to long-term erosion, it is also critical to establish the future design storm profile. For example, it may be prudent to design a coastal structure to withstand a 100-year condition after 15 years of long-term erosion (Headland, 1992). Such an approach requires that an estimate of the equilibrium shoreline profile be translated landward a distance corresponding to 15 years of erosion at the long-term erosion rate for the area. A cross-shore erosion model such as SBEACH or EDUNE can then be used to evaluate the design storm profile for the translated profile. A more systematic optimization approach has been developed to address this problem (Headland and Kotulak, 1996) and (Alfageme, Headland, and Kotulak, 1999) and is discussed in Section 8.5.6 of this chapter.

The bottom elevation is clearly a function of the location of the structure on the active beach profile. The bottom elevation is also a function of the beach profile response during a severe storm, taking into account long-term shoreline erosion. In other words, an estimate of the eroded profile during a severe storm is necessary in order to define the local bottom elevation, which, in turn, can be used with the design storm tide elevation to establish the design wave height.

While it is the penultimate issue governing the design of any coastal structure, determining the eroded storm profile fronting a coastal structure is not an easy matter. The best available method for estimating the design profile is to adopt results from cross-shore profile models such as SBEACH or EDUNE. Some of these models include the ability to simulate a sea wall, although not all of them consider the sloping/overtopped geometries that are built in everyday practice. One of the approaches used by the authors has been to use the vertical sea wall feature of SBEACH or EDUNE as a conservative estimate for storm depths at the toe of a structure. A general model for establishing depths fronting coastal structures has yet to be established for routine engineering applications.

An example estimate of an eroded profile fronting a buried sea wall is provided in Figure 8.9. The results presented in Figure 8.9 were developed using the EDUNE model. The example shows the prestorm and eroded profile and denotes the controlling still-water depth at the structure toe. Once this depth has been established it is possible to estimate the design wave heights that will reach the structure using the Goda model described above. The design wave conditions can then be used to establish the remaining geometry of the structure.

8.4.1.3. Structure Crest Elevation. The crest elevation of a coastal structure depends on its function. The crest elevation of sea walls, revetments, dunes, and dikes is normally selected so as to reduce wave overtopping and/or flooding to an acceptable level. The crest elevation of a coastal structure can be selected to reduce hinterland flooding to a prescribed level. More often than not, however, the crest elevation is selected to prevent extensive damage to the structure or dune itself. In other words, it is often the case that the level of protection necessary will be provided as long as the structure itself remains intact during the design storm. In this connection, it should be recognized that one of classic modes of failure for coastal protection is for the relevant structure to be eroded away at the crest by excessive overtopping.

The following formula has been used by the Dutch government (TAW, 1984) to establish the minimum remnant dune profile (see Figure 8.16) that must remain during a design storm event.

$$h_0 = \text{DSWL} + 0.12 T_p \sqrt{H_{os}(m)} \geq 2.5(m)$$

where h_0 = dune crest elevation
DSWL = design still water level (including astronomical and storm tide)
T_p = peak spectral wave period
H_{os} = deepwater significant wave height

The remnant dune profile is designed to prevent heavy wave overtopping which, if permitted to occur, could lead to dune breaching. The above approach can be used to make a quick estimate of the crest elevation required for a coastal dune. The crest elevation required for a coastal structure is best evaluated using the concept of allowable overtopping. This approach is detailed later in this chapter.

The crest elevation for the containment structure on a perched beach should be selected so that it prevents the offshore movement of the beach profile during storms and/or so that the structure reduces the volume needed in a beach nourishment project. A cross-shore erosion model can be very useful in determining both the toe and crest elevations of such a structure.

8.4.2. Structures to Reduce Longshore Sediment Transport

This section considers functional design of the structure types discussed above.

8.4.2.1. Groins. As previously stated, groins have been misused in many applications and have a tarnished reputation as a result. Nonetheless, groins can be a very effective means for increasing the longevity of a beach fill. They can be applied responsibly by filling the groin compartments, by providing an adequately designed transition to downdrift beaches, and by ensuring timely beach nourishment of both the groin field and downdrift beaches. The following paragraphs discuss the basics of functional groin design.

Groins interrupt longshore sediment transport and result in sand accretion on the up-

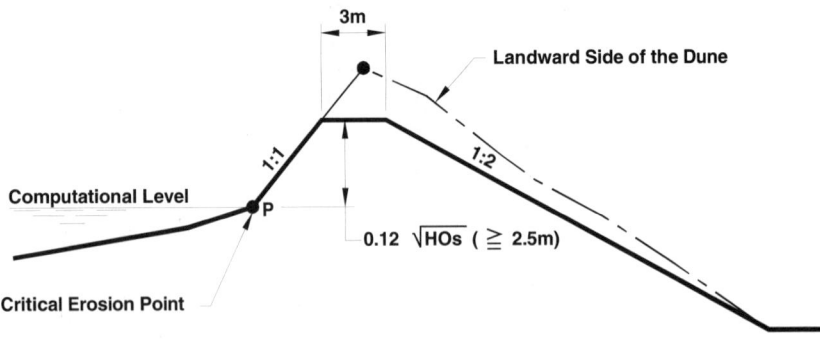

FIGURE 8.16. Minimum remnant dune profile (after TAW, 1984).

drift side of the groin in proportion to the length and height of the structure. The downdrift side of the groin is deprived of the accreted sand and usually erodes. Eventually, the shoreline up and downdrift of a groin will assume an equilibrium shape that is aligned to the predominant wave direction. In the case of a series of groins, the shoreline between the groins (groin compartment) will align itself in a similar equilibrium geometry. The shoreline shape is not straight, however. The central area of the groin compartment tends to be parallel to the predominant wave crests but the areas near the groins tend to be arcuate and correspond to the changes in wave direction that result as waves diffract around the groin. This arcuate shoreline is particularly distinct on the downdrift side of a groin.

Sand will tend to accumulate on the updrift side of a groin in proportion to the geometric profile of the groin. Specifically, the height of the impermeable section of a groin (inner core of a rubble mound groin) will act as a template or "form" that traps sand. Accordingly, the geometry of the beach profile can be controlled, to considerable degree, by the shape of the groin. Higher groins, for example, will tend to accrete sand at high levels in excess of the natural beach. Low-crested groins, on the other hand, may provide a template that is closer to the natural beach profile.

One of the factors that may reduce the efficacy of groins is the potential for rip currents (offshore-directed currents) that form adjacent to the groins. Wave-driven currents tend to travel parallel to shore. When these currents are interrupted by a groin, they may be trained offshore by the groins.

It is difficult to estimate the extent to which a groin field will retard longshore sediment transport. The following "rule of thumb" guidance has been generalized from the *Shore Protection Manual* (1984):

- 100% of the longshore transport will be interrupted if the groin is high and extends to a depth that is 167% of the normal or average breaker depth (i.e., 1.67 d_b)
- 75% of the longshore transport will be interrupted if the groin is low and extends to ≥ 1.67 d_b or if the groin is high and extends to a depth between 0.4 to 1.67 times d_b.
- 50% of the longshore transport will be interrupted if a high groin extends to ≤ 0.4 d_b, or if a low groin extends to <1.67 d_b.

Using the nomenclature described in the *Shore Protection Manual* (1984), a groin consists of three sections, as shown in Figure 8.17, namely: (1) the horizontal shore section (HSS), (2) the intermediate sloped section (ISS), and (3) the outer section (OS).

The HSS should extend a sufficient distance landward to prevent groin flanking (i.e., breaching of the beach area landward of the groin). The height of the core of the HSS is

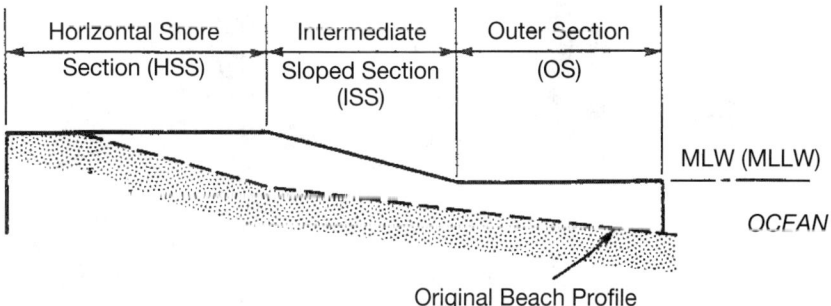

FIGURE 8.17. Typical groin (after *SPM*, 1984).

often set at the height of the natural berm. The ISS extends between the HSS and the OS and should parallel the natural foreshore slope. The crest elevation of the OS should be as low as possible, although groins are often built from land, which means that the groin must rise high enough out of the water to allow access for construction equipment. Additionally, for safety, the crest elevation should be visible under normal tide conditions.

A groin system should be designed to achieve a desired shoreline geometry. Normally, the intent of a groin field is to hold a desired protective beach in place. Particular attention should be paid to the shore alignment (i.e., berm position) in the vicinity of the groin. The original shoreline alignment and the shoreline orientation at nearby coastal structures can be very helpful in estimating the shoreline that will result after groin construction.

As shown in Figure 8.18, the berm crest on the updrift side of a groin will accrete to the seaward end of the HSS section. When evaluating the geometry of beaches adjacent to rubble mound groins it is important to focus on the geometry of the core of the groin rather than the armor, inasmuch as sand will often flow through the voids in the armor layer. The shoreline updrift of the first groin will extend from the seaward end of the HSS to the original shoreline. The sand impounded updrift of the groin is known as a fillet. The shape of the updrift shoreline can be estimated using a numerical (e.g., GENESIS) or analytical shoreline evolution model. The Pelnard-Considere model is summarized in the equations below and an example application is presented in Figure 8.19.

The following equation defines the shoreline position updrift of a complete littoral barrier from time 0 to the time when bypassing of the structure occurs.

$$y(x, t) = \left\{ \sqrt{\frac{4\varepsilon t}{\pi}} e^{-x^2/4\varepsilon t} - \text{erfc}\left(\frac{x}{\sqrt{4\varepsilon t}}\right) x \right\} \tan(\alpha)$$

where $y(x, t)$ = shoreline position at longitudinal distance x and time t
α = breaking wave angle
erfc( ) = complimentary error function = $1 - \text{erf}( )$
erf( ) = error function

$$\varepsilon = \frac{Q}{h\alpha}$$

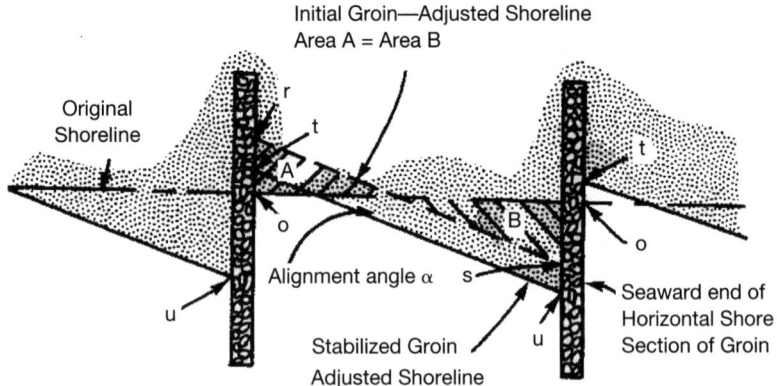

FIGURE 8.18. Shoreline alignment in a groin compartment (after *SPM*, 1984).

The time when bypassing occurs is equal to

$$t_{BP} = \frac{\pi L^2 h}{4Q\alpha}$$

Once bypassing begins, the shoreline position is governed by the following equation:

$$y(x, t) = \mathrm{erfc}\left(\frac{x}{\sqrt{4\varepsilon t}}\right) L$$

Matching time between the pre- and post-bypassing periods is done with the following equations:

$$t' = t - \left(1 - \frac{\pi^2}{16}\right) t_{BP}$$

$$Q_{BP} = Q\left[1 - \frac{2}{\pi\sqrt{\left(\frac{t}{t_{BP}} - 0.383\right)}}\right]$$

The example summarized in Figure 8.19 corresponds to a net longshore sediment transport rate of 157,000 cy/yr, a groin length of 100 feet and a wave angle of 2 degrees. The Figure shows that the shoreline grows to the seaward end of the updrift side of the groin and retreats an equal distance immediately downdrift of the groin. The groin begins to bypass sand after 1.2 years

The shoreline within a groin compartment can be estimated by extending a line from the seaward end of the HSS of the downdrift groin to the updrift groin. This line should be parallel to the predominant wave crest in the area. As shown in Figure 8.18, the new shoreline will accrete seaward of the original shoreline at the downdrift groin. There will be concomitant erosion relative to the original shoreline immediately downdrift of the updrift groin. The volumes of erosion and accretion will be roughly equal. This condition can result in excessive recession downdrift of the updrift groin, which can pose the risk of groin flanking. Downdrift erosion is an important consideration in determining the landward extent of the each groin. The erosion can be mitigated by adding sand to the groin compartment and by adjusting the groin lengths and/or spacing. As emphasized earlier, it is imperative to fill the groin compartments with sand in order to minimize erosion downdrift of the groin field. Such erosion will occur, in the absence of sand fill, because an unfilled groin field will trap sand until the compartments are filled to equilibrium. As a result, filling of the compartments to address erosion downdrift of the groin fill will also address the erosion on the downdrift side of each groin. It is important to realize, however, that it is highly likely that the groin field will still continue to reduce longshore sediment transport even after the compartments have been filled and there will be a need to provide a long-term means for mitigating the erosion downdrift of the groin field.

An analytical method for establishing the shoreline within a groin compartment has been advanced by Dean and Dalrymple (1993) and is presented below. This method can be used along with the above simple approach to evaluate shore geometries within the groin field.

$$y(x, t) = L - W\left(1 - \frac{x}{W}\right)\tan\alpha_b + \frac{\tan\alpha_b}{W}\sum_{n=0}^{\infty}\left\{\left[\frac{2W}{(2n+1)\pi}\right]^2 e^{-\varepsilon(2n+1)^2\pi^2 t/4W^2}\cos\left[\frac{(2n+1)\pi x}{W}\right]\right\}$$

where L = groin length
       W = groin compartment width

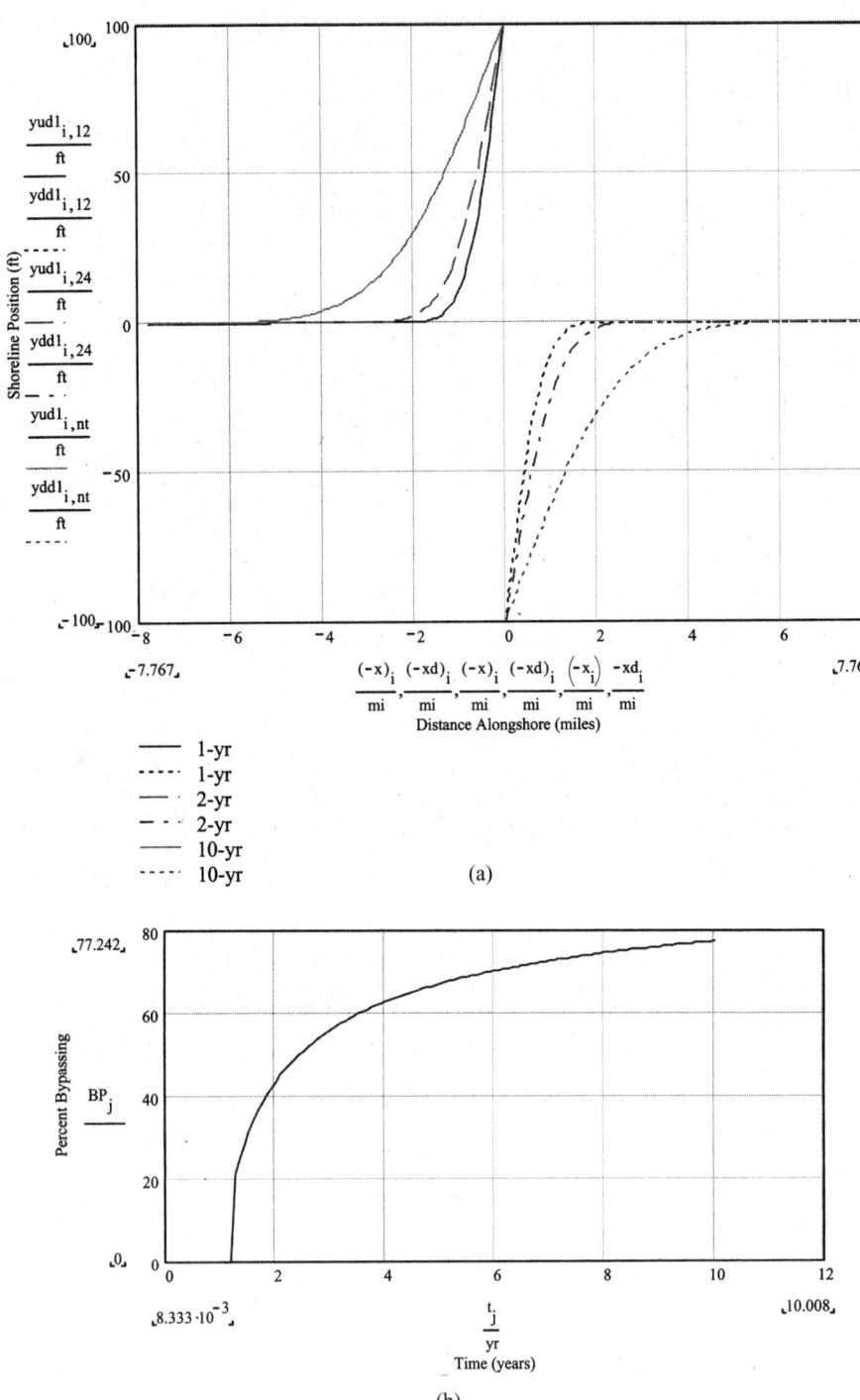

FIGURE 8.19. (a) Shoreline changes in the vicinity of groin. (b) Percent bypassing.

An application of the above equation is shown in Figure 8.20. Figure 8.20 corresponds to a net longshore transport rate of 157,000 cy/yr, groin lengths of 100 feet, and a groin compartment width of 300 feet. Figure 8.20 shows that the groin compartment shoreline aligns to the wave angle almost immediately. The model results also show that the shoreline grows to the seaward edge of the downdrift groin and shows that the shoreline immediately downdrift of the updrift groin does not retreat significantly landward. Numerical shoreline evolution models can also be used within groin fields, although care must be exercised to develop plausible answers.

The shoreline geometry downdrift of the last groin in a groin field is critical. Inasmuch as the area downdrift of a groin field can experience accelerated and detrimental erosion, it is inadvisable to use a groin field unless the terminal or last structure in the groin field is located at the boundary of a littoral cell (e.g., a headland or inlet) or unless a groin transition is provided. Even if such conditions exist, the designer should rigorously evaluate the long-term need for beach nourishment.

The shoreline geometry downdrift of a terminal groin can be estimated using numerical or analytical models. An example analytical model for a shoreline downdrift of a terminal groin is shown in Figure 8.21 and is based on the following equation developed by Larson et al. (1987):

$$y_{i,j}^w = \alpha_m \frac{\gamma}{2} \left\{ -\frac{4}{\gamma} \sqrt{\varepsilon_w \frac{t_j}{\pi}} e^{-(x_i^2/4\varepsilon_w t_j)} + 2\frac{x_i}{\gamma}\left[1 - \mathrm{erf}\left(\frac{x_i}{2\sqrt{\varepsilon_w t_j}}\right)\right] + \frac{1}{\gamma^2} e^{-\gamma x_i + \varepsilon_w t_j \gamma^2} \right.$$

$$\times \left[1 - \mathrm{erf}\left(\frac{x_i}{2\sqrt{\varepsilon_w t_j}}\right) - \gamma\sqrt{\varepsilon_w t_j}\right] - \left(\frac{1}{\gamma^2} e^{x_i + \varepsilon_w t_j \gamma^2}\right)\left[1 - \mathrm{erf}\left(\frac{x_i}{2\sqrt{\varepsilon_w t_j}} - \gamma\sqrt{\varepsilon_w t_j}\right)\right]$$

$$\left. + \frac{\alpha_m}{\gamma} e^{-\gamma x_i}(1 - e^{-\gamma^2 \varepsilon_w t_j}) \right\}$$

where $y_{i,j}$ = shoreline position at distance i and time j
x_i = shoreline position at point i
α_m = wave angle at groin

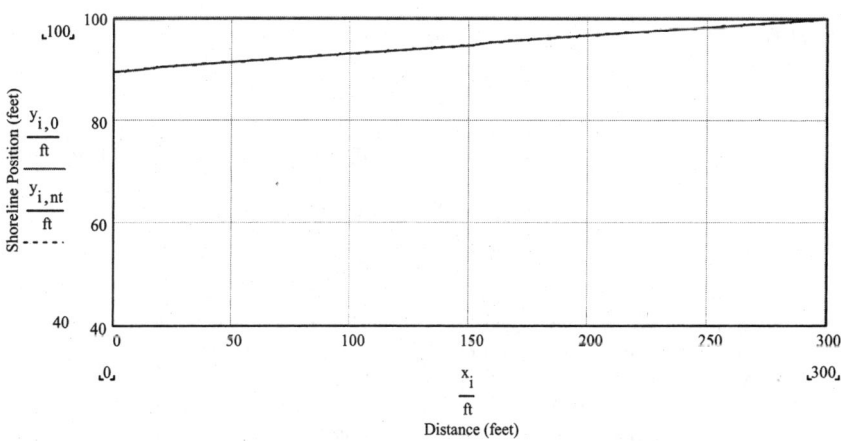

FIGURE 8.20. Analytical shoreline model results within a groin compartment.

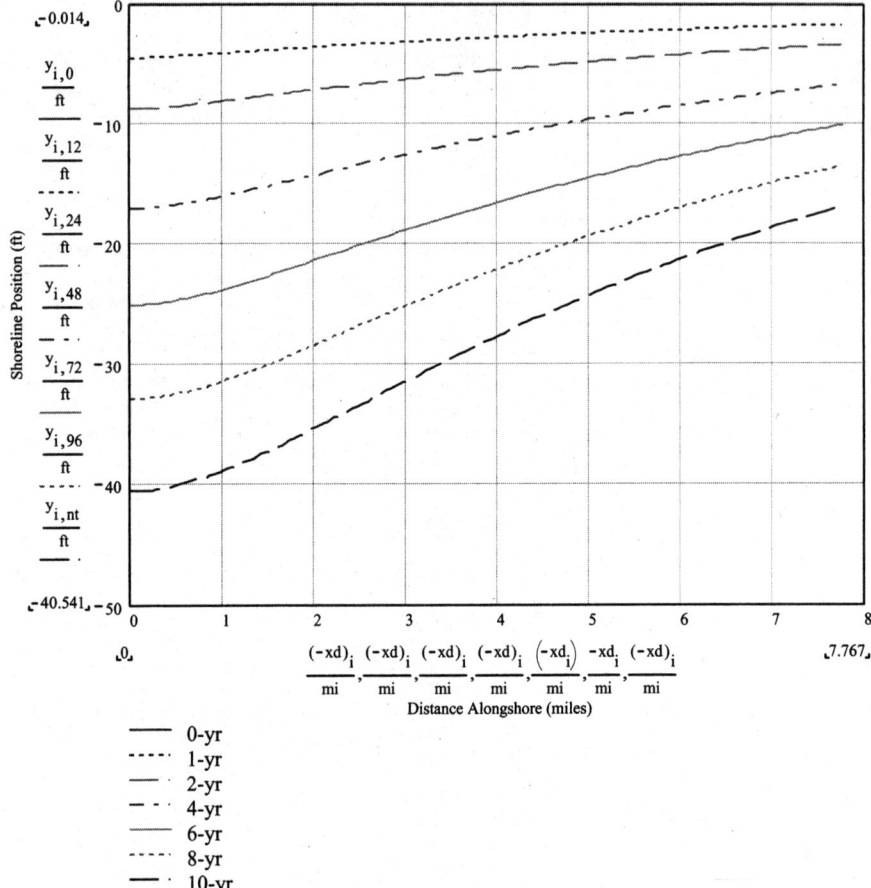

FIGURE 8.21. Analytical shoreline evolution model (including diffraction) downdrift of a groin.

γ = decay factor that accounts for the diffractive effect of the groin (units of 1/length)
ε_w = shoreline diffusivity for transport to the west = $2\,Q_w/D$
Q_w = longshore sediment transport rate
D = active depth of profile
erf = error function

The above equation is recommended in lieu of the Pelnard-Considere formula in view of the fact that the latter does not include the effects of wave diffraction.

Finally, many areas are characterized by seasonal reversals in the direction of longshore transport. Accordingly, the designer should assess the swings in shoreline position that can occur with changes in wave direction to assure that the swings will not flank the groins.

The updrift berm crest will grow to the seaward limit of the HSS. This position of the groin can be moved shoreward or seaward in order to achieve the desired beach width. The seaward extent of the OS is adjusted in accordance with the desired reduction in longshore transport, as discussed above. Estimates of the shore alignment within the groin field will normally dictate the groin spacing. As a general rule, the *Shore Protection Manual* (1984) recommends that the groin spacing be 2 to 3 times the groin length (measured from the berm crest to the seaward end of the OS).

A transition (i.e., groins of gradually reduced lengths) should be used to tie the groin field into the existing downdrift beach. Transitions should be used at both ends of the system where there are strong reversals in the direction of longshore littoral transport.

Kressner (1928) found that only three or four groins need to be shortened at the downdrift end of the system; see Figure 8.22. Kressner (1928) and Brunn (1952) both found that the transition is most effective if a line connecting the seaward ends of the shortened groins and the last full-length groin meets the natural shore alignment at an angle of about 6°. The authors' firm has designed a groin transition with the U.S. Army Corps of Engineers, New York District, at Westhampton Beach on Long Island using a transition angle of 3.45°, which is a milder angle than the 6° suggested by Kressner (Moffatt & Nichol Engineers, 1994a). The length of a groin, l, is measured from the crest of the beach berm to the seaward end. (The actual groin length extends shoreward of the berm.) The limit of the shortening is a matter of judgment; however, in the case of coastal tidal areas, it is suggested that the last transitional groin extend no farther seaward than the MLLW line.

The groin length and spacing for a transition must be selected so as to achieve a specified transition angle. The intent of a groin transition design is to provide a groin length to spacing ratio within the transition that is equal to length/spacing ratio within the groin field proper. The equations used to select groin lengths and groin spacing in a transition are as follows:

$$s_n = \frac{R_{sl}}{\left(1 + \frac{R_{sl}}{2}\tan(\theta)\right)} l_{n-1}$$

$$l_n = \frac{\left(1 - \frac{R_{sl}}{2}\tan(\theta)\right)}{\left(1 + \frac{R_{sl}}{2}\tan(\theta)\right)} l_{n-1}$$

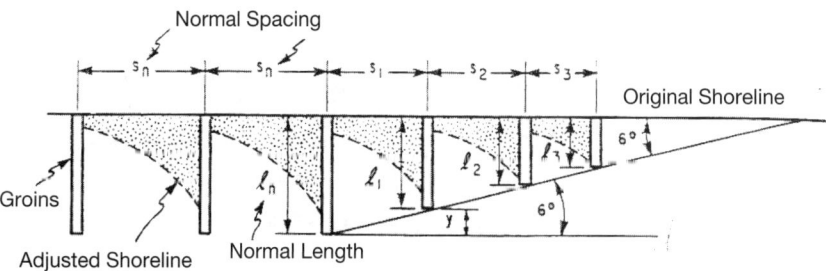

FIGURE 8.22. Schematic of groin-shortening procedure (after *SPM*, 1984).

where R_{sl} = ratio of groin spacing to groin length in the groin field
  l_n = length of the *n*th groin in the transition measured from berm crest to seaward end
  s_n = spacing between the groin *n* and groin *n* − 1 in the transition
  θ = transition angle
  l_{n-1} = length of groin *n* − 1

The above formulae are applied successively to evaluate the lengths and spacing of groins throughout the transition. The two variables that dictate the groin geometry are the groin spacing to groin length ratio, R_{sl}, and the taper angle, θ.

Sand will accrete to a higher elevation on the updrift side of groin than on the downdrift side of the groin. Estimating the beach profile on either side of the groin is an important element of structural design and can be used to estimate soil loads on groins comprised of vertical sheetpiles. The *Shore Protection Manual* (1984) suggests that the sand profile adjacent to a groin can be estimated by assuming that profile will extend from the seaward edge of the berm crest elevation on the HSS to the low water line. The appropriate point on the HSS is the seaward end of the HSS on the updrift side of the groin and landward of the HSS on the downdrift side in accordance with the groin compartment shoreline position. The slope of the profile will be the same as the native beach slope above low water. The profile below low water can be estimated by extending a line from the low water line to a point at the seaward end of the groin that lies on the native beach profile. A schematic profile is illustrated in Figure 8.23.

8.4.2.2. Offshore Breakwaters (also discussed in Chapter 5). Offshore breakwaters reduce longshore sediment transport by reducing incident wave heights in the regions protected by the structures. These devices can be used to build a wider beach (salient) in the lee of the structures or they can be designed to assure connection of the shoreline to the breakwaters. The latter formation is known as a tombolo.

Once a salient lee forms, longshore transport is likely to continue behind the breakwaters. If the transport continues, then the impacts to the downdrift shoreline will be minimized. A tombolo may substantially reduce or completely interrupt longshore transport. Depending on the situation, tombolo formation can result in detrimental impacts to the downdrift shoreline in much the same fashion as a groin field. The tombolo case is useful, however, in the design of artificial headlands and pocket beaches.

Numerous methods have been published for the functional design of offshore breakwaters (Dean-Rosati, 1990; Pilarczyk and Ziedler, 1996). A few of the methods are described below.

Figure 8.15 summarizes the geometry relevant to offshore breakwaters. In general, shoreline response to single and multiple offshore breakwaters can be estimated from the

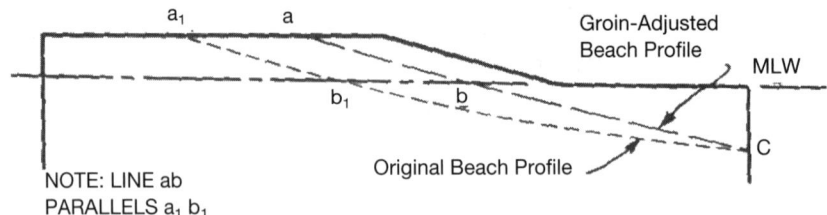

FIGURE 8.23. Determination of a beach profile adjacent to a groin (after *SPM*, 1984).

COASTAL PROTECTION METHODS   8.39

TABLE 8.1. Conditions for the formation of tombolos

Condition	Comments	References
$L_3/y > 2.0$		*SPM* (1984)
$L_3/y > 2.0$	Double tombolo	Gourley (1981)
$L_3/y > 0.67$ to 1.0	Tombolo shallow water	Gourley (1981)
$L_3/y > 2.5$	Periodic tombolo	Ahrens and Cox (1990)
$L_3/y > 1.5$ to 2.0	Tombolo	Daily and Pope (1986)
$L_3/y > 1.5$	Tombolo (multiple breakwaters)	Daily and Pope (1986)
$L_3/y > 1.0$	Tombolo (single breakwaters)	Suh and Dalrymple (1987)
$L_3/y > 2b/L_3$	Tombolo (multiple breakwaters)	Suh and Dalrymple (1987)

ratio of the breakwater length, L_s, to the distance offshore, X, see Tables 8.1 to 8.3. As shown in Tables 8.1 to 8.3, there is some overlap in the L_s/X values that correspond to tombolo and salient regions. Accordingly, care should be taken in applications.

Hsu and Silvester (1990) give the following equation for the shore-perpendicular length of a salient behind a single offshore breakwater.

$$\frac{Y_s}{Y_B} = 1 - 0.678\left(\frac{Y_B}{L_B}\right)^{0.215}$$

where Y_s = distance from the original shoreline to the seaward tip of the salient
Y_B = distance from the original shoreline to the breakwater
L_B = breakwater length

Suh and Dalrymple (1987) developed the following equation based on physical model test and prototype data:

$$X_s = 14.8\, XL_g \frac{X}{L_s^2} e^{-2.83\sqrt{L_g(X/L_s^2)}}$$

where X_s = distance from original shoreline to the seaward tip of the salient
X = breakwater distance from shore
L_s = breakwater length
L_g = breakwater gap distance

The above equation gives the distance from the original shoreline to the seaward tip of the accretion landward of the breakwater. The equation has been found to overpredict the sea-

TABLE 8.2. Conditions for the formation of salients

Condition	Comments	References
$L_3/y < 1.0$	No tombolo	*SPM* (1984)
$L_3/y < 0.4$ to 0.5	Salient	Gourley (1981)
$L_3/y < 0.5$ to 0.67	Salient	Daily and Pope (1986)
$L_3/y < 1.0$	No tombolo (single breakwater)	Suh and Dalrymple (1987)
$L_3/y < 2b/L_3$	No tombolo (multiple breakwaters)	Suh and Dalrymple (1987)
$L_3/y < 1.5$	Well-developed salient	Ahrens and Cox (1990)
$L_3/y < 0.8$ to 1.5	Subdued salient	Ahrens and Cox (1990)

TABLE 8.3. Conditions for minimal shoreline response

Condition	Comments	References
$L_3/y \geq 0.17$ to 0.33	No response	Inman and Frautschy (1966)
$L_3/y \geq 0.27$	No sinuosity	Ahrens and Cox (1990)
$L_3/y \geq 0.5$	No deposition	Nir (1982)
$L_3/y \geq 0125$	uniform protection	Daily and Pope (1986)
$L_3/y \geq 0.17$	Minimal impact	Noble 1978

ward tip of the salient for some prototype applications, but appears to be accurate for periodic tombolos and pocket beaches.

Suh and Dalrymple (1987) also suggest the following criteria for the formation of tombolos:

$$\frac{L_s}{X} \geq 1 \quad \text{Single Breakwaters}$$

$$\frac{L_g X}{L_s^2} \approx 0.5 \quad \text{Multiple Breakwaters}$$

Pope and Dean (1986) defined shoreline response as: (1) permanent tombolo, (2) periodic tombolo, (3) well-developed salient, (4) subdued salient, and (5) no sinuosity. Figure 8.24 provides design guidance in terms of L_s/L_g and X/d_s, where d_s is the depth at the structure.

Ahrens and Cox (1990) developed the following equation:

$$I_s = e^{1.72 - .41(L_s/X)}$$

where $I_s = 1$ (Permanent tombolo formation)
$I_s = 2$ (Periodic tombolo formation)
$I_s = 3$ (Well-developed salients)
$I_s = 4$ (Subdued salient)
$I_s = 5$ (No sinuosity)

Sedji et al. (1987) addressed the potential for shoreline retreat (relative to the original shoreline) opposite a multiple breakwater gap. They found that no erosion would occur for $L_s/X < 0.8$ while certain erosion would occur for $L_s/X > 1.3$. Erosion was possible between these two limits.

Hallermeir (1983) recommends that breakwaters be located in a depth d_{sa} to avoid tombolo formation. This depth is defined as

$$d_{sa} = \frac{2.9 H_e}{\sqrt{(S-1)}} - \frac{110 H_e^2}{(S-1) g T_e^2}$$

where d_{sa} = annual seaward limit of the littoral zone
H_e = deepwater waveheight exceeded 12 hours per year
T_e = wave period corresponding to the wave height
S = ratio of sediment to fluid density
g = acceleration of gravity

Hallermeir suggests a depth of $d_{sa}/3$ for tombolo formation and headland structures.

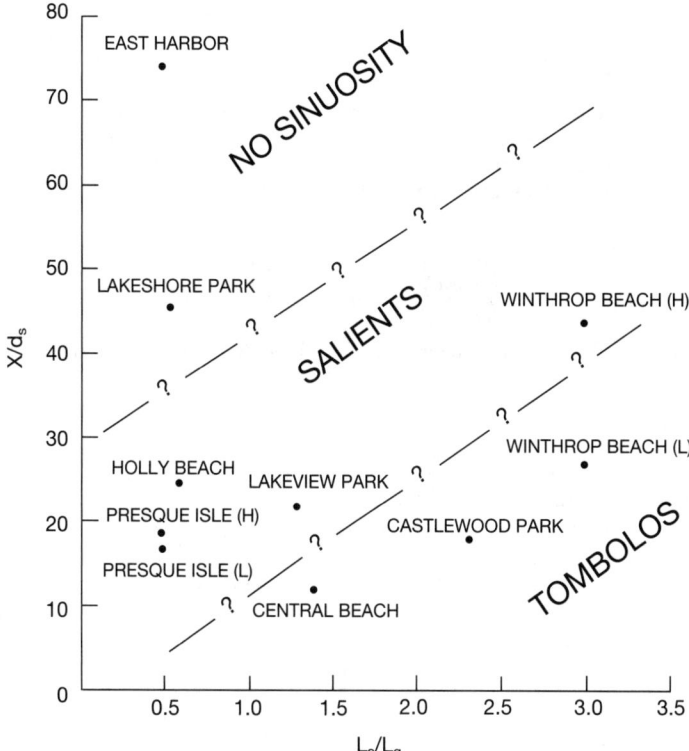

FIGURE 8.24. Dimensionless plot of United States segmented breakwater projects relative to configuration (from Pope and Dean, 1986).

8.4.2.3. Artificial Headlands and T-groins. A number of investigators have found that coastal embayments adjacent to headlands are of a crenulate shape. Beaches having this alignment may be in a state of static or dynamic equilibrium.

Silvester has promoted the use of artificial headlands as a means of coastal protection; see Figure 8.25. The intent of artificial headlands is to provide a shore-connected structure that mimics the crenulate-shaped embayment adjacent to headlands in nature. Artificial headlands can be viewed as offshore breakwaters with tombolos. The principle challenge for the designer is to establish the shape of the shoreline: (1) between adjacent headlands and (2) between a headland and the coastline downdrift of the headland.

Krumbein (1944), Yasso (1965), and Silvester (1960) have used a logarithmic spiral to describe the geometry of a crenulate bay. The logarithmic spiral is defined as follows:

$$R = R_0 e^{\theta \cot \alpha}$$

where R = length of the radius vector for a point P measured from the pole O
θ = angle from an arbitrary origin of angle measurement to the radius vector of the point P

8.42 HANDBOOK OF COASTAL ENGINEERING

R_0 = length of radius to arbitrary origin of angle measurement
α = characteristic constant angle between the tangent to the curve and radius at any point along the spiral

Silvester used results from model tests to develop Figure 8.25, which provides guidance for the value of α and the ratio a/b (a is the maximum indenture of the bay and b is the distance or gap between headlands).

Silvester and Hsu (1993) conducted further research and developed an alternative parabolic equation:

$$\frac{R}{R_0} = C_0 + C_1 \frac{\beta}{\theta} + C_2 \left[\frac{\beta}{\theta}\right]^2$$

Curve fits of C_0, C_1, and C_2 have been developed by the present authors and are as follows:

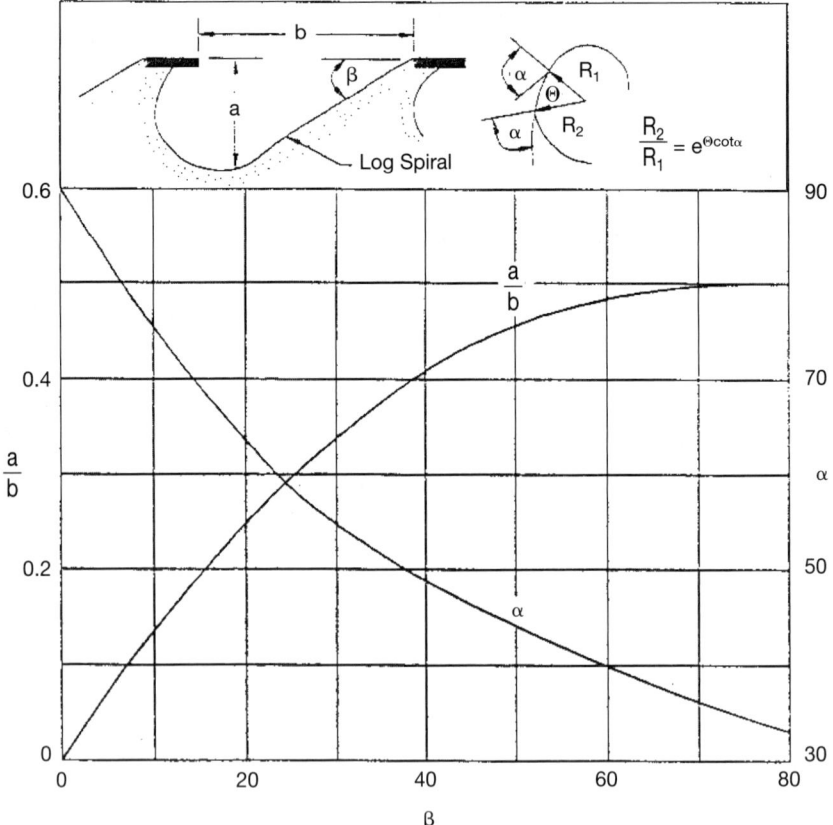

FIGURE 8.25. Parameters relating to bays in static equilibrium (after Silvester and Hsu, 1993).

$C_0 = 0.0000000479\beta^4 - 0.0000087963\beta^3 + 0.0003521878\beta^2 + 0.0047891887\beta + 0.0715244255$

$C_1 = -0.0000001281625\beta^4 + 0.0000181988465\beta^3 - 0.0004865839195\beta^2 + 0.0077130700611\beta + 0.9551247533875$

$C_2 = 0.00000011262\beta^4 - 0.00001561035\beta^3 + 0.00055939061\beta^2 - 0.01497707408\beta + 0.0859531142$

The geometry pertaining to the parabolic bay formulae are presented in Figure 8.26.

The parabolic method provides a better fit to crenulate shorelines than the logarithmic spiral in many cases and can be used to estimate shoreline positions between headlands.

The authors of the present chapter have noted that the parabolic bay methods of Silvester seem to work better for angles exceeding 15 to 20°. Many practical applications, however, involve relatively small wave angles. Accordingly, the design of artificial headlands, pocket beaches, and T-groins remains a matter of judgment, although some of the methods presented herein have been found to be useful.

The above equations can be used to estimate the shoreline shape that results from a series of artificial headlands at a given longshore spacing and structure orientation.

Combinations of coastal structures may be used in areas of relatively high long-term erosion rates. Such areas are often immediately downdrift of a littoral barrier (e.g., coastal inlet, harbor, submarine canyon, existing jetty, or coastal structure). Another example would be where a beach has been created in an area that is otherwise devoid of sand. Recent trends have been to combine the attractive features of groins with those of offshore breakwaters or artificial headlands. Groins are very effective in forming compartments of stabilized beaches but are vulnerable to rip currents and flanking. Offshore breakwaters (with tombolos) and artificial headlands can be used to compartmentalize/stabilize an eroding beach but risks of flanking those structures remain. Accordingly, some have advanced concepts involving the combined use of groins and offshore breakwaters/artificial headlands. The combined structures have been called T-groins and have been used to create pocket beaches. Sometimes, T-groins are used in combination with offshore breakwaters. An example T-groin application is shown in Figures 8.15. It should be noted that T-groins and pocket beaches tend to lock-in a beach segment. As a result, it is normally necessary to both fill the stabilized area during construction and provide a suitable transition to the downdrift shoreline in order to avoid erosion adjacent to the stabilized area.

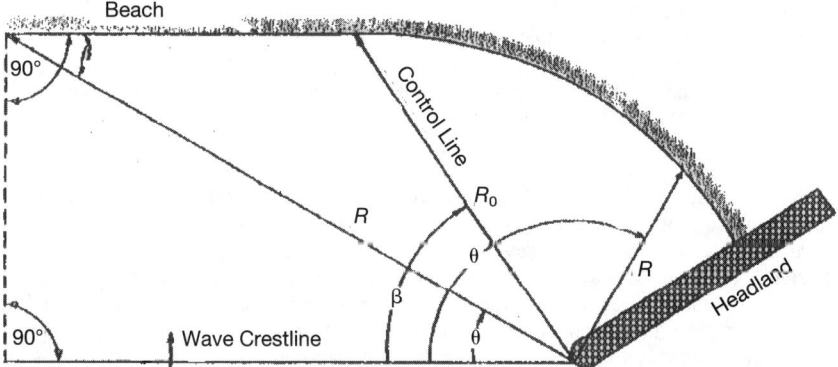

FIGURE 8.26. Definition sketch of the parabolic shape (after Silvester and Hsu, 1993).

The functional design of T-groins and pocket beaches involves a number of the above methodologies. One can start by locating offshore breakwaters or headlands sufficiently close to the shore to assure the formation of a tombolo. The above formulae for offshore breakwaters can be used to estimate gap spacing and lengths of the heads (i.e., "T's") of the structure. The trunks or groin sections of the structure can then be extended landward from the head a distance sufficient to avoid risk of flanking the groin. Next, one needs to establish the shape of the shoreline within the compartments between the structures. It is important here to establish the landward indentation of the shoreline from the gap or "control line" that extends between the breakwater/headland structures, see Figure 8.26. This indentation establishes the minimum beach width along the stabilized section and tends to dictate the level of protection offered by the pocket beach to hinterland areas. There are several methods for establishing the shoreline shape and resulting indenture.

Silvester's method, as summarized in Figures 8.25 and 8.26, has been used to establish the shoreline indenture and overall shape. Silvester's more recent parabolic shoreline equations presented above can also be used to estimate the shoreline between the structure heads. Bodge (1998) has advocated the use of a simple "⅓ rule" wherein the shoreline indenture (distance from the gap control line to the shoreline) is ⅓ of the gap distance. Bodge suggests that the shoreline between the heads of the breakwater structures is parallel to the control line (which is parallel to the principle wave direction) and is located a distance ⅓ of the gap length landward of the control line. Bodge suggests that the shoreline in the lee of the structure head can be estimated with a circle having a radius of ⅓ of the gap length. Finally, Bodge recommends that the final shoreline for design purposes can be taken as a composite average between the shoreline established using the ⅓ rule and the shoreline developed using a logarithmic spiral with its origin at the center of the head of the "T."

Berenguer and Enriquez (1988) offer design guidance for pocket beaches based on experience in Spain. They developed an equation for the maximum indenture from the gap control line based on empirical results from a number of constructed projects. The equation is given below:

$$A_1 = 25(m) + 0.85L_g$$

The above equation has the disadvantage of not giving a limit of zero indenture for zero gap length. Accordingly, the present authors reexamined the data of Berenguer and Enriquez and developed the following equation (which has a squared correlation coefficient of 0.76):

$$A_1 = -0.003685166L_g^2 + 1.151633229L_g$$

The above equation uses data for grain sizes less than 0.3 mm and neglects the high and low fill level data referenced in the original Berenguer and Enriquez paper. Berenguer and Enriquez also found that that $A_0 = 2A_1$, where A_0 is the maximum lateral distance of the compartment shoreline.

Hardaway and Gunn (1999) present a summary of design methods for headland pocket beaches and suggest the following ratio:

$$A_1 = 0.606L_g$$

Overall, the Berenguer and Enriquez method provides the largest estimate of the maximum indenture of the shoreline, followed by the Hardaway and Gunn method. This is particularly true for the small wave angles that are common in practice. Berenguer and Enriquez recommend that the compartment shoreline be fit with a logarithmic spiral. Specifically, they recommend that the center of the spiral be located at the midpoint of the breakwater gap, with a characteristic angle between 80 and 90 degrees. The pocket beach that results from this procedure tends to be relatively close to a circle and differs substantially from the shape produced by Silvester or Bodge's analyses. Nonetheless, the shape is consistent with the behavior of some pocket beaches and is certainly applicable to the

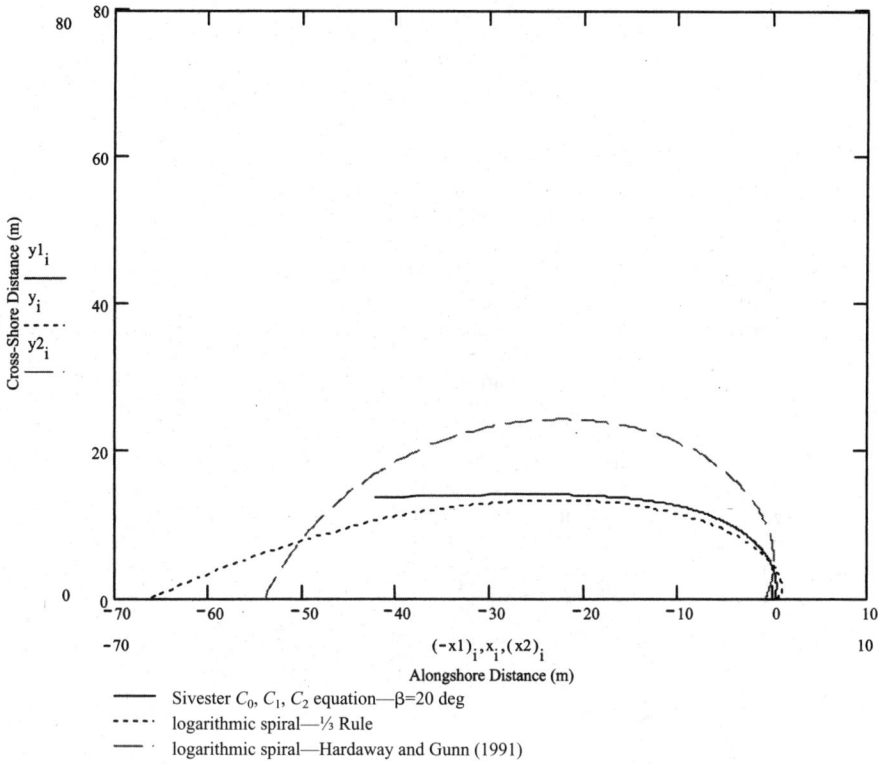

FIGURE 8.27. Example comparison of shorelines within a pocket beach using alternative methods.

Spanish cases presented by Berenguer and Enriquez. A comparison of the above methods for a particular case is shown in Figure 8.27. Overall, it is judged that a reasonable design can be developed using the Hardaway and Gunn ratio for the maximum indenture between headlands/offshore breakwaters. The Bodge, Silvester, and Berenguer and Enriquez methods can be used to bracket the possible range of shoreline geometries.

As shown in Figure 8.27, a variety of shoreline geometries can be estimated for a proposed headland/T-groin pocket beach, depending on the method adapted. Accordingly, it is normally necessary to exercise engineering judgement based on experience in order to develop a sound functional design.

8.5. STRUCTURAL DESIGN (DETERMINISTIC DESIGN, DESIGN OPTIMIZATION, AND PROBABILISTIC DESIGN)

This section addresses structural design of rubble mound coastal protection works. Methods pertaining to the coastal engineering aspects of vertical and monolithic structures can be found in Goda (1985) and the *SPM* (1984). Standard references such as DM-7 and Bowles (1977) provide useful information regarding structural and geotechnical design of vertical and monolithic structures.

8.5.1. Design Conditions

The design life selected for a coastal structure is an important factor in the overall planning.

Design guidance for the US Army Corps of Engineers (USACE) has required a project life of 50 years (ER-1110-2-1407 "Hydraulic Design of Coastal Shore Protection Projects"). This guidance has now been superseded by a revised ER-1110-2-1407 (November 1990), which dictates that a wider range of project lives be considered in order to account for differences in cost of repair, periodic replacements, and rehabilitation. The 50-year project life is consistent with the nature of routine coastal and hydraulic engineering projects that are designed to protect large areas of rural and urban infrastructure against flooding and/or wave-induced damages. Such projects are normally justified on the basis of a rigorous and codified economic analysis that assures that project benefits exceed project costs. The most rational means for selecting the project design life, however, is on the basis of economics (i.e., project costs and cost effectiveness).

A coastal protection structure must be designed for a given level of hydrodynamic design conditions including winds, waves, water levels, and currents. Design conditions can be stated in terms of risk levels and/or in terms of statistical return period. These two factors are related to one another and the project life through the following formula:

$$R = 1 - 1\left(1 - \frac{1}{RP}\right)^L$$

where R = risk or probability that a given condition will be equaled or exceeded
L = project life in years
RP = return period in years

Stated simply, the return period is the average time intervals between events of a similar magnitude. For example, a 50-year design wave would correspond to a wave that occurs an average of once every 50 years. A common USACE criteria stipulates that project should be designed for an event that has a 50% risk during a 50 year project life. Manipulation of the above formula will show that this criterion corresponds to a return period of 73 years.

Large coastal engineering and marine construction projects are often designed on the basis of an optimized approach in which a balance is obtained between initial construction costs and the maintenance costs associated with storm-induced damages. With regard to rubble mound coastal structures, two of the most important factors are the primary armor layer size and the crest elevation, although other aspects of the structure design can also be evaluated using an optimization methodology. Figure 8.28 presents a schematic design optimization plot. The x-axis is a measure of storm intensity (e.g. wave height or storm tide return period), whereas average annual costs are presented on the y-axis. A series of computations must be made to develop annualized construction costs for the various structure geometries that correspond to a given range of return periods. These computations result in the curve labeled *Construction Costs* in Figure 8.28 and shows that first cost increases with return period. Additional computations are made to estimate the damages (and corresponding maintenance and repair costs) that can occur to a cross-section designed for a particular return period under the full range of possible return periods. These computations result in the curve labeled *Maintenance & Repair Costs* in Figure 8.28. These two curves are then added to provide a curve labeled *Total Cost*. The location

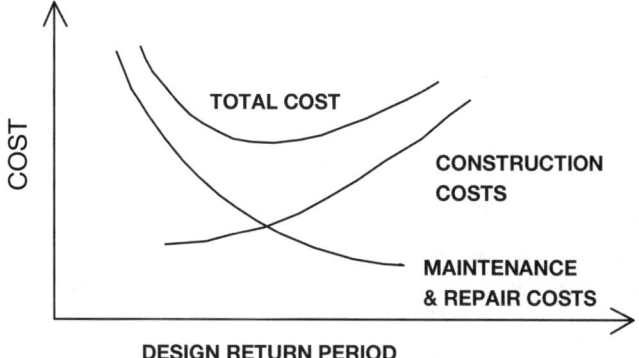

FIGURE 8.28. Schematic design optimization plot.

of the minimum point on the *Total Cost* curve corresponds to the optimal design condition. The approach summarized in Figure 8.28 provides a quantitative and objective means for assessing design conditions for coastal protection structures.

8.5.2. Wave Runup and Overtopping

One of the primary functions of most shore-parallel coastal structures (i.e., sea walls, revetments, and dikes) is to protect hinterland areas from the adverse effects of high water and waves. If a high level of protection is required, the structure should have a height above the maximum level of wave runup during storms. Typically, a high crest elevation will result if zero wave overtopping is allowed. If some overtopping is allowed based on the nature of the structure, however, the design requirement can be evaluated in terms of allowable overtopping.

Wave runup, and more importantly, overtopping computations allow an objective means for evaluating the level of protection (i.e., allowable overtopping) offered by various structure heights and armor protection combinations. In addition, wave overtopping computations provide a rational means for evaluating the relative risk of structure breaching and subsequent failure.

Wave runup is commonly evaluated on the basis of the composite-slope runup method outlined in the *Shore Protection Manual* (1984). This approach has been critically reviewed by the Federal Emergency Management Agency (FEMA) (1988). It was found that the composite-slope method provides a valid approach for estimating the *mean* runup value in random waves, but was lacking in its ability to predict *extreme* values of wave runup. Low or insignificant wave overtopping discharge values are normally computed on the basis of the mean wave runup values.

European engineers have long appreciated the need to consider wave runup levels higher than the mean values in design applications and have generally used the 2% exceedence runup value to select the heights of dunes and coastal dikes. Van der Meer (1998) published the following formulae for computing the 2% runup for sea walls and dikes:

$$R_{2\%} = 1.6 \gamma_b \gamma_f \gamma_B \xi_p H_s$$

$$\text{Maximum} = 3.2 \gamma_f \gamma_B$$

$$\xi_p = \frac{\tan \alpha}{\sqrt{S_p}}$$

$$S_p = \frac{H_s}{\frac{g}{2\pi}T_p^2}$$

where $R_{2\%}$ = 2% wave runup (wave runup exceeded only 2% of the time during a storm)
  γ_f = reduction factor for roughness
  γ_B = reduction factor for oblique wave attack
  ξ_p = breaker parameter (surf similarity parameter) based on equivalent slope
  α = angle of beach and or structure slope
  S_p = wave steepness
  H_s = significant wave height, average of highest one-third
  g = acceleration of gravity
  T_p = peak spectral wave period

Van der Meer's formulae are based on an extensive series of physical model tests including several full scale tests.

The influence of roughness based on van der Meer (1992) is summarized as follows:

Surface covering	Influence factor (γ_f)
One layer of rock	0.55–0.60
Two or more rock layers	0.50–0.55

More information on γ_f can be found in van der Meer (1998).

Wave runup is an important overall indicator of the protection offered by coastal structures. As previously mentioned, however, wave overtopping is judged to be a more objective and rationale method for estimating level of wave protection. Van der Meer (1998) presents the following formula for estimating the mean wave overtopping on sloping coastal structures subject to random waves:

$$\gamma_b = 1 - \frac{B}{L_{\text{berm}}}\left[1 - 0.5\left(\frac{d_h}{H_s}\right)^2\right]$$

with $0.6 \leq \gamma_b \leq 1$ and $-1 \leq d_h/H_s \leq 1$, and where

B = berm width
L_{berm} = berm length
d_h = wave depth over the berm

The reduction factor γ_B for short waves is computed as:

$$\gamma_B = 1 - 0.0022\,\beta \quad [2\% \text{ runup}]$$

$$\gamma_B = 1 - 0.0033\,\beta \quad [\text{overtopping}]$$

where β is the wave angle with respect to shore normal in degrees. Van der Meer (1998) provides additional details on each of the above reduction factors. That reference also provides guidance for examining the reliability of the above formulae.

$$\frac{q}{\sqrt{gH_s^3}} = \frac{0.06}{\sqrt{\tan \alpha}}\, \gamma_b \xi_p \exp\left[-4.7 \frac{R_c}{H_s} \frac{1}{\xi_p \gamma_b \gamma_f \gamma_B}\right]$$

with the following as a maximum:

$$\frac{q}{\sqrt{gH_s^3}} = 0.2 \exp\left[-2.3 \frac{R_c}{H_s} \frac{1}{\gamma_f \gamma_B}\right]$$

where q = mean wave overtopping discharge per unit width
R_c = dike crest freeboard (height of structure above still water)

The reliability of the above equation can be given by assuming that $\log(q/gH_s^3)$ has a normal distribution with a variation coefficient (the ratio of the standard deviation to the mean value) of 0.11. Reliability bands can then be calculated for various practical values of mean overtopping discharges.

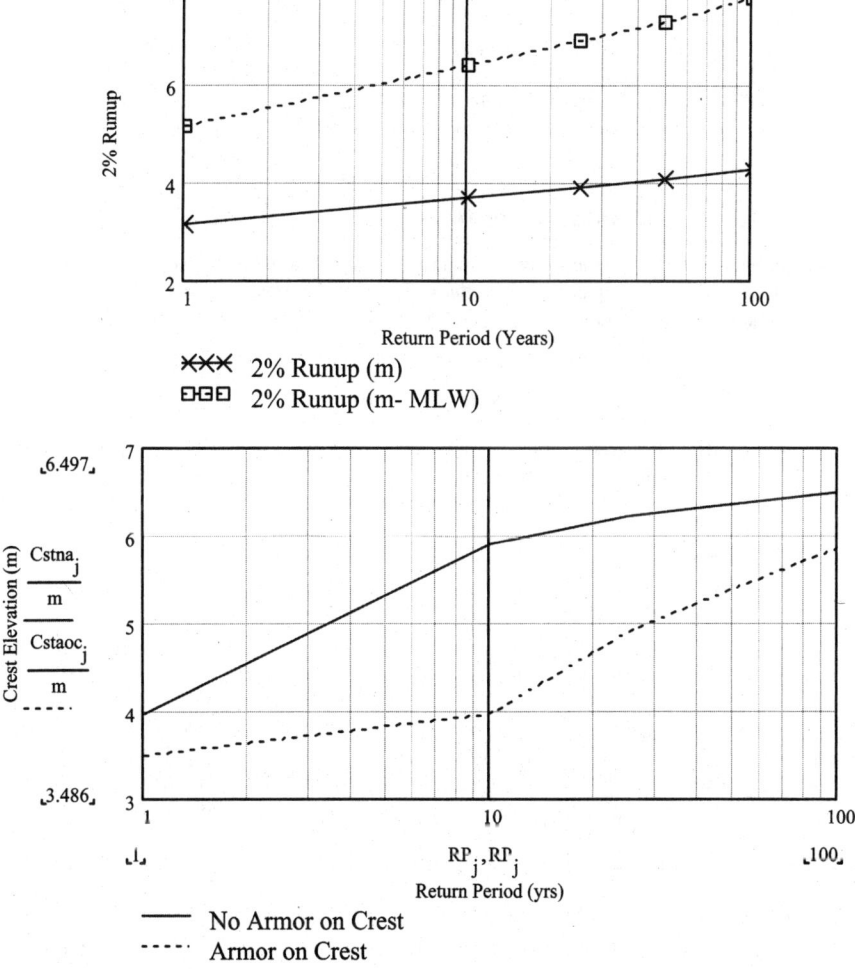

FIGURE 8.29. Example runup and overtopping analysis (Moffatt & Nichol Engineers, 1999b).

The above overtopping formula provides a means for computing wave overtopping on dikes of various geometries (i.e., structure slopes, slope breaks, and crest elevations). In order to evaluate the level of protection offered by a given structure configuration, it is necessary to establish limiting values for allowable overtopping. Critical or allowable overtopping discharges have been published by the United Kingdom (U.K.) Construction Industry Research and Information Association (CIRIA) and the Netherlands Centre for Civil Engineering Research and Codes (CUR) (CIRIA/CUR, 1991). Goda (1985) has also published similar values. The Goda allowable overtopping values have been used in this study and are summarized below:

Structure type	Surface armoring	Overtopping rate (Liters/m/s)
Type I: Coastal dike	Armor on front slope, soil on crown and back slope	5
Type II: Coastal dike	Armor on front slope and crown, soil on back slope	20
Type III: Coastal dike	Armor on front slope, crown and back	50
Type IV: Revetment	No pavement on ground	50
Type V: Revetment	Pavement on ground	200

8.5.3. Rubble Mound Armor Design

There are a number of methodologies available for determining armor stone requirements for coastal structures subject to wave attack. The most commonly used method is based on the Hudson equation, published in the *Shore Protection Manual* (1984):

$$W = \frac{\gamma_r^3 H}{K_D (S_r - 1)^3 \cot(\theta)}$$

where W = weight of armor stone
 γ_r = unit weight of the armor rock
 H = wave height to which the structure is exposed
 K_D = stability coefficient
 $S_r = \gamma_r/\gamma_w$
 γ_w = unit weight of water
 θ = angle of structure slope

Typical coastal protection structures will be located in relatively shallow water and will be exposed to a wave spectrum characterized by breaking waves as emphasized earlier in this chapter. There may also be cases however, when the structure will only be subjected to nonbreaking waves. The wave height used in the above equation depends on whether one is evaluating breaking or nonbreaking waves. According to the *SPM* (1984), an H_{10} wave height, which is equal to 1.27 times the significant wave height (H_s), is used for the nonbreaking wave height, whereas the maximum depth-limited wave height is used for breaking waves. The Hudson equation for breaking waves can result in overly conservative estimates of required armor stone sizes. This is especially true given the latest *Shore Protection Manual* guidance for the breaking wave stability coefficient, which is much lower than values published in previous editions of the manual. Stability coefficients recommended in the current and previous editions of *Shore Protection Manual* for rough angular stone are as follows:

COASTAL PROTECTION METHODS 8.51

Structure trunk	K_D Values	
	SPM (1984)	SPM (1977)
(Random placement, nonbreaking waves)	4.0	4.0
(Random placement, breaking waves)	2.0	3.5
Structure Head		
(Random placement, breaking waves)	1.9 for cot(θ) = 1.5	
(Random placement, breaking waves)	1.6 for cot(θ) = 2.0	
(Random placement, breaking waves)	1.3 for cot(θ) = 3.0	
(Random placement, nonbreaking waves)	3.2 for cot(θ) = 1.5	
(Random placement, nonbreaking waves)	3.8 for cot(θ) = 2.0	
(Random placement, nonbreaking waves)	2.3 for cot(θ) = 3.0	

With regard to geometry, a distinction is made between the seaward tip, or head, of the structure and the middle portion of the structure, known as the trunk. The head section is generally less stable than the trunk; consequently, a lower stability coefficient is normally specified for the structure head as indicated above.

Rock sizes using the newer criteria for breaking waves are 1.6 times larger (i.e., 3.5/2) than those computed using the older published criteria for the structure trunk.

An alternative method for computing armor requirements has been developed by van der Meer (1988). Van der Meer's equations for sizing armor stone subject to shallow-water random waves are as follows:

For plunging waves:

$$\frac{H_{2\%}}{\Delta} = 8.7 \, P^{0.18} \left(\frac{S}{\sqrt{N}} \right)^{0.2} \xi_m^{0.5}$$

For surging waves:

$$\frac{H_{2\%}}{\Delta} D_{n50} = 1.4 \, P^{-0.13} \left(\frac{S}{\sqrt{N}} \right)^{0.2} \xi_m^P$$

The waves are of the surging types when

$$\xi_{mc} \geq 6.2 \, P^{31} \sqrt{\tan} \, \theta^{1/(P+0.5)}$$

where $H_{2\%}$ = 2% exceedence of wave height
$\Delta = S_r$
D_{n50} = mean nominal diameter of the stone = $(W/S_r)^{1/3}$
S = structural damage level taken as 2 for 0–5% damage
N = number of waves in the storm (a value of 7000 was used)
P = structure permeability
ξ_m = surf similarity parameter

The surf similarity parameter is defined as

$$\xi_m = \frac{\tan(\theta)}{\sqrt{\frac{2\pi H_s}{gT_p^2}}}$$

where H_s = significant wave height
T_p = peak spectral wave period

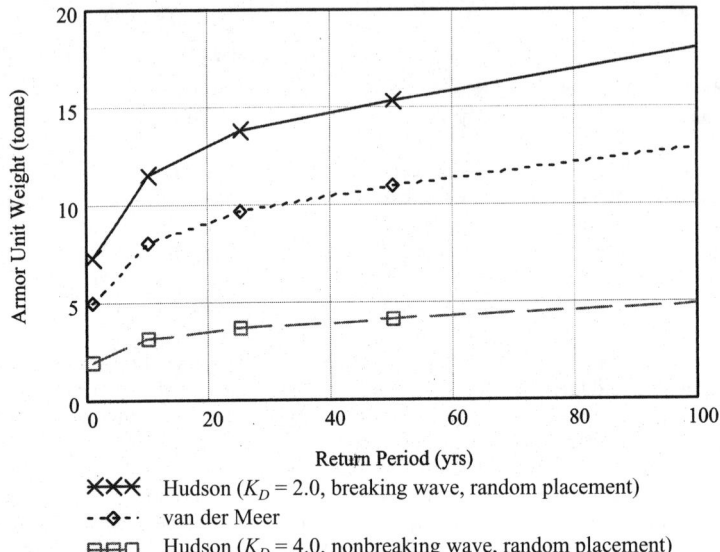

FIGURE 8.30. Example armor size computation results (Moffatt & Nichol Engineers, 1999b).

The methodology presented by van der Meer is often preferable because it is based on random wave conditions, whereas the Hudson formula is based on regular wave conditions.

Stone requirements for groins, headlands, offshore breakwaters and other structures in the surf zone require special consideration. Since the these structures are located in relatively shallow water on the open coast, the structures will be exposed to breaking waves. The wave height for design is the maximum depth-limited height that can reach the structures for a given storm tide water level and bottom elevation.

The outer sections of typical surf zone structures are relatively low, with crest elevations on the order of 0 to 1 m above low water. Accordingly, one can expect that the groins and breakwaters/headlands will be overtopped during attack by storm tides and severe waves. Some designers have taken an approach where the depth-limited waves assumed capable of causing damage on an overtopped structure are limited to those associated with water levels that exceed the crest of the structure by less than 1 m. Under such conditions it has been reasoned that waves associated with water levels exceeding the crest of the structure by more than 1 m do not cause as much damage as waves at lower water levels, since much of the wave crest passes over the structure without impinging on its rock face. This basic phenomenon can be examined using the following equation for submerged breakwaters developed by Van der Meer:

$$N_s^* = \frac{H}{\Delta D_{n50}} s_p^{-1/3}$$

where h_c = height of structure crest
       h = water depth at the toe of the structure
       $s_p = 2\pi H/(gT_p^2)$

Armor stone computations should be made for a range of water levels and bottom elevations. The general objective is to provide a structure that can sustain the design storm; however, the largest armor stones on any particular section of the groin or surf zone structure may not result from the waves and water levels attending the design event. This stems from the previously mentioned fact that waves overtop the structure when the water levels rise above the relatively low portions of the structure. As a result, the appropriate stone size requirements for the entire structure can only be determined through an examination of numerous combinations of water levels, bottom elevations, and structure heights. This is particularly important for groins where the bottom elevation and structure crest elevation vary significantly over the length of the structure.

The first step in examining stone size requirements for a given bottom elevation and water level is to compute the total water depth from which the maximum breaking wave height can be determined. This breaker depth, h_b, is the sum of the selected water elevation and the bottom elevation. The maximum breaker height that can be supported in the resulting water depth can be estimated using the Goda model presented earlier in this chapter. The resulting maximum breaker height can be used in the Hudson equation. The Van der Meer equations, on the other hand, require an estimate of the $H_{2\%}$ wave height, which Van der Meer (1990) states can be equated to the $H_{1/250}$ computed using the Goda model. Example results are presented in Figure 8.31.

It should be noted, however, that none of the above methods for sizing armor stone account for the condition of a single layer of armor stone that is common to groins and

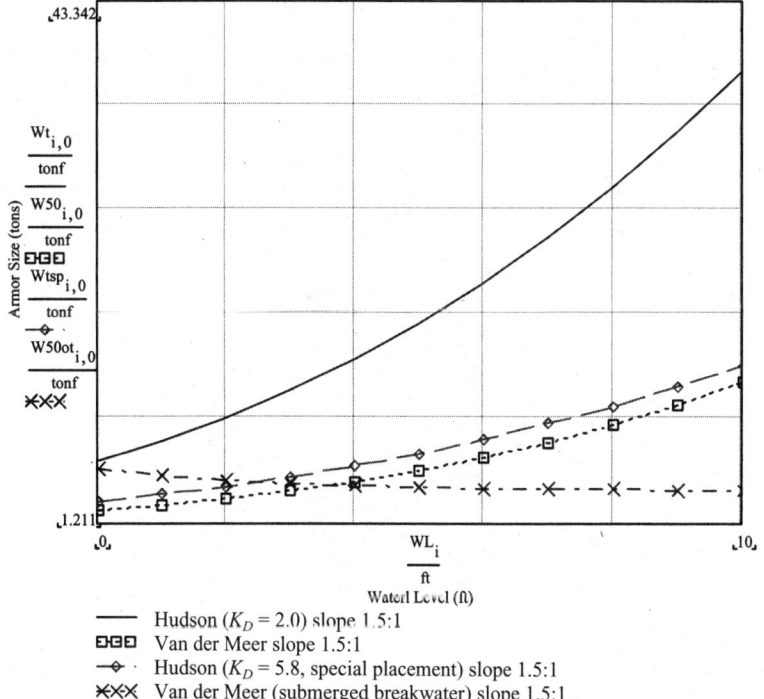

—— Hudson ($K_D = 2.0$) slope 1.5:1
ⒺⒽⒺ Van der Meer slope 1.5:1
—◆— Hudson ($K_D = 5.8$, special placement) slope 1.5:1
✖✖✖ Van der Meer (submerged breakwater) slope 1.5:1

FIGURE 8.31. Example groin armor size analysis (Moffatt & Nichol Engineers, 1994).

other coastal structures. In fact, it is stated in the *Shore Protection Manual* that single armor layers are not recommended for breaking wave conditions. Notwithstanding this recommendation, groins are commonly designed using a single layer of armor, since it is very difficult to fit two layers of armor in a groin without building a massive structure. Furthermore, two layers of armor result in a permeable structure that is a relatively inefficient barrier to longshore transport. Judgement must be exercised in actual applications to account for the decreased stability that might be associated with a single layer of armor.

8.5.4. Scour Protection

Supplemental armor is often provided at the toe of an armored slope in order to prevent wave energy from scouring and undercutting the slope. Factors that affect the severity of toe scour include wave breaking, wave runup and rundown, wave reflection and grain size distribution of the beach or bottom materials. Toe stability is essential because failure of the toe will generally lead to failure throughout the entire structure. Toe scour is a complex process and specific design guidance has not been developed. Some general guidelines, however, have been suggested and are summarized in Figure 8.32.

8.5.5. Underlayers and Filters

Coastal structures are normally constructed with an armor layer and one or more underlayers. Many structures have two layers of armor and one or two underlayers overlying a geotextile built upon a core of sand or clay. Small particles beneath the geotextile should not be washed through the fabric and the underlayer stones should not be washed through the armor. Geotechnical filter rules should be considered; see *SPM* (1984).

The *SPM* (1984) recommends that underlayer stone range from 1/10 to 1/15 of the armor weight. This results in a relatively large underlayer, which has two advantages. First, a large underlayer permits surface interlocking with the armor. Second, a large underlayer produces a more permeable structure and therefore has an influence on the stability of the armor layer.

8.5.6. Optimal and Probabilistic Design

The purpose of optimization analysis is to select the design return period that corresponds to an optimal balance between initial capital cost and maintenance/repair costs as discussed above. The following section considers two damage modes for rubble mound construction: (1) armor layer damage from direct wave attack, and (2) armor layer damage due to wave overtopping. It is important to note that the following optimization analysis only considers damage to the primary armor layer. Underlayer and structure core damage have not been included. This approach is often taken in practice because it is difficult to quantify underlayer/core damage and because the armor layer normally dictates optimal design. Specifically, damage to the underlayer and core can only occur after relatively severe damage to the armor layer and it is seldom optimal to allow severe damage to the armor layer.

8.5.6.1. Armor Damage from Direct Wave Attack. Smith (1986) outlined an optimization methodology using estimates of wave-induced armor damage based on the Hudson formula. The basic optimization procedure that is used by Smith (1986) is retained here.

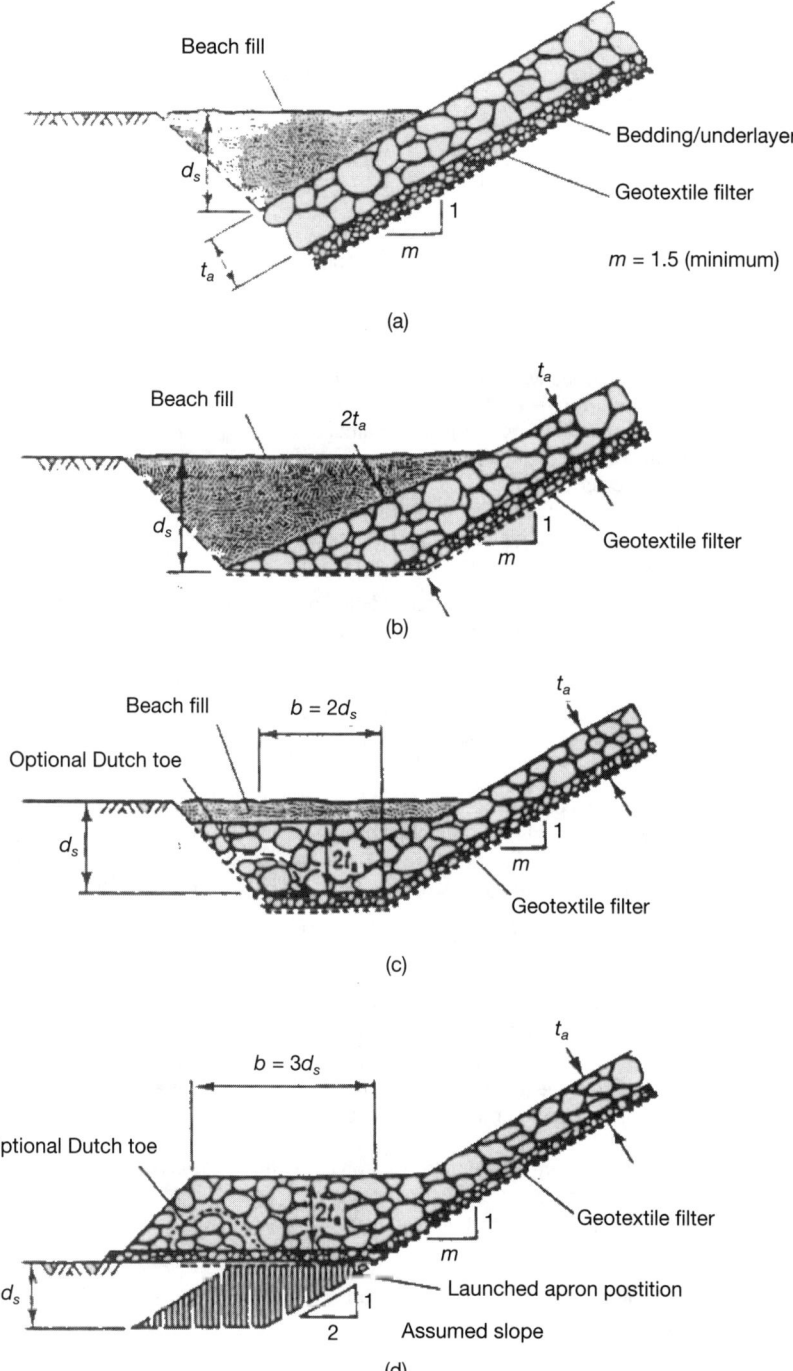

FIGURE 8.32. Alternative toe geometry (*SPM*, 1984).

However, primary armor damage is estimated using the following van der Meer equations for plunging waves:

$$S = \sqrt{N \left| \frac{H_{2\%} \sqrt{\xi_m}}{\Delta D_{n50} 8.7 P^{0.18}} \right|}$$

$$\%D = S \frac{(D_{n50})^2}{V} 100$$

where S = damage level
  N = number of waves
  $H_{2\%}$ = 2% exceedence wave height (defined as the design wave height)
  ξ_m = surf similarity parameter
  Δ = relative mass density of the armor units = $\gamma_{rock}/\gamma_{water} - 1 = 1.6$
  D_{n50} = mean nominal diameter of the stone
  P = structure permeability
  $\%D$ = damage level defined as a function of the eroded area, normalized by the square of the nominal armor unit diameter
  V = volume of armor unit per unit length of dike

The present approach accounts for the dependency of armor damage on structure slope through the surf similarity parameter (ξ_m) defined as

$$\xi_m = \frac{\tan \theta}{\sqrt{\frac{2 \pi H_s}{g T^2}}}$$

Overall, the van der Meer approach indicates that flatter slopes have more reserve strength and are less vulnerable to accelerated damage than steeper slopes.

Expected annual damage, E, is computed from the damage function $\%D$ presented above and a cumulative probability distribution for wave heights, $F(h)$, as follows:

$$E\left[\frac{\%D}{yr}\right] = \lambda \int_{H_d}^{H_\infty} \%D \left[\frac{dF(H)}{dH}\right] dH$$

The Poisson parameter, λ, is the average number of extreme events represented by VH values per year. H_∞ is the significant wave height with an annual probability of exceedence of 0.0000001. The above formulation assumes that the number of storms per year is a random variable and can be represented by a mean value. The approach further assumes that this number is independent of the H values which represent the intensity of individual storms.

The Extreme Type I (Gumbel) cumulative probability distribution (cpd) is used for explanation purposes, other statistical distributions can be used where appropriate. Additional data required to perform the optimization calculations include:

- Crest height
- Cross-sectional area per unit length of structure
- Total length of structure
- Initial armor cost per ton
- Design life (i.e., length of time to be considered for repair events)

- Present worth costs for each construction mobilization
- Interest rates to calculate present worth
- Annualization and escalation for repair
- Percent damage allowable before each repair event (5% damage)

The following equations are used to determine the parameters for the Extremal Type I distribution:

$$RP = \frac{1}{\lambda}[1 - P(H)]$$

where RP = return period
$P(H)$ = probability of exceedence
λ = number of storms per year

The Extremal Type I distribution is given by

$$-\ln\{-\ln[P(H)]\} = \alpha H - \alpha u$$

where H = design wave height per return period (here the $H_{2\%}$ wave height)
α = slope of linear regression of H and $-\ln\{-\ln[F(H)]\}$
αu = y-intercept of linear regression

For Type I distribution, mean wave height (H_m) and variance (σ^2) are related to α and u by

$$H_m = u + \frac{0.5772}{\alpha} \qquad \sigma^2 = \frac{\pi^2}{6\alpha^2}$$

The probability density function is defined as follows:

$$f(H) = \alpha \left[e^{(-\alpha(H-u) - e^{(-\alpha(H-u))})} \right]$$

The cpd is given by

$$F(H) = e^{[-e^{-\alpha(H-u)}]}$$

Calculation of the expected annual damage is performed by approximating the indefinite integral using Simpson's rule.

The average repair interval is determined by dividing the percent damage prior to repair by E. The estimated number of repairs that will be conducted during the design life of the structure is then calculated by dividing the design life by the average repair interval. The return period that is estimated to cause 5% damage is calculated using the following equations:

$$RP = \frac{1}{\lambda[-F(H_{5\%})]}$$

Present worth costs for future repairs consist of both mobilization and repair costs. The following equation was used to calculate the present worth values:

$$PW = C\left[\frac{1+i}{1+p}\right]^{nrt}$$

where PW = present worth value for a single repair and/or mobilization
C = present cost of repair and/or mobilization

i = escalation rate of repair cost (inflation)
p = prime interest rate
nr = number of repairs during life of the structure
t = average time interval (years) between repairs

The sum total for all repairs and mobilizations is calculated using the following equation:

$$PW_{\text{total}} = \sum_0^{nr} PW$$

Annualized repair costs are calculated as follows:

$$A = PW \frac{r(1+r)^{\text{life}}}{(1+r)^{\text{life}} - 1}$$

where A = annualized cost (either initial, repair or mobilization)
  PW = present worth cost (either initial capital cost, or present worth of repair and/or mobilization)
  r = interest rate to calculate annualization
  life = design life of structure, i.e., the number of years for which repair costs are calculated

8.5.6.2. Armor Damage from Wave Overtopping. Overtopping damage is estimated in a manner similar to that used for wave-induced armor damage. This damage mechanism is relevant to sea walls, revetments, dikes, and other structures that are used to protect an earthen embankment. Overtopping damage is not a factor in the design of offshore breakwaters, headlands, groins, and similar structures that are designed for overtopping. The primary differences between optimization computations for direct wave attack and overtopping are: (1) water level elevation is used in place of wave height in the cpd, and (2) damage functions are developed from de Waal and van der Meer's (1992) runup and overtopping equations. It should be noted that the design wave for breaking wave conditions is obtained directly from the design water levels and the input bottom elevation.

There are presently no quantitative methods for evaluating the level of damage that occurs when the overtopping rate exceeds the allowable value. In general, however, it is known that excessive overtopping will erode the area immediately landward of the revetment crest. If the overtopping is sufficiently excessive, then the structure may collapse completely.

Smith and Chapman (1982) present a detailed description of coastal revetments that have collapsed due to excessive overtopping. In the absence of specific guidance that relates overtopping discharge to damage it is necessary to make certain assumptions. For example, Headland and Kotulak (1996) assumed that a Type I coastal revetment would suffer 100% armor damage if the overtopping discharge exceeded 250 liters per meter per second. This value is 50 times larger than the allowable value. Similar assumptions could be made with other structure types. Notwithstanding the approximate nature of this approach, it would be imprudent to ignore the costs associated with this failure mechanism.

The damage estimates for overtopping are used with the cpd for water levels to calculate expected annual damage in a manner similar to that described for primary armor damage.

8.5.6.3. Results. Typical results are presented in Figure 8.33, which presents initial costs, maintenance costs, and total costs. The location of the minimum point on the *Total Present Worth* curve corresponds to the optimal design condition defined by the return

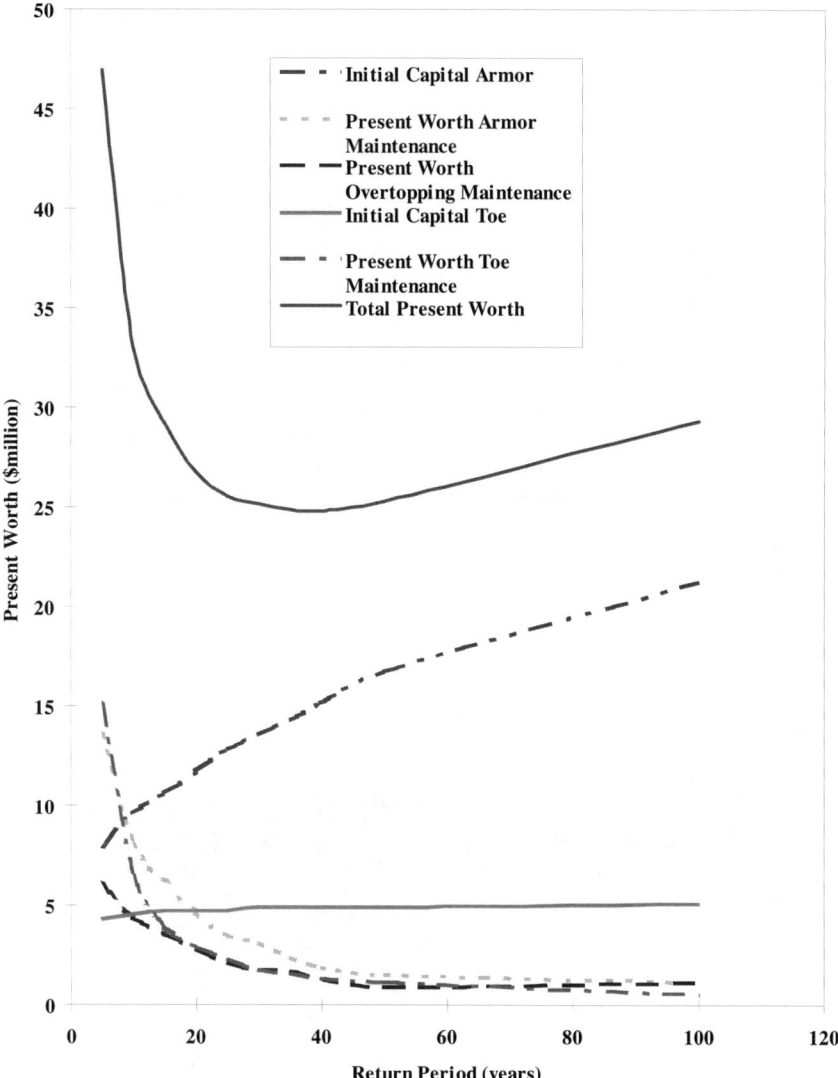

FIGURE 8.33. Optimization results (Moffatt & Nichol Engineers, 1996).

period. For the case presented, the return period for optimized design is approximately 25–40 years.

It is also possible to examine the optimal design for a retreating beach. In this case, the design wave height increases with time as the depth fronting a coastal structure increases due to shoreline retreat. Results from such analyses produce an optimization surface that can be used to select the optimal return period and the optimal year in the future for which

8.60   HANDBOOK OF COASTAL ENGINEERING

a structure should be designed. An example result from Headland and Kotulak (1996) is presented in Figure 8.34.

8.5.7. Probabilistic Design

An important consideration for design is that there is a low yet finite risk of dike damage and failure during the life of the structure. A reliability or probability analysis can be performed to provide an assessment of this risk. This section addresses detailed reliability analyses for armor design of coastal structures. Reliability analyses provide a probability-based means for evaluating the risk of damage to coastal structures throughout the life of the particular structure.

8.5.7.1. Methodology. The basic methodology, terminology and nomenclature used for this analyses is described in Van der Meer (1988). The first step in the risk analysis of the dike armor is to define the mechanisms, hazards, and consequence. Hazards are the independent (input) variables, whereas the mechanisms are the manners in which the structure responds to the hazards. Consequences are the results of structure damage. The combination of hazards and mechanisms leads, with a given probability, to the failure and/or

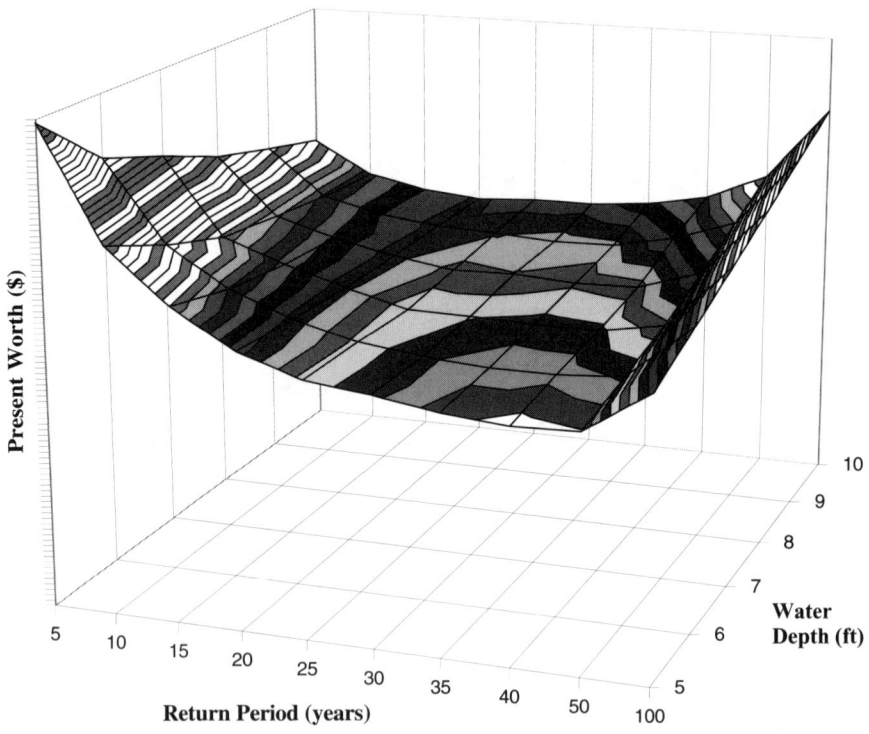

FIGURE 8.34. Long-term erosion results (Headland and Kotulak, 1996).

collapse of the structure. Failure or damage is associated with a particular probability of exceedence for a given time interval of structure life. The following analyses only consider the failure of the main armor of a coastal structure due to waves. Waves and water levels during storms constitute the hazard, whereas removal of the armor is the mechanism.

A probabilistic estimate of failure probability can be computed using the failure or reliability function Z associated with a given limit state of design. The Z function is simply equal to the load (i.e., waves) minus the strength (i.e., rock size) of the armor. Negative values of Z (i.e., load exceeds strength) correspond to armor failure. The probability of failure is designated $P(Z < 0)$. For armor layer design, the damage function S is equivalent to the failure criteria for the reliability function, allowing probabilities to be calculated for varying damage levels.

Risk-based computations of the failure probability were performed using the following reliability function which assumes plunging waves:

$$Z = S^{0.2} 8.7 P^{0.18} (\cot \theta)^{0.5} \Delta D_{n50} - (H_s + FH_s) \left(\frac{H_s}{L_z} \right)^{-0.25} N^{0.1}$$

The input variables used in the reliability analyses are the wave height at the toe of the structure for a given water depth (H_s), the nominal median rock diameter (D_{n50}), the relative mass density of the rock ($\Delta = SG_{rock}/SG_{water} - 1$), structure slope (as cot θ), structure permeability coefficient factor (P), the number of waves (N) for the given storm duration, the uncertainty parameter of the long-term distribution of the wave height (FH_s), wave steepness (H_s/L_z), and the experimentally derived coefficients in Van der Meer's stability equations (i.e., $a = 8.7$ and $b = 1.0$).

The above equation will provide an estimate of the probability of exceedence of damage level S for a 1 year period. Computation of the same probability for different structure lifetimes (X) is computed as follows:

$$P[Z < 0; X \text{ yr}] = 1 - (1 - P[Z < 0, 1 \text{ yr}])^X$$

8.5.7.2. Probability Distributions for Input Variables. The following table summarizes the input variables used in an example reliability computation. Probability distributions and related factors are consistent with the approach and values reported in Van der Meer (1988).

Input Variables for Reliability Computations

Parameter	Distribution type	Mean value	Standard deviation
H_s (ft)	Extreme value	4.98 (mode)	0.92 (scale)
D_{n50} (ft)	Normal	2.36	0.07
Δ	Normal	1.60	0.05
cot θ	Normal	3.00	0.15
P	Normal	0.10	0.01
N	Normal	7,000	1,500
FH_s (ft)	Normal	0.00	0.25
H_s/L_z	Normal	0.04	0.01
a	Normal	8.70	0.56
b	Normal	1.00	0.08

8.5.7.3. Monte Carlo Simulations. Computations were performed using a Monte Carlo simulation technique. This technique consists of a system that uses random numbers to measure the uncertainty in a model. Forecasts of uncertainty are made by randomly selecting numbers from the probability distributions for each variable and making a specific computation. A large number of random-number computations or trails are made in order to obtain a probability distribution for the computed value, which in this case is the damage function S.

8.5.7.4. Results. Results from the above analyses are presented in Figure 8.35. The results are provided in the form of a plot of probability of exceedence (ordinate axis) and armor damage S (abscissa axis). For example, Figure 8.35 shows that during any one year there is a 100 percent (i.e. 1 on y-axis) chance that the armor damage function S will exceed 0. This is to be expected insofar as the armor has been designed for an S value of 2, which corresponds to "no damage." The structure examined in Figure 8.32 corresponds to a 25-year return period. As a result, Figure 8.35 shows that there is about a 70% chance that a S value of 2 will be exceeded over a 20 year period. Similarly, Figure 8.35 indicates that there is a 90% chance that the dike will suffer a damage level corresponding to an S value of 3.5 over a 100-year period.

4.5.8. Construction Materials

Economical coastal structure designs cannot be developed without a detailed understanding of the materials available for construction as well as the long-term performance of various material types in the marine environments that characterize a particular site. The important materials involved in most applications are:

- Rock—including small quarry run sized stone for the breakwater core as well as large pieces for the underlayer and breakwater armor

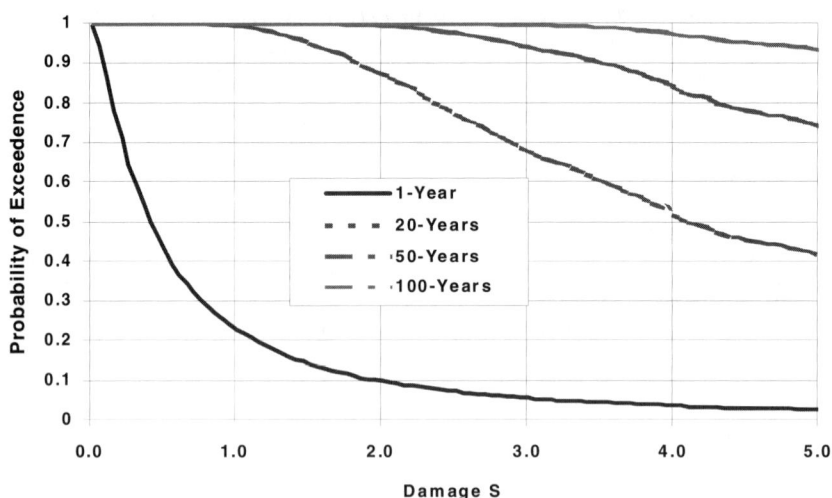

FIGURE 8.35. Probability of exceedence of damage level S (Moffatt & Nichol Engineers, 1996).

- Concrete—which may be used for the armor layers if suitably large rock armor cannot be produced by area quarries. Concrete may also be used to form a concrete wall, concrete sheet pile wall, or parapet wall for the crest of a rubble mound structure
- Geotextiles—which may be used as part of the structure foundation as a filter fabric or fascine mattress
- Steel—steel sheet piles are often used in coastal structures. Steel H-piles are also used in some applications
- Timber—timber sheet piling and timber piles can be used in both sea wall and groin applications

For rubble mound structures, special attention should be given to the quality and gradation of stone that can be obtained from local quarries. This evaluation should consider:

- Availability of stone
- Sizes produced by the quarry
- Transportation to the site
- Stockpiling
- Amount needed
- Cost

A rubble mound structure may be armored with either large quarry stone or concrete armor units. Concrete armor units are generally more cost-effective than natural stone when design waves are large and the quarry cannot economically produce sufficient quantities of large armor stone. Factors which much be considered in the design of structures armored by concrete units include:

- Availability of concrete forms
- Quality of concrete that can be produced
- Shape to be used (e.g. Accropodes, cubes, Core-Loc, Stabits, blocks, Dolos, Tribars, etc.)
- Number of armor unit layers
- Transportation
- Stockpiling
- Royalty costs for using a particular unit
- Placement procedures and tolerances
- Need to incorporate a concrete cap or parapet wall

8.5.9. Construction Methods

No coastal structure can be designed or estimated without a good understanding of the likely methods that will be used to construct the structure. Generally, there are two methods that can be used: (1) building the structure from land or (2) building the structure from the water. In some cases, it may also be feasible to use a combination of the two techniques. When building the structure from land, the core of the structure generally serves as the roadway for trucks carrying stone out from shore. The height of the core should be selected such that trucks can travel safely along the breakwater. Depending on the size of the armor and underlayer stone or concrete armor unit sizes, these units can be placed with a crane (larger sizes) or a backhoe (smaller sizes). When building from the water, the designer should recognize the potential for substantial downtime during construction.

REFERENCES

Ahrens, J. P. and Cox, J. (1990). "Design and Performance of Reef Breakwaters," *Journal of Coastal Research, 7:*61–75.
Alfageme, S., Headland, J. R., and Kotulak, P. (1999). "Dike Design Optimization and Reliability Analysis," Coastal Structures, Spain.
Berenguer, J. M., and Enriquez Fernandez, J. (1988). "Design of Pocket Beaches: The Spanish Case," Proceedings 21st ICCE, Malaga, Spain.
Bodge, K. R. (1998). "Beach Fill Stabilization with Tuned Structures: Experience in the Southeastern USA and Caribbean," FSBPA.
Bowles, J. E. (1977). *Foundation Analysis and Design,* McGraw-Hill.
Bruun, P. (1962). "Sea Level Rise as a Cause of Shore Erosion," *Journal of the Waterways and Horbor Division,* ASCE, Vol. 88, February.
Danish Hydraulics Institute. (1998). Mike 21 User's Guides and Reference Manuals.
Dean, R. G. and Dalrymple, R. A. (1993). "Coastal Processes with Engineering Applications," University of Florida Course Notes.
Dean, R. G. and Dalrymple, R. A. (1984). *Water Wave Mechanics for Engineers and Scientists,* Prentice-Hall.
Dean-Rosati, J. (1990). "Functional Design of Breakwaters for Shore Protection: Empirical Methods," U.S. Army Corps of Engineers, CERC-90-5.
de Waal, J. P. and van de Meer, J. W. (1992). Wave Runup and Overtopping on Coastal Structures. *Proceedings of the Twenty-third International Conference on Coastal Engineering.* American Society of Civil engineers (ASCE), pp. 1758–1771.
Federal Emergency Management Agency (FEMA). (1988). Investigation and Improvement of Capabilities of the FEMA Wave Runup Model (Technical Documentation for Version 2.0). Washington, DC.
Goda, Y. (1985). *Random Seas and Design of Maritime Structures.* University of Tokyo Press.
Hallermeier, R. J. (1983). "Sand Transport Limits in Coastal Structure Design," Coastal Structures '83, ASCE, Arlington, VA.
Hanson, H. and Kraus, N. C. (1991). "Genesis: Generalized Model for Simulating Shoreline Change," Report 1, USACE.
Hardaway, C. S. and Gunn, J. R. (1991). "Headland Breakwaters in the Chesapeake Bay," Coastal Zone 1267–1281.
Headland, J. R. (1992). "Design of Protective Dunes at Dam Neck, Virginia," ASCE Coastal Engineering Practice Conference, Long Beach, CA.
Headland, J. R. and Kotulak, P. (1996). "Probabilistic and Optimal Design of Coastal Dikes," 26th International Conference on Coastal Engineering, Orlando, FL.
Herbich, J. B. (1990–1993). *Handbook of Coastal and Ocean Engineering* Vols. 1–3, Gulf Publishing Co.
Hsu, J. R. C. and Silvester, R. (1990). "Accretion Behind Single Offshore Breakwaters," *Journal of WPC&O Engineering,* ASCE, 116(3).
Kirby, J. T. and Dalrymple, R. A. (1993). "REF/DIF 1 Combined Refraction/Diffraction Model," University of Delaware.
Kriebel, D. L., and Dean, R. G. (1985). "Numerical Simulation of Time-Dependent Beach and Dune Erosion," *Coastal Engineering,* Vol, 9.
Krumbein, W. C. (1944). "Shore processes and beach characteristics," Beach Erosion Board, Tech Memo. No. 3, U.S. Army Corps of Engineeers, Washington, D.C.
Larson, M., Hanson, H., and Kraus, N. C. (1987). "Analytical Solutions of the One-Line Model of Shoreline Change. " CERC-87-15, USACE-WES.
Larson, M. and Kraus, N. C. (1989). "SBEACH: Numerical Model for Simulating Storm-Induced Beach Change," Report 1, CERC, USACE.

Moffatt & Nichol Engineers. (1994a). "Rikers Island, New York, Section 14, Emergency Shoreline Protection Project," prepared for USACE, New York District.
Moffatt & Nichol Engineers. (1994b). "Fire Island to Montauk Point, New York, Moriches to Shinnecock Reach," Interim Plan for Storm Damage Protection, Technical Support Document, prepared for USACE, New York District.
Moffatt & Nichol Engineers. (1995). "North Shore of Long Island, New York, Storm Damage Protection and Beach Erosion Control Reconnaissance Study," prepared for USACE, New York District.
Moffatt & Nichol Engineers (in joint venture with Gahagan and Bryant, Associates). (1996). "Poplar Island Restoration Project, Hydrodynamic and Coastal Engineering Report, Environmental Study and Engineering Design," prepared for Maryland Port Administration, Chesapeake Bay, MD.
Moffatt & Nichol Engineers. (1998). "Orote Landfill Shore Protection Alternative Analysis and Basis of Design 100% Submittal," prepared for Ogden Environmental and Emergency Services.
Moffatt & Nichol Engineers. (1998b). "Sea Gate Reach of Coney Island," Field Data Gathering, Project Performance Analysis and Design Alternative Solutions to Improve Sandfill Retention, Final Report, Vols. 1–3 prepared for USACE, New York District.
Moffatt & Nichol Engineers. (1999a). "Preliminary Engineering Inlet Studies For Bolsa Chica Wetlands Restoration, 2nd Draft Report," prepared for Coastal Conservancy.
Moffatt & Nichol Engineers. (1999b). "Port of Manta Expansion Study," Manta Ecuador, prepared for U.S. Trade and Development Agency.
Naval Facilities Engineering Command. (1986). "Foundations and Earth Structures," Design Manual 7.02.
Pelnard-Considere, R. (1954). "Essai de theorie d l'evolution des formes de rivages en plages de sable et de galets," *IVme journees de l'Hydraulique,* Paris.
Pilarczyk, K. W. (Editor). 1990. *Coastal Protection.* A. A. Balkema, Rotterdam.
Pilarczyk, K. W. and Zeidler, R. B. (1996). *Offshore Breakwaters and Shore Evolution Control,* A. A. Balkem, Rotterdam/Brookfield.
Pope, J. and Dean, J. L. (1986). "Development of Design Criteria for Segmented Breakwaters," 20th ICCE, pp. 2144–2158.
Sedji, M. T., Ude, T., and Tanaka, S. (1987). "Statistical Study on the Effect and Stablitiy of Detached Breakwaters," *Coastal Engineering in Japan,* 30(1).
Silvester, R. and Hsu, J. R. C. (1993). *Coastal Stabilization: Innovative Concepts,* Prentice Hall.
Smith, A. W. and Chapman, D. M. (1982). "The Behaviour of Prototype Boulder Revetment Walls," 18th Coastal Engineering Conference, Capetown, Vol. III: pp. 1914–1928.
Smith, O. P. (1986). "Cost-Effective Optimization of Rubble-Mound Breakwater Cross Sections. Technical Report CERC-86-2." U.S. Army Corps of Engineers (USACE), Waterways Experiment Station (WES), Coastal Engineering Research Center (CERC). Final Report.
Suh, K. and Dalrymple, R. A. (1987). "Offshore Breakwaters in Laboratory and Field," *JWPCO,* 113(2).
Technical Advisory Committee on Water-Defences (TAW). (1984). *Guidelines for the Evaluation of the Safety of Dunes as Coastal Defense.*
United Kingdom (U.K.) Construction Industry Research and Information Association and the Netherlands Centre for Civil Engineering Research and Codes (CIRIA/CUR). (1991). Manual on the Use of Rock in Coastal and Shoreline Engineering. CIRIA Special Publication 83, London; CUR Report 154, The Netherlands.
U.S. Army Corps of Engineers (USACE), Waterways Experiment Station (WES), Coastal Engineering Research Center (CERC). (1977). *Shore Protection Manual,* Vols. I and II. Third Edition. U.S. Government Printing Office.
U.S. Army Corps of Engineers (USACE), Waterways Experiment Station (WES), Coastal Engineering Research Center(CERC). (1984). *Shore Protection Manual,* Vols. I and II. Fourth Edition. U.S. Government Printing Office.
U.S. Army Corps of Engineers (USACE), Waterways Experiment Station (WES), Coastal Engineering Research Center (CERC). (1987). *Geotechnical Engineering in the Coastal Zone. Final Report.* U.S. Government Printing Office.
U.S. Army Corps of Engineers (USACE), Waterways Experiment Station (WES), Coastal Engineer-

ing Research Center (CERC). (1989). *Study of Breakwaters Constructed with One Layer of Armor Stone Detroit District. Final Report*. U.S. Government Printing Office.

U.S. Army Corps of Engineers (USACE), Waterways Experiment Station (WES), Coastal Engineering Research Center (CERC). (1993). "SBEACH: Numerical Model for Simulating Storm-Induce Beach Change," Instruction Report CERC-93-2 Report 3, User's Manual.

U.S. Navy, Naval Facilities Engineering Command, DM-7. (1986). "Foundations and Earth Structures."

van der Meer, J. W. (1988). *Rock Slopes and Gravel Beaches Under Wave Attack*. Doctoral Thesis, Delft University of Technology.

van der Meer, J. W. (1992). "Wave Runup and Overtopping on Coastal Structures." *Journal of Coastal Engineering*.

van der Meer, J. W. and Stam, C. J. M. (1992). "Wave Runup on Smooth and Rock Slopes of Coastal Structures," *Journal of Waterway, Port, Coastal and Ocean Engineering*, 118(5).

van der Meer, J. W. (1998). "Wave Runup and Overtopping." In *Dikes and Revetments*, ed. K. W. Pilarczyk, Balkema, Rotterdam.

Van Rijn, L. C. (1993). *Principles of Sediment Transport in Rivers, Estuaries and Coastal Seas*, Aqua Publications, Amsterdam.

Yasso, W. E. (1965). "Plan Geometry of Headland Bay Beaches," *J. Geology, 73*.

CHAPTER 9
SHORELINE PROTECTION METHODS—JAPANESE EXPERIENCE

JOHN R. C. HSU
Department of Environmental Engineering,
University of Western Australia,
Nedlands, Western Australia, Australia

TAKAAKI UDA
Public Works Research Institute,
Ministry of Construction,
Tsukuba, Ibaraki, Japan

RICHARD SILVESTER
Department of Civil Engineering,
University of Western Australia,
Nedlands, Western Australia, Australia

1. Introduction	9.2
2. State of Beach Erosion in Japan	9.4
2.1. Physiography of Japan	9.4
2.2. Causes of Beach Erosion	9.6
2.3. State of Beach Erosion	9.7
2.4. National Budget for Coastal Preservation	9.8
3. Brief History of Coastal Engineering in Japan	9.10
3.1. Committee on Coastal Engineering, JSCE	9.10
3.2. Evolution of Coastal Engineering Activities	9.11
3.3. Coastal Engineering Research In Japan	9.12
4. Shoreline Protection Methods	9.13
4.1. Conventional Methods	9.14
4.2. Nonconventional Methods	9.20
5. Recent Developments	9.24
5.1. Artificial Reefs	9.24
5.2. Coastal Community Zone (CCZ) Projects	9.24
5.3. Integrated Shore Protection System (ISPS)	9.28
5.4. Control of Beach Groundwater Table	9.29
6. Case Studies	9.30
6.1. Amano-Hashidate	9.30

6.2. Ishizaki 9.32
6.3. Iwafune 9.33
6.4. Kaike 9.35
6.5. Katase East 9.41
6.6. Niigata 9.42
6.7. Oarai 9.46
6.8. Ohno-Kashima 9.50
6.9. Seppu 9.51
6.10. Yaizu 9.53
7. Engineering Applications of Coastal Geomorphology 9.55
7.1. An Empirical Bay Shape Equation 9.56
7.2. Bayed Beaches in Japan 9.58
7.3. Effect of Breakwater Construction 9.61
8. What Can We Learn from Japanese Experience? 9.69
Acknowledgments 9.72
References 9.72

1. INTRODUCTION

Japan, where the East meets West, and the past greets the present, is located at the fringe of the northeastern Asiatic continent. The Japanese landform is dominated by mountainous terrain in the center and narrow, low-lying sandy strips scattered along its edges (Figure 9.1). Traditionally, the Japanese economy has depended heavily on agriculture, for which rice paddy fields were created from reclaimed tidal flats plus coastal lagoons and low-lying river deltas. Since the Meiji era (1868–1912), modernization has seen industrial activities expand throughout the entire country, with factories and population crowding the narrow coastal strips (Koike, 1988).

The island nation of Japan has been subjected to frequent attacks of natural disasters, including typhoons, storm surges, tsunamis, and earthquakes. The low-lying coastal areas are susceptible to coastal flooding, caused by various meteorological, geomorphological, and oceanographical conditions. One of the worst natural disasters was the storm surge produced by Ise Bay typhoon, which killed 5,041 people on September 26, 1959 (Iwagaki, 1994). As such, Hom-ma and Horikawa (1960) once stated about coastal protection works and related problems in Japan that: "Japan has been utilizing the coastal areas to an utmost extent for survival as well as prosperity, and . . . this trend is destined to persist also in the future in order to keep pace with her economical development."

It is fair to say that the present status quo of the Japanese coastline has improved in the forty years since the observations of Hom-ma and Horikawa, but it has also worsened in terms of the total length under artificial protection. About 34% of its coastline is now covered by concrete structures, such as sea walls, groins, and detached breakwaters (Fukuya et al., 1994; Kawata and Shibayama, 1998; MOC, 1997).

The natural disasters occurring along the coastal strip in Japan, caused by meteorological or seismological factors, have been compounded by the effect of human interference, from damming of rivers for various purposes to the construction of breakwaters for harbors and protection. Japanese engineers have been striving to overcome the problems of beach erosion in order to preserve the valuable coast, of which the nation cannot afford to lose any more (Hom-ma and Horikawa, 1960).

Bill Emmott, a former bureau chief of *The Economist*, once wrote (Emmott, 1989): "Four decades ago, Japan was written off as a contender for political and economical

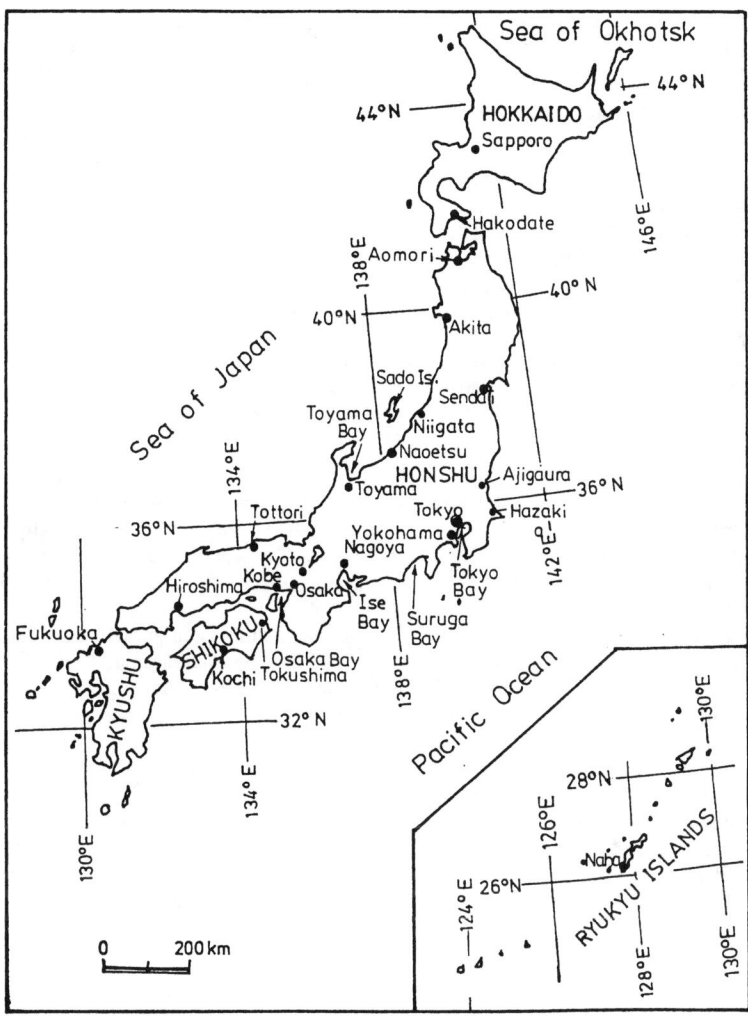

FIGURE 9.1. Location of Japan, showing the four main islands and those of Ryukyu, plus major cities and bays.

power; two decades ago, it was no more than a voice offstage; now it is a leading player in virtually every arena of international endeavor." Since World War II, Japan's influence is not strictly limited to economic power, but also in modern technology, with their products, such as motor vehicles, cameras, TV sets, camcorders, etc., found ubiquitously around the world. Ironically, Japan may also be considered as number one in shoreline protection, in terms of the total length of coastline protected by artificial structures and the money so invested (see Table 9.1 and Figure 9.2).

In this chapter the physiography of the Japanese coastline is introduced, followed by

TABLE 9.1. Statistics of beach erosion in Japan and changes in beach protection works between 1965 and 1996 (Toyoshima, 1982, 1984; Kawata and Shibayama, 1998)

Year	Sea walls (km)	Sea dikes (km)	Groins (units)	Groins (km)	Detached breakwaters (units)	Detached breakwaters (km)	Coastline eroded (km)	Coastline protected (km)
1962†	2,686	1,818	5,448		205			
1965‡	3,743	2,086	6,781	273	347	28	10,701	6,100
1972†	4,752	2,566	8,747		544			
1981*	5,579	2,838	10,043		2,305			
1982†	5,665	2,889	9,790		2,831			
1985‡	5,806	2,836	9,630	387	3,732	347	15,958	9,008
1996‡	6,005	2,922	9,387	384	7,371	837	19,762	9,466

Note: total length of coastline 34,608 km.
Sources: * Toyoshima (1982); †Toyoshima (1984); ‡Kawata and Shibayama (1998).

the main causes and state of beach erosion. The history of coastal engineering in Japan, its evolution, and the major coastal research organizations are described. The efforts and contributions of their coastal engineers to shoreline protection are then reviewed, followed by recent developments. Case studies are then discussed, including applications of coastal geomorphology.

The material forming the bulk of this chapter has mainly been published by Japanese engineers. These originate from the Proceedings of the International Conference on Coastal Engineering (ICCE) between 1956 to 1996, published by the American Society of Civil Engineers; papers selected from *Coastal Engineering in Japan* (referred to as *Coastal Engineering Journal*, JSCE from 1998), which is an English language journal published by the Japan Society of Civil Engineers; as well as from some limited research papers written in English by Japanese coastal scientists and engineers.

The present authors were the founding contact persons participating in the coastal engineering theme of the bilateral Japan/Australia Science and Technology Agreement in 1989. Research collaboration has been maintained since that time. They wish, through their combined efforts in coauthoring this chapter to share the views and experiences of coastal engineers in Japan with professionals in other countries.

2. STATE OF BEACH EROSION IN JAPAN

Many environmental factors have caused coastal disasters in Japan, among which are storm surge, beach erosion, extreme tropical cyclones, tsunamis, winter storm waves, land subsidence, and earthquakes (MOC, 1997). It is such natural phenomena that have caused vast investments in coastal structures, much greater than would be sanctioned by western governments. Only beach erosion on sandy coasts will be discussed in this section.

2.1. Physiography of Japan

Japan is located on the northeastern edge of the Asian continent. It consists of four main islands, namely Hokkaido, Honshu, Shikoku, and Kyushu, from north to south, extending from 129° to 146° East of Greenwich, and from 31° to 46° Northern Latitude (Figure 9.1), and about 4000 small islands of various dimensions, including the Ryukyu Islands extending southward of Kyushu (Horikawa, 1996; Katoh et al., 1994; Silvester, 1974).

The total land area of Japan is 337,815 square kilometers (Bureau of Statistics, 1988), with a total coastline of approximately 34,600 km (MOC, 1997), of which about 14,600 km (42%) surrounds the four main islands. The length of coastline per person in Japan is about 30 cm (Katoh et al., 1994) which is more than twice that estimated for the world population along the 440,000 km world shoreline, including the Arctic and Antarctic (Komar, 1998). Its 200 mile exclusive economic zone is about 3.86 million square kilometers, which is almost half that of the United States (Shapiro, 1989). In commenting on the coastline length versus land area, it has been stated that: "The length of shoreline per land area of island nations such as Japan and the United Kingdom is relatively longer than that of peninsula nations (Korea and Italy) and continental nations (USA, France, Canada and the former USSR). Japan has more than 90 kilometers shoreline per 1000 square kilometers" (MOT, 1998)

Because the mountainous area occupies about 70% of the total area, Katoh et al. (1994) once commented: "For this reason, the available space for economic and social activities are limited and concentrated along the coastal region." However, the environmental conditions around Japan are very severe, because of the natural disasters mentioned previously. Besides these, man has added to the problem, as noted by Goda (1983), who stated that the beach erosion is "generally severe due to decrease of sediment discharge from rivers by construction of many flood control dams and other structures." Beaches immediately downcoast of fishing harbors and commercial ports have badly eroded due to littoral drift being intercepted by breakwater construction.

2.1.1. Geography. The geography of the Japanese islands are dominated by relatively high mountains in the center and narrow strips of coastal plain, which consist of the foot of mountains (4% of the total land), terrace (12%), and low-lying land (13%) (Horikawa, 1996). Rivers connecting them are relatively short and steep, with the Shinano River at Niigata being the longest (367 km). Due to the topography, the watersheds in Japan are generally diminutive, but some of the embayments and coastal plains at their downstream ends are relatively large in size. The proportion of reasonably level plain is in the order of a quarter (Smith and Good, 1943), or 21% of habitable land (Goda, 1983). The coastal outline features prominent peninsulas and rocky headlands, with many minor promontories making up an altogether rugged shoreline (Silvester, 1966).

The Environmental Agency of Japan (1982; see Koike, 1988), has classified the status of Japanese coastlines into natural (59%), seminatural (13.5%), artificial (26.7%), and river mouth (0.8%). For natural and seminatural coasts, the statistics also indicate the types of coastline on the four major islands as sandy beach (25%), cliffed coast (24%), shingle beach (15%), and muddy coast (1%), whereas on the small islands they are sandy (17%), cliffed (39%), shingly (26%), and muddy (2%), respectively. Apparently, small islands have more cliffed and shingly coasts than their mainland counterparts. By 1997, about 34% of its total coastline was covered by protective concrete structures (Fukuya et al., 1994; Kawata, 1989; Kawata and Shibayama, 1998; MOC, 1997).

Traditionally, Japanese society has been heavily dependent on agriculture and fishing, which necessitates extensive reclamation around its estuaries and low-lying coastal areas. The coastal zone is fully utilized for marine transport. There are about 1000 commercial ports, 3000 fishing harbors, and many large-scale industrial areas and modern cities along its narrow coastal plains (Fukuya et al., 1994; Nagao, 1993; Horikawa, 1996). Active reclamation has produced offshore islands for large-scale utilization and development projects, most of which are for airports, situated so as to reduce noise levels over concentrated living areas.

There are 47 prefectures (states) in Japan, of which only eight are not adjacent to the sea, the latter accounting for 14% of the total land area and 15% of the total population. Of the remaining 85% of the population, about 39% is concentrated around three large bayed areas, namely Tokyo Bay, Ise Bay, and Osaka Bay. These, in turn, have populations of 25.1 million (20%), 8.4 million (7%), and 15 million (12%), respectively. Ecosociologically speaking, many large commercial ports are located within these three large bays; they handle 75% of the country's total exports and 60% of the total imports (Iwagaki, 1994). Unfortunately, many natural disasters, notably storm surges, have occurred at these three bays, where population and industries are heavily concentrated (Iwagaki, 1994). Coastal dikes and sea walls have been constructed to protect property and life from coastal flooding; protection against storm surges has been afforded by installing gates around Tokyo Bay, Ise Bay and Osaka port.

2.1.2. Wave climate and sediment movement. In general terms, physiography of a coast is not only affected by its geology but also by meteorological and oceanographical factors. The Japanese coast is located along the earthquake zone fringing the western Pacific Ocean. Crustal movement and earthquakes have repeatedly caused tsunami flooding, liquefaction of foundations, and destruction of coastal structures. Storm surge associated with violent winds and waves from typhoons (tropical cyclones) has often inundated low-lying areas, particularly in large sheltered bays. The most frequent natural disasters affecting the Japanese coast have been typhoons. These are by far the most destructive meteorological events affecting most of the Japanese coast. According to the statistics available (Hom-ma and Horikawa, 1960), these can attack the Pacific and Sea of Japan coasts, from May to November, on average four times per year. These typhoons form in the western Pacific at about 15°N and travel to 35°N latitude, affecting a wide area including the Philippines, Taiwan, Japan, and the Chinese mainland (Silvester, 1966). Besides these intrusions, the Sea of Japan suffers from northerly violent wind waves in winter, due to the passage of prevailing low-pressure centers across the northern area. The waves are of storm type, lasting only as long as the winds generating them, but they are almost continual in winter months.

In a similar manner, Silvester (1966) also broadly classified the coastlines of Japan into three zones in respect to wave climate (swell and storm), resulting from meteorological and oceanographical conditions. The major sector is the coast boarding the Pacific Ocean, which comprises the southern and eastern shorelines of all four main islands. The second zone is the section boarding the Seas of Japan and Okhotsk, generally facing north or west. The third zone is the coastal margins of the Seto Inland Sea, comprising the northern coastlines of Kyushu and Shikoku and the southern half of Honshu island. Prevailing directions of swell and storm waves for each zone were discussed, based on the distinct wind systems affecting each zone.

2.2. Causes of Beach Erosion

Beaches are unique features on coastlines around the world. They are removed and replaced almost annually and hence no vegetation occurs on them. They are relatively flat ribbons of sand between the sea and the hinterland, regions where continuous conflict takes place between nature and man, mainly due to the lack of knowledge of the processes involved. With swell waves, nature helps move sediment landward to build up the berm; thus the beach reappears. On the other hand, when storm waves arrive, a large quantity of sand is transported offshore to form a bar; in a relatively short period of time, the beach disappears. If waves approach a beach obliquely, a longshore current is induced, so promoting longshore sediment transport, or *littoral drift* as it is called. Where coastal structures are installed on a beach, which they should not be because of instability of the beach, as noted above, waves will reflect obliquely and cause short-crested waves. These enhance downcoast transport of sediment (Silvester, 1977).

The causes of beach erosion can be best summarized by Bruun (1972), who stated with a time-honored Dutch proverb: "Water shall not be compelled by any 'force,' or it will return that force unto you." He broadly classified the main causes of beach erosion into those due to nature and man. Silvester and Hsu (1993, 1997) also commented: "It must be accepted that wave obliquity, storm attack, imbalance of sediment transport, and perhaps sea level rise, are the primary factors. However, there are many ways to cause sand loss from a beach, including removal alongshore, blown inland, swept into inlets or lagoons, lost to deep sea, and extraction from the beach for construction purposes."

Although beach erosion has occurred in many countries around the world, due to nature or man himself, the problem is much more serious in Japan, because most activities there are concentrated in narrow coastal strips. The problem of beach erosion in Japan has existed since it was first reported in 1833 (Toyoshima, 1974). It became very severe in many coastal areas around the 1950s, on both sandy coasts and soft-cliff sections. The main factors affecting beach erosion in Japan were first identified by Hom-ma and Horikawa in 1960. They classified these into four categories, namely (1) geomorphological, (2) meteorological and oceanographical, (3) sedimentary, and (4) human interference. Their identifications were rather general and implicit, supported by evidence of beach erosion that occurred in the period from the end of the World War II to 1960, a period of 15 years.

Human intervention in sediment supply and interception has been considered one of the most detrimental factors causing beach erosion throughout Japan. This may be attributed to intervention with rivers (by damming or flood control and power generation), extraction of sediment from rivers and estuaries for construction purposes, as well as stopping littoral drift by various coastal structures. More recently, breakwaters constructed for fishing harbors and commercial ports have also caused beach erosion downcoast when located on a sandy coast or at a river mouth. Uda (1997) has reiterated the beach erosion mechanisms, and considers that they are due to (1) disturbance to the continuity of littoral drift, (2) wave sheltering by coastal structures, (3) offshore movement of sediment, (4) reduction in sediment supply, (5) dredging or extraction of sand and gravel for construction, and (6) land subsidence. He also discussed the erosive condition on cliffed coasts in Japan (Uda, 1997).

2.3. State of Beach Erosion

Although beach erosion has existed at many places in Japan from the early 1900's, most has occurred naturally, due to oblique wave incidence and during storm attack. Shorelines might have receded drastically as a result of complex natural forces, but it has been observed that the situation has worsened more recently. Toyoshima (1984) once commented that: "In most cases, shorelines were slightly advancing until 1950, although there are some exceptional coasts which have been under continuous erosion for more than a thousand years. However, in the last thirty years, beaches began to be eroded one after another, and the countermeasures against beach erosion has become one of the most important problems in the national land preservation." This implied that the problem, exacerbated after the World War II, was caused mainly by human intervention in the supply of sediment and its movement along the coast.

Toyoshima (1982, 1984), Kawata (1989), and Kawata and Shibayama (1998) collected statistics on beach erosion in Japan and the changes in major beach protection works between 1962 and 1996. The results, given in Table 9.1, show that by 1965 there were 10,701 km or 31% of coastline in Japan suffering from erosion. In the following 20 years from 1965 to 1985, an additional 5,257 km of shoreline was further eroded. And between 1985 and 1996, another 3,804 km of beach was disappearing, totalling 19,762 km by 1996. This shows 57% of the total coastline of 34,608 km in Japan (see Section 9.2.1) was affected by erosion, whereas only 9,466 km or 48% of those eroded were protected by various defense measures.

In addition to the above general statistics above, Uda (1997) has provided specific information for 41 locations where erosion has occurred and classified them as follows:

- Disturbance to the continuity of littoral drift (27% of the 41 sites)
- Reduction in sediment supply (20%)
- Wave sheltering by harbor structures (16%)
- Dredging or extraction for construction (16%)
- Offshore movement of sediment (5%)
- Land subsidence (1%).

The numbers in the parentheses above indicate the percentages of occurrence due to each cause of erosion. Most of the erosion reported was produced by multiple factors (Uda, 1997).

2.4. National Budget for Coastal Preservation

According to the Sea Coast Act, approved by the Japanese Diet (Parliament) on 12 May 1956, the national government is responsible for the preservation and protection of the coast against tsunamis, storm surges, stormy sea waves, land subsidence, and sea level change (Horikawa, 1996; MOC, 1997; MOT, 1998). This specific responsibility is delegated to three major Ministries, of which the Ministry of Agriculture, Forestry and Fisheries (MOAFF) is in charge of the agricultural land and areas adjacent to fishing harbors, the Ministry of Transport (MOT) is responsible for the coastal strips incorporating commercial and industrial harbors, and the Ministry of Construction (MOC) is responsible for the remainder of the coast.

The Governor of a prefectural government, as a designated coastal protection manager, is responsible for maintaining and preserving the coast within his prefecture, and should devise a coastal preservation plan for coastal regions endangered by natural disasters. The national government, through the Minister of each of the three ministries, is obliged to provide up to 50% of this expenditure. However, these three Ministries can also directly initiate coastal protection work of national significance, or expensive projects requiring modern technology and heavy machinery, or projects requiring mutual cooperation of two or more prefectural governments. The Act also stipulates uniform standards for the design and execution of coastal structures.

Under the Sea Coast Act enacted in 1956, plus amendments over the years, the Cabinet has requested the national government to allocate sufficient funds for coastal preservation and protection work. This national budget includes four different categories, namely, linear projects designated by the responsible Ministry, subsidiary projects of national and local significance, local individual works, and postdisaster and remedial cleanup. The national budget for coastal preservation and protection in Japan, from 1950 to 1996, is summarized in Figure 9.2. The Cabinet also introduced the so-called "five-year fiscal

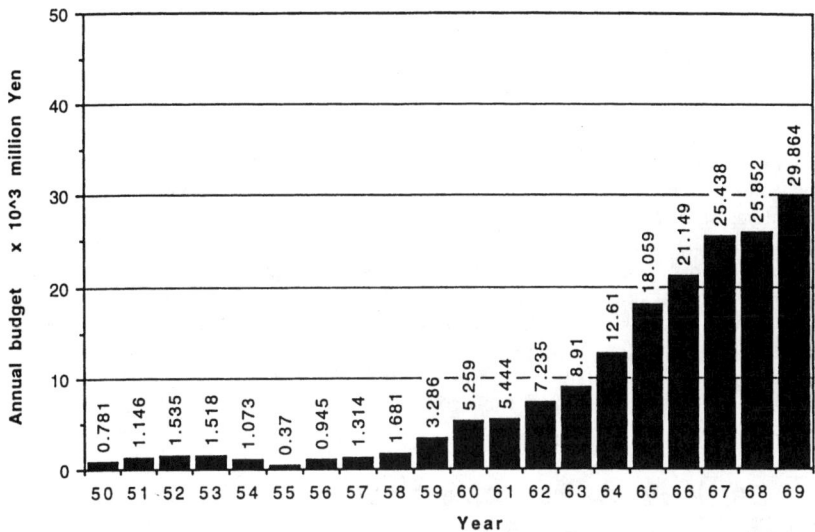

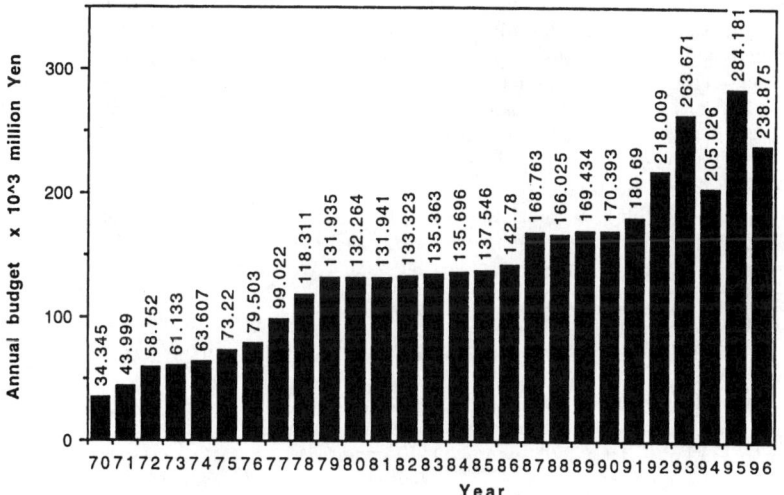

FIGURE 9.2. National budgets for coastal preservation works in Japan from 1950 to 1969 and 1970 to 1996. (Data source: MOC, 1997.)

plan" for coastal preservation and protection, in order to implement the work efficiently and effectively. The actual investment for each of the six five-year plans since 1970 are:

- First Plan (1970–1974)   261.84 × 10^3 million yen
- Annual budget for 1975   73.22 × 10^3 million yen
- Second Plan (1976–1980)   561.04 × 10^3 million yen

- Third Plan (1981–1985)   673.87 × 10³ million yen
- Fourth Plan (1986–1990)   817.40 × 10³ million yen
- Fifth Plan (1991–1995)   1151.58 × 10³ million yen
- Sixth Plan (1996–2002)   1340.0 × 10³ million yen

It is noted that the Sixth five-year plan was approved in December 1996 but was extended to a seven-year plan in January 1998, due to the recession of Japanese economy in recent years. This Sixth plan has a total budget of ¥1340 billion, or 11.2 billion U.S. dollars (taking U.S. $1.00 = ¥120 in 1996). Overall, the budget may be broadly divided into protection work against beach erosion (35%), storm surges (60%), and miscellaneous (5%). The average breakdown of the funds available for the three major Ministries is 25% for MOAFF, 25% for MOC, and 30% for MOT, although their shares in lengths of coastline are 8,089 km (23%, MOAFF), 16,273 km (47%, MOC) and 6,292 km (25%, MOT), respectively. This implies that the Ministry of Construction (MOC) maintains as much coastline as the other two for the same cost (MOC, 1997).

3. BRIEF HISTORY OF COASTAL ENGINEERING IN JAPAN

In historical terms, Japanese engineers have practiced reclamation and coastal protection using coastal dikes and sea walls since the Edo era (1603–1867), to meet the demand of expanding rice paddy fields and protecting the populace. However, the modern chapter of coastal engineering in Japan may be considered to start from the first Japanese Conference on Coastal Engineering (JCCE) held in Kobe in 1954, after the Ise Bay storm surge disaster in September 1953. This was four years after the first International Conference on Coastal Engineering (ICCE), which took place in 1950 at Long Beach, California. The JCCE has been held annually since then.

Horikawa (1996), in reporting the history of coastal engineering in Japan, has commented: "The storm surge damage at Ise Bay in 1953 opened the dawn of coastal engineering in Japan.... Before that, coastal related works in Japan were executed as a branch of river engineering and/or harbor engineering by government agencies, and the related research activities in this field were in an early stage of development." Since the end of the World War II, the Japanese coast has been governed by four government agencies, namely the Ministry of Construction (MOC), the Ministry of Transport (MOT), the Ministry of Agriculture, Forestry and Fisheries (MOAFF), and the Fishery Agency (FA) (Horikawa, 1996). The total lengths of coastline controlled by the three largest agencies are 16,273 km (MOC), 8,089 km (MOAFF) and 6,292 km (MOT), respectively (MOC, 1997), as mentioned in section 2.4.

3.1. Committee on Coastal Engineering, JSCE

The Committee on Coastal Engineering was organized within the Japan Society of Civil Engineers (JSCE) in April 1955. The Committee was charged with the publication of a *Design Manual for Coastal Structures* in August 1957. Based on this design manual, the Standard for Design and Execution of Coastal Structures was issued by the Japanese Government in December 1958. This standard has since been revised in 1969 and 1987, to incorporate the new knowledge available.

The Committee on Coastal Engineering, under the chairmanship of Professor Masashi

Hom-ma of Tokyo University, was responsible for the establishment of the now annual Japanese Conference on Coastal Engineering (JCCE) in 1955. From the statistics given by Horikawa (1996), a total of 876 papers were presented at the JSCE to 1993 (40 th conference). Among these, the three most popular topics were (1) wave theory (260 papers, or 30%), (2) wave action on structures (221 papers, 25%), and (3) coastal sediment (179 papers, 20%). It is worth noting that the total number of research papers (about 260) published in the 42nd JCCE in 1995 was almost the same as that in the 24th ICCE, held in Kobe in 1994.

An English-language journal *Coastal Engineering in Japan* (CEJ) has been published since 1958, aimed to promote coastal engineering research activities in Japan. This journal (volumes 1–39) was printed in Japan by the Japan Society of Civil Engineers until 1998 when a new name *Coastal Engineering Journal* (the new CEJ), was adopted and the World Scientific Publishing Company in Singapore was chosen as its publisher. It is worth noting that, for the first time, the new editorial committee includes two non-Japanese members.

3.2. Evolution of Coastal Engineering Activities

According to Iwagaki (1994), the topics studied by coastal engineers in Japan have evolved from coastal disasters associated with waves, storm surges, and tsunamis in the 1950s, to water quality, tidal currents, and heated water exchange in 1970s, and to environmental problems and sea level rise in the 1990s. The previous emphasis of coastal activities on beach protection has been shifted from protection to utilization and development and environmental preservation, focused on planning and management of the coastal zone.

Horikawa (1996) has produced a chronological review of coastal activities in Japan spanning the five decades from 1950 to the present. Only beach erosion and protective measures are highlighted in the following sections.

The First Decade (1950–1960). The main concern in this decade was coastal disaster prevention, since serious beach erosion had occurred at several locations along the Sea of Japan, including Niigata, Kaike, and Toyama. Investigative committees were set up by relevant government agencies to determine their causes and suggest preventive measures. During this period, the storm surge disaster at Ise Bay due to Typhoon 13 in September 1953 provided a good cause of concern for coastal protection in Japan. The inaugural First Japanese Conference on Coastal Engineering (JCCE) was held in Kobe in 1955. The Sea Coast Act was issued by the Japanese Government in 1956, and the first issue of *Coastal Engineering in Japan* was published in 1958.

The Second Decade (1960–1970). Vigorous economic development took place in Japan in this decade. As a result, active land reclamation for industrial sites created new land—27,738 hectares overall. However, Horikawa (1996) commented that the Japanese people had "little consideration on natural environment including ecological aspects," because "people could not afford the time, or the money, to take coastal environmental changes into consideration." Consequently, marine pollution contaminated seabed sediment and heated water discharges from power plants caused ecological consequences. Research activities were concentrated on wave action on structures, evaluation of storm surges and tsunamis, and mechanisms of beach erosion. During this period, the 10th International Conference on Coastal Engineering (ICCE) took place in Tokyo in 1966. Protection from natural disasters remained the top priority in this decade.

The Third Decade (1970–1980). A great deal of research effort was concentrated on reclaiming the marine and coastal environments from contamination. Topics were related to water quality, water pollution, seabed sediment pollution, and ecological chain reaction. Studies on coastal sediment extended to the preservation of the coastal environment.

The Fourth Decade (1980–1990). Strong demand for recreational facilities and urban development in the coastal zone created a new conflict between conservation and development. However, coastal protection works continued to receive large government funding.

The Fifth Decade (1990–present). The impact of sea level rise on human activities has become of great concern in this decade. The demand for protection of beaches from erosion remains high. Japan was fortunate to host the 24th ICCE held at Port Island at Kobe in 1994.

3.3. Coastal Engineering Research in Japan

Prior to 1953, a number of research papers relating to the present activities of coastal engineering were published by engineering scientists in the discipline of civil engineering, especially on harbor construction and public works. Horikawa (1996) classified the period before 1953 as the "pre-dawn period" in the development of coastal engineering in Japan, and that after 1954 as "post-dawn period." In presenting the history of coastal engineering in Japan, he has given a detailed outline on the research activities carried out by Japanese researchers in government agencies, universities, and the private sector, with the Committee on Coastal Engineering, JSCE, performing an effective liaison between the research scientists and practicing engineers.

Three of the four governmental agencies previously mentioned have research facilities, as follows: The Ministry of Construction (MOC) has its Public Works Research Institute (PWRI); the Ministry of Transport (MOT) has its Port and Harbour Research Institute (PHRI); and the Fishery Agency (FA) has its Fishery Engineering Research Institute (FERI). The PWRI has a field observation station with a 200 m pier at Ajigaura facing the Pacific Ocean, while the PHRI operates its own 427 m long pier at Hazaki, called Hazaki Oceanographical Research Facility (HORF), in close proximity to Kashima Port, also facing the Pacific Ocean.

Horikawa (1996) further mentioned that there are about 100 national universities, 50 public universities, and some 400 private universities in Japan. Among these, active research activities on coastal engineering can be found in about 50 universities (about 10%). Usually, the number of staff members so engaged is rather small. However, special mention must be given to two research institutes with substantial support (over 400 staff) working with well-equipped laboratory facilities and field observation stations. These are the Disaster Prevention Research Institute (DPRI) of Kyoto University and the Applied Mechanics Research Institute (AMRI) of Kyushu University. Moreover, the Nearshore Environment Research Center (NERC) operated a multiinstitutional research program from 1975 to 1980 at Tokyo University. It accumulated field data on nearshore waves, currents, and sediment transport. A reference book entitled *Nearshore Dynamics and Coastal Processes* was produced to mark the achievement of this special research program (Horikawa, 1988).

In Japan, there are several technical research institutes supported by general contractors and private consulting firms. One example is the Central Research Institute for Electric Power Industry (CRIEPI), financially supported by nine electric power companies in

Japan. The Institute installed a large-wave flume of 6 × 3.5 × 205 m in 1984. Beach profile changes and other related projects have been investigated in this large flume, believed to be one of four in the world using waves of 1 to 2 m and periods between 5 and 10 seconds. Examples of private sector supported research institutes include the Institute of Technology, supported by Shimizu Corporation, and the technical research institute of Kashima Construction.

Research subjects in Japan have revolved around practical problems in each of the five decades discussed in Horikawa (1996). It is believed that the papers presented at the annual Japanese Conferences of Coastal Engineering (JCCE) reflect the most popular topics undertaken by coastal researchers in Japan. The three most popular topics have been on waves, coastal sediment, and effects of waves on structures. Numerous wave theories from deep to shallow water and in transmission and reflection have been studied in the laboratory and field, as well as analytically and numerically. Mechanics of coastal sediment movement have also been studied intensively, in relation to both beach erosion and siltation. Methodology of beach erosion control has also been promoted from time to time by a number of prominent researchers, especially Professor Yoshito Tsuchiya and his coworkers at DPRI, Kyoto University. The interaction between waves and structures has received great attention, especially on reflectivity and stability, which have been investigated under laboratory conditions. Overall, although the number of papers on coastal sediment occupied about 25% of the total papers in 35th and 40th JCCE (Horikawa, 1996), the number of papers on the practical problem of beach erosion control was rather limited.

4. SHORELINE PROTECTION METHODS

Coastal protection works in Japan vary according to specific purpose, whether for defense against storm surge or tsunami, beach erosion on sandy beaches or cliff coasts, siltation at harbor entrances or river mouths, or preservation of recreational beaches. To appreciate the prodigious amount of taxpayers' money which has been channeled into beach preservation, or more explicitly protection of human lives and public and private assets adjacent to stretches of shoreline, the daunting task of Japanese coastal engineers in facing the seemingly impossible battle against nature must be realized. In a review of the changes of Japanese coastal environment and beach erosion control works between 1965 and 1996, Kawata and Shibayama (1998) produced alarming statistics, as given in Table 9.1. The progression in percentage of eroded beaches in Japan was 31%, 46%, and 57% of the total national coastline of 34,608 km in the years of 1965, 1985, and 1996, respectively. By 1996, the total length of the eroded beach was 19,762 km. The total length of coastline protected by various control works was 9,466 km, or 28% of the national coastline, approximately 48% of the eroded beaches. The lengths of beach protection works in 1996 were sea walls 6,005 km, seadikes 2,922 km, groins 384 km (or 9,387 units), and detached breakwaters 837 km (or 7,371 units).

The history of the Japanese approach to shoreline protection is unique. It began with sea walls, then groins were introduced, followed by detached breakwaters. These conventional protection methods will be treated first in this section. Beach renourishment was also used to create artificial beaches for a partially protected environment. However, at this point in time, sand bypassing has never been adopted as a preferred option, although two cases have been reported (Suyama et al., 1986; Irie et al., 1990), while some recent developments using mild-sloped sea walls, artificial reefs, ISPS (integrated shore protection system), and CCZ (coastal community zone) projects have become popular. Gravity drainage and control of beach ground water level for coastal stabilization are also receiving research attention.

4.1. Conventional Methods

It is not known by these authors which came first, the sea wall or the groin, but the most obvious for protecting property could well have been the former. Later, observation of longshore movement of sand or littoral current could have aided the introduction of groins. The concept of detached (offshore) breakwaters is more recent, supposedly inhibiting offshore transport of sand.

4.1.1. Sea Walls and Sea Dikes. This family of structures includes bulkheads, revetments, sea dikes, and normal sea walls. They are constructed parallel to the beach to halt shoreline erosion by receiving the impact of waves, or to prevent coastal flooding due to storm surges. In Japanese terminology, sea walls are used for beach erosion control, whereas sea dikes are for flood prevention. Silvester and Hsu (1993, 1997) have given a general description of sea walls. Despite their various locations of construction in relation to local beach and surf, sea walls are designed for physically blocking wave energy. The face fronting the wave action may be vertical or inclined, with or without armor protection. Bulkheads are normally vertical walls built on the upper reach of a beach, whereas revetments have sloping surfaces armored with various kinds of larger stones or concrete blocks. Normal sea walls are often constructed with limited width of fronting beach or even without, and so may receive direct wave action on their faces when wave height or water level increases, generally during storm action.

Sea walls and sea dikes are likely the very first structures ever constructed on a coast for protection against beach erosion and coastal flooding. Japan was no exception. Their cross-sections have evolved over time and have been dependent on material available at the time. They began as earth embankment and rubble mound dikes used for land reclamation and as protection against storm surge, and evolved into simple vertical masonry structures with concrete facing, with or without drainage at the back, after the World War II (Figure 9.3).

Although there were claims about the effectiveness of sea walls in physically blocking wave attack and reflecting energy back to the sea, most engineers did not realize that the reflecting waves also transport sediment deposited in front of the sea wall offshore, and also downcoast if wave approach is oblique. It may thus scour the bottom sediment supporting the toe of the sea wall and even cause the subsidence and collapse of the structure itself (Silvester, 1986).

Many attempts have been made in Japan to retard upward motion on vertical or sloping concrete walls. The first was to insert protruding blocks. The second was known as a "lattice sea wall" made of large blocks with crevices filled with cobbles, small blocks, or armor units (Toyoshima, 1978). Gentle slopes and permeable surfaces were meant to reduce reflection, which is now seen to be rather doubtful. Toyoshima (1984), who was involved in most of these developments, then invented the *Lotus-Uni* block of hexagonal shape with a hole at the center. Asakawa et al. (1992) termed these *crab* blocks, which were used on 1:5 slopes. Later, armor units were introduced in front of the toes of walls to reduce wave impact and reflection (Toyoshima, 1988; see Figure 9.4). However, these became ineffective when storm surge was present with storm waves.

From 1962 to 1981, sea walls and sea dikes were added at the rates of 145 and 51 km annually (Toyoshima, 1982). Other statistics from 1981 to 1985 (Table 9.1) have shown an addition of 57 km of sea walls each year (Kawata and Shibayama, 1998); see Figure

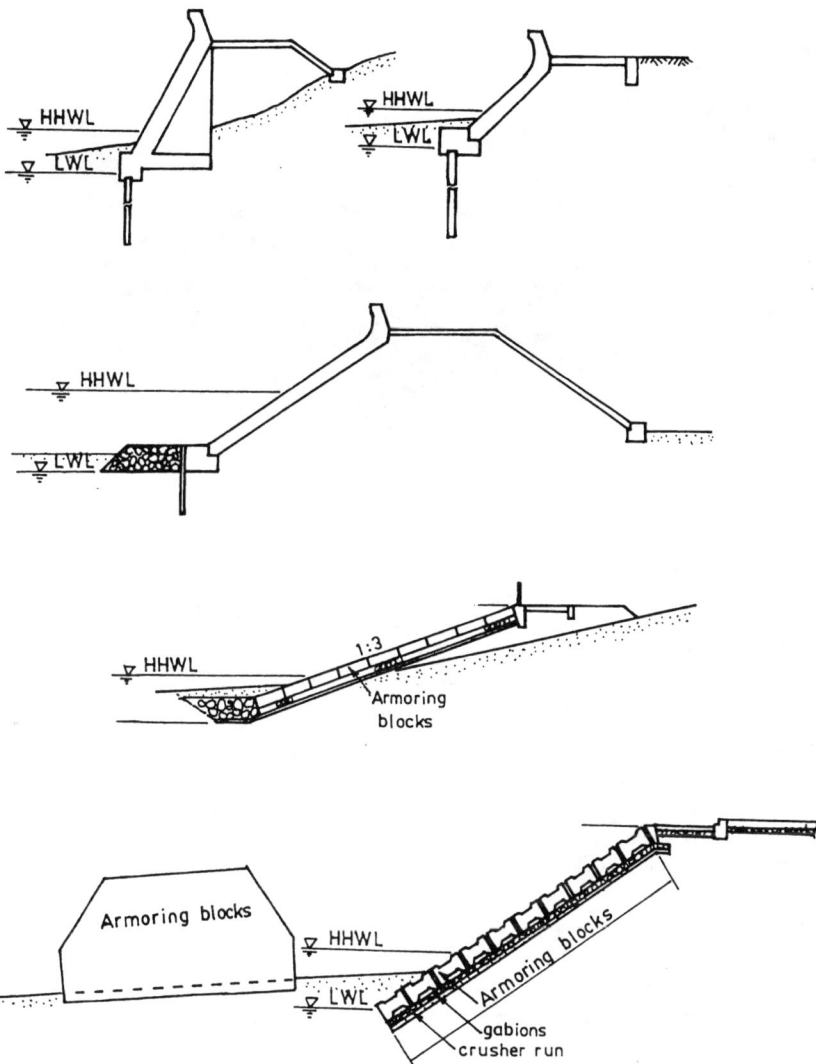

FIGURE 9.3. Evolution of sea wall cross-sections from vertical to inclined type, with toe protection and wave-dissipating blocks (Toyoshima, 1978).

9.5. The number was reduced to 18 km annually between 1985 and 1996 for sea walls, and to 8 km per year for sea dikes. This implies that a saturation point is being reached.

4.1.2. Groins. These structures were used to retard longshore sediment movement. They usually run normal to the coast out to the limit of breaking swell waves. Their length and height vary depending on the purpose they are meant to serve (Silvester and

FIGURE 9.4. Vertical-type sea wall with toe protection, wave-dissipating works, and detached breakwaters at Kurobe coast, Honshu (Toyoshima, 1988).

Hsu, 1993, 1997). Construction materials including timber, rubble mounds and concrete have been employed. They certainly are not effective if littoral drift is zero.

Groins have been used extensively in Japan since World War II, with 5,450 units installed up to 1962. From then until 1981, a further 4,600 were added, an average rate of 230 per annum. Few other western countries could claim such a figure despite their long shorelines. As seen in Table 9.1, the number of groins was reduced from 9,790 in 1982 to 9,630 in 1988, and further to 9,387 in 1996, indicating a disenchantment with this type of protection. Despite the decline in groin use, some representative installations should be noted. For example, a T-shaped groin system of 25 units was spread over an 8 km length in front of a sea wall on the Tokushima Coast, Shikoku (see Figure 9.6) during the period 1950–1960 (Ishihara and Sawaragi, 1964). Another application was at the Yaizu Coast, on the western side of Suruga Bay (see Figure 9.1), where groins were constructed at 75 m intervals in front of a sea wall of 1600 m (Seo and Fukuchi, 1966), to be discussed in Section 6.10.

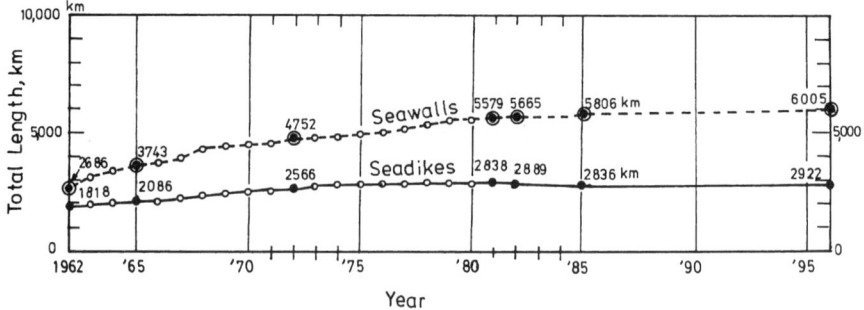

FIGURE 9.5. Total length of sea walls and seadikes in Japan between 1962 and 1996 (Data source: Toyoshima, 1982, 1984; Kawata and Shibayama, 1998.)

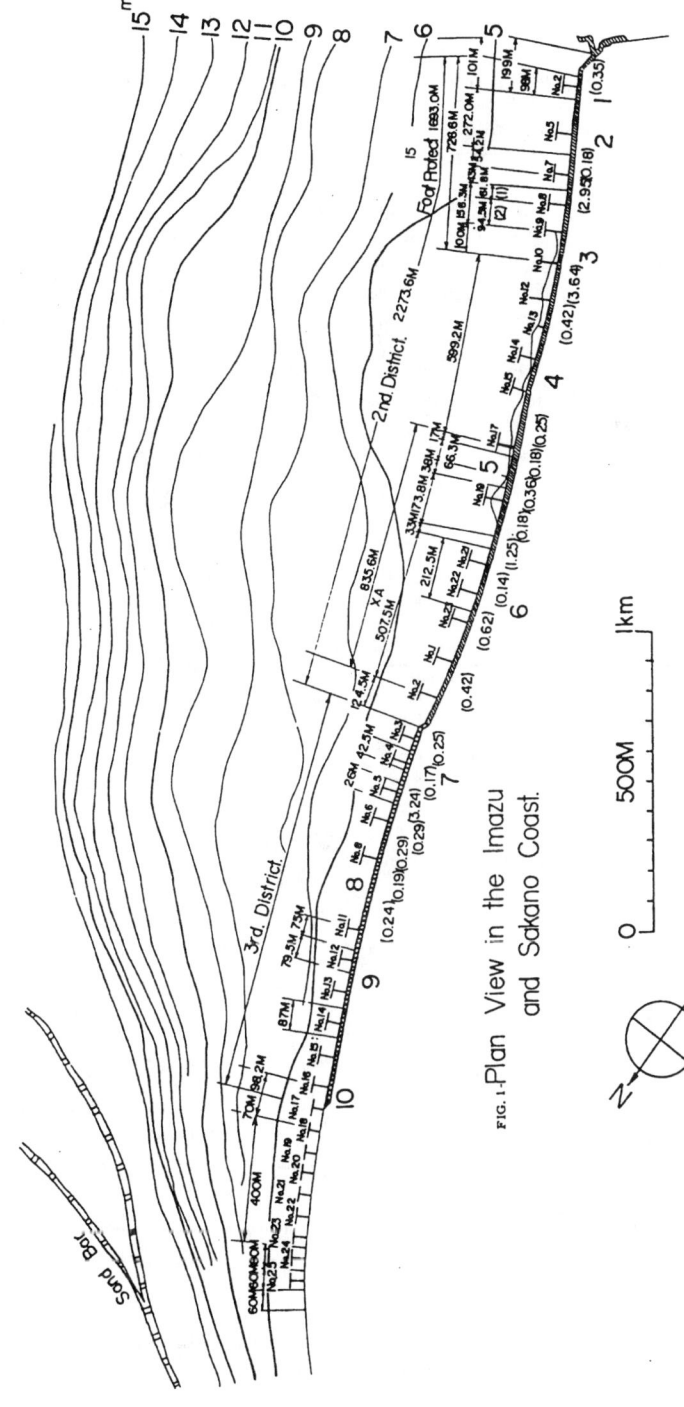

FIGURE 9.6. Imazu and Sakano coasts in Tokushima, Shikoku, protected by 25 groins in the 1960s (Ishihara and Sawaragi, 1964).

Silvester and Hsu (1993, 1997) have discussed groins thoroughly but a final word might be quoted from Magoon and Edge (1978): "Although thousands of groins have been built, it is unfortunately true that the technology of groin design has not been investigated adequately either in the laboratory or in the field. With this in mind, it is easy to understand why groins have so often not only failed but have created worse conditions than they were designed to prevent."

4.1.3. Detached Breakwaters. For some thirty years after the end of the World War II, structures built to combat the incessant problems of beach erosion in Japan were dominated by sea walls and groins, until the early 1970s, when the detached breakwater systems suddenly became the primary countermeasure. Despite some earlier construction in Japan, detached breakwaters (or offshore breakwaters as they are called) have been used almost indiscriminately since their successful application at Ishizaki, Hokkaido in 1966 (see Section 6.2). This can be found from the statistics (Figure 9.7 and Table 9.1) on the number of detached breakwaters constructed since their inception (Toyoshima, 1982; Uda, 1988). Within the 20 year period from 1962 to 1981, a total of 2100 new detached breakwaters were built, or a rate of 105 units per annum. Although they were widely used

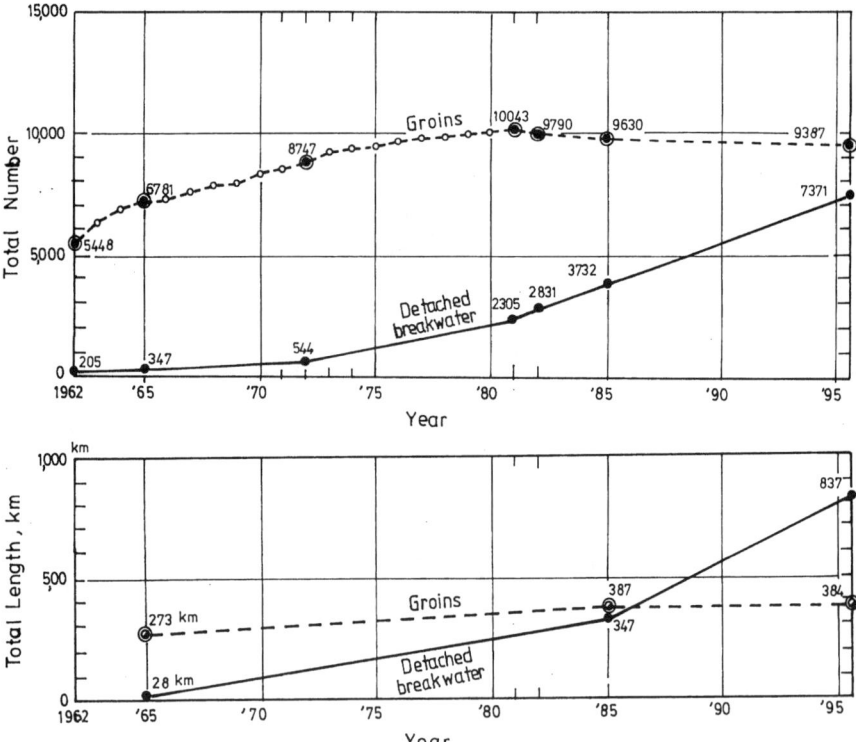

FIGURE 9.7. The total number and lengths of groins and detached breakwaters in Japan between 1962 and 1996, showing the sharp increase of detached breakwaters. (Data source: Toyoshima, 1982, 1984; Kawata and Shibayama, 1998.)

in the early 1960s, no systematic and scientific research on salient/tombolo formation was available until Shinohara and Tsubaki (1966) published their experimental results of a series of model studies on the change in sandy shoreline behind a single detached breakwater. Temporal development of salient plan forms and the volumetric sand accumulation behind the structure were reported for four geometric ratios of breakwater length to distance offshore and two wave steepnesses. However, prediction methods were not proposed for the apex location and the shape of the salient plan form.

Toyoshima (1974) believed sea walls and groins constructed after the World War II in Japan were mostly ineffective, and in some cases accelerated beach erosion. As the main protagonist of the detached breakwaters, he had found initial success at Ishizaki in 1966 (see Section 6.2) and then at Kaike in 1971 (see Section 6.4). From the 217 units constructed at 86 locations by 1969, totalling 35 km of shoreline, he attempted to find useful correlations between breakwater length, distance offshore, water depth, and gap and crown widths. The only design criterion he derived was water depth, with a structure installed in a depth less than 1 m to prevent sea wall scouring, and in 2 to 4 m (in the surf zone) and greater than 4 m (beyond the surf zone) for wave dissipation. Information on breakwater length and gap was specified.

Despite the widespread use of detached breakwaters in Japan since 1966 (see Table 9.1), there was no statistical analysis until that of Seiji et al. (1987) and Uda (1988), who produced the results of a survey from the more than 2,000 units installed by the MOC to 1981. They gave the frequency distributions of various dimensions, such as unit length, water depth, crown elevation above the MSL, offshore distance, crown width, and weight of the concrete blocks used (see Figure 9.8).

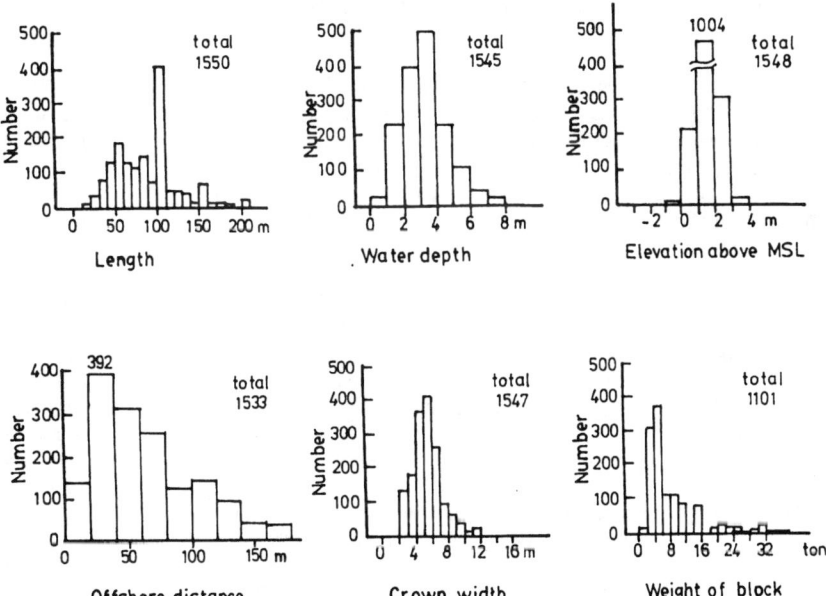

FIGURE 9.8. Frequency distributions of the various dimensions for detached breakwaters used in Japan (Uda, 1988).

4.2. Nonconventional Methods

Some shoreline protection methods used in Japan are not traditional, but many of them have achieved satisfactory results. These methods include mild-sloped sea walls, headland control, integrated shore protection system (ISPS), coastal community zone (CCZ) projects, and artificial reefs. The first two are discussed in this section and the last three are discussed in Sections 5.1 to 5.3.

4.2.1. Mild-sloped Sea Walls. Since the conventional sea wall cross-sections were rather steep, wave reflection coefficients were high, even if protected by concrete armoring blocks. The consequence of high wave reflectivity is toe scour. With the community attitude toward the continual use of hard concrete structures changing in the late 1970s, a new focus was on the aesthetic harmony with the surrounding environment and human accessibility to the beach for recreation. Sea walls with a gentler slope and without wave-dissipating blocks were then proposed. A mild-sloped sea wall using the Lotus-Uni blocks (or crab blocks) on a front face of 1 to 5 was first successfully carried out at Kurobe, about 5 km downcoast of the Miyazaki fishing harbor facing Toyama Bay on the northern Honshu (see Figure 9.1; also see Toyoshima, 1984).

Asakawa et al. (1992) believed "a structure which resembles the natural beach itself" is being chosen over the conventional types and significant changes in the planning of shore protection have resulted. They reported the development of a novel block called "Terrace Block" to increase comfort and safety afforded by existing blocks. The modified Terrace Blocks offer several advantages, including easy beach access, low crest elevation to blend in with the natural landscape, reduction in wave run-up, improved unit stability, plus flexibility for application to curved alignments. Combined with the ISPS (integrated shore protection system, to be discussed in Section 5.3), which incorporates nourishment, the mild-sloped sea walls or revetments (Figure 9.9) have been well accepted by Japanese communities.

4.2.2. Headland Control. For conventional beach protection methods (such as sea walls, sea dikes, and detached breakwaters) to be effective, they must cover an extended distance along the coastline. These may be considered as "linear protection." Detached breakwaters are used in front of sea walls or groins to dissipate incoming waves and so serve as toe protection. This combination covers a wider area, which may be classified as "plane protection." Both linear and plane protection strategies (as seen in Figure 9.10) have been promoted in Japan (MOC, 1997), with variable degrees of success and misfortune.

As well as the changing attitude toward the uses of hard structures, a new concept started in Japan in the late 1970s. This was the so-called "headland control." In promoting this concept, which was new in Japan at that time, Professor Yoshito Tsuchiya of the Disaster Prevention Research Institute, Kyoto University once quoted (Tsuchiya, 1987) the statement of Bruun (1978) that "Nature's engineering does not always satisfy man's ambitions. Man, however, learned from nature. Combining nature and man's efforts, practical solutions have been obtained." He echoed the suggestion of Silvester (1979) on "how to copy nature."

The concept of headland control for beach preservation was first applied at a pocket beach called Shirarahama in Kanayama Bay, Wakayama Prefecture (located at the southeastern corner of Osaka Bay), facing the Pacific Ocean. The project commenced in 1983 by constructing a submerged breakwater of 40 m long extended out from an existing headland. Then in 1987 a T-shaped groin was installed at about 400 m to the south of the headland, followed by nourishment of 35,000 m^3 over the entire beach in 1990. Since

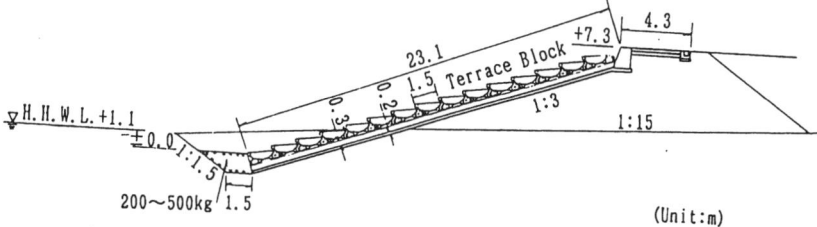

FIGURE 9.9. A mild-sloped revetment incorporating terrace block. (a) Typical cross-section. (b) A field application (Asakawa et al., 1992).

then the pocket beach at Shirarahama has been successfully preserved (Tsuchiya et al., 1992).

In addition, this new methodology has been implemented at two other locations in Japan—at the Ohno-Kashima and Joetsu Coasts. Both sites are fully exposed to ocean waves and the extent of coastline affected by headland control is much longer than that at Shirarahama. The essence of the headland control is to space structural components at large intervals, thus allowing waves to carve a stable sandy beach (Silvester et al., 1980). As such, this new concept may be referred to as "headland control" or "point protection," as opposed to the linear and plane protection mentioned above, as only minimum fixed components are used compared with the conventional protection methods.

The fundamental concept of headland control is to produce a condition in which the shoreline is reoriented parallel to the crests of the incoming waves, thus minimizing or eliminating totally the sediment transport alongshore (Silvester and Hsu, 1993, 1997). This condition is seen to have been provided by nature in her ability to sculpture crenulate or J-shaped bays downcoast of headlands, or close to symmetrical shapes between headlands. The curvature of the beach planform, whether symmetric or asymmetric, will depend on the wave obliquity. A suitably sized structure has to be installed, which serves as a headland, from which the position of the stable waterline can be calculated or designed using an empirical formula (e.g., Hsu and Evans, 1989; Silvester and Hsu 1993, 1997).

The headland control approach at the Ohno-Kashima Coast (see Figure 9.40), the

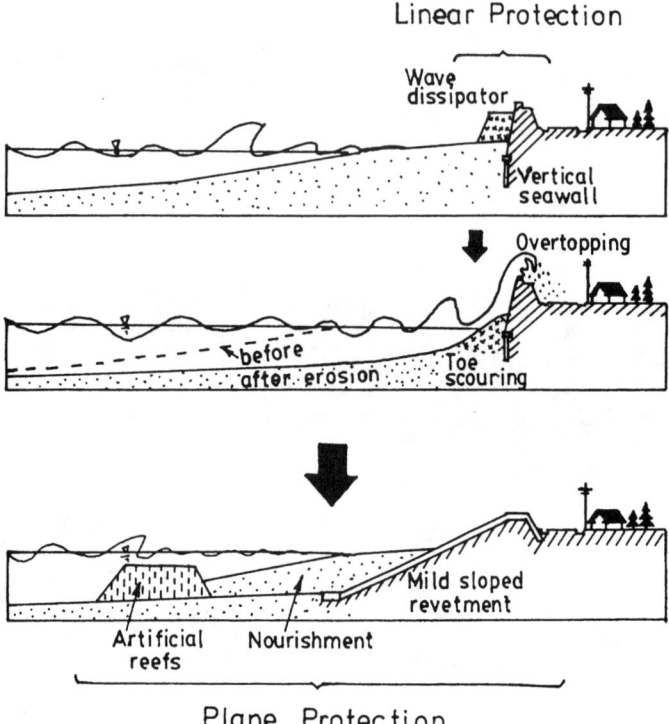

FIGURE 9.10. Beach erosion control strategy from linear to plane protection (MOC, 1997).

first large-scale field application on an open coast in Japan, was proposed by the Ministry of Construction (Uda et al., 1988) in conjunction with the Ibaraki Prefecture Government. The scheme installed groins at approximately 1 km intervals on the coast north of Kashima Port, after beach erosion was found several years after the port was constructed (see Section 6.8 for details). A total of eleven headlands were planned and constructed in stages by 1988, with two additional headlands to follow (Uda et al., 1988). The crown width of each headland was 6 m; length was 100 m, with tip at 2 m of water depth. Dynamically stable beaches of symmetrical shape were formed between headlands. It was found that the groin-type headland system could reduce the periodic changes in shoreline configuration due to seasonal variations in wave direction (Uda et al., 1988). Stable beaches have formed as expected. Later the straight groins were modified to assume an anchor shape (see Figure 9.11). This unique shape helps reduce the rip current and seaward sediment transport, thus reducing sediment loss offshore. A total of 40 modified headlands have been constructed on the Kashimanada coast (Saito et al., 1996).

The second large-scale application of headland control in Japan can be found at the Joetsu Coast, facing the Sea of Japan, where the beach was eroded severely following the completion of the main breakwaters for Naoetsu harbor (see Figure 9.1) in 1974. The conventional beach protection methods, using sea walls and detached breakwaters,

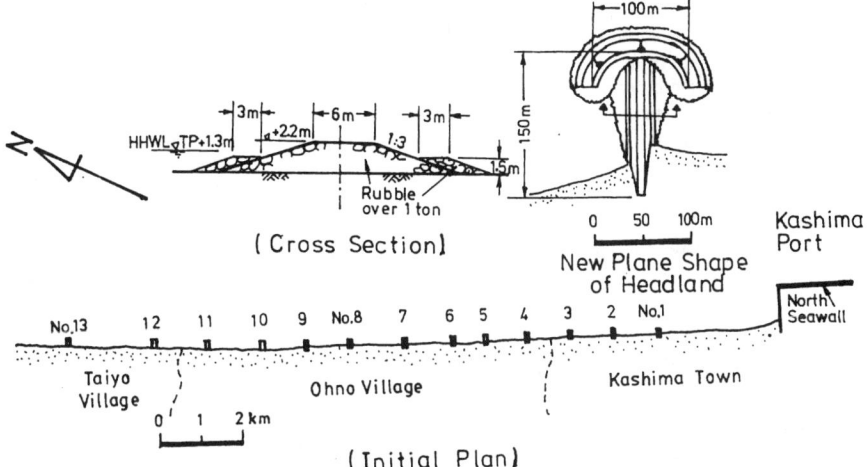

FIGURE 9.11. Headland control at Ohno-Kashima coast, Ibaraki Prefecture (Uda et al., 1988; Saito et al., 1996).

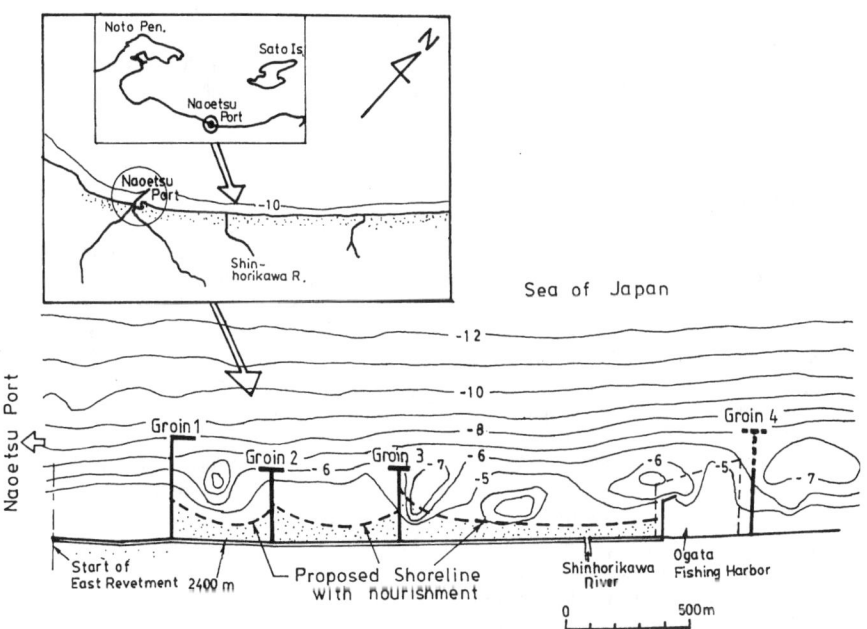

FIGURE 9.12. Headland control at Joetsu coast, downcoast of Naoetsu Harbor, Niigata Prefecture (courtesy of Takao Yamashita 1998).

had been implemented, which adversely transformed the beach profile character from dissipative to reflective, hence accelerating the erosion process. The headland control approach incorporating four groins with circular heads was then proposed (Tsuchiya et al., 1990). But the final design uses a series of four T-shaped groins installed beyond wave-breaking depth, with individual distance offshore and intervals between units reduced in a cascading manner (as seen in Figure 9.12). The work, comprising four headlands in front of a mildly sloped revetment of 2,400 m long and beach nourishment, was commenced in 1998. In addition to the above, Tsuchiya (1987, 1994) also proposed headland control for a beach downcoast of a river mouth, where sediment supply was interrupted by damming, and for an eroding beach downcoast of a large-scale coastal structure.

5. RECENT DEVELOPMENTS

With the economic success that started in Japan in the mid-1970s, there was a community movement demanding a better coastal living environment and full utilization of the coastal zones. But it was not until the 1980s that engineers began examining the aesthetic effects of the conventional protection structures. Several alternative methods were then proposed to improve the existing facilities. First, artificial reefs or submerged offshore breakwaters attracted the attention of coastal scientists and practicing engineers, with many field applications being carried out in Japan. Second, two similar beach protection schemes incorporating hard structures and stable beaches were introduced in succession by different Japanese Ministries in the 1980s. These were the CCZ (coastal community zone) projects administered by the Ministry of Construction (MOC) and the ISPS (integrated shore protection system) under the Ministry of Transport (MOT). More recently, the positive effect of lowering beach groundwater levels for mitigating beach erosion has also been investigated.

5.1. Artificial Reefs

Artificial reefs and submerged detached breakwaters are wave breaking structures. They share the same function—dissipating wave energy and promoting sediment accumulation in their lees if positioned properly. As such, they are used as beach protection structures and for creating recreational beaches. If installed in front of a sea wall, they will also reduce wave runup and the overtopping rate of the sea wall. They can also reduce the longshore drift. However, artificial reefs differ from conventional submerged offshore breakwaters in having a wider crown, usually in the range of 40 to 50 m, and the ability to inhibit offshore sediment transport.

Beginning in the 1980s, artificial reefs have gradually replaced detached breakwaters, in order to enhance the aesthetic appearance of local coastal environments. A typical cross-section is shown in Figure 9.13. By 1992, there were 70 artificial reefs installed in Japan (Asakawa et al., 1992). The variability in structural dimensions, such as crown width and submergence, as well as water depth and armor block weight, are given in Figure 9.14, where the water depth is measured from the low water level. There are no rules of thumb for the dimensions governing the installation of artificial reefs. However, a typical reef structure may have an average crest width between 40 to 50 m and a submergence of less than 2 m in water between 3 to 5 m. Concrete blocks weighting less than 2 tons are commonly used.

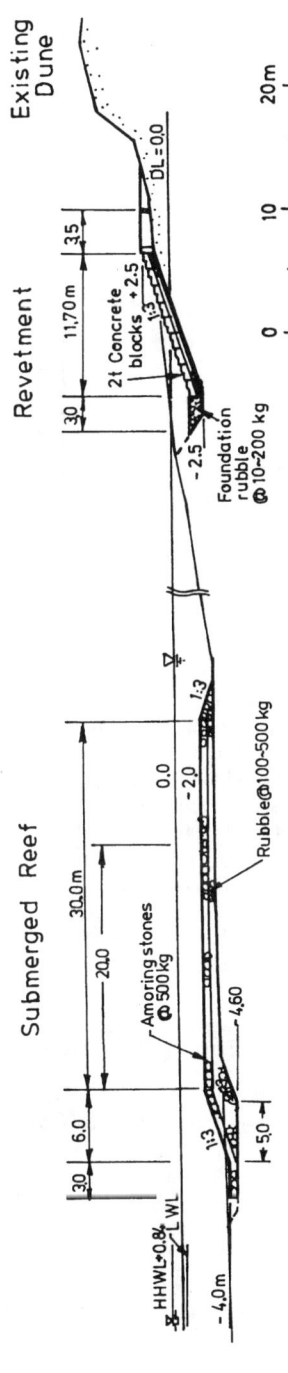

FIGURE 9.13. A typical cross-section of an artificial reef (courtesy of Tottori Prefectural Government 1992).

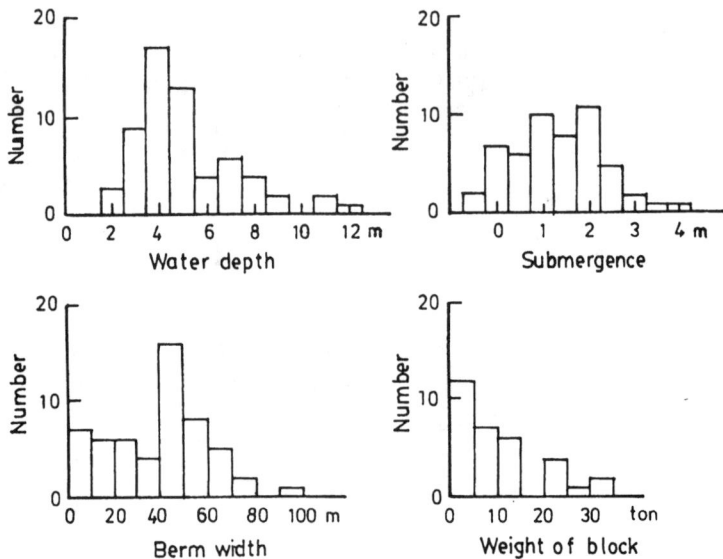

FIGURE 9.14. Frequency distributions for four parameters of artificial reefs (Asakawa et al., 1992).

Although reduction of wave heights by submerged breakwaters have been found in laboratory studies, there was no systematic design guideline available for artificial reefs. Practical problems associated with the construction of artificial reefs include their low stability, threat to public safety, which requires provision of warning signs using buoy systems, and need for deeper submergence to allow small craft to navigate. Yoshioka et al. (1993) have reported that the Ministry of Construction has systematically organized results of model experiments and data obtained through local observations on the effectiveness and stability of artificial reefs, and compiled a *Design Manual for Artificial Reefs*. The *Manual* also gives suggestions for the cross-section and plan shape of an artificial reef unit. It was claimed by Yoshioka et al. (1993) that this was the world's first manual that systematically explained the design of artificial reefs.

5.2. Coastal Community Zone (CCZ) Projects

In the 1980s, the Ministry of Construction also promoted the alternative concept of creating a calm seashore and protection zone. Aoki et al. (1989) reflectioned on this concept, stating that "In cities, there has been a growing demand for development of the waterfront as a leisure area for urban dwellers, and in rural areas, where the natural environment is still intact, people have started to develop the seashore by building resorts." According to the Sea Coast Act, enacted in 1956, the national government has an obligation to support local protection projects. However, prodigious sums of taxpayers' money were not available to meet the demand of many public projects, in addition to the routine and urgent beach protection works around the country. It was not until mid-1980s, when the Japanese government used the profits from selling Nippon Telephone and Telegraph shares, that funds were available to support local development projects (Aoki et al., 1989). From this, the CCZ projects, administered by the Office for Ocean Development within the Ministry of Construction, were implemented in 1987. The objectives of these projects were to integrate public works and private sector investment to enhance local coastal facilities and to advance development and utilization of coastal zones.

The principle concept of the CCZ projects was to improve the designated coastal community zones through a wide range of carefully designed activities, such as the maintenance, development, integration, or expansion of coastal areas (Aoki et al., 1989). Each CCZ improvement plan was drawn up by the local community, be it city, town, or village, and was to be implemented after receiving authorization from the Minister of Construction. The length of coastline for each project was limited to 2 or 3 km. Under a five-year plan, "public works under the jurisdiction of the Ministry of Construction, such as public works of seacoast, public parks, roads, sewage facilities and others, shall preferentially be enforced. At the same time, the vitality of private enterprise shall be positively introduced to build facilities that offer public convenience and entertainment" (Aoki et al., 1989). Initially, a total of 12 CCZ projects were authorized in the fiscal year of 1987. The total project cost to be invested over 5 years was ¥77 billion, of which approximately ¥52 billion was to be invested in public works and the other ¥25 billion in private works. It was estimated that, upon the completion of the infrastructure for each project, it would improve the local economies and hence the so-called "production-inducing effect," which was estimated at ¥170 billion. An additional eight CCZ projects were added in 1988. This was followed by five more in 1989, three in 1990, five in 1991, four in 1992, two in 1993, one in 1994, and one in 1996 (MOC, 1988, 1997). An example showing the completed CCZ project at Atami, Shizuoka, Kanagawa Prefecture is shown in Figure 9.15 (MOC,

FIGURE 9.15. An example of a CCZ project at Atami Coast, Shizuoka Prefecture, Honshu (MOC, 1995).

1995), which has transformed a waterfront concrete jungle into a popular recreational beach.

5.3. Integrated Shore Protection System (ISPS)

Most beach protection works in Japan were built in the 1950s to 1960s. From the data collected by Katoh et al. (1994) on the age of these structures, it may be estimated that, by the year 2000, 30% of the sea walls in Japan will be between 30 and 40 years old and 25% of them more than 40 years, whereas 50% of the groins will be between 30 and 40 years old with 30% of them exceeding 40 years. Instead of reinforcing the shore protection facilities, it was suggested to develop a new protection system that satisfies two basic functions. The first is to prevent a coastal disaster during storms and the second is to provide a beach with calm water under swell conditions. The ISPS was devised to satisfy these two principles.

According to Katoh et al. (1994), the ISPS is a "combined system of an offshore structure, a sandy beach, and a sea wall that brings each protective function into full play as a whole." Under this system, storm wave energy is reduced in three stages, using low-crest or submerged detached breakwaters to dissipate waves and also to help retain beach material within the system, then a natural or artificial beach to further reduce wave energy, and finally a low-crested mild-slope or step-type revetment to provide easy access to the beach. An example showing the conventional protection and the completed ISPS at Tsuda (see Figure 9.18), Kagawa Prefecture on the northeastern Shikoku is given in Figure 9.16. In some ways, this scheme is similar to the existing practice of creating artificial beaches. However, the usual artificial beaches are designed for recreational purposes, often created by artificial nourishment with or without protective structures, and they exclude a step-type revetment in the background. Therefore the conventional artificial beaches may not protect the hinterland from storm wave attack.

This integrated protection scheme has been promoted by the Ministry of Transport in Japan since the mid 1980s, and has gradually become popular in Japan since then, despite the cost involved. A rendering of the ISPS (MOT, 1988) is shown in Figure 9.17. Note that trees are planted behind the low crest of the step-type revetment to prevent sea spray. While describing the ISPS, Katoh et al. (1994) also warned that the fundamental mechanism of sudden beach profile changes due to infragravity waves and groundwater level in stormy conditions should be studied further.

FIGURE 9.16. Comparison between the conventional beach protection and the ISPS at Tsuda Port, Kagawa Prefecture, Shikoku (MOT, 1998).

FIGURE 9.17. A rendering of the ISPS, showing detached breakwaters, sandy beaches with salients, and low-crested step-type revetment with tree planting (MOT, 1988).

5.4. Control of Beach Groundwater Table

Kawata and Tsuchiya (1986) and Katoh and Yanagishima (1996) have recognized the effect of beach groundwater level on the diminution of a beach profile. It was observed that the groundwater table rose when the foreshore was eroded in a storm. By lowering it, beach material may be prevented from moving seaward during a storm attack. Different terminologies have been used for this procedure, such as subsand system (Kawata and Tsuchiya, 1986) and gravity drainage system (Katoh et al., 1994). The concept has been tested in laboratories (Kawata and Tsuchiya, 1986; Katayama et al., 1992; Kanazawa et al., 1996; Sato et al., 1996), as well as in the field during storms (Yanagishima et al., 1991; Katoh and Yanagishima, 1996). The potential benefit of using a gravity drainage system for beach stabilization has been confirmed from field tests in Japan; the system is identical to the beach dewatering systems implemented in United States (Curtis et al., 1996) and elsewhere.

The methodology behind the subsand filter system is to increase the inflow velocity into the beach through control of the sediment–fluid boundary, thus accelerating the accretion of sediment on the foreshore. In laboratory experiments reported by Kawata and Tsuchiya (1986), piping beneath the model beach was connected to a pump that controlled the flow rate out of the beach during tests. Wave run-up was reduced, thus resulting in the accumulation of sediment on the model beach, regardless of swell or storm waves. They concluded that: "As the results clearly show, the subsand filter system has high applicability to beach erosion control work."

Although the subsand filter system is capable of promoting sediment accumulation on beaches, it requires continuous pumping. Katoh et al. (1994) regarded this as "a large problem in this method," since high operational costs may result. Consequently, an alternative concept, called the gravitational drainage system, was proposed, in which the beach "groundwater is to be naturally drained offshore through a porous layer set up under the beach" (Katoh et al., 1994). A laboratory test was conducted in a two-dimensional wave flume, using a drainage layer 10 cm thick consisting of crushed stones of 13 to 20 mm in diameter (Katayama et al., 1992; Kanazawa et al., 1996). The result, after 5 hours of wave action, showed that the model beach without the drainage system had formed an offshore bar with consequent berm erosion, whereas accretion occurred on the beach with a porous layer. A subsequent field experiment on a gravity drainage system was successfully carried out at the HORF (the Hazaki Oceanographic Research Facilities, Ministry of Transport) during a storm in 1994 (Katoh and Yanagishima, 1996). It was found that a gravity drainage system was capable of decreasing the speed of foreshore erosion in a storm, and an eroded foreshore recovered quicker on a drained beach after a storm event. Thus the benefit of controlling or lowering the beach groundwater table is apparent.

6. CASE STUDIES

Most of Japan's economic and social activities are concentrated on its relatively narrow coastal plains, which suffer from natural disasters. Losses of land are calamitous and hence beach protection has received a high priority in terms of research funds and field experiments. One development not directly associated with the coast but affecting it closely has been the construction of dams on rivers for water supply, power generation, and flood control, mainly after World War II. Another has been the multitude of fishing ports that were expanded to large harbors between 1950 and 1960. The coastal damage caused by these developments has instigated many prevention schemes, which will now be briefly described. A locality map showing the erosion sites discussed in this section and others is given in Figure 9.18.

6.1. Amano-Hashidate

This sandy spit, some 3.6 km in length, separates Miyazu Bay from Asokai (an inland sea). Two narrow waterways at its southern end connect the two water bodies. The width of the spit varies from 20 to 170 m and in historic times it was covered with pine trees to make it one of the three most picturesque landscapes, including Matsushima and Miyajima, in ancient Japan. As seen in Figure 9.19, there are three rivers debouching into the Sea of Japan on the eastern margin of the Tango Peninsula, on the northern shoreline of Honshu, Kyoto Prefecture. To the south of these are two fishing harbors, Hioki and Ejiri. The relative shallowness of Miyazu Bay is associated with mildly sloped beaches because of the fine sand.

It is obvious from Figure 9.19 that longshore drift is southward, with the main sediment supply coming historically from the three rivers mentioned above. Prior to 1950, erosion was evident near their mouths due to reduction in sediment discharge, necessitating coastal defenses. Later, around 1970, dredging of the entrance channels to the two ports further intercepted this sand transit to the sand spit (Yajima et al., 1983; Irie et al., 1990). This removal was sporadic, dependent as it was on wave conditions within Miyazu Bay.

As early as in 1950, groins of 15 m long and spaced 50 m apart were inserted but found to be ineffective. Each fourth groin was doubled in length around 1970. Although erosion was considered to have been stopped, the natural scenic beauty had not been restored. Detached and submerged breakwaters were contemplated but the demand for large volumes of sand could not be met (Irie et al., 1990). Assisted by the Ministry of Trans-

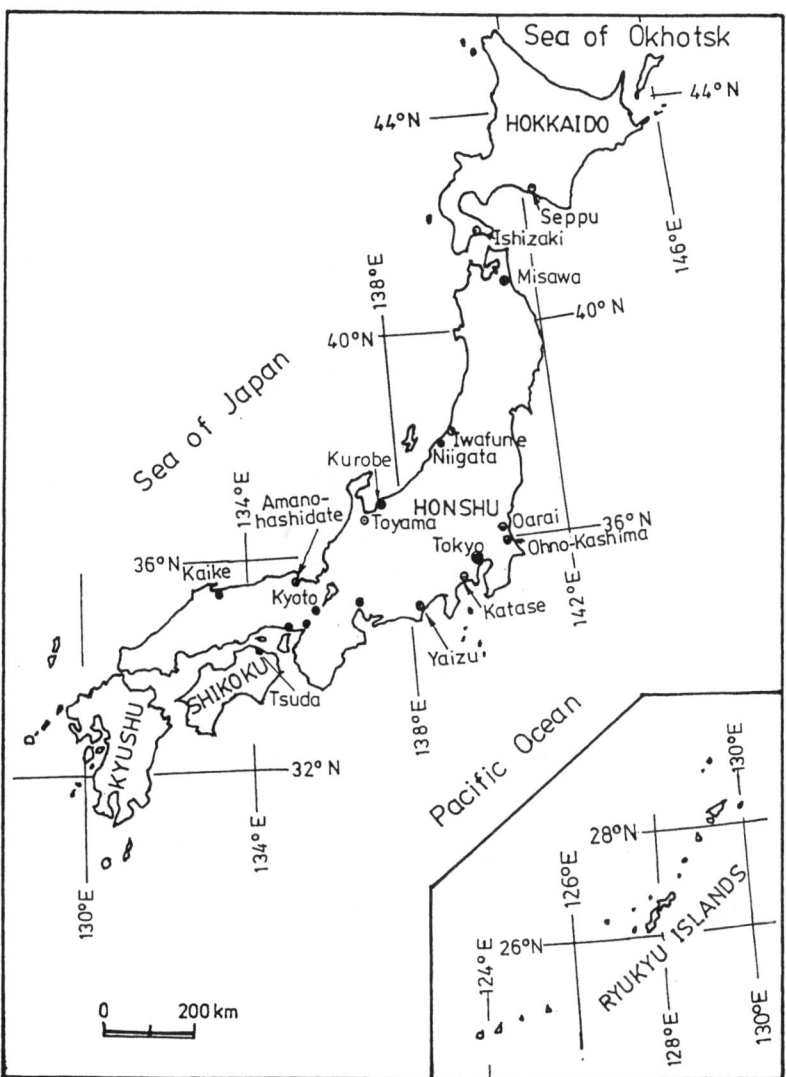

FIGURE 9.18. Locality map showing the erosion sites discussed in Section 6.

port, Kyoto Prefecture Government therefore proposed in 1982 to undertake a sand bypassing program. The annual average of 4,000 m^3 dredged from Hioki and Ejiri was to be transported by barge and deposited on the northern edge of the spit. The beach was smoothed in order for swell waves to carry the sand past the groins to the downcoast end, where it was recycled as indicated in Figure 9.20. Yajima et al. (1983) stated that this was the first bypassing initiative to be used in Japan. Irie et al. (1990) reported other attempts, but these were mainly channel maintenance procedures.

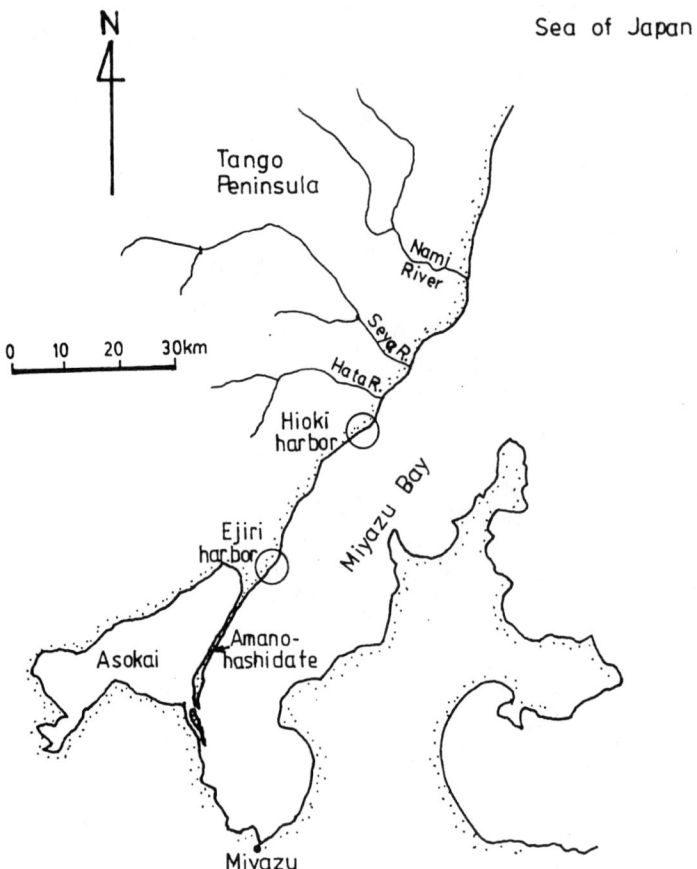

FIGURE 9.19. Location of Amano-hashidate in Miyazu Bay, showing rivers and harbors on Tango Peninsula to its north, and the Asokai to its left (Irie et al., 1990).

Surveys over many years have proven that the sand bypassing scheme at Amano-hashidate has not only preserved the spit from further erosion but also enhanced the aesthetic value of the unique landscape. Amano-hashidate has since become a popular beach for bathing and leisure. Chin (1993) and Chin et al. (1994) have reported that the combination of groins, beach nourishment and sand bypassing at Amano-hashidate has created a series of dynamically stable sandy beaches between groins (see Figure 9.21).

6.2. Ishizaki

The first detached breakwater system in Japan was constructed in 1966 at Zenikamezawa Beach on the Ishizaki Coast, in Hokkaido east of Hakodate City (see Figure 9.1). Toyoshima (1974) reported on the erosion of this beach: "Before 1960, there were wide

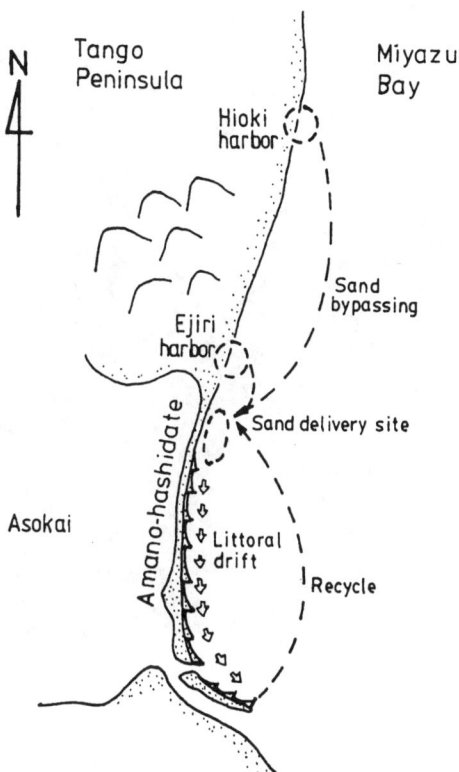

FIGURE 9.20. Sand bypassing scheme at Amano-hashidate Coast (Irie et al., 1990).

sandy beaches with 1 to 2 meters thickness of sand on the rocky seabed. After finishing the sea wall construction works, which were carried out early in the 1960's, the sandy beaches have vanished." It was Toyoshima who suggested the construction of a detached breakwater at the site, some 60 m in length, parallel to the existing wall and 35 m distant from it. A tombolo was formed from sand on the rocky seabed at a depth of 4 m. This prompted construction of five more units by 1971 (Figure 9.22) and five more by 1974. The problem of beach erosion has since been overcome. The tombolo has been instrumental in growing new seaweed on the rocky seabed, which has helped the local fishing industries (Uda and Igarashi 1992).

6.3. Iwafune

Iwafune fishing harbor (see Figure 9.18) is located on the mouth of the Ishi River, some 45 km north of Niigata and 5 km south of the Miomote River, as seen in Figure 9.23. The direction of most winter storm and swell waves is NW to WNW, whereas the shoreline up to the Miomote River is almost NNE and becomes NW–SE for the local beach at the river mouth. The littoral drift was estimated at about 40,000 m^3 per year from the rivers to

FIGURE 9.21. An oblique aerial view of Amano-hashidate spit in 1993 (Chin, 1993).

FIGURE 9.22. Five detached breakwaters on Zenikamezawa Beach, Ishizaki, near Hakodate in September 1971, showing tombolos and the narrow strip of land protected by coastal structures (sea wall, groins, and detached breakwaters) in front of a high cliff (Uda and Igarashi, 1992).

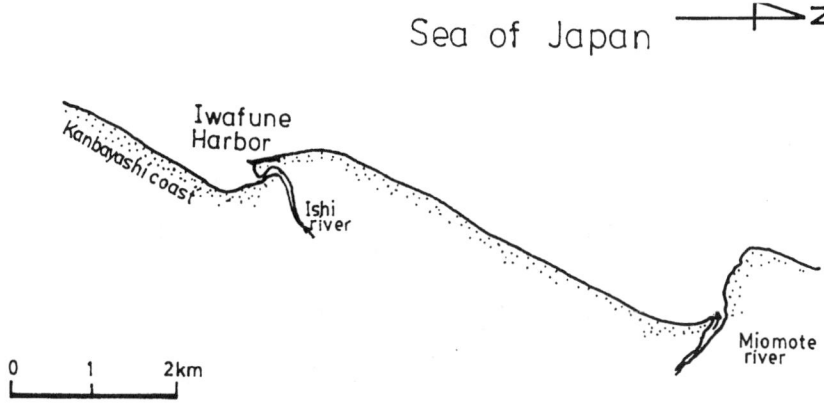

FIGURE 9.23. Location of Iwafune fishing harbor in Niigata Prefecture, showing local shoreline orientations (Haruta, 1961).

the north of the harbor (Haruta, 1961). This results in continuous siltation at the river mouth, even with the construction over many decades of breakwaters, groins, and sea walls, as depicted in Figure 9.24. Haruta (1961) has commented on this incessant shoaling. Hom-ma and Horikawa (1960) have also stated: "We should say that this fishing harbor embodies the history of the Japanese engineers who fought coastal sediments with unswerving perseverance." Koh (1966) has surveyed these developments from 1913 to 1957, noting the volume of dredging required. As seen in Figure 9.24, siltation was the eternal problem, but by 1965 dam construction on the Miomote River had reduced this sediment supply. This may be a benefit of dam construction, as distinct from the adverse effect of causing beach erosion.

In 1965, the shape of the shoreline south of the harbor was crenulate. In this same year, an extension was started to the northern breakwater. Further work was done in 1976, 1979, and 1980, and continued until 1984, with even an island extension in that year (see Figure 9.25). These involved seaward protuberances of almost 1 km (Uda and Noguchi, 1991), which effected new diffraction points for waves and hence new shapes for the bayed beach downcoast. The accretion toward these salients, behind the extended northern breakwater, are seen in Figure 9.25. The salient growth along the two survey lines S_0 and N_2 of Figure 9.25 is exhibited in Figure 9.26. In this figure, the correlation between the curves for salient line S_0 and the outer breakwater length is very definitive, but note the constancy of the N_2 axis upcoast of the harbor. It is seen in Figure 9.25 that a long groin has been needed in the harbor basin to prevent siltation, but as is evident in this figure, much more accretion could occur if material is available either upcoast or downcoast. This has occasioned the insertion of groins and detached breakwaters on the southern Kanbayashi Coast to prevent this supply, and for beach protection (Figure 9.27).

6.4. Kaike

This is another section of coast that has suffered significant erosion and has been protected by groins, sea walls, and detached breakwaters, in that order. It is also the most

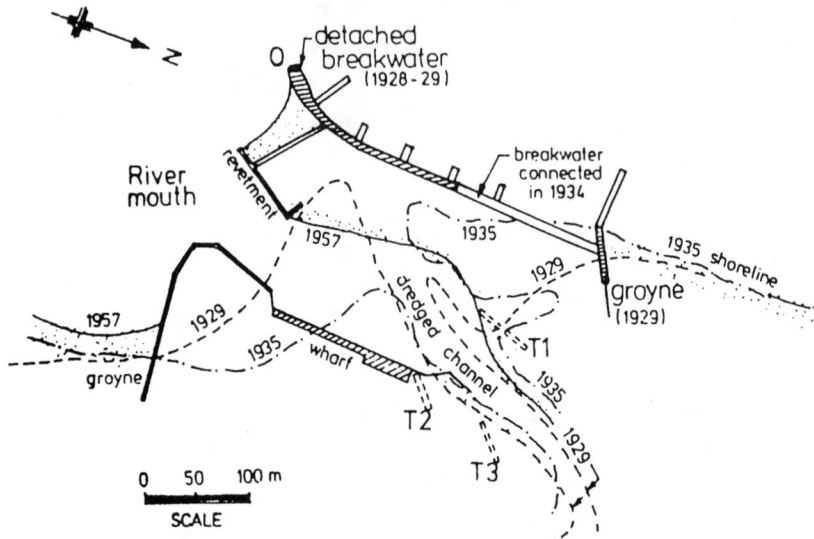

FIGURE 9.24. A series development at Iwafune, showing the first river-mouth jetty T3, two subsequent jetties T1 and T2, groin, and detached breakwater in 1928–1929, the main breakwater connected in 1934, and river-mouth revetment and harbor wharf in 1957 (Hsu et al., 1993, modified from Haruta, 1961).

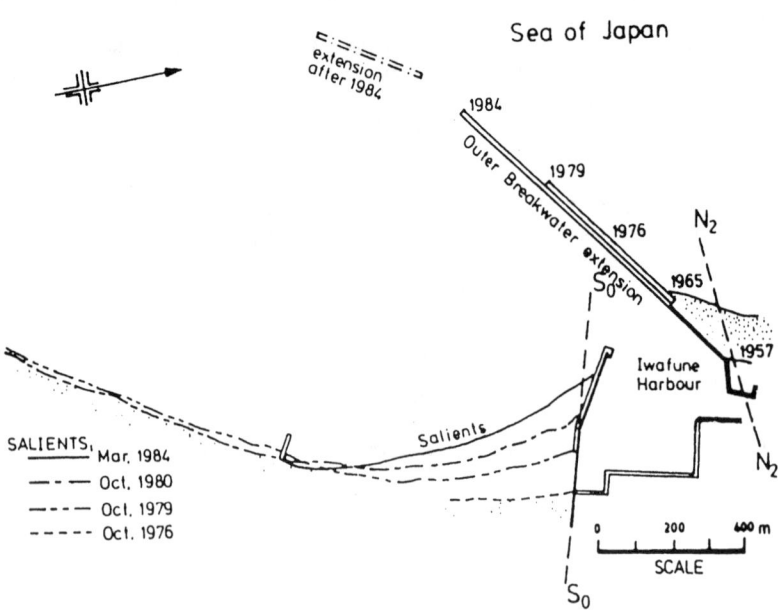

FIGURE 9.25. Outer breakwater extension at Iwafune fishing harbor from 1965 to 1984, showing potential salients in the lee of the main breakwater (Uda and Noguchi, 1991; Hsu et al., 1993).

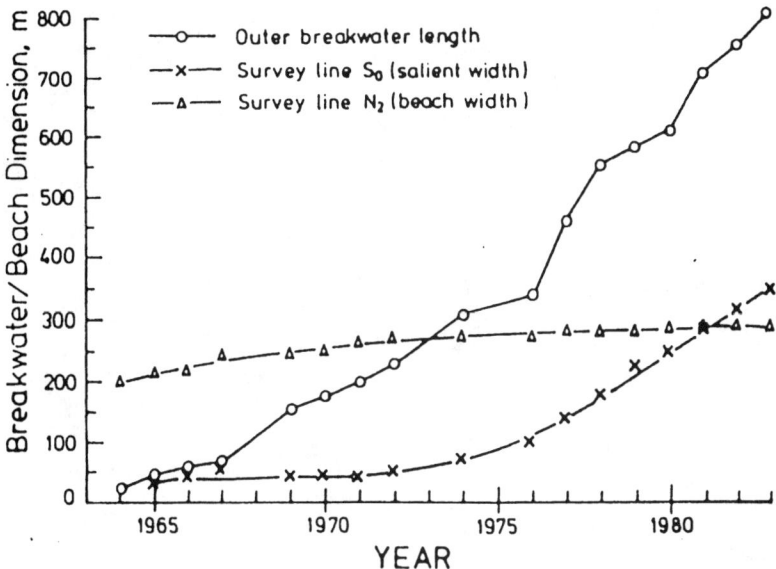

FIGURE 9.26. Relationship between the extension of the outer breakwater at Iwafune and the shoreline advancement along two survey lines S_0 and N_2 in Figure 9.25, from data of Uda and Noguchi (1991) (Hsu et al., 1993).

publicized example of beach protection employing detached breakwaters. As seen in Figure 9.28, the Kaike Coast can be found at the western part of Tottori Prefecture on the northern shoreline of Honshu (see Figure 9.1). It is located at the root of a major sand spit (called Yumiga-hama), virtually in the lee of a large headland (Shimane Peninsula), and supplied by sediment from the Hino River (Hom-ma and Horikawa, 1960; Toyoshima, 1974, 1976, 1982; Uda and Igarashi 1992). This spit also contains hot springs, making it a much desired resort. The curved beach has been shaped by the NE and ENE waves of the Sea of Japan, as they diffract at the eastern tip of the Shimane Peninsula into Miho Bay. Prior to 1910, the beach was accreting and was widened some 200 m. By 1920, erosion commenced due to closure of a sand mine up the Hino River, where dam construction was also underway. Between 1920 and 1970, the Kaike beaches receded 200 m (Toyoshima, 1974, 1976) with many buildings falling prey to nature.

The losses created a national alarm, with protection measures taking various forms as materials and technology became available. During World War II, no cement could be used for such activities, so wooden truss groins were employed, being promptly destroyed in storms. In 1947, rubble stone versions were erected, which also subsided upon completion. The Tottori Prefecture Government then established a beach erosion research committee, comprised of engineers from universities and agencies of the local and national governments. This was an unprecedented action in the history of beach management in Japan (Toyoshima, 1976). Their deliberations led to construction of five more rubble stone groins in 1950, in front of the Kaike Spa, with some initial accretion appearing to ensue. This led to construction of eight more groins from 1951–1954, but the complete section of beach disappeared in a typhoon in 1955.

(a)

(b)

FIGURE 9.27. Iwafune Harbor and beach downcoast (a) in 1965 and (b) in 1988, showing breakwater extension plus groins and detached breakwaters used. (Courtesy of Ministry of Construction.)

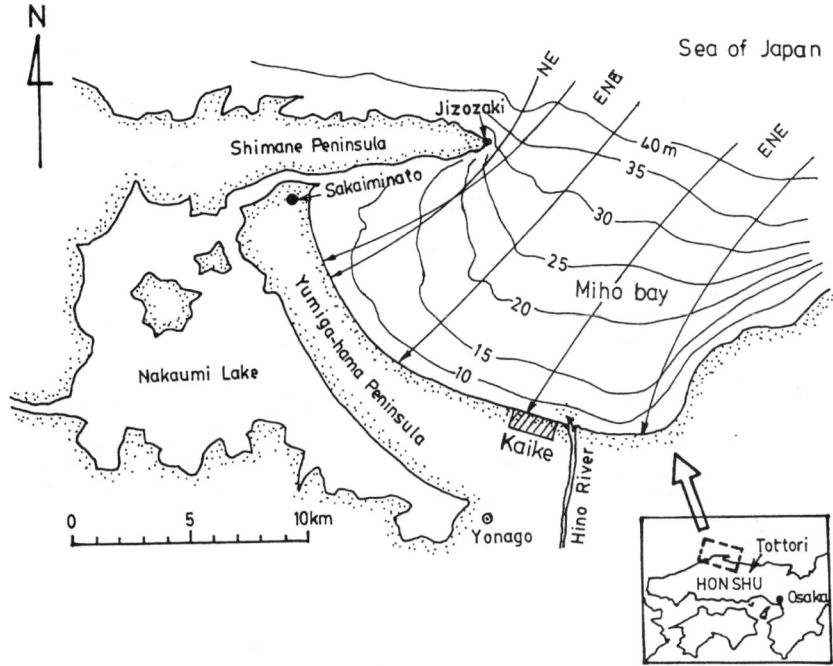

FIGURE 9.28. Location of the Kaike Coast, showing directions of the prevailing waves within Miho Bay, facing the Sea of Japan (Toyoshima, 1976).

The ferocity of the wave attack occasioned the construction of a sea wall with concrete steps some 1.3 km in length over six years, starting in 1955. The beach berm actually built up to the soffit (Figure 9.29), but in 1961 (on completion of the wall), storm waves had removed the sand and reflection took place with resulting scour and collapse of this major structure. A stronger but inclined wall followed but still required toe protection with what were known as hexa-leg blocks. By this time no beach existed in front of the sea wall, and maintenance was difficult.

A change in direction was then initiated by Toyoshima (1974), who was influential in the Ministry of Construction and who had had some success with detached breakwaters at Ishizaki Beach, Hokkaido, in 1966 and at 20 other locations. A single unit was installed at Kaike in 1971 using rubble stone armored by 16 tonne tetrapods (Figure 9.30). These structures were 150 m long and located 110 m offshore in 5 m water. The sand trapped in the tombolo was said to have come from 12 m depths offshore. One detached breakwater unit was added toward the west each year for the next decade. The gap between units was 50 m, with salients or tombolos providing calm beaches for swimmers. The only drawback has been the scour on the seaward edge due to the oblique reflection of all waves, storm and swell. Further westward of the protected region, a concrete sea wall is still used, with severe dune erosion beyond this. On the eastern side of the Hino River mouth, detached breakwaters have also been employed effectively, as seen in Figure 9.31.

FIGURE 9.29. Sandy beach of moderate width on the Kaike Coast in 1957, showing berm elevation about the soffit of the sea wall (Uda and Igarashi, 1992).

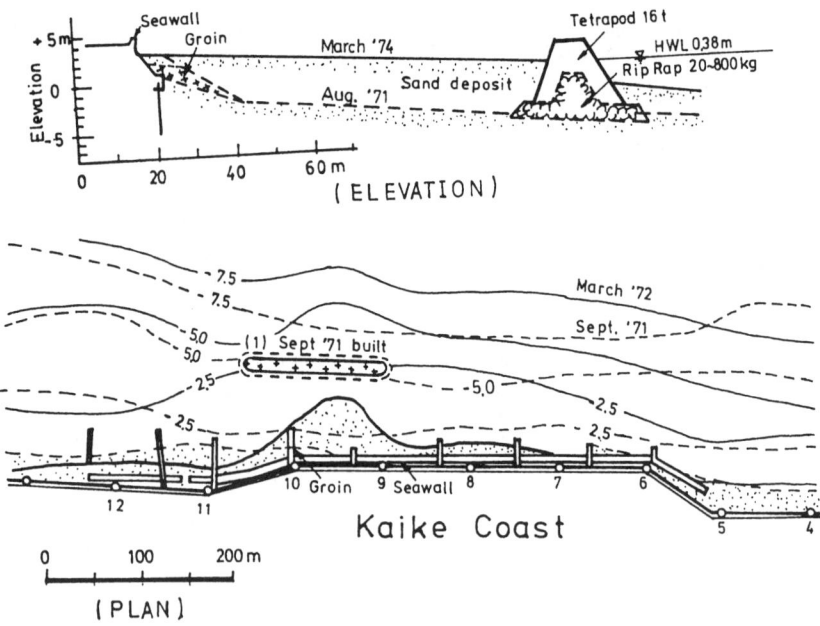

FIGURE 9.30. Cross-section of the first detached breakwater at Kaike in 1971 and the change of shoreline after its completion (Toyoshima, 1974).

FIGURE 9.31. The beach protection scheme at the Kaike Coast in 1991, including east of the Hino River. (Courtesy of Ministry of Construction.)

6.5. Katase East

This beach, also known as Enoshima Beach because it is in the lee of an island by that name. It is on the eastern side of a tombolo that was formed due to a man-made extension to Enoshima Island during the construction of a marina. This previous salient, at the mouth of the Katase River, is located at the NE corner of Sagami Bay, as seen in Figure 9.32 (Hom-ma et al. 1960). Prior to the tombolo formation, sand from the Sagami River moved eastward to feed the salient on both sides, despite the Katase River debouching on the western beach. This eastern littoral drift is clearly shown by the deflection of the Hikichi mouth in that direction.

The Katase East Beach is limited in length (0.9 km) because of a rocky promontory (Point Koyurugi) where, as seen in Figure 9.33, another marina has been built adjacent to a small river outlet. It is one of the most popular beaches for the vast population of Tokyo, 50 km to the east, as exhibited by Figure 9.34 on a hot summer day. Okajima et al. (1993) stated that: "As many as 10,000,000 people visit the East and West beaches of Katase, the largest bathing resort in Japan." Its popularity is enhanced by the mild waves, caused by the large diffraction around the eastern tip of Enoshima and its enormous marina.

Progressive shoreline recession has occurred over the entire length of the coast east of the Sagami River, partly because of the reduction in sediment discharge due to the construction of two dams upstream of the Sagami river in the 1940s, for flood control and to supply water to Tokyo and Yokohama cities (Hom-ma et al., 1960). Extraction of large quantities of sand and gravel from the river bed for construction purposes further accelerated the recession of the beaches within Sagami Bay, including the Katase East Beach. Hom-ma et al. (1960) have reported severe erosion in this area.

The erosion of Katase East beach today can be abated with the natural supply of sand or renourishment by man. The only material available for the elongated length to Enoshima Island itself has to come from the curved eastern segment adjacent to the mainland marina at Point Koyurugi. Previous causes of erosion have included the construction of a highway on the seaside edge of the sand dunes for the complete length of this beach (Figure 9.35). This was probably the cheapest way of providing motor access because houses

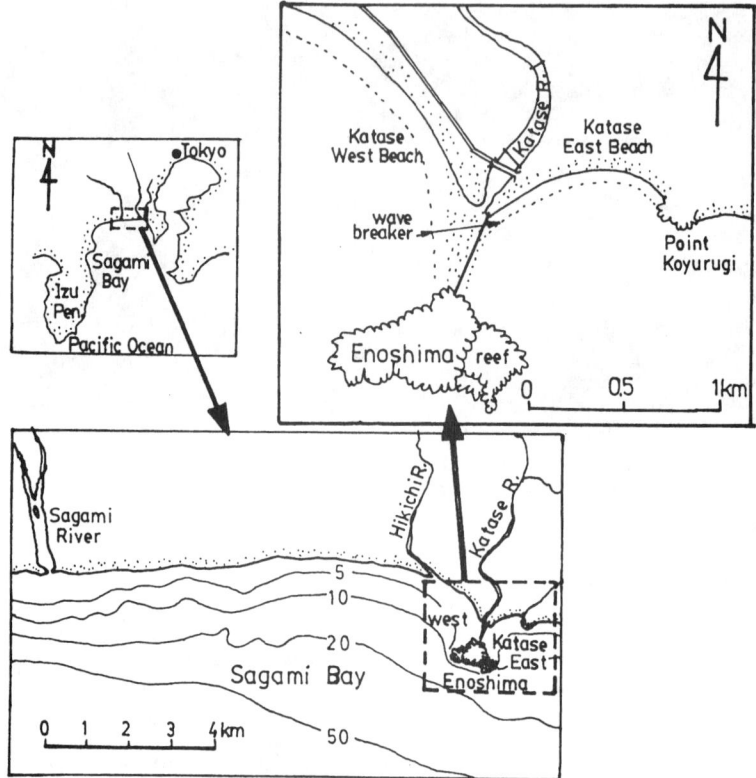

FIGURE 9.32. Location of Katase East Beach in Sagami Bay (modified from Hom-ma et al., 1960).

and commercial buildings edged the original beaches. The road had to be protected from the sea, so revetments were built on its seaward edge. During storms, with storm surge levels, waves reflected from these revetments with concomitant severe scour. Sediment so moved offshore has not been readily returned to the berm because of the small swell waves (Hom-ma et al., 1960).

This coast has major storm centers and typhoons passing eastward, both north and south of this coast. It suffers winds from NE and SWS, neither of which have much influence on Katase East Beach, except as small diffracted waves. However, typhoons can create strong southerly gales, accompanied by storm surge, which can effect erosion swiftly and drastically. The eastern edge will be influenced more than the western margin, but there the width is small and could disappear if the salient reformed.

6.6. Niigata

The coast to the SW of Niigata City (Figures 9.1 and 9.36) is a salient formed in the lee of Sado Island, some 75 km offshore but 130 km in length. For many centuries, the Shinano

SHORELINE PROTECTION METHODS—JAPANESE EXPERIENCE 9.43

FIGURE 9.33. Aerial photograph showing Enoshima Island and marinas in 1980. (Courtesy of Ministry of Construction.)

FIGURE 9.34. The Katase East Beach on a hot summer day (Okajima et al., 1993).

9.44 HANDBOOK OF COASTAL ENGINEERING

FIGURE 9.35. Highway Route 134 run through the berm of Katase East Beach in 1968 (Uda and Igarashi, 1992).

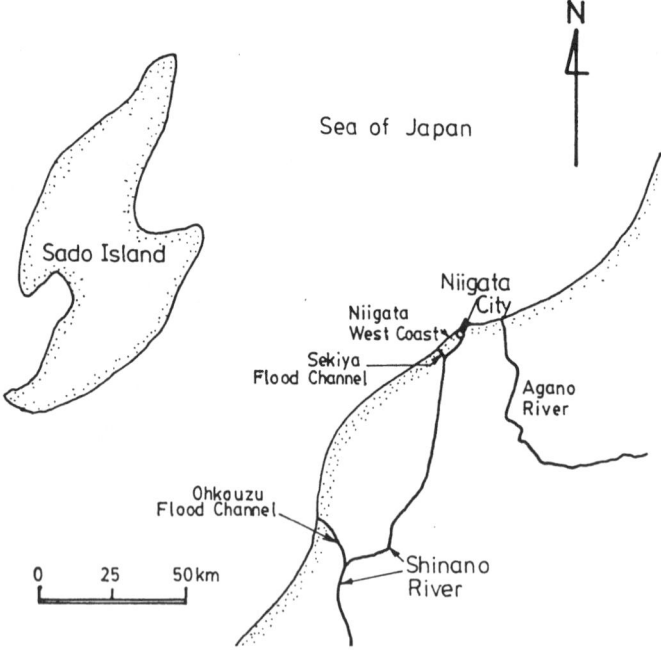

FIGURE 9.36. Location of the Niigata Coast and Shinano River on Honshu (Irie et al., 1990).

SHORELINE PROTECTION METHODS—JAPANESE EXPERIENCE   9.45

River (the longest in Japan) has debouched northward into the Sea of Japan. It has discharged tremendous volumes of sand that has been transported to the SW by waves from the NW quadrant. To overcome siltation at the port mouth and reduce flooding of the city, an alternate exit was dredged between 1909 and 1922, known as the Ohkouzu Flood Channel, which, as seen in Figure 9.36, is some 70 km upstream of the original river mouth. The original discharge was reduced by 95%, which meant that sediment to the west coast of Niigata city became negligible.

The resulting erosion was excessive on both sides of the original mouth but particularly toward to the west. Hom-ma and Horikawa (1960) considered this recession as the most prominent example in Japan of human interference with natural processes. Irie et al. (1990) surveyed this west coast situation, which involved the losses of sand dunes and some hundreds of meters of beach width, as indicated in Figure 9.37, from 1889 to 1984. The greatest loss was near the river mouth because material was still bypassing points to the SW. Irie et al. (1990) referred to this recession as a "very rare case in Japan."

Protection measures were first proposed in 1945, but an extensive report was issued in 1951, and construction commenced in 1957 (Hom-ma and Horikawa, 1960). This involved a sea wall, groins, and submerged breakwaters. These latter structures suffered from scour at the seaward toe that required continuous additions. Subsidence of the subsoil due to natural gas extraction since 1949 also exacerbated the problem, resulting in a breakwater cross-section shown in Figure 9.38. To improve the efficiency of these submerged structures, they were broadened and designed for lower transmission and reflection coefficients. Figure 9.39 shows the coastal structures installed at Niigata in 1973.

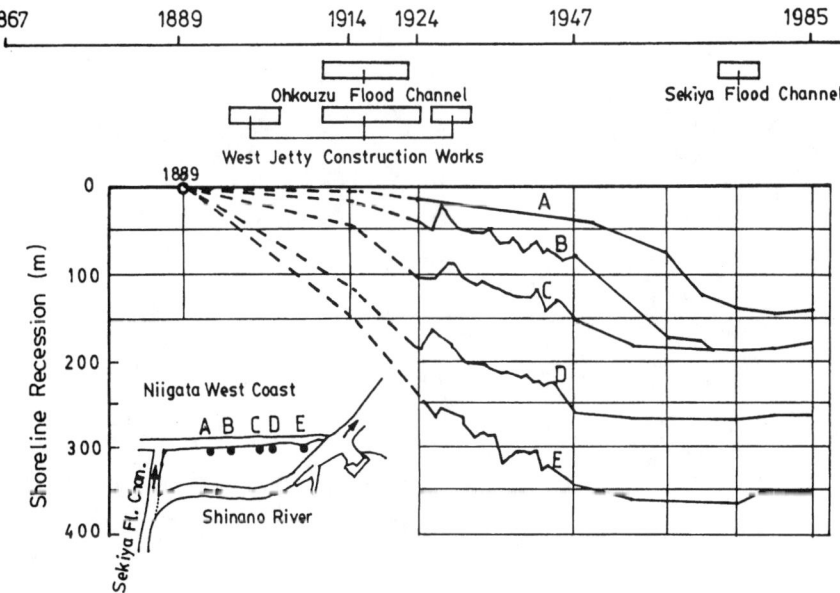

FIGURE 9.37. Relationship between Shinano River improvement works and shoreline changes at Niigata West Coast (Irie et al., 1990).

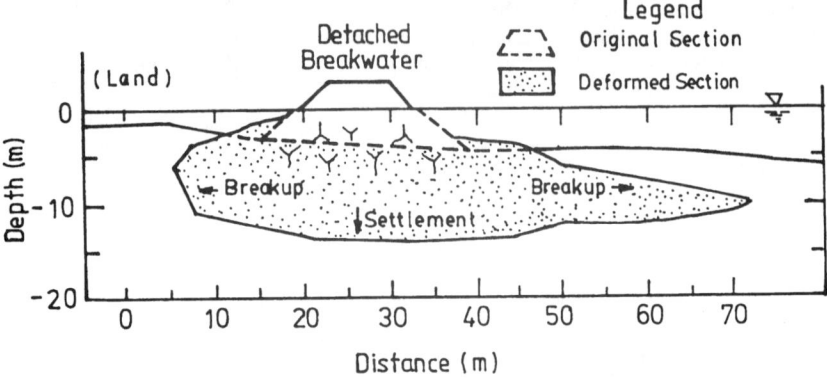

FIGURE 9.38. Deformed shape of detached breakwater at Niigata Coast (Irie at al., 1990).

6.7. Oarai

This port is located north of the Kashimanada Coast (see Figure 9.40), which is a bow-shaped sandy beach about 70 km long between the Naka River to the north and the Tone River in the south. The predominant wave direction at Oarai is southward (Kraus et al., 1984; Uda et al., 1986). Sediment discharge from the Naka River was transported past the Oarai headland. Originally, Isohama Harbor was built, as seen in Figure 9.41, but it silted

FIGURE 9.39. Aerial photograph showing beach protective structures installed at Niigata West and East Coasts in 1973. (Courtesy of Ministry of Construction.)

SHORELINE PROTECTION METHODS—JAPANESE EXPERIENCE 9.47

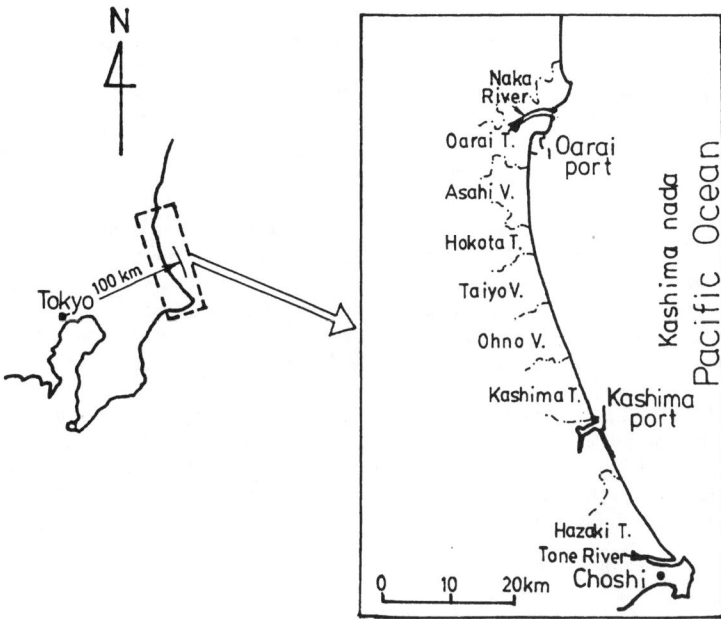

FIGURE 9.40. Locality map of the Kashimanada Coast, showing Oarai and Kashima ports and the towns (T.) and villages (V.) between them (Uda et al., 1988).

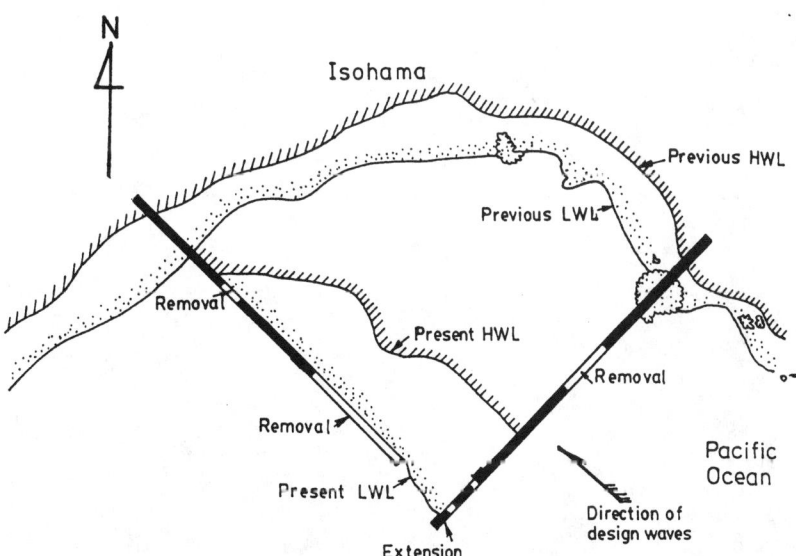

FIGURE 9.41. A small fishing harbor at Isohama (presently Oarai) completely disabled by severe sedimentation in 1911–1916 (Hom-ma and Horikawa, 1960).

up before it was completed in 1916 (Hom-ma and Horikawa, 1960). No development took place between 1916 and 1961. The modern history of Oarai port has resumed since 1977 (Uda 1997), when the old fishing harbor was converted to a ferry and cargo port, involving the construction of a long breakwater. This certainly overcame the siltation problem.

As seen in Figure 9.42, an L-shaped breakwater was constructed between 1977 and 1981 to deflect littoral drift seaward. In 1981 the breakwater had a 45° arm added, which caused beach erosion downcoast. An offshore structure was added between 1981 and 1985 (Figure 9.43), creating erosion further downcoast (not indicated in this figure). As seen in this figure, the offshore contours had been moving seaward, indicating local aggradation and beach degradation further downcoast. As can be seen by the existing beach alignment in 1983, harbor siltation was anticipated, requiring a longer groin or breakwater extension in 1985. The material for these required additions was not coming from the north but was being supplied by beaches some 1 to 3 km to the south, which began to erode (Uda et al., 1986; Uda, 1997). Sea walls and concrete block revetments were installed (Mizumura, 1980; Kraus et al., 1984). Even other groins were inserted to try and prevent this accretion, to no avail. The beach profiles steepened and the walls finally failed, as seen in Figure 9.44 (Ibaraki Prefecture Government, 1991; Uda, 1997). Thus the port of Oarai has suffered both siltation and erosion. An aerial view of the Oarai port in 1988 is depicted in Figure 9.45.

FIGURE 9.42. Aerial photograph showing shoreline at Oarai in 1976. (Courtesy of Ministry of Construction).

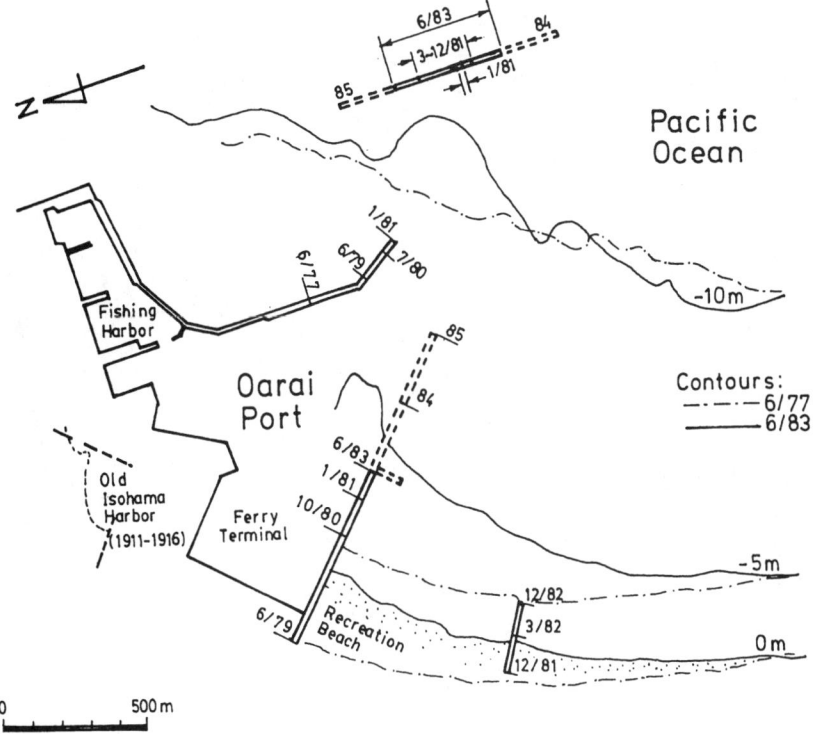

FIGURE 9.43. Progressive construction of breakwaters between 1961 and 1985 and contour changes between 1977 and 1983 at Oarai (Kraus et al., 1984).

FIGURE 9.44. Sea wall on the eroding beach south of the Oarai port in 1985, showing rows of concrete blocks for toe protection (Uda, 1997).

FIGURE 9.45. Aerial photograph showing shoreline at Oarai in 1988, as well as the breakwaters and groins. (Courtesy of Ministry of Construction.)

6.8. Ohno-Kashima

The Kashimanada Coast is located on the eastern margin of Honshu, as seen in Figure 9.40. It is a mildly indented bay shape at the north and almost straight toward the south (Sato and Tanaka, 1966). The distance from Kashima Port to Ohno Beach is around 15 km. The longshore transport of sand reverses seasonally but has a southward predominance, although this is in dispute (Uda et al., 1988).

The Ministry of Transport was responsible for the construction of the port, which commenced in 1963 and was completed in 1975. The main southern breakwater runs 3500 m seaward at about 45° to the shoreline, after a shore-normal segment of some 1000 m, as shown in Figure 9.46. A northern breakwater, normal to the beach, is 1050 m long. To the south, 6.75 km of sea wall with submerged detached breakwaters of the same length protects the coast nourished by the dredged soil.

After the construction of the main breakwater for Kashima Port to the water depth of 16 m, the beach at Ohno Village some 15 km to its north began to erode, possibly due to littoral drift being intercepted by the breakwater. Shoreline changes surveyed in October 1985 and November 1987 (Uda, 1997) indicated that beach retreat of about 30 m was experienced at some sections of Ohno Village. The severe erosion occasioned the construction of headlands, the first large-scale attempt in Japan of the method known as "headland control" (Uda et al., 1988; Silvester, 1976). Initially, a total of eleven headlands in the form of straight groins were constructed at an intervals of approximately 1 km, commencing in 1988. The crown width of each headland is 6 m and the length 100 m, with its tip at 2 m of water depth. Dynamically stable beaches of symmetrical shape have formed between headlands.

A follow-up report by Saito et al. (1996) has confirmed the success of the headland

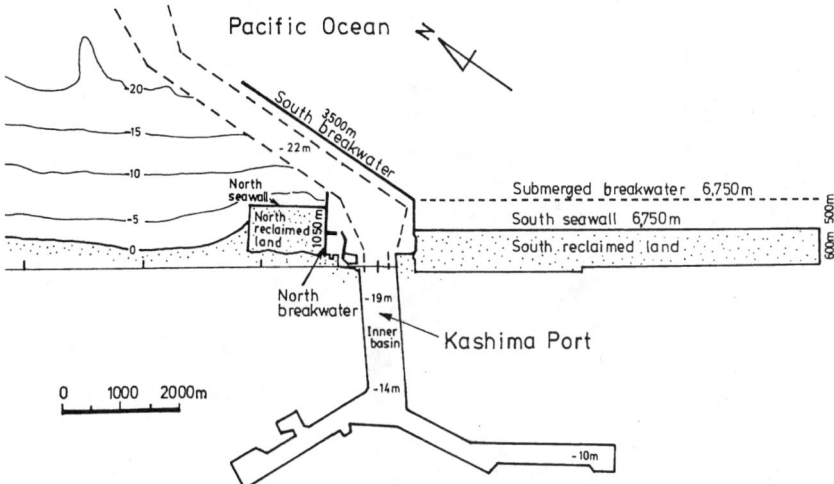

FIGURE 9.46. Layout of Kashima Port, including contours in its north in 1973 (Tanaka and Sato, 1976).

control approach at the Ohno-Kashima Coast. However, the headlands are modified into an anchor shape (Figure 9.47), 150 m in length. The semicircular shape for the headland has a projected length of 100 m and a mild slope of 1:3, designed to reduce wave impact. The anchor shape helps reduce the rip currents and seaward sediment transport. Saito et al. (1996) have also reported that the construction of 40 modified headlands has been underway on the Kashimanada Coast. In this way, stable beaches have been maintained whose orientations change from season to season. New shapes and dimensions have been employed, mainly to reduce reflection from the semicircular heads.

6.9. Seppu

There are approximately 3000 fishing harbors in Japan, an average of one harbor for every 10 kilometers of coastline (Fukuya et al., 1994). Many of them have suffered siltation. Dredging and/or structural options have been employed as countermeasures. Groin fields were first used to intercept longshore transport, without much success. Detached breakwaters were installed, which succeeded in preventing shoaling of harbor entrances. A typical application at Seppu fishing harbor on Hokkaido is discussed in this section (see Figure 9.18).

The Hidaka Coast, where Seppu fishing harbor is located, is a 130 km long coastline with a gently curved arc at the southeastern corner of Hokkaido (Figure 9.48). The coast, stretching approximately in a SE–NW alignment from Point Erimo to Tomakomai Harbor, receives swell and storm waves from the northern Pacific Ocean. Waves in the winter months sweeping from the NW direction are said to cause more serious beach erosion than the waves in the summer, which are essentially from the SE. Ozaki (1964) suggested that the best way to combat the potential blocking of the harbor entrance at Seppu was to construct an offshore breakwater outside the main breakwater. The longshore drift, which was transported down to the tip of the main breakwater previously, could then be diverted

9.52 HANDBOOK OF COASTAL ENGINEERING

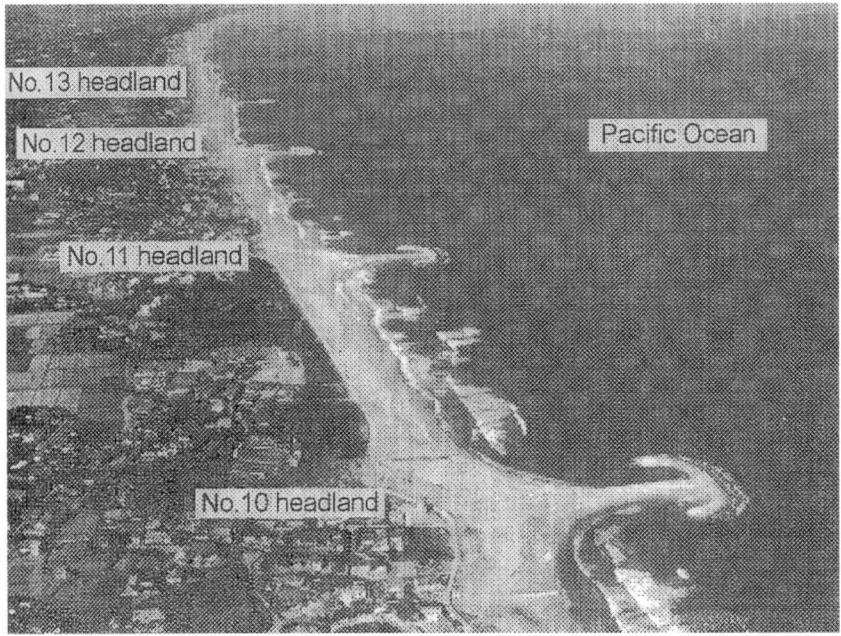

FIGURE 9.47. Headlands installed on the Kashimanada coast in December 1993 (Saito et al., 1996).

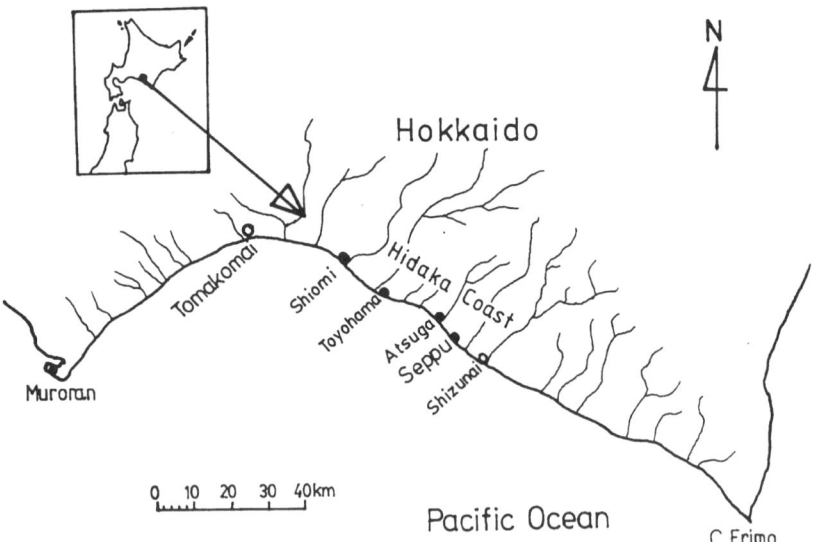

FIGURE 9.48. Location of the Hidaka Coast on Hokkaido, showing Seppu fishing harbor and others (Ozaki, 1964).

to construct a salient or tombolo behind the offshore breakwater. An view of Seppu fishing harbor is given in Figure 9.49. Similar examples can also be found at Atsuga, Shiomi, and Toyohama harbors on the Hidaka Coast (Uda, 1997). Since then, Japanese engineers seem to have developed a model for preventing harbor entrances from being blocked by the accumulation of littoral drift around the tip of a breakwater.

6.10. Yaizu

This fishing port is on the western shoreline of Suruga Bay (Figure 9.1), one of the largest on the central southern coast of Honshu (see Figure 9.50). The bay is extremely deep—greater than 2,000 m near the center—even though it has been fed with sediment by four short rivers. Its sandy shorelines are also steep—1:10 and 1:5 to the 5 m depth. Seo and Fukuchi (1966) regarded Yaizu as "one of the greatest fishing ports in Japan," as it was developed "from ancient days, and today it has grown into an important base of pelagic fishing." The city stretches along the narrow coastline with houses, fish-processing plants, shipyards, and other facilities. A mammoth sea dike with high crown elevation and groin system runs 1.6 km southward, as seen in Figure 9.51.

Waves arrive essentially from the S or SE, with large significant heights of 4.5 m and period 8.4 s. On some stormy days, breaking takes place close to the wall causing spray 20 m in height. During these events, offshore bars are not formed due to the lack of sand and the steepness of the nearshore area. These conditions have called for shore protection works since 1772, because shoreline retreat was on the order of 100 m until up to 1870 and an additional 30 m was lost in front of the revetment up to 1935. The sea dikes have been built and destroyed repeatedly several times since the end of the 18th century (Seo and Fukuchi, 1966).

The erosion at Yaizu has accelerated since 1900 for several reasons. First, the Oi River to the south (see Figure 9.50) has been dammed for flood control and power generation, thus reducing sediment supply. Second, breakwaters at the main harbor have inter-

FIGURE 9.49. Field application of single detached breakwater at Seppu fishing harbor (Uda, 1997).

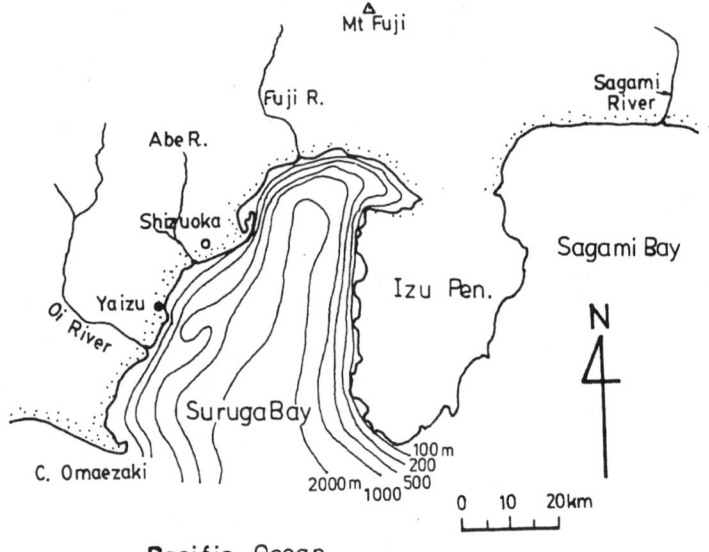

FIGURE 9.50. Location of Yaizu fishing port in Suruga Bay (Seo and Fukuchi, 1966).

FIGURE 9.51. Aerial view of the Yaizu Coast viewed from its south, showing the groin system and the Yaizu fishing port at the far end (Seo and Fukuchi, 1966).

cepted what little littoral drift the groins allowed to pass and the steepness of the foreshore has added to the problem. The sea dikes have collapsed in 1948, 1954 and 1959, causing losses of hundreds of houses and industrial properties. The replacement structures became more massive, but bigger is not better if foundation conditions are not improved.

7. ENGINEERING APPLICATIONS OF COASTAL GEOMORPHOLOGY

The strip of boundary between the land and the sea may be classified in terms of landform materials, such as cliff, gravel, sand, or mud. The shoreline along this boundary may be either straight or curved. It may remain straight over a large length or may be separated by headlands, natural or man-made (Figure 9.52). Spits may form at turning points of coastline. The coast may be fully exposed to the waves, affecting its stability. It may be partially sheltered or even fully protected by headlands or structures. This is indeed a complex environment. Coastal geomorphology is a tool used to study coastal landforms, their evolution, the processes acting on them, and the changes taking place (Bird, 1984). Traditionally, geologists have classified coasts as submergent or emergent, from a genetic point of view, and more recently in relation to sea level rise. However, a practical approach for coastal engineers is to examine the effects of wave climate on beach processes. This implies that landforms in each of their own environmental settings respond dynamically to the local conditions of waves, tides, and currents, which interplay with various landform materials. The changes are not often peaceful, but very volatile, and sometimes catastrophic. This results in erosion and accretion occurring either continuously or in se-

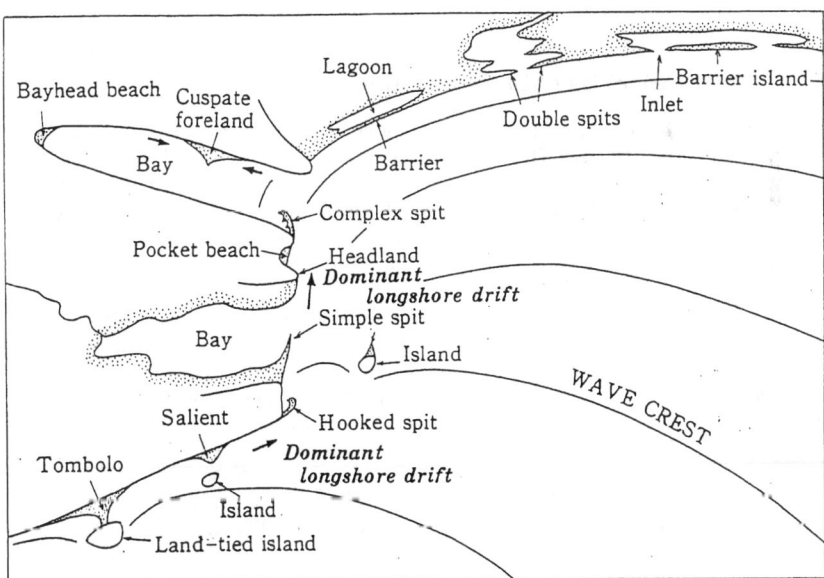

FIGURE 9.52. Typical depositional landforms along the coast with a predominant swell wave direction (Horikawa, 1988).

quence along a limited length of coast, due mainly to the action of waves and other meteorological and oceanographical factors.

A coastline is rarely straight; some segments may curve gently in plan and others are indented or of even more intricate configuration, with various shapes, sizes, and names (Silvester and Hsu, 1973, 1977). Even without professional training, one can easily identify a curved shoreline between headlands on the coast from aerial photographs, maps, and travel magazines. There are bays or sandy bay beaches, which have existed over long geological periods. Most of these are aesthetically beautiful and, surprisingly, much more stable than their straight counterparts. Silvester and Hsu (1993) reported that this is "because the orientation of these bays is an excellent indication of the direction of the net sediment movement along the coast. By observing these shapes along large reaches of shoreline it can be shown that littoral drift is in the same direction for many kilometers or even hundreds of kilometers. This knowledge can help man predict future trends in shoreline movement over large lengths of time." However, many new bays created by human activities are not so fortunate, as they are nowhere near the static equilibrium shape.

Although coastal geomorphology has not been systematically studied in Japan, local understanding has already produced practical engineering works for disaster prevention in various locations. Based on the research publications available, it seems that the concept of coastal geomorphology, especially the bay shape, has only been considered by a few workers in Japan (Tsuchiya, 1987, 1994), often without referring to it. The majority of the coastal scientists and engineers are not aware of the usefulness of coastal geomorphology for examining beach processes and application to beach erosion control. One of the notable geomorphological features is the formation of salients in the lees of breakwaters constructed for harbors, which have caused both siltation and beach erosion in unsuspected areas. Hsu and Silvester (1996) commented that "In terms of the beach erosion downcoast of harbors, it is not so much a need to look back on our achievements but a need to observe natural processes in geomorphological terms and apply them."

7.1. An Empirical Bay Shape Equation

Bay beaches are found downcoast of protruding headlands, either natural or man-made, such as groins or breakwaters. These shapes also occur as salients in the lee of offshore islands and detached breakwaters. Despite having a great variety of indentations, the stability of these curved shapes can now be predicted. This is always related to the swell-built profile, since it is under these conditions that beaches are widest. Also, swell waves are the most persistent waves on any coast, since storm sequences are very sporadic and last for short periods of time. This predicted curvature applies to the condition known as *static equilibrium,* for which littoral drift has ceased. When it still ensues, the crenulate shape is said to be in *dynamic equilibrium* because further indentation can take place until the stable limit is reached, but no further. Many bays can remain in dynamic equilibrium for years, decades, or centuries, as long as the supply of sediment to them balances the removal by waves at the downcoast end. A third case can exist, termed *dynamically unstable,* where a beach is actively eroding toward the static equilibrium state.

Although a logarithmic bay shape equation was available (Yasso, 1965), it was found difficult to use (Silvester and Hsu, 1993). This was because the diffraction point was not recognized. As a consequence, an empirical parabolic equation was derived for bayed beaches in static equilibrium (Hsu and Evans, 1989),

$$\frac{R}{R_0} = C_0 + C_1\left(\frac{\beta}{\theta}\right) + C_2\left(\frac{\beta}{\theta}\right)^2 \tag{1}$$

Application of this equation has been presented (Hsu et al. 1993; Silvester and Hsu, 1993, 1997; Hsu and Silvester, 1996). The main basic variable β, as illustrated in Figure 9.53, is the angle of wave obliquity or that between the incident wave crests (assumed linear) and the *control line,* joining the upcoast diffraction point to the near-straight downcoast beach. The control line of length R_0 is also angled β to the tangent at its beach end; in fact, that is how it is determined from maps or vertical aerial photographs. The radius R to any beach point around the bay is angled θ from the wave crest line. The other variables C_0, C_1, and C_2 in Equation (1) vary directly with β, as seen in Figure 9.54, as obtained from many bays known to be in static equilibrium, such as on islands or along large promontories where rivers are no longer adding sediment to the system.

Values of radii R can then be calculated for each θ value, once β and R_0 are determined. Values of the nondimensional ratios R/R_0 versus increments of 2° of β from 20° to 80° can be prepared (see Table 9.2), which eases applications greatly (Silvester and Hsu, 1993, 1997). When values of R at angle θ are drawn on the existing plan shape, the nearness to static equilibrium can be gauged. If the predicted curve is landward of the existing beach, it can be considered to be in dynamic equilibrium, or even may be dynamically unstable or eroding swiftly. It can certainly degrade or aggrade as sand supply decreases or increases.

Besides predicting bays in the lee of headlands, Equation (1) can also be applied to salients behind detached breakwaters (Hsu and Silvester, 1990), as seen in Figure 9.55. These accretions assume a parabolic form to an apex whose position can also be determined. This accretion is at the expense of material beyond the extremities of the detached breakwater and hence erosion can occur there. Similar salients form in the lee of harbor breakwaters, causing silting of the refuge basin.

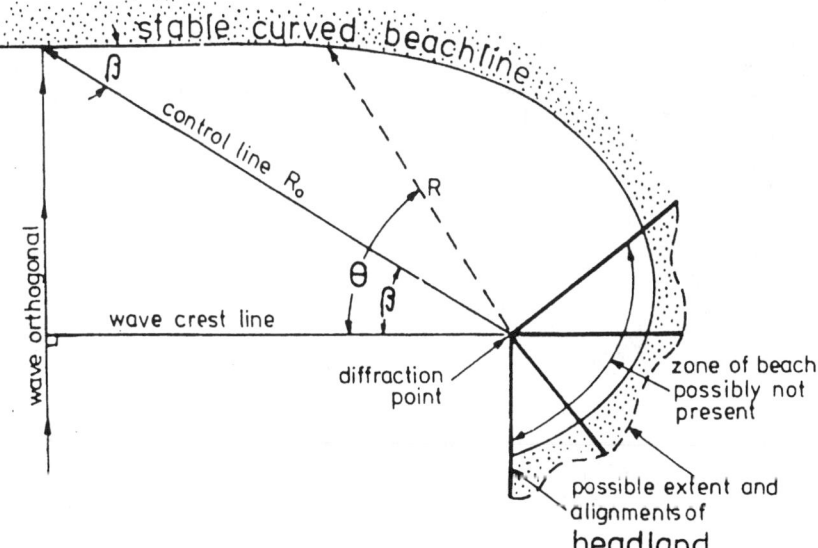

FIGURE 9.53. Definition sketch of a bayed beach in static equilibrium (Hsu and Evans, 1989; Silvester and Hsu, 1993, 1997).

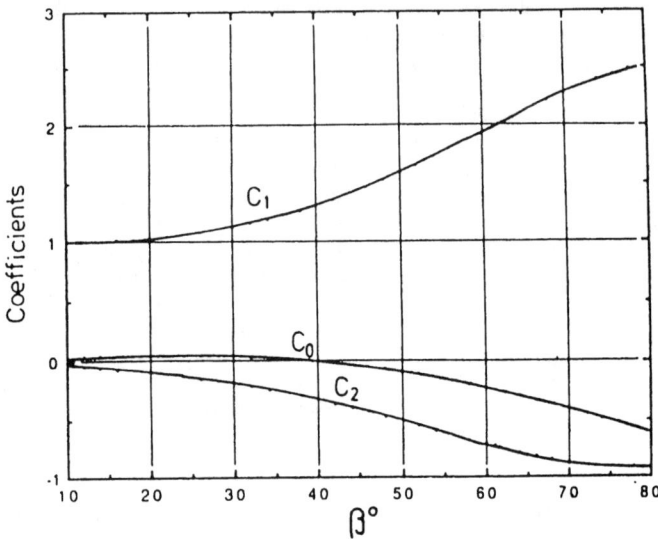

FIGURE 9.54. The three nondimensional C coefficients in Equation (1) versus wave obliquity angle β (Hsu and Evans, 1989; Silvester and Hsu, 1993, 1997).

7.2. Bayed Beaches in Japan

Bayed beaches are readily identified in aerial photographs, of which thousands are taken annually around the Japanese coast for many purposes. They stand out as white or yellow strips between the green inland and blue sea, both appearing as darker color in black and white. The width of beach can be determined, which indicates the strength of the storm waves in removing the berm almost annually. The seaward edge during the nonstorm season is the one to be used in testing the shape against the predicted stable form.

Having many natural promontories and offshore islands, Japan has a multitude of sedimentary beaches between them. Some are very localized and do not feed material into adjoining bays. However, those in one locality will all be oriented in the same direction, as the offshore wave climate is common for extensive longshore distances. Any bayed beach is notable for a strong curvature in the lee of the upcoast headland, with the beach straightening out toward the downcoast end, hence the likening to a logarithmic spiral in early studies (Yasso, 1965; Silvester and Ho, 1972; Silvester, 1976). The sizes of these embayment features can vary from tens of meters to many kilometers.

It must be stressed that the static equilibrium shape predicted by Equation (1) refers only to the swell-built profile or the widest width of berm. This can be narrowed very swiftly or disappear altogether during a severe storm. Hence the beach is not an area on which to construct buildings, facilities, or even protective sea walls. At some stage, waves will reflect from vertical or sloping surfaces of such structures and promote scouring and hence disappearance of beaches. Many papers have been published on this topic (e.g., Nishimura et al. 1978). Japan is notable for sea walls and sea dikes constructed for protection against waves, storm surges, and tsunamis. Coastal defense has to account for a great variety of situations.

SHORELINE PROTECTION METHODS—JAPANESE EXPERIENCE

TABLE 9.2. Radii ratios R/R_0 versus wave obliquity angle β for a range of value of θ in Equation (1) for bayed beaches in static equilibrium (Silvester and Hsu, 1993, 1997)

						Radii ratios R/R_0 for $\theta =$				
β	C_0	C_1	C_2	30	45	60	75	90	105	120	135	150
20	0.05392	1.03954	−0.09413	0.7051	0.4973	0.39	0.324	0.28	0.249	0.225	0.2059	0.1909
22	0.05421	1.05284	−0.10889	0.7677	0.5429	0.426	0.354	0.305	0.27	0.244	0.2229	0.2063
24	0.05377	1.06887	−0.12531	0.8287	0.5882	0.461	0.383	0.33	0.292	0.263	0.2398	0.2216
26	0.05241	1.08792	−0.14356	0.8874	0.6331	0.497	0.412	0.355	0.313	0.281	0.2566	0.2367
28	0.04992	1.11028	−0.16375	0.9435	0.6774	0.532	0.442	0.379	0.334	0.3	0.2732	0.2515
30	0.04615	1.13613	−0.18597	0.9963	0.7209	0.568	0.471	0.404	0.356	0.319	0.2894	0.2659
32	0.04092	1.16565	−0.21027		0.7635	0.603	0.5	0.429	0.377	0.337	0.3054	0.28
34	0.03411	1.19894	−0.23665		0.8049	0.638	0.529	0.453	0.398	0.355	0.3211	0.2937
36	0.0256	1.23606	−0.2651		0.8448	0.672	0.558	0.478	0.418	0.373	0.3364	0.307
38	0.01528	1.27704	−0.29554		0.8829	0.706	0.586	0.502	0.439	0.39	0.3513	0.3198
40	0.00308	1.32182	−0.3279		0.919	0.739	0.615	0.526	0.459	0.407	0.3659	0.3323
42	−0.0111	1.37034	−0.36203		0.9526	0.771	0.643	0.55	0.479	0.424	0.3802	0.3442
44	−0.0272	1.42244	−0.39777		0.9833	0.802	0.67	0.573	0.499	0.441	0.3942	0.3558
46	−0.0454	1.47796	−0.43491			0.832	0.698	0.596	0.519	0.457	0.4077	0.367
48	−0.0655	1.53664	−0.47322			0.861	0.724	0.619	0.538	0.473	0.421	0.3777
50	−0.0877	1.59823	−0.51241			0.888	0.75	0.642	0.557	0.489	0.4339	0.3881
52	−0.1119	1.66237	−0.55219			0.914	0.775	0.664	0.576	0.505	0.4465	0.398
54	−0.1381	1.72869	−0.59219			0.938	0.8	0.686	0.594	0.52	0.4587	0.4075
56	−0.1661	1.79676	−0.63205			0.96	0.823	0.707	0.612	0.535	0.4705	0.4166
58	−0.1959	1.8661	−0.67133			0.981	0.846	0.728	0.63	0.549	0.482	0.4253
60	−0.2274	1.93618	−0.70959			0.999	0.867	0.748	0.647	0.563	0.493	0.4336
62	−0.2604	2.00643	−0.74633				0.888	0.768	0.664	0.577	0.5036	0.4414
64	−0.2949	2.07622	−0.78103				0.908	0.787	0.68	0.59	0.5138	0.4488
66	−0.3307	2.14487	−0.81313				0.927	0.805	0.696	0.603	0.5236	0.4556
68	−0.3676	2.21166	−0.84202				0.945	0.823	0.712	0.615	0.5328	0.462
70	−0.4054	2.27581	−0.86708				0.963	0.84	0.726	0.627	0.5416	0.4678
72	−0.4439	2.3365	−0.88763				0.981	0.857	0.741	0.638	0.5498	0.4731
74	−0.4828	2.39286	−0.90296				0.999	0.874	0.755	0.649	0.5575	0.4779
76	−0.522	2.44396	−0.91235					0.891	0.769	0.66	0.5647	0.4821
78	−0.5611	2.48883	−0.915					0.909	0.783	0.67	0.5714	0.4856
80	−0.6	2.52646	−0.91011					0.927	0.797	0.68	0.5776	0.4886

Many obvious examples of bay shapes can be identified in Japan. For example, there are bayed beaches either side of but leeward of the Oga Hanto (Peninsula) on northwestern Honshu, located at 140°E and 40°N facing the Sea of Japan (see Figure 9.56). In this case, the natural headland at Oga Peninsula behaves like a single detached breakwater. The second example is the large bay shape along the Yumigahama Peninsula, where the Kaike Coast is located (see Section 6.4). As seen in Figure 9.28, the Shimane Peninsula may be considered as a headland, with its tip at Jizozaki as the upcoast control point. The waves approaching the beach from the Sea of Japan diffract and refract within Miho Bay. The stability of the bay beach on the Yumigahama Peninsula facing Miho Bay is dependent upon the availability of sediment supply from the Hino River. With the reduction of supply due to damming of the river, beach erosion was inevitable at Kaike. Moreover, three other examples cited in the case studies above have involved

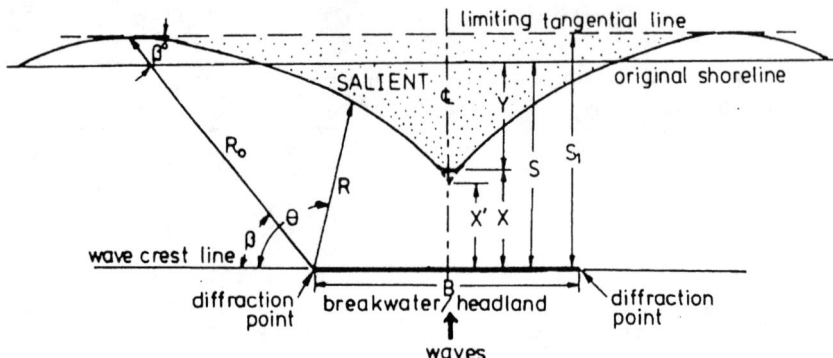

FIGURE 9.55. Definition sketch of a salient formed in the lee of a detached breakwater, with normal wave approach, based on the static bay shape equation (Hsu and Silvester, 1990; Silvester and Hsu, 1993).

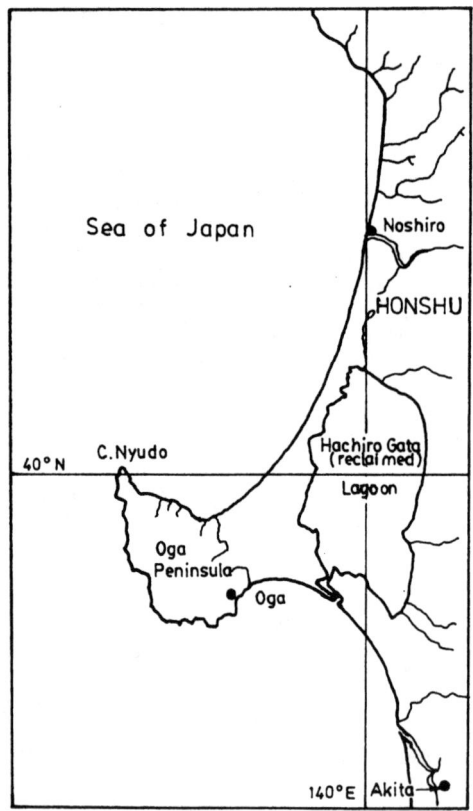

FIGURE 9.56. Tombolo formed leeward of Oga Peninsula located at 140°E and 40°N facing the Sea of Japan.

bayed beaches; for example, at Katase East, Iwafune, and Oarai. Many more bay shapes can be readily identified.

7.3. Effect of Breakwater Construction

Many harbors built on sandy coasts, where natural shelter is not available, may require long breakwaters to inhibit siltation and entry by storm or swell waves. These structures are often extended some kilometers to afford greater protection and are generally angled around 45° to the shore. Their mammoth lengths can often interrupt littoral drift completely and so, apparently, solve the siltation problem. However, the beach downcoast may be severely eroded, hence requiring protective structures to be installed. An example can be found at Misawa fishing harbor, on the Pacific coast of Aomori Prefecture, as seen in Figure 9.57 (see Figure 9.18 for location). Alternatively, a salient will be formed in its lee, thus bringing sand in from further downcoast and causing both a siltation and erosion problem. This material must come from beaches beyond the jurisdiction of the port authority and hence conflicts of interest are involved with at least two different government agencies.

The above problem is being discussed in this chapter on Japan, but it occurs worldwide in projects designed by engineers who have not heard the message of coastal geomorphology. Herron and Harris (1966) have commented that "harbor works are the principal offenders" in interrupting the balance of natural littoral drift that has existed for hundreds or thousands of years. Komar (1983) also has observed that "many occurrences of destructive coastal erosion have resulted directly from the construction of jetties,

FIGURE 9.57. Accretion upcoast and erosion downcoast in October 1989, resulting from the breakwater construction at Misawa fishing harbor, Aomori Prefecture (Uda, 1997).

breakwaters, and other engineering structures." Moutzouris (1990) attributes some coastal erosion in Greece to harbor construction.

The salient so formed can be predicted by Equation (1) since the shoreline will build out until zero littoral drift is approached. With no material available around the tip of the main breakwater, all material for the silting salient must come upcoast from beaches on the downcoast area of the resulting bay shape. Despite the spate of papers on wave kinematics and sediment motion, the answer appears to involve a macroscopic view of the coast or a geomorphological approach. The unique curvature of these beaches is the result of short- plus long-term processes that are bound to happen as night follows day.

In Japan, the effect of harbor construction on sandy beaches has been reported by many researchers, such as Hiroi (1921). Sato and Irie (1970) and Tanaka (1983) have examined seabed variations from breakwaters. Tanaka and Sato (1976) have cited examples at Kashima, Kashiwazaki, Oarai, Kochi, and Hitachi, and Uda et al. (1986) have reported beach erosion downcoast from Ohtsu, Hitachi, Oarai, and Kashima, all facing the Pacific Ocean. Tsuchiya et al. (1990) have studied the erosion in the vicinity of Naoetsu harbor facing the Sea of Japan. It has been a common practice that no preventive countermeasures were planned prior to breakwater construction, due perhaps to the fact that harbor construction assumed the highest priority.

In the examples of beach changes noted above, the concept of stable bays has not been mentioned, despite the clear curvature of the beaches involved. This is understandable because most of this activity occurred prior to the 1980s when knowledge of this static equilibrium shape was not well known, the first paper being that of Silvester (1970) and more recently Hsu et al. (1993), Silvester and Hsu (1993, 1997), and Hsu and Silvester (1996). However, many papers and books have been available on beach erosion downcoast of harbors for three decades and hence the phenomenon should be verifiable, even if its relationship with the bay shape has not been mentioned explicitly. The headlands, if required to prevent erosion in downcoast areas, should be installed before the construction of the main harbor structure that will cause the problem because removal is swift. The sheltered zone that will be silted by the salient should also be saved by a groin or a secondary breakwater, whose length is predictable as soon as the layout of the main breakwater is known, from which the wave diffraction tip can be determined (Silvester and Hsu, 1993, 1997).

7.3.1. Straight Shoreline. Silvester and Hsu (1993, 1997) have observed that "A breakwater running out to sea acts like a headland from which the prevailing waves will diffract and refract to the downcoast beaches. This must result in the formation of a bay with concomitant erosion of the beach some distance from the heel of the structure. This deleterious denudation is overcome at great cost by the construction of a sea wall in the most affected area, followed possibly by groins and even offshore breakwaters." A typical case is depicted in Figure 9.58, where the main breakwater is angled to the shore and intercepts a large volume of littoral drift. The secondary structure at A provides shelter from the diffracting waves and also from storm waves that can arrive from many directions.

Some accretion can be expected at A as the bay commences to form. Sediment that may naturally bypass the main breakwater could follow the path shown to be deposited on the coast at point C, some distance from A. As seen in the figure, the area from B to C could be eroded in the continued formation of the resulting bay. It is in this zone that headland control (as seen in the figure) is preferred, preferably installed prior to the construction of the main breakwater to prevent the swift erosion that will occur. The bays so formed will be in static equilibrium, as predicted by Equation (1), since littoral drift will be cut off.

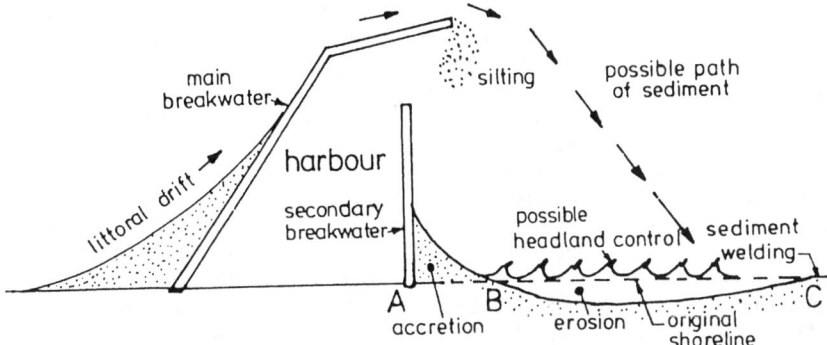

FIGURE 9.58. Effect of harbor construction on a straight sandy coast, showing potential erosion, accretion and headland control (Silvester and Hsu, 1993, 1997).

7.3.2. Bay in Dynamic Equilibrium. If a harbor is constructed in a bay with small indentation, or a modest harbor is built, littoral drift may only be partially interrupted. In this case, the bay or the beach downcoast of the harbor is said to be in the dynamic condition, because the predicted static bay shape (dashed line in Figure 9.59) is landward of the existing bay (solid line). Should reduction in littoral drift from upcoast occur after breakwater construction, the beach downcoast will recede, thus resulting in beach erosion. The conventional approach would be to build sea walls, groins, or even detached breakwaters, one after another. All of these have not been proven effective, as shown in some of case studies mentioned.

One alternative is to reorient the shoreline into static equilibrium. This can be seen from the effect of positioning a breakwater in a bay in dynamic equilibrium in Figure 9.59. In this case, the downcoast tangent would not change, based on the same persistent swell, or the shoreline boundary further downcoast remains in the same direction as the

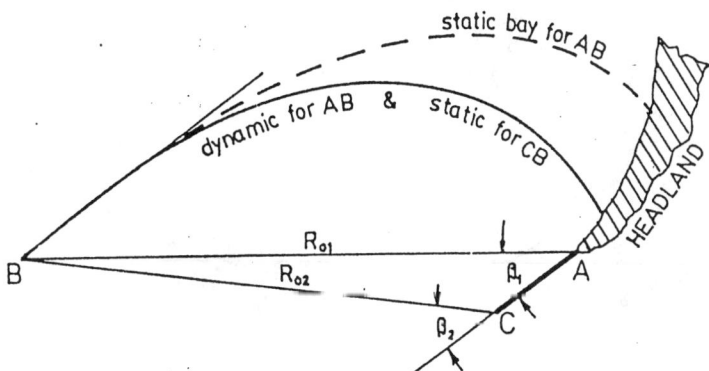

FIGURE 9.59. Effect of positioning a breakwater in a bay in dynamic equilibrium (Hsu and Silvester, 1996).

persistent swell breaking crests. Once the downcoast limit at B is located, control line AB for the original bay is angled to the downcoast tangent by wave obliquity β_1. When breakwater AC is built, the new wave diffraction point is C, which gives a new control line CB and new obliquity β_2. This results in a static bay shape; the bay is in stable condition even if upcoast sediment supply is stopped altogether. Ideally, no hard structure is ever required on the bayed beach, except a wide beach berm to safeguard against storm attack.

Silvester and Hsu (1993, 1997) have shown that similar-sized breakwaters placed in slightly different locations or extended in different orientations can have a beneficial or at least not deleterious effect on a downcoast beach area. This is due to the new diffraction and refraction pattern causing waves to arrive near normal to the existing shoreline of a bay in dynamic equilibrium, which could erode if sediment supply to it is terminated by natural processes or man's intervention. This extends the concept discussed in Figure 9.59 into a multiple-choice approach for protecting a beach downcoast.

An example of harbor construction with breakwater extension is now given to explain the procedure of converting a bay beach downcoast of a harbor from dynamic to static equilibrium (Silvester and Hsu, 1993, 1997), which is relevant to many harbors in Japan and elsewhere. Consider a modest port complex, in the form of breakwaters AB and CD in Figure 9.60 built at a natural headland upcoast of bay AE, which is in dynamic equilibrium. By transferring the tangent at E across to B or A, the wave crest line is angled to the control line AE at 32°, as shown. The stable bay shape for no port structures can thus be drawn as "static bay for AE." This illustrates the erosion that could take place should littoral drift from upcoast of the natural headland at A cease altogether. However, the existing shoreline, as shown, indicates that sufficient sediment is bypassing point A to maintain it in a state of dynamic equilibrium.

When breakwater AB is constructed, the new diffraction point (B) causes a stable shape further seaward, marked as "bay for BE." This again is for a situation of no littoral drift. An alternative breakwater FG provides a new value of $\beta = 38°$ and R_0, from which the static bay, denoted as "bay for GE," forms. This, it is seen, matches the existing shoreline fairly well, with slight accretion to the shoreline adjacent to CD and slight erosion

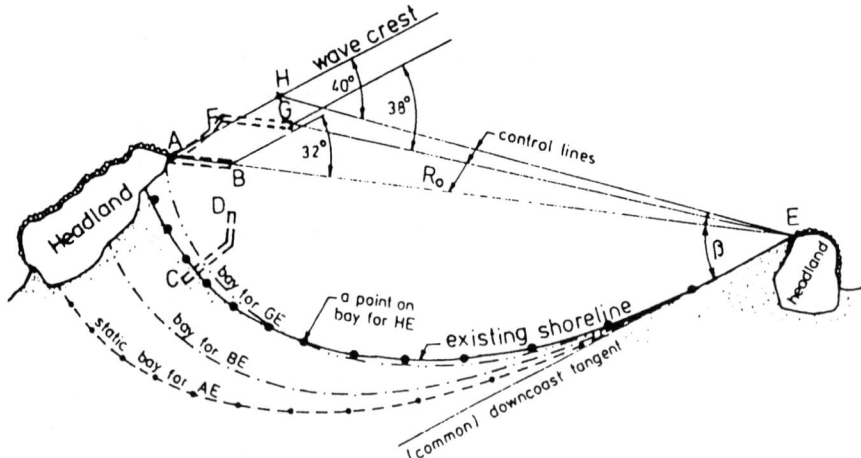

FIGURE 9.60. Converting a bay beach from a dynamic condition to static equilibrium by positioning the tip of a breakwater at an appropriate location (Silvester and Hsu 1993, 1997).

downcoast of it. It should be remembered that in time sediment would be bypassing point G so that the bay would again be in dynamic equilibrium, resulting in a shoreline (GE) seaward of this static equilibrium outline. However, it can be seen that the initial construction of FG instead of AB has almost solved any erosion problems on the bayed coast. These two structures are of similar length and, except for differing depths, should cost much the same. Its obliquity to the wave crests will aid transmission of sediment beyond G without forming shoals. The curved structure between A and F could serve to retain sand and build a straight beach between A and F, oriented parallel to the wave crest line. This structure will not be attacked by waves, so it need rise only to MSL and its units could be modest in size, just to retain the sand. In fact, sand could just accrete in this zone and provide natural protection (Zwemmer and Van't Hoff, 1982).

Should it be desired that the coast from A to E be in static equilibrium, because of imminent interception of littoral drift upcoast from headland A, due possibly to an access channel being dredged for a port in that region, then point H could be chosen as the diffraction point. As seen, $\beta = 40°$ and with the new R_0 of HE, the static bay, as shown by larger dots, results. This matches the existing shoreline almost exactly, inferring that permanent erosion would not be experienced at any point around its periphery, even with no sediment input.

An actual example of such shoreline evolution is given for Oarai Harbor, discussed in Section 6.7. As seen in Figure 9.61, two angled breakwaters formed a harbor at Isohama

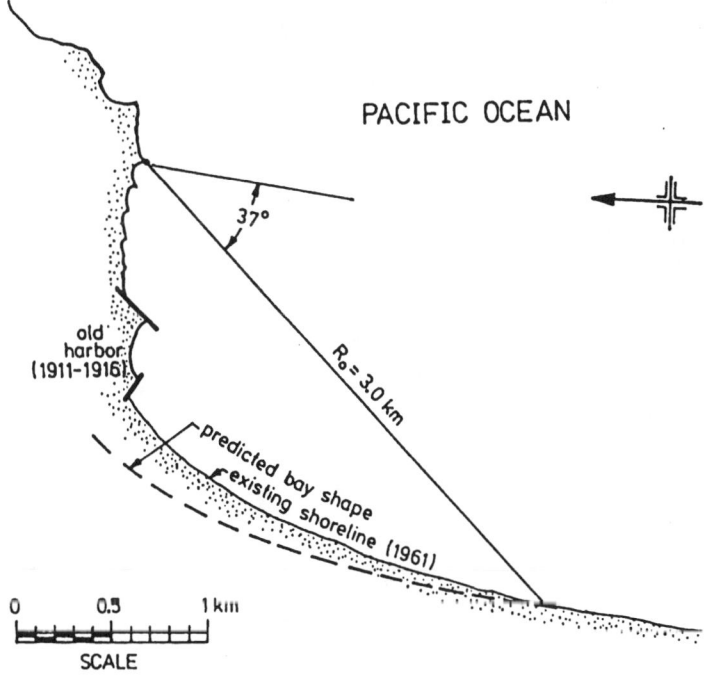

FIGURE 9.61. Shoreline downcoast of Oarai Harbor, showing the once-buried harbor at Isohama and a dynamic bay shape in 1961 (Silvester and Hsu, 1993, 1997).

during 1911–1916, which quickly silted up. The stable shoreline south of the site, for no further input of sediment, is shown for $\beta = 37°$ and $R_0 = 3.0$ km, using an upcoast control on Oarai Point. However, because of littoral drift the shoreline was seaward of this, implying the existing beach was in dynamic equilibrium. An extension to the harbor, as in Figure 9.62, established a new diffraction point at A, with $\beta = 45°$ and $R_0 = 1.92$ km, resulting in the predicted shoreline, which matches the existing one very well.

A further extension of this breakwater from A to B was completed in 1981. And seen in Figure 9.63, it gives $\beta = 35°$ and $R_0 = 3.48$ km. The new static equilibrium shape is shown dotted, together with the 1976 shoreline. During 1983 to 1985, a detached breakwater CD was constructed, for which a new stable bay shape for CE is depicted, plus the existing shoreline of 1988 (Mizumura, 1980; Kraus et al., 1984). It is seen that the shoreline in 1988, even with two long groins inserted, had not accreted out to this new limit for zero littoral drift. Material collected as a salient in the lee of breakwater CD has meant that sand for this accretion has had to come from downcoast, resulting in erosion from 1 to 3 km south of the harbor. This has initiated the installation of sea walls, groins, and concrete block revetments. If and when normal drift conditions are established in the future, the bay shape for CE could eventuate.

The changeability of a beach shape from dynamic equilibrium to static, if used with proper planning, may protect a beach downcoast from further erosion. For example, extension of breakwaters to a certain length may be carried out in a suitable direction, or an existing breakwater system modified, which will swiftly change the present salient of dynamic condition to a static bay shape, similar to that introduced in Figure 9.60. The approach of headland control, without the need of nourishing the beach but simply installing a suitable upcoast control point, deserves more study.

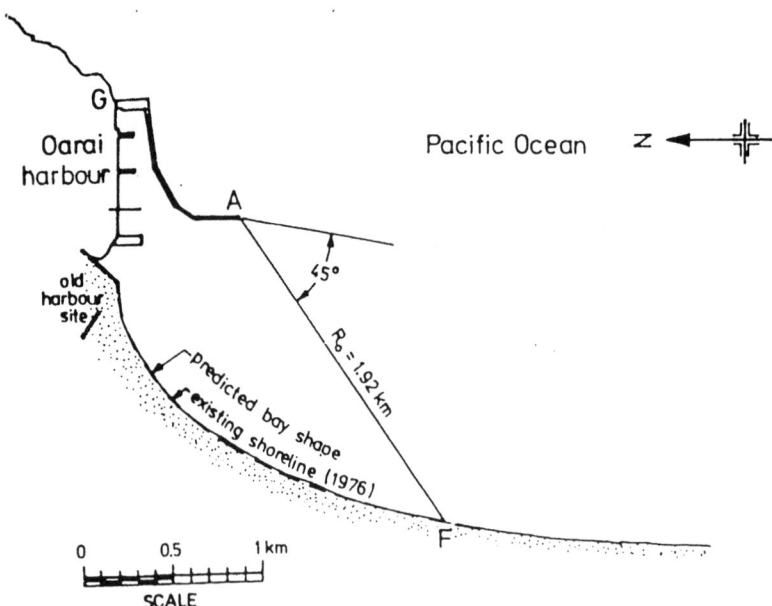

FIGURE 9.62. Shoreline downcoast of Oarai Harbor, showing a static bay shape in 1976 (Silvester and Hsu, 1993, 1997).

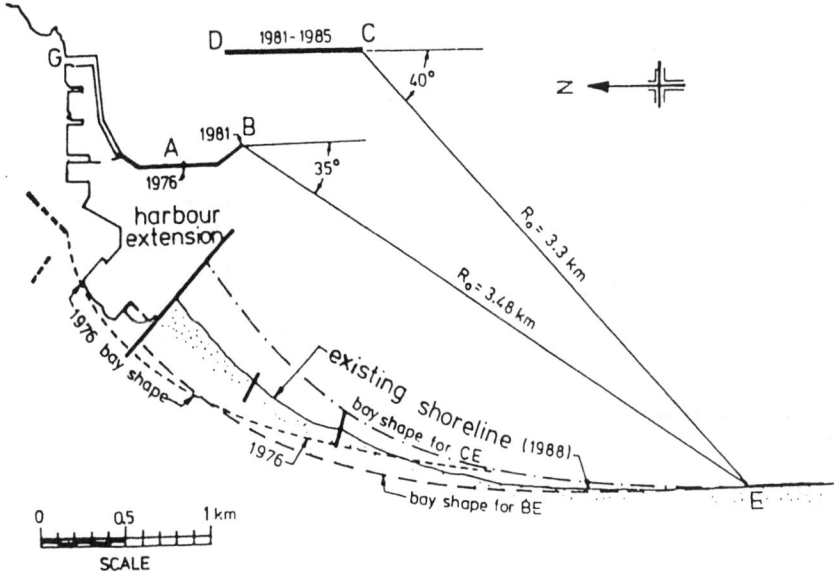

FIGURE 9.63. Shoreline downcoast of Oarai Harbor in 1988, showing salient positions in relationship to breakwater extensions (Silvester and Hsu 1993, 1997).

7.3.3. Bay in Static Equilibrium. Although, as shown above, a bay in dynamic equilibrium can be made stable by placing breakwaters at the upcoast headland in a suitable position, this is not necessarily the case for a bay already in static equilibrium. The shoreline in the vicinity of Iwafune Harbor on the Niigata Coast facing the Sea of Japan was discussed in Section 6.3. It was originally a small river delta, with a 50 m wide mouth, receiving sediment from the river itself and also a river 5 km to the north. As seen in Figure 9.24, an upcoast groin and detached breakwater were completed in 1929 to prevent siltation. These were connected in 1934, with an inner revetment and wharf added in 1957 (Haruta, 1961), which became the outer breakwater for this river mouth harbor. Reports on the beach erosion downcoast at Iwafune can be found in Uda and Noguchi (1991), Hsu et al. (1993), and Silvester and Hsu (1993, 1997).

Aerial photographs made in 1965 show reasonably wide beaches upcoast of the main breakwater, since most sediment comes from the north. As seen in Figure 9.64, some downcoast position had to be located as the limit of the bay formed south of the harbor at point B. The tip of the breakwater (A) and wave-crest line, parallel to the tangent at B, provides β at 30° and $R_0 = 1.35$ km. The static equilibrium shape fits the actual shoreline as of 1965 and hence no further erosion could be expected.

Mammoth breakwater extensions were carried out, as shown in Figure 9.65, resulting in the formation of new salients at each stage of the extensions, which caused shoaling of the channel from the port, with massive dredging required. In this case two points D and E along the same sandy beach were selected, with similar β values of 40° to the two breakwater tips B and C, to obtain the static bay shapes. The predicted static bay shape shorelines are seen to be well seaward of the 1988 shoreline with its many groins. Material for this addition must come from downcoast, which instigated the groin construction

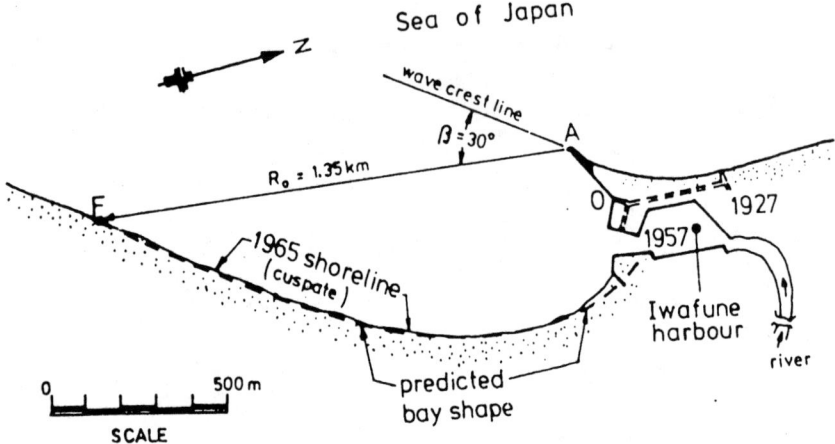

FIGURE 9.64. Shoreline downcoast of Iwafune Harbor in 1965, showing a stable bay shape in static equilibrium (Silvester and Hsu, 1993, 1997).

and even the offshore breakwaters beyond the downcoast control point. These will not prevent the recession necessary for nature to reshape the beach line to that applicable to the boundary conditions.

Where a bay downcoast of a harbor complex is already stable, any modifications to it, by diffraction points being established from breakwater extensions, must involve erosion of one segment to feed the other, in order to produce the salient predicted, using the em-

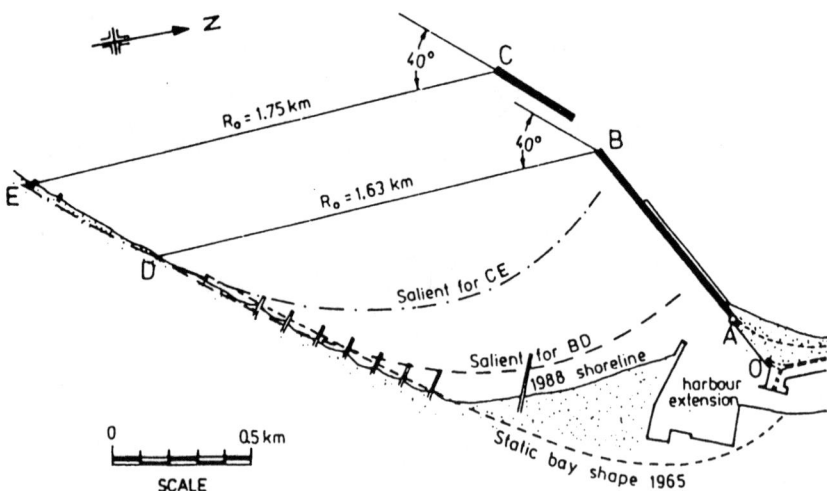

FIGURE 9.65. Shoreline downcoast of Iwafune Harbor in 1988, showing salient growth in the lee and beach erosion downcoast, with protection works (Silvester and Hsu, 1993, 1997).

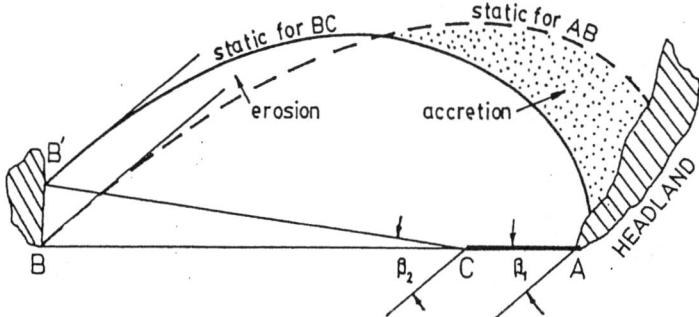

FIGURE 9.66. Effect of positioning a breakwater tip in a bay in static equilibrium for B'C, for harbor planning and beach erosion control (Hsu and Silvester, 1996).

pirical bay shape equation for static equilibrium. This concept is illustrated in Figure 9.66, in which the dotted curve represents the static equilibrium shape for control line AB, which is the existing beach. When breakwater AC is added, the new control line CB' creates a new static beach, which involves massive accretion in the lee of new headland at C. The material for this has to be supplied from the region of B because the beach was originally stable. This causes a new downcoast tangent to operate that determines a new wave obliquity β_2 for Equation (1), given in Section 7.1.

8. WHAT CAN WE LEARN FROM JAPANESE EXPERIENCE?

The Japanese are essentially said to be a "coastal zone race" (Nagao, 1993), because the majority of its population inhabits and conducts its social and economic activities in the narrow strips of coastland. However, because of the country's unique geographical location and geological condition, it has also suffered from severe natural disasters. Consequently, protection works along its boundary with the sea has had to account for many variables. Nagao, of Kyoto University, who edited the special volume of *Coastlines of Japan II* in 1993, observed:

> Foreign visitors shall be surprised by finding many artificial structures covering the Japanese coastal zone when they take a walk on the coast. There are gigantic reclaimed lands near big cities, big harbors, many fishing ports, seaside roads connecting points, and shore protection, embankments, breakwaters, etc. It is a scenery like the "Great Walls" of the sea. Its total extension today almost reaches 10,000 km.

This was echoed by Kawata and Shibayama (1998) on reviewing the direction of coastal sedimentation problems in Japan. Nevertheless, coastal scientists around the world will appreciate the severe environmental factors affecting the coastal zone in Japan, and at the same time admire the courage of Japanese engineers in striving to overcome these problems.

On the other hand, different opinions have also emerged. For example, Shapiro, of Osaka University, concluded that "Japan cannot be called a sea-oriented nation" (Shapiro,

1989). His conclusion was based on a judgment about "the pathetic condition of its coast and nearshore sea." In his paper called "Japan, a Sea-Oriented Nation?" he tried "to look beneath the surface of economic and geographic determinism in an attempt to identify some of the cultural roots, in its literacy tradition, . . . , as well as recent signs of hope for the future." It is understandable that, for example, water quality and preservation of the coastal landscape have become a concern not only nearshore (Iwagaki, 1994; Horikawa, 1996) but also in some popular bathing places (Okajima et al., 1993).

Whatever the perception or conclusion might be, coastal scientists and engineers in Japan have gone through a very difficult time over many decades, facing simple yet intricate tasks of protecting beaches from erosion and other natural calamities. One cannot help but conjecture about the problems they have tried so hard to overcome, which are not created by themselves but are the consequence of nature and other members of the engineering profession, such as those who build dams and ports around the country. This touches a fundamental issue of mutual dependence in modern society—we are not alone but dependent on each other. A single and isolated effort attempting to solve the coastal problem would not be fruitful. Mutual cooperation is essential for an ultimately successful outcome.

Of the Japanese history of disaster prevention, such as for coastal flooding and beach erosion, Ishihara (1979) has stated that:

> In the past, research activities relating to protection from natural disasters and mitigation of their damages have been such that the existing fields of study—e.g., earth science, civil, structural, and agricultural engineering, economics, and psychology—have proceeded independently of each other. This lack of coordination among the efforts to alleviate disasters has prevented the individual disciplines from fulfilling the objectives common to them all.

He proposed knowledge-sharing with "the researchers in Japan and outside it," as he believed "the welfare of the societies of the world depends on harmony with nature." A nationwide coordination on mutual research has culminated in the publication of a new *Journal of Natural Disaster Science* since 1979, partly due to the effort of Tsuchiya of the Disaster Prevention Research Institute, Kyoto University (Ishihara, 1979).

Further, on the issue of mutual cooperation for protection and preservation of the coastal environment, Horikawa (1996) observed that the Japanese coast, as a public property, "has in practice been separately controlled by four government agencies, namely the Ministry of Construction, the Ministry of Transport, the Ministry of Agriculture, Forestry and Fisheries, and the Fishery Agency, for particular coastal regions, depending on the designation and usage of the region." Similar to the practice in the United States, little mutual cooperation in research activities and beach protection among these four government agencies can be found from the published literature.

One question to ask is "What can we learn from the Japanese experience?" The experience of Japanese engineers has been much the same as in other countries—a case of trial and error and learn from past mistakes. However, their environment is unique and their approaches have been strongly supported by substantial sums of the taxpayers' money. Despite the plethora of papers on wave theory and resulting sediment movement from hydrodynamics, not enough thought has been given to the macrostructure of the coast rather than the microstructure of a river mouth, an individual structure, or part of a structure. If a beach is going to silt or to erode because of a general abundance or dearth of material, then no marine structure is going to prevent it. This large-scale view is known as *geomorphology*; it has been discounted by engineers who are trained to solve technical problems with detailed mathematics and physics. This geomorphic knowledge, originated by geographers, has been extended by some coastal engineers to cover the evolution and stability of bayed beaches behind coastal structures (Hsu et al., 1993; Silvester and Hsu, 1993, 1997; Hsu and Silvester, 1996), selected applications of which have been illustrated in this chapter. There is a need to observe the coast from a macroscopic point of view and to look downward from a high altitude before applying theory.

Japanese knowledge of coastal engineering has matched that of elsewhere around the world, commencing just after World War II and facing similar problems of decreased sediment supply and use of bigger and larger structures. Having inherited the traditional protection devices—sea walls and sea dikes—that reflect wave energy back to sea, Japanese engineers have slowly replaced these with mild-sloped sections, but with the knowledge that this alternative also exacerbates most problems. When engineers analyzed coastal concerns only with regard to uniform or long-crested waves, both in theory and models, they were omitting the most prominent phenomenon of angled or short-crested waves, which has only recently been analyzed (Hsu, 1990).

It has been seen how the Japanese experience has involved employing sea walls, groins, detached breakwaters, and submerged artificial reefs, almost in that order, for solving many coastal problems. In respect to beaches, these conventional structures need rethinking and headland control also requires attention. Expensive solutions are sometimes applied that may not solve the problem fully, or last for lengthy periods. Various types of precast concrete blocks, in excessive quantities, have been placed on sloping surfaces, at the toes of structures, and on frontal dunes to dissipate wave energy. Wave reflection has worsened in most cases, with disappearance of beaches and scour at the toes. Despite the introduction of roughened surfaces and milder-sloped and perforated blocks, reflection has still damaged sea walls and dikes. In many cases this was due to the extra-high water levels caused by storm surges. However, the basic issue of liquefaction had not been recognized by most practicing engineers, even though soil scientists had published many papers on this topic. Where waves steepen due to shoaling and becoming partially standing due to reflection, they will interact with sediment sublayers and create a situation of nonload bearing. Structures will then subside to depths equal to the liquefaction limit.

Another phenomenon not widely appreciated by Japanese engineers, in company with colleagues around the world, is that of rip currents that carry sediment offshore. This can occur at groins during storm sequences, so assisting beach degradation, even though protective structures have been provided. Strong seaward currents can also form between gaps of detached breakwaters when being overtopped by storm waves. In this way, much of the salient formed in their lee can be deposited offshore when short-crested waves due to obliquity of reflected swell waves carries this material downcoast.

The scour adjacent to marine structures is a continuous problem, caused by fluctuating water velocities and pressures. It is mostly associated with reflected waves, whose heights are almost doubled and wave lengths halved, or steepnesses multiplied by four. This can happen with submerged structures as much as with those rising above sea level. It also takes place with rubble mound structures as the crevices between units are filled with water and so provide a virtually continuous surface for reflection. The influence of waves on scour is omitted in many flume studies because structures are sited on concrete floors rather than sedimentary beds.

Thus we have much to learn from the Japanese experience because they have been willing to experiment full-scale or construct prototype situations in order to collect data, sometimes over a period of several decades. As outside observers, it may be easy to criticize mistakes, but similar misfortunes are still being experienced in all countries, perhaps because we will not learn from someone else's mistakes.

The financing of coastal protection works is generally by government agencies because they are so costly. As with road construction, expenditure can be spread over a number of years—part of a highway this year, more in following years—as a budget is constructed. This is not the case with the seashore, which must be seen macroscopically or in terms of many kilometers within a cell of sediment movement. Projects must be planned years ahead, even before a major port expansion is carried out that is likely to affect downcoast beaches. The needed protective structures, perhaps with the addition of sand, are required before the real reason for their purpose is obvious to the populace or even to some politicians. An integrated but expensive plane protection strategy, incorporating artificial reefs, beach nourishment, and mild-sloped revetment, has been implemented by both the Ministries of Construction and Transport in Japan.

In conclusion, we quote Nagao (1993): "The Japanese thought they were one of the races that believed that the human being should exist together with the nature, thus . . . the Japanese of the present age are at a loss either to select a way to control nature and so raise living standards or to select a way to understand nature's providence and seek to survive harmoniously with nature." This philosophy of living was echoed by Iwagaki (1994) of Kyoto University and Horikawa (1996) of the University of Tokyo. In fact, there has been a fresh view of coastal protection in Japan in recent years, as can be found from the mission statement of the Fifth and Sixth Five-Year Plans of Coastal Protection—that the major aims of coastal endeavors are to create stable and safe beaches, to coexist with nature, and to provide easy access for all beach users (MOC, 1997). This integrated concept highlights a careful balance among the conflicting requirements of protection, development, and conservation, with planning and management fulfilling the central role. The single task of coastal protection is seemingly to be gradually replaced by the preservation of the coastal environment. The authors of this chapter wonder whether most developing countries would adopt this approach instead of employing the conventional hard structures that Japanese engineers have used, with such mixed results, in the past.

ACKNOWLEDGMENTS

The authors wish to acknowledge the provision of references included in this chapter from Professor Yoshiaki Kawata and Associate Professor Takao Yamashita of the Disaster Prevention Research Institute, Kyoto University; Professor Isao Irie of Kyushu University; and Dr. Yoshiaki Kuriyama of the Port and Harbor Research Institute, Ministry of Transport.

REFERENCES

Aoki, H., K. Kumagai, and S. Takazawa. 1989. Coastal community zone project (CCZ). *Proceedings of Coastal Zone '89*, Vol. 5, pp. 4213–4219. New York: American Society of Civil Engineers.

Asakawa, T., M. Hasegawa, H. Sato, and N. Hamaguchi. 1992. Recent developments on shore protection in Japan. In *Coastal Structures and Breakwaters*. The Institution of Civil Engineers, UK, pp. 409–422. London: Thomas Telford Ltd.

Bird, E. C. F. 1984. *Coasts: An Introduction to Coastal Geomorphology*, 3rd ed. Canberra: Australian National University Press.

Bruun, P. 1972. The history and philosophy of shore protection. *Proceedings of the 13th International Conference on Coastal Engineering*, Vol. 1, pp. 33–74. New York: American Society of Civil Engineers.

Bruun, P. 1978. *Stability of Tidal Inlets*. Elsevier: Amsterdam.

Bureau of Statistics. 1988. *Statistical Handbook of Japan*. Statistics Bureau Management and Coordination Agency, Tokyo.
Chin, I. 1993. *Research on Barrier Spit Coast: Application to Amano-hashidate Coast*. Doctor of Engineering Thesis, Kyoto University, Japan. (In Japanese)
Chin, I., M. Yamada, and Y. Tsuchiya. 1994. Formation of dynamically stable sandy beaches on the Amanohashidate coast by sand bypassing. *Proceedings of the 24th International Conference on Coastal Engineering*, Vol. 3, pp. 3478–3490. New York: American Society of Civil Engineers.
Curtis, W. R., J. E. Davis, and I. L. Turner. 1996. Evaluation of a beach dewatering system: Nantucket, USA. *Proceedings of the 25th International Conference on Coastal Engineering*, Vol. 3, pp. 2677–2690. New York: American Society of Civil Engineers.
Emmott, B. 1989. *The Sun also Set : The Limits to Japan's Economic Power*. New York: Simon & Schuster, 292 pp.
Environmental Agency of Japan. 1982. *The Natural Environment of Japan*. Tokyo: Nature Conservation Bureau, Environmental Agency of Japan. 249 pp. (In Japanese with English abstract)
Fukuya, M., N. Takaki, K. Ota, S. Harikai, and M. Ikeda. 1994. Littoral drift in fishing ports and approach channels; Problems and countermeasures. In *Hydro-Port '94, Proceedings of the International Conference on Hydro-Technical Engineering for Port and Harbor Construction*, Vol. 2, pp. 1059–1076. Yokosuka, Japan: Port and Harbour Research Institute.
Goda, Y. 1983. Facts and data on coastal engineering study in Japan. Tokyo: Port and Harbour Research Institute, Ministry of Transport, Japan. 13 pp.
Haruta, T. 1961. Recent coastal processes in Niigata Prefecture. *Coastal Engineering in Japan*, 4: 73–83. Tokyo: Japan Society of Civil Engineers.
Herron, W. J., and R. L. Harris. 1966. Littoral bypassing and beach restoration in the vicinity of Port Hueneme, California. *Proceedings of the 10th International Conference on Coastal Engineering*, Vol. 1, pp. 651–675. New York: American Society of Civil Engineers.
Hiroi, I. 1921. On the nature of drifting sand as affecting harbor construction on sandy coasts. *Journal College of Engineering*, 11 (3): 47–82. Tokyo: Tokyo Imperial University.
Hom-ma, M., and K. Horikawa. 1960. Coastal protection works and related problems in Japan. *Proceedings of the 7th International Conference on Coastal Engineering*, pp. 904–930. New York: American Society of Civil Engineers.
Hom-ma, M., K. Horikawa, and C. Sonu. 1960. A study on beach erosion at the sheltered beaches of Katase and Kamakura, Japan. *Coastal Engineering in Japan*, 3: 101–122. Tokyo: Japan Society of Civil Engineers.
Horikawa, K. (ed.). 1988. *Nearshore Dynamics and Coastal Processes: Theory, Measurement, and Predictive Methods*. Tokyo: University of Tokyo Press. 522 pp.
Horikawa, K. 1996. History of coastal engineering in Japan. In *History and Heritage of Coastal Engineering*, (ed. N. C. Kraus), pp. 336–374. New York: American Society of Civil Engineers.
Hsu, J. R. C. 1990. Short-crested waves. Chapter 3 in *Handbook of Coastal and Ocean Engineering*, (ed. J. B. Herbich). Vol. 1, pp. 95–174. Houston, TX: Gulf Publishing.
Hsu, J. R. C., and C. Evans. 1989. Parabolic bay shapes and applications. *Proceedings Institution of Civil Engineers*, Part 2, 87: 557–570. London: Thomas Telford Ltd.
Hsu, J. R. C., and R. Silvester. 1990. Accretion behind single offshore breakwater. *J. Waterway, Port, Coastal and Ocean Engineering*, 116 (3): 362–380. New York, NY: American Society of Civil Engineers.
Hsu, J. R. C., and R. Silvester. 1996. Stabilizing beaches downcoast of harbor extension. *Proceedings of the 25th International Conference on Coastal Engineering*, Vol. 4, pp. 3986–3999. New York: American Society of Civil Engineers.
Hsu, J. R. C., T. Uda, and R. Silvester. 1993. Beaches downcoast of harbours in bays. *Coastal Engineering*, 19(1, 2): 163–181. Amsterdam: Elsevier.
Ibaraki Prefectural Government. 1991. Master plan on coastal management: Evaluation of beach changes. Report, Civil Engineering Department, 38 pp. (In Japanese)
Irie, I., M. Sasaki, T. Yagyu, and K. Katoh. 1990. Development of shore protection engineering in Japan. *Proceedings 27th PIANC*, Issue S-2-1, 29–45.
Ishihara, Y. 1979. Preface. *Journal of Natural Disaster Science*, Vol. 1 (1), pp. 1. Kyoto: Dohosha Co.
Ishihara, T., and T. Sawaragi. 1964. Stability of beaches using groins. *Proceedings of the 9th International Conference on Coastal Engineering*, pp. 299–309. New York: American Society of Civil Engineers.

Iwagaki, Y. 1994. The present and future of coastal engineering in Japan. *Proceedings of the 24th International Conference on Coastal Engineering*, Vol. 1, pp. 1–11. New York: American Society of Civil Engineers.

Kanazawa, H., F. Matsukawa, K. Katoh, and I. Hasegawa. 1996. Experimental study on the effect of gravity drainage system on beach stabilization. *Proceedings of the 25th International Conference on Coastal Engineering*, Vol. 3, pp. 2640–2653. New York: American Society of Civil Engineers.

Katayama, T., M. Kurokawa, S. Yanagishima, K. Katoh, and I. Hasegawa. 1992. Lowering of water table with drainage layer under the foreshore. *Proceedings of the 39th Japanese Conference on Coastal Engineering*, pp. 871–875. Tokyo: Japane Society of Civil Engineers. (In Japanese)

Katoh, K., and S. Yanagishima. 1996. Field experiment on the gravity drainage system on beach stabilization. *Proceedings of the 25th International Conference on Coastal Engineering*, Vol. 3, pp. 2654–2665. New York: American Society of Civil Engineers.

Katoh, K., S. Yanagishima, S. Nakamura, and M. Fukuta. 1994. Stabilization of beach in Integrated Shore Protection System. In *Hydro-Port '94, Proceedings of the International Conference on Hydro-Technical Engineering for Port and Harbor Construction*, Vol. 2, pp. 1077–1096. Yokosuka, Japan: Port and Harbour Research Institute.

Kawata, Y. 1989. Methodology of beach erosion control and its application. *Coastal Engineering in Japan*, 32 (1): 113–132. Tokyo: Japan Society of Civil Engineers.

Kawata, Y., and T. Shibayama (eds.). 1998. *Guideline of Creating Sediment Transport Environment.* Current Status Review sub-committee, Committee on Coastal Engineering, JSCE. Tokyo: Japan Society of Civil Engineers.

Kawata, Y., and Y. Tsuchiya. 1986. Application of sub-sand system to beach erosion control. *Proceedings of the 20th International Conference on Coastal Engineering*, Vol. 2, pp. 1255–1267. New York: American Society of Civil Engineers.

Koh, R. 1966. Littoral drift along Iwafune port. *Coastal Engineering in Japan*, 39: 127–136. Tokyo: Japan Society of Civil Engineers.

Koike, K. 1988. Japan. In *Artificial Structures and Shorelines,* ed. H. J. Walker, pp. 317–330. Dordrecht: Kluwer.

Komar, P. D. 1983. Coastal erosion in response to the construction of jetties and breakwaters. Chapter 9 in *CRC Handbook of Coastal Processes and Erosion,* ed. P. D. Komar. Boca Raton, FL: CRC Press, pp. 191–204.

Komar, P. D. 1998. *Beach Processes and Sedimentation,* 2nd ed. Englewood Cliffs, NJ: Prentice Hall. 544 pp.

Kraus, N. C., H. Hanson, and S. Harikai. 1984. Shoreline change at Oarai beach: Past, present and future. *Proceedings of the 19th International Conference on Coastal Engineering*. Vol. 2, pp. 2107–2123. New York: American Society of Civil Engineers.

Magoon, O., and B. L. Edge. 1978. Stabilization of shorelines by use of artificial headlands and enclosed beaches. *Proceedings Coastal Zone '78,* Vol. 2, pp. 1367–1370. New York: American Society of Civil Engineers.

MOC. 1988. 20 Coastal Community Zones. Tokyo: Ministry of Construction. 34 pp. (In Japanese)

MOC. 1995. The coastal projects in our country. Tokyo: River Bureau, Ministry of Construction. 34 pp. (In Japanese)

MOC. 1997. *1997 Coast Handbook.* Tokyo: National Coast Association. 210 pp. (In Japanese)

MOT. 1988. A brochure on coast. Tokyo: Port and Harbour Bureau, Ministry of Transport. (In Japanese)

MOT. 1998. *Coastal protection in Japan.* Tokyo: Port and Harbour Bureau, Ministry of Transport. 24 pp.

Mizumura, K. 1980. Littoral drift of sand near port of Oarai. *Proceedings of the 17th International Conference on Coastal Engineering*, Vol. 3, pp. 2159–2173. New York: American Society of Civil Engineers.

Moutzouris, C. 1990. Effect of harbour works in the morphology of three Greek coasts. *Proceedings of the 27th International Navigation Congress*, pp. 17–21.

Nagao, Y. 1993. Preface. In *Coastlines of Japan II,* ed. Y. Nagao, pp. v–viii. New York: American Society of Civil Engineers.

Nishimura, H., A. Watanabe, and K. Horikawa. 1978. Scouring at the toe of a seawall due to tsunami. *Proceedings of the 16th International Conference on Coastal Engineering*, Vol. 3, pp. 2541–2547. New York: American Society of Civil Engineers.

Okajima, Y., Y. Oh-Hashi, and N. Ohshima. 1993. The present condition and future prospects of Shonan Nagisa Plan. In *Coastlines of Japan II*, ed. Y. Nagao, pp. 54–68. New York: American Society of Civil Engineers.
Ozaki, A. 1964. On the effect of an offshore breakwater on the maintenance of a harbor constructed on a sandy beach. *Proceedings of the 9th International Conference on Coastal Engineering*, pp. 323–345. New York: American Society of Civil Engineers.
Saito, K., T. Uda, K. Yokota, S. Ohara, Y. Kawanakajima, and K. Uchida. 1996. Observation of nearshore currents and beach changes around headlands built on the Kashimanada coast, Japan. *Proceedings of the 25th International Conference on Coastal Engineering*, Vol. 3, pp. 4000–4013. New York: American Society of Civil Engineers.
Sato, M., T. Fukushima, R. Nishi, and M. Fukunaga. 1996. On the change of velocity field in nearshore zone due to coastal drain and the consequent beach transformation. *Proceedings of the 25th International Conference on Coastal Engineering*, Vol. 3, pp. 2666–2676. New York: American Society of Civil Engineers.
Sato, S., and I. Irie. 1970. Variation of topography of sea-bed caused by the construction of breakwaters. *Proceedings of the 12th International Conference on Coastal Engineering*, Vol. 2, pp. 1301–1320. New York: American Society of Civil Engineers.
Sato, S., and N. Tanaka. 1966. Field investigation on sand drift at Port Kashima facing the Pacific Ocean. *Proceedings of the 10th International Conference on Coastal Engineering*, Vol. 1, pp. 595–613. New York: American Society of Civil Engineers.
Seiji, M., T. Uda, and S. Tanaka. 1987. Statistical study on the effect and stability of detached breakwaters. *Coastal Engineering in Japan*, 30(1): 131–141. Tokyo: Japan Society of Civil Engineers.
Seo, G., and T. Fukuchi. 1966. Shore protection on the coast of "Yaizu." *Proceedings of the 10th International Conference on Coastal Engineering*, Vol. 2, pp. 1183–1200. New York: American Society of Civil Engineers.
Shapiro, H. A. 1989. Japan, a sea-oriented nation? *Proceedings of Coastal Zone '89*, Vol. 4, pp. 3502–3513. New York: American Society of Civil Engineers.
Shinohara, K., and T. Tsubaki. 1966. Model study on the change of shoreline of sandy beach. *Proceedings of the 10th International Conference on Coastal Engineering*, Vol. 1, pp. 550–563. New York: American Society of Civil Engineers.
Silvester, R. 1966. Sediment transport and accretion around the coastlines of Japan. *Proceedings of the 10th International Conference on Coastal Engineering*, Vol. 1, pp. 469–488. New York: American Society of Civil Engineers.
Silvester, R. 1970. Development of crenulate shaped bays to equilibrium. *J. Waterways and Harbors Division*, 96 (WW2): 275–287. New York: American Society of Civil Engineers.
Silvester, R. 1974. *Coastal Engineering*, II. Amsterdam: Elsevier.
Silvester, R. 1976. Headland defense for coasts. *Proceedings of the 15th International Conference on Coastal Engineering*, Vol. 2, pp. 1347–1365. New York: American Society of Civil Engineers.
Silvester, R. 1977. The role of wave reflection in coastal processes. *Proceedings Coastal Sediments '77*, pp. 639–654. New York: American Society of Civil Engineers.
Silvester, R. 1979. A new look at beach erosion control. *Annual,* Disaster Prevention Research Institute, Kyoto University, Vol. 22 (A), pp. 19–31.
Silvester, R. 1986. The influence of oblique reflection on breakwaters. *Proceedings of the 20th International Conference on Coastal Engineering*, Vol. 3, pp. 2253–2267. New York: American Society of Civil Engineers.
Silvester, R., and S. K. Ho. 1972. Use of crenulate shaped bays to stabilize coasts. *Proceedings of the 13th International Conference on Coastal Engineering*, Vol. 2, pp. 1394–1406. New York: American Society of Civil Engineers.
Silvester, R., and J. R. C. Hsu. 1993. *Coastal Stabilization: Innovative Concepts.* Englewood Cliffs, NJ.: Prentice Hall. 578 pp.
Silvester, R., and J. R. C. Hsu. 1997. *Coastal Stabilization.* Advanced Series on Ocean Engineering, No. 14. Singapore: World Scientific Publ. Co. 578 pp. (Reprint of Silvester and Hsu, 1993)
Silvester, R., Y. Tsuchiya, and T. Shibano. 1980. Zeta bays, pocket beaches and headland control. *Proceedings of the 17th International Conference on Coastal Engineering.* Vol. 2, pp. 1306–1319. New York: American Society Civil Engineers.
Smith, G. H. and Good, D. 1943. *Japan—A Geographical Review*. Special Publication, No. 28. Washington: American Geographical Society.

Suyama, H., T. Uda, and T. Yoshimura. 1986. Beach change around detached breakwaters due to artificial nourishment of by-passed sand. *Proceedings of the 20th International Conference on Coastal Engineering*, Vol. 2, pp. 1565–1575. New York: American Society of Civil Engineers.
Tanaka, N. 1983. A study on characteristics of littoral drift along the coast of Japan and topographic change resulted from construction of harbors on sandy beach. *Technical Notes* No. 453. Yokosuka: Port and Harbour Research Institute. 148 pp. (In Japanese)
Tanaka, N., and S. Sato. 1976. Topographic change resulting from construction of a harbor on a sandy beach: Kashima Port. *Proceedings of the 15th International Conference on Coastal Engineering*, Vol. 2, pp. 1824–1843. New York: American Society of Civil Engineers.
Toyoshima, O. 1974. Design of a detached breakwater system. *Proceedings of the 14th International Conference on Coastal Engineering*, Vol. 2, pp. 1419–1431. New York: American Society of Civil Engineers.
Toyoshima, O. 1976. Changes of sea bed due to detached breakwaters. *Proceedings of the 15th International Conference on Coastal Engineering*, Vol. 2, pp. 1572–1589. New York: American Society of Civil Engineers.
Toyoshima, O. 1978. Effectiveness of seadikes with rough slope. *Proceedings of the 16th International Conference on Coastal Engineering*, Vol. 3, pp. 2528–2539. New York: American Society of Civil Engineers.
Toyoshima, O. 1982. Variation of foreshore due to detached breakwaters. *Proceedings of the 18th International Conference on Coastal Engineering*, Vol. 2, pp. 1873–1892. New York: American Society of Civil Engineers.
Toyoshima, O. 1984. New type block for seawall slope protection. *Proceedings of the 19th International Conference on Coastal Engineering*, Vol. 3, pp. 2536–2545. New York: American Society of Civil Engineers.
Toyoshima, O. 1988. Gentle slope seawalls covered with armour units. *Proceedings of the 21st International Conference on Coastal Engineering*, Vol. 3, pp. 1983–1996. New York: American Society of Civil Engineers.
Tsuchiya, Y. 1987. Beach erosion control. *Proceedings Hydraulic and Sanitary Engineering*, JSCE, No. 387, II-8, pp. 11–23. Tokyo: Japan Society of Civil Engineering. (In Japanese)
Tsuchiya, Y., T. Yamashita, and R. Silvester. 1990. Beach erosion due to large coastal structure and its control. *Proceedings of the 22nd International Conference on Coastal Engineering*, Vol. 3, pp. 2726–2739. New York: American Society of Civil Engineers.
Tsuchiya, Y., Y. Kawata, T. Yamashita, T. Shibano, M. Kawasaki, and S. Habara. 1992. Sandy beach stabilization: Preservation of Shirarahama beach, Wakayama. *Proceedings of the 23rd International Conference on Coastal Engineering*, Vol. 3, pp. 3426–3439. New York: American Society of Civil Engineers.
Tsuchiya, Y., T. Yamashita, and T. Izumi. 1994. Erosion control by considering large scale coastal behavior. *Proceedings of the 24th International Conference on Coastal Engineering*, Vol. 3, pp. 3378–3392. New York: American Society of Civil Engineers.
Tsuchiya, Y. 1994. Formation of stable sandy beaches and beach erosion control: A methodology for beach erosion control using headlands and its applications. *Bulletin,* Disaster Prevention Research Institute, Vol. 44, Part 3, pp. 139–173. Kyoto: Disaster Prevention Research Institute, Kyoto University.
Uda, T. 1988. Statistical analysis of detached breakwaters in Japan. *Proceedings of the 21st International Conference on Coastal Engineering*, Vol. 2, pp. 2028–2042. New York: American Society of Civil Engineers.
Uda, T. 1997. *Beach Erosion in Japan.* Tokyo: Sankai Do Publishing Co. 442 pp. (In Japanese)
Uda, T., and T. Igarashi. 1992. *Portraits of Coastal Disasters:* The slides collected by Dr. Osamu Toyoshima. Document No. 3075. Tsukuba: Public Works Research Institute, Ministry of Construction, Japan. (In Japanese)
Uda, T., and K. Noguchi. 1991. Beach changes around Iwafune port in northern Niigata Prefecture. *Technical Notes in Civil Engineering,* Vol. 33 (11), pp. 28–33. Tokyo: Japan Society of Civil Engineers. (In Japanese)
Uda, T., M. Sumiya, and Y. Kobayashi. 1986. Analysis of beach erosion around large-scale coastal structures. *Proceedings of the 20th International Conference on Coastal Engineering*, Vol. 3, pp. 2329–2343. New York: American Society of Civil Engineers.
Uda, T., M. Sumiya, and H. Sakuramoto. 1988. Stabilization of coast by construction of headlands

on the Kashimanada coast, Japan. *Proceedings of the 21st International Conference on Coastal Engineering*, Vol. 3, pp. 2791–2805. New York: American Society of Civil Engineers.

Yajima, M., A. Uezono, T. Yauchi, and F. Yamada. 1983. Application of sand bypassing to Amanohashidate beach. *Coastal Engineering in Japan*, 26 (1): 151–162. Tokyo: Japan Society of Civil Engineers.

Yanagishima, S., K. Katoh, T. Katayama, and H. Murakami. 1991. Effects of depressing water table on change of foreshore profile. *Proceedings of the 38th Conference on Coastal Engineering*, pp. 266–270. Tokyo: Japan Society of Civil Engineers. (In Japanese)

Yasso, W. E. 1965. Plan geometry of headland breakwaters. *J. Geology,* 73: 702–714.

Yoshioka, K., T. Kawakami, S. Tanaka, M. Koarai, and T. Uda. 1993. Design manual for artificial reefs. In *Coastlines of Japan II*, ed. Y. Nagao, pp. 93–107. New York: American Society of Civil Engineers.

Zwemmer, D. and J. Van't Hoff. 1982. Spending beach breakwater at Saldanha bay. *Proceedings of the 18th International Conference on Coastal Engineering*, Vol. 2, pp. 1248–1267. New York: American Society of Civil Engineers.

CHAPTER 10
BEACH NOURISHMENT DESIGN

BILLY L. EDGE
Civil Engineering Department
Ocean Engineering Program
Texas A&M University
College Station, TX

Introduction	10.1
Does Beach Nourishment Work?	10.2
How Should Success Be Measured?	10.3
Are Beach Nourishment Projects Economically Justified?	10.3
How Can Beach Nourishment Projects Be Improved?	10.4
What is the Role of Fixed Structures?	10.4
What is the Role of Nourishment in Flood Protection?	10.4
Findings of the Marine Board Study on Beach Nourishment	10.4
Beach Nourishment Design	10.6
Lessons from Experience	10.6
Changes in Design Methodology	10.6
Construction Practice	10.9
Overview of the Beach Nourishment Design Process	10.11
Equilibrium Beach Profiles	10.12
Volume of Fill	10.16
Planform Models	10.19
Beach Nourishment with Coastal Structures	10.23
Requirements for Renourishment	10.25
Monitoring and Maintenance	10.25
Hot Spots	10.26
Benefits of Beach Nourishment	10.27
Summary	10.29
References	10.30

INTRODUCTION

Maintenance of beaches is fundamental to their recreational value and economic worth to the tourist economy. International visitors were expected to spend $77 billion in the United States during 1995, which will make this country have the largest tourist income of all nations—more than twice the tourist income of France, the second-ranked nation. (Lekic, 1995) The World Tourism Organization estimates that by the end of the century the number of people traveling internationally for vacations will double the 1985 levels. A similar increase is expected for the national population. The majority of the visits to this country include a trip to the beaches as a part of the vacation. Therefore it is imperative for the stability of the tourism industry as well as the stability of the United States coastline that the beaches must be maintained. Maintenance of the nation's beaches requires a commitment on behalf of all levels of government, especially the Federal government, which has a national perspective. Without a Federal commitment, the efforts will not be consolidated and coordinated. This commitment should take the form of continuing efforts by the Army Corps of Engineers in designing and constructing these projects for the people of the country and by the consideration of the true value of the recreational aspects of the beaches.

Beach nourishment has been ongoing in the United States since 1922, when the Coney Island project was constructed. Since then there have been numerous projects, small and large, and as more knowledge has been gained through experience and monitoring, understanding of coastal processes has increased and design and construction of the projects have become more successful. Technology has moved from a "simple" pragmatic approach, to an empirical one, to the definition of alongshore and cross-shore transport within the framework of existing knowledge of these processes. Much still remains to be learned about beach nourishment, but the fact is clear that it is the best solution that currently exists to provide stability to an eroding shoreline while preserving access and recreational opportunities.

In addition to the value of beaches for recreation, they offer several other important features. First, beaches provide habitat for various flora and fauna, not the least of which is the sea turtle, which uses sandy beaches for nesting. There are also many species of birds that use beaches for nesting habitat. The beaches provide protection role to upland areas and in the case of barrier islands, protection to the mainland and the waters between.

There are many forms of beach protection—from sea walls to breakwaters—but it is clear that at present, beach nourishment is by far the best single available device for nearly all applications. Moreover, beach nourishment in combination with fixed structures offers a viable solution to control erosion of a beach where "hot spots" appear in a nourishment project.

The National Academy of Sciences/National Academy of Engineering acting through its National Research Council commissioned a study by the Marine Board (1995) to address six questions about using nourishment for beach preservation:

1. Does beach nourishment work?
2. How is success measured in nourishment projects?
3. Is beach nourishment feasible?
4. How can the process be improved?
5. Will fixed (hard) structures help improve the performance of the nourishment?
6. What is its effect in storm surge protection?

The answers to the six questions posed to the Marine Board are summarized below.

Does Beach Nourishment Work?

The Marine Board study addressed this question and determined that the effectiveness of beach nourishment will depend upon the following requirements:

1. Erosion rates must be effectively incorporated into project design.
2. State-of-the-art engineering standards must be used for planning, design, and construction.
3. It is necessary that nourishment projects be maintained according to design specifications and that when renourishment is required, it must be applied (Figure 10.1).

FIGURE 10.1. Beach replenishment from offshore sources at Pompano Beach, Florida (Herbich, 1992).

How Should Success Be Measured?

It was determined that different stakeholders measure success differently. For instance, the local community will guage success by the width of the dry beach, whereas the engineer may view a successful project as one that erodes at the predicted rate with most of the lost sediment remaining within the beach profile. There is no single measure of success. However, measures of performance may include:

1. Dry beach width
2. Total sand volume remaining
3. Poststorm damage assessments
4. Residual protection capacity

Are Beach Nourishment Projects Economically Justified?

Federal and nonfederal sponsors use different criteria to address economic justification. All projects that are supported by the Corps of Engineers requires conformation to the NED (National Economic Development) Plan. The NED Plan does not account for regional or recreational benefits to justify the project. Rather, the NED Plan requires economic justification on flood or storm protection. On the other hand, the local sponsor is vitally concerned that the project inject an economic boost to the local economy. Therefore, the project is economically viable if local economic expectations are met and, for Federal participation, the NED Plan criteria are met. One of the important recommendations of the Marine Board study suggested that the full range of costs and benefits need to be accounted for in any economic valuation, which is not currently done.

How Can Beach Nourishment Projects Be Improved?

Project planners need to expand public involvement in the planning process to include design expectations. The planners need to obtain long-term commitments for sand sources. The project planners and engineers can improve the project design and benefits by broadening the project boundaries to include the entire littoral cell. Typically, nearshore sources of sand are sought for borrow material; however, when offshore sources are selected, the borrow sites for sand should be beyond the depth of closure. The project planners should ensure that the 50-year design life of the project is matched with a 50-year funding plan. Moreover, improvement in the planning, design, and construction of beach nourishment projects will only occur with adequate physical and biological monitoring.

What is The Role of Fixed Structures?

Permanent structures can enhance the performance of beach nourishment by reducing end losses, reducing local wave intensity, and preventing flooding where dunes do not exist. Structures do not increase the amount of sand in the system, they simply rearrange or control the flow. Unconventional approaches to structural enhancement of the nourishment project often trap sand by robbing neighboring beaches. If the beach nourishment project is not maintained, structures built to support the project should be removed.

What is the Role of Nourishment in Flood Protection?

It is clear that beach nourishment provides storm damage reduction for upland buildings and infrastructure. It also provides a reduction in flooding risk, which merits a reduction in flood insurance premiums. The study group also found that uncertainties in continued financing for future renourishment efforts and a permanent source of future sand supply require physical setbacks even where a healthy nourishment project has been constructed. The building standards behind a nourishment project should reflect the level of risk caused by the diminution or maintenance of the protection offered by the nourishment through time.

Findings of the Marine Board Study on Beach Nourishment

The Marine Board report (1995) demonstrates that beach nourishment is a viable engineering alternative for storm protection, shoreline protection, and maintenance of a recreational beach. There are specific project performance criteria that can measure the success of a project. The formula developed for cost sharing between the stakeholders should match the benefits of the beach nourishment project. Federal procedures for calculating benefits are overly restrictive. The Marine Board recommended that postconstruction economic evaluations should be made to identify and measure actual project benefits and costs (see Table 10.1).

TABLE 10.1. Recommendations for U.S. Army Corps of Engineers (Marine Board, 1995)

Engineering
   Publish detailed and comprehensive state-of-the-art engineering guidance on design of beach nourishment.
   Develop consistent methodology for design.
   Develop performance-based specifications for nontraditional devices.
Planning
   Incorporate the true social costs and benefits for the entire region.
   Couple projects with local land-use plans to increase net benefits.
   Establish updated guidelines for measuring benefits and consistently apply the guidelines in all Corps district offices.
Management
   Conduct postconstruction economic evaluation to identify and measure actual costs and benefits.
   Require consideration of the economic value of beach quality sand dredged from navigation projects.
   Undertake a cooperative program with NOAA and USGS to establish standardized decadal rates of erosion and accretion.

There were a number of issues that arose in the Marine Board study that remain unresolved. Only further monitoring of completed nourishment projects will provide the information for some of the unresolved issues. Other issues will require changes in Federal policies and programs. The remaining issues are noted below:

- Should benefits from sand transported to adjacent beaches be included in the economic evaluation?
- Should navigation projects be "charged" for sand deficits on adjacent beaches?
- Publish detailed and comprehensive state-of-the-art engineering guidance on beach nourishment design and monitoring.
- Accreditation of beach nourishment for the U.S. Federal Flood Insurance Program.
- Postconstruction assessment of costs and economic benefits, and effects of economic development.
- Define "engineered beach" to include technical criteria and monitoring.
- Remove restrictions on hard structures that improve project performance.
- Include public involvement in determining criteria for success.

The Marine Board study did conclude that large short-term fluctuations mask the effects of long-term sea level changes and this must be considered in project evaluation and design. A lack of public involvement in the planning and selection of performance criteria has heightened the controversy over beach nourishment success. The apportionment of costs as mandated by the Federal government does not reflect the distribution of benefits that are either planned or realized. The current Federal policy on offshore disposal of beach quality sand does not recognize the economic value of sand and is counterproductive to littoral processes that would support a beach nourishment project. Lastly, the report recognizes that beach nourishment reduces the risk of flooding and supports a reduction in premiums for flood insurance.

BEACH NOURISHMENT DESIGN

Lessons from Experience

There are approximately fifty reasonably large beach nourishment and shore protection projects that have been constructed since the 1950s by the U.S. Army Corps of Engineers (Sudar et al., 1995). There also have been several large projects constructed by others, including a very large 4,100,000 m^3 beach nourishment constructed by the U.S. Navy at Gulf Islands National Seashore at Perdido Key, Florida. And, of course, there have been many smaller projects of 10,000 to 1,000,000 m^3 constructed where there was no Federal participation due to lack of "Federal interest." A number of beach nourishment projects have been summarized in terms of volumes, funding, and renourishment by the Institute for Water Resources (USACE, 1996) for Federal projects, Trembanis and Pilkey (1998) for Gulf of Mexico beaches, and Pilkey and Clayton (1989) for the Atlantic beaches. Over the course of the construction of these projects, many lessons have been learned. Unfortunately, it was many years after the initial projects were constructed before detailed monitoring began to be used routinely. As a result of the monitoring, there is now a much better understanding of the processes, construction methods, design procedures, and project expectations (Rijkswaterstaat, 1987; Dean and Yoo, 1992; Wise and Kraus, 1993; USACE, 1994).

For example, the monitoring and analysis reported by Wise and Kraus (1993) demonstrate the recovery process of nourished beaches after significant winter storms. Wiegel (1992, 1994) provides a very complete account of the experiences of beach nourishment projects on the Dade County, Florida and the U.S. Pacific coasts, respectively. Wiegel points out that the Dade County project has been an extremely successful beach nourishment project with many benefits to the Miami Beach area. Yet, Weggel (1995) indicated that it is impossible to properly monitor a beach nourishment project unless directional waves and profiles to the depth of closure are included in the plan. As the understanding of the coastal processes has evolved, so has the understanding of beach fills. This understanding has lead to improvements in both the design and construction of beach construction. Nevertheless, to continue improving the design procedures, it is imperative that completed projects continue to be monitored.

There are many other projects that have been successfully constructed around the world—in Spain (Gomez-Pina and Ramirez, 1994), Denmark (Laustrup and Madsen, 1994), and The Netherlands (Kroon et al., 1994). The success of the projects in Spain has in fact pushed that country from the third-largest tourist destination spot in the world (in numbers of tourists) to second, raising it above the United States (*Houston Chronicle*, 1996). The United States remains the largest country in terms of income from tourism even though the number of tourists are below those of Spain and France.

Changes in Design Methodology

The first beach nourishment project in the United States began at Coney Island, New York, in 1922 (Dornhelm, 1995), with placement of approximately 1,300,000 m^3 of sand. The initial design template used for project construction was less than optimal and subsequent projects varied in shape and design until guidance was published by the Corps of Engineers, which suggested using the overfill factor proposed by James (1975) and based upon Krumbein and James (1965) to quantify the amount of sand required. This information was then included in Corps of Engineers report T.R. 4, "Shore Protection, Planning, and Design," which was subsequently replaced with the *Shore Protection*

Manual (USACE, 1984). The equations and curves provided by these reports have been the basis for the design of beach nourishment projects for nearly twenty years.

James presented an overfill factor, R_A, which is the ratio of fill material required for a given borrow site compared to that required using the existing beach sediments. The overfill factor has been used to define the actual quantity of borrow material that will be required for a project fill based upon the desired construction template. The template is typically given in terms of what the contractor can establish as opposed to the final wave-worked profile. Thus the overfill factor takes into consideration the mean grain size and distribution of the borrow and native materials and provides an indication of the loss of material that will occur as a result of the differing sediment distributions.

The overfill factor is shown in Figure 10.2. In this figure,

R_A = the overfill factor

$$\sigma_\phi = \frac{(\phi_{84} + \phi_{16})}{2}$$

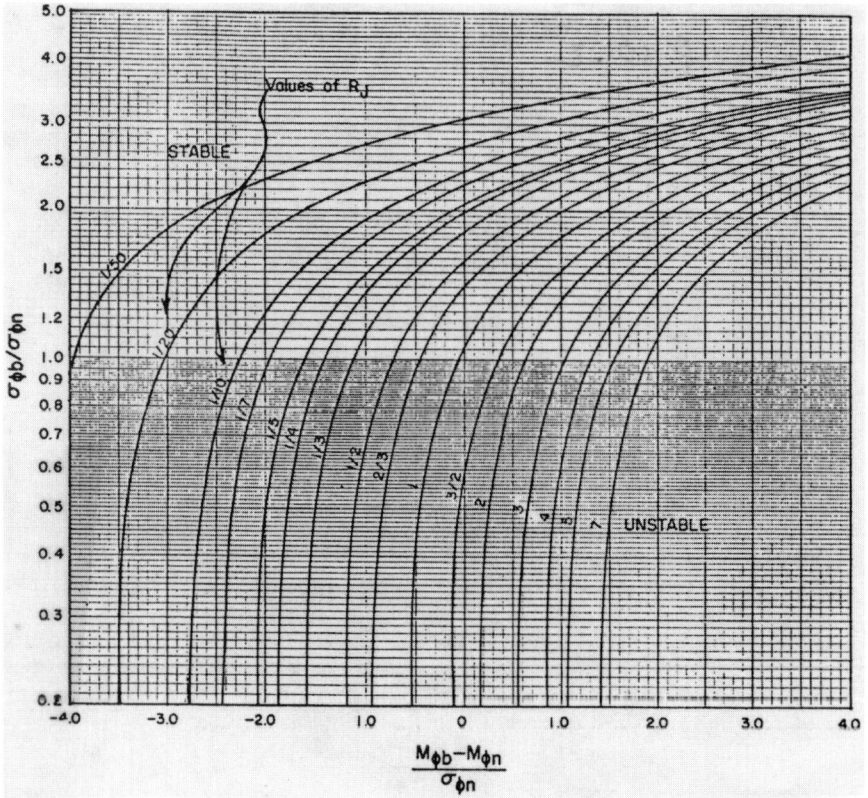

FIGURE 10.2. Beach fill adjustment factor, R_A. (After USACE, 1984.)

$$M_\phi = \frac{(\phi_{84} + \phi_{16})}{2}$$

ϕ_{84} = 84th percentile in phi units

ϕ_{16} = 16th percentile in phi units

The subscripts b and n refer to borrow and native material, respectively. The parameter σ_ϕ is the standard deviation of the sediment sample and M_ϕ is the mean diameter of grain size. Both of these parameters are based upon the sample size distribution, being given in phi units. Phi units are defined as $\phi = -\log_2 d_{mm}$. Where d_{mm} is the equivalent sediment diameter in mm. James (1975) also presented the renourishment factor, R_J, which is the ratio of the rate at which the borrow material will erode as compared with the native material. The renourishment factor is show in Figure 10.3. It too is seen to be a function of the standard deviation and the mean diameter of the sediments. However, this procedure neglects several important aspects of specific beach projects, such as storm effects, dependency of profile on sediment size distribution across the beach profile, length of project, coastal structures, and alongshore transport.

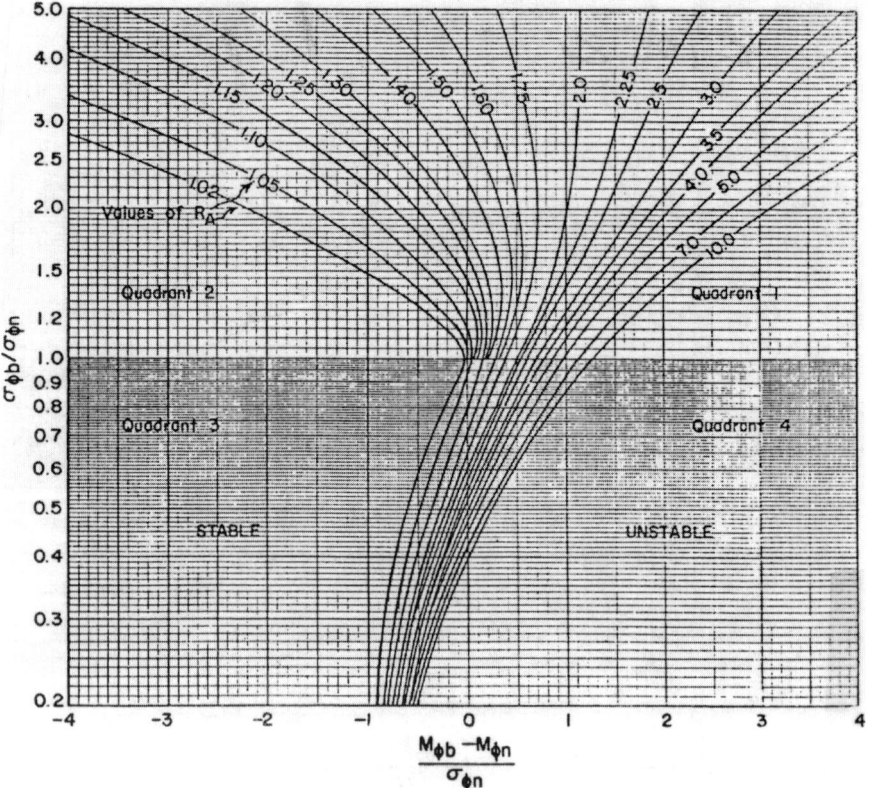

FIGURE 10.3. Renourishment factor, R_J. (After USACE, 1984.)

The *Manual on Artificial Beach Nourishment* by Rijkswaterstaat (1987) was the first design manual to be published that represented a major change in the design procedure for a beach nourishment project. Subsequently, the Corps of Engineers (USACE, 1995) presented an alternative to the simple use of the overfill factor with a standard construction shape as proposed in the *Shore Protection Manual* (USACE, 1984). The alternative included consideration of cross-shore transport and alongshore spreading of the fill material. Moreover, it suggested the use of one-line models in evaluating the influence of coastal structures on the response of the fill to wave conditions at the site. In the future, the design of beach nourishment projects will shift even further away from the use of the overfill factor and will be based upon the use of more modern or basic tools.

Construction Practice

Beach nourishment is most commonly achieved by placement of hydraulic fill and less commonly by placement of dry fill. The selection of either approach typically dictates the shape of the design or construction fill template. One of the results of monitoring has been the modification of contract specifications (Edge et al., 1994) to take advantage of the natural shaping of the beach that takes place as a result of normal wave activity. This basically constitutes a recognition that the contractor only has control over the finished lines and grades above the waterline and that natural processes will shape the fill below the waterline. The Corps of Engineers (USACE, 1995) suggest several ideas for template design but the basic one in use in the United States is a berm plus dune. A variety of construction templates are shown in Figure 10.4. The choice of the appropriate template for the project will depend upon the project objectives, construction methods available, and available borrow material. For example, in some countries, the preferred method of placement of borrow material is from an offshore hopper, spraying the material onto the active beach, whereas in other locations, the site conditions are better met with an offshore submerged berm. The method used to properly construct any of these templates depends largely upon the availability of adequate borrow sites and the method of placement. For example, an underwater berm was created at Mobile off the Alabama Coast (Hands, 1991) that has shown some success in providing protection and migrating shoreward (McLellan, 1990; McLellan and Kraus, 1991). In Corpus Christi Bay, where the wave climate is limited by the available fetch in the Bay, a uniform layer of material with a large amount of fines was placed and then overlain with upland sand with success (Kieslich and Brunt, 1989). The success of this project is related to the limited wave conditions in the Bay.

More typically, a design template similar to that shown in Figure 10.5 has been used in beach construction in the United States. This is basically a lateral, seaward shift of the existing profile a distance of Y, with the addition of material in the construction profile to account for the underwater material that will be washed seaward during natural processes once the material is placed upon the beach. The distance X represents the width of the additional construction profile to account for the loss of material to offshore as the profile adjusts.

There are no limitations on the capability of the dredging industry to support needed beach nourishment in most countries. In fact there are dredges specifically designed and built for beach nourishment projects. If the time window of construction is not severely constrained by weather and environmental conditions, then dredging companies with smaller dredges can be competitive for beach nourishment projects. But this continues to be a major problem as the "windows" of construction time are often defined without ade-

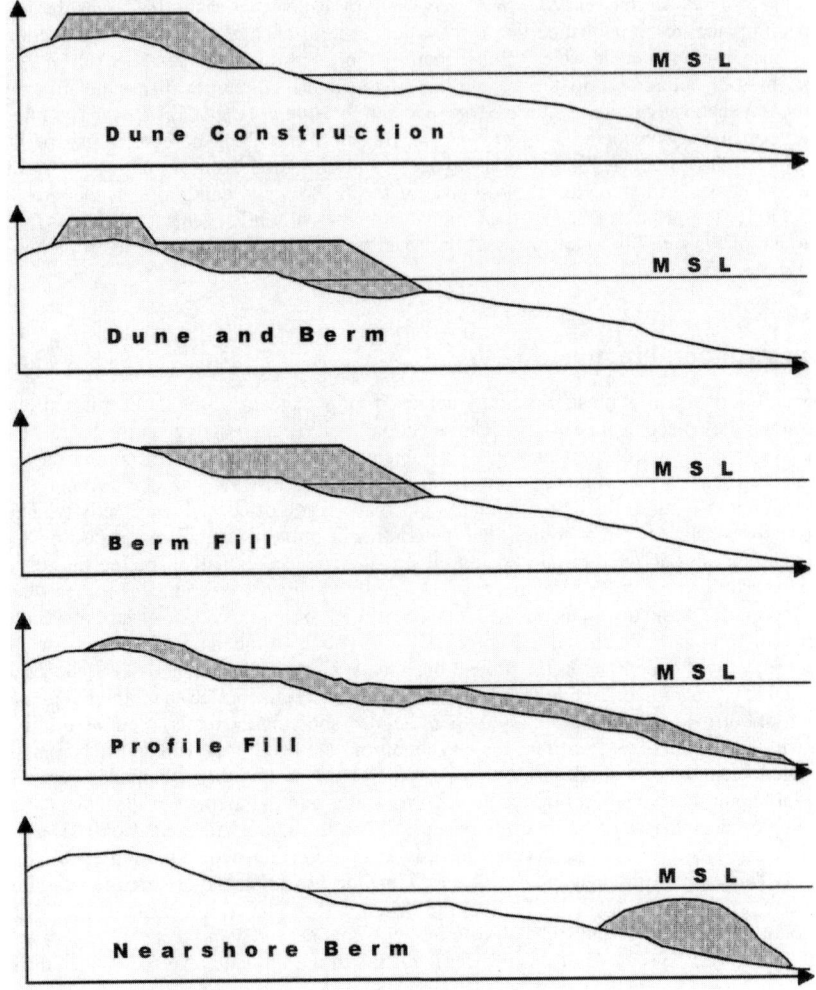

FIGURE 10.4. Example of various construction templates. (After USACE, 1995.)

quate data or understanding of the interference of the construction with the biological systems that form the basis for the closure windows.

One real challenge to the viability of beach nourishment projects is the narrowing window of the available construction season because of various constraints, the primary one being environmental. Because the turtle nesting season coincides with the most suitable dredging period, much of the coastline must be nourished during the period of energetic wave conditions. In addition to the turtle nesting season, the nesting season of some endangered or threatened shorebirds and the spawning or migration season of commercially important fish further serve to reduce the available dredging window to a narrow band of opportunity in the middle of the winter season. Dredging during the winter season not

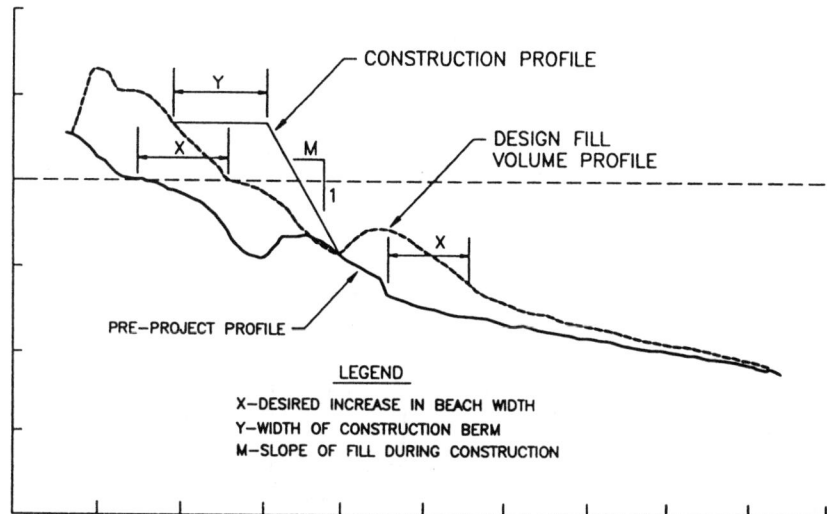

FIGURE 10.5. Typical approach to determination of construction profile and volume of material to be placed. (After USACE, 1995.)

only brings the potential for bad weather and downtime, but it also creates the possibility of significant losses of pay material from the constructed section. Work is needed to clarify the importance of the closure periods on the species involved so the controversy can reach closure. Additionally, work is needed to identify the additional costs of nourishment projects caused by construction during the winter season. Studies to answer these questions have been proposed by the Marine Board and are being conducted by the USACE.

On the other hand, there are locations and times when the temporal limitation to dredging is controlled by the tourist season. Beach communities will not want to disrupt the tourist season with heavy equipment on the beach and the concomitant risks and disruption to the tourists. In spite of the best efforts by dredging contractors, tourists are attracted to the dredge operations on the beach during a nourishment project. The tourists often gather to inspect the work of the equipment operators and surveyors and the discharge from the pipeline.

Overview of the Beach Nourishment Design Process

The components of the design will vary according to requirements of the project and the the organization performing it. As shown in Figure 10.6, the process begins after identification of the problem, with a definition of objectives that can be met, which are a combination of local desires and site characteristics. Once the preliminary engineering advances to the stage where the economics of the project can be adequately developed, additional considerations are incorporated into the design and economics. These additional considerations include minimum environmental aspects and national and regional funding and policy issues. Once the issues are fully developed and the project economics are finally

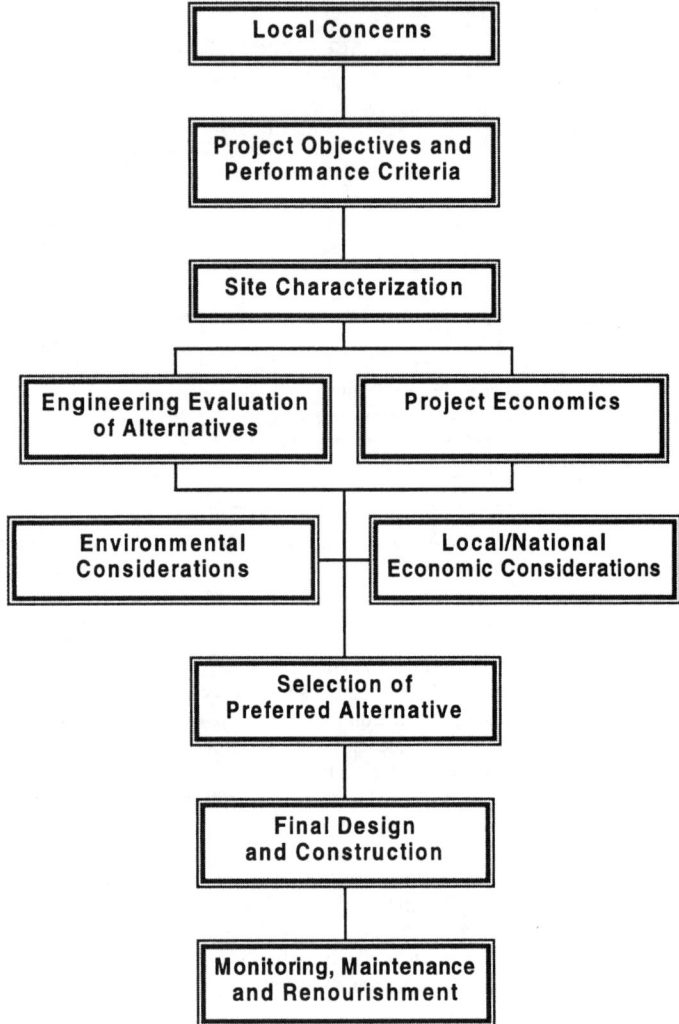

FIGURE 10.6. General outline of project progress from conception to future renourishment. (After USACE, 1995.)

developed, the preferred alternative is selected through discussion with the local sponsor, designer, and regulatory and other officials. When the beach nourishment project proceeds without input from the local communities, the project has less of a chance of satisfying the expectations of the local residents and stakeholders.

When a beach nourishment project is developed in the United States with Federal participation (funding), there are specific criteria used in defining the project's economic viability. Currently, as noted above, the project must be justified for flood protection before any recreational or other benefits can be considered. Moreover, by Federal law, the project must be shown to optimize the National Economic Development benefits, which are prescribed in detail by the Federal Government. The project can then proceed to final design, preparation of contract documents, selection of contractor, and construction. Once the initial nourishment is placed, monitoring should begin immediately to confirm the design considerations and to establish a base for the maintenance of the project. Preproject physical and environmental baseline surveys of the site and the borrow area should be completed as part of the planning and design process and these will become a part of the project monitoring data base. As a minimum, the design process for beach nourishment projects should include the components shown in Table 10.2.

The last two steps in the design process were added to Table 10.2 to ensure that they are not overlooked in the planning for the project. The monitoring and maintenance programs are very important in keeping the local communities involved in the project and fully aware of the realistic expectations for the performance of the project.

The first two components in Table 10.2 have been included in project plans since the earliest designs for beach nourishment were started. The design process is somewhat more complex now that significantly more wave and storm condition information are available than previously, and it is in more complex form. The list in Table 10.2 does not include environmental conditions that will control a large part of the design, from the selection of the borrow source, to the available window for construction, to spatial limits of project construction.

Equilibrium Beach Profiles

The most significant advances in the design process stem from the models that have been developed and verified in recent years for cross-shore and alongshore transport. The cross-shore transport models are based upon the equilibrium beach profile concept proposed by Bruun (1962) in his discussions on the effects of sea level rise on coastal erosion and further discussed by Dean (1991, 1998). The equilibrium beach will have a profile that is stable over the long term for the existing wave climate at the site and it can be described as

$$h(y) = Ay^{2/3} \qquad (1)$$

where $h(y)$ is the depth of the beach profile, y is the distance offshore, and A is a parameter related to sediment fall velocity. The sediment parameter A is frequently given by

$$A = 0.067w^{0.44} \qquad (2)$$

TABLE 10.2. Steps in beach nourishment design (after Campbell et al., 1990)

1. Assessment of existing and historical beach conditions
2. Definition of wave climate and storm conditions
3. Identification of permanent borrow source(s) and characterization of its quality and quantity
4. Evaluation of existing and proposed structures
5. Sand budget of the "reach"
6. Design of section for recreation and/or storm protection
7. Effect of spreading losses on design section
8. Define postproject monitoring program
9. Definition of maintenance and renourishment requirements

where w is the sediment fall velocity in m/s and was shown by Hallermeier (1981) to be related to sediment diameter d (m) as

$$w = 14d^{1.1} \tag{3}$$

The parameter A was determined by laboratory experiments and field measurements to take the form shown in Figure 10.7. Development and utilization of the equilibrium profile is extensively discussed in Dean (1983, 1991). The equilibrium profile concept can be used to predict the response of the shore profile to the combination of storm surges and waves. The equilibrium beach profile is approximately valid to the depth of closure, which is the seaward limit at which the bottom sediments are effectively moved by the waves. The closure depth is given by Birkemeier (1985) as

$$h^* = 1.75 H_{s0.137} - 57.9\left(\frac{H_{s0.137}}{gT_s^2}\right) \tag{4}$$

where

h^* = depth of closure
$H_{s0.137}$ = "extreme" nearshore significant wave height exceeded 12 hour/year (m)
T_s = period of waves associated with $H_{s0.137}$

Houston (1996) shows that h^* can be shown to be related to the mean annual significant wave height, H_s, as

$$h^* = 1.5 H_{s0.137} \approx 6.75 H_s \tag{5}$$

where

$$H_{s0.137} = H_s + 5.6\sigma \tag{6}$$

In the above equation, σ represents the significant wave height standard deviation, which should be obtained from actual wave height records or from a hindcast data base similar to the Wave Information Study (WIS) maintained by the USACE.

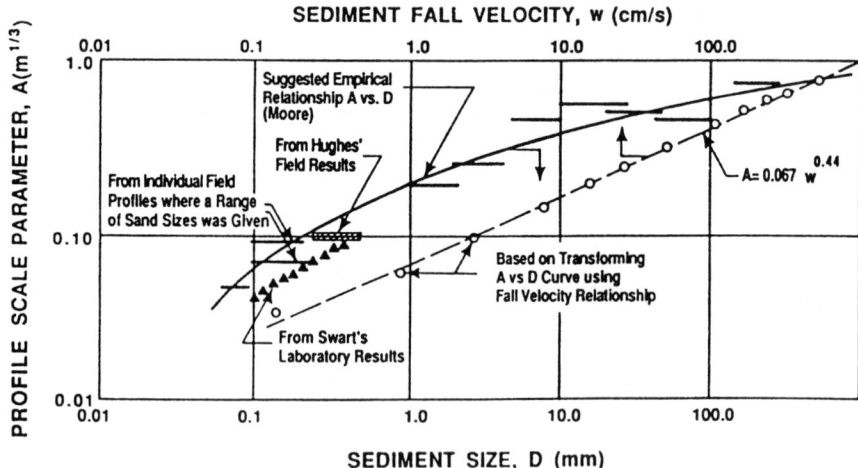

FIGURE 10.7. Beach profile factor, A, as a function of diameter, d, and fall velocity, w. (After Dean, 1991; modified from Moore, 1982.)

The equilibrium beach concept forms the basic foundation of the commonly used dune erosion models in the United States: the SBEACH model (Larson and Kraus, 1989, 1990, 1991) and the EDUNE model (Kriebel and Dean, 1985). The primary difference between SBEACH and EDUNE is that SBEACH divides the cross-shore into four zones that have different transport properties. SBEACH will allow simultaneous erosion from the foreshore and onshore movement from offshore, resulting in the development of an offshore bar, as shown in Figure 10.8. Wise et al. (1996) provide further validation of the SBEACH model and provide a method of extending the model to handle irregular waves. The basic SBEACH model provides the transport rate relations for each region as given in the following set of equations.

Region I:
$$q = q_b e^{-\lambda 1(x-x_b)} \qquad x_b < x \qquad (7)$$

Region II:
$$q = q_b e^{-\lambda 2(x-x_b)} \qquad x_p < x \leq x_b \qquad (8)$$

Region III:
$$q = \begin{cases} K\left(D - D_{eq} + \dfrac{\varepsilon}{K}\dfrac{dh}{dx}\right) & \text{for } D > \left(D_{eq} - \dfrac{\varepsilon}{K}\dfrac{dh}{dx}\right) \\ 0 & \text{for } D \leq \left(D_{eq} - \dfrac{\varepsilon}{K}\dfrac{dh}{dx}\right) \end{cases} \quad x_z \leq x \leq x_p \qquad (9)$$

Region IV:
$$q = q_z\left(\dfrac{x - x_r}{x_z - x_r}\right) \qquad x_r < x < x_z \qquad (10)$$

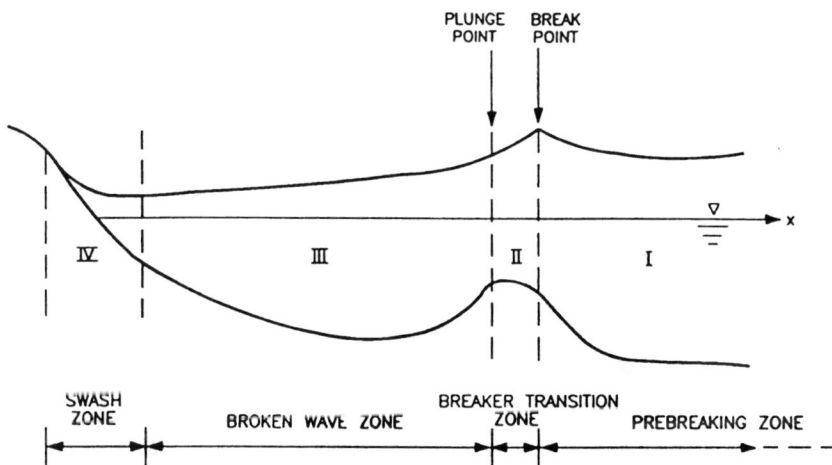

FIGURE 10.8. Definition sketch of SBEACH model regions of Applicability. (After Larson and Kraus, 1989.)

where

q = net cross-shore sand transport rate (m³/m-sec)
$\lambda_{1,2}$ = spatial decay coefficients in Regions I and II, (1/m)
x = cross-shore coordinate directed positive offshore (m)
K = sand transport rate coefficient (m⁴/N)
D = wave energy dissipation per unit water volume (N-m/m³-sec)
D_{eq} = equilibrium wave energy dissipation per unit volume of water (N-m/m³-sec)
ε = slope-related sand transport rate coefficient (m²/sec)
h = still water depth (m)

The subscripts b, p, r, and z refer to the related quantities at break point, plunge point, run-up limit and landward limit, of the surf zone. The values of the spatial decay coefficients are given as

$$\lambda_1 = 0.4 \left(\frac{d_{50}}{H_b} \right)^{0.47} \tag{11}$$

$$\lambda_2 = 0.2\lambda_1 \tag{12}$$

in which d_{50} is the median grain size (mm) and H_b is the breaking wave height (m). Implementation of the SBEACH model is well described in the reports from the Coastal Engineering Research Center listed in the references (Larson and Kraus, 1989, 1990, 1991).

The use of both EDUNE and SBEACH is for determination of short-term, event-driven erosion of the nourished beach profile. These models serve to define the amount of protection provided by the beach nourishment project to the design storm levels. The results of the erosion response determined by the dune erosion model is often used in defining the benefits to be assigned to the project from the reduction in flood damages due to storms. Determination of the alongshore and offshore loss of material due to long-term shoreline evolution is best defined with a shoreline evolution model such as GENESIS, as described below.

Volume of Fill

To determine the amount of fill required in a unit width of beach cross-section, consider the typical section shown in Figure 10.9. Assuming that the fill material is identical to the native material, the total volume of fill required is simply given as

$$V = (B + h^*)Y \tag{13}$$

where

B = design berm height (m)
h^* = depth of closure (m)
Y = desired seaward translation of the berm (m)

The material added to the construction profile is equivalent to the material in the submerged profile seaward of the toe of the construction profile. In general, if there is a difference between the native and the fill materials, then adjustments must be made using the techniques presented in the previous sections.

Ideally, some amount of the fill material placed below the additional construction profile could be of lesser-quality material than the native material if there is a definite commitment to renourish the beach before the underlayer material becomes exposed. This would mean a layer of finer material being placed before the beach-quality material is

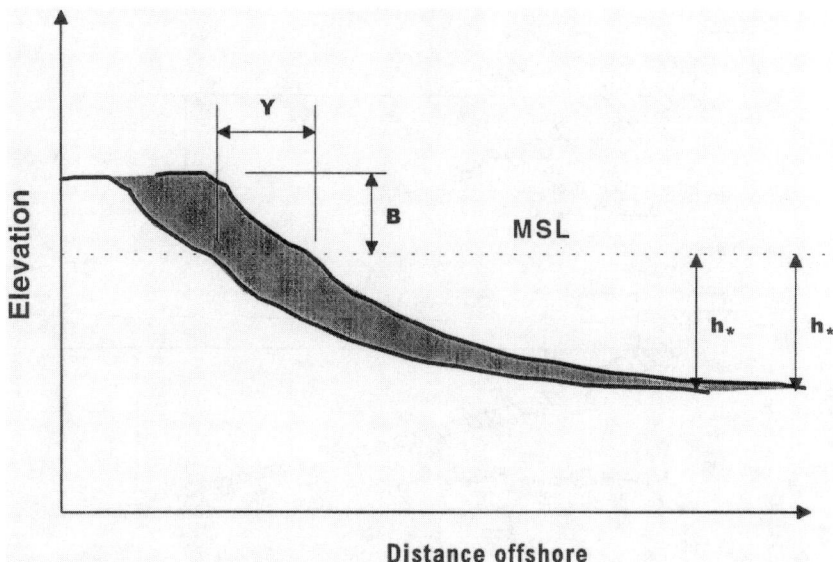

FIGURE 10.9. Schematic of the idealized nourishment profile with a seaward translation of Y to the berm height, B, and the depth of closure, h.

placed, similar to the successful nourishment project at the beach at Corpus Christi Bay, Texas (Kieslich and Brunt, 1989). However, this is infrequently done and is not recommended except for very special cases.

An important use of the equilibrium beach profile concept is developing a more realistic estimate of the volume of fill required. Dean (1991) showed that the volume calculations depend upon three different types of possible profiles:

Profile classification	Sediment characteristic	Profile characteristic
Intersecting	$A_F > A_N$	Nourished profile intersects landward of depth of closure
Nonintersecting	$A_F \leq A_N$	Nourished profile does not intersect before depth of closure
Submerged nonintersecting	$A_F < A_N$	Nourished profile does not intersect native profile

An illustration of each of these three cases is given in Figure 10.10. In this figure it is seen that for the same volume of material, the width of the berm continues to shrink with decreasing values of A_F and the position of the toe of the fill moves seaward. More specifically, Dean shows that the criteria for intersecting and nonintersecting profiles are given by:

Intersecting:

$$Y\left(\frac{A_N}{h^*}\right)^{3/2} + \left(\frac{A_N}{A_F}\right)^{3/2} < 1 \tag{14}$$

10.18 HANDBOOK OF COASTAL ENGINEERING

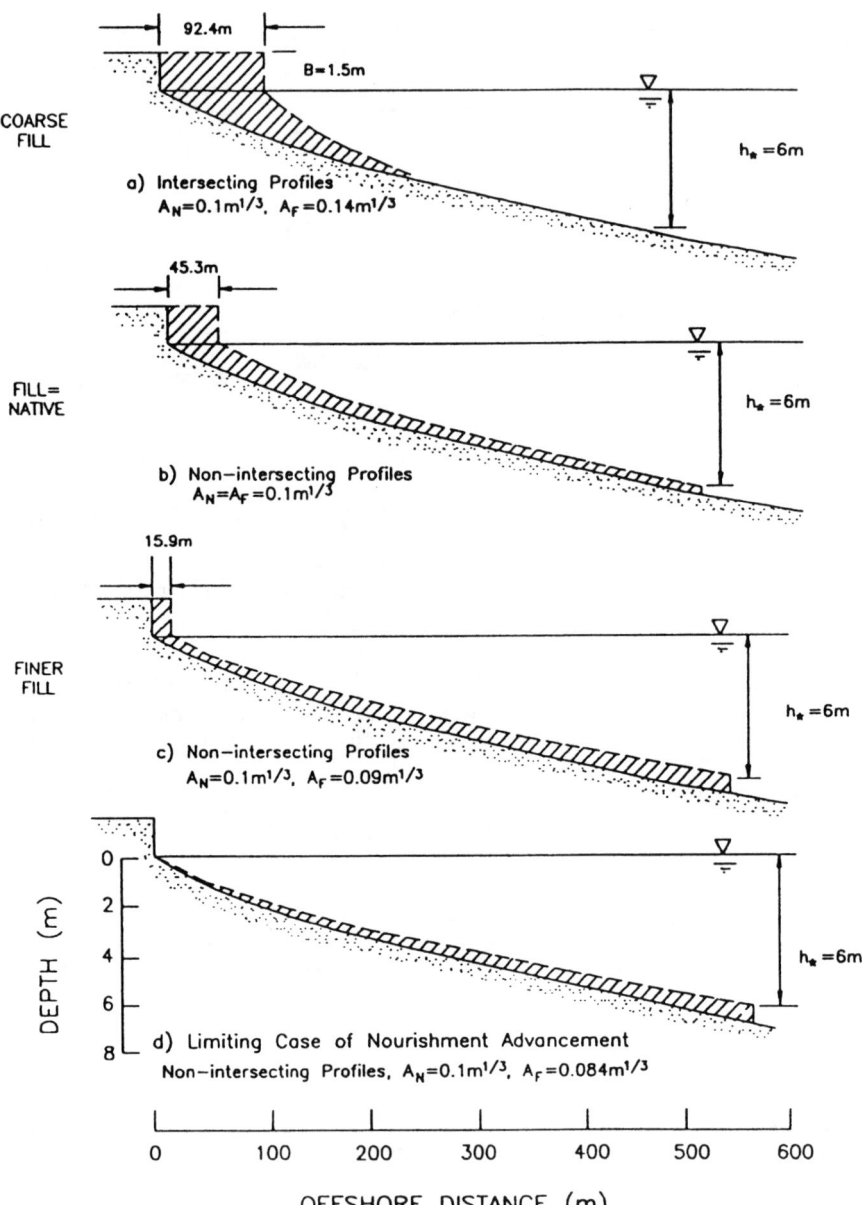

FIGURE 10.10. Response of the nourished beach profile to the ratio A_N/A_F. (After Dean, 1991.)

Nonintersecting:

$$Y\left(\frac{A_N}{h^*}\right)^{3/2} + \left(\frac{A_N}{A_F}\right)^{3/2} > 1 \tag{15}$$

The calculation of the volume of material required for a nonintersecting nourished profile for the condition of a subaerial beach with a movement of Y seaward and berm height B is given by Houston (1996), based upon Dean (1991), as

$$V = YB + \frac{3}{5}h^{*5/2}\left[A_N\left(\frac{Y}{h^{*3/2}} + A_F^{-3/2}\right)^{5/3} - A_F^{-3/2}\right] \tag{16}$$

The corresponding volume for the condition of an intersecting nourished profile is given by

$$V = YB + \frac{3}{5}A_N Y^{5/3}\left[1 - \left(\frac{A_N}{A_F}\right)^{3/2}\right]^{-2/3} \tag{17}$$

This method does not provide an allowance for the variation in distribution of the borrow and native materials. For example, no consideration is given to the loss of fines from the nourished beach if there are substantial amounts in the material that is placed on the beachface. These fines will be quickly washed away if the wave energy is sufficient. In this case it is appropriate to apply the beach-fill adjustment factor, R_A, as given in Figure 10.1, to modify the amounts given above.

In the case where the fill material is finer than the existing material, the following relationship can be used to determine the minimum amount of sand with A_F that must be placed before the sand will emerge above the waterline:

$$V = \frac{3}{5}h^{*5/2}(A_F)^{-3/2}\left(\frac{A_N}{A_F} - 1\right) \tag{18}$$

If less than this amount is placed, all of the fill material will lie below the existing waterline. An additional limitation of the approach is the limitation to a single value of A_N. Methods to include a variation of grain size across the beach profile are available but are more cumbersome.

Planform Models

The second set of models that are available to assist with the design provide tools for evaluating the alongshore transport and the planform performance of the beach fill. The most general of the widely used numerical models is GENESIS. An application of this model is well described by Hanson and Kraus (1989, 1991, and 1993). This model allows the incorporation of hard coastal structures into the analysis and is able to accept the offshore wave conditions directly from the WIS (wave information study) to provide realistic conditions with a high degree of definition of the local conditions. The user's manual is by Gravens and Kraus (1991); it is the second of a two-volume series on the GENESIS shoreline evolution model. The evolution model discussed here is a one-line model (e.g., mean high tide), the simplest form; N-line, or multiple line models will, when they become available, offer a much better treatment of the three-dimensional processes (Dean, 1998).

The GENESIS model is based upon the conservation of sand volume. It simulates shoreline change produced by spatial and temporal differences in longshore transport. The model is intended to provide the user with a realistic expectation of the evolution of the shoreline due to specific storms, wave climate, riverine influence, nourishment, and various types of structures. The model does not include any cross-shore variation—assuming that the beach is and remains in equilibrium. A definition of the conservation system is given is Figure 10.11. From the figure, the following governing equation for the rate of change of shoreline position can be obtained (Hanson and Kraus, 1989).[1]

$$\frac{\partial y}{\partial t} + \frac{1}{(B + h^*)}\left(\frac{\partial Q}{\partial x} - q\right) = 0 \qquad (19)$$

where

y = distance measured positive offshore (m)
x = distance measured alongshore (m)
Q = alongshore transport (m³/sec), and
q = source or sink of sand (m³/m/sec)

In GENESIS the longshore transport rate is determined from

$$Q = (H^2 C_g)_b \left(a_1 \sin 2\alpha_b - a_2 \cos \alpha_b \frac{\partial H}{\partial x}\right)_b \qquad (20)$$

where

α_b = angle of breaking waves at the local shoreline
C_g = wave group speed (linear wave theory)

The subscript b represents conditions at breaking. The nondimensional parameters a_1 and a_2 are given by

$$a_1 = \frac{K_1}{16\left(\frac{\rho_s}{\rho} - 1\right)(1 - p)(1.416)^{5/2}} \qquad (21)$$

$$a_2 = \frac{K_2}{8\left(\frac{\rho_s}{\rho} - 1\right)(1 - p)\tan \beta (1.416)^{7/2}} \qquad (22)$$

where

K_1, K_2 = empirical coefficients used for calibration
ρ_s = density of sand (kg/m³)
ρ = density of water (kg/m³)
p = porosity of sand
$\tan \beta$ = bottom slope from the shoreline to the depth of active longshore transport

The value of K_1 is recommended as 0.77 in the *Shore Protection Manual* (1984) and the value of K_2 is determined from testing with historical shoreline data. Some reports have indicated better results with values of K_1 less than 0.77. Some of the author's experience has been with values of 0.4 to 0.6, to provide correlation with historical shoreline data. GENESIS is able to represent groins, detached and attached breakwaters, and terminal groins. The boundary conditions are able to accommodate a known shoreline position or a known flux of sediment.

Dean and Yoo (1992 and 1994) provide an analytical method and a simplified numerical method to treat the planform performance of the beach fill. They show that the linearized form of the governing equations reduces to the general diffusion equation

---

[1] Note that the distance measured offshore is y instead of the value of x used previously.

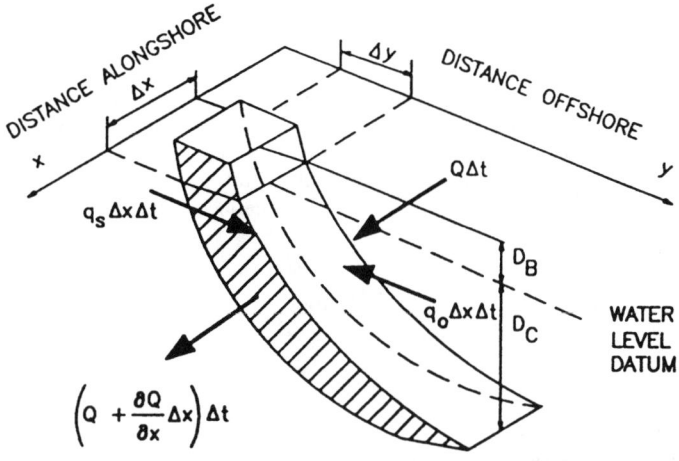

a. Cross-section view

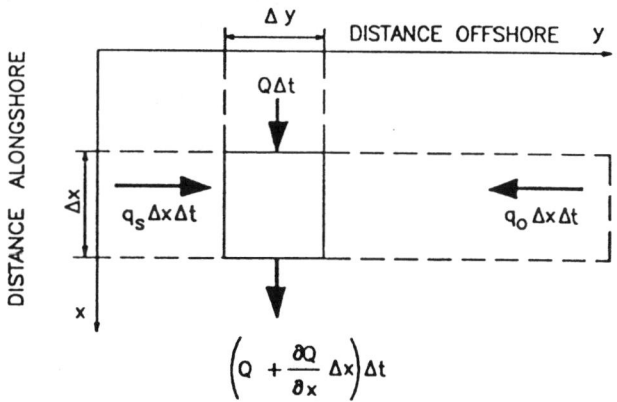

b. Plan view

FIGURE 10.11. Definition of parameters upon which the GENESIS model is developed. (After Hanson and Kraus, 1989.)

$$\frac{\partial y}{\partial t} = G \frac{\partial^2 y}{\partial x^2} \qquad (23)$$

where y is the distance offshore and x is the distance alongshore and G is the longshore diffusivity and is related to wave and sediment conditions. This immediately points to the upcoast and downcoast benefits (Dean, 1988) that will be effected by a project and it helps to define the length of time that the material will remain in the project area.

In Equation (23), the longshore diffusivity can be defined as (Dean, 1998):

$$G = \frac{KH_b^{5/2}\sqrt{\frac{g}{\kappa}}}{8(s-1)(1-p)(h^*+B)} \qquad (24)$$

where

K = sediment transport coefficient
H_b = representative breaking wave height
g = gravity
κ = ratio of breaking wave height to depth
s = specific gravity of sediment
p = sediment porosity

The value of K is often taken as 0.77, as noted in Equation 21, and is frequently used as a calibration to field observations; κ is often taken as 0.78. The solution to Equation (23) is relatively easy to obtain for most initial and boundary conditions. For an initially rectangular beach fill of length ℓ, extending offshore a distance Y, the solution is

$$y(x,t) = \frac{Y}{2}\left\{\text{erf}\left[\frac{\ell}{4\sqrt{Gt}}\left(\frac{2x}{\ell}+1\right)\right] - \text{erf}\left[\frac{\ell}{4\sqrt{Gt}}\left(\frac{2x}{\ell}-1\right)\right]\right\} \qquad (25)$$

Consider the following conditions as an example:

ℓ = 10 km
H_b = 1.5 m
B = 1 m
h^* = 6 m
s = 2.56
p = 0.35
g = 9.81 m/s^2

The solution for five subsequent periods of time—1 month, and 1, 3, 5, and 10 years—is given in Figure 10.12. The solution as presented does not include any background erosion; it only gives the change from the initial condition to which a background erosion rate would need to be added if it existed. Moreover, it is clear that the longer the project, ℓ, the slower the loss of material will be as a result of losses alongshore. Equation (23) can also be solved with $G(\chi)$, which could arise from variations in breaking wave height alongshore, sediment characteristics, or berm height. Much can be done with Equation (23) to analyze the characteristics of a beach fill, as suggested by Dean (1983, 1998). For example, the amount of material remaining in the reach of length ℓ at any given time will be

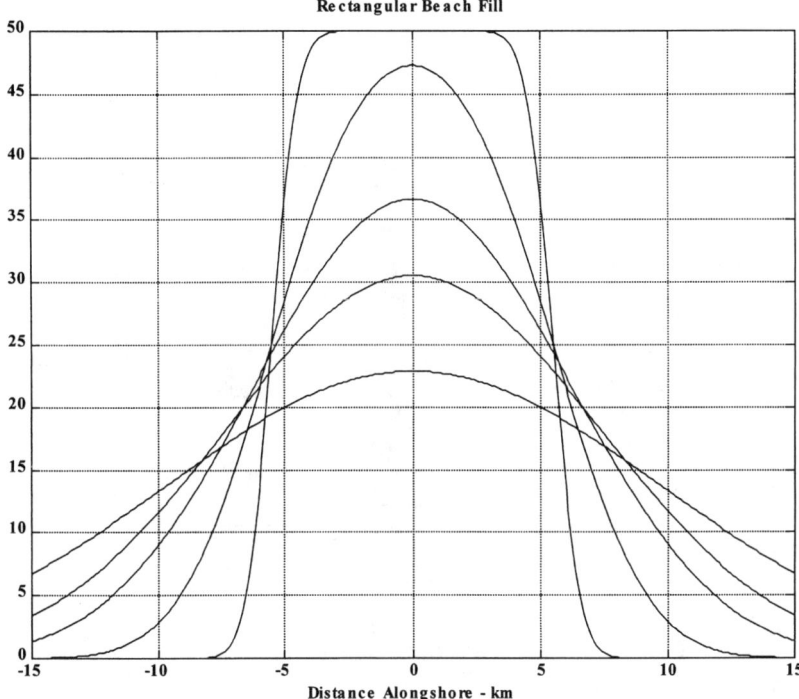

FIGURE 10.12. Plan view of a 10-km rectangular nourishment at intervals of 1 month and 1, 3, 5, and 10 years.

$$M(t) = \frac{\sqrt{4Gt}}{\ell\sqrt{\pi}} \exp\left[-\left(\frac{\ell}{\sqrt{4Gt}}\right)^2\right] + \text{erf}\left(\frac{\ell}{\sqrt{4Gt}}\right) \qquad (26)$$

This basically is a two-dimensional result of the plan area remaining of the original fill. Recognizing the assumption of an equilibrium beach profile, the result in equation (26) is equally valid in determining the volume of material remaining in the limits of the project.

Beach Nourishment with Coastal Structures

Ebersole et al. (1996) show the advantages of incorporating coastal structures into a shore protection project where the primary method is beach nourishment. At Folly Beach, South Carolina, there are several areas of higher erosion rates. Traditionally these areas have been "protected" by groins. The design of the project considered the higher longshore transport rates in these localized areas and determined that the longevity of the nourishment project would be enhanced by the reconstruction of groins in these high erosion areas. The plan-view of Folly Beach (Figure 10.13), shows the location of the existing groin field before the nourishment project and the location where the groins where reconstructed. The selection of the groins for reconstruction was made through the application of the GENESIS model. The GENESIS model provided the shoreline evolution in the critical areas over the project

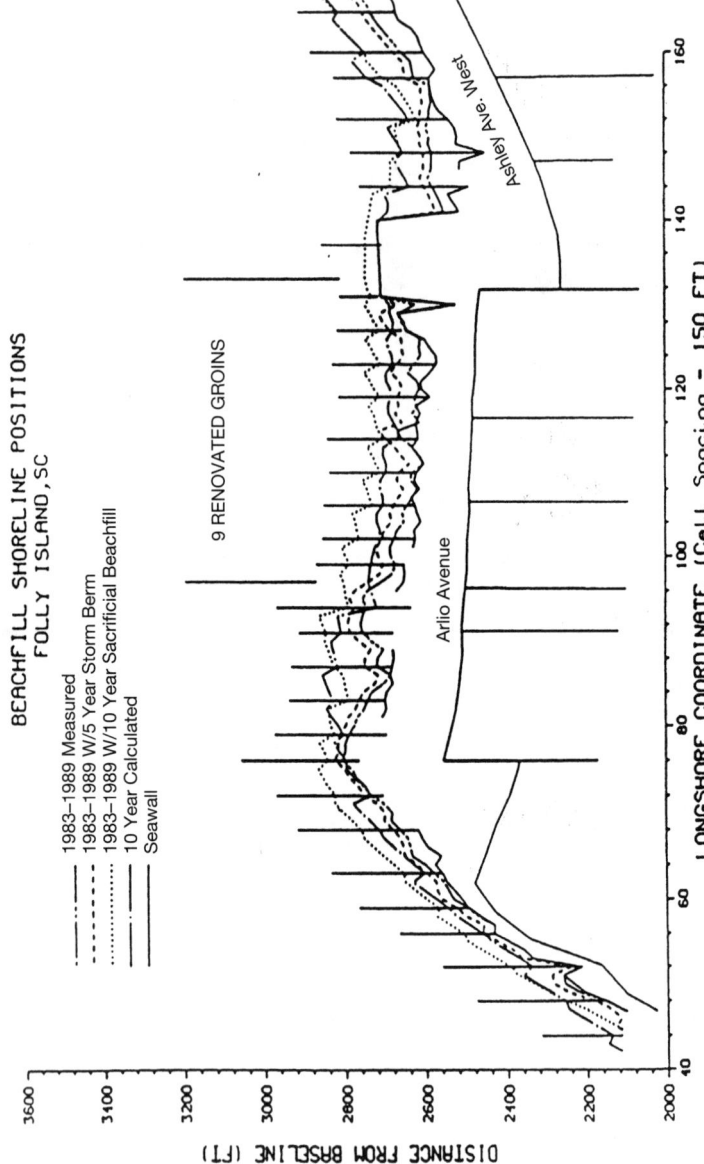

FIGURE 10.13. Example of GENESIS model showing the location of groin field at Folly Beach, South Carolina, and the groins used in the final construction

life. The SBEACH model was used to determine the amount of flooding that would result at particular times based upon the recession of the shoreline as shown by GENESIS.

Additional considerations for incorporating structures with a nourishment project are given by Tondello et al. (1998). They used a physical model to evaluate alternative structure–nourishment solutions at three sites along the Italian coastline. Their alternatives considered both groins and submerged breakwaters.

Requirements for Renourishment

The frequency and amount of material required has been addressed by Dette et al. (1994). They developed a two-dimensional analytical model to investigate the rate and volume of renourishment that would optimize the costs of renourishment while maintaining the design requirements of the project. Their results showed that a smaller amount of sand would be required over the life of the project if a more frequent application of renourishment were followed. They further showed that, based upon field results, the initial erosion rate of the newly placed material approaches the background rate of the region as the nourishment project matures.

A model for determining the optimal nourishment cycles was developed by van Noortwijk and Peerbolte (1995). They developed a decision model using the discrete renewal theorem to individual reach renourishment decisions whose expected costs are minimal with respect to the available information, including the probability distribution of the limiting average rate on ongoing erosion. Their results show that the method of discounted costs and equivalent average costs, when applied over an unbounded time span, are appropriate to obtain optimal sand nourishment decisions.

Monitoring and Maintenance

Monitoring of beach nourishment projects is essential to adequately determine the performance of the engineered project. Monitoring allows comparison of the actual performance of the beach fill to the predicted performance and provides the basis for future modifications and refinement of future renourishment of the project. It is common to include both physical and biological based monitoring of the nourished project and the borrow site. The monitoring should begin immediately after construction is complete and it should be a supplement to and a continuation of the surveys taken before construction begins. In the beginning of the postproject monitoring, extensive surveys—often quarterly—are conducted to see the rapid changes following installation. As the project demonstrates a performance similar to that predicted, the rate of monitoring can be scaled back to semiannual surveys.

Routine monitoring in the initial stages of the project will help to quantitatively define "hot spots" that may occur. These are areas where the newly placed sand erodes at a much higher rate than the average measured rate for the project. Hot spots are a frequent concern in nourishment projects and likely hot spots can often be identified in the planning and design phase. They are often associated with offshore conditions, including the submerged profile if it is irregularly placed, causing a focusing of wave energy through refraction. The response to hot spots that occur is to add more sand to the site either from the original source or by means of relocating the sand from other more stable areas of the project. The origin of and response to hot spots remains an interesting topic deserving of more study.

Typically the physical monitoring will consist of beach profiles out to the depth of closure, sediment samples, and aerial and shoreline photography of the nourished area and areas both updrift and downdrift. Additionally there should be information available on water levels and wave conditions. Wise (1995) gives a detailed account of the pre- and postproject monitoring plan for nourishment projects. Biological monitoring is frequently related to species identification and quantification at the borrow site (for biological recovery) and at the nourished site (for biological adaptation to the new environment). One of the common concerns on the Gulf and Atlantic coasts is the turtles that nest on the beach. Other endangered species, such as the piping plover, are frequently a concern because their desired habitat conflicts with the location of possible borrow sites or even portions of the project site. Detailed information on biological monitoring is given by Stauble (1991).

Maintenance of the beach nourishment project takes two forms. The first and most obvious is the renourishment when the volume of sand in the beach profile or the location of the MHW (mean high water) line reaches a predetermined value. Since the financing, planning, design, and construction may take considerable lead time, this should be considered when the predetermined values for residual volume or position of the MHW is selected. The second type of maintenance is associated with day-to-day care of the nourished project, which may consist of replanting vegetation, maintenance of sand fences, repairing and replacing signage, maintaining dune walkovers, and repair of any coastal structures associated with the project, such as groins or buried outfalls. Moreover, it is important to avoid other construction and activities that may have a deleterious effect on the associated vegetation or the new beach sand.

Hot Spots

These are areas of a nourishment project that lose volume and beach width much more rapidly than adjacent areas. Occasionally, these areas exhibit a problem before the project is complete and the contractor can solve the problem by adding more sediment to the localized area. When the accelerated loss does not occur until after the project is built and the contractor is no longer on site, the remedy becomes much more costly. The causes of hot spots is not always clearly obvious. It is clear that analysis of historical shoreline positions will lead to a better understanding of where they might be expected.

Hamilton et al. (1996) demonstrate the use of numerical methods to predict hot spots at Jupiter Island, Florida. At Jupiter Island, the hot spot erosion rates have been measured between 10 and 30 ft per year. The analysis shows that these erosion areas are largely influenced by wave energy focused by offshore natural ridges and rock outcrops as well as nearshore dredge holes. These features result in a significant spatial variation in longshore transport. At this site, the proposed solution was to construct headlands in combination with overnourishment in areas predicted to have hot spots.

A detailed analysis by Stauble (1991) uses the nourishment efforts at Ocean City, Maryland, to identify the location and effects of hot spots on beach fill performance. Stauble suggests that offsetting the effects of hot spots requires a thorough understanding of the three-dimensional nature of postnourishment redistribution of material under the prevailing coastal processes. Clearly, prediction is preferable to postproject response or correcting deficiencies as they occur.

In the following example, a hot spot occurs in a 10-km beach fill and its effect of creating erosion is essentially instantaneous. This is demonstrated in the one-line evolution model suggested for this problem by Dalrymple (1999). For this example, the following parameters were selected and used with Equation (25):

$Y = 50$ m
$p = 37\%$
$S = 2.65$
$K = 0.77$
$\kappa = 0.78$

$h^* = 6$ m

$B = 1$ m

$H_b = 1.5$ m and 2.0 m in the center of the project for a distance of 1000 m

As shown, within a month a significant "erosional" problem exists with the new nourishment in the region of the hot spot (Figure 10.14). Immediately adjacent to the hot spot, the effect is a temporary period of accretion. However, after a period of a year, the effect of the hot spot is to induce erosion along a much larger area of the planform. Moreover, it seems that the long-term effect is to cause an even greater diffusivity along the entire beach. Thus a positive loss from the nourished beach would occur due to a hot spot in the center of the project covering only 10 % of the project length. It is easy to expect that similar changes in porosity, specific gravity (such as from a different nourishment source, including much shell) would generate such a modified performance of the shoreline.

BENEFITS OF BEACH NOURISHMENT

There are currently several initiatives originating from the public and governmental sectors to promote beach reconstruction and continued nourishment. It is clear that the na-

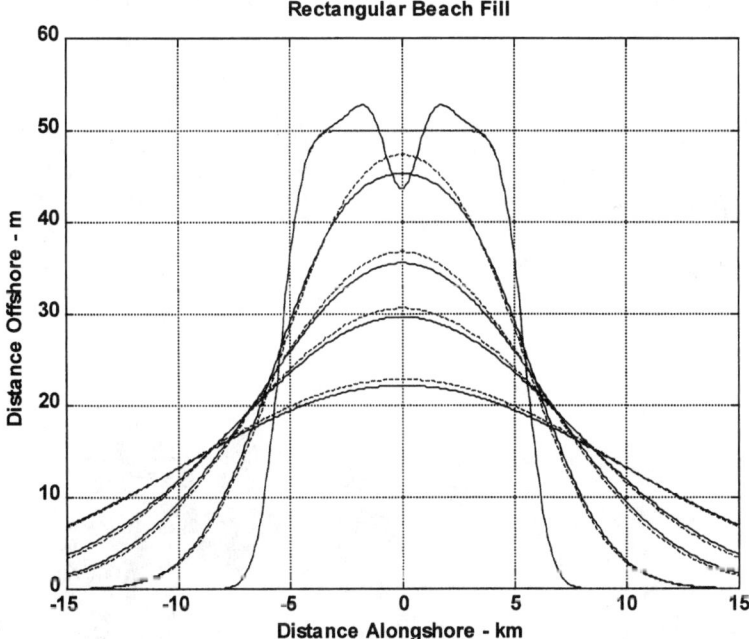

FIGURE 10.14. Plan view of beach with a 10-km nourishment project in which a "hot spot" occurs from a larger wave condition. The dashed line represents a constant wave condition and the solid line represents the nourishment with the hot spot. The line pairs represent 1 month, and 1, 3, 5, and 10 years.

tional beach restoration and maintenance programs in Spain and Germany are very successful. Although beach nourishment projects have been undertaken for more than 44 years, the subject has seen much debate in the United States in the last 15 years. Much of the debate is based upon qualitative considerations rather than sound facts. Foremost in the issue is the need for recreational beaches in the coastal United States. Travel and tourism is the single largest industry in the United States and in the world, with worldwide revenues estimated at $2.9 trillion according to Houston (1995). This represents a very large expenditure, considering that the United States is the only country in the world with an annual gross national product exceeding $2.9 trillion. In the United States, travel and tourism is the largest employer, with more than 11,000,000 people, compared with only 18,000,000 workers in all of manufacturing (Houston, 1995).

Beach nourishment projects (Figure 10.15) are not just for the benefit of the local residents; beaches serve the needs of visitors from throughout the home state, the region, other parts of the nation and, frequently, other countries. The economic impact of beach usage and the enhancement of property values is analyzed by Stronge (1991 and 1995). Beach nourishment projects are not unlike most other public works projects in that they are constructed for the public benefit. However, each project must satisfy a specific set of criteria to determine whether it meets the requirements of the governmental unit funding the project. One of the most common forms of analysis uses the benefit versus cost approach. The set of procedures used in the benefit–cost method will depend upon the governmental unit involved. Those used in the Federal government must follow the specific criteria in the USACE (1991) *National Economic Development Procedures Manual for Coastal Storm Damage and Erosion and the Economic and Environmental Principles* and *Guidelines for Water and Related Land Resources Implementation Studies* (Water Resources Council, 1983).

FIGURE 10.15. Construction of a berm and discharge from the pipeline at Folly Beach, South Carolina.

In determining the benefits of a beach nourishment project it would be appropriate to include those benefits identified in Table 10.3. As can be seen, these benefits are not all easy to identify much less to quantify. The costs of a project may include those listed in Table 10.2.

The costs, like the benefits, are very difficult to accurately quantify in every instance, especially for new projects for which there are only modeled effects of the project without local ground truth. Dean (1988) clearly demonstrates the downstream benefits (outside of the project boundaries) that always occur for storm damage reduction and recreation.

SUMMARY

The opponents of spending additional moneys on enhancing and restoring recreational beaches in the United States cite the ineffectiveness of the works as their major point of contention. See, for example, Dixon and Pilkey (1991), Leonard, Clayton and Pilkey (1990), Leonard, Dixon and Pilkey (1990), and Pilkey and Clayton (1989). They point out that design procedures used in the projects they cite are generally lacking and are ineffective for the proposed solutions. Notwithstanding environmental improvements, there has been an evolution of beach fill design so that it is now be based more on science than the empirical methods used in the past. There are cross-shore and planform evolution models that give consistently reasonable results and which should be employed in the design of beach fills.

By appropriate application of the equilibrium beach theory, the previously employed empirical overfill factor can be replaced and the question then becomes one of defining the proper depth of closure (Hanson and Lillycrop, 1988) and defining sand sources that will optimize the performance of the nourishment. A combination of initial dredging costs and renourishment costs is also required. Through proper design, not only will the projects be more cost effective but the resulting nourished beach will receive less opposition and criticism than is now the case. This in turn will fuel interest in pursuing more beach restoration projects as the overall costs decrease and project performance improves through better planning and design.

The damage reduction potential resulting from beach nourishment projects should be recognized by the insurance industry, thus adding stimulus for beach nourishment projects that may not otherwise meet the benefits–cost requirements. With the current initia-

TABLE 10.3. Benefits and costs of beach nourishment projects

Benefits	Costs
Increase in property values	Construction costs
Storm damage reduction	Lost opportunity costs
Increase in willingness to pay by users	Increase in infrastructure demand and consequent costs
Increase in esthetic value	Ecological and downstream impacts
Increase in commercial profits	Negative effects on quality of life
Regional and downstream benefits	
Ecological benefits	
Positive effects on quality of life	
Creation of employment opportunities	
Increase of foreign income	

tives, it is very likely that the tourism industry will be successful in obtaining a larger commitment to maintaining the nation's beaches, which will surely call upon the resources of the dredging industry. Moreover, governments in coastal areas must rally to support the ongoing initiatives for continuing beach restoration and nourishment.

REFERENCES

Bruun, P. 1962. Sea Level Rise as a Cause of Erosion. *Journal of the Waterways and Harbors Division*. ASCE, 88:117–133.
Bruun, P. 1996. Discussion of Interdependence of Beach Fill Volumes and Repetition Intervals. *Journal of Waterway, Port, Coastal and Ocean Engineering*, ASCE, 122:103–104.
Campbell, T. J., Dean, R. G., Mehta, A. J., and Wang, H. 1990. Short Course on Principles and Applications of Beach Nourishment. Organized by Florida Shore and Beach Preservation Association and Coastal and Oceanographic Engineering Department, University of Florida.
Dalrymple, R. A. 1999. Personal Communication.
Dean, R. G. 1983. Principles of Beach Nourishment. In *Handbook of Coastal Processes and Erosion*. Boca Raton, Florida: CRC Press, pp. 217–232.
Dean, R. G. 1988. Realistic Economic Benefits from Beach Nourishment. *Proceedings 21st International Coastal Engineering Conference*, ASCE, pp. 1558–1572.
Dean, R. G. 1991. Equilibrium Beach Profiles: Characteristics and Applications. *Journal of Coastal Research* 7(1):53–84.
Dean, R. G. and Yoo, C-H. 1992 Beach Nourishment Performance Predictions. *Journal of Waterway, Port, Coastal and Ocean Engineering*, ASCE. 118(6):557–586.
Dean, R. G. and Yoo, C-H. 1994. Beach Nourishment in the Presence of Seawalls. *Journal of Waterway, Port, Coastal and Ocean Engineering*, ASCE, 120(3):302–316.
Dean, R. G. 1998. Beach Nourishment: A Limited Review and Some Recent Results. *Proceedings of the 27th International Conference on Coastal Engineering*, Copenhagen, ASCE.
Dette, H-H., Fuehrboeter, and Raudkivi, A. J. 1994. Interdependence of Beach Fill Volumes and Repetition Intervals. *Journal of Waterway, Port, Coastal and Ocean Engineering*, ASCE, 118:580–593.
Dixon, K. and Pilkey, O. H. 1991. Summary of Beach Replenishment on the U.S. Gulf of Mexico Shoreline. *Journal of Coastal Research*, 7:249–256.
Ebersole, B. A., Neilans, P. J. and Dowd, M. W. 1996. Beach-Fill Performance at Folly Beach, South Carolina (1 Year After Construction) and Evaluation of Design Methods. *Journal of the American Shore and Beach Preservation Association*, 64:11–26.
Edge, B. L., Johnson, P., and Granger, C. W. 1994. Beach Nourishment at Folly Beach, SC. *Proceedings 15th Annual Conference of the Western Dredging Association and 27th Annual Texas A&M Dredging Seminar*, San Diego, CA.
Edge, B. L., Magoon, O. T., Converse, H. and Tobin, L. T. 1995. Recreation Values of Urban Beaches. *Abstracts, Coastal Zone '95*, Tampa, Florida, ASCE.
Gomez-Pina, G. and Ramirez, J. L. 1994. The Complementary Interaction Between Beach Nourishment and Harbour Management: Four Cases in Spain. *Proceedings of the 23rd International Conference on Coastal Engineering*, Kobe, ASCE, 3:3507–3521.
Gravens, M. B. and Kraus, N. C. 1991. GENESIS: Generalized Model for Simulating Shoreline Change. Report 2: Workbook and System User's Manual. Technical Report No. CERC-89-19. Vicksburg, MS: Coastal Engineering Research Center, U.S. Army Engineer Waterways Experiment Station.
Hallermeier, R. J. 1978. Uses for a Calculated Limit Depth to Beach Erosion. *Proceedings of the Sixteenth International Conference on Coastal Engineering*. New York, NY: ASCE, 1493–1512.
Hallermeier, R. J. 1981. A Profile Zonation for Seasonal Sand Beaches from Wave Climate. *Coastal Engineering*, 4:253–277.
Hamilton, R. P., Ramsey, J. S., and Aubrey, D. G. 1996. Numerical Predictions of Erosional "Hot-Spots" at Jupiter Island, Florida. *Proceedings of the 9th National Conference on Beach Preservation Technology*. St. Petersburg, FL. Florida Shore and Beach Preservation Association.

Hands, E. B. 1991. Unprecedented Migration of a Submerged Mound off the Alabama Coast. *Proceedings 12th Annual Conference of the Western Dredging Association and 24th Annual Texas A&M Dredging Seminar.* Las Vegas.
Hanson, H. and Kraus, N. C. 1989. GENESIS: Generalized Model for Simulating Shoreline Change. Report 1: Reference Manual and Users Guide. Technical Report No. CERC-89-19. Vicksburg, MS: Coastal Engineering Research Center, U.S. Army Engineer Waterways Experiment Station.
Hanson, H. and Kraus, N.C. 1991. Numerical Simulation of Shoreline Change at Lorain, Ohio. *Journal of Waterway, Port, Coastal and Ocean Engineering*, ASCE, *117:*1–18.
Hanson, H. and Kraus, N. C. 1993. Optimization of Beach Fill Transitions. In D. K. Stauble and N.C. Kraus, eds., *Volume on Beach Nourishment Engineering and Management Considerations, Proceedings of Coastal Zone '93*. New York, NY: ASCE, pp. 103–117.
Hanson, M. and Lillycrop, W. J. 1988. Evaluation of Closure Depth and Its Role in Estimating Beach Fill Volumes. *Proceedings Beach Preservation Technology '88*. Florida Shore and Beach Preservation Association, Tallahasse, FL., pp. 107–114.
Herbich, J. B. 1992. *Handbook of Dredging Engineering,* New York, McGraw-Hill.
Houston, J. R. 1995. Beach Nourishment, *Shore and Beach.* 63:21–24.
Houston, J. R. 1996. Simplified Dean's Method for Beach-Fill Design. *Journal of Waterway, Port, Coastal and Ocean Engineering*, ASCE, *122:*143–146.
Houston Chronicle. Spain Tops Uncle Sam as No. 2 Destination. Feb. 11.
James, W. R. 1975. Techniques in Evaluating Suitability of Borrow Material for Beach Nourishment. Technical Manual No. 60. Ft. Belvoir, VA: U.S. Army Coastal Engineering Research Center.
Kieslich, J. M. and Brunt, D. H. III. 1989. Assessment of a Two-Layer Beach Fill at Corpus Christi Beach, Texas. *Proceedings, 6th Symposium on Coastal and Ocean Management, Coastal Zone '89.* New York, ASCE, pp. 3975–3984.
Kraus, N. C. and Wise, R. A. 1993. Simulation of January 4, 1992 Storm Erosion at Ocean City, Maryland. *Shore and Beach, 61*(1):34–41.
Kriebel, D. L. and Dean, R. G. 1985. Numerical Simulation of Time Dependent Beach and Dune Erosion. *Coastal Engineering, 9:*221–245.
Kroon, A., Hoekstra, P., Houwman, K., and Ruessink, G. 1994. Morphological Monitoring of a Shoreface Nourishment NOURTEC Experiment at Terschelling, The Netherlands. *Proceedings of the 24th International Conference on Coastal Engineering,* Kobe, ASCE, *2:*2222–2236.
Krumbein, W. C. and James, W. R. 1965. A Lognormal Size Distribution Model for Estimating Stability of Beach Fill Material. Technical Memorandum No. 16. Washington, DC: U.S. Army Coastal Engineering Research Center.
Larson, M. and Kraus, N. C. 1989. SBEACH: Numerical Model for Simulating Storm-Induced Beach Change. Report 1: Empirical Foundation and Model Development. Technical Report No. CERC-89-9. Coastal Engineering Research Center, U.S. Army Engineer Waterways Experiment Station, Vicksburg, MS.
Larson, M. and Kraus, N. C. 1990. SBEACH: Numerical Model for Simulating Storm-Induced Beach Change. Report 2, Numerical Foundation and Model Tests. Technical Report CERC-89-9. Vicksburg, MS: Coastal Engineering Research Center, U.S. Army Waterways Experiment Station.
Larson, M. and Kraus, N. C. 1991. Mathematical Modeling of the Fate of Beach Fill. *Coastal Engineering, 16:*83–114.
Larson, M. and Kraus, N. C. 1994. Prediction of Cross-Shore Sediment Transport at Different Spatial and Temporal Scales. *Marine Geology, 126:*111–127.
Laustrup, C. and Holger T. M. 1994. Design of Breakwaters and Beach Nourishment. *Proceedings of the 24th International Conference on Coastal Engineering,* Kobe, ASCE, *2:*1359–1372.
Lee, C-E., Kim, M-H., and Edge, B. L. 1996. Numerical Model for On-Offshore Sediment Transport with Moving Boundaries. *Journal of Waterway, Port, Coastal and Ocean Engineering*, ASCE, *122:*84–92.
Lekic, S. 1995. U.S. Not Tapping Full Tourism Potential. *Houston Chronicle.* July 31, p. 1B.
Leonard, L. A., Clayton, T. D. and Pilkey, O. H. 1990. An Analysis of Replenished Beach Design Parameters on U.S. East Coast Barrier Islands. *Journal of Coastal Research*, Special Issue No. 6:15–36.

Leonard, L. A., Dixon, K. L., and Pilkey, O. H. 1990. A Comparison of Beach Replenishment on the U.S. Atlantic, Pacific and Gulf Coasts. *Journal of Coastal Research*, Special Issue 6:127–140.
Marine Board. 1995. *Beach Nourishment and Protection.* National Research Council. Washington, DC: National Academy Press, 334 p.
McLellan, T. N. 1990. Nearshore Mound Construction Using Dredged Material. *Journal of Coastal Research*, Special Issue No. 7:99–107.
McLellan, T. N. and Kraus, N. C. 1991. Design Guidance for Nearshore Berm Construction. *Proceedings of Coastal Sediments '91.* New York, NY: ASCE, pp. 2000–2011.
Moore, B. D. 1982. Beach Profile Evolution in Response to Changes in Water Level and Wave Height, MCE Thesis, Department of Civil Engineering, University of Delaware, 162 pp.
National Research Council. 1995. *Beach Nourishment and Protection.* Marine Board, Committee on Beach Nourishment and Protection, Commission on Engineering and Technical Systems, Washington, DC: National Academy Press.
Pilkey, O. H. and Clayton, T. D. 1989. Summary of Beach Replenishment Experience on U.S. East Coast Barrier Islands. *Journal of Coastal Research*, 5(1), 147–159.
Rijkswaterstaat. 1987. *Manual on Artificial Beach Nourishment.* Report 130. Center for Civil Engineering Research, Codes and Specifications. Delft, The Netherlands: Delft Hydraulics.
Stauble, D. K. 1991. Elements of Beach Nourishment Project Monitoring, CETN II-26. Coastal Engineering Research Center, U.S. Army Engineer Waterways Experiment Station, Vicksburg, MS.
Stronge, W. B. 1991. Recreational Benefits of Barrier Island Beaches: Anna Maria, Captiva, and Marco, A Comparative Analysis. Preserving and Enhancing Our Beach Environment, *Proceeding of the 4th Annual National Conference,* Florida Shore and Beach Preservation Association, Tallahassee, FL.
Stronge, W. B. 1995. The Economics of Government Funding for Beach Nourishment Projects: The Florida Case. *Shore and Beach, 63:3*
Sudar, R. A., Pope, J., Hillyer, T., and Crumm, J. 1995. Shore Protection Projects of the U.S. Army Corps of Engineers. *Shore and Beach, 63:*3–16.
Swart, D. H. 1974. Offshore Sediment Transport and Equilibrium Beach Profiles. Publication No. 131. Delft, The Netherlands: Delft Hydraulics Laboratory.
Tondello, M., Ruol, P., Sclavo, M., and Capobianco, M. 1998. Model Tests for Evaluating Beach Nourishment Performance. *Proceedings of the 27th International Conference on Coastal Engineering,* Copenhagen, ASCE.
Trembanis, A. C. and Pilkey, O. H. 1998. Summary of Beach Nourishment Along the U.S. Gulf of Mexico Shoreline. *Journal of Coastal Research, 14*(2), 407–417.
USACE. 1984. *Shore Protection Manual* (4th Ed.) (2 Volumes). Prepared by Coastal Engineering Research Center, U.S. Army Corps of Engineers (USACE). Publication No. 008-002-00218-9. Washington, DC: U.S. Government Printing Office.
USACE. 1986. *Storm Surge Analysis.* Engineer Manual No. EM 1110-2-1412. Prepared by U.S. Army Corps of Engineers (USACE) Washington, DC: U.S. Government Printing Office.
USACE. 1989. *Water Level and Wave Heights for Coastal Engineering Design.* Engineer Manual 1110-2-1414. Prepared by U.S. Army Corps of Engineers (USACE). Washington, DC: U.S. Government Printing Office.
USACE. 1991. National Economic Development Procedures Manual for Coastal Storm Damage and Erosion. Institute of Water Resources Report No. 91-R-6. Fort Belvoir, VA.
USACE. 1994. *Design of Beach Fills.* Engineer Manual No. EM 1110-2-3301. Prepared by U.S. Army Corps of Engineers (USACE) Washington, DC: U.S. Government Printing Office.
USACE. 1995. Design Aspects of Corps Beach Nourishment Projects. Coastal Engineering Technical Note I-61. Prepared by U.S. Army Corps of Engineers (USACE). Washington, DC: U.S. Government Printing Office.
USACE. 1996. Shoreline Protection and Beach Erosion Control Study, Final Report: An Analysis of the U.S. Army Corps of Engineers Shore Protection Program. IWR Report 96-PS-1, pp. 362.
van Noortwijk, J. M. and Peerbolte, E. B. 1995. Optimal Sand Nourishment Decisions. Report No. 493. Delft, The Netherlands: Delft Hydraulics.
Water Resources Council. 1983. *Economic and Environmental Principles and Guidelines for Water and Related Land Resources Implementation Studies.* Washington, DC: U.S. Government Printing Office.

Weggel, J. R. 1995. A Primer on Monitoring Beach Nourishment Projects. *Shore and Beach, 63:*3
Wiegel, R. L. 1992. Dade County, Florida, Beach Nourishment and Hurricane Surge Protection. *Shore and Beach 60*(4):2–28.
Wiegel, R. L. 1994. Ocean Beach Nourishment on the USA Pacific Coast. *Shore and Beach 62*(1):11–36.
Wise, R. A. and Kraus, N. C. 1993. Simulation of Beach Fill Response to Multiple Storms, Ocean City, Maryland. *Proceedings of Coastal Zone '93.* New York, NY: ASCE, pp. 133–147.
Wise, R. A. 1995. Recommended Base-Level Physical Monitoring of Beach Fills. CETN II-35. Coastal Engineering Research Center, U.S. Army Engineer Waterways Experiment Station, Vicksburg, MS.
Wise, R. A., Smith, S. J., and Larson, M. 1996. SBEACH: Numerical Model for Simulating Storm-Induced Beach Change, Report 4—Cross-Shore Transport under Random Waves and Model Validation with SUPERTANK and Field Data. T.R. CERC-89-9. 140 pp.

CHAPTER 11
NAVIGATION CHANNELS— DESIGN AND OPERATION

JOOP A. ZWAMBORN
Coastal and Hydraulic Engineer
Stellenbosch, South Africa

Notation	11.2
1. Introduction	11.4
2. Principles and Definitions	11.4
2.1. Types of Harbors	11.4
2.2. Harbor and Channel Design Features	11.4
2.3. Main Parameters and Definitions	11.5
3. Design Process	11.6
3.1. Design Approach	11.6
3.2. Preliminary Desk Design	11.7
3.3. Detailed Design	11.7
4. Physical Environmental Data	11.7
4.1. Data Requirements	11.7
4.2. Bathymetry and Geotechnics	11.8
4.3. Morphological Changes and Siltation	11.8
4.4. Wave Data	11.8
4.5. Wind and Visibility Data	11.9
4.6. Tide and Water Levels	11.9
4.7. Water Temperature and Salinity	11.9
4.8. Currents	11.10
5. Design Ships	11.10
5.1. Choice of Design Ships	11.10
5.2. Type of Ships and Propulsion	11.11
5.3. Ship Dimensions	11.12
5.4. Loading Conditions	11.12
5.5. Auxiliary Craft	11.13
6. Channel Layout	11.13
6.1. Design Principles	11.13
6.2. Channel Alignment	11.15
6.3. Channel Bends	11.15
6.4. Navigational Aspects	11.18
7. Channel Length	11.18
7.1. Problem Statement	11.18
7.2. Stopping Distances	11.20
7.3. Entry Maneuver	11.23

8. Channel Width	11.23
8.1. Design Factors	11.23
8.2. Empirical Formulae	11.24
8.3. Bend Width	11.25
8.4. Physical Model and Prototype Data	11.26
8.5. Maneuvering Simulations	11.26
9. Channel Depth	11.30
9.1. Design Principles	11.30
9.2. Squat Formulae and Studies	11.32
9.3. General Data on Wave-Induced Ship Motions	11.32
9.4. Detailed Data on Wave-Induced Ship Motions	11.34
9.5. Additional Data on Wave-Induced Ship Motions	11.42
9.6. Determination of Channel Depth	11.42
10. Aids to Navigation	11.43
10.1. Definitions and Principles	11.43
10.2. Leading Lights	11.44
10.3. Additional Aids	11.45
11. Harbor Channel Operation	11.45
11.1. Operational Principles	11.45
11.2. Operational Procedures	11.47
11.3. Underkeel Allowance Criteria	11.47
11.4. Port Accessibility	11.49
12. Examples of Harbor Channel Design	11.56
12.1. Port of Richards Bay	11.56
12.2. Port of Haifa	11.61
Acknowledgments	11.67
References	11.67

NOTATION

A	wave-induced vertical ship motion amplitude (m)
A^1	ship's submerged cross-sectional area (m^2)
A_{max}	maximum overdraught (m)
$A_{max(p)}$	maximum amplitude with probability p (m)
A_s	significant vertical ship motion amplitude (m)
A_w	channel cross-sectional area (m^2)
B	beam (m)
C_B	block coefficient (–)
D	draught (midship) (m)
D_{ap}	aft perpendicular draught (m)
D_{fp}	forward perpendicular draught (m)
D_p	midship port draught (m)
D_s	midship starboard draught (m)
ΔD	sinkage due to squat (m)
d	channel depth (m)
f	function
f_p	spectral peak frequency (Hz)
g	acceleration due to gravity (m/s^2)
H_{max}	maximum wave height (m)

H_{mo}	characteristic wave height (m)
H_s	significant wave height (m)
h	heel (m)
k	channel lifetime number of bottom contacts
L	ship length (m)
L_{oa}	length overall (m)
L_{pp}	length between perpendiculars (m)
l	channel length (m)
m_0	zeroth moment of wave spectrum (m^2)
N	number of waves
p	probability of bottom contact
p_1	lifetime probability of bottom contact
R	radius of channel curvature (m)
$S(f)$	wave energy density spectrum (m^2/Hz)
$s(f \cdot \theta)$	directional wave spreading [(m^2/Hz)/°]
sh	sagging and hogging (m)
T_1	channel lifetime (years)
t	stationary trim (m)
t_1	recurrence interval (years)
T_p	spectral peak period (s)
T_{pE}	encounter wave period (s)
T_s	significant wave period (s)
T_z	zeroth-crossing wave period (s)
UKA	underkeel allowance (m or % of D)
UKC	underkeel clearance (m or % of D)
V	ship forward speed (m/s) or V_k (knots)
w	channel bottom width (m)
w_{BM}	basic maneuvering lane width (m)
w_{Bg}	starboard (green) bank clearance (m)
w_{Br}	port (red) bank clearance (m)
w_i	additional widths required (m)
w_p	passing distance (m)
w_s	maneuvering lane width (m)
Z_{max}	maximum overdraught (m)
Z_0	tide level (m)
ΔZ	channel siltation (m)
α	vector angle of wave propagation relative to ship (°)
α^1	angle between ship speed and cross-current vectors (°)
β	bend angle (°)
γ	Jonswap enhancement factor
θ	mean vector direction of wave propagation (°)
λ_p	peak period wave length (m)
ρ	density of fresh water (kg/m^3)
ρ_s	density of seawater (kg/m^3)
ρ_w	density (sea) water (t/m^3)
∇	displacement (m^3)
∇_m	mass displacement (t)
CD	chart datum (reference plane)
LSD	land survey datum level or MSL (reference plane)
$LWOST$	low water ordinary spring tide (m)
ML	mean water level (m)

1. INTRODUCTION

The design of the approach, entrance, and internal channels is an important part of the design of new ports and of extensions to existing ports. Obviously, nautical aspects must play a major role in the layout design, whereas the channel cross-sections are largely dependant on design ship characteristics and the prevailing environmental conditions.

In the following, the emphasis will be on the design of the channel alignment, length, width, and depth vis à vis the local environmental conditions, the design ship(s), and navigational requirements. In addition, optimization of harbor channel operation will be briefly discussed. The design should allow for the maximum possible degree of flexibility in view of possible changes in the expected future traffic and ship developments.

Although the design process discussed below could be applied, in principle, to any size port, the data and examples included refer mainly to major commercial ports, including bulk ports.

Sedimentation and maintenance dredging of harbor channels is covered in Chapter 12 (refer also to PIANC, 1989).

2. PRINCIPLES AND DEFINITIONS

2.1. Types of Harbors

The four main types of harbors are:

1. **River ports.** The harbor development is inland along the banks of a river and the river mouth acts as harbor entrance. The river mouth may have to be improved by dredging an entrance channel, which will be protected against siltation by a set of jetties or breakwaters. Depending on depth requirements, further channel dredging may be required inside and groins may be necessary to concentrate the river flow to maintain the dredged channel depth.
2. **Lagoon ports.** In this case, harbor basins are dredged in a shallow lagoon and/or low-lying inland area. An entrance channel from the sea must be dredged and protected with, normally, two breakwaters or jetties. Depending on the length of these breakwaters, an approach channel in the open sea may also have to be dredged.
3. **Ports in natural bays.** Ideally, port development should take place in coastal bays that provide some natural protection, thus reducing the need for expensive breakwaters. If the port area is too shallow, the harbor basins will need dredging and, depending on the available depths, approach and entrance channels will also have to be excavated.
4. **Open coastal ports.** These are ports built out into the sea along a featureless (almost straight) open coastline (with little or no natural protection). The harbor basins will probably need dredging, whereas the quay and other dry working areas will have to be reclaimed from the sea. These ports require major breakwaters to enclose and protect both the harbor basins and the quay areas. Depending on the depth requirements, an approach and/or entrance channel has to be dredged for these ports.

2.2. Harbor and Channel Design Features

The design of harbors and the associated navigational channels is, nowadays, normally preceded by a strategic environmental assessment (SEA) study, including public participation. The design involves several or all of the following design features, depending on the type of harbor.

- **Site selection.** The choice of the harbor site is influenced by many factors, including economical, political, geographical, demographical, and environmental aspects.
- **Design ships.** No doubt, the most important feature for the design process is a clear understanding of the sizes and types of ships for which the port will be designed, both initially and allowing for possible future requirements.
- **Environmental data.** Detailed data must be collected at the site, including wave, wind, ocean currents, tidal levels and currents, river flow, visibility, sedimentology, presence of contaminated sediments, and ecological data. These data must be evaluated and interpreted with respect to the planned harbor development.
- **Subbottom and seismic data.** Data on the subbottom conditions, particularly on the dredgability of the subbottom layers, and on possible seismic activities, are essential for the design process.
- **Nautical requirements.** The channel and maneuvering basins' layouts and sections are largely dependant on the design ship(s) and navigational and safety requirements.
- **Wave protection.** To satisfy accepted wave agitation and mooring criteria in the port, wave agitation studies must be done. These must reflect the effect of wave refraction due to a possible dredged approach channel and wave diffraction due to the protecting breakwaters.
- **Sedimentology.** Both breakwaters or jetties and dredged channels will cause changes in the local morphology. The effects of these structures and the possibility of channel siltation must be minimized by detailed sedimentological design studies. These must include the effects of waves, wind, tides (storm surges), currents, and river flow.
- **Channel/basin dredging.** The design process must include a phase whereby dredging versus reclamation for quay areas and/or coastal nourishment are optimized.

2.3. Main Parameters and Definitions

The main parameters related to shipping channel design and operation are:
(a) **Ship-related parameters**
- Deadweight tonnage, dwt (tons), is the collective term for the mass of everything except the mass of the ship itself, thus the total carrying capacity of the ship (cargo, fuel, crew, consumables, etc.)
- Displacement, $\nabla = C_B \cdot L_{pp} \cdot B \cdot D$ (m³) is the total volume of water displaced by the ship; the mass displacement is $\nabla_m = \rho_w \cdot \nabla$ (tons)
- Block coefficient, C_B (–)
- Ship length, L (m)
- Length between perpendiculars, L_{pp} (m)
- Length overall, L_{oa} (m)
- Beam, B (m)
- Mean or midship draught, D (m)
- Area of ship cross-section (submerged), $A^1 \simeq B \cdot D$ (m²)
- Stationary trim, $t = D_{ap} - D_{fp}$ (m), where D_{ap} and D_{fp} are the draughts at the aft and forward perpendiculars, respectively
- Heel, $h = D_s - D_p$ (m), where D_s and D_p are the midship's draughts at the starboard and port sides of the ship
- Sagging and hogging, $sh = D - \frac{1}{2}(D_{ap} + D_{fp})$

- Bilge keels fitted to the bottom corners of the ship's hull significantly reduce ship roll; they are normally fitted to bulk carriers over the center third of the ship and extend about 0.5 m from the hull

(b) **Channel-related parameters**
  - Channel direction
  - Channel curvature, radius, R (m) and bend angle, β (°)
  - Bottom width, w (m)
  - Channel length, l (m)
  - Channel depth, d (m)

(c) **Ship-in-channel-related parameters**
  - Ship forward speed, V (m/s)
  - Vector angle of wave propagation relative to ship, α (°)
  - Basic maneuvering lane, w_{BM} (m) is the maneuvering width required for the design ship under favorable conditions; also referred to as swept path
  - Starboard (green) bank clearance, w_{Bg} (m) and port side (red) bank clearance, w_{Br} (m)
  - Additional width for adverse environmental and operational conditions, w_i (m)
  - Passing distance in the case of two-way traffic, w_p (m)
  - Underkeel allowance, $UKA = d - D$ (m), normally expressed as a percentage or ratio of the ship's draught, D
  - Ship sinkage due to squat, ΔD (m)
  - Amplitude of vertical ship movements due to waves, A (m) with significant amplitude, A_s, and maximum amplitude, A_{max}
  - Maximum overdraught, $Z_{max} = sh + \Delta D + A_{max}$ (m)
  - Tide level at the time of the maneuver, Z_0 (m) relative to chart datum, CD
  - Underkeel clearance, $UKC = d - Z_0 - D - sh - \Delta D - A_{max}$ (m)

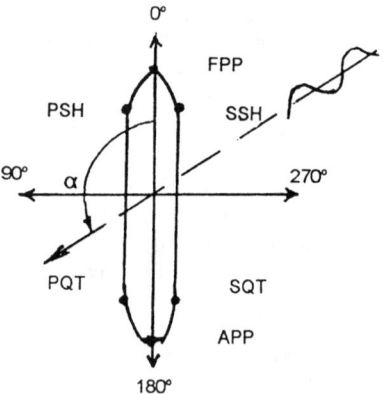

3. DESIGN PROCESS

3.1. Design Approach

As mentioned in Section 2.2, the site selection of a port and the related navigation channels must be based on a proper evaluation of economical, geographical, demographical, environmental, and political factors and considerations. Once the basic site has been selected, the design aims will be to:

- Minimize restrictions to navigation due to channel limitations and adverse environmental conditions
- Optimize the capital vis à vis maintenance costs
- Ensure acceptable safety of navigation with minimal tug and aids to navigation requirements
- Develop an environmentally acceptable scheme.

To achieve these aims, the design approach consists of two stages—the preliminary desk design followed by the detailed design.

3.2. Preliminary Desk Design

As a first step, a survey has to be done of the existing and/or the expected traffic, including the expected type and size of ships that will use the facilities, types of cargo, and shipping intensity.

The preliminary channel design is then usually based on available hydrographic and environmental data. Channel layout and dimensions are based on empirical data obtained from the literature and from similar existing projects. Experience also plays a major role in this design phase.

3.3. Detailed Design

For the detailed channel design phase, comprehensive environmental data are required, which, in most cases, have to be collected and measured specially for the scheme.

These data are then used in physical and/or numerical models to determine, in more detail, the wave, current, tide (storm surge), and sedimentological conditions in the channel areas. These design studies may also have to include ship motion studies to determine the required channel depths and the risk of bottom contact more accurately.

In addition, ship maneuvering studies have to be done, preferably on a full-mission, real-time ship maneuvering simulator, to confirm the channel layout and to determine operational limits of the design (PIANC, 1992). Provided sufficient repeat runs are made on the simulator, the risk of exceeding the channel boundaries can also be estimated on the basis of the simulation studies.

The detailed design must also include an optimization phase and cost estimates of the final design.

4. PHYSICAL ENVIRONMENTAL DATA

4.1. Data Requirements

The quality of a port design, in general, and of navigation channels, in particular, is largely dependant on the extent and quality of the physical environmental data available for the design. Normally, the initial design can be based on available (historic) data, but for the detailed design reliable statistical data must be available, collected over extensive time periods. The lengths of these periods depend on the type of data; for instance, astronomical tidal predictions can be made on the basis of about one year of records. For the prediction of storm surges, a much longer recording period is needed—at least about 10 years—whereas sea level rise, related to global warming, is difficult to predict reliably even when "long-term" records are available.

A brief discussion of the required data is given below. A detailed description of the data collection methods falls outside the scope of this chapter; the reader is referred to Zwamborn et al. (1972) for low-cost measuring procedures and PIANC (1989, 1997) for more up-to-date methods. A description of environmental design data specifically aimed at the determination of the forces exerted on a moving vessel are included in PIANC (1979).

4.2. Bathymetry and Geotechnics

A depth chart of the channel area must be available, preferably with 0.5 m interval contours. Since land levels are normally given to mean sea level (MSL), whereas hydrographic charts are relative to chart datum (CD) or a similar low-water reference level, care must be taken in combining these data and preferably one datum level, such as CD, should be used for the entire scheme. Particularly in the case of a tidal lagoon, exact concurrence of datum levels is essential.

Channel location and cost are very dependant on the subbottom conditions. A seismic subbottom survey will have to be available, combined with bottom samples and borehole results, to be able to identify the presence of rock or other hard layers that would be difficult to dredge.

4.3. Morphological Changes and Siltation

Bottom and/or bank changes may occur due to waves, currents, and river discharge. Available survey data must be compared to determine any morphological changes in the channel area. These would then be used to estimate the expected future changes, including the possibility and extent of channel siltation.

Survey data going as far as possible back in time should be used, but for the detailed design, special repeat surveys should be made over a period of at least three years. Detailed data collection methods are included in the PIANC publication on channel maintenance (PIANC, 1989).

4.4. Wave Data

Wave data are probably the most important input for approach and entrance channel design. Statistical data of wave height, wave period, and direction are needed for the preliminary design. These could be recorded data, made either visually or with the aid of instruments, or hindcast data if sufficient wind data are available. If no local data are available, a first estimate of the wave climate can be obtained from Hogben and Lumb's (1967) worldwide VOS-based *Ocean Wave Statistics* (1953–1961) and/or by using voluntary observing ship (VOS) data directly, which can be accessed from the relevant weather bureaus (for example, KNMI, de Bilt, the Netherlands).

Proper instrument-recorded local wave data based on wave energy spectra are required for the detailed design. Data must be collected at regular intervals (e.g., 3 or 6 hourly). The recording period should be as long as possible with a minimum of, say, three years. If, for practical reasons, the recording period is shorter, the recorded data should be correlated with other longer-term data to improve their statistical reliability.

The wave data must be converted from deep sea or the local recording point to the areas relevant for the channel design using appropriate refraction and diffraction models, as described, inter alia, in CERC (1984).

The wave parameters relevant for channel design are as follows (refer also to the "List of Sea State Parameters," PIANC, 1986):

- Wave height (m): maximum (for the recording period, usually 20–30 min) H_{max}; significant (average of the highest one-third of the wave heights) $H_s \simeq H_{mo} = 4\sqrt{m_0}$, where m_0 (m^2) is the zeroth moment of the wave spectrum

- Wave period (s): significant (average of the highest one-third waves) $T_s \simeq T_z$, the ze-

roth crossing period; spectral peak period $T_p = 1/f_p$, the inverse of the spectral peak frequency, f_p (Hz)
- Wavelength, related to the peak period and the water depth, λ_p (m)
- Shape of wave energy density spectrum, $S(f)$ (m²/Hz)
- Wave direction (°): mean vector direction θ of wave propagation (direction into which waves propagate) and directional spreading, $s(f \cdot \theta)$

For optimum channel design, joint distributions (percentages of occurrence) of wave height, period and direction are required using, for instance, 0.5 m height, 1 s period, and 10° direction intervals.

4.5. Wind and Visibility Data

Normally, wind data are available at or near the site and can be obtained from the relevant weather offices. Joint distributions (percentages of occurrence) of wind direction and strength (velocities) are needed, particularly for channel operation aspects. Wind velocities should be converted to the standard height of 10 m above (mean) sea level (see CERC, 1984).

For the detailed design, local wind data should be collected over a period of not less than one year. These data must then be correlated with data from a "long-term" station to improve statistical reliability.

Again, for port operation (downtime), the percentages of occurrence of heavy fog, the particular seasons, months, and times of the main occurrences, and the duration of fog events are needed. These data can also be obtained from relevant weather offices.

4.6. Tide and Water Levels

Astronomical tidal data can usually be obtained from available tide tables. For channel design, both the mean water level (ML) and the low water ordinary spring tide (LWOST) level are needed. These levels must be related to CD, that is, the reference level of the hydrographic surveys. Also, the ML is not necessarily the same as the land survey datum level (LSD), in South Africa referred to as mean sea level (MSL). Particularly in the case of tidal rivers and lagoons, differences in these reference levels can cause serious problems.

If no tidal predictions are available for the site, a tide recorder should be installed near the site. Tide records over a period of at least one year should be collected, particularly if the records are also used to abstract long-wave data such as wave and wind setup during storms and possible (basin) resonances. These data are needed for optimization of the channel and quay wall (mooring conditions) operations.

4.7. Water Temperature and Salinity

Ship draught, D, is dependent on the density of the water which, in turn, depends on the water temperature and salinity. The following density values can normally be used in the design:
- fresh water, $\rho = 1000$ kg/m³ (salinity 0, temperature +4° C)
- seawater, $\rho_s = 1025$ kg/m³ (salinity 35.5 ‰, temperature +20° C).

In the detailed design, attention must be given to possible deviations from these values, particularly in the case of fresh–seawater interaction in river mouths or lagoons (PIANC, 1997).

An important secondary effect may occur when water temperatures are rather high, such as in the Mediterranean Sea and the Red Sea. Also, when power station cooling water is discharged into a harbor basin in these areas, the ambient water temperature may rise above the 32° normally used for ship design, causing a loss of efficiency in the ship's generators and severe limitations (loss of power) to the tug's main engines.

4.8. Currents

Records of ocean currents, tidal velocities (storm-surge-generated velocities), and river flows must be available for the design. These data are inherently more difficult to collect because of the spatial variation in current patterns.

For the preliminary design, available data can be used, which may include references to the area concerned in local or international "sailing directions," as issued by the various hydrographic offices. In addition, some simple measurements using floats, dye bombs, moored current buoys, or jelly bottles can be made (Zwamborn et al., 1972).

For the detailed design both the vertical and the horizontal distribution of the currents must be known as well as their variations in time. The only practical way is to install appropriate recording equipment at a few critical points and to use the data to calibrate a suitable numerical flow model, which will then provide the more comprehensive coverage required. Initial runs with the model can also be used to design the measuring program, that is, to indicate the most relevant recording points.

The duration of the measurements depends on the local conditions but, generally, should be at least one year.

5. DESIGN SHIPS

5.1. Choice of Design Ships

The choice of the design ships must be based on thorough traffic analyses. This must include the type and volumes of cargo both for the present and for predicted future conditions (see Halber et al., 1985).

For a single-purpose port, such as an oil terminal or a bulk coal/ore handling facility, the choice of design ship is fairly simple; normally the largest bulk carrier will be determinative for the design of the channels. In the case of a mixed port, however, more than one ship will have to be considered in the design process. For instance, if the port has to provide for bulk and container traffic, no doubt the bulk carrier will be determinative for the channel depth but the container vessel, with its large windage area, may require more maneuvering space and greater channel widths in areas with dominantly strong winds.

Thus, in the final choice of the design ships various factors must be considered, including ship size (dwt), ship dimensions (L_{oa}, B, D), windage areas, ship propulsion, maneuvering characteristics and type of cargo (hazardous cargo or not). Care must be taken to choose ships that are representative for the expected ship mix. Although it would not be realistic to choose a design ship with the worst maneuvering characteristics, because ship designs can be expected to improve with time, the choice must definitely not fall on the best-performing vessel in its class, but rather on an average-performing ship.

Because of the inherent inaccuracies in predictions of future traffic and ship developments (Vickerman, 1992), the design of the channels should allow for a maximum degree of flexibility so that adjustment can be made relatively easily to satisfy future requirements.

5.2. Type of Ships and Propulsion

The following ship types, classified according to load, make use of commercial ports:

- tankers (ULCC), 350–500,000 dwt
- tankers (VLCC), 200–300,000 dwt
- tankers, 60–175,000 dwt
- chemical tankers, 3–50,000 dwt
- bulk carriers, 10–400,000 dwt
- container ships (post-Panamax), 55–70,000 dwt
- container ships (Panamax), 10–60,000 dwt
- ro-ro ships (freight), 5–50,000 dwt
- cargo vessels, 2.5–40,000 dwt
- cruise liners, 35–80,000 dwt
- ferries, 15–50,000 dwt

Typical deadweight tonnages, lengths, beams, draughts, and block coefficients of some eighty of the above types of ships are given in PIANC (1997).

There were five types of main propulsion machinery used in large ships (bulk carriers and tankers) in 1975, namely (Williamson, 1975):

- steam reciprocating (only in old ships)
- turbo-electric (only in old ships)
- geared steam turbine (5% of bulk carriers)
- direct-drive slow-speed diesel (89% of bulk carriers)
- geared medium-speed diesel (6% of bulk carriers)

Presently, most ships have diesel engines, because fuel consumption is significantly less than for steam turbine engines.

The steam turbine ships can use lower revolutions and can thus move slower than the motor ships because the propeller can be left to rotate (windmill) when the engines are stopped. This improves manoeuvrability at low speeds (better rudder action). However, their reaction time is very slow; it takes about 2 min for a change in revolutions. Also, the astern power is much less, down to 50%, of the forward power.

In comparison, the direct-drive motor ships cannot move very slowly (5 knots for dead slow compared to 3 knots for turbine ships) and their steering ability reduces because the propeller stops when the engines are stopped. The reaction time of the motor ships is, however, much shorter and they have almost full (85–100%) astern power. In the case of the geared medium-speed diesel engines, low propeller speeds and windmilling are possible, eliminating the low-speed disadvantage compared to the turbine ships. In addition, bow and stern thrusters usually installed in the large container vessels make these ships highly maneuverable at low speeds.

A factor that also affects the design of channels, particularly the layout, is the fact that almost all large ships have right-turning screws, which results in a clockwise tendency to turn when astern power is used in the stopping maneuver.

5.3. Ship Dimensions

For the larger ships of all kinds, there is generally a fairly good correlation between ship length, beam, draught, and deadweight tonnage. This is clear from the typical ship dimensions included in Appendix B of PIANC (1997). For bulk carriers, the following approximate dimensions, based on detailed analyses by Williamson (1975), can be used for the initial design:

Deadweight tonnage	L_{oa} (m)	L_{pp} (m)	B (m)	D (m)
20,000	162	152	22	9.5
40,000	202	190	28	11.5
60,000	230	220	32	12.5
80,000	248	238	35	13.5
100,000	258	246	39	15
120,000	266	254	41	16
140,000	280	268	43	17
160,000	294	284	45	17.5
180,000	304	294	46.5	18.5
200,000	314	302	48	19
250,000	330	316	52	20.5

Block coefficients, C_B, vary between 0.75 and 0.85 (Williamson, 1975) but bulk coal carriers calling at Richards Bay (South Africa) had C_B values from 0.82 to 0.85, with a mean of 0.84 (17 ships).

For the detailed design, up-to-date records of ship sizes, particularly of the ships-on-order, should be obtained from Lloyds, who publish an annual update on "World Fleet Statistics." Refer also to PIANC (1997).

5.4. Loading Conditions

For the channel design, the fully laden (maximum draught) design ships are normally determinative. However, conditions for a ship in ballast must also be checked, particularly regarding the port layout, since the maneuvering characteristics of ships in ballast are significantly different from those for the fully laden ships.

A minimum load factor at sea of about 0.3 (30% of the maximum deadweight) is the accepted rule to ensure adequate stability, moderate ship motions, and full immersion of the propeller (Williamson, 1975). This means, for example, that for a 130,000 dwt bulk carrier with a loaded draught of 16.7 m the mean draught in ballast will be 7.2 m, compared with the light draught of 2.8 m. Under these ballast conditions, the ship will normally be trimmed several meters by the stern to ensure submergence of the screw and most of the rudder area.

In the case of bulk carriers, the vessels normally enter full and leave empty, or vice versa, which simplifies the design process. For general cargo and particularly for container vessels, it is difficult to predict with what draught these ships may have on entering or leaving. Again, the fully laden (maximum draught) condition is normally considered but the design should also be checked for a semiladen container vessel that, for instance, would experience larger wind forces. In addition, trial data of a 80,000 dwt container vessel show a 40% increase in tactical turning diameter and transfer for the ship in ballast compared with the full ship, although these results apply to "deep" water. Nevertheless, it would be prudent to check the channel layout for at least two loading conditions in the case of container vessels.

5.5. Auxiliary Craft

The design of the maneuvering areas, in general, and the stopping distances, in particular, are largely dependant on the type, bollard pull, and number of available tugs. Thus, even at the stage of the preliminary design, a decision must be reached, in principle, on the extent of tug assistance to be provided, and detailed information is required in the final stages of the design, that is, for the ship-simulation studies.

There are two main types of tugs: conventional tugs with a normal stern propulsion/rudder arrangement and tractor tugs with propulsion units that enable them to move in any direction.

Most ports, and definitely all new ports, now employ tractor tugs which, although much more expensive than the normal or Kort-nozzle (ducted propellers) tugs, are more versatile and are preferred by the nautical staff, e.g., pilots (see Figure 11.1). The Schottel and Voith–Schneider tugs have twin propulsion units underneath the hull that are used for both propulsion and steerage (Atkinson, 1980). A Japanese development of the Schottel principle is the Z-peller tug, which has the drive units under the stern, thereby reducing the draught. Z-peller tugs are therefor particularly useful when operating in "shallow" water (typically, the draught of a Schottel tug is 5.5 m and of the Z-peller tug 4.5 m).

Whereas a normal propulsion tug attached by a line to a ship can only open up (start to assist) when the ship's speed has reduced to 2 to 3 knots, tractor tugs can make fast at virtually any speed and they can start to assist with steerage at 4 to 5 knots with waves up to H_s = 1.5 to 1.8 m (H_{max} = 2.5 to 3 m). Thus the use of tractor tugs can result in a "cheaper" harbor layout with a shorter stopping length.

Normally, two to three tugs with bollard pulls of 30 to 45 t are used in major ports.

6. CHANNEL LAYOUT

6.1. Design Principles

The channel layout must be designed in accordance with the following general engineering requirements:

- Shortest route between deep water, the mooring areas, and between areas to be connected by navigation channels
- Bottom material should be easy to dredge
- Away from areas with major river flow and littoral sand movements
- Location should have moderate environmental conditions (waves, wind, and currents)
- Provide a low-risk entry/approach for ships carrying dangerous cargoes

Kort nozzle—Richards Bay

Schottel/Z-peller—Port Elizabeth

Voith–Schneider—Saldanha Bay

FIGURE 11.1. Different types of tugs.

In addition, the design must take cognizance of the following navigational requirements:

- Channels should preferably be straight or consist of straight sections connected with appropriate bends
- Avoid (shallow) obstacles in the approach route/channel and travelling "long" distances along a lee shore
- The "point-of-no-return" should be as close as possible to the harbor entrance (exit zones)
- Shouldering or heading and beam waves are preferable to following to quartering waves, which can influence the alignment of the approach and entrance channels
- Cross-channel currents and strong winds should be avoided, if possible

The preliminary design can be based on available data, empirical information, and experience, but the suitability of the layout, including possible alternatives, must be checked in the detailed design phase using appropriate maneuvering simulation facilities (PIANC, 1992).

6.2. Channel Alignment

Channel alignment is largely influenced by topography and subbottom conditions (presence of rock), and some information on the latter must therefore be available at the early stages of the design process. Obviously, where possible, use will be made of any naturally deep areas and/or subbottom "channels" in the case of the presence of rock.

If there is no subbottom restriction in the locationing of the channel, different options must be considered when weighing the requirements in Section 6.1 against overall cost and environmental considerations. The shortest-route, normal-to-shore channel alignment, which is often also the main wave-propagation direction, is generally unfavorable from a wave protection point of view. However, this proved not to be a problem in the case of a dredged outer channel that was found to refract the directly incoming waves away from the entrance, resulting in low wave penetration for this direction (Zwamborn and Grieve, 1974).

An advantage of a channel alignment in the main wave-propagation direction (θ) is that the vertical ship motions are much smaller than for beam or quartering waves. This will result in a shallower and thus cheaper channel. On the other hand, crosscurrents and/or side wind will increase the required lane width for shipping, resulting in increased costs. These advantages and disadvantages will have to be carefully weighed in the detailed design phase.

6.3. Channel Bends

Ideally, the use of bends should be limited as much as possible for ease of maneuvering. However, bends cannot always be avoided, sometimes even just before the harbor entrance as, for example, is the case for the ports of Port Elizabeth (South Africa) and Haifa (Israel) (see Figures 11.2 and 11.3). Such bends must be properly designed (radius and width) and adequately marked by aids to navigation, such as intersecting leading lights, different color sector lights, and/or (lighted) marking buoys.

It is important that for unassisted navigation (no tugs), the curvature of bends in the approach and entrance channels and between two straight sections of shipping channel not be too sharp but also not too gentle. In calm water, ships can maneuver reasonably well around a bend with a radius of 1.8 to 2 times the ship length ($R = 1.8$ to $2\,L$). Howev-

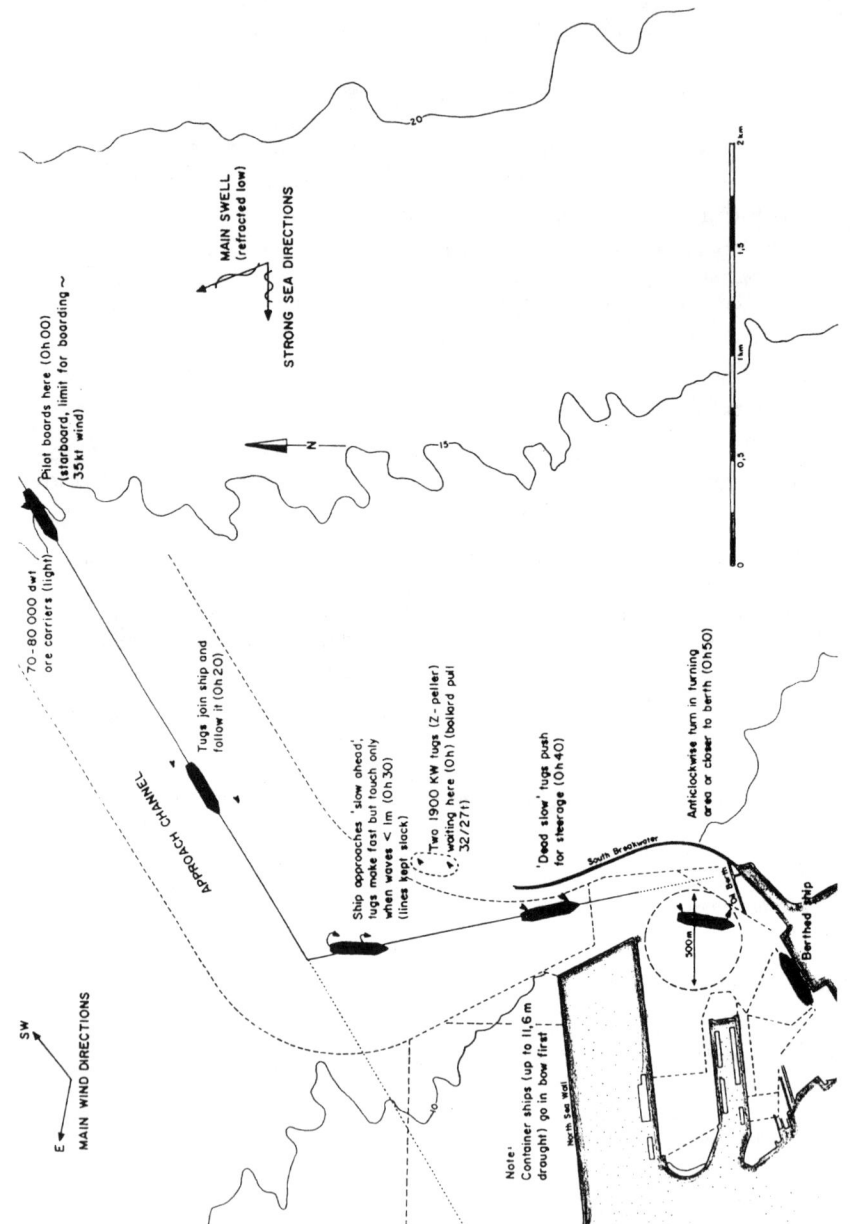

FIGURE 11.2. Harbor approach at Port Elizabeth (South Africa).

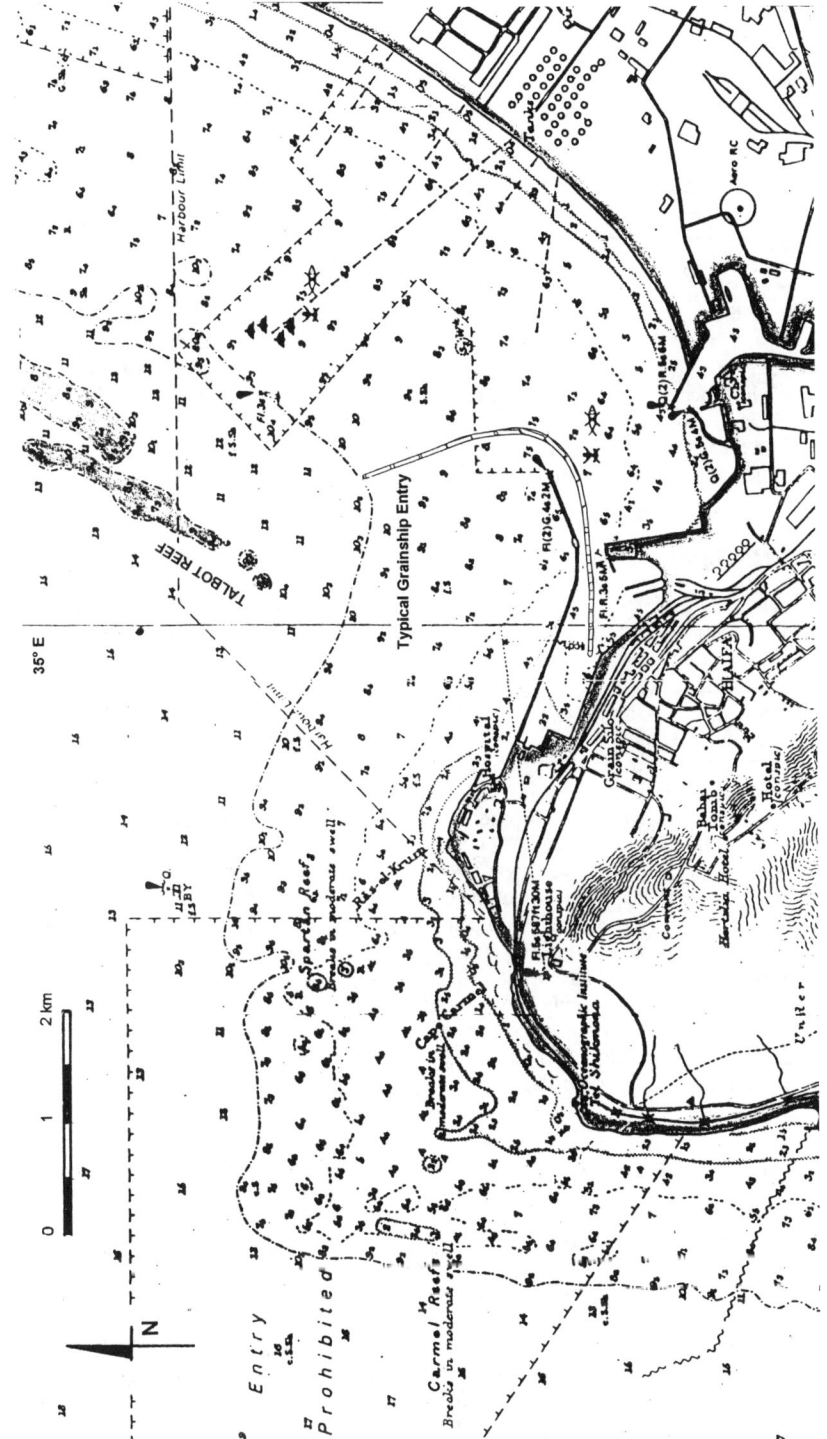

FIGURE 11.3. Harbor approach at Haifa (Israel).

er, this applies to deep water conditions ($d \geq 2\,D$). For small underkeel clearances, larger radii apply, namely, according to PIANC (1997):

d/D	∞	1.5	1.3	1.2	1.15	1.1
R/L_{pp}	3.1	3.9	4.8	5.7	6.6	8.6

Rudder angle 20°; change of heading, 90°.

However, these small underkeel clearances generally occur only inside the port, where tug assistance is available, making these large radii less relevant. In the case of a very gentle curve ($R > 5\,L$), navigation becomes difficult due to the need for almost continuous course adjustments. Also, positioning in a long gentle curve becomes problematic unless the bend is clearly marked with a large number of (numbered) buoys.

In the area where tug assistance is available, the ship can obviously negotiate much sharper bends, as is required in the berthing maneuvers.

6.4. Navigational Aspects

Ship handling can become rather difficult for near-following to quartering waves, but only when the wave period is close to the ship's pitch and roll periods, as in the case of T_p = 10 to 14 s (long swell). With these quartering waves, the ship tends to "lock" into the waves, that is, it tends to turn parallel with the wave crests ("broaching"). This tendency was observed in the 1 in 100 scale physical model tests of the Port of Richards Bay (Campbell and Zwamborn, 1977) and in the Port of Saldanha during entering of 200,000 dwt ore carriers. From a ship handling point of view, beam to shouldering waves are therefore preferable (IAPH, Section 1, 1979).

The effects that environmental conditions can have on a vessel in an approach channel are summarized in an example in Figure 11.4. The effects of waves (drift forces), wind, and currents are indicated by arrows. The relative importance of the various forces depends on the direction and severity of the conditions relative to the channel direction.

Finally, although the channel alignment can be chosen in such a way that the disturbing forces act in opposite directions, in an attempt to balance them out, full-mission maneuvering simulations will have to be done for the detailed design to check the overall acceptability of the adopted channel layout. Acceptability is based on certain safety criteria, namely, no exceedances of channel boundaries, the use of half-ahead and half-astern as 'maximum' telegraph settings to be used, and rudder angles not exceeding 15 to 20° for any length of time, leaving sufficient margins for emergencies.

7. CHANNEL LENGTH

7.1. Problem Statement

Channel length, l, is particularly relevant in the case of harbor-approach and entrance channels.

Approaching ships must come into the protected port area at a safe maneuvering speed which depends on the environmental conditions. The minimum maneuvering speeds under calm conditions are 3 knots for turbine ships and geared motor vessels and 5 knots for direct-drive diesel engine ships, which are most common. This means that for safe entry

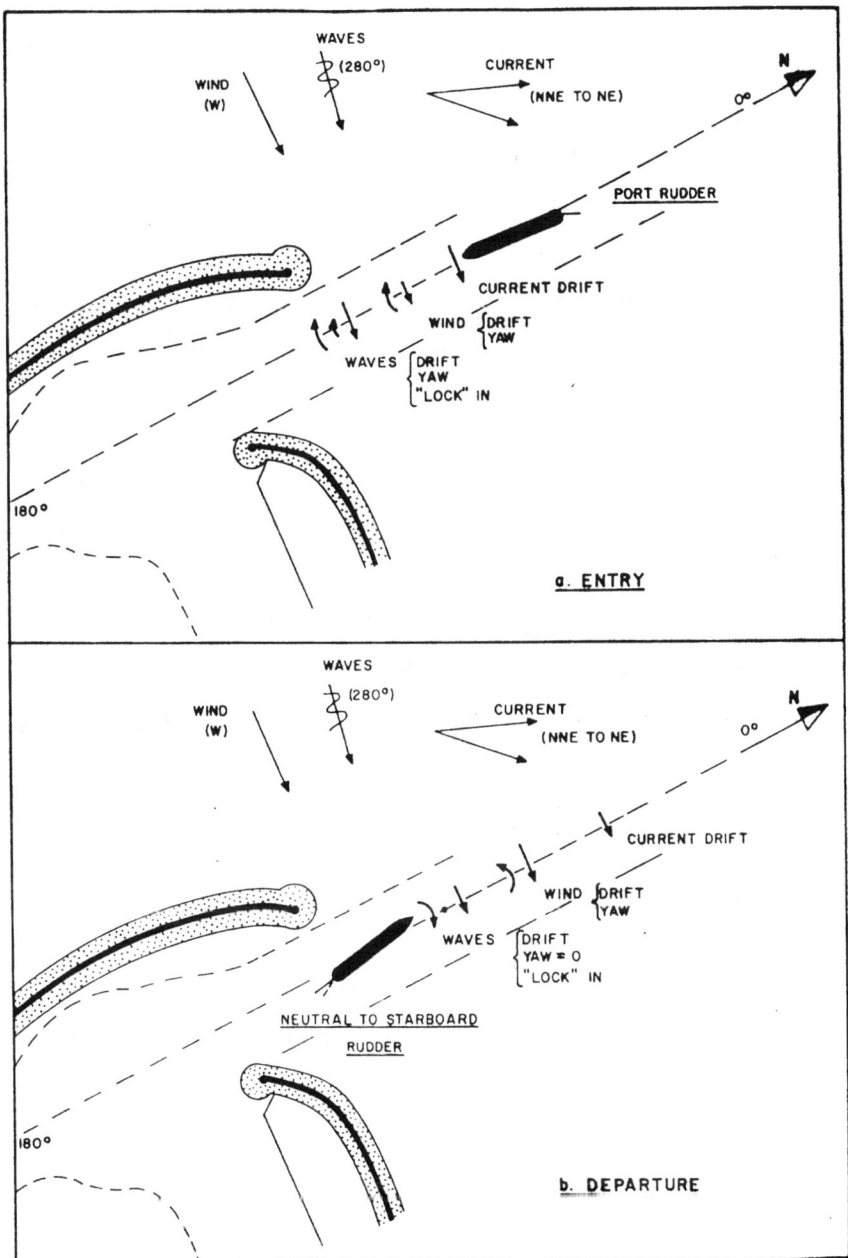

FIGURE 11.4. Effects of waves, wind, and currents on ship handling.

of ships under less favorable conditions, entry velocities must be between 5 to 7 knots (V = 2.5 to 3.5 m/s). Even using tugs, it takes a considerable channel length, the so called stopping distance, to gradually reduce this speed to zero.

Vertical ship motions due to waves are a stochastic process and the maximum expected vertical sinkage due to wave action is dependent on the number of waves, N, encountered in the channel and thus the channel length, l. Accepting a Rayleigh distribution for irregular sea waves, the maximum motion amplitude is found from:

$$A_{max} = \frac{1}{\sqrt{2}} (\ln N)^{1/2} A_s$$

and, since N is proportional with l, A_{max} will increase with l, which means that long approach/entry channels will have to be significantly deeper for the same nonbottom-touching safety criterion. For instance, comparing a 1 km with a 10 km approach channel and assuming $V = 3$ m/s and a wave encounter period of 13 s, A_{max} (10 km) = 1.3 A_{max} (1 km).

7.2. Stopping Distances

Crash stops in the open sea can be made from full ahead to full astern with the rudder in mid-position or by putting the rudder hard over, say to starboard. The stopping distance for a 150 to 200,000 dwt vessel will be about 4 to 5 km in the first case and the head reach about 1 km in the second case. In both situations, however, there is a considerable transfer of up to 1 km, which is unpredictable regarding direction and distance. Thus, such uncontrolled crash stops cannot be made in confined waters such as a harbor entrance.

Some recorded and calculated ship tracks for crash stops from full ahead are shown in Figure 11.5. Figure 11.6 shows examples of crash stops for the same 191,000 dwt tanker from half ahead, which result in stopping distances not exceeding 1 km, thus much smaller than the above 4 to 5 km (Crane, 1973). Similar results based on sea trials were found for a 80,000 dwt post-Panamax container vessel, namely a crash stopping distance of 5 km from full ahead to full astern and about 1.2 km from half ahead.

Data on maneuvering trials carried out on fully laden 20,000, 50,000, 100,000, and 200,000 dwt ships starting from a maneuvering speed of 5 knots are included in the Working Group No. 2 Report of PIANC (1973). The data show stopping distances relative to the ship lengths (l/L) of 2.3, 1.9, 2.3, and 2.7 for full astern (which is possible with tug assistance) and 3.7, 4.0, 3.8, and 4.1 for half astern (which is preferable inside a harbor), respectively, for the four ship sizes.

The stopping maneuver in the restricted waters of a harbor channel must be fully controlled. This can only be achieved when the engines are going ahead (slow to dead slow) so that there is positive rudder action, or, when the ship is kept on course by two or more tugs, in which case the engine can be put into reverse. The actual stopping distance for a ship entering the port depends on:

Ship-related factors	External factors
Ship size	Wave conditions
Load condition	Wind conditions
Entry speed	Current conditions
Type of propulsion	Available water depth
Engine power	Channel width
Number and type of crew	Aids to navigation
(for making fast the tugs)	(particularly for distance)
Other human factors	Type and number of tugs

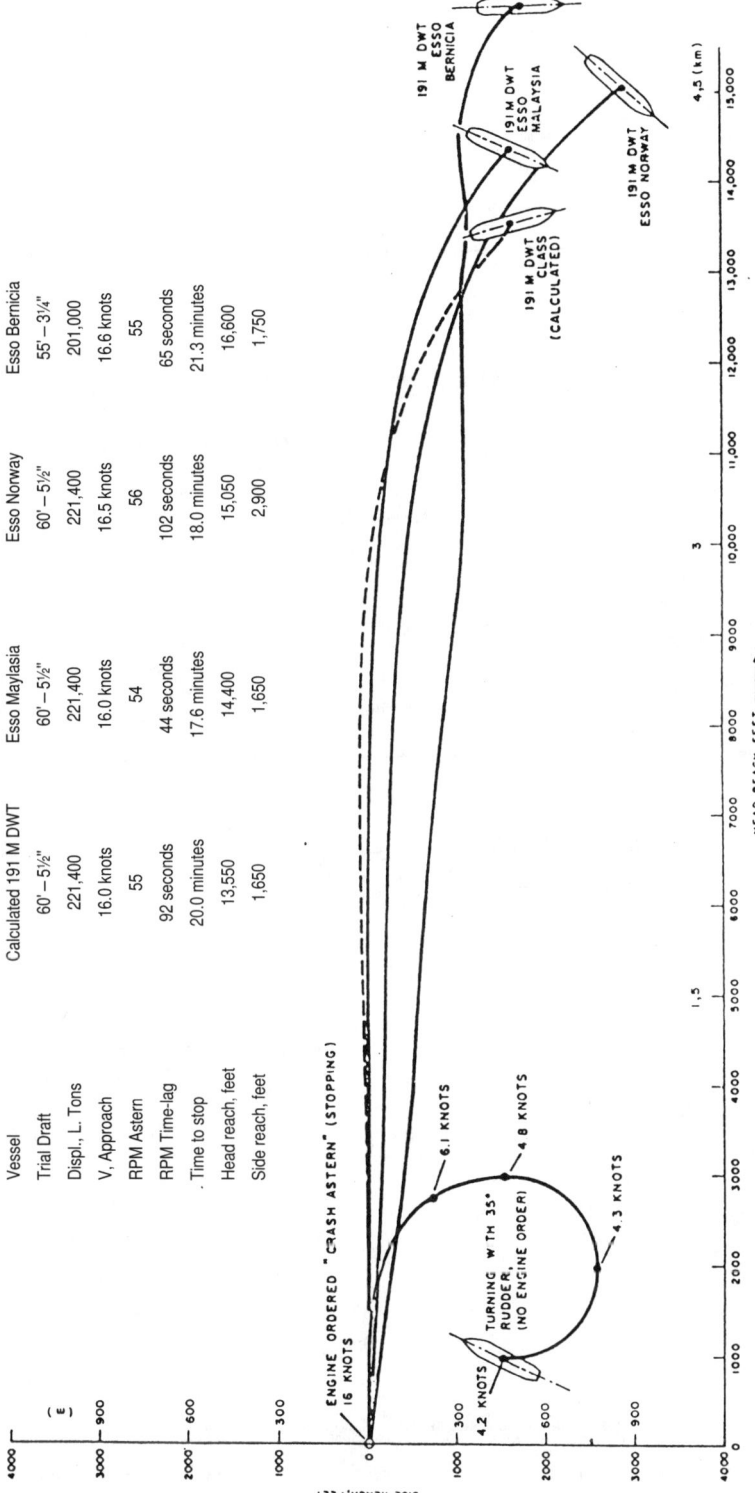

FIGURE 11.5. Comparison of full-scale trial results with calculated crash astern maneuver for 191,000 dwt tanker. (Reproduced from Crane, 1973.)

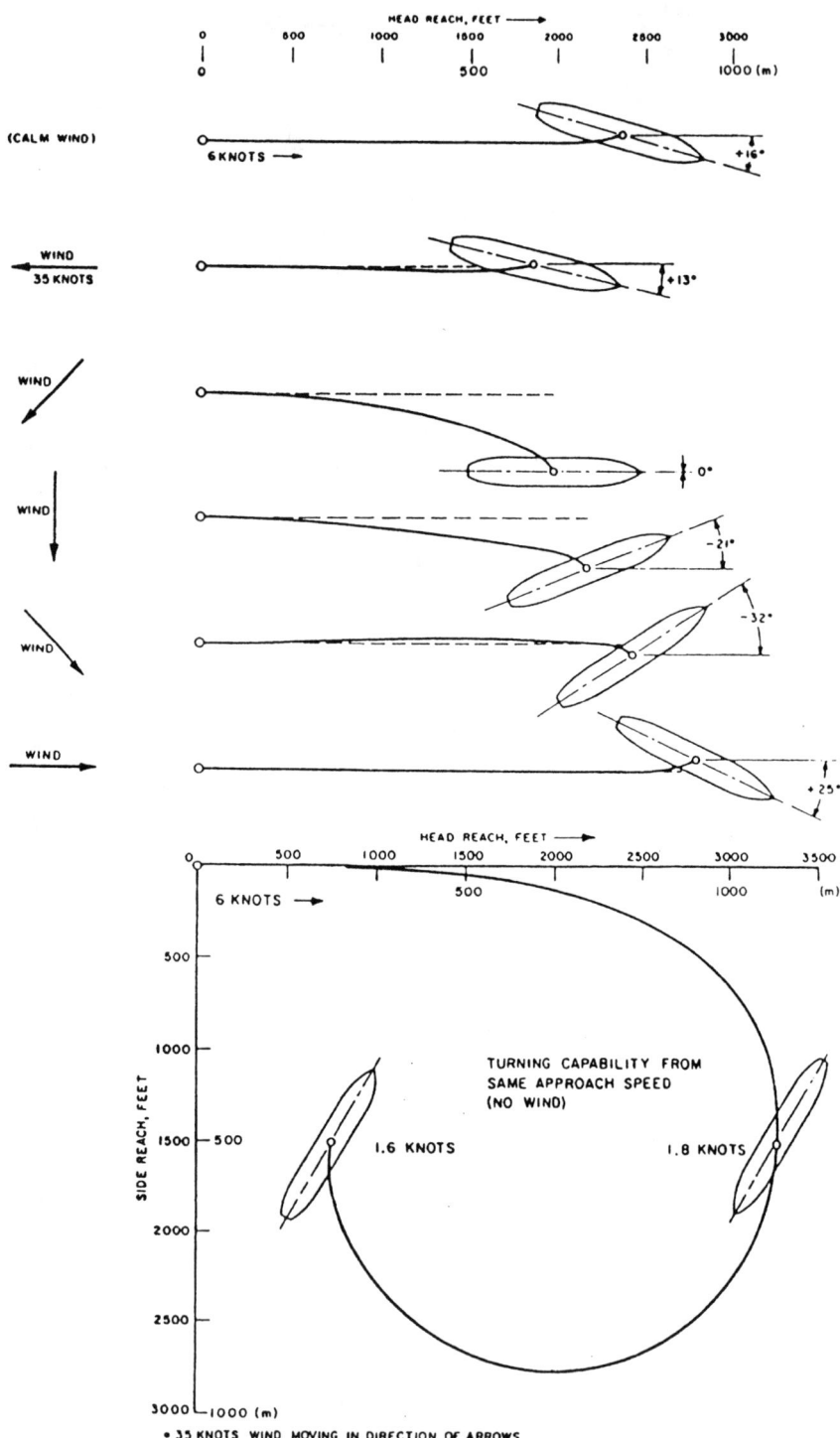

FIGURE 11.6. Calculated crash astern maneuvers (from 6 knots) of 191,000 dwt tanker in ballast. (Reproduced from Crane, 1973.)

The most important factors are the ship's dead weight tonnage, the entry speed, and the tug assistance.

Thus, for the preliminary design, a minimum stopping length, $l = 3$ to $4 L$, must be allowed for, whereas during the detailed design phase, the layout and stopping area should be checked on a full-mission maneuvering simulator. Such simulations for the Ashdod port extensions (Israel) using two to three tugs showed stopping distances between 600 m and 900 m for calm to rough conditions, respectively, or 2 to 3 L_{pp}. Allowing about one ship length as safety margin, this confirms the correctness of the above value of $l = 3$ to $4 L$.

7.3. Entry Maneuver

Tug assistance—the type, bollard pull, and number of tugs—play a major role in ship maneuvering in port entrances. Particularly the ability of tractor tugs to make fast at virtually any speed and to start assisting at a ship speed of 4 to 5 knots results in significant savings in port entrance layout due to the shorter stopping lengths required for a fully controlled tractor-tug-assisted entry (stopping) maneuver. The following procedures can be followed:

- The ship approaches at a safe maneuvering speed (3 to 6 knots, depending on the type of ship and the environmental conditions). Tugs make fast outside the port entrance and assist in the stopping and entry maneuver. This procedure, which is applied at the port of Haifa, Israel, results in the shortest possible stopping distance but can only be applied when wave heights outside the port entrance are always low, $H_s \leq 1.5$ m.
- The ship approaches and tugs make fast in a similar way but no assistance is provided by the tugs outside the port (tugs on loose lines). However, soon after entry (depending on the wave protection provided by the breakwaters, which must ensure $H_s < 1.5$ to 1.8 m), the tugs can start to assist and the ship can put its engines in reverse. This procedure, which is applied at the port of Ashdod, Israel, results in an intermediate stopping distance.
- The ship approaches in a similar way but the tugs only make fast inside the protection of the breakwaters, where they can start to assist as soon as they have made fast. This procedure, which is applied at the port of Richards Bay, South Africa, is necessary when the waves are dominantly high, $H_s \geq 1.5$ m, and it results in a long stopping distance.

The choice of which procedure to use in the design and thus what entrance channel length to allow for depends on the local conditions. Maneuvering runs on a real-time ship simulator must be done as part of the detailed design to confirm the suitability of the adopted procedure and adequacy of the stopping distance/channel length.

8. CHANNEL WIDTH

8.1. Design Factors

The required navigation channel width depends on many factors, the more important being:

- One- or two-way traffic (traffic density)
- Ship dimensions, particularly length (L) and beam (B)

- Wave, wind, and current conditions
- Ship speed (V), steering characteristics, and tug assistance, which affect the width of the swept-path or maneuvering lane (w_{BM})
- Channel cross-section
- Channel length (l) and aids to navigation (leading lights)
- Degree of curvature of the channel
- Human factors (pilot service)

Many of the above factors are interrelated. For instance, in the case of a short stopping length inside the port, the entry speed may have to be low and, due to dominant adverse wave and wind conditions, a relatively wide channel may be necessary.

Due to the complexity of this issue, no theoretical formulae are available for the determination of the required channel width and the preliminary design has to be based on empirical data. For the detailed design, suitable maneuvering simulations are required, which could be done on fast-time or real-time full-mission simulator facilities or both (PIANC, 1992).

8.2. Empirical Formulae

In PIANC (1973) it is proposed to calculate the channel width, w, from

$$w = 2B + L \sin \alpha'$$

where α' is the angle between the ship speed and cross-current vectors. For example, for a current velocity of 0.5 m/s (1 knot) and $V = 2$ m/s (4 knots), $\alpha' = 14°$ and $L \sin \alpha' = 0.24L \approx 1.6B$ (for $L = 6.7B$) or

$$w \approx 3.6B$$

In PIANC (1980), a minimum channel width of $5B$ is suggested but the effect of current must be added, thus for the above example:

$$w = 5B + L \sin \alpha' \approx 6.6B$$

In addition, allowance must be made for bend effects, that is $L^2/8R$ must be added, where R is the radius of the bend (for example, for $R = 1200$ m, $L = 300$ m, the additional width becomes about 10 m).

Herbich et al. (1981) proposed the following channel widths:

$$w_{inner} = 5B \qquad w_{outer} = 6B$$

Thus, the above data suggest straight channel widths varying from about $4B$ to $6.6B$.

In PIANC (1995, 1997) a more sophisticated approach is followed to estimate the channel width. The bottom width, w, of a one-way straight navigation channel is given as follows (see Figure 11.7):

$$w = w_{BM} + \sum_{i=1}^{n} w_i + w_{Br} + w_{Bg}$$

and for a two-way channel:

$$w = 2w_{BM} + 2\sum_{i=1}^{n} w_i + w_{Br} + w_{Bg} + w_p$$

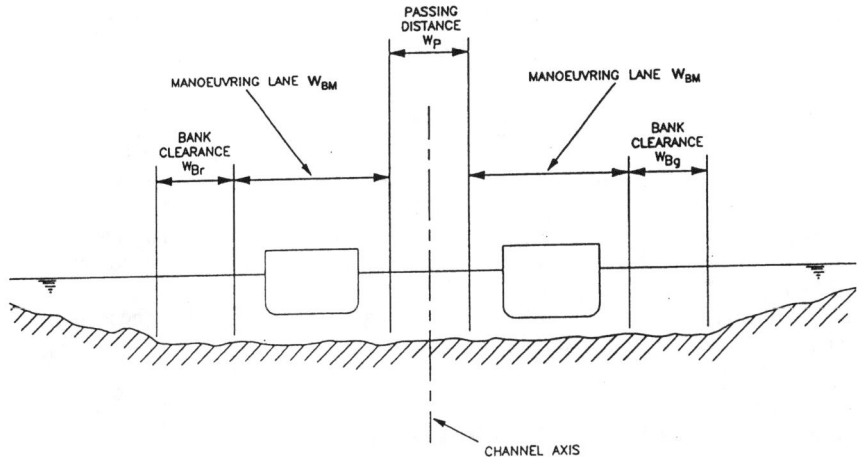

FIGURE 11.7. Two-way channel width elements.

where

w_{BM} = width of the maneuvering lane, which depends on the maneuverability of the vessel

w_i = additional widths to allow for vessel speed, cross wind, cross currents, longitudinal currents, wave height and length, aids to navigation, bottom surface, depth of channel, and cargo hazard level

w_{Br} and w_{Bg} = bank clearances on the red and green side, depending on the vessel speed

w_p = passing distance, depending on inner/outer channel and vessel speed.

Detailed values for w_{BM}, w_i, w_{Br}, w_{Bg}, and w_p are included in Tables 5.1, 5.2, 5.3, and 5.4 of PIANC (1997). Based on these, a realistic first estimate can be made of the required channel widths, both for single-lane and double-lane traffic, although correct judgment of the values of the various parameters need a reasonable degree of experience, otherwise, unrealistic values could be found. In this regard, it is interesting to consider the extremes, namely,

outer channel $w_{\min} = 1.9B$ and $w_{\max} = 12.1B$

inner channel $w_{\min} = 1.9B$ and $w_{\max} = 8.0B$

Here $w_{\min}$ represents a one-way channel under the most ideal conditions and $w_{\max}$ under the worst possible combination of negative effects. It is clear that the range is very large and that even with this improved method, it cannot be expected to derive exact channel width values.

8.3. Bend Width

The increase in bend width according to the earlier PIANC guidelines, namely $L^2/8R$, is quite small and probably negligible.

In PIANC (1995, 1997), very little guidance is given for the required bend width increase. The data provided apply to a single-screw, single-rudder container ship and they give the width of the swept track or maneuvering lane width (w_s) versus d/D and rudder angle. The following is a summary of these data for rudder angle of 20°:

d/D	∞	1.5	1.3	1.2	1.15	1.1
w_s/B	1.8	1.5	1.4	1.3	1.25	1.1

Compared with this, the basic maneuvering lane widths for a straight channel are w_{BM} = 1.3B, 1.5B, and 1.8B for a good, moderate, and poor maneuvering ship, respectively (PIANC,1997).

To find the overall bend width, w_s or w_{BM}, whichever is the largest, must be used in the above equations for the overall channel width, w (see Section 8.2).

8.4. Physical Model and Prototype Data

Physical model tests with self-propelled model ships have been used in detailed design studies for port entrances (Campbell and Zwamborn, 1977). These studies at a scale of 1:100 clearly showed the relationship between ship speed (150,000 dwt bulk carrier) and channel width. The safe entry speed for a 300 m wide channel was found to be 4.8 m/s (9.6 knots) and for a 400 to 500 m channel 3.5 to 4 m/s (7 to 8 knots). However, since human reactions cannot be scaled (they would have to be 10 times faster when steering the model ship), these physical model results can really only be accepted to yield relative results.

Confirmation of the preliminary design width can also be obtained by comparison with actual prototype measurements at a port that is similar, both in layout and with regard to environmental conditions, to the port to be designed. Such measurements were done at the port of Richards Bay and the results are shown in Figure 11.8. They represent the measured tracks of 50 loaded 100 to 150 000 dwt coal carriers leaving the port of Richards Bay, South Africa (Van Wyk, 1982). It is clear that virtually all ships remained within the 300 to 500 m bottom width channel, regardless of the weather and sea conditions, with recorded departure speeds between 7 to 10 knots. It must be noted that sea conditions off Richards Bay are generally rough, with H_s = 1 to 2.5 m and T_p = 9 to 13 s being the dominant swell condition (about 80% occurrence). Thus the widths of the approach channel at Richards Bay have been found to be adequate.

Thevenot (1992) reported on channel width studies carried out at WES in Vicksburg, Mississippi. The channel widths based on physical model tests, using the usual limiting criterion of 20° rudder angle, were found to result in up to 40% wider channels. However, if the criteria used for the physical model tests are somewhat relaxed (30° instead of 20°), good agreement was found between channel width requirements based on the physical model tests and those based on computer simulations.

8.5. Maneuvering Simulations

Mini- and micromaneuvering simulators can be used effectively in the preliminary design phase but, no doubt, the best method available today to confirm the design of channel widths is the use of real-time, full-mission simulators, because of the interactions between the many design factors, listed in Section 8.1 (PIANC, 1992, 1997). The simulator runs must be made by pilots having experience with the type of design ship used and, if possible, with the local conditions. The simulator bridge should, preferably, have all the equip-

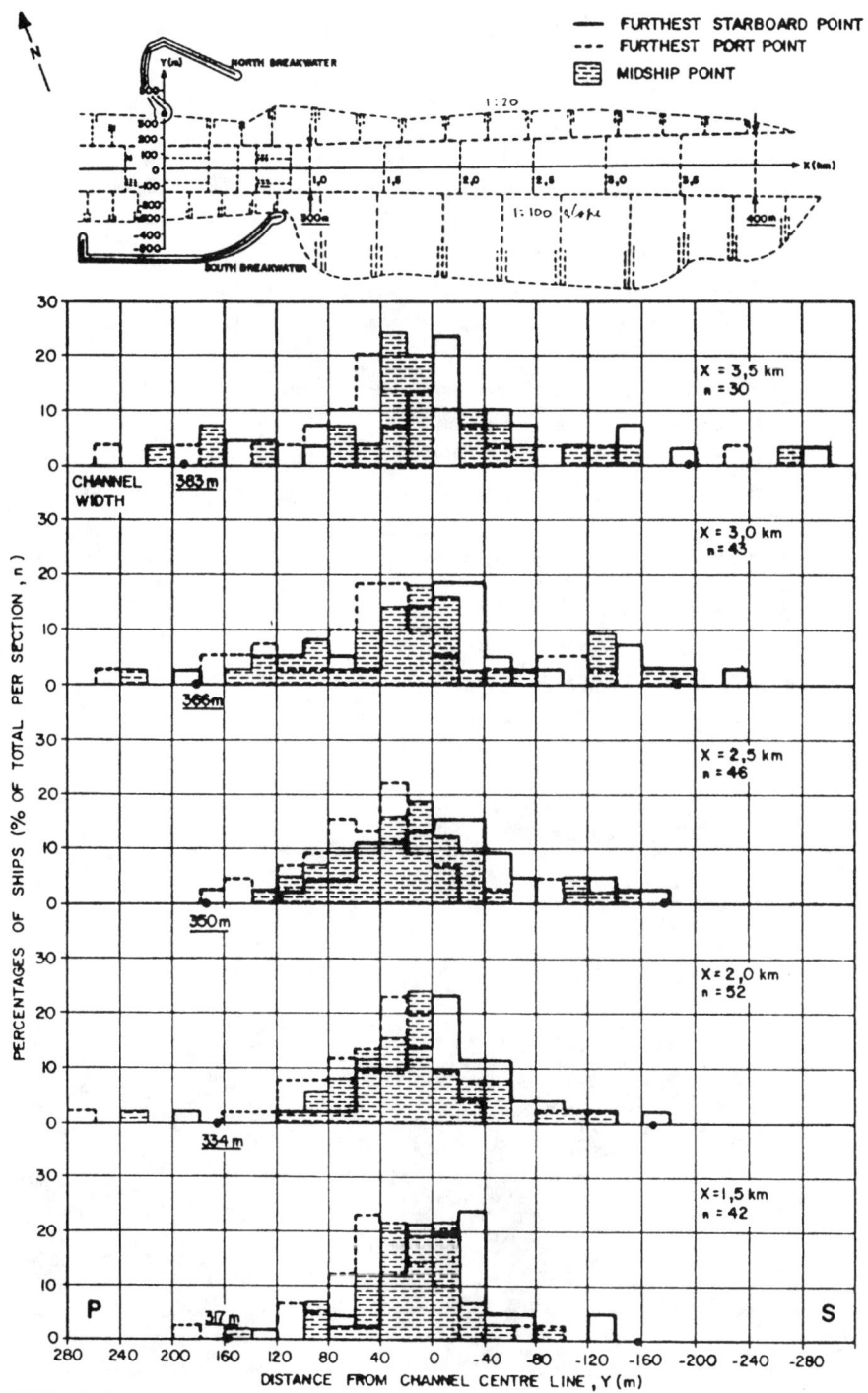

FIGURE 11.8. Swept paths of coal carriers in the Richards Bay entrance channel (South Africa).

ment normally found on a modern ship; for example, see Figure 11.9 (if the design ship does not have a certain piece of equipment, this can be switched off). A description of the various simulation models is included in PIANC (1992).

Typically, the scope of a simulation study would be to check the preliminary layout, make adjustments if necessary, and determine the limiting conditions for safe operations. This could include confirmation of:

- Entrance channel width
- Available stopping distance and area
- Extent and location of the turning areas
- Width of the mooring basins
- Minimum required tug assistance to ensure save maneuvering
- Need and effectiveness of aids to navigation
- Need for special precautions, if any, to be taken by the pilots
- Practicability and safety of the complete entry, turning, mooring, and departure maneuvers.

As an example, in the case of proposed extensions to the port of Ashdod, Israel (Halber et al., 1985), the following environmental test conditions were specified for the simulation studies (see Figure 11.10):

- Wave directions, W and NW
- Wave heights, H_s = 1.5, 2.5, and 3.5 m

FIGURE 11.9. Full-mission, real-time simulator. (Courtesy of MARIN, Wageningen, The Netherlands.)

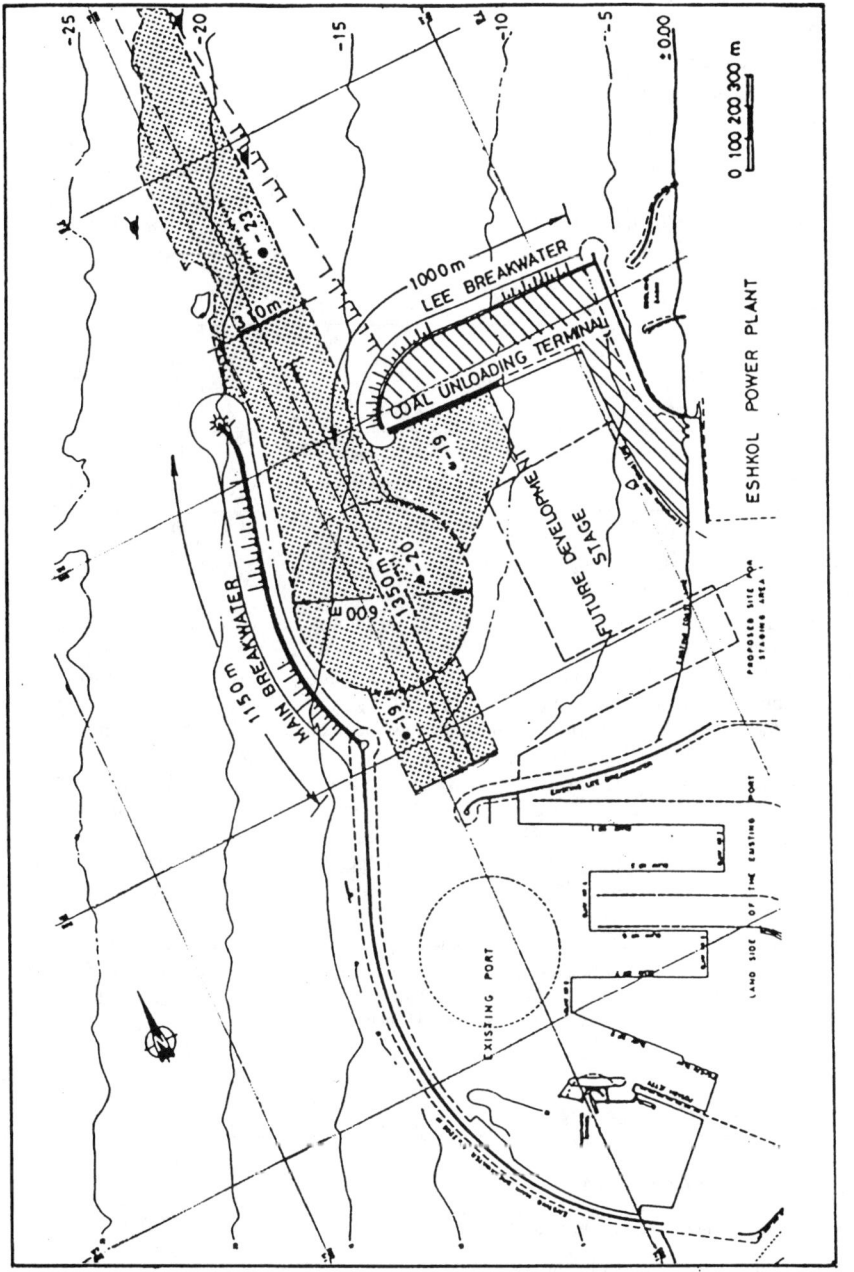

FIGURE 11.10. Ashdod port layout used for maneuvering simulations.

- Wave periods, $T_z = 7$, 8 and 10 s
- Wind directions, WSW, NW, and N
- Wind speeds, 5, 10, and 20 m/s
- Current directions, NNE and SSW
- Current speeds, 0.25 and 0.5 m/s
- Water level, + 0.1 m CD

Based on the above and taking into account practical limitations (number of runs versus time and cost), the following test run program was used for the Ashdod example which involved an approximately 80,000 dwt post-Panamax container vessel and a 150,000 dwt bulk carrier as design ships:

Swell	Direction	W						NW		
	H_s (m)	1.5		2.5		3.5		1.5	2.5	
	T_z (s)	7		8		10		7	8	
Wind	Direction	WSW						N	NW	N
	Speed (m/s)	0	5	5	10	10	20	10	20	10
Current	Direction	—	—	—	NNE	NNE	NNE	—	SSW	SSW
	Speed (m/s)	0	0	0	0.5	0.5	0.5	0	0.25	0.25
Starting Speed (knots)		4	5	5	6	6	7	4	6	5

A minimum of four repeat runs were done per condition, normally two runs by a local (Ashdod) pilot and two runs by a pilot experienced with the design ships. More runs were done for the more extreme conditions, so that a total of about 80–100 runs were involved. If a change is made to the layout (for instance a change in breakwater length) more repeat runs using mainly the most critical conditions may be needed. Some of the runs may also have to be done during night conditions, depending on the port procedures.

In the case of the Ashdod example, it was found that the wave drift forces played an important role in the channel design, and these forces could definitely not be omitted from the studies (Halber et al., 1985). The effect of waves and currents is shown by the track plots and entry speeds in Figure 11.11. The entry speed is seen to increase from 4 to 6 knots for $H_s = 1$ to 3 m. In addition, the mean rudder angle was found to increase from 5 to 10° for $H_s = 1$ m, to 20° for $H_s = 2$ m, and to 28° for $H_s = 3$ m.

These real-time simulation studies significantly increase the confidence of the channel design, as does involvement of the local nautical staff (harbor masters and pilots). A good description of the application of ship maneuvering simulation results for the design of the Avilés channel in Spain is given by Iribarren (1999).

9. CHANNEL DEPTH

9.1. Design Principles

With regard to depth requirements, the following channel sections can be differentiated:
- The approach or outer channel, which is the part of an entrance channel outside the protection of breakwaters

NAVIGATION CHANNELS—DESIGN AND OPERATION **11.31**

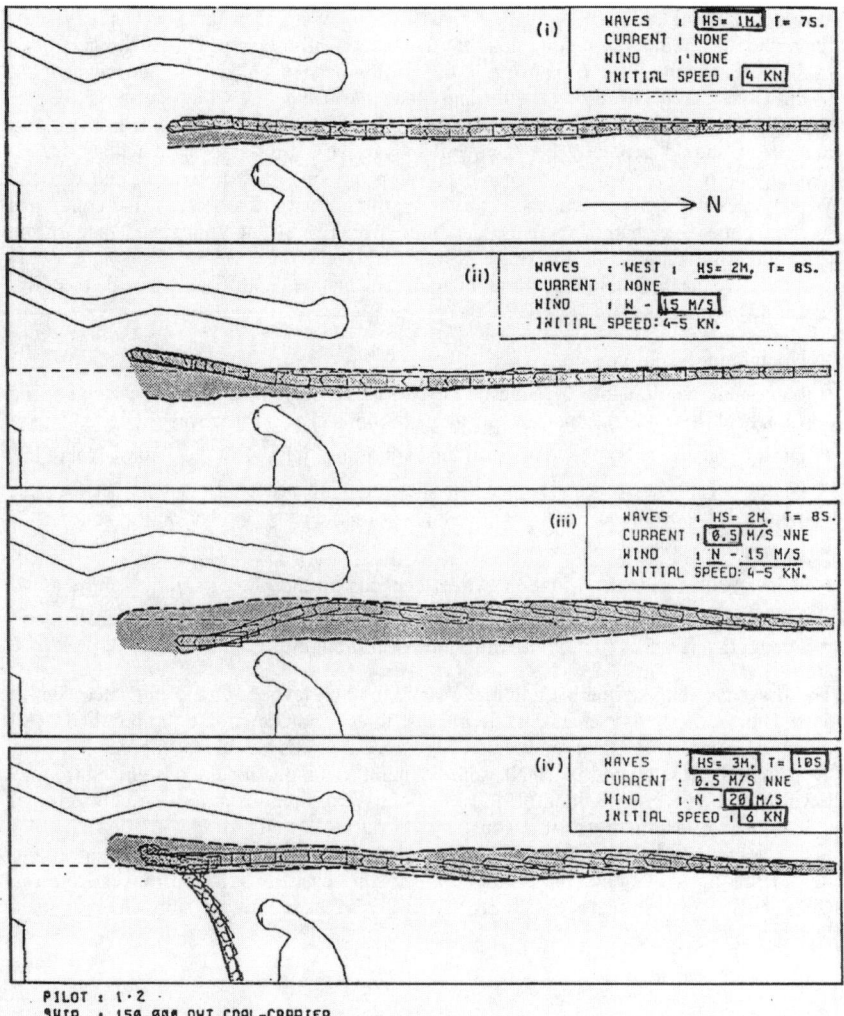

FIGURE 11.11. Envelopes of maneuvering simulation runs under various environmental conditions.

- The inner channel, which is inside the protection of breakwaters
- The turning basin area
- The berthing area or mooring basins
- Interconnecting channels between basins and/or different port areas

The outer channel is normally exposed to waves or swell from various directions as well as to crosscurrents and wind. The wave exposure in the inner channel is much less severe and, in addition, will generally consist of following or heading waves only. Currents will also be mainly longitudinal but the cross wind problem will be similar to that in the outer channel. Wave exposure in the turning basin should be further reduced, but beam waves can occur during the turning maneuver. Changing wind (and possibly current) directions during turning must also be taken into account. Conditions in the mooring basins should generally be calm, but long-wave action associated with some remaining short waves will result in moored ship motions, both horizontal and vertical. Wind forces will be a critical factor in the berthing maneuvers. Finally, the conditions in interconnecting channels will be determined mainly by the ship speed (squat) and possible cross winds.

There are many factors that affect the required channel depth. These include (see Figure 11.12):

- Draught of the design ships (D)
- Maximum overdraught (Z_{max}) during the passage through the channel consisting of sag (sh), heel (h), trim (t), squat (ΔD), and maximum sinkage due to waves (A_{max})
- Design water level(s), based on local tidal variations, relative to the datum level (CD)
- Underkeel clearance (UKC), which is the minimum accepted distance between the nominal seabed and any point on the ship's bottom during passage
- Accuracy of hydrographic survey
- Allowance for possible siltation, which depends on the rate of siltation and maintenance dredging frequency
- Dredging tolerances (bottom irregularities after dredging)

The first four items define the nominal channel bottom level or the promulgated depths, as shown on the harbor charts, while the last three items determine the required overdredging to ensure that the promulgated depths are always available to shipping.

Although these should be small, some draught variations for the design ship(s) may have to be considered. Kimon (1982) mentions a possible maximum draught error of 0.3 m due to uncertainty in fuel consumption during a long voyage. A proper survey of the vessel before departure should, however, minimize or eliminate such uncertainties. A draught variation of up to about 0.5 m can occur when a ship moves from salt (sea) water into fresh water. This must obviously be taken into account, where applicable.

9.2. Squat Formulae and Studies

Over the years, considerable attention has been given by international commissions to the problem of squat—the mean sinkage of a ship due to its speed—and the associated dynamical trim, normally for full-bodied ships towards the bow (negative trim). Relevant references are PIANC (1973, 1980, 1985, 1995, 1997). These efforts are mainly concerned with internal shipping channels, where squat plays a major role in the design. Also, approximate theoretical treatment of this problem is feasible although empirical and/or model data are required to substantiate the theoretical results. Even with these considerable efforts, PIANC (1979) states "... it is apparent that the use of the different (but appropriate) formulae can give widely varying values for squat . . ." (up to a factor of 2) and "... experience and judgement are necessary for their application"; also, "... specific research may be necessary."

For the preliminary design, PIANC (1997) suggest the use of the formulae from Huus-

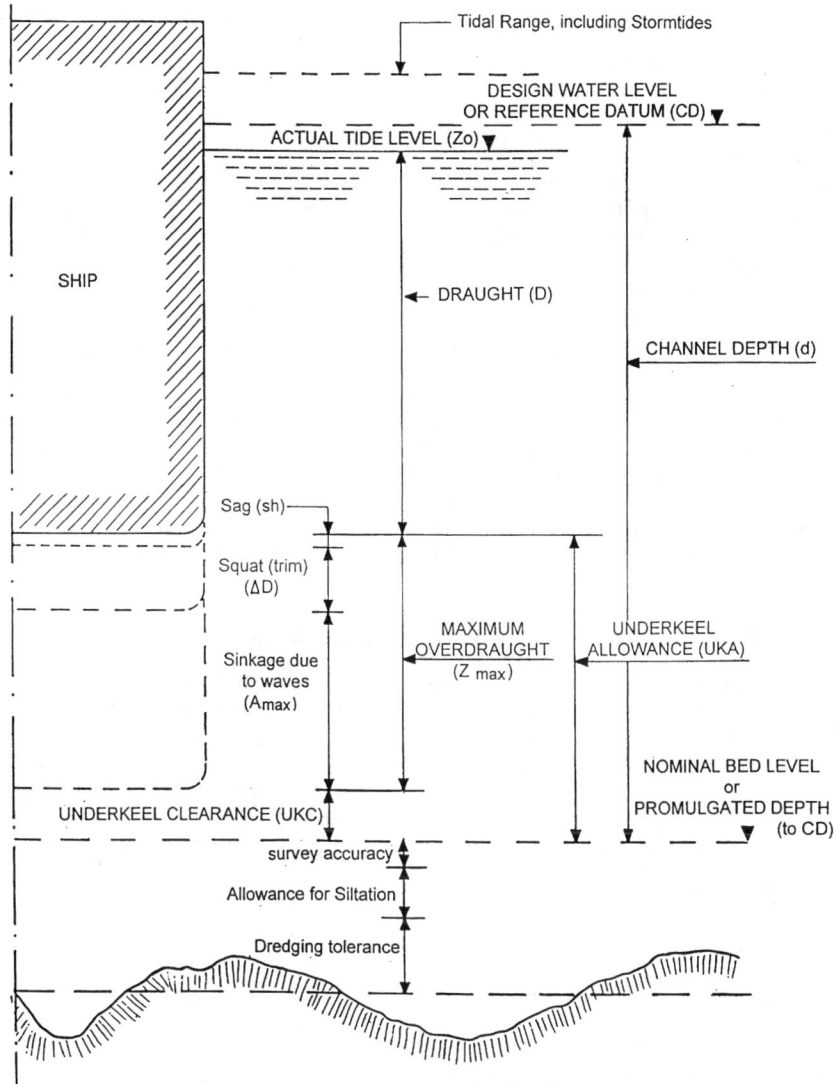

FIGURE 11.12. Factors affecting channel depth.

ka and Gullev, Barrass II, and Eryuzlu et al. (see references in PIANC, 1997) Of these, the formula for squat, from Barrass, which is the easiest to apply, is given here:

$$\Delta D = 1/30\ C_B[A^1/(A_w - A^1)]^{2/3} V_k^{2.08}$$

with A_w being the channel cross sectional area and V_k the ship speed in knots.

This equation applies to both open water and restricted channel conditions for d/D ratios ranging from 1.1 to 1.5, C_B from 0.5 to 0.9, and $V/\sqrt{gd} \leq 0.7$.

In Barrass (1979), a further simplification of the above equation is given:

$$\Delta D = C_B V_k^2 / 100$$

applicable to open water conditions (outer channel), and

$$\Delta D = 2 C_B V_k^2 / 100 = C_B V_k^2 / 50$$

for confined channel conditions.

The Barrass simplified equation is also given here as a basis for the preliminary design because its results compared well with prototype measurements carried out at Richards Bay (Zwamborn and Van Wyk, 1981; Van Wyk, 1982; CSIR, 1985). In fact, based on a comparison with nine different theories it was concluded in CSIR (1985) that the Richards Bay entrance channel squat data are best represented by the formulae of Barrass, Huuska and Eryuzlu, and Hausser, the same as those suggested for use in PIANC (1997).

For the detailed design of outer and inner entrance channels, the above squat formulae and/or those included in Appendix C of PIANC (1997) are considered sufficiently accurate, particularly, in the presence of significant wave action, when the vertical motions due to waves are much larger than the squat. In the turning and mooring areas, the ship speed is, generally, too low to cause significant squat (only a few centimetres). However, in the case of interconnecting shipping channels, special research, including tests with self-propelled model ships, may have to be done to optimize the channel depth design.

9.3. General Data on Wave-Induced Ship Motions

Compared with the data provided on squat, PIANC guidelines include very few quantitative data on vertical motions due to waves. This is rather surprising because this is one of the main factors in entrance channel design of major seaports.

According to PIANC (1973), the net underkeel clearance (UKC, see Figure 11.12) "... moving at planned passage speed under the action of the most severe planned tolerable wind and wave conditions" must be 0.5 m for sandy bottoms and 1.0 m for rocky bottoms. Some results on specific model tests on a 500,000 dwt tanker model are given but these cannot be applied to other designs.

In PIANC (1980), some 500,000 to 600,000 dwt model test data for the port of Antifer are given. The maximum "increase in draught," $Z_{max} = 0.39\ H_s$, is reported for $T_p = 10$ s. The importance of wave period is mentioned in a reference to low-amplitude oscillations, for which $A_{max} = 1.0\ H_s$. Preliminary guidelines are given for channel depth design, namely:

- $d = 1.2\ D$ for open seaways exposed to strong long stern or quartering swell
- $d = 1.15\ D$ for waiting areas exposed to strong long swell
- $d = 1.1\ D$ for less-exposed channels (inner channels)
- $d = 1.07\ D$ for turning and mooring areas

Even less attention is given to wave-induced ship motions in PIANC (1985), which deals specifically with "underkeel clearance for large ships." It is mentioned that the longer wave periods will have the greatest effect on ship motions, whereas a small underkeel clearance will reduce these motions, except for roll under roll-resonance conditions. A minimum maneuvering margin of 1.0 m is suggested.

In PIANC (1995) no data on vertical-wave-induced motions are included but design UKA values are given, as follows:

- 0.1 D for sheltered waters
- 0.3 D for waves up to 1 m in height
- 0.5 D for higher waves with unfavorable periods and directions

These values are significantly larger than the PIANC (1980) ones, presumably in an attempt to make better allowance for wave motions. Although they could be used for preliminary design, they are a very rough guide only.

The most recent approach-channel design guide (PIANC, 1997) maintains the above *UKA* values. For the detailed design, only general comments are made about wave-induced ship motions and using linear theory to determine vertical ship response. Also, physical model tests are recommended to determine shallow water ship motions.

PIANC (1997) does contain a detailed description of channels in muddy areas or the nautical depth concept. The nautical depth is the water depth relative to a level near the bottom where the density of the water/mud mixture is about 1200 kg/m³. Model test results on the ship behavior in muddy areas (ship speed, squat, forces, and maneuverability) are included in Appendix D of PIANC (1997). However, apart from the effect on squat, which is usually less for muddy conditions, again no data are provided on the effect on the wave-induced motions.

Several mathematical models of ship response in waves have been developed, for instance, Spencer et al.'s (1990) report on "UNDERKEEL" which is a frequency domain linear potential theory wave response model. This model can calculate, by linear superposition, the response to a multidirectional wave spectrum. The model is reported to give realistic results that are particularly valuable during the feasibility stage of the design. The model, however, cannot correctly reproduce roll response and, can therefore not be used reliably in the case of beam or near-beam waves, which often occur in outer harbor channels. According to the authors, in the detailed design stage, the results of UNDERKEEL should be checked in a physical model, which is absolutely essential in the case of beam waves.

9.4. Detailed Data on Wave-Induced Ship Motions

Comprehensive ship dynamics studies were undertaken at the South African Council for Scientific and Industrial Research (CSIR) laboratories in Stellenbosch over a period of some 10 years (1978 to 1988). These studies were initiated by the South African Ports Authority, which at the time considered the use of deeper-draught ships, particularly for the increased export of coal (presently about 67 million tons per annum) (Campbell and Zwamborn, 1977; CSIR, 1991). These studies included (see also Table 11.1):

- Field monitoring using shore-based equipment. Vertical motions (including squat) and horizontal tracks of over 200 loaded ships in excess of 100,000 dwt were recorded over a 2 km long section of the Richards Bay entrance channel between 1978 and 1984. See Figures 11.13 and 11.14 (Zwamborn and Van Wyk, 1981; CSIR, 1991).
- Physical model tests with self-propelled models of a 150,000 dwt and a 270,000 dwt bulk carrier. Some 3,000 test runs were made, both in a flat-bottom basin (basic tests) and in a 1 in 100 scale model of the Richards Bay port entrance (specific tests), using different wave directions, height and period distributions, and recording both the vertical motions and the horizontal tracks (van Wyk and Zwamborn, 1984, 1988).
- Numerical model runs for 100,000 dwt, 150,000 dwt, and 270,000 dwt bulk carriers using the three-dimensional source technique model VESDYN (OEC, 1983). The

TABLE 11.1. Three-legged ship motions study approach

```
                          ┌─────────────────┐
                          │ Ship dynamics   │
                          │ study approach  │
                          └────────┬────────┘
          ┌────────────────────────┼────────────────────────┐
          │                        │                        │
┌─────────────────┐      ┌─────────────────┐      ┌─────────────────┐
│   Prototype     │      │  Mathematical   │      │    Physical     │
│  measurements   │      │  model studies  │      │  model studies  │
└────────┬────────┘      └────────┬────────┘      └────────┬────────┘
         │                        │                        │
┌─────────────────┐      ┌─────────────────┐      ┌─────────────────┐
│ Development of  │      │      Model      │      │ Development of  │
│ monitoring      │      │   development   │      │ test facilities │
│ techniques      │      └────────┬────────┘      └────────┬────────┘
└────────┬────────┘               │                        │
         │              ┌─────────────────┐      ┌─────────────────┐
         │              │ Free-moving     │      │ Irregular wave  │
         │              │ ships, shallow  │      │ basin, ship     │
         │              │ water, irregular│      │ models,         │
         │              │ waves, time     │      │ monitoring eqpt │
         │              │ domain, viscous │      │ recording eqpt  │
         │              │ damping         │      │ data handling   │
         │              └─────────────────┘      │ procedure       │
         │                                       └─────────────────┘
┌─────────────────┐                                        │
│  Free-moving    │                              ┌─────────────────┐
│     ships       │                              │  Calibration    │
└────────┬────────┘                              │     tests       │
         │                                       └────────┬────────┘
┌─────────────────┐                              ┌─────────────────┐
│ Motion recording│                              │ Prototype data, │
│ roll, pitch,    │                              │ mathematical    │
│ heave, squat    │                              │ model data      │
│ position,       │                              └─────────────────┘
│ heading,        │                                       │
│ stopping        │      ┌─────────────────┐     ┌─────────────────┐
│ distances, tug  │      │     Model       │     │   Model tests   │
│ assistance, and │──────│  calibration    │─────└────────┬────────┘
│ environmental   │      └────────┬────────┘              │
│ data            │               │              ┌─────────────────┐
└─────────────────┘      ┌─────────────────┐     │ Underkeel       │
                         │ Prototype data, │     │ clearance,      │
                         │ physical model  │     │ channel         │
                         │ data            │     │ deepening,      │
                         └─────────────────┘     │ channel         │
                                  │              │ widening,       │
                         ┌─────────────────┐     │ acceptable wave │
                         │     Model       │     │ penetration     │
                         │  application    │     └─────────────────┘
                         └────────┬────────┘
                                  │
                         ┌─────────────────┐
                         │ Design of harbor│
                         │    entrances    │
                         └─────────────────┘
                                  │
                         ┌─────────────────┐
                         │ Port operation, │
                         │  emergencies,   │
                         │ safety          │
                         │ regulations     │
                         └─────────────────┘
```

FIGURE 11.13. Ship motion monitoring at Richards Bay.

model assumes that the wave–ship interaction is linear and it applies only to small motions and UKC exceeding about 0.2 D; after calibration against the physical model and prototype data, it was mainly used for sensitivity tests.

The detailed results of these studies are included in a series of laboratory reports and are summarized in the various publications mentioned. The data have been inter-correlated, as shown in Table 11.1, and the results are therefore considered reliable and, if correctly applied, should provide safe design and operational criteria.

The results of the basic physical model tests, carried out in the flat-bottom basin, can be used directly for the preliminary design, and in certain cases even for the detailed design (depending on the layout and the conditions) of the channel depths. Figure 11.15 shows the effect of wave direction for three ship draughts (D = 17, 19, and 21 m) for d/D = 1,3, V = 4 m/s, and T_p = 14 s. This figure provides the basic results. Figure 11.16 shows the effect of reduced UKC and Figure 11.17 shows the influence of wave period (CSIR, 1991). In the case of wave period, the encounter wave period must be used:

$$T_{pE} = T_p/(1 - VT_p \cos \alpha/\lambda_p)$$

To find the "corrected" significant response for a specific case, the relevant A_s/H_s values of Figure 11.15, $f(A_s/H_s)_\alpha$, are simply multiplied with the "correction" for draught, $f(d/D)$, and that for wave period, $f(T_{pE})$ or

$$A_s/H_s = f(A_s/H_s)_\alpha \times f(d/D) \times f(T_{pE})$$

For example, a ship with D = 19 m, V = 2.5 m/s (5 knots), wave incidence α = 135° (starboard shouldering), d = 24 m, d/D = 1.26, H_s = 3 m, and T_p = 14s or T_{pE} = 12.4s will have a significant motion response of:

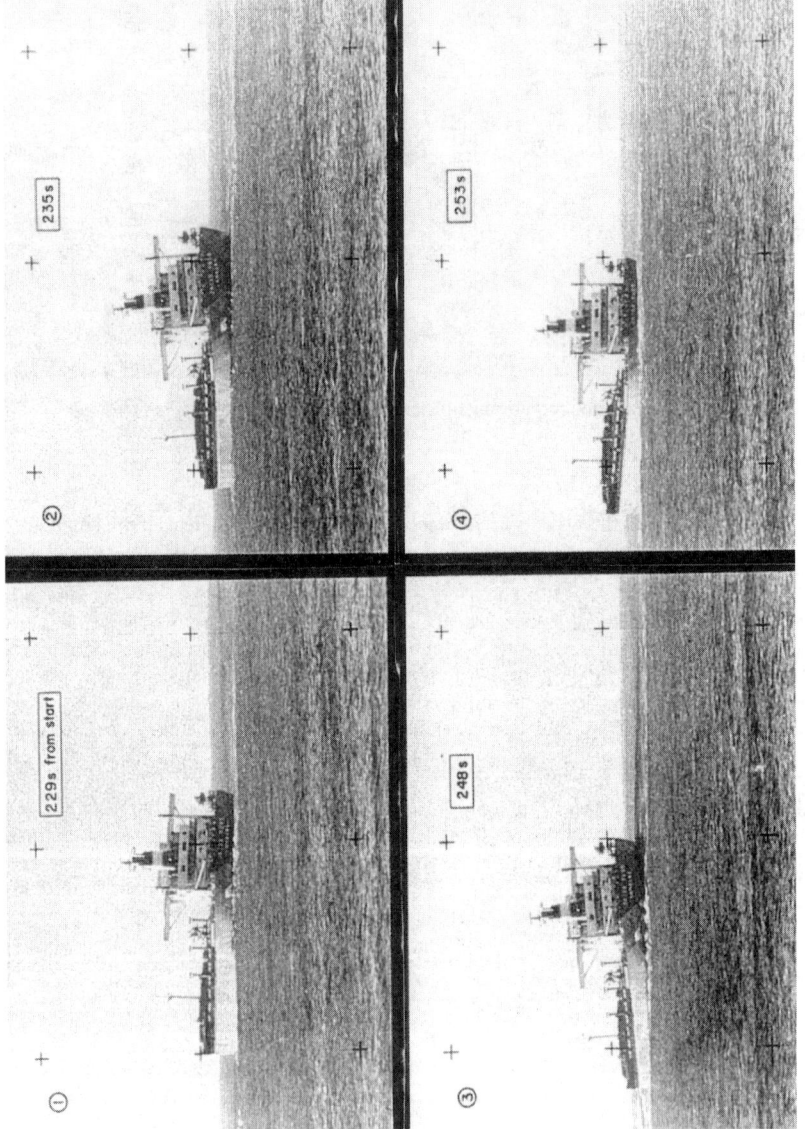

FIGURE 11.14. Recorded extreme roll, pitch, and heave motions.

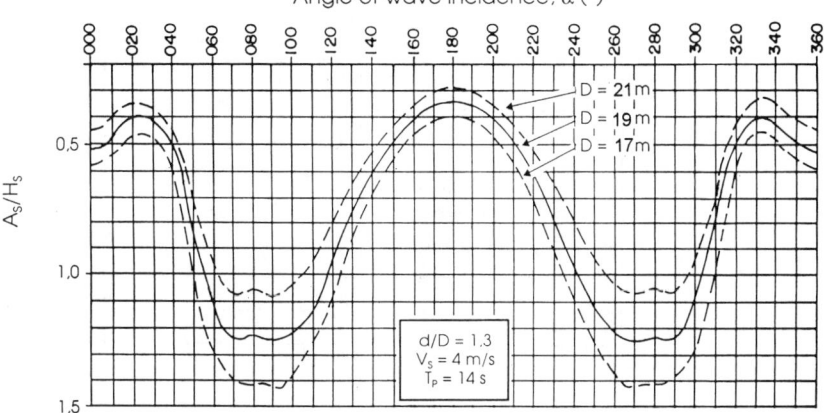

FIGURE 11.15. Significant hull motion response as function of wave direction and ship draught. (Reproduced from CSIR, 1991.)

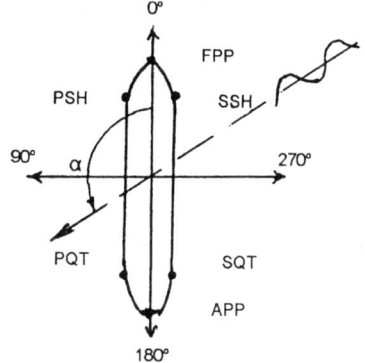

$$A_s/H_s = 0.65 \times 0.9 \times 1.3 = 0.76$$

or with $H_s = 3$ m, $A_s = 2{,}3$ m.

For the same ship, but with $\alpha = 90°$ ($T_{pE} = 14$ s):

$$A_s/H_s = 1.24 \times 0.9 \times 1.0 = 1.12$$

or with $H_s = 3$ m, $A_s = 3.4$ m.

The tests results also showed the following:

- Ship motions increase linearly for $H_s = 1$ to 5 m and $d/D > 1.1$.
- Wave spectral shape has little influence on the maximum response. Most tests were conducted with the normalized Richards Bay spectrum (with $\gamma = 2.74$), the narrow band Jonswap spectrum ($\gamma = 3.3$) showed 10 to 20% larger extreme motions (a

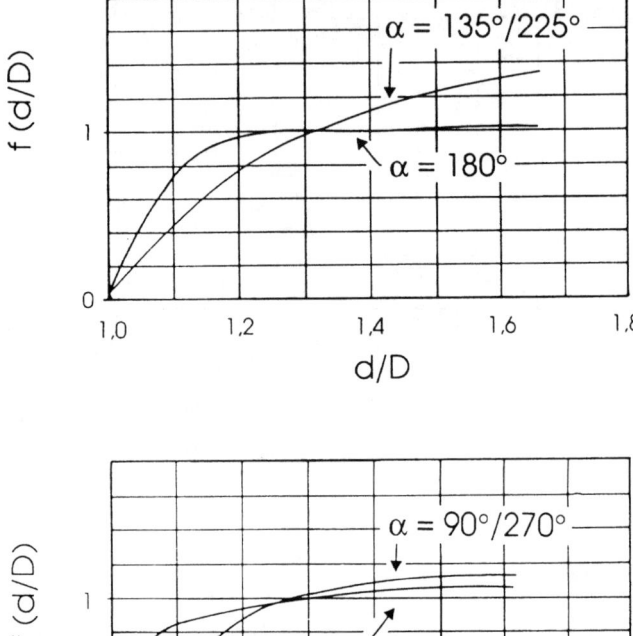

FIGURE 11.16. Hull motion response correction for underkeel clearance.

safety margin of 10 to 20% could be included if the spectral shape is thought to be very narrow).
- Most of the tests were conducted for ships with bilge keels. A limited number of tests without bilge keels showed much larger roll motions, up to twice as high for beam waves.
- The influence of directional spreading was found to be insignificant for near-beam (beam to quartering) waves and a somewhat larger increase was found for near-following and near-heading waves (the 10 to 20% safety margin for spectral shape would also compensate for any directional spreading effects).

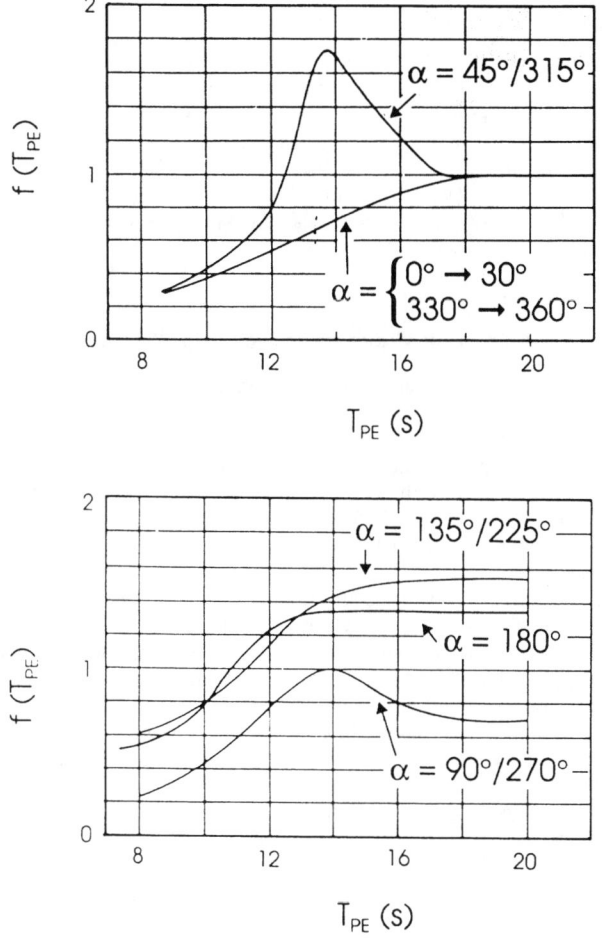

FIGURE 11.17. Hull motion response correction for encounter period.

The vertical motions were also found to be Rayleigh distributed, thus the maximum response can be calculated from

$$A_{max} = \frac{1}{\sqrt{2}} (\ln N)^{1/2} A_s$$

where N is the number of oscillations during passage.

Thus, for a 2 km channel, N for the above examples will be 64.5 and 57.1, respectively, and the maximum expected motion amplitudes (A_{max}) become

$\alpha = 135° \rightarrow A_{max} = 1.44\,A_s$ or with $A_s/H_s = 0.76$, $A_{max}/H_s = 1.09$,   thus $A_{max} = 3.3$ m

$\alpha = 90° \rightarrow A_{max} = 1.42\,A_s$ or with $A_s/H_s = 1.12$, $A_{max}/H_s = 1.59$,   thus $A_{max} = 4.8$ m

In addition, assume a narrow peaked spectrum plus directional wave spreading, which requires, say, a 20% increase so that for the shouldering wave case $A'_{max} = 3.9$ m and for the beam waves $A'_{max} = 5.7$ m. In addition, the squat must be added, namely $\Delta D = C_B V_k^2 / 100 = 0.83 \times 5^2/100 = 0.2$ m.

Since $d = 24$ m and $D = 19$ m, the shouldering wave case is in order ($D + \Delta D + A'_{max} = 23.1$ m) but the ship cannot enter or leave under the 3 m beam wave condition ($D + \Delta D + A'_{max} = 24.9$ m).

9.5. Additional Data on Wave-Induced Ship Motions

After the completion of the Richards Bay ship motion studies, the CSIR continued with basic vertical ship motion studies to cover the various ship sizes used in the South African ports. It was found from the Richards Bay results that the effect of ship speed ($V < 5$ m/s) on ship motions could be represented by using the encounter period, T_{pE}. The additional basic tests were therefore carried out at zero speed. Models of 270,000 dwt, 150,000 dwt, 110,000 dwt, and 65,000 dwt bulk carriers were tested at a scale of 1 in 100. Wave directions from 0° to 180° (45° steps), $d/D = 1.1$ to 2.0, $H_s = 3$ m, and $T_p = 10$ s were used.

The results of these extensive tests are included in the form of tables (A_s/H_s for eight different d/D values) in a series of four CSIR research reports. The results are very valuable, particularly for comparison with mathematical model data. However, since most of the tests were only conducted with one value of $T_p = 10$ s, the results must be properly interpreted before application. For instance, some sensitivity tests for the 150,000 dwt model ship using $T_p = 12$ s showed about three times greater A_s/H_s values. Thus, significant corrections would have to be made to these results if the design wave period deviates significantly from $T_p = 10$ s.

Finally, further physical model tests at zero speed were conducted for a 35,000 dwt model ship, the results of which were used to further calibrate the VESDYN model. It was found that this model cannot be used effectively for $d/D < 1.3$ without the introduction of viscous roll damping.

9.6. Determination of Channel Depth

There are two approaches for the determination of the required channel depth: the deterministic and the probabilistic approach (PIANC, 1985). In the deterministic approach, fixed allowances are made for the various uncertainties. Normally, maximum values are used throughout, which could result in a too conservative design depth. This can be avoided by selecting more realistic values in terms of their likelihood of occurring simultaneously. The probabilistic approach is based on mean values plus the possible variation (variance) for each factor. A fully probabilistic approach is generally not possible because of the lack of reliable data on the various variances. Therefore, a semiprobabilistic approach that includes risk of bottom contact is normally used.

Deterministic Approach. The following factors are involved (also see Section 2.3 and Figure 11.12):

- Overdredging 0.5 to 1.0 m, which covers the survey errors (0.2 m), dredging tolerance (0.3 to 0.5 m), and siltation
- Ship draught variation, including stationary sag, heel, and trim, about 0.3 m
- Overdraught, $Z_{max} = \Delta D + A_{max}$ (+0.3 m), which is the sum of sinkage due to squat and waves (and, where relevant, allowance for sag, heel, and trim)
- Minimum maneuvering margin for outer channels of 1.0 m. In the presence of waves, the minimum maneuvering margin is 1.0 m or A_{max}, whichever is the largest, because the vertical motions of the ship may penetrate the maneuvering margin (thus UKC $\geq$ 0; see PIANC, 1985).
- Reduced minimum maneuvering margins (*UKC*) of 0.7 and 0.5 m in the turning areas and inner channels

Thus, the channel depth relative to the design water level becomes:

$$d = D + \Delta D + 1.0 \text{ m} (+0.3 \text{ m}) \quad \text{or} \quad d = D + \Delta D + A_{max} (+0.3 \text{ m})$$

whichever is the largest. This is the depth relative to the accepted design or reference level and the nominal or promulgated bed level.

Semiprobabilistic Approach. Again, a minimum maneuvering margin of 1.0 m will be assumed for the outer channel. With regard to possible bottom contact, it is then accepted that the probability of nominal bottom contact during a particular maneuver under design conditions may not exceed a value p, where $p = 0.01$ was used for Europoort (van de Kaa, 1984). The value of $p = 0.01$ is equivalent to saying that there is a 99% guarantee of no bottom contact during any passage of a design ship during extreme (design) conditions.

The corresponding maximum vertical ship motion amplitude now follows from:

$$A_{max(p)} = \frac{1}{\sqrt{2}} (\ln N/p)^{\frac{1}{2}} A_s$$

and the required water depth becomes:

$$d = D + \Delta D + A_{max(p)} (+0.3 \text{ m})$$

Thus, for the example in Section 9.4 with $\alpha = 135°$, $A_{max(p)} = 2.09 \ A_s$ (instead of 1.44 A_s) $= 2.09 \times 0.76 \ H_s = 1.59 \times 3 = 4.77$ m or $d = 19 + 0.2 + 4.77 = 24.0$ m, thus only just sufficient (if no allowance is made for draught variation).

10. AIDS TO NAVIGATION

10.1. Definitions and Principles

According to the International Association of Ports and Harbours (IAPH, Section 4, 1979), the following definitions should be adhered to:

- A navigation(al) aid is any device or system used for ship maneuvering on board ship under control of the ship master.
- Aid to navigation is a shore-bound ship guidance system or device (operated by shore staff).
- Co-operative aid to navigation is normally operated by a shore authority but requires on-board equipment (e.g., Decca Navigator or a special harbor entrance electronic positioning system).

The aids to navigation are the responsibility of the port authority and they serve a very important purpose in harbor operation. They can form part of a vessel traffic services (VTS) system, which is used in many modern ports today to reduce operational and environmental risks in marine transportation (Young, 1994). The harbor (entrance) channel design must therefore include the evaluation and design of such systems. The preliminary aids design can be based on empirical data but the confirmation and optimization of the systems for the detailed design can best be done on a full-mission, real-time simulator where the pilots can check their efficiency.

The following aids can be considered:

- Leading lights and sector lights
- Navigation buoys (red, green, lighted)
- Fixed lights on prominant points (e.g., breakwater heads, corners of quays)
- Radar and special electronic positioning systems

Of course, existing natural or artificial markers, such as a prominent hill or escarpment, chimneys, or high buildings, can also be used for guidance, particularly in the approach to the port (see Section 12.2).

Approximate accuracies of some systems are given in IAPH, Section 4 (1979):

Type	Accuracies	Remarks
Buoys, light and radar reflector	30 to 40 m	Possibility of shifting, high maintenance cost
Leading-lights	8 m at 7 km* 140 m at 10 km 280 m at 14 km	Effectiveness depends on visibility
Radar	40 to 80 m at 5 km	Depending on type
Decca Hi-Fix	3 m at 5 km 4 m at 10 km	Cannot be used during electric storms

* Europort data.

10.2. Leading Lights

Leading lights provide an excellent guiding system that can be used by all ships, large and small, provided there is reasonable to good visibility. The strength of the lights, and thus their visibility, depends on the harbor entrance layout, but a range of at least 10 km is suggested.

As an example, details of leading lights at the port of Richards Bay are as follows:

			Distance (km)		Strength (10^3 CD)	
Light	Color	Height (m to CD)	Harbor Entrance	Entrance Approach Channel	Day	Night
Front	Red	27	6.7	10.2	4,800	57 (flashing)
Rear	White	50	9.2	12.7	3,600	57 (flashing)

The "correct" separation and height of the leading lights may be determined from the following IALA recommendations for use in simulation studies (PIANC, 1997):

$$H_{rl} = D_{fl}/650 + H_{fl}$$

$$D_{fl-rl} = KD_{fl}(H_{rl} - H_{fl})/w$$

where

H_{rl} = height of rear light above MHW
H_{fl} = height of front light above MHW
D_{fl} = distance from front light to limit of useful range
D_{fl-rl} = distance from front to rear light
K = coefficient of lateral sensitivity, optimum 2.5 but use of 1.5 is suggested for design
w = approach channel width

For the Richards Bay lights, with H_{rl} = 50 m and H_{fl} = 27 m, D_{fl} becomes 15 km (thus 5 km seaward of the entrance of the approach channel). With w = 300 m and D_{fl-rl} = 2,500 m, K = 2.2, which falls between the values of 2.5 and 1.5 quoted.

Recently, sector lights have become popular, particularly in close-to-port maneuvering situations. These lights consist of two or more different color lights that cover adjoining sectors, say three 10° sectors, total 30 °. These lights generally have a smaller range but they can be used effectively if there is a no-go area (red sector). These lights are being used in ports in Europe and they are being considered at present for improved approach safety in the ports of Ashdod and Haifa in Israel.

10.3. Additional Aids

Navigation buoys are of particular value in marking shipping hazards and shallow areas. They are also very effective in marking bends, particularly the start and end of bends.

Radar scanning equipment is essential when visibility is bad. Electronic position-fixing equipment, for instance using DGPS, can be very useful for large ships in narrow approach and entrance channels. However, this equipment must be taken on board by the pilot, which is a drawback.

11. HARBOR CHANNEL OPERATION

11.1. Operational Principles

For most navigation channels, particularly those directly connected to the open sea, there are certain operational restrictions, such as too low tidal levels and excessive wave action, or generally adverse weather conditions (strong wind, bad visibility, etc.). For optimum and yet safe operation it will be necessary to clearly define the limiting conditions and to accurately record the environmental conditions at the time of the maneuver, which must be transmitted to port control where, after proper interpretation, the master or pilot is advised whether or when the maneuver can take place. A flow chart of these logistics is included in Figure 11.18.

In principle, the following are required for a successful port operation system:

- Good communication system between port control and the vessel

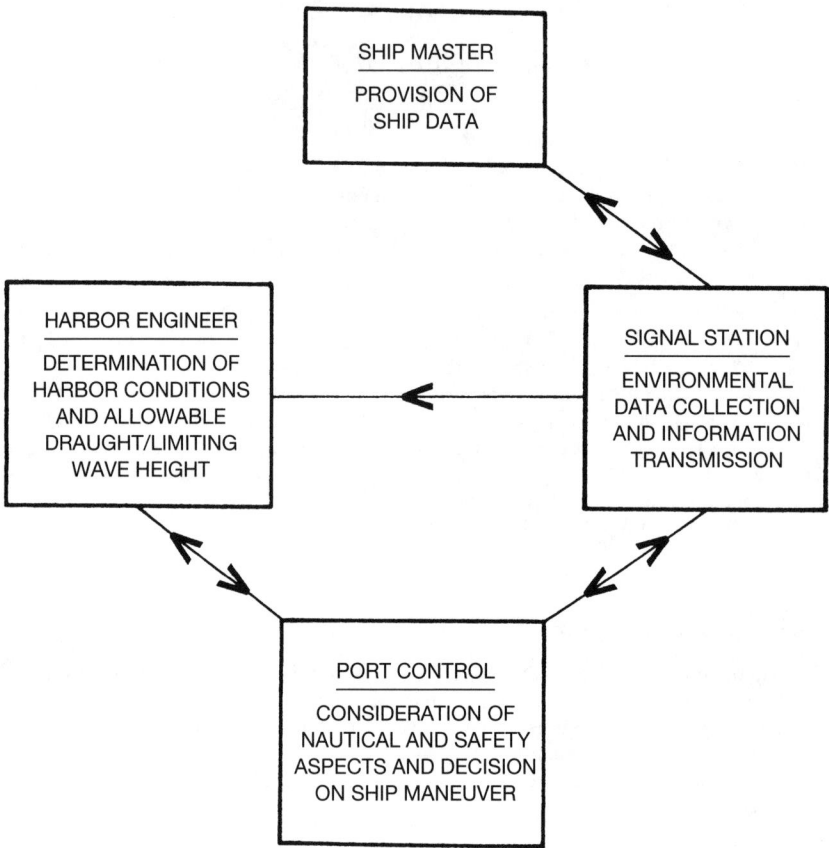

FIGURE 11.18. Example of port operation logistics.

- Accurate data on the ship and its loading conditions
- On-line data collection system at port control for tide levels, wave directions, heights, and periods, wind direction, speed, and visibility
- Accurate data on the channels with regard to available water depths (siltation)
- Operational charts or computer model at port control to determine the passage limitations (limiting water depth and/or wave conditions)

Standard equipment can be used for the communication links and the environmental data monitoring. The operational charts or computer model for the determination of the allowance criteria must, however, be specially developed based on detailed ship motion data.

NAVIGATION CHANNELS—DESIGN AND OPERATION 11.47

11.2. Operational Procedures

Sophisticated traffic guidance and control systems, including on-line data collection systems, are in use in a number of world ports, for example, at Europort in the Netherlands. The port of Richards Bay in South Africa, completed in 1976, was initially designed for the export of 12 million tons of coking coal in up to 150,000 dwt ($D = 17$ m) bulk carriers. The export capacity was doubled in 1978, increased to 44 million tons in 1984, further increased to about 50 million tons in 1990 (Dunn et al., 1985), and is presently well in excess of 60 million tons per annum.

In 1978, the South African Port Authority therefore initiated a major study program, carried out by the South African CSIR, to assist in coping with this enormous growth of coal export. The terms of reference were, briefly, to determine accessibility criteria for deeper-draught ships ($D > 17$ m) to use the existing channels (–24 m CD outer and –19.5 m inner channels) and to determine minimum dredging requirements to make the port accessible to ships with $D = 21$ m with maximum 1% downtime.

Based on these studies, an easy-to-use initial Port Operational Manual, Mark I, was prepared and was used at Richards Bay as early as in 1981 (Zwamborn and Cox, 1982; Campbell and Zwamborn, 1984). An updated guide, Mark II, which includes the results of extensive ship motion studies, refered to in Section 9.4, replaced Mark I in 1989. This manual was later computerized and improvements have been made, on a regular basis, to the data acquisition system. A flowchart of the operational procedures of this system is shown in Figure 11.19.

A proposed "ultimate operational scheme" prepared for the port of Richards Bay, including a detailed description of the various aspects, is shown in Tables 11.2 and 11.3. This includes a vertical (underkeel clearance) and a horizontal (maneuvering) mode.

11.3. Underkeel Allowance Criteria

Similar to the channel depth design discussed in Section 9.6, the operational criteria with regard to vertical ship motions should be based on the following:

- Nominal or promulgated channel depth
- Allowance for possible siltation in the channels, which will depend on the maintenance dredging program (ΔZ)
- Actual tidal level(s) at the time of the maneuver (Z_0)
- Allowance for squat (ΔD)
- Allowance for vertical ship motions due to waves, based on a probability of, say, less than 1% ($p = 0.01$) for touching bottom during one channel passage under extreme wave conditions ($A_{\max(p)}$)
- Minimum maneuvering allowance of 1 m in the outer channel, 0.7 m in the inner channels, and 0.5 m in the mooring areas (as discussed in Section 9.6)

Detailed data are needed for the determination of the expected vertical motions due to waves. The "basic" data, included in Sections 9.4 and 9.5, can be used for this if the local conditions are similar to those used in the model tests, otherwise site-specific studies for the relevant types of ships will have to be made. Numerical models could also be used, provided they have been properly calibrated, particularly regarding viscous roll motions.

Examples of draught allowance criteria in use at the port of Richards Bay are shown in Figures 11.20 and 11.21. The examples apply to the most critical wave period ranges,

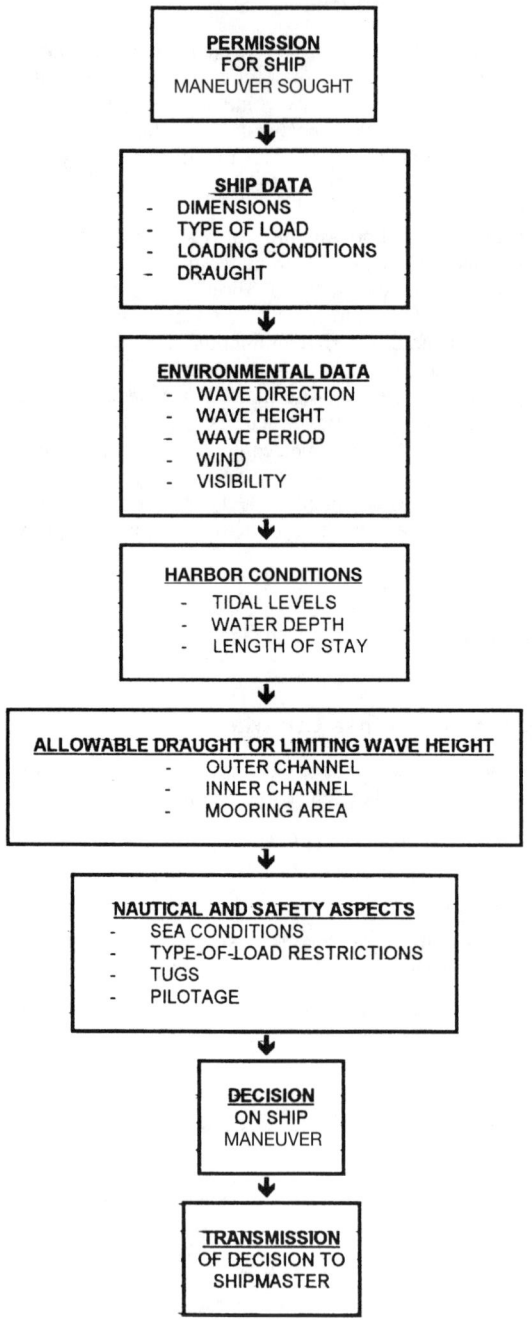

FIGURE 11.19. Example of port operation procedures.

TABLE 11.2. Underkeel clearance mode of port operation system

INPUT	TRANSMISSION	PROCESSING
Ship data Characteristic dimensions, draughts (d_{Fpp}, d_M, d_{APP}), postition of center of gravity, natural roll period, type of load	Preferably by telex from ship's captain or agents Alternatively, using normal radio links	Information to be entered into the control computer
Wave data Wave directions, wave heights, wave periods (applicable to wave recorder site)	By special radio link, which forms part of the wave recording system Results to be fed into control computer	Determine dominant direction, characteristic wave height (H_{mo}), characteristic wave period (T_p), directional spreading, wave energy spectrum, transfer of wave data to outer and inner channels, and mooring area
Harbor conditions Actual tide levels, predicted tide levels, channel depths	Tide records by overland cable Channel depths by existing telecommunication links from harbor engineer	Compare actual and predicted tides, determine tide level for time maneuver will take place, determine net water depths in different harbor areas
Underkeel clearance Amplitude response functions for different ships, load, water depth, and wave and/or simplified hydrodynamic ship motion model equation to determine squat	Stored in control computer	Determine maximum expected vertical sinkage, including squat for different harbor areas, determine net underkeel clearance in different areas, if $\geq$ accepted minimum value, maneuver can take place

$T_p \geq 11.4$ s for entry and $T_p \geq 14.5$ s for departure (see CSIR, 1991). There are similar charts for other period ranges because, apart from the wave direction, the wave period has a major influence on the vertical ship motions. These charts have now been converted into a user-friendly computer model, directly linked to the wave recording system. Typical outputs from this model are shown in Tables 11.4 and 11.5 (hypothetical case).

11.4. Port Accessibility

Port accessibility is determined by the type and size of ship, its loading condition, the occurrence of limiting wave and tide conditions with respect to the available channel depth,

TABLE 11.3. Maneuvering mode of port operation system

INPUT	TRANSMISSION	PROCESSING
Ship data — Characteristic dimensions, loading conditions, details of superstructure (windage), draughts	Preferably by telex from ship's captain or agents. Alternatively, using normal radio links	Information to be entered into the control computer
Wind data — Wind directions, wind velocities	By special radio link, data to be fed into control computer	Determine mean wind direction and speed over maneuvering period (also gusts)
Current data — Current directions, current velocities, extent of current field	By special cable or radio link	Determine average cross-current velocities in the entrance area (perpendicular to the channel axis)
Wave characteristics — Dominant direction, characteristic height (H_{mo}), characteristic period (T_p)	Available in the control computer from the underkeel clearance mode	
Maneuvering — Safe maneuvering speeds as functions of wave conditions and ship size for the given channel width and length	Stored in control computer	Determine expected drift angles due to wind and current effects, determine safe maneuvering speed for given ship and waves, transmit information to pilot

limiting sea, wind and current conditions with respect to pilotage and the use of tugs, and the occurrence of fog (bad visibility).

In the deterministic approach, channel downtime is based on the maximum expected vertical motion during a passage (A_{max}) of the design ship under low (conservative) or mean (more realistic) tide conditions. The limiting maximum vertical motion for a channel depth d, relative to the chosen tidal level, is found from

$$A_{max} = d - D - \Delta D$$

and the limiting significant motion amplitude follows from

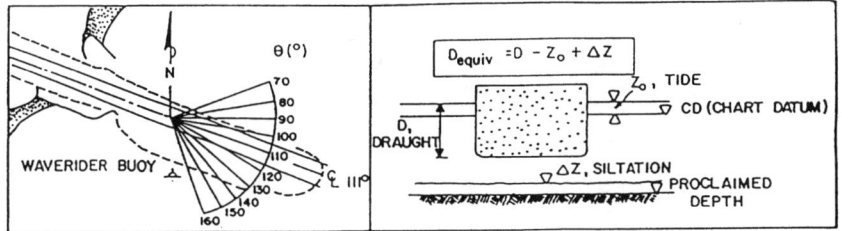

Reproduced from: CSIR (1991)

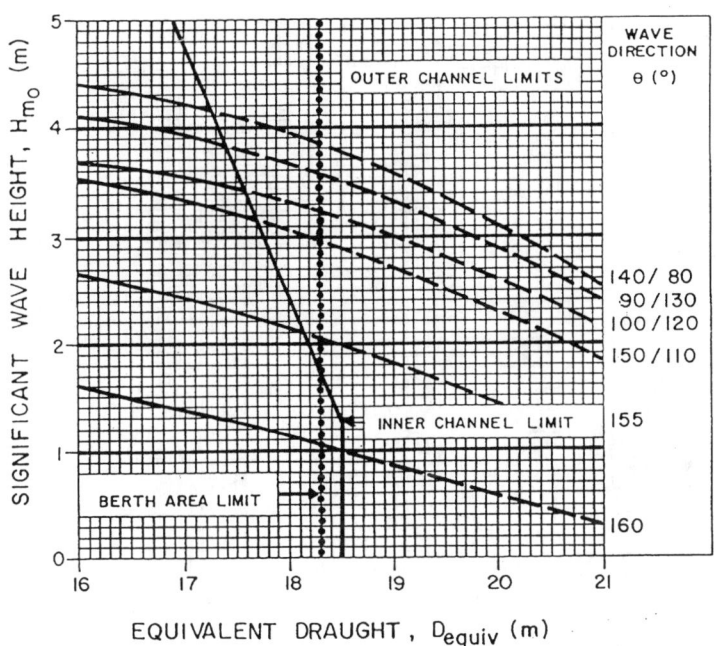

FIGURE 11.20. Draught allowances for bulk carriers entering Richards Bay Harbor ($T_p \geq$ 11.4 s).

$$A_s = A_{max} [\sqrt{2}/(\ln N)^{1/2}]$$

Using the joint occurrence distribution of wave direction, height and period and ship motion data such as given in Section 9.4 and shown in Figures 11.15 to 11.17, the percentage downtime can be found as the sum of all wave conditions that result in ship motions of the design ship exceeding the above limiting value. Ideally, this downtime should not be more than a few percent.

Alternatively, a more detailed probabilistic approach can be followed in which the various channel-depth-determining factors are integrated in such a way that all possible

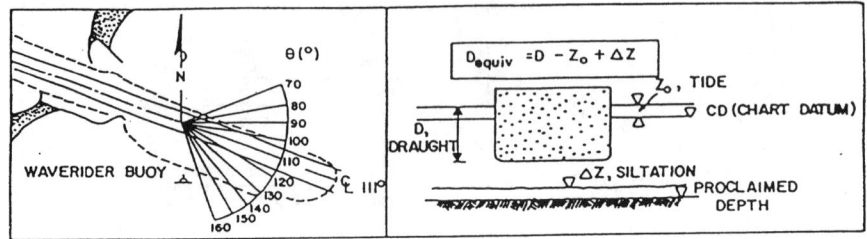

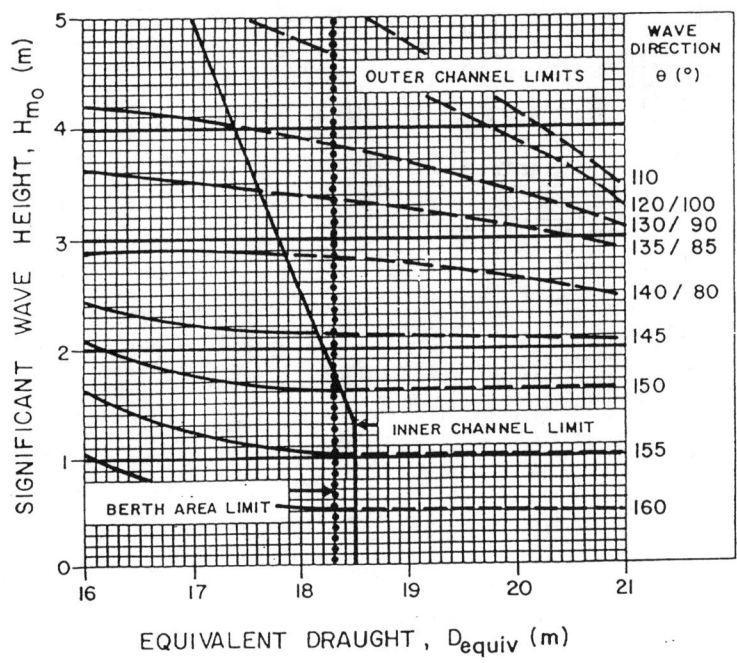

FIGURE 11.21. Draught allowances for bulk carriers leaving Richards Bay Harbor ($T_p \geq$ 14.5 s).

circumstances during the expected lifetime of the channel contribute to the downtime estimation in proportion to their respective probabilities of occurrence (PIANC, 1985; Strating et al., 1982; Savenÿe, 1996).

For this method, the following are needed:

- Joint frequencies of occurrence of extreme wave and tide conditions
- Expected average channel use by near-design ships over the design channel life (number of passages)

TABLE 11.4. Richards Bay computer system—input data

RICHARDS BAY DECISION SUPPORT		09h23:56
		WAVE
		F1-Height
SHIP MANOEUVRING AND		F2-Period
MOORING ANALYSIS		F3-Direction
		WIND
Del-Delete	PgDn-Browse	F4-Speed
		F5-Direction
SHIP NAME.........: KONKAR DIOS_		**AIR**
SHIP DRAUGHT (m)...: 18.00		F6-Temp.
SILTATION LEVELS:-		F7-Barometer
INNER CHANNEL....: 0.00		**TIDE**
OUTER CHANNEL....: 0.00		F8-Sea Level
		CURRENT
		F9-Speed
		F10-Direct.
ENVIRONMENTAL DATA	MEASURED	AGE (hours)
WAVE HEIGHT (Hmo)(m).: 3.00	0.00	856006.7
WAVE PERIOD (Tp) (s).: 12.00	0.0	856006.7
WAVE DIRECTION (deg.): 70.00	0	856006.7
TIDE (m)............: 0.50	0.00	856006.7
		LATEST DATA
		Alt+F1-View
Esc-Process		**MANOEUVRE**
		Alt+F2-Ship

TABLE 11.5. Richards Bay computer system—allowance criteria

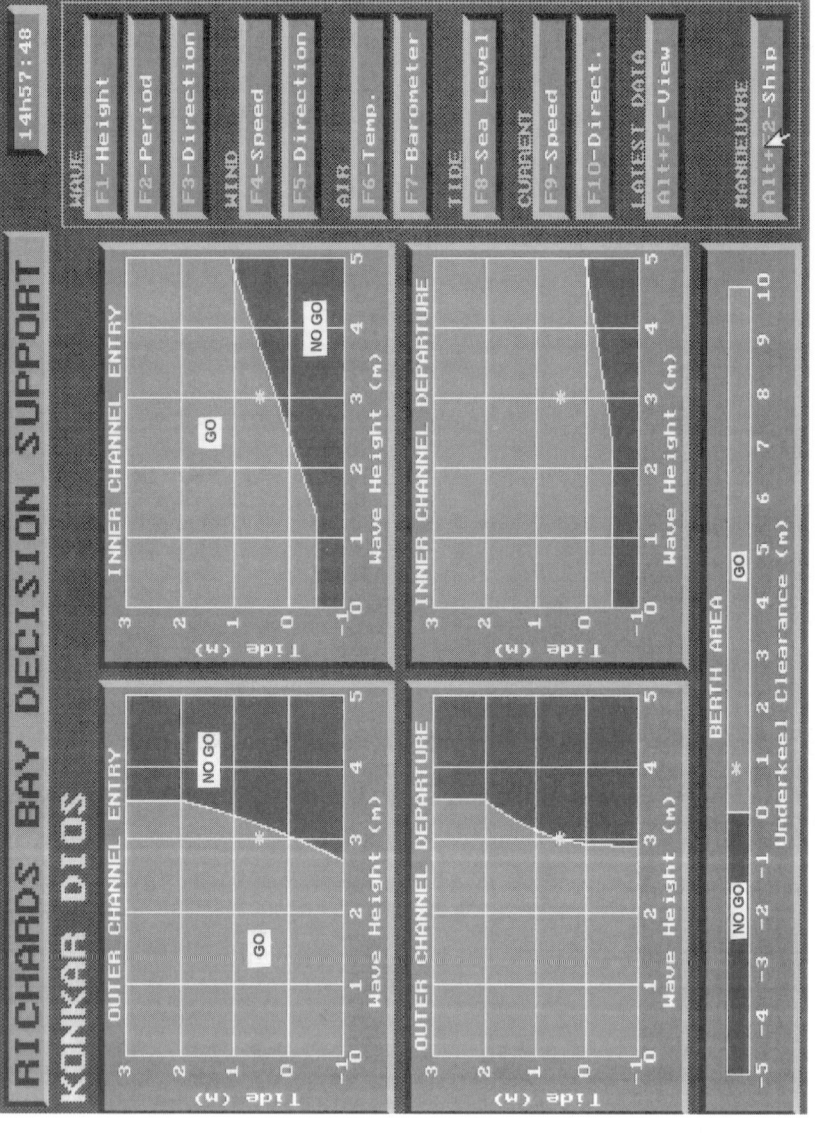

- Definition of the lifetime safety criterion, that is, the probability that the number of bottom contacts during the channel lifetime does not exceed a given number k, must be limited to a small value p_1 (this means that one can be $(1 - p_1) \times 100\%$ sure that no more than k bottom contacts will occur during the lifetime of the channel)
- Expected maximum overdraught, Z_{max}, during one channel passage with a probability of nonexceedance of p (e.g., $p = 0.01$)

For example, for the design of the Europort channel, a probability of exceedance of 10% ($p_1 = 0.1$) was accepted for a 25-year lifetime of the channel. The Poisson distribution can be used to relate the occurrence of a rare event to the channel lifetime, namely,

$$p_1 = 1 - \left(1 - \frac{1}{t_1}\right)^{T_l}$$

or

$$\frac{1}{t_1} = 1 - (1 - p_1)^{1/T_l}$$

where

p_1 = probability of bottom contact during the channel lifetime
T_l = channel lifetime (years)
t_1 = recurrence interval of event, i.e., bottom contact (years)

Thus, with $T_l = 25$ years, the expected frequency of bottom contact becomes

$$\frac{1}{t_1} = 1 - (1 - 0.1)^{1/25} = 0.0042$$

and thus $t_1 = 237$ years. Alternatively, the number of bottom contacts, k, for the design life of the channel would be $k = 25/237 = 0.105$.

Sensitivity studies for the port of Ashdod (Israel) using values for $k = 1$, 1, and 3 combined with $p_1 = 0.05$, 0.1, and 0.1, respectively, showed very little difference in the results, thus the choice of these values does not appear to be critical, although the variation in p_1 and k values was not very large.

For a particular port, the choice of probabilities should be linked to risk, which is the product of probability and consequence (degree of expected damage to the ship, spillage of oil, etc). The probabilistic method is described in detail in Strating et al. (1982) and Savenÿe (1996).

Although the conditions causing channel depth downtime will also often cause nautical problems (ship handling, use of tugs, pilotage), some additional downtime could occur as a result of nautical restrictions, for instance, very strong winds and heavy fog. Experience in South Africa's major ports has led to the following approximate limiting ship handling criteria:

- Problems with pilot boarding for $H_{max} > 5$ m or $H_s > 3$ m
- Ship handling problems for wind speeds exceeding 30 to 40 knots
- Problems with visibility when less than 1 mile

Downtimes as a result of the nautical restrictions can thus be estimated from wave, wind, and visibility records. However, if these occur concurrently with the channel restrictions, they need not be added to the total downtime.

12. EXAMPLES OF HARBOR CHANNEL DESIGN

12.1. Port of Richards Bay

The Richards Bay coal port was designed in the period 1969 to 1972 (Campbell and Zwamborn, 1977). The design of the entrance channel was based on empirical data and limited physical model tests with a 150,000 dwt model ship at a scale of 1 in 100, using regular waves only. The construction of the port was completed in 1976 and the detailed ship motion studies involving prototype measurements, physical model tests with irregular waves, and numerical modeling only started in 1978, as described in Section 9.4.

Since the data on which the channel design was based were very limited, the basic test results, described in Section 9.4, will be used to carry out a preliminary design check, whereafter a more detailed evaluation of the design will be discussed, based on the prototype measurements at Richards Bay and the results of the site-specific model tests of the Richards Bay entrance channel.

Preliminary Design. The following design conditions were used in the original design (see Campbell and Zwamborn, 1977):

- Design ship—150,000 dwt, $L = 300$ m, $B = 45$ m, $D = 17$ m (possible future design ship 250,000 dwt, $L = 350$ m, $B = 52$ m, $D = 20$ m)
- Waves—median wave period 12 s

Wave directions (deep sea)	Occurrence (%)	Entrance area direction (°)	1% wave height (m)
SSE (157.5°)	26.7	145	2.7
SSW (202.5°)	8.7	166	2.5

- Tides—0.5 m neap tide, 2.1 m springtide
- Currents—seldom exceeding 0.5 m/s (parallel to the shore)
- Wind—dominant directions, NNE and SW, with the latter being the stronger

The harbor layout, together with the relevant data are shown in Figure 11.22.

The channel alignment was fixed by the presence of a subbottom channel, which could be easily dredged.

The length of the channel was based on the fact that entry velocities of 6 to 8 knots are needed for safe maneuvering under design conditions. Also, tugs cannot make fast outside the protection of the breakwaters because of the dominant 1.5 to 2 m high swell and the high ship speed. Tugs were assumed to make fast when the ship speed is reduced to 3 to 4 knots, which means well inside the inner channel. Experience has shown that the available 4.6 km inside straight channel length is sufficient for the stopping maneuvers but, in view of a possible future design ship of 250,000 dwt, a total length of 6.1 km was provided for (Campbell and Zwamborn, 1977).

The channel width was based on experience and the initial physical model tests. The 300 to 400 m width for the outer channel shown in Figure 11.22, represents 6.7 to 8.9 B for the 150,000 dwt ship (5.8 to 7.7 B for the 250,000 dwt future ship). Using the PIANC (1997) approach discussed in Section 8.2 for single-lane traffic, the required channel width becomes, for $w_a = 0.0\ B$, $w_b = 0.4\ B$, $w_c = 0.7\ B$, $w_d = 0.0\ B$, $w_e = 1.0\ B$, $w_f = 0.1\ B$, $w_g = 0.1\ B$, $w_h = 0.1\ B$, $w_i = 0.0\ B$ or $w_{i(total)} = 2.4\ B$ and thus

$$w = w_{BM} + w_{i(\text{total})} + w_{Br} + w_{Bg} = (1.5 + 2.4 + 0.5 + 0.5)B = 4.9\,B$$

This is significantly less than the available width at Richards Bay. Ships do, however pass at times in the 300 m wide inner channel, indicating that the channel width may be overdesigned, but this is only done under favorable conditions. The PIANC method, therefore, does appear to be nonconservative. This is also borne out by the prototype monitoring at Richards Bay of ship tracks, the results being shown in Figure 11.8.

Much attention was given to the required channel depth for the port of Richards Bay. Based on tests with the 150,000 dwt design model ship at a scale of 1 in 100, the maximum vertical motion in the outer channel was found to be 6.0 m for the most critical SSW waves with 2.5 m height and 11.6 s wave period. This was for the model ship without bilge keels and thus conservative (with bilge keels, the maximum motion was about 5 m). Based on this, the channel depth was −24 m CD, leaving 24 − 17 − 6 = 1 m UKC at low water.

Using the basic flat-bottom detailed test results discussed in Section 9.4, the following is found, using the original design conditions given above; ship speed of 3.5 m/s, wave length in 24 m depth, $\lambda_p = 160$ m, an outer channel length of 4.5 km and a channel direction of 111° [for the critical SSW deepsea waves, $\alpha = 180° - (166° - 111°) = 125°$ for a departing ship]:

$$T_{pE} = \frac{12}{1 - 3.5 \times 12 \times \cos 125°/160} = 10.4\,\text{s}$$

thus, by using the data given in Figures 11.15 to 11.17 ($D = 17$ m and $d = 24$ m):

$$A_s/H_s = 1.0 \times 1.1 \times 0.8 = 0.9$$

and with $N = 4500/10.4 \times 3.5 = 124$

$$A_{\max}/A_s = \frac{1}{\sqrt{2}}\,(\ln 124)^{\frac{1}{2}} = 1.55$$

thus

$$A_{\max} = 1.55 \times 0.9\,H_s = 1.4\,H_s$$

With $H_s = 2.5$ m (1% SSW wave height), $A_{\max} = 3.5$ m and the required channel depth becomes

$$d_{\text{req}} = D + \Delta D + A_{\max} = 20.5 + \Delta D$$

Assuming semiconfined channel conditions, according to Section 9.2

$$\Delta D = 1.5 C_B V_k^2/100 = 1.5 \times 0.83 \times 7^2/100 = 0.6\,\text{m}$$

and thus

$$d_{\text{req}} = 20.5 + 0.6 = 21.1\,\text{m}$$

Even allowing for possible draft variation, sagging, etc., say 0.3 m, the required depth is still only 21.4 m, thus 2.6 m less than the available depth.

Detailed Design. The apparent overdesign of the channel depth came somewhat as a surprise at the time. However, some 200 shipping events had been monitored at Richards Bay and the recorded maximum vertical motions ($A_{\max}$) of 70 of the more important events are correlated in Figure 11.23 with the significant wave height (H_s). These results showed that $A_{\max}/H_s$ ratios of slightly above 2.0 occurred in real life, thus significantly more than the "maximum" value of 1.4 found above. Thus, for SSW waves with $H_s = 2.5$m, $A_{\max}$ would become 5.0 m instead of the 3.5 m found above.

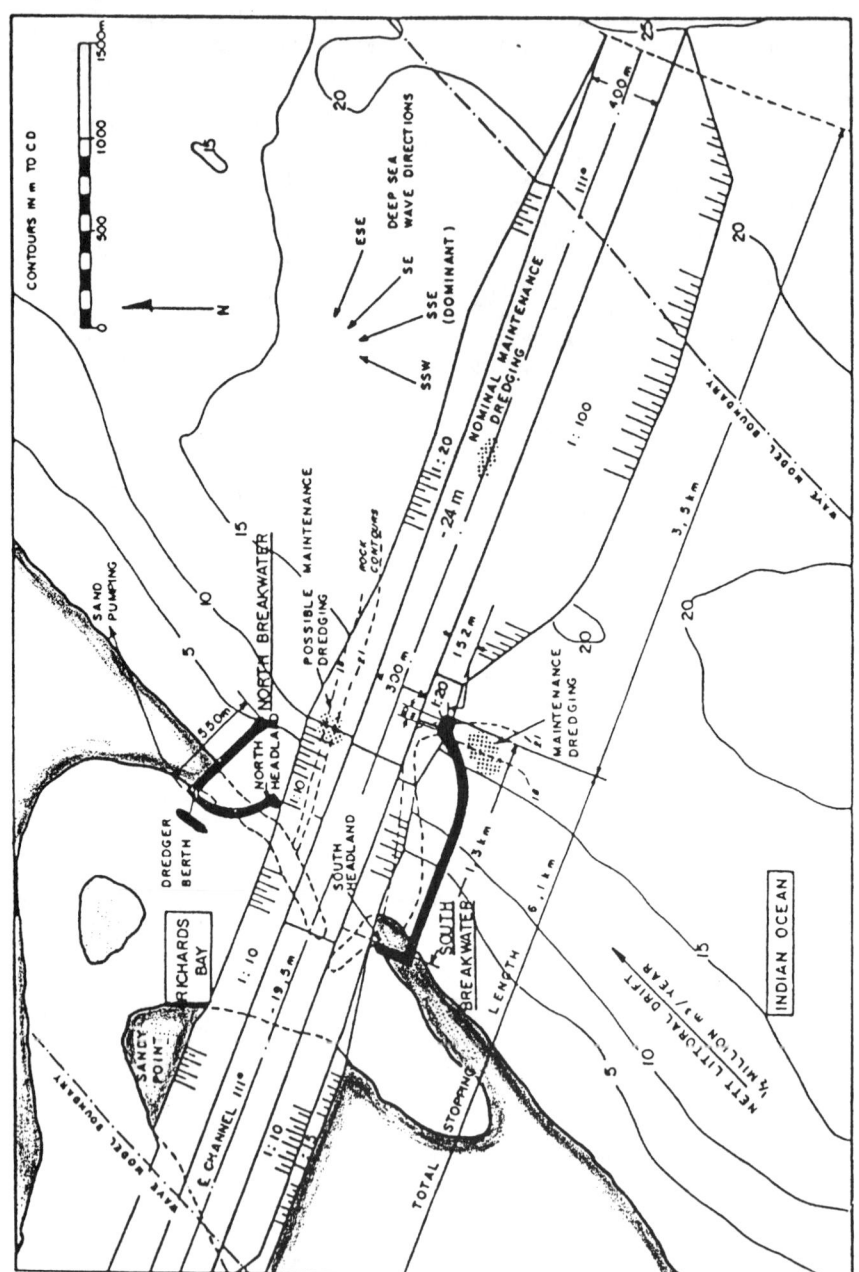

FIGURE 11.22. Richards Bay entrance channel design for 150,000 dwt ships.

However, detailed tests were done in a 1 in 100 scale model of the Richards Bay entrance channel to record the actual vertical motions for design wave conditions from the principal deep sea direction sector, namely E to SSW. For the input conditions at the model boundary (Figure 11.22) due allowance was made for selective wave refraction for waves of different periods, thus introducing a directional spreading at the model boundary. This resulted in significantly different motions in the entrance channel than found from the unidirectional basic tests, particularly for the departing loaded vessels. This becomes clear from Figure 11.24 which shows the Richards Bay model data, the

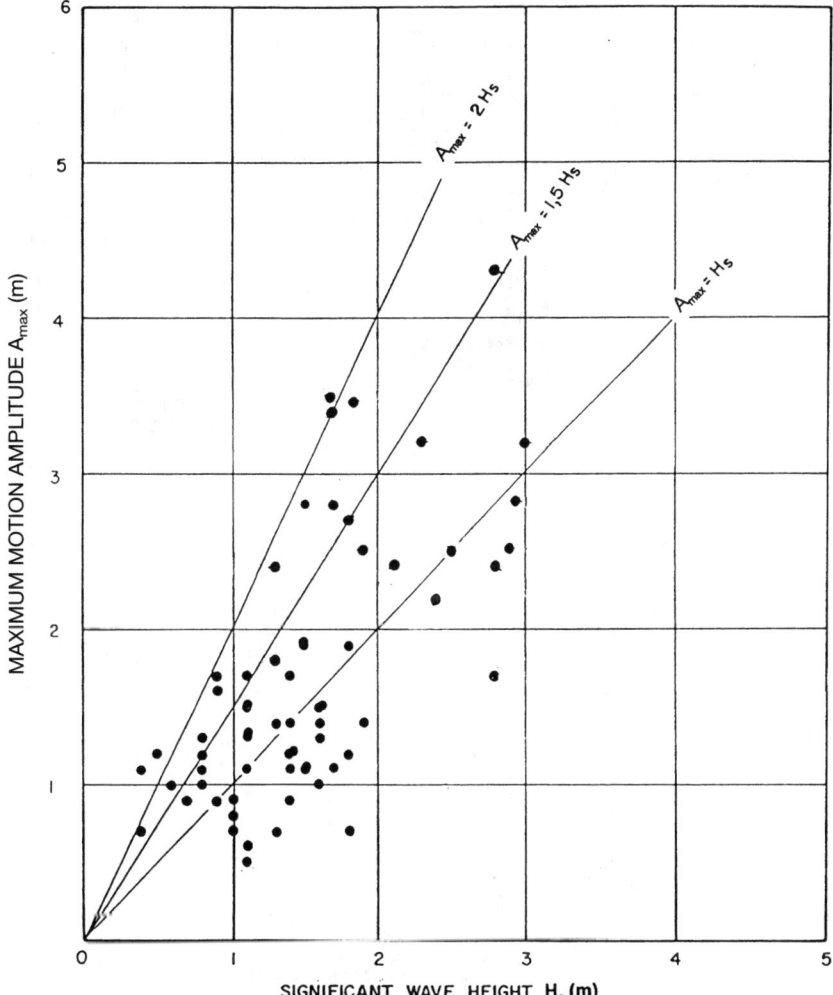

FIGURE 11.23. Recorded maximum wave-induced vertical ship motions at Richards Bay (CSIR, 1991).

prototype data (which compare very well), and an example of the basic test results for $T_p = 14$ s. The Richards Bay model data for $T_p = 14$ s (with directional spreading) shows A_s/H_s values up to twice as much as the basic test results. This was attributed to a shift of wave energy, due to refraction, toward near-to-beam wave directions (refer A_s/H_s values in Figure 11.15 for $\alpha < 125°$). Thus, in the case of significant refraction effects as experienced at Richards Bay, the basic flat-bottom test results can give incorrect results, which, in this case, would have resulted in an underdesign of the channel depth.

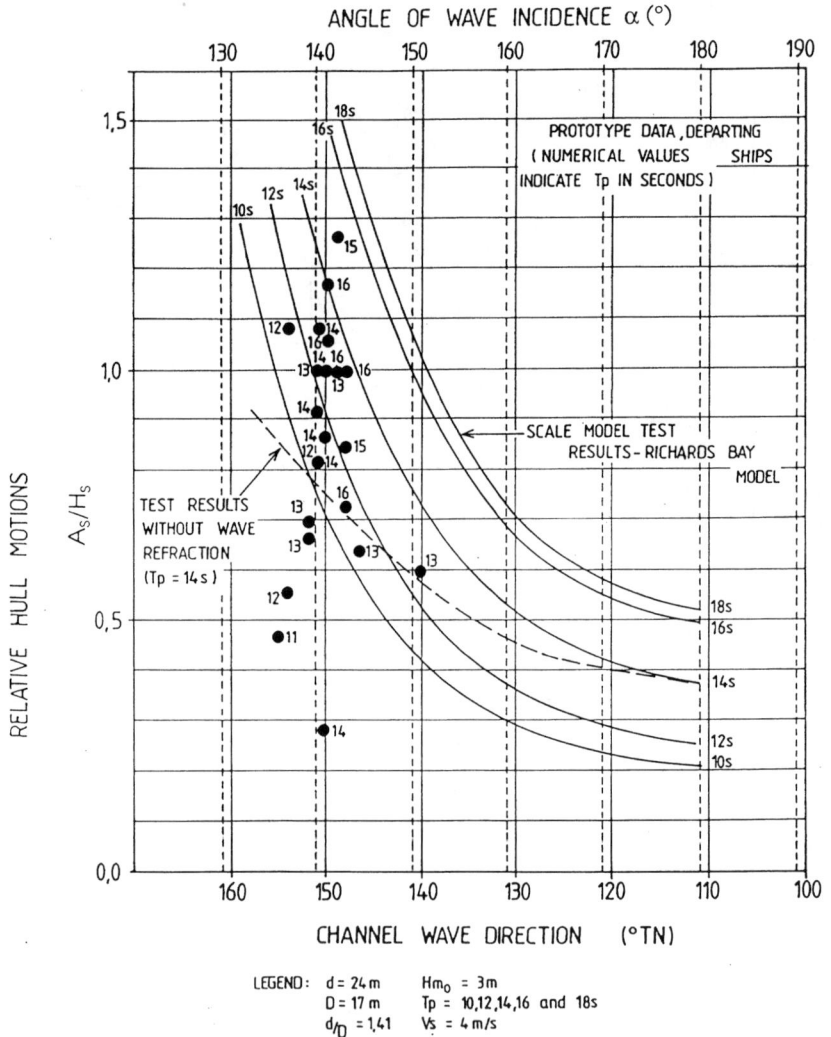

FIGURE 11.24. Comparison of basic and Richards Bay model hull motions with prototype data (CSIR, 1991).

Figure 11.24 gives a value of $A_s/H_s \simeq 1.6$ for the 12 s design wave and $\alpha = 130°$ (close to SSW). Thus, the "corrected" A_{max} value becomes:

$$A_{max} = 1.55 \times 1.6\, H_s = 2.5\, H_s$$

or, for $H_s = 2.5$ m

$$A_{max} = 6.25 \text{ m}$$

and

$$d_{req} = 17 + 0.6 + 6.25 = 23.85 \text{ m}$$

leaving 0.15 m for possible draft variations. Thus, the detailed studies, in principle, confirmed the channel depth of –24 m CD.

Based a probabilistic approach it was also concluded from these studies that ships up to 18.3 m draught can use the present port safely with only 1% downtime and another 1 to 2% waiting time (due to tide).

12.2. Port of Haifa

The existing port of Haifa is situated at the southern end of Haifa Bay and, due to the presence of the Carmel headland, is semiprotected from the dominant W to WNW'ly waves, as can be seen in Figure 11.3. Major extensions are planned for the port in three stages, B, C, and D, as shown in Figure 11.25. Only Stage B will be constructed in the foreseeable future and the main design effort is therefor concentrated on Stage B. However, Stages C and D are also considered in the design to ensure their easy implimentation when the need arises at some future date. A description of the nautical design aspects of the Haifa port extensions is included in Zwamborn et al. (1998).

Preliminary Design. The following are the design conditions for the Haifa B extensions:

- Design ships—post Panamax (PP) container vessel, 84,000 dwt, $L = 318$ m, $B = 42.8$ m, $D = 14$ m; bulk carrier, 150,000 dwt, $L = 295$ m, $B = 44$ m, $D = 17$ m
- Waves—refracted design-wave data in the entrance area are as follows (in %):

	$H_{mo} > 1.5$ m	
Wave direction	Long-term data, 1958–1996	Directional waverider, 1993–1996
NW	3.43	0.77
NNW	0.53	0.10
All directions	3.96	0.87

- Tides—spring tides –0.1 m to +0.3 m LSD; neap tides 0.0 m to +0.2 m LSD (LSD = land survey datum $\simeq$ – 0.1 m to mean water level)
- Currents—there are no current measurements near the entrance area but currents are known to be weak in Haifa Bay (an easterly current of 0.3 m/s, mean over depth, was calculated numerically in the entrance area for a 40 knot W'ly wind)
- Wind—dominant directions W, NW, and SE; strongest winds from W, E, and SE; wind speeds > 10 m/s (20 knots) occur 2.9% (all directions) and > 15 m/s (30 knots) occur less than 3 days/year (< 0.8%)

It is clear from these data that the conditions in the approach to the port are generally calm and since tugs can make fast in waves up to $H_{mo} = 1.5$ to 1.8 m (refer to Section 5.5), at least 96% of the time tugs can assist ships with the entry/departure maneuvers, if required, in the open bay area (outside the breakwater protection).

The entrance channel alignment was the subject of extensive studies and various options were considered. The approach route from the open sea is virtually fixed due to the presence of extensive reefs (e.g., the Talbot Reef) and the proposed 500 m main breakwater extension should go as far easterly as possible to provide better protection against wave penetration. The alignment shown in Figure 11.25 with a 53° single-radius (1,100 m) bend was eventually adopted in the initial design phase. The radius of the bend is about 3.5 L (longest ship), which agrees with the criterion of > 1.8 to 2 L of Section 6.3, although it is not considered very relevant in this case because of the possibility of tug assistance in the bend area.

Since the ships can enter at low maneuvering speeds (< 4 to 5 knots) with tugs already attached, if required, a relatively short stopping length, l, is needed. According to Section 7.2, $l = 3$ to $4 L$ or 945 to 1,260 m. The minimum available stopping distance from the new breakwater head to the southern end of the turning circle is about 1,100 m. However, the PP container vessel must berth at Quay no. 2 and the bulk carrier at Quay no. 3 so that both ships, on entry, can continue on the 1,100 m radius curve, providing additional stopping length.

The channel width can be estimated according to the PIANC (1997) approach discussed in Section 8.2, assuming single-lane traffic. For the open channel case $w_a = 0.0\ B$, $w_b = 0.5\ B$, $w_c = 0.3\ B$, $w_d = 0.0\ B$, $w_e = 0.5\ B$, $w_f = 0.1\ B$, $w_g = 0.1\ B$, $w_h = 0.2\ B$, $w_i = 0.0\ B$, or $w_i(\text{total}) = 1.7\ B$ and thus

$$w = (1.5 + 1.7 + 0.3 + 0.3)B = 3.8\ B$$

Thus for $B = 42.8$ m, $w = 162.6$ m.

Assuming $d/D = 1.15$ to 1.2, the bend width $w_s = 1.25$ to 1.3 B, according to Section 8.3. Since $w_{BM} = 1.5$ was used in the above (moderate ship manoeuvrability), no extra widening would be needed, according to PIANC (1997). Nevertheless, a minimum channel width of 300 m was adopted at the new breakwater head and about 330 m in the bend for the preliminary design, thus $w = 7.0$ to $7.7\ B$ (which is similar to the Richards Bay example).

The outer channel depth can again be determined using the basic test results discussed in Section 9.4. The more critical long-period waves are NW'ly (315°) in the approach and entrance area, thus with the channel direction being 126°, $\alpha = 351°$ (port side near-following waves) for an entering ship. However, in the bend area, just before passing behind the breakwater extension, the ship is heading due south or 180°. For the same NW'ly waves, $\alpha = 45°$ (starboard quartering waves). For the design of the channel depth, both these two cases must be investigated. The 1% occurrence H_s and T_p values in the entrance area are between 1.5 m and 2.0 m (waverider and long-term data) and 11 s (waverider), respectively. For the determination of channel depth, $H = 2$ m will be used with $T_p = 11$ s, a wavelength in 18 m depth, which is the general depth in the area concerned, $\lambda_p = 132$ m, and a ship entry speed of $V = 2$ m/s (4 knots). The bulk carrier with $D = 17$ m is determinative for the channel depth. They will come in loaded and depart light, thus entry is the critical case.

The two cases give the following results: $\alpha = 351°$ (approximate channel length 2 km)

$$T_{pE} = \frac{11}{1 - 2 \times 11 \times \cos 351°/132} = 13.2\ s$$

Using Figures 11.15 to 11.17 with (assumed) $d = 20$ m and thus $d/D = 1.18$:

$$A_s/H_s = 0.56 \times 0.9 \times 0.65 = 0.33$$

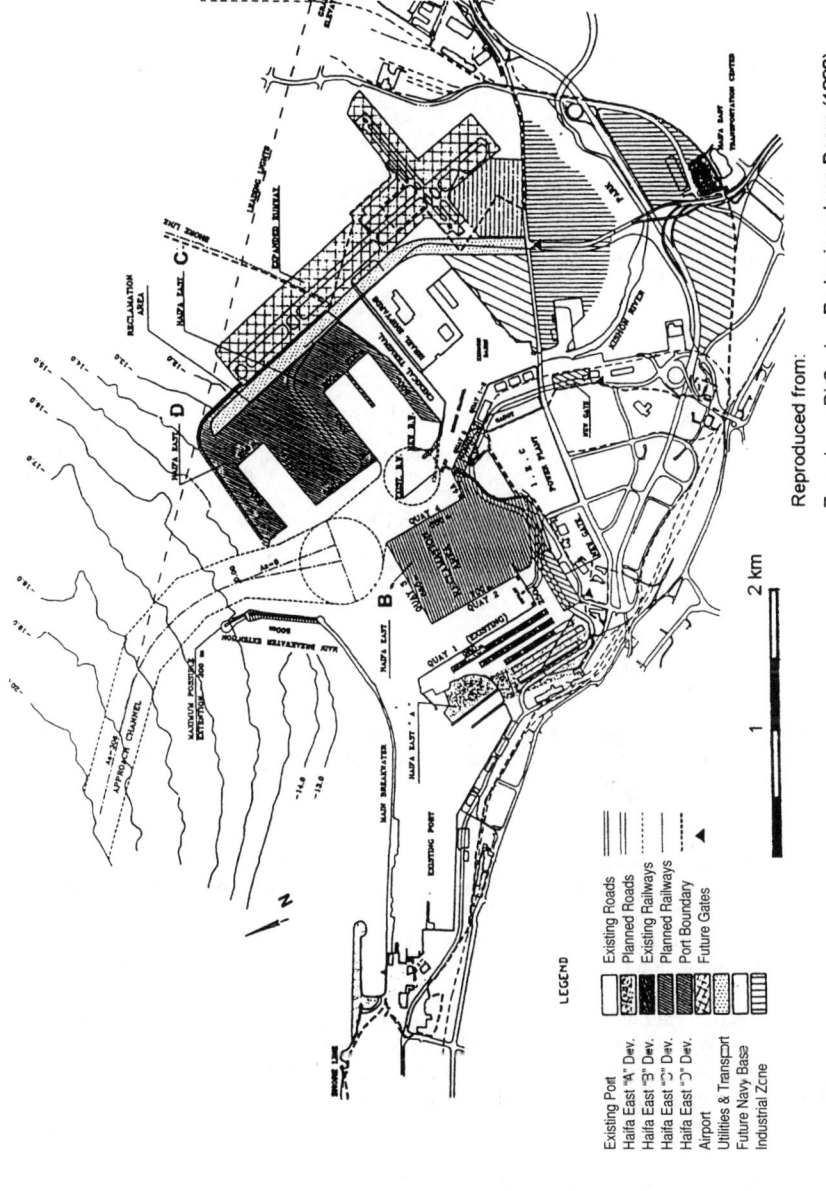

FIGURE 11.25. Port of Haifa development stages B, C, and D (Zwamborn et al., 1998).

with $N = 2000/13.2 \times 2 = 75$

$$A_{max}/A_s = \frac{1}{\sqrt{2}} (\ln 75)^{1/2} = 1.47$$

thus

$$A_{max} = 1.47 \times 0.33 \, H_s = 0.5 \, H_s$$

$\alpha = 45°$ (approximate channel length 750 m)

$$T_{pE} = \frac{11}{1 - 2 \times 11 \times \cos 45°/132} = 12.5 \text{ s}$$

With Figures 11.15 to 11.17:

$$A_s/H_s = 0.8 \times 0.9 \times 1.1 = 0.79$$

with $N = 750/12.5 \times 2 = 30$

$$A_{max}/A_s = \frac{1}{\sqrt{2}} (\ln 30)^{1/2} = 1.30$$

thus

$$A_{max} = 1.30 \times 0.79 \, H_s = 1.0 \, H_s$$

Thus the last case is much more critical for the channel depth. With $H_s = 2.0$ m, $A_{max} = 2.0$ m and the required channel depth becomes:

$$D_{req} = D + \Delta D + A_{max} = 19 + \Delta D$$

Assuming again semiconfined channel conditions, according to Section 9.2

$$\Delta D = 1.5 \times 0.83 \times 4^2/100 = 0.2 \text{ m}$$

and thus

$$d_{req} = 19 + 0.2 = 19.2 \text{ m}$$

A depth of (–19.5 to) –20 m LSD was accepted for the preliminary design, which allows 0.8 m for draught variations (0.3 m) and somewhat longer wave periods and/or directional spreading effects in the bend area (0.5 m).

Detailed Design. In the case of the Haifa port extensions, the detailed design of the layout was checked on a full-mission, real-time simulator, using Haifa pilots. These studies are described fully in Zwamborn et al. (1998).

A simulation program similar to that described in Section 8.5 was used and, overall, 102 real-time simulation runs were made. Some of the results for the PP container vessel and the bulk carrier are shown in Figure 11.26a and b. Some minor channel boundary transgressions occurred during extreme conditions but these were not serious (sufficient depth still available) and can almost certainly be avoided when experience is gained in the operation of the port. It is clear, however, that the PIANC (1997) channel widths calculated above (162.6 m in channel and bend) are insufficient and the previous PIANC (1980) recommended minimum channel width of $w = 5 \, B$ appears to be more appropriate (see Section 8.2).

Conclusions from the detailed simulation studies were:

- The approach and entrance channel are satisfactory for safe operation of the design

——— : 10% Exc. Line
— ·— : 1% Exc. Line

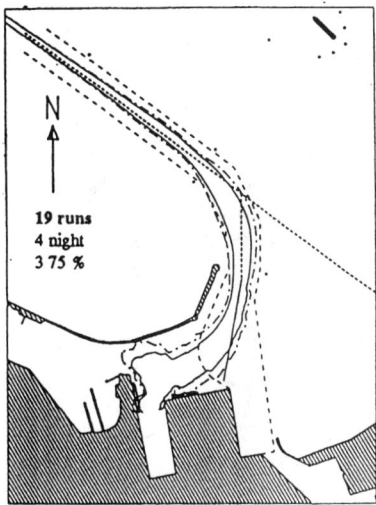
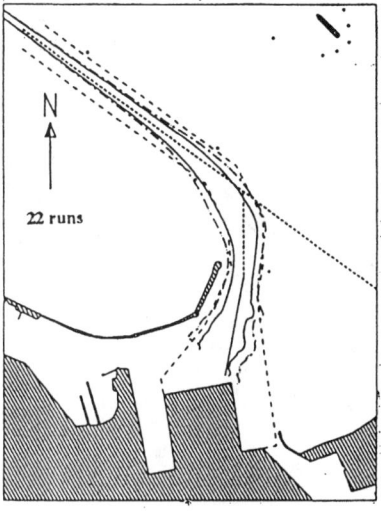

Ship	: Post Panamax Container		Ship	: 150,000 dwt Bulkcarrier
Alternative	: 3c		Alternative	: 3c. Depth 20m
Sailing Dir	: Inbound		Sailing Dir	: Inbound
Wind	: W. 10–15 m/s		Wind	: W. 10–20 m/s
Waves	: NW, Hs: 1.0/2.0m		Waves	: NW, Hs: 1.0/2.0m
Current	: 0.2–0.4m/s		Current	: 0.2–0.4m/s

a) Container Vessel (80,000 dwt)   b) Bulk Carriers (150,000 dwt)

FIGURE 11.26. Maneuvering simulations for Haifa B port extensions.

ships for waves up to $H_s = 2$ m and 20 to 30 knots wind for the PP container vessel and 40 knots for the bulk carrier.

- The limiting conditions for operation will be bringing aboard pilots and the use of service craft.
- Normally, two 30 t bollard pull tractor tugs were found to be sufficient to assist the ships.
- The optimized aids to navigation (leading lights, channel buoys and markers on the breakwater, and quay wall corner lights) were found to be satisfactory and are shown in Figure 11.27.

No site-specific model tests with a model bulk carrier or PP container ship were done, but since most of the time, in the approach channel, the waves are near following, possible directional wave spreading will have little effect on the depth requirements. The adopted channel depth for the simulations of –20 m LSD (0.5 m deeper than the minimum required depth of –19.5 m LSD) also proved to be acceptable for negotiating the 1100 m radius bend during the simulation runs.

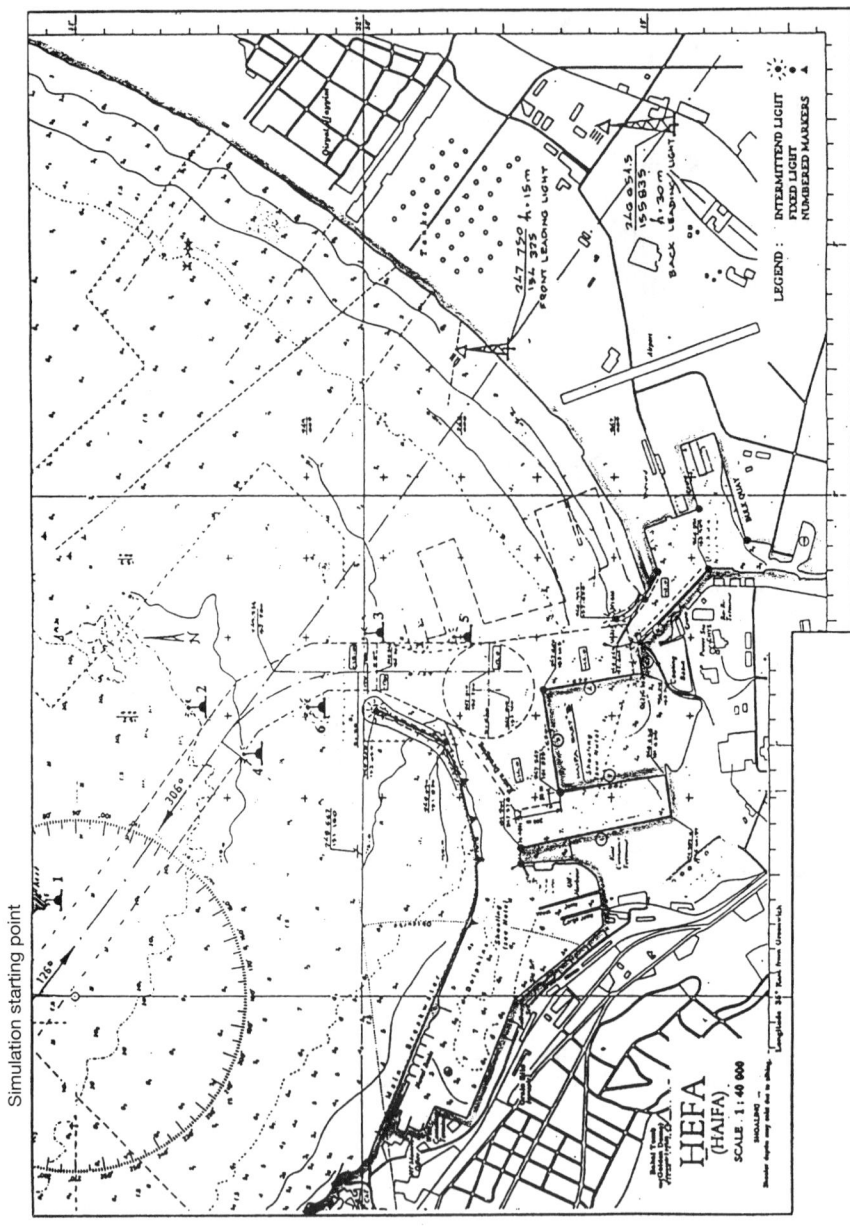

FIGURE 11.27. Haifa B port extensions—channel layout and aids to navigation.

ACKNOWLEDGMENTS

Much of the work on vertical ship motions carried out by the South African Council for Scientific and Industrial Research in Stellenbosch was done by A. (André) C. Van Wyk, who worked under the author's supervision. André tragically lost his life in December 1994 while on a well-earned holiday in Mauritius. His major contribution to this work is hereby acknowledged.

Thanks are also due to J. (Hans) Moes, Project Leader, Coastal Development and Marine Resources Programme, CSIR, Stellenbosch for constructive comments and detailed proofreading.

REFERENCES

Atkinson, J. A. (1980). Tugs as Aids to Safety of Harbours and Ports. *Proc. Ports '80*, ASCE, Norfolk, Virginia, USA.
Barrass, C. B. (1979). The Phenomenon of Ship Squat. *Terra et Aqua, 18,* The Netherlands.
Campbell, N. P. and Zwamborn, J. A. (1977). Special Features in the Design and Construction of the New Harbour for Bulk Cargoes at Richards Bay. *Proc. 24th PIANC Congress,* Leningrad, USSR.
Campbell, N. P. and Zwamborn, J. A. (1984). Richards Bay Harbour, Port Operation Manual, Mark I. *PIANC Bulletin, 8*(45), Brussels, Belgium.
CERC (1984). *Shore Protection Manual.* Coastal Engineering Research Center, WES, Vicksburg, Mississippi, USA.
Crane, L. C. (1973). Manoeuvring Safety of Large Tankers: Stopping, Turning and Speed Selections. *Trans. Society of Naval Architects and Marine Eng., 81,* USA.
CSIR (1985). Richards Bay Harbour Squat Study. CSIR Report C/Sea 8527, Stellenbosch, South Africa.
CSIR (1991). Ship Motion Studies, Optimum Use of the Port of Richards Bay by Large Ships. CSIR Report EMA-C 90158, Stellenbosch, South Africa.
Dunn, M. B., Campbell, N. P., and Zwamborn, J. A. (1985). Richards Bay, South Africa's Major Export Port. *Proc. 26th PIANC Congress,* Brussels, Belgium.
Halber, D., Stadler, L., and Zwamborn, J. A. (1985). Design of Ashdod Port Extension, Coal Unloading Terminal for 150,000 dwt Bulk Carriers. *Proc. 26th PIANC Congress,* Brussels, Belgium.
Herbich, J. B. et al. (1981). Factors in the Determination of a Cost Effective Dredging Cycle. *Proc. 25th PIANC Congress,* Edinburgh, Scotland, UK.
Hogben, N. and Lumb, F. E. (1967). *S.E. Ocean Wave Statistics.* Ministry of Technology, National Physical Laboratory, London, UK.
IAPH, Section 1 (1979). Ships Characteristics and Manoeuvrability. *Proc. 11th Conference International Association on Ports and Harbours,* Vol. 2, Deauville, Le Havre, France.
IAPH, Section 4 (1979). Aids to Navigation. *Proc. 11th Conference International Association on Ports and Harbours,* Vol. 2, Deauville, Le Havre, France.
Iribarren, J. R. (1999). Determining the Horizontal Dimensions of Ship Manoeuvring Areas, General Recommendations and Simulator Studies. *PIANC Bulletin,* No. 100–1999, Brussels, Belgium.
Kimon, P. M. (1982). Underkeel Clearance in Ports. Report no. E II, TM 82, Exxon International Company, USA.
OEC (1983). Software Package VESDYN. Ocean Engineering Consultants, Honolulu, Hawaii, USA.
PIANC (1973). Big Tankers and their Reception. PIANC Excerpt of Bulletin no. 16, Brussels, Belgium.
PIANC (1979). International Commission for the Reception of Large Ships. PIANC supplement to Bulletin no. 32, Brussels, Belgium.
PIANC (1980). International Commision for the Reception of Large Ships. PIANC Supplement to Bulletin no. 35, Brussels, Belgium.

PIANC (1985). Underkeel Clearance for Large Ships in Maritime Fairways with Hard Bottom. PIANC Supplement to Bulletin no. 51, Brussels, Belgium.
PIANC (1986). List of Sea State Parameters, January 1986. PIANC Supplement to Bulletin no. 52, Brussels, Belgium.
PIANC (1989). Economic Methods of Channel Maintenance. PIANC Supplement to Bulletin no. 67, Brussels, Belgium.
PIANC (1992). Capability of Ship Manoeuvring Simulation Models for Approach Channels and Fairways in Harbours. PIANC Supplement to Bulletin no. 77, Brussels, Belgium.
PIANC (1995). Approach Channels, Preliminary Guidelines. PIANC Supplement to Bulletin no. 87, Brussels, Belgium.
PIANC (1997). Approach Channels, A Guide for Design. PIANC Supplement to Bulletin no. 95, Brussels, Belgium.
Savenÿe, P. Ph. A. C. (1996). Probabilistic Admittance Policy for Deep Draught Vessels. PIANC Bulletion no. 91, Brussels, Belgium.
Spencer, J. M. A., Bowers, E. C. and Lean, G. H. (1990). Safe Underkeel Allowances for Vessels in Navigation Channels. *Proc. 24th ICCE,* Delft, The Netherlands.
Strating, J., Schilperoort, T. and Blaauw, H. G. (1982). Optimisation of Depths of Channels. Delft Publication no. 278, Delft, The Netherlands.
Thevenot, M. M. (1992). Evaluation of Width Criteria for Deep-Draft Channels Using Computer Simulation. *PIANC Bulletin* nos. 78–79, Brussels, Belgium.
Van de Kaa, E. J. (1984). Safety Criteria for Channel Depth Design. Contribution to WG III, Navigation Channels, Charleston, South Carolina, USA.
Van Wyk, A. C. (1982). Wave-induced Ship Motions in Harbour Entrances—A Field Study. *Proc. 18th ICCE,* Cape Town, South Africa.
Van Wyk, A. C. and Zwamborn, J. A. (1984). The Effect of Wave Direction on Ship Motions in a Harbour Entrance Channel—Model Study Approach. *Proc. 19th ICCE,* Houston, USA.
Van Wyk, A. C. and Zwamborn, J. A. (1988). Wave Induced Ship Motions in Harbour Approach Channels. *Proc. 21st ICCE,* Malaga, Spain.
Vickerman, M. J. (1992). The Influence of Future Vessel Design on Ports. *Proc. Ports '92,* ASCE, Seattle, USA.
Williamson, G. A. (1975). The Principle Dimensions and Operating Draughts of Bulk Carriers. Report Marine Transport Centre, University of Liverpool, Liverpool, UK.
Young, W. (1994). What are Vessel Traffic Services and What Can They Do? *Journal of the Institute of Navigation, 41*(1), USA.
Zwamborn, J. A., Russell, K. S. and Nicholson, J. (1972). Coastal Engineering Measurements. *Proc. 13th ICCE,* Vancouver, Canada.
Zwamborn, J. A. and Grieve, G. (1974). Wave Alternation and Concentration with Harbour Approach Channels. *Proc. 14th ICCE,* Copenhagen, Denmark.
Zwamborn, J. A. and Van Wyk, A. C. (1981). Monitoring of Ship Movements in the Richards Bay Entrance Channel. *Proc. 25th PIANC Congress,* Edinburgh, Scotland, UK.
Zwamborn, J. A. and Cox, P. J. (1982). Operational Procedures, Richards Bay Harbour. *Proc. 18th ICCE,* Cape Town, South Africa.
Zwamborn, J. A., Di Castro, F., Radomir, M., and van Doorn, J. T. M. (1998). Nautical Design Studies, Haifa Port Extensions. *Proc. 26th ICCE,* Copenhagen, Denmark.

CHAPTER 12
MAINTENANCE DREDGING IN CHANNELS AND HARBORS

JOHN HEADLAND
Moffatt & Nichol Engineers
New York, NY

PETER KOTULAK
Moffatt & Nichol Engineers
Baltimore MD

SANTIAGO ALFAGEME
Moffatt & Nichol Engineers
New York, NY

1. Introduction   12.1
2. Sedimentation in Dredged Channels   12.2
   2.1. Methodology   12.2
   2.2. Example Results   12.5
3. Sedimentation in Semienclosed Harbor Basins   12.7
   3.1. Hydraulic Exchange   12.7
   3.2. Sedimentation   12.14
   3.3. Example Results   12.15
4. Example Numerical Model Results   12.18
References   12.18

1. INTRODUCTION

The following paragraphs present methods for estimating sedimentation in dredging channels and harbor basins. Emphasis is placed on the methods used to compute the sedimentation that results in a need for maintenance dredging. Methods and dredging equipment necessary to accomplish maintenance dredging can be found in other references (e.g., Herbich, 1992). The methods discussed below have been used extensively by the authors for projects throughout the world. Recent important applications have been made for New York Harbor (Moffatt & Nichol Engineers, 1996; Moffatt & Nichol Engineers, 1998a), Colombia South America (Moffatt & Nichol Engineers, 1996) and Thailand (Moffatt & Nichol Engineers, 1998b). The methods have also been used in connection with subchannel placement cells for contaminated dredged material (Headland and Kotulak, 1999).

It should be noted that there are numerous more sophisticated numerical models for sedimentation in channels and harbors. Examples include two-dimensional (horizontal) models such as MIKE-21, which is a finite-difference model (Danish Hydraulics Institute, 1999) and the SED2D finite-element model (Araithurai and Krone, 1977), which is part of the FAST-TABS numerical modeling system. These models have been used extensively by the authors and example applications are presented at the end of this section. The following paragraphs, however, focus on the application of simpler analytical models. These models can be easily applied to practical problems and have the advantage of being easily extended to include a probabilistic approach. In practice, the authors tend to use both the analytical models presented in this chapter and the referenced numerical models. Example projects include Moffatt & Nichol Engineers (1996) and Moffatt & Nichol Engineers (1999).

2. SEDIMENTATION IN DREDGED CHANNELS

Dredged channels are often prone to sedimentation because currents within the deepened areas are weaker than the currents that existed prior to channel dredging. The weaker currents are often incapable of preventing sediments from depositing. The following paragraphs summarize the mechanisms responsible for sedimentation in channels. The principal cause and effect relationships responsible for sedimentation within channels are summarized along with methods for making quantitative predictions. The methodology used to estimate sedimentation in channels has been adapted from Eysink & Vermaas (1983). This model simulates sediment deposition in a way that lends itself to any abrupt change in depth.

2.1. Methodology

Predicting the rate of sedimentation in dredged channels is difficult and inexact due to the complex nature of sediment transport processes. The important variables to consider include: hydrodynamic characteristics such as velocity, tidal fluctuations, and salinity, and sediment characteristics such as suspended sediment concentration, particle size, fall velocity, and sediment type (i.e., sand, silt, clay, etc.).

Most methods for predicting the rate of sedimentation in dredged channels assume that the channel is aligned perpendicular to the flow direction. This assumption simplifies the situation but also restricts the applicability of these prediction methods.

Recent formulae have been developed for conditions with flows crossing a dredged channel at an arbitrary angle. These formulae account for: (1) local sediment transport, (2) proposed channel width and depth, (3) existing water depth, (4) sediment settling velocity, (5) currents and wave conditions (Mayor-Mora et al., 1976; Lean, 1980). These methods assume that channel deepening disturbs an existing condition of dynamic equilibrium. The change in sediment transport rate within the channel can be estimated on the basis of the changes in flow conditions. Sedimentation within the channel is estimated using the change in sediment transport rates upstream and within the channel. The referenced formulae yield fairly good predictions when the angle between channel and flow direction is 30° to 90°. Under such conditions, flow velocity in the channel satisfies the following equation of continuity:

$$u_2 = u_1 h_1 / h_2 \qquad (1)$$

where

u_2 = flow velocity in the channel
u_1 = flow velocity above the undisturbed bed just upstream of the channel
h_1 = undisturbed water depth
h_2 = (increased) water depth in the channel

Channels nearly parallel to existing flows, however, tend to attract flow due to the increased cross section of the channel and the lower bottom shear stresses in the channel relative to the original bed. In many cases, flow velocity in the deepened channel can exceed that in the predredged channel. Many of the existing formulae for channel sedimentation will predict erosion for such cases, although the channel may actually experience sedimentation.

A realistic assessment of sedimentation in dredged channels requires a good understanding of the two-dimensional flow field and attendant sediment transport capacities per unit of width (the latter values depend on local current velocities). In principle, the sediment transport capacities must be based on the same discharge. Deposition occurs when sediment flux into the channel exceeds sediment flux exiting the channel, whereas erosion occurs when sediment flux exiting the channel exceeds sediment flux entering the channel. This approach is illustrated in Figure 12.1, which uses the same nomenclature as Eysink and Vermaas (1983). Erosion in the channel occurs when

$$fT_{1e} < T_{2e} \tag{2}$$

where

$$f = \frac{u_{2g}(\Delta F_1 + \Delta F)}{u_1(\Delta F_1)} = \frac{u_{2g}h_2}{u_1 h_1} \tag{3}$$

The relation between the equilibrium transports T_{1e} and T_{2e} generally can be written as

$$T_{2e} = (u_{2g}^*/u_1^*)^n T_{1e} = \left(\frac{u_{2g} R_c}{u_1 G}\right)^n T_{1e} \tag{4}$$

where

$$R_C = C_1/C_2 = 1 + \frac{18}{C_2} \log \frac{h_1}{h_2} \tag{5}$$

$$G = \left\{\frac{1 + 1/2\left(\xi_1 \frac{u_{01}}{u_1}\right)^2}{1 + 1/2\left(\xi_2 \frac{u_{02}}{u_{02}}\right)^2}\right\}^{1/2} \tag{6}$$

where

u^* = shear stress velocity
n = exponent varying between 3 and 5
C = roughness coefficient of Chézy
ξ = coefficient according to Bijker (1971)
u_0 = orbital wave velocity at the bottom

Manipulation of the above equations gives

$$\frac{G^3}{R_C}\left[1 + \frac{B}{B^t}\left\{\left(\frac{h_2}{h_1}\right)^{3/2} \frac{1}{R_C} - 1\right\}\right]^2 < 1 \tag{7}$$

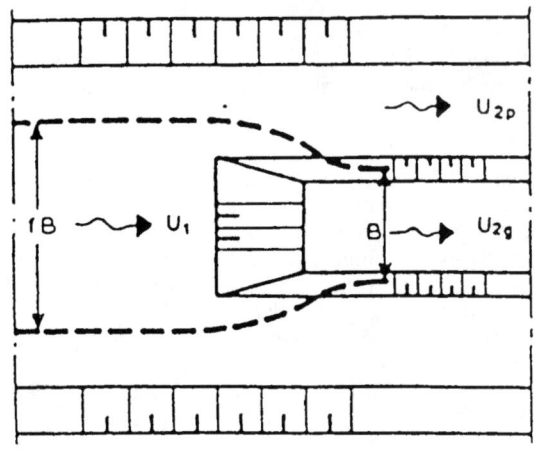

TOP VIEW

sediment influx into channel : $(x=0)$; $T_{in} = fBT_{1e}$
minimum sediment flux out of channel : $(x=\infty)$ $T_{out} = BT_{2e}$
if $f = 1$ than $U_{2g} = U_2$

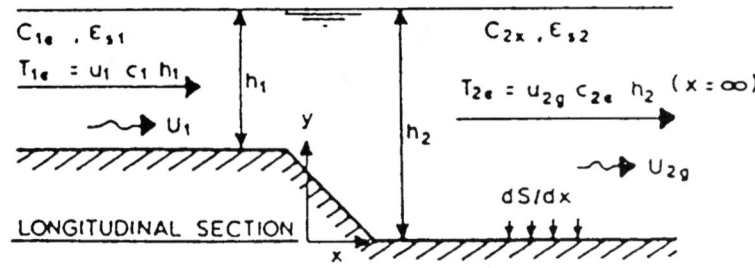

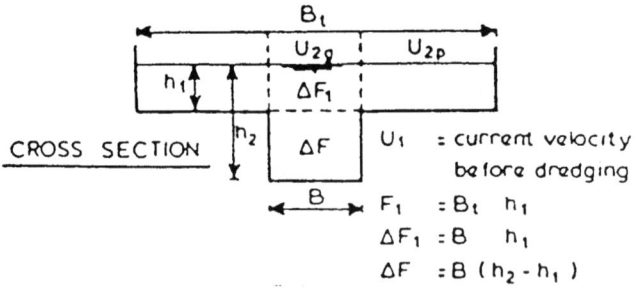

U_1 = current velocity before dredging
$F_1 = B_t \; h_1$
$\Delta F_1 = B \; h_1$
$\Delta F = B(h_2 - h_1)$

FIGURE 12.1. Basic quantities of sedimentation in dredged channel, after Eysink and Vermaas (1983).

For an artificial deepening it holds that

$$h_2/h_1 > 1$$
$$R_C < 1$$

and

$$G \geq 1$$

Hence, the left-hand side of the above equation is greater than one. This means that the dredged channel would be subject to sedimentation even though it attracts flow.

In order to estimate sedimentation (or erosion) within a deepened channel, it is necessary to estimate the vertical sediment concentration distribution at the upstream side of the channel. Eysink (1989) estimated this concentration distribution on basis of the SURTRENCH numerical model. SURTRENCH is a two-dimensional model of sediment transport in the vertical plane and has been verified using physical models and field data. Experience from the SURTRENCH model indicates that local sedimentation can be described by

$$\frac{dS}{dX} = -h_2 u_{2g} \frac{d\bar{c}_{2x}}{dx} = A \frac{\bar{c}_{2x} - \bar{c}_{2e}}{\bar{c}_{2e}} \frac{T_{2e}}{h_2} \tag{8}$$

where

A = dimensionless constant depending on sediment characteristics and the hydraulic conditions in the channel; $f(w/u^*, k/h)$
k = roughness of channel bottom

Integration of the above equation with the proper boundary conditions yields

$$\bar{c}_{2x} = (\bar{c}_1 - \bar{c}_{2e}) \exp\left(-A \frac{x}{h_2}\right) + \bar{c}_{2e} \tag{9}$$

The total sedimentation per unit of width between $x = 0$ and x is approximately

$$S(x) = T_{in} - T_{2x} = \bar{c}_1 u_2 h_2 - \bar{c}_{2x} u_2 h_2 \tag{10}$$

Manipulation of the above two equations yields

$$S(x) = (T_{in} - T_{2e})\left\{1 - \exp\left(-A \frac{x}{h_2}\right)\right\} \tag{11}$$

The constant A depends only on the sediment characteristics and the hydraulic conditions in the channel. The above equation is valid for suspended sediment transport where w/u^* is less than 0.3 to 0.4.

The flow velocity in the channel may be determined using the above equation based on simple flow continuity or a numerical flow model.

2.2. Example Results

Example applications of the channel sedimentation model are shown in Figures 12.2 and 12.3. Figure 12.2 is for a dredged channel subject to cohesive sedimentation (i.e., mud) whereas Figure 12.3 pertains to a channel dredged in sand. The model applications have been applied using a deterministic and probabilistic approach. Input parameters are

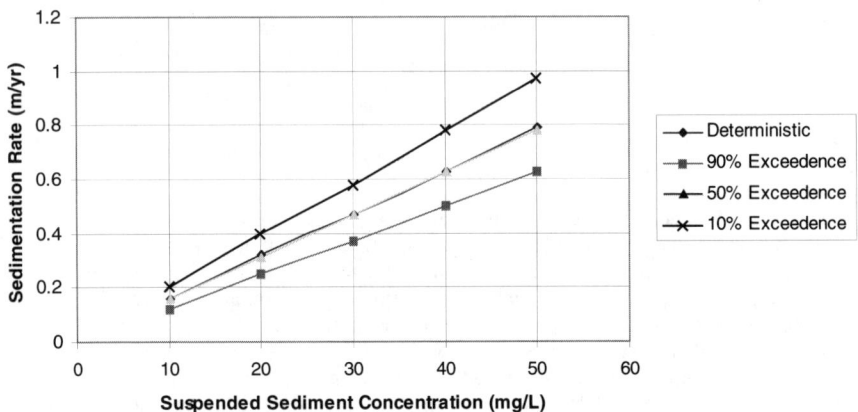

FIGURE 12.2. Example results from channel sedimentation model for mud.

fixed for the former case, whereas probability distributions are used as input for the probabilistic analysis. The probabilistic analysis is performed using the Monte Carlo method.

Input parameters for the model results in Figure 12.2 include: (1) $u_1 = 0.35$ m/s, (2) $h_1 = 6$ m, (3) $h_2 = 8$ m, (4) width = 50 m, (5) current angle = 0°, (6) tide = 0.5 m, and (7) various upstream suspended sediment concentrations. The probabilistic runs were made using a normal distribution for each input variable with a mean value equal to the deterministic value and a standard deviation equal to 10% of the mean value.

Input parameters for the sand model results in Figure 12.3 include: (1) $u_1 = 0.1$ m/s, (2) $h_1 = 7.25$ m, (3) $h_2 = 8$ m, (4) width = 50 m, (5) current angle = 0°, (6) tide = 0.5 m, (7) $H = 0.4$ m, (8) $T = 5$ sec, and (9) various sediment sizes. The probabilistic runs

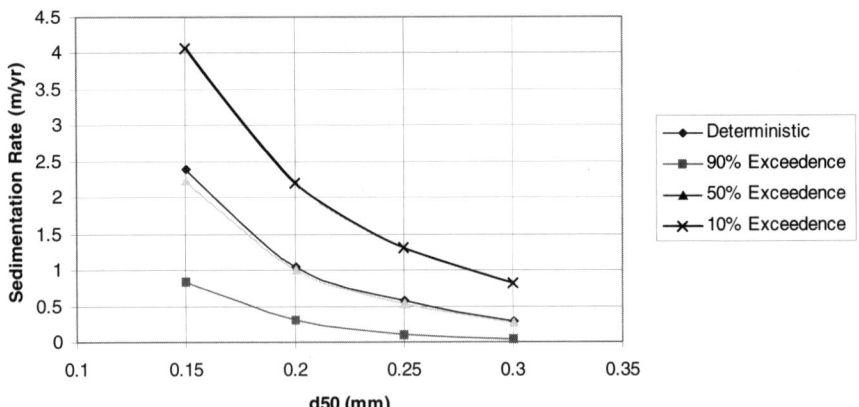

FIGURE 12.3. Example results from the channel sedimentation model for sand.

were made using a normal distribution for each input variable with a mean value equal to the deterministic value and a standard deviation equal to 10% of the mean value.

3. SEDIMENTATION IN SEMIENCLOSED HARBOR BASINS

The sedimentation rate of a semienclosed harbor depends on: (1) the hydraulic exchange rate of the harbor and (2) the sediment trapping efficiency of the harbor. The method for computing sedimentation in a harbor basin is based on the work of Eysink (1989) and is presented below.

The hydraulic exchange rate of a harbor has been shown to be a result of three exchange mechanisms (Roelfzema and Van Os, 1977; Eysink and Vermaas, 1983; and Eysink, 1989):

- Tidal prism exchange: exchange resulting from the rise and fall of the tide
- Horizontal eddy exchange: exchange resulting from an eddy in the harbor entrance
- Density current exchange: exchange resulting from differences in the density of the harbor water and that of the adjoining water body

Currents associated with tidal prism and horizontal eddy exchange are often weak and velocities diminish quickly with distance from the harbor entrance. In contrast, density current velocities are relatively high provided there is a density difference on the order of 1 kg/m^3 or more. Furthermore, density currents extend throughout the harbor and are not confined to the harbor entrance area. When evaluating sedimentation at a specific harbor it is important to establish the relative magnitude of each exchange mechanism in order to determine the merits of remedial measures to reduce sedimentation.

3.1. Hydraulic Exchange

As previously stated, the hydraulic exchange of a harbor basin is comprised of tidal prism exchange, horizontal eddy exchange, and density current exchange.

1.2.1.1. Tidal Prism Exchange. The volume of water exchanged by the tide during a single tidal period is equal to the tidal prism:

$$V_t = 2\eta A \tag{12}$$

where

V_t = tidal prism of harbor basin
η = tidal amplitude
A = storage area of harbor basin

Tidal prism exchange is shown graphically in Figure 12.2.

1.2.1.2. Horizontal Eddy Exchange. There is a strong lateral gradient in flow velocity when water flows swiftly past the entrance to a relatively quiescent harbor basin. This gradient gives rise to an eddy as shown in Figure 12.4. This phenomenon results in an increase in the harbor exchange rate near the harbor entrance. The hydraulic exchange rate

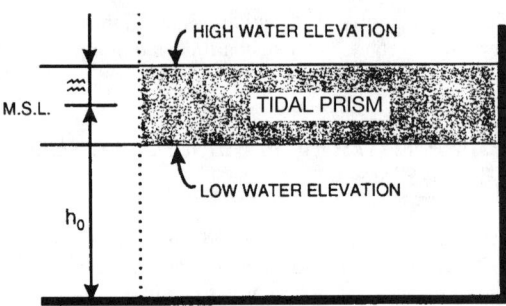

LONGITUDINAL SECTION

TIDAL PRISM EXCHANGE

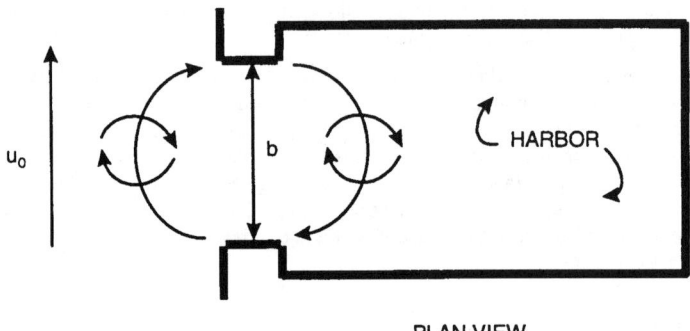

PLAN VIEW

HORIZONTAL EDDY EXCHANGE

FIGURE 12.4. Hydraulic exchange within a harbor, after Eysink (1989).

from this mechanism depends on the flow velocity outside the harbor entrance and geometry of the harbor entrance. Graaf and Reinalda (1977) have presented the following formula for approximating the horizontal exchange rate:

$$Q_h = f_1 h b u_0 - f_2 Q_t \tag{13}$$

where

Q_h = rate of horizontal volume exchange
f_1, f_2 = empirical coefficients
h = entrance depth
b = entrance width
u_0 = main flow velocity in front of the harbor
Q_t = filling flow rate due to tide

According to Graaf and Reinalda (1977), there is negligible horizontal exchange during ebb tide. Substituting assumed sinusoidal variations in water depth, h, and the main flow velocity, u_0, into the above formula and integrating over the flood portion of the tide yields the following expression:

$$V_h = f_1 h_0 b \frac{u_{0,\max}}{\pi} T - f_2 V_t \tag{14}$$

where

V_h = volume of horizontal eddy water exchange per tide
h_0 = depth in the entrance relative to MSL
T = tidal period
V_t = tidal prism of the harbor basin

The coefficients f_1 and f_2 vary according to harbor geometry.

1.2.1.3. Hydraulic Exchange From Density Currents. Hydraulic exchange occurs when the water density outside a harbor differs from that inside the harbor. Density differences can result from differences in water temperature or salinity (more common). Two aspects of density-driven exchange make it a very important factor in harbor sedimentation. First, relatively large currents are generated by relatively small variations in density (1 kg/m^3) for typical harbor entrance depths. Often, these density currents are much higher than tidal filling or horizontal exchange currents. Second, density currents are maintained and attendant volume exchange occurs throughout the harbor basin. In contrast, horizontal eddy velocities are negligible and tidal prism velocities are very weak at locations distant from the harbor entrance.

Figure 12.5 depicts typical density current patterns for a harbor entrance. Density currents can be computed using the following formula (Eysink and Vermaas, 1983; Eysink, 1989):

$$u_{d0} = f_3 \left(\frac{\Delta \rho}{\rho} gh \right)^{1/2} \tag{15}$$

where

u_{d0} = density current excluding tidal influence
ρ = water density
$\Delta \rho$ = characteristic density difference
f_3 = empirical coefficient
g = acceleration due to gravity
h = water depth

The rate of exchange by density currents is computed as follows:

$$Q_d = \frac{1}{2}(u_{d0} - u_t)hb \tag{16}$$

where

u_t = tidal velocity

Figure 12.6 summarizes the mutual relationships of the important parameters including:

- Variation in water level outside the harbor (h)
- Variation in river velocity outside the harbor (u_0)

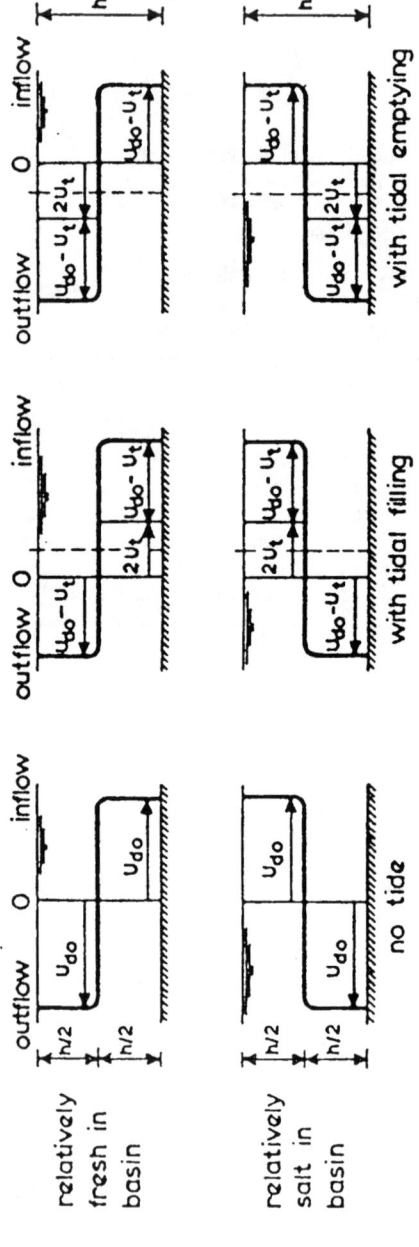

FIGURE 12.5. Typical density current patterns, after Eysink (1989).

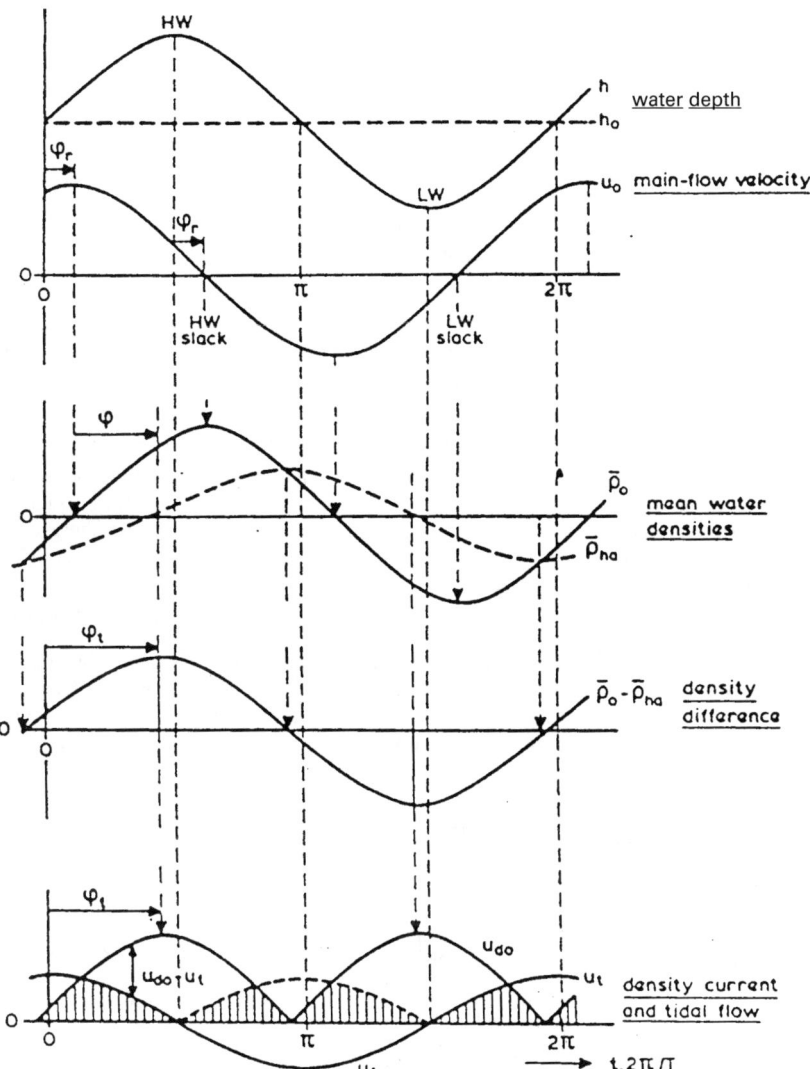

FIGURE 12.6. Relations of parameters for hydraulic exchange, after Eysink (1989).

- Depth averaged density variation outside the harbor ($\bar{\rho}_o$)
- Depth averaged density variation within the harbor ($\bar{\rho}_{ha}$)
- Time varying difference in density of outside and harbor waters
- Tidal current variations in harbor entrance (u_t)

Assuming linear harmonic relationships, the density induced exchange flow rate can be integrated over a tidal cycle which yields the following expressions:

$$V_d = f_4 h_0 b \left(\frac{\Delta \rho_{max}}{\rho} g h_0 \right)^{1/2} T - f_5 V_t \quad (17)$$

with

$$\Delta \rho_{max} = \frac{1}{2} (\overline{\rho}_{0,max} - \overline{\rho}_{0,min}) \quad (18)$$

$$V_{d0} = f_{4,max} h_0 b \left(\frac{\Delta \rho_{max}}{\rho} g h_0 \right)^{1/2} \quad (19)$$

where

V_d = exchange volume per tide due to density currents
f_4 = empirical coefficient depending on V_{d0}/V_{ha}, with V_{ha} being the harbor volume below MSL
f_5 = empirical coefficient depending on V_{d0}/V_t and phase lag φ between u_0 and u_t

In principle, the average water density of a relatively small harbor will follow the average water density fluctuation outside the harbor. This results in a reduction in the characteristic density difference between the river and the harbor. On the other hand, the average density of a relatively large harbor will not correspond closely to the river density. This effect has been included in the coefficient f_4, which decreases with increasing V_{d0}/V_{ha} as shown in Figure 12.7. The coefficient f_5 is given in Figure 12.8 as a function of V_{d0}/V_t and phase lag φ.

Eysink (1989) provided a summary of data used to determine f_4 based on 21 field experiments and 26 model investigations. These data represent a wide range of hydraulic conditions. For example, the boundary conditions in front of the harbor varied from a well-mixed system to a stratified system with fluctuations in depth-averaged density over

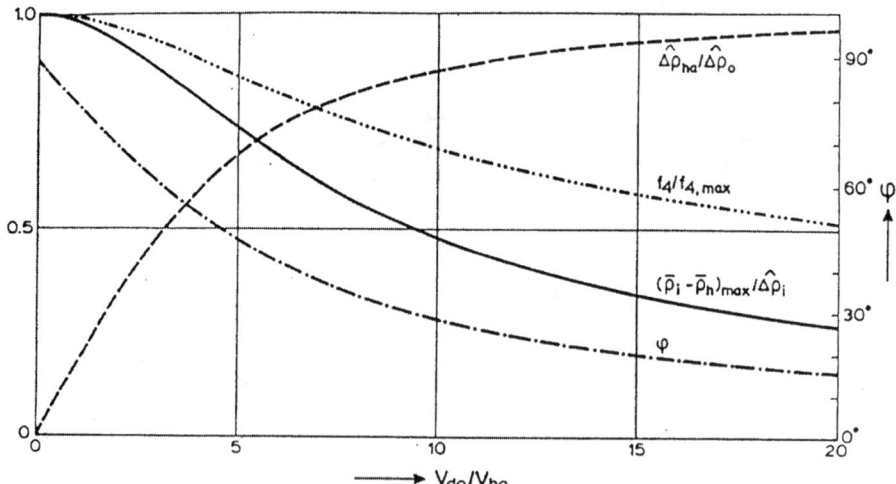

FIGURE 12.7. Coefficient f_4, after Eysink (1989).

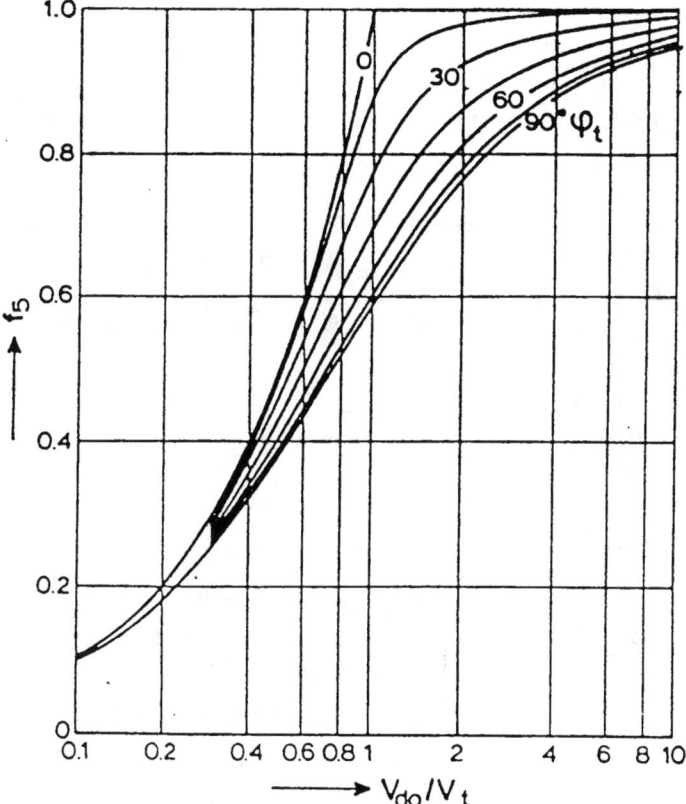

FIGURE 12.8. Coefficient f_5, after Eysink (1989).

the tide ranging from 0.2 to 18.2 kg/m^3. The harbor entrances included openings that were perpendicular and oblique to the river flow, single and composite entrances, and entrances with a sill. The areas of the basins ranged from 0.52 to 8.3 km^2 and the ratios of V_{d0}/V_t and V_{d0}/V_{ha} from 1.1 to 7.7 and 0.1 to 1.35, respectively. In spite of the considerable variations in these quantities, the results derived for $f_{4,\max}$ generally are within a fair range around the average value and seem to be independent of the various controlling parameters, see Figure 12.9. Headland (1991) used hydraulic model tests and field data to confirm these coefficients. Although there are some uncertainties regarding the data (e.g., in the discharge measurements and the density data) the majority of the $f_{4,\max}$ lies within a range of 20% of the average value of $f_{4,\max}$.

1.2.1.4. Total Hydraulic Exchange. The total water exchange volume per tide, V_e, is equal to the sum of the tidal prism exchange, V_t, horizontal eddy exchange, V_h, and density current exchange, V_d:

$$V_e = V_t + V_h + V_d \qquad (20)$$

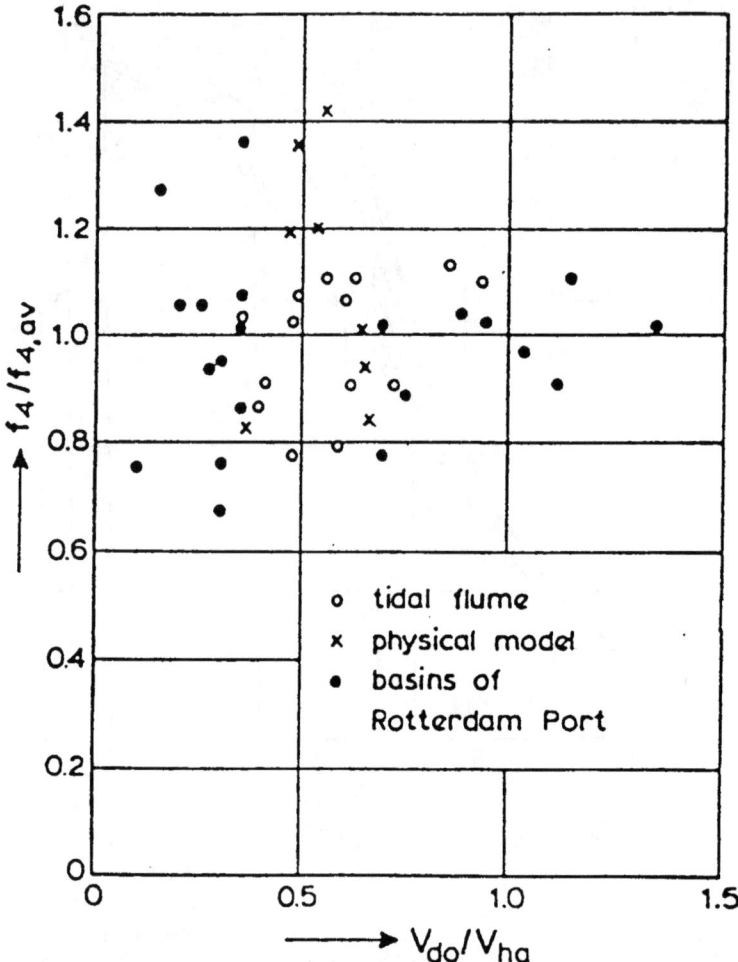

FIGURE 12.9. Physical model and field data used to determine f_4, after Eysink (1989).

3.2. Sedimentation

Eysink's method computes the sedimentation rate as follows:

$$\Delta T_s = p \frac{T_s}{\rho_d} \qquad (21)$$

where

ΔT_s = volume rate of sedimentation
p = percent of the penetrating suspended sediment that settles in the harbor
T_s = mass sediment influx per tide
ρ_d = dry bulk density of the sediment deposit

The factor p depends on the ratio of: (1) the harbor water depth, h, to the effective sediment fall velocity, w_{eff}, which is a measure of the time required for complete settling and (2) the retention time, T_r, of the harbor basin. The effective sediment fall velocity is used to account for the fact that hydraulic turbulence may reduce the actual sediment fall velocity, w. The effective fall velocity is defined as follows:

$$w_{eff} = w\left(1 - \frac{u^2}{u_c^2}\right)_{basin} \tag{22}$$

where

u = flow velocity in the basin and subscript c denotes the critical value below which sedimentation occurs

The expression for the portion of incoming sediment trapped in the harbor at each tide is then:

$$p = 1 - \exp\left[-\frac{w}{h}\left(1 - \frac{u^2}{u_c^2}\right)_{basin} T_r\right] \tag{23}$$

The hydraulic residence time of the harbor, which is the time required to replace the average harbor volume at the tidally averaged exchange rate, can be expressed as follows:

$$T_r = \frac{V_{ha}}{V_e} \tag{24}$$

where

T_r = harbor residence time

Finally, the mass of sediment deposited during a tidal cycle is

$$T_s = \overline{c_{ent}} V_e \tag{25}$$

where

$\overline{c_{ent}}$ = tidal mean suspended sediment concentration of water entering the harbor

3.3. Example Results

Example application of the harbor sedimentation model is shown in Figure 12.10. The model application has been applied using a deterministic and probabilistic approach. In-

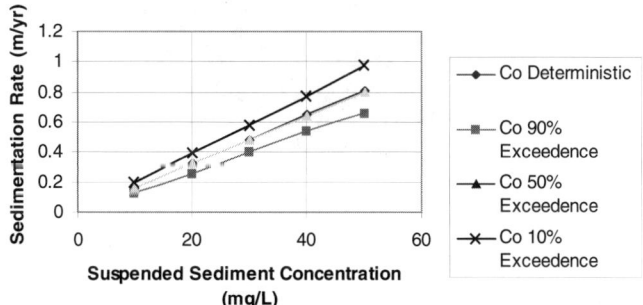

FIGURE 12.10. Basin sedimentation model results.

put parameters are fixed for the former case, whereas probability distributions are used as input for the probabilistic analysis. The probabilistic analysis is performed using the Monte Carlo method, as was the case for the probabilistic implementation of the channel sedimentation model. A normal distribution with a mean equal to the deterministic value and a standard deviation equal to 10% of the mean was used for the following variables: (1) salinity gradient, (2) river velocity, (3) fall velocity, (4) suspended sediment concentration, and (5) sediment dry density.

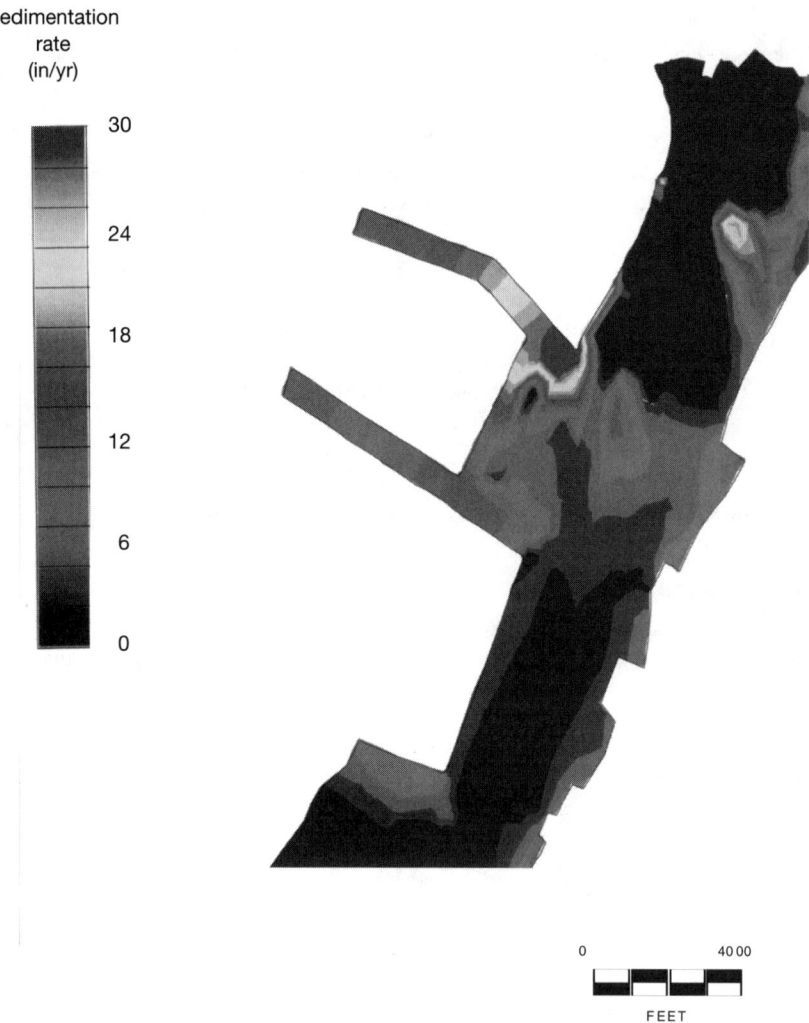

FIGURE 12.11. FAST-TABS modeling results for Newark Bay, New York, existing conditions.

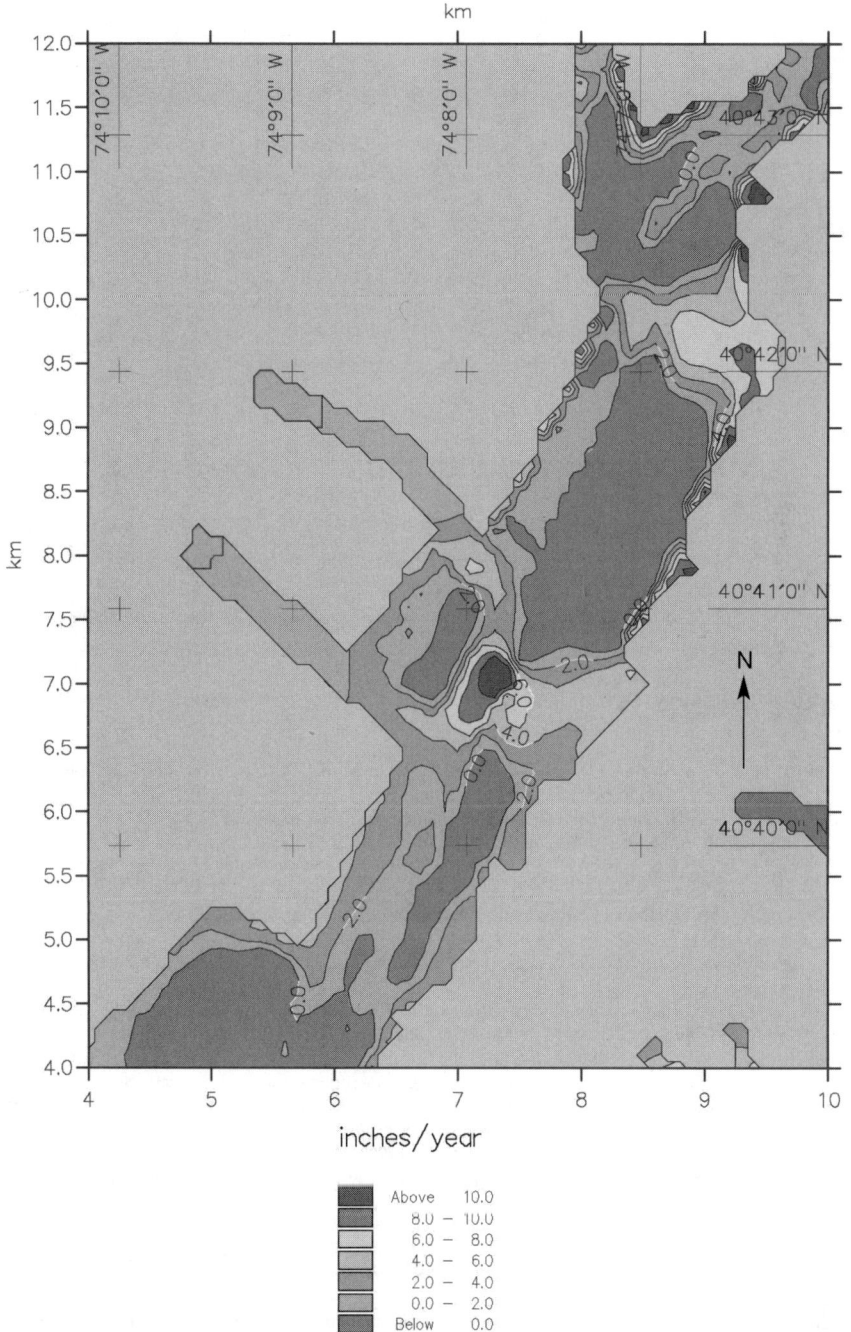

FIGURE 12.12. MIKE 21 MT modeling tesults for Newark Bay, New York, project conditions.

Input Variables	
r	1.4 m
h	3.4 m
b	36.6 m
s_{max}	25 mg/L
s_{min}	21 mg/L
$u_{0,max}$	0.61 m/s
T	12.4 hrs
ρ	1024 kg/L
w	0.0003 m/sec
c_{in}	50 mg/L
ρ dry	300 kg/L

4. EXAMPLE NUMERICAL MODEL RESULTS

Figures 12.11 and 12.12 present the types of results that can be obtained from the FAST-TABS and MIKE 21 models mentioned above. Figure 12.11 is FAST-TABS plot for Newark Bay in New York Harbor and Figure 12.12 is a MIKE 21 plot for the same area.

REFERENCES

Araithurai, R. and Krone, R. B. (1977). Finite Element Model for Cohesive Sediment Transport. *Journal of Hydraulic Division, ASCE, 102,* No. HY3.
Bijker, E. W. (1971). Longshore Transport Computations. *JWHCE, 97,* No. WW4.
Danish Hydraulics Institute. (1999). *Mike 21 MT and Mike 21 ST Model Documentation.*
Eysink, W. D and Vermaas, H. (1983). *Computational Methods to Estimate the Sedimentation in Dredged Channels and Harbor Basins in Estuarine Environments.* Delft Hydraulics Publication No. 307. Delft, The Netherlands: Delft Hydraulics. 12 pp.
Eysink, W. D. (1989). *Sedimentation in Harbour Basins. Small Density Differences May Cause Serious Effects.* Delft Hydraulics Publication No. 417. Delft, The Netherlands: Delft Hydraulics. 13 pp.
Graaf, J. v. d. and Reinalda, R. (1977). *Historical Exchange in Distorted Scale Models.* Delft Hydraulics Laboratory, Research Report S 61. Delft, The Netherlands: Delft Hydraulics.
Headland, J.R. 1991. An Engineering Evaluation of Fine Sedimentation at the Mayport Naval Basin, Mayport, Florida. Coastal Sediments '91, American Society of Civil Engineers, Seattle.
Headland, J. R., and Kotulak, P. (1999). "Design of Sub-Channel Placement Cells (SPC) for Dredged Material." WEDA Conference, Louisville, Kentucky.
Herbich, J. B. (1992). *Handbook of Dredging Engineering.* New York, McGraw-Hill.
Lean, G. H. (1980). *Estimation of Maintenance Dredging for Navigation Channels.* Wallingford, UK: HR Wallingford.
Mayor-Mora, D., Mortensen, P., and Fredsoe, J. (1976). Sedimentation Studies on the Niger River Delta. *Proceedings International Conference of Coastal Engineering, Honolulu,* pp. 2151–2169, New York, ASCE.
Moffatt & Nichol Engineers. (1996). Access Channel Design—Buenaventura, Colombia.
Moffatt & Nichol Engineers. (1997). Sedimentation Reduction/Mitigation Methods. Prepared for the U.S. Army Corps of Engineers, New York District.

Moffatt & Nichol Engineers. (1998a). Sub-Channel Placement and CAD Cells Design, prepared for the U.S. Army Corps of Engineers, New York District.

Moffatt & Nichol Engineers. (1998b). Consulting Services of the Feasibility Study on the Southern Seaboard Port and Industrial Complex Development Project. Prepared for U.S. Trade and Development Agency.

Roelfzema, A. and Van Os, A. G. (1977). Effect of Harbours on Salt Intrusion in Estuaries. *Proceedings 15th Coastal Engineering, Copenhagen, Denmark.* New York, ASCE..

CHAPTER 13
SUBAQUEOUS CAPPING OF CONTAMINATED SEDIMENT

MICHAEL R. PALERMO, PH.D., P.E.
U.S. Army Engineer Waterways Experiment Station
Vicksburg, Mississippi

Introduction	13.2
Background	13.2
Purpose and Scope	13.2
Capping Defined	13.3
Design Issues for Capping	13.4
Viability of Capping as an Alternative	13.4
Design Sequence for Capping	13.5
Sediment Characterization and Site Selection	13.5
Sediment Characterization	13.5
General Considerations for Site Selection	13.7
Equipment and Placement Techniques	13.9
Considerations for Contaminated Material Dredging and Placement	13.9
Considerations for Capping Material Placement	13.10
Geotechnical Considerations	13.10
Surface Discharge Using Conventional Equipment	13.11
Spreading by Barge Movement	13.11
Spreading by Hopper Dredges	13.12
Pipeline with Baffle Plate or Sand Box	13.12
Submerged Discharge	13.13
Submerged Diffuser	13.14
Sand Spreader Barge	13.15
Gravity-fed Downpipe (Trémie)	13.16
Hopper Dredge Pump-down	13.16
Compatibility of Operations	13.16
Navigation and Positioning	13.17
Exposure Time between Placement of Contaminated Material and Cap	13.17
Dispersion and Mound Development	13.18
Evaluation of Dispersion during Placement	13.18
Models for Prediction of Short-Term Fate during Placement	13.19
Evaluation of Spread and Mounding	13.19
Typical Contaminated Mound Geometry	13.20
Geometry for CAD Projects	13.21
Cap Design	13.21

Chemical Isolation	13.22
Bioturbation	13.23
Erosion	13.23
Consolidation	13.24
Variation in Cap Thickness	13.24
Required Areal Coverage of the Cap	13.24
Evaluation of Long-Term Contaminant Flux	13.24
Significance of Flux	13.27
Considerations for Interim Caps	13.27
Long-Term Cap Stability	13.27
Evaluation of Consolidation	13.28
Evaluation of Erosion Potential	13.28
Cap Monitoring	13.29
Designating Management Actions	13.30
Summary	13.30
Acknowledgments	13.31
References	13.31

INTRODUCTION

Background

Potential for water column and benthic effects related to sediment contamination must be evaluated when considering open-water disposal of dredged material. Management options aimed at reducing the release of contaminants to the water column during disposal and/or subsequent isolation of the material from benthic organisms may be considered to control potential contaminant effects. Such options include operational modifications, use of subaqueous discharge points, diffusers, subaqueous lateral confinement of material, or capping of contaminated material with suitable material (Francingues and Palermo, 1984; USACE/EPA, 1992).

Subaqueous dredged material capping is the controlled, accurate placement of contaminated dredged material at an appropriately selected open-water disposal site, followed by a covering or cap of suitable isolating material. Capping of contaminated dredged material in open-water sites began in the late 1970s, and a number of capping operations under a variety of placement conditions have been accomplished. Conventional disposal equipment and techniques are frequently used for a capping project, but these practices must be controlled more precisely than for conventional disposal (see also Chapter 17).

Purpose and Scope

This chapter provides an overview of methods for evaluation of subaqueous dredged material capping projects. Design requirements, a design sequence, site selection, equipment and placement techniques, geotechnical considerations, mixing and dispersion during placement, required capping sediment thickness, material spread and mounding during placement, cap stability, and monitoring are described.

More details on capping and related evaluation procedures are found in technical guidance documents prepared jointly by the U.S. Army Corps of Engineers (USACE) and the Environmental Protection Agency (EPA) (Palermo et al., 1998).

Capping Defined

Level-bottom capping (LBC) is defined as the placement of a contaminated material in a mounded configuration and the subsequent covering of the mound with clean sediment. Contained aquatic disposal (CAD) is similar to LBC but with the additional provision of some form of lateral confinement (e.g., placement in bottom depressions or behind subaqueous berms) to minimize spread of the materials on the bottom. An illustration of LBC and CAD is shown in Figure 13.1.

The objective of level-bottom capping is to place a discrete mound of contaminated material on an existing flat or very gently sloping natural bottom. A cap is then applied over the mound by one of several techniques, but usually in a series of disposal sequences to ensure adequate coverage. CAD is generally used where the mechanical properties of the contaminated material and/or bottom conditions (e.g., slopes) require positive lateral control measures during placement. Use of CAD can also reduce the required quantity of cap material and thus the costs. Options might include the use of an existing natural or excavated depression, preexcavation of a disposal pit, or construction of one or more submerged dikes for confinement (Truitt et al., 1989).

Capping is also a potential alternative for remediation of contaminated sediments in place or in situ. However, a clear distinction should be made between navigation project dredged material capping and capping in the remediation context. For dredged material capping associated with navigation projects, the sediment of concern would normally require capping because it may exhibit potential for toxicity or significant bioaccumulation in benthic organisms. Often, these sediments are only marginally contaminated in comparison to other sediments in the area. The objective of capping in this context is to effectively eliminate direct exposure of benthic organisms to the contaminated sediments and thus reduce potential benthic toxicity or bioaccumulation. For in-situ capping in the remediation context, the sediments of concern are sufficiently contaminated to warrant some sort of cleanup action. The objective of capping in the remediation context may involve objectives over and above isolation of the sediment from the benthic environment. Guidance for in-situ capping for sediment remediation has been developed by the U.S. Environmental Protection Agency under its Assessment and Remediation of Contaminated Sediments (ARCS) Program (Palermo et al., 1996).

Capping field experiences for a number of projects have been documented. For example, a harbor basin in the Port of Rotterdam, the Netherlands, contained heavily contaminated material. Several options (upland, open water, dredged pits, and confined behind a sheet-piled dam) were considered for disposing of the contaminated material, as described

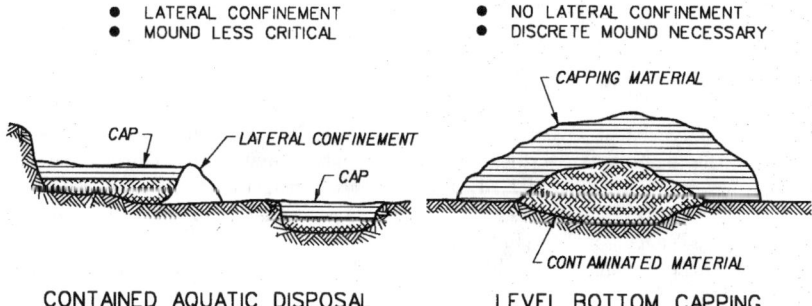

FIGURE 13.1. Schematic illustrating level bottom capping (LBC) and contained aquatic disposal (CAD) (after Palermo, 1991a).

by Kleinbloesem and van der Weijde (1983). The alternative finally selected was a CAD project, which consisted of excavating pits in the basin, dredging the contaminated material, disposing of it in the pits, and capping and lining the pit with clean material. Hiroshima Bay in the Inland Sea of Japan was the site of bottom-sediment improvement testing using a special barge unloader–sand spreader (Kikegawa, 1983). A conveyor barge with eighteen hopper bins was used in Hiroshima Bay for another sand spreading test (Togashi, 1983). In the United States, level-bottom capping is routinely used in the Long Island Sound (SAIC, 1995), and several in-situ capping projects have been completed in Puget Sound in which sand caps were placed by incremental release from barges (Sumeri, 1995).

Design Issues for Capping

Capping is a contaminant control measure to prevent impacts. However, dredged material capping requires initial placement of contaminated material at an open-water site. There are several issues that therefore must be carefully considered within the context of a capping project design. These include:

a) Potential water column impacts during placement—assessment should consider evaluation of potential release of contaminants to the water column, evaluation of potential water column toxicity, and evaluation of initial mixing. Elutriate test procedures for water quality, water column bioassay tests, and computer models for dispersion and mixing are available to address these requirements. The mass loss of contaminants during placement (fraction dispersed off-site and remaining uncapped) may also be predicted using these same tests and models.

b) Efficacy of cap placement—assessment should consider available capping materials, methods for dredging and placement of both contaminated material and cap material, and compatibility of site conditions, material physical properties, and dredging and placement techniques. Guidance on selection of appropriate methods, compatibility with site conditions and material properties, and computer models for predicting mound development and spreading behavior are available.

c) Long-term cap integrity—assessment should consider the physical isolation of contaminants, potential bioturbation of the cap by benthos, consolidation of the sediments, long-term contaminant flux due to advection/diffusion, and potential for physical disturbance or erosion of the cap by currents, waves, and other forces such as anchors, ship traffic, ice, etc. Test procedures for contaminant isolation and consolidation, and computer models for evaluation of long-term contaminant flux and resistance to erosion are described below.

Each of these issues must be appropriately addressed by the project design.

Viability of Capping as an Alternative

Capping is only one of several alternatives that may be considered for contaminated dredged material and requires isolation from the benthic environment if open-water disposal is proposed. If the issues described above can be satisfactorily addressed in the project design for the specific set of sediment, site, and operational conditions under consideration, capping is a technically viable option.

Capping is not a technically viable option for a specific set of sediment, site, and operational conditions when:

a) The contaminant release and dispersion behavior of the contaminated material (even with consideration of controls) results in unacceptable water column impacts during placement.

b) The spreading or mounding behavior of the contaminated material or cap material (even with consideration of controls) indicate that the required cap cannot be effectively placed.
c) The energy conditions or operational conditions at the site are such that the required cap thickness cannot be effectively maintained in the long term.
d) The institutional constraints do not provide the ability to commit to the long-term monitoring and management requirements.

Under such circumstances, other options for disposal of the contaminated sediments must be considered.

Design Sequence for Capping

Capping must not be viewed merely as a form of unrestricted open-water placement. A capping operation should be treated as an engineered project with carefully considered design, construction, and monitoring to ensure that the design is adequate. The basic criteria for a successful capping operation is that the cap thickness required to isolate the contaminated material from the environment be successfully placed and maintained.

The design requirements for a LBC or CAD project include characterization of both contaminated and capping sediments, selection of an appropriate site, selection of compatible equipment and placement techniques, prediction of mixing and dispersion during placement, determination of the required capping sediment thickness, prediction of material spread and mounding during placement, evaluation of cap stability against erosion and bioturbation, and development of a monitoring program (Palermo, 1992).

The flowchart shown in Figure 13.2 illustrates the major design requirements for a capping project and the sequence in which the design requirements should be considered (Palermo, 1991a). There is a strong interdependence between all components of design for a capping project. For example, the initial consideration of a capping site and placement techniques for both the contaminated and capping materials strongly influence all subsequent evaluations, and these initial choices must also be compatible for a successful project (Shields and Montgomery, 1984). Each step in the process must be clearly identified and documented before a decision can be made to proceed.

When an efficient sequence of activities for the design of a capping project is followed, unnecessary data collection and evaluations can be avoided. General descriptions of the various design requirements are given below, corresponding to the recommended design sequence.

SEDIMENT CHARACTERIZATION AND SITE SELECTION

Sediment Characterization

Characterization of both the contaminated sediment and potential capping sediments is necessary for evaluation of the environmental acceptability of sediments for open-water placement and to determine physical and engineering properties necessary for prediction of both short- and long-term behavior of the sediments. Some characterization data may have been obtained as a part of a more general investigation of disposal alternatives prior to consideration of capping.

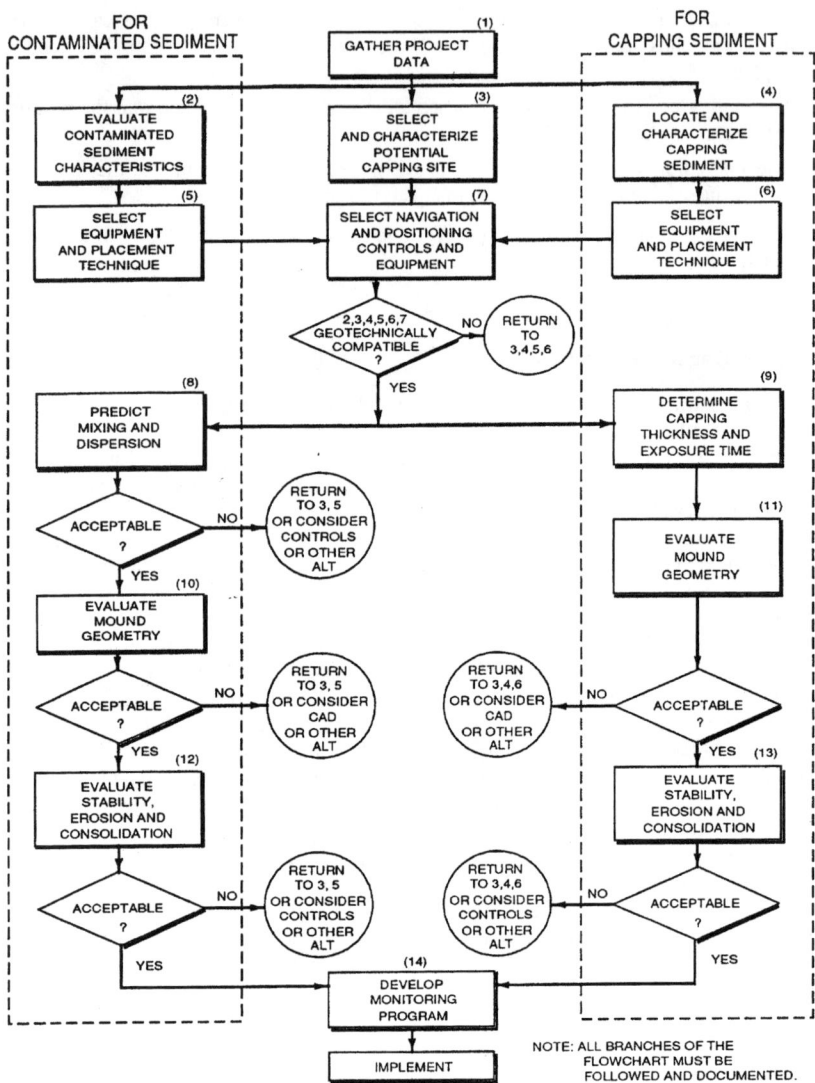

FIGURE 13.2. Flowchart illustrating design sequence for dredged material capping projects (after Palermo, 1991a).

The contaminated sediments to be capped are likely to have been characterized to some degree prior to consideration of capping. In any event, the contaminated sediment must be characterized from physical, chemical, and biological standpoints.

Several sediments may be available for selection as the capping sediment to be used for a particular capping project. For economic reasons, a capping sediment is usually taken from an area that also requires dredging. If this is the case, there may be a choice between multiple dredging projects; the schedule for dredging will then be an important consideration. In other cases, removal of bottom sediments from areas adjacent to the capping site may be considered. If CAD is under consideration, material removed to create CAD cells may be stockpiled and used later in the capping operation (Sumeri, 1995). Such materials must be characterized and determined to be suitable for open-water placement.

Previous studies have shown that both fine-grained materials and sandy materials can be effective capping materials. However, the physical characteristics of the capping sediment should be compatible with the contaminated sediment, considering the dredging and placement techniques used for both.

General Considerations for Site Selection

The selection of an appropriate site is a critical requirement for any capping operation. Since the cap must provide long-term isolation of the contaminated material, capping sites should generally be characterized as nondispersive sites, where material is intended to remain in a stable deposit. Therefore, the considerations for site selection for a conventional nondispersive open-water disposal site also apply to capping sites (Palermo, 1991b).

The selection of a potential site for capping is subject to the same constraints and tradeoffs as any other nondispersive open-water disposal site. However, beyond the normal considerations, the capping site should be in a relatively low-energy environment with little potential for erosion of the cap. While capping a low-energy site is desirable, such sites are not always available. Higher-energy sites can be considered for dredged material capping, but a detailed study of erosion potential is required, and increases in cap thickness to account for potential erosion may be required. Monitoring and maintenance costs may also be higher for higher-energy sites.

Special consideration of site bathymetry, currents, water depths, bottom sediment characteristics, and operational requirements such as distance, sea state, etc. are required in screening or selecting sites for capping (Truitt et al., 1989).

Bathymetry. The bathymetry of the site has an influence on the degree of spread during placement of both contaminated and capping material. The flatter the bottom slope, the more desirable it is for LBC projects, especially if material is to be placed by hopper dredge. If the bottom in a disposal area is not horizontal, a component of the gravity force influences the energy balance of the bottom surge (lateral movement of the disposed material as it impacts the sea bottom) and density flows due to slope following impact of the discharge with the bottom. It is difficult to estimate the effects of slope alone, since bottom roughness plays an equally important role in the mechanics of the spreading process. Placement of material on steep bottom slopes (more than a few degrees) should generally be avoided for a level-bottom capping project (Truitt et al., 1989). Bathymetry forming a natural depression tends to confine the material, resulting in a CAD project. This is the most desirable type of site bathymetry for a capping project.

Currents. Water column currents affect the degree of dispersion during placement and mound location with respect to the point of discharge. Of more importance are bottom currents, which could potentially cause resuspension and erosion of the mound and cap. The effects of storm-induced waves on bottom current velocities must also be considered. Capping sites need to have current and wave climate characteristics that result in long-term stability of the capped mound or deposit.

Collection of basic current information is necessary at prospective disposal sites to identify site-specific conditions. Storm-induced currents are also of interest in the long-term stability of the site. However, disposal operations are not conducted during storms, so the designer doesn't need to consider storm-induced currents during disposal. Measured current data can be supplemented by estimates for external events using standard techniques; for example, see the *Shore Protection Manual* (Coastal Engineering Research Center, 1984). Selection of a nondispersive site in a relatively low-energy environment normally results in a site with low bottom current velocity and little potential for erosion. However, if the material is hydraulically placed, a thorough analysis of the potential for resuspension and erosion is necessary. Conventional methods for analysis of sediment transport can be used to evaluate erosion potential (Teeter, 1988; Dortch et al., 1990). These methods can range from simple analytical techniques to numerical modeling (Scheffner et al., 1995). In the analysis of erosion, the effects of self-armoring due to the winnowing away of finer particles should be considered.

Predicted erosion values due to currents and storm waves should be used as a criterion in screening and/or selecting sites. For projects in which no subsequent capping is anticipated (e.g., the final cap for a site) or for which materials for cap nourishment are not easily obtained, net cap erosion should not exceed one foot over a period of 20 years of normal current/wave energies, or one foot erosion for a 100 year extreme event. For projects in which subsequent capping is planned or for which materials for cap nourishment can be easily obtained, higher erosion rates may be considered. In areas where available capping materials and current and wave conditions are severe, a coarse-grained layer of material may be incorporated into the cap design to provide protection against erosive currents at the site. An armoring layer for resistance to erosion can also be considered in the cap design (Palermo et al., 1996).

Average Water Depths. Case studies have indicated that water depth is of particular interest in evaluating the potential suitability of a site for capping operations (Palermo, 1989). The deepest water depth for which a capping project has been executed (as of 1995) is approximately 100 feet. Greater water depths generally provide more stable bottom conditions with less potential for erosion. However, the greater the average water depth at the site, the greater the potential for water entrainment and dispersion during placement. The expense and difficulty in monitoring is also increased with a greater water depth.

As water depth increases, both the contaminated and clean material must descend through a greater water column depth. More material is released to the water column during deep-water placement as compared to shallower-water placement, all other factors being equal. Therefore, the fraction of the contaminated material that is not finally capped is greater.

Entrainment of ambient water causes the descending material to become more buoyant; therefore, the effect of density stratification in the water column needs to be evaluated. Although density stratification in the water column may be encountered at some deep-water sites, stratification is not likely to prevent the descent of the dredged material mass during placement. The very cohesive fraction of mechanically dredged material (clods or clumps) attains terminal speed quickly after release from a barge and does not accelerate further with depth.

The increased water entrainment with deep-water placement may also result in a greater spread of the more fluid material on the bottom, but entrainment reduces the overall potential energy at bottom impact. Field studies indicate that the bottom surge does not spread at a faster rate than that occurring in shallower depths, although because of additional entrainment, the initial thickness of the surge has been shown to be a function of water depth (Bokuniewicz et al., 1978). Greater care in control of placement may therefore be required as water depth increases, to develop a discrete mound of contaminated material and adequate coverage of the mound with capping material.

The use of a deep-water site for capping generally holds an advantage over a shallower site from the standpoint of cap stability from erosive forces. Deep water acts as a buffer from wave action, and the resulting wave-induced currents from storm events are less than in shallow water. Therefore, deep-water sites are usually quiescent, near-bottom, low-energy environments that are better suited to capping from the standpoint of cap stability, but this must be balanced against material loss during placement. Generally, greater water depth at a site has more favorable influence on long-term cap stability than unfavorable influence on dispersion during the placement process (Truitt et al., 1989).

Operational Requirements. Among the operational criteria that need to be considered in evaluating potential capping sites are: site volumetric capacity, nearby obstructions or structures, haul distances, bottom shear due to ship traffic (in addition to natural currents), location of available cap material, potential use of bottom drag-fishing equipment, and ice influences. The effects of shipping are especially important, since bottom stresses due to anchoring, propeller wash, and direct hull contact at shallow sites are typically of a greater magnitude than the combined effects of waves and other currents (Truitt et al., 1989).

EQUIPMENT AND PLACEMENT TECHNIQUES

Equipment and techniques applicable to placement of contaminated material to be capped and clean material used for capping include conventional discharge from barges, hopper dredges, and pipelines; diffusers and trémie approaches for submerged discharge; and spreading techniques for cap placement (Palermo, 1991c; Palermo, 1994).

Considerations for Contaminated Material Dredging and Placement

Placement of contaminated material for a capping project should be accomplished so that the resulting deposit can be defined by monitoring and effectively capped. Therefore, the equipment and techniques for dredging, transport, and placement must be compatible with that of the capping material. Since capping is a contaminant-control measure for potential benthic effects, the contaminated material should be placed such that the exposure of the material prior to capping is minimized. In most cases, the water column dispersion and bottom spread occurring during placement should also be reduced to the greatest possible extent. This minimizes the release of contaminants during placement and provides for easier capping. If the placement of the contaminated sediment has potentially unacceptable water column impacts, controls to specifically reduce water column dispersion (for example submerged discharge) may be required.

For level-bottom capping, the dredging equipment and placement technique for contaminated sediment must result in a tight, compact mound that is easily capped. Compact mounds generally result when the material is dredged and placed at or near its in-situ density prior to dredging. This is most easily accomplished with mechanical dredging techniques and precision point discharges from barges.

For contained aquatic disposal projects, the provision for lateral containment in the form of a bottom depression or other feature defines and limits the extent of bottom spread. For this reason, either mechanical dredging or hydraulic placement of the contaminated material may be acceptable for CAD. If the contaminated material is placed hydraulically, a suitable time period (usually a few weeks) must be allowed for settling and consolidation to occur prior to placement of the capping material to avoid potential mixing of the materials, unless capped by slow sprinkling of sand.

Considerations for Capping Material Placement

Placement of capping material is accomplished so that the deposit forms a layer of the required thickness over the contaminated material. For most projects, the surface area of the contaminated material to be capped may be several hundred feet or more in diameter. Placement of a cap of required thickness over such an area may require spreading the material to some degree to achieve coverage.

The equipment and placement technique is selected and the rate of application of capping material is controlled to avoid displacement or mixing with the previously placed contaminated material to the extent possible. Placement of capping material at equal or lesser density than the contaminated material generally meets this requirement. However, sand caps have been successfully placed over fine-grained contaminated material. Since capping materials are not contaminated, water column dispersion of capping material is not usually of concern (except for loss when slowly placing a sand cap), the use of submerged discharge for capping placement need only be considered from the standpoint of placement control.

Geotechnical Considerations

Geotechnical considerations are important in capping because of the fact that most contaminated sediments are fine-grained silts and clays and usually have high water contents and low shear strengths in situ. Once sediments are dredged and placed at a subaqueous site, the water contents may be higher and the shear strengths lower than in situ.

Capping involves the placement of a layer of clean sediment of perhaps three feet or more in thickness over such low-shear-strength material. Field monitoring data has definitively shown that contaminated sediments with low strength have been successfully capped with slow placement of sandy material. The geotechnical considerations involved can be described in terms of the ability of a capped deposit with given shear strength to support a cap from the standpoint of slope stability and/or bearing capacity (Ling et al., 1996).

Only very limited geotechnical evaluations have been considered in past capping projects. In virtually all of past capping projects, the design was empirical, i.e., prior field experience showed that it worked, but actual geotechnical design calculations were not conducted. Research on this topic is now underway and more detailed guidance on this aspect of capping design will be provided in the future. Additional research is also planned to define geotechnical design for bearing capacity, slope failure, loading rate, impact penetration, etc. At present, geotechnical aspects of capping project design are limited to the evaluation of compatibility of equipment and placement technique for contaminated and capping sediments with sediment properties (Palermo, 1994; Palermo et al., 1998).

Surface Discharge Using Conventional Equipment

The behavior of a discharge of dredged material into open water is dependent on a number of factors including the physical characteristics of the material, the site conditions, and the method of dredging and placement (that is, from barges, hopper dredges, or direct pipeline). Dredged material released at the water's surface using conventional equipment tends to descend rapidly to the bottom as a dense jet with minimal short-term losses to the overlying water column (Bokuniewicz et al., 1978). Thus, the use of conventional equipment can be considered for placement of both contaminated and capping material if the bottom spread and water column dispersion resulting from such a discharge are acceptable.

The surface release of mechanically dredged material from barges results in a faster descent, tighter mound, and less water column dispersion as compared to surface discharge of hydraulically dredged material from a pipeline. Placement characteristics resulting from surface release of hydraulically dredged material from a hopper dredge fall between the characteristics resulting from surface release of hydraulically dredged material from barges and from surface discharge of hydraulically dredged material from a pipeline; that is, the descent is slower than the former but faster than the latter, the mound is looser than the former but tighter than the latter, and more water column dispersion results from the former than from the latter.

Field experiences with LBC operations in Long Island Sound and the New York Bight have shown that mechanically dredged silt and clay released from barges tends to remain in clumps during descent and forms nonflowing discrete mounds on the bottom, which can be effectively capped. Such mounds have been capped with both mechanically dredged material released from barges and with material released from hopper dredges (O'Connor and O'Connor, 1983; Morton, 1983, 1987). In fact, mechanically dredged cohesive sediments often remain in a clumped condition, reflecting the shape of the dredge bucket. Mounds of such material are very stable, resist displacement during capping operations, and present conditions ideal for subsequent LBC (Sanderson and McKnight, 1986). However, these mounds may experience initial surface erosion due to irregular surface geometry and higher friction coefficients.

Spreading by Barge Movement

A layer of capping material can be spread or gradually built up using bottom-dump barges if provisions are made for controlled opening or movement of the barges. This can be accomplished by slowly opening a conventional split-hull barge over a period of tens of minutes, depending on the size of the barge and site conditions. Such techniques have been successfully used for controlled placement of predominantly coarse-grained, sandy capping materials (Sumeri, 1995). The gradual opening of the split-hull or multicompartmented barges allows the material to be released slowly from the barge in a sprinkling manner. If tugs are used to slowly move the barge during the release, the material can be spread in a thin layer over a large area (Figure 13.3). Multiple barge loads are necessary to cap larger areas in an overlapping manner. The gradual release of mechanically dredged fine-grained silts and clays from barges may not be possible due to potential "bridging" action; that is, the cohesion of such materials may cause the entire bargeload to "bridge" the split-hull opening until a critical point is reached, at which time the entire bargeload is released. If the water content of fine-grained material is high, the material exits the barge in a matter of seconds as a dense slurry, even though the barge is only partially opened.

FIGURE 13.3. Spreading technique for capping by barge movement.

Spreading of thin layers of cap material over large areas can also be accomplished by gradually opening a conventional split-hull barge while underway by tow. These techniques were used for in-situ capping operations at Eagle Harbor, Washington (Sumeri, 1995).

Spreading by Hopper Dredges

Hopper dredges can also be used to spread a sand cap. During the summer and fall of 1993, the Port Newark/Elizabeth capping project in New York Bight used hopper dredges to spread a sand cap over 580,000 yd^3 of dioxin-contaminated sediments. To facilitate spreading the cap in a thin layer (6 inches) to quickly isolate the contaminants and to lower the potential for resuspension of the contaminated material, conventional point dumping was not done. Instead, a split-hull dredge cracked the hull open 1 ft and released its load over a 20–30 minute period while sailing at 1–2 knots. Also, as an alternative means of placing the cap, another dredge used pump-out over the side of the vessel through twin vertical pipes with end plates to force the slurry in the direction the vessel was traveling. As with the cracked-hull method described above, injecting the slurry in the direction of travel of the vessel increased turbulence, reducing the downward velocity of the slurry particles and thus the potential for resuspension of the contaminated sediments (Randall et al., 1994).

Pipeline with Baffle Plate or Sand Box

Spreading placement for capping operations can be easily accomplished with surface discharge from a pipeline aided by an energy dissipating device such as a baffle plate or sand box attached to the end of the pipeline. Hydraulic placement is well suited to placement of thin layers over large surface areas.

A baffle plate (Figure 13.4), sometimes called an impingement or momentum plate, serves two functions. First, as the pipeline discharge strikes the plate, the discharge is sprayed in a radial fashion and allowed to fall vertically into the water column. The decrease in velocity reduces the potential of the discharge to erode material already in place. Second, the angle of the plate can be adjusted so that the momentum of the discharge exerts a force that can be used to swing the end of the floating pipeline in an arc. Such plates are commonly used in river dredging operations, where material is deposited in thin layers in areas adjacent to the dredged channel (Elliot, 1932). Such equipment can be used in capping operations to spread very thin layers of material over a large area, thereby gradually building up the required capping thickness.

A device called a "sand box" (Figure 13.5) serves a similar function. This device acts as a diffuser box with baffles and sideboards to dissipate the energy of the discharge. The bottom and sides of the box are constructed as an open grid or with a pattern of holes so that the discharge is released through the entire box. The box is mounted on the end of a spud barge so that it can be swung about the spud using anchor lines (Sumeri, 1995).

Submerged Discharge

If the placement of the contaminated sediment with surface discharge results in unacceptable water column impacts, or if the anticipated degree of spreading and water column

FIGURE 13.4. Spreader plate for hydraulic pipeline discharge.

FIGURE 13.5. Spreader box or "sand box" for hydraulic pipeline discharge.

dispersion for either the contaminated or capping material is unacceptable, submerged discharge is a potential control measure.

In the case of contaminated dredged material, submerged discharge serves to isolate the material from the water column during at least part of its descent. This isolation can minimize potential chemical releases due to water column dispersion and significantly reduce entrainment of site water, thereby reducing bottom spread and the area and volume to be capped. In the case of capping material, the use of submerged discharge provides additional control and accuracy during placement, thereby potentially reducing the volume of capping material required. Several equipment alternatives are available for submerged discharge (Palermo, 1994) and are described in the following paragraphs.

Submerged Diffuser

A submerged diffuser (Figure 13.6) can be used to provide additional control for submerged pipeline discharge. The diffuser consists of conical and radial sections joined to form the diffuser assembly, which is mounted to the end of the discharge pipeline. A small discharge barge is required to position the diffuser and pipeline vertically in the water column. By positioning the diffuser several feet above the bottom, the discharge is isolated from the upper water column. The diffuser design allows material to be radially discharged parallel to the bottom and with a reduced velocity. Movement of the discharge barge can serve to spread the discharge to cap larger areas. The diffuser can also be used with any hydraulic pipeline operation, including hydraulic pipeline dredges, pump-out from hopper dredges, and reslurried pump-out from barges.

A design for a submerged diffuser system was developed by JBF Corporation as a part of the USACE Dredged Material Research Program (DMRP). This design consists of a funnel-shaped diffuser oriented vertically at the end of a submerged pipeline section, which discharges the slurry radially. The diffuser and pipe section is attached to a pivot

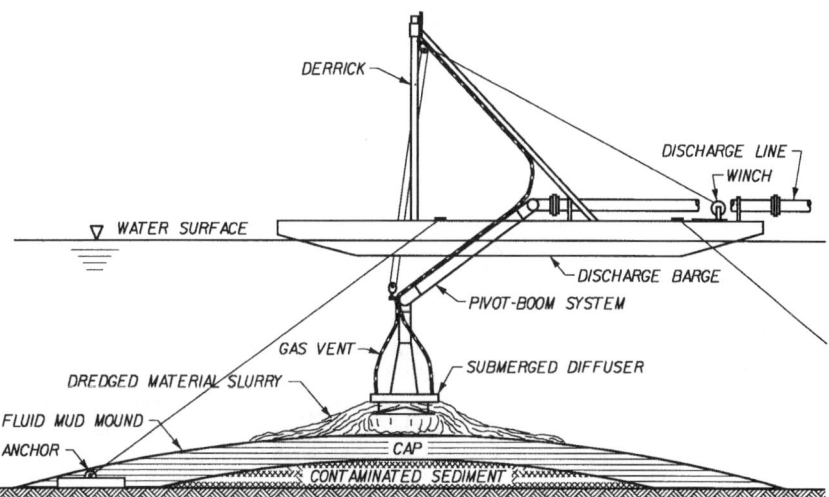

FIGURE 13.6. Submerged diffuser system, including the diffuser and discharge barge.

boom system on a discharge barge. Design specifications for this submerged diffuser system are available (Neal et al., 1978).

A variation of the DMRP diffuser design was used in an equipment demonstration at Calumet Harbor, Illinois. Although not constructed to the DMRP specifications, this diffuser significantly reduced pipeline exit velocity, confined the discharged material to the lower portion of the water column, and reduced suspended solids in the upper portion of the water column (McLellan and Truitt, 1986). Diffusers have been constructed using the DMRP design and used at a habitat creation project in the Chesapeake Bay (Earhart et al., 1988), and at a Superfund pilot dredging project at New Bedford Harbor, Massachusetts involving subaqueous capping (USACE, New England Division, 1990). At the Chesapeake Bay site, the diffuser was used to effectively achieve dredged material mounding prior to placement of a layer of oyster shell to provide substrate for attachment of oyster spat. At the New Bedford site, the diffuser was used to place contaminated sediment in an excavated subaqueous cell and was effective in reducing sediment resuspension and in controlling placement of contaminated sediment. However, capping operations were started immediately and positioning of the diffuser within two feet of the contaminated sediment layer resulted in mixing of cap sediment with contaminated sediment. These results indicate the need for a high degree of control when capping newly placed slurry with a diffuser, and the need for adequate time to allow for some self-weight consolidation of slurry material prior to capping. Diffusers have also been successfully used to place and cap contaminated sediments at projects in Rotterdam Harbor in the Netherlands (d'Angremond et al., 1984) and in Antwerp Harbor in Belgium (Van Wijck and Smits, 1991).

Sand Spreader Barge

Specialized equipment for hydraulic spreading of sand for capping has been used by the Japanese (Kikegawa, 1981; Sanderson and McKnight, 1986). This equipment employs the basic features of a hydraulic dredge with submerged discharge. Material is brought to the spreader by barge, where water is added to slurry the sand. The spreader then pumps the slurried sand through a submerged pipeline. A winch and anchoring system is used to swing the spreader from side to side and forward, thereby capping a large area.

Gravity-fed Downpipe (Trémie)

Trémie equipment can be used for submerged discharge of either mechanically or hydraulically dredged material. The equipment consists of a large-diameter conduit extending vertically from the surface through the water column to some point near or above the bottom. The conduit provides the desired isolation of the discharge from the upper water column and improved placement accuracy. However, because the conduit is a large-diameter straight vertical section, there is little reduction in momentum or impact energy over conventional surface discharge. The weight and rigid nature of the conduit requires a sound structural design and consideration of the forces due to currents and waves.

The Japanese have used trémie technology in the design of specialized conveyor barges for capping operations (Togashi, 1983; Sanderson and McKnight 1986). This equipment consists of a trémie conduit attached to a barge equipped with a conveyor. The material is initially placed in the barge mechanically. The conveyor then mechanically feeds the material to the trémie conduit. A telescoping feature of the trémie allows placement at depths of up to approximately 40 feet. Anchor and winch systems are used to swing the barge from side to side and forward so that larger areas can be capped, similar to the sand spreader barge.

Hopper Dredge Pump-down

Some hopper dredges have pump-out capability by which material from the hoppers is discharged as from a conventional hydraulic pipeline dredge. In addition, some have further modifications that allow pumps to be reversed so that material is pumped down through the dredge's extended drag arms. Because of the expansion at the drag head, the result is similar to using a diffuser section. Pump-out depth is limited, however, to the maximum dredging depth, typically about 60–70 ft.

Compatibility of Operations

An acceptable match of equipment and placement techniques for contaminated and capping material is essential to avoid displacement of the previously placed contaminated material or excessive mixing of capping and contaminated material. The availability of certain types of equipment and the distance between dredging and placement sites may also influence selection of compatible equipment types. The nature of the materials (cohesive versus noncohesive), the dredging method (mechanical versus hydraulic), the method of discharge (instantaneous dump from hopper dredge or barge versus continuous pipeline), the location of discharge (surface or submerged), frequency and scheduling of discharges, physical characteristics of discharge material, and other factors influence the tendency of the material to mound or flow and the tendency to displace or mix with material already placed.

Navigation and Positioning

Controlled and accurate placement of both the contaminated and capping material is crucial for a successful capping project. Once the dredging equipment and placement techniques and potential capping site have been selected, the needs for navigation and positioning equipment and controls can be addressed. The objective here is to place both the contaminated and capping materials (whether by the bargeload, hopperload, or pipeline) at the desired location in a consistently accurate manner so that adequate coverage by the cap is attained.

For pipeline placement in shallow water, the desired positioning of the pipeline discharge can be maintained with little difficulty. Accurate navigation to the placement site and precise positioning during material placement by bottom-dump barge or hopper dredge is more difficult, especially for sites well offshore. State-of-the-art equipment (i.e.,differential global positioning system (DGPS) or microwave systems) and techniques must be employed to assure accurate point placement to the extent deemed necessary. Taut-moored buoys, mooring barges, various acoustical positioning devices, and computer-assisted, real-time helmsman's aids should be considered. In all cases, barges or scows must be required to release the material within a prescribed radius of the designated point of placement. A positioning system for the disposal barges must be specified that has sufficient accuracy to ensure placement within the minimum specified radius of placement. Diligent inspection of operations to insure compliance with specifications is essential.

Exposure Time between Placement of Contaminated Material and Cap

Scheduling of the contaminated material placement and capping operation must satisfy environmental and engineering/operational constraints. Following the placement of contaminated material, there is necessarily some time lag prior to completion of the capping operation. This results in some exposure time for benthic organisms to colonize a portion of the contaminated material deposit. Placement of the cap material must begin as soon as practicable following completion of the placement of contaminated material to minimize this exposure time. However, a delay of one to two weeks is desirable from an engineering standpoint to allow initial consolidation of the contaminated material to occur, with an accompanying increase in shear strength, prior to placement of the cap.

Factors to consider in arriving at an appropriate exposure time are:

1. Potential effects due to exposure prior to capping
2. Estimates of time required for initial colonization of the site by benthic organisms
3. Estimates of time required for initial consolidation of the contaminated material due to self-weight
4. Monitoring requirements prior to cap placement

The process of recolonization by opportunistic species may begin almost as soon as contaminated material placement operations are completed (Rhoads and Boyer, 1982; Rhoads and Germano, 1982). However, it is probably several weeks before any significant recolonization by organisms occurs (Scott et al., 1987). Bioaccumulation by opportunistic species can occur during this time, but they will be buried and thus physically isolated by placement of the cap. It is therefore unlikely that any significant bioaccumulation will result in unacceptable effects during this period of time.

Some delay between completion of contaminated material placement and initiation of capping is desirable from an engineering standpoint. Consolidation of the contaminated material and a corresponding increase in density and strength occurs due to the weight of the material as it is placed in the deposit. This process is called self-weight consolidation. The contaminated material should be allowed to undergo initial self-weight consolidation prior to capping to increase its stability and resistance to displacement during cap placement. This is especially important for slurried materials placed by pipeline or by hopper dredge. For slurried materials, a large portion of the self-weight consolidation occurs within a few weeks of placement. Mechanically dredged materials placed by barge release are initially deposited at essentially the same density at which they were dredged, and the potential degree of self-weight consolidation is less than for slurried materials.

Monitoring is required to determine the areal extent of the contaminated deposit prior to capping. Surveys and other sampling and monitoring activities may require several weeks to complete. An appropriate delay between contaminated material placement and capping must balance environmental exposure with the engineering requirements of stability and scheduling constraints for monitoring and dredging required for capping. Considering the times required for recolonization and consolidation, a time lag of three to four weeks between completion of contaminated material placement and initiation of capping is generally considered acceptable.

DISPERSION AND MOUND DEVELOPMENT

Evaluation of Dispersion during Placement

The physical behavior of a dredged material discharge depends on the type of dredging and disposal operation used, nature of the material (physical characteristics), and hydrodynamics of the disposal site. For capping operations, it is essential to determine beforehand the nature of the discharge for both contaminated and capping material. The degree of dispersion and associated water column contaminant release dictates whether a given discharge is acceptable from the standpoint of water column impacts. The geometry of the subaqueous deposit or mound dictates the required area to be capped and cap configuration.

A knowledge of the short-term physical fate of both the contaminated material and capping material is necessary to determine the acceptability of the equipment and placement operation under consideration. Short-term fate is defined as the behavior exhibited by the material during and immediately following discharge. The dispersion of material released into the water column and the deposition of the material on the bottom are also of interest. These processes occur over a time period of a few minutes to several hours for a single release from a barge or hopper dredge.

In addition to physical dispersion of suspended material, an evaluation of water column mixing of released contaminants or suspended dredged material is necessary whenever potential water column contaminant effects are of concern. Such an evaluation may involve comparison of predicted water column contaminant concentrations with water quality criteria (or standards) or predicted suspended dredged material concentrations with bioassay test results. Water column effects measured in the field on actual projects (see Truitt, 1986) may be valuable in quantifying water quality effects. For capping operations, such evaluations are normally required for the contaminated material to determine if water column control measures (i.e., submerged discharge) are necessary during placement. In addition, the prediction indicates what portion of the contaminated material is dispersed during placement and is not capped.

The physical development of a mound or deposit on the bottom due to a number of barge or hopper releases or prolonged discharge from a pipeline is also of interest. Such information can be used to define the areal extent of the mound or deposit for the contaminated material. This dictates the required volume of capping material.

Models for Prediction of Short-Term Fate during Placement

A computer model is available for evaluating the short-term fate of dredged material discharges in open water from hoppers or barges. The model is called the Short Term FATE (STFATE) model (Johnson et al., 1993; Johnson and Fong, 1995) and can be run on a personal computer (PC). The model is available as a part of the Automated Dredging and Disposal Alternatives Management System (ADDAMS) (Schroeder and Palermo, 1990). Versions of the model are also included in the USACE ocean and inland testing manuals (USACE/EPA, 1991; USACE/EPA, in preparation).

Input data required to run the models include: 1) description of the disposal operation, 2) description of the disposal site, 3) description of the dredged material, 4) model coefficients, and 5) controls for input, execution, and output. More detailed descriptions and guidance for selection of values for many of the parameters is provided directly on-line in the system software, or default values may be used.

Model output includes a time history of the descent and collapse phases of the discharge and suspended sediment concentrations for various particle size ranges as a function of depth and time. At the conclusion of the model simulation, the thickness of the deposited material on the bottom is given. This allows an estimate of the areal extent or "footprint" of contaminated material as deposited on the bottom for a single disposal operation (i.e., a single barge or hopper load of material). The STFATE model has been verified under a range of site conditions, and has been shown to accurately simulate the processes involved in the discharge of material from hoppers or barges.

Evaluation of Spread and Mounding

The mound geometry, including contaminated material mound and cap, will influence the design of the cap and volume of capping material required. The smaller the footprint of the contaminated material as placed, the less volume of capping material will be required to achieve a given cap thickness.

For LBC sites, the geometry of the contaminated material mound depends on the physical characteristics of the material (grain size and cohesion) and the placement technique used (hydraulic placement will result in greater spread than mechanical placement). Assuming that the material from multiple barge loads or pipeline can be accurately placed at a single point, the angle of repose taken by the material and the total volume placed will dictate the mound spread.

However, few data are available on the volume changes resulting from entrainment of water during open-water placement or the shear strengths of dredged material initially deposited in open-water sites. For these reasons, a priori estimates of mound spread made to date have been made based on the observed characteristics of previous mounds created with similar placement techniques and similar sediments (Palermo et al., 1989). Models have been developed that will account for the development of mounds due to a number of barge or hopper discharges (Moritz and Randall, 1995; Wiley, 1994).

The USACE mound building model that models Multiple Disposals from barges and hopper dredges and their FATE (MDFATE), is a modification of the STFATE model. In the MDFATE model, a streamlined version of the STFATE model is run for each barge disposal. Thus the input requirements for MDFATE are similar to those for STFATE. In

MDFATE, the program keeps track of the mound thickness in each grid cell, then algebraically adds the thickness from subsequent disposals. MDFATE allows a number of typical disposal patterns to be automated, it allows for moving barges and can import actual site bathymetry in real-world coordinates. MDFATE also allows interaction with the Long Term FATE of Dredged Material (LTFATE) model (Scheffner et al., 1995). This allows the mound created in MDFATE to be eroded by waves and currents during mound creations that may last months. Similar to the output from STFATE, output from the MDFATE model includes the volume of material on the bottom and contour and cross-section plots of mound bathymetry. One limitation of MDFATE is that it has been verified on only one actual project to date (Moritz and Randall, 1995). A second model developed for the USACE New England Division (Wiley, 1994) is also based on the STFATE model. Although it does not consider moving vessels or erosion by waves and currents, it has the advantage of having been verified for a number of mounds constructed by the New England Division (NED) in Long Island Sound.

Typical Contaminated Mound Geometry

As noted in the previous chapter, for LBC projects, virtually all of the mounds created have been constructed using mechanical dredging with transportation and placement by bottom dump barges. The resulting mounds created have had reasonably consistent geometries. Most mounds have been round or elliptical in shape, with a defined crest that is relatively flat, a main mound side slope (also termed the inner flank), sometimes an outer flank, and a thin outer apron. The following paragraphs describe each of the mound features in more detail.

Mound Crest. Most contaminated mounds to date have had main mound crest elevations of 1–2 m, though some contaminated mounds with elevations of 3+ m have been constructed. Higher mounds have been constructed from noncontaminated material. For point-dumped projects in NED, mound crests have generally been circles or ellipses 100 to 200 m in diameter, reflecting good control of the disposal process around a taut-moored buoy (disposal within about 25 m of a buoy). They have been moderate sized projects, generally 20,000 to 100,000 yd^3. The 1993 Port Newark/Elizabeth project used disposal lanes 150 m in width and 300–420 m long to create a triangular-shaped mound, approximately 630 by 645 m, with peak elevations of 1.5 to 2.4 m.

Inner Flank. At the edge of the main mound, the inner flank of the mounds slope downward at a slope of approximately 1:35 to 1:70, with most of the mound slopes between 1:40 and 1:50. For the Port Newark/Elizabeth mound, the inner flank extended from the mound crest down to an elevation of about 1.5 m above the preplacement bottom.

Outer Flank. For the Port Newark/Elizabeth project, a break in slope generally occurred at the 0.5 m elevation, the outer flank then sloped down to an elevation of about 0.15 m at a slope of about 1:400. Data from New England Division projects has not been examined in sufficient detail to determine if a similar feature exists for those mounds.

Apron. During the dynamic collapse phase (when the energy of the vertically descending jet of material disposed from a barge or hopper dredge is converted to horizontal velocity), some portion of the low-shear-strength, fine-grained material with high water contents may be transported a considerable distance from the disposal point. At the completion of the contaminated material placement, an apron of fine-grained material, typically 1–15 cm in thickness but extending up to several hundreds of meters beyond the main mound flanks, has occurred on almost all LBC projects. Defining the apron as the material less than 15 cm is done because 15–20 cm is the resolution limit for high-quality bathymetry in water depths of 25 m or less. A sediment profiling camera (SPC) can reliably measure apron thickness from 1–2 cm up to 20 cm. Thus, the outer limit of the apron should be defined as the point at which the apron can no longer be conclusively distinguished by the SPC, a thickness of 1–2 cm. Some contaminated material extends beyond the apron edge as defined by the 1–2 cm SPC limit; however, the percentage of the total volume is likely to be extremely small.

Geometry for CAD Projects

The geometry of the deposit for CAD sites is largely controlled by the geometry of the depression or subaqueous berms that form the lateral containment. If hydraulic methods are used to dredge the contaminated materials going into the CAD site, and if the site has a relatively small surface area, the materials will tend to spread in a layer of even thickness over the entire area. If the site has a large surface area, or if the contaminated material is mechanically dredged and placed by barges, the material may tend to form a mound within the site, not covering the entire surface area. If this is the case, methods for intentionally spreading the contaminated material within the CAD site boundaries may be appropriate. Contaminated materials should be placed in CAD sites as a layer of uniform thickness, so that the required thickness of cap material can be placed using a minimum volume of cap material.

Cap geometry for CAD sites should be developed as the design cap thickness placed uniformly over the entire contaminated deposit. Assuming the contaminated material has been placed as a fairly uniform layer, the cap would essentially be placed from bank to bank within a depression, pit, or contained area formed by subaqueous berms.

The same tools as described above for LBC projects can be used for evaluation of deposit geometry for CAD sites. The major consideration for CAD geometry is the placement of both contaminated and cap layers in a uniform and level configuration.

Bulking is an important consideration for CAD geometry. The volume of contaminated material and cap and associated bulking must be closely estimated to ensure that all the material and cap can be placed within the available contained volume. For mechanically dredged sediments, bulking of 10–20% (Herbich, 1992) is reasonable. For hydraulically dredged sediments, dredged and placed by hopper or pipeline, much of the excess water will be expelled as the material is placed within the CAD site, but the volume occupied during the placement operation must be closely estimated. A project-specific investigation of the expected increase in volume for a particular dredging/placement method and sediment is warranted. Sedimentation analysis to determine a volume occupied by hydraulic pipeline placement to a CAD site has been conducted using procedures developed for diked confined disposal facilities (Averett et al., 1989). Procedures for such an analysis are outlined in detail in the USACE Engineer Manual 1110-2-5027, *Confined Disposal of Dredged Material* (USACE, 1987).

CAP DESIGN

The composition and dimensions (thickness) of the components of a cap can be referred to as the cap design. Dredged material caps are typically constructed with a single layer of "clean" sediments because sediments from other dredging projects are often available as cap materials and disposal/capping sites with low potential for erosion do not normally require an armoring layer.

Determining the minimum required cap thickness depends on the physical and chemical properties of the contaminated and capping sediments and the potential for bioturbation of the cap by aquatic organisms. The thickness for chemical isolation plus the thickness for bioturbation is considered the minimum required cap thickness. This minimum required thickness must be maintained to ensure long-term integrity of the cap. The integrity of the cap from the standpoint of physical changes in cap thickness and long-term migration of contaminants through the cap should also be considered. The potential for a physical reduction in cap thickness due to the effects of consolidation and erosion can be evaluated once the overall size and configuration of the capped mound is determined. The design cap thickness can then be adjusted such that the minimum required cap thickness is maintained. A schematic of the cap thickness components is shown in Figure 13.7.

Chemical Isolation

The basic function of the required sediment cap is that associated with physical and/or chemical isolation. For sediment cap materials, the thickness that provides an effective

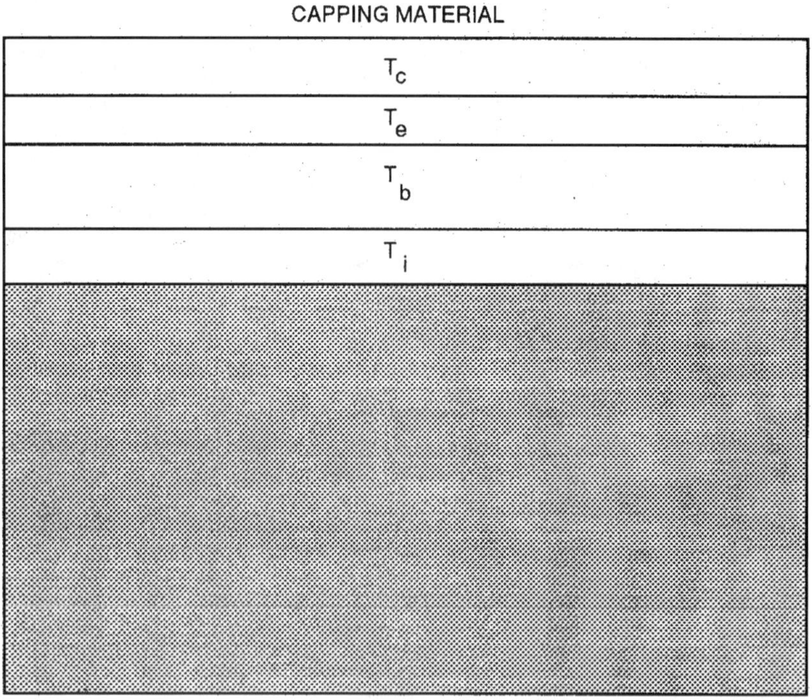

FIGURE 13.7. Schematic of the individual cap thickness components for a sediment cap.

physical/chemical barrier may be defined as T_i. A primary function of the cap is physical isolation from benthic organisms, and the isolation component provides a buffer between the organisms at their burrowing depth and the contaminated materials.

Chemical isolation tests conducted to date (Sturgis and Gunnison, 1988) have shown the minimum required cap thickness for chemical isolation to be on the order of 30 cm (1 ft) for most sediments. A thickness of one foot of capping material for this purpose is therefore considered conservative for isolation of nutrient-rich sediments or sediments with relatively low levels of contamination. For sediments with relatively high levels of contamination, a more involved analysis, including capping effectiveness testing and modeling of potential long term flux (see following discussion), would be required. In this case, a value of one foot may be considered as a trial value for the isolation component for purposes of the modeling effort.

Bioturbation

Thickness required for bioturbation, T_b, is equivalent to the depth to which the deepest burrowing organism likely to colonize the site in significant numbers can reach. The importance of bioturbation by burrowing aquatic organisms to the mobility of contaminants cannot be overestimated. In addition to the disruption (breaching) of a thin cap that can result when organisms actively work the surface sediments, there is the problem of the direct exposure of the burrowing organisms to the underlying contaminated sediment. Bioturbation depths are highly variable, but have been on the order of 30–60 cm (1–2 ft) for most organisms that populate marine sites in great numbers. Consulting with experts on bioturbation in the region of the disposal site location is desirable. The thickness needed to prevent breaching of cap integrity through bioturbation can be determined indirectly from other information sources. For example, the benthic biota of U.S. coastal and freshwater areas have been fairly well examined, and estimates of the depth to which benthic animals burrow should be available from regional authorities.

Erosion

The total cap thickness should include a thickness for erosion, T_e, which may occur primarily due to long-term continuous processes (i.e., tidal currents and normal wave activity) or episodic events such as storms. This portion of the total thickness can be lost after many years of normal levels of wave and current activity, after an abnormally severe storm season, or in a few days during extreme events. Monitoring activities should result in detecting the loss of cap followed by a management decision to place additional material to bring the cap back to its design thickness.

Detailed methods for predicted erosion values must be used to determine the final value for the erosion thickness component. For projects in which no subsequent capping is anticipated (e.g., the final cap for a site) or for which materials for cap nourishment are not easily obtained, cap erosion thickness should be equivalent to the greater of erosion calculated for period of 20 years of normal current/wave levels and management efforts or a 100 year return period erosion event. For projects in which subsequent capping is planned or for which materials for cap nourishment can be easily obtained, higher erosion rates may be considered. In areas where available capping materials and current and wave conditions are severe, a coarse-grained layer of material may be incorporated into the cap design to provide protection against erosive currents at the site.

Consolidation

The total cap thickness should include a thickness for consolidation of the cap, T_c, so that the minimum required cap thickness is maintained. Such consolidation occurs over a period of time following cap placement, but does not occur more than once. Therefore the total mound thickness can be reduced due to consolidation of the contaminated layer without the need to nourish the cap. If a fine-grained cap will be used, consolidation of the cap will also have to be considered. Also, the potential for consolidation of the native material should also be considered and predicted. Thus a prediction of the value of T_c (from the cap, contaminated material, and native bottom) is also important in interpreting monitoring data to differentiate between changes in mound thickness due to consolidation as opposed to those potentially due to erosion.

Variation in Cap Thickness

The sum of the cap design components for chemical isolation, bioturbation, consolidation, and erosion comprise the design cap thickness. However, the placement process will likely result in some unevenness of the cap thickness. This unevenness should be considered in calculation of the volume of capping material required. Additional discussion of this topic is provided later in this chapter in the Placement Options, Restrictions, and Tolerances Section. Monitoring will define the areas where minimum cap thickness is not achieved, and additional cap material may be required for those areas.

Required Areal Coverage of the Cap

For a capping operation to be successful, the required cap thickness must be placed over the deposit of contaminated material. However, it is not possible or necessary to cap every particle of contaminated material with the full design cap thickness. Within this context, the contaminated material deposit is considered that which can be detected. Typically, the edge of the contaminated mound will be detected with a sediment profiling camera (SPC), which can reliably detect contaminated layers thickness of 1–2 cm.

A small percentage [field data suggests 1 to 5 percent (Truitt, 1986)] of the contaminated material is dispersed in the water column and transported away from the site. The dispersed material settles to the bottom in ever-decreasing thicknesses with increasing distance from the point of discharge. The capping material is similarly dispersed, especially if the grain size and placement methods are similar. Both contaminated and capping materials can therefore be considered to form a "thin veneer" over an area surrounding the identifiable mound or deposit, but this thin layer would quickly be diluted by mixing with the existing sediment profile by hydrodynamic processes and bioturbation, effectively offsetting any potential adverse effects.

Evaluation of Long-Term Contaminant Flux

Depending on the levels of contaminants, types of contaminants, site and operational conditions, and other factors, a detailed assessment of the long-term effectiveness of the cap in controlling contaminant flux may be performed. For example, if the reason for capping is to isolate a sediment that is nontoxic to benthic organisms, and exhibits bioaccumulation only marginally above that for a reference sediment, an isolation thickness of one foot is likely to provide sufficient control, and there is little reason to conduct a detailed assessment. Conversely, if the sediment to be capped has exhibited toxicity to benthic organisms, a detailed assessment of long-term effectiveness would be advisable.

Once processes such as consolidation and erosion have been assessed and an overall required cap thickness for the various components has been determined (at least as a trial value), a detailed assessment of the flux can be conducted. Some components for cap thickness should not be considered in evaluating long-term flux. For example, the depth of overturning due to bioturbation can be assumed to be a totally mixed layer that will offer no resistance to long-term flux. The component for erosion may be assumed to be absent for short periods of time (assuming the eroded layer would be replenished). Components for operational considerations, such as an added thickness to ensure uniform placement, would provide long-term resistance to flux. The changes in void ratio or density of the cap layer due to consolidation should be considered in the flux assessment.

Any detailed assessment of flux must be based on modeling, since the processes involved are potentially very long term. Laboratory testing to more precisely determine parameters for the available models may also be conducted.

Long-Term Processes. Long-term migration of contaminants upward into the cap can potentially occur due to advection (pore water movement) and diffusion. The most common process causing advection is consolidation of the contaminated sediment and/or underlying native sediments. Groundwater flow upward through near-shore sediments is also a potential advection process. The effect of long-term diffusion on the design cap thickness is normally negligible, because long-term diffusion of contaminants through a cap is an extremely slow process and contaminants are likely to adsorb to the clean cap material particles. Migration of pore water upward due to consolidation of the contaminated material is also a slow process and contaminants in the pore water similarly adsorb to the clean cap material particles.

Properly placed capping material acts as a filter layer against any migration of contaminated sediment particulates. There is essentially no driving force that would cause any long-term migration of sediment particles upward into a cap layer. Most contaminants of concern also tend to remain tightly bound to sediment particles. However, the potential movement of contaminants by advection upward into the cap or by molecular diffusion over extremely long time periods is possible.

Advection could occur as an essentially continuous process if there is an upward groundwater gradient acting below the capped deposit. It could also occur as a result of compression or consolidation of the contaminated sediment layer or other layers of underlying sediment. Movement of pore water due to consolidation would be a finite, short-term phenomenon, in that the consolidation process slows as time progresses and the magnitude of consolidation is a function of the loading placed on the compressible layer. The weight of the cap will "squeeze" the sediments, and as the pore water from the sediments moves upward, it displaces pore water in the cap. The result is that contaminants can move part or all the way through the cap in a short period of time. This advective movement can cause a short-term loss, or it can reduce the breakthrough time for long-term advective/diffusive loss.

Even if contaminants were released to the pore water, the cap would act as both a filter and buffer during advection and diffusion. As pore waters move up into the relatively uncontaminated cap, the cap sediments can be expected to scavenge contaminants so that any pore water that traveled completely through the cap theoretically would carry a relatively small contaminant load to the water column. Furthermore, through-cap transport can be minimized by using a cap that has sufficient thickness to contain the entire volume of pore water that leaves the contaminated deposit during consolidation. For example, Bokuniewicz (1989) has estimated that the pore water front coming from a consolidating two-meter-thick mud layer would only advance 24 cm into an overlying sand cap (Sumeri et al., 1991).

Diffusion is a molecular process in which chemical movement occurs from material with higher chemical concentration to material with lower concentration. Diffusion results in extremely slow but steady movement of contaminants. The effect of long-term diffusion on the design cap thickness is normally negligible, because long-term diffusion of contaminants through a cap is an extremely slow process and contaminants are likely to adsorb to the clean cap material particles.

Analytical or Modeling Approaches. A model has been developed by EPA to predict long-term movement of contaminants into or through caps due to advection and diffusion processes. This model has been developed based on accepted scientific principles and observed diffusion behavior in laboratory studies (Bosworth and Thibodeaux, 1990; Thoma et al., 1993; Myers et al., 1996). The model considers the thickness of sediment layers, physical properties of the sediments, concentrations of contaminants in the sediments, and other parameters. The results generated by the model include flux rates, breakthrough times, and pore water concentrations at breakthrough. Such results can be compared to applicable water quality criteria, or interpreted in terms of a mass loss of contaminants as a function of time.

The diffusion relationships used in the model have been verified against laboratory data. However, no field verification studies for the model have been conducted. There is a need for a comprehensive and field verified predictive tool for capping effectiveness, and additional research on this topic is planned. But in absence of such a tool, analytical approaches should be used in calculating long term contaminant flux for capped deposits as long as conservative assumptions are used in the calculations.

Field Data on Long-Term Effectiveness. Some field studies have been conducted on long-term effectiveness of caps. Sequences of cores have been taken at capped dredged material sites in which contaminant concentrations were measured over time periods of up to 15 years (Fredette et al., 1992' Brannon and Poindexter-Rollings, 1990, Sumeri et al. 1994). Core samples taken from capped sites in Long Island Sound, the New York Bight, and Puget Sound exhibit sharp concentration shifts at the cap/contaminated layer interface. For the Puget Sound sites, these results showed no change in vertical contaminant distribution in 5 years of monitoring with 18 mo and 5 yr vibracore samples taken in close proximity to each other. In the New York Bight and Long Island Sound sites, respectively, cores were taken from capped disposal mounds created approximately 3 and 11 years prior to sampling. Visual observations of the transition from cap to contaminated sediment closely correlated with the sharp changes in the sediment chemistry profiles. The lack of diminishing concentration gradients away from the contaminated sediments strongly suggests that there has been minimal long-term transport of contaminants up into the caps. Additional sampling for longer time intervals is planned.

These results confirm that no gross movement of contaminated sediments or contaminants occurs with a properly placed cap, that only porewater advection and molecular diffusion would act to move contaminants into a cap over the long term, that such processes move contaminants at extremely slow rates, and therefore contaminants are effectively isolated from the aquatic environment for extremely long time periods (Brannon and Poindexter-Rollings, 1990).

Laboratory Tests for Flux Evaluation. Results of laboratory tests conducted with samples of the contaminated sediments to be capped and the proposed capping sediments should yield sediment-specific and capping-material-specific values of diffusion coefficients, partitioning coefficients, and other parameters needed to model long-term cap effectiveness. Model predictions of long-term effectiveness using the laboratory-derived parameters should be more reliable than predictions bases on so-called default parameters.

At present, there are several tests that have been applied for this purpose. The USACE developed a first-generation capping effectiveness test in the mid 1980s as part of the initial examination of capping as a dredged material disposal alternative (Sturgis and Gunnison, 1988). Louisiana State University has conducted laboratory tests to assess diffusion rates for specific contaminated sediments to be capped and materials proposed for caps (Wang et al., 1991). Environment Canada has performed tank tests on sediments to investigate the interaction of capping sand and compressible sediments, and additional tests are planned in which migration of contaminants due to consolidation-induced advective flow will be evaluated (Zeman, 1993). The USACE has also developed leach tests to assess the quality of water moving through a contaminated sediment layer into groundwater in a confined disposal-facility environment (Myers and Brannon, 1991). This test is being applied to similarly assess the quality of water potentially moving upward into a cap due to advective forces.

Significance of Flux

The significance of contaminant flux in evaluation of a design cap thickness will be project-specific. As mentioned above, flux should not be of concern for contaminated sediments exhibiting bioaccumulation only marginally above reference sediments. For contaminated sediments where flux is of concern, calculated flux rates, breakthrough times, and pore water concentrations at breakthrough can be compared to a water quality standard or criteria in much the same way as water column contaminant releases during the placement process. Compliance of the flux concentrations at the boundary of the site or edge of an established mixing zone would be appropriate. In this way, the cap thickness component for isolation required to meet the water quality standards can be determined.

Considerations for Interim Caps

Some capping projects could be designed in the context of anticipated multiuse or multiuser applications. In such cases, one site (e.g., a subaqueous borrow pit) could be selected for placement of contaminated sediments from several projects. If several placements of contaminated sediments are to be placed with such frequency that the site could not effectively recolonize, there would be no pathway for bioaccumulation or benthic toxicity. Also, if the site were located in a sheltered area, or the energy from low-frequency events would not cause significant erosion, no placement of cap material or placement of an interim cap with lesser thickness than the full design cap could be considered. Studies are planned to develop guidance on appropriate intervals and thicknesses of interim caps.

LONG-TERM CAP STABILITY

When contaminated material is isolated from the environment through a dredged material capping operation, it is essential that not only the precision and thoroughness of initial cap placement be considered but also the long-term integrity, or stability, of the capped deposit be evaluated on a regular basis. The critical element in successful performance of a cap is preservation of an adequate thickness of this clean material to prevent escape of contaminants from or intrusion of biological organisms into the contaminated sediment. In evaluating long-term cap performance, factors that must be addressed include:

1. Potential for erosion (considering the wave and current conditions at the disposal site and dredged material particle size and cohesion)
2. Possible consolidation (of capping material, contaminated sediment, and foundation material) for effect on long-term site capacity, differentiation from erosion, and quantification of contaminated pore water volume expelled
3. Migration of chemical contaminants (out of the contaminated sediment, through the cap, and/or into underlying foundation sediments)

Each of these factors is important and must be evaluated. However, assessment of consolidation and chemical migration are mathematically tractable, whereas the very stochastic nature of erosion makes it much more complicated to predict.

If any one of these factors (erosion, consolidation, and chemical migration) causes the cap to be too thin to effectively isolate the contaminated material from the surrounding environment, then remedial actions will be required to reestablish cap integrity. These issues are discussed in the following paragraphs, along with recommended techniques and computer models available for analysis.

Evaluation of Consolidation

The process of consolidation occurs as soil particles are pressed together under load. Consolidation may occur in the capping material, in the contaminated sediment, and/or in the native (foundation) sediments. In the capping material, consolidation will result from self-weight of the material, whereas in the contaminated sediment, consolidation will occur first from self-weight and second as the result of the load imposed on it by the capping material. Consolidation of the natural bottom underlying the recently constructed capped mound or deposit will occur as the result of surcharge loading caused by the mound. Consolidation will generally occur in most fine-grained soils, although the amount can vary greatly depending on a number of factors, including the particle type (clay versus silt, high versus low plasticity), moisture content/density, and permeability of the deposit, combined with the loading conditions and thickness of the compressible layers. All of the above factors interact to significantly affect the compressibility of sediment layers. Coarse-grained sediments will not consolidate appreciably.

Evaluation of Erosion Potential

If practical, capping should normally be conducted in predominantly nondispersive sites with relatively little potential for erosion. However, existing sites with potential for erosion can be used for capping projects after completing studies of the frequency of erosion of a specific capping material (considering grain size and cohesion) for expected wave and current conditions (to include storms) over time predicted in the area. The results from such a study will provide data that can be used to predict the expected cumulative amount of erosion over time along with confidence intervals on the answers. These numbers can then be used to define the design cap thickness for erosion protection required for a given length of time (say, 20 to 100 years). Obviously, periodic monitoring is required to measure existing cap thickness, along with plans that describe at what reduced cap thickness contingency plans for placement of additional material will be required.

The deposit of contaminated dredged material must also be stable against excessive erosion and resuspension of material before placement of the cap. The potential for resuspension and erosion depends on bottom current velocity, potential for wave-induced currents, sediment particle size, and sediment cohesion. Site selection criteria as described above would normally result in a site with low bottom current velocity and little potential for erosion during the window for placement of the contaminated sediments and cap. However, if the contaminated sediment is hydraulically dredged and placed fine-grained material, a thorough analysis of the potential for resuspension and erosion needs to be performed.

Conventional methods for analysis of sediment transport are available to evaluate erosion potential (Teeter, 1988; Dortch et al., 1990). These methods can range from simple analytical techniques to numerical modeling. In the analysis of erosion, the effects of self-armoring due to the winnowing away of finer particles should be considered. If erosion is considered to be a problem, armoring and geotextiles may be considered as engineering approaches to overcome or protect against this problem.

The first level of investigation of cap stability against erosion involves examination of the normal wave and current regime to determine if these cause measurable amounts of erosion. However, sites where day-to-day waves and currents cause measurable amounts of erosion would be very poor sites for capping projects.

CAP MONITORING

Monitoring of capped disposal projects is required to ensure that capping acts as an effective control measure (Palermo et al., 1992). Monitoring is therefore required before, during, and following placement of the contaminated and capping material to insure that an effective cap has been constructed. (This effort also may be defined as construction monitoring.) Monitoring may also be required to ensure that the cap as constructed will be effective in isolating the contaminants and that long-term integrity of the cap is maintained. (This activity also may be defined as long-term monitoring.)

The monitoring discussed here does not focus on water column processes or the water column contaminant pathway during the placement of contaminated material prior to capping. If a determination is made that the contaminated material has potential for unacceptable water column impacts during placement, other control measures to offset those impacts (i.e., silt curtain, diffuser, trémie pipe, etc.) and additional monitoring of water column processes as necessary. Also, the monitoring discussed here does not focus on those aspects of open water site monitoring pertaining to site designation or on the direct physical effects of disposal. Any such monitoring would be considered in the context of the overall site selection process and site monitoring plan (Palermo, 1991b).

Appropriate objectives for a capping monitoring program/plan may include the following:

1. Determine bathymetry, organisms, and sediment type at capping site
2. Determine currents for evaluating erosion and dispersion potential
3. Define areal extent and thickness of contaminated material deposit (to include the apron thickness) to guide cap placement

4. Define areal extent and thickness of the cap
5. Determine that desired capping thickness is maintained
6. Determine cap effectiveness in isolating contaminated material from benthic environment
7. Determine extent of recolonization of biology and bioturbation potential

Selection of the types of samples or observations to be made, the equipment to be used, the number of samples or observations, etc. is highly project-dependent. Fredette et al. (1990b) contains guidelines on available equipment and techniques. Physical measurements may include bathymetry, cap thickness, sediment physical properties (e.g. grain size distribution and density), wave and current conditions, etc. Depth sounders, side-scan sonar and subbottom profilers, sediment sampling and coring devices, sediment profiling cameras, and instruments for measuring engineering properties of the sediment are required to make these physical measurements.

Designating Management Actions

When any acceptable threshold values are exceed, some type(s) of management action(s) are required. The appropriate management actions should be determined/defined early in the disposal planning process; they should not be determined after the threshold value(s) have been exceeded.

Management options in early tiers could include increasing the level of monitoring to the next tier, the addition of more sediment to form a thicker cap, or stopping use of the site. Management options in later tiers could include stopping use of the site, changing the cap material, or the addition of a less-porous material in cases where contaminant transport due to biological or physical processes is occurring. For caps that are experiencing erosion, additional cap can also be added, although it may be advisable to choose a coarser material (coarse sand or gravel) to provide armoring. In cases where extreme problems are encountered, removal of the contaminated material and placement at another site could be considered.

SUMMARY

This chapter presents an overview of guidance for subaqueous dredged material capping. Capping is the controlled accurate placement of contaminated material at an open water disposal site, followed by a covering or cap of clean isolating material. A capping operation must be treated as an engineered project with carefully considered design, construction, and monitoring to ensure that the design is adequate.

Capping of contaminated material in open water sites began in the late 1970's, and a number of capping operations under a variety of disposal conditions have been accomplished. Field experience with these projects has shown the capping concept is technically and operationally feasible.

The cost of capping is generally lower than alternatives involving confined (diked) disposal facilities. The geochemical environment for subaqueous capping favors long-term stability of contaminants as compared to the upland environment where geochemical changes may favor increased mobility of contaminants. Capping is therefore an attractive alternative for disposal of contaminated sediments from both the economic and environmental standpoints.

ACKNOWLEDGMENTS

The information presented in this chapter is a summary of detailed guidance prepared by the U.S. Army Corps of Engineers (USACE). The contributions of Mr. Jim Clausner, U.S. Army Engineer Waterways Experiment Station (WES); Dr. Robert E. Randall, Texas A&M University; and Dr. Thomas Fredette, New England Division, USACE are acknowledged. Permission was granted by the Chief of Engineers to publish this information.

REFERENCES

Averett, D. E., Palermo, M. R., Otis, M. J. and Rubinoff, P. B. 1989. "New Bedford Harbor Superfund Project, Acushnet River Estuary Engineering Feasibility Study of Alternatives for Dredging and Dredged Material Disposal, Report No. 11, Evaluation of Conceptual Dredging and Disposal Alternatives," Technical Report EL-88-15, Report 11, U. S. Army Engineer Waterways Experiment Station, Vicksburg, MS.

Bokuniewicz, H. J., Gerbert, J., Gordon, R. B., Higgins, J. L., Kaminsky, P., Pilbeam, C. C., Reed, M., and Tuttle, C. 1978. "Field Study of the Mechanics of the Placement of Dredged Material at Open-Water Disposal Sites," Technical Report D-78-7, U. S. Army Engineer Waterways Experiment Station, Vicksburg, MS.

Bosworth, W. S. and Thibodeaux, L. J. 1990. "Bioturbation: A Facilitator of Contaminant Transport in Bed Sediment," *Environmental Progress,* Vol. 9, No. 4.

Brannon, J., and Poindexter-Rollings, M. E. 1990. "Consolidation and Contaminant Migration in a Capped Dredged Material Deposit," *The Science of the Total Environment,* Vol. 91, pp 115–126.

Coastal Engineering Research Center. 1984. *Shore Protection Manual,* 4th ed., 2 Vols, prepared by the US Army Engineer Waterways Experiment Station, available for purchase from the Superintendent of Documents, US Government Printing Office, Washington, DC.

d'Angremond, K., de Jong, A. J., and De Waard, C. -P. 1986. "Dredging of Polluted Sediment in the 1st Petroleum Harbor, Rotterdam" (in preparation), Proceedings of the Third United States/The Netherlands Meeting on Dredging and Related Technology, prepared for the U.S. Army Corps of Engineers Water Resources Support Center by the U.S. Army Engineer Waterways Experiment Station, Vicksburg, MS.

Dortch, M. S. et al. 1990. "Methods of Determining the Long-Term Fate of Dredged Material for Aquatic Disposal Sites," Technical Report D-90-1, U.S. Army Engineer Waterways Experiment Station, Vicksburg, MS.

Earhart, G., Clarke, D., and Shipley, J. 1988. "Beneficial Uses of Dredged Material in Shallow Coastal Waters; Chesapeake Bay Demonstrations," Information Exchange Bulletin D-88-6, U. S. Army Engineer Waterways Experiment Station, Vicksburg, MS.

Elliot, D. O. 1932. *The Improvement of the Lower Mississippi River for Flood Control and Navigation,* Vol. II, U.S. Army Engineer Waterways Experiment Station, Vicksburg, MS.

Francingues, N. R. and Palermo, M. R. 1984. "Management Strategy for the Disposal of Dredged Material," *Proceedings of the Specialty Conference Dredging '84,* American Society of Civil Engineers, Clearwater, FL.

Fredette, T. J., Nelson, D. A., Clausner, J. E., and Anders, F. J. 1990a. "Guidelines for Physical and Biological Monitoring of Aquatic Dredged Material Disposal Sites," Technical Report D-90-12, U.S. Army Engineer Waterways Experiment Station, Vicksburg, MS.

Fredette, T. J., Nelson, D. A., Miller-Way, T., Adair, J. A., Sotler, V. A., Clausner, J. E., Hands, E. B., and Anders, F. J. 1990b. "Selected Tools and Techniques for Physical and Biological Monitoring of Aquatic Dredged Material Disposal Sites," Technical Report D-90-11, U.S. Army Engineer Waterways Experiment Station, Vicksburg, MS.

Fredette, T. J., Germano, J. D., Kullberg, P. G., Carey, D. A. and Murray, P. 1992. "Chemical Stability of Capped Dredged Material Disposal Mounds in Long Island Sound, USA. " 1st International Ocean Pollution Symoposium, Mayaguez, Puerto Rico. Chemistry and Ecology.

Gunnison, D., et al. 1987. "Development of a Simplified Column Test for Evaluation of Thickness of Capping Material Required to Isolate Contaminated Dredged Material," Miscellaneous Paper D-87-2, U.S. Army Engineer Waterways Experiment Station, Vicksburg, MS.
Herbich, J. B. 1992. *Handbook of Dredging Engineering.* McGraw Hill, New York.
Johnson, B. H. and Fong, M. T. 1995. "Development and Verification of Numerical Models for Predicting the Initial Fate of Dredged Material Disposed in Open Water; Report 2, Theoretical Developments and Verification Results." Prepared for U.S. Army Corps of Engineer Waterways Experiment Station, Vicksburg, MS. Technical Report DRP-93-1, NTIS No. AD-A292 040.
Kikegawa, K. 1983. "Sand Overlying for Bottom Sediment Improvement by Sand Spreader," Management of Bottom Sediments Containing Toxic Substances: Proceedings of the 7th U.S./Japan Experts Meeting, pp. 79–103, prepared for the U.S. Army Corps of Engineers Water Resources Support Center by the U.S. Army Engineer Waterways Experiment Station, Vicksburg, MS.
Kleinboesem, W. C. H. and van der Weijde, R. W. 1983. "A Special Way of Dredging and Disposing of Heavily Polluted Silt in Rotterdam," World Dredging Congress 1983, Paper M2, Singapore, pp. 509–525.
Ling, H. I., Leshchinsky, D., Gilbert, P. A., and Palermo, M. R. 1996. "In-Situ Capping of Contaminated Sediments: Geotechnical Considerations," Second International Congress on Environmental Geotechnics, Osaka, Japan.
McLellan, T. N. and Truitt, C. L. 1986. "Demonstration of a Submerged Diffuser for Dredged Material Disposal," *Proceedings of the Conference Oceans '86,* Washington, DC.
Moritz, H. R. and Randall, R. E. 1995. "Simulating dredged material placement at open water disposal sites," *Journal of Waterway, Port, Coastal, and Ocean Engineering,* ASCE, Vol. 121, No 1.
Morton, R. W. 1983. "Precision Bathymetric Study of Dredged Material Capping Experiment in Long Island Sound," In: Kester, D. R., Duedall, I. W., Ketchum, B. H., and Parks, P. K. (editors). *Wastes in the Ocean,* Volume II. *Dredged Material Disposal.* Wiley, New York, pp. 99–121.
Morton, R. W. 1987. "Recent Studies Concerning the Capping of Contaminated Dredged Material," Proceedings of the United States/Japan Experts Meeting on Management of Bottom Sediments Containing Toxic Substances, November 3–6, 1987, Baltimore, MD.
Myers, T. E. and Brannon, J. M. 1991. "Technical Considerations for Application of Leach Tests to Sediments and Dredged Material," Environmental Effects of Dredging Program Technical Note EEDP-02-15, U.S. Army Engineer Waterways Experiment Station: Vicksburg, MS.
Myers, T. E., M. R. Palermo, T. J. Olin, D. E. Averett, D. D. Reible, J. L. Martin and S. C. McCutcheon. 1996. "Estimating Contaminant Losses from Components of Remediation Alternatives for Contaminated Sediments," prepared for the U.S. Environmental Protection Agency, Great Lakes National Program Office, Chicago, IL.
Neal, R. W., Henry, G., and Greene, S. H. 1978. "Evaluation of the Submerged Discharge of Dredged Material Slurry During Pipeline Dredged Operations," Technical Report D-78-44, by JBF Scientific Corp. for U.S. Army Engineer Waterways Experiment Station, Vicksburg, MS.
O'Connor, J. M. and O'Connor, S. G. 1983. "Evaluation of the 1980 Capping Operations at the Experimental Mud Dump Site, New York Bight Apex," Technical Report D-83-3, U.S. Army Engineer Waterways Experiment Station, Vicksburg, MS.
Palermo, M. R. 1989. "Capping Contaminated Dredged Material in Deep Water," *Proceedings of the Specialty Conference Ports 89,* American Society of Civil Engineers, Boston, MA.
Palermo, M. R. 1991a. "Design Requirements for Capping," Dredging Research Technical Note DRP-5-03, U.S. Army Engineer Waterways Experiment Station, Vicksburg, MS.
Palermo, M. R. 1991b. "Site Selection Considerations for Capping," Technical Note DRP-5-04, U.S. Army Engineer Waterways Experiment Station, Vicksburg, MS.
Palermo, M. R. 1991c. "Equipment and Placement Techniques for Capping," Technical Note DRP-5-05, U.S. Army Engineer Waterways Experiment Station, Vicksburg, MS.
Palermo, M. R. 1994. "Options for Submerged Discharge of Dredged Material," *Proceedings of the 25th Dredging Seminar and Western Dredging Association XIII Annual Meeting,* May 18–20, 1994, San Diego, CA.
Palermo, M. R., 1994. "Placement Techniques for Capping Contaminated Sediments," *Proceedings of Dredging '94, The Second International Conference and Exhibition on Dredging and Dredged Material Placement,* American Society of Civil Engineers, Lake Buena Vista, Fl.

Palermo, M. R., et al. 1989. "Evaluation of Dredged Material Disposal Alternatives for U.S. Navy Homeport at Everett, Washington," Technical Report EL-89-1, U.S. Army Engineer Waterways Experiment Station, Vicksburg, MS.

Palermo, M. R., Fredette, T. J. and Randall, R. E., 1992. "Monitoring Considerations for Capping," Dredging Research Technical Notes, DRP-5-07, U.S. Army Engineer Waterways Experiment Station, Vicksburg, MS.

Palermo, M. R. 1992. "Design Procedure for Capping Contaminated Sediments," *Proceedings of the Twenty-Fifth Annual Dredging Seminar and Western Dredging Association XIII Annual Meeting,* May 26–29, 1992, Mobile AL, CDS Report No. 323, Center for Dredging Studies, Texas A&M University, College Station, TX.

Palermo, M. R., Miller, J., Maynord, S., and Reible, D. 1996. "Guidance for In-Situ Subaqueous Capping of Contaminated Sediments," EPA 905-B96-004, Great Lakes National Program Office, U.S. Environmental Protection Agency, Chicago, IL.

Palermo, M. R., Clausner, J. E., Rolllings, M. P., Williams, G. L., Myers, T. E., Fredette, T. J., and Randall, R. E. 1998. "Guidance for Subaqueous Dredged Material Capping," Technical Report DOER-1, U.S. Army Engineer Waterways Experiment Station, Vicksburg, MS, http://www.wes.army.mil/el/dots/doer/pdf/doer-1.pdf.

Pequegnat, W. E., Gallaway, B. J., and Wright, T. D. 1990. "Revised Procedural Guide for Designation Surveys of Ocean Dredged Material Disposal Sites," Technical Report D-90-8, US Army Engineer Waterways Experiment Station, Vicksburg, MS.

Randall, R. E., Clausner, J. E., and Johnson, B. H. 1994. "Modeling Cap Placement at New York Mud Dump Site," *Proceedings of Dredging '94, The Second International Conference and Exhibition on Dredging and Dredged Material Placement,* American Society of Civil Engineers, Lake Buena Vista, FL.

Rhoads, D. C. and Boyer, L. F. 1982. "The Effect of Marine Benthos on Physical Properties of Sediments; A Successional Perspective," Chapter 1 in *Organism-Sediment Relations; Biogenic Alteration of Sediments.* Eds. P. L. McCall and M. J. S. Tevesz, pp. 3–52. Plenum Press, New York.

Rhoads, D. C. and Germano, J. D. 1982. "Characterization of Organism–Sediment Relations Using Sediment Profile Imaging: An Efficient Method of Remote Ecological Monitoring of the Sea Floor (REMOTS System)," *Marine Ecology Progress Series,* Vol 8, pp. 115–128.

Sanderson, W. H., and McKnight, A. L. 1986. "Survey of Equipment and Construction Techniques for Capping Dredged Material," Miscellaneous Paper D-86-6, US Army Engineer Waterways Experiment Station, Vicksburg, MS.

Scheffner, N. W., Thevenot, M. M., Tallent, J. R, and Mason, J. M. 1995. "LTFATE: A model to investigate the long-term fate and stability of dredged material disposal sites," Technical Report DRP-94-xx, U.S. Army Engineer Waterways Experiment Station, Vicksburg, MS.

Schroeder, P. R. and Palermo, M. R. 1990. "The Automated Dredging and Disposal Alternatives Management System (ADDAMS)," Technical Note EEDP-06-12, U.S. Army Engineer Waterways Experiment Station, Vicksburg, MS.

Science Applications International Corporation (SAIC). 1995. Sediment Capping of Subaqueous Dredged Material Disposal Mounds: An Overview of the New England Experience, DAMOS Contribution #95. SAIC Report # SAIC-90/7573&C84. U.S. Army Corps of Engineers, New England Division, Waltham, MA.

Scott, K. J. Rhoads, D., Rosen, J., Pratt, S., and Gentile, J. 1987. "Impact of Open Water Disposal of Black Rock Harbor Dredged Material on Benthic Recolonization at the FVP Site," Technical Report D-87-4, U.S. Army Engineer Waterways Experiment Station, Vicksburg, MS.

Shields, F. D. and Montgomery, R. L. 1984. "Fundamentals of Capping Contaminated Dredged Material," *Proceedings of the Conference Dredging '84,* American Society of Civil Engineers, Vol 1, pp. 446–460.

Sturgis, T., and Gunnison, D. 1988. "A Procedure for Determining Cap Thickness for Capping Subaqueous Dredged Material Deposits," Technical Note EEDP-0109, U.S. Army Engineer Waterways Experiment Station, Vicksburg, MS.

Sumeri, A. 1995. "Dredged Material is Not a Spoil: A status on the Use of Dredged Material in Puget Sound to Isolate Contaminated Sediments," *Proceedings of the 14th World Dredging Congress,* World Organization of Dredging Associations, Amsterdam, The Netherlands, November, 1995.

Sumeri, A., Fredette, T. J., Kullberg, P. G., Germano, J. D., Carey, D. A., 1991 "Sediment Chemistry

Profiles of Capped In-Situ and Dredged Sediment Deposits: Results from Three U.S. Army Corps of Engineers Offices," WEDA XII Western Dredging Association and Twenty fourth Annual Dredging Seminar, Center for Dredging Studies, Texas A&M University, College Station, TX.

Sumeri, A., Fredette, T. J., Kullberg, P. G., Germano, J. D, Carey, D. A., and Pechko, P. 1994. "Sediment Chemistry Profiles of Capped Dredged Material Deposits Taken 3 to 11 Years after Capping," Dredging Research Technical Note DR-5-09, U.S. Army Engineer Waterways Experiment Station, Vicksburg, MS.

Teeter, A. M. 1988. "New Bedford Harbor Superfund Project, Acushnet River Estuary Engineering Feasibility Study of Dredging and Dredged Material Disposal Alternatives, Report 2, Sediment and Contaminant Hydraulic Transport Investigations," Technical Report EL-88-15, Report 2, U.S. Army Engineer Waterways Experiment Station, Vicksburg, MS.

Thoma, G. J., Reible, D. D., Valsaraj, K. T., and Thibodeaux, L. J. 1993. "Efficiency of Capping Contaminated Sediments in-Situ: 2. Mathematics of Diffusion-Adsorption in the Capping Layer," *Environ. Sci. Technol.*

Togashi, H. 1983. "Sand Overlaying for Sea Bottom Sediment Improvement by Conveyor Barge," Management of Bottom Sediments Containing Toxic Substances: Proceedings of the 7th U.S./Japan Experts Meeting, pp 59–78, prepared for the U.S. Army Corps of Engineers Water Resources Support Center by the U.S. Army Engineer Waterways Experiment Station, Vicksburg, MS.

Truitt, C. L. 1986. "Fate of Dredged Material During Open Water Disposal," Environmental Effects of Dredging Programs Technical Note EEDP-01-2, U.S. Army Engineer Waterways Experiment Station, Vicksburg, MS.

Truitt, Clifford L. 1988. "Dredged Material Behavior During Open-Water Disposal," *Journal of Coastal Research*, Volume 4. Number 2.

Truitt, C. L., Clausner, J. E., and McLellan, T. N. 1989. "Considerations for Capping Subaqueous Dredged Material Deposits," *Journal of Waterway, Port, Coastal, and Ocean Engineering*, Vol. 115, No. 6.

U.S. Army Corps of Engineers (USACE). 1987. "Confined Disposal of Dredged Material," Engineer Manual 1110-2-5027, Washington, DC.

U.S. Army Corps of Engineers/ Environmental Protection Agency (USACE and EPA). 1992. "Evaluating Environmental Effects of Dredged Material Management Alternatives—A Technical Framework," EPA842-B-92-008, Environmental Protection Agency, Office of Marine and Estuarine Protection, Washington, DC.

U.S. Army Corps of Engineers/Environmental Protection Agency (USACE and EPA). 1991. "Evaluation of Dredged Material Proposed for Ocean Disposal—Testing Manual," EPA-503-8-91/001, Environmental Protection Agency Office of Marine and Estuarine Protection, Washington, DC.

U.S. Army Corps of Engineers/Environmental Protection Agency (USACE and EPA). In Preparation. "Evaluation of Dredged Material Proposed for Inland and Near Coastal Disposal—Testing Manual," EPA-xxx-x-/xxx, Environmental Protection Agency, Office of Science and Technology, Washington, DC.

U.S. Army Corps of Engineers, New England Division. 1990. "New Bedford Harbor Superfund Pilot Study, Evaluation of Dredging and Dredged Material Disposal," U.S. Army Corps of Engineers, New England Division, Waltham, MA., May 1990.

Van Wijck, J. and Smits, J. 1991. "Underwater Disposal of Dredged Material: A Viable Solution for the Maintenance of Dredging Works in the River Scheldt," Bulletin No. 73, Permanent International Association of Navigation Congresses.

Wang, X. Q., Thibodeaux, L. J., Valsaraj, K. T., and Reible, D. D. 1991. "Efficiency of Capping Contaminated Bed Sediments in Situ. 1. Laboratory-Scale Experiments on Diffusion-Adsorption in the Capping Layer," *Environmental Science and Technology*, Vol. 25, No. 9.

Wiley, M. B. 1994. "DAMOS Capping Model Verification," Contribution #89, Science Applications International Corporation, submitted to U.S. Army Corps of Engineers, New England Division, Waltham, MA.

Zeman, A. J. 1993. "Subaqueous Capping of Very Soft Contaminated Sediments," 4th Canadian Conference on Marine Geotechnical Engineering, St. John's, Newfoundland.

CHAPTER 14
MODELING THE PHYSICAL AND CHEMICAL STABILITY OF UNDERWATER CAPS IN RIVERS AND HARBORS

RAM K. MOHAN, P.E., PH.D.
Gahagan & Bryant Associates, Inc.
Baltimore, MD

Notation	14.2
Introduction	14.4
Typical Cap Components	14.5
Cap Design Criteria	14.6
Cap Placement Analysis	14.7
Impact Velocity of Sand	14.7
Thickness of the Slug	14.7
Horizontal Placement of the Slug	14.8
Required Number of Lifts	14.8
Geotechnical Stability Analysis	14.8
Settlement Analysis	14.8
Bearing Capacity Analysis	14.9
Slope Stability Analysis	14.10
Chemical Stability Analysis	14.11
Contaminant Release by Diffusion	14.11
Contaminant Release by Advection/Dispersion	14.12
Contaminant Release by Pore Water Release During Sediment Compression	14.12
Physical Stability Analysis	14.13
Protection from River Flow	14.13
Protection from Wave Action	14.13
Protection from Propeller Wash Forces	14.14
Protection from Ice Loading	14.14
Erosion of Contaminated Sediments Due to Pore Water Release during Sediment Compression	14.15
Hydraulic Filtering Analysis	14.15
Filter Design Criteria	14.15
Recommended Thickness and Gradation	14.16
A General Modeling Approach—The CAPSTABL Model	14.17

Example Model Application		14.19
Summary		14.24
References		14.24

NOTATION

A_c	Arial extent of the cap
B	Total width of the cap
c'	Effective cohesive strength
C	Sand–armor interface chemical concentration
C_1	Constant (0.86 for high turbulence and 1.2 for low turbulence)
C_2	Constant (0.22 for nonducted propeller)
C_3	Constant (4.4 for nonducted propeller)
C_{bs}	Contaminant concentration in the bottom 2.54 cm of the sand cap
C_c	Compression index of the sediments
C_0	Sediment–sand interface chemical concentration
C_s	Depth-averaged sediment chemical concentration
C_t	Average concentration of the contaminant in the native sediment zone
c_u	Undrained shear strength
$c_{u,n}$	Undrained shear strength of the native sediment sublayers
C_v	Coefficient of consolidation
C_α	Coefficient of secondary compression
d_{15}	15% passing size
$d_{15(A)}$	Diameter through which 15% of the armor passes
$d_{15(B)}$	Diameter through which 15% of the base material passes
$D_{15(F)}$	Diameter through which 15% of the filter passes
d_{50}	50% passing size
$d_{50(A)}$	Diameter through which 50% of the armor passes
$d_{50(B)}$	Diameter through which 50% of the base material passes
$D_{50(F)}$	Diameter through which 50% of the filter passes
d_{85}	85% passing size
$d_{85(B)}$	Diameter through which 85% of the base material passes
$d_{85(F)}$	Diameter through which 85% of the filter passes
D_e	Effective diffusivity
D_H	Hydrodynamic dispersion coefficient
D_P	Propeller diameter
d_{sl}	Distance the slug travels in water before it hits the bottom
D_t	Transient transport effective diffusion coefficient
D_w	Chemical diffusivity
e_0	Initial void ratio
erfc	Error function
F_b	Total negative buoyancy of the slug
F_c	Contaminant flux per area
F_l	Long-term flux
f_{occ}	Organic carbon content of the cap
f_{ocs}	Organic carbon content of the native sediments
FS	Factor of safety
F_{ss}	Steady-state flux
g	Acceleration due to gravity

G_{sc}	Specific gravity of the cap layer
G_{ss}	Specific gravity of the native sediments
H_c	Height (or thickness) of the cap layer
$H_{c(max)}$	Maximum allowable thickness of the cap
H_{dr}	Drainage path length for pore water expulsion
H_f	Height of the fill
H_P	Distance from propeller shaft to channel bottom
H_s	Height (or thickness) of the native sediment layer
$H_{s,n}$	Height (or thickness) of the native sediment sublayers
h_{sl}	Thickness of the slug as it reaches the bottom
H_{SSL}	Actual thickness of the sand sorptive layer
i	Angle between native sediments and the edge of the cap
K_{oc}	Chemical distribution coefficient
K_p	Partitioning coefficient
n	random integer
n_c	Porosity of the cap layer
NGVD	National Geodetic Vertical Datum
N_{hl}	Required number of horizontal lifts of the cap
n_s	Porosity of the native sediment layer
n_{sl}	Porosity of the slug
N_{vl}	Required number of vertical lifts for the cap
p_0	Initial vertical effective stress
PC	Personal computer
PCB	Poly-Chloro-Biphenyl
P_d	Engine power
P_0	Average pore water chemical concentration
P_w	Water column chemical concentration
q	Surcharge
q_{ult}	Ultimate bearing capacity of the sediments
r	Radial coordinate
R	Retardation factor
R_0	Original radius of the slug
R_f	Final radius of the slug
S_D	Standard deviation of the solid particles within the slug
S_{rl}	Riprap layer thickness
S_{sl}	Overlap factor for the slug
S_0	Primary consolidation settlement
S_T	Total consolidation settlement
S_α	Secondary consolidation settlement
t	Time period
t_0	Beginning of secondary compression
T_A	Thickness of the armor layer
t_b	Breakthrough time
t_d	End of design time frame
t_e	Length of time elapsed following cap placement
T_F	Thickness of the filter layer
t_{sl}	Time of travel of the slug before it hits the bottom
t_{ss}	Steady-state time
T_v	Dimensionless time factor
U	Degree of consolidation
U_0	Jet velocity exiting the propeller

USACE   U.S. Army Corps of Engineers
U_w Ambient flow or current velocity of the water
V_b Bottom velocity
V_{bp} Maximum bottom velocity due to propeller wash forces
V_f Mean flow velocity above the armor layer
V_g Groundwater velocity
V_i Impact velocity of sand
V_{pw} Velocity of release of the pore water
V_{sl} Volume of the slug
x_{sl} Horizontal displacement of the slug
Δp Additional vertical effective stress due to effective cap weight
α Empirical constant
β Cap placement function
ϕ'_c Effective friction angle of the cap material
ϕ_c Friction angle of the cap material
ϕ'_s Effective friction angle of the native sediments
γ_a Unit weight of the armor stone
γ_c Unit weight of the cap
γ'_c Effective unit weight of the cap
γ_f Unit weight of the fill material
γ'_s Effective unit weight of the native sediments
γ_w Unit weight of water
μ_w Kinematic viscosity of water
ρ_{sl} Density of the slug
ρ_{bc} Bulk density of the cap
ρ_{bs} Bulk density of the native sediments
σ' Effective normal stress
τ_r Required shear strength

INTRODUCTION

Rapid, largely underregulated industrialization of the past has caused contamination of rivers, waterways, and harbors all over the world. Regardless of the source of contamination (industrial discharges, storm sewers, wastewater, landfill runoff and leachate), they severely impair the ecological and recreational functions of the affected water body. Typically, the released contaminants adhere to the finest fraction of the suspended sediments in the water column and get deposited in relatively quiescent areas of the water body. Once in place, contaminated sediments can exert a significant oxygen demand, support a poor diversity of benthic organisms, and adversely affect local (overlying) and downstream water quality. Removal of the contaminated sediments can be expensive because a high degree of efficiency and reliability is required in such operations. Subaqueous capping is an attractive, nonintrusive and cost-effective method of remediating contaminated sediments in rivers and harbors where draft restriction is not a major concern. The same physiochemical properties and hydraulic conditions that favored the initial adsorption to and deposition of the contaminated sediments typically favor successful containment by capping. Since capping is an in-situ technique, it can often be accomplished at approximately 20 to 30% of the cost of a remedial dredging project.

Successful design of an underwater cap requires the proper application of hydraulic (armor and filter equations), chemical (diffusive and advective/dispersive transport equations), and geotechnical (settlement and stability equations) engineering principles. An underwater cap should be able to withstand the worst design case physical and chemical events at the site in order to ensure sufficient protectiveness and isolation of the contaminants from the surrounding environment. The major physical destabilizing forces include extreme water flows, storm waves, and tidal currents. Molecular diffusion, groundwater-induced advection, hydrodynamic dispersion, and pore water flux due to consolidation are the major chemical destabilizing forces. A base sand isolation layer provides protection from chemical events and an armor/filter layer (usually riprap and gravel/cobbles) provides protection from physical forces. Bioturbation (sediment processing by benthic organisms) is another major destabilizing force causing mixing of the water/sediment interface, thereby releasing contaminants to the water column. Therefore, the design thickness of the cap should account for physical and chemical isolation, as well as bioturbation. Since the likelihood and type of organisms that may colonize the cap would vary from site to site, a site-specific survey of benthic organisms is often required. This could also be estimated by chemical and isotopic analysis of thinly sectioned sediment cores for sediment mixing depths.

Since the early 1980s, several pilot-scale and full-scale projects have been conducted to evaluate the effectiveness of capping under a variety of site conditions [Kikegawa (1983); Strugis and Gunnison (1985); Palermo (1991); Sumeri et al. (1991); Wang et al. (1991); Fredette et al. (1992); Zeeman (1993); Averett and Francingues (1994); Nelson et al. (1994); Randall et al. (1994); Thibodeaux et al. (1994); GeoEngineers (1995); Wright and Kim (1995); Environment Canada (1996); Ling et al. (1996); Hull et al. (1998); Laboyrie and Flach (1998); Li et al. (1998); Lillycrop and Clausner (1998); Shaw et al. (1998); and Mohan et al. (1999a)]. While these studies established the validity of the environmental isolation provided by the capping process, design of underwater caps is still a complicated process requiring specialized skills in hydraulic, chemical, and geotechnical engineering aspects. This chapter outlines the theoretical considerations for modeling the physical and chemical stability of underwater caps in river and harbor environments, and presents a generic setup of a cap evaluation model (CAPSTABL). In addition, the various factors directly affecting the long-term stability of underwater caps in the marine environment are identified through an example application of the model in a project case study.

TYPICAL CAP COMPONENTS

Capping involves placement of one or more of the following above the contaminated sediments (see Figure 14.1):

1. **Base stabilizing layer** (usually geotextile or granular soil) to provide additional bearing capacity to the in-situ sediments
2. **Base isolation layer** (usually sands) to provide environmental (physical and chemical) isolation of the contaminated materials
3. **Filter layer** (usually gravels or geotextiles) to provide protection to the isolation layer from hydraulic erosive forces that may act through the voids in the armor layer
4. **Armor layer** (usually rocks) to provide physical (hydraulic) protection to the cap layer against erosive forces

The success of a capping operation relies on sound engineering design and proper selection of the cap placement techniques.

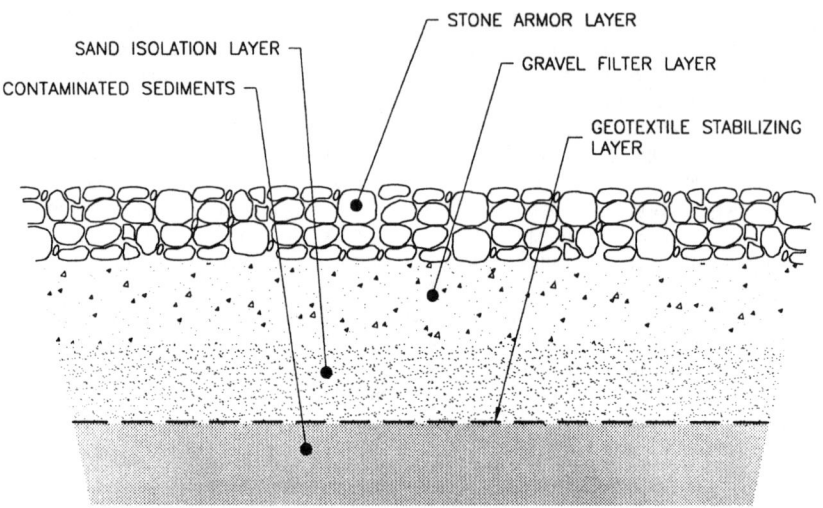

FIGURE 14.1. General schematic of a cap system.

CAP DESIGN CRITERIA

The general design principles of the cap system have been fairly well developed recently through the efforts of several investigators [Palermo (1991); Silva et al. (1991); USACE (1991); Wang et al. (1991); Fredette et al. (1992); Maynord and Oswalt (1993); Thoma et al. (1993); Randall et al. (1994); Thibodeaux et al. (1994); Ling et al. (1996); Mohan (1997); Lillycrop and Clausner (1998); Mohan et al. (1999a; 1999b); and Palermo (2000)]. Special factors of interest to cap design/evaluation include cap behavior during placement, settlement of the cap and native sediments following placement, potential contaminant release mechanisms, and stability of the cap in the long term.

Design considerations for an underwater cap can be classified into six categories:

1. **Arial extent of the cap.** This is usually determined based on mapping the horizontal distribution of the contaminants. Ideally, the cap should cover the entire horizontal extent of the contaminated sediments. However, in special cases where sediment contamination is vast and widely distributed spatially within the water body, the extent of the cap should be determined based on human health exposure risks and isolation of the bulk of the contaminant mass to the food chain (Mohan et al., 1999a).

2. **Cap behavior during placement.** Several mechanisms are of interest as the cap is being placed. These include the impact velocity of sand, thickness of the slug during placement, horizontal displacement of the slug, and required number of lifts for attaining the final desired thickness of the cap.

3. **Geotechnical stability.** Once in place, the geotechnical stability of the cap is of major concern. This includes the potential for settlement (including differential settlement) of the cap and native sediments, bearing capacity of the cap/native sediments, and slope stability.

4. **Chemical stability.** Major contaminant release mechanisms of concern include diffusion, advection/dispersion, and pore water release during sediment compression. The

effectiveness of the cap against these effects is typically measured in terms of the rate of release of the contaminants with and without the cap ("flux"), and the time associated with the release of the bulk (typically set at 95%) of the contaminants to the water column ("breakthrough time").

5. **Physical stability.** As with any coastal structure, an underwater cap/armor must be designed for a specific level of protection. This implies that the cap/armor should be able to withstand the impact forces resulting from a certain return period event (typically the 100-year event for moderately contaminated sediments). Considerations include protection from river flow, wave action, propeller wash forces, ice loads, and forces resulting from pore water release mechanisms.

6. **Hydraulic filtering.** The cap/armor should also be designed to withstand the impacts of the hydraulic filtering forces. There are two ways of providing this protection: (i) by installing a separate filter layer between the armor and cap layers, and (ii) by installing a "graded" armor layer (i.e., with a wider grain size range) above the cap layer (Wright and Kim, 1995).

CAP PLACEMENT ANALYSIS

Underwater caps may be placed using several techniques [Herbich (1992b); Mohan (1997); Palermo (2000)]. These include surface placement from barges, clamshells, or pipelines, placement using spreader boxes, trémie pipes, and underwater placement using clamshells. One of the main concerns during placement of the cap is the behavior of the slug (velocity and properties) as it reaches the ocean or river bed. Theoretical considerations about these aspects are discussed in this section.

Impact Velocity of Sand

Krishnappan (1975) proposed the following relationship for the impact velocity of sand (V_i in m/s) as it reaches the bottom:

$$V_i = \beta F_b^{0.5}/d_{sl} \tag{1}$$

where β varies as a function of $\gamma_c \rho_{sl} d_{50}^3/\mu_w^2$ (β varies linearly from 1.2 when $\gamma_c \rho_{sl} d_{50}^3/\mu_w^2 = 1$, to $\beta = 2.1$ when $\gamma_c \rho_{sl} d_{50}^3/\mu_w^2 = 1000$), γ_c is the unit weight of the sand particles (kg/m^3), ρ_{sl} is the density of the slug (kg/m^3), d_{50} is the median grain size diameter of the sand particles (m), μ_w is the kinematic viscosity of water (m^2/s), F_b is the total negative buoyancy of the slug [given by $F_b = (\gamma_w - \gamma_c)V_{sl}$], γ_w is the unit weight of water (kg/m^3), V_{sl} is the volume of the slug (m^3), and d_{sl} is the distance the slug travels in water before it hits the bottom (m).

Thickness of the Slug

The thickness of the slug as it reaches the bottom (h_{sl} in m) can be evaluated based on the work of Krishnappan (1975) as follows:

$$h_{sl} = V_{sl}/[2\pi(1 - n_{sl})S_D^2] \tag{2}$$

where, V_{sl} is the volume of the slug (m^3), n_{sl} is the porosity of the slug, and S_D is the standard deviation of the solid particles within the slug given by

$$S_D = R_f/2 = R_0 + \alpha d_{sl} \qquad (3)$$

where, R_f is the final radius of the slug (m), R_0 is the original radius of the slug (m), α is an empirical constant determined by Krishnappan (1975), and d_{sl} is the distance the slug travels in water before it hits the bottom (m).

Horizontal Displacement of the Slug

The flow velocity and underwater currents will induce a horizontal displacement to the slug as it moves down through the water column. Graham and Zeeman (1992) suggest the following relationship for determining the horizontal displacement of the slug (x_{sl} in m).

$$x_{sl} = h_{sl} \exp[-(r - U_W t_{sl})^2/(2S_D^2)] \qquad (4)$$

where r is the radial coordinate (m), U_W is the ambient flow or current velocity of the water (m/s), and t_{sl} is the time of travel of the slug before it hits the bottom (sec).

Required Number of Lifts

The required number of vertical lifts for the cap (N_{vl}) can be estimated for a given desired thickness of the cap (H_c in m), as follows:

$$N_{vl} = H_c/h_{sl} \qquad (5)$$

The required number of horizontal lifts of the cap (N_{hl}) can then be estimated as follows:

$$N_{hl} = [A_c \cdot H_c]/[\pi (R_f)^2 (h_{sl} \cdot S_{sl})] \qquad (6)$$

where A_c is the aerial extent of the cap (m²), H_c is the required final thickness of the cap (m), R_f is the radius of the slug (m), h_{sl} is the thickness of the slug (m), and S_{sl} is the overlap factor for the slug. A value of overlap equal to the radius of the slug is typically recommended.

GEOTECHNICAL STABILITY ANALYSIS

Construction of the sand cap over soft native sediments may cause stability concerns due to the increased stresses in those layers. The primary stability concern during cap placement is the potential for the sediments to fail under the increased stresses, and to form "mudwaves," which could be trapped in subsequent layers of sand cap material. This concern is particularly valid since contaminated sediments typically have very high water contents and low shear strengths, resulting from the lack of sediment consolidation. Cap stability analysis involves consideration of three major factors: (i) settlement of the cap and foundation material, (ii) bearing capacity of the cap, and (iii) slope stability of the cap and foundation material.

Settlement Analysis

The settlement of the native sediments due to the load from the cap system can be evaluated using Terzaghi's small-strain consolidation theory (Terzaghi and Peck, 1967). The average consolidation settlement of the native sediment layer can be modeled assuming that the total settlement would be comprised of primary settlement due to expulsion of pore water from the sediments and secondary compression due to time-dependent creep of the sediments. Primary consolidation settlement (S_0 in m) of the native sediments can be evaluated using (Das, 1984)

$$S_0 = [C_c/(1 + e_0)][H_s \log_{10}\{(p_0 + \Delta p)/(p_0)\}] \tag{7a}$$

$$p_0 = \gamma_s'(H_s/2) \tag{7b}$$

$$\Delta p = \gamma_c' H_c \tag{7c}$$

$$\gamma_s' = \gamma_w(G_{ss} - 1)/(1 + e_0) \tag{7d}$$

where, C_c is the compression index of the native (in-place) sediments, e_0 is the initial void ratio of the sediment layer, H_s is the height (or thickness) of the native sediment layer (m), H_c is the height (or thickness) of the cap layer (m), p_0 is the initial vertical effective stress (kg/m^2), Δp is the additional vertical effective stress due to effective cap weight (kg/m^2), γ_s' is the effective unit weight of the native sediments (kg/m^3), γ_c' is the effective unit weight of the cap (kg/m^3), G_{ss} is the specific gravity of the native sediments, and γ_w is the unit weight of water (kg/m^3).

Secondary consolidation settlement (S_α in m) is due to time dependent deformation (creep) of the native sediments under a relatively constant load. It can be evaluated using

$$S_\alpha = H_s C_\alpha \log_{10}(t_d/t_0) \tag{8}$$

where, C_α is the coefficient of secondary compression, t_d is the end of design time frame (sec) and t_0 is the beginning of secondary compression (sec).

The total settlement (S_T in m) of the native sediments due to stresses induced by the proposed cap can then be computed as the sum of primary and secondary settlements as shown below:

$$S_T = S_0 + S_\alpha \tag{9}$$

The length of time elapsed following cap placement (t_e, in sec) and the degree of consolidation (U, in %) can be expressed as

$$t_e = H_{dr}^2 T_v/C_v \tag{10}$$

$$U(\%) = f(T_v) \tag{11}$$

where T_v is the dimensionless time factor given by standard geotechnical tables (Sivaram and Swamee; 1977) and plots, H_{dr} is the length of the drainage path for the pore water ($H_{dr} = H_s$ for single drainage; $H_{dr} = 0.5H_s$ for double drainage) and C_v is the coefficient of consolidation (in m^2/s).

Note that for evaluating the settlement of hydraulically placed dredged material caps, the finite strain consolidation theory (Cargill, 1983) should be used.

Bearing Capacity Analysis

For this analysis, the cap is considered as a footing acting over the native sediments with a surcharge (q in kg/m^2) given by (Terzaghi and Peck, 1967)

$$q = \gamma_c' H_c \tag{12}$$

where γ_c' is the effective unit weight of the cap (kg/m^3), and H_c is the thickness of the cap (m).

Assuming that the sediments will fail in an undrained local shear failure mode (due to the soft nature of the native sediments), the ultimate bearing capacity of the sediments (q_{ult} in kg/m^2) can be expressed as (Ling et al., 1996)

$$q_{ult} = 3.43\, c_u \tag{13}$$

where c_u is the undrained shear strength of the sediments (kg/m^2).

Assuming a design factor of safety of 3.0, the maximum allowable thickness of the cap [$H_{c(max)}$ in m] due to bearing capacity considerations can be expressed as (Ling et al., 1996)

$$H_{c(max)} = 1.14\, c_u/\gamma'_c \tag{14}$$

Slope Stability Analysis

It is well known that slopes may fail either slowly or suddenly, with or without the presence of an apparent provocative force [Terzaghi and Peck (1967); USBR (1974); Das (1984); CGS (1992); Abramson et al. (1996)]. Slope failures can occur along the edges (side slopes) of the cap layer because of one or more factors. These include removal of soil mass by erosion, natural slope movements, overloading of the soil, transient effects (earthquakes), increase in lateral pressure, changes in soil structure, and effect of pore pressure changes. In general, slope failures can be classified into five types: (i) translational failure, (ii) planar or wedge failure, (iii) circular failure, (iv) noncircular failure, and (v) a combination mode of failures, consisting of one or more of the above types.

The protection of a slope against potential failure is typically expressed in terms of the factor of safety (*FS*), which is expressed as follows:

$$FS = c_u/\tau_r \quad \text{(for short-term analysis)} \tag{15}$$

$$FS = (c' + \sigma' \tan \phi')/\tau_r \quad \text{(for long-term analysis)} \tag{16}$$

where c_u is the total available undrained shear strength (kg/m^2), τ_r is the required shear strength (kg/m^2), c' is the effective cohesive strength (kg/m^2), ϕ' is the effective friction angle, and σ' is the effective normal stress (kg/m^2).

The slope stability of an underwater cap may be evaluated using the method described by Leschinsky and Smith (1989). This method assumes a log-spiral type of failure in the sand cap, and a circular failure in the native sediments, as shown in Figure 14.2. Typically, for long-term analysis, the drained parameters are used (for short-term analysis on cohesive soils, the undrained analysis should be used). The cap and sediment properties may be defined as several individual layers [with a corresponding thickness of cap (H_c in m), effective unit weight of the cap material (γ'_c in kg/m^3), effective friction angle of the cap material (ϕ'), thickness of native sediments (H_s in m), and undrained shear strength of the native sediment layer ($c_{u,n}$ in kg/m^2)]. Due to the symmetry of the cap cross-section, it is required to conduct the analysis for only half of the cap cross section ($B/2$, where B is the total width of the cap in ft). Once the material parameters are known, the factor of safety can be computed for a variety of failure assumptions using standard theories described above. Typically, a factor of safety of at least 1.5 in the short term (i.e., immediately following construction) and 2.0 in the long term is desirable.

Some researchers argue that the method presented above is complicated due to the use of the log spiral. If there is such concern, it is also permissible to use a circular arc or straight vertical line in the sand layer. Also, it may be appropriate to give zero strength to the sand layer to increase the conservative nature of the analysis. Details of such methods are presented in Abramson et al. (1996).

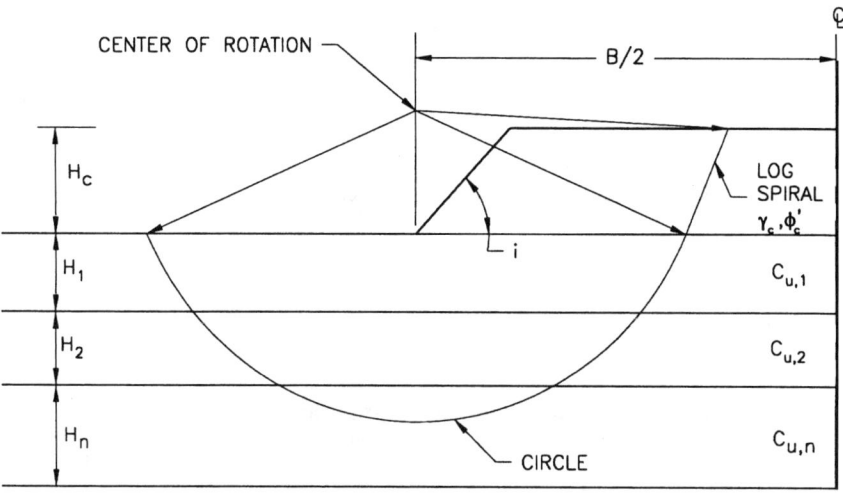

FIGURE 14.2. Cap stability analysis.

CHEMICAL STABILITY ANALYSIS

There are four major mechanisms that could potentially release contaminants through the cap system: (i) diffusion, in the absence of groundwater flow, (ii) advection and dispersion, if groundwater flow through the cap is present, (iii) release of pore water due to sediment compression, and (iv) erosion of contaminated sediments due to pore water release during sediment compression.

Contaminant Release by Diffusion

Molecular diffusion in the absence of groundwater flow can cause the sediment-bound contaminants to be released into the overlying water column. The time for 5% of the contaminants to break through the upper surface of the cap into the water column (defined as the breakthrough time, t_b, in sec) can be evaluated as follows (Wang et al., 1991; Thoma et al., 1993):

$$0.95 = -2 \sum_{N=1}^{N=6} (-1)^N \exp\{(-D_t N^2 \pi^2 t_b)/H_c^2\} \tag{17}$$

$$D_t = D_e/(n_c + \rho_{bc} K_{oc} f_{occ}) \tag{18}$$

where H_c is the cap thickness (m), D_t is the effective transient transport diffusion coefficient (m²/s), D_e is the effective diffusivity (m²/s), n_c is the cap porosity, K_{oc} is the chemical distribution coefficient, f_{occ} is the cap organic carbon content (expressed as a fraction), and ρ_{bc} is the bulk density of the cap (kg/m³).

Steady-state time (t_{ss} in sec) is defined as the time for the contaminant release to reach 95% of the maximum release rate (flux) and is expressed as

$$t_{ss} = (3.69\, H_c^2)/(D_t \pi^2) \tag{19}$$

The steady-state flux (F_{ss} in kg/m²/s) can be computed by

$$F_{ss} = D_e (P_0 - P_w)/H_c \qquad (20)$$

$$P_0 = C_s/K_p \qquad (21)$$

where P_w is the chemical concentration in the water column (mg/L), P_0 is the average chemical concentration in the pore water (mg/L), K_p is the organic carbon partition coefficient ($K_p = K_{oc} \cdot f_{oc}$), and C_s is the depth averaged sediment chemical concentration (mg/kg).

The long-term flux (F_l in kg/m²/s) can then be estimated from

$$F_l = P_0 \{(D_e R)/(\pi t)\}^{0.5} \qquad (22)$$

where R is the retardation factor ($R = n_c + \rho_{bc} K_p$) and t is the time period (sec).

Contaminant Release by Advection/Dispersion

The contaminant release at the sand–armor interface due to the processes of advection and dispersion (expressed as C, in mg/kg) can be evaluated based on the works by Ogata and Bank (1961), Bedient et al. (1985), Tchobangolous and Schroeder (1987), and Fetter (1993):

$$C = 0.5 \, C_0 \, \mathrm{erfc}[(RH_c - V_g t)/\{2 \, (RD_H t)^{0.5}\}] \qquad (23)$$

where C_0 is the sediment–sand (lower) interface chemical concentration (mg/kg), H_c is the cap thickness (m), V_g is the groundwater velocity (m/s), t is the time (sec), D_H is the hydrodynamic dispersion coefficient, and erfc is the error function given by

$$\mathrm{erfc}(x) = 1 - [\{2/(\pi)^{0.5}\}\{x - (x^3/3) + (x^{2n-1}/\{(2n-1)[(n-1)!]\}) - \ldots \}] \qquad (24)$$

Contaminant Release by Pore Water Release during Sediment Compression

Chemical migration into the sand cap layer due to pore water release during sediment compression can be estimated based on the consolidation properties of the native sediment layer. Assuming that pore water expulsion during sediment compression will result in a short-term flux of dissolved contaminants into the cap system, the flux per area (F_c in mg) can be estimated as

$$F_c = 10 \cdot S_T \cdot C_t / (K_{oc} \cdot f_{ocs}) \qquad (25)$$

where S_T is the total settlement of the native sediments (cm), C_t is the average concentration of the contaminant in the native sediment zone (mg/L), K_{oc} is the chemical distribution coefficient, and f_{ocs} is the total organic carbon content of the native sediment layer.

Now, assuming that the contaminants in the pore water are sorbed into the bottom 2.54 cm (1 inch) of the sand cap layer (based on experimental results of Gunnison et al., 1987), the concentration of contaminant in the bottom 2.54 cm (1 inch) layer of the sand cap (C_{bs} in mg/kg) can be estimated as

$$C_{bs} = F_c / \{25.4 \cdot [G_{sc} (1 - n_c)]\} \qquad (26)$$

where n_c is the porosity of the cap layer and G_{sc} is the specific gravity of the cap layer.

The actual thickness of the sand sorptive layer (H_{SSL} in mm) can then be estimated as

$$H_{SSL} = 25.4 \cdot (C_{bs}/C_t) \qquad (27)$$

PHYSICAL STABILITY ANALYSIS

An armor layer is typically provided to protect the sand isolation layer from being eroded away due to hydraulic and environmental forces. Typically, bottom velocities exceeding about 0.3 m/s (1 fps) can initiate sediment erosion (Vanoni, 1977). In general, protection against four potential erosive forces should be considered. These include: (i) river flow, (ii) wave action, (iii) propeller wash forces, and (iv) ice scour.

Protection from River Flow

For design purposes, the maximum expected river flow velocity for the design event (usually the worst-case flood event, 100-year, or even 500-year flood) is first computed. Hydraulic numerical models may be used to compute the worst-case flood velocities (US-ACE, 1985). Once this is obtained, the Isbach equation (USACE, 1991) may be used to determine the stable stone size based on expected bottom velocities as shown below:

$$V_b = C_1[2g\{(\gamma_a/\gamma_w) - 1\}]^{0.5}(d_{50(A)})^{0.5} \tag{28}$$

where $d_{50(A)}$ is the median diameter of the armor stone (m), γ_w is the unit weight of water (kg/m^3), γ_a is the unit weight of the armor stone (kg/m^3), g is the acceleration due to gravity (m/s^2), C_1 is a constant (0.86 for high turbulence and 1.2 for low turbulence), and V_b is the bottom velocity (m/s).

Figure 14.3 illustrates a design chart for determining armor stone sizes based on maximum expected velocities (USACE, 1991).

Protection from Wave Action

The erosive velocities resulting from the design wave event (usually the worst-case storm event, 100-year, or even 500-year storm) is computed. A design wave analysis should

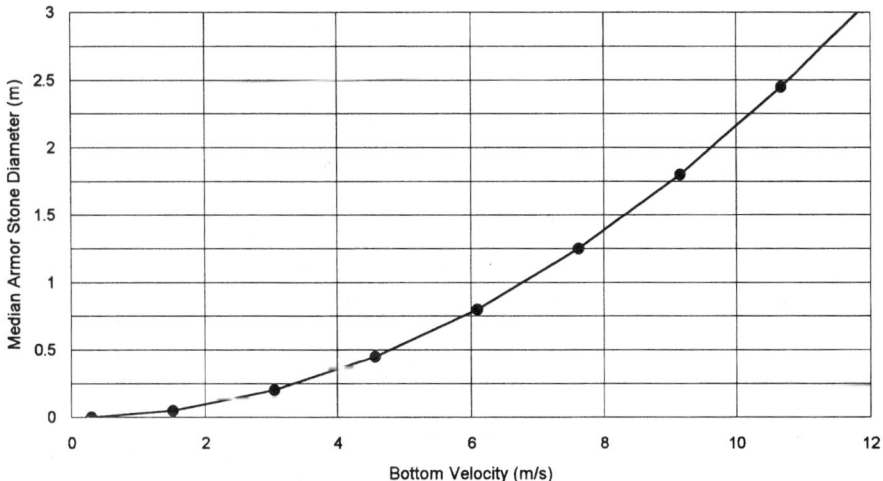

FIGURE 14.3. Relationship between median armor stone size and bottom velocity.

first be performed to define the maximum wave characteristics for the desired recurrence interval. Based on the generated wave characteristics (wave height, wave length, water depth, and wave period), the best-suited wave theory should be selected. For transitional and deep waters, the horizontal orbital velocity should be estimated from standard wave theories [USACE (1984); Pilarczyk (1990); Herbich (1992a); ASCE (1999)]. For shallow-water cases, the velocity of the wave breaking on the shoreline should also be considered. The worst storm-induced wave velocities may also be computed using coastal numerical models (USACE, 1992). Once this is obtained, the Isbach equation (USACE, 1991) may be used to determine the stable stone size based on expected maximum velocities (Equation 28, or Figure 14.3).

Protection from Propeller Wash Forces

In some cases, the effects of bottom stresses induced by navigational craft can be much greater than those associated with river flows and waves. The U.S. Army Corps of Engineers (Palermo, 1991) have demonstrated that the erosional effects of propeller wash forces from maneuvering vessels can be much greater than from vessels that are under constant motion.

The following equation may be used to determine the maximum bottom velocity due to propeller wash forces from a navigating vessel (Blaauw and van de Kaa, 1978):

$$V_{bp} = C_2 U_0 D_P / H_P \tag{29}$$

where V_{bp} is the maximum bottom velocity (m/s), C_2 is a constant (0.22 for nonducted propellers), U_0 is the jet velocity exiting the propeller (m/s), D_P is the propeller diameter (m), and H_P is the distance from propeller shaft to channel bottom (m).

The jet velocity exiting the propeller is given as follows (Blaauw and van de Kaa, 1978):

$$U_0 = C_3 [P_d / (D_P)^2]^{0.333} \tag{30}$$

where U_0 is the jet velocity exiting the propeller (m/s), P_d is the engine power (HP), D_P is the propeller diameter (m), and C_3 is a constant (4.4 for nonducted propellers).

Knowing the maximum design velocities from these equations, the armor stone size can be determined from the Isbach equation (Equation 28 or Figure 14.3).

Protection from Ice Loading

Ice scour becomes a potential concern during armor design if the possibility exists for ice being piled or ridged to the total depth (or a significant part) of the water column. The *Shore Protection Manual* (USACE, 1984) suggests that, in general, ice piled on shore by wind and wave action would not cause serious damage to riprap, but would provide additional protection against severe winter waves. Field observations and/or conversations with local residents may be a good method to determine whether significant ice flows or ice damming have been observed at the site. Where there is a significant concern for ice scour, the U.S. Army Corps of Engineers (USACE, 1984) recommends that an impact of 28.15 kg/cm^2 for ice crushing and 10.56 kg/cm^2 for blows from floating ice be accounted for during selection of the armor materials (ASTM, 1993).

Erosion of Contaminated Sediments Due to Pore Water Release during Sediment Compression

The potential for erosion of contaminated sediment particles due to pore water release during sediment compression can be evaluated based on the results of the geotechnical settlement analysis discussed above. Knowing the total settlement and the time to achieve that from the above equations, the equivalent velocity of release of the pore water (V_{pw} in m/s) can be computed as

$$V_{pw} = S_T/t_e \tag{31}$$

This velocity can then be used to compute the particle size of the native sediments that would potentially be eroded by the pore water flux into the sand cap and the associated erosion loss volumes (Vanoni, 1997).

HYDRAULIC FILTERING ANALYSIS

Hydraulic forces such as eddies (which are transmitted to the bottom through the voids in a protective layer), seepage forces (developed by percolating water), or a combination of both may initiate movement of bed material beneath an armor layer. A filter layer provides: (i) protection of the base material from direct attack by such hydraulic forces, and (ii) ability to distribute the loads due to the armor layer evenly to the foundation soils. In this section, the need for a filter layer and its design criteria are evaluated.

Filter Design Criteria

Several investigators have studied the filtering process of granular soils [De Abreu-Lima and Morgan (1951); Jetter (1951); Manamperi (1952); Sherman (1953); Karpoff (1955); Zweck and Davidenkoff (1957); Sherard et al. (1984a,b); Kenney and Lau (1985); Kenney et al. (1985); La Fleur et al. (1989); Aberg (1993)]. The Terzaghi–Vicksburg criteria [Terzaghi (1949); USACE (1949)] is often used as a design guideline to investigate the need for a filter layer and to design a filter layer. It essentially consists of three criteria that a filter material must satisfy in order to provide hydraulic stability to the base material.

$$d_{15(F)} < 5 \, d_{85(B)} \tag{32}$$

$$20 \, d_{15(B)} > d_{15(F)} > 5 \, d_{15(B)} \tag{33}$$

$$d_{50(F)} < 25 \, d_{50(B)} \tag{34}$$

where $d_{15(F)}$ is the diameter through which 15% of the filter passes (cm), $d_{50(F)}$ is the diameter through which 50% of the filter passes (cm), $d_{15(B)}$ is the diameter through which 15% of the base material passes (cm), $d_{50(B)}$ is the diameter through which 50% of the base material passes (cm), and $d_{85(B)}$ is the diameter through which 85% of the base material passes (cm). Note that these diameters are based on percent passing by weight.

The first criterion (Equation 32) prevents the largest base material grains from being washed into the pores of the filter material. Washout of the smaller grains will then be prevented by means of internal formation of filters in the base material (self-filtration). The second criterion (Equation 33) improves the drainage characteristics of the filter.

Worman (1989) defines stability of the protective layer by the following relationship:

$$(V_f^2/gS_{rl}) = 6\ (d_{85(F)}/d_{15(A)}) \qquad (35)$$

where V_f is the mean flow velocity above the armor layer (m/s), g is the acceleration due to gravity (m/s^2), S_{rl} is the riprap layer thickness (m), $d_{85(F)}$ is the 85% passing size of the filter material (cm), and $d_{15(A)}$ is the 15% passing size of the armor layer (cm).

Some researchers [Kenney and Lau (1985); Maynord (1985); Worman (1989); CGS (1992); Aberg (1993)] have suggested that if the pore spaces of the armor layer were packed with smaller-sized stones, the density and hydraulic stability of the layer would increase due to the reduction in void space. It was found that, in general, a 20 to 80% by weight mixture of filter and armor stones packed uniformly provides a level of protection equivalent to that provided by a layer of armor and a layer of filter placed underneath (Wright and Kim, 1995). However, it should be noted that the most reliable method for determining cap stability is to perform model tests simulating the actual external forces at the site.

Recommended Thickness and Gradation

The U.S. Army Corps of Engineers [USACE (1984, 1992)] recommends the following relationship for thickness of a filter layer:

$$T_F = T_A/4 \qquad (36)$$

where T_F is the thickness of the filter layer (m) and T_A is the thickness of the armor layer (m). Gradation curves for the filter layer are presented in Figures 14.4–14.8.

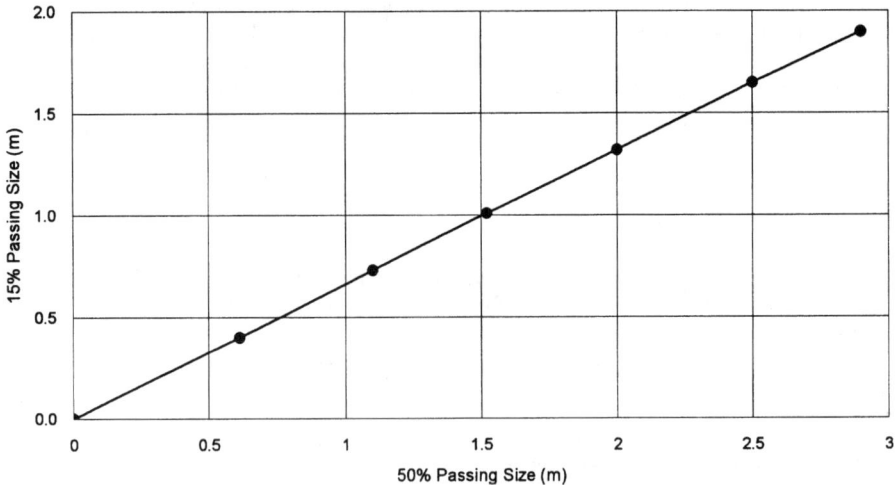

FIGURE 14.4. Relationship between 15% passing size and 50% passing size for filters.

MODELING THE PHYSICAL AND CHEMICAL STABILITY OF UNDERWATER CAPS **14.17**

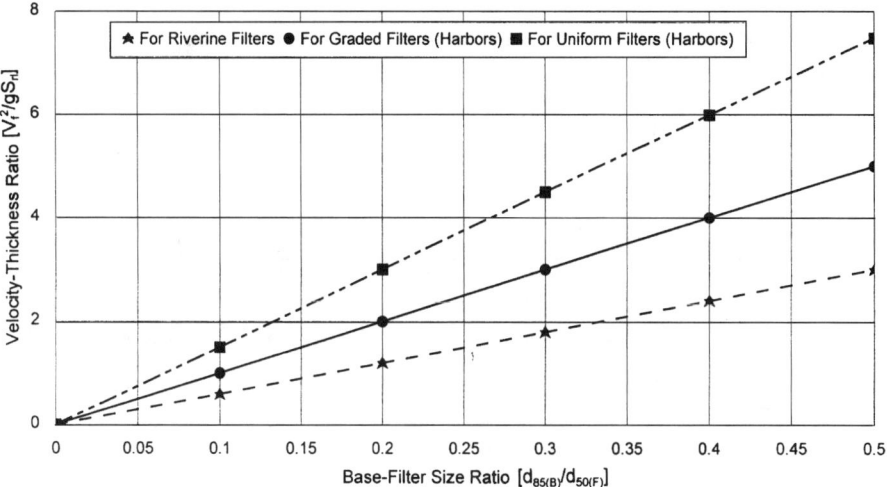

FIGURE 14.5. Relationship between velocity–thickness ratio and base–filter size ratio.

A GENERAL MODELING APPROACH— THE CAPSTABL MODEL

Design of underwater caps can be quite a challenging task, given all the variables involved in the design. A computer model for cap design would offer the designer an opportunity to quickly analyze the effect of cap thickness, material type, carbon content, and

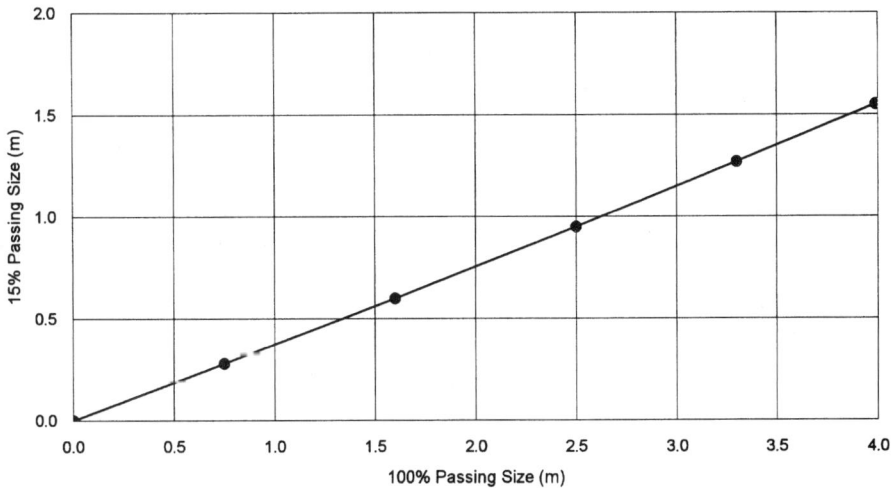

FIGURE 14.6. Relationship between 15% passing size and 100% passing size for filters.

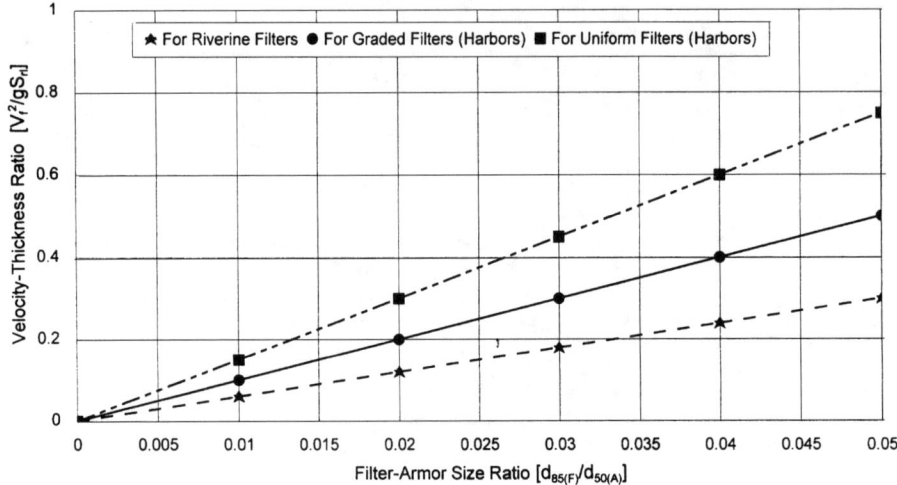

FIGURE 14.7. Relationship between velocity–thickness ratio and filter–armor size ratio.

other variables on cap effectiveness for the particular site. The author has used the theoretical formulations in this chapter to develop a spreadsheet-based computer model—CAPSTABL—that can be executed on a personal computer (PC).

The model consists of the following five major modules (Figure 14.9):

1. **PLACE,** which simulates the behavior of the cap during placement
2. **GEOTEC,** which simulates the geotechnical behavior of the cap and native sediments following placement (i.e., settlement, bearing capacity, and slope stability)

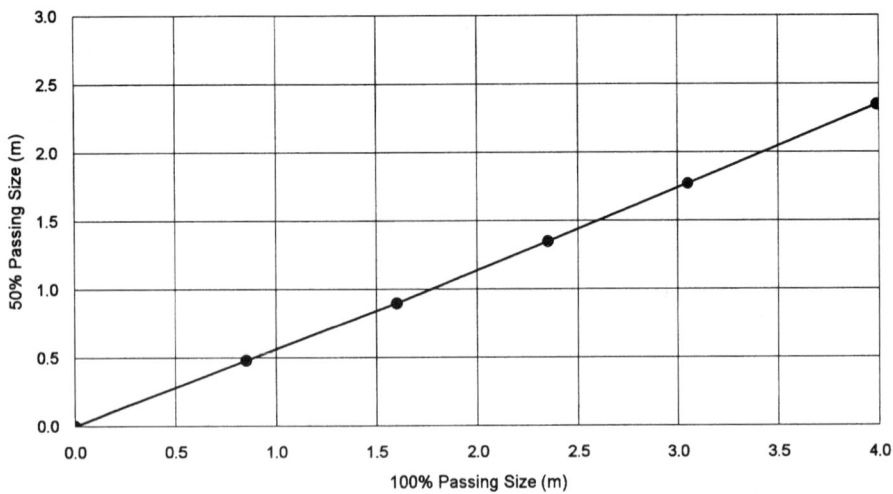

FIGURE 14.8. Relationship between 50% passing size and 100% passing size for filters.

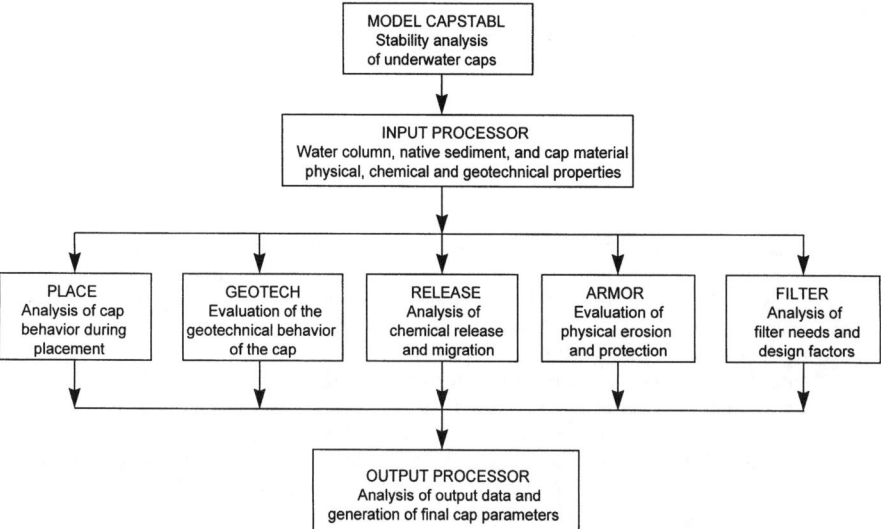

FIGURE 14.9. Schematic representation of computer model CAPSTABL.

3. **RELEASE,** which simulates the chemical migration patterns and contaminant release mechanisms in the cap (by accounting for the effects of diffusion, advection/dispersion, and sediment compression)
4. **ARMOR,** which estimates the required gradation of the armor to yield sufficient erosive protection to the sand cap against forces from river flow, waves, propeller wash, ice loads, and pore water release mechanisms
5. **FILTER,** which evaluates the need for a filter layer based on the sand cap and armor material gradations.

Note that these modules can be run either jointly or individually for evaluating various parameters of the cap.

EXAMPLE MODEL APPLICATION

An example application of the CAPSTABL model is presented here. The case involves execution of the RELEASE module of the model to evaluate the effectiveness of a sand cap at a confidential contaminated site. Water conditions at the site were primarily saline, with water depths varying from 1 to 3 m. The 100-year storm event at the site was expected to generate a maximum river velocity of 1.2 m/s, with a corresponding maximum flow rate of 225 m³/s. Mean high water and mean low water at the site were estimated to be 1.5 m-NGVD and −0.5 m-NGVD, respectively. The primary contaminant was PCB, with an average concentration of 100 mg/kg. Native sediments at the site were comprised of a top layer of silty sediments (varying thickness; typically 0.5 to 1.5 m) underlain by a clay layer (varying thickness; typically 3 to 5 m). This layer was underlain by a bedrock layer.

In order to remediate the contaminated sediments, a 0.25-acre area of the river was proposed to be remediated by in-situ capping, providing isolation of over 95% of the total

TABLE 14.1. Effect of cap and contaminant properties on diffusive releases

Property	Diffusive time	Diffusive flux
Cap thickness (H_c)	$1/H_c^2$	$1/H_c$
Cap organic carbon content (f_{occ})	f_{occ}	no effect
Distribution coefficient (K_{oc})	K_{oc}	$1/K_{oc}$

PCB mass present at the site. A design sand cap layer thickness of 23 cm (~9 inch) with a median grain size of 2 mm and organic carbon content of 0.5% was assumed for preliminary analysis. The cap layer would be overlain by a 15 cm (~6 inch) thick graded armor layer comprised of gravel and cobbles (median grain size of 5 cm). The armor layer will provide the required isolation from bioturbation.

A sensitivity analysis was first performed on the RELEASE submodel, the results of which are summarized in Tables 14.1 and 14.2. Basically, the effectiveness of the cap increases with increasing total organic carbon content of the cap (f_{occ}), thickness of the cap (H_c), and chemical distribution coefficient (K_{oc}). Similarly, cap effectiveness also increases with decreasing cap porosity (n_c) and chemical diffusivity in water (D_w). Breakthrough and steady-state times are proportional to H_c^2 for diffusion and H_c for advection. Steady-state flux is inversely proportional to H_c for diffusion and independent of H_c for advection. Finally, breakthrough and steady-state times for both diffusion and advection are proportional to total organic carbon content of the cap material.

The proposed thickness of the sand cap was then evaluated separately for protection from chemical migration using the RELEASE module (see Tables 14.3 and 14.4). It was estimated that placement of a 23 cm (~9 inch) sand cap over the sediments would considerably enhance the isolation of the contaminants from the water column (on the order of several hundred years) when compared to the existing conditions for both diffusion and advection assumptions. Several cap design parameters were modeled including varying thickness and organic carbon contents. Diffusive breakthrough times of PCB through the cap were estimated to be approximately 600 and 2,300 years for total organic carbon contents of 0.5 and 2%, respectively, for the 23 cm cap. The calculated steady-state flux of PCB was only 9 gm/year/acre. This is in comparison to 623 gm/year/acre from the existing sediment surface (neglecting the releases associated with scour). For consideration of natural sedimentation within the armor layer over the sand cap, the breakthrough times increased to over 6,400 years.

In order to estimate the potential contribution of groundwater movement on PCB release, the hydraulic conductivity and regional groundwater gradient were estimated from available data. The resulting advective breakthrough times for the 0.5 and 2% total organic carbon sand caps (23 cm thick) were approximately 112 and 440 years, respectively. As in the case of diffusive transport, assumption of natural sedimentation over the sand cap increases this breakthrough time even further by severalfold.

TABLE 14.2. Effect of cap and contaminant properties on advective and dispersive releases

Property	Advective/dispersive time	Advective/dispersive flux
Cap thickness (H_c)	$1/H_c^2$	no effect
Cap organic carbon content (f_{occ})	f_{occ}	no effect
Distribution coefficient (K_{oc})	K_{oc}	$1/K_{oc}$

TABLE 14.3. Summary input data from CAPSTABL RELEASE module for the example application

Input data	Units	Value
Chemical of concern	n/a	PCB
Chemical partitioning coefficient	n/a	15,000
Native sediment chemical concentration	mg/kg	100
Water column chemical concentration	mg/L	0
Native sediment density	gm/cm^3	1.5
Native sediment porosity	n/a	0.5
Native sediment organic carbon content	%	5
Estimated amount of native sediment compression	cm	15
Ground water velocity	cm/day	0.01
Pore water velocity	cm/day	0.1
Surface area of cap	acres	0.25
Capping material	n/a	sand
Thickness of cap	cm	23
Cap organic carbon content	%	0.5
Cap porosity	n/a	0.4
Effective diffusivity	cm^2/s	1.18E-006
Hydrodynamic dispersion coefficient	cm^2/s	1.00E-006

TABLE 14.4. Summary output data from CAPSTABL RELEASE module for the example application

Output data	Units	Value
Diffusion Parameters		
Maximum flux (without cap)	gm/yr/acre	623
Maximum flux (with cap)	gm/yr/acre	9
Percent reduction in flux provided by cap	%	98.5
Breakthrough time (5% of maximum flux)	years	90
Steady-state time (95% of maximum flux)	year	600
Advection/Dispersion Parameters		
Maximum flux (without cap)	gm/yr/acre	197
Maximum flux (with cap)	gm/yr/acre	78
Percent reduction in flux provided by cap	%	60
Breakthrough time (5% of maximum flux)	years	47
Steady-state time (95% of maximum flux)	year	112
Compression Parameters		
Volume of water released from native sediments	liters	154,000
Volume of water released from cap	liters	92,000
Mass of chemicals released from native sediments	gm	20
Mass of chemicals released from cap	gm	12

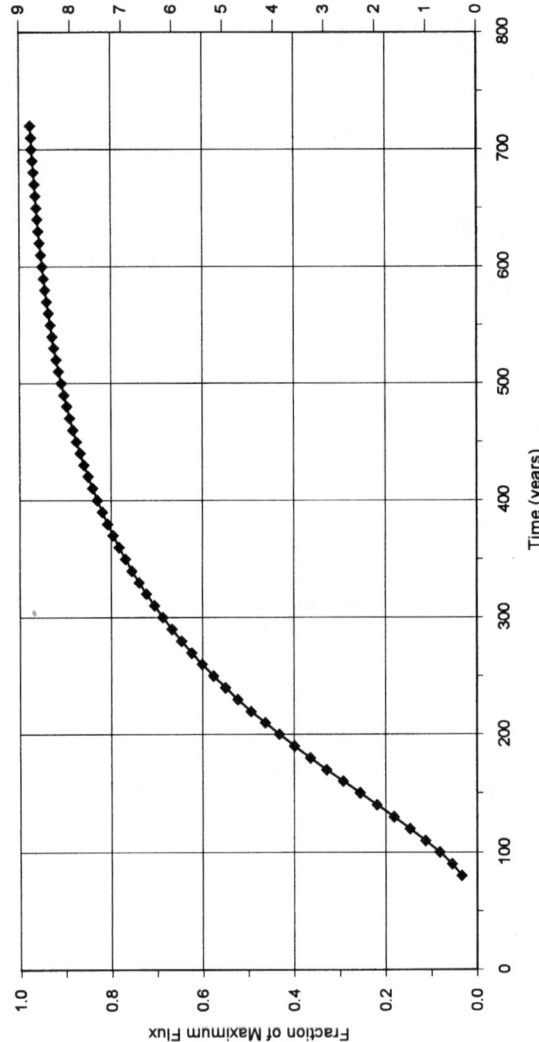

FIGURE 14.10. Diffusive breakthrough curve for PCB.

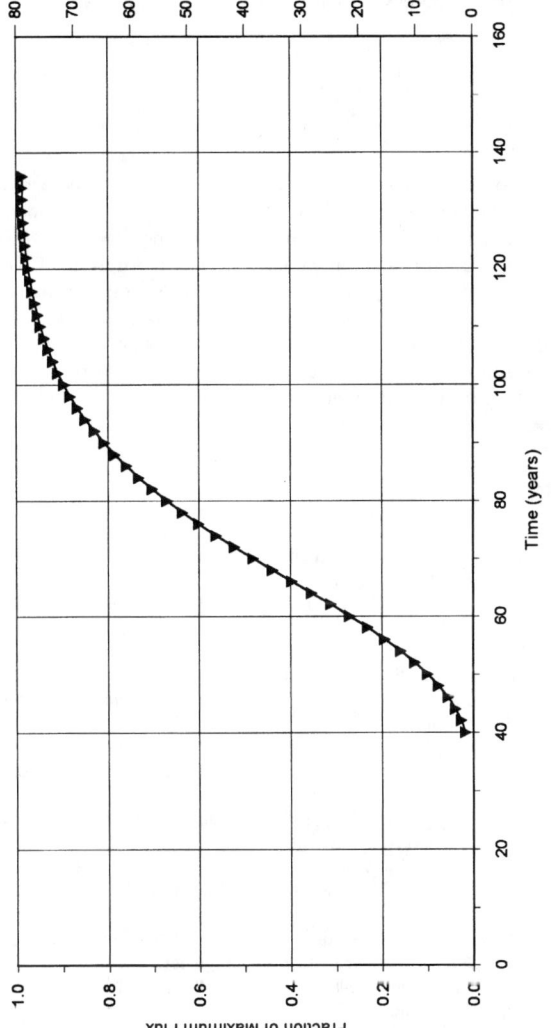

FIGURE 14.11. Advective/dispersive breakthough curve for PCB.

The contaminant breakthrough curves are shown in Figures 14.10 and 14.11, which indicate that a 23 cm sand cap would provide sufficient isolation of the PCB contaminated sediments on the order of 112 years (advection/dispersion) to 600 years (diffusion). Even after that time, the amount of contaminants leaving the cap would be reduced by as much as 60 to 98% (for advection and dispersion, respectively) when compared to the existing condition. Thus, a cap consisting of a 23 cm sand layer and a 15 cm graded armor/filter layer would yield sufficient chemical, environmental, and hydraulic protection to the in-situ contaminants, thereby providing isolation of the contaminants from the water column.

SUMMARY

Theoretical formulation for modeling the placement behavior, geotechnical stability, chemical stability, physical stability, and hydraulic filter stability of underwater caps in rivers and harbors was presented in this chapter. Development of a general spreadsheet-based computer model using the governing equations was also described. The application of the model was illustrated using a project case study. The computer model described in this chapter and the theoretical formulations can be used to predict the performance of capping projects in the field, during the project planning/feasibility study stage. Finally, since this chapter presents the entire cap design equations and theory in one document, it is hoped that this will be of use to designers and planners.

REFERENCES

Aberg, B. (1993). "Washout of Grains from Filtered Sand and Gravel Materials," *ASCE Journal of Geotechnical Engineering,* Vol. 119, No. 1.

Abramson, L. W., Lee, T. J., Sharma, S., and Boyce, G. M. (1996). *Slope Stability and Stabilization Methods,* Wiley, New York, NY.

ASCE (1999). *ASCE Standards for Shore Protection Systems,* Committee on Standards, American Society of Civil Engineers, Reston, VA (in press).

ASTM. (1993). *Annual Book of Standards: Soil & Rock; Dimension Stone; Geosynthetics,* Volume 04. 08, American Society for Testing and Materials, New York, NY.

Averett, D. E., and Francingues, N. R. (1994). "Sediment Remediation: An International Review," Proceedings Dredging-94 Conference, American Society of Civil Engineers, Lake Buena Vista, FL.

Bedient, P. B., Borden, R. C., and Leib, D. I. (1985). "Basic Concepts for Groundwater Transport Modeling," *Groundwater Quality,* Wiley, New York, NY.

Blaauw, H. G., and Van de Kaa, E. J. (1978). "Erosion of Bottom and Banks Caused by the Screw Race of Maneuvering Ships," DHL Publication No. 202, Delft Hydraulics Laboratory, Delft, The Netherlands.

Cargill, K. W. (1983). "Procedures for the Prediction of Consolidation in Soft Fine-Grained Dredged Material," Report No. D-83-1, U.S. Army Corps of Engineers, Washington, DC.

CGS. (1992). *Canadian Foundation Engineering Manual.* 3rd Edition, Canadian Geotechnical Society, Richmond, British Columbia.

Das, B. M. (1984). *Principles of Geotechnical Engineering,* PWS-Kent Publishing Company, Boston, MA.

De Abreu-Lima, C. and Morgan, S. (1951). "Protection of Earth Dams by Riprap of Uniform Size," Thesis submitted to the State University of Iowa, Des Moines, Iowa.

Environment Canada. (1996). "Hamilton Harbor Pilot Capping Project—Fact Sheet," Environment Canada, Toronto, Canada.

Fetter, C. W. (1993). "Mass Transport in Saturated Media," In *Contaminant Hydrogeology,* Macmillan, New York, NY.
Fredette, T. J., Germano, J. D., Carey, D. A., Murray, P. M., and Kullberg, P. G. (1992). "Chemical Stability of Capped Dredged Material Disposal Mounds in Long Island Sound," *Chemistry and Ecology,* Vol. 7, No. 2.
GeoEngineers, Inc. (1995). "Geotechnical Study for Manistique River and Harbor Sediment Cap," Technical Report to Blasland, Bouck & Lee, Inc. by GeoEngineers, Inc., Richmond, WA.
Graham, J. E., and Zeeman, A. J. (1992). "Distribution of Sand on Harbor Bottom during Cap Placement," Technical Report No. 92-52, National Water Research Institute, Burlington, Ontario, Canada.
Gunnison, D., Brannon, J. M., Sturgis, T. C., and Smith, I. (1987). "Development of a Simplified Column Test for Evaluation of Thickness of Capping Material Required to Isolate Contaminated Dredged Material," Paper No. D-87-2, Corps of Engineers, Vicksburg, MS.
Herbich, J. B. (1992a). *Handbook of Ocean and Coastal Engineering,* Gulf Publishing, Houston, Texas.
Herbich, J. B. (1992b). *Handbook of Dredging Engineering,* McGraw-Hill, New York, NY.
Hull, J., Jersaka, J., Pochop, P., and Cummings, J. (1998). "Evaluating a New In-Situ Capping Technology for Mitigating Contaminated Sediments," Proceedings Wodcon-XV Conference, World Dredging Association, Las Vegas, NV.
Jetter, K. (1951). "Tests on Sand Dikes Protected against Erosion by Overflowing Water," Thesis submitted to the State University of Iowa, Des Moines, Iowa.
Karpoff, K. P. (1955). "The Use of Laboratory Tests to Develop Design Criteria for Protective Filters," *Proceedings of the American Society for Testing and Materials,* Vol. 55, No. 4.
Kenney, T. C., and Lau, D. (1985). "Internal Stability of Granular Filters." *Canadian Geotechnical Journal,* Vol. 22, No. 3.
Kenney, T. C., Chahal, R., Chiu, E., Ofoegbu, G. I., Omange, G. N., and Ume, C. A. (1985). "Controlling Constriction Sizes of Granular Filters," *Canadian Geotechnical Journal,* Vol. 22, No. 3.
Kikegawa, K. (1983). "A Sand Overlaying for Bottom Sediment Improvement by Sand Spreade," Proceedings of the 7th U.S./Japan Experts Meeting, Ft. Belvoir, VA.
Krishnappan, V. (1975). "Dispersion of Granular Material Dumped in Deep Water," Technical Report No. 55, National Water Research Institute, Burlington, Ontario, Canada.
Laboyrie, H., and Flach, B. (1998). "The Handling of Contaminated Dredged Material in the Netherlands," Proceedings Wodcon-XV Conference, World Dredging Association, Las Vegas, NV.
LaFleur, J., Mlynarek, J., and Rollin, A. L. (1989). "Filtration of Broadly Graded Cohesionless Soils," *Journal of Geotechnical Engineering,* Vol. 115, No. 12.
Leschinsky, D., and Smith, D. S. (1989). "Deep Seated Failure of a Granular Embankment Over Clay: Stability Analysis." *Soils and Foundations,* Vol. 29, No. 3.
Li, W., Chian, C., and Walker, J. (1998). "Contaminated Dredged Material Disposal: Approaches and Experience," Proceedings Ports-98 Conference, American Society of Civil Engineers, Long Beach, CA.
Lillycrop, L., and Clausner, J. (1998). "Numerical Design of the 1997 Capping Project at the Mud Dump Site," Proceedings Wodcon-XV Conference, World Dredging Association, Las Vegas, NV.
Ling, H. I., Leshinsky, D., Gilbert, P. A., and Palermo, M. R. (1996). "Geotechnical Considerations Related to In-Situ Capping of Contaminated Submarine Sediments," Proceedings of the 29th TAMU & WEDA XVI Technical Conference, Western Dredging Association, New Orleans, LA.
Manamperi, H. D. S. (1952). "Tests of Graded Riprap for Protection of Erosible Material," Thesis submitted to the State University of Iowa, Des Moines, Iowa.
Maynord, S. T. (1995). "Manistique Intermediate Rock Layer," Internal Memorandum, U.S. Army Corps of Engineers, Waterways Experiment Station, Vicksburg, MS.
Maynord, S. T., and Oswalt, R. (1993). "Design Considerations for Capping and Armoring Contaminated Sediments In-Place," U.S. Army Corps of Engineers, Vicksburg, MS.
Mohan, R. K. (1997). "Design and Construction of Subaqueous Caps for Restoring Contaminated Coastal Areas," Proceedings, Coastal Zone '97 Conference, Boston, MA.
Mohan, R. K., D'Hollander, R. D., Johnson, A. N., Brozowski, P., D'Ambrosio, K. T., and Jerome, J.

(1999a). "Remediation of Contaminated Sediments by In-Place Capping—A Case Study of Rahway River, NJ," *Journal of Marine Environmental Engineering,* Vol. 5, No. 2.
Mohan, R. K., Mageau, D. M., and Brown, M. P. (1999b) "Geo-Environmental Modeling of Native Sediment Disturbance during Aquatic Sediment Cap Construction," *Marine Technology Society Journal,* Vol. 33, No. 3.
Nelson, E. E., Vanderheiden, A. L., and Schuldt, A. D. (1994). "Eagle Harbor Superfund Project," Proceedings Dredging-94 Conference, American Society of Civil Engineers, Lake Buena Vista, FL.
Ogata, A., and Bank, R. B. (1961). "A Solution to the Differential Equation of Longitudinal Dispersion in Porous Media," Paper No. 411-A, U.S. Geological Survey, Washington, DC.
Palermo, M. R. (1991). "Design Requirements for Capping," Dredging Research Technical Notes, U.S. Army Corps of Engineers, Waterways Experiment Station, Vicksburg, MS.
Palermo, M. R. (2000). "Subaqueous Capping of Contaminated Sediment," in *Handbook of Coastal Engineering* (J. B. Herbich, Ed.), McGraw-Hill, New York, NY.
Pilarczyk, K. W. (1990). *Coastal Protection,* A. A. Balkema, Rotterdam, The Netherlands.
Randall, R. E., Clausner, J. E., and Johnson, B. H. (1994). "Modeling Cap Placement at New York Mud Dump Site," Proceedings Dredging-94 Conference, American Society of Civil Engineers, Lake Buena Vista, FL.
Shaw, J., Whiteside, P., and Ng, K. (1998). "Contaminated Mud in Hong Kong: A Case Study of Contained Seabed Disposal," Proceedings Wodcon-XV Conference, World Dredging Association, Las Vegas, NV.
Sherard, J. L., Dunnigan, L. P., and Talbot, J. R. (1984a). "Filters for Silts and Clays," *ASCE Journal of Geotechnical Engineering,* Vol. 110, No. 6.
Sherard, J. L., Dunnigan, L. P., and Talbot, J. R. (1984b). "Basic Properties of Sand and Gravel Filters," Journal of Geotechnical Engineering. Vol. 110, No. 6.
Sherman, W. C. (1953). "Filter Experiments and Design Criteria," U.S. Army Waterways Experiment Station. Vicksburg, MS.
Silva, A. J., Brandes, H. G., Tian, W. M., and Uchtyil, C. J. (1991). "Damos Program Cap Site Study," Technical Report, University of Rhode Island, Narragansett, RI.
Sivaram, B., and Swamee, A. (1977). "A Computational Method for Consolidation Coefficient," *Soils and Foundations,* Vol. 17, No. 2.
Strugis, T. C., and Gunnison, D. (1985). "New Bedford Harbor Superfund Report No. 6," WES Technical Report No. EL-88-15, U.S. Army Corps of Engineers, Vicksburg, MS.
Sumeri, A., Fredette, T. J., Kullberg, P. G., Germano, J. D., Carey, D. A., and Pechko, P. (1991). "Sediment Chemistry Profiles of Capped In-Situ and Dredged Sediment Deposits: Results from Three U.S. Army Corps of Engineers Offices," Proceedings, 24th Annual Dredging Seminar, Texas A&M University, College Station, TX.
Tchobangolous, G., and Schroeder, E. D. (1987). *Water Quality: Characteristics, Modeling, Modification,* McGraw-Hill, New York, NY.
Terzaghi, K. (1949). *Theoretical Soil Mechanics,* Wiley, New York.
Terzaghi, K., and Peck, R. B. (1967). *Soil Mechanics in Engineering Practice,* Wiley, New York, NY.
Thibodeaux, L. J., Valsaraj, K. T., and Reible, D. D. (1994). "Capping Contaminated Sediments—The Theoretical Basis and Laboratory Evidence for Chemical Containment," Proceedings Dredging-94 Conference, American Society of Civil Engineers, Lake Buena Vista, FL.
Thoma, G. J., Reible, D. D., Valsaraj, K. T., and Thibodeaux, L. J. (1993). "Efficiency of Capping Contaminated Sediments In-Situ: Mathematics of Diffusion–Adsorption in the Capping Layer," *Environmental Science and Technology,* Vol. 27, No. 11.
USACE (1949). "Preliminary Report on Slope Protection for Earth Dams," U.S. Army Corps of Engineers, Waterways Experiment Station, Vicksburg, MS.
USACE (1984). *Shore Protection Manual,* Volumes 1 & 2, U.S. Army Corps of Engineers, Vicksburg, MS.
USACE (1985). *Users Manual for HEC-2 Computer Program,* U.S. Army Corps of Engineers, San Francisco, CA.
USACE (1991). "Hydraulic Design of Flood Control Channels," EM-1110-2-1601, U.S. Army Corps of Engineers, Vicksburg, MS.
USACE (1992). *Automated Coastal Engineering System Users Manual,* U.S. Army Corps of Engineers, Vicksburg, MS.

USBR. (1974). *Earth Manual,* 2nd Edition, U.S. Bureau of Reclamation, Reston, VA.

Vanoni, V. A. (1977). "Sedimentation Engineering," Manuals and Reports on Engineering Practice, No. 54, American Society of Civil Engineers, New York, NY.

Wang, X. Q., Thibodeaux, L. J., Valsaraj, K. T., and Reible, D. D. (1991). "Efficiency of Capping Contaminated Sediments In-Situ: Laboratory-Scale Experiments on Diffusion–Adsorption in the Capping Layer," *Environmental Science and Technology,* Vol. 25, No. 9.

Worman, A. (1989). "Riprap Protection Without Filter Layers," *Journal of Hydraulic Engineering,* Vol. 115, No. 2.

Wright, S. J., and Kim, C. (1995). "Armor Layer for Protective Cap: Manistique River and Harbor, Michigan," Report UMCEE 95-16, Department of Civil and Environmental Engineering, University of Michigan, Ann Arbor, MI.

Zeeman, A. J. (1993). "Subaqueous Capping of Contaminated Sediments," *Canadian Geotechnical Journal,* Vol. 31, No. 4.

Zweck, H., and Davidenkoff, R. (1957). "Etude Experimentale des Filtres de Granulometrie Uniforme," Proceedings, Fourth International Conference on Soil Mechanics & Foundation Engineering, American Society of Civil Engineers, New York, NY.

CHAPTER 15
THE USACE DREDGING OPERATIONS AND ENVIRONMENTAL RESEARCH (DOER) PROGRAM

CLARK McNAIR
LYNDELL HALES
U.S. Army Engineer Waterways Experiment Station
Vicksburg, Mississippi

Introduction	15.3
Contaminated Sediment Characterization and Management	15.4
Rapid Screens for Chlorinated Hydrocarbons (Dioxins) in Contaminated Sediments	15.6
Feasibility of Reclamation and Reuse of Contaminated Dredged Material from CDFs	15.7
Design of Confined Disposal Facility Containment Features	15.8
CDF-Based Intrinsic Aerobic Bioremediation	15.8
CDF Management for Solids Separation	15.9
Screening Procedures for CDF Leachate	15.9
Vegetation Management for CDFs	15.10
Exposure/Effects-Based Testing for CDFs	15.10
Biomarker-Based Analyses for Contaminants in Sediments/Soils	15.11
Field Verification of Simplified Surface Runoff Test	15.11
Field Demonstration and Validation of Chronic Sublethal Bioassays	15.12
Field Verification of Volatile Losses from Dredged Material	15.12
Improved Effectiveness of Subaqueous Caps	15.13
Field Operations for Manufactured Soil at CDFs	15.13
Treatment Technologies for CDF Effluent, Runoff, and Leachate	15.13
CDF Filling Operations to Enhance Bioremediation	15.14
Predicting Trophic Transfer Potential for Sediment-Associated Contaminants	15.14
Contaminant Pathway Screening for CDFs	15.15
Validation of Biomarkers for Effects Assessments	15.15
Effects-Based Testing of Reclaimed Materials	15.16
Anaerobic Bioremediation in CDFs	15.17
Prediction and Control of Contaminant Losses during Dredging	15.17
Environmental Windows for Dredging Operations	15.18

Research and Operational Measures to Reduce the Need for
   Environmental Windows   15.18
Technical Evaluation of Hydraulic Entrainment Effects on
   Biological Resources   15.20
Technical Evaluation of Environmental Windows to Protect Migratory Fishes   15.21
Numerical Modeling of Suspended Sediment and Turbidity Plumes   15.21
Physical Modeling of Suspended Sediment/Turbidity Plumes in the
   Water Column   15.22
Technical Evaluation of Suspended Sediment Effects on Biological Resources   15.22
Physical Monitoring of Near-Bottom Suspended Sediment/Turbidity Plumes   15.23
Technical Evaluation of Sedimentation Effects on Biological Resources   15.23
Instrumentation for Dredge and Site Monitoring   15.24
   Silent Inspector for Pipeline Dredges   15.25
   Dredge Contract Payment Using Tons Dry Solids (TDS) Method   15.26
   Silent Inspector for Dump Scows and Mechanical Dredges   15.26
   Improved Dredging Site Contaminant Characterization   15.26
Nearshore/Aquatic Placement of Dredged Materials   15.27
   Evaluation and Design of Nearshore/Aquatic Placement of Mixed Sediments   15.28
   Geotechnical Design of Contained Aquatic Disposal (CAD) Sites   15.28
   Near-Field Modeling of Subaqueous Dredged Material Placements   15.30
   Comprehensive Open Water Site Management Software   15.30
   Modeling Dispersion from Pipeline Disposals   15.31
   Estimating Vertical Variation in Erosion and Deposition Rates of
      Mixed Sediments   15.31
   Environmental Effects of Placing Dredged Material in the Active
      Nearshore Zone   15.32
   Far-Field Fate Predictions for Subaqueous Dredged Material Placement   15.32
   Near-Field Prediction of Bottom Surges Due to the Placement of
      Dredged Material   15.33
   Entrainment and Transport of Sediment Mixtures Placed in the Nearshore   15.33
   Fate of Dredged Material Placed on Channel Margins   15.34
   Open-Water Alternatives for Beneficial Uses of Dredged Material   15.34
   Geotechnical Properties of Dredged Material   15.35
   Tools for Estimating Sediment Fate in Estuaries   15.35
   Thin-Layer Disposal of Dredged Material   15.36
   Dredged Material Decision Support System   15.36
   Mass Balance in Dredging and Dredged Material Management   15.37
   Criteria for Construction of Subaqueous Dredged Material Deposits   15.37
Innovative Dredging Technology   15.38
   Demonstrations of Innovative Dredging Technologies   15.39
Ecological Risk Management for Dredging and Disposal Projects   15.40
   Risk Assessment Framework for Management of Dredged Material   15.41
   Risk Characterization of Dredged Material   15.41
   Decision Support Tools for Environmental Risk Assessment   15.42
   Integrating Ecological Risk Assessment and Economic Factors   15.42
Benefits of the DOER Program   15.43
Acknowledgments   15.43
Appendix A: Genesis of the Dredging Operations and Environmental
   Research (DOER) Program   15.43
   Dredged Material Research Program (DMRP)   15.44
   Dredging Operations Technical Support (DOTS) Program   15.45
   Field Verification Program (FVP)   15.46

Long-Term Effects of Dredging Operations (LEDO) Program   15.47
Dredging Research Program (DRP)   15.47
Synopsis   15.48
Appendix B: DOER Technical Reports and Technical Notes   15.49
 Contaminated Sediment Characterization and Management   15.49
 Environmental Windows for Dredging Operations   15.51
 Instrumentation for Dredge and Site Monitoring   15.52
 Nearshore/Aquatic Placement of Dredged Materials   15.53
 Innovative Dredging Technology   15.54
 Ecological Risk Management for Dredging and Disposal Projects   15.55

INTRODUCTION

Waterborne transportation is the most efficient and economical means for national and international commerce, and continued maintenance and development is necessary to ensure continuing benefits. Dredging is a viable option for developing and maintaining ports, harbors, coastal channels, and inland waterways. Protection and enhancement of the environment associated with waterway infrastructure operation and maintenance (O&M) is a substantial priority. The U.S. Army Corps of Engineers (USACE) is faced with the challenge of maintaining a viable navigation system through dredging while providing necessary environmental protection or enhancement. Thus, navigation dredging and environmental requirements are inseparable. The USACE is involved in virtually every navigation dredging operation performed in the United States. The Corps' navigation mission entails maintenance and improvement of about 40,000 km of navigable channels serving about 400 ports, including 130 of the nation's 150 largest cities. U.S. dredging costs to maintain viable navigation now approximate $600 million annually and are increasing. These costs are borne by individual dredging projects, and are increased by Federal and state statutory requirements for environmental sustainment.

Research and development are integral components of managing the USACE dredging program to assure an efficient and environmentally sustainable navigation system. Integration of operational and environmental aspects of dredging and disposal must be accomplished within a climate of increasing dredging workload needs and environmental constraints with decreasing fiscal and manpower resources. The Dredging Operations and Environmental Research (DOER) program addresses documented requirements of the primary Corps users (field operating Division and District offices). Requirements identified by the field offices are categorized into six specific applied research focus areas, each with research tasks describing objectives, research methodologies, user products, and time/cost schedules. The 8-year, $48-million DOER program was initiated in October 1996, and is scheduled for completion at the end of September 2004. A chronology of events giving rise to the DOER program is presented in Appendix A.

Objectives of the DOER program being conducted by the U.S. Army Engineer Waterways Experiment Station (WES) are to develop technologies, methodologies, and techniques which, in concert with an aggressive technology transfer and interagency coordination effort, will assure that the operational and environmental requirements of the USACE dredging program are adequately and efficiently met. Each research task of the six focus areas is addressing a specific deficiency, and will be completed within (usually) 3 or 4 years, culminating in a fieldable product consisting of a Technical Report (TR) or Technical Note (TN) pertaining to a documented solution of that specific deficiency. TRs are comprehensive documents containing detailed information pertaining to the problem that was investigated, analysis techniques, and conclusions reached from the study. TNs deal with one specific aspect of dredging technology, and present in a succinct format certain data emerging from the research program. All DOER TRs and TNs are being made available via the Internet as soon as they are completed, and may be found at http://www.wes.army.mil/el/dots/doer in portable document format (.pdf) at this Website, along with other information about USACE dredging research. Website visitors may save the materials in electronic form or may print copies in color or black and white.

DOER is addressing and solving issues in six focus areas. (1) Contaminated Sediment Characterization and Management will reduce cost and improve the reliability and acceptability of assessing, dredging, placing, managing, and controlling contaminated dredged material. (2) Environmental Windows for Dredging Operations will minimize cost of dredging operations attributable to compliance with environmental windows that are determined to be overrestrictive, inconsistent, or technically unjustified. (3) Instrumentation for Dredge and Site Monitoring will provide automated dredge inspection and innovative dredge contract payment methods, remotely delineate sediment characteristics, and improve the effectiveness of monitoring. (4) Nearshore/Aquatic Placement of Dredged Materials will predict the movement, cost-effectiveness, and beneficial aspects of dredged material placed in the nearshore environment. (5) Innovative Dredging Technology will document and facilitate demonstrations of emerging dredging and disposal technologies in cooperation with field offices. (6) Ecological Risk Management for Dredging and Disposal Projects will develop a technically sound approach for characterizing and managing risk that makes use of existing guidance and proven tools for conducting risk-based evaluations of dredged material that are consistent with the U.S. National Academy of Science/U.S. Environmental Protection Agency (USEPA) paradigms for risk assessment.

All research tasks of the DOER program are under continuous review by the user Field Review Group (FRG) and Corps Headquarters (HQUSACE). A research task may be modified or terminated by the FRG and/or HQUSACE, depending on the assigned priority and annual funding. Likewise, since field user needs are not static, new research tasks may be initiated as appropriate. TRs and TNs presently on the Internet, and those that are scheduled for preparation during the remainder of the research program, are listed in Appendix B. As the program proceeds from its initial years and products from the research increasingly mature, an accelerated number of TRs and TNs will be available on the Internet. A summary of the DOER research tasks by focus area follows.

CONTAMINATED SEDIMENT CHARACTERIZATION AND MANAGEMENT

The presence of contaminated sediments within the boundaries of federal navigation projects that are located in many industrial and urbanized harbors and waterways contributes to environmental degradation and inhibits the ability of the Corps to dredge, transport, and relocate sediments in performance of its navigation mission. The presence of chlorinated hydrocarbons such as dioxins is especially viewed as a potential threat to the environment and human health, often resulting in significant project delays and management cost increases. Although problems are severe only in some areas, the public perception associated with contaminated sediments affects the entire navigation program. Currently, contaminated sediments unsuitable for conventional disposal may be confined, contained, treated, or simply not dredged.

Most commonly considered alternatives for management of contaminated sediments are (a) placement in confined disposal facilities (CDFs) (Figure 15.1), and (b) capping, an option for containment in subaqueous sites. CDFs are located on land or in areas of relatively sheltered water. Many CDFs are near closure and future locations may include nontraditional areas such as offshore. Treatment to reclaim CDF capacity may be promising for certain sites. Capping has significant potential as a disposal alternative, but issues related to its long-term effectiveness and potential application to deeper waters or high-energy environments require additional engineering and environmental investigation. DOER will address high priority research needs aimed at reducing costs associated with screening and assessing potential impacts associated with contaminants and increasing the reliability and acceptability of CDFs and capping options for management of contaminated sediments.

DOER is developing low-cost, rapid, and interpretable biological screening methods for chlorinated hydrocarbons and other contaminants. These methods will reduce the number and cost of chemical analyses and quickly identify contaminated sediments and marginally contaminated dredged material in existing CDFs that can be reused. Tiered screening tools will be developed for estimating contaminant losses from CDFs and capped sites in order to reduce the need for more expensive environmental testing. Techniques requiring minimal data, such as bulk sediment chemistry, will be emphasized and developed for implementation on desktop computers.

Risk-based assessments for contaminated dredged materials are being developed for both open water and CDF placement options. The risk assessment process for contaminated sediments includes effects and exposure assessment (e.g., contaminant pathway testing). Results of risk-based assessments facilitate risk management which, for contaminated dredged material, may include identification of design requirements for contaminant controls and treatment. Effects assessments for dioxin contaminated dredged material will tie directly into the overall environmental risk assessment framework developed under

FIGURE 15.1. Confined disposal facility for contaminated sediments, Kewaunee, Wisconsin (photo courtesy of Michael R. Palermo, WES).

this research focus area. Effects data from laboratory testing will be compared with field measurements of effects on populations of organisms in areas where sediments are contaminated with chlorinated hydrocarbons (dioxins). Cost-effective laboratory test procedures and predictive tools for exposure assessment will address CDF groundwater leachate, surface runoff, and volatile pathways.

CDF research is developing and validating contaminant controls, treatment methods, and management techniques. Design of CDFs as treatment structures, groundwater and surface water protection, and overall contaminant retention will be emphasized. Design criteria for treatment and/or control of toxic contaminants are being developed, including low-cost effective methods for CDF management to meet state water quality certification requirements. Research on filtration treatment structures and enhanced biodegradation of contaminants in CDFs is receiving the highest priority. Techniques for reclamation of CDF capacity will be developed for sites with materials marginally contaminated with chlorinated hydrocarbons. Tools for predicting capped material chemical migration are being refined and used as a basis for more cost-effective capping designs. Research on environmental aspects of capping and CDFs is being integrated with research on physical aspects under the DOER Nearshore/Aquatic Placement of Dredged Materials focus area to provide comprehensive guidance for these management options. Laboratory studies and technical assessments of control and containment technologies for open water disposal other than capping also are being conducted. Table 15.1 shows the planned time line for conducting the research tasks of this focus area. Brief descriptions of these research tasks follow.

Rapid Screens for Chlorinated Hydrocarbons (Dioxins) in Contaminated Sediments

With recent publication of the USEPA's Dioxin Risk Reassessment Study, dioxin has reemerged as the most feared of all environmental contaminants. New analytical advances have pushed the detection limits for dioxin in sediments and organisms down into the parts-per-quadrillion range. The demand for high-resolution dioxin chemical analysis now frequently made of permittees has driven the cost of some dredging projects to exorbitant heights. This research is using existing biotechnology to develop rapid, low-cost, in vitro biomarker assays for detecting and quantifying high-profile contaminants [dioxins/furans, polycyclic aromatic hydrocarbons (PAHs), and coplanar polychlorinated biphenyls (PCBs)] in soils and sediments. The H4IIE cell line is being developed as a first-generation screening assay by the refinement of methodology and the application of accelerated solvent extraction techniques. The P450 Reporter Gene System (P450 RGS) cell line is being developed as a dioxin assay, and testing against the H4IIE assay is being performed to determine relative sensitivity, selectivity for dioxins, and cost and ease of performance.

The costs of monitoring dioxin contamination in high-risk sediments will be greatly reduced by the availability of a screening assay. Cost per assay can be expected to be about 1/10th of that of analytical chemical methods of gas chromatography/electron capture detection (GC/ECD) or gas chromatography/mass spectrometry (GC/MS). Analysis time per sample will be reduced to 1 week or less, enabling faster response to environmental needs. This research will integrate the effects of mixtures of dioxin-like chemicals that may be additive or antagonistic, thus providing a more toxicologically relevant endpoint than is possible with analytical chemistry.

TABLE 15.1. Contaminated sediment characterization and management research tasks schedule

	1997	1998	1999	2000	2001	2002	2003	2004
Rapid screens for chlorinated hydrocarbons (dioxins) in contaminated sediments	****	****	****					
Feasibility of reclamation and reuse of contaminated dredged material from CDFs	****	****	****					
Design of CDF containment features			****	****				
CDF-based intrinsic aerobic bioremediation		****	****	****	****			
CDF management for solids separation			****	****	****			
Screening procedures for CDF leachate			****	****	****			
Vegetation management for CDFs			****	****	****			
Exposure/effects-based testing for CDFs				****	****			
Biomarker-based analyses for contaminants in sediments/soils				****	****	****		
Field verification of simplified surface runoff test				****	****	****		
Field demonstration and validation of chronic sublethal bioassays				****	****	****		
Field verification of volatile losses from dredged material				****	****	****		
Improved effectiveness of subaqueous caps					****	****		
Field operations for manufactured soil at CDFs				****	****	****		
Treatment technologies for CDF effluent, runoff, and leachate				****	****	****		
CDF filling operations to enhance bioremediation				****	****	****		
Predicting trophic transfer potential for sediment-associated contaminants				****	****	****		
Contaminant pathway screening for CDFs						****	****	****
Validation of biomarkers for effects assessments						****	****	****
Effects-based testing of reclaimed materials						****	****	****
Anaerobic bioremediation in CDFs						****	****	****
Prediction and control of contaminant losses during dredging						****	****	****

Feasibility of Reclamation and Reuse of Contaminated Dredged Material from CDFs

Disposal sites for future dredging projects are becoming scarce or are not available. Existing CDFs are being filled. Contaminants in dredged material in CDFs reduces both the potential or beneficial reuse of the dredged material and the removal of dredged material from CDFs to provide additional storage space for future dredged material. There is a need to determine how many CDFs contain contamination, how much dredged material is contaminated, what contaminants are present, what levels of contaminants are present, and the location and accessibility of the CDFs so alternative management strategies can be formulated. This research is developing techniques to beneficially reclaim/reuse dredged material from existing CDFs to increase storage capacity at minimum cost.

New dredged material disposal sites will not be required if existing CDFs can be emptied and used for future dredging projects. Revenue will be generated through reuse/recovery opportunities for dredged material, CDF costs will be reduced by recovery of storage capacity, greater regulatory and public acceptance of CDFs will occur, beneficial reuse of marginally contaminated material will happen, and avoidance of more expensive options will be realized.

Design of Confined Disposal Facility Containment Features

Confined disposal facilities are widely used for disposal of dredged material unsuitable for open water disposal. Containment measures to minimize contaminant losses via leachate, surface runoff, and volatilization are often considered to meet stringent release standards or to manage risk for contaminated dredged material. Available containment features include surface covers to reduce infiltration of precipitation (leachate generation), to reduce losses of volatile contaminants, to prevent uptake by plants and animals, and to avoid contaminated surface runoff. For leachate control, reducing permeability of dikes and the CDF bottom is often suggested as a means for control. There is no technical guidance for design and construction of these systems for dredged material containment areas. The fact that many CDFs are constructed in open water further complicates application of these systems for contaminated dredged material.

Objectives of this research are to develop design procedures, construction methods, and cost estimates for covers, dikes, and liners appropriate for CDFs receiving high-risk dredged material. Application of techniques for containing contaminants in CDFs will be significantly improved. The capability to design cost-effective containment systems will reduce exposure concentrations and subsequent risks.

CDF-Based Intrinsic Aerobic Bioremediation

Research on soil remediation indicates that aerobic, intrinsic bioremediation technology could be applied to the unsaturated zone in CDFs. Soils contaminated with soluble, readily biodegradable organics (i.e., benzene and toluene) have been bioremediated at many land sites. The problem organics in dredged material, however, are usually not very soluble or biodegradable. Application of bioremediation technology to large volumes of dredged material without the intensive equipment, energy, and other requirements associated with most bioremediation technologies presents unique challenges. This research is developing environmental engineering components for design, operation, and management of the unsaturated zone in CDFs to decontaminate dredged material using intrinsic bioremediation technology to allow for beneficial use (e.g., manufactured soil).

The work is focused on the more easily biodegradable contaminants [petroleum hydrocarbons, 2- and 3-ring PAHs and PCBs, and polychlorinated dibenzo-p-dioxins (PCDDs)/polychlorinated dibenzofurans (PCDFs) with low chlorination numbers] through pilot and field-scale studies. This research will bridge the technology gap between simple storage of contaminated dredged material in CDFs and expensive high-technology treatment alternatives. The technology that is being developed (CDF-based bioremediation) will reduce the size of future CDFs and provide revenue generation through reuse/recovery opportunities, greater regulatory and public acceptance of CDFs and biotechnology, and avoidance of more expensive high-tech alternatives.

CDF Management for Solids Separation

Exposure/effects evaluation and interpretive guidance are required for effective management of contaminants to minimize unacceptable adverse impacts in CDFs, and for beneficial use projects using contaminated dredged material. These procedures will determine where management is necessary and will provide strategies prior to dredging so that placement, site design, and any remedial action can be cost-effectively implemented. This research is taking advantage of the most recent developments in effects-testing procedures, and will supply required data for risk assessment at CDFs and beneficial use projects where contaminated dredged material is used.

Objectives of this research are to (a) develop bench-scale tools for evaluating the physical and chemical feasibility of separating contaminated from noncontaminated sediment fractions in existing CDFs, (b) develop guidance for evaluation of existing CDFs to determine the potential volume of material suitable for physical separation and reuse, and (c) develop guidance for evaluating the practicality of physical separation as a contaminated sediment management strategy from a cost/benefit perspective. Products of this research can be directly applied in the field to cost-effective evaluation of the materials recovery potential of CDFs, and recovery of CDF storage capacity. The potential benefits of CDF capacity recovery include operations cost savings, production of material suitable for beneficial use, revenues associated with production of a marketable material, and obvious environmental benefits as the need to develop new sites is diminished.

Screening Procedures for CDF Leachate

Confined dredged material disposal facilities produce leachate containing contaminants. The evaluation of leachate and flow to groundwater can be the most complex and expensive pathway evaluation for CDFs. Presently there are no standard procedures for screening or evaluating the significance of these releases from CDFs. Further, complex, sophisticated, site-specific groundwater modeling or expensive leachate testing may not be needed, depending on the dredged material characteristics, foundation soils, and groundwater flow conditions. This research is developing a leachate screening evaluation methodology for estimating the potential for adverse impacts of contaminated leachate on resources and provide guidance for applying the methodology.

At the completion of this research, a leachate screening methodology will be available for estimating the potential for adverse impacts of contaminated seepage on ecological resources at all confined disposal facilities. The screening methodology will eliminate the need for leachate testing, site characterization, and complex modeling of facilities having negligible or little potential for contamination of groundwater resources. Screening models will be used to determine the need for leachate testing, site characterization, and the need for complex, sophisticated, site-specific groundwater modeling. This research will provide savings in analyses, site characterization and modeling, and facilitate rapid preparation of environmental impact statements relating to leachate impacts.

Vegetation Management for CDFs

CDFs usually become colonized by vegetation. Undesirable vegetation needs to be controlled and more desirable vegetation to detoxify and decontaminate dredged material needs to be established. Large CDFs of metal-contaminated and/or organic-contaminated sediments/dredged material require reclamation before the dredged material can be used in a beneficial way. Application of physical and chemical treatment technologies to large areas and massive amounts of dredged material would be an enormous cost to society. Recent scientific discoveries in the fields of plant and soil sciences present two opportunities never before available: (a) the development of metal hyperaccumulator plant species to phytoremediate and decontaminate dredged material containing excessive heavy metals, and (b) the development of plant/root rhizobia associations that can phytoremediate and degrade organic contaminants in dredged material. These phytoreclamation technologies, when perfected, could have an enormous dual application to Corps CDFs as well as to Department of Defense (DOD) installations and Superfund sites.

Objectives of this research are to identify, develop, and test vegetation management techniques that control and eliminate undesirable vegetation, and to establish and utilize plant species and associated plant rhizosphere microbes to hyperaccumulate metals or degrade organic contaminants in dredged material in CDFs. This research is taking advantage of the most recent discoveries of metal hyperaccumulator plants and will advance the state-of-the-art in genetic engineering of "super" metal-accumulator plants that will reduce the amount of time required to clean up metal contaminated dredged material. Likewise, the state-of-the-art in phytoreclamation of organic contaminated dredged material will be advanced. The resultant phytoreclaimed soil will cost less than current treatment technologies, will be a permanent solution, and will allow the reuse of the dredged material in manufactured soil production. Metals accumulated in these phytoreclamation crops can be recovered and used to offset the cost of this phytoreclamation technology, and greatly reduce the economic cost to society. This technology can be applied to Superfund sites and will also substantially improve Department of Defense capability to clean up metal and/or organic contaminated sediment/soil at many installations.

Exposure/Effects-Based Testing for CDFs

Plants and animals colonizing existing and future CDFs are exposed to contaminants that can move from dredged material into food webs, and can result in unacceptable risks outside the CDF. Exposure/effects-based assessments of contaminant pathways are required prior to dredging to evaluate impacts to plants and animals where terrestrial placement is selected as a disposal alternative. As in aquatic pathways, there are no specific guidelines for contaminant concentrations in plants and animals. Risk assessments are currently based on exposure/effects-based evaluations using index species test data that may not necessarily provide an accurate indication of the fate and effects of contaminants in the species that may actually colonize the CDF. Uncertainties need to be resolved and guidance developed to standardize the interpretation of test results used in risk assessments to determine effective long-term management strategies for CDFs. The objective of this research is to evaluate effects-based testing of CDF dredged material contaminants that will interrelate the WES plant and earthworm bioassay procedures to key ecosystem components and processes as part of an overall risk assessment.

Exposure/effects evaluation and interpretive guidance will be provided for effective management of contaminants to minimize unacceptable risks in CDFs, and for beneficial use projects using contaminated dredged material. These procedures will determine where management is necessary, and will provide strategies prior to dredging so that placement, site design, and any remedial action can be cost-effectively implemented. This research will take advantage of the most recent developments in effects testing procedures and will supply required data for risk assessments at CDFs and beneficial use projects where contaminated dredged material is used.

Biomarker-Based Analyses for Contaminants in Sediments/Soils

Dredged sediment evaluations and contaminant monitoring for remediation, confinement, and treatment of high-risk dredged material involves expensive and time-consuming chemical analyses and bioassay/bioaccumulation testing. Although a number of rapid biological tests for toxic chemicals exist, these have never been developed and applied systematically to screening contaminated sediments. At present, there are few means available for quickly and inexpensively discriminating among possibly contaminated sediments. Consequently, disposal or treatment options are most often decided upon only after sediment characterization by analytical chemistry and/or lengthy biological studies. A significant portion of the high cost of testing could be eliminated if faster and cheaper screening tests were available to identify and characterize the more contaminated sediments, thereby eliminating them from inclusion in an extensive testing effort. The objective of this research is to extend the application of biomarker-based assays begun in other DOER research tasks for dioxins and related compounds to additional sediment and soil contaminants.

New techniques developed in the Corps' Long-term Effects of Dredging Operations (LEDO) research program's genotoxicity research will be incorporated into this research to develop a suite of rapid, low-cost assays. Rapid screening methods will identify sediments that can be eliminated from further testing. High-risk sediments will be identified quickly, thus facilitating decision making. The nonextractive immunoassays will be capable of being applied directly in the field to give real-time results at probable costs of $5 to $10/sample.

Field Verification of Simplified Surface Runoff Test

Surface runoff from CDFs is a potential contaminant loss pathway that must be considered in risk assessments and in managing CDFs. Prediction of surface runoff water quality has been accomplished using the WES rainfall simulator/soil bed lysimeter. This accurate procedure has been field verified, but requires large quantities of dredged material and 6 months to complete a prediction. A simplified laboratory test has been developed under the LEDO research program, but this new procedure requires field verification before it can be recommended nationwide. The objective of this research is to field verify the newly developed simplified surface runoff test at selected CDFs

Simplified surface runoff test results will be verified under field conditions at selected CDFs. Field verification will be conducted at CDFs containing either freshwater or estuarine dredged material. The WES rainfall simulator will be used to determine surface runoff water quality on freshly placed dredged material and then after the dredged material has dried and oxidized under field conditions. Field tests results will be used to verify laboratory results of the simplified surface runoff test. Contaminants will include metals, PAHs, PCBs, pesticides, and Dioxins. Reduced cost and time requirement for predicting surface runoff water quality from CDFs will be realized. Test results can be used in the Corps/EPA technical framework to determine the need for restrictions or treatment controls for surface runoff, and can provide exposure data for use in risk assessments.

Field Demonstration and Validation of Chronic Sublethal Bioassays

Federal regulations governing the disposal of dredged material require the Corps and EPA to evaluate the potential for disposal to result in unacceptable long-term/chronic environmental effects. Toxicity tests currently used to evaluate the suitability of dredged material only measure the potential for short-term/acute effects. Chronic sublethal bioassays developed by the Corps and EPA will soon be available for use nationwide. Prior to implementing chronic tests the quality of the predictions these tests provide must be evaluated. The predictive quality of acute tests has been verified by a number of studies conducted by the Corps, EPA, and others. However, no such effort has been undertaken for chronic sublethal bioassays. The objective of this research is to field demonstrate and validate chronic sublethal sediment bioassays as predictive tools for evaluating dredged material.

The selection of a particular dredged material disposal option (e.g., open water, open water with capping, upland confined disposal, etc.) is based to a large degree on the results of biological tests (e.g., toxicity bioassays) performed during the evaluation process. Because of the large cost differences among disposal alternatives, it is necessary to ensure that the test predictions driving the management decisions are accurate. The results of this research will provide a quantitative description of the relationship between chronic bioassay results and population/community-level impacts. The credibility of the Corps among other agencies and the public will be significantly enhanced by demonstrating the quality of test predictions, resulting in less controversy, fewer conflicts, and fewer project delays.

Field Verification of Volatile Losses from Dredged Material

Loss of volatile contaminants, many of which can be present in high concentrations in dredged sediments, is increasingly recognized as a potential environmental problem. Laboratory studies show that volatile emissions from dredged material are affected by the stage of filling and management of a CDF. Volatile emissions are particularly sensitive to the amount of exposed sediment, sediment contaminant concentration, sediment water content, humidity, wind speed, and vegetation cover. Predictive techniques to determine volatile losses from exposed sediment and from sediment during dredging and disposal have been developed using bench scale flux chambers under the LEDO research program. The laboratory results were used to evaluate and modify the chemodynamic equations developed from mass transport theory. Field verification of the equations, and evaluation of the effects of vegetation on volatile losses (a scenario best evaluated in field simulations) have not been conducted. The objective of this research is to verify predictive methodologies for describing the loss of volatile organic compounds from dredged material under field conditions (to include during dredging and disposal) from exposed sediment and from sediment covered by vegetation.

Active dredging projects will be sampled and controlled field simulation studies will be conducted to evaluate volatile losses in the field environment. Chemodynamic models will be applied to the field data to determine if direct extrapolation of the laboratory data to full-scale field systems is feasible or if modification of the models are needed for certain field situations. Currently, projects wherein volatile losses are of potential concern must either use unverified models or commission costly site-specific studies to address potential problems. Knowledge gained from this research will allow preproject evaluation of volatile losses to be accurately assessed from the initial dredging and disposal operation to the final stages of project development.

Improved Effectiveness of Subaqueous Caps

Subaqueous capping of contaminated dredged material is an accepted disposal alternative in the New England Division and New York and Seattle Districts, and results in significant cost savings over other alternatives. The capping alternative is now being considered in the Great Lakes Districts and other Districts for differing disposal conditions and material types. However, capping is often a controversial alternative because there is no proven technical guidance for important aspects of capping project design related to the ability of caps to contain contaminants in the long term. Because of the lack of solid technical design guidance, inappropriate capping projects are being proposed, requiring difficult regulatory decisions by Districts. Research is needed to refine existing design guidance for long-term capping effectiveness under a range of sediment, capping material, equipment, and site conditions. The objective of this research is to develop improved guidance for evaluating chemical isolation effectiveness and refine overall technical guidance for design of capping projects.

Existing guidance on capping project design will be refined and expanded. Additional research will be conducted to develop improved tools for capping evaluations, emphasizing development of sound engineering approaches. Predictive tools for evaluation of long-term cap integrity, considering chemical migration via diffusion, advection, and bioturbation will be refined using both analytical and modeling approaches. Batch equilibrium tests for cap material will be developed to provide coefficients for predicting contaminant movement through cap material by advection and diffusion. Monitoring programs will field verify the improved predictive tools. Development of improved design capability will allow consideration of capping on the basis of technical and economic merit, thereby realizing significant cost savings over other alternatives.

Field Operations for Manufactured Soil at CDFs

There are opportunities for reuse of marginally contaminated materials as manufactured soil that could provide significant recovery of storage capacity. Before this can be realized, there is a need to field demonstrate and verify the reclamation and reuse of contaminated dredged material as manufactured soil. The objective of this research is to develop guidance on beneficially reusing dredged material as manufactured soil from existing CDFs to increase storage capacity at minimum cost.

The potential for utilizing dredged material beneficially as manufactured soil will be demonstrated and verified at selected CDFs. Candidate screening tests to evaluate potential reuse and the process of manufacturing soil from dredged material will be demonstrated under field conditions. Undesirable vegetation will be controlled by appropriate measures, dredged material will be collected from the CDFs, blended with appropriate cellulose and biosolids to manufacture topsoil, and will be used at selected landscaping demonstration sites. Guidance for manufacturing soil will encourage application of this approach, resulting in reclamation of storage capacity. Revenue will be generated through reuse/recovery opportunities for dredged material. CDF costs will be reduced by recovery of storage capacity and avoidance of more expensive treatment options or construction of new CDFs.

Treatment Technologies for CDF Effluent, Runoff, and Leachate

CDFs are widely used for disposal of dredged material unsuitable for open water disposal. Waterborne contaminants may render effluent, runoff, and leachate from these disposal areas unsuitable for discharge or release. Clarification and filtration systems may provide sufficient particulate removal to meet effluent suspended solids requirements, but dredging and disposal of contaminated sediments may generate contaminant concentrations requiring improved particulate removal and removal or destruction of dissolved contaminants. For high-risk contaminated dredged material, treatment technologies for particulate-bound and dissolved contaminants are needed to meet increasingly stringent State 401 Water Quality Certification requirements. The nature of CDF operations requires short-term, high-volume technologies that can be easily mobilized or incorporated into CDF design features. The objectives of this research are to develop design procedures, laboratory methods for obtaining design parameters, and implementation concepts for treatment of waterborne contaminant releases from CDFs.

The functional capabilities and performance characteristics of filtering systems, adsorption materials, and chemical treatment to remove contaminants in dredged material effluent, runoff, and leachate will be investigated in a series of laboratory and field studies. Various filtration or chemically reactive media, adsorptive materials, packing densities, and flow modes (upflow, downflow, and horizontal flow) will be the focus of laboratory investigations. Advanced oxidation techniques will be considered for removal of trace-level organics. Engineering capability for contaminant removal from effluents, runoff, and leachates from CDFs will be significantly improved. Capability to design and operate cost-effective treatment and containment systems will reduce exposure concentrations and subsequent risks.

CDF Filling Operations to Enhance Bioremediation

The large volumes of contaminated dredged material from navigation projects not suitable for open water disposal require innovative dredged material management. The unique properties of dredged material (large volumes, fine grain, high oxygen demand) present significant technical challenges to implementing bioremediation technology at CDFs. CDF filling practices that are consistent with and enhance environmental engineering design will be required to maximize the cleanup potential of a CDF-based bioremediation technology. Presently, there is no information or guidance on CDF filling practices that maximize the bioremediation potential for dredged material in CDFs. The objective of this research is to develop filling and operation practices that enhance intrinsic biodegradation in CDFs and the feasibility of reclaiming dredged material for beneficial use(s) (e.g., manufactured soil).

This work will focus on identifying, characterizing, and developing best filling practices for enhancing intrinsic biodegradation of organic compounds in CDFs. Hydraulic versus mechanical filling, impact of lift thickness, benefits of rotation among cells, and optimal filling frequencies will be evaluated. It is anticipated that the relative benefits of various filling practices will be tested at full or pilot scale in field studies. Guidance will be developed for selecting filling practices that minimize oxygen transfer limitations and maximize intrinsic biodegradation. Significant cost savings will be realized through maximizing the amount of material that can be bioremediated and converted to beneficial use, and minimizing equipment and energy costs associated with bioremediation processing of material.

Predicting Trophic Transfer Potential for Sediment-Associated Contaminants

Concern over the environmental impacts of dredged material disposal are commonly focused on high trophic level organisms (e.g., fish). Adverse effects caused by the presence of contaminants in sediment can result from direct exposure to contaminated sediment, or through indirect exposure via the ingestion of contaminated prey (i.e., trophic transfer). Persistent, bioaccumulative contaminants will move through food chains as higher trophic level organisms feed on lower trophic organisms in direct contact with contaminated media. No guidance currently exists that describes how such high trophic level assessments should be performed. The objective of this research is to develop a database of trophic transfer coefficients for sediment-associated contaminants to be used in modeling exposure to higher trophic level receptors, thereby reducing reliance on overly conservative default assumptions.

This research will develop a data base of trophic transfer coefficients for sediment-associated contaminants and a modeling framework to accommodate their use in projecting exposure to higher trophic level organisms. A trophic transfer modeling system will be developed to allow users to make projections about likely exposure to fish and other valued aquatic receptors. These models will be constructed to provide information about the most likely exposure concentrations (i.e., the mean) and the uncertainty associated with these exposure estimates. Modeled estimates of trophic transfer exposure, used in combination with the Environmental Residue–Effects Database, will provide the means to answer questions about the potential for high trophic level impacts at dredged material disposal sites. The ability to provide defendable guidance for evaluating risk resulting from exposure through trophic transfer will support the Corps' environmental mission by ensuring that dredged material evaluations are properly focused and technically sound.

Contaminant Pathway Screening for CDFs

Evaluation of dredged material disposal alternatives presently requires extensive testing to provide the information needed for evaluation of potential releases of contaminants by effluent, surface runoff, leachate, uptake, and volatilization pathways. Extensive testing without prescreening does not facilitate cost-effective or timely decision making. Desk-top screening tools for providing planning level estimates of exposure concentrations and contaminant mass losses are needed for preliminary evaluation of alternatives, to guide resource allocation for detailed pathway testing, and to reduce the costs of testing. The objectives of this research are to develop tiered screening tools for estimating contaminant losses from in-water, nearshore, and upland confined disposal facilities, and dredged material treatment facilities, with and without restrictions.

Available a priori techniques for predicting contaminant release for various disposal alternatives will be identified and developed for implementation on desktop computers. Effects of restrictions on key contaminant migration pathways will be identified for analysis. Rapid and inexpensive methods for estimating contaminant pathway release concentrations will eliminate unnecessary testing and will allow available resources to be focused on those pathways of most concern. This research will result in more cost-effective exposure assessments, rapid analysis of contaminant losses for selected disposal and treatment alternatives, and computerized tools suitable for widespread application.

Validation of Biomarkers for Effects Assessments

Rapid cost-effective screening tools have been developed to assess the potential for effects/exposure of contaminants in dredged material. Typically, these tools take advantage of normal subcellular biochemical processes (biomarkers) involved in metabolism of contaminants. Many of these tools have been developed as in vitro procedures in which tissue extracts containing the pertinent subcellular constituents are exposed to an extract of a sediment, and a response is measured (e.g., fluorescence in a test tube). While these subcellular responses are generally rapid, induced by very low levels of contaminants, and offer some degree of specificity (i.e., class of contaminant), they may or may not indicate the potential for whole organism effects. Therefore, before these tools can be used as predictive screens for sediment associated contaminant effects, the relationship between biomarker response and whole organism response must be validated. The objective of this research is to establish the validity of biomarker responses in predicting potential effects in whole organisms and to characterize the uncertainty associated with these predictive screening tools.

Validated biomarker screening-level technologies will allow for rapid, cost-effective screening of a large number of samples for potential contaminant effects. The use of such procedures will significantly reduce and may in some instances even eliminate the need for bulk sediment chemistry and whole organism bioassay tests, resulting in substantial cost savings. Because these tests are generally inexpensive and rapid, a larger number of samples can be processed, allowing for better spatial characterization of potentially contaminated reaches. As a result of enhanced spatial characterization, smaller volumes of material can be isolated and identified for further testing or application of special management techniques, again resulting in substantial cost savings.

Effects-Based Testing of Reclaimed Materials

Because of the lack of available land for siting and increasing construction costs, it is unlikely that many new CDFS will be constructed to address the Corps' future dredged material disposal needs. One potential solution to this dilemma is to reclaim material from these CDFs for beneficial use and/or aquatic disposal. Contamination in the surface layers of many CDFs has been naturally attenuated via biotic and abiotic processes. Consequently, material that was once deemed unsuitable for aquatic disposal may now be suitable. Also, application of special management techniques such as solids separation, bioremediation, phytoremediation, etc. to these sites may allow material to be reclaimed and/or made suitable for aquatic disposal. Although numerous tests exist for evaluating suitability of sediment for aquatic disposal, the direct application of such tests to dewatered dredged material reclaimed from a CDF may not be appropriate. Placing such material back into an aquatic environment will have profound effects on the physical/geochemical nature of the material, potentially affecting bioavailability/toxicity of sediment-associated contaminants. The objective of this research is to evaluate and adapt the design of currently used effects-based bioassays so that they may be used in a technically sound manner to assess the potential toxicity of material reclaimed from CDFs and establish the suitability of this material for aquatic disposal.

A range of contaminated and uncontaminated materials (based on bulk chemical analysis) will be collected from CDFs. Materials will be rehydrated with either fresh water or seawater to simulate aquatic disposal. Toxicity of this material will be evaluated over time at representative ambient conditions using selected standard effects-based tests (e.g., 10-day amphipod toxicity, 28-day bioaccumulation, etc.). Demonstrating suitability of reclaimed material for aquatic disposal will lead to substantial increases in the storage capacity and the operational life spans of existing CDFs. This will result in major cost savings via avoidance of other, more costly measures (e.g., treatment, siting, and construction of new CDFs, etc.) to meet the Corps' future dredged material disposal needs.

Anaerobic Bioremediation in CDFs

The saturated zone in existing CDFs and future facilities designed to enhance aerobic bioremediation presents major obstacles to dredged material cleanup using natural processes. The high sediment oxygen demand and fine-grain nature of contaminated dredged material make it impractical to dewater and aerate these materials in situ. Soil bioremediation research indicates that anaerobic biodegradation of organic compounds probably occurs in the unsaturated zone in CDFs, but intrinsic degradation rates may be too slow for effective remediation. Contaminants in the saturated zone of CDFs are sometimes viewed by resource agencies as perpetual environmental threats, and practical alternatives to perpetual storage are not always available. The objective of this research is to develop approaches for anaerobic biodegradation of contaminated dredged material and management of the saturated zone in confined disposal facilities.

Applied research will be conducted to demonstrate/validate anaerobic bioremediation of the more easily biodegradable contaminants (petroleum hydrocarbons, 2- and 3- ring PAHs, and PCBs and PCDDs/PCDFs with low chlorination numbers) in the saturated zone through laboratory and field-scale studies. Laboratory studies will be conducted to determine the optimum conditions (denitrifying, iron reducing, sulfate reducing, or methanogenic) for anaerobic biodegradation of PAHs, PCBs, and PCDDs/PCDFs. Knowledge gained from this research will provide the technical basis for improved CDF operation and management with greater potential for regulatory and public acceptance of CDFs and biotechnology, and avoidance of more expensive cleanup options.

Prediction and Control of Contaminant Losses during Dredging

Evaluation of the environmental risks of a contaminated sediment dredging operation requires estimates of the sediment and contaminant concentrations released to the water column. Improved techniques for making these estimates will allow the consideration of operational controls and alternative dredges for reducing potential contaminant losses. One of the control measures often suggested or required by regulatory agencies is the deployment of silt curtains around the dredging site. Silt curtains are not suitable for all sites; they present operational problems to the dredging operation, they are expensive, and their benefits are sometimes oversold. The Corps, other government agencies, and port authorities would benefit from updated guidance on the use, design, and limitations of silt curtains and other management approaches. The objectives of this research are to develop predictive techniques for sediment and contaminant releases to the water column when dredging contaminated sediments, and to develop guidance for field use on control measures for sediment resuspension.

Laboratory and field data are needed to identify critical operations parameters for hydraulic and mechanical dredges. These data will be used to improve the predictive techniques for broad application to contaminated sediment dredging. Alternatives for reducing risks associated with contaminated sediment dredging can be evaluated with field-verified techniques. Case studies for resuspension control measures will be analyzed to determine lessons learned and causes of failure and success. Experiences from the Superfund program will also be reviewed. The guidance developed on control measures will include design impacts on dredging operations, and an assessment of effectiveness in controlling and managing resuspended sediments for various scenarios. Application of improved methods of evaluating contaminated sediment dredging environmental effects will avoid the costs associated with implementing controls that are not needed.

ENVIRONMENTAL WINDOWS FOR DREDGING OPERATIONS

Environmental windows are routinely recommended by resource agencies with the intent of protecting sensitive biological resources or their habitats from potential detrimental effects of dredging operations (Figure 15.2). About 80% of all USACE dredging is subject to environmental windows. However, many inconsistencies exist in the application of windows and in the technical bases used to justify windows. Compliance with requests for windows can result in reduced options for contracting dredge plant and equipment, severely constrain mobilization/demobilization schedules, limit contingencies for repairs and severe weather shutdowns, create hazardous working conditions, and ultimately increase dredging project costs.

This focus area research will resolve longstanding controversial issues that underlie recommendations for restrictive windows. A detailed analysis is being performed to ascertain the economic effects of compliance with windows so that DOER program funding can be directed toward research of highest priority with greatest potential return on investment. Investigations are required to fill gaps in the state of knowledge concerning dredging related effects of suspended sediments, turbidity, sedimentation, and entrainment on aquatic organisms. Coordination and collaboration of research activities with appropriate resource agencies is critical. Salient results are being published in peer-reviewed literature. Guidance documents on effective operational measures to reduce or eliminate the need for windows will be disseminated. Table 15.2 shows the planned time line for conducting the research tasks of this focus area. Brief descriptions of these research tasks follow.

Research and Operational Measures to Reduce the Need for Environmental Windows

Compliance with environmental windows can increase dredging cost in numerous ways (e.g., contracting complications, reduced dredge plant and equipment options, reduced flexibility for mobilization/demobilization, limited contingencies for unanticipated repairs and weather delays, and increased safety hazards). These costs have not been rigorously quantified. Optimal return on research can only be achieved through focused effort on technically resolvable issues associated with windows having the greatest impact on the national dredging program. This research task is quantifying actual and potential monetary costs of environmental windows within and between Districts by a survey of the planning, operations, and regulatory elements. Incremental costs per volume of dredged material for all classes of windows and types of dredging projects is being quantified.

Multidisciplinary (scientists, dredging industry experts, and District and resource agency personnel) working groups are being convened to integrate prioritized research with ongoing dredging projects. A database of economic effects of windows on various dredging project types is being compiled. Data will be shared so that dredging and resource experts can work together to determine acceptable and cost-effective engineering solutions. Practical operational guidelines will be merged with objectively based window start/end dates. The first detailed information on the overall economic impact of environmental windows on the national dredging program is being derived. The validity of certain windows is being resolved. Removal of ambiguous concerns and inconsistent applications of windows will greatly enhance the dredging project manager's ability to meet channel and harbor maintenance schedules. A rational balance will be

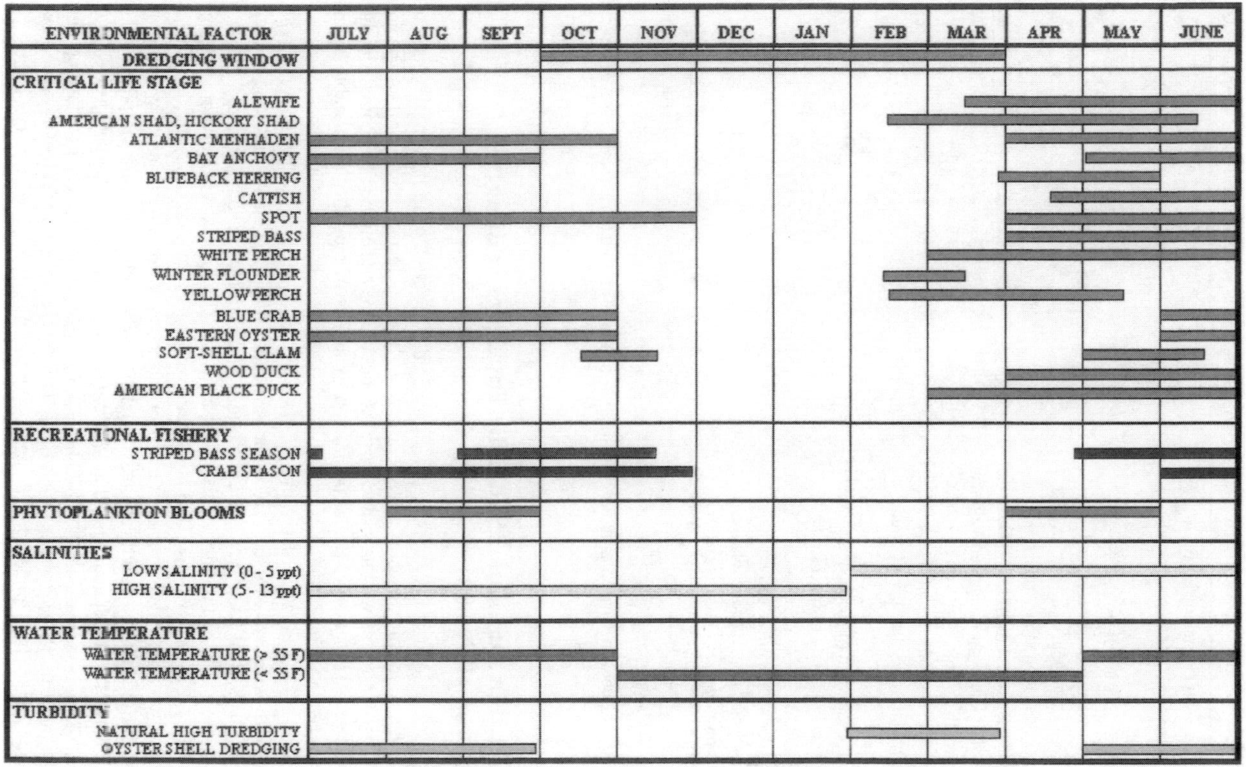

FIGURE 15.2. Composite environmental windows of various species in U.S. Army Engineer District, Wilmington (figure courtesy of Douglas G. Clarke, WES).

TABLE 15.2. Environmental windows for dredging operations research tasks schedule

	1997	1998	1999	2000	2001	2002	2003	2004
Research and operational measures to reduce the need for environmental windows	****	****	****	****	****	****	****	****
Technical evaluation of hydraulic entrainment effects on biological resources			****	****	****			
Technical evaluation of environmental windows to protect migratory fishes		****	****	****				
Numerical modeling of suspended sediment and turbidity plumes		****	****	****				
Physical modeling of suspended sediment/turbidity plumes in the water column		****	****	****				
Technical evaluation of suspended sediment effects on biological resources					****	****	****	
Physical monitoring of near-bottom suspended sediment/turbidity plumes					****	****	****	
Technical evaluation of sedimentation effects on biological resources					****	****	****	

achieved between justifiable natural resource protection measures and cost-effective dredging operations.

Technical Evaluation of Hydraulic Entrainment Effects on Biological Resources

Restrictive windows are commonly placed on dredging activities to prevent the entrainment of sensitive life stages of many commercially or recreationally important species (e.g., oyster larvae) as well as species listed as threatened/endangered (e.g., juvenile chinook salmon). The magnitude and population dynamics of mortality attributable to entrainment remain speculative. Attempts to protect hypothetically susceptible resources in the absence of quantitative data have resulted in frequent requests for dredging windows. These windows can severely impair dredging project planning and, due to their ambiguous technical basis, become contentious issues in the coordination process.

This research is evaluating the technical basis for dredging windows where entrainment of larval or juvenile fish and shellfish is of concern. Operational dredging techniques that will minimize entrainment impact and the need for imposition of windows is being identified. This research requires a modeling approach, with field sampling to define the temporal and spatial distribution of larval and juvenile fishes and invertebrates to estimate their relative risk of entrainment. These data are being used to develop and verify models to estimate actual risk of population-level consequences of observed entrainment. Alternative models are being tested and adapted to provide specificity for various dredging scenarios. This work will provide a comprehensive evaluation of the technical validity of dredging windows related to entrainment of larval and juvenile fish and shellfish. Conclusive demonstration of the lack of need for particular windows, or means to avoid significant risk during dredging operations, will result in substantial cost savings.

Technical Evaluation of Environmental Windows to Protect Migratory Fishes

It is a concern of resource agencies that dredging-induced turbidity plumes effectively block or delay the upstream or downstream migration of fishes. This is particularly true for commercially and recreationally important anadromous fishes such as striped bass, salmon, sturgeon, herring, and shad that have suffered dramatic population declines in recent years. Migratory runs of several species have been designated as threatened or endangered. Dredging projects have become highly controversial, although no conclusive evidence has linked dredging to blockage of migratory stocks. This research is investigating the behavioral response of key migratory species to the presence of turbidity plumes surrounding dredging operations. The degrees of avoidance, if any, of turbidity plumes on various temporal and spatial scales is being documented.

Population dynamics of observed responses of key species to the presence of turbidity plumes is being estimated through coordinated field and laboratory investigations. Responses to actual dredging conditions are being analyzed to formulate guidance for operational measures, including objective windows, to protect target species. Quantifying the effects of turbidity plumes on fish and shellfish movements will remove ambiguity from windows required to adequately protect these resources. This knowledge will be used to develop alternative dredging strategies to minimize impacts on susceptible stocks. This information will place observed dredging effects into perspective with both natural and other anthropogenic events to which migratory species are exposed. Increased flexibility resulting from less restrictive windows, and identification of operational modifications that will remove the need for windows, will realize significant cost savings.

Numerical Modeling of Suspended Sediment and Turbidity Plumes

Interpreting biological impacts due to exposures to dredging-related turbidity and suspended sediment plumes requires quantification of the spatial and temporal characteristics of dredging/disposal plumes. Other simultaneous sources of background turbidity and suspended sediment concentrations may also be significant (e.g., storms, trawling, ship traffic). When the relationships between biological effects, turbidity, suspended sediment concentration, and sedimentation exposures are quantified, numerical models to accurately predict these physical parameters will be developed to minimize the need for site-specific empirical studies. This research is developing a field-tested numerical model for predicting the generation and subsequent fate of both ambient and dredging-related turbidity and suspended sediment plumes at appropriate, realistic temporal and spatial resolutions required to predict relevant biological effects.

A multidimensional model for predicting the fate of ambient and dredging/disposal sediment plumes is being developed with an efficient and accurate transport/diffusion algorithm fed by source modules for the introduction of suspended sediment into the water column. The model will simulate a variety of sources of resuspension, including bottom shear due to ship traffic and storms. The model will run in either stochastic or receptor modes to yield probabilities of sediment plumes reaching environmentally sensitive areas. Verification is being accomplished by field data collection under actual dredging operations by other research tasks of this focus area. This research will allow the expansion and probable elimination of windows from many projects. This PC-based numerical modeling technology, peer-reviewed by resource agencies, will produce greater credibility for the Corps' environmentally sensitive national dredging program.

Physical Modeling of Suspended Sediment/Turbidity Plumes in the Water Column

Dredging and disposal operations resuspend sediments into the water column, creating plumes to which aquatic organisms are exposed. Interpretation of biological responses to various degrees of exposure requires quantification of physical conditions within and surrounding the plumes. It is also required that dredging-induced exposures be placed into correct context with other sources of sediment resuspension (e.g., storms, trawling, river discharges, ship traffic). The current state of knowledge regarding natural, dredging/disposal, and other sources of sediment resuspension is inadequate to develop environmental windows for resource protection. Objectives of this research are to provide synoptic physical and biological parameter monitoring during field investigations for other focus area research tasks.

Quantitative descriptions are being obtained of the spatial distributions and temporal variability of all sources of suspended sediment turbidity. Field verification data are being obtained for full development of numerical simulation models. Site-specific synoptic measurements of turbidity and suspended sediment concentrations are being obtained in coordination with biological field studies. Monitoring of the three-dimensional water column is being performed to characterize the spatial and temporal variability of the plumes. These data will be used to relate background suspended sediment concentrations and turbidity with those present during dredging and disposal operations. This research will verify simulations of plume dynamics generated by numerical models presently under development. These data will minimize the need to collect extensive, expensive plume characterization data during future dredging and disposal operations, thereby accruing savings and monitoring costs through distant future events.

Technical Evaluation of Suspended Sediment Effects on Biological Resources

Laboratory studies show that suspended sediment concentrations that result from dredging and disposal plumes can reduce viability of fish eggs, growth and survival of larval and early juvenile fish, growth and survival of bivalves, and movement of larval fish and shellfish into nursery areas from offshore waters. These effects can occur in several ways, most notably from deposited sediments smothering animals, suspended sediments reducing feeding abilities, or suspended sediments forcing nektonic animals to remain inactive on the bottom for extended time periods. The applicability of these observations to the field is unclear, yet because relevant field studies have not been done, resource agencies have acted conservatively by restricting dredging. Objectives of this research are to (a) characterize the distribution, growth, and survival of shellfish and larval fish in the field in the presence and absence of dredging and disposal plumes to demonstrate whether environmental windows for dredging are necessary to protect these resources, and (b) to characterize the transport of larval and juvenile fishes through and around dredged material plumes to determine if plumes act as a barriers or significant sources of stress to these organisms.

Active dredging projects in areas where environmental windows are most restrictive will be examined to compare the abundance and species composition of larval fish and shellfish within and outside dredging-induced plumes. Daily growth increments of larval fish and juvenile shellfish from these areas will be determined and growth rates in the presence and absence of dredging, vessel traffic, and storm-induced plumes will be compared. Currently, environmental windows are imposed upon dredging based on incomplete laboratory studies and scant anecdotal field observations. Knowledge gained from this research will allow the need for environmental windows to be assessed objectively based upon quantitative field evidence by clearly demonstrating the effects, if any, of dredged material plumes on the growth and survival of shellfish and larval fishes.

Physical Monitoring of Near-Bottom Suspended Sediment/Turbidity Plumes

Assessments of effects of dredging and disposal operations on bottom-dwelling biota cannot be performed without detailed information on the temporal and spatial aspects of near-bottom turbidity and suspended sediments. Current knowledge is inadequate to relate tolerance of sensitive biota (e.g., oyster reefs) to both natural and dredging-induced conditions. Consequently, environmental windows to protect these resources tend to assume worst-case conditions. Accurate near-bottom physical data are therefore required to establish meaningful, defensible guidance on issues relating to protection of bottom-dwelling biota. Objectives of this research are to (a) develop quantitative descriptions of the spatial and temporal distribution characteristics of near-bottom suspended sediment and turbidity conditions surrounding dredging and disposal operations, (b) generate characterizations of near-bottom phenomena associated with natural and anthropogenic resuspension events for comparison with dredging-induced sediment resuspension, and (c) provide field data required to verify numerical modeling tools presently being developed.

This research will address existing gaps in the state of knowledge regarding spatial and temporal scales of near-bottom sediment resuspension caused by dredging and disposal operations. These quantitative data will remove subjective judgments and speculative assumptions from the decision process in establishing appropriate environmental windows. More complete information on natural resuspension events will assist field offices in assessment of the degree of actual disturbance entailed in dredging and disposal operations. Potential projects for which windows are requested could be screened and perhaps resolved without resort to expensive monitoring efforts.

Technical Evaluation of Sedimentation Effects on Biological Resources

Temporal and spatial dredging windows have routinely been requested to minimize adverse effects of sedimentation on sensitive benthic resources such as oyster bars or beds of submerged aquatic vegetation, or on critical life history stages of both fish and shellfish (e.g., effects on demersal eggs of winter flounder). While these resources are particularly vulnerable to smothering effects of sediment plumes because of their relative immobility, similarities between dredging and storm-induced sedimentation rates suggest that these organisms have adaptations to deal with some sedimentation regimes. The information necessary for regulators to formulate technically sound windows is generally inaccessible or unavailable and has led to restricted project flexibility, increased costs, and inconsistent application of windows among Corps Districts or between resource agencies. Objectives of this research are to (a) quantify the spatial and temporal scales of bottom disturbance (i.e., elevated turbidity and sedimentation) attributable to dredging operations, and the thresholds of disturbance that result in detrimental impacts to bottom fauna and flora, (b) examine the technical merits of existing environmental windows in light of knowledge regarding such disturbances, and (c) develop guidance for environmental windows that balance dredging project requirements with adequate safeguards against adverse impacts.

This research will provide technically sound guidance for acceptance and implementation of environmental windows dealing with sedimentation effects on sensitive benthic habitats. This research also will provide the quantitative field evidence needed to determine if seasonally adjusted windows on dredging are needed for projects that occur in the vicinity of sensitive fixed benthic resources. Not only will increased project flexibility (and lower costs) result for project managers, windows also will have sounder scientific bases, and there will be increased protection to natural resources.

INSTRUMENTATION FOR DREDGE AND SITE MONITORING

Improvements in instrumented measurements are needed to meet increasingly stringent environmental monitoring requirements and to expand the Corps' automated operational monitoring and bottom characterization capabilities (Figure 15.3). The DOER program will bridge the gap between existing Corps instrumentation needs and the tendency of commercial firms to direct their product development toward broader, nondredging market segments.

FIGURE 15.3. Silent Inspector monitoring instrumentation aboard hopper dredge Columbus (photo courtesy James Rosati, WES).

TABLE 15.3. Instrumentation for dredge and site monitoring—research tasks schedule

	1997	1998	1999	2000	2001	2002	2003	2004
Silent Inspector for pipeline dredges	****	****	****	****				
Dredge contract payment using tons dry solids (TDS) method	****	****	****	****				
Silent Inspector for dump scows and mechanical dredges					****	****	****	
Improved dredging site contaminant characterization						****	****	****

This focus area research is augmenting commercial products where feasible and is supporting specialized product applications. By developing system standards and specifications and demonstrating prototypes, an overall goal of product implementation and support through the commercial sector is being achieved. Research is developing systems and standards to (a) monitor contractor pipeline and mechanical dredges and dump scows, (b) pay hopper-dredge contracts on a dry-weight basis, (c) precisely locate dredging, transport, and disposal, (d) use cable array technology to monitor cap thickness and erosion potential, and (e) improve dredging site contaminant characterization technology so that more accurate and cost-effective core and bed material sampling plans can be developed to delineate contaminated materials. Instrument and system development is being preceded by a thorough requirement analysis to ensure that quantified performance goals are based on verifiable need. Dredge monitoring and data management technology is being incorporated into Corps business practices through a process action team composed of both Corps and industry. Table 15.3 shows the planned time line for conducting the research tasks of this focus area. Brief descriptions of these research tasks follow.

Silent Inspector for Pipeline Dredges

The USACE depends almost completely on inspectors for quality control and performance monitoring of hydraulic pipeline dredging. Limitations on the number of available inspectors has led to development of automated tools to assist in maintaining continuous and effective inspection. Claim cost reduction and more accessible dredging production records are required. There is a vital need to continuously know where dredging, transport, and disposal occur. This research is providing a quality assurance tool for monitoring compliance of Corps pipeline dredging contracts. Commercially implementable standards and requirements for monitoring and reporting dredge data for environmental compliance and production verification are being developed. Dredge data compatibility throughout the Corps and standardized contractor data reporting will result from this research.

This research will provide the technical basis for defining the needed quality assurance steps to be taken by the Corps to provide accurate and reliable pipeline dredge contract data. The number of contractor claims and the cost of responding to them will be significantly reduced as a result of the products arising from this research. The system will provide data needed to assure compliance with environmental restrictions. The system also will provide a means of inspection and payment for contract dredging where Corps contract managers have a constrained inspector presence, as well as a significantly improved means for managing the acquired dredge data. The technology to implement this tool is being demonstrated in a prototype system. Existing standards are being used where feasible and Silent Inspector specific standards are being developed as needed.

Dredge Contract Payment Using Tons Dry Solids (TDS) Method

The dry weight of material in the hopper or tons dry solid (TDS) method is an accurate and repeatable measure of hopper dredge production. Presently, payment is based on volume dredged as determined by pre- and postdredging surveys. The TDS method provides an objective means to determine hopper dredge production in dredge and haul work. Payment based on each load of material removed from the dredge prism provides an incentive to the contractor to maximize his efficiency. TDS can be used to describe the mass quantity of sediment removed. This research is providing a technically and scientifically defensible basis for the Corps to implement the TDS method of payment for hopper dredging. Uncertainty analysis is being applied to determine the overall accuracy of the system. Standards and quality assurance procedures are being developed upon which the TDS system can be defended during disputes or claims.

This research will assure that District personnel can easily use TDS data to generate reports, graphs, and other products for use in production verification, project reporting, project planning, and avoidance or defense of claims that may arise from a project. TDS will provide the dredging community with a fair and accurate alternative for dredge contract payment. This research will provide the scientific background, guidance, and specifications that are essential for effective implementation of an automated and technically defensible TDS pay system. The TDS method also provides a positive incentive for the contractor to shorten project completion time.

Silent Inspector for Dump Scows and Mechanical Dredges

As with pipeline dredging, the USACE depends almost completely on inspectors for quality control and performance monitoring of mechanical dredging. Limitations on the number of available inspectors has led to improved productivity and development of automated tools to assist inspectors in maintaining continuous and effective inspection. Claim cost reduction and more accessible dredging production records are required. There is also a need to precisely locate where dredging, transport, and disposal occur. The objective of this research is to provide a quality assurance tool for monitoring compliance of Corps dredging contracts that utilize mechanical dredging and dump scows for transporting the dredged material to the placement site.

Dump scow monitoring technology will be optimized to meet Corps requirements at minimum cost. A scow monitoring system composed of commercial off-the-shelf components will be developed for ready implementation by field operating authorities or dredging contractors. This research will reduce the number of contractor claims and the cost of responding to them. The system will provide data analogous to that required of pipeline dredging, which is needed to assure compliance with environmental restrictions.

Improved Dredging Site Contaminant Characterization

The costs associated with both new work and maintenance dredging for some projects are elevated because of the sampling plans necessary to define materials and possible contaminants. Since contaminants tend to attach to certain soil types, the problem is to identify these soil types and determine which of the soil areas should be sampled before dredging occurs. If the number of sample locations can be minimized while providing the same scientific characterization benefit, then the Corps realizes a cost savings for predredging surveys. The objective of this research is to improve an existing bottom classification system so that more accurate and cost-effective core and bed material sampling plans can be developed.

The present bottom classification system uses acoustic signals from several different frequency transducers to measure bottom characteristics. These data are then processed to create coverage maps delineating material types throughout the survey area. Through a detailed study in the laboratory, a data base can be developed correlating the acoustic signatures with bottom physical properties. The bottom properties can then be correlated with possible contaminants. This data base of information will assist in improving the analysis of the acquired acoustic data. By placing the acoustic data into a Geographical Information System (GIS), the improved analysis techniques can be applied to the data to create a better coverage map of the system. The last portion of the study will be to interrogate the new coverage map and lay out a sampling plan to best define the areas to be dredged. The immediate benefits of this research will be to produce better coring plans to define the regions to be dredged. This will eliminate unnecessary core samples and costly lab analysis. The Corps will have a much better understanding of the material types throughout the system being managed.

NEARSHORE/AQUATIC PLACEMENT OF DREDGED MATERIALS

Nearshore locations present a variety of challenges and opportunities for cost-effective, environmentally acceptable placement of dredged material. Conventional disposal practices are costly, and usually remove sandy and silty materials that build and maintain beaches, barrier islands, and other land features. Estuarine placement options of fine-grained sediments are becoming limited due to environmental concerns. On some riverine dredging projects, near-bank placement options are constrained because of concerns about adjacent backwaters. Many aquatic placement options for dredged material are presently constrained by a lack of predictive tools for sediment transport, which translates to inadequate assessment capabilities and operational guidance for the physical aspects of placement/environmental interactions.

This focus area research is predicting time-dependent movement of disposed dredged material, including sand/silt mixtures placed nearshore on the open coast and in estuaries, all materials placed in estuaries and offshore, and sand/silt mixtures in rivers (Figure 15.4). Similar predictions for estuarine placement of fine-grained material near the dredging site and near-bank placement in rivers is allowing expanded placement options at reduced cost while quantifying risks to resources. For nearshore open ocean, estuarine, and riverine placement, detailed monitoring and evaluation of prototype demonstrations are being combined with physical model testing and sediment transport numerical model enhancements to produce well-documented design procedures and predictions of environmental impacts. Geotechnical aspects crucial to sediment transport predictions and design of underwater features are being researched. Integration of nearshore dredged material placement design aspects into a comprehensive computer-based system called Dredged Material Spatial Management, Analysis, and Record Tool (DMSMART) will facilitate planning, engineering, and operational aspects of dredged material management.

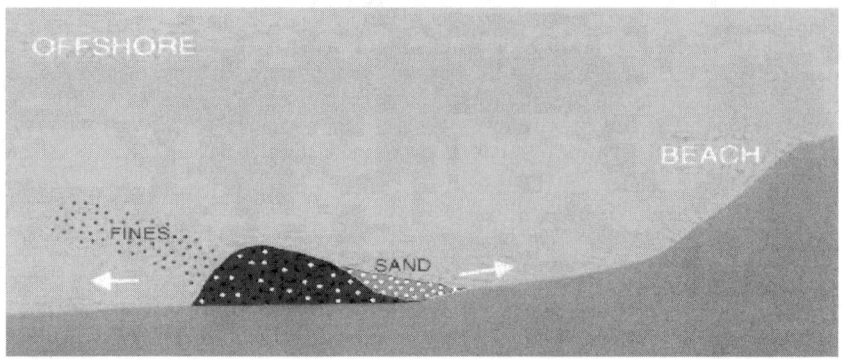

FIGURE 15.4. Proper design of sand/silt sediment mixtures placed in nearshore waters provides cost-effective beach-building material (sketch courtesy James E. Clausner, WES).

Table 15.4 shows the planned time line for conducting the research tasks of this focus area. Brief descriptions of these research tasks follow.

Evaluation and Design of Nearshore/Aquatic Placement of Mixed Sediments

USACE dredges large amounts of sand/silt mixtures that cannot be beneficially placed nearshore due to regulatory and public perceptions of negative environmental impacts. To gain acceptance for beneficially placing these composites nearshore, knowledge of the long-term history of material placed in the nearshore is required. Objectives of this research are to develop numerical modeling tools to assist planners, engineers, and operations personnel in predicting a placement option's ability to enhance the nearshore profile; to site, design, and monitor a project; and to assess physical and environmental benefits/impacts of such Corps projects.

Field data collection from placement of mixed sediments is being obtained so verification of numerical models presently under development can be facilitated. Modeling tools are being incorporated into guidance as development is completed. Numerical modeling and data collection of physical parameters for the nearshore is being integrated into the Comprehensive Open Water Site Management Software research task. This umbrella research task will continue through the life of the focus area. Technically based numerical modeling tools will provide comprehensive guidance for nearshore/aquatic placement of dredged material. Cost savings will result from increasing the acceptance of placing mixed sediments in the nearshore zone, reducing haul distance, and reducing shoreline erosion.

Geotechnical Design of Contained Aquatic Disposal (CAD) Sites

Many Corps Districts are now, or soon will be, using contained aquatic disposal (CAD) sites for placement of contaminated sediments (e.g., New England, New York, Los Angeles, Seattle). Existing or newly constructed borrow pits within the water body are used to laterally confine the contaminated dredged material below navigation depths. These pits range in depth from a few feet to 30 to 40 ft. The horizontal dimensions of the pits vary

TABLE 15.4. Nearshore/aquatic placement of dredged material—research tasks schedule

	1997	1998	1999	2000	2001	2002	2003	2004
Evaluation and design of nearshore/aquatic placement of mixed sediments	****	****	****	****	****	****	****	****
Geotechnical design of contained aquatic disposal (CAD) sites			****					
Near-field modeling of subaqueous dredged material placements	****	****		****				
Comprehensive open water site management software	****	****	****	****				
Modeling dispersion from pipeline disposals			****	****	****			
Estimating vertical variation in erosion and deposition rates of mixed sediments			****	****	****	****		
Environmental effects of placing dredged material in the active nearshore zone			****	****	****	****		
Far-field fate predictions for subaqueous dredged material placement				****	****	****		
Near-field prediction of bottom surges due to the placement of dredged material				****	****	****		
Entrainment and transport of sediment mixtures placed in the nearshore				****	****	****		
Fate of dredged material placed on channel margins				****	****	****		
Open-water alternatives for beneficial uses of dredged material				****	****	****		
Geotechnical properties of dredged material				****	****	****	****	
Tools for estimating sediment fate in estuaries					****	****	****	
Thin-layer disposal of dredged material					****	****	****	
Dredged material decision support system					****	****	****	****
Mass balance in dredging and dredged material management						****	****	****
Criteria for construction of subaqueous dredged material deposits						****	****	****

considerably, but some of the larger pits are on the order of 1,500 ft wide by 5,000 ft long. For many of the larger sites, the pits will be filled segmentally; that is, a smaller portion (an end or a corner) will be filled with contaminated sediments and capped. Subsequently other sections of the pit will be utilized for CAD development. With this sequential plan of filling, it is essential that geotechnical issues be addressed in the design of these CAD projects. This research is developing geotechnical design criteria for contained aquatic disposal projects.

When CAD pits have been filled without geotechnical designs, difficulties have been experienced. For example, recent problems with determining cap thickness due to rapid consolidation of the contaminated sediments have been reported in the demonstration

CAD project in Boston Harbor; this has led to reports in the public domain that the contaminated sediments were not successfully capped. It is essential that geotechnical design guidance be developed for these contained aquatic disposal projects so that they function as sound engineering structures with behavior that can be quantified and accurately predicted. This work will provide guidance for geotechnical design of contained aquatic disposal pits to provide the most efficient and stable capped disposal sites. This guidance will allow construction of space-efficient, cost-effective deposits of improved long-term integrity, and will provide means of quantifying and more accurately assessing observed field performance.

Near-Field Modeling of Subaqueous Dredged Material Placements

Numerical models exist for forecasting the fate of noncohesive, single grain size offshore placement mounds. However, frequently the sediments to be placed are cohesive, mixed cohesive and noncohesive, or multiple grain size noncohesive. Nearshore/aquatic placement at locations such as estuaries or wave-impacted areas in need of beach nourishment is often of interest. Exiting models need to be enhanced and extended such that predictions of disposal site stability can be made for a wide range of sediments and wave/current/depth conditions.

This research is enhancing the Long-Term FATE (LTFATE) of dredged material model to include multiple grain sizes (silts and clays, sand mixed with silts and clays, multiple grain size sands), armoring, winnowing for noncohesive sediments, and a layered sediment bed model. LTFATE also is being expanded to include processes and characteristics relevant to nearshore regions. The enhanced LTFATE numerical model will be integrated into the Comprehensive Open Water Site Management Software research task. An LTFATE model that includes cohesive and noncohesive sediments, armoring, winnowing, and consolidation will provide the District with a forecasting tool that accurately predicts the fate of disposal mounds placed in estuaries, nearshore protected areas, or just outside the surf zone for the purpose of cost-effective disposal or beach nourishment.

Comprehensive Open Water Site Management Software

Effective open water placement site management for multiuser sites is becoming more difficult. To maximize efficient use of existing sites, particular those with multiple users, a variety of material types, and those with capping projects, an integrated system/set of tools/software package is needed. This research is developing DMSMART, an integrated suite of numerical tools for managing open water placement sites.

DMSMART will include the PC-based fate models Short-Term FATE (STFATE), Multiple Disposals FATE (MDFATE), and LTFATE, and applicable programs from the Automated Dredging and Disposal Alternatives Management System (ADDAMS) family. DMSMART will provide site managers with easily accessible or modeled information (e.g., historical site bathymetry, disposal logs, volume changes over time, predicted mound geometry from proposed dredging projects including capping, source of material, core logs, sediment camera profiles, sediment chemistry, biological testing, waves, and currents). DMSMART will be critical for managing day-to-day activities at disposal sites. Benefits include much reduced effort associated with providing routine and unusual requests for data (e.g., the report to the London Convention), and more efficient use of limited, existing placement sites, thereby saving time and expense with new disposal site designations.

Modeling Dispersion from Pipeline Disposals

Open water disposal from a pipeline dredge represents a proven economic alternative for disposal of dredged material from inland channels. Many Corps Districts rely on this type of disposal operation for economic disposal of channel maintenance material. The fate of sediment plumes generated by pipeline disposal operations in shallow water is not well understood or quantified. Pipeline disposal in open water is frequently utilized, but often opposed because the sediment plume fate cannot be reliably predicted. Numerical methods need to be developed for predicting the advective and diffusive sediment fluxes that describe plume behavior for a pipeline discharge. If open water disposal operations are suspended because of regulatory concerns, the cost of dredging operations will substantially increase, placing a financial burden on the O&M functions of the affected Districts. Objectives of this research are to develop methods for describing the near-field suspended sediment distribution and concentration for a pipeline point source disposal into shallow water of 15 ft or less.

Numerical models will be developed for predicting (a) discharge sediment concentrations for various discharge configurations and sediment types, and (b) the fate of the sediment plume in shallow water. The models will predict a mixing zone, suspended sediment, fluidized sediment layer, and bed deposit characteristics. This research will provide methods of predicting sediment plumes from pipeline disposal operations, thus meeting regulatory concerns about water column plume impacts. Near-field results will provide mixing zone definition and input for the far-field sediment plume model being developed under the Environmental Windows focus area.

Estimating Vertical Variation in Erosion and Deposition Rates of Mixed Sediments

Storm-induced erosion of dredged sediments placed in open waters is a major concern. This problem is of special concern when contaminated sediments are involved because of potential environmental impact. Presently, erosion rate equations exist for cohesive and mixed sediments, but the parameters required by these equations are for low-stress conditions (i.e., nonstorms), are site specific, vary significantly with depth, and may vary over orders of magnitude between sites. Estimating these parameters under high shear conditions for a specific site is time consuming, expensive, and has hindered past efforts in estimating the erosion potential of existing mounds. If algorithms relating the erosion parameters to readily measured sediment characteristics (e.g., density, grain size, etc) are developed, less site-specific erosion rate analysis will be required for each project. This is critical to Districts, where time and money often limit the present expensive analysis necessary for each site. Objectives of this research are to quantify as a function of sediment characteristics, depth of burial, and shear stress, the erosion and deposition rates of cohesive and mixed cohesive/noncohesive sediments under high-shear (storm) conditions.

Estimating erosion and deposition rates of cohesive and mixed sediments is essential for accurate sediment transport and fate modeling. Presently, the LTFATE model has been successfully used to estimate noncohesive sediment mound movement. However, many sediments placed in offshore mounds are cohesive or mixed. Algorithms developed from this research will be incorporated into near-field (LTFATE) and far-field Advanced CIRCulation (ADCIRC) sediment transport models enhancements. The erosion and deposition rate algorithms developed in this research task, when incorporated into these models, will provide the District with the ability to predict (a) the stability of sediment mounds during storm conditions, and (b) the ultimate fate of sediments that erode from these mounds. Therefore, the District will have a tool that can estimate the near and far-field impacts of open-water sediment placement.

Environmental Effects of Placing Dredged Material in the Active Nearshore Zone

Resource agencies have raised concerns that dredged material placement in the nearshore active zone could detrimentally affect use of nearshore habitats by fishery resources. Hypothetical impacts may include loss of critical nursery habitat due to avoidance responses to the presence of turbidity plumes, blockage of migratory corridors, and effects of suspended sediments on growth and survival of larval and juvenile fish and shellfish. Also, sensitive habitats might be buried when dredged material is placed in the active zone or as the material adjusts to the shore face. The validity of these concerns remains uncertain, but the concerns must be addressed during the coordination process. For resource agencies to accept nearshore/aquatic placement of dredged material as a viable management technique, it must be clearly demonstrated that potential detriments are offset by the physical benefits derived. Objectives of this research are to (a) document responses of fishery resources to turbidity plumes and sediment resuspension in the nearshore zone, (b) measure effects of project alterations on growth and survival of target species, and (c) provide guidance on siting, timing, and mode of disposal that would minimize disturbance of nearshore habitats critical to fishery resources.

Opposition by resource agencies to placement of dredged material in the nearshore zone can lead to substantial project delays and increased monitoring requirements. Acquisition of technically defensible data on the temporal and spatial scales of impacts to fishery resources, if any, will facilitate the coordination process. Knowledge of the use of the nearshore zone and tolerances of key species will enable project managers to develop and employ disposal techniques that minimize any impacts that might occur.

Far-Field Fate Predictions for Subaqueous Dredged Material Placement

Locating and designating acceptable sites for the offshore placement of dredged material from navigation projects is a serious problem for many Corps District offices. This problem is especially true for fine-grained sediments with adsorbed contaminants, because of the possibility of impact to regions far removed from the disposal site. The inability to predict the fate of plumes associated with fine-grained sediments placed in deep water has also been a problem at some sites. The numerical model LTFATE can be used to predict the near-field (1–5 km) fate of sediments eroded from the bottom; however, a larger domain model will have to be developed to predict the far-field (5–500 km) impact of sediment eroded and transported from the disposal site or sediment plumes associated with placement of fines in deep water. This far-field impact may become an issue on which site designation is based. Without some method of quantifying far-field effects, the site designation processes may be delayed or terminated. Objectives of this research are to (a) construct and make available to the District a PC- and Workstation-based large domain hydrodynamic and sediment transport model, and (b) to predict the fate of eroded sediments from specified disposal sites or the fate of plumes associated with deep-water disposal.

The long-wave model Advanced CIRCulation (ADCIRC), will provide the basic hydrodynamics for the far-field model. A sediment transport capability will be coupled to the model such that erosion and transport (or transport only for plumes) of fine-grained sediments from a specific location can be computed over large domains and for long periods of time. Transport and deposition of material on the order of hundreds of kilometers from the source will be practical using ADCIRC. The model will be applicable to both offshore disposal sites as well as nearshore sites, providing that the nearshore sites are not located within the active surf zone. Development of this far-field sediment fate model will provide the District with a numerical tool that can accurately predict the fate of fine-grained sediments from dispersive disposal sites.

Near-field Prediction of Bottom Surges Due to the Placement of Dredged Material

During the placement of dredged material in depressions, two issues must be addressed: (a) water column effects, and (b) whether energy possessed by the falling material will be dissipated during the bottom surge such that the surge does not run up the side of the depression and deposit material outside the depression. Predicting surge run-up for various pit depths for disposal at various locations in the pit is critical during the design of disposal pits. Likewise, during the placement of material on top of existing mounds or berms, an accurate computation of the final extent of the resulting bottom surge, which can be greatly increased due to the side slope of the mound, is needed for predicting the extent of mound aprons. Such predictions are critical in determining volumes of clean material required for capping mounds. The objective of this research is to provide a fully three-dimensional, near-field model for computing the bottom surge and deposition of material resulting from the placement of dredged material by various disposal methods in open water on arbitrary bottom bathymetries.

An existing, highly accurate, 3-D near-field hydrodynamic numerical model will be modified to remove certain inherent limitations such as the flat-bottom assumption. The resulting model will be entirely physics-based (e.g., 3-D turbulence, allow entrainment of ambient water into a heavier sediment/water mixture, etc.), and will not be constrained by assumptions concerning the shape of the bottom surge or the bottom bathymetry. Various methods of disposal (e.g., disposal from split-hull barges, hopper dredges, trémie tubes, and diffusers placed in the water column) will be allowed. Small-scale laboratory experiments will be conducted to collect verification data in existing flumes. This research will provide a highly accurate predictive tool to increase acceptance of both nearshore and offshore placement of dredged material by regulatory agencies. This will be especially true for the disposal of contaminated material in pits followed by capping operations.

Entrainment and Transport of Sediment Mixtures Placed in the Nearshore

Material dredged from navigation channels typically is a sediment mixture comprised of some combination of noncohesive sediment, silt, clay, and organic material. If the dredged material contains a significant portion of beach-quality sand, then offshore placement effectively eliminates this sand from the beach system. If placed in the nearshore, some of the beach-quality sand may migrate toward the beach and benefit the beach system by reducing erosion of the downdrift shore. Resource agencies are reluctant to approve nearshore/aquatic placement of mixed sediments due to concerns about the fine-grained fractions. To benefit from nearshore/aquatic placement of sediment mixtures, it is necessary to understand how sediment mixtures are transported by waves and currents, how selective sorting of grain sizes occurs, and how the silt, clay and organic components are entrained into the water column and transported away from the disposal site. The objective of this research is to devise useful quasiempirical formulations from theoretical considerations and laboratory measurements to predict entrainment and transport of placed sediment mixtures as a function of the sediment properties and the predominant wave and current hydrodynamic parameters.

Waves propagating over placed sediment mixtures will induce cyclic shear stresses that effectively mobilize sediment for transport by currents and entrain fine-grained particles into suspension. Laboratory tests in a movable-bed flume with waves and currents will be performed to investigate response of different sediment mixtures. Appropriate sediment transport and sediment resuspension theoretical expressions will be adopted and extended by including parameters that adequately characterize sediment mixtures. Laboratory and field data will be used to validate the theoretical expressions. This research will provide physical insight into the processes that dictate the ultimate fate of sand and fine-grained fractions of dredged material placed in the nearshore zone. The work will improve the ability to predict the fate of sediment mixtures in very complex hydrodynamic environments and will provide theoretical and empirical predictive expressions for entrainment and transport of sediment mixtures needed by numerical models that predict the time-history of dredged material placed in the nearshore.

Fate of Dredged Material Placed on Channel Margins

Dredged material placement along channel margins is a common practice along several inland navigation channels. The fate of this material, however, is unknown. Dredged material may be transported during high flows back into the navigation channel or into backchannels where it may impact upon habitat areas. A methodology is needed for quantifying the magnitude of sediment redistributions and for identifying low-impact disposal. Until recently, the only reliable method to track the fate of sediment was through costly field or experimental studies. Fortunately, existing multidimensional, multi-grain-size modeling capabilities now allow for rapid calculation of sediment transport, both as bed load and suspended load. These models can be used to develop guidance for identifying low-impact disposal sites and for determining the magnitude of sediment redistribution for a range of typical project operating conditions. Using existing multidimensional, multi-grain-size modeling techniques, a rational procedure to compute sediment transport from channel margin disposal sites into the navigation channel and into backwater zones will be developed.

The methodology for computing sediment transport from channel margin disposal sites will be demonstrated for a variety of existing diversion sites. Field data will be collected to verify numerical model predictions. Using the verified numerical model, sediment diversion to backwater zones will be developed for a range of geometric and hydrodynamic conditions with a range of sediment sizes. Project impacts can be rapidly assessed, and appropriate dredge material placement sites can be rapidly identified.

Open-Water Alternatives for Beneficial Uses of Dredged Material

A large majority of beneficial use projects implemented to date involve either emergent marsh creation, restoration to provide habitat for wildlife and fishery resources, or island creation for avian habitat. Although many other types of potential beneficial uses for habitat development exist (e.g., intertidal mudflats, oyster reefs, artificial reefs), few have actually been constructed and their probabilities of long-term success has not been established. In many cases, beneficial use projects offer opportunities to place dredged material at reduced cost relative to other modes of disposal, and derive public relations and education benefits as well. These benefits are not currently being pursued to the fullest. Expanding available options for beneficial uses will require sufficient documentation of planning requirements, design and performance criteria, engineering and construction constraints, and monitoring approaches. This research will demonstrate the economic and ecological feasibility of new alternatives for beneficial use of dredged material, and provide guidance for implementing such projects.

Results of this research will increase the acceptability of open-water dredged material disposal by resource agencies and the general public. Maintaining viable placement alternatives during a period of increasing pressure to forego open-water disposal in favor of more expensive modes (e.g., upland containment areas) can contribute to substantial cost savings when these projects are given prioritized consideration. Several project types may provide long-term solutions for disposal requirements as habitat benefits are accrued incrementally over multiple dredging cycles.

Geotechnical Properties of Dredged Material

Increasingly large amounts of dredged materials are being used to construct geotechnical structures (e.g., subaqueous mounds, berms, CAD pits) or for beneficial uses (e.g., wetland creation, fill for airport expansions). It becomes increasingly important that the materials' engineering properties be determined so engineering behavior can be predicted. The variety of sediment types, in situ conditions, and complex depositional stratigraphy encountered in dredging projects makes predicting the postdisposal geotechnical properties of dredged material deposits extremely difficult. However, quantitative descriptions of the temporal and spatial variability of the geotechnical properties including classification, compressibility, shear strength, and density are critical to analyses of consolidation, stability, and/or erosion of constructed structures and facilities. Because a lack of pertinent geotechnical data has previously prevented successful analysis of dredged material behavior even for high-profile, well-funded disposal operations, it is essential that research be conducted to determine the geotechnical properties of dredged materials. Objectives of this research are to (a) develop quantitative descriptions of the temporal and spatial distribution of geotechnical properties in dredged material deposits,(b) provide techniques for assessing salient geotechnical characteristics, and (c) provide geotechnical input parameters relevant to erosion of and design/construction with sediments and dredged material.

This research will provide definitive information on the range and variability of dredged material properties encountered throughout the United States. It will provide a framework for determination of geotechnical properties, to include expedient testing and correlation of material properties to engineering behavior. This framework will provide a rational basis for selecting pertinent laboratory testing, eliminate unnecessary testing, expedite material property determination, and reduce time and costs for material characterization.

Tools for Estimating Sediment Fate in Estuaries

Many dredging projects encounter the engineering and environmental issue of the fate of sediments in an estuarine setting. The issue arises during dredging activity, for all methods of excavation, and for the ultimate fate of in-water placement, whether confined or open water placement. The existing tools for tracking of sediments are limited. Most techniques available for the analysis of the movement of sediments are based on a budget type of approach. These methods are unable to track individual sediment particles. Alternative approaches could be developed if a truly multiple grain size estuarine sediment model were available to address the full range of sediment classes found in estuarine waters, from fine clays to coarse sand, with the appropriate bed structure/exchange model. Such a model does not currently exist. Sediment tracking examples include turbidity plumes from dredging activity, mixing zone analyses for effluent from CDFs, migration of material placed in open water disposal areas, identification of sources of sediment to shoals in navigation channels, and potential impacts from contaminated sediments. The objective of this research is to provide a USACE tool that will permit predicting the fate of sediments resuspended by vessels or from open water disposal sites in estuarine settings.

A multiple-grain-size model over the full range of estuarine sediments will be developed with a bed layering model that will address armoring, cementation of sands by cohesive minerals, and the interaction of all sediment classes in the bed and water column. The model will be demonstrated on a monitored field site.

Thin-Layer Disposal of Dredged Material

Although thin-layer disposal of dredged material has been identified as a potentially environmentally acceptable alternative for over a decade, this concept has been demonstrated only in the Mobile District. Thin-layer disposal represents a viable option for a variety of dredging projects, with application to unconfined open-water and marsh disposal sites. Results of projects in the Mobile District have been very encouraging, but thin-layer techniques have not been given serious consideration in other regions. This may be due to reluctance on the part of resource agencies to accept untested methods of disposal or to concur with results of monitoring studies not conducted in their specific region. Negative perceptions of conventional modes of open-water disposal tend to constrain dredging project managers to comparatively expensive upland options. This research will demonstrate the economic and ecological feasibility of thin-layer disposal technology on a regional basis.

The initial phase of this work will involve packaging existing information on the various forms of thin-layer disposal, emphasizing the positive results of the Mobile District experience, and disseminating this information to Corps field offices. Workshops will be held to familiarize dredging project managers and appropriate resource agency personnel with the potential advantages of thin-layer disposal. Candidate projects will be identified for collaborative field demonstrations, with this work supporting critical monitoring studies to continue to build a strong technical case for further applications of this disposal technique. Remaining gaps in the knowledge base regarding thin-layer techniques will be addressed.

Dredged Material Decision Support System

When DMSMART and the Silent Inspector are endorsed as effective tools for dredging management, then their capabilities should be expanded to incorporate improved predictive and measurement tools being generated within DOER. The inability of DMSMART to incorporate real-time disposal and removal data from a variety of dredge types will be a major limitation of the usefulness of the system for site and project management. As these tools are more widely used, the ability to easily extend them to meet unique District requirements becomes increasingly important. This research will integrate and extend a suite of tools for managing open water placement sites and dredging areas. These prediction tools will access and assimilate disposal and performance data from dredges, and improve both dredge material estimates and the Corps' ability to assure contract monitoring and compliance with EPA regulations.

The Dredge Material Decision Support System (DMDSS) will incorporate updated fate models and will have the ability to assimilate measured dredge data (performance and disposal) from the Silent Inspector system on Corps and contractor dredges and dump scows as well as survey data. The DMDSS will provide Corps' Districts with an easily useable package of tools that will serve the diverse needs of the many disciplines associated with dredging and open water placement activities with a common set of readily available information. This will be invaluable for managing the day-to-day activities at disposal sites, providing planning and design information for future more demanding placements (e.g., increased volumes associated with channel deepening, capping), and allow ready access and presentation of data to regulators (EPA and states) and environmental groups.

Mass Balance in Dredging and Dredged Material Management

Recent environmental and political pressures placed on Corps' dredging projects have raised concerns about "cradle to grave" tracking of all dredged material excavated from a location during a particular project. These concerns are particularly notable for suspended sediments and/or contaminated sediments where a certain size fraction that is closely associated with contaminants may be more susceptible to loss. These losses are problematic because a larger percentage of contaminants may be lost with the finer grain sizes than that assumed based on gross mass losses. Dredged material losses can differ depending on the dredging method (hopper, clamshell, cutterhead, etc.) and can occur in all phases of the dredging process: disturbed in-situ material, overflow from hopper dredges, pipeline leaks, barge leaks, resuspension during submerged placement or from upland runoff and/or long-term redistribution (erosion/consolidation) at the placement site. Objectives of this research are to (a) quantify the total mass and size fraction lost from dredged material excavation, transportation, and placement by understanding better the physical processes involved, (b) develop instrumentation to assist in mass accounting, thus focusing on those phases of the dredging/disposal process with the greatest uncertainty in mass losses, and (c) examine mass losses of sediments prior to and after leaving a dredge, thus allowing this research to compliment the Tons Dry Solid research, which is primarily concerned with mass accounting within a dredge.

Understanding mass balance results in understanding dredged material losses. Mass balance of dredged material provides knowledge of actual quantities of dredged material that are lost, which is important for tracking actual amounts of contaminants associated with particular fractions of "lost" sediments. This research will also identify new instrument needs and/or new applications of existing instruments that may be applicable to installation on dredges for tracking the total mass of dredged material. This research also will serve as a vehicle for coordination of mass balance concerns with Federal and state resource agencies.

Criteria for Construction of Subaqueous Dredged Material Deposits

Offshore placement of contaminated dredged material and capping of these deposits with clean material is an attractive alternative for disposal of sediments removed from navigation channels. However, existing geotechnical design procedures are very limited and little is known about the geotechnical behavior of these materials during or after subaqueous construction. While regulatory concerns have driven extensive research in chemical and biological characterization of sediments, most physical studies have dealt simply with hydraulic concerns. The variety of material types and in situ conditions encountered in dredging significantly affects the postdredging geotechnical behavior of capped deposits and should be considered in any consolidation, stability, and/or erosion analysis at nearshore and offshore placement locations. The objective of this research is to develop geotechnical criteria for underwater mound construction and performance evaluation by analyzing the interdependent soil mechanics processes that control mound behavior during construction, immediately after construction (often the most critical time for geotechnical materials), and in the long term.

This research will provide geotechnical design and construction criteria for dredged material deposits constructed of typical types of dredged material. The criteria will provide guidance on maximum rates of material placement to maintain stability and bearing capacity, and thickness limitations dictated by type and consistency of construction materials and site conditions. This research will also address slope angles required to maintain stability both during and after construction. When dredged material deposit behavior is quantified, timing of various phases of construction can be optimized, and deposits can likely be constructed more cost-effectively, with less equipment, time, and manpower, and with increased long- and short-term integrity.

INNOVATIVE DREDGING TECHNOLOGY

In the past, there has not been a programmatic approach to insure the adequate demonstration, monitoring, evaluation, and reporting of new innovative dredging technology applications as they are identified. This focus area is responding to important targets of opportunity to demonstrate new and innovative equipment and process technologies to potential users (Figure 15.5). Technologies to be demonstrated are not limited to dredge equipment enhancements for removing material, but may also include physical mechanisms for handling and treating materials that require special attention due to their composition or chemical content. These new technologies may be applicable to the dredging site, in the transport and/or treatment process, or at the ultimate disposal and/or storage containment region. Additionally, the new techniques may arise from either government or private sources, and may consist of an activity being utilized only in a specific location but which might have national application with appropriate technology transfer and dissemination. (The telescoping weir of Figure 15.5 presently being used by the Corps' Norfolk District has been demonstrated to Corps dredging operations personnel from around the nation for potential adoption in other locations.) Innovative dredging equipment and operations procedures developed by both domestic and foreign dredging interests are being identified for possible demonstration.

Technologies with the highest cost savings potential are being evaluated by a review committee composed of USACE operations and maintenance personnel, research and development personnel, and Corps field representatives. The highest priority equipment and techniques are being demonstrated. Dredging technology demonstration ideas received from USACE field operating agencies and HQUSACE are being catalogued and will be prioritized via a ranking process for implementation. Table 15.5 indicates that the one research task of this focus area will continue for the life of the program. A brief description of this research task follows.

FIGURE 15.5. Telescoping weir, innovative dredging technology used by the U.S. Army Engineer District, Norfolk, to dewater a confined disposal facility (photo courtesy Elka Briuer, WES).

Demonstrations of Innovative Dredging Technologies

During the course of a research program designed to extend over eight years, many technological opportunities will be encountered that are outside the scope of work as finalized. A program that can readily respond to new equipment, ideas, and techniques should be implemented to demonstrate these technologies to potential users. The objectives of this effort are to (a) identify and evaluate dredging technologies currently not in wide use by the Corps, (b) conduct demonstrations of attractive dredging technologies or operations in cooperation with USACE field offices or sister agencies, and (c) provide documentation of the results in written reports and readily accessible technology resource data bases.

Dredging technology demonstration ideas will be evaluated for consideration as demonstration projects. Technologies with the highest potential for cost savings will be evaluated by a review group composed of HQUSACE O&M and R&D personnel and

TABLE 15.5. Innovative dredging technology—research task schedule

	1997	1998	1999	2000	2001	2002	2003	2004
Demonstrations of innovative dredging technologies[a]	****	****	****	****	****	****	****	****

[a]Demonstrations will be scheduled and conducted throughout the life of the program as opportunities arise.

ECOLOGICAL RISK MANAGEMENT FOR DREDGING AND DISPOSAL PROJECTS

The specter of adverse ecological impacts has become an increasingly important consideration in the dredging manager's decision-making process, and is often an intractable impediment. Ecological impacts most often cited include (a) effects of sediment-associated contaminants on aquatic, wetland, and terrestrial organisms (Figure 15.6), (b) human consumption of (dioxin-) contaminated fish and shellfish, (c) turbidity effects on anadromous fish, oysters, etc., (d) entrainment of valued species (Dungeness crabs, sea turtles, etc.), (e) habitat destruction/disturbance (nesting shorebirds), and (f) diminished water quality (high ammonia and low dissolved oxygen). Research is providing tools for the dredging manager to quantify the probability or likelihood of these purported impacts.

Existing guidance is being evaluated for incorporation into a broader environmental risk assessment framework. Input is being obtained from experts in the areas of contaminated sediment management and environmental risk assessment. Field demonstrations of risk-based tools and newly developed methods are being conducted to verify their utility

FIGURE 15.6. Sediment-associated contaminant effects on the food chain (illustration courtesy of Todd Bridges, WES).

TABLE 15.6. Ecological risk management for dredging and disposal projects—research tasks schedule

	1997	1998	1999	2000	2001	2002	2003	2004
Risk assessment framework for management of dredged material	****	****	****	****	****			
Risk characterization of dredged material		****	****	****	****			
Decision support tools for environmental risk assessment				****	****	****	****	
Integrating ecological risk assessment and economic factors					****	****	****	

for assisting the dredging manager in making difficult, cost-effective decisions on controversial projects. Existing information (e.g., residue-effects data) and assessment tools (e.g., trophic transfer models and risk-based decision support tools) are being enhanced to facilitate risk-based assessments. Output from risk assessment will lead to risk management where incremental reductions in risk for a range of actions/alternatives (i.e., CDFs, capping, etc.) can be evaluated in terms of dollars expended. Table 15.6 shows the planned time line for conducting the research tasks of this focus area. Brief descriptions of these research tasks follow.

Risk Assessment Framework for Management of Dredged Material

To optimize available funding, a dredging manager evaluates critical factors (e.g., suitability of material, cost, availability of equipment, windows, disposal options, consequences of not dredging), and makes a decision regarding which project to dredge. This is risk management. However, the dredging manager often lacks repeatable, defendable tools to quantify the process; the information is not consistent with existing frameworks (Environmental Protection Agency/National Research Council paradigms). This research is developing an overall framework for applying risk assessment to the Corps' dredging and disposal program. Existing guidance is being identified and evaluated. Information gaps are being identified and prioritized to ensure that future R&D is directed toward areas of greatest need. The framework under development will demonstrate the usefulness of the risk assessment process for assisting Corps dredging managers in making difficult decisions on controversial, closely ranked projects.

The ability to provide defendable, quantifiable support for decisions will (a) improve the execution of dredging and disposal at controversial projects (e.g., those involving contaminated sediments, conflicting disposal options, marginal projects), (b) reduce disagreements with resource agencies, local port authorities, states, Congressional staff, etc., and (c) increase the Corps' credibility with other agencies that embrace risk management techniques (e.g., EPA).

Risk Characterization of Dredged Material

Characterizing the environmental risk posed by dredged material requires information on the likelihood that organisms will be exposed to contaminants and the probability that such exposure will produce adverse effects in the environmental receptors (organisms) of concern. Uncertainty associated with these estimated risks has been high. This high uncertainty has led to disputes with other regulatory and resource agencies. Objectives of this research are to develop quantitative tools for integrating exposure and effects data during characterization of risk that will reduce the level of uncertainty associated with estimated risks.

Those elements of the dredged material evaluation process that contribute to this uncertainty are being identified and prioritized by their degree of contribution to the uncertainty; from this, quantifiable solutions are being developed. Techniques for describing both the spatial and temporal aspects of contaminant exposure at dredged material disposal sites will be available. Modeling techniques will define the potential for contaminant trophic transfer resulting in adverse effects on higher trophic level organisms. The ability to provide defendable, quantifiable support for dredged material management decisions will improve the execution (fewer delays and lower cost) of dredging and disposal. Disagreements with resource agencies, local port authorities, and state agencies will be reduced, and the Corps' credibility with other agencies that embrace the risk paradigm (e.g., EPA) will be increased. Reducing uncertainty will benefit the Corps' national dredging program by reducing costs and delays.

Decision Support Tools for Environmental Risk Assessment

Environmental risk assessment can be a complex process requiring multidisciplinary expertise. To facilitate initial screening-level assessments, PC-based risk assessment decision support tools have been developed and applied to numerous land sites for the evaluation of both human and ecological risks from exposures to hazardous and radioactive wastes. Although these decision support tools have proven successful in providing site-specific risk estimates for human health and potential ecological impacts at superfund sites, they have not been adapted for use in evaluating the potential impacts from dredging operations. These tools will require modification before they can be adapted for estimating risks associated with dredging/disposal operations. The objective of this research is to develop a PC-based risk assessment decision support tool for evaluating the effects of dredging and disposal operations on human and ecological resources through modification of existing technology.

Potential risk assessment decision support tools include those that have been developed to meet the remedial investigation/risk assessment/feasibility study (RI/RA/FS) process under the Comprehensive Environmental Response Compensation and Liability Act (CERCLA) of 1980 and Superfund Amendments and Reauthorization Act of 1986. The PC-based tools developed for the RI/RA/FS process were designed primarily for evaluating risks associated with cleanup activities at hazardous waste sites. Two existing tools that may potentially serve as building blocks include the Multimedia Environmental Pollutant Assessment System (MEPAS) and Remedial Action Assessment System (RAAS). At the completion of this research, a risk assessment decision support tool will be available that can be used to conduct screening-level risk assessments for the evaluation of dredge/disposal management options. The final product will aid in the design and planning process by providing quick, cost-effective, technically sound, screening-level estimates of environmental risk.

Integrating Ecological Risk Assessment and Economic Factors

Risk assessment provides the dredged material manager with a logical, repeatable, and technically defensible approach for evaluating the ecological risk associated with a range of potential dredged material management options. This information must then be integrated along with other factors impacting on the manager's decision making environment. Such factors include economic costs/benefits associated with selected management options. Often these costs/benefits are not explicit, are difficult to quantify (e.g., cost recovery via beneficial use, avoidance of litigation, etc.), and weigh against incremental differences in ecological risk. Objectives of this research are to identify and incorporate appropriate economic cost/benefit forecasting tools into the risk management decision making process. The costs/benefits associated with a particular management option will be explicitly defined and include measures of uncertainty around projected estimates.

Selected forecasting tools will be modified as necessary to integrate with ecological risk such that dollars saved/expended are expressed in terms of incremental reductions in risk. Once fully developed, a PC-based risk management decision support tool will be demonstrated on selected dredging projects. The ability to provide defendable, quantifiable support for decisions will improve the execution of dredging and disposal on controversial projects (e.g., those involving contaminated sediments, conflicting disposal options, marginal projects), reduce disagreements with resource agencies, local port authorities, and states, and increase the Corps' credibility with other agencies that embrace risk management techniques (e.g., EPA).

BENEFITS OF THE DOER PROGRAM

Benefits of the DOER program include cost-effective practices for dredging and dredge material disposal, environmental protection enhancement through application of effective environmental windows, compliance with environmental statutes for identifying and managing contaminated sediments, reduction of costs of disposal of dredged material by beneficial placement in the nearshore zone, greater flexibility in dredging in sensitive ecological areas, and expanded options for beneficial uses of contaminated and noncontaminated dredged materials. This research is being conducted with full coordination and cooperation of other appropriate agencies, including the EPA, the U.S. National Marine Fisheries Service, and the U.S. Fish and Wildlife Service. The DOER program includes an aggressive technology transfer mechanism to ensure rapid implementation of research products into Corps projects.

ACKNOWLEDGMENTS

Permission to publish this paper was granted by the Chief of Engineers, Headquarters, U.S. Army Corps of Engineeers, Washington D.C.

APPENDIX A: GENESIS OF THE DREDGING OPERATIONS AND ENVIRONMENTAL RESEARCH (DOER) PROGRAM

In the 1960s, as sediments in many waterways and harbors became polluted, concern developed that dredging and disposal of this material would adversely affect water quality or aquatic organisms. A number of localized studies were conducted to investigate the environmental impact of specific disposal practices and to explore alternative disposal methods. However, these studies did not provide sufficient definitive information on the environmental impact of then-current disposal practices nor did they fully investigate alternative disposal methods. Significant changes occurred in the conduct of U.S. dredging operations, and the coordination of such dredging with environmental protection agencies, as a result of the National Environmental Policy Act of 1969. The Corps of Engineers was authorized by Congress in the 1970 River and Harbor Act to initiate a comprehensive nationwide study to provide more definitive information on the environmental impact of dredging and dredged material disposal operations, and to develop new or improved dredged material disposal practices. That study, conducted during the period June 1971 through April 1972, concluded that a broad-based program of research was needed to develop the widest possible choice of technically satisfactory, environmentally compatible, and economically feasible disposal practices. Thus, the Dredged Material Research Program (DMRP) was born in 1973 and, for the first time in dredging research, definitive knowledge was developed pertaining to the impacts of dredged material disposal and methods to ameliorate any resulting adverse effects.

Dredged Material Research Program (DMRP)

A basic premise of the DMRP, a 6-year $30 million program initiated by WES in 1973 and completed in 1978, was that dredging of large volumes of material would continue because of a lack of alternatives in creating and maintaining navigable waterways for the nation's economic development. Another premise was that concern over the environmental impact of man's activities would result in significant controls over methods used to dredge and dispose of that material. Each potential disposal practice would be developed to the point where it could be demonstrated and documented as a working system.

Much of the concern over potential adverse environmental impacts of disposal operations related to the practice of open water disposal of "polluted" material, and it was acknowledged that there must be constraints on the type of material that could be disposed in open water. However, there were strong reservations within the scientific community over the then-current tendency to use the chemical composition of dredged sediments as the sole indicator of pollution status. Research was needed to develop a better understanding of the nature and magnitude of the effects on water quality and aquatic organisms due to varying methods of disposal and types of material deposited. Constraints relating to open water disposal should be based on sound technical information, and should allow for periodic review and update based on the natural regional variability found throughout the nation's marine and fresh waters.

Many lines of evidence indicated that dredged material could be considered a manageable resource. Numerous cases of beneficial environmental effects of disposal operations could be documented, and these were examined from the standpoint of identifying those aspects needing research to permit widespread application. Research was conducted into the regional practicality and environmental effects of beneficial open water disposal practices such as marsh creation, island development for wildlife habitat, and beach nourishment. Deep water disposal was an attractive alternative disposal method but should first be the subject of research to answer basic questions regarding biologic effects of pollutants in a largely unknown environment.

As a result of the concern over open water disposal of polluted material, it was believed there would be more land disposal of dredged material in future years. Most of the material to be disposed on land would come from highly developed areas (e.g., ports, harbors) where land disposal sites were and would increasingly be more difficult to obtain. Expanded land disposal would probably require barge or pipeline systems that could transport the material long distances, and/or the development of techniques to permit the use of coastal disposal areas (confined disposal facilities). Considerations such as the environmental impact of land disposal and technical problems related to design, construction, operation, and utilization of land disposal sites were addressed by the DMRP.

The DMRP was designed for national application, including research on all major types of dredging activities, all regions of the country, and all types of environmental settings. Results were first-generation procedures for evaluating the physical, chemical, and biological impacts of a variety of disposal alternatives in water, on land, and in wetland areas. Research resulted in cost-effective methods and guidelines for assessing and minimizing the impacts of conventional disposal alternatives. At the same time, DMRP activities demonstrated viability and limits of new disposal alternatives, including the beneficial use of dredged material in areas such as wetlands. New or improved procedures for designing, constructing, and managing confined disposal areas to maximize service life and minimize adverse environmental impact were developed. Procedures to predict and minimize turbidity from dredging material disposal operations, and methods to predict movement of dredged material in the aquatic systems, also were products of the DMRP.

The DMRP determined that no single disposal alternative was most suited for a region or a type of project. Conversely, there was no single disposal alternative that could be dismissed as environmentally unsatisfactory due to potential impacts. From a technical standpoint, it was found there was no inherent effect or characteristic of an alternative disposal method that precludes it consideration before specific site assessment. That conclusion held for ocean disposal, confined disposal, or any other alternative. To address a variety of environmental factors and considerations adequately, long-range planning was required for effective disposal of dredged material. Through use of disposal management plans that consider project types, dredged material characteristics, disposal alternatives, and other factors, the best opportunity would exist for maximum environmental protection at an acceptable cost.

Dredging Operations Technical Support (DOTS) Program

The Dredging Operations Technical Support (DOTS) Program was established at WES in 1978 initially to provide comprehensive and interdisciplinary technology transfer/technology application of DMRP products to all Corps dredging operations projects. Subsequently and to the present time, DOTS has been a mechanism for providing support to Corps headquarters and field offices in applying state-of-the-art dredging technology toward implementation of the Clean Water Act (CWA); Marine Protection, Research, and Sanctuaries Act (MPRSA); Water Resources Development Acts (WRDA); Resource Conservation and Recovery Act (RCRA); and Toxic Substances Control Act (TSCA). DOTS is managed from a centralized program to maximize cost effectiveness and help implement national policies, laws, and complex technical requirements on a consistent basis. Additionally, short-term applied research efforts are frequently developed to address technical dredging and dredged material disposal problems.

DOTS provides training to Corps staff that includes the latest environmental and engineering techniques associated with dredging and dredged material disposal management, water quality and related aquatic environmental issues, and technology transfer of new and emerging techniques used to determine compliance with environmental protection statutes regarding management of dredged material. DOTS personnel have contributed to the development and preparation of manuals that implement inland and ocean disposal regulations. DOTS also provides for the (a) transfer of technology developed by the Corps and others that deal with the treatment and decontamination of highly contaminated sediments, (b) application and transfer of innovative risk-based technologies that identify, assess, and manage sediments contaminated with high-profile chemicals such as dioxins, polyaromatic hydrocarbons, and polychlorinated biphenyls, and (c) assessment and management protocols for beneficial uses of dredged material, protection of threatened and endangered species, and general environmental management guidance. Generic technology is presented by DOTS in the form of field-ready decision support systems. Data bases, research information, and publications can be located through the DOTS web site at http://www.wes.army.mil/el/dots/.

Field Verification Program (FVP)

The Corps of Engineers and the Environmental Protection Agency (EPA) Interagency Field Verification of Testing and Predictive Methodologies for Dredged Material Disposal Alternatives Program, referred to as the Field Verification Program (FVP), investigated methods for predictive evaluation of dredged material disposal alternatives. Research focused on comprehensive evaluation and comparison of environmental effects of highly contaminated dredged material placed in upland, wetland, and aquatic environments. The Corps and the EPA conducted the $7 million FVP over the 6-year period 1982–1987.

Samples of sediment for use in laboratory studies were collected from the industrialized Black Rock Harbor channel, Bridgeport, CT, prior to dredging. The channel was then dredged, and portions of the material were (a) placed in an upland disposal site, (b) placed in an aquatic disposal site, and (c) used to create a wetland. Predisposal studies were conducted in the laboratory. These laboratory results were then compared with results of the same techniques applied in the field after disposal. These studies provided a basis for determining (a) the reproducibility of the test using dredged material in the laboratory, (b) the ability of the laboratory test methods to predict effects in the field, and (c) the comparative effects of the same contaminated dredged material in upland, aquatic, and wetland environments. The test methods had previously been developed by the Corps, EPA, and the European Economic Commission, but had not been evaluated for dredged material predictive accuracy.

Results showed that laboratory methods for predicting effluent, surface water quality, and plant toxicity in upland disposal sites compared well with field data. The techniques for predicting effluent and surface water quality were shown to have good utility for predisposal evaluations of dredged material proposed for upland disposal. Methods for testing toxicity and bioaccumulation in plants in the wetland environment showed good predictive ability. However, optimum utility for predictive evaluations for animal bioassays requires further confirmation of the reproducibility. Techniques with good utility for predisposal evaluation of dredged material proposed for aquatic disposal include toxicity, bioaccumulation, intrinsic rate of population increase, and scope of growth.

Methods for predicting aquatic impacts were shown to have good utility for predisposal evaluations. In general, the effects of aquatic disposal predicted in the laboratory and observed in the field were less persistent than in the other two environments. Wetland creation showed greater effects than aquatic disposal. Upland disposal produced the greatest and most persistent impacts. This is compatible with expectations based on the physicochemical behavior of contaminated dredged material in the three environments.

The same ranking of effects in the upland, wetland, and aquatic environments can be expected in similar situations although the relative magnitude of effects may be different.

Long-Term Effects of Dredging Operations (LEDO) Program

The LEDO Program focuses on cost-effective, environmentally responsible techniques for dredging and dredged material disposal in aquatic, wetland, and upland environments. The long-term effects of contaminated dredged material disposal are not well known, and assessment techniques are either undeveloped or in early stages of development. Environmental, legislative, and public interest groups' concerns with regard to Corps dredging programs have become more intense, and have resulted in state and federal laws and regulations to address these concerns. LEDO was established at WES in 1982. Current research emphasizes risk-based procedures for effects assessment, exposure assessment, and risk characterization. The LEDO objective is to provide the latest proven technologies for identifying, quantifying, and managing contaminated sediments in support of the Corps's congressionally mandated navigation mission.

The primary LEDO benefit is a more timely, complete, and cost-effective execution of the Corps' responsibilities under the CWA, MPRSA, WRDA, RCRA, and TSCA. LEDO is providing significant cost avoidances for achieving restoration goals for contaminated dredged material disposal as mandated by law. Recent major LEDO products include a chronic sublethal genotoxic assay that meets national regulatory requirements, and the Environmental Residue- Effects Database (ERED), which is the first Internet-accessible data base of bioaccumulation effects to improve the accuracy and defensibility of environmental impact predictions while significantly reducing evaluation costs. Recent costs for the disposal of contaminated dredged material from New York Harbor could have been reduced by 90% if the ERED had been then available. The ERED can be found on the Internet at http://www.wes.army.mil/el/ered/index.html.

Dredging Research Program (DRP)

A long period in which the Corps' dredging activities were almost totally concerned with maintaining existing waterways and harbors ended with passage of the Water Resources Development Act of 1986. That legislation authorized major deepening and widening of existing navigation projects. Future changes in dredging were not expected to be any less dramatic than those that had occurred in recent years. It was believed the Corps would continually be challenged in pursuing optimal means of performing it's dredging activities. Implementation of an applied R&D program to meet demands of changing conditions, and generation of significant technology to be adopted by all dredging interests, would be means of reducing the cost of dredging the nation's waterways and harbors. The 7-year, $35 million Dredging Research Program (DRP) was initiated at WES in 1988 and successfully concluded in 1994, having achieved all major goals established by field users during the program development process.

The DRP had as major objectives the development of equipment, instrumentation, software, and operational monitoring and management procedures to significantly enhance the Corps' dredging activities. The DRP's purpose was to reduce the cost of dredging to a minimum consistent with Corps mission requirements and environmental responsibility. The DRP consisted of five technical areas from which distinct products could be developed, and where annual and one-time direct and indirect benefits could be quantified. The five technical areas of the DRP included (1) analysis of dredged material disposed in open water, (2) material properties related to navigation and dredging, (3) dredge plant equipment and equipment processes, (4) vessel positioning, survey controls, and dredge monitoring systems, and (5) management of dredging projects.

DRP Technical Area 1 results included (a) better understanding of boundary-layer properties of both cohesive and noncohesive sediments, (b) development of instrumentation for monitoring suspended-sediment plumes and for determining resuspension from disposal mounds, (c) formulation and verification of short- and long-term numerical models for determining fate of material disposed in open water, and (d) improved field techniques for collecting long-term data pertaining to movement of nearshore berms designed as feeder beach material or wave-attenuation devices. Technical Area 2 (a) developed descriptors for bottom sediments to be dredged, (b) devised new drilling parameter recorder technology, (c) designed and field-tested a methodology for surveying navigation channels containing fluid mud, (d) and developed acoustic impedance techniques for performing rapid measurements of characteristics of consolidated sediments. Technical Area 3 (a) developed new technologies and enhanced existing dredge systems to achieve economic hopper dredge load during dredging operations, (b) improved eductors, (c) developed a single-point mooring system for direct pumpout of hopper dredges onto beaches, (d) enhanced technology for dredge production and process monitoring, and (e) modified designs for hopper-dredge dragheads to increase production. Technical Area 4 (a) addressed the need for reporting offshore onsite water levels, (b) enhanced a vessel-heave system for dredging purposes, (c) developed real-time, on-the-fly GPS navigation for surveying and dredging, (d) evaluated dredge production meters, and (e) developed the operational aspects of a silent inspector for monitoring contract hopper dredges. Technical Area 5 (a) developed a framework for comprehensive site management of the open-water placement and monitoring of dredged sediments, (b) designed criteria for level-bottom capping and contained aquatic disposal of contaminated dredged materials, and (c) provided engineering design guidance for nearshore berms constructed of clean sediments.

The DRP concentrated primarily on the physical aspects of handling dredged material. It became apparent during the conduct of the program that significant and substantial uncertainties pertaining to the chemistry and biology of dredged material management should be pursued by a formally structured R&D program. This understanding resulted in the development and execution of the 8-year, $48 million Dredging Operations and Environmental Research (DOER) Program at WES in October 1996.

Synopsis

Since 1972, the Corps of Engineers has invested over $150 million through the Waterways Experiment Station in dredging and dredged material disposal research. This research has focused on nearly all aspects of the physical, chemical, and biological systems involved. WES dredging research has played a vital role in identifying environmentally appropriate dredged disposal alternatives. Research has been directed at all facets of sediment management, from beneficial uses (e.g., wetlands establishment, manufactured soil development, etc.), to design and management of contained disposal facilities, to the toxicology and management of contaminated sediments, to equipment selection for speciality dredging, to dredging instrumentation for environmental purposes, etc. Much of this research has provided the scientific basis for the Corps and the Environmental Protection Agency to classify sediments according to contamination potential and to regulate dredged material in a cost-effective and environmentally responsible manner. The ultimate objective of this research is to provide meaningful scientific information that will help decision-makers choose among alternatives and make better-informed and more scientifically based decisions.

APPENDIX B: DOER TECHNICAL REPORTS AND TECHNICAL NOTES

Contaminated Sediment Characterization and Management

"Guidance for Subaqueous Dredged Material Capping," Technical Report DOER-1, June 1998, M. R. Palermo, J. E. Clausner, M. P. Rollings, G. L. Williams, and T. E. Myers, U.S. Army Engineer Waterways Experiment Station, Vicksburg, MS; T. J. Fredette, U.S. Army Engineer Division, New England; and R. E. Randall, Texas A&M University, College Station, TX.
"Guidance for Performance of the H4IIE Dioxin Screening Assay," Technical Note DOER-C1, February 1998, V. A. McFarland, U.S. Army Engineer Waterways Experiment Station, Vicksburg, MS; and D. D. McCant and L. S. Inouye, AScI Corporation, Vicksburg, MS.
"Dredged Material Screening Tests for Beneficial Use Suitability," Technical Note DOER-C2, February 1998, L. E. Winfield and C. R. Lee, U.S. Army Engineer Waterways Experiment Station, Vicksburg, MS.

The following Technical Reports and Technical Notes are scheduled to be prepared during the DOER program:

"Determining Material Recovery Potential at CDFs," Technical Report.
"Cost/Benefit Analysis in Solids Separation," Technical Report.
"Bench Scale Characterization Protocol for Contaminated Sediments," Technical Report.
"Comparison of Two Cell-based Assays for Screening Dioxin and Dioxin-like Compounds in Sediments," Technical Note.
"Evaluation of Dredged Material for Phytoreclamation Suitability," Technical Note.
"Screening Test for Assessing the Bioreclamation of Dredged Material," Technical Note.
"Bioremediation of PAH-Contaminated Dredged Material at the Jones Island CDF: Materials, Equipment, and Initial Operations," Technical Note.
"Design Concepts for Bioremediation in CDFs," Technical Note.
"Surface Management Practices for Enhancing Natural Decontamination at CDFs," Technical Note.
"Methods for Enhancing Intrinsic, Aerobic Biodegradation at CDFs," Technical Note.
"Field Operating Experiences with CDF Containment Measures," Technical Note.
"Design and Construction Guidance for CDF Containment Measures," Technical Note.
"Considerations for Leachate Screening Evaluation," Technical Note.
"Methodology for Leachate Screening for Confined Disposal Facilities," Technical Note.
"Guidance for Leachate Screening for Confined Disposal Facilities," Technical Note.
"Potential for Phytoreclamation of Contaminated Dredged Material, Results of a Working Group," Technical Note.
"Vegetation Management for Desirable Plant Species," Technical Note.
"Soil Conditions that Optimize Metal Uptake," Technical Note.
"Soil Conditions that Optimize Organic Degradation," Technical Note.
"Genetically Engineered Zn/Cd Hyper-accumulator Plants," Technical Note.
"Plant/Rhizome Associations that Degrade PAH Contaminants," Technical Note.

"Guidance for Conduct of Cell-based and Microbial Toxicity Screening Assays," Technical Note.
"Non-extractive ELISAs for Screening Sediment Contaminants," Technical Note.
"Field Validation of Sediment Screening Assays," Technical Note.
"A priori Estimation of Effluent Losses—CDFs," Technical Note.
"A priori Estimation of Runoff Losses—CDFs," Technical Note.
"A priori Estimation of Leachate Losses—CDFs," Technical Note.
"A priori Estimation of Volatile Losses—CDFs," Technical Note.
"A priori Estimation of CAD Losses," Technical Note.
"A priori Estimation of Dredging Losses—CDFs," Technical Note.
"Application of Plant and Animal Bioassays in an Upland Risk Assessment Framework," Technical Note.
"Field Validation of Efficacy of Using a Bioassay Organism to Predict Contaminant Uptake by Naturally Colonizing Organisms," Technical Note.
"Field Verification at Selected Freshwater CDFs," Technical Note.
"Field Verification of Estuarine Dredged Material at Selected CDFs," Technical Note.
"Field Verification Implementation Guidance for Predicting Surface Runoff Water Quality from CDFs," Technical Note.
"Testing Dredged Materials with Chronic Sublethal Bioassays," Technical Note.
"Field Demonstration and Validation of Chronic Sublethal Bioassays," Technical Note.
"Predictive Quality and Interpretive Guidance for Chronic Sublethal Bioassays," Technical Note.
"Field Evaluation of Volatile Losses from Exposed Sediment," Technical Note.
"Field Evaluation of Volatile Losses During Dredging and Disposal," Technical Note.
"Evaluation of the Effects of Vegetative Cover on Volatile Losses," Technical Note.
"Guidance for Selecting Biomarkers of Effect Screening Tools," Technical Note.
"Efficacy of Biomarkers of Effect in Predicting Whole Organism Response," Technical Note.
"Guidance for Application and Interpretation of Biomarkers of Effect in Dredged Material Evaluations," Technical Note.
"Guidance for Evaluating Suitability of Dredged Material in CDFs for Aquatic Disposal: Freshwater," Technical Note.
"Guidance for Evaluating Suitability of Dredged Material in CDFs for Aquatic Disposal: Marine," Technical Note.
"Suitability of Reclaimed Material for Aquatic Disposal," Technical Note.
"Estimation of Bioturbation Depths," Technical Note.
"Factors in Long-Term Cap Integrity," Technical Note.
"Improved Laboratory Tests for Cap Design," Technical Note.
"Improved Models for Cap Design," Technical Note.
"Laboratory Verification of Cap Design," Technical Note.
"Field Verification of Cap Design," Technical Note.
"Field Demonstration and Verification of Manufactured Soil from Selected Freshwater CDFs," Technical Note.
"Field Demonstration and Verification of Manufactured Soil from Selected Estuarine CDFs," Technical Note.
"Field Verification Implementation Guidance for Reclamation and Reuse of Contaminated Dredged Material as Manufactured Soil," Technical Note.
"Performance Data for Filtration Systems at CDFs," Technical Note.
"Media for Contaminant Removal from CDFs," Technical Note.
"Guidance for Designing Contaminant Removal Systems at CDFs, Technical Note.
"Impacts of Filling Practices on Intrinsic Bioremediation at CFS," Technical Note.

"Relative Benefits of Optimized Filling Practices at CDFs," Technical Note.
"Best Filling Practices for Enhancing Bioremediation at CDFs," Technical Note.
"Potential for Anaerobic Biodegradation in CDFs," Technical Note.
"Methods for Accelerating Anaerobic Biodegradation in CDFs," Technical Note.
"Cost Effective Management Practices for Enhancing Intrinsic Anaerobic Biodegradation in CDFs," Technical Note.
"Refined Computer Programs for Resuspension Predictions," Technical Note.
"Guidance on Use of Silt Curtains during Dredging," Technical Note.
"Development of Control Measures for Contaminant Losses during Dredging," Technical Note.
"Field Verification Results for Control Measures for Contaminant Losses during Dredging," Technical Note.
"Tropic Transfer Evaluations for Aquatic Disposal," Technical Note.
"Tropic Transfer Modeling during Evaluations of Sediment-associated Contaminants in Dredged Material," Technical Note.
"Guidance for Evaluating Trophic Transfer using Models and the ERED," Technical Note.

Environmental Windows for Dredging Operations

"Entrainment by Hydraulic Dredges—A Review of Potential Impacts," Technical Note DOER-E1, October 1998, K. J. Reine and D. G. Clarke, U.S. Army Engineer Waterways Experiment Station, Vicksburg, MS.
"Environmental Windows Associated with Dredging Operations," Technical Note DOER-E2, December 1998, K. J. Reine, D. D. Dickerson, and D. G. Clarke, U.S. Army Engineer Waterways Experiment Station, Vicksburg, MS.
"Economic Impact of Environmental Windows Associated with Dredging Operations," Technical Note DOER-E3, December 1998, D. D. Dickerson, K. J. Reine, and D. G. Clarke, U.S. Army Engineer Waterways Experiment Station, Vicksburg, MS.
"FISHFATE: Population Dynamics Models to Assess Risks of Hydraulic Entrainment by Dredges," Technical Note DOER-E4, December 1998, J. S. Ault and K. C. Lindeman, Rosenstiel School of Marine and Atmospheric Science, University of Miami, Miami, FL; and D. G. Clarke, U.S. Army Engineer Waterways Experiment Station, Vicksburg, MS.
"Evaluation of Dredged Material Plumes—Physical Monitoring Techniques," Technical Note DOER-E5, December 1998, P. T. Puckett, General Engineering Laboratories, Charleston, SC
"Estimating Dredging Sediment Resuspension Sources," Technical Note DOER-E6, March 1999, B. H. Johnson and T. M. Pachure, U.S. Army Engineer Waterways Experiment Station, Vicksburg, MS.

The following Technical Reports and Technical Notes are scheduled to be prepared during the DOER program:

"Technical Framework and Operational Guidelines for Evaluating and Minimizing Windows," Technical Report.
"Development and Verification of the Numerical Modeling System of Suspended Sediment and Turbidity Plumes," Technical Report.
"Sediment Resuspension During Dredging Operations—A Review of Potential Impacts to Fish and Shellfish," Technical Note.
"Preliminary Applications of FISHFATE," Technical Note.

"Entrainment of Larval Fish and Shellfish," Technical Note.
"Entrainment of Juvenile Salmonids," Technical Note.
"Entrainment of Juvenile Sturgeon," Technical Note.
"Sampling Protocols for Determining Fish Behavior in Relation to Turbidity Plumes," Technical Note.
"Responses of Anadromous Fishes to Turbidity Plumes—Regional Demonstration I," Technical Note.
"Responses of Anadromous Fishes to Turbidity Plumes—Regional Demonstration II," Technical Note.
"Responses of Anadromous Fishes to Turbidity Plumes—Regional Demonstration III," Technical Note.
"Description of the System for Numerical Modeling of Suspended Sediment and Turbidity Plumes," Technical Note.
"Demonstration of the System for Numerical Modeling of Suspended Sediment and Turbidity Plumes," Technical Note.
"Verification of the Numerical Model of Suspended Sediment and Turbidity Plumes," Technical Note.
"Monitoring of Ambient and Dredging-related Suspended Sediment and Turbidity Plumes during Migratory Studies," Technical Note.
"Monitoring of Storm, River Discharge, Trawling, and Ship Traffic Sources of Sediment Resuspension—Results of Field Studies," Technical Note.
"Monitoring of Ambient and Dredging-related Suspended Sediment and Turbidity Plumes during Investigations of Effects on Early Life History Stages," Technical Note.
"Monitoring Methods for Determination of Near-bottom Background and Dredging-related Suspended Sediment Concentrations and Turbidity," Technical Note.
"Effects of Natural and Anthropogenic Sources of Sedimentation of Aquatic Communities," Technical Note.
"Differential Seasonal Susceptibility on Aquatic Biota to Adverse Effects of Sedimentation," Technical Note.
"Effects of Dredged Material Placement on Benthic Primary Producers," Technical Note.
"Effects of Dredged Material Placement on Benthic Fishery Resources," Technical Note.
"Differential Growth as a Measure of Suspended Sediment Exposure Impact," Technical Note.
"Interactions of Dredging Conditions and Growth of Critical Life History Stages," Technical Note.
"Effects of Dredging Projects on Growth and Survival of Fishery Resources," Technical Note.
"Population Dynamics Consequences of Differential Growth Rates," Technical Note.

Instrumentation for Dredge and Site Monitoring

The following Technical Reports and Technical Notes are scheduled to be prepared during the DOER program:

"Use and Capabilities of the Mechanical Dredge Monitoring System," Technical Report.
"Corps Experience with Pipeline Dredge Monitoring," Technical Note.
"Field Experience with the Pipeline Dredge Silent Inspector," Technical Note.
"Preliminary Results of the Tons Dry Solids (TDS) Investigation," Technical Note.
"Initial Field Experience with the Mechanical Dredge Silent Inspector," Technical Note.

Nearshore/Aquatic Placement of Dredged Materials

"LTFATE Cohesive Sediment Transport Model," Technical Note DOER-N1, April 1998, J. Z. Gailani, U.S. Army Engineer Waterways Experiment Station, Vicksburg, MS.

"Dredged Material Spatial Management, Analysis, and Record Tool (DMSMART)," Technical Note DOER-N2, April 1998, J. E. Clausner, U.S. Army Engineer Waterways Experiment Station, Vicksburg, MS.

"Planning Considerations for Nearshore/Aquatic Placement of Mixed Dredged Sediments," Technical Note DOER-N3, March 1998, G. L. Williams and T. L. Prickett, U.S. Army Engineer Waterways Experiment Station, Vicksburg, MS.

The following Technical Reports and Technical Notes are scheduled to be prepared during the DOER program:

"Siting, Designing, and Evaluation of Nearshore/Aquatic Placement," Technical Report.
"Algorithms for Estimating Vertical Variation in Erosion and Deposition Rates of Mixed Sediments," Technical Report.
"Model Development, Laboratory Experiments, Verification, and Simulation Results of Near-field Bottom Surge Due to Placement of Dredged Material," Technical Report.
"Sediment Mixture Flume Test Results and Conclusions," Technical Report.
"Shear Strength of Dredged Material," Technical Report.
"Parametric Analysis of Various Deposits," Technical Report.
"Design Criteria for Dredged Material Deposits," Technical Report.
"Framework for Geotechnical Testing of Dredged Materials," Technical Report.
"Bed Model for Estimating Sediment Fate in Estuaries," Technical Report.
"Enhancement of Modeling Tools for Multiple-constituent Transport," Technical Report.
"Fate of Dredged Material Placed on Channel Margins," Technical Report.
"Interim Tools for Siting and Evaluation of Nearshore/Aquatic Placement," Technical Note.
"Numerical Model for Grain Size Depth of Residence," Technical Note.
"Nearshore/Aquatic Placement Monitoring at Site 1," Technical Note.
"Updated Nearshore/Aquatic Placement Database," Technical Note.
"Updated Interim Siting and Evaluation," Technical Note.
"Final Nearshore/Aquatic Placement Database," Technical Note.
"Improved Shear-stress Model and Layered Sediment Bed Model for LTFATE," Technical Note.
"Nearshore Version of LTFATE," Technical Note.
"Winnowing Processes, Flocculation, and Deposition Rates," Technical Note.
"Algorithms for Cohesive Sediment Erosion/Deposition in LTFATE," Technical Note.
"District Requirements for DMSMART," Technical Note.
"Guidance for District Database Development," Technical Note.
"Results of Mobile Site Analysis," Technical Note.
"Results of First Site Analysis," Technical Note.
"Preliminary Erosion and Deposition Rate Algorithms that are less Site-specific," Technical Note.
"Results of Second Site Analysis," Technical Note.
"Results of Third Site Analysis," Technical Note.
"Sampling Design for Assessment of Nearshore/Aquatic Placement Impacts," Technical Note.
"Patterns of Use of Nearshore Habitats by Fishery Resources," Technical Note.
"Sensitive Habitat Susceptibility to Nearshore/Aquatic Placement Impacts," Technical Note.

"Environmental Assessment of Nearshore/Aquatic Placement Impacts," Technical Note.
"Effects of Chronic Turbidity on Fishery Resources in the Nearshore Zone," Technical Note.
"Design Guidance for Contained Aquatic Disposal," Technical Note.
"Pipeline Plume Behavior," Technical Note.
"Near-field Module for Dispersion from Pipeline Disposals," Technical Note.
"Verification of Near-field Simulation using Field Data," Technical Note.
"Mass Balance Instrument Development," Technical Note.
"Mass Balance Instrument Deployment Test Results," Technical Note.
"Preliminary Results from Demonstration of Far-field Fate Predictions for Subaqueous Dredged Material Placement," Technical Note.
"Application Demonstration of Offshore Model of Far-field Fate Predictions for Subaqueous Dredged Material Placement," Technical Note.
"Far-field Fate Predictions on the East and West Coasts, and the Great Lakes," Technical Note.
"Results from Laboratory Tests of Near-field Prediction of Bottom Surges Due to Placement of Dredged Material," Technical Note.
"Technical Workshop Summary and Recommendations for Entrainment and Transport of Sediment Mixtures Placed Nearshore," Technical Note.
"Validation Flume Tests Description and Data for Entrainment and Transport of Mixed Sediments," Technical Note.
"1-D Consolidation Model Evaluation of Subaqueous Dredged Material Deposits," Technical Note.
"Interim Design Criteria for Subaqueous Deposits," Technical Note.
"Construction Rates for Subaqueous Dredged Material Deposits," Technical Note.
"Analysis Procedure for Subaqueous Deposit Designs," Technical Note.
"In situ Expedient Test Methods of Geotechnical Properties of Dredged Materials," Technical Note.
"Variability of Geotechnical Properties," Technical Note.
"DMSMART Applications in the Dredged Material Decision Support System," Technical Note.
"New Programs in the Dredged Material Decision Support System," Technical Note.
"Dredged Material Decision Support System Applications," Technical Note.
"Application of Silent Inspector Data in the Dredged Material Decision Support System," Technical Note.
"Additional Dredged Material Decision Support System Applications," Technical Note.
"Demonstration of Open Water Beneficial Use Alternative I," Technical Note.
"Demonstration of Open Water Beneficial Use Alternative II," Technical Note.
"Demonstration of Open Water Beneficial Use Alternative III," Technical Note.
"Thin-layer Disposal of Dredged Material, Regional Demonstration I," Technical Note.
"Thin-layer Disposal of Dredged Material, Regional Demonstration II," Technical Note.

Innovative Dredging Technology

The following Technical Reports and Technical Notes are scheduled to be prepared during the DOER program:

"Innovative Dredging Technologies—Purpose and Scope," Technical Note.
"Developments in the European Dredging and Channel Maintenance Program," Technical Note.
"Innovations within the Corps of Engineers," Technical Note.

"Demonstration of Innovative Telescoping Weir for Confined Disposal of Dredged Material," Technical Note.

Ecological Risk Management for Dredging and Disposal Projects

"Environmental Risk Assessment and Dredged Material Management: Issues and Application—February 1998 Workshop Proceedings," Technical Report DOER-2, December 1998, D. W. Moore, T. S. Bridges, and C. Ruiz, U.S. Army Engineer Waterways Experiment Station, Vicksburg, MS.; J. Cara, S. K. Driscoll, and D. Vorhees, Menzie-Cara and Associates, Inc., Chelmsford, MA.; and Dick Peddicord, Dick Peddicord & Co., Inc., Parkton, MD.

"Improving Dredged Material Management Decisions with Uncertainty Analysis," Technical Report DOER-3, December 1998, D. J. Vorhees, S. B. K. Driscoll, and K. von Stackelberg, Menzie-Cara and Associates, Inc., Chelmsford, MA.; and T. S. Bridges, U.S. Army Engineer Waterways Experiment Station, Vicksburg, MS.

"Dredging/Dredged Material Management Risk Assessment," Technical Note DOER-R1, September 1998, T. S. Bridges, U.S. Army Engineer Waterways Experiment Station, Vicksburg, MS; D. W. Moore, MEC Analytical Systems, Carlsbad, CA; and J. Cara, Menzie-Cara & Associates, Inc., Chelmsford, MA.

The following Technical Reports and Technical Notes are scheduled to be prepared during the DOER program:

"Improving Dredged Material Management Decisions with Uncertainty Analysis," Technical Report.
"Use of Risk Assessment in Dredging and Dredged Material Management," Technical Note.
"Application of Risk Assessment to a Dredging Project (Field Demonstration)," Technical Note.
"Guidance on Using Uncertainty Analysis in Decision Making," Technical Note.
"Application of Uncertainty Analysis for Dredged Material Evaluations," Technical Note.
"Guidance for Considering Economic Factors in Risk Management," Technical Note.
McNair, DOER, 2 June 1999

CHAPTER 16
NUMERICAL MODELS FOR PREDICTING THE FATE OF DREDGED MATERIAL PLACED IN OPEN WATER

HANS R. MORITZ
*U.S. Army Corps of Engineers, Portland District,
Portland, Oregon*

**BILLY H. JOHNSON
NORMAN W. SCHEFFNER**
*U.S. Army Engineer Waterways Experiment Station,
Vicksburg, Mississippi*

Introduction	16.2
Processes Affecting Dredged Material Placed in Open Water	16.3
Short-Term Fate	16.3
Long-Term Fate	16.3
Numerical Modeling of Dredged Material Placed in Open Water	16.4
STFATE Concepts	16.5
Convective Descent	16.6
Dynamic Collapse in Water Column	16.7
Dynamic Collapse on Bottom	16.8
Transport–Diffusion Phase	16.8
Other Processes	16.9
Model Use	16.10
LTFATE Concepts	16.10
Data Bases	16.10
Hydrodynamics Module	16.11
Sediment Transport	16.12
Bathymetry Change	16.12
Model Use	16.13
MDFATE Concepts	16.14
Discretizing an Ocean Dredged Material Disposal Site	16.14
Simulating a Dredged Material Disposal Operation	16.16
Example MDFATE Application	16.16
Problem	16.16
Input Data	16.16

MDFATE Results 16.18
Summary 16.24
References 16.25

INTRODUCTION

The option of placing dredged material within the open water, although often more cost effective than upland disposal, introduces additional concerns to the overall management of dredged material disposal. The management of an open water dredged material disposal site (ODMDS) includes environmental requirements such as obtaining and maintaining regulatory approval for site use and operational constraints concerning navigation issues and disposal efficiency. To satisfy environmental requirements and operational constraints, the project manager of an ODMDS will likely address one or more of the following site management issues:

- Demonstrate that dredged material placed in open water does not violate applicable water quality standards.
- Ensure that dredged material placed in open water does not accumulate in a fashion that would pose a navigational hazard.
- If required, demonstrate that placed dredged material stays within the boundaries of a designated ODMDS.
- Design an ODMDS with areal boundaries commensurate with dredging disposal requirements and attain maximum utilization of ODMDS volumetric capacity.
- In the case of an ODMDS sited within the littoral zone, ensure that placed dredged material is transported alongshore away from the point of dredging or accretes toward a terrestrial resource.

The common management link between operational constraints, environmental requirements, and efficient site management is the ability to accurately predict and track the placement of dredged material within a given ODMDS. A major consideration is whether a given ODMDS is either dispersive or nondispersive. At nondispersive sites, approximately 95 to 99% of dredged material placed within the ODMDS reaches the immediate benthos (Tavolaro, 1984; Truitt, 1988). However, at a dispersive ODMDS most of the placed dredged material is transported along the bottom or dispersed through the water column by ambient currents and waves. Without proper evaluation, dredged material disposal could inadvertently impact sensitive resources or the ODMDS capacity may be exceeded sooner than expected. ODMDS capacity can roughly be defined as that quantity of material that can be placed within the legally designated disposal site without extending beyond the site boundaries or interfering with navigation (Poindexter-Rollings, 1990).

Depending on local statutes, obtaining regulatory approval for a new permanent ODMDS may require considerable time and resource expenditures, e.g., $500,000 or more per disposal site in the United States. The key to a successful ODMDS designation is knowing in advance (or reliably predicting) the fate of dredged material placed at the disposal site. In addition, proper life-cycle management of an ODMDS requires an approach to quantitatively predict and assess the behavior of dredged material placed in open water. This requires a knowledge of the physical processes acting on placed dredged material and the subsequent modeling of those processes to enable the quantitative prediction of the placed material.

PROCESSES AFFECTING DREDGED MATERIAL PLACED IN OPEN WATER

In the simplest of terms, the physical forces affecting dredged material placed in open water include gravity and forcing due to waves, and currents. Field evaluations by Bokuniewicz et. al. (1978) and laboratory tests by Johnson et al. (1993) have shown that open water disposal of dredged material conforms to a three-step process:

1. Convective descent, during which the material falls through the water column under the influence of gravity followed by
2. Dynamic collapse during which the descending plume impacts the bottom or arrives at a level of neutral buoyancy, and finally
3. Passive transport–dispersion commencing when material transport is governed more by ambient processes than by the dynamics of the disposal operation.

Apart from gravity, water column currents are the dominating factor in terms of the environmental forces that act on dredged material when placed in open water. Currents generally result from the combined actions of several components, e.g., large-scale ocean/coastal current regimes due to tidal circulation and/or storm surge propagation, locally generated wind-stress generated currents, short- and long-wave induced currents, inertial currents, and estuarine/riverine plume effects.

Short-Term Fate

At the point of release from the disposal vessel, dredged material convects through the water column under the influence of gravity while entraining ambient water, advects and diffuses laterally, and eventually comes to rest on the seafloor. This scenario characterizes the short-term fate of dredged material placed in open water. Depending upon the speed of the disposal vessel, water depth, water column current and density, ambient bathymetry, and dredged material type, the dredged material can spread out on the seabed to varying degrees. In some cases, dredged material placed in open water can be dispersed a considerable distance (thousands of meters) within the water column with little deposition near the point of disposal. This occurs at dispersive ODMDSs and is referred to as far-field dispersion. ODMDS siting and management considerations will be significantly different for dispersive disposal sites as compared to nondispersive sites. Dredged material dispersion can also occur after the placed material has come to rest on the receiving benthos if environmental forcings result in erosion of the deposited material.

Long-Term Fate

After dredged material has come to rest on the seabed, it can be eroded and transported by waves and/or currents. Furthermore, if the dredged material is cohesive, it can experience self-consolidation due to gravity. In addition, if many loads of dredged material are placed one on top of another such that a steep aggregate mound develops on the ambient benthos, the mound will avalanche and material will be transported downslope as a function of gravity and material characteristics. The combination of these processes define the long-term fate of dredged material placed in open water. In addition to sediment characteristics, water depth, wave activity, and current regime are the primary factors that contribute to the long-term fate of dredged material placed at a given ODMDS.

Depending upon the type of dredged material to be placed at an ODMDS and surrounding resources, long-term dispersiveness may or may not be a desirable aspect. If the placed dredged material is expected to remain within an ODMDS (due to incompatibility issues with ambient or adjacent resouces), a site should be selected that minimizes long-term dispersiveness. If the goal of the ODMDS is to facilitate re-introduction of dredged material into the littoral system, then an ODMDS should be selected that maximizes long-term dispersiveness. The degree of long-term stability for dredged material placed at an ODMDS is an important factor that dictates the amount of dynamic site capacity for a given location.

An ODMDS can exhibit little dispersiveness during dredged material disposal, while having a high degree of long-term dispersion during moderate to severe wave or current activity. To predict the fate of dredged material placed in open water, the dispersiveness of an existing or new candidate ODMDS must be fully considered for both short-term and long-term aspects. Numerical models are required for these predictions.

NUMERICAL MODELING OF DREDGED MATERIAL PLACED IN OPEN WATER

Until recently, estimating the fate of dredged material placed in open water in terms of the resulting distribution through the water column and on the ambient bathymetry was performed by:

1. Direct point measurement of sediment dispersion through field monitoring
2. Estimated point prediction of sediment transport potential by empirical methods
3. Comparison of consecutive hydrographic surveys for bathymetric accumulation and transport

All of the above sediment fate prediction methods possess a high degree of uncertainty when applied to complex large-scale conditions and do not provide the flexibility to address the variations associated with "what if" scenarios. Point measurement/estimation methods do not address the need for examining the (sub)areal distribution of placed dredged material within the water column and on the receiving benthos required for total site management. In addition, point estimate methods assume equal transport potential throughout a given area. Comparison of predisposal to postdisposal bathymetry provides a "hind-cast" for dredged material fate, but precludes a predictive means for managing dredged material.

Under the United States Army Corps of Engineers (USACE) Dredging Research Program (DRP), three deterministic sediment fate numerical models were developed or enhanced to improve the reliability of site management for an ODMDS. The numerical models incorporate state-of-the-art techniques for predicting short- and long-term behavior of dredged material placed in open water, and account for a variety of disposal operations and environmental settings. The models were developed at USACE, Waterways Experiment Station (WES) and are known as STFATE (Short-Term FATE), LTFATE (Long-Term FATE), and MDFATE (Multiple-Dump FATE). Each is summarized below and subsequently discussed later in the chapter.

Short-Term FATE: Simulates the areal distribution of dredged material within the water column and the resulting bathymetric distribution of dredged material after it has passed through the water column on an individual "dump" (disposal vessel load) basis.

Long-Term FATE: Simulates the bathymetric change of an existing mound due to self-weight consolidation and sediment transport arising from the interaction of waves and currents.

Multiple-Dump FATE: Simulates the change in bathymetry at an ODMDS resulting from a series of "dumps" as the resulting mound is simultaneously subjected to long-term transport processes. MDFATE uses modified components of STFATE and LTFATE to simulate a disposal operation that could extend over a year and consist of hundreds of "dumps."

In addition to these models, under the DRP other models such as an empirical-based model called EBERM (Empirical BERM) (Hands and, Allison, 1991), which determines if a bathymetric feature (berm or mound) will remain stationary or move due to erosive forces and COSED1V (COhesive SEDiment 1-Dimensional Vertical) (Chou et al., 1998), which provides guidance on whether cohesive material on the sea bottom will be eroded were developed. Neither EBERM nor COSED1V predict the physical fate of deposited dredged material and thus are not covered here. Details of these models can be found in the referenced reports.

The primary restriction of the DRP predictive sediment fate models is that reliable environmental input data are required to properly simulate dredged material behavior at an ODMDS. Fate predictions of dredged material placed in open water can only be as good as the poorest estimate for the forcing environment, i.e. waves and currents. In many locations, prototype environmental data may not exist.

To minimize the collection of prototype data, numerical models [ADCIRC ADvanced CIRCulation), HPDPRE (Height, Period, Direction PRE-Processor), and HPDSIM (Height, Period, Direction SIMulation)] were developed at WES to facilitate the generation of "synthetic" environmental data to be used in place of prototype data in the DRP sediment fate models. The suitability of synthetic versus prototype data in terms of sediment fate prediction has not been thoroughly resolved, although synthetic environmental data have been successfully used at several locations (USACE, 1995; Scheffner, 1992; and Moritz and Randall, 1995).

The programs HPDPRE and HPDSIM are used to simulate a time series representation for wave height, period, and direction for any location where a Wave Information Study (WIS) database has been produced (Borgman and Scheffner, 1991). The ADCIRC model is used to simulate elevations and currents for the equilibrium Newtonian tidal signal, as well as for tropical and extratropical storm events, for any location along the Gulf, Caribbean, NW Atlantic, and NE Pacific Coasts (Hench et al., 1994). Details concerning these programs are not given here. The interested reader should refer to the referenced reports.

The remainder of the chapter will focus on conceptual discussions of STFATE, LTFATE, and MDFATE, and will conclude with an example application of MDFATE. Although both STFATE and LTFATE are stand-alone models, MDFATE will likely be the model of choice in the management of an ODMS, since it simulates the simultaneous building and transport of sediment mounds. If water column issues are important, STFATE should be used; whereas, if the interest is in analyzing the long-term fate of an existing mound without further disposals in the ODMDS, LTFATE should be employed.

STFATE CONCEPTS

As previously noted, field evaluations by Bokuniewicz et. al. (1978) and laboratory tests by Johnson et al. (1993) have shown that the placement of dredged material generally follows a three-step process: (1) convective descent, during which the material falls under the influence of gravity, (2) dynamic collapse, occurring when the descending cloud or jet either impacts the bottom or arrives at a level of neutral buoyancy, in which case the descent is retarded and horizontal spreading dominates, and (c) passive transport–dispersion, commencing when the material transport and spreading are determined more by ambient currents and turbulence than by the dynamics of the disposal operation. Figure 16.1 illustrates these phases.

Model development in this area was initiated in the early 1970s with the work of Koh and Chang (1973) and was continued with developments by Brandsma and Divoky (1976) and Johnson (1990). However, deficiencies remained. Research in the DRP, which resulted in the STFATE model, was directed at removing many of these deficiencies, e.g., inadequate representation of disposal from hopper dredges, the inability to represent the nonhomogenity of disposal material, the inability to model disposals at dispersive sites, inadequate representation of the bottom collapse phase, and inability to model disposal over bottom mounds. Basic concepts employed in STFATE are presented below. Details can be found in Johnson and Fong (1995).

Convective Descent

In STFATE, multiple convecting sediment clouds that maintain a hemispherical shape during convective descent are assumed to be released. Figure 16.2 illustrates the basic concept. By representing the disposal operation as a sequence of convecting clouds, both split-hull barge disposal as well as disposal from a hopper dredge can be modeled. For example, the material in each hopper might be contained in one cloud. This concept also allows for a more realistic representation of the disposal material when consolidation in the disposal vessel has occurred. Denser consolidated material might be represented by one cloud and the less dense more fluid-like fraction overlying the consolidated material represented by a separate cloud. In addition, the use of multiple convecting clouds allows for a more realistic representation of disposal from a moving vessel where the disposal oper-

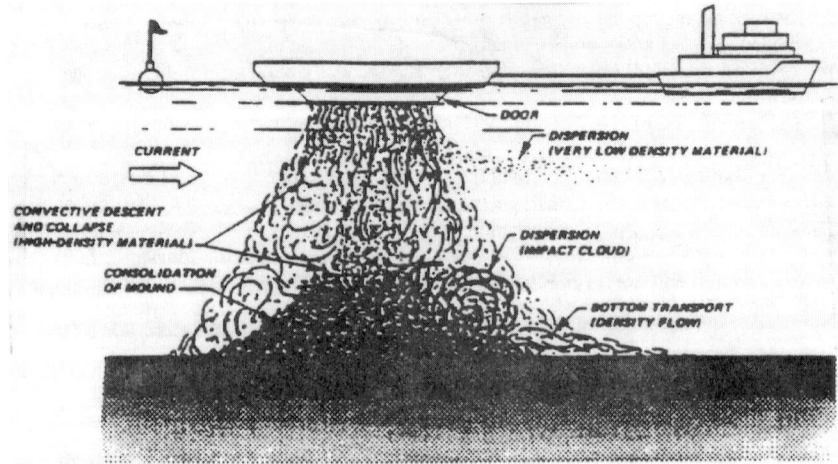

FIGURE 16.1. Illustration of placement processes.

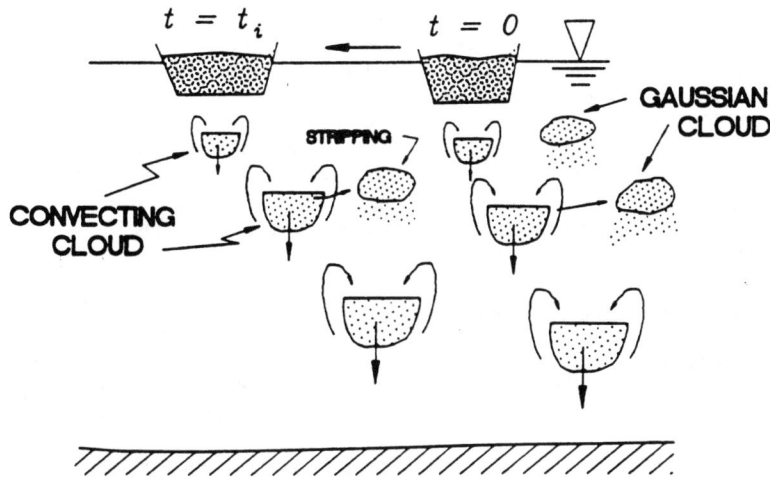

FIGURE 16.2. Multiple convecting clouds.

ation typically requires several seconds to perhaps 1–2 minutes for completion. As each small convecting cloud descends through the water column, material can be stripped from the cloud to settle with its particle settling velocity. Movement of the stripped material as small Gaussian clouds is discussed later.

Other than the concept of multiple convecting clouds and the stripping of material from those clouds, treatment of the convective descent phase for each of the convecting clouds is the same as initially developed by Koh and Chang (1973).

The equations governing the motion are those for the conservation of mass, momentum, buoyancy, solid particles, and vorticity. These equations are ordinary differential equations with time as the independent variable. Basic dependent variables are the cloud's radius, velocity, and density along with the concentration of each solid fraction.

Given a set of initial conditions, e.g., the total volume of disposal material, sediment concentrations, etc., along with ambient conditions such as the water column stratification and current, the governing system of equations are solved using numerical techniques. In STFATE, the forward Euler scheme is used. Given the idealized representation of the disposal operation and the uncertainty of initial conditions, the need for higher-order numerical schemes is questionable.

Dynamic Collapse in Water Column

As the disposal cloud goes through the convective descent phase, its mass and momentum change due to entrainment of surrounding water. The horizontal velocity of the cloud tends to approach that of the ambient fluid. Coincidentally, the disposed material concentration is greatly reduced and the cloud's vorticity becomes insignificant because of dissipation by ambient stratification and turbulence. If the cloud reaches the depth of neutral buoyancy, its momentum will tend to make it overshoot the neutrally buoyant point while the buoyancy force will tend to bring it back to the neutrally buoyant position. The combined action of these forces will make the cloud undergo a decaying vertical oscillation. As the vertical motion of the cloud is being suppressed, the cloud tends to collapse vertically and spread out horizontally, seeking hydrostatic equilibrium within the stratified ambient fluid. As the cloud collapses, its cross section becomes elongated in the horizontal. The idealized shape that the cloud is assumed to take is that of an oblate spheroid.

With the exception of vorticity, the conservation equations used for convective descent still hold. Any differences are due to the additional dimension used to describe the cloud. Final results from the convective descent phase such as cloud size, position, and velocity become the initial conditions for the solution of these equations.

Dynamic Collapse on Bottom

As previously noted, the convective descent and water column collapse phases follow the development of Koh and Chang (1973). However, the manner in which collapse of the disposal cloud is treated when the bottom is encountered during convective descent is different.

Bottom collapse in STFATE is computed from a conservation of energy concept. When the cloud strikes the bottom, it possesses a certain amount of potential energy that can be computed since the mass of the cloud and the location of its centroid are known. In addition, the kinetic energy of the impacting cloud can be computed since its velocity and mass are known. Thus, the total energy of the cloud at the moment of impact is known. This energy is then available to drive the resulting bottom collapse or surge.

A basic assumption is that the bottom collapsing cloud is one-half of an ellipsoid. If the bottom is flat, the cloud becomes one-half of an oblate spheroid. As illustrated by Johnson and Fong (1995), expressions for the time rate of change of the three cloud dimensions can be derived that are dependent on the cloud's kinetic energy.

Given that the change in the total energy of the cloud, which is the sum of the kinetic and potential energies, is equal to work done by the cloud, the temporal variation of the kinetic energy can be determined. The potential energy of the cloud is easily computed and changes as the cloud density decreases due to entrainment of ambient fluid, the loss of sediment, and to the decrease of the height of the centroid of the cloud as the cloud spreads over the bottom. The kinetic energy of the cloud is then known since the change in kinetic energy can be computed once the work done by the collapsing cloud is quantified. Work must be done to overcome bottom friction, drag, the production of internal turbulence, and the setting of the ambient fluid in motion. Given the kinetic energy of the cloud, the cloud shape can be determined from the expressions derived by Johnson and Fong (1995) relating the change in the cloud axis to the cloud's kinetic energy.

Transport–Diffusion Phase

At most disposal sites, the convective descent and dynamic collapse phases only last on the order of a few minutes. When the rate of spreading of the collapsing cloud becomes less than an estimated rate of spreading due to turbulent diffusion, the collapse phase is terminated and the "longer" term transport–diffusion phase is initiated. In this phase, material in suspension is transported and diffused by the ambient current while undergoing settling. Any nonsediment constituents being modeled are also transported and diffused.

A basic assumption in the transport–diffusion phase is that material (or constituent) concentrations can be determined from a superposition of small clouds characterized by a normal or Gaussian distribution. These clouds are formed as material is stripped away during the descent of the convecting clouds as well as during the collapse phase. For collapse in the water column, small clouds are formed as material settles from the collapsing cloud. However, during collapse on the bottom, laboratory experiments by Johnson et al. (1993) as well as field data collected at Mobile, Alabama by Kraus (1991), imply that fine material is also lost to the water column at the top of the collapsing cloud.

At the end of each time step, each cloud is advected horizontally by the input velocity field. In addition to the advection or transport of the cloud, the cloud grows both horizontally and vertically as a result of turbulent diffusion. If long-term output is desired at the end of a particular time step, the concentration of each solid type is given at each grid point at a particular vertical location by summing the contribution from individual clouds. This approach for the transport–diffusion phase follows the work of Brandsma and Sauer (1983). The surface and all solid boundaries except the bottom are handled by assuming reflection from the boundaries.

In addition to the horizontal advection and diffusion of material, settling of the suspended solids also occurs. If a solid fraction is specified as being cohesive, the settling velocity is computed as a function of the suspended sediment concentration of that solid type. Therefore, at each net point the amount of solid material deposited on the bottom and a corresponding thickness are also determined. Since a normal distribution is assumed for material in the small clouds, deposited material is also assumed to take such a distribution on the bottom. A basic assumption in the model is that once material is deposited on the bottom it remains there; i.e., neither erosion nor bed-load movement of material is allowed. However, deposition is prohibited if the computed bottom shear stress exceeds a specified critical shear stress for deposition for each solid fraction.

The discussion presented above for the transport–diffusion of solids also applies to the disposed fluid with its dissolved constituents. The constituents are assumed to be conservative with no further adsorption on or desorption from the solids in the water column or those deposited on the bottom. Computing the resultant time history of constituent concentration provides information on the dilution that can be expected over a period of time at the disposal site and enables the computation of mixing zones in water column evaluations.

Other Processes

Bottom Slope. The influence of a bottom slope is to increase the spread of the bottom surge in a downslope direction and to decrease the spread in an upslope direction. This effect is simulated in STFATE by using the bottom slopes at the centroids of the axis of the bottom footprint to compute either an increase or decrease of the cloud's axis by first computing the movement of the centroids.

Disposal at Dispersive Sites. In the earlier disposal model work, there was no mechanism for keeping material from depositing, and once material was deposited on the bottom it remained there. Thus, at highly dispersive sites, model results were questionable. In STFATE, a critical shear stress for deposition is specified for each sediment fraction. A computed bottom shear stress is then compared to the critical value to determine if suspended material can be initially deposited.

If the bottom shear stress is affected only by currents and no waves are present, its computation is based upon the assumption of a log velocity profile. If waves are present at the disposal site, the approach derived by Madsen and Wikramanayake (1991) is used. With this approach, the ambient current angle, wave amplitude, wave period, and an initial value for the wave–current angle must be inputted.

Model Use

Generally, STFATE is used to address environmental issues concerning individual disposal operations, e.g., the determination of mixing zones. However, a streamlined version of STFATE comprises the short-term computations for individual disposals in MDFATE. In that version, many of the computations concerned with water column concentrations have been removed since they are not required to address mound formation and the long-term fate of those mounds.

Basic data required to operate STFATE are descriptions of the disposal material, disposal vessel, and the ambient environment at the disposal site. Therefore, information such as the bulk density of the material, the number of solid fractions and their concentrations, the barge length and width, and the ambient current and water column depth and stratification must be known. Basic output are the time history of water column concentrations in horizontal planes at particular depths and the initial deposition of material on the bottom.

LTFATE CONCEPTS

The discussion given below describes concepts employed in the site evaluation model known as LTFATE. LTFATE provides a systematic and quantifiable means of accurately predicting the evolution of in-place dredged material mounds over long periods of time. Time scales of interest range from months to years in the case of normal wave and current conditions and on the order of days in the case of tropical or extratropical storm induced erosion. The PC-based code is intended for application in protected nearshore applications such as placement sites within estuaries or in the vicinity of barrier islands as well as for unprotected sites in the proximity of navigation channels.

LTFATE is composed of three modules: hydrodynamic, sediment transport, and bathymetry change. However, before describing each of these modules, it should be noted that a critical requirement for the successful application of LTFATE is the development of a data base of information that can be accessed by the user to define design wave and current boundary conditions that are realistically representative of conditions at the disposal site. This data base of wave and current information is available at the alongshore spatial density of the Phase II WIS data base, i.e., a spacing of approximately 20 miles. This density is considered sufficient for defining environmental conditions near any nearshore area subject to the disposal of dredged material. LTFATE then utilizes this data base to provide boundary conditions to drive the hydrodynamics module.

Data Bases

The erosion and transport of sediment from a disposal site is primarily a function of the combined action of waves and currents. To realistically model long-term changes in sediment transport, and thus quantify site stability, realistic boundary conditions for input to the hydrodynamic module must be defined. These boundary conditions represent time series of both long-term fluctuations of waves and currents and the shorter-term effects of storm events. Both stochastic and deterministic analyses are required to adequately define these conditions for inclusion into the data base. The stochastic components represent those data that can only be specified according to their statistical patterns of frequency, magnitude, etc. Waves and storm events fall into this category. Neither can be quantitatively predicted; however, each can be approximated according to their historical patterns of occurrence.

The deterministic component of the data base represents those contributions of the overall wave and current regime which are predictable and can therefore be determined. For example, tidal elevations and currents at any location can be accurately computed using known harmonic constituent boundary conditions. Similarly, if storm parameters are specified, the resulting surge height and current distribution throughout a computational grid can be computed. Both the tidal and storm surge contributions to the data base are computed via a highly adaptive large-scale, two-dimensional (2D) hydrodynamic numerical model developed specifically to provide boundary conditions for LTFATE applications. Documentation of the theory and implementation of the 2D code is reported by Luettich et al. (1992).

Waves. A methodology for generating time series of wave height, period, and direction that simulate the WIS 20-year hindcast time series data base for any specific WIS station is reported by Borgman and Scheffner (1991). The simulation technique does not require access to the hindcast data base but instead requires access to a precomputed matrix of coefficient multipliers that contain the statistical correlations of the wave field parameters for each WIS station. The procedure results in the generation of arbitrarily long sequences of wave data that are statistically similar to those of the entire WIS data base, preserving both seasonality and wave sequencing. Example use of the PC version of the code in the generation and analysis of the synthetic time series is demonstrated in Scheffner and Tallent (1994).

Storms. Storm events can be described according to several descriptive parameters, including accompanying wave fields, duration, and some measure of intensity. Both tropical (hurricanes) and extratropical (northeasters) storms are necessary for analyzing the long-term fate of dredged material placed in an ODMDS. Since the characteristics of each storm type are different, separate analysis techniques are used to predict parameters descriptive of the storm event.

The data base of storm events were generated by first identifying a representative number of tropical and extratropical events that have impacted the U.S. coast. Each event was then input to the ADCIRC model to generate a location-indexed data base of elevation and current hydrographs for each of the alongshore Phase II WIS stations. This data base of event hydrographs is available for use as input to LTFATE for investigating storm-related effects on the dispersive characteristics of a disposal site.

Tides. The tidal component of the data base is provided in the form of harmonic constituents for tidal elevation and currents. Similar to the generation of wave sequences, a postprocessing program accesses the data base and generates an arbitrarily long time sequence of tidal data. The spatial density of these components are specified by the user, therefore, time series can be generated to correspond to the WIS data station spacing adopted for the wave simulation program and used for the storm-event archiving.

Hydrodynamics Module

The hydrodynamic modeling component of LTFATE solves the depth-averaged Navier–Stokes and continuity equations with depth-averaged currents specified at the computational boundaries. The equations that govern the conservation of linear momentum and continuity are well known, and, thus, are not given here. The solution scheme is an explicit, central difference approximation of the governing equations computed on a fixed space orthogonal grid. The model incorporates a radiation boundary condition such that transient waves are not trapped within the small computational domain.

The purpose of the hydrodynamic module is to compute the current distribution around an arbitrarily shaped disposal feature as a function of the local boundary conditions described above. Because the temporal input for the LTFATE site analysis model is limited in time by the frequency of wave data (provided at a 3-hour time spacing), the hydrodynamic module was developed to represent a "steady-state" current distribution corresponding to every 3 hours.

Sediment Transport

LTFATE allows for the dredged material disposal mounds to be composed of either noncohesive or cohesive sediments. The equations reported by Ackers and White (1973) were selected as the basis for the noncohesive transport modeling component. These relationships predict sediment transport as a primary function of sediment grain size, depth, and depth-averaged velocity. The equations are applicable to uniformly graded noncohesive sediment with a grain diameter in the range of 0.04 mm to 4.0 mm (White, 1972).

Because many disposal sites are located in relatively shallow water, a modification of the Ackers–White equations was incorporated to reflect an increase in the transport rate when ambient currents are accompanied by surface waves. The modification, in the form of an effective increase in the depth-averaged velocity, is based on the concepts developed by Bijker (1971) and implemented by Swart (1976). The modified transport equations have been shown to be well behaved within the stated limits of the equations, i.e., noncohesive sediment with a grain size between 0.04 and 4.0 mm.

Fine-grained sediments are hydraulically transported almost entirely in suspension rather than as bed load. Therefore, the governing differential equation for fine-grained sediment transport is a transport–diffusion equation containing a settling term and terms representing rates of deposition and erosion based upon critical shear stresses. The terms utilized in LTFATE were developed by Krone (1962) and Partheniades (1962).

Bathymetry Change

In LTFATE, the bathymetry of an initial mound must be specified with no additional disposal allowed during the simulation. The time change in the bathymetry of the mound is then determined from a sediment continuity equation relating the change in local depth with respect to time as a function of the spatial gradients of the sediment transport provided by the sediment transport module.

Avalanching. The concept of slope failure is incorporated in LTFATE to insure stability of the dredged mound configuration. Thus, a routine is included to account for transport induced by slope failure (Allen, 1970; Larson et al., 1990). Allen recognized two different limiting profile slopes: the angle of initial yield (angle of repose) and the residual angle after shearing.

In the model, if the local slope exceeds some initial yield angle, ϕ_{iy}, material will be redistributed in neighboring cells such that the slope in each cell of redistribution does not exceed the residual angle, ϕ_{ra}. Computationally, this is done by checking the local slope, ϕ, at each time step along the grid, and, if $\phi > \phi_{iy}$, avalanching is again initiated. Redistribution of the sand following avalanching is governed by two conditions: (1) mass must be conserved, and (2) the residual angle after shearing is equal to ϕ_{ra}.

Consolidation. A second component of the bathymetry change module relates to the consolidation of the mound. Consolidation of dredged material on the sea floor occurs due to large initial void ratios in the newly formed mound following the rapid disposal process. Consolidation is primarily confined to the silt and clay component of the dredged material. The consolidation process occurs gradually due to the low permeability of these fine soils.

Consolidation calculations used in this model are based on finite strain theory described by Gibson et al., 1981, and later refined and modeled by Cargill (1982, 1985). This theory is well suited for the prediction of consolidation in cases of thick deposits of fine-grained material because it provides for the effect of self-weight, permeability varying with void ratio, a nonlinear void ratio–effective stress relationship, and large strains. Furthermore, the predictive capability of this relationship has been verified in several field studies conducted by Poindexter-Rollings (1990).

At the end of each 3-hour simulation, assuming steady-state hydrodynamics, the new configuration of the mound is determined by summing the change in mound thickness based upon mass conservation, the change in mound thickness due to avalanching, and the decrease in mound thickness due to consolidation. The new configuration then becomes the initial state of the sea floor for the next computational period.

Model Use

The discussion above presents the concepts employed in LTFATE. A limitation of LTFATE is that additional material cannot be added to the initial mound. Thus, the MDFATE model discussed next, which combines features of both STFATE and LTFATE to allow for the simultaneous building and transport of sediment mounds, is the more general model for use in site management studies. However, if one is only interested in simulating the long-term fate of an existing mound, LTFATE would be the model of choice.

As previously noted, input to the LTFATE model includes local wave and/or current conditions for the site of interest. These data can be obtained from the World Wide Web home page for the Coastal and Hydraulics Laboratory in Vicksburg, Mississippi (http://bigfoot.wes.army.mil/c205.html). Additional information required in an application of LTFATE includes local bathymetry, disposal feature geometry, and sediment characteristics. Use of these data for input to and application of the model are described in detail in Scheffner et al. (1995).

MDFATE CONCEPTS

The PC-driven numerical model MDFATE was developed for the quantitative assessment of open water disposal activities with regard to both short- and long-term morphology of dredged material placed on the seafloor. The model spatially accounts for bathymetric changes based upon LTFATE concepts and will generally be the model of choice to assist in planning and managing the use of an ODMDS. The MDFATE model has evolved from several earlier concepts (Moritz and Randall, 1992, 1995).

MDFATE defines an ODMDS in terms of a numerical grid and, as previously noted, incorporates modified versions of STFATE and LTFATE to predict the ODMDS bathymetry resulting from a series of disposal cycles or "dumps." Execution of MDFATE is controlled by an easy to follow menu and prompt–input interface.

Discretizing an Ocean Dredged Material Disposal Site

As a first step in simulating the life-cycle for a given ODMDS, MDFATE is used to produce a discretized representation (rectangular digital elevation model) of the ODMDS of interest. Since the ODMDS is rectangular, all that is required from the user are the ODMDS corner coordinates and the desired grid interval with which to descretize the ODMDS. Horizontal control (x, y) is manifested in terms of the coordinate system used to describe the site in the prototype scale. State plane and geographic (latitude–longitude) coordinate systems are supported. Up to 40,000 grid points can be used to represent a given ODMDS in terms of a MDFATE grid. This is sufficient to represent a 10,000 by 10,000 ft disposal site with a 50-foot grid interval. Bathymetric (z) data are represented in terms of a specific vertical datum. Within MDFATE, subsequent modification of ODMDS bathymetry is performed with respect to the vertical datum established during the creation of the disposal area grid. MDFATE can either automatically generate the ODMDS grid bathymetry (flat or sloping), or adapt survey data $(x, y, z$ ASCII format) consistent with the actual sites's coordinate system. Survey data are fitted to the MDFATE digital elevation model (DEM) by a multipoint polynomial interpolant scheme.

MDFATE can produce cross-sectional, 2-D contour, and 3-D surface renderings of an ODMDS grid of interest. At the conclusion of every MDFATE activity, two data files are produced for postprocessing and informational purposes. An ASCII (x, y, z) DEM data file is produced for additional postprocessing of the ODMDS bathymetry. This file is generic and can be used by a variety of 2-D or 3-D software packages. The second file serves as a schematic representation of the ODMDS and contains parameters describing the site and grid characteristics.

Simulating a Dredged Material Disposal Operation

Once a particular ODMDS grid has been created, MDFATE can be used to simulate a given disposal operation, which may extend over one year and consist of hundreds of disposal cycles or "dumps." A "dump" consists of one load of dredged material released into open water from either a barge, scow, or hopper dredge. The entire disposal operation is divided into separate week-long episodes over which long-term fate processes governing dredged material behavior on the seafloor are simulated using a modified version of the LTFATE model. Results are modeled in a cumulative manner. As previously discussed, long-term processes include self-weight consolidation, sediment transport by waves and currents, and mound avalanching.

Within each episode, a modified version of STFATE simulates the short-term fate processes which govern each "dump" occurring inside the ODMDS of interest. Short-term processes are those that influence placed dredged material up to the point at which all momentum imparted to the material from the "dump" activity is expended through convection, diffusion, and bottom friction. As in an application of LTFATE, HPDSIM, and TIDEPRED (TIDE PREDiction) are utilized to provide wave and tidal information for every 3-hour interval during the disposal operation. This information is also utilized in the short-term computations to simulate wave-current affects acting upon each "dump" as dredged material passes through the water column and comes to rest on the seafloor. The short-term computations are made using the actual ODMDS bathymetry to simulate cumulative mound distribution arising from each "dump."

A prime assumption for the MDFATE model is that other than the "placed" dredged material, no sedimentary material enters the ODMDS. This assumption may be invalid at times of high estuary discharge when sediment enters an ODMDS from estuarine transport.

When applying MDFATE, specification of the disposal operation is performed through a menu-driven format in which the user specifies basic data. These include:

- Dredging disposal vessel parameterization

    Vessel type (split-hull or hopper), dimensions, and volumetric capacity

    Placement duration per load (time to place each load)

    Vessel speed and heading during placement

    Total volume of dredged material to place at the candidate ODMDS during a given disposal operation, or the number of individual "dumps"

- Method of disposal vessel control during the disposal operation (i.e. how each dump is placed). Four options are available; namely:

    1. Within a specified radial distance of a predetermined geographic location (i.e., coordinates defining a disposal buoy location). Dumps are placed in a random manner and are weighted in the direction of disposal vessel approach.
    2. Along a predetermined transect line based on beginning and ending coordinates
    3. Each dump location defined by the user entering coordinates of each location
    4. Dump locations based upon prerecorded coordinates for each load. Coordinates (x, y) are contained in an ASCII data file queued by MDFATE.

- Dredged material parameters (density, grain size, solids concentration, void ratio, critical shear stress)
- Existing bathymetry at the ODMDS
- Time series data for tidal elevation and tidal induced depth-averaged currents
- Time series data for wind-driven surface waves
- Ambient depth-averaged residual currents present at the ODMDS

For modeling purposes, the annual dredging and disposal season for an ODMDS management is usually broken into two discrete time periods. Dredging disposal at a given ODMDS normally begins during the Spring time frame and continues through the summer until fall. After the fall season, the ODMDS is not used but is affected by the energetic winter environment until the following spring time frame when dredged disposal again commences. The proposed schematic shown below describes how the MDFATE model would be applied to simulate dredged material disposal at a candidate ODMDS to account for the seasonality of ODMDS management.

◄─────── DUMPING ───────► Simulate Disposal Season	◄─ Simulate Remaining Part of ─► the Year: Long-term Fate Only
(1) Model a series of 1 week dump episodes for short-term fate processes (2) Follow-on with long-term fate calculations for 1 week after each dump episode	Model bathymetric change due to sediment transport
April September	April

Schematic Timeline for a Dredged Material Disposal Cycle at an ODMDS

Using the above schematic timeline as an example, during any given year at an ODMDS, short-term and long-term fate processes are simulated simultaneously at the disposal site for a 5-month period (April–September). Only long-term fate processes are then simulated for a 7-month period, until the following year when the annual cycle begins again. Semiannual ODMDS bathymetric surveys are typically taken during April and September, which makes the above disposal cycle discretization desirable in terms of comparing before and after surveys to simulated results.

Following the simulated disposal operation, MDFATE generates a brief report documenting disposal equipment used, operation duration, dredged material parameters, and each dump location. Also described within each MDFATE disposal report are ODMDS locations where dredged material accumulation has exceeded navigable depth, dumps that exceeded lateral boundary tolerances, and remaining site capacity.

MDFATE may also be used to edit/update/manipulate the bathymetry within an ODMDS grid to generate "differences" between predisposal and postdisposal conditions and to obtain volume estimates. Options within MDFATE allow for assessing sediment transport rates based upon variations in water depth, wave height and period, current speed, and sediment characteristics. A variety of bathymetric features (berms and mounds) can be generated for purposes of modeling long-term morphological behavior.

EXAMPLE MDFATE APPLICATION

Problem

Assume that the site manager for a major port and waterway is considering limiting dredged material disposal at most of the available ODMDSs for upcoming maintenance dredging operations. This is due to potentially adverse navigation impacts associated with placing dredged material at these ODMDSs. Bar pilots and other navigational interests are concerned with hopper dredges crossing the entrance channel and interfering with the shipping lane traffic when dumping at locations other than ODMDS "B" (Figure 16.3). To avoid placement of dredged material at the other ODMDSs, approximately 4.3 million cubic yards of sandy dredged material are assumed to be placed at site "B" during the upcoming dredging cycle.

Placement of dredged material at ODMDS "B" is to be restricted to the western half of the site (2,000 ft × 6,000 ft) due to excessive mounding in the eastern half. The feasibility of operating site "B" to accommodate disposal of 4.3 million y^3/year in the site's western half for a 3-year period is investigated using MDFATE. Development of model input data for the ODMDS "B" assessment and the results obtained from the MDFATE application are presented below.

Input Data

An annualized wave environment was simulated for ODMDS "B" using the HPDSIM program (Borgman and Scheffner, 1991). Based upon the location of ODMDS "B," the tidal environment at ODMDS "B" was simulated using the five primary tidal constituents generated from the ADCIRC-derived data base (Hench, et al., 1994). The residual current was estimated to be 0.5 ft/sec @ 280° (T). To properly characterize the total effective current at ODMDS "B," the simulated tidal current data were combined within MDFATE with the estimated residual current to simulate the effective water column current at the site.

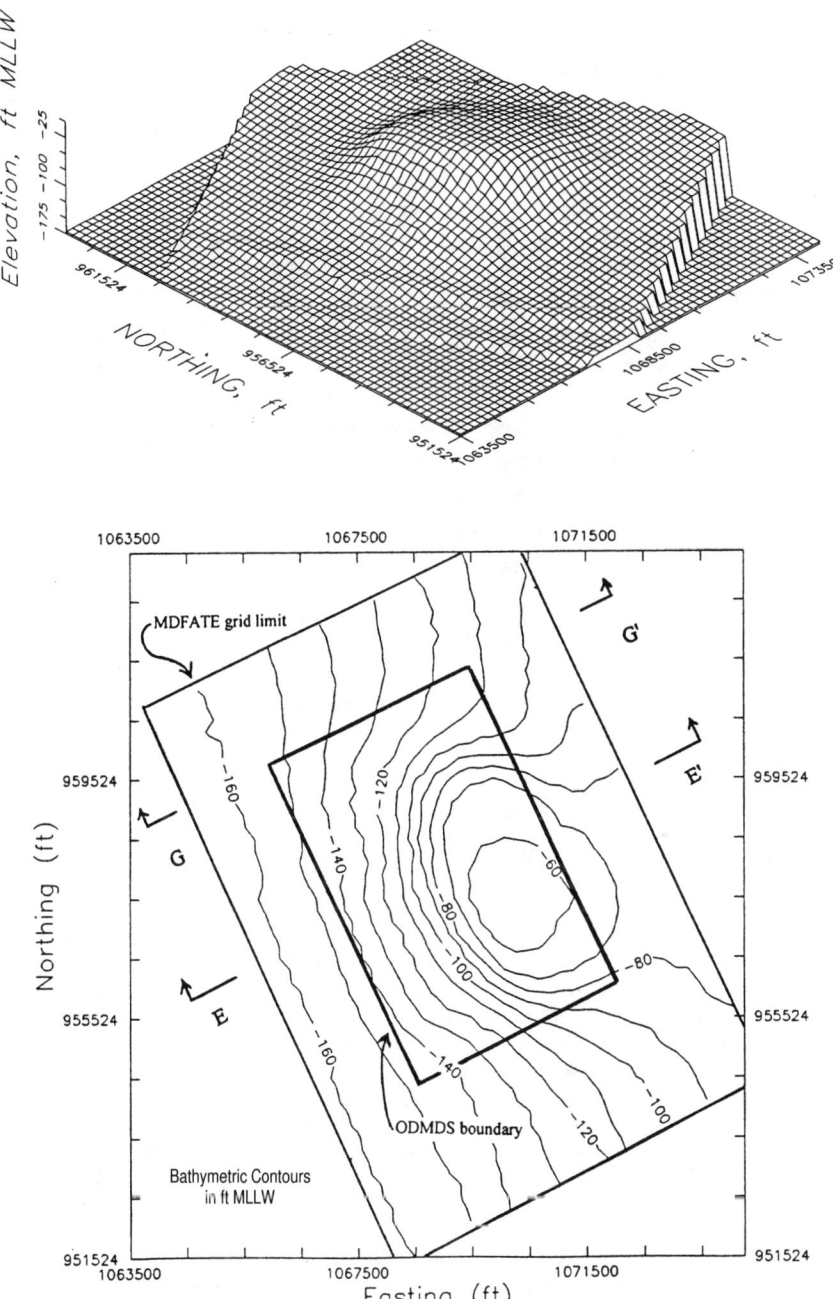

FIGURE 16.3. Predisposal bathymetry at ODMDS "B."

To minimize the vertical accumulation of dredged material placed at ODMDS "B," two disposal scenarios were considered for upcoming maintenance dredging. Both disposal scenarios were assessed using MDFATE:

1. The western half of site "B" was divided into 3 cells, each 2,000 ft × 2,000 ft. Dredged material was placed randomly about the centroid of each cell based upon a 1,400 ft radius. Random placement was weighted towards the hopper dredge's approach direction, during disposal.
2. The western half of site "B" was divided into 12 cells, each 1,000 ft × 1,000 ft. Dredged material was placed randomly about the centroid of each cell based upon a 800 ft radius. Random placement was weighted towards the hopper dredge's approach direction, during disposal.

For both disposal scenarios, an ocean-going hopper dredge was assumed to be the vessel used for the disposal at ODMDS "B." Disposal vessel and dredged material characteristics are described below:

- 4,500 y^3/dump capacity (total of 957 dumps per year = 4.3 million yd^3)
- Approaching the disposal site from the southeast
- Dumping while underway at 3.5 ft/sec
- Duration of active disposal per dump estimated at 4 minutes
- Dredged material parameters were:
    Dredged material type = fine sand
    D50 = 0.20 mm
    S.G. = 2.71
    Concentration of solids by volume in the disposal vessel = 0.50
    e_d = depositional void ratio = 0.71
    ϕ_s = subaqueous shearing angle = 1.8°
    ϕ_{ps} = subaqueous postshearing angle = 1.5°

MDFATE Results

As noted, the predisposal configuration for ODMDS "B" is shown in Figure 16.3. The bold interior line defines the boundaries for ODMDS "B." The cross-sections for the predisposal configuration are shown in Figures 16.4a and 16.5a. Note the prominent dredged material mound at the center of the disposal area (section E–E'). The northern part of the disposal site (section G–G') is similar to the ambient bathymetry. Ambient water depths at ODMDS "B" are about 140 ft.

For both disposal scenarios (12-point and 3-point), overall bathymetric impacts at ODMDS "B" due to the proposed disposal operation for year 1 are greater than expected. The largest predicted increase in bathymetric relief for the 3-point dump is 23 ft, near the northwest end of ODMDS "B." The largest predicted bathymetric increase for the 12-point dump is 16 ft, near the west center of site "B." Results for both dump scenarios are shown in Figures 16.4–16.8.

Based on the above MDFATE results, it would not be advisable to continue the proposed disposal operation at ODMDS "B" for more than one year: Mounding will be rapid and may worsen wave conditions at the site. As long as disposal is limited to the western half of site B and material is evenly distributed, the "problem area" (high point) associated with the existing mound at site B should not be made worse by the proposed disposal

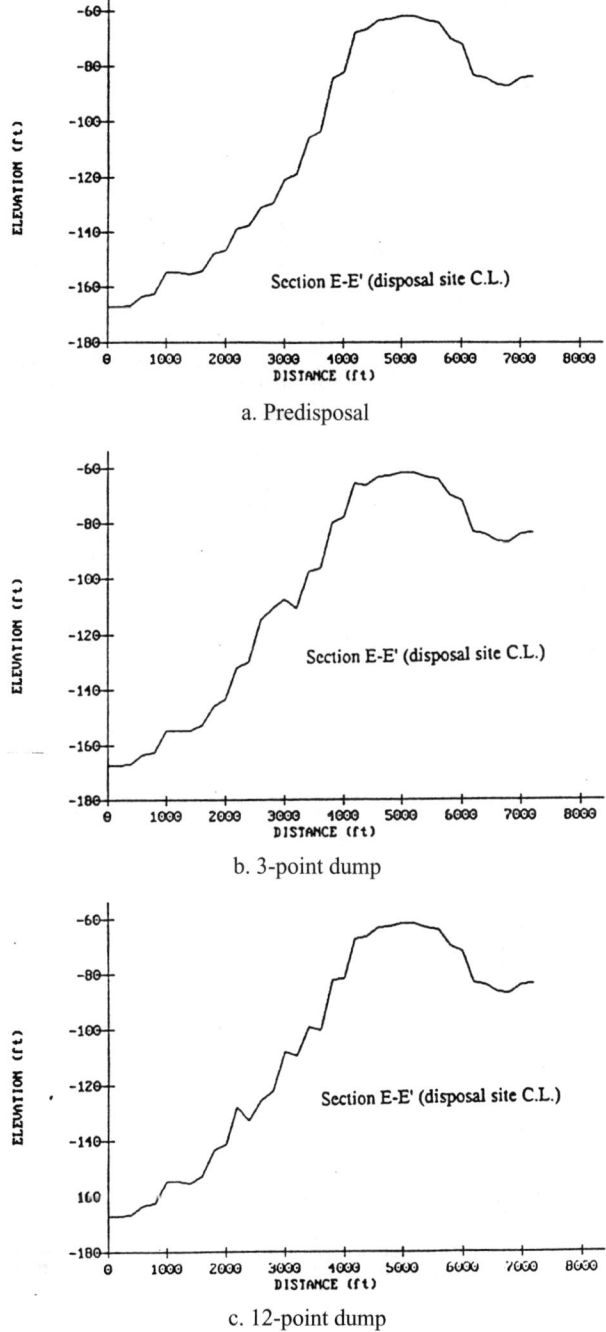

FIGURE 16.4. Onshore–offshore cross-sections at ODMDS B, Section E–E′.

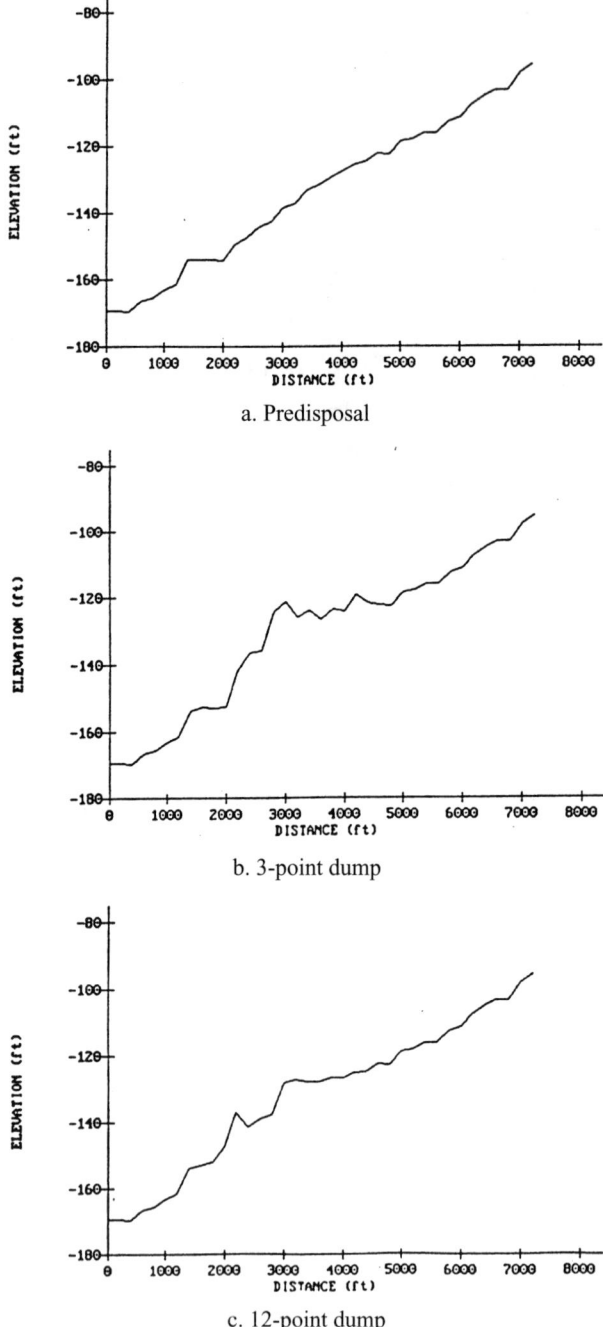

a. Predisposal

b. 3-point dump

c. 12-point dump

FIGURE 16.5. Onshore–offshore cross-sections at ODMDS B, Section G–G′.

PREDICTING THE FATE OF DREDGED MATERIAL PLACED IN OPEN WATER **16.21**

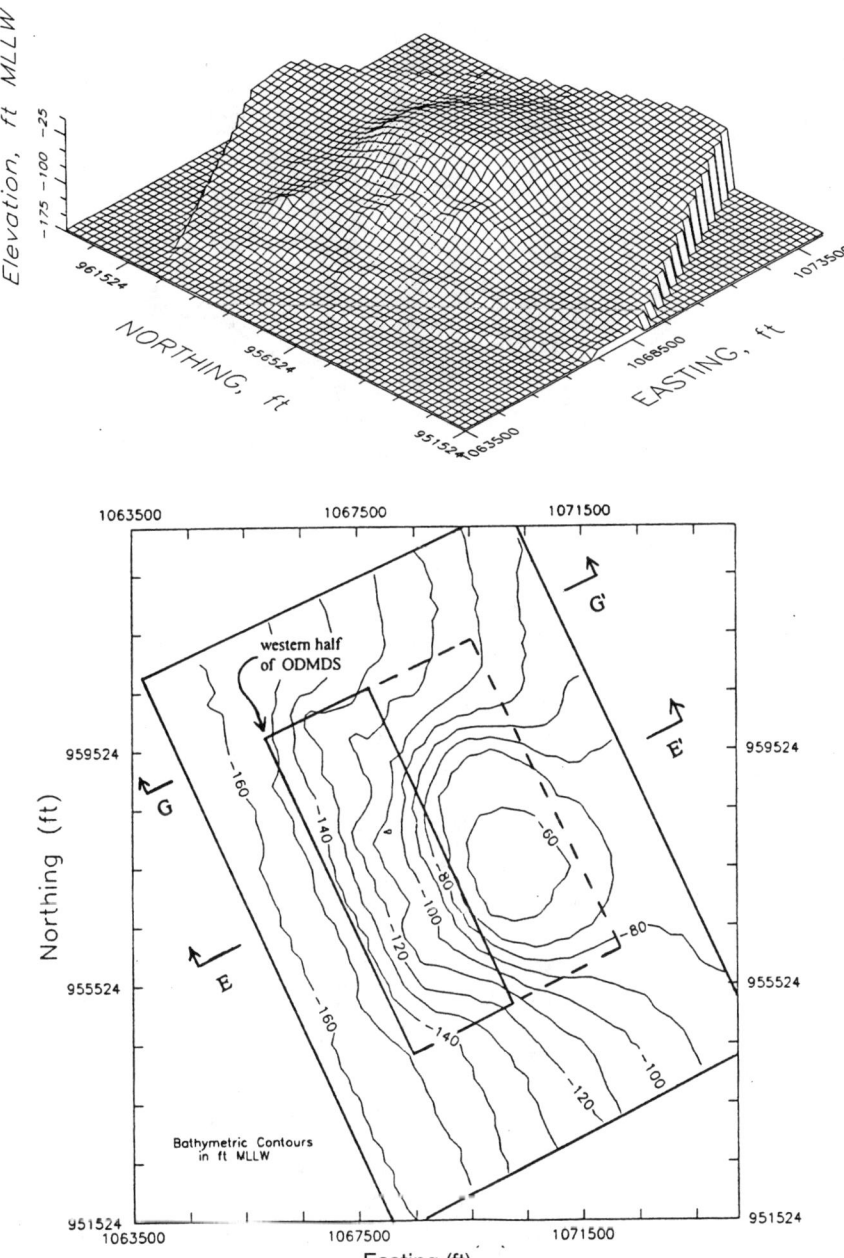

FIGURE 16.6. Predicted Postdisposal bathymetry for 3-point disposal.

16.22 HANDBOOK OF COASTAL ENGINEERING

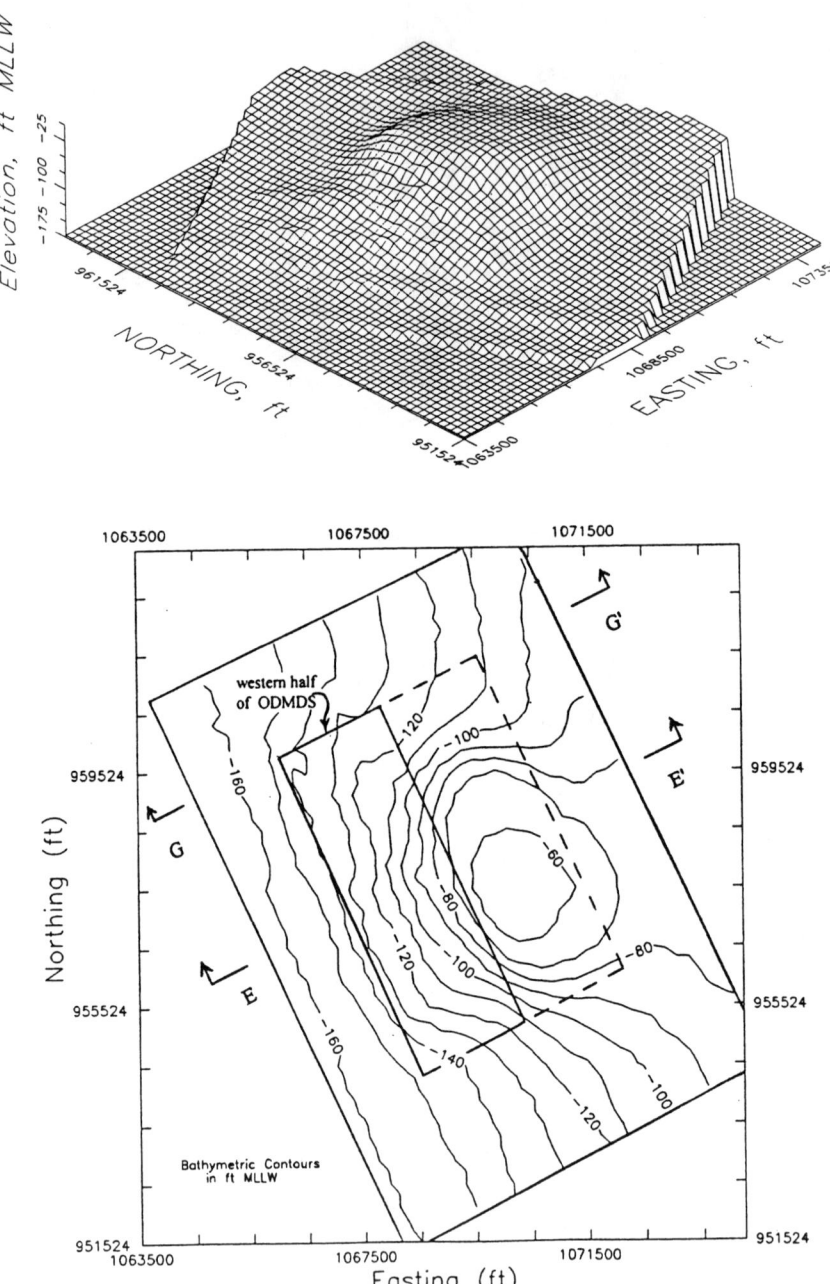

FIGURE 16.7. Predicted Postdisposal bathymetry for 12-point disposal.

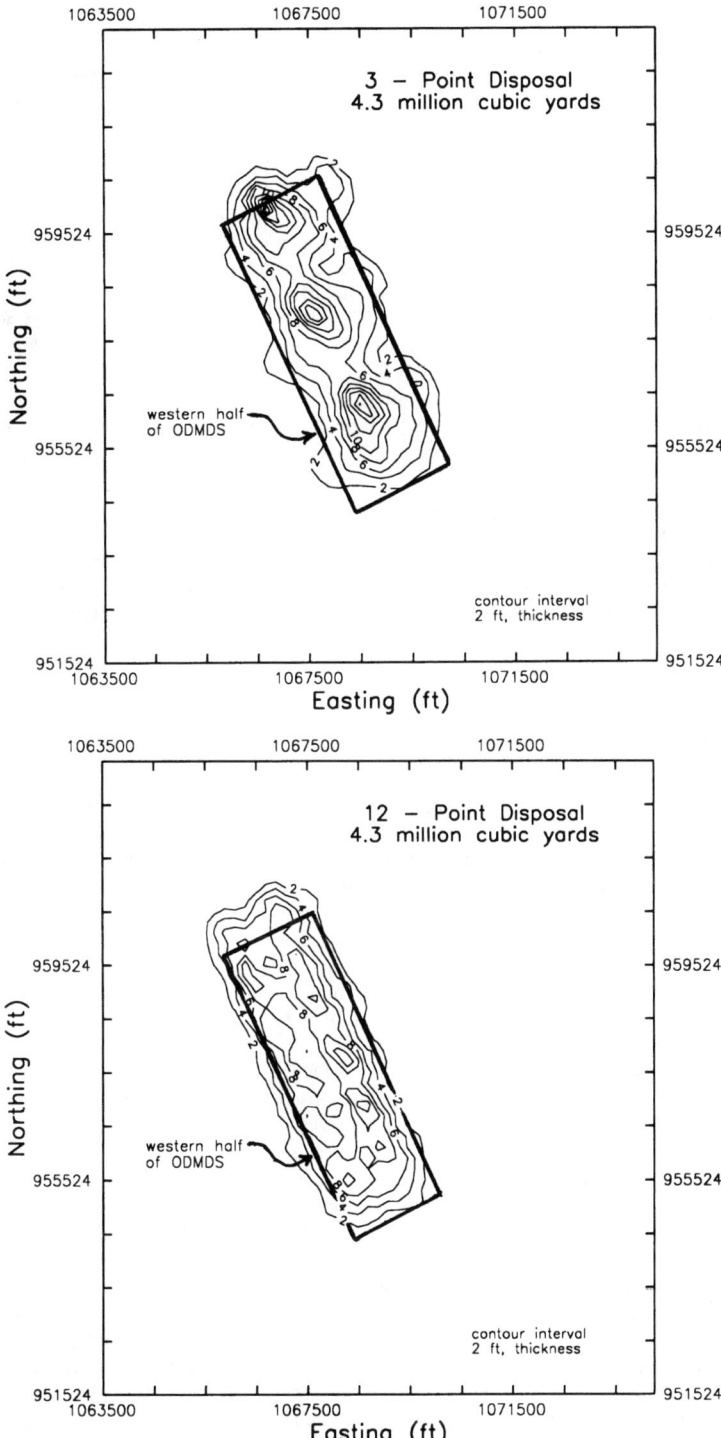

FIGURE 16.8. Contour plot of predicted bathymetry difference surface.

operation. According to the MDFATE results, the existing mound at ODMDS "B" will experience additional "growth" along the western and northwestern areas due to the proposed disposal operation. While difficult to predict, if the proposed 4.3 million yd^3 disposal operation were to have any additional negative effect upon the wave environment, it would be at the northern end of site "B" (Figures 16.4 and 16.5, section G–G').

The 12-point dump scenario should minimize potential wave effects due to the lower resultant mounding compared to the 3-point dump scenario. Overall, the 12-point scenario provides for more uniform dispersal of dredged material along the western half of site B than does the 3-point dump scenario. The difference between predisposal and simulated postdisposal conditions is shown in Figure 16.8. Upon disposal from the hopper dredge, both dump scenarios result in 5–10% of placed dredged material leaving the formal boundaries of site "B."

It is believed that disposal direction plays an important role in confining dredged material disposal within the formal site boundaries. The southeast (SE) approach direction assumed for this example produced the best case for keeping material within ODMDS "B." Temporary acquisition of a placement "buffer" extending 750 ft beyond the existing site boundaries would prevent dredged material from being placed outside of the site.

Based on MDFATE results, the dredged material placed in the western half of ODMDS "B" did not experience any significant movement of material except during severe storm events; what is placed, stays. This is due to the greater water depths along the western part of site "B" as compared to the depth at the top of the existing mound (50–60 ft).

SUMMARY

The manager of an ODMDS must simultaneously address the needs of obtaining regulatory approval for open water disposal of dredged sediments, ensure environmental compliance, and maximize disposal site efficiency. To satisfy these constraints, the ODMDS manager must ensure that dredged material placed offshore does not accumulate in a fashion that would pose a navigational hazard, must demonstrate that unacceptable adverse impacts to significant resources do not occur, and must attain maximum utilization of site volumetric capacity. Given these operational constraints, the proper management of many disposal sites requires a physically consistent methodology to quantitatively predict and assess the behavior of dredged material placed in an ODMDS.

Three numerical models developed under the U.S. Army Corps of Engineers Dredging Research Program have been discussed. STFATE predicts the fate of dredged material from a single disposal operation and is primarily used to address water column issues associated with regulatory approval. LTFATE is used to predict the long-term movement of an existing disposal mound without additional placement of material on the mound. Several auxiliary models for generating wave and current data required for the long-term simulation of disposal mounds were developed during the development of LTFATE.

MDFATE was developed to bridge the gap between the modeling of individual disposal events or "dumps" and tracking a myriad of disposals that occur within a disposal area over the duration of a disposal site's operative cycle. Thus MDFATE should prove to be a valuable tool to aid in the managing of a typical ODMDS. Other potential uses for MDFATE include addressing fate issues associated with subaqueous capping of contaminated sediments and the assessment of feeder berms/nearshore beach nourishment.

Advantages for using the numerical models presented in an ODMDS designation process and the subsequent management of the site include:

1. Streamlining decision making and interagency interaction by verifying ODMDS evaluation methods that increase the reliability of sediment fate estimates and minimize the need for redesignation efforts.
2. Increasing year-to-year dredged material placement options for ODMDS project management and the dredge contractor by reliably determining site capacity.
3. Reducing monitoring and prototype data collection costs as verified models increase confidence in regulatory adherence.
4. Reducing liability related to alleged navigation hazards associated with ODMDSs as verified models help avoid future adverse wave effects due to mounding dredged material.

The STFATE, LTFATE, and MDFATE models discussed can be obtained by contacting Clark McNair of the U.S. Army Engineer Waterways Experiment Station. Mr. McNair was the Program Manager of the DRP. His address is:

CE-ERD-HV-CD
3909 Halls Ferry Road
Vicksburg, MS 39180-6199

REFERENCES

Ackers, P. and White, R. W. 1973. "Sediment Transport: New Approach and Analysis," ASCE *Journal of the Hydraulics Division,* Vol. 99, No. HY11, pp. 2041–2060.

Allen, J. R. L. 1970. "The Avalanching of Granular Solids on Dune and Similar Slopes," *Journal of Geology,* Vol. 78, No. 3, pp. 326–351.

Bijker, E. W. 1971. "Longshore Transport Computations," ASCE *Journal of the Waterways, Harbors, and Coastal Engineering Division,* No. WW4, pp. 687–701.

Bokuniewicz, H. J., Gilbert, J., Gordon, R. B., Higgins, J. L., Kaminsky, P., Pilbeam, C. C., Reed, M., and Tuttle, C. 1978. "Field study of the mechanics of the placement of dredged material at open-water disposal sites. Vol. 1: Main text and appendices A–I," Technical Report D-78-7, U.S. Army Engineer Waterways Experiment Station, Vicksburg, MS.

Borgman, L. E. and Scheffner, N. W. 1991. "The Simulation of Time Sequences of Wave Height, Period, and Direction," Technical Report DRP-91-2, Coastal Engineering Research Center, U.S. Army Engineer Waterways Experiment Station, Vicksburg, MS.

Brandsma, M. G., and Divoky, D. J. 1976. "Development of models for prediction of short-term fate of dredged material discharged in the estuarine environment," Contract Report D-76-5, U.S. Army Engineer Waterways Experiment Station, Vicksburg, MS; prepared by Tetra Tech, Inc., Pasadena, CA.

Brandsma, M. G., and Sauer, T. C., Jr. 1983. "Mud discharge model: report and user's guide," Exxon Production Research Co., Houston, TX.

Chou, L. et al. 1998. "Dredged Material Management Plan (DMRP) for the Ports of New York and New Jersey—Modeling Studies to Support Island CDF and Constructed CAD Pit Design." Draft Report, U.S. Army Coastal and Hydraulic Laboratory, Waterways Experiment Station, Vicksburg, MS.

Cargrill, K. W. 1982. "Consolidation of Soft Layers by Finite Strain Theory," Miscellaneous Paper GL-82-3, Geotechnical Laboratory, U.S. Army Engineer Waterways Experiment Station, Vicksburg, MS.

Cargrill, K. W. 1985. "Mathematical Model of the Consolidation/Dessication Process in Dredged Material," Technical Report D-85-4, Geotechnical Laboratory, U.S. Army Engineer Waterways Experiment Station, Vicksburg, MS.

Gibson, R. E., England, G. L., and Cargrill, K. W. 1981. "Theory of One-Dimensional Consolidation

of Saturated Clays. 1. Finite Nonlinear Consolidation of Thin Homogeneous Layers," *Geotechnique,* Vol. 17, No. 3, pp. 261–273.
Hands, E. B. and Allison, M. C. 1991. "Mound Migration in Deeper Water and Methods of Catagorizing Active and Stable Mounds", Proceedings, Coastal Sediments '91, Seattle, Washington.
Hench, J. L., Luettich, R. A., Westerink, J. J., and Scheffner, N. W. 1994. "ADCIRC: An advanced three-dimensional circulation model for shelves, coasts, and estuaries," Technical Report DRP-92-6, U.S. Army Engineer Waterways Experiment Station, Vicksburg, MS.
Johnson, B. H. 1990. "User's guide for models of dredged material disposal in open water," Technical Report D-90-5, U.S. Army Engineer Waterways Experiment Station, Vicksburg, MS.
Johnson, B. H., McComas, D. N., McVan, D. C., and Trawle, M. J. 1993. "Development and verification of numerical models for predicting the initial fate of dredged material disposed in open water. Report 1—Physical model tests of dredged material disposal from a split-hull barge and a multiple bin vessel," Technical Report DRP-93-1, U.S. Army Engineer Waterways Experiment Station, Vicksburg, MS.
Johnson, B. H. and Fong, M. 1995. "Development and verification of numerical models for predicting the initial fate of dredged material disposed in open water, Report 2—Theoretical developments and verification," Technical Report DRP-93-1, U.S. Army Engineer Waterways Experiment Station, Vicksburg, MS.
Koh, R. C. Y., and Chang, Y. C. 1973. "Mathematical model for barged ocean disposal of waste," Technical Series EPA 660/2-73-029, U.S. Environment Protection Agency, Washington, DC.
Kraus, N. C. 1991. "Mobile Alabama, field data collection project, 18 August—2 September 1989—Report 1—Dredged material plume survey data report," Technical Report DRP-91-3, U.S. Army Engineer Waterways Experiment Station, Vicksburg, MS.
Krone, R. B., 1962, "Flume Studies of the Transport of Sediment in Estuarial Shoaling Processes," Technical Report, Hydraulic Engineering Laboratory, University of California, Berkely, CA.
Larson, M., Kraus, N. C., and Byrnes, M. R. 1990. "SBEACH: Numerical Model for Simulating Storm Induced Beach Change, Report 2," Technical Report CERC-89-9, Coastal Engineering Research Center, U.S. Army Engineer Waterways Experiment Station, Vicksburg, MS.
Luettich, R. A., Westerink, J. J., and Scheffner, N. W., 1992. "ADCIRC: An Advanced Three-Dimensional Circulation Model for Shelves, Coasts and Estuaries Report 1: Theory and Methodology of ADCIRC-2DDI and ADCIRC-3DL," Technical Report DRP-92-6, Report 1 of a Series, U.S. Army Engineer Waterways Experiment Station, Vicksburg, MS, November 1992.
Madsen, O. S. and Wikramanayake, P. N. 1991. "Simple models for turbulent wave-current bottom boundary layer flow," Contract Report DRP-91-1, U.S. Army Engineer Waterways Experiment Station, Vicksburg, MS.
Moritz, H. R. and Randall, R. E. 1992. "Users Guide for the Open Water Disposal Area Management Simulation," Contract Number DACW39-90-k-0015, Final Report to the U.S. Army Engineer Waterways Experiment Station, Submitted through the Texas A&M Research Foundation, College Station, TX.
Moritz, H. R. and Randall, R. E. 1995. "Simulating dredged material placement at open water disposal sites," ASCE *Journal of Waterway, Port, Coastal, and Ocean Engineering,* Vol. 121, No. 1.
Partheniades, E., 1962, "A Study of Erosion and Deposition of Cohesive Soils in Salt Water," Doctoral Dissertation, University of California, Berkely, CA.
Poindexter-Rollings, M. E., 1990. "Methodology for Analysis of Subaqueous Sediment Mounds," Technical Report D-90-2, US Army Engineer Waterways Experiment Station, Vicksburg, MS.
Scheffner, N. W. 1992. "Dispersion analysis of Humboldt Bay, California, Interim Offshore Disposal Site," Technical Report DRP 92-1, Coastal Engineering Research Center, U.S. Army Waterways Experiment Station, Vicksburg, MS.
Scheffner, N. W. and Tallent, J. R. 1994. "Dispersion Analysis of Charleston, South Carolina, Ocean Dredged Material Disposal Site," Miscellaneous Paper DRP-94-1, U.S. Army Engineer Waterways Experiment Station, Vicksburg, MS, March 1994.
Scheffner, N. W., Thevenot, M. M., Tallent, J. R., and Mason, J. M. 1995. "LTFATE: A model to investigate fate and stability of dredged material disposal sites," Technical Report DRP 95-1,

Coastal Engineering Research Center, U.S. Army Waterways Experiment Station, Vicksburg, MS.
Swart, D. H. 1976. "Predictive Equations Regarding Coastal Transport," in *Proceedings of the 15th Coastal Engineering Conference,* ASCE, pp. 1113–1132.
Tavolaro, J. F. 1984. "A Sediment Budget Study of Clamshell Dredging and Ocean Disposal Activities in the New York Bight," *Environmental Geology and Water Science,* Vol. 6, No. 3, pp. 133–140.
Truitt, L. C. 1988. "Dredged Material Behavior During Open-Water Disposal," *Journal of Coastal Research,* Vol. 4, No. 3, Summer 1988.
U.S. Army Engineer District, Portland 1995. "Simulation of dredged material disposal at Coos Bay ocean dredged material disposal site "F." Department of the Army, Corps of Engineers, Portland District, Portland, OR.
White, P. 1972. "Sediment Transport in Channels: A General Function," Report Int 104, Hydraulics Research Station, Wallingford, England.

# CHAPTER 17
REMOVAL OF CONTAMINATED SEDIMENT BY DREDGING

JOHN B. HERBICH, PH.D., P.E.
W. H. Bauer Professor Emeritus
Ocean Engineering Program
Civil Engineering Department
Texas A&M University
College Station, TX

Introduction	17.1
Extent of Contamination	17.5
Classification Methodologies	17.7
Evaluation Approach	17.10
Interim and Long-Term Controls	17.11
Interim Controls	17.11
Long-Term Controls and Technologies	17.12
Comparative Analysis of Controls and Technologies	17.14
Containment	17.15
Confined Disposal	17.15
Comparative Analysis	17.18
Risk Assessment	17.19
Dredging Equipment for Handling Contaminated Sediment	17.20
Mechanical Type Dredges	17.22
Enclosed Clamshell	17.22
Cable Arm Clamshell Bucket	17.22
Instrumented Backhoe	17.23
Mechanical–Hydraulic Dredges	17.23
Mud Cat	17.23
"Cleanup" System	17.24
"Dozer" Dredge	17.26
Hydraulic Suction Dredges	17.28
"Refresher" Dredge	17.29
"Matchbox" Dredge	17.30
Case Studies	17.30
Domestic	17.30
Baltimore, Maryland	17.30
Bayou Bonfouca, Louisiana	17.33
Cold Spring, New York	17.35
Commencement Bay, Washington	17.40

Eagle Harbor, Washington	17.50
Grand Calumet River, Indiana	17.50
Manistique River, Michigan	17.55
Massena, New York, General Motors Site	17.62
Massena, New York, Reynolds Metals Facility	17.71
New Bedford Harbor, Massachusetts	17.76
Waukegan, Illinois	17.86
Foreign	17.89
Antwerp, Belgium	17.89
Ketelmeer, the Netherlands	17.91
Minamata Bay, Japan	17.95
Penetanguishene, Canada	17.97
Scarborough Bluffs, Canada	17.98
Southampton, United Kingdom	17.99
Zeebrugge, Belgium	17.99
Summary	17.100
Appendix 1. Canadian Sediment Quality Criteria Guidelines	17.103
Appendix 2. The Use of Sediment Quality Guidelines as Benchmarks	17.103
References	17.115

INTRODUCTION

The American Society of Civil Engineers (ASCE) lists in its 1998 Report Card for America's Infrastructure a D– grade for hazardous waste. "More than 530 million tons of municipal and hazardous waste is generated in the U.S. each year. Since 1980, only 423 (32 percent) of the 1,200 Superfund Sites on the National Priorities List have been cleaned up. The National Priorities List is expected to grow to 2,000 in the next several years. The price tag for Superfund and related cleanup programs is an estimated $750 billion and could rise to $1 trillion over the next 30 years."

In the early 1980s it became apparent that the coastal and river sediments might be contaminated, but the actual extent of the contamination was not very well known. In the middle 1980s the problem of contaminated marine sediments emerged as an environmental issue of national importance. The pervasive and widespread nature of the problem resulted from decades of using coastal waters intentionally or unintentionally for waste disposal. Harbor areas in particular have been found to contain high levels of contaminants in bottom sediments due to wastes from urban, industrial, and riverine sources, as well as navigation.

Legislative authority for the management of contaminated marine sediments falls largely under three statutes: the Comprehensive Environmental Response, Compensation, and Liability Act of 1980 (CERCLA), the Marine Protection, Research and Sanctuaries Act of 1972 (MPRSA), and the Clean Water Act (CWA). The Comprehensive Environmental Response Compensation and Liability Act of 1980, as amended by the Superfund amendments and Reauthorization Act (SARA) of 1986, is aimed at the cleanup and remediation of inactive or abandoned hazardous waste sites, regardless of location. Superfund sites are currently ranked by the Environmental Protection Agency (EPA) based on the hazard they may pose to human health and the environment via releases of contaminants to groundwater, surface water, and air. Underwater accumulations of hazardous wastes in marine environments are unlikely to threaten human health except by way of food chain exposure. Under the 1986 Superfund amendments, however, EPA was required to modify its Hazard Ranking System to address "the damage to natural resources which may affect the human food chain and which is associated with any release (of a hazardous substance)" (Section 105(a)(2)). It is likely, therefore, that once this amendment is accepted, there will be a significant increase in the number of "underwater Superfund sites" in both coastal and inland areas.

The 1987 amendments to the Clean Water Act further required EPA to study and conduct projects related to the removal of toxic pollutants from Great Lakes bottom sediments (Section 118(c)(3)), and to identify and implement individual control strategies to reduce toxic pollutant inputs into contaminated waterway segments (Section 304(1)).

In response to Title II of the Marine Protection, Research, and Sanctuaries Act of 1972 (PL 92-532) and the National Ocean Pollution Planning Act, the National Oceanic and Atmospheric Administration (NOAA) Office of Marine Pollution Assessment conducts comprehensive interdisciplinary assessments of the effects of human activities on estuarine and coastal environments. Among these assessment activities is the National Status and Trends Program (NST), which attempts to create, maintain, and assess a long-term record of contaminant concentrations and biological responses to contamination in the coastal and estuarine waters of the United States. This assessment provides some insight into the extent of contamination nationally.

As a result of legislative responsibility and programmatic interests, a wide variety of federal agencies have shown active interest in this subject. EPA's responsibilities under Superfund and the CWA are the source of its interests in water quality concerns and remediation of uncontrolled hazardous waste sites. The U.S. Army Corps of Engineers (USACE) is involved because of its responsibility to dredge and maintain navigable rivers and harbors. The USACE also assists in the design and implementation of remedial cleanup actions under Superfund. NOAA has the responsibility for assessing the potential threat of Superfund sites to coastal marine resources as a natural resource trustee as well as under its NST program. The U.S. Fish and Wildlife Service has legal authority for various endangered coastal species, food chain relationships, and habitat considerations, all of which are potentially impacted by contaminated sediments. The Navy has had experience in assessing contaminated sediments and now must grapple with such problems in maintaining home ports for Navy vessels.

In response to this emerging problem, the National Research Council convened the Committee on Contaminated Marine Sediments. A three-year study culminated in an extensive report entitled *Contaminated Marine Sediments—Assessment and Remediation* (Marine Board, 1989).

The U.S. Army Corps of Engineers and the U.S. Environmental Protection Agency (EPA) share the responsibility of regulating dredged material management alternatives under the Marine Protection, Research, and Sanctuaries Act (MPRSA) and the Federal Water Pollution Control Act Amendments of 1972, also called the Clean Water Act (CWA). Such management alternatives must also comply with the applicable requirements of the National Environmental Policy Act (NEPA).

The Technical Framework is intended to serve as a consistent "road map" for USACE and EPA personnel to follow in evaluating the environmental acceptability of dredged material management alternatives. Specifically, its major objectives are to provide:

- A general technical framework for evaluating the environmental acceptability of the full continuum of dredged material management alternatives (open-water placement, confined (diked) placement, and beneficial uses applications)

- Additional technical guidance to supplement present implementation and testing manuals for addressing the environmental acceptability of available management options for the discharge of dredged material in both open-water and confined sites

- Enhanced consistency and coordination in USACE and EPA decision making in accordance with Federal environmental statutes regulating dredged material management

The Technical Framework can be applied nationwide and is relatively general, but comprehensive. Because the Technical Framework provides national guidance, flexibility is necessary. It should not be followed rigidly; rather, it should be used as a technical guide to evaluate the commonly important factors to be considered in managing dredged material in an environmentally acceptable manner. A sequence of broad steps is followed, as shown in Figure 17.1.

For both open-water and confined placement, alternatives include the following broad steps:

- Determining the characteristics of disposal sites
- Evaluating direct physical impacts and site capacity
- Evaluating contaminant pathways of concern
- Evaluating control measures
- Retaining environmentally acceptable alternatives

For beneficial use applications, the detailed assessment includes:

- Determining beneficial use needs and opportunities
- Evaluating physical suitability of material for proposed uses.

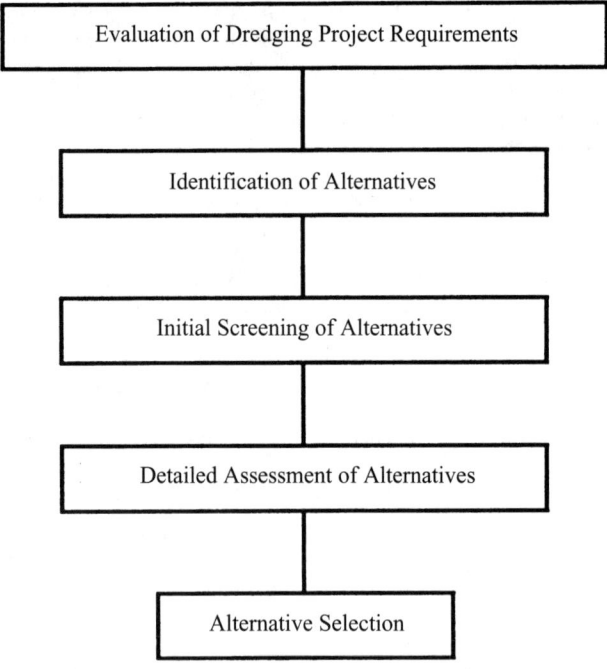

FIGURE 17.1. Initial steps in technical framework.

- Evaluating logistical and management requirements
- Evaluating environmental suitability.
- Retaining environmentally acceptable alternatives (Palermo, 1993)

EXTENT OF CONTAMINATION

Before evaluating the extent of sediment contamination in U.S. estuaries and rivers, one needs to define what contamination is, and what is the threshold of contamination.

The Marine Board (1989) definition of contaminated sediments is "Contaminated sediments are those that contain chemical substances at concentrations which pose a known or suspected environmental or human health threat."

Many marine sites are known to contain sediments with high levels of anthropogenic chemicals or to have altered biological characteristics. However, there are no generally accepted definitions of contamination that trigger consideration of remedial action. The sites that require the most urgent attention are those reservoirs of contamination that affect regions or that have the most severe impacts on health and the environment.

Many contaminated marine sediments are located along all coasts of the contiguous United States, both in local "hot spots" and distributed over large areas. Some of these sites (but not many) have been well characterized. Existing data on individual sites and their contamination vary widely in content and organization. Assessments using available data have been conducted on the national extent of contamination and have identified a partial picture of the total contaminated sediment problem. These studies have shown that a wide variety of contaminants are found in sediments, including heavy metals, polychlorinated biphenyls (PCBs), DDT, and polynuclear aromatic hydrocarbons (PAHs).

The location and extent of contaminated marine sediments were not at that time comprehensively assessed on a national basis to identify site-specific remediation targets. In regions of concern, or in areas of known hot spots, special attention should be directed to identifying and characterizing specific contaminated sites. The search for new sites or the reclassification of known sites should proceed concurrently with remedial action.

A major effort was undertaken by the United States Environmental Protection Agency (USEPA) as required by the Water Resources Development act of 1992 (WRDA). The report describes the accumulation of chemical contaminants in river, lake, ocean, and estuary bottoms and includes a screening assessment of the potential for associated adverse effects to human and environmental health. The U.S. Army Corps of Engineers and the National Oceanic and Atmospheric Administration (NOAA) were consulted in compiling the data and preparing the report (USEPA, 1997b). EPA studied available data from the 65% of the 2,111 watersheds in the continental United States and identified 96 watersheds containing "areas of probable concern" (Figure 17.2). In portions of these watersheds, environmental conditions may be unsuitable for bottom-dwelling creatures, and fish that live in these waters may contain chemicals at levels that make them unsafe for regular consumption. Areas of probable concern are located in regions affected by urban and agricultural runoff, municipal and industrial waste discharge, and other pollution sources.

The USEPA estimates that approximately 10% of the sediment underlying our nation's surface water is sufficiently contaminated with toxic pollutants to pose potential risks to fish and to humans and wildlife that consume fish. This represents about 1.2 billion cubic yards of contaminated sediment out of the approximately 12 billion cubic yards of total surface sediments (upper 5 centimeters) where many bottom dwelling organisms live and where the primary exchange processes between the sediment and overlying surface water occur. Approximately 300 million cubic yards of sediments are dredged from

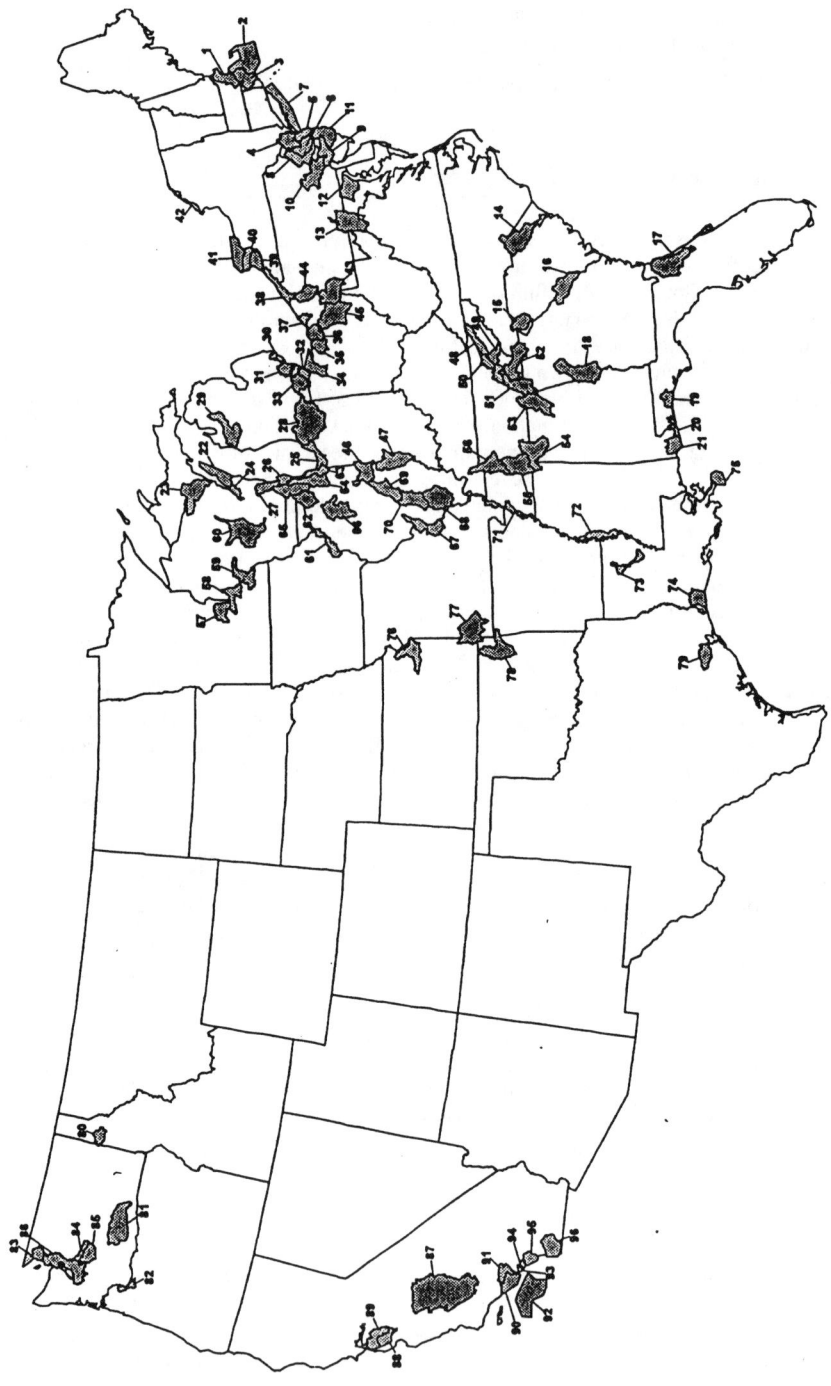

FIGURE 17.2. Watersheds containing areas of probable concern in sediment contamination (APCs) (EPA, 1997).

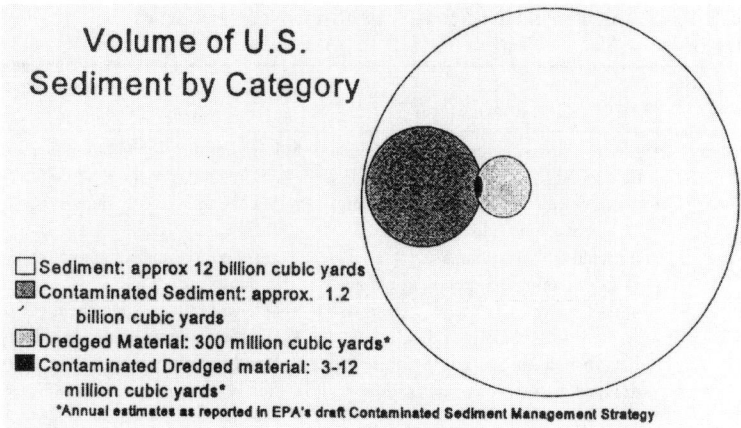

FIGURE 17.3. Volume of U.S. sediment by category.

harbors and shipping channels annually to maintain commerce, and about 3–12 million cubic yards of those are sufficiently contaminated to require special handling and disposal. These amounts are graphically illustrated in Figure 17.3.

USEPA has studied data from 1,372 of the 2,111 watersheds in the continental United States. Of these, USEPA has identified 96 watersheds that contain "areas of probable concern" where potential adverse effects of sediment contamination are more likely to be found. These areas, identified in Figure 17.2, are on the Atlantic, Gulf of Mexico, Great Lakes, and Pacific coasts, as well as in inland waterways in regions affected by urban and agricultural runoff, municipal and industrial waste discharges, and other pollution sources. Some of these areas have been studied extensively, and now have appropriate management actions in place. However, others may require further evaluation to confirm that environmental effects are occurring.

CLASSIFICATION METHODOLOGIES

A variety of biological and chemical sediment classification methods are available. Individually or in combination, they attempt to systematically characterize marine sediments with elevated levels of contaminants, and correlate such concentration increases with adverse biological effects. With one possible exception (the acute amphipod bioassay), none of these techniques are routinely used and each has its limitations. Indeed the cost and complexity of a number of these tests virtually ensures they will be used routinely only at large sites.

The committee examined several contaminated sediment classification techniques: sediment bioassays, sediment quality triad approach, apparent effects threshold technique, and equilibrium partitioning. Some of the advantages and disadvantages of each for remedial action screening and sediment quality criteria development are presented in Table 17.1.

From a remedial cleanup standpoint, the most useful sediment testing and classification procedures would be those that are simple and inexpensive, with rapidly available

TABLE 17.1. Assessment of sediment classification methodologies (Marine Board, 1989)

Classification method	Advantages	Disadvantages
• Bioassay	• Follows toxicological methods developed for water quality criteria • A direct measure of sediment toxicity • Does not require identification of individual contaminants • Does not assume a specific route of uptake • Acute results available quickly • Established test procedures in use for dredged material characterization	• Requires development of standard chronic bioassay methodologies • May be more costly than some chemical analyses • Difficult to translate laboratory results to natural conditions • Difficult to determine chemical effects • Does not address human health impacts • Results of chronic tests may not be timely • May not identify causative contaminants
Sediment quality triad	• Based on combination of laboratory and field data indicating effects of actual contaminated sediments • Based on observed biological effects • Does not assume a specific route of chemical uptake • Applicable to complex mixtures	• Limited by the availability of existing data or by the ability to collect large amounts of new data • Available data may be of highly variable quality • Difficult to translate laboratory results to natural conditions • Does not address human health impacts • May not identify causative contaminants • Indicators are not independent; covary with grain size and organic carbon content • Potentially not comparable between geographic locations • Does not consider chemical bioavailability from site to site
Apparent effects threshold	• Uses existing data (from field and laboratory; e.g., Sediment quality triad) • Applicable to all chemicals and all biological effects • Most useful for prioritizing contaminated areas within a large site • Based on observed biological effects • Does not assume a specific route of chemical uptake • Applicable to complex mixtures	• Limited by the availability and quality of existing data • Varies with choice of biological effects indicator • Relies on correlations/may not identify causative contaminants • Potentially not comparable between geographic locations • May be both over- and under-protective • Difficult to translate laboratory results to natural conditions • Does not address human health impacts • Multicompound interactions not accounted for • Indicators are not independent; covary with grain size and organic content

TABLE 17.1. *Continued*

Classification method	Advantages	Disadvantages
Equilibrium partitioning	• Provides a chemical specific criterion • Utilizes large toxicological data base incorporated in water quality criteria and other toxicological endpoints • Relies on well-developed partitioning theory • Accounts for the bioavailability of the chemical interest • Provides a standard basis for comparison within and among sites • Where data are available allows quick and inexpensive characterization • Incorporates a built-in "how clean is clean" standard • Is a direct measurement of sediment characteristics • Can be readily incorporated into existing regulatory frameworks	• Does not consider complex mixtures and chemical interactions • Currently limited to hydrophobic neutral organic compounds • Does not address human health impacts • Limited to contaminant for which both water quality criteria (or other suitable toxicological endpoints) and sediment-water partitioning coefficients are available • Relies on K_{oc}^{a}* measurements, which are often variable • Does not account for contaminant uptake by ingestion of particles or direct absorption/adsorption from sediments • Sediment and water may not be at equilibrium with respect to contaminant concentration • Does not use toxicological data derived from the sediment of interest • Assumption of constant bioaccumulation factor for various contaminants and organisms is questionable

*K_{oc}^{a} = carbon normalized sediment–water partition coefficient.

test results. If sediment quality criteria methodologies are adopted by USEPA, a routine basis for establishing the presence of unacceptable high levels of sediment contaminants may be available. The design and implementation of remedial action for contaminated sediments are likely to be delayed and frustrated unless one can readily determine "how clean is clean." Development of an interim working methodology to establish such a criterion would alleviate the delay (Marine Board, 1989).

The USEPA (1997) report was sent to Congress in four volumes:

- Volume 1: *National Sediment Quality Survey*—Screening analysis to qualitatively assess the probability of associated adverse human or ecological effects based on a weight-of-evidence evaluation
- Volume 2: *Data Summaries for Areas of Probable Concern (APCs)*—Sampling station location maps and chemical and biological summary data for watersheds containing APCs
- Volume 3: *National Sediment Contaminant Point Source Inventory*—Screening analysis to identify probable point source contributors of sediment pollutants

- Volume 4: *National Sediment Contaminant Nonpoint Source Inventory*—Screening analysis to identify probable nonpoint source contributors of sediment pollutants (in preparation for subsequent biennial reports)

The NSI is the largest set of sediment chemistry and related biological data ever compiled by USEPA. It includes approximately two million records for more than 21,000 monitoring stations across the country. To efficiently collect usable information for inclusion in the NSI, USEPA sought data that were available in electronic format, represented broad geographic coverage, and represented specific sampling locations identified by latitude and longitude coordinates. The minimum data requirements for inclusion of computerized data in the NSI were monitoring program, sampling date, latitude and longitude coordinates, and measured units. Additional data fields such as sampling method and other quality assurance/quality control information were retained in the NSI if available, but were not required for a data set to be included in the NSI.

The NSI includes data from the following data storage systems and monitoring programs:

- Selected data from USEPA's Storage and Retrieval System (STORET)
- NOAA's Coastal Sediment Inventory (COSED)
- USEPA's Ocean Data Evaluation System (ODES)
- USEPA Region 4's Sediment Quality Inventory
- Gulf of Mexico Program's Contaminated Sediment Inventory
- USEPA Region 10/USACE Seattle District's Sediment Inventory
- USEPA Region 9's Dredged Material Tracking System (DMATS)
- USEPA's Great Lakes Sediment Inventory
- USEPA's Environmental Monitoring and Assessment Program (EMAP)
- United States Geological Survey (Massachusetts Bay) Data

In addition to sediment chemistry data, the NSI includes tissue residue, toxicity, benthic abundance, histopathology, and fish abundance data. The sediment chemistry, tissue residue, and toxicity data were evaluated for this report to Congress. Data from 1980 to 1993 were used in the National Sediment Inventory (NSI) data evaluation, but older data also are maintained in the NSI.

EVALUATION APPROACH

The WRDA defines contaminated sediment as "aquatic sediment that contains chemical substances in excess of appropriate geochemical, toxicological, or sediment quality criteria or measures, or is otherwise considered to pose a threat to human health or the environment." The approach used to evaluate the NSI data focuses on the risk to benthic organisms exposed directly to contaminated sediments, and the risk to human consumers of organisms exposed to sediment contaminants. EPA evaluated sediment chemistry data, chemical residue levels in edible tissue of aquatic organisms, and sediment toxicity data taken at the same sampling station (where available) using a variety of assessment methods.

In the 1997 report, EPA associates sampling stations with their "probability of adverse effects." Each sampling station falls into one of three categories, or tiers:

- Tier 1: associated adverse effects are probable
- Tier 2: associated adverse effects are possible, but expected infrequently

- Tier 3: no indication of associated adverse effects (any sampling station not classified as Tier 1 or Tier 2; includes sampling stations for which substantial data were available, as well as sampling stations for which limited data were available) (USEPA, 1997).

Based on the evaluation, sediment contamination exists at levels where associated adverse effects are probable (Tier 1) in some locations within each region and state of the country. The water bodies affected include streams, lakes, harbors, nearshore areas, and oceans. A number of specific areas in the United States had large numbers of sampling stations where associated adverse effects are probable. Puget Sound, Boston Harbor, the Detroit River, San Diego Bay, and portions of the Tennessee River were among those locations. Several U.S. harbors (e.g., Boston Harbor, Puget Sound, Los Angeles, Chicago, Detroit) appear to have some of the most severely contaminated sediments in the country. This finding is not surprising since major U.S. harbors have been affected throughout the years by large volumes of boat traffic, contaminant loadings from upstream sources, and many local point and nonpoint sources.

Thousands of other water bodies in hundreds of watersheds throughout the country contain sampling stations classified as Tier 1. Many of these sampling stations may represent isolated "hot spots" rather than widespread sediment contamination, although insufficient data were available in the NSI to make such a determination. EPA's River Reach File 1 (RF1) delineates the nation's rivers and waterways into segments, or reaches, of approximately 1 to 10 miles in length. Based on RF1, approximately 11% of all river reaches in the United States contained NSI sampling stations. More than 5,000 sampling stations in approximately 2,400 river reaches across the country (4% of all reaches) were classified as Tier 1. Another 10,000 sampling stations were classified as Tier 2. In total, over 5,000 river reaches in the United States—approximately 8% of all river reaches—include at least one Tier 1 or Tier 2 station.

It appears that further investigation and assessment of contaminated sediment is warranted.

The EPA's recommendations relate to five activities of information needs:

1. Further investigate conditions in the 96 targeted watersheds
2. Coordinate efforts to address sediment quality through watershed management programs
3. Incorporate a weight-of-evidence approach and measures of chemical bioavailability into sediment monitoring programs
4. Evaluate the National Sediment Inventory's (NSI's) coverage and capabilities and provide better access to information in the NSI
5. Develop better monitoring and assessment tools

INTERIM AND LONG-TERM CONTROLS

The Marine Board (1997) discussed the interim and long-term controls and technologies currently available in the United States.

Interim Controls

Interim controls may prove helpful when sediment contamination poses an imminent hazard. Identification of an imminent hazard is usually a matter of judgment, but in general an imminent hazard exists when contamination levels exceed by a significant amount the sum of a defined threshold level plus the associated uncertainty. Administrative interim controls (e.g., signs, health advisories) have been used a number of times. Only two applications of structural interim approaches (e.g., thin caps) were identified, but additional structural approaches, such as the use of confined disposal facilities (CDFs) for temporary storage, appear promising. Few data are available concerning the effectiveness of interim controls because to date they have not been used often or evaluated in detail (Marine Board, 1997).

Long-Term Controls and Technologies

Technologies for remediating contaminated sediments are at various stages of development. Sediment-handling technologies are the most advanced, although benefits can be realized from improvements in the precision of dredging (and, concurrently, site characterization). The state of practice for in situ controls ranges from immature (e.g., bioremediation) to evolving (e.g., capping). Ex situ containment is commonplace. A number of existing ex situ treatment technologies can probably be applied successfully to treating contaminated sediments, but full-scale demonstrations are needed to determine their effectiveness. But these technologies are expensive, and it is not clear whether unit costs would drop significantly in full-scale implementation.

The cost of cleanup depends on the number of steps involved—the more handling required, the higher the cost—and the type of approach used. The costs of removing and transporting contaminated sediments (generally less than \$15 to \$20/yd^3) tend to be higher than costs of conventional navigational dredging (seldom more than \$5/yd^3) but much lower than the costs of treatment (usually more than \$100/yd^3). Volume reduction (i.e., removing only sediments that require treatment and entraining as little water as possible) will mean greater cost savings than increased production rates; improved site characterization coupled with precision dredging techniques hold particular promise for reducing volume. Treatment costs may also be reduced through pretreatment.

In situ management offers the potential advantage of avoiding the costs and potential material losses associated with the excavation and relocation of sediments. Among the inherent disadvantages of in situ management measures is that they are seldom feasible in navigation channels that are subject to routine maintenance dredging. In addition, monitoring needs to be an integral part of any in situ approach to ensure effectiveness over the long term.

Natural recovery is a viable alternative under some circumstances and offers the advantages of low cost and, in certain situations, the lowest risk of human and ecosystem exposure to sediment contamination. Natural recovery is most likely to be effective where surficial concentrations of contaminants are low, where surface contamination is covered over rapidly by cleaner sediments, or where natural processes destroy or modify the contaminants, so that contaminant releases to the environment decrease over time. A disadvantage of natural recovery is that the sediment bed is subject to resuspension by storms or anthropogenic processes. For natural recovery to be pursued with confidence, the physical, chemical, and hydrological processes at a site need to be understood adequately; however, no capability currently exists for completely quantifying chemical movements. Extensive site-specific studies may be required.

In situ management promotes chemical isolation and may protect the underlying contaminated sediments from resuspension until naturally occurring biological degradation of contaminants has occurred. The original bed must be able to support the cap, suitable capping materials must be available to create the cap, and suitable hydraulic conditions (including water depth) must exist to permit placement of the cap and to avoid compromising the integrity of the cap. Changes in the local substrate, the benthic community structure, or the bathymetry at a depositional site may subject the cap to erosion. Improved long-term monitoring methods are needed. A regulatory barrier to the use of capping is the language of Superfund legislation (Section 121[b]), which gives preference to "permanent" controls. Capping is not considered by regulators to be a permanent control, but available evidence suggests that properly managed caps can be effective, and many capping projects have been successful.

Neither in situ immobilization nor chemical treatment of contaminated sediments has been demonstrated successfully in the marine environment, although both concepts are attractive because they do not require sediment removal. Their application would be complicated by the need to isolate sediments from the water column during treatment, by inaccuracies in reagent placement, and by the need for long-term follow-up monitoring. Other constituents (e.g., natural organic matter, oil and grease, metal sulfide precipitates) could interfere with chemical oxidation. Immobilization techniques may not be applicable to fine-grained sediments with a high water content.

Biodegradation has been observed in soils, in groundwater, and along shorelines contaminated by a variety of organic compounds (e.g., petroleum products, PCBs, polyaromatic hydrocarbons, pesticides). However, the use of biodegradation in subaqueous and especially marine environments presents unresolved microbial, geochemical, and hydrological issues and has yet to be demonstrated (USEPA, 1997).

When sediments must be moved for ex situ remediation or confinement, efficient hydraulic and mechanical methods are available for removal and transportation. Most dredging technologies can be used successfully to remove contaminated sediments; however, they have been designed for large-volume navigational dredging rather than for the precise removal of hot spots (surgical dredging). Promising technologies offering precision control include electronically positioned dredge heads and bottom-crawling hydraulic dredges. The latter may also have the capability to dredge in depths beyond the standard maximum operating capacity. The cost effectiveness of dredging innovations can best be judged by side-by-side comparison to technologies in current use. Dredging equipment for surgical dredging has been developed in other countries; however, this equipment may not be deployed in the United States because of the Jones Law (unless the hull of the vessel has been constructed in a U.S. shipyard).

Containment technologies, particularly confined disposal facilities (CDFs), have been used successfully in numerous projects. A CDF can be effective for long-term containment if it is well designed to contain sediment particles and contaminants and if a suitable site can be found. A CDF can also be a valuable treatment or interim storage facility, allowing the separation of sediments for varying levels of treatment and, in some cases, beneficial reuse. Costs are reasonable; in some parts of the country it may be cheaper to reuse CDFs than to build new ones. Disadvantages of this technology include the imperfect methods for controlling contaminant release pathways. There is also a need for improved long-term monitoring methods.

Contained aquatic disposal (CAD) is applicable particularly to contaminated sites in shallow waters where in situ capping is not possible and to the disposal and containment of slightly contaminated material from navigation dredging. Although the methodology has been developed, CAD has not been widely used. Among the advantages of a CAD are that it can be filled with conventional dredging equipment and the chemical environment surrounding the cap remains unchanged. Disadvantages include the possible loss of contaminated sediments during placement operations. Improved tools are needed for the design of sediment caps and armor layers and for the valuation of their long-term stability and effectiveness.

Scores of ex situ treatment technologies have been bench tested and pilot tested, and some warrant larger-scale testing in marine systems, depending on their applicability to particular problems. Chemical separation, thermal desorption, and immobilization technologies have been used successfully but are expensive, complicated, and only effective for treating certain types of sediments. Similarly, because of extraordinarily high unit costs, thermal and chemical destruction techniques do not appear to be near-term, cost-effective approaches for the remediation of large volumes of contaminated dredged sediment.

Ex situ bioremediation, which is not as far along in development as are other ex situ treatment approaches, presents so many technical problems that its application to contaminated sediments would be expensive. If these technical problems can be resolved, however, ex situ bioremediation has the potential, over the long term, for the cost-effective remediation of large volumes of sediments. Ex situ bioremediation is much more promising than in situ bioremediation because conditions can be controlled more effectively in a contained facility. The approach has been demonstrated on a pilot scale with some success, but complex questions remain concerning how to engineer the system.

Comparative Analysis of Controls and Technologies

Table 17.2 summarizes the overall assessment of the feasibility, effectiveness, practicality, and costs of controls and technologies. For each control and technology, the four characteristics were rated separately on a scale of 0 to 4, with 4 representing the best available (not necessarily the best theoretically possible) features. The effectiveness rating is an estimate of contaminant reduction or isolation and removal efficiency; scores represent a

TABLE 17.2. Comparative analysis of technology categories (Marine Board, 1997)

Approach	Feasibility	Effectiveness	Practicality	Cost
Interim control				
Administrative	0	4	2	4
Technological	1	3	1	3
Long-term control				
In situ				
Natural recovery	0	4	1	4
Capping	2	3	3	3
Treatment	1	1	2	2
Sediment removal and transport	2	4	3	2
Ex situ Treatment				
Physical	1	4	4	1
Chemical	1	2	4	1
Thermal	4	4	3	0
Biological	0	1	4	1
Ex situ containment	2	4	2	2
Scoring				
0	<90%	Concept	Not acceptable, very uncertain	$1,000/yd^3
1	90%	Bench		$100/yd^3
2	99%	Pilot		$10/yd^3
3	99.9%	Field		$1/yd^3
4	99.99%	Commercial	Acceptable, certain	<$1/yd^3

range of less than 90% to nearly 100%. The feasibility rating represents the extent of technology development, with 0 for a concept that has not been verified experimentally and 4 for a technology that has been commercialized. The practicality ranking reflects public acceptance; 0 means no tolerance for an activity and 4 represents widespread acceptance. The cost ranking is inversely related to the cost of using the control of technology (not including expenses associated with monitoring, environmental resource damage, or the loss of use of public facilities).

The overall pattern of the ratings underscores the need for trade-offs in the selection of technologies. No single approach emerges with the highest scores across the board, and each control or technology has at least one low or moderate ranking. In general, interim controls and in situ approaches are feasible and low in cost but less effective than the most practical ex situ approaches, which tend to be high in cost and complexity. Decisions about which approach is the most appropriate must be made on a project-by-project basis.

CONTAINMENT

Containment is a common approach to the ex situ management of contaminated sediments that have been dredged and transported. Ex situ containment has been widely used, at perhaps several hundred sites in North America. Containment technologies can be implemented in various ways. Figure 17.4 is an illustration depicting containment technologies, in situ capping, and deep-ocean dumping. The illustration highlights the distinctions among the different types of containment structures, particularly in terms of transport and isolating barriers. The subsections that follow assess confined disposal facilities (CDF) and contained aquatic disposal (CAD).

CONFINED DISPOSAL

Confined disposal involves the placement of dredged material within diked near-shore, island, or land-based CDFs. Confinement or retention dikes or structures in a CDF enclose the disposal area above any adjacent water surface, isolating the dredged material from

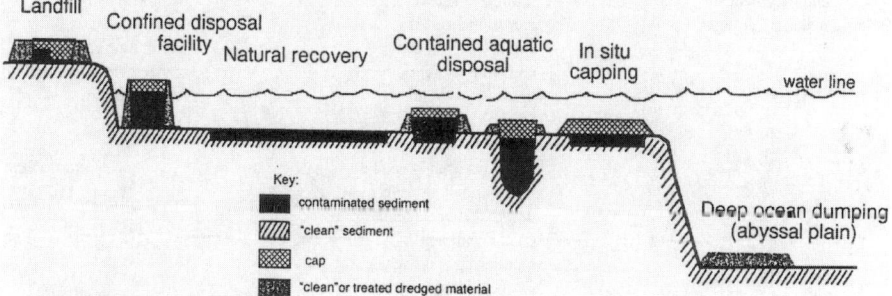

FIGURE 17.4. Conceptual illustration of containment, disposal, and natural recovery technologies. Dumping contaminated sediments in water anywhere but in the open ocean is not permitted under the Marine Protection, Research and Sanctuaries Act. (*Source:* Marine Board, 1997).

adjacent waters during placement. The enclosed disposal area of CDFs distinguishes this disposal method from other methods, such as disposal on confined land, placement on wetland, or CAD, which is a form of subaqueous capping (USACE and EPA, 1992). The placement of dredged material in CDFs differs from the placement of waste materials in licensed solid-waste landfills.

The two objectives in the design and operation of CDFs that are used for contaminated sediments are to provide adequate storage capacity to meet dredging requirements and to maximize efficiency in controlling contaminant releases. Possible migration pathways of contaminants from CDFs include effluent discharges to surface water during filling operations and subsequent settling and dewatering, rainfall-generated runoff, leaching into groundwater, volatilization to the atmosphere, and direct uptake. Direct uptake includes plant uptake subsequent cycling through food chains, and direct uptake by animals. Effects on surface water quality, groundwater quality, air quality, plants, and animals depend on the characteristics of the dredged material, the management and operation of the site during and after filling, and the proximity of the CDF to potential receptors of the contaminants. If evaluations of contaminant pathways indicate that impacts will be unacceptable, special or additional management and contaminant control measures can be considered, including modification to the dredging operation or site; treatment of effluent, runoff, or leachate; treatment of dredged material solids; and site controls, such as surface covers or liners. Techniques for evaluating pathways have been developed (USACE/EPA, 1992; Myers and Zappi, 1992). Key considerations are summarized in Table 17.3.

TABLE 17.3. Confined disposal facility

State of practice (system maturity, known pilot studies, etc.)	Applicability	Advantages/effectiveness	Limitations	Research needs
(a) The most commonly used placement alternative for contaminated sediments; (b) hundreds of sites nationwide for navigation dredging projects; (c) often used for pretreatment prior to final placement or as final sediment placement site for remediation projects.	Applicable to a wide variety of sediment types and project conditions.	(a) Low cost compared to ex situ treatment; (b) compatible with a variety of dredging techniques, especially direct placement by hydraulic pipeline; (c) proper design results in high retention of suspended sediments and associated contaminants; (d) engineering for basic containment normally involves conventional technology; (e) controls for contaminant pathways usually can be incorporated into site design and management; (f) conventional monitoring approaches can be used; (g) site can be used for beneficial purposes following closure, with proper safeguards.	(a) Does not destroy or detoxify contaminants unless combined with treatment; (b) control of some contaminant-loss pathways may be expensive.	(a) Design approaches, such as covers and liners, needed for low-cost containment controls; (b) design criteria for treatment of releases or control strategies for high-profile contaminant; (c) methods for site management to allow restoration of site capacity and potential use of treated materials.

The cost of using CDF's to contain contaminated sediments ranges from $15 to $50/yd³, plus the operation and maintenance costs associated with closed CDFs (USEPA, 1993). Thus, storage in a CDF can be less expensive than landfill disposal, which can cost $20 to $120/yd³ (USEPA, 1994). The design, construction, and operation of CDFs require conventional engineering approaches that have been used successfully for numerous other projects (USACE, 1987). A CDF can foster harbor development in urban areas; however, near-shore space may be difficult to find if wetlands must be preserved. In some cases, it may be difficult to find an area and construct dikes in deep water to accommodate large volumes of material. If a freshwater CDF is located above an aquifer, controls may be required to prevent groundwater contamination or oxygenation of the sediment by rainfall, because the acids formed may cause the release of metals to groundwater cleanup projects.

The state of practice of CAD for restoring bottom sediment is not well advanced (Table 17.4). Like in situ capping, a successful CAD operation requires only that the cap that isolates the contaminated material be accurately placed and well maintained. It is important that CAD be carried out in areas where erosion is minimal or controllable. The USACE has developed guidelines for planning CAD projects (Truitt, 1987a,b), determining the required capping thickness (Sturgis and Gunnison, 1984), determining design requirements (Palermo, 1991a), selecting sites (Palermo, 1991b), evaluating equipment and placement techniques (Palermo, 1991c), and evaluating monitoring systems (Palermo et al., 1991c). In cooperation with the EPA, the USACE has also developed guidelines for in-place capping for restoration purposes (Palermo and Miller, 1995). A joint USACE and EPA technical document for the subaqueous capping of dredged material is also in preparation.

TABLE 17.4. Contained aquatic disposal (Marine Board, 1997)

State of practice (system maturity, known pilot studies, etc.)	Applicability	Advantages/ effectiveness	Limitations	Research needs
Limited application. Reviews exist concerning (a) necessary data, equipment, and procedures; (b) engineering considerations; (c) guidelines for cap armoring design; (d) predicting chemical containment effectiveness	(a) Costs and environmental effects of relocation are factors; (b) suitable types and quantities of cap material are available; (c) hydrologic conditions will not compromise the cap; (d) cap can be supported by original bed; (e) appropriate for sites where excavation is problematic or removal efficiency is low, (f) cap material is compatible with existing aquatic environment.	(a) Eliminates need to remove contaminated sediments; (b) cost effective for sites with large surface areas; (c) effective in containing contaminants by reducing bioaccessibility; (d) promotes in situ chemical or biological degradation; (e) maintains stable geochemical and geohydraulic conditions, minimizing containment release to surface water, groundwater, and air.	(a) Laboratory and field validation of capping procedures and tools; (b) analysis of data from existing and ongoing field demonstrations to support capping effectiveness; (c) test for chemical release during bed placement and consolidation; (d) tests to evaluate and simulate the effects of cap penetration by deep burrowing organisms; (e) simulate and evaluate consequences of mixing; (f) potential loss of contaminants to the water column may require controls during placement.	(a) Design criteria for treatment of releases or control strategies for high-profile contaminants; (b) improved methods for evaluation of potential contaminant release pathways; (c) develop reliable cost estimates.

A major advantage of CAD is that it can be performed with conventional dredging equipment, although the equipment may have to be operated in special ways. Also, unlike the CDF option, the chemical environment surrounding the contaminated material remains virtually unchanged because the sediment remains in the waters of its origin. A major consideration is the potential loss of contaminated sediments during placement operations. Controls comparable to the ones used with CDF technology must be applied to minimize such losses. Research is needed to improve control capabilities, to determine the effects of losses on the ecosystem, and to assess the associated risks. Research on the long-term effectiveness of various types of capping, including CAD, is also needed. Resolution of these issues would probably enhance the acceptability of this technology for restoring contaminated sediment sites. Accurate data on the actual costs of CAD are not available.

Another possible approach to subaqueous offshore containment, at least for small volumes of material, might be to encase contaminated sediments in woven or nonwoven permeable synthetic fabrics. The casings could be expected to eliminate losses during placement and to contain the contaminated sediment on the seafloor. Fabric has been used for some 30 years to make various types of receptacles, such as air bags, geotextile tubing, and geotextile containers (see Fowler et al., 1994; Pilarczyk, 1994, and references therein). This approach was demonstrated with contaminated materials dredged from Marina Del Rey in California, where the use of geotextile containers added more than \$50/yd^3 to the cost of the project (Clausner, 1996). Because most contaminants are sorbed to sediments and would not seep through the fabric, placement of filled geotextile bags in the water might be environmentally safe and would eliminate the need for land-based disposal sites. However, no data are available about the environmental effects of this approach (Clausner, 1996). A collection of bags could be capped, if necessary.

In addition to their utility in civil engineering projects and in the dewatering of dredged sediments, geotextile containers could provide a unique system for demonstrating emerging ex situ bioremediation technologies for certain contaminants. As disposal sites become increasingly difficult to find, the treatment of contaminated sediments in constructed cells, CDFs, or geotextile containers could be ways of reusing scarce sites. No long-term information is available on survival of geotextile containers.

Another idea that has received some attention is the placement of contained wastes on the abyssal plain [roughly 4,500 m (14,440 ft) deep] in the ocean. This idea was recently examined in a U.S. Department of Defense-sponsored study of ways to place and monitor clean dredged material, sewage sludge, and combustion fly ash (Valent and Young, 1995). The most attractive technique involves the use of fabric-like containers to isolate wastes from the water column during deployment from the transport ship or barge. Although this proposal has technical merit, legal barriers (dumping of contaminated sediments in the open ocean is not allowed) and environmental uncertainties must be still investigated. Because of the expense of the long-distance ocean transport of contaminated sediment, the cost-effectiveness of this idea needs to be closely examined.

COMPARATIVE ANALYSIS

The 1997 Marine Board Committee summarized its evaluation of remediation on a qualitative basis as shown in Table 17.5.

Natural recovery is the cheapest method, but the main difficulty in the selection process is the difficulty of estimating the rate of the natural recovery.

TABLE 17.5. Qualitative comparison of the state of the art in remediation technologies (Marine Board, 1997)[a]

Feature technology	State-of-design guidance	Number of times used	Scale of application	Cost (per yd^3)	Limitations[b]
Natural recovery	Nonexistent	2	Full scale	Low	Source control, sedimentation storms
In-place containment	Developing rapidly	<10	Full scale	<$20	Limited technical guidance, legal/regulation uncertainty
In-place treatment	Nonexistent	=2	Pilot scale	Unknown	Technical problems, few proponents, need to treat entire volume
Excavation and containment	Substantial and well developed	Several hundred	Full scale	$20 to $100	Site availability, public assistance
Excavation and treatment	Limited and extrapolated from soil	<10	Full scale	$50 to $1,000	High cost, inefficient for low concentration, residue toxic, need for treatment train

[a]Estimates for North America.
[b]See Table 17.4 for further details.

Excavation and treatment is the most expensive method; however, several Superfund projects (Waukegan Harbor, IL, Cold Spring, NY, Bayou Bonfouca, LA, etc.) were completed successfully.

RISK ASSESSMENT

Decisions about remediating contaminated sediment sites are made on the basis of human health risks. The health risk is defined as expected frequency of occurrence of an unacceptable health response in a population exposed to hazardous substances (Elliott, 1992).

CERCLA and the Superfund program were developed to overcome the further deterioration of the environment and to reduce the threat to human health. Individual health risk was the parameter chosen to measure the severity of a health hazard posed by a contaminated site. The USEPA follows a deterministic logic that quantifies individual health risk for "reasonable" maximum exposures (USEPA, 1990).

An alternative method is a probabilistic approach to calculate individual health risk (Elliott, 1992).

A more recent research effort conducted by the Corps of Engineers is the Dredging Operations and Environmental Research Program (DOER; see also Chapter 15, this volume). Part of this program is the development of risk-based effects assessments for contaminated dredged material that include contaminant control, treatment and removal technology, and reuse of marginally contaminated sediments from existing CDFs.

The risk process for contaminated sediments includes hazard assessment, contaminated pathway testing, exposure assessment, and identification of contaminated controls and treatment (Hummer, 1998; also see Chapter 15, this volume).

DREDGING EQUIPMENT FOR HANDLING CONTAMINATED SEDIMENT

Conventional dredging equipment typically handles large volumes of material in maintaining or deepening navigational channels. Such equipment may be operated in a "modified" procedure to handle relatively small volumes of contaminated material. However, in some cases, it may be more appropriate to use "special purpose" dredges (either specially developed or adapted) which are more suitable for handling contaminated sediments.

Several "special purpose" dredges are described below and their capabilities discussed (see Tables 17.6–17.9).

The selection of proper dredging equipment for any project is important to achieve an efficient operation. In case of contaminated sediment, it is even more important, since any additional contamination generated during dredging must be avoided. Selection depends on a number of factors, including:

1. Characteristics of sediments
2. Quantity of sediments to be removed
3. Degree of contamination
4. Toxicity of contaminants
5. Location
6. Environmental conditions at the site—waves, currents, tides, etc.
7. Distance to the disposal site

TABLE 17.6. Summary table—mechanical dredges

Type	Production	Depth limitation	Resuspension of Sediment	Comments
Open clamshell bucket	Low	30–40 ft	High	
Water-tight clamshell bucket	Low	30–40 ft	Low	Experiments conducted in the St. John's River

TABLE 17.7. Summary table—mechanical–hydraulic dredges

Type	Production	Depth limitation	Resuspension of Sediment	Comments
"Mud Cat"	Moderate	15 ft	Low to moderate	Extensively used
Remotely controlled "Mud Cat"	Low	15 ft	Low to moderate	New development
"Clean-up" system	Moderate	70 ft	Low to moderate	Extensively used in Japan

TABLE 17.8. Summary table—hydraulic suction dredges

Type	Production	Depth limitation	Resuspension of Sediment	Comments
"Refresher"	Moderate to high	60–115 ft	Low	Extensively used in Japan
"Waterless"	Moderate		Low	Limited experience
"Matchbox"	Moderate to high	85 ft	Low	Experiments conducted in Calumet Harbor
"Wide Sweeper"	Moderate	100 ft	Low	Used in Japan

8. Type of disposal
9. Availability of particular equipment

There are several types of dredges for conventional operations, designed principally for moving large volumes of material efficiently. Conventional equipment operated in a "modified" procedure can be effective, as reported by Hayes et al. (1992). However, in some cases, it may be more appropriate to use "special purpose" dredges (either specially developed or adapted) suitable for handling contaminated sediments.

There are several dredges that may be placed in a "special purpose" category, listed below under mechanical, mechanical–hydraulic, hydraulic, and pneumatic types (Herbich, 1990–1995).

1. Mechanical type
   a) Enclosed clamshell bucket
   b) Cable arm clamshell bucket
   c) Instrumented backhoe
2. Mechanical–hydraulic type
   a) Cutterhead dredge
   b) "Mud Cat"
   c) Remotely controlled "Mud Cat"
   d) "Cleanup" system
   e) High-density dredge
   f) Scraper dredge
3. Hydraulic type
   a) "Refresher"
   b) "Matchbox"

TABLE 17.9. Summary table—pneumatic pumps (dredges)

Type	Production	Depth limitation	Resuspension of sediment	Comments
"Pneumatic"	Low to moderate	+100 ft	Low	Evaluated by USAE Waterways Experiment Station
"Oozer"	Moderate to high	59 ft	Low	Used extensively in Japan

c) "Wide Sweeper" cutter-less dredge
d) Vibrating auger–positive displacement pump
4. Pneumatic
   a) "Pneuma"
   b) "Oozer"
   c) Agitation and emulsion dredge

In addition, an encapsulation method may be considered that provides for a dredge to inject material into silt deposits.

Mechanical Type Dredges

Enclosed Clamshell. The Japanese have developed a "watertight" clamshell for use with grab bucket dredges. An evaluation of the "watertight" bucket was made by the U.S. Army Engineer Waterways Experiment Station in 1982 (Figure 17.5). Experiments conducted at the Jacksonville District indicated that the "watertight" bucket significantly reduced water column turbidity and did not reduce production.

Figure 17.6 shows the benefit of using an enclosed bucket. The operation of the dredge can be modified slightly to reduce sediment resuspension by slowing the raising and lowering of the bucket through the water column. It must be noted that this operation modification reduces the production rate of the dredge, and generally high unit costs are associated with this type of mechanical dredging.

Cable Arm Clamshell Bucket. The design of the cable arm clamshell bucket differs from the conventional clamshell buckets, as the sweep of the bucket is controlled by the employment of cables (Figure 17.7). One main cable controls the descent of the bucket, four spreader cables control the opening of the shell, and another cable closes the shell and lifts the bucket. To reduce resuspension of the sediment, a venting system is constructed to allow water and air to pass through the bucket during its descent. The cable arm bucket was deployed by Environment Canada (Buchberger et al., 1993) during the Toronto Harbor demonstration. An improvement in the design was made possible through

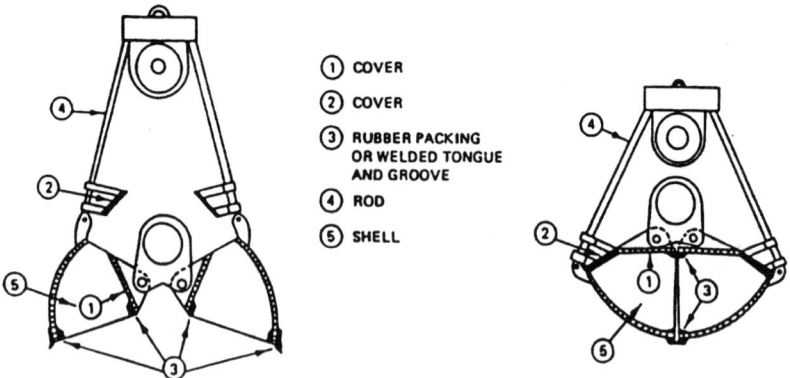

FIGURE 17.5. Open and closed positions of the watertight clamshell bucket.

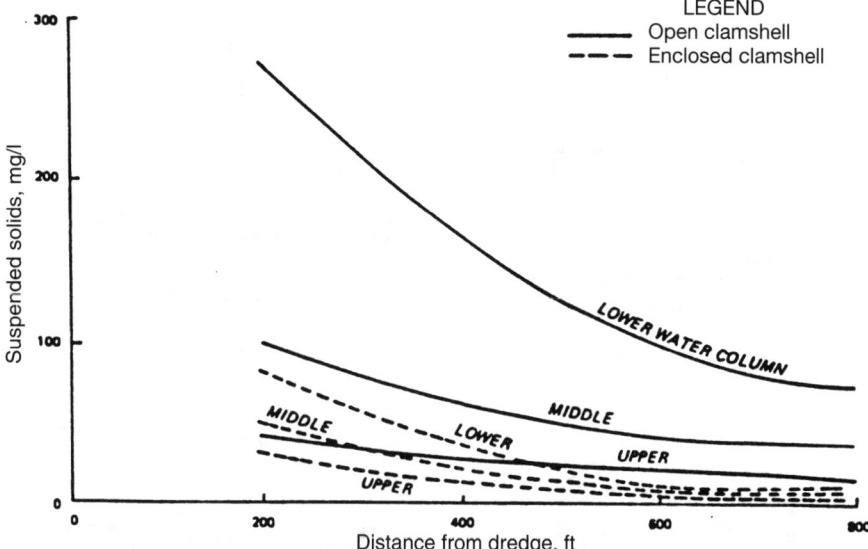

FIGURE 17.6. Resuspended sediment levels from open and enclosed clamshell dredge operations in the St. John's River.

the installation of an inner deflecting plate that restricted lateral movement; a greater quantity of sediment was retained within the bucket as it closed.

Instrumented Backhoe. A backhoe bucket was specially designed for the Bayou Bonfouca, Louisiana project to remove PAH-contaminated sediment. The bucket has a 5.2 yd^3 capacity and is installed on a barge. This excavator was completely instrumented with sensors and controls to achieve a 3-in dredging accuracy (Figure 17.8).

Mechanical–Hydraulic Dredges

"Mud Cat." The Mud Cat has a horizontal cutterhead equipped with knives and spiral augers that cut the material and move it laterally toward the center of the augers where it is picked up by suction. The dredge can remove sediments in a 2.4 m (8 ft) width in water depths of up to 10.1 m (33 ft). The dredge operates on anchors and cables; the manufacturer claims that the dredge leaves the bottom of the dredged area flat and free of the windrows that are characteristic of a typical cutterhead and hopper dredge operation (Figure 17.9).

By covering the cutter–auger combination with a retractable mud shield, the amount of turbidity generated by the Mud Cat's operation can be reduced.

The Mud Cat was developed in the United States and has been used extensively in removing sediments from reservoirs and disposal ponds. Production is relatively low and depth limitation is 4.6 m (15 ft). Resuspension of sediment may be classified as low to

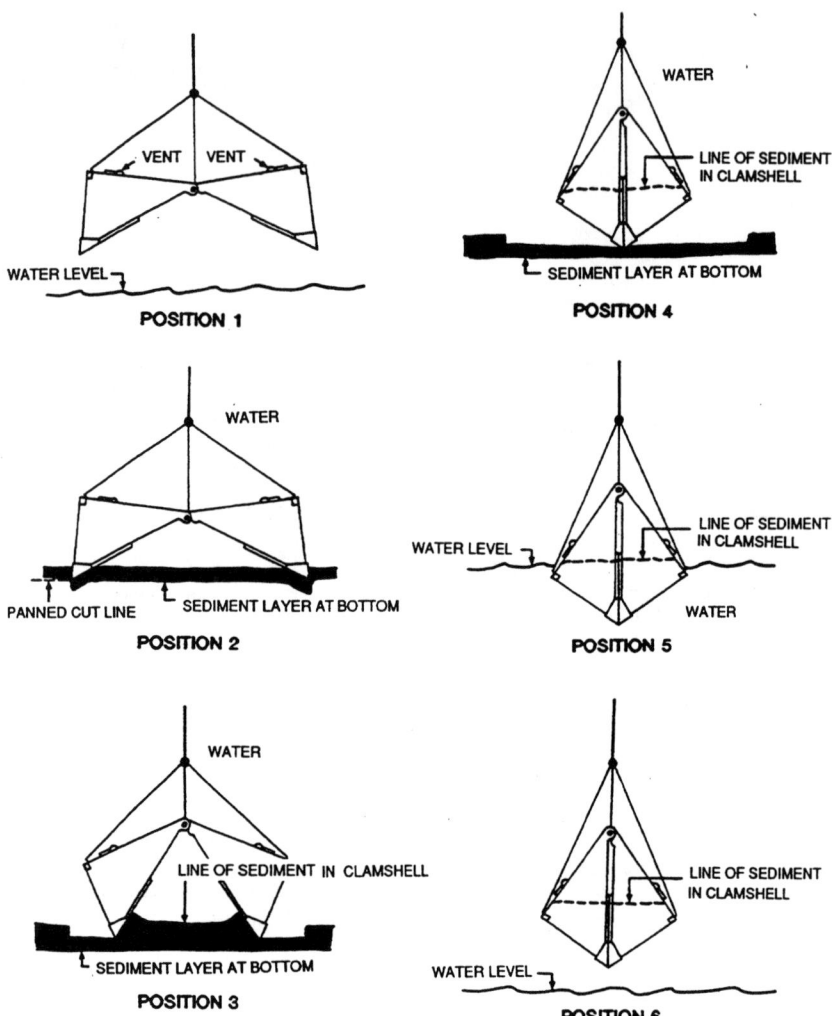

FIGURE 17.7. Cable arm environmental clamshell. Typical dredging cycle. (Courtesy of Cable Arm™ Clamshell.)

moderate; field test results are presented in a later section. A remotely controlled unit is also available.

"Cleanup" System. To reduce or minimize resuspension of the sediment, Toa Harbor Works, Japan, has developed a unique Cleanup system for dredging highly contaminated sediment (Sato, 1984). The Cleanup head consists of a shielded auger that collects sediment when the dredge swings back and forth; the auger guides the sediment toward the suction of a submerged centrifugal pump (Figure 17.10). To minimize sediment resuspension, the auger is shielded and a movable wing covers the sediment as it is being

FIGURE 17.8. Dredge configuration—"The Bonacavor." (Courtesy of Bean Stuyvesant, L.L.C.)

collected by the auger. Sonar devices indicate the elevation of the bottom. An underwater television camera also indicates the amount of material being resuspended during a particular operation. Figure 17.11 shows details of a shielded auger. Fairly large volumes (2,200,000 m^3 up to 1981) have been excavated by Cleanup dredges in soft muds and sand containing various contaminants such as mercury, cadmium, PCB, and oily, organic substances.

Cleanup dredges have been used extensively in Japan. In 1981 there were four such dredges having centrifugal dredge pumps; one dredge was equipped with a "Pneuma" pump. The resuspension of the sediment may be classified between low and moderate. Several dredges developed specifically for cleanup of contaminated sediments are manufactured outside of the United States and thus are unavailable under the Jones Act.

FIGURE 17.9. "Mud Cat" dredge (shield over the auger head raised to show the augers). (Courtesy of Ellicott Machine Corp., Baltimore, Maryland.)

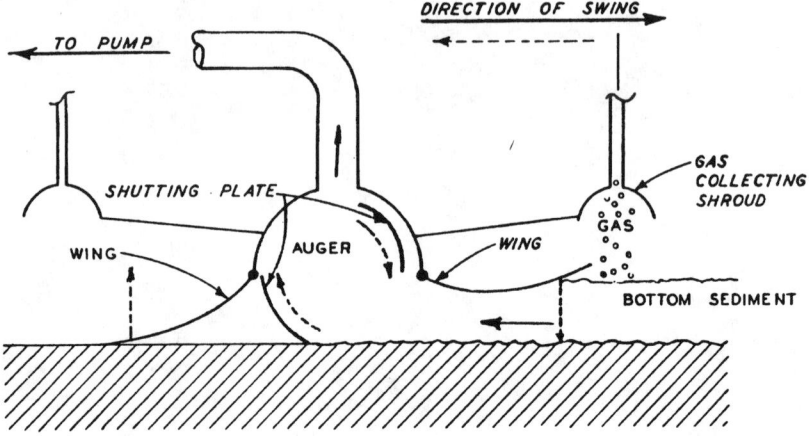

FIGURE 17.10. The "Clean-up" system.

"Oozer" Dredge. The Oozer pump was developed by Toyo Construction Company, Japan. The pump operates in a manner similar to that of the Pneuma pump system; however, there are two cylinders (instead of three) and vacuum is applied during the cylinder-filling state when the hydrostatic pressure is not sufficient to rapidly fill the cylinders. The pump is usually mounted at the end of a ladder and equipped with special suction heads and cutter units depending on the type of material being dredged.

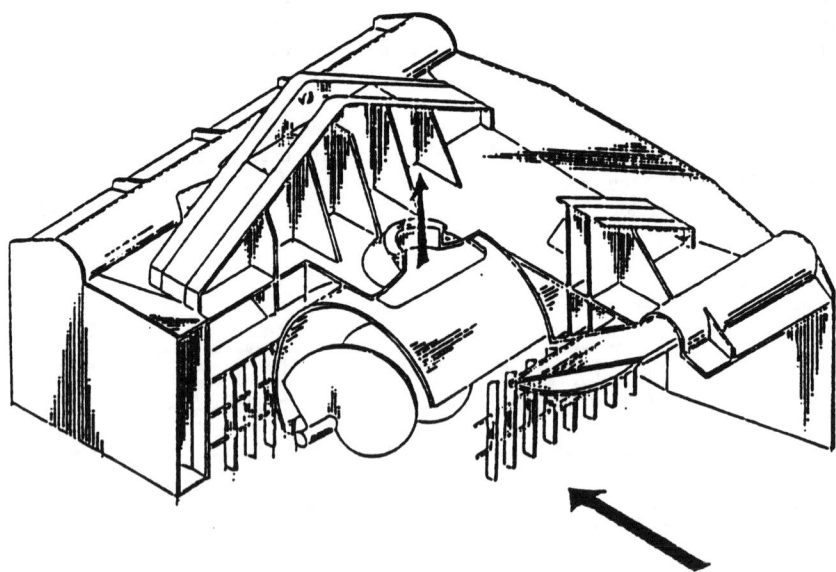

FIGURE 17.11. "Clean-up" shielded auger head.

The conditions around the dredging system such as the thickness of the sediment being dredged, the bottom elevation after dredging, as well as the amount of resuspension, are monitored by high-frequency acoustic sensors and an underwater television camera. A large Oozer pump has a dredging capacity ranging from 300 to 500 m^3/hr. During one dredging operation, suspended solids levels within 3 m of the dredging head were all within background concentrations of less than 6 mg/l. Figure 17.12 is a sketch of the Oozer dredge and Figure 17.13 shows the OOZER dredging system, "DREX," consisting of a suction mouth and a device that permits a back and forth movement of the suction mouth. This modified system is said to increase solids concentration up to 60%.

The main features of the "Oozer" dredge are as follows:

1. The dredge can effectively remove contaminated sediments from a maximum depth of 18 m (59 ft).
2. Because the swing speed can be adjusted from 0 to 20 m/min (0–65.5 ft/min), the dredge can be effective in removing suspended sediments.
3. Five acoustic sediment sensors can measure the bearing pressure of sediment to be removed and the thickness of various sediment layers.
4. Underwater TV cameras monitor the presence of turbidity near the suction intake.
5. Toxic gases released during dredging pass through gas scrubbers to remove toxic content before gases are released to the atmosphere.
6. A screen is provided at the suction mouth to prevent large objects from entering. Double-suction valves and electrically controlled check valves provide secondary protection.
7. An Oozer dredge can, under ideal conditions, pump sediments at in situ density.
8. Different cutters and suction heads are available for dredging sediments ranging from clay to sand.

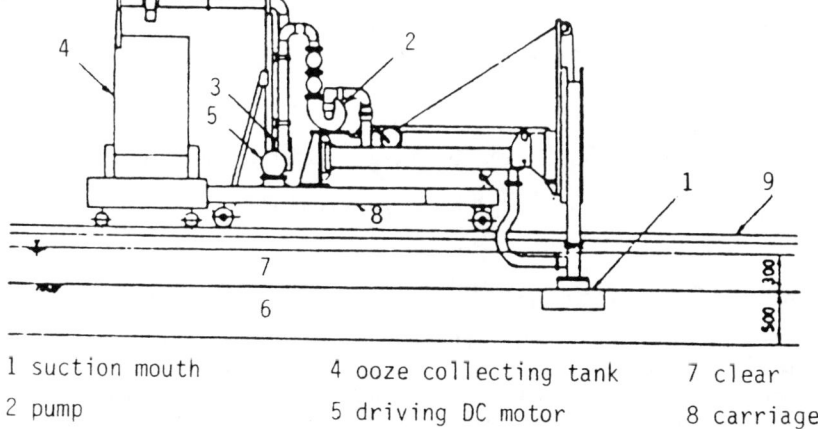

1 suction mouth	4 ooze collecting tank	7 clear
2 pump	5 driving DC motor	8 carriage
3 magnetic flow meter	6 test soil	9 rail

FIGURE 17.12. Outline of an "Oozer" dredge.

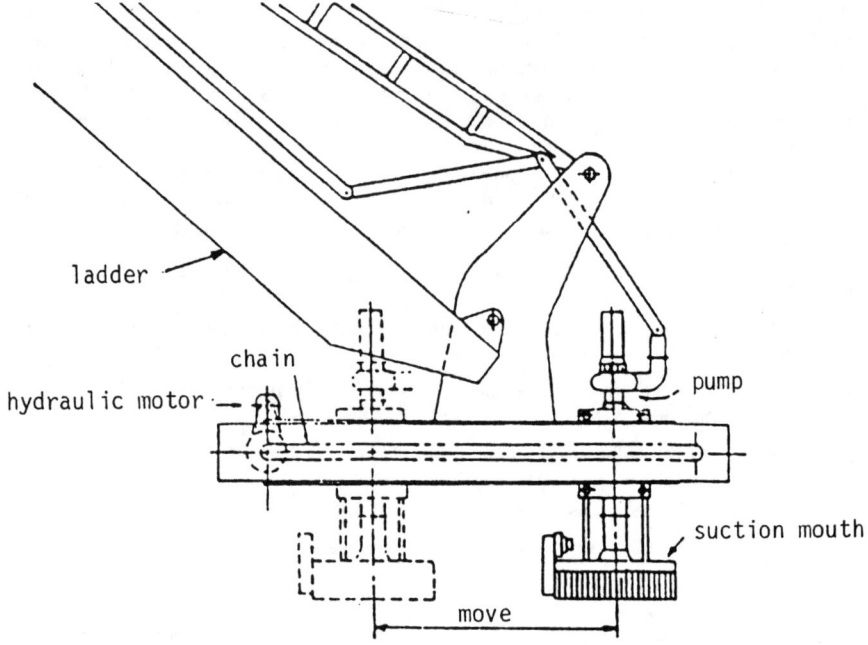

FIGURE 17.13. Sketch of "Oozer" dredging system.

Specifications for the Oozer dredge *Taian Maru* are as follows:

Hull:
Overall length	37 m	(121 ft)
Breadth	12 m	(39 ft)
Depth	3 m	(10 ft)
Draft	2.2 m	(7 ft)
Dredging depth	17 m	(55 ft)

Engine:
Oozer pump   1
Type: cylindrical twin-barrel, negative pressure suction and positive pressure discharge.
Dredging capacities (Pressure intensity): 7 kg/cm^2 (100 psi)

Hydraulic Suction Dredges

"Refresher" Dredge. The "Refresher" dredge was purposely developed for removal of contaminated materials by Penta-Ocean Construction Company, Ltd. The dredge material is confined by a specially designed flexible enclosure that completely covers the cutter, preventing escape of sediments to the outside of the immediate dredging area (Figures 17.14 and 17.15). The working open section is always on the swing side of the cutterhead. A gas removal system is also installed and can be activated as needed to prevent gas from

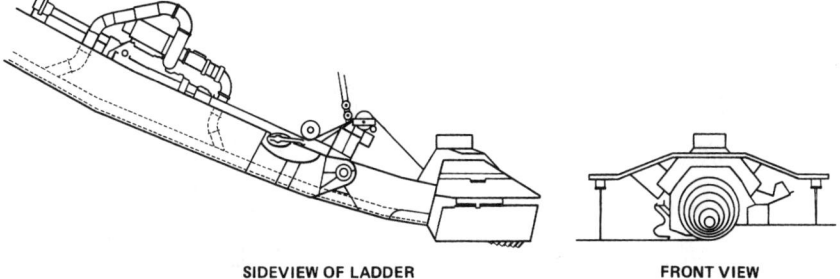

SIDEVIEW OF LADDER   FRONT VIEW

FIGURE 17.14. "Refresher" dredge.

moving up the suction pipe. The flexible enclosure of the cutterhead is automatically adjusted to the bottom contours.

The "Refresher" dredge is equipped with a main pump and an additional pump on the ladder to provide a high level of protection. Automatic valves in the suction pipe prevent the sediment–water mixture from flowing back in case of power failure (Figure 17.15).

"Matchbox" Dredge. A special suction head was developed by a dredging contractor in the Netherlands (Figure 17.16) to replace the traditional cutterhead (Herbich, 1992). The main design points are as follows:

1. A large plate covers the top of the dredge head to avoid inflow of water and escape of gas bubbles.
2. An adjustable angle between the drag head and the ladder creates the optimum position of the drag head independent of the dredging depth.
3. There are openings in both sides of the drag head to improve dredging efficiency. During swinging action, the leeward side is closed to prevent water inflow.
4. Dimensions of the head must be carefully designed for the average flow rate and swing rate.

A diffuser may be installed at the submerged end of the discharge pipe to reduce the dispersion of fine sediment in the water column (Figure 17.17). By gradually widening the cross section, the flow can decelerate to an acceptable velocity to reduce turbulence. Outflow velocities are designed to be between 0.2 to 0.3 m/sec (0.6 to 1 ft/sec); however, it is unlikely that contaminated material would be discharged in open water. A possible application may be to employ such a diffuser in a containment area. A degassing system is also installed to prevent or reduce the amount of gas moving up the suction pipe.

A direct comparison between a "Matchbox" suction head and a conventional cutterhead was made by the Waterways Experiment Station in Calumet Harbor. The "Matchbox" was specifically designed to be fitted on the ladder of the U.S. Army Corps of Engineers dredge *Dubuque*. The Calumet Harbor demonstration project indicated that the clamshell dredge generated the largest suspended sediment plume affecting the entire water column. The cutterhead (when operated properly) and the "Matchbox" dredges were able to limit the sediment resuspension to the lower portion of the water column. The cutterhead slightly outperformed the "Matchbox" dredge.

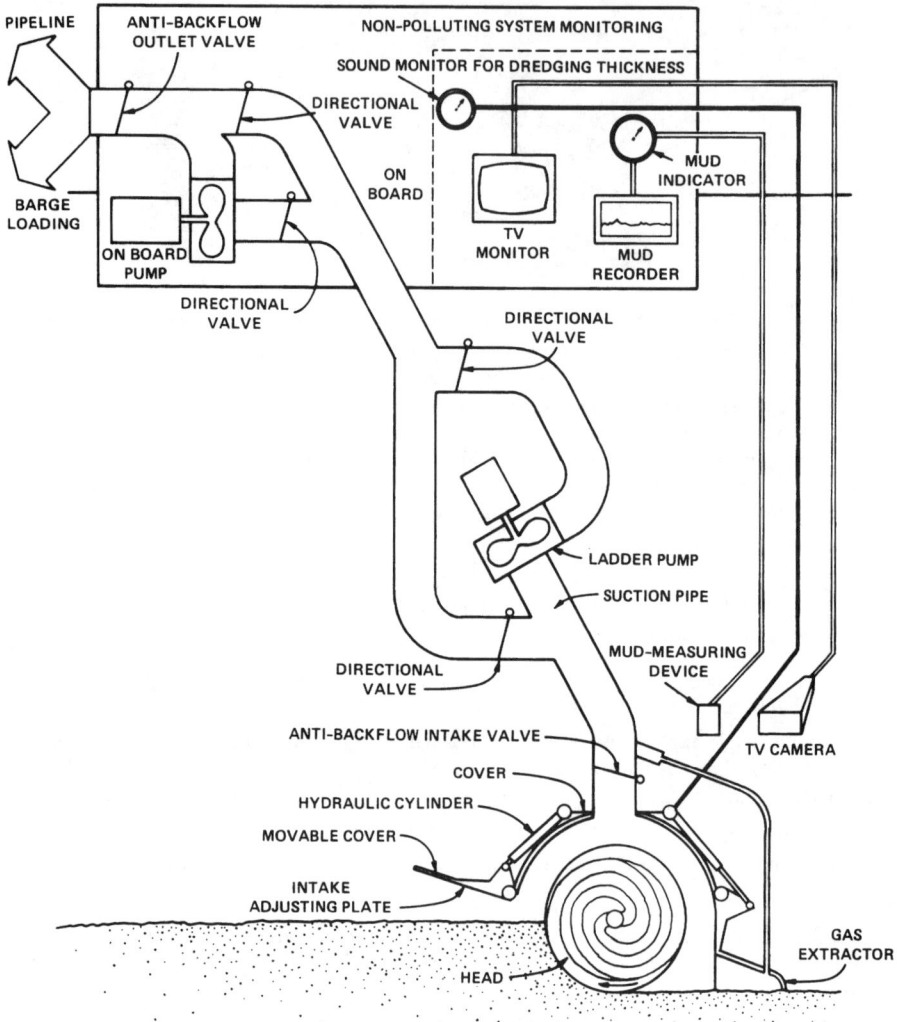

FIGURE 17.15. Description of a "Refresher" dredge.

CASE STUDIES

Domestic

Baltimore, Maryland

Introduction. The Baltimore Inner Harbor has been a center of trade since the early 1600s. In the 1840s, Baltimore Works began processing raw chrome ore into various

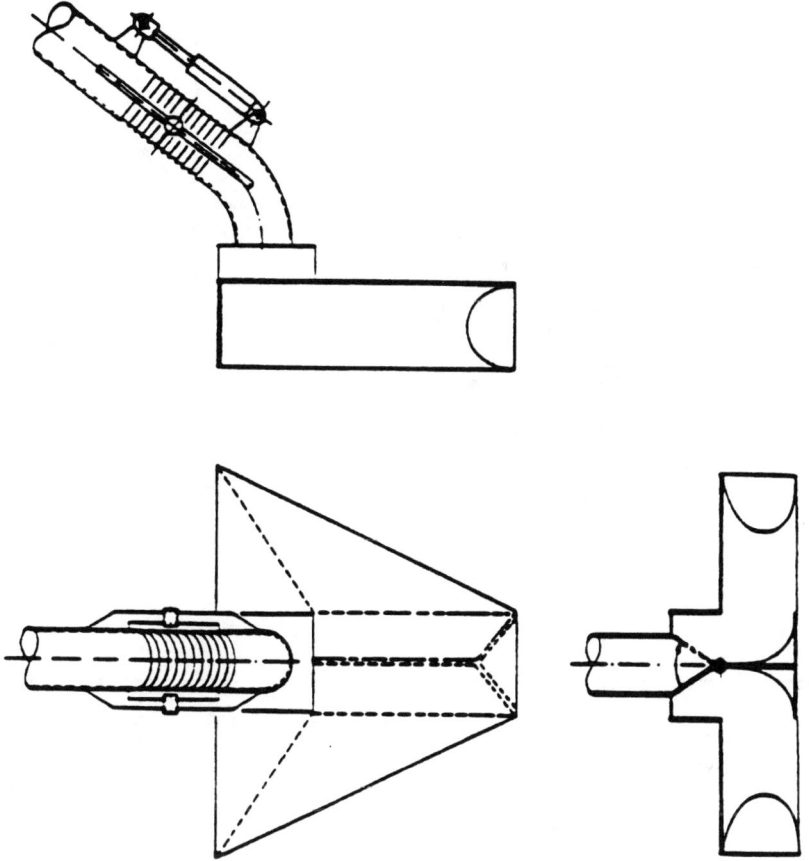

FIGURE 17.16. "Matchbox" suction head.

chemicals, consequently resulting in a great deal of waste being discharged into the harbor. The processing of chrome was abandoned in 1985 and remediation began.

A 15-acre site was contaminated with hexavalent chromium, which also migrated to the surrounding sediments (Figure 17.18). Predredging surveys showed chromium content between 240 and 36,000 ppm (average 339 ppm), copper content between 76 and 1,000 ppm (average 152 ppm), and lead content between 240 and 1,600 ppm (average 185 ppm), as well as oil and grease.

Consent Order. Remediation was conducted under a consent order issued by USEPA negotiated with Allied Signal and the State of Maryland. Remediation consisted of: (a) placing perimeter hydraulic barrier within the outboard embankment; (b) installing a groundwater pumping system inside the containment areas; (c) placing a site-wide cap, and (d) conducting long-term monitoring (Snyder et al., 1997).

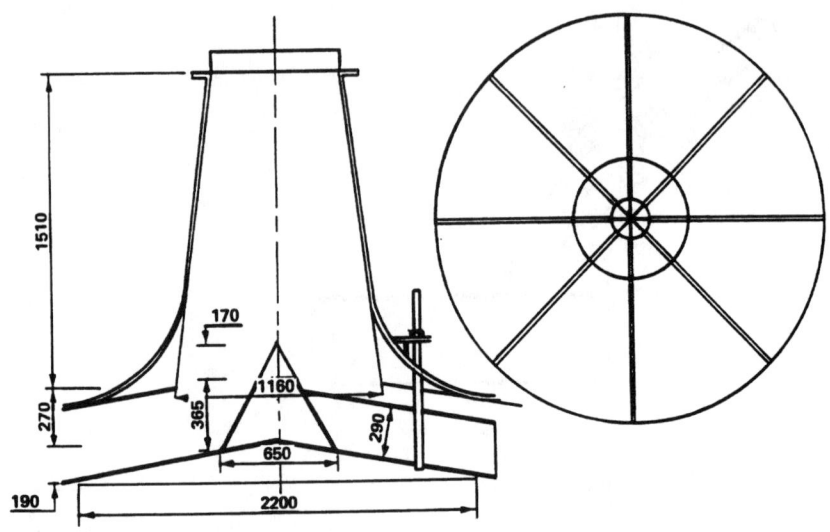

FIGURE 17.17. Schematic of a diffuser.

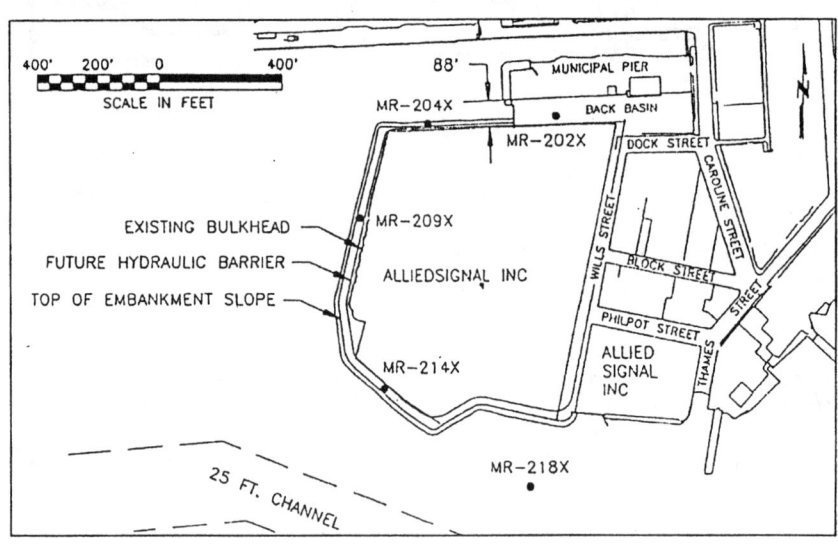

FIGURE 17.18. Site plan, Baltimore, Maryland.

Dredging and Disposal. Because of environmental concerns, regulatory constraints were imposed by the Federal 404 permit, State of Maryland Water Quality Certification and wetlands permit, Maryland Port Authority construction permit, and dredging and disposal permit.

During the summer of 1990, approximately 150,000 cubic yards of sediment were dredged from the Inner Harbor. The dredged area was 100 ft wide and 2,000 ft long with a water depth of 29 ft. Initially, 12 yd^3 and 26 yd^3 environmental buckets were used; later 26 yd^3 and 12 yd^3 conventional buckets were allowed. Dredged material was placed in 6,000 yd^3 scows and transported to Hart Miller Island for disposal (Table 17.10). Only sediments meeting the maximum contaminant concentration levels were allowed to be placed at the disposal area (Figure 17.19).

The dredged material was pumped hydraulically from the scows. The discharge pipes were moved frequently to spread the sediment evenly. A Mudcat dredge was deployed in the Hart Miller disposal area to mix the contaminated dredged material with the clean material placed from a channel-deepening navigation project.

The Back Basin, north of the site was capped. The procedure was as follows: (a) the basin was isolated with steel sheeting: (b) wide drains were used to consolidate the sediment basin; (c) the basin was covered with geomembrane; (d) the basin was covered with 3 ft of bentonite-augmented soil, and finally (e) 16 ft of surcharge was placed to consolidate underlying soft sediments.

Bayou Bonfouca, Louisiana

Site History. From 1892–1970 a creosote plant operated on Bayou Bonfouca, Louisiana. In 1970, the plant caught fire releasing large volumes of creosote; hundreds of creosote-treated timbers found their way into the Bayou. This resulted in 169,000 yd^3 of contaminated soil spread over 55 acres.

In 1982 the site was declared eligible for Superfund cleanup. Nine years later, Bean Dredging Company was awarded a dredging contract. The removal concentration goal was 1,300 ppm of creosote (PAH).

Dredging Equipment and Positioning. Because of the nature of the contaminant and the location and accuracy required, special equipment had to be developed. The dredging equipment selected was a hydraulic backhoe with a 5 yd^3 bucket called the Bonacavor (Figure 17.20) (Taylor, 1999).

The backhoe-type bucket was installed on a barge 140 ft long and 45 ft wide. The barge was equipped with a spud system to insure stability. Innovative sensors and controls were installed, including a laser positioning system, computerized excavation con-

TABLE 17.10. Summary of dredged material

	Scow volume m^3 (yd^3)	Total chromium mg/kg	Lead mg/kg	Copper mg/kg
Minimum value	1,040 (1,360)	75	21	20
Maximum value	3,601 (4,709)	16,000	1,900	1,300
Average value	2,233 (2,920)	3,473	753	482
MCL[a]	NA	12,000	2,500	1,500

Note: Minimum, maximum, and average values based upon 74 individual scow samples.
[a]Permit maximum concentration level.
Source: B&V, 1993.

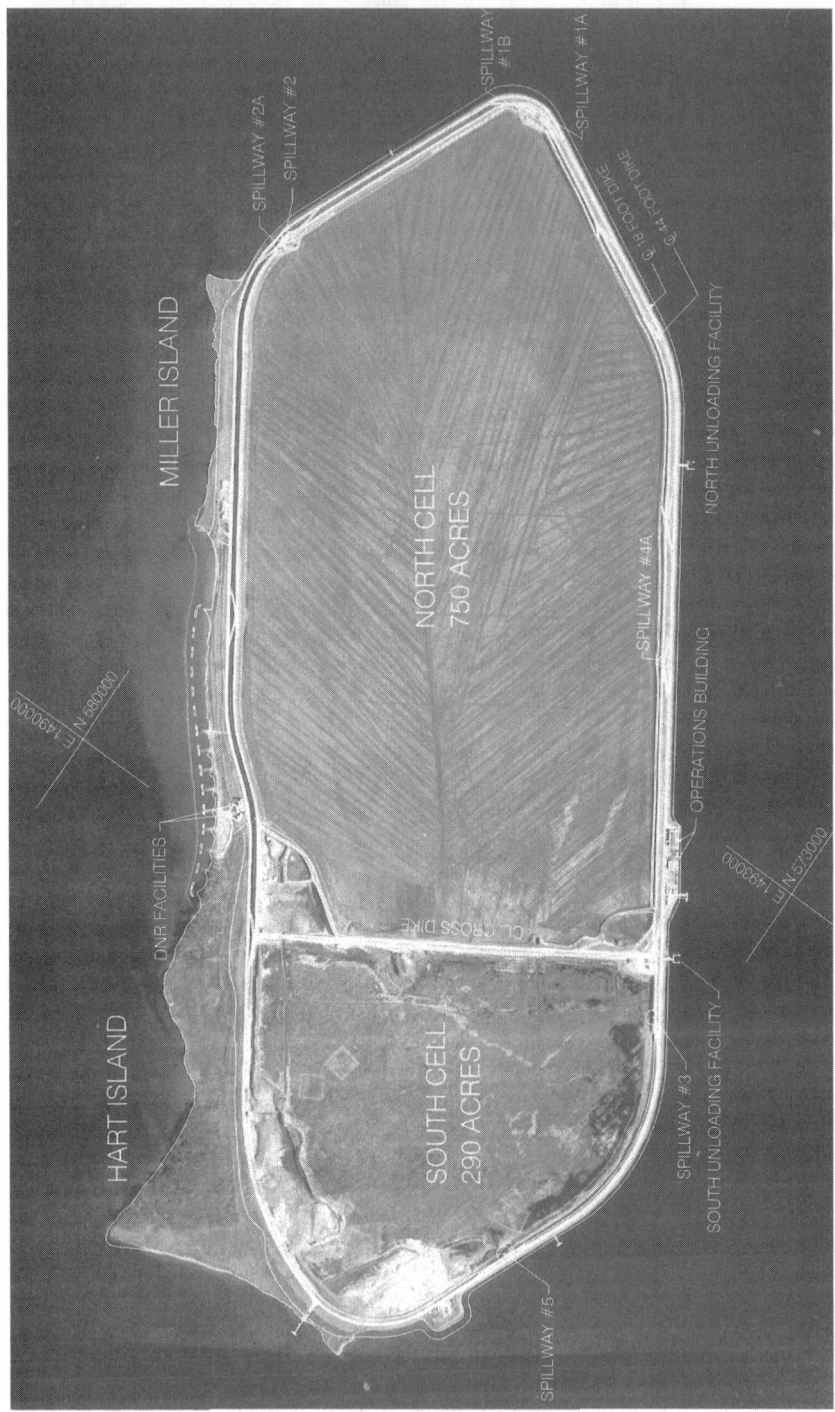

FIGURE 17.19. Hart Miller Island. Comparison of pre- and postbathymetric survey. (Courtesy of Kevin Wikar, Maryland Port Administration, Maryland Environmental Services.)

FIGURE 17.20. Hydraulic backhoe—"The Bonacavor." (Courtesy of Bean Stuyvesant, L.L.C.)

trol tied to the positioning system, and a real-time telemetry link with the corporate headquarters. The goal was to achieve a 3-in dredging accuracy.

One of the challenges in precision (surgical) dredging is to secure the position of the dredge. Wind, waves, currents, and tides make vessel positioning difficult. For tracking the vessel position onshore, lasers allow real-time monitoring. A three-dimensional model of the seafloor is highly desirable. There are a number of survey techniques available on the market.

A digital model can drive positioning systems and also drive the excavator control system. Accurate monitoring of the excavator position was achieved by: (a) position sensors at every joint in the excavator; (b) tracking turret rotation, arm angle, bucket angle, bucket depth, etc.; (c) real-time display of excavator position, relative to vessel and sea bottom; (d) simultaneous monitoring at the corporate headquarters (Figure 17.21)

Additional features included a requirement for the personnel to wear protective clothing. The transport pipeline was completely encased in a second pipeline to prevent any possible leakage.

Five layers of silt curtains were deployed, two near the dredge and three placed in succession away from the dredge. Dredging was limited to an 8-hour day, 5-day week. The quantity of sediment removed, transported and treated was 162,246 yd^3. The tolerances achieved were: topside +0 in, bottom –6 in; the average overdredge dimension was 0.17 ft. It took 21 months to complete the project. The total cost is estimated at $115 million.

Cold Spring, New York*

Dredging Operations. The dredge equipment consisted of a custom-built horizontal auger dredge similar in configuration to an Ellicott Machine Corporation Mud Cat. The 30.5 cm (12 in) diameter horizontal auger was 2.4 m (8 ft) wide. The cutterhead was

---

*Adapted from P. F. Marano and T. W. Russell, "Contaminated Sediment Dredging in a Tidal Estuary," Geoenvironmental 2000 Conference, February 1995.

FIGURE 17.21. Display on the monitor. (Courtesy of Bean Stuyvesant, L.L.C.)

connected to a 25.4 cm (10 in) diameter suction line. The dredge contained a 20.3 cm × 20.3 cm (8 in × 8 in) Denver-Thomas pump powered by a 167 metric hp (165 hp) Detroit Diesel engine. The discharge line was 20.3 cm (8 in) diameter, butt-fused, high-density polyethylene (HDPE).

The draft of the dredge was 61 cm (24 in). For this project, the draft was decreased to 45.7 cm (18 in) by the addition of flotation tanks. The auger and suction head can operate efficiently in as little as 35.6 cm (14 in) of water.

Prior to dredging, the vegetation in the cove and pond was removed. It was initially intended to use a "hydrorake" for this task. This equipment simultaneously removes the growth as well as its roots and rhizomes. However, it was believed that this device would cause excessive disturbance of contaminated sediments. As a result, a weed harvester was used. This unit contains a sickle bar that cuts the growth approximately 15 cm (6 in) above the cove/pond bottom. It also collects and stores the cut vegetation in an on-board hopper. The remaining roots and rhizomes were removed by the auger during dredging and were screened out at the discharge outfall.

The dredging pattern consisted of 2.4 m (8 ft) wide straight-line runs. Generally, two passes were needed to complete a 30.5 cm (12 in) cut; the first pass usually removed 20 to 23 cm (8 to 9 in) and the second pass was used to clean up the remainder. The removal cannot be efficiently performed with one pass; some of the contaminated sediment that becomes suspended during the first pass will settle out after a short delay, resulting in the need to redredge to achieve cleanup criteria.

The auger/suction line can accept debris up to 25.4 cm (10 in) in diameter. A "rock box" within the suction line collects debris greater than 10.2 cm (4 in), as the pump cannot accept particles larger than this diameter. Boulders and debris larger than 25.4 cm (10 in) are removed by mechanical or manual lift methods.

Horizontal control was accomplished by means of a winch and cable system. The winch also provides propulsion for the dredge. The cables are secured to shoreline features (trees, poles, etc.) and have flotation collars to minimize disturbance to contaminated bottom sediments. Control was also maintained through the use of a dredge-mounted Lietz Red Mini 2 electronic distance measuring device (EDM) in conjunction with short-stationed prisms. The accuracy of this system is ±6 cm (0.2 ft).

Vertical control was accomplished through a combination of tidal gauges, cutterhead sensors, fathometers, manual sounding rods, and proprietary methods. The accuracy of this system is ±3 cm (0.1 ft). This removal accuracy has been maintained during the dredging to date. However, additional care was needed to maintain vertical control in very soft sediments and peat-like material.

Quality control procedures include surveying to verify dredge depth and sampling/laboratory analyses to determine the decrease in contaminant levels. A predredge topographic survey was performed to establish original elevation. Within 12 hours after an area had been dredged, a postdredge survey was conducted on a 7.6 m (25 ft) grid to establish the actual dredge cut.

In addition, postdredge sediment samples were obtained within 24 hours of dredging. These samples were collected on a 30.5 m (100 ft) grid and analyzed for concentrations of total cadmium. The sampling and analyses were subject to standard quality assurance procedures.

Production/Output. The dredge pump typically operated at a flow rate between 75.7 to 94.6 l/sec (1,200 to 1,500 gpm). The flow rate was limited by the velocity control required for the relatively short (152 to 457 m, or 500 to 1,500 ft) discharge lines. The flow meters showed a typical range of 15–20% solids, although the first pass in very soft or peat-like materials had been as great as 30% solids, and a second (cleanup) pass in most materials had been as low as 10% solids.

The average production in the early stages had been 47.4 in situ m^3 (62 yd^3) per hour. This rate includes downtime for dredge set-up, repositioning, removal of obstructions, etc. The typical operating range was on the order of 45.9 to 65 m^3 (60 to 85 yd^3) per hour. Surprisingly, the variability of gradation (sands versus silts/clays) can be accommodated by a skilled dredge operator through adjustments to flow rates and auger speeds, resulting in no significant change in production rate. There was, however, a wide variation of production rate that resulted from the hardness of the sediments; these rates can drop to as low as 15.3 m^3 (20 yd^3) per hour. Obviously, the greater the amount of obstructions encountered, the lower the production rates.

Dredged Material Handling. The dredged material was discharged into a settling basin for initial treatment. Originally, the discharge fed directly to the landside mechanical dewatering facility. However, changes in grain size (sands versus silts/clays) and fluctuations in flow rates were too rapid to allow for dewatering plant adjustments. The use of the settling basin assures a smoother, more uniform input to the dewatering plant.

Once the majority of the solids settled out, the free water was pumped off and passed through sand filters, or if significantly turbid, through plate and frame filter presses. The treated water was tested and if it achieved New York State Department of Environmental

Control (NYSDEC) discharge effluent criteria, it was pumped back into East Foundry Cove. Treated water that did not achieve this criteria was retreated and retested.

The solids were removed from the dewatering basin with backhoes, and fixated in a pugmill using a Maectite process. If the solids were too wet, they were passed through the filter presses prior to fixation. The fixated solids were allowed to cure (typically 12 hours) and tested with the Toxicity Characteristic Leaching Procedure (TCLP) for cadmium and lead. If the treated material met TCLP criteria, it was loaded into railcars and shipped off-site for disposal. If the material failed TCLP criteria, it was retreated and retested.

Environmental Concerns. There are two major environmental concerns with respect to the dredging of contaminated sediments: removal efficiency and sediment resuspension. Removal efficiency is an obvious major concern, as the original intent of the dredging is to lower the existing contamination level within the sediment. However, this requirement is also balanced by economic concerns. Whereas it may be desirable to dredge deeper than necessary to assure complete removal, each extra cubic yard of unnecessary removal incurs an additional remediation and disposal cost of over $160/yd^3.

Another major concern is sediment resuspension that can result in contamination of adjacent areas. This is especially critical in the areas adjacent to a wildlife sanctuary. The dredging operation itself purposely disturbs the sediment so that the suction line can collect the resuspended particles. However, the suction line is not 100% effective in collecting these particles. In addition, the complementary activities such as repositioning the dredge, back-flushing of the dredge line caused by pump shut-down, and the action of the work boat propellers in shallow water can increase sediment resuspension.

Several dredging techniques are used to minimize sediment resuspension. The dredge operator minimizes the auger speed as conditions allow to reduce resuspension. To prevent back-flushing of the dredge line during pump shut-down (which results in additional bottom disturbance), a shut-off valve is typically installed in the dredge line. However, this valve reduces efficiency and is prone to clogging, especially when dredging sands. The dredging contractor at this site used a proprietary procedure to eliminate back-flushing in lieu of a shut-off valve.

To minimize the dispersion of resuspended sediments, silt curtains were deployed. These barriers were placed around the perimeters of East Foundry Pond and the dredge zones in East Foundry Cove. Silt curtains were also installed along the East Foundry Cove channel by the railroad embankment causeway; however, the flow velocity through this channel was greater than 0.49 mps (1.6 fps), resulting in bowing, submergence of the curtain, and disruption of the barrier.

In addition to the preventative measures, the resuspension of sediment was monitored. The most effective way to monitor this phenomenon is through the measurement of total suspended solids (TSS). However, real-time measurement of TSS was not possible. Therefore, a correlation was developed prior to dredging between turbidity (which can be measured in real time) and TSS. This correlation was obtained by collecting turbidity and TSS data at various locations during high and low tide for varying conditions of precipitation. These measurements also considered the effects of the inflow from Foundry Brook. This site-specific correlation is shown in Figure 17.22.

These measurements also established a background level of TSS (and, through the correlations, a corresponding background value for turbidity). During dredging operations, turbidity measurements were obtained with a Hach 2100P portable turbidimeter (accurate to 0.01 NTU) at regular intervals and compared to the correlated background value. If the measurement exceeded 1.5 times the background value, dredging was stopped until corrections could be effected to reduce turbidity to acceptable levels. Some

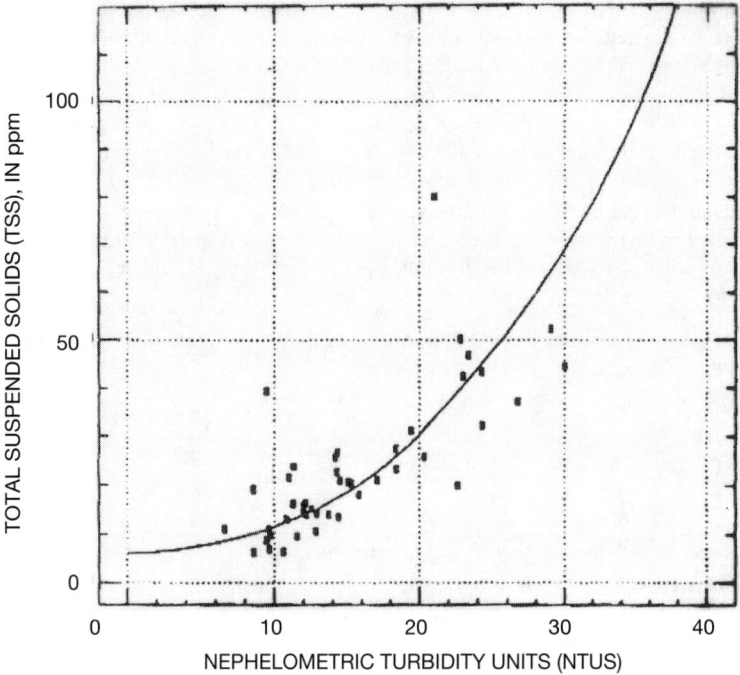

FIGURE 17.22. TSS/Turbidity correlation.

of the turbidity samples were analyzed at regular intervals for TSS and for total concentrations of cadmium, lead, cobalt, nickel, and zinc. These analyses served as a verification of the original correlation.

Results/Conclusions. The following are the preliminary conclusions based upon dredging performed.

a) Preliminary data indicated that a significant reduction in cadmium levels had been achieved.

b) A critical factor in gauging the effectiveness of dredging is an accurate characterization of the cadmium levels in the sediments. The Marathon sediments typically have less than 50% solids and contain organics. Whereas the analytical results of split samples from EFC/EFP were fairly consistent (±30%), these physical properties could significantly affect the variability, and thus reliability, of confirmatory analysis on future dredging remediation projects.

c) The resuspension and resulting redistribution of sediment during dredging was not as far-reaching as initially believed. Resuspension appears to be very localized, to within a few feet of the original disturbance.

d) The efficiency of dredging is dependent upon the cleanup levels required. The majority of the areas have required more than the initially planned two passes. To date, 79% of the dredged areas have required at least one additional dredge pass, and of these, 21% have required more than one additional pass.

e) Multiple passes of a dredge are more effective than a single, deeper pass for removal of contaminated sediment. It is suspected that multiple passes result in less overall removed material for the same level of removal efficiency.

f) Although numerous factors can affect dredge productivity, the biggest factors are delays in water/solids treatment coupled with a lack of dredged material storage. The treatment facilities must be able to keep pace with the dredge output in order to maintain an economical, efficient removal operation.

g) Silt curtains are ineffective in flow velocities greater than 0.46 mps (1.5 fps), as documented in technical publications. Control of sediment dispersion at the source (i.e., controlling the speed and depth of cut and other dredging techniques) is more effective.

All dredging was completed in 1995. In situ sediment volumes removed by dredging were as follows:

East Foundry Cove	40,674 m^3	(53,200 yd^3)
East Foundry Pond	11,009 m^3	(14,400 yd^3)
Cold Spring Pier Area	7,339 m^3	(9,600 yd^3)
Total	59,022 m^3	(77,200 yd^3)

*Commencement Bay Nearshore/Tideflats, Tacoma, WA**

Introduction. *Site Location.* The Commencement Bay Nearshore/Tideflats (CB/NT) Superfund site is located in Tacoma, Washington at the southern end of the main basin in Puget Sound (Figure 17.23). The site encompasses an active commercial

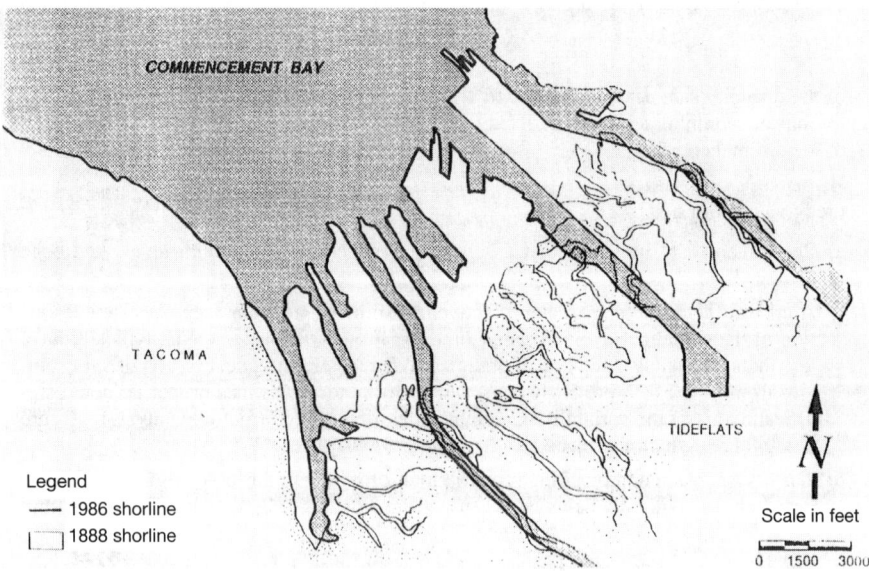

FIGURE 17.23. Commencement Bay tideflats and shoreline modifications as of 1986.

*Adapted from Gilmur and Saathoff (1994).

seaport and includes 10–12 square miles of shallow water, shoreline, and adjacent land, most of which is highly developed and industrialized. The upland boundaries of the site are defined according to the contours of localized drainage basins that flow into the marine waters. The marine boundary of the site is limited to the shoreline, intertidal areas, bottom sediments, and water of depths less than 60 feet below mean lower low water. The nearshore portion of the site is defined as the area along the Ruston shoreline from the mouth of City Waterway to Pt. Defiance. The tideflats portion of the site includes the Hylebos, Blair, Sitcum, Milwaukee, St. Paul, Middle, Wheeler-Osgood, and City Waterways; the Puyallup River upstream to the Interstate-5 bridge; and the adjacent land areas. Because the landward boundary of the B/NT site is defined by drainage pathways rather than political boundaries, the precise landward extent of the site may be adjusted as new information regarding surface water and groundwater flow patterns is developed.

Land Use (1989). The CB/NT site is located within the city of Tacoma, which has a population of 162,100. The land, water, and shoreline within the study area are owned by various parties, including the state of Washington, the Port of Tacoma, the city of Tacoma, Pierce County, the Puyallup Tribe of Indians, and numerous private entities. Much of the publicly owned land is leased to private enterprises. Within the site boundaries, land use is chiefly industrial and commercial.

The Port of Tacoma owns approximately 35–40% of the 2,700 acres that make up the port and industrial areas within the CB/NT site. The port operates many cargo handling and storage facilities along the waterways and leases other properties to large and small industrial, manufacturing, and commercial tenants. Many of the remaining properties within the port and industrial area were under port ownership at one time, but have since been sold. Major private landowners include lumber, chemical, and petroleum companies. Property along the Hylebos Waterway is owned almost exclusively by private companies, and there are several privately owned parcels along the Blair Waterway. Other privately owned parcels are found predominantly at the landward end of the port and industrial area.

A large portion of the tideland and offshore areas of the CB/NT site is either owned outright by the state or is designated as state-owned harbor areas. The Port of Tacoma owns tidelands and bottom sediments in several areas including the head of Hylebos Waterway, the head of Blair Waterway, and Milwaukee and Sitcum waterways. The St. Paul and Wheeler-Osgood waterways are privately owned. Private ownership of shorelines and intertidal areas in many portions of the site generally corresponds with ownership of the adjacent upland property parcels.

The Puyallup Tribe of Indians has asserted title to land in the Tacoma tideflats area, including former Puyallup River bottomland and filled tidelands adjacent to the Puyallup Reservation. Negotiations among the Puyallup Tribe of Indians, the federal government, the state of Washington, the Port of Tacoma, and other affected parties were completed during the summer of 1988 to resolve various land ownership issues. The settlement agreement was approved on 27 August 1988 by tribal members and by federal, state, and local governments. On 21 June 1989, the Puyallup Tribe of Indians Settlement Act of 1989 (PSA) was signed into law by the President, incorporating the August 1988 settlement agreement and technical documents. Efforts are underway to implement the terms of the agreement, which adds to the tribe's land base and provides for substantial restoration and enhancement of fisheries resources. Several large parcels of property within the CB/NT site boundaries that are slated for environmental cleanup by the Port of Tacoma will be transferred to the tribe within the next few years.

Contaminants in the CB/NT area originate from both point and nonpoint sources. Industrial surveys conducted by the Tacoma–Pierce County Health Department (TPCHD) and the Port of Tacoma indicate there are more than 281 active industrial facilities in the

CB/NT area. Approximately 34 of these facilities are National Pollutant Discharge Elimination System (NPDES)-permitted dischargers, including two sewage treatment plants. Nonpoint sources include two creeks, the Puyallup River, numerous storm drains, seeps, and open channels, groundwater seepage, atmospheric deposition, and spills. The TPCHD has identified approximately 480 point and nonpoint sources that empty into Commencement Bay (Rogers et al., 1983).

Environmental Setting. Commencement Bay is a large, deepwater embayment of about 9 square miles in southern Puget Sound. Puget Sound was designated in March 1987 by USEPA as an estuary of national significance. Several waterways, including the Puyallup River adjoin Commencement Bay. The drainage area for the Puyallup River, is approximately 950 square miles.

Site History. At the time of urban and industrial development in the late 1800s, the south end of Commencement Bay was composed largely of tideflats formed by the Puyallup River delta. Dredge and fill activities have significantly altered the estuarine nature of the bay since the 1920s. Intertidal areas were covered and meandering streams and rivers were channelized (Figure 17.23). Numerous industrial and commercial operations have located in the filled areas of the bay, including shipbuilding, chemical manufacturing, ore smelting, oil refining, food preserving, and transportation facilities (Figure 17.24).

With industrialization, the release of hazardous substances and waste materials into the environment has resulted in alterations to the chemical quality of waters and sediments in many areas of the bay. Contaminants found in the area include arsenic, lead, zinc, cadmium, copper, mercury, and various organic compounds such as polychlorinated biphenyls (PCBs), and polycyclic aromatic hydrocarbons (PAHs).

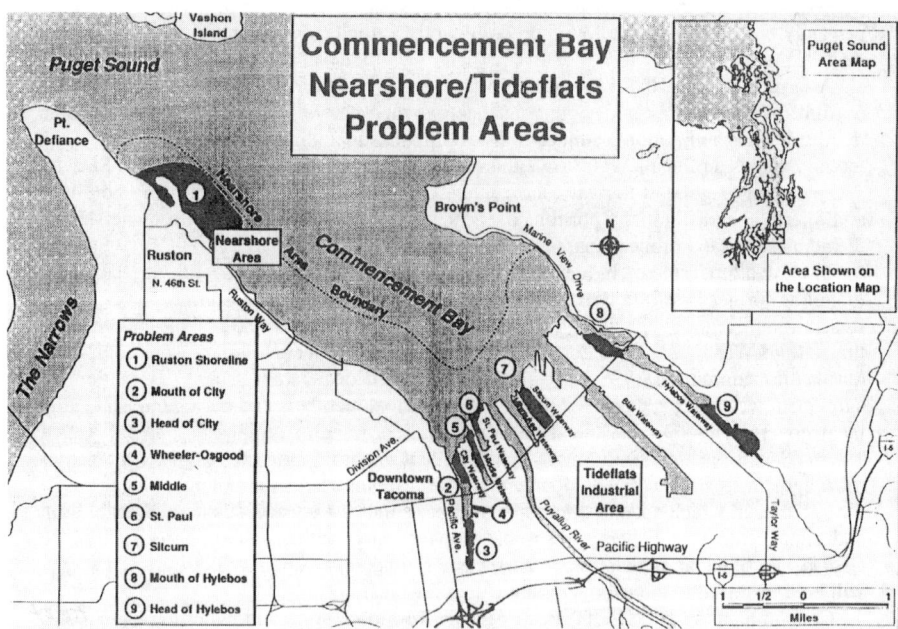

FIGURE 17.24. Commencement Bay Nearshore/Tideflats Problem Areas.

Commencement Bay was placed on a national interim list of 115 highest-priority hazardous waste sites on October 23, 1981. Initially, the Commencement Bay site was divided into four areas: deepwater, nearshore, tideflats/industrial, and south Tacoma channel. The National Priorities List promulgated on September 8, 1983 designated the CB/NT area and the Commencement Bay South Tacoma Channel (CB/STC) as separate National Priorities List sites. The deepwater portion of the bay was eliminated from the list at that time because water quality studies indicated there was minimal contamination in the area.

On 13 April 1983, USEPA announced that a cooperative agreement had been reached with Washington State Department of Ecology to conduct a remedial investigation/feasibility study on the nature and extent of contamination in the CB/NT site. Under the agreement, Ecology was designated as the lead agency for the investigation. The Commencement Bay Nearshore/Tideflats Remedial Investigation (Tetra Tech, 1985), completed in August 1985, characterized the nature and extent of contamination at the site. The commencement Bay Nearshore/Tideflats Feasibility Study (Tetra Tech 1988a) was completed in December 1988 and described feasible alternatives for sediment remedial action at the site. The feasibility study included an integrated action plan (PTI, 1988) to coordinate ongoing source control efforts and sediment remedial alternatives, and a sediment quality goals document (PTI, 1989) to develop sediment quality objectives.

Description of Alternatives. The feasibility study described cleanup objectives for the site and then presented a range of alternatives that offered viable means of achieving those objectives.

Ten candidate alternatives were identified in the CB/NT feasibility study:

1. No action
2. Institutional controls
3. In situ capping
4. Removal/confined aquatic disposal
5. Removal/nearshore disposal
6. Removal/upland disposal
7. Removal/solidification/upland disposal
8. Removal/incineration/upland disposal
9. Removal/solvent extraction/upland disposal
10. Removal/land treatment

USEPA Selected Remedy (1989). In the proposed plan for the feasibility study (the agencies' study), the agencies recommended that a performance-based remedy incorporating multiple sediment remedial options would be preferable to one that limited remedial action to a single specific technology. The recommendation was based on evaluations in the feasibility study indicating that all four confinement options offered similarly feasible and cost-effective means of achieving the project cleanup objectives.

However, in the CB/NT feasibility study, a preferred remedy was identified for each problem area that included a specific confinement option (e.g., nearshore disposal was preferred for the Head of Hylebos Waterway). The decision to define a generalized confinement element for sediment remediation instead of the specific confinement options identified in the feasibility study or a performance-based remedy as recommended in the feasibility study was based on comments received during the public comment period, and additional technical and administrative review conducted by USEPA and Ecology. This decision affects only the sediment remedial action element of the remedy. Source control and natural recovery remain key elements of each problem area remedy.

The preferred alternative identified in the CB/NT feasibility study and the selected remedy are summarized in Table 17.11. The remedy selected for the St. Paul Waterway problem area represents one of the four confinement options: in situ capping. For the Mouth of Hylebos Waterway, Head of City Waterway, and Wheeler-Osgood problem areas, open-water confined aquatic disposal was identified as the preferred alternative in the feasibility study. Nearshore disposal was identified in the feasibility study as the preferred alternative for Head of Hylebos, Sitcum, and Middle Waterways problem areas. Institutional control (including natural recovery) was selected as the preferred alternative for the Mouth of City Waterway problem area.

In 1990, USEPA requested that the Port consider including Sitcum sediment remediation as part of a long-standing nearshore fill development project, the Milwaukee Waterway nearshore fill and Blair Waterway dredging project. The Port evaluated the request as discussed at the end of the following subsection, and determined that combining the Sitcum Superfund cleanup with the Milwaukee/Blair project was feasible. Subsequently, the Port initiated a voluntary cleanup action for Sitcum Waterway. However, since the Sitcum is a Superfund site, USEPA required that the work be conducted under formal USEPA order in accordance with the ROD and Superfund regulations. The Sitcum Waterway Remediation Project is the first Superfund sediment remediation action initiated in Commencement Bay since the issuance of the ROD.

Milwaukee Waterway Nearshore Fill and Blair Waterway Dredging Project. Milwaukee Waterway is between the Puyallup River and Sitcum Waterway (see Figures 17.23 and 17.24). Created by dredge and fill operations between 1910 and 1913, Milwaukee Waterway contains approximately 30 acres of water surface area and ranges in depth from approximately –30 to –40 feet mean lower low water (MLLW). Milwaukee Waterway originally served as a marine terminal for the Milwaukee Railroad and remained in active use until the 1960s. Milwaukee Waterway has not been used actively by the Port for over two decades because of its functional obsolescence as a shipping waterway. It is obsolete in terms of depth, width, and available backup land for terminal or pier facilities. It is also an obstacle to existing Sea-Land terminal operations, which are located on the two peninsulas of land that it separates. The waterway sediments are contaminated from historical uses, although not to a level that warranted its designation as a Superfund problem area (Gilmur and Saathoff, 1994).

The project became an integral part of the Puyallup Indians Land Claims lawsuit. As

TABLE 17.11. Sediment remedies selected in the feasibility study and record of decision

Problem area	Feasibility study	Record of decision
Head of Hylebos	Nearshore disposal	Confined disposal[a]
Mouth of Hylebos	Confined aquatic disposal	Confined disposal[a]
Sitcum	Nearshore disposal	Confined disposal[a]
St. Paul	In situ capping	In situ capping
Middle	Nearshore disposal	Confined disposal[a]
Head of City	Confined aquatic disposal	Confined disposal[a]
Wheeler–Osgood	Confined aquatic disposal	Confined disposal[a]
Mouth of City[b]	Institutional controls	Confined disposal[a]

[a]In situ capping confined aquatic disposal, nearshore disposal, upland disposal.
[b]Predicted to recover following source controls.

such, the project was substantially revised and reduced in scope. The Port refined its proposal through analysis of the minimum operational needs necessary to keep the Milwaukee/Blair project viable. The Port's intent was to design the fill project to avoid and minimize, and then to mitigate, adverse effects while still fulfilling the project purpose and need. The Port and the Puyallup Tribe, as well as the United States of America, state of Washington, Pierce County, City of Tacoma, City of Fife, City of Puyallup, and various private parties resolved the Land Claims lawsuit in August 1988 when the Puyallup Tribe accepted the PSA. In order to become effective, however, the PSA had to be formally approved by the various parties. This took until March 1990.

As part of the PSA, the Port is obligated to dredge the Blair Waterway for navigational and environmental improvements. The Blair Waterway, which is now maintained at a depth of –40 feet MLLW, will be deepened to a minimum of –45 feet with a majority of the waterway at –48 feet MLLW. Further, placement of the Blair dredge material into the Milwaukee Waterway fill is specified in the PSA and its implementing legislation, Puyallup Tribe of Indians Settlement Act of 1989, Public Law 101-41. The PSA did not obviate the need for the Port to obtain necessary project permits.

On March 24, 1990, the PSA became effective, and the Port reactivated the Milwaukee/Blair project with the intent of completion as soon as possible. The project now included filling 72.5% of Milwaukee Waterway (approximately 2,400 lineal feet) to create approximately 23.7 acres of upland for the expansion of the existing Sea-Land marine container terminal, providing habitation mitigation at and beyond the mouth of the Milwaukee Waterway, and dredging approximately 2.4 million cubic yards from Blair Waterway and using the materials for filling the Milwaukee Waterway and constructing the mitigation project (see Figure 17.25).

Blair Waterway contains sediments that are, for project purposes, identified as clean sediments and sediments designated for confinement. The clean sediments are those which are equal to or below the Puget Sound Dredge Disposal Analysis (PSDDA, 1988) screening level (SL) criteria. These sediments are deemed appropriate for use in the closure berm and habitat mitigation fill, and/or as capping material for the nearshore confined disposal fill. Blair Waterway sediments that exceed the PSDDA SL criteria are des-

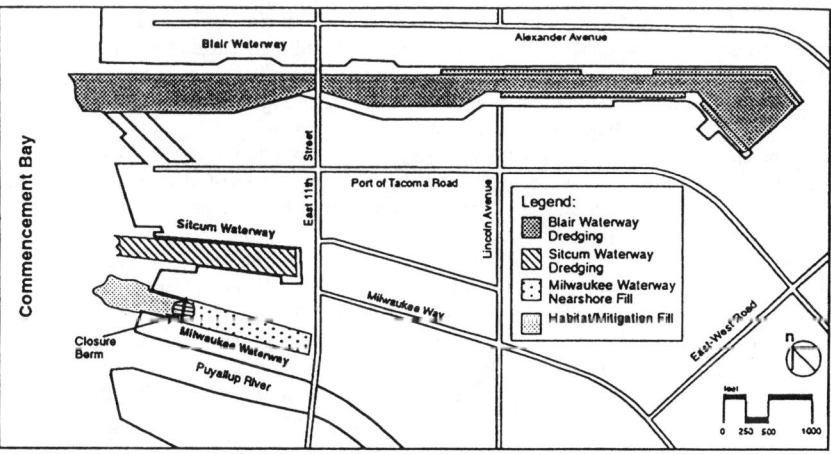

FIGURE 17.25. Sitcum Waterway Remediation Project Areas.

ignated for confinement and require placement in the confined disposal fill below +9 feet MLLW, which is below the groundwater elevation.

The Port tested a portion of the Blair Waterway (approximately 594,000 yd^3) sediments for open water disposal. As this material passed the PSDDA criteria for open water disposal, the Port was able to determine that USEPA's request was feasible.

Administrative Order on Consent. The Port elected to pursue remediation of Sitcum Waterway as a voluntary cleanup action under USEPA Superfund authority. As a voluntary action, USEPA and the Port negotiated an Administrative Order on Consent (Order) (USEPA, 1991a), which became effective in March 1991. It should be noted that the federal, state and tribal Natural Resource Trustees (Trustees) also participated in the negotiations to ensure that Trustee interests were considered. The trustees for the CB/NT site are: (1) the national Oceanic and Atmospheric Administration of the U.S. Department of Commerce, (2) the U.S. Department of Interior, (3) the Washington Department of Ecology (on behalf of the Washington Department of Fisheries, the Washington Department of Natural Resources, and the Washington Department of Wildlife), (4) the Puyallup Tribe of Indians, and (5) the Muckleshoot Indian Tribe. The Trustees were not signatory to the Order.

Sitcum Waterway Remediation Project. *Design Strategy and Basis* (Gilmur and Saathof, 1994). The approach to establishing sediment volume requiring remediation included determining the area and depth of Sediment Quality Objectives (SQO) exceedence, determining navigational requirements of the waterway, determining limitations for dredging techniques and equipment, and determining implementation issues such as the potential for contamination resulting from sediment resuspension or slough along slopes. In consideration of these issues, an approach was established that achieves the objectives of Superfund sediment remediation and maintains the navigational purposes and characteristics of the waterway, including pier and side slope integrity and stability.

Summary of Project Design. The nearshore confined disposal site is in the Milwaukee Waterway, positioned between the Sitcum Waterway and the Puyallup River (see Figure 17.25). A closure berm was located approximately 2,400 feet out from the inland limit of the existing Milwaukee Waterway and was constructed using approximately 170,700 yd^3 of clean Blair Waterway sediments (includes an outside toe buttress), and approximately 21,500 yd^3 of clean imported material. The closure berm provided final confinement for Sitcum Waterway sediments (Phase 1 and Phase 2), Blair Waterway sediments designated for confinement, and clean Blair sediments used for capping. Blair Waterway sediments used for closure berm construction will be dredged by mechanical dredge, and bottom dump barges used to transport the dredged sediment from Blair Waterway to Milwaukee Waterway for placement.

The closure berm was constructed from an elevation of –40 to 0 feet MLLW with side slopes of 6 horizontal to 1 vertical (6H:1V) by position release of clean Blair sediments from bottom dumped barges. The closure berm has a top elevation of +12 ft MLLW.

Habitat impacts from the project were compensated by providing approximately 20.6 acres of intertidal (–12 feet to –4 feet MLLW) and shallow subtidal (–4 feet to –10 feet MLLW) habitat at and beyond the mouth of Milwaukee Waterway, approximately 1.6 acres of upland habitat on the edges of the peninsulas and closure berm abutting the intertidal habitat, and an additional 9.5 acres of restored wetland/aquatic habitat at another site located on a tributary to the lower Puyallup river. Clean sediments from Blair Waterway were used to build the habitat mitigation area and beyond the mouth of Milwaukee Waterway. The sediments were placed by bottom dump barge and hydraulic pipeline.

A summary of remedial options evaluation is presented in Table 17.12 and a summary of costs are given in Table 17.13.

TABLE 17.12. Summary of remedial option evaluation

Criteria	Confined aquatic disposal	Nearshore disposal	Upland disposal
Overall protection of human health and the environment	Yes	Yes	Yes
Compliance with ARARs and TBCs	Yes	Yes	Yes
Long-term effectiveness	Moderate	High	Moderate
Reduction in toxicity, mobility, or volume by treatment	Low	Low	Low
Short-term effectiveness	Low	Moderate	Low
Implementability	Moderate	High	Moderate
Cost	High	Low	High

Federal Consent Decree. As part of the Superfund process, USEPA authorized implementation of the Sitcum Waterway Remediation Project under a Federal Consent Decree, which was negotiated by the Port and USEPA, and entered into federal court on October 8, 1993. The Consent Decree is a legal document that contains all of the requirements for the cleanup action. The Consent Decree includes agreement by the Port to pay past and future USEPA oversight and response costs, implement the remedial action in accordance with the approved remedial design, meet performance standards, conduct construction and postconstruction monitoring, and implement contingency actions or plans as necessary.

The Consent Decree also incorporates the terms of a settlement of the Port's Natural Resource Damages liability. This settlement resolves Trustee claims for damages due to injury to, destruction of, or loss of natural resources caused by releases of hazardous substances from property owned, managed or operated by the Port within Commencement Bay. The primary component of the settlement is payment to the Trustees of $12,000,000 over 6 years. It is intended that these funds be used to benefit natural resources of Commencement Bay in Pierce County injured as a result of releases of hazardous substances.

The USEPA Superfund process as applied to Sitcum Waterway will enable significant port cleanup and development projects to be accomplished concurrently under a Federal Consent Decree. The Sitcum Waterway Remediation Project has been designed to accomplish the following goals: (1) provide Superfund cleanup of Sitcum Waterway channel and nearshore areas, (2) provide sufficient depth for navigation in Sitcum Waterway, (3) provide cleanup and navigation improvements in Blair Waterway, (4) fulfill certain obligations of the PSA and its implementing legislation, the Puyallup Tribe of Indians Settlement act of 1989, (5) expand the existing Sea-Land marine container terminal facility, (6) provide ecologically acceptable disposal for Blair Waterway sediments, (7) minimize en-

TABLE 17.13. Summary of costs

Option	Estimated cost
Confined aquatic disposal	$11,523,861
Nearshore disposal	$3,978,235
Upland disposal	$20,555,599

vironmental impacts by combining project actions, (8) facilitate habitat restoration by the prompt resolution of all claims for natural resource damages, and (9) promote economic benefits for citizens of Pierce County and Washington State.

Project Implementation. The Port of Tacoma received bids on June 30, 1993; the total base bid was $18,136,541.50 for the combined Blair (not Superfund) and Sitcum (Superfund) Projects. Unit dredging costs varied from $1.50 per cubic yard (dredging of Sitcum Waterway and depositing dredged material in the Milwaukee Waterway) to $25.00 per cubic yard (dredging of Sitcum Waterway side slopes under piers and depositing dredged material in the Milwaukee Waterway).

Dredging was completed at the end of 1994. On August 28, 1996 the EPA stated that

> Cleanup progress has been made in all areas of the Commencement Bay Nearshore/Tideflats (CB/NT) site, and cleanup actions have eliminated the threat to public health or the environment at some portions of the site. EPA is recognizing these areas as clean and proposing to remove them from the National Priority List (NPL).

> This proposal pertains only to St. Paul Waterway sediments, Blair Waterway sediments, sources draining to the St. Paul and Blair Waterways, and to four properties which were transferred to the Puyallup Tribe under the Puyallup Land Claims Settlement Act of 1989.

USEPA Progress on the Contaminated Sediment Management Strategy. *Purpose of the Strategy.* USEPA proposed its Contaminated Sediment Management Strategy in August 1994.

> The purpose of the Environmental Protection Agency's (EPA's) Contaminated Sediment Management Strategy is: to describe EPA's understanding of the extent and severity of sediment contamination, including uncertainties about the dimension of the problem; to describe the Agency's cross-program policy framework in which EPA intends to promote consideration and reduction of ecological and human health risks posed by sediment contamination; and to describe actions EPA believes are needed to bring about consideration and reduction of risks posed by contaminated sediments.

Additionally, the fourth goal of the Strategy is to develop and consistently apply methodologies for analyzing contaminated sediments.

The Strategy is an Agency work plan issued in support of USEPA's regulatory and policy initiatives and is for Agency guidance only. The Strategy does not propose new regulations.

Progress on Responses and Response to Public Comments Document. Nearly 500 pages of comments were received from 126 organizations on the proposed Contaminated Sediment Management Strategy. USEPA's Office of Science and Technology (OST) within the Office of Water (OW) has drafted a 375-page "Comment and Response Document," which is being reviewed by four USEPA workgroups that developed the Strategy. The chapters in the "Comment and Response Document" are organized to reflect the chapter organization of the "Contaminated Sediment Management Strategy." The Comment Response Document addresses a wide range of comments concerning contaminated sediment assessment, prevention, abatement and control, remediation, dredged material management, research, and outreach.

Public Comments on the Strategy. The comments address legal, policy, and technical issues. Major technical concerns include the following:

- Uncertainties regarding the extent and severity of the contaminated sediment problem
- The availability of methods to assess contaminated sediment
- Understanding of the sources of sediment contamination
- The validity and use of numerical sediment quality criteria
- The use of fate and transport models, particularly for permitting programs (404 permits, NPDES permits)
- The need for cost/benefit studies of regulatory alternatives
- Selection of appropriate remediation technologies and development of new ones (especially for decontamination)
- Appropriate handling and disposal of contaminated dredged material
- Additional research including Quantitative Structure Activity Relationships (QSARs), assessment methods, fate and transport models, and causes of toxicity

After completing dredging of Sitcum and Blair Waterways, the EPA has been involved with planning the remediation of other waterways.

Hylebos Waterway. Remediation is to include (a) sediment sampling to define cleanup areas, (b) cleanup at the Occidental Plant, and (c) Wood Debris Pilot Project.

Sediment Sampling Continues to Define the Cleanup Area. The Hylebos Cleanup Committee (HCC), a group of five companies and the Port of Tacoma, continues its work to better define the cleanup area. Sediment samples were taken in the Hylebos Waterway in the spring and summer of 1998 to define the edges of cleanup areas. This information will be used to develop the HCC's proposed cleanup areas and disposal plan, which they submitted to USEPA at the end of 1998.

In the meantime, some property owners on the Hylebos Waterway are taking steps to clean up contaminated intertidal sediments on their properties:

- The intertidal areas of the former Tacoma Boat property are now clean, thanks to cleanup efforts by the new property owner, Ace Tank.
- General Metals will clean up intertidal areas in front of their property as part of a project to upgrade their wharf and scrap loading facilities.
- Automobile shredder residue (shredded car interiors) was removed from the upland and intertidal portions of the Mather Auctioneers property.

Cleanup Progress at the Former Occidental Plant. In November 1998, USEPA and Occidental Chemical Corporation agreed to address the embankment area and the adjacent offshore area in front of the former Occidental facility. Sampling of the embankment area is now complete and sampling of the offshore area continues. These areas contain the highest concentrations of contaminants found anywhere in the Hylebos Waterway.

Wood Debris Pilot Project. USEPA, Ecology and the National Oceanic and Atmospheric Administration (NOAA) are reviewing the studies submitted by the Hylebos Wood Debris Group (WDG). The studies determine where wood waste accumulates and where cleanup is needed in the head of the Hylebos Waterway.

A pilot project will pick up whole logs from the waterway and determine whether they could be reused or recycled. Further sampling and analysis will also be conducted to determine the best disposal option for the materials removed from the waterway. In addition to preparing for cleanup of wood debris, the WDG is developing an operation, maintenance, and monitoring plan that will address prevention of future accumulations of wood debris within the waterway.

Disposal sites for Thea Foss Contaminated Sediments. USEPA has been working with parties potentially responsible for the cleanup of contaminated sediments in the bay, as well as interested citizens and state and federal agencies, to come up with a plan for disposing of contaminated sediments from the waterways in Commencement Bay.

EPA is reviewing the City's report recommending a cleanup plan and a disposal site for contaminated sediments in the Thea Foss and Wheeler Osgood Waterways. The report evaluates options, including upland disposal and confined aquatic disposal.

In addition, discussions have taken place concerning a proposal by Simpson Tacoma Kraft to use the St. Paul Waterway as a disposal site for contaminated sediment dredged from the Thea Foss and Middle Waterways. Several issues face this proposal, including whether the area is big enough to hold the contaminated sediments needing disposal. Issues are also arising out of the recent proposed listing of Puget Sound Chinook salmon as a threatened species under the Endangered Species Act and need resolution.

Investigation of Dioxin at the Olympic View Restoration Site. The Department of Natural Resources (DNR) was notified by the City of Tacoma that high concentrations of dioxin were found at the head of the peninsula between Thea Foss and Middle Waterways. This area, known as the Olympic View Restoration Site, is slated for a habitat restoration project under an agreement between the City and the National Resources Trustees.

The City of Tacoma is developing a sampling and analysis plan to investigate this contamination further and the Department of Ecology is coordinating the review of the City's plan with USEPA and the Trustees.

Eagle Harbor, Washington

Introduction. In the early 1980s it was discovered that an inordinately high percentage of fish caught in a small harbor on the western shore of Puget Sound had liver disease and tumors.

In 1985 USEPA confirmed that the sediments in the eastern portion of Eagle Harbor were heavily contaminated with polynuclear aromatic hydrocarbons (PAHs). Mercury was also found in sediments in the western part of the Harbor.

In 1987, Eagle Harbor and the wood treatment facility were designated a Superfund site by USEPA.

Capping. Several alternatives were considered including: (a) removal of mercury hot spots, and placing a thick layer of clean sediment over the remaining contaminated sediment (West Operable Unit); (b) dredging the contaminated sediment and placing it in a confined aquatic disposal (CAD) site, or in a waste landfill, or treating it (East Operable Area). The estimated costs of these alternatives ranged from $14.1 and $23.8 million (Nelson et al., 1984).

In 1993, USEPA requested the Corps of Engineers to use some of the dredged material from Snohomish River for capping. This sediment provided a 0.9 meter cap over a 220,000 square meter contaminated sediment area in Eagle Harbor. Placement quantities for Area 1 were estimated at 104,700 m^3 and 107,000 m^3 for Area 2 (Figure 17.26). The cost of placing the cap was $1.45 per m^3 for Area 1 and 3.17 per m^3 in Area 2. Thus, the total cost of capping was about $446,000. Extensive monitoring of the site has been planned.

Grand Calumet River, Indiana

1. Introduction. USEPA Region 5 considered responses to comments from the public concerning a USX Corporation proposal to dredge a portion of the Grand Calumet River and dispose of PCB-contaminated sediments in a disposal facility to be constructed on USX property in Gary, Indiana.

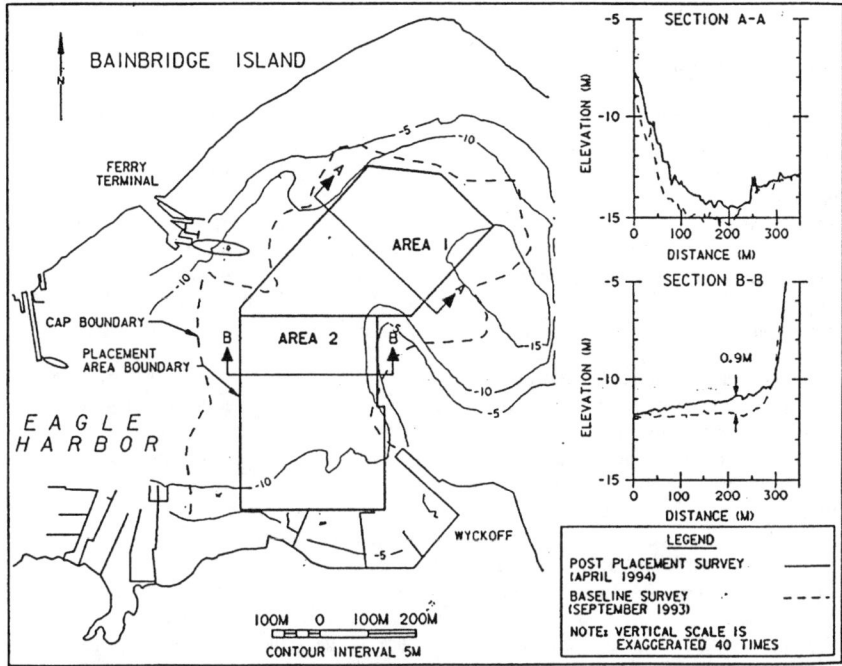

FIGURE 17.26. Comparison of pre- and postbathymetric surveys.

USX proposes to remove about 687,000 yd^3 of contaminated sediment from the upper 5 miles of the East Branch of the river, adjacent to the USX steel production facility known as the Gary Works. Some 125,000 yd^3 of sediment are contaminated with polychlorinated biphenyls (PCBs). A USX sediment study completed in 1993 shows that the river contains heavy metals iron, lead, zinc, cadmium, and chromium; oil and greases; PCBs; polycyclic aromated hydrocarbons; benzene; cyanide; and other pollutants.

The project was proposed by USX in cooperation with USEPA and the Indiana Department of Environmental Management. It will be implemented according to all applicable state and federal environmental laws. USX has submitted a plan to EPA that is being reviewed to ensure that it adheres to those laws. The Agency has asked the public to comment on the disposal facility and on the manner in which some of the sediments will be managed.

2. Sediment Dredging Proposed. Under a Resource Conservation and Recovery Act (RCRA) corrective action order, sediments will be dredged using a hydraulic suction pump. Such pumps minimize resuspension of disturbed sediments, but moving any contaminated sediments has the potential to resuspend contaminated material in the water column. Consequently, short-term exposure of aquatic life to contaminated sediments may increase, and contaminated material may be transported to cleaner areas. USX proposes to minimize these impacts through sound management practices, including total containment of the most contaminated areas during dredging.

Dredged sediments are to be disposed of in a 40-acre, double-lined corrective action management unit (CAMU) to be built on USX property. This is the safest way of handling sediments because they will be taken out of the river completely. The CAMU will be constructed to contain all dredged sediments and prevent future releases of hazardous constituents. It will be sited near the project area to minimize sediment transport, in an area formerly used to dispose of dredged sediments.

After all dredged sediments are placed in the CAMU, the CAMU will be dewatered and the dredged solids will be consolidated. Wastewater from this process will be collected and treated in a project-specific treatment system. The oil skimmed from the CAMU will be treated and disposed of in accordance with applicable state and federal law. Although not a complete treatment process, this step will reduce some of the sediment toxicity. Once the dewatering has been completed, USX will install a temporary vegetative cover over the CAMU to control dust and minimize erosion.

3. Excess Disposal Capacity. The CAMU will be designed with excess capacity that may be used to dispose of additional wastes from the site cleanup. By law, only wastes generated during the implementation of the corrective action may be disposed of in the CAMU. It will not be used for waste disposal from other site cleanups. The excess capacity would be used only if USEPA approves, and if the physical and chemical makeup of the additional waste is compatible with the dredged sediments.

USX must prepare a groundwater monitoring plan to ensure that contaminants are contained by the CAMU, which must be approved by USEPA. After the activities specified in the consent order are completed, USX must submit a closure plan, which also must be approved by USEPA. After closure, USX will routinely inspect the CAMU. The dredging will be an interim measure under a corrective action order pursuant to RCRA section 3008(h).

The cleanup is expected to take about 5 years to complete. This period includes 2 years for engineering studies, design, and permit application preparation; 1 year for constructing the CAMU and wastewater treatment system; and 2 years to dredge the sediments and place them in the CAMU.

The USX proposal goes beyond the requirements of the settlement of a 1990 federal lawsuit brought under the Clean Water Act. Under the consent decree that resulted from the suit, the company agreed to improve its wastewater treatment equipment and its water delivery and sewer systems; spend up to $2.5 million to investigate pollution in a 13-mile stretch of the Grand Calumet River (GCR) and up to $5 million to clean the upper 5 miles of river adjacent to its property; and pay a $1.6 million fine.

4. Summary Description of the Remediation Plan. USX agreed to perform the activities set forth in the SOW. The following is a summary of the activities USX will perform pursuant to the SOW; they are not intended to relieve USX of any obligation which may be described in more detail elsewhere.

- Submit complete Workplan(s) and Engineering and Design Reports.
- Submit to the appropriate governmental agencies complete applications for permits and approvals, and remediate the GCR in accordance with those permits, as well as this SOW and all applicable Federal, State and local regulations.
- Construct and manage an on-site disposal area approvable as a Corrective Action Management Unit (CAMU) under RCRA and as an alternative disposal method under the Toxic Substances Control Act (TSCA).
- Provide verification of nonnative sediment removal through comparison between pre- and postdredge sediment surveys.

REMOVAL OF CONTAMINATED SEDIMENT BY DREDGING   **17.53**

- Remove nonnative sediment as measured in the predredge survey in Transects 1 through 11 from within river isolation cells formed by the installation of bulkheads upstream and downstream and impounding, diverting, or bypassing flow and placing dredged sediment in a discrete disposal cell within the CAMU.
- Remove nonnative sediment as measured in the predredge survey downstream of Transect 11 (USX Outfall 018) during open flow conditions in the river channel (i.e., no isolation cells required) and place in the CAMU.
- Remove nonnative sediment as measured in the predredge survey from Transect 17, Horizon 1 during open flow conditions in the river channel and place in a discrete disposal cell within the CAMU.
- Provide wastewater treatment for the dredge water generated during sediment removal and monitor prior to conveyance to the Terminal Lagoons for subsequent discharge to the GCR through a permitted National Pollutant Discharge Elimination System (NPDES) outfall.
- Perform remediation plan work (i.e., facilities construction and completion of dredging program) within 3 years or less following receipt of necessary permits and other approvals.
- To the extent practicable, include 6 inches of over-dredging in all the areas of the GCR to which the SOW applies.

In addition, USX will:

- Conduct a USEPA-approved postremediation monitoring program 3 and 6 years after completion of the Grand Calumet River (GCR) dredging.
- Estimate the annual potential for air emissions from the sediment remediation project and the CAMU, provide for ambient air monitoring during the duration of the project, conduct bench-scale tests and air modeling to establish site-specific air emissions action levels and operational standards for the project, and submit the proposed air emissions action levels and operational standards to USEPA for approval.
- Prepare and implement a health and safety plan consistent with the RCRA Corrective Action Order.
- Prepare and submit a summary of all the dredging activities, volume/sediment verification studies and findings to the USEPA and IDEM within 90 days of the completion of the GCR dredging activities.
- Prepare and submit the findings of the postremediation monitoring program within 90 days of completion of the third and sixth year monitoring events, respectively.

The specific activities and methodologies that comprise the remediation plan are summarized below.

Sediment Removal: Sediment from Transects 1 through 36 will be removed by hydraulic dredging methods and delivered, via slurry pipeline, to the CAMU. Dredging activities will be confined to the river channel. Nonnative sediment will be removed from the river channel with allowances provided for incidental sloughing from "soft-side" areas and overdredging. The estimated total quantity of sediment to be dredged (i.e., approximately 687,000 yd^3 consists of: 559,000 yd^3 of nonnative sediment from the river channel; 38,000 yd^3 of native river bottom material due to incidental overdredging (i.e., an estimated six inches); 90,000 yd^3 due to incidental sloughing from soft-side areas.

For purposes of this SOW and the subsequent verification of nonnative sediment removal, the pre- and postdredge surveys defined the sediment as follows:

17.54 HANDBOOK OF COASTAL ENGINEERING

- Nonnative sediment is soft, dark or black in color, and typically consists of fine sands, silts, and clayey silts and can be readily differentiated from native sediment on the basis of color, grain size, resistance to penetration, organic content, and odor.
- Within the river channel.
- Bounded on the top by the water column and on the bottom by the interface with the native river bottom, which is material that is gray and consists of very dense beach and dune sands.
- Bounded by the interconnection of elevation points (spaced nominally at 10-foot intervals) with straight lines outlining the end area template (see Figure 17.27).

Sediment removed from Transects 1 through 11 will be dredged from within river isolation cells formed by the installation of upstream and downstream bulkheads. Dredged sediment will be delivered to the CAMU via pipeline in slurry form for disposal. Outfalls discharging to the GCR along the cell being dredged will be intercepted and pumped to the GCR downstream of the cell. Upstream flow will be impounded or bypassed around the cell being dredged. Provision for managing stormwater runoff is included in the design.

Hydraulic dredging of Transects 17, Horizon 1 will be performed during open-channel flow conditions in the GCR and will precede the dredging of other reaches. Transects 12 through 36 will be hydraulically dredged during open-channel flow conditions following

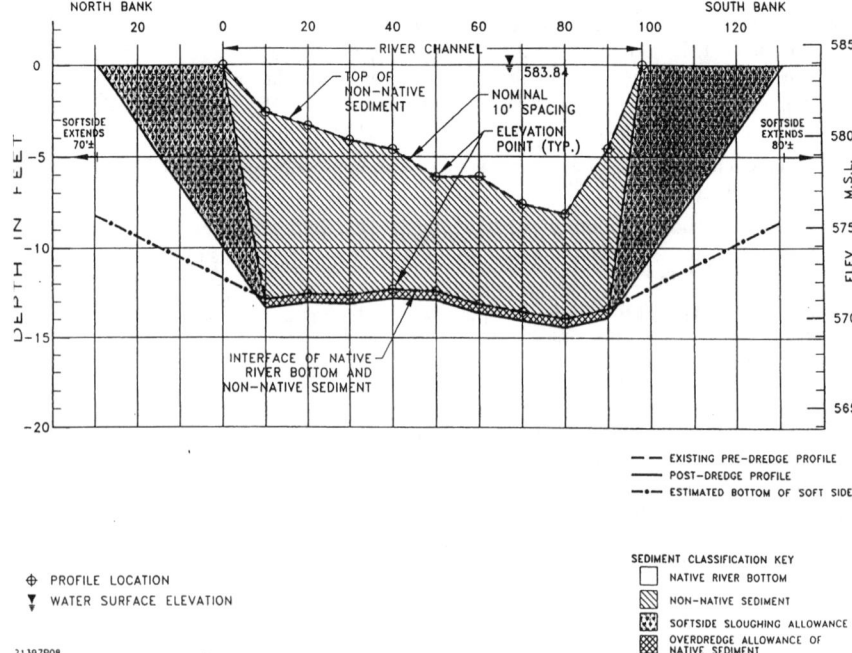

FIGURE 17.27. Typical Grand Calumet River cross-section (end area template).

completion of Transect 17, Horizon 1 and concurrent with the dredging of Transects 1 through 11. USX shall perform field surveys of the river channel prior to and following completion of the dredging to verify that the nonnative sediment has been removed.

USX will employ silt screens, oil booms, and/or other equivalent suspended solids collection devices during the open channel dredging operation of Transects 12–36 to minimize the resuspension and downstream discharge of oil and/or other materials. The isolation cells of Transects 1–11 will minimize the downstream migration of resuspended materials. In addition, USX will employ oil booms within the cells to control floating oils.

Management of Dredged Sediment Slurry: Dredged sediment from the project area (Transects 1 through 36) will be passively dewatered within the CAMU. Sediment from Transects 1 through 11 and Transect 17, Horizon 1 will be passively dewatered within a discrete disposal cell in the CAMU.

Sediment Disposal: Dredged sediment will be disposed of in the CAMU in accordance with the terms and conditions hereof. The liner, leachate collection, and cover systems shall be designed in accordance with RCRA Subtitle C (Hazardous waste) landfills (see Figures 17.28–17.31).

Management of Dredge Waters: Water from Transects 1 through 11 and Transect 17, Horizon 1 will be processed through a project-specific wastewater treatment plant, monitored and then conveyed to the Terminal Lagoons prior to discharge to the GCR in accordance with IDEM-approved modifications to USX's NPDES permit. The remaining dredge waters from Transects 12 through 36 will be treated and monitored before conveyance to the Terminal Lagoons and discharged to the GCR in accordance with IDEM-approved modifications to USX's NPDES permit.

Manistique River, Michigan

1. Location. The Manistique River and Harbor site is located in Manistique, Schoolcraft County, Michigan, on the southern shore of Michigan's Upper Peninsula, where the Manistique River empties into Lake Michigan. Since the 1970s, a number of federal and state agencies have conducted sampling at the site. The principal contaminants identified in the sampling were polychlorinated biphenyls (PCBs) in the sediments of the river and harbor. In 1987, the site was designated as one of 43 Areas of Concern in the Great Lakes because of the PCB contamination in the river and harbor. Several area companies whose operating practices may have contributed to the contamination have been identified as potentially responsible parties (PRPs). This provides an update on the actions being taken by the USEPA to address site contamination.

2. Site Background. The Manistique River and Harbor site is bounded on the east and west by the Manistique River, on the south by Lake Michigan, and on the north by a dam. Throughout its history, waste disposal in the Manistique River included paper and wood wastes as well as various chemicals and oil wastes from industrial and paper milling operations. In addition, storm water and sanitary wastes were discharged directly into the river prior to the construction of a municipal wastewater treatment plant in 1959.

A report prepared in 1985 on Great Lakes water quality identified several problem areas in Manistique Harbor. The report found a reduction in the number and variety of aquatic life as a result of the presence of PCB contamination. Fish from the harbor were found to contain PCBs above the action levels of 2 parts per million (ppm) set by the Food and Drug Administration (FDA). The Michigan Department of Public Health has issued a public health advisory warning people against eating carp caught in the harbor. Sampling conducted by various federal and state agencies since the 1970s has confirmed the presence of PCB contamination in the river and harbor.

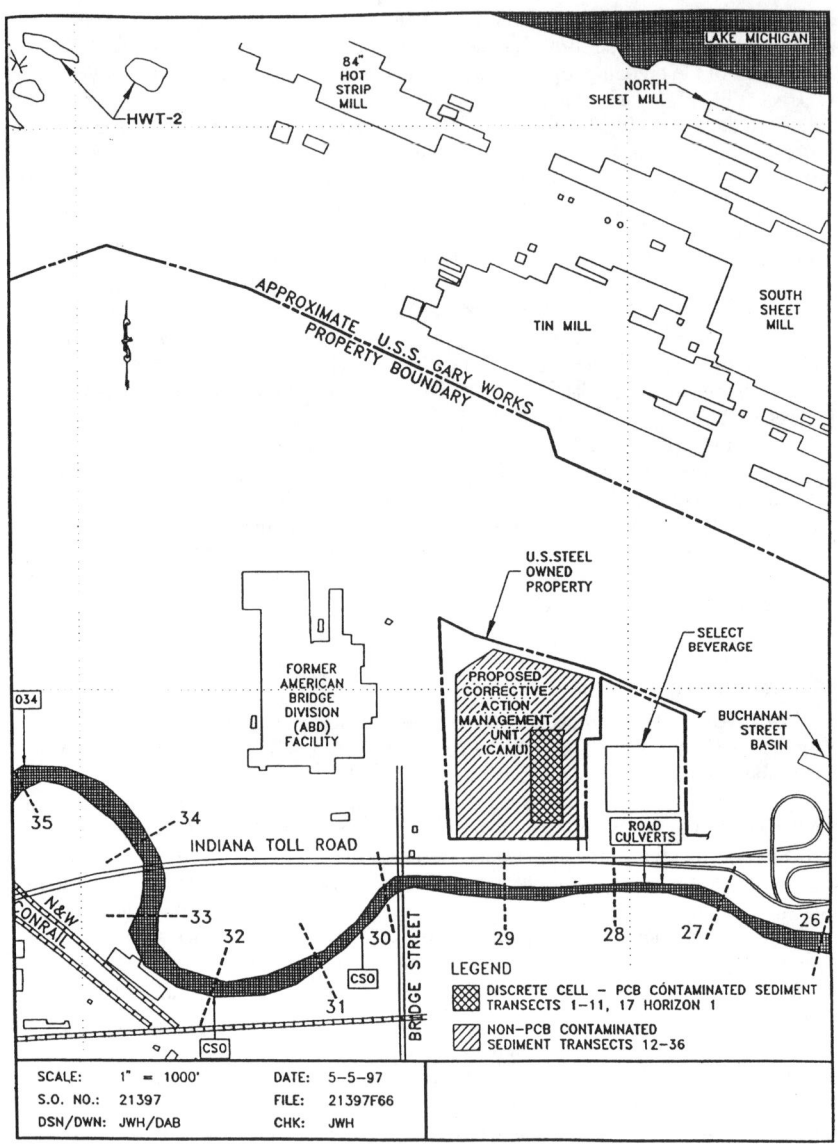

FIGURE 17.28. Sediment disposal facilities—location plan.

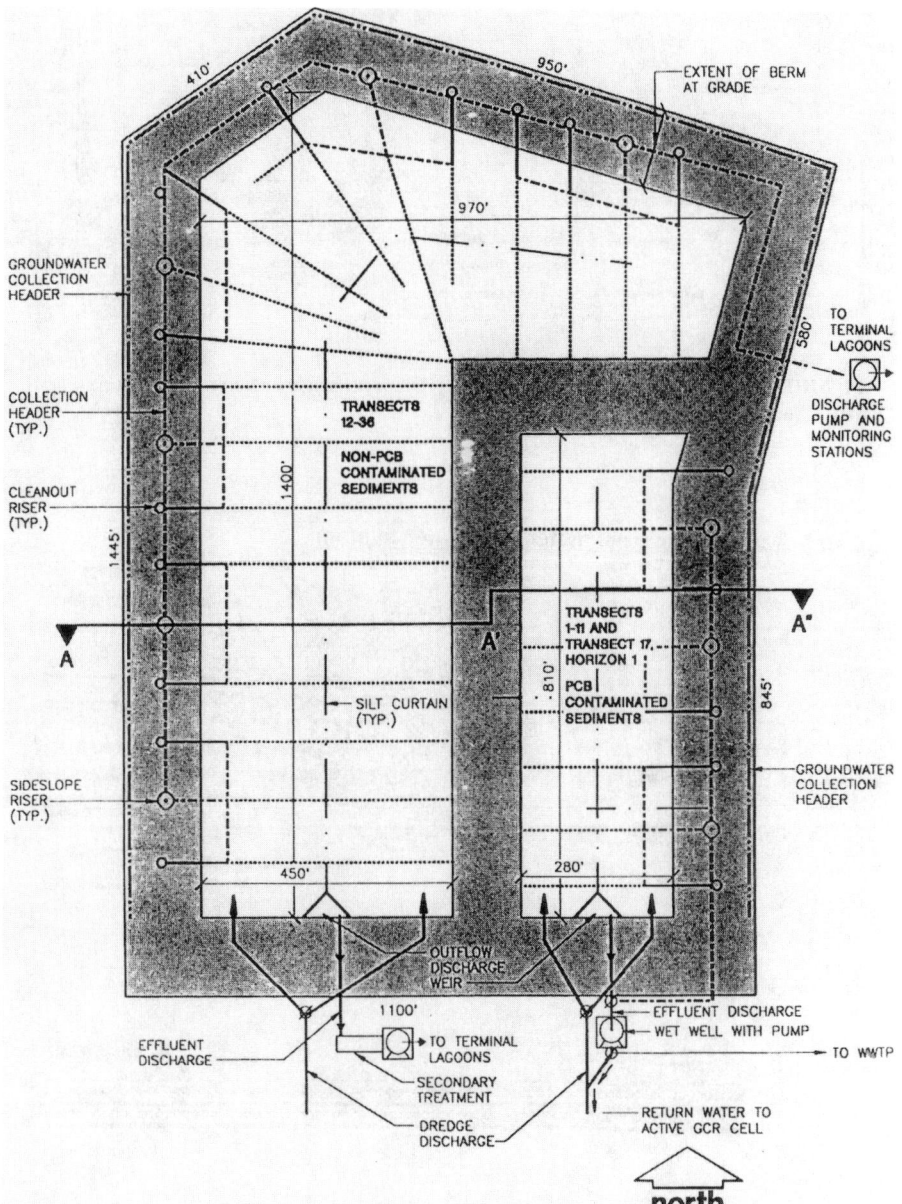

FIGURE 17.29. Plan view—corrective action management unit, U.S. Steel, Gary Works (not to scale).

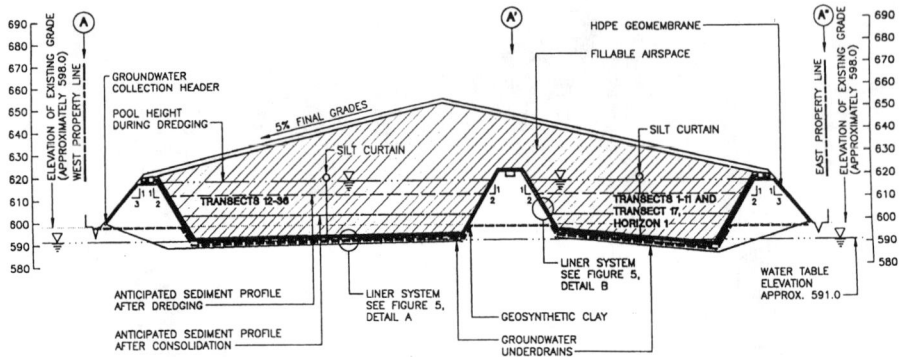

FIGURE 17.30. Section A-A'-A''—corrective action management unit, U.S. Steel, Gary Works (not to scale).

Detail A Base and perimeter berm liner system

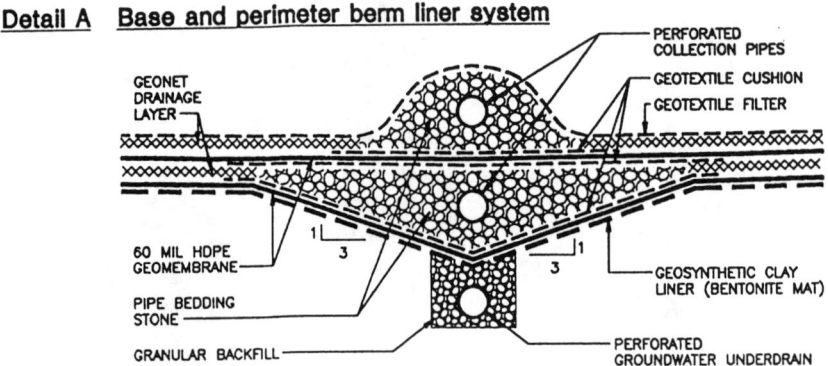

Detail B Interior berm liner system

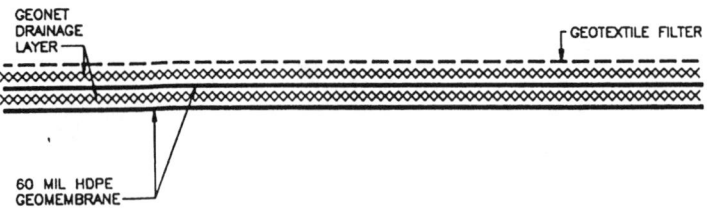

FIGURE 17.31. Proposed liner system detail, corrective action management unit, U.S. Steel, Gary Works (not to scale).

USEPA has identified several parties potentially responsible for contributing to the Manistique River and Harbor site. In June 1993, two of the PRPs—Manistique Papers, Inc. (Manistique Papers), and Edison Sault Electric Company (Edison Sault)—conducted sediment surveys in cooperation with USEPA to assess the distribution of PCB contamination in the river and harbor. In November 1993, USEPA began a time-critical response action at the site. The response action consisted of placing a temporary cap over a 2-acre area of Manistique Harbor bottom where surface sediments contain the highest concentrations of PCBs and river currents are the strongest.

3. USEPA Selects Response Actions. Under a USEPA Consent Order, Manistique Paper and Edison Sault also conducted an engineering evaluation/cost analysis (EE/CA) to evaluate several possible technological alternatives for addressing the PCB contamination in the river and harbor. USEPA reviewed the alternatives and selected an approach that is being implemented as a non-time-critical response. The remedy consists of a combination of two technologies: (1) capping, to contain PCB contamination in areas of the river and harbor, and (2) dredging, to remove contaminated sediment from an upstream area of the river. USEPA is requiring the response action to reduce PCB contamination to a concentration of 10 parts per million (ppm) or less in surface sediments.

4. Capping Activities. Capping to address most of the contaminated sediments in the river and harbor is conducted by the PRPs with USEPA oversight. Capping activities will take place in two areas of the Manistique River and Harbor: a 2-acre area that includes a temporary cap previously placed by USEPA and a 15-acre area downstream in the harbor area bordering Lake Michigan. The caps for both areas will consist of a 20-inch-thick layer of sand and a 12-inch-thick layer of stone armor. An additional layer of activated carbon will be added to the sand cap material over areas with higher PCB concentrations. To ensure that the caps will continue to contain the PCB contamination, USEPA requires the PRPs to implement a continual monitoring program consisting of visual inspections, cap coring, and sampling of the water column, groundwater, sediments and aquatic life.

5. Dredging Activities. USEPA has also chosen to dredge a portion of the Manistique River and Harbor site to test several dredging removal techniques and containment technologies. Funding for the dredging activities is being provided by USEPA under the Superfund program.

Sediment will be dredged from an area near the western shore of the Manistique River, upstream of the U.S. Route 2 bridge. This is the area where sediment surveys have found the highest concentrations of PCBs in the sediments. The area of the river to be dredged has been isolated to prevent disturbance and the spread of the contaminated sediments (Figure 17.32). USEPA will monitor the dredging to ensure that the PCB contamination remains confined to the dredging area. Sediment that is dredged from the river will be dewatered and transported for proper disposal at a commercial PCB disposal facility in Model City, New York. Following the dredging, USEPA will conduct sampling to confirm that the dredging has reduced the contamination to acceptable levels. If necessary, areas containing PCB concentrations that exceed acceptable levels may also be capped.

The project consists of capping a 15-acre area in the harbor and another 2-acre area upstream in the river. The cap cross-section will consist of a 20-inch thick "enhanced" (with carbon) base sand isolation layer overlain by a 12-inch thick "well graded" (with a filter component) armor stone layer. This area will not be dredged.

The USEPA employed a vacuum diver-assisted dredge on a 2-acre isolated area in the river during the Fall 1995. The aim of the project was to remove all sediments in that area with PCB concentrations exceeding 10 ppm. However, as of December 1995 only about

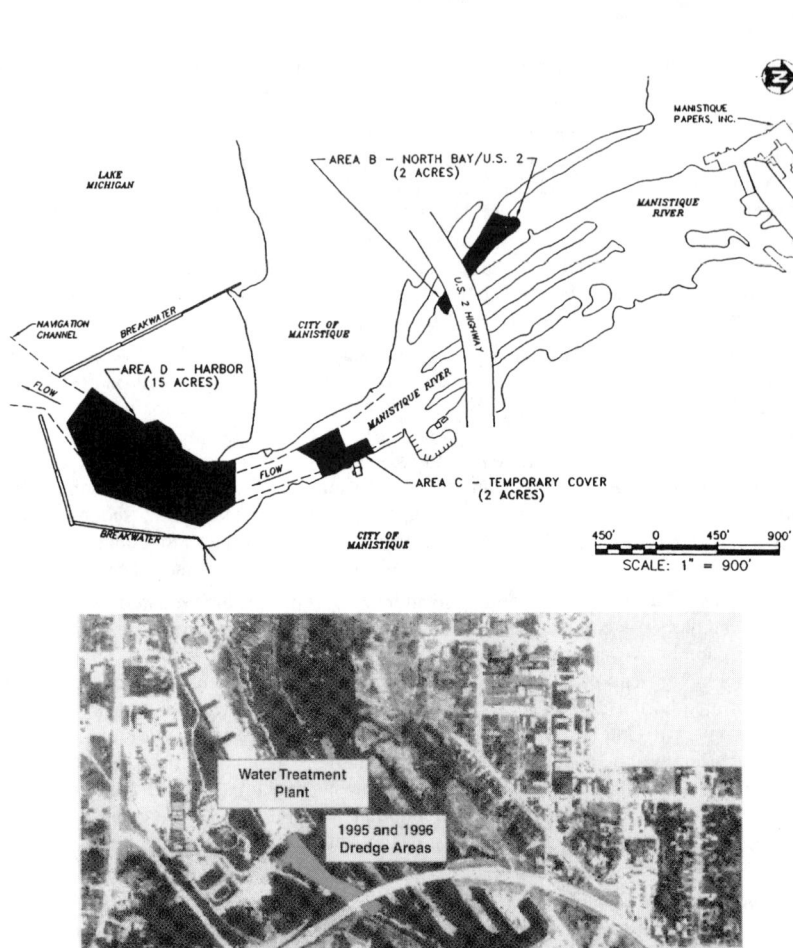

FIGURE 17.32. Location of Manistique site dredging areas.

half of the target volume was removed (10,000 yd^3 of sediments at a rate of approximately 15 yd^3/hr).

6. USEPA Final Decision on Dredging Proposal. On September 10, 1996, USEPA finalized an action memorandum that documents its decision to dredge all contaminated sediment in the river and harbor (see Figure 17.4). Prior to this decision, USEPA planned to cap some areas. Before making its decision, USEPA announced the dredging proposal and held a public comment period, during which individuals within the community were encouraged to voice their opinions about the proposal. The public comment period, which was extended twice, began on April 1, 1996 and ended on July 21, 1996. USEPA carefully reviewed and responded to all comments received before making its final decision.

In addition to documenting USEPA's decision, the recent action memorandum approved the following activities:

- Removing a temporary cover in Area C (see Figure 17.32).
- Installing a silt curtain to prevent sediments in the dredging area from migrating to other areas.
- Using hydraulic dredging techniques in Areas C and D to remove sediments with polychlorinated biphenyl (PCB) concentrations greater than 10 parts per million.
- Conducting a demonstration and evaluation of physical separation techniques to segregate contaminated sediments from noncontaminated sediments. If the techniques are demonstrated to be practical and cost-effective, they will be used during dredging activities.
- Treating all PCB-contaminated water and returning treated, clean water to the river.
- Disposing all PCB-contaminated sediments at approved off-site facilities.
- Returning clean sediments to the river or harbor (clean sediments may also be disposed of or reused during construction activity).
- Covering with clean sand any residual PCB-contaminated sediments (with concentrations greater than 10 parts per million) that remain in the dredged areas.

7. Site Activities. On July 10, 1996, dredging began in North Bay with a floating hydraulic auger dredge. This technique is being used to clear areas that contain mostly wood chips contaminated with PCBs. That summer, approximately 8,000 cubic yards of contaminated sediments, including wood chips and silty sediments, have been dredged and pumped for treatment. Samples collected for screening and analysis have confirmed that PCB concentrations in the remaining sediments along the eastern side of North Bay are less than 10 parts per million. Several small areas in the center of the bay are still covered with wood chips; these areas will be redredged at a later date. All other areas of North Bay have been cleared to bedrock.

During the Summer of 1996, approximately 28 million gallons of dredge water were treated, confirmed clean, and discharged into the river. Approximately 500 yd^3 of PCB-contaminated material have been shipped to a PCB disposal facility, and about 700 yd^3 of sediments with relatively low PCB concentrations have been sent to a local landfill. The dredging rate has been increased with the addition of new pumps and modified carbon and sand filters. A new barrier was also installed to enclose the dredge area under and immediately south of the bridge on US Highway 2.

A new access road south of US Highway 2 on the west side of the river was completed, as well as a new diver and dredge station.

8. The Next Steps. EPA expects to complete dredging activities near the US

Highway 2 bridge by the end of November 1996. Next year EPA will remove the temporary cover at Area C and will dredge that area. EPA will then begin dredging the navigation channel in the harbor area and will attempt to complete dredging in this area during 1997. If dredging activities in the harbor navigation channel are not completed during 1997, EPA will complete those dredging activities during the 1998 construction season (May to October).

EPA was scheduled to resume dredging in Spring 1999. EPA anticipated completing dredging activities in the river and harbor by winter of 1998. However, dredging activities during the Fall were unexpectedly delayed due to rock and wood debris at the bottom of the harbor. This hard material significantly slowed the dredging process because it is more difficult to dredge it.

An average of 124 yd^3 of sediment was removed each day of the dredging from May 11 to October 14, 1998. The sediment treatment plant was in operation 12 hours a day. During 1998, about 31,100 yd^3 of wood chips, sawdust, and other solid material were dredged. In addition, about 12,600 tons of waste has been shipped offsite. About 1,525 yd^3 of sand containing less than 1 ppm of PCBs was collected year-to-date during the dredging process.

On October 14, 1998, dredging activities in the river and harbor ended for the year. Site shut-down activities began on October 15, 1998 and were completed November 2, 1998. Shut-down activities consisted of (1) cleaning out settling chambers used to remove solids from the fine sediments, (2) emptying sand and carbon filters used to filter sediments from the water, (3) decontaminating dredging equipment, (4) decontaminating the concrete pad used to hold solid materials collected during dredging activities, and (5) demobilizing equipment.

As of November, 1999, about 25,049 yd^3 of solid material was dredged. About 205 million gallons of water had been treated, and about 13,880 total tons of waste had been shipped off site. The average values of PCBs before dredging were 30.2 ppm; after dredging the values were between 13.8 and 17.9 ppm (USEPA, 2000).

9. Comments. The EPA requirements on maximum PCB contamination allowed varies from project to project, from 1 to 50 ppm. On this project the limit is 10 ppm.

Massena, New York General Motors Site

1. Introduction. Cleanup of the General Motors (GM) Superfund Site in Massena, New York began in 1993. The goal of the cleanup is to remove polychlorinated biphenyls (PCBs) from around the GM facility and surrounding areas off GM property. In the past, PCBs were used in hydraulic fluids at the plant. USEPA banned the sale and use of PCBs in 1976. PCBs are no longer used at the plant; however, PCBs remain in the soil, sediments and water around the plant as a result of past disposal practices. The cleanup site is shown in Figure 17.33.

The cleanup is being conducted by GM with oversight from USEPA and technical assistance from the U.S. Army Corps of Engineers and Environment Canada officials. The New York State Department of Environmental Conservation and the St. Regis Mohawk Tribe provide support to USEPA in overseeing the cleanup. PCB contamination is shown in Figure 17.34.

The site cleanup includes five components, each a key element of the program:

- Site characterization activities
- St. Lawrence River cleanup
- St. Regis Mohawk Reservation and Raquette River cleanup
- Treatment of affected material
- Containment of affected material

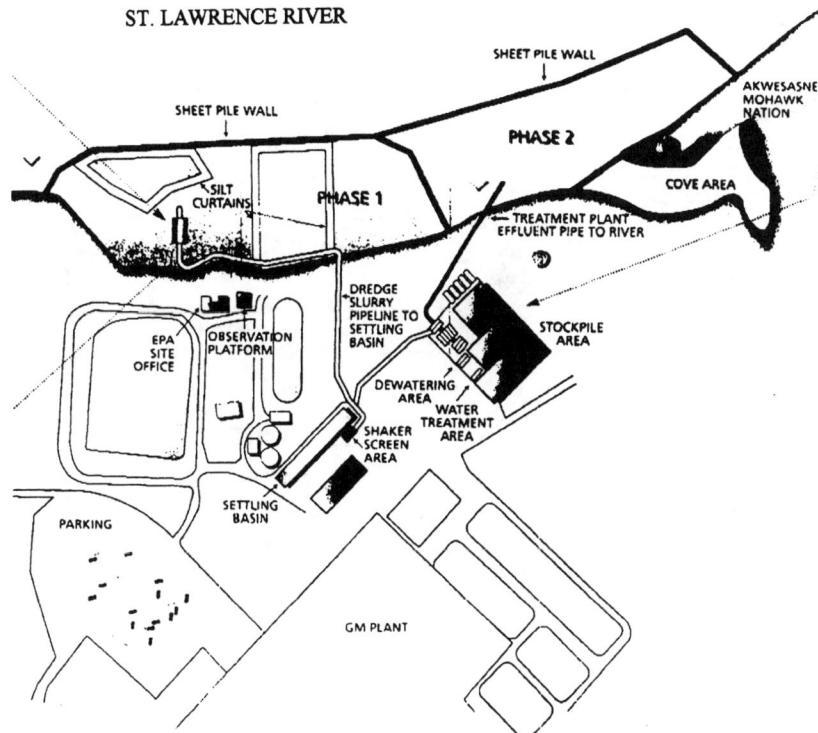

FIGURE 17.33. Cleanup site.

As a result of these tests, USEPA and GM believe that thermal extraction is the most appropriate technology for treating PCBs from the site. This is because thermal extraction was able to reduce PCB concentrations below 10 parts per million (ppm) in a cost-effective manner.

EPA's original cleanup plans for the site included treatment of all soils, sediments, and sludges with PCB concentrations above 10 ppm. Based on data collected in 1993, GM has requested that USEPA only treat materials with high PCB concentrations (above 550 ppm) and contain materials with lower PCB concentrations in a secured area on the site.

2. Site Characterization Activities. During 1993 and 1994, GM and their consultants collected and analyzed additional data required to complete the site cleanup. The data collected were used by designers and engineers to prepare cleanup plans. In the summer of 1994, GM's consultants also collected field data to determine which areas of the site are wetlands or floodplains and determined whether archeologically or historically significant remnants are likely to be uncovered during cleanup.

3. Silt Curtains. In July 1994, site work began with the construction of dewatering, sediment stockpiling, and water treatment facilities. During September, problems were encountered during the installation of silt curtains in the river. Silt curtains were made of high-strength, impermeable materials that are installed vertically in the water to reduce

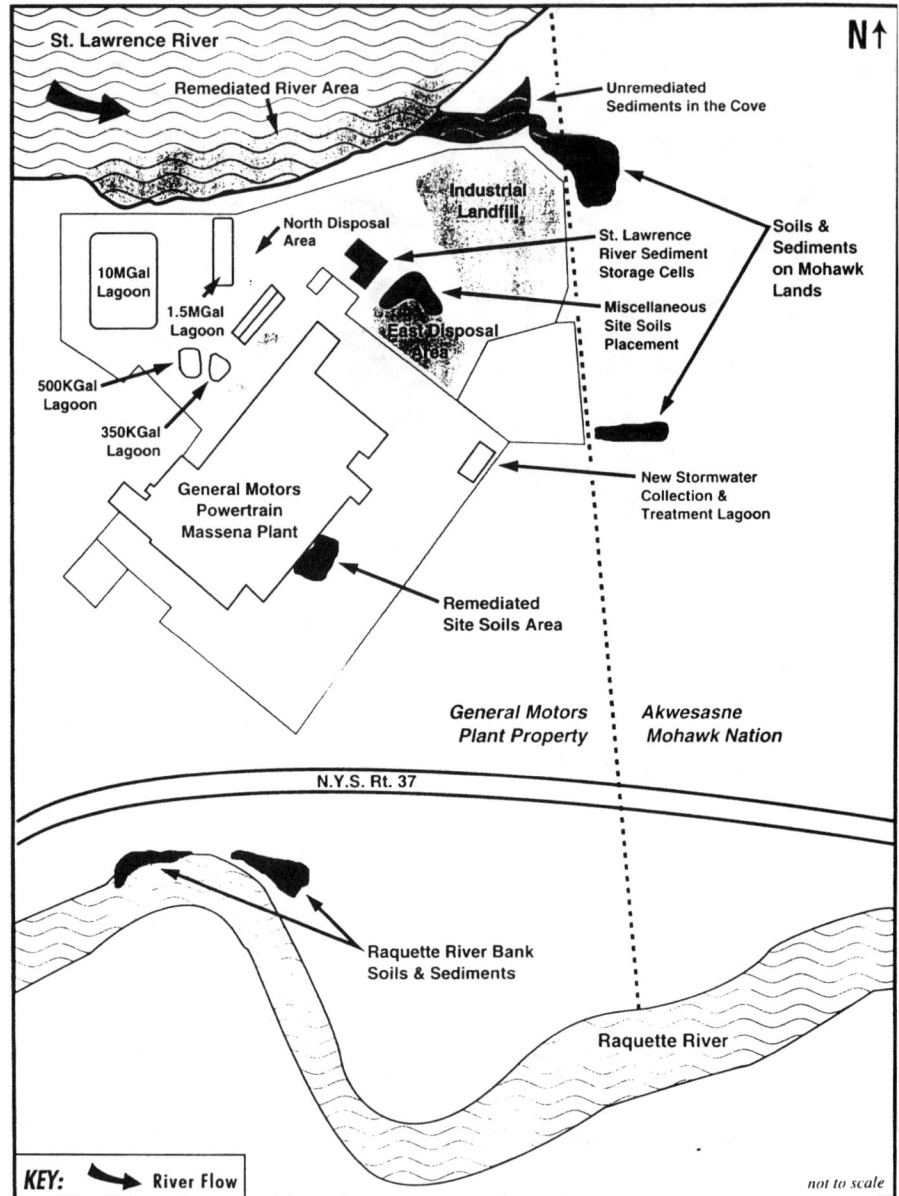

FIGURE 17.34. PCB contamination at the General Motors site.

currents and minimize movement of suspended particles. Site contractors had difficulty installing the curtains. In addition, the curtains were not the same as those called for by the design engineers and required modification before installation. During installation problems with anchoring and adjusting the curtains above the river bottom were encountered. Debris on the river bottom caused significant damage to the curtains. The contractor was dismissed from the project.

GM's consultants made adjustments to the curtains and their alignment during October, but could not align the curtains adequately to achieve effective sediment control. With the winter season approaching, USEPA and GM agreed to postpone river work until 1995.

4. Sheet Pile Wall. USEPA and GM have established a dredging plan that minimizes the potential for contaminated sediments to be stirred up (resuspended) as a result of dredging. The dredging plan is also designed to limit the potential for downstream movement of sediments.

The most important measure for preventing downstream movement is the installation of a sheet pile wall around the dredging area. The wall consists of interlocking steel sheets, approximately 2 feet wide by 30 feet long, driven 1 to 2 feet into the river bottom. The wall forms a barrier that prevents resuspended sediments from leaving the dredging area and moving downstream.

A new contractor was hired to conduct dredging early in 1995.

The sheet pile wall has been placed around all areas to be dredged; a second, inner barrier has been placed around areas with high PCB concentrations. This secondary barrier is made of strong plastic material, commonly referred to as "silt curtains."

In addition, air water and turbidity in water were monitored daily.

5. Dredging. Approximately 25,000 yd^3 of sediments have been removed using a horizontal auger hydraulic dredge and a "long stick" hydraulic hoe installed on a barge. Dredging started in July 1995. The dredging process began on the west end and is proceeding toward the east. A steel sheet pile wall was constructed around the dredging area and impermeable "silt curtains" made of plastic fabric were installed within the sheet pile wall, encapsulating areas where PCB concentrations exceed 500 parts per million (ppm). The containment system will prevent the downstream movement of sediments if any are stirred up (resuspended) during dredging.

Dredging of contaminated sediments was completed in November 1995.

6. Transfer of Sediments. The dredged sediments are transferred from the dredge via a floating pipeline to an onshore booster pump. The booster pump delivers the sediments up the river embankment to a concrete settling basin. At this stage, a submersible pump sends the sediments from the settling basin to the dewatering and water treatment system.

7. Dewatering/Water Treatment. The sediments are pumped through filter presses to remove the water from the solids. The dewatered sediments are then placed in temporary stockpile areas until a PCB treatment system is operational. The sediment stockpiles have been lined with an impermeable geomembrane liner and have a sand drainage layer and collection sumps to retrieve any additional water that may drain from the sediments. Air monitoring is performed around the stockpile area.

Water removed from the sediments is treated in an on-site plant that was set up specifically to remove any residual PCBs before the water is discharged back to the river (Figure 17.35).

8. Water and Air Monitoring. Monitoring stations were located in the river around the perimeter of the work area to measure the turbidity and level of PCBs in the water during dredging. The sensors alerted USEPA of the need to take any corrective measures, including stopping work, if necessary.

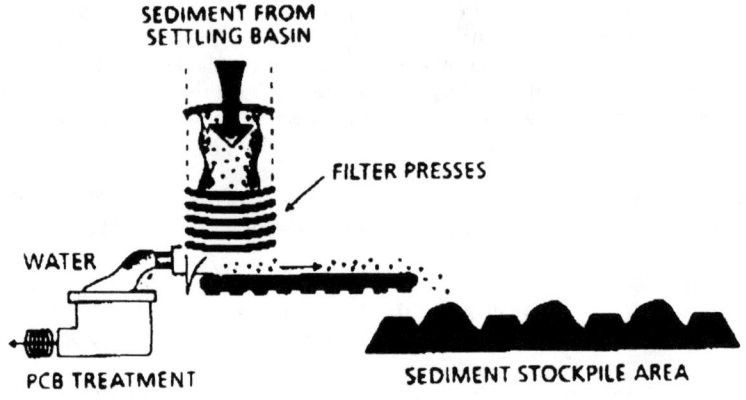

FIGURE 17.35. On-site plant.

Air monitoring is was carried out at several locations to ensure that no PCBs are being released into the environment.

9. Postdecision Proposed USEPA Plan (June 1995).
This Postdecision Proposed Plan describes proposed changes to the cleanup plan selected by USEPA in the December 17, 1990 Record of Decision (ROD) for General Motors Corporation—Central Foundry Division Superfund Site (Site). The 1990 ROD was issued by EPA, as lead agency, and concurred on by the New York State Department of Environmental Conservation (NYSDEC) and the St. Regis Mohawk Tribe.

The 1990 ROD cleanup remedy included: excavation and dredging of polychlorinated biphenyl (PCB) contaminated sediments, sludges, soils, and debris from the GM facility, the St. Regis Mohawk Reservation, and the surrounding river system followed by treatment of dredged/excavated material with PCB concentrations above 10 parts per million (ppm); interim surface runoff control to prevent migration of contamination from the East Disposal Area on the site, and recovery and treatment of contaminated groundwater from the site with discharge of treated groundwater to the St. Lawrence River.

This Postdecision Proposed Plan recommends that all sludge, as well as dredged/excavated material with PCB concentrations above 500 ppm, be treated by thermal desorption (an innovative treatment technology) and that dredged/excavated material with PCB concentrations above 10 ppm but less than 500 ppm be contained on-site, rather than treated. USEPA is, in this proposed plan, in effect recommending that the PCB treatment threshold for soil on the GM site be changed from 10 ppm to 500 ppm. In addition, USEPA proposed that a multilayer cap be used to cover treated materials and untreated materials with PCB concentrations below 500 ppm. Originally, a soil cover was selected in the 1990 ROD. USEPA is not proposing any changes to those portions of the 1990 cleanup plan that pertain to the East Disposal Area or to groundwater. However, USEPA announced its decision that thermal desorption be used to treat material on-site. USEPA is soliciting public comment on the siting and use of a thermal desorber on the GM facility. Finally, USEPA is proposing to designate a portion of the site as a corrective action management unit to facilitate placement of remediation wastes from the facility in this area.

USEPA is proposing to reduce the amount of material that would be treated on the site from approximately 171,000 yd^3 to approximately 54,000 yd^3. The total present cost of the remedy with the proposed changes described below is estimated by USEPA to be approximately $59 million.

All PCB-contaminated sediments in the St. Lawrence River and Raquette River with PCB concentrations above 1 ppm would be dredged. In addition, sediments with PCB concentrations above 0.1 ppm in Turtle Creek would also be dredged or excavated. Excavated sediments would be dewatered, as necessary. Decanted water would be treated and discharged to the St. Lawrence River. All dewatered sediments with PCB concentrations above 500 ppm would be treated using thermal desorption to reduce PCB concentrations below 10 ppm. Treated sediments and untreated dewatered sediments would be disposed on the GM facility in the East Disposal Area and covered with a multilayer composite cap. EPA is proposing to designate the East Disposal Area as a corrective action management unit.

Standing water would be drained from the Industrial Lagoons, treated, and discharged to the St. Lawrence River. Soil and sludge in the Industrial Lagoons, the North Disposal Area, and soil in miscellaneous areas on GM property with PCB concentrations above 10 ppm would be excavated. Excavated soil with PCB concentrations above 500 ppm, and all excavated sludge material (regardless of concentration), would be treated using thermal desorption to levels below 10 ppm. Treated and untreated material would be backfilled on GM property and covered with a multilayer composite cap. The two active lagoons would be remediated once taken out of service by GM.

Soil on the St. Regis Mohawk Reservation with PCB concentrations above 1 ppm would be excavated. Because the soils have PCB concentrations below 500 ppm (in fact, below 50 ppm), they would not be treated. These materials would be backfilled on GM property in the East Disposal Area and covered with a multilayer composite cap.

The estimated capital cost is $56,000,000; the estimated O&M cost: $243,000–$435,144/year; the estimated 30-year present worth cost: $59,000,000. Time to implement: 3 years (includes 1½ years of on-site treatment).

10. St. Regis Mohawk Reservation and Raquette River Cleanup. USEPA, GM, and the St. Regis Mohawk Tribe plan to excavate PCB-containing soil and sediment from areas on the St. Regis Mohawk Reservation and from the Raquette River. Final plans for soil and sediment excavation will be made after GM completes and USEPA reviews the wetlands and cultural resources assessment final report. Excavation and restoration will begin in 1996, after the site containment area is prepared to accept PCB-containing materials.

11. USEPA Record of Decision. The USEPA Record of Decision specified a 1 ppm PCB cleanup goal and a 10 ppm treatment goal in the St. Lawrence and Raquette Rivers. A cleanup level of 0.1 ppm was set for Turtle Creek (Mohawk Tribal Lands). The actual cleanup level was difficult to achieve, as measurements taken after dredging indicated. After dredging was completed at Massena, "an analysis of the samples indicated that PCB concentrations in all areas except one averaged less than 3 ppm." The area adjacent to the GM plant outfall, however, continued to show elevated PCB levels during dredging. Despite repeated dredging attempts, final sampling results showed an average of 25 ppm. With the exception of one sample at 6,000 ppm out of a total of 113 different sample locations in the river, more samples tested above 100 ppm. By early November 1995 EPA determined that continued dredging was not likely to result in further PCB reduction.

12. Comments

a) Containment of the proposed dredging area by reinforced silt curtains failed, delaying the project.

b) The sheet-pile wall was successfully deployed to prevent resuspended contaminated sediments from reaching the St. Lawrence River. The sheet pile was removed in December 1995.

c) The threshold level of PCB concentration was discussed over the proposed remediation cycle; values from 0.1 to 500 ppm were considered. EPA's original cleanup plans for the site included treatment of all soils and sediments with PCB concentrations above 10 ppm. The post-decision proposed plan recommends that all dredged/excavated material with PCB concentrations above 500 ppm be treated by the thermal desorption method. All sediments in the Raquette River with PCB concentration above 1 ppm will be dredged. In addition, sediments with PCB concentrations above 0.1 ppm in Turtle Creek would also be dredged or excavated. 0.1 and 1.0 ppm concentrations appear to be an unrealistic goal.

d) Sampling results discussed in Section 11 indicate that USEPA has set unrealistic cleanup goals.

13. 1998 Proposed Plan. The proposed plan recommends changes to some aspects of the 1990 Record of Decision (ROD). The remedy for the St. Lawrence River had three major components: dredging sediments greater than 1 ppm PCBs, treatment of dredged materials with PCB concentrations greater than 10 ppm, and on-site containment of untreated and treated sediments with concentrations of PCBs less than and equal to 10 ppm (Table 11.14).

The proposed changes do not suggest a change in any of the site-specific cleanup levels but deal only with how the sediments/soils are handled after they are excavated or dredged. The proposed changes recommend that soils and sediments with PCB concentrations greater than 10 ppm, which have been removed from the St. Lawrence River and will be removed from the Raquette River, and soils excavated during the installation of site-wide groundwater controls, be disposed off-site rather than treated on-site.

a) St. Lawrence River. The only change to the 1990 remedy for the St. Lawrence River would be the elimination of on-site treatment for dredged materials. Instead, the dredged materials with concentrations of PCBs greater than 10 ppm would be disposed off-site in a secure facility.

During the processing of the sediments after dredging, sediments from areas of high contaminant levels were mixed with sediments from areas with lower concentrations. The sediments were pumped from the river into a settling pond. From that settlement pond, the water was sent to the treatment system to further remove PCBs. During this processing, the materials were mixed and, as a result, all St. Lawrence River sediments that were dredged, processed, and stored on-site in 1995 have an average PCB concentration of 200 ppm. Therefore, all of the stockpiled St. Lawrence River sediments would be shipped off-site for disposal to a Resource Conservation and Recovery Act (RCRA)- and TSCA-approved facility. The estimated cost for the off-site disposal of the approximately 10,230 yd^3 of sediments dredged from the St. Lawrence River is $2.3 million.

b) Raquette River. The only change to the remedy for the Raquette River selected in

TABLE 17.14. Extent of PCB contamination (in yd^3)

	Levels	
	<10 ppm PCBs	>10 ppm PCBs
St. Lawrence River	n/a	10,230
Raquette River	1,400	2,600
Soils from site-wide groundwater controls	17,600	5,100

1990 would be the elimination of treatment for the dredged/excavated sediments and soils. The cleanup level for the Raquette River remains the same. However, instead of on-site treatment, the excavated/dredged materials with PCB concentrations greater than 10 ppm (2,600 yd³) would be disposed off-site in a RCRA- and TSCA-approved facility. The remaining 1,400 yd³ of materials with PCB concentrations of 1–10 ppm would be contained on-site and covered with a vegetated soil cap meeting New York State and TSCA requirements for chemical waste landfill. The estimated cost for the off-site disposal of approximately 2,600 yd³ of soils/sediments with PCB concentrations greater than 10 ppm is $0.7 million. This cost represents only the off-site disposal cost of the Raquette River materials that have PCB concentrations greater than 10 ppm and does not include the costs for excavation/dredging of the sediments/soils, which are fixed.

c) Groundwater Control System Soils. The only change to the remedy selected in 1990 for soils excavated during the construction of the site-wide groundwater control system would be the elimination of treatment for the excavated soils. Any soils with PCB concentrations greater than 10 ppm excavated during the installation of the groundwater control system would be shipped off-site for disposal at a RCRA- and TSCA-approved facility. This includes an area of contamination at the toe of the landfill slope. The remaining soil with PCB concentrations less than or equal to 10 ppm (approximately 17,600 yd³) would be contained on-site under a soil cap meeting New York State and TSCA requirements for a chemical waste landfill. The estimated cost for the off-site disposal of the estimated 5,100 yd³ of soils with PCB concentrations greater than 10 ppm to be removed during the installation of site-wide groundwater controls is $1.4 million (see discussion above regarding volume estimates). This cost reflects only the cost for off-site disposal of the excavated soils and does not include the costs of the installation of a groundwater control system.

The total approximate cost for the off-site disposal of sediments and soils with PCB concentrations greater than 10 ppm removed from the St. Lawrence River, Raquette River, and soils excavated during the construction of site-wide groundwater controls is $4.4 million (Table 17.15). Comparison of original (1990) remedy and proposed changes are listed in Table 17.16.

The proposed changes to the remedy are more effective in the short term than the 1990 remedy because they can be implemented more quickly. The proposed remedy would be implemented in approximately 6 months to a year, rather than the 2–3 years originally planned for the procurement, mobilization, and operation of the thermal desorption treatment system. These time estimates only reflect the time needed for off-site disposal or on-site treatment of materials after they are excavated or dredged and do not include the time needed for excavating or dredging, which remains the same.

Although the treatment unit would be operated in compliance with applicable regula-

TABLE 17.15. Treatment and disposal costs

Location and volume >10 ppm PCB (yd³)	Treatment costs ($M) (as per 1990 ROD)	Disposal cost ($M) (as per proposed change)
St. Lawrence River (10,230 yd³)	$4.6	$2.3
Raquette River (2,600 yd³)	$1.2	$0.7
Site-wide groundwater controls (5,100 yd³)	$2.4	$1.4
Total	$8.2	$4.4

TABLE 17.16. Proposed changes to the 1990 remedy

1990 Remedy	Proposed changes
Dredge the St. Lawrence River to cleanup goals	No change
Treat dredged St. Lawrence River sediments with >10 ppm PCBs	Off-site disposal of dredged sediments with >10 ppm PCBs
Dredge Raquette River and excavate riverbank soils to cleanup goals	No change
Treat Raquette River sediments with >10 ppm PCBs	Off-site disposal of dredged sediments with >10 ppm PCBs
Downgradient groundwater recovery and treatment	No change
Treatment of soils >10 ppm PCBs excavated during installation of site-wide groundwater controls	Off-site disposal of excavated soils with >10 ppm PCBs

tions, potential air quality impacts from the operation of the thermal desorber are possible. These risks would be eliminated by using off-site disposal. Further, potential risks to on-site workers would be lessened by reducing the materials handling requirements needed for on-site treatment. The potential short-term risks associated with transporting PCB-contaminated sediments to an off-site landfill would increase. However, these risks are estimated to be minimal due to the short duration of off-site disposal activities.

14. Costs. capital costs for the existing and proposed changes are presented in Table 17.17. In the case of existing and proposed changes to the remedy, the present-worth analysis is not applicable since there is only a one-time capital investment. There are no long-term monitoring costs since the waste would either be destroyed or sent for off-site disposal. Any materials left on site would have PCB concentrations less than 10 ppm. Although the property would be monitored as long as contaminants remain on site, the monitoring would not be specifically for the materials with PCB concentrations less than or equal to 10 ppm, but for the balance of the site. Since the costs of monitoring apply to the entire site, they are not affected by this change and therefore not included in this comparison analysis.

The capital cost for the relevant portions of the original remedy, which includes the on-site treatment of materials dredged or excavated from the St. Lawrence River and Raquette River and site-wide groundwater controls, is approximately $8.2 million. The

TABLE 17.17. Comparison of the 1993 and current estimates of sediment volume to be dredged from the St. Lawrence River

PCB concentration range (ppm)	1993 Volume estimate (yd^3)	Current volume estimate (yd^3)
Between 1 and 25	37,000	38,700
Greater than 25	14,500	
Between 25 and 50		4,700*
50 to 500		29,700
Greater than 500		4,500
Total	51,500	77,600

*It is noted that the 4,700 yd^3 combined with the 38,700 yd^3 having PCB levels between 1 and 25 ppm results in a total of 43,400 yd^3 of sediment with PCB levels of less than 500 ppm that would be landfilled on site under this plan.

capital cost for the off-site disposal of those materials at an approved facility is $4.4 million. This represents a decrease of $3.8 million. Based on these estimates, off-site disposal is significantly more cost-effective.

15. Summary. Based upon an evaluation of the two remedies, USEPA recommends that the remedy as selected in the 1990 Record of Decision be changed to allow for the off-site disposal, rather than on-site treatment, of materials excavated or dredged from the St. Lawrence River and Raquette River and soils excavated during the installation of site-wide groundwater controls. All dredged/excavated materials with PCB concentrations above 10 ppm would be transported off site to a RCRA- and TSCA-approved landfill. All dredged/excavated materials with PCB concentrations less than or equal to 10 ppm would be contained on the GM site, in keeping with the original 1990 Record of Decision. The cleanup goals set by the 1990 OUI ROD are not changed.

Massena, New York, Reynolds Metals Facility

1. Background. The Reynolds Facility is an active aluminum production plant located on 1,600 acres in the Town of Massena, New York. It is bordered on the north by the Grasse and St. Lawrence Rivers, on the east by the New York Central Railroad, on the west by Haverstock Road (south Grasse River Road), and on the south by the Raquette River. The facility is located off Route 37 near the Massena–Cornwall International Bridge, and directly upriver of the General Motors Powertrain Division Plant, a federal Superfund site being remediated under the direct oversight of USEPA. St. Regis Mohawk tribal lands, known as Akwesasne, are within one-half mile of the Facility. The Aluminum Company of America (Alcoa) manufacturing facility is located 8 miles west and upriver of the facility (Figure 17.36).

The cleanup of the Reynolds Facility and surrounding areas is divided into two phases

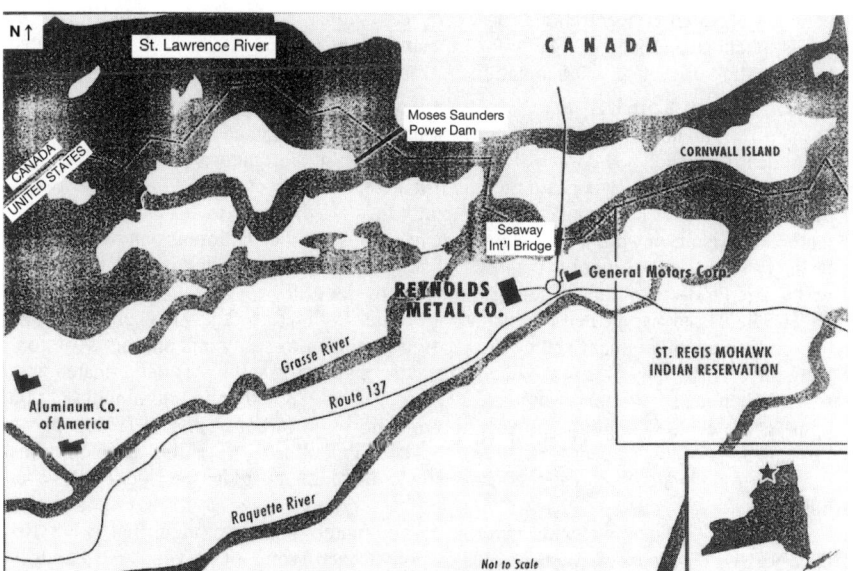

FIGURE 17.36. Site location map.

that are being overseen separately by state and federal regulatory authorities; NYSDEC oversees the land-based portion of the cleanup, and USEPA oversees the cleanup of the river system adjacent to the Reynolds facility, designated as the "Reynolds Study Area" (the Site). The site includes those portions of the St. Lawrence, Grasse, and Raquette Rivers, their tributaries, and any wetlands that are adjacent to the Reynolds Facility.

EPA oversees all three projects—General Motors (Massena Plant), Alcoa (Massena Plant), and Reynolds Metals Facility.

The Reynolds Facility was constructed in 1958 for the production of aluminum. The plant buildings occupy about 7%t or 112 acres of Reynolds' property. The remaining portion of the property consists primarily of forested areas and wetlands.

The aluminum is produced in individual metal pots which are lined with potliner, a material composed of a mixture of carbon compounds, including coal tar pitch and coke. As a result of plant operations, various types of industrial waste, including hazardous substances, were generated and disposed of at the Facility. Major areas of contamination include the Black Mud Pond, the Industrial Landfill, a former Potliner Storage Area, a wetland, and the North Yard. Other areas of contamination include the plant outfalls (open drainage ditches), which discharge wastewater and surface-water runoff from the Facility to the St. Lawrence River (Figure 17.37). The Facility has been placed on NYSDEC's Registry of Inactive Hazardous Waste sites.

2. Black Mud Pond. Black Mud Pond was an unlined disposal pit constructed in 1973 along the western side of the Facility. Its purpose was to hold settling carbon solids produced as a by-product of plant operations. The Pond had a surface area of approximately six acres and a volume of approximately 20 million gallons prior to its closure by NYSDEC in 1996. It is estimated that the Pond contained 165,000 yd^3 of black mud, underlain by approximately 22,000 yd^3 of contaminated soil. The waste materials consisted primarily of alumina (30–40%), carbon (35–45%) and fluoride (2–5%). Cyanide and PCBs were also present at concentrations up to 61 ppm and 8 ppm, respectively. The groundwater beneath the Pond, which occurs at depths ranging from a few feet to 15 feet, contains elevated concentrations of several contaminants, including cyanide, fluoride, PCBs, phenols, and sulfate. The low permeabilies of the underlying till have prevented the migration of groundwater contaminant from the Pond area.

3. Industrial Landfill and Former Potliner Storage Area. The Industrial Landfill is an unlined landfill that covers an area of approximately 11.5 acres and is located near the southwest corner of the Reynolds Facility. Groundwater is encountered in the fill materials directly beneath the Landfill at depths between five and eight feet below the ground surface. The groundwater generally flows to the south to discharge to the wetland. There is an upward vertical component of flow in the shallow groundwater zone. From 1957 to 1990, Reynolds disposed of solid waste, industrial debris, spent potlining waste, and PCB-contaminated sewage sludge at the landfill.

The Industrial Landfill and the adjacent Former Potliner Storage Area can be characterized as one contaminant source area, based on their proximity and similarity of contaminants. Waste materials at the Landfill and the underlying soil were contaminated with several chemicals including cyanide (300 ppm), fluoride (8,500 ppm), aluminum (87,000 ppm), sulfate (13,000 ppm), polycyclic aromatic hydrocarbons (PAHs) (2,200 ppm), PCBs (906 ppm), and phenols (21 ppm). Several contaminants were also detected in the shallow groundwater at elevated concentrations, including cyanide, fluoride, PCBs, phenols, and sulfate.

A tract of regulated wetlands covering approximately 170 acres (identified as No. Rr-6 by NYSDEC) occurs on the Reynolds property directly south of the Industrial Landfill. It is one of the three largest wetlands in the Town of Massena. The wetland is a groundwater discharge area for the southern portion of the property. Drainage from the wetland

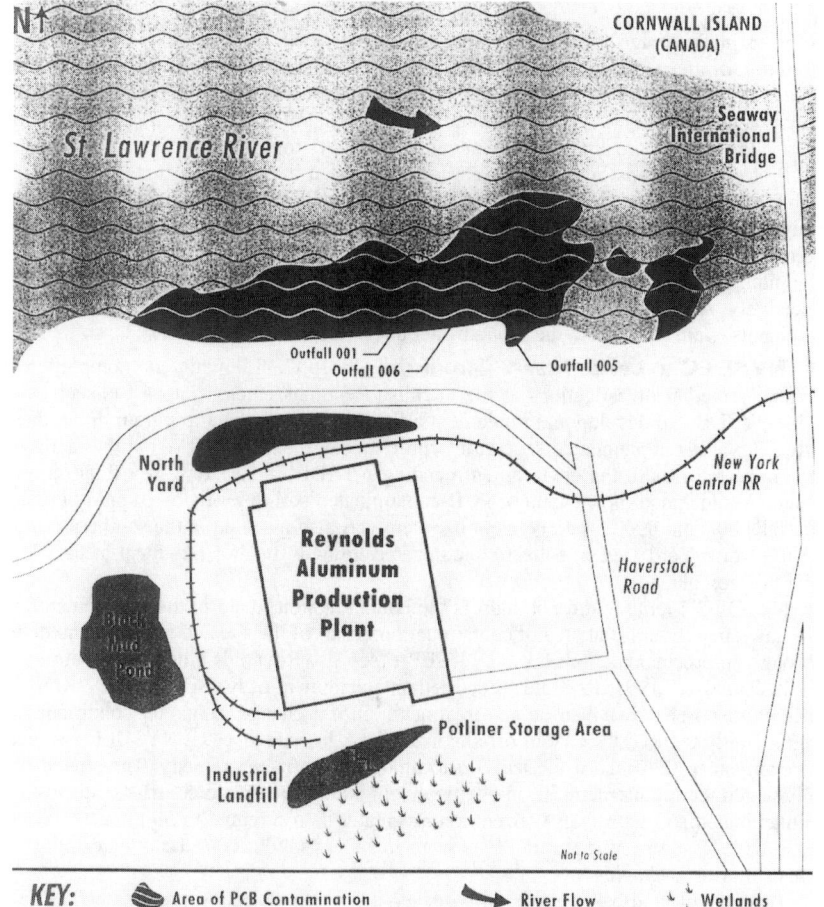

FIGURE 17.37. PCB contamination in the Reynolds study area.

flows southward via two streams and discharges into the Raquette River. Prior to the construction of the partial surface-water runoff controls and leachate collection system for the Landfill area, the leachate, groundwater, and surface water from the Landfill area discharged directly to the wetland. The wetland also received contaminated sediment and surface water that drained from other areas of the Facility.

Several contaminants were found in the wetland sediments at elevated concentrations, including aluminum, arsenic, cyanide, fluoride, sulfate, and phenols. PCBs were also detected in the sediments at concentrations ranging up to 19 ppm. Contaminants found in surface-water samples collected from the wetland included PCBs, chrysene, fluoride, and bio(2-ethylhexyl)phthalate.

The North Yard is central to the majority of production activities. All raw materials and finished products enter and leave the Facility through this area. Other plant facilities at the North Yard include the unloading shed for receiving alumina, coke, soda ash, and fluoride, pitch storage tanks, and the truck unloading dock.

Groundwater is present in the fill materials (reworked till) beneath the North Yard area at depths ranging from 2 to 15 feet and flows northward into the St. Lawrence River. Man-made structures (i.e., utility trenches and a drain collection system) in the southern part of the North Yard area act as preferential pathways for groundwater flow.

Past investigations in the North Yard area revealed the presence of several contaminants in the soil and groundwater. PCBs, polychlorinated dibenzofurans (PCDFs), and polychlorinated dibenzo-p-dioxins (PCDDs) were detected in the soil. PCB levels were found as high as 89,000 ppm. PCDDs and PCDFs were detected in soils primarily near the pitch pump house at concentrations of 9.92 parts per billion (ppb) and 9.35 ppb, respectively. Aluminum, arsenic, cyanide, fluoride, phenols, and PCBs, among other contaminants, were detected in the groundwater at elevated concentrations.

4. NYSDEC's Land-Based Cleanup. In 1987, following the completion of several remedial investigations at the Facility, Reynolds entered into a Consent Order with NYSDEC to develop and implement a facility-wide remedial program. Soon thereafter, Reynolds implemented several interim remedial measures (IRMs) at the facility to minimize releases to the environment pending NYSDEC's selection of a final remedy. The IRMs included the cleanup of PCB-contaminated soil to levels of 10 ppm or less at Outfalls 002 and 006. A rain collection system was also installed in the southern portion of the North Yard area to collect contaminated groundwater for treatment at the North Yard carbon plant.

NYSDEC's remedy did not address the contamination found in the sediments of the St. Lawrence River. That area of contamination is part of the site and is being remediated through enforcement actions taken by USEPA (see also Reynolds Study Area, below).

In January of 1995, Reynolds requested an amendment of NYSDEC's 1992 ROD for off-site disposal rather than on-site treatment, since a change in market conditions had significantly reduced the cost of off-site landfilling. In June of 1995, NYSDEC issued an amendment to the Record of Decision to modify the selected remedy. The amendment eliminated the requirement for on-site treatment and instead required off-site disposal of contaminated soil with PCB concentrations greater than or equal to 50 ppm. Excavated soils with less than 50 ppm of PCBs were to be consolidated in the on-site Industrial Landfill prior to capping.

To date, all land-based remedial work has been completed. Approximately 135,300 yd^3 of excavated soil with PCB concentrations less than 50 ppm have been consolidated into the Industrial Landfill. Capping will be completed following placement into the landfill of the additional sediments to be dredged from the St. Lawrence River as part of USEPA's remedial efforts.

5. Reynolds Study Area. In 1988 and 1989, Reynolds performed an initial study of the sediments in the St. Lawrence River adjacent to its facility. A total of 67 sediment samples were collected at 35 sampling locations along the riverbed. The analytical results showed PCBs were present in 17 sediment samples at concentrations ranging from 10 ppm to greater than 1,000 ppm. The highest concentrations were found in samples collected from within 250 feet of the facility outfalls, primarily Outfall 001.

Using the data from Reynolds' Assessment and Remediation of Sediments (ARS), USEPA performed baseline human health and ecological risk assessments to evaluate the potential risks to human health and the environment associated with this site. The results of those assessments showed that the contaminated sediments at the Site posed a threat to human health and the environment. The greatest health risk was associated with the ingestion of PCB-contaminated fish caught in the St. Lawrence River by residents and fishermen. Other health threats were associated with the direct contact or ingestion of contaminated sediments.

1998 Plan. The original remedy selected in 1993 addressed contaminated sediments in that portion of the St. Lawrence River located in the vicinity of the Reynolds Facility. The major components of that remedy were: (1) dredging of river sediments contaminated with polychlorinated biphenyls (PCBs) and other chemicals; (2) on-site treatment of sediments with PCB concentrations greater than 25 parts per million (ppm); and (3) on-site disposal of treatment residuals and sediments having PCB levels less than or equal to 25 ppm in the Black Mud Pond (a former disposal pit at the Facility). The remedy did not address contamination present on the land-based portion of the Facility. The land-based cleanup was performed under the direct oversight of NYSDEC (Table 17.18).

In 1998 EPA proposed changes to the original remedy which was equally protective of human health and the environment, but significantly less expensive. Under this Plan, the more contaminated river sediments (PCB levels equal to 50 ppm or greater) would be disposed of at an approved off-site facility; sediments with PCB levels exceeding 500 ppm would be treated off-site, whereas sediments with PCB levels between 50 and 500 ppm would be landfilled off-site. The lower level of contaminated sediments (less than 50 ppm of PCBs) would be disposed of at the Industrial Landfill, another disposal area located at the Facility. The cost savings are due to a change in market conditions that significantly improved the cost-effectiveness of off-site disposal as compared to on-site treatment. Additionally, there has been an increase in the estimated volume of contaminated sediment requiring treatment.

The proposed changes would also be consistent with NYSDEC's land-based cleanup, where approximately 135,300 yd^3 of soil with low levels of PCBs (less than 50 ppm) have been consolidated in the Industrial landfill. If the actions described in this Plan are implemented, it is estimated that an additional 43,400 yd^3 of sediment would be disposed therein. Following such disposal, the Industrial Landfill would be capped as part of NYSDEC's program.

The proposed changes to the 1993 remedy are as follows:

- Eliminate the on-site thermal desorption treatment component of the remedy
- Landfill all dredged and dewatered sediments with PCB levels between 50 and 500 ppm (approximately 29,700 yd^3 at an approved off-site facility)
- Treat all dredged and dewatered sediments with PCB levels exceeding 500 ppm (approximately 4,500 yd^3) at an approved off-site facility
- Consolidate all dredged and dewatered sediments with less than 50 ppm of PCBs (approximately 43,400 yd^3) in the on-site Industrial Landfill, which will be capped in compliance with NYSDEC's 1992 ROD and 1993 Consent Order for the land-based cleanup; black Mud Pond was capped in 1996 as part of NYSDEC's cleanup and, therefore, is not available for sediment disposal

TABLE 17.18. Cost estimate with 1993 remedy

Capital cost	$72 million (originally estimated to be $34.7 million based on 1993 volume estimates)
O & M cost	$400,000 for postremediation monitoring over a 5 year period (does not include O&M costs for long-term management of Black Mud Pond, which would be conducted under NYSDEC's land-based remedial program)
Present-worth cost	$72.4 million (originally estimated to be $35.1 million based on 1993 volume estimates)
Time to implement	Approximately 4 years

TABLE 17.19. Capital cost for the proposed remedy

Capital cost	$62.8 million
O&M cost	$400,000 for postremediation monitoring over a 5-year period (does not include O&M costs for long-term management of the Industrial Landfill, which would be conducted under NYSDEC's land-based remedial program)
Present-worth cost	$63.2 million
Time to implement	Approximately 1–2 years

The 43,400 yd^3 of sediment to be landfilled on-site consists of 38,700 yd^3 with PCB levels less than 25 ppm and 4,700 yd^3 with PCB levels between 25 and 50 ppm.

The costs associated with the 1998 Plan are shown in Table 17.19. The other capital costs are primarily associated with dredging of sediments. The time estimated to complete the work is 1 to 2 years. Table 17.20 presents a comparison of the 1993 Plan with the 1998 Plan proposed changes. The evaluation criteria considered in the 1998 Plan are summarized in Table 17.21. The final USEPA decision was made after public comments were received.

New Bedford Harbor, Massachusetts

1. Volume to be Dredged. A decision to dredge approximately 10,000 yd^3 of contaminated sediments was made on February 10, 1994. At that time incineration was considered as the best alternative for treating the "hot spot" sediments.

2. Equipment. A small cutterhead dredge was selected. A diffuser was attached to the discharge pipe to reduce turbulence when the contaminated dredged material was placed in the containment disposal facility (CDF) approximately one mile from the dredged site.

3. Dredging. Generally, dredging was performed during high tide over a 4-hour period. The original estimates indicated that the entire project should be completed by early September 1994. The actual dredging was completed on September 6, 1995, or a year later than anticipated. One area (B) was not dredged because of complications with sub-

TABLE 17.20. Comparison of proposed changes in the remedy

1993 Remedy	Proposed changes
Dredge the St. Lawrence River	No change
Treat dredged sediment with PCB levels ≥ 25 ppm	Dispose of dredged sediments with PCB levels between 50 and 500 ppm in an off-site landfill; treat dredged sediments with PCB levels >500 ppm at an off-site facility
Contain sediments with PCB levels ≤ 25 ppm on site in the Black Mud Pond	Contain sediment with PCBs <50 ppm on site in the Industrial Landfill
Black Mud Pond will be capped as part of NYSDEC's land-based cleanup	Industrial Landfill will be capped as part of NYSDEC's land-based cleanup
Monitor St. Lawrence River	No change
Present-worth cost: $72.4 million (current estimate)	Present-worth cost: $63.2 million

TABLE 17.21. Summary

- *Overall Protection of Human Health and the Environment* addresses whether or not a remedy provides adequate protection and describes how risks are eliminated, reduced, or controlled through treatment, engineering controls, or institutional controls.
- *Compliance with ARARs* addresses whether or not a remedy will meet all of the applicable or relevant and appropriate requirements of other environmental statutes and requirements or provide grounds for invoking a waiver.
- *Long-Term Effectiveness* refers to the ability of a remedy to maintain protection of human health and the environment once cleanup goals have been met.
- *Reduction of Toxicity, Mobility, or Volume through Treatment* is the anticipated performance of the treatment technologies a remedy may employ.
- *Short-Term Effectiveness* addresses the period of time needed to achieve protection and any adverse impacts on human health and the environment that may be posed during the construction and implementation period until cleanup goals are achieved.
- *Implementability* is the technical and administrative feasibility of a remedy, including the availability of materials and services needed to implement a particular option.
- *Cost* includes estimated capital and operation and maintenance costs, and net present-worth costs.
- *State Acceptance* indicates whether, based on its review of this plan, the state concurs, opposes, or has no comment on the preferred alternative.
- *Community Acceptance* will be assessed in the ROD amendment following a review of the public comments received on this plan.

merged high voltage power lines. It is anticipated that the dredging of area B will be addressed as part of the second phase of dredging.

4. Dredging and Incineration Contract. The contracts for the "hot spot" remedial action involved the following components:

- Dredging
- Modification of the confined disposal facility
- Construction and operation of a wastewater treatment facility
- Incineration of the sediments with the associated sediment dewatering, material and ash handling
- Construction of a permanent cap over the disposal facility
- Extensive monitoring (air and water quality)

The original contract price was $19,357,720; however, numerous contract modifications have increased the contract to $21,165,436.

Several developments significantly affected the cost of this project. The first and most significant was USEPA's decision not to incinerate the sediments that required termination of this portion of the contract. Dredging also proceeded at a much slower pace than originally anticipated due somewhat to the extensive air monitoring effort that was required.

It was anticipated that the total contract costs would be approximately $18 million, which would not involve the treatment of any sediment.

5. Interim Storage of Dredged Sediments. Due to the need for evaluating treatment technologies other than incineration, the dredged contaminated sediments are being stored until the final treatment technology is actually implemented.

6. Air Monitoring. Extensive air monitoring of PCBs was conducted. Air monitoring samples for airborne PCBs were collected from six locations around the dredging areas on each day that dredging occurred. Collection of air samples around the Confined Disposal Facility occurred at least twice weekly during normal operations. During 1994, ranges of air samples were observed as shown in Table 17.22.

7. Controlling Dredging Operations. Control of dredging operations was based on the PCBs content in the air. There are three standards based on PCBs in the air that are used to control dredging operations of the "hot spots": a *shut-down level*, an *action level* and a *notice level*. The shut-down level has never been exceeded during the first four weeks of operation. The action level and the notice level both were exceeded during the first week of dredging, but there have been no exceedances since changes were made to dredging operations at the start of the second week of dredging.

a) The shut-down level is based on the National Institute of Occupational Safety and Health (NIOSH) standard for toxic substances in the workplace, which is 1000 ng/m^3. NIOSH considers it safe for an individual to be exposed to 1000 ng/m^3 of PCBs for 40 hours per week, every week for 50 years. If any one air sample exceeds the shut-down level, dredging will be stopped until the source of the PCBs can be identified and corrective action taken.

b) The action level is one-half the NIOSH standard, or 500 ng/m^3. If any one air sample exceeds the action level, the dredging contractor must make changes in the dredging process to reduce the levels of PCBs emitted. These changes may include reducing the period of dredging each day or operating the dredge at lower speeds.

c) The notice level is based upon measured background PCB levels in vicinity of the "hot spot." The background levels are the PCB levels determined during times when there is no dredging. The notice level is based on the average of the four previous background sampling events plus 30 ng/m^3 that represents an increased risk of less than 3 in 10,000,000 for the total dredging period. The background level is determined from air samplers in the dredging area.

If the average daily sample exceeds the "hot spot" notice level, the contractor must no-

TABLE 17.22. Air monitoring results

	"Hot spots"		CDF Area	
	Range (ng/m^3)[a]	Average (ng/m^3)	Range (ng/m^3)	Average (ng/m^3)
Before dredging started	3–335	54.2	3–41	11.6
During dredging				
1994	21–190	—	23–1032	—
1995	11–639	—	6–1113	—

[a] ng/m^3 = one billionth of a gram/m^3

tify the government and identify a corrective action. If the notice level is exceeded in two consecutive average samples, the government can direct the contractor to implement the corrective action. If the notice level is exceeded in 5 of 10 consecutive sampling events, the contractor is required to either implement corrective actions or shut down the dredge.

8. Preliminary Comments

a) Original estimates to complete dredging and treatment were 5 months. The actual dredging was from April 26, 1994 until September 6, 1995. Considering winter shutdowns, the dredging was completed in about 15 months. Treatment of contaminated sediment is on hold.

b) Extensive air monitoring contributed to dredging delays.

c) The original contract price was $19,357,720. Modifications increased the cost to $21,165,436. Since the treatment contract was canceled, the anticipated total costs were approximately $18,000,000. The original dredging costs were estimated at $651,400. Although the final figures are not available, the dredging costs have undoubtedly increased considerably.

d) Caution should be taken on new projects not to underestimate the cost and time of remediation projects.

9. The Proposed USEPA Cleanup Plan (1996).

On October 30, 1996 USEPA proposed a major cleanup plan for Upper and Lower New Bedford Harbor.

USEPA, in cooperation with the Massachusetts Department of Environmental Protection and the New Bedford Harbor Superfund Site Community Forum, is proposing to dredge PCB-contaminated sediments in upper and lower New Bedford Harbor and isolate the sediment in four shoreline confined disposal facilities. The key elements of the proposal are:

- Dredge approximately 450,000 yd^3 of PCB-contaminated sediment spread over approximately 170 acres of the harbor. In the upper harbor north of Coggeshall Street, sediments above 10 parts per million (ppm) PCBs will be dredged; in the lower harbor and in salt marshes, sediments above 50 ppm will be dredged.

- Construct four shoreline confined disposal facilities, or CDFs, to contain and isolate the dredge sediments from the public and the marine ecosystem (Figure 17.38).

- Drain water from the sediments once they are placed in the CDFs and treat the water to remove contaminants before returning the water to the harbor (Figure 17.39).

- Construct an impermeable cover or cap, once the sediments have sufficiently settled, on top of the CDFs (approximately three years after final placement in the CDF).

- Develop a long-term monitoring and maintenance program for the CDFs.

- Explore potential reuse of the completed CDFs.

This "Phase II" cleanup proposal evolved after an extensive process of studying the harbor and developing community, state, and federal agencies to arrive at a consensus for a solution. USEPA originally issued a proposed plan and addendum for the upper and lower harbors in January and May 1992, respectively. The recent proposal reflects the USEPA's consideration of the comments received on the two documents, as well as extensive dialogue with the local Forum. The record of decision for the upper and lower harbors will contain the EPA's response to all comments previously received as well as any new comments received on this proposal.

According to the Deputy Director of the Massachusetts Office of Dispute Resolution, Facilitator of the New Bedford Harbor Forum:

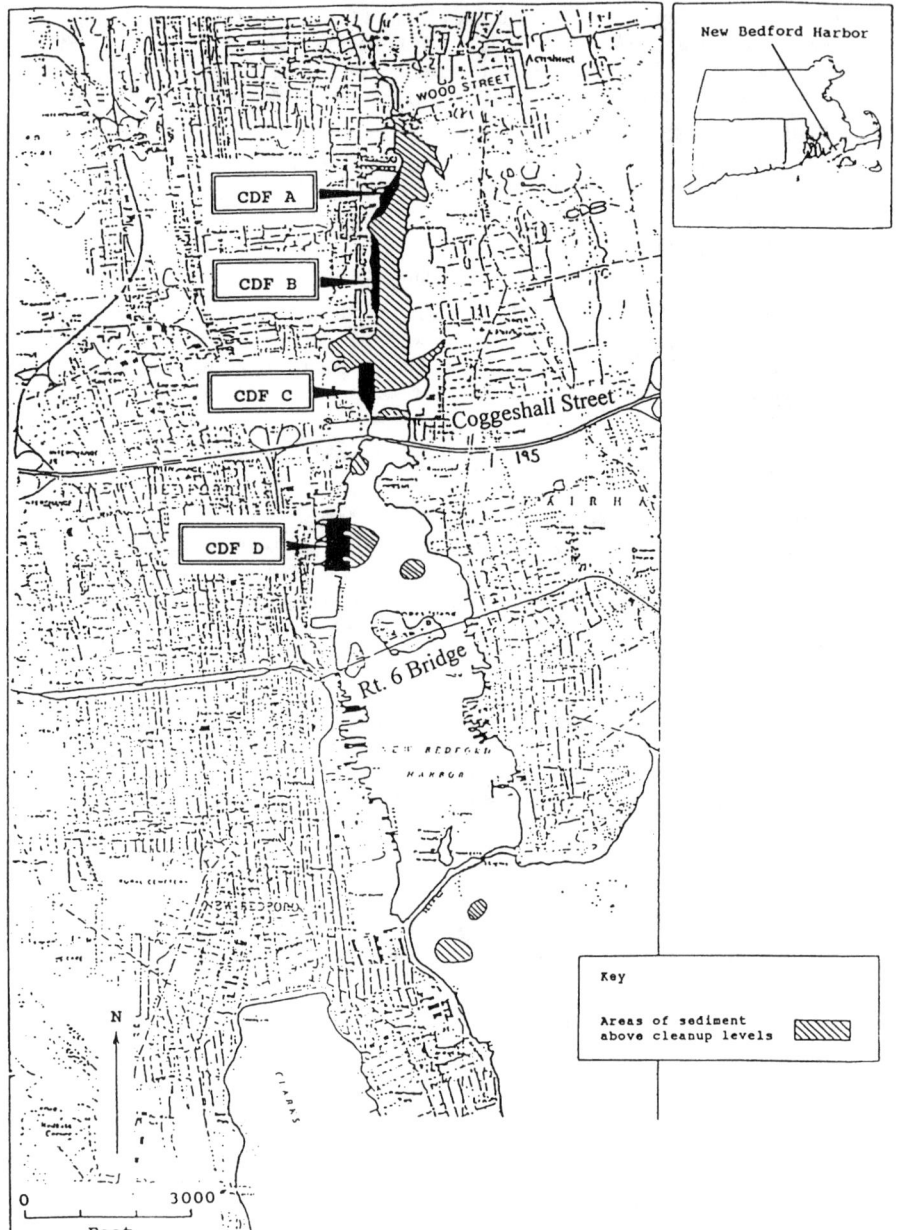

FIGURE 17.38. New Bedford Harbor ROD II proposed dredging areas and CDFs.

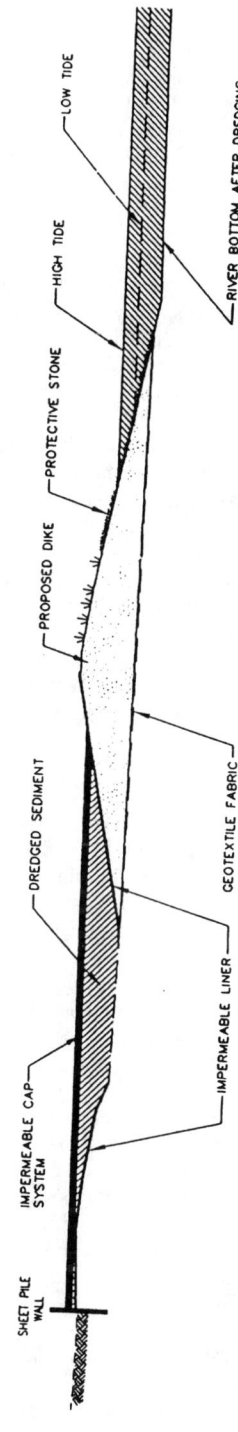

FIGURE 17.39. Typical CDF cross-section.

The plan includes dredging and isolating contaminated sediments. This represents an important milestone toward cleaning up New Bedford Harbor.

The members of the New Bedford Harbor Forum have demonstrated extraordinary commitment throughout this very demanding process. The results, both in sophistication regarding technical issues and in building a strong community–government working relationship, provide a model of excellence which will no doubt be studied and replicated.

The New Bedford Harbor Superfund site encompasses 18,000 acres of New Bedford Harbor and parts of Buzzards Bay. The primary contaminants at the site are the result of past disposal of PCB-contaminated process wastes from industries along the Acushnet River and Buzzards Bay. In 1977, testing of edible fish tissue revealed PCB levels in excess of U.S. Food and Drug Administration guidelines. As a result, the Massachusetts Department of Public Health has restricted fishing in areas of New Bedford Harbor and Buzzards Bay.

In 1983, USEPA added the site to the National Priorities List of hazardous waste sites eligible for Superfund cleanup funds. Massachusetts has designated the New Bedford Harbor site as its top priority Federal Superfund site.

USEPA has conducted and continues to conduct extensive studies of the nature and extent of PCB contamination in the harbor. In 1990, USEPA signed a Record of Decision to dredge the most contaminated hot spot sediment and treat the sediment by on-site incineration (Phase I). However, due to public disapproval, USEPA is reevaluating this remedy and currently conducting on-site treatability studies for an alternative treatment technology.

Once treatability tests are completed, USEPA will issue a separate proposed plan for treatment of the Phase I sediments to solicit public comments on the preferred technology. The hot spots were dredged in 1994–1995 and are currently being contained in an on-site confined disposal facility.

1. A Closer Look at USEPA's Proposal. The proposed cleanup consists of a dredging and sediment containment approach using the cleanup levels discussed previously combined with treatment of the seawater drained from the CDFs. The remedy can be divided into the six different implementation components described below.

1. *Construction of CDFs and Water Treatment Facilities.* The first step in the cleanup process is to design and construct the CDFs and associated water treatment facilities. Dredged contaminated sediments will then be piped into the CDFs and associated water treatment facilities and passively dewatered. Dredging need not wait until all these facilities are constructed, however, since the process can be staged so that dredging begins once the first CDF is ready to accept sediments.

Locating the four CDFs in contaminated areas avoids dredging approximately 126,000 yd^3 of underlying contaminated sediment. The side walls of these CDFs will be lined with a synthetic impermeable material, but not the bottoms of the CDFs, since (a) the existing sediments in these areas are very low in permeability, (b) the integrity of a man-made impermeable liner constructed in saturated conditions cannot be guaranteed, and (c) the dredged sediments themselves will form into a highly impermeable product. Computer modeling indicates that long-term worst-case PCB leakage rates from the CDFs would be at least 98% lower than the current rate of PCBs in the water fluxing seaward at the Coggeshall Street bridge. Each CDF will include perimeter groundwater monitoring wells in order to verify that they are operating safely.

2. *Dredging of Sediments with PCB Levels Above Cleanup Levels.* Sediments above

the target cleanup levels will be removed from the river bottom by a cutterhead dredge, a type of dredge proven to be environmentally safe. The sediments will then be pumped by the dredge to one of the four CDFs. A few areas of contaminated sediments in deep water and in the Upper Harbor salt marshes may have to be removed by a less intrusive method such as clamshell bucket and transported separately to the CDFs. The sediments near submerged high-voltage power cables in the Upper Harbor will not be removed for safety reasons. However, USEPA proposes to reduce the overall width of this undredged cable corridor as much as possible through collaborative efforts with the ComElectric power utility. Other potential remedies for this cable corridor will also be considered, including capping, bioremediation, and reconstruction of the power lines in a way that allow complete dredging.

To ensure that residential areas neighboring the dredging areas are not impacted by airborne PCBs, air monitoring in these neighborhoods will be performed throughout the dredging process. Similarly, the water column will be sampled during dredging operations to ensure that sediment resuspension will be below preestablished safe levels.

3. *Operation of the CDFs and Water Treatment.* Once the dredge starts pumping contaminated sediment into the CDFs, a process of continually draining off and treating the seawater brought in along with the sediments will occur. This water will be treated by physical and chemical processes to reduce PCBs and heavy metals to very low levels before discharge back into the Harbor.

To insure the safety of workers at the CDFs, as well as other workers and residents in areas surrounding the CDFs, air monitoring will be performed at all CDFs in addition to the off-site areas. To help control airborne PCB emissions from the CDFs, a minimum of two feet of water will be maintained above the dredged sediment during dredging operations. Other emission control methods will be employed as necessary.

4. *Preliminary Capping and Sediment Consolidation.* Once the CDFs have been filled with sediment, a preliminary cap will be installed to prevent escape of PCB dust and allow for precipitation runoff while the underlying contaminated sediment consolidates. This consolidation or settling process is required to establish appropriate foundation conditions prior to construction of a final impermeable cap. Where manageable (e.g., CDFD), and if scheduling permits, cleaner dredged sediments from Harbor shipping channels will be used for the preliminary cap during this settling period. It is anticipated that approximately 3 years of sediment consideration will be required before final capping can be initiated.

Perimeter air and groundwater monitoring around the CDFs will continue during this interim time frame, although the air monitoring will be on a more limited basis than during full operation of the CDFs.

5. *Final Capping of the CDFs, Monitoring, and Long-Term Maintenance.* Once the dredged sediment has sufficiently consolidated, the final cap or cover on the CDFs will be constructed. This cap will consist of a multilayered system that will prevent water infiltration into and promote surface drainage away from the underlying sediments. Once capped, a long-term monitoring and maintenance program will begin to keep the CDFs in good repair. Monitoring to prevent groundwater contamination and airborne PCB emissions will also be conducted. Figure 17.2 shows a cross-sectional view of a final CDF.

USEPA will continue to work with the local communities to develop appropriate plans for beneficial reuse of each CDF. For example, the City of New Bedford has expressed an interest in the reuse of CDFD as a commercial marine facility. As a result, the conceptual design of this CDF includes sheet pile walls (rather than earthen dikes) on the seaward side of the CDF to promote docking, and an area that will accommodate future boat hauling activities. Design accommodations also can be made to the other CDFs provided that the ultimate land use is developed in advance in conjunction with the surrounding communities.

6. *Review of the Completed Remedy.* Because contaminated sediments will remain in CDFs at the Site, USEPA will review the cleanup, as required by the Superfund statute, no less than every five years after the cleanup begins. The five year review is performed to ensure that human health and the environment are protected by the cleanup action. In addition, as agreed to in the New Bedford Harbor Superfund Community Forum agreement for ROD 2, USEPA will conduct an ongoing literature review of treatment alternatives for potential use on materials in the CDFs until the CDFs are capped. Once capped, the technology review will continue at least every five years.

2. Proposed Enhancement of the Remedy to Include Navigational Dredging.

In addition to this Superfund cleanup, the Commonwealth of Massachusetts has requested an enhancement of the remedy, as allowed by Superfund regulations. The enhancement would include dredging and disposal of an additional one million cubic yards of sediments generated from the maintenance dredging of navigational channels. The Commonwealth's proposal to implement this remedy enhancement is expressly dependent on its ability to receive state bond funding for navigational dredging of "designated port areas" of the Commonwealth such as New Bedford Harbor. Although these "navigational" sediments fall below the proposed target cleanup levels for PCBs (and thus do not overlap with the sediments slated for Superfund dredging), they are still contaminated with metals and lower levels of PCBs. As a result, disposal options are limited, and an alternative disposal plan is required if the Harbor shipping channels are to be maintained to their originally approved depths.

Although this enhancement of the remedy would be administered and funded entirely by the Commonwealth of Massachusetts, and would be implemented in a way that would not significantly delay the Superfund cleanup, the USEPA is soliciting comments on the concept of performing the remedial and navigational dredging simultaneously as an enhanced remedy. From USEPA's and the Commonwealth's perspectives the benefits of such a linkage would primarily stem from a streamlined permitting process for navigational sediment disposal facilities, as well as the possibility of using navigational sediments for preliminary cap material (especially for CDFD). Also, the preferred alternative would enjoy an added degree of protectiveness as a result of the enhanced remedy since the navigational sediment contains PCBs up to 50 ppm and heavy metals. The navigational dredging works in concert with the City's plans for developing the public and economic uses of the Harbor. The MA Office of Coastal Zone Management intends to conduct a separate public comment process on the specific details of navigational dredging projects in the State, including New Bedford Harbor.

If this enhancement occurs, the 28,000 yd^3 of sediment from the two areas of remedial dredging just south of the hurricane barrier would most likely be disposal of in large proposed "navigational" CDF just north of this barrier on the New Bedford shore. Implementation of this enhancement is expressly dependent on available Commonwealth funding.

3. The Proposed USEPA Cleanup Plan (1997).

USEPA proposes to design and build four shoreline CDFs and associated water treatment facilities. The CDFs would be built in contaminated areas to avoid dredging approximately 126,000 yd^3 of underlying contaminated sediment. Dredged contaminated sediments would be piped into the CDFs and passively dewatered. Groundwater monitoring wells would be installed around each CDF to verify that it is operating safely.

Once construction of the first CDF is complete, dredging would commence. Approximately 450,000 yd^3 of PCB-contaminated sediment are to be dredged: in the Upper Harbor, sediments above 10 parts per million (ppm) PCBs, and in the Lower Harbor and salt marsh areas, sediments containing more than 50 ppm PCBs would be dredged. Sediments above the target cleanup levels would be removed from the river bottom by a cutterhead dredge, a type of dredge proven in the pilot studies to be environmentally safe. The sediments would then be pumped by the dredge to one of the four CDFs. Other dredging methods may be used for deep water or salt marshes.

The air quality in nearby residential areas would be monitored throughout the dredging process, and a minimum of 2 feet of water would be maintained above the sediment during dredging operations to control airborne PCB emissions. Similarly, the water column would be sampled during dredging to ensure that sediment resuspension is below preestablished safe levels. During dredging, seawater would be drained from the sediments and treated physically and chemically to reduce levels of PCBs and heavy metals before discharge back into the harbor.

After the CDFs are filled with sediment, a preliminary cap would be installed to prevent escape of PCB dust and to allow for precipitation runoff while the underlying contaminated sediment consolidates. This consolidation process, which is expected to take approximately 3 years, is necessary to establish appropriate foundation conditions prior to construction of a final impermeable cap. When the dredged sediment has sufficiently consolidated, a multilayered cap would be constructed to prevent water infiltration into, and promote surface drainage away from the underlying sediments.

4. Proposed Remedy Enhancement to Include Navigational Dredging.
The Commonwealth of Massachusetts has requested an enhancement of the Superfund remedy to include dredging and disposal of an additional 1,000,000 yd^3 of sediments generated from the maintenance dredging of navigational channels. Although these "navigational" sediments fall below the proposed target cleanup levels for PCBs, and thus do not overlap with the sediments slated for Superfund dredging, they are still contaminated with metals and low levels of PCBs. As a result, disposal options are limited, and an alternative disposal plan is required if the harbor shipping channels are to be maintained at their originally approved depths.

This enhancement could entail removing 28,000 yd^3 of sediment from two areas for disposal in a large proposed "navigational" CDF. The benefits of this action would be the possibility of using navigational sediments as preliminary cap material, the removal of additional PCBs and heavy metals in the navigational sediments, and streamlined permitting procedures. The navigational dredging would also work in concert with the City's plans for developing the public and economic uses of the harbor. If the proposed enhancement is accepted, its implementation would be contingent on appropriate state funding and would be directed by the Commonwealth and the Army Corps of Engineers, rather than the federal Superfund program.

5. 1998 (October) EPA Cleanup Plan for the Second Phase.
This is one of the largest projects of its kind in the country. It is a 10-year, $120 million cleanup plan which includes the dredging and containment of one half million cubic yards of PCB-contaminated sediment from 170 acres of the harbor floor. Design of the plan will begin immediately. The key elements of the cleanup are:

- Approximately 450,000 yd^3 of sediment contaminated with polychlorinated biphenyls (PCBs) will be removed.

- In certain popular beachcombing shoreline areas, sediments between the high and low tide levels will be removed if above 25 ppm PCBs. In areas where homes directly abut the harbor and where contact with sediment is expected, sediments between the high and low tide levels will be removed if above 1 ppm PCBs.

- Four shoreline confined Disposal Facilities (CDF) will be constructed on 44 acres to contain and isolate the dredged sediments. Three of these facilities will be in the Upper Harbor, and one will be in the Lower Harbor. Archaeological surveys will be performed prior to construction of the CDFs and before dredging is started.

- The large volumes of water that are brought in by the dredging process will be decanted and treated before being discharged back to the Harbor. The sediments will be stored in the CDFs.

- A cap will be constructed at each CDF, and where possible, cleaner sediment from the harbor's navigational channels will be used as part of the interim caps.

- The CDF will be available for beneficial reuse as shoreline open space, parks or, in the case of the lower harbor CDF, a commercial marine facility.

- The capped confined disposal facilities will be monitored and maintained over the long term to ensure their integrity.

The decision was achieved through close collaboration and consensus building with the New Bedford Harbor Superfund Site Community Forum, community and neighborhood leaders, and local, state, and federal officials.

Waukegan, Illinois
The Outboard Marine Corporation (OMC) site is located on Sea Horse Drive on the west shore of Lake Michigan in Waukegan, Illinois, about 30 miles north of Chicago and 10 miles south of the Wisconsin border.

Polychlorinated biphenyls (PCBs) have been found in Waukegan Harbor, the North Ditch, the Parking Lot Area, and in the North Ditch area.

Waukegan Harbor is an irregularly shaped harbor about 37 acres in area. The harbor has been divided into three general areas of PCB contamination: Slip No. 3, with PCB concentrations in excess of 500 parts per million (ppm); the Upper Harbor, with PCB concentrations from 50 to 500 ppm; and the Lower Harbor, with PCB concentrations from 10 to 50 ppm. The Feasibility Study (FS) focused on the PCB contamination above 50 ppm. The two areas with PCB contamination above 50 ppm are shown in Figure 17.40. Water depths in the harbor generally vary from 14 to 25 feet (ft), with some shallower depths in parts of Slip No. 3.

The harbor sediments consist of 1 to 7 ft of very soft organic silt (muck) overlying typically 4 ft of medium dense, fine to coarse sand. A very stiff silt (glacial till) that typically ranges from 50 to more than 100 ft thick underlies the sand. The entire harbor is bordered by 20- to 25-ft long steel sheet piling. The sheet piles are believed to generally extend into the sand layer above the glacial till.

The North Ditch is a small tributary of Lake Michigan that drains surface runoff from about 0.11 square mile (mi^2) of OMC and North Shore Sanitary District property. The U.S. Department of the Interior measured the mean daily discharge of the ditch between March and September 1979 as 1.8 cubic ft per second (cfs), with a maximum discharge of 5.3 cfs. They calculated the 5-year storm event to be 23 cfs.

The Parking Lot Area is located north of OMC's Plant No. 2 and is about 9 acres in area. The generalized subsurface conditions in the North Ditch/Parking Lot Area consist of typically 30 ft of medium dense, very fine to fine sand overlying a stiff to very stiff silt (glacial till). The thickness of the glacial till ranges from 50 to more than 100 ft.

USEPA Suit. The presence of high levels of PCBs in Waukegan Harbor sediment and in the soil adjacent to the OMC plant was discovered in 1975. The USEPA and the State of Illinois encouraged OMC to cease PCB discharges and to control the PCB-contaminated sediment and soil. A suit was filed by USEPA on March 1, 1978 against OMC to cleanup the PCB contamination (USEPA, 1984).

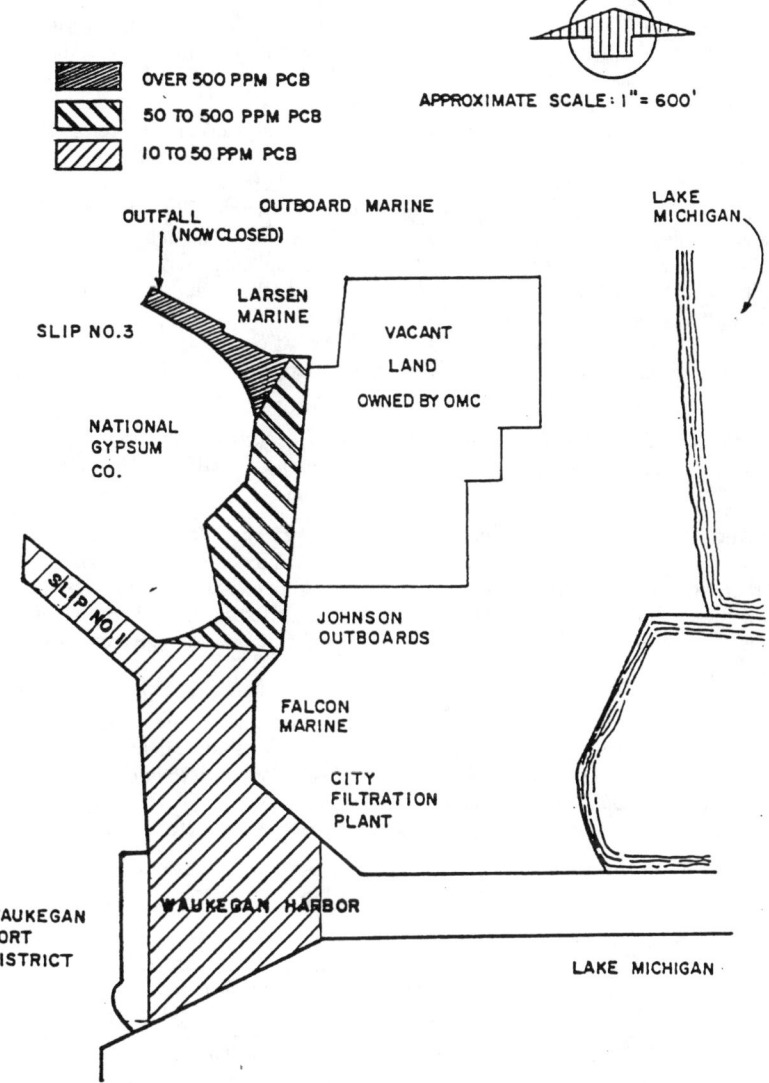

FIGURE 17.40. Waukegan.

Harbor Navigation

Waterborne commerce. The major portion of waterborne commerce in Waukegan Harbor is building cement and gypsum received by Gold Bond Building Products and Huron Cement Company, both divisions of National Gypsum Company. In 1982, 114,000 tons of building cement were received and in 1981, 130,000 tons of building cement and 81,000 tons of gypsum were received. A commercial fishing fleet of eight active boats also operates out of the harbor. Thirty-six tons of fresh fish were unloaded at the harbor in 1982 and twenty-five tons were unloaded in 1981. The Port of Waukegan is also homesite to a number of small- and large-scale industries, including OMC Johnson and Outboard Marine Corporation, together employing over 2,000 persons. Other industries include Falcon Marine and a marine contractor.

Recreational boating. Another key use of the Port of Waukegan is recreational boating. In 1994, the Waukegan Port District operated 158 slips and moorings as well as 103 dry dock spaces. Directly to the north of Slip 3, Larson Marine service housed approximately 300 small pleasure craft for storage and repair. Since the mid 1970s the Waukegan area has been recognized as one of the major cohoe and salmon fishing areas on Lake Michigan. The recreational use of the Waukegan Harbor has grown significantly over the past twenty years and has served as the stimulus for the construction of new harbor facilities. The new facilities will include 761 new slips for small pleasure craft. This expansion will also increase the number of charter fishing boats from 35 in 1983 to a projected 60 charter boats operating out of the Waukegan area (USACE, 1987).

Extent of the Problem. According to 1982 reports (Mason and Hanger, 1981, 1982) about 42,900 yd^3 of upper sediments (muck) and 3,700 yd^3 of deep sediments (sand and silt) would have to be dredged to clean up areas of contamination (Table 17.23).

Project Execution. The goal was to remove sediments with PCB concentrations greater than 50 ppm. A ten-inch cutterhead dredge was employed in dredging the Upper Harbor. Silt curtains anchored to the bottom were deployed at the lower part of the Upper Harbor and at the entrance to Slip No. 4. 32,000 yd^3 of sediment were removed from the Upper Harbor and 18,000 yd^3 were removed from the ditches, storage area, oval lagoon, and parking lot.

Slip No. 3 became a confined disposal area by constructing a cofferdam between Slip No. 3 and the Upper Harbor. Because Slip No. 3 contained highly contaminated sediments, such sediments were removed by a small Mud Cat dredge, treated and returned to the Slip. All dredged sediments from the Upper Harbor were placed in Slip No. 3, which was eventually capped (after 3 years).

Waukegan remediation project took about 13 years to complete at a cost of about $21 million.

TABLE 17.23. Extent of contamination

Location	Volume to be removed (yd^3)	PCB contamination (ppm)	Sediment
Slip No. 3	2,000	10,000	Muck
Slip No. 3	3,700	10,000	Sand and silt
Slip No. 3	2,200	1,000	Muck
Slip No. 3	3,000	6500–1,000	Muck
North Ditch/Parking Area			
Crescent Ditch north of the dike storage area	28,900	5,000–38,000	Soil
North of Storage Area	2,300	200	Soil
Oval Lagoon	14,600	26,000	Soil
North Ditch	25,000	500–5,000	Soil
Parking Lot	68,000	50–500	Soil

Foreign

Antwerp, Belgium

1. Introduction. Vacuum consolidation of contaminated silt can accelerate the natural process by a factor of 100. Breakthroughs in horizontal drainage systems could transform the economics of contaminated silt disposal.

The new vacuum consolidation system is being used for the first time to increase the capacity of an underwater silt reception pit in Antwerp. This involves installing a network of horizontal drains within the pit.

This technology can accelerate the natural process of consolidation by a factor of 100 and is said to have a major impact on policies for the storage, disposal and use of contaminated silt. The system (for which a patent is pending) has been developed by the University of Ghent's, Department of Soil Mechanics and Foundation Engineering together with Dredging International (DI).

The company's Operating Manager said: "This technology is a major step forward in the drive to find environmentally acceptable and cost-effective disposal solutions for contaminated silt. Many research projects have demonstrated the environmental advantages of underwater disposal. We can now make optimum use of these facilities—vacuum consolidation, over a relatively short period, significantly increases the space available. Furthermore, the efficiency of the new system is such that, in many reclamation applications, it will be possible to leave the silt in-situ and reclaim the area with clean sand."

2. Antwerp Scheme. The Antwerp pilot project was conducted by Combinatie Kallo under a contract with the Flemish Ministry of Environment and Public Works. This scheme has its origins in requirements introduced in 1992 to greatly reduce the migration of polluted silt from the Belgian Schelde to the Dutch border. This requires the removal of 300,000 tons of contaminated silt every year from Antwerp's Kallo Lock, a major site of natural accumulation. This silt has been placed in a series of 10 large pits within the nearby Left Bank Harbor. These pits were excavated many years ago, and the excavated sand was employed for land reclamation.

These large pits were dredged to a depth of 25 m, to the clay level, but the demand for disposal space for polluted silt is now so high that they have become virtually full in a very short period. But, if DI's vacuum consolidation trials are successful, similar horizontal drain systems could be installed in the other pits and this would provide extra space for a further two years' underwater disposal.

The Project Manager, said: "Without this new technology, the Left Bank Harbor disposal pits will close within 12 months, forcing the authorities to find alternative space on land—which is very limited and extremely expensive. A two-year life extension for the Left Bank Harbor disposal pits would produce an important saving."

Dredging in the Kallo Lock is being undertaken by the DI vessel Brabo, equipped with "scoopdredger" technology specifically developed for use at environmentally sensitive locations. When silt is dumped in Left Bank Harbor pits, it has a water content of 90% and would take 100 years for natural consolidation. Vacuum consolidation can reduce this period to 12 months.

The system harnesses a combination of water pressure and atmospheric pressure to force water out of the silt. Work in the Left Bank Harbor began in early January and installation of the horizontal drains was completed by the end of February. The system will operate for a six-month period, with full technical evaluation scheduled at the end of the experiment.

The novel engineering aspect is the ability to achieve a highly accurate installation of the horizontal drainage system. The use of vertical drains is commonplace on reclamation sites, yet horizontal drains are far more efficient. In the past, however, attempts to employ horizontal drainage have run into problems during the installation phase. Great accuracy is essential if a horizontal drainage system is to perform efficiently.

These problems have been overcome with a laybarge-mounted system that plows, places four drains simultaneously, and then backfills. The pilot project involves the installation of 150 km drains at 2 m horizontal and vertical spacings.

The drains are held on a main spool mounted on the laybarge deck. The drainage consists of plastic composite with a geotextile sheath for maximum protection. The plow features a twin-bladed head equipped with a high-pressure water jet system. The equipment can install drains to depths up to 8 m. The Project Manager stated:

> This system has important implications for the management of polluted silt in many areas of the world, including the Far East. Accelerated consolidation transforms the economics of contaminated silt disposal. In many cases, it will now be possible to leave the silt in-situ, consolidate it with this technology and then reclaim with sand. (Figure 17.41)
>
> The system also has obvious applications in other sectors, such as the highly accurate installation of cables and pipes. It is ideal for shallow water crossings, to water depths of up to 40 m in the present configuration. The major advantages include speed (laying up to 1 km per day) and simultaneous installation of groups of

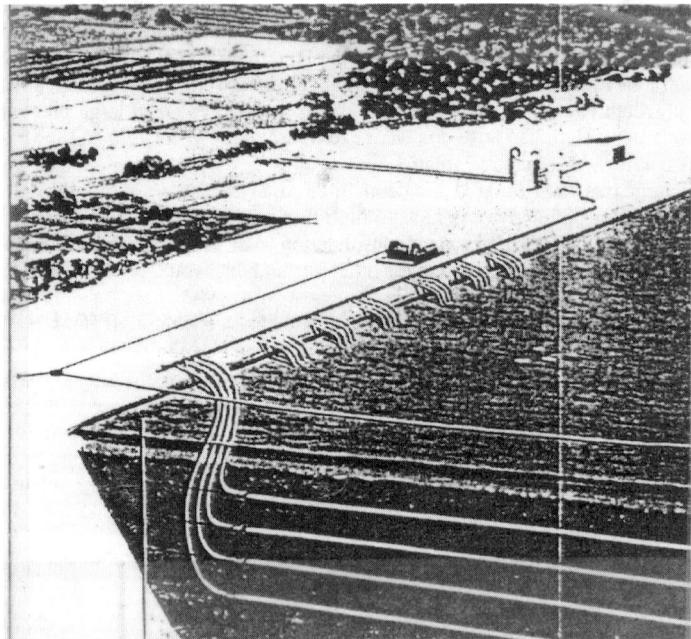

FIGURE 17.41. A cross section of the Left Bank Harbor disposal pit, Antwerp, showing a group of four horizontal drains connected to a primary line running to the pumping station.

cables or pipes with extreme accuracy. Discussions with telecommunication groups and other utilities were held prior to the trials.

Ketelmeer, the Netherlands

1. Introduction. The Dutch Ministry of Public Works (Rijkswaterstaat) awarded one of the largest environmental dredging contracts in Western Europe (the Ketelmeer Project). The 350 hectare (845 acres) Ketelmeer Lake lies to the east of Ejsselmeer, the northern part of the Netherlands. For decades, silt deposited in Ketelmeer by the Ijssel River has spread across the sandy bottom of the 12 km wide lake. This sediment is contaminated with heavy metals, PACs, PCBs, and other hazardous pollutants.

Contaminated silt has four classifications in the Netherlands. Ketelmeer's silt is rated as Class 4, the category reserved for the most heavily contaminated material, which requires disposal in fully contained special waste facilities.

Four dredging contractors were selected by the Ministry to participate in a series of three-week trials. Each has a distinct and innovative solution based on either a newly developed technique or the modification of a conventional technique to suit the demands of environmental dredging (Figure 17.42).

2. The Ministry of Public Works. The Ministry of Public Works needs to arrive at an optimum specification that offers the best balance between cost and performance. It is seeking to gather information through field trials on the best methods by which the eventual work might be carried out. Under the Development Program Treatment

FIGURE 17.42. The Sweephead mounted on dredge "Vlaanderen XV."

Processes for Polluted Sediments (POSW), data are being collected to compare different dredging techniques in a similar location, eliminating variations in hydrodynamic conditions and sediment characteristics.

3. Accuracy Pays. Accurate cutting of ultra-thin layers of sediment is the crucial issue at Ketelmeer, as the contaminated silt has been deposited in a layer that is between 20 and 40 cm thick, with occasional pockets that are thicker. In a conventional dredging contract where 2 to 3 m thick layers are removed, an accuracy of 30 cm would be considered normal. A greater accuracy of 10 cm may also be possible; however, a greater accuracy would result in lower production. At Ketelmeer, a 10 cm accuracy would not be acceptable, because it would necessitate the removal of an additional 5 million cubic meters of sediment.

Each centimeter of contaminated layer would add 280,000 m³ of removed sediment, resulting in an additional disposal cost of $2.8 million (at $10 per m³). Such a high cost gives ample reason to investigate state of the art dredging equipment and survey methods. The stakes are high. Some 20 million cubic meters of contaminated material must be dredged and pumped to a special disposal facility now under construction in the center of the lake. This consists of a 1 km diameter disposal site 45 m deep. It will be ringed by eight dikes and other separating structures (Figure 17.43).

The project commenced in the late 1990s and will continue throughout the next decade. The project costs are expected to exceed $110 million, although savings may re-

FIGURE 17.43. Dredge "Vlaanderen XV" pictured here during the Sweephead trials on Ketelmeer. The sheet-piled area visible to the left marks the perimeter of what will be Ketelmeer's permanent silt disposal facility.

sult from the trials themselves. Total cost will relate directly to the amount of material ultimately dredged and discoveries born of these trials could reduce the final amount.

4. Ongoing Trials. In pursuit of this most lucrative contract are Baggerbedrijf de Boer (Dutch Dredging), HAM, Boskalis, and Dredging International. Each is required to deposit trial dredged material in a temporary confined disposal site situated next to the dredging area in order to minimize transport cost. Each trial is to be completed within a three-week period with the amount of dredged material not to exceed 15,000 m^3. Each participating company engages in a separate trial. The HAM and Boskalis trials took place in the summer and autumn of last year. Dredging International's trial was conducted in June 1996 and Dutch Dredging conduted their trial later.

To achieve the high levels of accuracy required of the trials, the newly developed kinematic DGPS-Kart was chosen as the main positioning system for both vertical and horizontal positions. A monitoring station was set up close to the dredging area to check accuracy and reliability. All bathymetric measurements prepared for the project and for continuous monitoring during the trials were made following strict procedures. Coring was used to establish the boundary between clean and contaminated sediments. Two digital terrain models with a grid size of 1 m were used in the trials, one each at the top and bottom of the contaminated layer.

Monitoring focuses on three main criteria: accuracy, sediment resuspension and turbidity. Overall accuracy is calculated by means of comparison between the desired profile and the actual dredged profile, that is determined by a high-precision echosounder system.

Sediment resuspension is defined as all material loosened by dredging and entering the water column transport system. Turbidity is a most important consideration at Ketelmeer, as contaminants can be attached to suspended solids causing contamination of clean sediments in the surrounding area. The turbidity measurement technique complies with the Dutch standard developed by the Rotterdam Harbor Authority in combination with the dredging contractors and the Ministry. It allows an objective comparison between different dredging techniques.

The teams at the trials work closely with a 45-person technical group from the Ministry and its consultants. The trials explore the optimum balance between high productivity, extreme accuracy, minimum sediment resuspension, and low turbidity. The participants work through a complex program that has taken six months to formulate and prepare.

Each day of the trials consists of a carefully orchestrated test schedule with several cycles of dredging, soundings, monitoring resuspended sediments, measurements, and other activities. The program explores three scenarios across a variety of horizontal and sloping profiles. These were: dredging a 20 cm layer of the silt surface, dredging a 40 cm layer of underlying sand, and finally, dredging at a greater depth to remove deeper deposits.

5. Four Concepts. The companies participating at Ketelmeer provided four very different environmental dredging solutions.

Dutch Dredging brought the Aalschover, an environmentally friendly bucket ladder dredge. The conventional 540 liter bucket dredger was converted from its normal specification by enclosing the upper section of the bucket ladder so that any overspill slid back down the chute to the digging point without affecting the ambient water.

After discharging their loads, the buckets are pressure washed in an enclosed part of the superstructure so that they reenter the water free of contaminants. The buckets are also fitted with valves to let the air out of the buckets on reentry to eliminate the risk of resuspension of sediment.

Crew accommodation is sealed to protect personnel from the sediments. Special washing equipment and protective clothing are also provided.

HAM has developed the auger dredge as an environmentally friendly unit with a patented silt screen mounted around the auger (Figure 17.44).

As used at Ketelmeer, an 8-m wide auger with a diameter of 1.25 m was mounted on the Willem Bever pontoon. Dredging was performed in 7 m wide tracks in one direction with 1 m overlap. To accurately haul the pontoon along the track, six constant-tension winches were used. This system combined with an automated hauling system achieved a hauling accuracy within 0.25 m. To minimize turbidity caused by anchor wires dragging along the lake bottom sediment, floating pontoons were used to keep the wires off the bottom of the lake.

As layers to a maximum thickness of 0.85 m were removed on each track, the flexible silt screen curtain automatically followed the bottom profile. A debris box in front of the suction mouth protected the system and resuspension of sediment was prevented by keeping the lower side of the auger's cutting blade at nearly the same level as the cutting edge of the auger blades.

Boskalis utilized the environmental disk cutter, derived from the disk bottom cutter and based on model tests conducted at the University of Delft.

FIGURE 17.44. New environmental suction dredger "Ham 291."

Past experience has shown the disk bottom cutter develops a flat profile with a low spillage percentage. At Ketelmeer, work focused on increasing the positional accuracy of the cutter, increasing the mixture concentration, and reducing turbidity around the cutter.

The soil was cut by means of a cylindrical cutter with a flat, closed bottom and vertical axis of rotation. A suction mouth, for the removal of cut material, was located inside the cutter. An adjustable screen around the cutter was used to prevent resuspension of sediment and turbidity; its height can be automatically corrected to the height of the soil being cut.

Dredging International deployed its Sweephead, a variant of the Scoophead, introduced for low-turbidity dredging tasks. The Sweepdredger combines the low turbidity advantages of the Scoopdredger with the latest silt draghead technology.

The Sweephead's effectiveness was first demonstrated during earlier trials in the Brussels Sea Canal. It was then deployed at the Njeuwpoort, Belgium access channel linking harbor to sea. It is mounted on the cutter suction dredger "Vlaanderen XV" in a conversion that took only a week to complete. The Sweephead blends high-production cutter technology and high-density silt dredging.

Whereas a conventional cutter might average 400 m^3 per hour while cutting an ultra-thin layer, the Sweephead equipped "Vlaanderen XV" was capable of up to 1,500 m^3 per hour, depending on prevailing conditions and turbidity restrictions. A primary aim at the Ketelmeer trials was to progress as close as possible toward 1,500 m^3 per hour without compromising compliance with the other key parameters.

At an optimal productivity of 1,000 m^3 per hour, the Sweephead-equipped "Vlaanderen XV" produced minimal turbidity and achieved a spillage target of less than 1 cm and an accuracy to within 5–10 cm.

The trials explored a range of concentration values from 30 to 80%, with the dredged material pumped from the Sweephead to the disposal facility via a floating pipeline The vessel's fully automated dredging system provided digitized terrain modeling of both the silt surface and underlying sand profiles, but during the trials more sand was encountered than expected. It was discovered that closely packed layers of coarse sand had a tendency to clog the pipeline.

After the trials, "Vlaanderen XV" returned to the new Brussels Sea Canal, dredging sand at Hingene, near Antwerp. The work serves as a further trial of a reduced pipeline diameter, is expected to increase flow and enhance performance. In addition, the positioning of the water jets in relation to the Sweephead cutter blades is being adjusted to increase the fluidity of the sand entering the dredge.

With such a large contract at stake, the companies involved are highly motivated to achieve optimal performance by way of fine tuning. There will be some anxious waiting over the winter by the four participants as the Rijkswaterstaat completes its full technical evaluation. It is not known whether the entire contract will go to one winner or whether the work will be apportioned in the interest of an earlier completion date.

By committing themselves so fully to the Ketelmeer trials, the participants have pioneered new technology and, in some cases, given themselves the capability to fulfill both conventional and environmental aspects of a dredging contract employing one vessel (Ketelmeer Feature, 1996). Whoever wins, both the cause and the practice of environmental dredging will have been advanced significantly.

Minamata Bay, Japan
Introduction
Waste from a chemical factory in Minamata City (located at the southern part of Kumamoto Prefecture in Southern Japan) containing methyl mercury was found in Minama-

ta Bay. Mercury accumulated in the Bay's fish and shellfish. These were eaten by people unaware of this. Over 2,000 people were affected by the mercury consumption, and some 900 died.

In 1974 the Minister for Transport, the Director General of the Environmental Agency, and the Director of Kumamoto Prefecture reached a basic agreement on the cleanup of Minamata Bay (Yoshinaga, 1995).

Basic Plan. The cleanup goal was to remove all contaminated sediment having a mercury content of 25 ppm or higher from an area of 2,090,000 m^2. The most contaminated area, 580,000 m^2, was enclosed with a watertight sheet pile cellular barrier.

Dredging. Mercury was contained in the upper thin layer of sediment in the Bay, whereas in the inner part of the Bay a 50-cm thick layer of sediment had to be removed by dredging (Figure 17.45). This indicated a need for precision dredging, as any overdredging would increase the volume required in the disposal site. Another requirement was to minimize the resuspension of sediment by dredging. The resuspension could cause mercury to dissolve.

A cutterless hydraulic dredge was selected for removal of the mercury-contaminated sediments. Four cutterless dredges were deployed. The suction mouth shape was modified to suit the soft mud in the Bay. A TV camera was installed to observe turbidity generated during dredging. Turbidity was continuously measured by turbidity meters attached to the suction head.

A considerable amount of data were obtained on the dredging layer thickness, bathymetry (before and after dredging), and dredged volume and concentration of the dredged

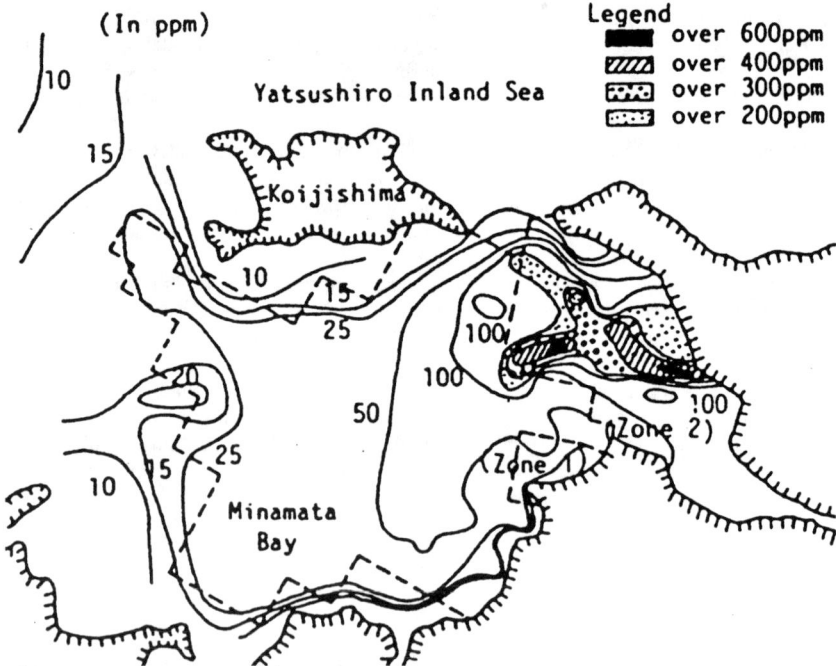

FIGURE 17.45. Mercury contamination distribution before dredging.

material. In addition, other instrumentation measured the suction head position, swing speed and swing direction. To prevent dredged material exposure to air and sunlight, the slurry was discharged underwater. The dredged material was discharged into the reclamation area up to sea level and covered with a geotextile. Lightweight volcanic ash and good quality soil was placed on the geotextile.

The effluent discharged back to sea could not contain more than 0.05 ppm of mercury; the maximum limit for turbidity was set at 40 ppm. Mercury concentration distribution after dredging is shown in Figure 17.46.

Conclusion. Dredging was successfully completed in December 1987, and the project, after 10 years, was completed in March 1990.

Penetanguishene, Canada
The Canadian Government, in accordance with the 1987 Canada–USA Water Quality Agreement, launched a $125-million Great Lakes Action Plan in 1989. As a result, $55

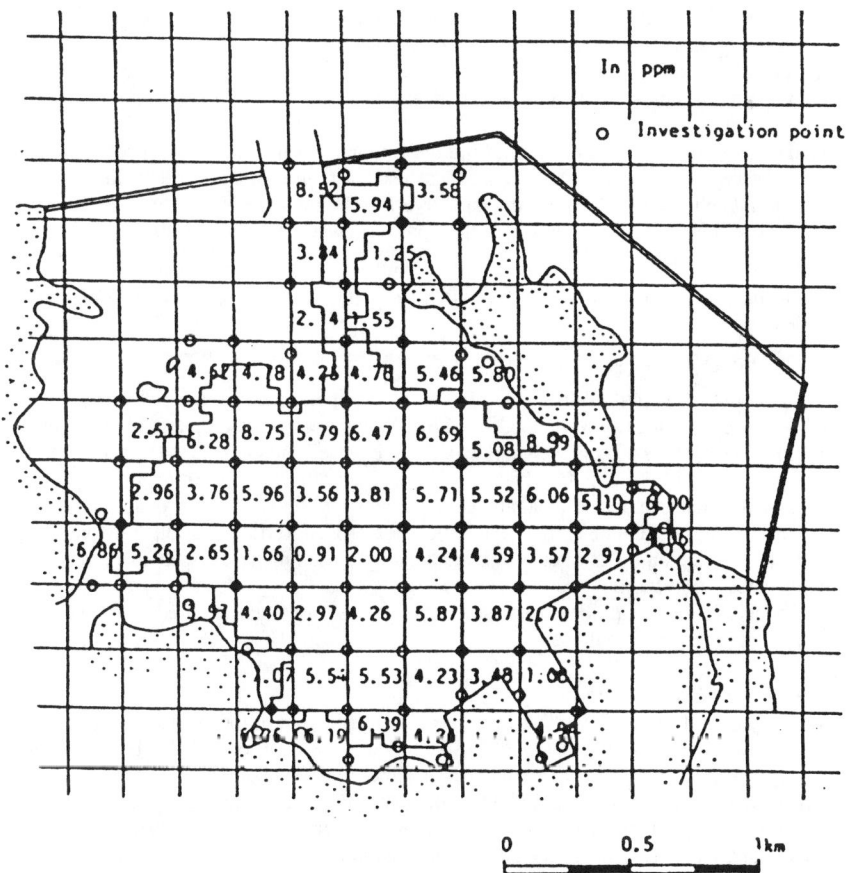

FIGURE 17.46. Mercury concentration distribution after dredging.

million was allocated to Environment Canada's Great Lakes 2000 Cleanup Fund, which created the Remediation Technologies Program in 1990. This program is designed to demonstrate and assist in the commercialization of innovative technologies for remediation of contaminated sediment (Environment Canada, 1990).

Severn Sound is located in the southeastern portion of Georgian Bay and consists of a group of bays including Penetang Bay, Midland Bay, Hog Bay, Sturgeon Bay, and Matchedash Bay. From the 1800s through the early 1960s, sawmills and logging companies flourished in Penetanguishene, located in Severn Sound. In addition to sawmill activities, transport of logs contributed to the degradation of Penetanguishene's waterfront.

In November 1992, a drought lowered the water level of Severn Sound by one meter. As a result, several hectares of degraded fish habitat were exposed (approximately 300 m × 200 m) in the shallow nearshore area of Penetang Bay. The area was covered with logs, wood slabs, and sawdust.

In early 1994, a partnership was established with the Town of Penetanguishene, the Severn Sound Remedial Action Plan, and Great Lakes 2000 Cleanup Fund's Remediation Technologies Program. Together they provided funding and expertise to help restore the valuable fish and wildlife habitat of a section of Penetang Bay.

This portion of the project had two components. First, the removal of large wood wastes using conventional technologies and methodologies. Second, the demonstration of an innovative sediment removal technology to assess its effectiveness in the removal of contaminated sediment in other areas of concern throughout the Brimley Road storm sewer.

In order to prevent park closure during the summer, an in-water holding cell was used as a primary settling basin. An excavator was used to remove the pumped sediment from the holding cell to an adjacent drying pad. Overflow water was directed to secondary and tertiary in-water settling basins, where finer particles were allowed to settle.

Results indicated:

- The average production rate achieved by the Amphibex dredge during this project was close to 50 m^3/hr.

- The average percentage of solids was between 25–35%.

Production rates and timing were affected by the presence of the active storm sewer and a significant quantity of debris. Since the Amphibex is portable, amphibious, versatile, and powerful, its use avoided park closure and permanent damage to the park. During the construction of the retention basin, the Amphibex attracted a great deal of public attention and acted as a promotional and educational agent for Greater Toronto.

Throughout this demonstration, the Amphibex proved to be one of the most versatile dredges available on the market. Additional testing on sediment resuspension and pumping distance was required.

Financially, the Amphibex was less expensive than conventional dredging, making it attractive to other potential users.

Scarborough Bluffs, Canada

Located just east of the City of Toronto, Scarborough is part of the Metropolitan Toronto and Region Area of Concern (AOC). Erosion from the Scarborough Bluffs is a major source of suspended solids in lake Ontario. Another problem associated with the Scarborough Bluffs is the storm sewer and occasional combined sewer overflows that discharge in Bluffers Park and ultimately in Lake Ontario. As part of the Water Quality Enhancement Strategy for the Brimley road drainage area (storm sewer source to Bluffers Park), a Dunkers Flow Balancing System was proposed within the Bluffers Park embayment. Construction of this facility requires dredging of sediment currently in the embayment.

The quality of the sediment was determined to identify acceptable disposal methods according to the Ontario Ministry of Environment and Energy Guidelines for the Protection and Management of Aquatic Sediment Quality.

Due to the nature of material to be removed, the shallowness of the embayment, and difficulties associated with dredging in a parkland, an innovative amphibious dredging technology (Amphibex) was selected. The production rates and timing were affected by the presence of the active storm sewer and a significant quantity of debris. Since the Amphibex is portable, amphibious, versatile and powerful, its use avoided park closure and permanent damage to the park. During the construction of the retention basin, the Amphibex attracted a great deal of public attention and acted as a promotion and educational agent for Greater Toronto.

Throughout this demonstration, the Amphibex proved to be one of the more versatile dredges available on the market. Additional testing on sediment resuspension and pumping distance was required.

Financially, the Amphibex was less expensive than conventional dredging, making it attractive to other potential users.

Southampton, United Kingdom

1. Geotechnical Data. A soil investigation showed silt layers with irregular thickness, mainly in the shallower areas of the port where it was impossible to work with a Trailing Suction Hopper (TH) dredge. The deposits were a mixture of green sand (a very fine glauconite sand) and clay. Since particles brought into suspension of this type of material settle very slowly, the use of a normal cutter suction dredge was not appropriate. Under normal dredging methods the silt layer would have been displaced and consequently would have resulted in displacement of the silt from one area to another. The silt had to be carried to a designated dump area, 50 km from the dredge site.

A modified draghead was developed by Jan De Nul (1998) called "sweephead," capable of removing thin layers of silt with great accuracy and minimum disturbance to the environment.

The "sweephead" has two inlets and works without additional mechanical movements. A hydraulic valve in the head opens the inlet toward the dredging direction while the shape of the contact surfaces ensures optimal sediment transport. Process water is minimal, which means that the material is pumped at almost in-situ density.

Additional sensors and instruments allow for better process monitoring and more accurate dredging. The survey campaign showed that the turbidity generated by the actual dredging operations was reduced to such a level that other sources such as normal shipping and maneuvering auxiliary vessels brought more solids in suspension than the dredging.

A cutterhead dredge "Dirk Martens" was also outfitted with the "sweephead" for dredging in the Port of Southampton.

Zeebrugge, Belgium

As part of a research project at Zeebrugge in combination with the Belgian authorities, different dredging and overflow techniques were compared to quantify the generation of turbidity and the mobilization of pollutants.

For this dredging campaign, the multi-functional TH dredge "Cristoforo Colombo" was selected.

This modern dredge is capable of several dredging methods: standard dredging without overflow; dredging with one or two suction pipes until the overflow level has reached the top of the hopper; standard dredging with overflow; continue loading the hopper barge with higher density material while process water is removed from the hopper by

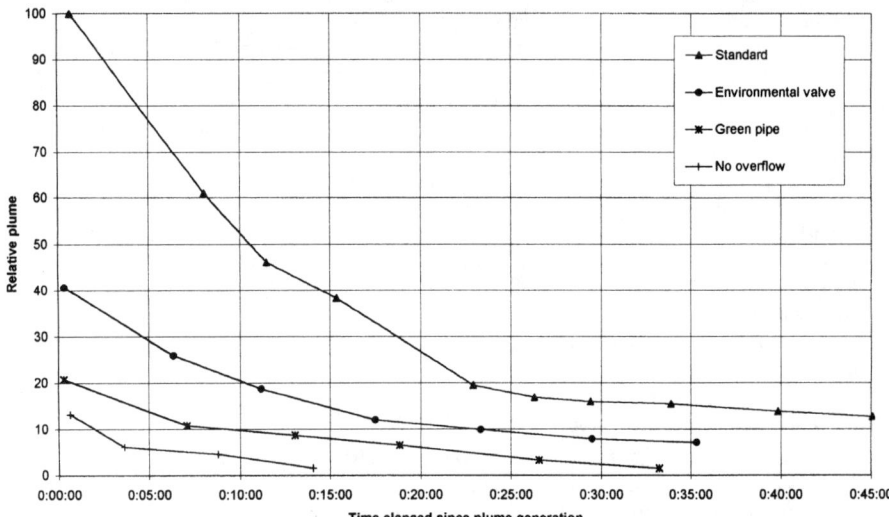

FIGURE 17.47. Graph showing turbidity levels relative to dredging techniques. (Courtesy of *World Dredging, Mining & Construction,* August, 1998.)

standard overflow; environmental valve in the overflow funnel—an adjustable valve in the overflow funnel chokes the flow in such a way that no air is entrained in the suspension leaving the hopper. The overflow suspension is pumped through another pipe (called the "green pipe"), mounted on top of the suction pipe, back to the suction head, where it is used as process water.

The purpose of the project was to determine how alternative dredging and overflow methods influence the generation and the behavior of particles brought into suspension.

Figure 17.47 demonstrates how turbidity was reduced by selecting different dredging and overflow techniques and by setting out the relative turbidity of the plume section against the time elapsed since the passage of the dredge.

SUMMARY

1. Each remediation project is very site-specific; experience on one project does not necessarily translate to another project.
2. Removal of contaminated sediment from a given site requires a lengthy process of negotiations between the regulatory agencies at federal and state levels and the owner of the site.
3. There are no single cleanup goals for the whole country; the range of PCBs cleanup goals, for example, varies from 0.1 ppm to 50 ppm. Goals, in some instances, are unrealistic and cannot be achieved through repeated passes of dredging equipment.

    On a number of projects it was not possible to remove the contaminated sediments to the desired goals. There were several reasons for this:

a) Predredging surveys may not have been accurate, and/or an insufficient number of borings may have been taken.
b) Dredging processes resuspend sediments that can be moved by currents or waves.
c) Dredging processes mix clean sediments with contaminated sediments, and repeated passes to achieve predredging goals may not be economical.

Table 17.24 shows that it is difficult to achieve low residual contaminant concentrations on a consistent basis (Romagnoli, 1998).

4. Remediation costs are enormous.
5. Lengths of projects were up to 10 years, or more.
6. There is a lack of specially designed equipment in the United States (with a few exceptions) to conduct "surgical" dredging. Dredges available in other countries cannot be used in the United States because of the Jones Law.
7. There are two principles for successful remediation dredging:
   a. Reduction, or elimination of resuspension of sediment during dredging
   b. Achievement of a low water content in the dredged material
8. Table 17.25 summarizes the more important, selected remediation projects in the United States. Table 17.26 shows the costs of remediation for selected projects.

TABLE 17.24. Results from completed PCB dredging projects (Romagnoli, 1998)

Site	Sediment volume removed (yd^3)	PCB Concentration Preremediation range (ppm)	PCB Concentration Postremediation range (ppm)
New Bedford Harbor, MA[a] (CRS, 1993)	10,000	150–580	0.5–270
Sheboygan River, WI (BBLES, January 1991 and BBLES, March 1992)	3,800	0.2–4,500	0.45–295
Ruck Pond/Cedar Creek, WI (BBL,[d] August 1995)	7,500	ND–2,500	9.2–320
St. Lawrence River, NY (BBL, June 1996)	13,250 (in situ), including approx. 1,800 yd^3 boulders	ND–7,120	ND–91[b]
Grasse River, NY (BBL, December 1995)	2,600 and 400 (boulders)	12–1,708	1.1–260
Shiawasse River, MI	2	Up to 520	Up to 7.2
Duwamish Waterway, WA	NA[c]	NA (although removal was in response to PCB spill)	Up to 50
Portland General Electric, OR (CH2M Hill, January 1991)	14	Up to 190	Up to 21

[a]Data from New Bedford Harbor Pilot Study, not hot-spot removal.
[b]An additional sediment sample collected following the postdredging sampling program indicated PCBs in excess of 6,000 ppm.
[c]NA = Not available
[d]BBL = Blasland, Bouck & Lee, Inc. (BBL).

TABLE 17.25. Summary of selected projects in the United States

Project	Contaminant levels	Contaminant goal levels	Volume removed yd^3
Alcoa (Massena, NY)	12–11,000 ppm PCBs	<10 ppm PCBs	3,100
Bayou Bonfouca, LA	N/A	1,300 PAHs	159,000
Black River, OH	PAHs and Metals	Remove to natural till	60,000
Cold Spring, NY	Cadmium	95% removal	77,200
Duwamish Waterway, WA	260 gallons PCBs (total) 760 ppm PCBs	92% recovery	27,000
GM (Massena, NY)	N/A	1 ppm PCBs	24,000
LTV Steel, IN	PAHs and oils	Remove to natural hardpan	114,000
Manistique, MI	Hot spots	10 ppm PCBs	87,000
New Bedford, MA	10–100,000 ppm	<4000 ppm	14,000
River Raisin, MI	Hot spots	10 ppm PCBs	29,000
Sheboygan, WI	Hot spots	10 ppm PCBs (?)	5,900
Waukegan, IL	Up to 500,000 ppm (hot spots)	50 ppm PCBs	32,000

Source: Summarized and modified from Cushing (1998).

TABLE 17.26. Approximate costs of remediation (modified from Cushing, 1998 and Romagnoli, 1998)

Location	Total costs ($ million)	Total unit cost ($/yd^3)	Dredging costs ($ million)	Dredging unit costs ($/yd^3)
Bayou Bonfouca, LA	115	71	2.1	120 or 132 (WODCON)
Black River, OH	5	80	—	—
Cold Spring, NY	9–11	115–140	—	35
Grass River	4.9	1,635	—	—
LTV Steel, IN	12	105	—	—
Manistique River, MI	15.5 (three years)	178	—	—
Massena, NY (ALCOA)	4.9	1,670	—	—
Massena, NY (General Motors)	10	714	—	—
New Bedford Harbor, MA	20.1	1,430	1.7	124
River Raisin, MI	5.4	186	—	—
Sheboygan River, WI	2.6 (without disposal)	684 (without disposal)	—	—
Waukegan Harbor, IL	21 (in water and on land)	420	—	36–40

APPENDIX 1. CANADIAN SEDIMENT QUALITY GUIDELINES

Environment Canada (Guidelines Division, Science Policy and Environmental Quality Branch) develops Canadian sediment quality guidelines for the protection of aquatic life as part of its obligations under the Canadian Environmental Protection Act (CEPA).

The Act dictates the Canadian government's responsibilities regarding monitoring activities, substance assessments, pollution prevention and control strategies, and regulatory activities (e.g., ocean disposal of dredged sediments). These national sediment quality guidelines are developed cooperatively with the provincial and territorial governments through the Water Quality Guidelines Task Group of the Canadian Council of Ministers of the Environment (CCME). CCME is a joint federal, provincial, and territorial council committed to intergovernmental cooperation on environmental matters in Canada.

National sediment quality guidelines for chemical substances, which are developed using toxicological information, represent concentrations of individual chemicals below which adverse biological effects are not expected. They are developed with the intention to be conservative, national benchmarks (i.e., reference points) to protect and sustain aquatic life. These resource-use-based guidelines provide scientifically defined measures to evaluate the status of, and progress toward, societal goals for the maintenance, protection, and remediation of environmental quality (Gaudet et al., 1995).

Although Canadian sediment quality guidelines provide a nationally consistent, scientific basis for management decisions, such as the development of substance-, site-, or issue-specific objectives or standards, they do not directly incorporate management considerations (e.g., cost and technological limitations) nor are they intended to serve directly as management objectives without due consideration of such factors. Therefore, effective implementation of national sediment quality guidelines requires that the distinction between generic guidelines and site-specific objectives be recognized within a broader decision-making framework.

In Canada, sediment quality guidelines are developed using a nationally approved protocol (CCME, 1995) to ensure consistency, transparency, and scientific defensibility in the process. Sediment quality guidelines technical documents for a number of individual chemicals and groups of substances are being developed by the Environment Canada Guidelines Division. The document *Canadian Sediment Quality Guidelines for Cadmium* is available. The draft Environment Canada document, *Proposed Interim Canadian Sediment Quality Guidelines for the Protection of Aquatic Life,* will be available once an internal department review is complete.

Excerpts from the *Protocol for the Derivation of Canadian Sediment Quality Guidelines for the Protection of Aquatic Life* (CCME, 1995) follow (Appendix 2).

APPENDIX 2. THE USE OF SEDIMENT QUALITY GUIDELINES AS BENCHMARKS

Glossary

ASTM American Society for Testing and Materials
BEDS Biological Effects Database for Sediments
CCREM Canadian Council of Resource and Environment Ministers
CCME Canadian Council of Ministers of the Environment
EC_{50} median effective concentration

ERL	effects range low
ERM	effects range median
ISQG	interim sediment quality guideline
LC_{50}	median lethal concentration
LOEL	lowest-observed-effect level
NC	no concordance
NE	no effect
NG	no gradient
NOEL	no-observed-effect level
NSTP	National Status and Trends Program
PEL	probable effect level
SG	small gradient
SQG	sediment quality guideline
SSTT	spiked-sediment toxicity test
TEL	threshold effect level
TOC	total organic carbon

Preface

Canadian sediment guidelines for the protection of aquatic life are being developed under the auspices of the Canadian Council of Ministers of the Environment (CCME). Sediment quality issues have become an important focus in the environmental assessment, protection, and management of aquatic ecosystems. Historically, water quality activities were motivated by concerns for human health (e.g., drinking water quality guidelines; Health and Welfare Canada, 1993), but attention has shifted in recent years towards the protection of other components of the ecosystem (e.g., sediments, soil) and other water uses. These water uses include freshwater and marine aquatic life, recreation and aesthetics, irrigation and livestock watering, and industrial water supplies. In Canada, acceptable water quality for the protection of these uses has been evaluated against the Canadian Water Quality Guidelines (CCREM, 1987). Sediment quality guidelines for the protection of aquatic life will be used in a complementary manner to evaluate sediment quality.

The Canadian water quality guidelines were adopted on the basis of a review and evaluation of existing water quality guidelines from other jurisdictions (CCREM, 1987). In some cases, scientific information was used to modify the guidelines so that they were applicable to Canadian conditions. Guidelines were not recommended for the parameters for which guidelines from other jurisdictions were deemed inappropriate or for which scientific data were lacking. Recently, a formal protocol has been developed for the consistent derivation of numerical water quality guidelines for the protection of freshwater and marine aquatic life (CCME, 1991a) and for the protection of agricultural water uses (CCME, 1993). Similarly, interim soil quality guidelines have been adopted for 60 substances, and a formal protocol for their refinement is currently being developed (CCME, 1991b, 1994; Environment Canada 1991).

Water quality guidelines play an important role in protecting water uses and in assessing the impact of environmental contaminants on the quality and uses of aquatic resources. Sediment quality guidelines are also important because sediments have a profound influence on the health of aquatic organisms, which may be exposed to chemicals through their immediate interactions with bed sediments. Therefore, the use of sediment quality guidelines for evaluating the toxicological significance of sediment-associated chemicals has become an important part of the protection and management of freshwater, estuarine, and marine ecosystems.

In 1989, the CCME Environmental Protection Committee mandated the responsibility of establishing Canadian sediment quality guidelines (SQGs) to the CCME Task Group on Water Quality Guidelines. The Evaluation and Interpretation Branch, Ecosystem Conservation Directorate, of Environment Canada provides scientific and technical support to the Task Group on the development of these guidelines. Sediment quality guidelines can be used to help set targets for sediment quality that will sustain aquatic ecosystem health for the long term, they are required to support the interpretation of sediment chemistry data and the overall assessment of sediment quality conditions within the context of specific water uses, and they support the development of site-specific objectives.

The use and interpretation of the terms criteria, guidelines, objectives, and standards vary among different agencies and countries. For the purposes of this document, these terms are defined as follows:

Criteria—The scientific data that are evaluated to derive sediment quality guidelines.

Guidelines—Numerical limits or narrative statements recommended to support and maintain designated uses of the aquatic environment.

Objectives—Numerical limits or narrative statements that have been established to protect and maintain designated uses of the aquatic environment at a particular site.

Standards—Sediment quality objectives that are recognized in enforceable environmental control laws of one or more levels of government.

These definitions are consistent with those used in the discussion of Canadian water quality guidelines (CCREM, 1987). The term sediment refers to the bottom deposits in aquatic environments that are composed of particulate material (of various sizes, shapes, mineralogy) from various sources (e.g., terrigenous, biogenic, authigenic).

In response to the identified need for SQGs in Canada, Environment Canada commissioned a study in 1988 (MacDonald et al., 1992) to review and evaluate the available approaches used to develop such guidelines. The document also provided an extensive compilation of existing sediment quality assessment values from around the world. The approaches reviewed included those of the

- Sediment background
- Spiked-sediment toxicity test
- Water quality guidelines
- Interstitial water toxicity
- Equilibrium partitioning
- Tissue residue
- Benthic community structure assessment
- Screening level concentration
- Sediment quality triad
- Apparent effect threshold
- International Joint Commission sediment assessment strategy
- National Status and Trends Program

MacDonald et al., (1992) provided a brief description of the methodology of each of the approaches, their major advantages and limitations, and their current uses. Numerous other reviews of these approaches have been published (Beak Consultants, 1987, 1988; Chapman, 1989; Sediment Criteria Subcommittee, 1989; Adams et al. 1992; Persaud et al., 1992; Lamberson and Swartz, 1992; Long and MacDonald, 1992).

A preliminary evaluation indicated at that time that no single approach was likely to fully support the immediate need for national, scientifically defensible guidelines, as well as the long-term need for guidelines that explicitly consider the factors that influence the toxicity of sediment-associated contaminants. Until a formalized protocol was developed, MacDonald et al. (1992) recommended that effect-based sediment quality assessment values from other jurisdictions be evaluated for their applicability to Canadian conditions, modified with existing scientific data (if necessary) to increase their applicability to Canadian conditions, and adopted as interim sediment quality guidelines if deemed suitable. This recommendation parallels the initial CCME strategy followed for adopting interim water quality and soil quality guidelines.

Further to the recommendations of MacDonald et al. (1992), a study was commissioned by Environment Canada to validate and update the National Status and Trends Program database (developed by the National Oceanic and Atmospheric Administration), which contained information on the biological effects of sediment-associated contaminants. The results of these initiatives provided a sound basis for developing a formal protocol (presented in Chapter 1*) for the development of national SQGs for use in Canada.

The formal protocol established for the derivation of numerical SQGs is applicable to the protection of both freshwater and marine (including estuarine) aquatic life associated with bed sediments (separate guidelines are derived for each of these systems). The protocol relies mainly on the National Status and Trends Program approach, with the complementary use of the spiked-sediment toxicity test approach in the future, once methodological concerns have been resolved. Because the development of SQGs relies on current scientific information, they will be refined as new and relevant data become available.

Sediment quality guidelines are developed from the available scientific information on the biological effects of sediment-associated chemicals. These tools are intended to provide guidance to provincial, federal, territorial, and nongovernmental agencies involved in the protection, assessment, and management of sediment quality. SQGs provide a scientific review of the existing toxicological information for a chemical, which can be used to support the establishment of sediment quality objectives to protect aquatic life, which are developed to reflect a number of site-specific considerations. The recommended SQGs may be employed as nationally consistent screening tools; however, variations in environmental conditions across Canada will affect sediment quality in different ways. Complementary information on background concentrations of natural substances is evaluated during the development of SQGs and should be considered during their implementation and the development of site-specific sediment quality objectives. Impairment of sediments of superior quality to national guideline concentrations should not be advocated.

This document focuses on the procedures to be used in deriving national SQGs for the protection of aquatic life. National SQGs are currently being developed on a chemical-by-chemical basis through the CCME Task Group on Water Quality Guidelines. This document also provides introductory' guidance on how these guidelines are intended to be used with other types of information, and will be followed by other documents providing national guidance specific to the implementation of SQGs and the development of site-specific sediment quality objectives.

The Use of Sediment Quality Guidelines as Benchmarks

Introduction. Canadians have begun to recognize that developmental activities represent potential hazards to the health and integrity of their aquatic ecosystems. Fisheries

---

*Cross-references refer to CCME, 1995.

closures on shellfish and advisories on finfish consumption in the vicinity of pulpmills in British Columbia provide graphic examples of the potential social and economic impacts that can result from the release of chemicals into the environment (M. Nassichuk, Department Fisheries and Oceans, Vancouver, B.C., personal communication, 1992). The conservation and protection of our aquatic resources has become a high priority goal, and management efforts are focusing on reducing inputs of toxic substances into the environment as well as cleaning up contaminated areas to restore the quality of degraded ecosystems (CEPA, 1988; Government of Canada, 1990).

Sediments are important in influencing the fate of chemicals in aquatic ecosystems and they provide habitat for many benthic and epibenthic organisms. Concerns regarding the protection and management of sediment quality have raised important questions about the toxicological significance of sediment-associated chemicals and their potential to impair the designated uses of aquatic environments. Therefore, SQGs for the protection of aquatic life are needed to provide relevant benchmarks to help address these concerns.

SQGs are only one of the many scientific tools available to help in the protection and management of sediment quality. They can be used to help interpret whether existing or predicted sediment quality conditions pose a threat to benthic organisms. The use of SQGs in exclusion of other information (such as background concentrations of naturally occurring substances, other assessment values such as the PEL, or biological tests) can lead to erroneous conclusions or predictions about sediment quality. Therefore, SQGs and all other relevant information should be considered, to support practical and informed decision-making regarding sediment quality. These considerations are equally important whether the focus is to maintain, protect, or improve sediment quality conditions at a particular site.

This chapter is intended to provide a brief generic example of the use of national SQGs along with other relevant information. The use of SQGs as national benchmarks provides a broadly applicable example for many potential users of these guidelines and does not imply the preclusion of other specific uses that have not been discussed (such as their use with site-specific information and contaminant transport models to help evaluate discharge limits, their use to help focus the cleanup of contaminated sites, or as the scientific basis for developing site-specific objectives). Potential application scenarios of SQGs, as well as guidance on when or how to proceed with various management options, will be provided in future documents.

The following guidance is given within the context of a general sediment assessment framework, which integrates the types of information that should be considered along with SQGs. Specific management options, such as defining site-specific objectives to maintain and protect sediment quality conditions or defining remediation objectives to improve sediment quality conditions, cannot be appropriately chosen without an understanding of ambient sediment quality. The framework provides environmental managers with a consistent process for using SQGs with other relevant information to assess sediment quality (i.e., to help answer the question of whether existing or predicted concentrations of chemicals in sediment pose a hazard to sediment-associated organisms).

Application of Sediment Quality Guidelines as Benchmarks. Sediment quality guidelines can be used as benchmarks for evaluating sediment chemistry information to identify situations that may be harmful to aquatic organisms associated with bed sediments. They can also be used as benchmarks to help set targets for sediment quality, within broader management strategies, that will sustain aquatic ecosystem health for the long term. The latter use will be discussed more fully in a future document. For sediments of superior quality, "impairment" to the national SQGs should not be advocated.

In using SQGs as benchmarks, adverse biological effects are not predicted when the measured concentrations of sediment-associated chemicals at a site are at or below the national SQGs. (Note that the term site in the context of this chapter is meant to be generic, whether it refers to a region, a basin, a specific site, or a given quantity of sediment.) Further investigation of sediment quality at the site is usually not necessary, but may be warranted under some circumstances (e.g., when sediments at the site have low levels of TOC, when other variables are suspected to be increasing the bioavailability of chemicals, or when SQGs do not exist for particular chemicals that are measured in the sediment). The potential for observing adverse biological effects is recognized when the concentration of one or more sediment-associated chemicals is greater than the national SQG, with the incidence and severity of these effects generally increasing with increasing chemical concentrations (Long et al., 1994; see also Appendix A).

A Generic Example. The context of a general sediment assessment is used in the following sections to illustrate how SQGs function as benchmarks and how they should be used with other complementary information. A general framework providing an example of this is outlined in Figure 17.A2 (note that this framework is not intended to replace accepted sediment-testing protocols or monitoring programs that have already been established). Depending on the goals of the assessment, various tools (e.g., SQGs, biological tests) and site-specific information (e.g., background concentrations of naturally occurring substances) can be integrated to achieve the desired goals. The components of this framework are briefly discussed in the following sections.

In a broad context, the initial phase of a sediment quality assessment should involve the identification of sediment quality issues and concerns for the areas that are primarily associated with (existing and/or potential) point and nonpoint sources of contaminants. The goal of the assessment also needs to be clearly defined, incuding its affirmation of the broader environmental management goal (of which sediment quality is only one component) that has been defined for the region. A regional assessment of sediment quality may be required to determine the relative conditions of ambient sediments at a number of sites, with the aim of prioritizing and focusing future activities. In contrast, the assessment of sediment quality at a specific site may need to characterize site conditions more comprehensively in order to implement specific management options (such as the maintenance and/or protection of ambient sediment quality conditions, the development of site-specific objectives, or the potential remediation of an area). Besides being very different in geographical scale, the complexity of regional and site-specific sediment quality assessments may vary depending on the situation. The types of tools and information used in the assessment is ultimately the choice of the environmental manager. The following sections provide a brief discussion on the various assessment tools available and the types of site-specific information that should be considered when assessing sediment quality.

Water and Land Use Activities. The first step in the framework involves reviewing available information on existing and potential land and water use for the site(s) under consideration. Information on past, present, and future industries and businesses in the area; the location of wastewater treatment plants; land use activities in upland areas; stormwater drainage systems; and residential developments are important. Such information provides a basis for identifying historical, current, and potential sources of contaminants to the aquatic ecosystem. Information on the existing and potential chemical composition of point and nonpoint source discharges of contaminants, on the water quality conditions (in many cases, based on comparisons with water quality guidelines), and on the physical/chemical properties of these substances provides a basis for identifying potential sediment chemistry concerns at the site(s). Subsequently, a list of substances of potential concern can be compiled. This site-specific information on the past, present, and future uses of the site(s) provides a basis for making decisions regarding the nature and

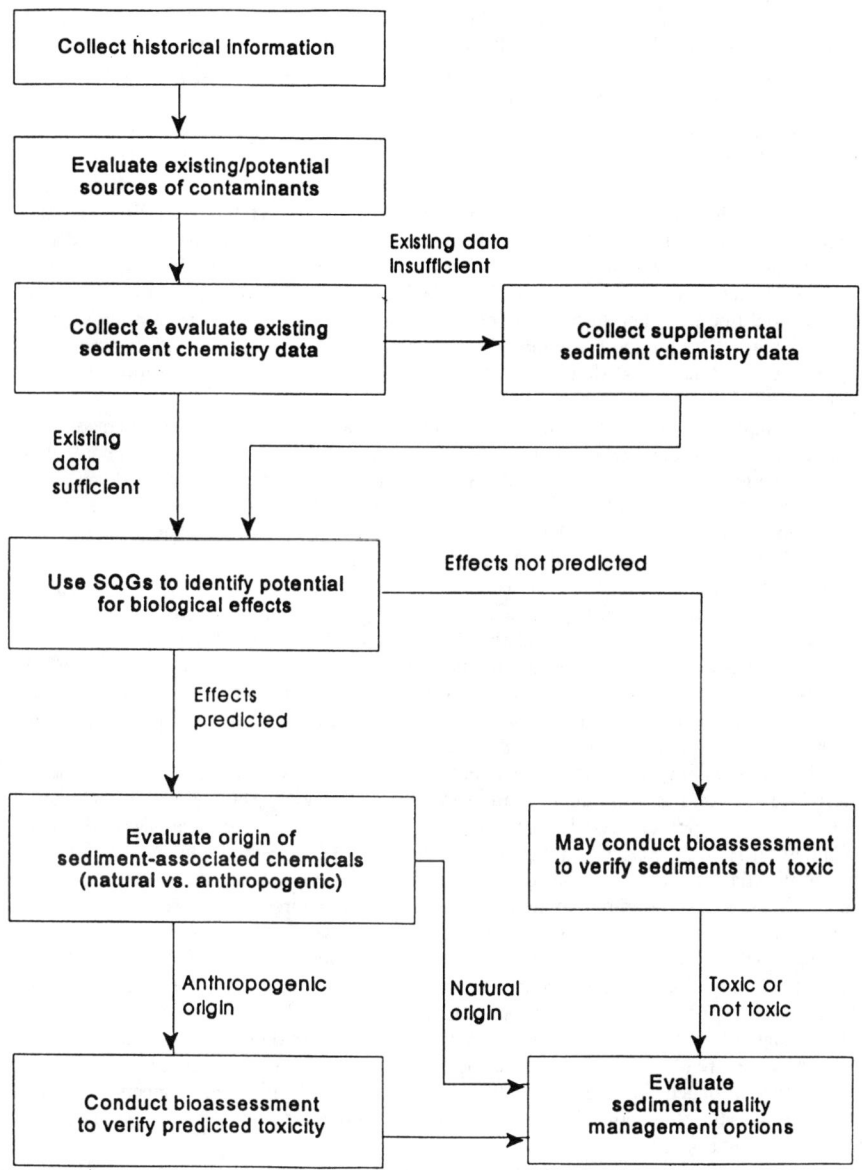

FIGURE 17.A.2. A framework for the assessment of sediment quality.

extent of the investigations that should be conducted. More detailed descriptions of the type of information that should be collected and how such data can be used to assess ambient sediment quality is reviewed elsewhere (Baudo and Muntau, 1990; Mudroch and MacKnight, 1991; MacDonald, 1993).

Existing Sediment Chemistry Data. All of the available sediment chemistry data for the site(s) under consideration should be assembled to initiate a preliminary assessment of sediment quality conditions. The applicability of these data must be fully evaluated to determine the overall quality of the existing data set, the degree to which the data are thought to represent the current conditions at the site(s), and the degree to which they address the issues and concerns identified. In addition, available information on other physical and chemical characteristics of the sediment (e.g., TOC, particle size distribution) should be compiled in conjunction with the data on chemical concentrations. Such factors may influence the bioavailability (and hence the toxicity) of sediment-associated chemicals, and the site(s) should be assessed to determine the distinct conditions that exist.

An important consideration in the evaluation of the quality of the available sediment chemistry data is the quality assurance/quality control measures that were implemented during collection, transport, and analysis of sediment samples. Conventional practices have recently been established that provide guidance on the field aspects of sediment sampling programs (ASTM, 1990a; U.S. EPA and ACE, 1991; Mudroch and MacKnight, 1991; Environment Canada, 1994a). A diversity of analytical procedures have been developed to quantify concentrations of chemicals in sediments, and acceptable methods have been reported (U.S. EPA and ACE, 1991). More novel analytical procedures may be evaluated based on the reported accuracy and precision of the technique (i.e., the results of analyses performed on standard reference materials, split samples, or spiked-sediment samples). The detection limits reported must be relevant to assessing the potential for biological effects at the site. The applicability of the detection limits may be assessed by comparing them to the SQGs recommended for that substance.

The applicability of existing sediment chemistry data should also be evaluated in terms of temporal and spatial variability in sediment quality. The age of the chemistry data is an important consideration. Natural degradative processes (Mosello and Calderoni, 1990), meterological events (such as storms) that result in the transport of sediments (Allan, 1986), industrial developments, and regulatory activities may alter the sources and composition of chemicals released into the environment. As a result of these processes, historical chemical data may not be representative of existing conditions. The list of variables for which chemical analyses are completed should incorporate knowledge of the potential contaminants from land and water use activities in the area. Since the chemistry of bed sediments may vary spatially (Mah et al., 1989; Long and Morgan, 1990; Hakanson, 1992), data from a number of stations are required to provide a representative picture of sediment quality conditions in an area. The actual number of stations required will depend on the size of the area under consideration, the concentrations of sediment-associated chemicals, the variability of chemical concentrations, and the overall goal of the assessment.

If existing sediment chemistry data are considered acceptable, the potential for observing adverse biological effects at the site(s) can be assessed. If the sediment chemistry data are considered to be of unacceptable quality or are not considered to adequately represent the site(s), additional sediment chemistry data should be collected.

Supplemental Sediment Chemistry Data. A focused, well-designed field survey must be implemented to collect the types of supplemental sediment chemistry data that will support the goal of the assessment. The initial list of substances of potential concern defined during the evaluation of the land and water use activities provides a defensible means of identifying substances for inclusion in a field survey. The field survey should be designed to delineate temporal and spatial variability in sediment quality and should explicitly outline the quality assurance/quality control measures that will be implemented. Collection, handling, and storage of sediment samples should follow established protocols (ASTM, 1990a; Environment Canada, 1994a). Analytical methods and detection limits should be appropriate for the substances under consideration.

The Potential for Adverse Biological Effects. The next step in the framework for assessing sediment quality involves determining the potential for observing adverse biological effects at the site(s) under consideration. Information on concentrations of chemicals in sediments provides essential data for evaluating the nature and spatial extent of sediment chemistry. However, these data alone do not provide a measure of adverse biological effects or an estimate of the potential for such effects. SQGs are used to determine whether sediment-associated chemicals are present at concentrations, with the potential to impair the aquatic life at that site(s).

As described in Chapter 1, a weight-of-evidence approach is used to derive national SQGs. A second sediment quality assessment value, the PEL, can also be derived using the guideline derivation tables (MacDonald, 1993). The PEL represents the lower limit of the range of chemical concentrations that are usually or always associated with adverse biological effects. The national SQG and the PEL are used to define three ranges of chemical concentrations for a particular chemical, those that are rarely (SQG), occasionally (between the SQG and the PEL), and frequently ($\geq$ PEL) associated with adverse biological effects (see Appendix A; MacDonald 1993; Long et al. 1994). The quantification of the incidence of biological effects within each of these concentration ranges provides a useful tool for estimating the probability of observing similar adverse effects within the defined concentration ranges of particular chemicals. Therefore, the frequency with which and degree to which measured sediment chemical concentrations at a site fall within each of these concentration ranges are useful to distinguish sites and chemicals of little toxicological concern, of potential toxicological concern, or significantly hazardous to exposed organisms.

Sediments with measured chemical concentrations that are equal to or lower than the national SQGs are considered to be of acceptable quality. In general, further investigations of these sediments would be of relatively low priority. Management options at these sites would focus on the protection of existing sediment quality conditions. However, in some cases, biological testing may be required for validation of this conclusion (for example, in sediments with low levels of TOC, when other variables are suspected to be increasing the bioavailability of sediment-associated chemicals, or when SQGs do not exist for particular chemicals that are measured in the sediment).

Sediments with measured chemical concentrations between the national SQG and the PEL are considered to represent potential hazards to exposed organisms. Although adverse biological effects are possible within this range of concentrations, their occurrence, nature, and severity are difficult to reliably predict on an a priori basis. Specific conditions at these sites are likely to control the expression of toxic effects. Further investigations on these sediments are needed to determine whether sediment-associated chemicals represent significant hazards to aquatic organisms. Such investigations may include the determination of background concentrations for naturally occurring substances and/or a series of biological tests designed to evaluate the toxicological significance of particular chemicals (with respect to key species of aquatic biota and factors at the site that may be influencing the bioavailability of the chemical).

Sediments with measured chemical concentrations equal to or greater than the PEL are considered to represent significant and immediate hazards to exposed organisms. Sediments with concentrations of one or more chemicals that fall within this range of concentrations should be considered to be of the highest priority for appropriate management actions to improve sediment quality and restore the desired level of protection, if necessary. Biological assessment is recommended at these sites to determine the nature and extent of effects that are being manifested as a result of the sediment-associated contaminants.

Background Concentrations of Natural Substances. The determination of background concentrations of naturally occurring substances is important when adverse biological effects have been predicted using SQGs (i.e., measured chemical concentrations at a site are > SQG). Information on background concentrations of natural substances is used to determine the extent to which human activities have contributed to the concentrations of sediment-associated chemicals measured at a site, and is particularly important for metals and certain organic substances that may be enriched through natural processes. Although natural levels of these substances may have an adverse effect on certain organisms, defensible management options should consider the contribution of natural processes in order to focus on the sites and chemicals that are primarily influenced by human activities.

An interpretive tool has been developed that provides an effective means of distinguishing the probable origin (i.e., natural vs. anthropogenic) of many metals in sediments (Loring, 1990, 1991; Schropp and Windom, 1988; Schropp et al., 1989; Schropp et al., 1990; Loring and Rantala, 1992). This method involves determining the ratio of metal concentrations to those of a reference element. Because such ratios are relatively constant in the earth's crust, they can be used to interpret the degree of anthropogenic enrichment of metals at other locations. This method has been demonstrated only for some marine sediments, however, and its applicability to freshwater sediments is unknown (but should be evaluated).

The development and use of this interpretive tool involves intensive sampling at a number of uncontaminated sites within a region. For example, data on sediment metal concentrations were collected from roughly 100 sites throughout Florida, sites thought to be representative of natural estuarine areas in the state (Schropp and Windom, 1988). These data were then used to develop correlations between log-transformed concentrations of various metals and those of a reference element (such as lithium, iron, or aluminum). Simple linear regressions were performed on these data and the 95% confidence limits were calculated. In Florida, significant correlations with aluminum concentrations were obtained for arsenic, cadmium, chromium, copper, lead, nickel, and zinc. These relationships provided the basis for interpreting data on the concentrations of metals in sediments at various sites, such that anthropogenic enrichment of metal levels was suspected when metal-to-aluminum ratios exceeded the upper 95% confidence limits. Subsequent evaluations have confirmed the effectiveness and utility of this interpretive tool at a variety of locations (in the St. Lawrence estuary and the Gulf of the St. Lawrence [Loring, 1991], and in the United States [Schropp et al., 1989; Schropp et al., 1990; MacDonald 1993]). Loring (1990, 1991) has established similar correlations between concentrations of lithium and several metals. Lithium appears to be most appropriate for the normalization of metal data from sediments derived from glacial erosion of crystalline rocks, which are common in Canada (Loring 1990, 1991).

Since synthetic organic contaminants are released into the environment only as a result of human activities, equivalent tools for distinguishing their probable origin (i.e., natural vs. anthropogenic) are needed for only those substances (such as polycyclic aromatic hydrocarbons) that have important natural sources. Concentrations of naturally occurring organic substances from essentially uncontaminated sediments at reference sites far from point-source discharges could be used as a guide to defining surrogate background concentrations of these substances. For freshwater sediments, alternative methods for evaluating the origin (i.e., natural vs. anthropogenic) of sediment-associated chemicals may include the choice of an appropriate reference area that is unaffected by point-source discharges or of the "pre-colonial" sediment horizon for determining the background concentrations of natural substances (Persaud et al., 1992).

Biological Assessment. Biological testing is an important component of a sediment quality assessment. The nature and extent of the available information on the effects of sediment-associated chemicals in North America is such that most of the data do not support the establishment of cause-and-effect relationships (most of the data compiled in BEDS support "associative" relationships only, and the quantity of spiked-sediment bioassay data for any given chemical is limited). The relationships between characteristics of the sediment and/or overlying water column (e.g., TOC, pH) and observed adverse effects also need to be defined to the extent that the relative importance of these modifiers of sediment toxicity are predictable under field conditions. Therefore, there is some level of uncertainty associated with predicting toxicological effects in the field (although this uncertainty can be evaluated to a great extent through the calculation of the incidence of effects within defined concentration ranges of chemicals; see Appendix A). Biological tests used in conjunction with chemical analyses of sediments can provide definitive information regarding the toxicity of sediment-associated chemicals under a wide variety of circumstances.

Further investigation involving biological testing is recommended to support a sediment quality assessment when the concentrations of one or more chemicals are higher than the national SQGs and established background concentrations of natural substances. Biological testing may also be used to assess the toxicity of sediments that may contain unmeasured chemicals or that have distinct physical characteristics (e.g., low levels of TOC). It may also be necessary to consider sensitive species representative of the site(s). Such studies can be used to assess the applicability of the national SQGs to the site conditions and will contribute site-specific toxicological information that will support the development of sediment quality objectives for the site. Biological testing should be performed at sites where sediment conditions are considered to be significant and immediate hazards to exposed organisms (i.e., with measured chemical concentrations equal to or greater than the PEL) to determine the nature and extent of effects that are being manifested at these sites.

A number of tests have been developed to evaluate the toxicological significance of sediment contamination. These tests range in complexity from spiked-sediment bioassays (which study a single contaminant and a single species) to microcosm studies (which investigate the long-term effects of chemical mixtures on ecosystem dynamics). In addition, tests may be designed to assess the toxicity of whole sediments (solid phase), suspended sediments, elutriates, sediment extracts, or pore (interstitial) water. The organisms that are routinely tested include microorganisms, algae, aquatic macrophytes, invertebrates, and fish (Schiewe et al., 1985; Burton and Stemmer, 1988; ASTM, 1990b,c, E.V.S. Consultants, 1990; U.S. EPA and ACE, 1991; Burton, 1991; Phipps et al., 1993). While requirements for biological tests differ among applications, sediment toxicity tests should follow established or approved methods (such as those established by provincial, federal, or national agencies). Such methods may be modified to assess the toxicity to resident species, toxicity over longer periods (i.e., to address chronic toxicity), or for different end points. However, the basic principles of accepted protocols should be followed.

Whole-sediment bioassays are the most relevant for assessing the effects of chemicals associated with bottom sediments. Environment Canada (1992a) has developed 10-day toxicity tests using six Canadian species of sediment-burrowing amphipods (*Amphiporeia virginiana, Corophium volutator, Eohaustorius estuarius, Foxiphalus xiximeus, Leptocheirus pinguis,* and *Rhepoxynius abronius*). Likewise, the American Society for Testing and Materials (ASTM, 1990b) has developed and approved four tests for assessing the toxicity of marine and estuarine sediments to four species of amphipod. These bioassays may be modified to assess toxicity to other benthic invertebrate species that occur in estuarine and marine environments, including other amphipods, other crustaceans, polychaetes, and bivalves (ASTM, 1990b). A 20-day sublethal test for polychaetes and a 10-day sublethal test for mussels in whole sediment are also under development by Environment Canada (J. Osborne, Office of Waste Management, Environment Canada, Ottawa, personal communication, 1993). The ASTM is also considering procedures for conducting sediment toxicity tests with polychaetes and echinoderms (Ingersoll, 1991). Similar techniques have also been developed to assess the toxicity of sediment-associated chemicals in freshwater (ASTM, 1990c; Burton, 1992; Burton et al., 1992). Environment Canada (R. Scroggins, Technology Development Branch, Environment Canada, Ottawa, personal communication, 1993) is developing growth inhibition/survival tests for freshwater amphipods (*Hyalella azteca*), chironomids (*Chironomus tentans/C. riparius*), and mayflies (*Hexagenia* sp.).

In addition to whole-sediment toxicity tests, various procedures are available for assessing the potential for adverse effects on aquatic organisms due to the resuspension of sediments or the partitioning of chemicals into the water column. The bacterial luminescence or Microtox® test is frequently used (Burton and Stemmer, 1988; Environment Canada, 1992b). Tests using algae, invertebrates, and fish have also been employed to assess the toxicity of the suspended and/or aqueous phases. Environment Canada (1992c) provides guidance on the use of an echinoderm fertilization assay for testing the toxicity of sediment pore water or elutriate and is developing a similar test for oyster larvae. The use of oyster and echinoderm embryos and larvae in sediment toxicity testing of marine sediments is currently under evaluation by ASTM (Ingersoll, 1991). In addition, procedures for conducting water column bioassays and bioaccumulation tests have been recommended by the U.S. EPA and ACE (1991) and Lee et al. (1989), and a document on sediment resuspension testing is under consideration by ASTM.

Other types of biological information may also be used in the sediment quality assessment process. For example, comparison of the diversity and abundance of benthic invertebrate communities at test sites with appropriate reference sites (e.g., sites with similar particle size distributions, TOC) provides a means of assessing the relative toxicity of test sediments (Diaz, 1992; La Point and Fairchild, 1992; Persaud et al., 1992; Reynoldson and Zarull, 1993). Various statistical procedures may be used to help identify the chemicals that are associated with the observed biological effects when adequate sediment chemistry data are available. In addition, spiked-sediment bioassays may be used to establish cause-and-effect relationships for specific substances or mixtures of chemicals (Swartz, 1987; Burton, 1991; U.S. EPA, 1992). Information on levels of chemicals in aquatic biota may also provide a basis for determining the significance of chemical levels in sediments relative to the protection of the health of wildlife consumers of aquatic organisms.

Management Options. Along with the information obtained through a preliminary sediment quality assessment, management options are evaluated against the sediment quality guidelines or objectives. A number of management options are possible whether the ultimate goal is to maintain, protect, or improve sediment quality. For sites where further investigation is not warranted, appropriate management options should be chosen to protect existing sediment quality at the site. Continued monitoring of sediment quality conditions and evaluation against national SQGs will provide a means of identifying changes in sediment quality that may lead to problems (which can then be addressed proactively). Other possibilities include continued monitoring for the assessment of trends or the development of sediment quality objectives that address distinct characteristics of the site. The identification of priority chemicals and sites of concern during the sediment quality assessment can focus further biological investigations on potential problem areas or Identified areas of immediate concern. At sites that are seriously contaminated, some remedial action (including source control measures) may be necessary to achieve environmental management goals. These remedial actions could include removal and treatment of toxic materials, isolation of contaminated sediments, or no action at all (i.e., permitting natural degradative and sedimentation processes to mitigate contaminant effects; Sullivan and Bixby, 1989).

Many other factors may also influence the sediment quality management strategies that are ultimately implemented at a site. These factors include the management goals for the site, the nature and severity of the contamination, the potential for exposure of aquatic organisms, availability and costs of remediation technologies, unique characteristics that should be preserved at the site, public expectations, and other political, social, and economic factors. The integration of such information challenges the environmental manager to formulate an effective management strategy that will address the sediment quality issues and concerns that have been identified.

REFERENCES AND SELECTED READINGS

Adams, W. J., R. A. Kimerle, and J. W. Barnett, Jr. 1992. Sediment quality and aquatic life assessment. *Environ. Sci. Technol.* 26:1865–1875.
Allan, R. J. 1986. The role of particulate matter in the fate of contaminants in aquatic ecosystems. Scientific Series No. 142. Inland Waters Directorate, National Water Research Institute, Burlington, Ontario. 128 pp.
Anonymous. 1998. Jan De Nul's Environmental Approach, *World Dredging, Mining, and Construction,* pp. 8–9, 25–26, August.
Anonymous. 1997. Remediation Technologies, Great Lakes 2000, Environment Canada, March.
ASTM (American Society for Testing and Materials). 1990a. Standard guide for collection, storage, characterization, and manipulation of sediments for toxicological testing. A5TM Designation: E 1391–90. 15 pp.
ASTM. 1990b. Standard guide for conducting 10-day static sediment toxicity tests with marine and estuarine amphipods. ASTM Designation: E 1367–90. 24 pp. A5TM. 1990c. Standard guide for conducting sediment toxicity tests with freshwater invertebrates. ASTM Designation E 1383–90.
Baudo, R. and H. Muntau. 1990. Lesser known in-place pollutants and diffuse source problems. In *Sediments: Chemistry and Toxicity of In-Place Pollutants,* ed. R. Baudo, J. Giesy, and H. Muntau, pp. 1–14. Boca Raton, Fla.: Lewis Publishers, Inc.
Beak Consultants Ltd. 1987. Development of sediment quality objectives: Phase I—options. Prepared for Ontario Ministry of Environment. Mississauga, Ont.
Beak Consultants Ltd. 1988. Development of sediment quality objectives: Phase II—guidelines development. Prepared for Ontario Ministry of Environment. Mississauga, Ont.
Buchberger, C., Santiago, R., and Orchard, I. 1993. Contaminated sediment removal program with a view to the future. Proc. 26th Dredging Seminar, Texas A&M University, College Station, TX.
Burton, G. A. 1991. Assessing the toxicity of freshwater sediments. *Environ. Toxicol. Chem.* 10:1585–1627.
Burton, G. A. 1992. Plankton, macrophyte, fish, and amphibian toxicity testing of freshwater sediments. In *Sediment Toxicity Assessment,* ed. G. A. Burton, pp. 167–182. Boca Raton, Fla.: Lewis Publishers, Inc.
Burton, G. A., and B. L. Stemmer. 1988. Evaluation of surrogate tests in toxicant impact assessments. *Toxic. Assess.* 3:255–269.
Burton, G. A., M. K. Nelson, and C. G. Ingersoll. 1992. Freshwater benthic toxicity tests. In *Sediment Toxicity Assessment,* ed. G. A. Burton, pp. 213–240. Boca Raton, Fla.: Lewis Publishers, Inc.
CCME (Canadian Council of Ministers of the Environment). 1991 a. A protocol for the derivation of water quality guidelines for the protection of aquatic life. In *Canadian Water Quality Guide-*

lines, App. 9. 1987. Prepared by the Task Force on Water Quality Guidelines, Ottawa, Ont. 8 pp.
CCME. 1991 b. Interim Canadian environmental quality criteria for contaminated sites. Report CCME EPC-C534. Prepared by the CCME Subcommittee on Environmental Quality Criteria for Contaminated Sites, Winnipeg, Man. 20 pp.
CCME. 1993. Protocols for the derivation of water quality guidelines for protection of agricultural water uses. Eco-Health Branch, Ecosystem Science and Evaluation Directorate, Environment Canada, Ottawa, Ont.
CCME. 1994. A protocol for the derivation of ecological effects-based and human health-based soil quality criteria for contaminated sites. CCME Subcommittee on Environmental Quality Criteria for Contaminated Sites, Winnipeg, Man. Second draft.
CCME (Canadian Council of Ministers of the Environment). 1995. *Protocol for the Derivation of Canadian Sediment Quality Guidelines for the Protection of Aquatic Life,* Report CCME EPC-298E. Prepared by the Technical Secretariat of the Water Quality Guidelines Task Group, Winnipeg, Manitoba, 38 p.
CCREM (Canadian Council of Resource and Environment Ministers). 1987. Canadian water quality guidelines. Task Force on Water Quality Guidelines, Ottawa, Ont.
CEPA (Canadian Environmental Protection Act), S. C. 1988. c. 22 (now R. S. C. 1985 [fourth supp], c. 16).
Chapman, P. M. 1989. Current approaches to developing sediment quality criteria. *Environ. Toxicol. Chem.* 8:589–599.
Clausner, J. E. 1996. Potential Application of Geosynthetic Fabric Container for Open Water Placement of Contaminated Dredged Material, Technical Note EEDP-01-39, U.S. Army Engineer Waterways Experiment Station, Vicksburg, MS.
Cushing, B. S. 1998. Remediation of Sediments by Dredging Methods and Case Histories, *Proceedings 15th World Dredging Congress,* World Organization of Dredging Associations, Las Vegas, NV, June/July.
Diaz, R. J. 1992. Ecosystem assessment using estuarine and marine benthic community structure. In *Sediment Toxicity Assessment,* ed. G. A. Burton, pp. 67–85. Boca Raton, Fla.: Lewis Publishers, Inc.
DPC. 1997. Solution for Contaminated Silt Disposal. April.
Elliott, G. M. 1992. Risk Assessment and Contaminated Sites, *Superfund Risk Assessment in Soil Contamination Studies,* ASTM STP 1158, K.B. Hoddinott (Ed.), American Society for Testing and Materials, Philadelphia, PA.
Environment Canada. 1991. Review and recommendations for Canadian interim environmental quality criteria for contaminated sites. Scientific Series No. 197. Inland Waters Directorate, Water Quality Branch. Ottawa, Ont.
Environment Canada. 1992a. Biological test method: acute test for sediment toxicity using marine or estuarine amphipods. Report EPS 1/RM/26. Environmental Protection, Conservation and Protection. Ottawa, Ont.
Environment Canada. 1992b. Toxicity test using luminescent bacteria (Photobacterium phosphoreum). Report EPS 1/RM/24. Environmental Protection, conservation and Protection. Ottawa, Ont.
Environment Canada. 1992c. Fertilization assay with echinoids (sea urchins and sand dollars). Report EPS 1/RM127. Environmental Protection, Conservation and Protection. Ottawa, Ont.
Environment Canada. 1994a. Guidance for the collection, handling, transport, storage and manipulation of sediments for chemical characterization and toxicity testing. Draft report. Environmental Protection, Conservation and Protection. Ottawa, Ont.
Environment Canada. 1994b. Guidance on control of test precision using a spiked control sediment toxicity test. Draft report. Environmental Protection, Conservation and Protection. Ottawa, Ont.
E.V.S. Consultants. 1990. Review of sediment monitoring techniques. Prepared for the Ministry of Energy, Mines and Petroleum Resources, B.C. Acid Mine Drainage Task Force. 87 pp.
Fowler, J., Sprague, D. J. and Toups, D. 1994. Dredged Material-Filled Geotextile Containers, *Environmental Effects of Dredging,* Technical Notes, U.S. Army Engineer Waterways Experiment Station, Vicksburg, MS.
Francingues, N. R. 1984. Management Strategy for Disposal of Dredged Material, *Environmental Effects of Dredging,* Vol. D-84-6, U.S. Army Engineer Waterways Experiment Station, Vicksburg, MS.

Francingues, N. R., Palermo, M. R., Peddicord, R. K. and Lee, C. R. 1985. Management Strategy for the Disposal of Dredged Material: Contaminant Testing and Controls, Miscellaneous Paper EL-85-1, U.S. Army Engineer Waterways Experiment Station, Vicksburg, MS.
Gaudet, C. L., K. A. Keenleyside, R. A. Kent, S. L. Smith, and M. P. Wong. 1995. How Should Numerical Criteria be Used? The Canadian Approach. *Human and Ecological Risk Assessment* $1(1)$:19–28.
Gilmur, R. C., and D. D. Saathoff. 1994. Sediment Remediation and Navigational Improvements Within the Port of Tacoma, Washington, *Proc. WEDA XV Technical Conference and 27th Annual Texas A&M Dredging Seminar,* J. B. Herbich (Ed.), San Diego, CA, May.
Government of Canada. 1990. Canadian Environmental Protection Act. Report for the period ending March 1990.
Hakanson, L. 1992. Sediment variability. In *Sediment Toxicity Assessment,* ed. G. A. Burton, pp. 19–35. Boca Raton, Fla.: Lewis Publishers, Inc.
Hart Crowser. 1987. Sediment Sampling Assessment, Final Report for Proposed Tacoma Terminals Pier Extension, Port of Tacoma, Washington.
Hart Crowser. 1990. Sediment Quality Assessment Final Report for Proposed Pier 7-D Extension.
Hayes, D. F., McLellan, T. N., and Truitt, C. L. 1988. Demonstration of innovative and conventional dredging equipment at Calumet Harbor, Illinois. MP-EL-88-1. U.S. Army Engineer Waterways Experiment Station, Vicksburg, MS.
Hayes, D. F. and Crockett, T. R. 1994. Modeling Near-Field Resuspensions in Cutterhead Suction Dredging Operations, 6th International Symposium on the Interactions Between Sediment and Water, Santa Barbara, CA, December 5–8, 1993.
Health and Welfare Canada. 1990. Biological safety factors in toxicological risk assessment. Environmental Health Directorate, Health Protection Branch.
Health and Welfare Canada. 1993. Guidelines for Canadian drinking water quality. 5th ed. Ottawa, Ont.
Herbich, J. B. (Ed.) 1990–1992. *Coastal and Ocean Engineering,* Vols. 1–3, McGraw-Hill, New York.
Herbich, J. B. 1992. *Handbook of Dredging Engineering,* McGraw Hill, New York.
Herbich, J. B. 1993a. Dredging Equipment for the Removal of contaminated Sediment State-of-the-Art, *6th International Symposium on the Interactions Between Sediments and Water,* Santa Barbara, CA, December 5–8.
Herbich, J. B. 1993b, Review of New Bedford Harbor Pilot Study—Removal of Contaminated Sediments, *Proceedings 26th Dredging Seminar,* Texas A&M University, pp. 25–43, May.
Herbich, J. B. 1995. Removal of Contaminated Sediments: Equipment and Recent Field Studies, in *Dredging, Remediation and Containment of Contaminated Sediments,* K. R. Demars, et al., Eds., ASTM Publication Code Number (PCN) 04-012930-38.
Hummer, C. W., Jr. 1998. DOER: A Major Dredging Research Programme. *Terra et Aqua,* No. 72, pp. 13–17, September.
Ingersoll, C. 1991. Sediment toxicity and bioaccumulation testing: E47. 03 develops standard guides for evaluating the toxicity and bioaccumulation of contaminants in sediment to aquatic organisms. Standardization News 19(4):28–33.
Iwaskai, M., Kuioka, K., Izumi, S., and Miyata, N. 1992. High Density Dredging and Pneumatic Conveying System, *Proceedings, XIIIth World Dredging Congress, WODCON XIII,* Bombay, India, pp. 773–792, April.
Ketelmeer Feature (1996), *Dredging and Port Construction.*
Lamberson, J. O., and R. C. Swartz. 1992. Spiked-sediment toxicity test approach. In *Sediment Classification Compendium.* EPA 823-R-92-006. U.S. EPA, Office of Water, Washington, D.C.
La Point, T. W., and J. F. Fairchild. 1992. Evaluation of sediment contaminant toxicity: the use of freshwater community structure. In *Sediment Toxicity Assessment,* ed. G. A. Burton, pp. 87 110. Boca Raton, Fla.: Lewis Publishers, Inc.
Lee, H., B. L. Boese, J. Pelletier, M. Windsor, D. T. Specht, and R. C. Randall. 1989. Guidance manual:, bedded sediment bioaccumulation tests. EPA/600/X-90/302. Environmental Protection Agency. Newport, Oreg. 232 pp.
Long, E. R., and D. D. MacDonald. 1992. National Status and Trends Program approach. In *Sediment Classification Methods Compendium.* EPA 823-R-92-006. U.S. EPA, Office of Water, Washington, D.C.
Long, E. R., and L. G. Morgan. 1990. The potential for biological effects of sediment-sorbed con-

taminants tested in the National Status and Trends Program. National Oceanic and Atmospheric Administration Tech. Memo. NOS OMA 52. Seattle, Wash. 175 pp. + app.

Long, E. R., D. D. MacDonald, S. L. Smith, and F. D. Calder. 1994. Incidence of adverse biological effects within ranges of chemical concentrations in marine and estuarine sediments. *Environ. Manage.* In press.

Loring, D. H. 1990. Lithium—a new approach for the granulometric normalization of trace metal data. *Mar. Chem. 29*:155–168.

Loring, D. H. 1991. Normalization of heavy-metal data from estuarine and coastal sediments. *ICES J. Mar. Sci. 48*:101–115.

Loring, D. H., and R. T. T. Rantala. 1992. Manual for the geochemical analysis of marine sediments and suspended particulate matter. *Earth-Sci Rev. 32*:235.

MacDonald, D. D. 1993. Development of an approach to the assessment of sediment quality in Florida coastal waters. Prepared for the Florida Department of Environmental Protection. MacDonald Environmental Sciences, Ltd., Ladysmith, B.C. Vol. 1, 128 pp. ; Vol. 2, 117 pp.

MacDonald, D. D, S. L. Smith, M. P. Wong, and P. Mudroch. 1992. The development of Canadian marine environmental quality guidelines. Report prepared for the Interdepartmental Working Group on Marine Environmental Quality Guidelines and the Canadian Council of Ministers of the Environment. Environment Canada, Ottawa, Ont. 50 pp. + app.

Mah, F. T. S., D. D. MacDonald, S. W. Sheehan, T. N. Tuominen, and D. Valiela. 1989. Dioxins and furans in sediments and fish from the vicinity of ten inland pulp mills in British Columbia. Water Quality Branch. Environment Canada. Vancouver, B.C. 77 pp.

Malins, D. C., B. B. McCain, D. W. Brown, A. K. Sparks, and H. O. Hodgins. 1980. Chemical Contaminants and Biological Abnormalities in Central and Southern Puget Sound, *NOAA Technical Memorandum.*

Malins, D. C., S. L. Chan, B. B. McCain, D. W. Brown, A. K. Sparks and H. O. Hodgins. 1981. Puget Sound Pollution and Its Effects on Marine Biota.

Marine Board. 1997. *Contaminated Sediments in Ports and Waterways,* National Research Council, National Academy Press, Washington, DC.

Marine Board. 1989. *Contaminated Marine Sediments—Assessment and Remediation,* National Research Council, National Academy Press, Washington, DC.

Mason & Hanger-Silas Mason, Inc., Lexington, Kentucky. 1981. An engineering study for the removal and disposition of PCB contamination in the Waukegan Harbor and North Ditch at Waukegan, Illinois. (Unpublished report, January)

McLellan, T. N., Havis, R. N., Hayes, D. F. and Raymond, G. L. 1989. Field Studies of Sediment Resuspension Characteristics of Selected Dredges, Technical Report HL-89-9, U.S. Army Engineer Waterways Experiment Station, Vicksburg, MS.

Mosello, R., and A. Calderoni. 1990. Pollution and recovery of Lake Orta (northern Italy). In *Sediments: Chemistry and Toxicity of In-Place Pollutants,* ed. R. Baudo, J. Geisy, and H. Muntau. Chelsea, Mich.: Lewis Publishers, Inc.

Mudroch, A., and S. D. MacKnight, eds. 1991. *CRC handbook of techniques for aquatic sediment sampling.* Boca Raton, Fla.: CRC Press.

Myers, T. E., and Zappi, M. E. 1992. Laboratory Evaluation of Stabilization/Solidification Technology for Reducing the Mobility of Heavy Metals in New Bedford Harbor Superfund Site Sediment, in *Stabilization and Solidification of Hazardous, Radioactive and Mixed Wastes,* Vol. 2, T. M. Gilliam and C. C. Wiles (Eds.) ASTM STP 1123, American Society of Testing and Materials, Philadelphia, PA.

Orchard, I. 1992. Toronto Harbour Contaminated Sediment Removal Demonstration, Environment Canada, September.

Palermo, M. R. 1991a. Design Requirements for Capping, Dredging Research Technical Notes, DRP 05-03, U.S. Army Engineer Waterways Experiment Station, Vicksburg, MS.

Palermo, M. R. 1991b. Site Selection Considerations for Capping, Dredging Research Technical Notes, DRP-5-04, U.S. Army Engineer Waterways Experiment Station, Vicksburg, MS.

Palermo, M. R. 1991c. Equipment and Placement Techniques for Capping, Dredging Research Technical Notes, DRP-5-05, U.S. Army Engineer Waterways Experiment Station, Vicksburg, MS.

Palermo, M. R. 1993. Technical Framework for Environmental Evaluation of Dredged Material Management Alternatives, *Environmental Effects of Dredging,* Vol. D-93-4, October.

Pelletier, J. P., and Leaney, A. 1992. Hamilton Harbor Removal and Treatment Demonstration, Envi-

ronmental Screening Document, Environmental Protection, Environment Canada, Toronto, Ontario, Canada.
Persaud, D., R. Jaagumagi, and A. Hayton. 1992. Guidelines for the protection and management of aquatic sediment quality in Ontario. Water Resources Branch, Ontario Ministry of the Environment, Toronto. 26 pp.
Phipps, G. L., G. T. Ankley, D. A. Benoit, and V. R. Mattson. 1993. Use of the aquatic oligochaete Lumbriculus variegatus for assessing the toxicity and bioaccumulation of sediment-associated contaminants. *Environ. Toxicol. Chem.* 12:269–279.
Pilarczyk, K. W. 1994. *Novel Systems in Coastal Engineering Geotextile System and Other Methods: An Overview,* Rijkwaterstaat, Delft, the Netherlands.
Port of Tacoma. 1992a. Sitcum Waterway Remediation Project, Phase 1 Pre-Remedial Design Evaluation and Phase 2 Preliminary Evaluation of Remedial Options Reports, Port of Tacoma, September.
Port of Tacoma. 1992b. Data Report, Sitcum Waterway Pre-Remedial Design, Phase 1 and Phase 2 Areas, Port of Tacoma.
Port of Tacoma. 1992c. Data Report, Sitcum Waterway Pre-Remedial Design, Natural Resource Data Report.
PSDDA. 1988. Evaluation Procedures Technical Appendix. Sampling, Testing, and Test Interpretation of Dredged Material Proposed for Unconfined, Open-Water Disposal in Central Puget Sound.
Reynoldson, T. B., and M. A. Zarull. 1993. An approach to the development of biological sediment guidelines. In *Ecological Integrity and the Management of Ecosystems,* ed. S. Woodley, J. Kay, and G. Francis, pp. 177–200. Delray Beach, Fla.: St. Lucie Press.
Richardson, T. W., Hite, J. E., Shafer, R. A., and Ethridge, J. D. 1982. Pumping Performance and turbidity Generation of Model 600/100 Pneuma Pump, TR HL-82-8, U.S. Army Engineer Waterways Experiment Station, Vicksburg, MS.
Romagnoli, R, Van Dewalker, H. M., Doody, J. P. and Anckner, W. H. 1998. The Future Challengers on Environmental Dredging, *Proceedings 15th World Dredging Congress,* Vols. 1 and 2, WODA, Las Vegas, NV, June 28–July 1.
Sato, E. 1984. Bottom Sediment Dredge CLEAN UP, *Proceedings, 8th U.S. Japan Experts Meeting, Principles and Results, Management of Bottom Sediments Containing Toxic Substances,* T. T. Patin, Ed., U.S. Army Engineer Waterways Experiment Station, Vicksburg, MS, pp. 403–418.
Schiewe, M. H., E. G. Hawk, D. I. Actor, and M. M. Krahn. 1985. Use of a bacterial bioluminescence assay to assess toxicity of contaminated marine sediments. *Can. J. Fish. Aquat. Sci.* 42:1244–1248.
Schropp, S. J., and H. L. Windom. 1988. A guide to interpretation of metal concentrations in estuarine sediments. Coastal Zone Management Section, Florida Department of Environmental Regulation. Tallahassee, Fla. 44 pp. + app.
Schropp, S. J., F. D. Calder, L. C. Burney, and H. L. Windom. 1989. A practical approach for assessing metals contamination in coastal sediments—an example from Tampa Bay. In Proc. Sixth Symposium on Coastal and Ocean Management, July 11–14, 1989. American Society of Civil Engineers. Charleston, S.C.
Schropp, S. J., F. G. Lewis, H. L Windom, J. D. Ryan, F. D. Calder, and L. C. Burney. 1990. Interpretation of metal concentrations in estuarine sediments of Florida using aluminum as a reference element. *Estuaries* 13(3):227–235.
Sediment Criteria Subcommittee. 1989. Review of the apparent effects threshold approach to setting sediment criteria. Report of the Science Advisory Board. U.S. EPA. Washington, D.C. 18 pp.
Snyder, G. W., Ponton, J. R. and Desring, P.W. 1997. Dredging with Environmental Controls, Baltimore, MD Inner Harbor, Dredging and Management of Dredged Material, Geotechnical Special Publication No. 65, ASCE, New York, NY.
Sturgis, T. and Gunnison, D. 1984. A Procedure for Determining the Cap Thickness for Capping Subaqueous Dredged Material Deposits, Technical Note EEDP-01-09, U.S. Army Engineer Waterways Experiment Station, Vicksburg, MS.
Sullivan, J., and A. Bixby. 1989. A citizen's guide: cleaning up contaminated sediment. Lake Michigan Federation, Ann Arbor, Mich. 32 pp.
Swain, L. G., and R. A. Nijman. 1991. An approach to the development of sediment quality objec-

tives for Burrard Inlet. Proc. 17th Annual Aquatic Toxicity Workshop: November 5–7, 1990, Vancouver, B.C., vol. 2, ed. P. Chapman, F. Bishay, E. Power, K. Hall, L. Harding, D. McLeay, M. Nassichuk, and W. Knapp. *Can. Tech. Rep. Fish. Aquat. Sci. 1774,* vol. 2. 12 pp.

Swartz, R. C. 1987. Toxicological methods for determining the effects of contaminated sediments on marine organisms. In *Fate and Effects of Sediment-Bound Chemicals in Aquatic Systems.,* ed. K. L. Dickson, A. W. Maki, and W. A. Brungs. Elmsford, N.Y.: Pergamon Press.

Taylor, A. 1999. Bean Dredging Corporation, personal communication.

Tetra Tech Co. 1985. Commencement Bay nearshore/tideflats remedial investigation. Prepared for Washington State Department of Ecology and USEPA.

Tetra Tech Co. 1988. Commencement Bay nearshore/tide flats feasibility studies. December.

Truitt, C. L. 1987a. Engineering Considerations for Capping Subaqueous Dredged Material Deposits—Background and Preliminary Planning, *Environmental Effects of Dredging,* Technical Note EEDP-01-3, U.S. Army Engineer Waterways Experiment Station, Vicksburg, MS.

Truitt, C. L. 1987b. Engineering Considerations for Capping Subaqueous Dredged Material Deposits—Design Concepts and Placement Techniques, *Environmental Effects of Dredging,* Technical Note EEDP-01-4, U.S. Army Engineer Waterways Experiment Station, Vicksburg, MS.

USACE. 1984. Waukegan Harbor, Illinois, Confined Dredge Disposal Facility, Chicago District, April.

USACE. 1987. Confined Disposal of Dredged Material—Engineering Manual, EM-1110-2-5027, U.S. Army Corps of Engineers, Washington, DC.

USACE/EPA. 1992. Evaluating Environmental Effects of Dredged Material Management Alternatives—A Technical Framework, EPA-842/B-92/008, Washington, DC.

USEPA. 1984. Conceptual Design. OMC Hazardous Waste Site, Waukegan, Illinois, EPA 13-5M28.0, September 14.

USEPA. 1989. Commencement Bay Nearshore/Tideflats Record of Decision.

USEPA. 1989. *Exposure Factors Handbook,* Office of Health and Environmental Assessment, Washington, DC.

USEPA. 1991a. Administrative Order on Consent for Remedial Design Study.

USEPA. 1991b. Statement of Work for the Remedial Design and Remedial Action for the Sitcum Waterway Problem Area and Other Areas of the Commencement Bay Nearshore/Tideflats Superfund Site.

USEPA. 1992. Review of sediment criteria development methodology for non-ionic organic contaminants. Prepared by the Sediment Quality Subcommittee of the Ecological Processes and Effects Committee. EPA-SAB-EPEC-93-002. U. S. EPA, Science Advisory Board, Washington, D.C. 12 pp.

USEPA and ACE (United States Environmental Protection Agency and Department of the Army, Army Corps of Engineers). 1991. Evaluation of dredged material proposed for ocean disposal. EPA-503/8-91/001. USEPA, Office of Water, Washington, D.C.

USEPA. 1993. Selecting Remediation Technique for Contaminated Sediment, Office of Water, EPA-823-b93-001, Environmental Protection Agency, Washington, DC.

USEPA. 1994. Assessment and Remediation of Contaminated Sediments (ARCS) Program, Remediation Guidance Document, Great Lakes National Program Office, EPA 905-R94-003, Environmental Protection Agency, Chicago, IL.

USEPA. 1997a. Contaminated Sediment News, EPA-823-N-97-006.

USEPA. 1997b. The Incidence and Severity of Sediment Contamination in Surface Waters of the United States, Vols. 1,2,3, EPA 23-R-97-006, 007 and 008, Washington, DC.

USEPA 2000. Manistique site update No. 8.

Valent, P. J. and Young, D. K. 1995. Technical and Economic Assessment of Storage of Industrial Waste on Abyssal Plain, paper presented to the Marine Board, John C. Stennis Space Center, MS.

Van Drimmellen, C. and Schut, T. 1992. New and Adapted Small Dredges for Remedial Dredging Operations, *WODCON XIII,* Bombay, India, pp. 156–169, April.

Yoshinaga, K. 1995. Mercury Contaminated Sludge Treatment by Dredging in Minamato Bay, *Dredging, Remediation and Contaminated Sediments,* K. R. Demars, et al., Eds., ASTM Publication Code Number (PCN) 04-012930-38.

CHAPTER 18
MARINE AGGREGATE DREDGING

HENRY BOKUNIEWICZ
Marine Science Research Center
State University of New York
Stony Brook, New York

Introduction	18.1
Methodology	18.2
Demand for Mined Aggregates	18.3
Environmental Issues	18.6
Management	18.7
References	18.8

INTRODUCTION

Offshore, submerged marine aggregates consist primarily of sand and gravel, but mined marine resources can include boulders, clays, marl, shell, and shingle. Marine aggregates have been mined in most countries of northern Europe, as well as in Canada, the United States, Japan, and the city of Hong Kong (Figure 18.3). Japan recovers 80 million tons (44 million m^3) per year, making her the leader in world production (Narumi, 1989, as cited in Smith and Collis, 1993). The United Kingdom and the Netherlands are other large producers. Over 15% of the demand for mined aggregates in United Kingdom are met with offshore resources. This can amount to almost 24 million tons (19 million m^3) per year. Denmark also obtains 10 to 15% of its needs from offshore resources (ICES, 1992).

The reasons for turning to marine aggregates can be both economic and environmental. In some regions, land-based deposits may coincide with good agricultural land but in many cases development pressures are incompatible with active land-based sand mining because of noise, dust, and transport traffic (Smith and Collis, 1993). Land-based operations unavoidably alter landscape, change terrestrial habitats, and can threaten groundwater. Marine aggregate mining, however, must also be concerned with alterations of the benthic habitat and with potential conflicts with fishing activity, navigation, and coastal development. In Sweden, for example, marine aggregates had been extracted at rates in excess of 500,000 m^3 per year but all mining has stopped since the deposits are now in a marine nature reserve around the Falsterbo Peninsula. In Japan, also, fisheries interests tend to restrict mining to areas beyond 5 or 10 km offshore. In Hiroshima, offshore mining recently has been prohibited. In Canada, a policy of "no net loss" of fisheries habitat discourages attempts to develop offshore mining. In the late 1970s and early 1980's, about 7 million m^3 were being mined annually from the Canadian Atlantic Region

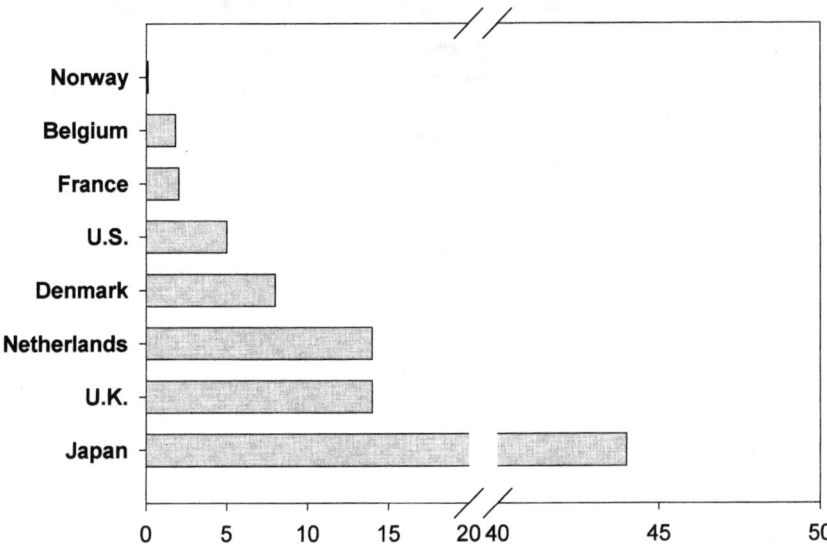

FIGURE 18.1. Estimated annual extraction of marine aggregates within the past decade (millions of m³). Values can fluctuate widely from year to year, in part because extractions are often used for large public works projects.

(deGroot, 1986); this extraction has ceased in recent years, although mining is still done in Arctic waters associated with oil and gas production.

The cost of starting a marine aggregate dredging operation may be high, but the economy of scale is favorable for several reasons. One reason is that marine sites can be worked at high capacity. This is an advantage especially for construction projects done at coastal locations or at sea, for which a rapid delivery rate is required in order to take advantage of short periods of favorable conditions and reduce the risks to construction during adverse weather. Water-borne transport of high-volume, bulk material, such as sand and gravel, is most economical, and marine sites can be close to the areas of heaviest demand. Most of the world's heavily developed areas are coastal. Thirteen of the world's 15 largest cities are on or near the coast.

METHODOLOGY

The mining of aggregates from the sea floor is commonly referred to as dredging because dredges are used for this purpose. Any type of dredge can, and has, been used for marine mining, including cutter–suction dredges, bucket dredges, dipper dredges, and clamshell dredges. The choice of equipment depends not only on the nature of the sediment but also on wind, wave, and current conditions, water depth, volume to be excavated, use of the aggregate, and the availability of equipment. All dredges are suitable for sand but dredgeability tends to decrease with more coarse materials. Self-propelled, self-contained plants can tolerate the widest range of wave conditions. A trailing suction hopper dredge, for example, can reasonably be operated in waves up to at least 2.5 m in height (Groothuizen,

1998). Dredges typically work deposits less than 30 m deep but can reach material in excess of 40 m deep.

Clamshell dredges have been used extensively in Japan and the United States. They are generally employed in sheltered shallow waters close to shore, but operations in water depths to 275 m have been reported (Packer, 1987). Since they are stationary when dredging, they may pose navigational problems, and operations using a clamshell dredge tend to leave behind an irregular, pitted sea floor.

Often, a suction dredge or a trailer dredge is used. A moving dredge seems most appropriate for water depths less than 25 m, whereas deeper dredging can usually be accommodated with a stationary dredge (Smith and Collis, 1993). A stationary suction dredge has been used to mine sand and gravel in water 80 m deep (Packer, 1987). The Dutch have recently employed a "pinpoint" dredge to extract sand from small areas in shallow water. The suction dredge is anchored over the deposits to be recovered and creates large pits in the sea floor as it extracts material. Borrow pits from a single anchor location may be 20 m in depth below the ambient sea floor and perhaps 75 m in diameter. Repeated anchoring, however, may result in a pit, or series of pits, covering hundreds of acres. In addition to extractions in the Netherlands, sand and gravel mining in New York and in Denmark has been done by this method (Bokuniewicz, 1988; ICES, 1992). Trailer suction dredges extract sand and gravel by suction while under way. Such an operation tends to form furrows on the sea floor about 0.3 m deep and 1–4 m across. They can operate continuously and at a high efficiency over large areas of a smooth sea floor. Much of the mining of marine sands and gravel in the United Kingdom is conducted in this way (Hitchcock and Drucker, 1996).

Except where environmental concerns prohibit it, overflow of the barge or hopper is allowed to wash out silt and clay from the load. Usually, however, most washing of the load occurs at the shoreside facility (Smith and Collis, 1993). On-board screening can also be carried out at sea as the aggregate is loaded, in order to maintain a specific product composition. The unwanted size fractions are returned to the sea floor on site.

Suction dredges will commonly unload through a hydraulic pipeline. This method has the advantage of a high discharge rate (2000 tons/hour) so that the unloading time is minimized and the dredge can spend most of its time recovering sand. If the receiving site is close by, sand may be sprayed or "rainbowed" directly to the desired location. Split-hull barges or bottom-dump barges loaded by dipper, bucket, or clamshell dredges may be emptied directly over submerged sites. Otherwise, barges may be unloaded by a hydraulic pump. Alternatively, grab cranes or a drag line may be used to unload the aggregate (Figure 18.2; Uren, 1989). Additional processing at a shoreside facility may include washing with fresh water to remove fines and salts. Although the solid aggregate does not contain salts, the sea water does, so that the salt content of unwashed aggregate may make it unacceptable for use in concrete production. The use of marine aggregates for building material in Norway, for example, has declined over the last 15 years to less than 100,000 m^3 per year due primarily to problems with salt. Washing is commonly done through polymer mats supported on screens (Smith and Collis, 1993). The product may then be screened and blended with other sand or crushed rock to achieve the desired specification.

DEMAND FOR MINED AGGREGATES

Marine aggregates, meeting the usual specifications, can be used in all engineering applications. These include needs for mortar sand, grout sand, cushion sand used for concrete block and slope paving, mineral filler in bituminous concrete mixtures, blasting sand, fill

FIGURE 18.2. Unloading the *Sandy Hook*, a trailing-suction hopper dredge, by means of drag-line buckets at a shoreside facility (courtesy of Amboy Aggregates, a subsidiary of Great Lakes Dredge and Dock Company).

sand for roadways, filter sand, and concrete aggregate. About 70% of the marine aggregate extracted in the United Kingdom is used for construction, mainly the production of concrete. Most of Japan's production is also used for concrete (Smith and Collis, 1993). The sole commercial operation in the United States using marine aggregate provides about one million m^3 per year, making it the leading supplier of construction aggregate in the New York metropolitan area. Much of this material is used for highway construction.

One of the earliest, and still the most extensive, uses of marine sands is for beach replenishment. Offshore sand having been exposed to wave reworking is especially suitable for constructing a beach that is "natural" in both texture and color. In the Netherlands, beach nourishment projects require an average of 5 million m^3 per year taken from the Dutch section of the North Sea. This is about one-third of the total sand extraction in the Netherlands. About three million m^3 of sand are used for beach nourishment in Denmark each year. In the United States, between two and six million m^3 typically are extracted annually from offshore sources for beach renourishment projects. These are primarily along the Atlantic and Gulf coasts, with some in southern California. Occasionally, very large projects can be added to these totals. A single project at Miami Beach in the United States required 13 million m^3 of sand. Nineteen million m^3, were placed in a single project in the Netherlands. The largest single beach nourishment project in the world to date is being done along the New Jersey coast, involving some 24 million m^3 of marine sand. It is interesting to note that, despite the great volume of marine aggregate mining done in Japan, beach nourishment projects there sometimes use imported sand, for example, from Australia, at great expense because of the limitations of national reserves.

Other forms of coastal reclamation also use marine aggregates. Gravel, sand, or even mud can be used as backfill behind sheet piling to create landfill for commercial, recreational, residential, or agricultural uses. This use of marine aggregate is often seen in large public works projects. The construction of Newark Airport in 1950 and its expansion in 1963 utilized over 13 million m³ of marine sand. (Bokuniewicz, 1988) More than 9 million m³ of sand fill and till were used in Denmark between 1989 and 1993 for the construction of the Great Belt bridge-and-tunnel project. In 1989, 7.7 million m³ were extracted for use in reclamation projects in Denmark. Each restoration project along the Jutland west coast typically requires one to one-and-a-half million m³ per year (ICES, 1992). Over nine million m³ were recovered for reclamation projects in the Netherlands in 1989. Offshore resources have been used for over 40 years in Hong Kong, primarily for major civil engineering projects and for reclamation. The Kia Tak Airport in Hong Kong (Figure 18.3) was constructed on 7.5 million cubic meters of marine fill, and 8 million m³ were used in the construction of a container terminal there (Guilford, 1988). The Kansai International Airport in Japan was also constructed largely from sand mined from the floor of the Seto Inlet Sea. Honolulu airport was constructed of marine sand.

Worldwide, marine aggregates have found uses in a wide variety of other types of reclamation projects. Marine sand and gravel mining in Canada, for example, is primarily confined to the Beaufort Sea for the construction of drilling islands for offshore oil and gas production. In the 1970's and 1980's, dozens of large islands were contructed, each requiring about one million m³ of sand and gravel. Artificial gravel islands also are constructed from marine aggregate for oil and gas production in Alaskan waters.

Artificial islands have been constructed for the containment of contaminated sediments. These can be relatively large. Craney Island in Virginia covers 2,500 acres and has

FIGURE 18.3. An artist's impression of Hong Kong's Kia Tak Airport. The dredging part of the contract is thought to be one of the largest operations ever carried out in the world, involving 240 million cubic meters (314 million cubic yards) of marine sediments and fill. The contract created a 1,248 hectare (3,083 acre) island stretching 3.5 km (2.8 miles) into the sea.

been in use for over 35 years for the disposal of dredged sediment. (Palermo et al., 1981) Hart-Miller Island in Chesapeake Bay near Baltimore covers 1,110 acres. Artificial islands are constructed for industrial uses in densely populated regions such as Hong Kong and Japan; an example is the industrial island in Kobe, Japan, which covers 940 acres. Other examples off the coast of Japan are used for ventilating undersea coal mines. Artificial reefs and breakwaters are also constructed not only for shoreline protection but also to promote fishing activity. In Scandinavia, boulders (glacial erratics) are extracted from the sea floor by special vessels called "stone fishers" expressly for use in the construction of breakwaters and revetments. Marine aggregate can also be used as a base for creating wetlands in mitigation for losses due to other development.

Carbonate material is mined for diverse uses. About 15% of the extraction of marine aggregate in France, or about 500,000 tonnes per year, is comprised of marl and calcareous sand from Brittany (ICES, 1992). In Norway, 100,000 to 200,000 tonnes of marine carbonate sands are extracted annually. This material is mainly used for fertilizer (STIL, 1995). In the Dutch Wadden Sea and western Scheldt, shell has been extracted for the construction, for example, of bicycle paths and drainage beds for houses (ICES, 1995). In the Bahamas, aragonite has been mined by cutter–suction dredge. Aragonite is used in the production of cement and glass, agricultural liming, and buffering of acid wastes.

ENVIRONMENTAL ISSUES

Environmental concerns and the avoidance of user conflicts can be as important as engineering considerations in sand mining operations. There are three principal physical considerations for sand mining operations. These are (1) the removal of substrate and the alteration of sea floor bathymetry, (2) the generation of excess turbidity, and (3) redeposition of unwanted sediment either from turbid plumes or screening. The relative importance of these impacts varies depending on the particular operation.

Anchored dredging creates large borrow pits on the sea floor. Since these features are not in equilibrium with the ambient sea floor, sedimentation tends to occur with the pits filling more or less rapidly with sediment of a finer grain size than that originally present. The type of benthic animals in an area depends, in part, on the sediment type, so that any change in substrate has direct consequences for the composition of benthic faunal communities. If muddy sediments accumulate, the change could also result in oxygen depletion. In addition, the presence of pits nearshore will alter wave patterns and could result in shoreline erosion. In some cases, the recommended procedure is to confine mining to water depths beyond which sand is not moved by waves (or beyond the "closure depth"), in order to avoid impacts on the shoreline due to the disturbance of wave patterns. This may be in excess of 30 m in the open ocean offshore, as, for example, off the coast of Japan, taxing the capabilities of the dredging equipment and limiting the extent of potentially suitable deposits. In shallower water where there is concern, forecasts of wave conditions over the projected, altered bathymetry may be required. In the United Kingdom, for example, this is done routinely using numerical, wave refraction, and diffraction models.

Trailing suction dredges tend to have less intense impacts over a greater area of the sea floor. The furrows produced by single passes of a trailing dredge are unlikely to alter the bathymetry enough to seriously alter wave conditions, although the cumulative impacts of repeated mining in one area may change wave refraction and diffraction patterns. Furrows left by trailing dredges can be obliterated in less than a year if ambient sediment transport is high. Where bottom sediments are not actively transported by waves and currents, however, dredged furrows may persist for many years, slowly filling with fine-grained sediment.

Turbidity is generated both at the dredge head and from hopper overflow or screening. The importance of turbidity plumes depends both on the amount of fine-grained sediment in the dredged deposit and on the ambient suspended sediment concentrations. Turbidity plumes, however, tend to be both local and transient. (i.e., Hitchcock and Drucker, 1996) Near the dredge, turbidity may increase 100-fold, but returns to background levels quickly when dredging ceases. Redeposition is concentrated initially in the dredging area, although resettled fine-grained sediment can be subject to further resuspension and dispersion until it ultimately accumulates in depressions where the current is weak or in areas of natural mud accumulation. The discharge of screened material could, in principle, convert a gravel substrate to one of mobile sand or, alternatively, armor the sea floor with immobile, coarser-grained material. The potential impacts of these activities on the marine ecosystem depends on many site-specified factors, such as the ability of currents to transport material, the sediment mobility, the ambient turbidity and bottom morphology, as well as on the extraction practice.

Chemical impacts due to aggregate dredging are likely to be minor, since the favored sediment deposits are coarse-grained with low organic content and only trace amounts of fines. The dredging operation, however, will liberate pore water to be mixed with the ambient seawater along with dissolved nutrients or contaminants. The resuspension of fine-grained particles and organic components during dredging, overflow of the hoppers, or on-site washing could potentially add to the water column dissolved oxygen demand. Studies of dredging operations in fine-grained sediments, a worse-case scenario, have shown that any such effects are both local and transient.

Excavation of marine aggregates unavoidably disturbs the benthic biological community. Organisms in the reach of the dredge are removed. A succession of new communities develops on the sea floor after the disturbance is past. Anchored dredging produces an intense, localized disturbance. The change in bathymetry, redeposition of substrate, and circulation in borrow pits when the operation is finished will likely result in the reestablishment of a benthic community that is substantially different from that which was removed. Trailer-suction dredging, on the other hand, affects a larger surface area, but the sea floor is left in a condition more similar to the ambient, undredged state. In this case, the dredged furrows would be recolonized by the ambient forms. The rate of recolonization in both cases depends on the season, the supply of larvae, and other conditions. It may be as short as a few months to as long as several years. In warmer climates the recolonization may be rapid, but, in some instances, the new community may never completely revert to the ambient composition. Commercial marine animals can be affected by aggregates mining operations because of damage to their breeding or spawning habitats. In European waters, these species include sand eels, herring, coregonid fishes, and crabs. It should be noted that fishing activity (trawling) is also a source of considerable turbidity (Churchill, 1989). Even in intensely mined areas of the North Sea, trawling disturbs the seabed much more extensively than all other activity combined, including aggregate extraction (ICES, 1992).

MANAGEMENT

The most extensively mined marine deposits are fluvio-glacial deposits submerged by the postglacial rise in sea level. These may be alluvial sands deposited within the last 10,000 years, when sea level was some 130 meters below the present sea level, as well as littoral sands or marine channel sands stranded during the last transgression (Choot, 1988). Sand deposits tend to be voluminous near the former margins of glacial ice. Near former ice margins, the size of the some types of minable deposits may be restricted (e.g., eskers or drumlins) but extensive, thick deposits may also be accessible, such as the outwash deposit of the northeast coast of the United States (Cruickshank and Hess, 1975), East Anglia (Harrison, 1988), and around Japan (Narumi, 1989, as cited in Smith and Collis, 1993). Mapping of these submerged resources is critical to their proper management. In many instances, mapping is done by government agencies and includes textural characteristics of the deposits, evidence for deposit thickness (cores, boring, or seismic reflection surveys) and may also include information on mobile or relict bedform (side-scan surveys, bathymetric surveys). A few countries have advanced mapping programs at scales more detailed than 1:100,000.

Submerged aggregate resources on the continental shelf or in territorial waters are predominantly public lands under government jurisdiction in which legislation is enacted to regulate mining activity. A system of leasing or licensing, such as that used in the regulation or oil and gas exploration and development, may be established and royalties imposed. The regulatory authority normally acts in consultation with other interests representing, for example, fisheries, defense, navigation, energy production, coastal protection, waste disposal, and various marine construction organizations, such as those maintaining underwater cables, pipelines, sewer outfalls, etc. A code of practice to promote sound management in the commercial extraction of marine sediments has been put forward by the International Council for the Exploration of the Sea (ICES, 1992).

REFERENCES

Bokuniewicz, H. J. 1988. Sand mining in New York Harbor. *Marine Mining* 7:7–19.
Choot, G.E. B. 1988. *Marine Sources of Fill.* Geotechnical Control Office, Civil Engineering Services Department, Hong Kong. Advisory Rpt. No. ADR 30/88, 132 pp.
Churchill, J. H. 1989. The effect of commercial trawling on sediment resuspension and transport over the Middle Atlantic Bight continental shelf. *Continental Shelf Research,* 9:841–864.
Cruickshank, M. J. and H. D. Hess. 1975. Marine sand and gravel mining. *Oceanus, 19*:2–44.
deGroot, S. J. 1986. Marine sand and gravel extraction in the North Atlantic and its potential environmental impact, with emphasis on the North Sea. *Ocean Management, 10*:21–36.
Groothuizen, B. 1998. Dredging methods and materials for reclamation, in *The sand and gravel resources of Hong Kong.* P. G. D Whiteside and N. Wragge-Morley, editors. Geological Society of Hong Kong: 83–99.
Guilford, C. M. 1988. Marine fill in Hong Kong—a 35 year resume. In *The Sand and Gravel Resources of Hong Kong,* P. G. D. Whiteside and N. Wragge-Marley, editors. Geological Society of Hong Kong, pp. 11–22.
Harrison, D. J. 1998. The marine sand and gravel resourses of Great Yarmouth and Southold, East Anglia. British Geological Survey, Technical Report WB/88/9.
Hitchcock D. R. and B. R. Drucker. 1996. Investigation of benthic and surface plumes associated with marine aggregates mining in the United Kingdom, *Proceedings of Oceanology '96,* pp. 221–234.
ICES (International Council for the Exploration of the Sea), 1992. Effects of extraction of marine sediments of fisheries, ICES Cooperative Research Report No. 182, 78 pp.
ICES, 1995. Report of the working group on the effects of extraction of marine sediments on the marine ecosystem. ICES CM 1995/E:5, 97 pp.
Narumi, Y. 1989. *Quarry Management,* March, 11–16.
Packer, T. 1987. *Survey of Environmental Effects of Marine Mining.* Ocean Mining Division, Mineral Policy Section, Canada. 56 pp.
Palermo, M. R., F. D. Shields, and D. F. Hayes. 1981. *Development of a Management Plan for Craney Island Disposal Area.* Technical Rpt. EL-91-11. Environmental Laboratory, U.S. Army Corps of Engineers, Waterways Experiment Station, Vicksburg, MS, 170 pp.

Smith, M. R. and L. Collis (editors). 1993. *Aggregates: Sand, Gravel and Crushed Rock Aggregates for Construction Purposes*. 2nd edition. Geological Society Engineering Geology Special Publication No. 9. London: 339 pp.

STIL, 1995. Statistics for the agricultural use of carbonates. National Agricultural Institue (STIL). Norway.

Uren, M., 1989. Supplying aggregates from the sea bed. Quarry Management, December, 19–25.

CHAPTER 19
METHODOLOGY FOR DELINEATION OF COASTAL HAZARD ZONES AND DEVELOPMENT SETBACK FOR OPEN DUNED COASTS

TERRY R. HEALY
Coastal Marine Group
Department of Earth Sciences
University of Waikato, Hamilton, New Zealand

ROBERT G. DEAN
Department of Coastal and Oceanographic Engineering
The University of Florida
Gainesville Florida

Notation	19.2
Introduction: Why the Need for a Development Setback?	19.3
Coastal Hazard Zones and Development Setback	19.5
Legislative Need for a CHZ and Development Setback	19.6
Some U.S. Coastal Setback Precedents	19.7
Hazards in the Coastal Zone	19.8
Role of the Frontal Dunes	19.9
Planning Horizon	19.9
Methodology for Establishment of Development Setback Incorporating the Coastal Hazard Zone	19.9
Long-Term Duneline Trend (R)	19.10
Maximum Short-Term Duneline Fluctuation ($F_{(max)}$)	19.10
Effect of Relative Sea Level Rise (Δy)	19.10
Dune Topographic Stability Factor (D)	19.15
Tests for Applying the CHZ Setback	19.15
Episodic Storm Cut and Sand Reservoir Considerations	19.16
Storm Surge Washover and Flooding	19.16
Possible Tsunami Hazard	19.19
Application to New Zealand's Bay of Plenty Coast	19.20
What is the Appropriate Projection for Sea Level Rise?	19.20

Other Functions of a CHZ and Development Setback		19.23
Concluding Issues		19.24
Acknowledgments		19.25
References		19.25

NOTATION

A	= profile scale parameter ($m^{1/3}$)
CHZ	= Coastal Hazard Zone as a linear distance from frontal dune toe (m)
d	= median sand grain diameter (m)
D	= dune topography stability factor = $E/\tan \alpha$
E	= dune elevation above datum (m)
E_f	= frontal dune elevation (m)
$F_{(max)}$	= decadal fluctuation of frontal dune toe (m)
g	= gravitational acceleration (m·sec^{-2})
h	= water depth, depth at any point on equilibrium beach profile (m)
H	= wave height (m)
h^*	= limiting depth of sand exchange for profile (m)
$h_{\Delta p}$	= barometric set-up (m)
h_1	= depth at shelf break (m)
h_2	= depth at shore (m)
h_b	= depth at breaking
H_b	= wave height at breaking (m)
HIL	= Hallermeier inner limit representing the day-to-day exchange of sand from the inner shelf and surf zone to the beach
H_0	= deep water wave height (m)
HOL	= Hallermeier outer limit water depth for annual significant diabathic sediment exchange
h_r	= wave run-up (m)
h_s	= wave set-up (m)
H_s	= significant wave height (m)
$H_{s0.137}$	= maximum nearshore significant wave height occurring for 12 hours per year
h_T	= height of predicted astronomical tide above datum (m)
h_{total}	= total predicted storm surge and wave run-up (m)
H_{ts}	= tsunami nearshore wave height
h_w	= wind set-up (m)
k	= wave number (m^{-1})
K	= constant
L_0	= deepwater wave length (m)
R	= long-term shoreline erosion (or accretion) rate per 100 years (m)
R_{ts}	= tsunami run-up inundation level (m)
T	= wave period (sec)
u	= wind speed (m·sec^{-1})
W	= width of continental shelf (m)
y	= distance from shoreline on equilibrium beach profile (m)
α	= natural angle of repose of dune sand
β	= angle of the offshore bar, beach slope
γ	= relative submerged density
Δs	= projected rate of local relative sea level rise (m/100 yr)
$\Delta s'$	= net local relative sea level rise effect (m/100 yr)

ΔS_h = historical rate of local relative sea level rise (m/100 yr)
ΔS_L = the long-term (100 year) projected sea level change rate (m/100 yr)
ΔT = the 100 year vertical tectonic movement of the shoreline
ΔG = any discernible change in the ocean geodynamic surface.
ΔV = volume per unit beach length (m^3·m^{-1})
$\Delta V_{(max)}$ = maximum volume of storm cut (m^3·m^{-1}) from frontal dune
Δy = dune line retreat due to relative sea level rise per 100 years (m)
ξ = horizontal wave orbital semiexcursion distance (m)
ℓ^* = length of offshore profile (m)
ρ = fluid density (kg·m^{-3})
ρ_s = sediment density (kg·m^{-3})
ω = wave radian frequency (radians·sec^{-1})

INTRODUCTION: WHY THE NEED FOR A DEVELOPMENT SETBACK?

Many cases of coastal erosion become manifest in spectacular episodic storm events (Figure 19.1a); others evolve and persist for a longer term, but are nonetheless emotive (Figure 19.1b,c). Many cases lead to demand from the beachfront property owners for government or local authority "action." Formerly, a favored response was to construct sea walls or groins to defend the properties against the attacks of the sea; or alternatively, the landowners took the matter into their own hands and constructed sea wall protection in front of their own property (Pilkey et al., 1982, 1983; Kaufman and Pilkey, 1983). In some parts of the world there has been strong centralized government influence to build protective structures against erosion, for example, the Ministry of Construction in Japan and the U.S. Army Corps of Engineers in the United States.

In recent years, however, "solid structure" responses have often been considered as inappropriate for cases of coastal erosion. Rather, partly related to the growth of environmental awareness by the global public, coastal management philosophy has veered toward soft solutions that attempt to harmonize with nature. Beach renourishment—especially in view of the potential impact from sea level rise—has become a major management option (Dean, 1983; Dean and Yoo, 1992).

There is also the realization that beach erosion problems are only part of the coastal management spectrum and that the solution for the coastal managers often requires interaction between coastal engineers, scientists, planners, landscape architects, and social scientists. Beach erosion problems may thus require additional focus on:

- Protection of the natural and intrinsic character of the coastal environment. This, after all, is what attracts people to the coast, and to smother the coastline in unabated development ultimately despoils the essential attraction of the coast.

- Practice of "sustainable management" of the coastal resources. This concept is enshrined in the 1993 Rio de Janeiro UNCED (United Nations Conference on the Environment and Development) Agenda 21, which recognizes that many people are economically reliant on the sustainable productivity of the coastal zone. In particular, coasts are important for tourism: people flock to Florida, the Spanish coasts, the Queensland Gold Coast, or Phuket in Thailand, for example, for the sunshine and to experience the coastline. With tourism such big global business, the issue becomes: what is the best way to manage the coastal resources for long-term sustainable management?

FIGURE 19.1. Storm cut of a frontal dune (a). When dwellings are constructed on the frontal dunes they become endangered during such events; (b) if there is a long-term sediment budget deficit leading to shoreline retreat; and (c) near tidal inlets that are prone to migrate. Such situations can be avoided by implementation of a sensible coastal hazard zone setback.

FIGURE 19.1. *Continued.*

- Allowing for cultural issues of indigenous peoples and their traditional links with the coast in matters of food supply and sites of cultural significance.

Beatley (1994) recently stated with regard to U.S. shorelines: "management of the coastline (particularly with respect to natural hazards) has been neither effective nor stringent," and as much of the coast is already developed, there is increasing justification to treat the remaining areas of coast in a more conservative manner.

Accordingly, in this chapter a method is presented that identifies a development setback and coastal hazard zone for open duned coasts based on physical principles of protection against coastal hazards, but which allows for the additional issues of retention of natural character and protection of dune ecosystems and sites of cultural interest. The methodology has evolved from applications on the barrier coasts of Florida and the barrier embayed coasts of New Zealand.

COASTAL HAZARD ZONES AND DEVELOPMENT SETBACK

Bruun (1964) defined a setback line as "an established survey line indicating the limits for certain types of developments," which will depend on technical as well as developmental and administrative aspects.

The concept of "coastal hazard zone" (CHZ) was introduced by Gibb (1981), who defined the CHZ as "an adequate width of land between any development and the beach."

By implication, the hazard zone was defined from assessment of the potential for identified hazards to "damage or destroy beachfront property and assets" along the beach (Gibb and Aburn, 1986).

A coastal hazard zone, then, may be defined as a sector of coastal terrain that is subject to hazards from the marine environment. The hazards are mainly manifested as storm wave erosion, storm surge and flooding, or tsunami wave washover.

Development "setback," as defined here, is subtly more general in that it relates to providing a buffer zone between the beach and developments, such as buildings or structures, including roads. However, setback may include considerations other than coastal hazard; for example, preservation of the "natural character" of the coastline, or protection of sites of special or cultural interest. A setback is usually imposed at the time when coastal developments are being planned, and may become incorporated into local government ordinances.

In this chapter, the term "setback" and "coastal hazard zone" are used more or less synonymously to mean that zone measured as a linear distance landwards from a reference feature, usually taken as the toe of the frontal dune, to a line on the ground that is subject to hazards from the marine environment, and which, on the balance of evidence and in the light of scientific knowledge of the moment, it would be prudent to restrict development.

The coastal reference feature from which the CHZ is measured may be variously identified as (i) the seaward toe of the frontal dune (Healy, 1981; National Research Council, 1990); (ii) the line of vegetation; (iii) the mean high water mark; or (iv) a specified elevation contour. Of these, the seaward toe of the frontal dune is the most easily identifiable in the field and least variable through time, and is thus the preferred reference feature from which to measure the CHZ (Healy, 1980b, 1981, 1982, 1988, 1993).

LEGISLATIVE NEED FOR A CHZ AND DEVELOPMENT SETBACK

The coastal hazard zone, as normally understood and applied, primarily serves to protect development from coastal hazards.

However, an important consideration of contemporary coastal management strategy is that cognizance should be paid to concepts of "sustainable management," safeguarding ecosystems and avoiding, mitigating, or remedying any adverse effects on the environment. Such ideas are manifest in sections 5, 6, and 7 of New Zealand's Resource Management Act of 1991, which states, inter alia, that the purpose of the act is:

> sustaining the potential of natural and physical resources . . . to meet the reasonably foreseeable needs of future generations; and . . . safeguarding the life-supporting capacity of air, water, soil, and ecosystems; and . . . avoiding, remedying, or mitigating any adverse effects of activities on the environment. (S5)

S6 of the Act outlines matters of national importance as:

> . . . managing the use, development and protection of natural and physical resources, shall recognize and provide for the following matters of national importance:
> (a) The preservation of the natural character of the coastal environment . . . and the protection of (it) from inappropriate subdivision, use and development;

(b) The protection of outstanding natural features and landscapes from inappropriate subdivision, use and development. . . .

S7 of the Act states that in achieving these aims particular regard shall be paid to:

(c) The maintenance and enhancement of amenity values
(d) Intrinsic values of ecosystems
. . .
(f) Maintenance and enhancement of the quality of the environment

Similar legislation is evident in Florida in four acts (Brindell, 1990). In respect of the Coastal Zone Protection Element, the "Planning Act" includes among its objectives (Brindell, 1990):

- Public right to reasonable access to beaches
- Protection of coastal resources and dune systems from adverse effects of development
- Prohibition of development and other activities that disturb coastal dune systems, and ensuring and promoting the restoration of coastal dune systems that are damaged
- Avoidance of expenditure of state funds that subsidize development in high-hazard coastal areas

In addition, the Coastal Zone Protection Element must address, inter alia,

- Principles to be used to control development to mitigate adverse impacts on coastal resources
- Hazard mitigation and protection against natural disasters
- Protection of existing beach and dune systems from human-induced erosion and restoration of altered beach and dune systems
- A plan to eliminate inappropriate and unsafe redevelopment

Clearly, although both New Zealand and the State of Florida have enacted different legislation in detail, both possess similar coastal management philosophies. The methodology presented in this chapter is one way to address the issues encapsulated within the respective legislations.

SOME U.S. COASTAL SETBACK PRECEDENTS

Various examples of setback applied along shorelines of the United States are summarized by the National Research Council (1990). State and local governments employ several techniques to transform recession rate data to CHZs, with varying regulations in different areas and latitude allowed by the local government authorities.

A fundamental component of the State of Florida regulatory program is the Coastal Construction Control Line (CCCL), which delineates the limit of severe fluctuations due to a 100 year storm event (Chiu and Dean, 1984; Kriebel and Dean, 1985). The CCCL is depicted on a series of aerial photographs prepared on a county-by-county basis for each of the 24 coastal counties in the CCCL program. Relevant is the fact that the Florida CCCL is the line of jurisdiction separating the seaward zones of state control from the more landward zones, where the local entities have regulatory responsibility. Thus, the CCCL is not a setback line but simply a jurisdictional line seaward of which the state requires permits for extensive modifications within this zone. The CCCL is established by a thorough program involving field measurements of dune–beach profiles, aerial photography for identification of changes in vegetation limits, and following this, an extensive numerical modeling program within which the characteristics of historical hurricanes are synthesized to establish the 100 year storm surge and the associated dune erosion. The location of the control line is the most landward limit of the following three positions: (1) the landward limit of dune erosion, (2) the landward limit of penetration of a 0.91 m (3 ft) wave in conjunction with the 100 year storm surge, and (3) the landward limit of overwash during the 100 year storm. A detailed description of the methodology employed in this program is available in Chiu and Dean (1984). The State of Florida also requires a setback from the seasonal high water contour (SHWC), being 30 times the average annual erosion rate for single-family structures and 60 times the average annual erosion rate for multifamily structures. The SHWC is the contour corresponding to the seasonal high water elevation (SHWE), which is defined as 1.5 times the mean tidal range above mean high water.

Of interest for open duned coasts are setback regulations from the State of North Carolina, where setback must be 30 times the annual erosion rate (60 times for large structures), a minimum of 18 m (36 m for large structures) from the vegetation line, and include the crest of a primary dune above the 100 year storm flood level.

The State of South Carolina utilizes a setback line in its coastal regulatory program. The setback line is located 6 m landward of the primary dune line and is a line of prohibition. The dune line is established through ground-based surveys and aerial photography. For those locations where sea walls preclude the presence of a dune crest, the effective dune line is estimated as the location where the dune crest would be if the sea wall were not present.

HAZARDS IN THE COASTAL ZONE

Hazards operating in the coastal hazard zone are several. Not all hazards will be evident at any one location or at any one time. In a two-dimensional beach profile sense they include:

- Beach erosion, i.e., the loss of sand from the beach and dune resulting in landward retreat of the beach system and dune face (Hattersley and Foster, 1968; Healy, 1980a,b; Dean and Maurmeyer, 1983; Kaufmann and Pilkey, 1983; Everts, 1985; van de Graaff, 1994)
- Storm surge induced flooding—the effect of storms, including onshore wind stress, barometric set-up, wave set-up and run-up, which combine to pile up water against the shoreline, and perhaps lead to flooding of low-lying coastal swales and dunes (Kriebel and Dean, 1985; van de Graaff, 1986, 1994)
- Blowouts in the frontal dune caused by strong onshore winds result in sand loss from the frontal dune system. This often migrates inland as a transgressive sheet (Sherman and Nordstrom, 1994)
- Tsunamis, which may surge over low-lying coastal dunes and structures causing damage from flooding as well as from the wave bore effect (Synolakis, 1991; Yeh, 1991; Camfield, 1994)

Other hazards may be human-induced, caused by the construction of groins, jetties, or wharves (Dean, 1988), or the removal of sand from the littoral system (Griggs and Fulton-Bennett, 1988; Finkl, 1994a,b; Griggs, 1994).

ROLE OF THE FRONTAL DUNES

The application of CHZ methodology presented here applies to duned coastlines facing the open sea, but the general nature of the coastline may also be exposed open coast or embayed. Typically, these contain a landform suite comprising a variety of morphodynamically interlinked subaqueous and subaerial landform features from the inner shelf/shore face to the offshore bar, trough, swash platform, beach face, berm, back berm, frontal dune, and swale.

The geomorphic beach types include barrier islands, barrier spits, bayhead beaches, and indeed any sandy beaches comprising noncohesive sandy sediments and containing the typical beach geomorphic elements. It is widely recognized that beaches, especially barrier beaches, are very changeable features in historical time (Leatherman, 1988; Dean, 1990).

The natural role of open coast frontal dunes acting as a reservoir of sand for rare but severe storms (Figure 19.1b) has been well-emphasized in the literature (Hattersly and Foster, 1968; Healy, 1974, 1980a, 1981; Healy et al., 1977; *Shore Protection Manual*, 1984; National Research Council, 1987; 1990). In storms, the beach is lowered as sand is removed offshore by erosive waves, and then the waves attack the frontal dune sand reservoir. After the storm, as the onshore wind stress and wave forcing decrease, and wave period increases, the waves tend to bring sand from the surf and nearshore zones back up onto the beach, forming a berm. In time, with onshore winds the sand is blown from the berm landward to reform the dune. Recognition of this process has been widely adopted as a cornerstone for dune management, protection, and enhancement programs. Thus a major underlying philosophy of the CHZ management methodology is to preserve the frontal dunes intact, in recognition of their role in the dynamic dune–beach–nearshore system.

PLANNING HORIZON

The temporal horizon for coastal planning has ranged between 30–100 years in various states. The advantage of the longer planning horizon is that projections for sea level change are usually given for up to the year 2100, and that coastal developments are rarely, if ever, made to be abandoned. Once developed, the coastal zone is highly likely to continue to be redeveloped and upgraded over time. Accordingly, it is better to allow for a long planning horizon, 100 years reasonably being regarded as the foreseeable future in the planning context.

METHODOLOGY FOR ESTABLISHMENT OF DEVELOPMENT SETBACK INCORPORATING THE COASTAL HAZARD ZONE

Kay et al. (1993) have undertaken a review of methods of coastal hazard zone delineation in New Zealand. They identified the most comprehensive method for coastal hazard zone determination as that evolved by Healy (1980b, 1981, 1988, 1993). The approach is based upon beach profile conceptualization, i.e., is two-dimensional, and may be expressed as:

$$CHZ = R + 2F_{(max)} + \Delta y + D \qquad (1)$$

where CHZ is a linear distance measured inland from a reference point, here taken as the toe of the frontal dune (or failing, that the vegetation line); R is the long-term shoreline erosion or accretion rate trend; $F_{(max)}$ is the decadal term duneline fluctuation, representing the maximum observed cyclical fluctuation of extreme storm cuts within recent decades, say 50 years; Δy is the dune line retreat consequent upon projected relative sea level rise; and D is the dune topography stability factor.

The various components required to be calculated for the CHZ are defined in the following sections.

Long-Term Duneline Trend (R)

The long-term trend of the shoreline change, with units in m/100 years, can be obtained from historical shoreline MHWM surveys dating from earlier centuries, or from alternative data such as repetitive dune/beach survey profiles where they exist, such as in Florida (National Research Council, 1990; Dean, 1994), or aerial photos. In practice, we take the extent of locational variation of the toe of the frontal dune, defined as the point where the base of the frontal dune intersects with the back beach, as measured off, for example, historical aerial photographs, which typically are available since the 1940s. Whereas in many cases the trend is for erosion, in some locations the trend may be for progradation, for example, parts of New Zealand's Bay of Plenty coast and much of the Florida coast (Dean, 1994). Thus R can be either positive (eroding) or negative (prograding).

Maximum Short-Term Duneline Fluctuation ($F_{(max)}$)

This measure of maximum duneline "cut and fill" from rare severe storms represents short-to-medium-term trends that are superimposed on the long term trends. Such duneline oscillations relate either to episodic erosive events or decadal trends. For New Zealand's Bay of Plenty, for example, $F_{(max)}$ was surveyed as a 20 m duneline retreat representing about 120 m^3 per meter of beach cut after a major storm (Healy, 1978a,b). This is consistent with model predictions for a peak surge of 3.5 m (Kriebel and Dean, 1985) applied to the Florida coast.

A safety factor may be applied to the $F_{(max)}$ parameter because few data are available either on the magnitude of the episodic storm cuts or on the long-term trends. For the data presented by Healy (1982), it was shown that a maximum likelihood statistical estimation of storm cuts was about twice the maximum cut known at the time. Thus, in the absence of long-term detailed beach profile survey data, including both pre- and poststorm profile surveys, from which the appropriate 1 in a 100 year storm event dune cutback can be calculated, a safety factor of 2 is in the meantime taken for the $F_{(max)}$ parameter.

Effect of Relative Sea Level Rise (Δy)

There is substantial evidence of historical sea level rise on a world scale of approximately 12 cm per century (Barnett, 1983; Barth and Titus, 1984; Pirazzoli, 1986; Bruun, 1986, 1987, 1990; Stewart et al., 1990; Gornitz, 1993), but the projected rate of future rise is an issue (Aubrey and Emery, 1993; Warrick, 1993a,b).

Clearly, there is wide international acceptance that "considerations of sea level changes should be incorporated into land use planning" (National Research Council, 1987, p. 3) and that "Long term planning and policy development should explicitly consider the high probability of future increased rates of sea level rise" (National Research Council, 1987, p. 7). The National Research Council (1987) explicitly warns that feasibility studies for coastal projects should consider the storm surge as well as the high probability of accelerated sea level rise, and that "obviously a rising relative sea level, with resultant higher storm tides and larger waves can only increase these hazards" (National Research Council, 1987, p. 34).

Assessment of future changes of sea level at any specific coastal site is a function of several factors; viz., the effect of local tectonic uplift or down-sinking, local sea surface geodynamic surface effects (Mörner, 1976; Tooley, 1993), steric effects (Stewart et al., 1990), any human-induced effects, for example alteration to the groundwater table or hydrocarbon pumping inducing local or regional relative sea level change (Boesch et al., 1994).

The relative sea level change at any one coastal sector is specific to that site, and may even be affected differentially by possible global changes of the ocean geodynamic surface (Mörner, 1976). However, relative variation of the ocean geodynamic surface, indicated by Mörner to be possibly of the order of 100 m on a global scale, has been shown by satellite altimeter measurements to be much less—more of the order of 2 m. Thus, changes in the earth geodynamic surface as a result of global warming effects are likely to be small over the time frame considered.

Of more importance may be tectonic processes, especially where rapid. Rapid uplift is known to be occurring near Santa Cruz in Monterey Bay, for example (Griggs, 1994). Where it is known that a given sector of coast is subject to measurable uplift or down sinking, then assessment of the projected local relative sea level change (Δs) should appropriately reflect these additional effects, i.e.

$$\Delta s = \Delta S_L \pm \Delta T \pm \Delta G \tag{2}$$

where ΔS_L is the long-term (100 year) projected sea level change rate, ΔT is the 100 year vertical tectonic movement of the shoreline, and ΔG is any discernible change in the ocean geodynamic surface.

One should be aware of the possibility of "double dipping" when applying a sea level rise projection (Δs) in conjunction with a long-term erosion rate (R) in Equation 1. This arises because the existing beaches and dunes are presumably already adjusted to the local historical rate of sea level rise, which is implicit in determination of a value for R based on historical shoreline measurements. Accordingly, it may be appropriate to subtract this effect from future projected local relative sea level rise, as advocated by Gibb and Aburn (1986). Thus:

$$\Delta s' = \Delta s - \Delta S_h \tag{3}$$

where $\Delta s'$ is the net local relative sea level rise effect, Δs is the local relative sea level change projection and ΔS_h is the historical local relative sea level change over, say, the last 50–100 years.

Bruun (1962, 1983, 1989) developed a two-dimensional model of the effect of sea level rise inducing erosion and shoreline retreat, in which the eroded material is redistributed over the inner shelf. Because a profile of equilibrium is assumed, it is proposed that the nearshore inner shelf is raised in direct proportion to the elevational increase in sea level, and the volume of sediment deposited must thus be equal to that eroded from the dunes (Schwartz, 1967, 1968, 1987; Komar et al., 1991).

The original model from Bruun (1962, 1989) (see Figure 19.2a) is expressed as

$$\Delta y = \frac{\Delta s \cdot \ell^*}{h^*} \tag{4}$$

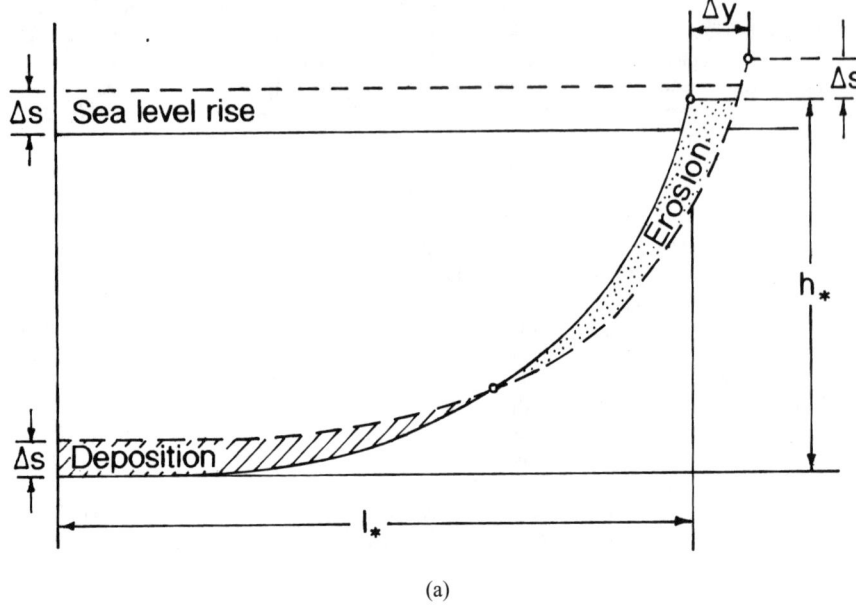

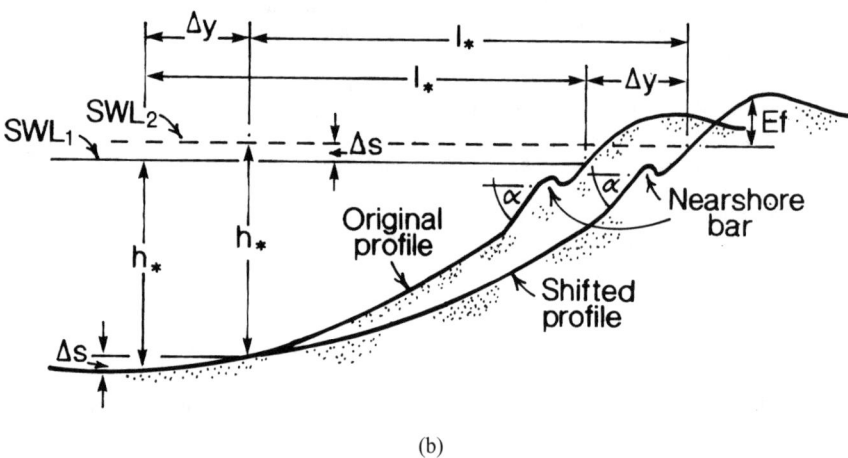

FIGURE 19.2. (a) Bruun model of shoreline adjustment. (b) Weggel modification of the Bruun model.

where ℓ^* is the length of the offshore profile out to a supposed depth, h^*, of the limit of material exchange from the beach and the offshore, Δy is the dune line erosion, and Δs is the rate of sea level rise.

Consideration of the height of the dunes is only implicit in the Bruun model, but Weggel (1979) explicitly includes the height of the dunes at the shoreline so that

$$\Delta y = \frac{\Delta s \cdot \ell^*}{h^* + E} \tag{5}$$

where E is the elevation (height) of the dunes and the other symbols are as previously defined (Figure 19.2b).

A simplification of the Bruun model was given by Dubois (1977) and may be expressed simply as

$$\Delta y = \Delta s \cdot \tan \beta \tag{6}$$

where Δy is the duneline retreat, Δs is the sea level rise rate, and $\tan \beta$ is the slope of the offshore bar.

Problems with applying the Bruun model include the difficulty of defining precisely the maximum and frequent depths from which there is exchange from the offshore to the beach. As depth of disturbance of bottom sediment by waves is a function of wave height, this is a statistical problem dependent upon the frequency of various wave heights and periods. This problem has been addressed by Hallermeier (1981) (see Figure 19.3). Hallermeier defined two limits based on the wave climate and grain size, viz., an inner limit (HIL) representing the day-to-day exchange of sand from the surf zone to the beach, and an outer limit (HOL) that conceptually defines sand transport outside the surf zone and represents the maximum offshore extent of wave shoaling sediment transport. It relates to bottom sand motion induced from the largest waves, which occur for less than 12 hours per year. The Hallermeier limits are expressed by:

$$(HIL)\ 0.03\ h \geq (\xi \omega)^2/\gamma g \geq 8\ d (HOL) \tag{7}$$

where h is the water depth (m), ω is the wave radian frequency (radian·sec^{-1}), d is the median sand diameter (m), g is gravitational acceleration (m·sec^{-2}), ξ is the horizontal wave orbital semiexcursion distance (m), given by linear wave theory as:

$$\xi = \frac{H}{2\ \sinh\ (kh)} \tag{8}$$

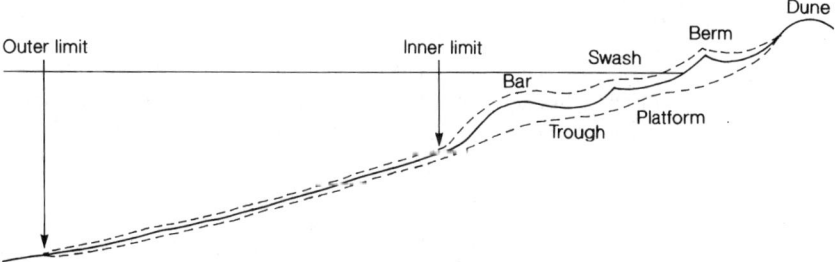

FIGURE 19.3. The concept of Hallermeier limits illustrating annual (outer) and daily (inner) closure depths.

where H is wave height (m), and k is the wave number (m^{-1}), and

$$\gamma = \frac{\rho_s - \rho}{\rho} \qquad (9)$$

where ρ_s is the sediment density (kg·m^{-3}), and ρ is the fluid density (kg·m^{-3}).

CUR (1987) note that for practical purposes the inner limit may be taken as about 1.75 $H_{s0.137}$ and that the outer limit is about twice that or 3.5 $H_{s0.137}$ where the $H_{s0.137}$ is the maximum nearshore significant wave height occurring for 12 hours per year.

Values for Hallermeier limits vary depending on wave climate, grain size, and inner shelf conditions. For example, Foster (1991), from detailed study of a nearshore beach renourishment, reviewed the Bay of Plenty, New Zealand, wave data for the Hallermeier limits. For that situation, the inner limits depicting day-to-day surf zone diabathic exchange were estimated as 6 m water depth. The outer limit, depicting seaward limits of annual profile exchange, was calculated as 13 m. These values compare to those estimated from data in Harray and Healy (1978), which result in an inner limit of 5.7 m and an outer limit of 9.4 m, whereas Gibb and Aburn (1986) suggest an inner limit of 6 m and de Lange and Healy (1994) give limits of 5.5 m and 10.2 m from the detailed data set from monitoring inner shelf processes and dredge spoil and beach renourishment off the Tauranga Harbour entrance (Dahm and Healy, 1980, 1985; Bradshaw, 1991; Bradshaw et al., 1991; Foster et al., 1991, 1994; Healy et al., 1991).

For the Florida shorelines, possessing a rather wider and shallower continental shelf, the effective depth for diabathic exchange is given as between 4–6 m (Dean and Grant, 1989) in relation to "closure depth" on the shore-normal profile. This relates to an effective deep water H on the order of 3–5 m and $T = 7$–9 sec (Chiu and Dean, 1984).

Bruun's model assumes that the nearshore profile will build up by precisely the same amount of sea level rise; thus the selection of a value for ℓ^* is critical for the amount of sand that is supposedly eroded from the dunes to build up the sea floor as required by the model (Healy, 1991).

The main problems with applying the Dubois simplification are that no cognizance is taken of the height of the dunes to be eroded (equivalent to the reservoir of sand), while the angle of the seaward slope of the offshore bar can vary considerably both temporally and alongshore.

Recent work by Dean (1994) has shown that in cases of advancing shorelines, such as much of the east coast of Florida, straightforward application of the Bruun rule is inappropriate. Rather, a model is presented that considers the deviation from the Bruun rule can be ascribed to an addition to, or removal of sediment volume from, the active profile. Extending the Bruun rule to include this source term:

$$\Delta y = \frac{1}{(h^* + E)} (\Delta s \ell^* + \Delta V) \qquad (10)$$

where ΔV is the volume per unit beach length added to the active profile to achieve the Bruun rule. The concept of the equilibrium beach profile (Dean, 1991; Dean et al., 1993) was used to estimate the active profile length ℓ^* from the depth of active motion h^*

$$h = Ay^{2/3} \qquad (11)$$

or

$$\ell^* = (h^*/A)^{3/2} \qquad (12)$$

where A is a profile scale parameter, which varies over a limited range for Florida beaches but has a representative value of 0.1 m$^{1/3}$.

Dean (1994) showed that applying Equations 9–11 with existing conditions for the Florida coast permits a shoreline progradation of about 8 cm per year; but with doubling the rate of sea level rise, the average shoreline advancement would change to a retreat of 5 cm per year.

The conclusion seems inescapable that if the system is two-dimensional, any accelerated sea level rise resulting from "greenhouse" global warming effects will cause accelerated erosion of sandy coastlines, and must be allowed for in planning for coastal developments and management of the coastal resources.

Dune Topographic Stability Factor (*D*)

The natural stability angle, or angle of repose of medium dune sand is about 30°. Allowance must be made for this in development planning and in setback, for should there be a severe dune cutback, structures developed within the zone of natural stability angle will be at risk from slope instability (Figure 19.4). Thus for every 1 m elevation of the dunes, one needs approximately 2 m linear distance to allow the dune sand to rest at its natural angle of repose. In the general sense, the dune stability factor, D (in meters), is given by

$$D = E/\tan \alpha \qquad (13)$$

where $\tan \alpha$ is the natural angle of repose of the dune sand, and E is the elevation of the dunes, above datum.

TESTS FOR APPLYING THE CHZ SETBACK

An integral part of applying CHZ methodology is that once the initial estimation of the CHZ is calculated as a summation of the four parameters, it is then subjected to three tests to ensure that the CHZ setback determined as a linear distance back from the reference point, normally taken as the toe of the frontal dune, is appropriate for the potential hazards for the particular dune geomorphology at that site. The three tests are episodic storm cut and sand reservoir considerations, storm surge washover and flooding, and possible tsunami hazard.

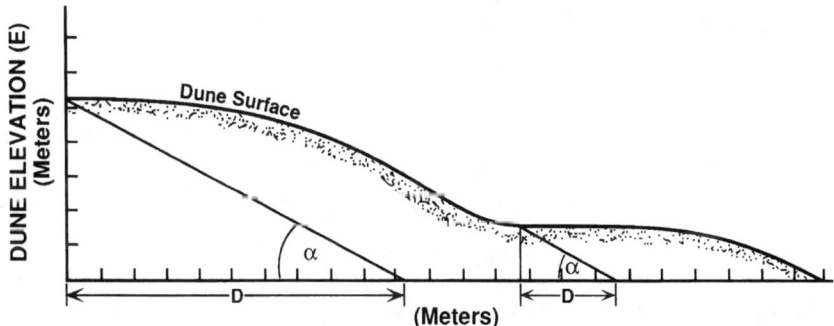

FIGURE 19.4. Concept of dune topographic stability factor, *D*, for both high and low dunes.

Episodic Storm Cut and Sand Reservoir Considerations

Apart from the medium-term trends shown by aerial photos, the critical question is: How much storm cut and dune recession can be expected in a severe storm, and particularly in a "worst case" or "1-in-100 year" erosion event?

At very few coastal sites are there long-term records available and suitable for undertaking dune cut and fill extreme-value statistical analysis. Where there are profile survey data, they are usually taken as regular monthly, seasonal, or annual surveys by the environmental or beach management organization. It is rare that they truly represent the maximum storm dune cut, because surveying teams are not normally available immediately prior to and at the height of the storm. Additionally, the required survey control may be destroyed by the storm.

In the absence of extreme-value storm dune cut data, it is acceptable to follow the standard adopted by Dutch engineers (van de Graaff, 1986; 1994; CUR, 1989) and to retain a sand reservoir of 400 m³·m⁻¹ within the CHZ above a mean sea level datum after the worst known storm cut. For such conditions, a storm surge and run-up of approximately 5 m is assumed.

Because the CHZ is given as a linear distance from a reference point, to convert storm profile "cut" to distance we must consider the average elevation of the dune field (E). Accordingly, when we apply this test

$$CHZ > (\Delta V_{(max)} + 400)/E \qquad (14)$$

where $\Delta V_{(max)}$ is the maximum storm cut volume in m³·m⁻¹ of beach, and E is the average dune elevation, and the 400 m³·m⁻¹ of beach is the Dutch recommended frontal dune sand reservoir (Figure 19.5).

Storm Surge Wave Washover and Flooding

Many sectors of the southeast U.S. coast have suffered from storm surge washover and flooding in severe storms. This is one of the major coastal hazards and coastal setback should be sufficient to avoid surge and flooding hazard. Accordingly, a design storm surge and run-up calculation may be undertaken, based on the five components illustrated in Figure 19.6. The implication is that if the general dune topography within the initially delineated CHZ is lower than the design surge and flood level, the CHZ must be extended landward until the terrain is sufficiently high. Details of the calculations to derive the total predicted storm wave run-up follow.

(i) *Tidal difference* (h_T) is the height of the predicted astronomical tide above datum:

$$h_T = \text{(highest predicted tide)} - \text{(datum)} \qquad (15)$$

(ii) *Barometric set-up* ($h_{\Delta p}$). Tidal predictions assume an average barometric pressure of 1014 hPa. A deviation of 1 hPa from this causes approximately a 10 mm change in sea level, so that

$$h_{\Delta p} = 0.010 \,(1014 - \text{storm barometric pressure}) \qquad (16)$$

where the pressure is in hectapascals (hPa) or millibars (mb).

(iii) *Wind set-up* (h_w). An onshore wind will result in an increase in the still water level along the coastline as a result of wind-induced shear stresses acting on the water surface. Silvester's (1974) relationship for wind set-up is expressed as:

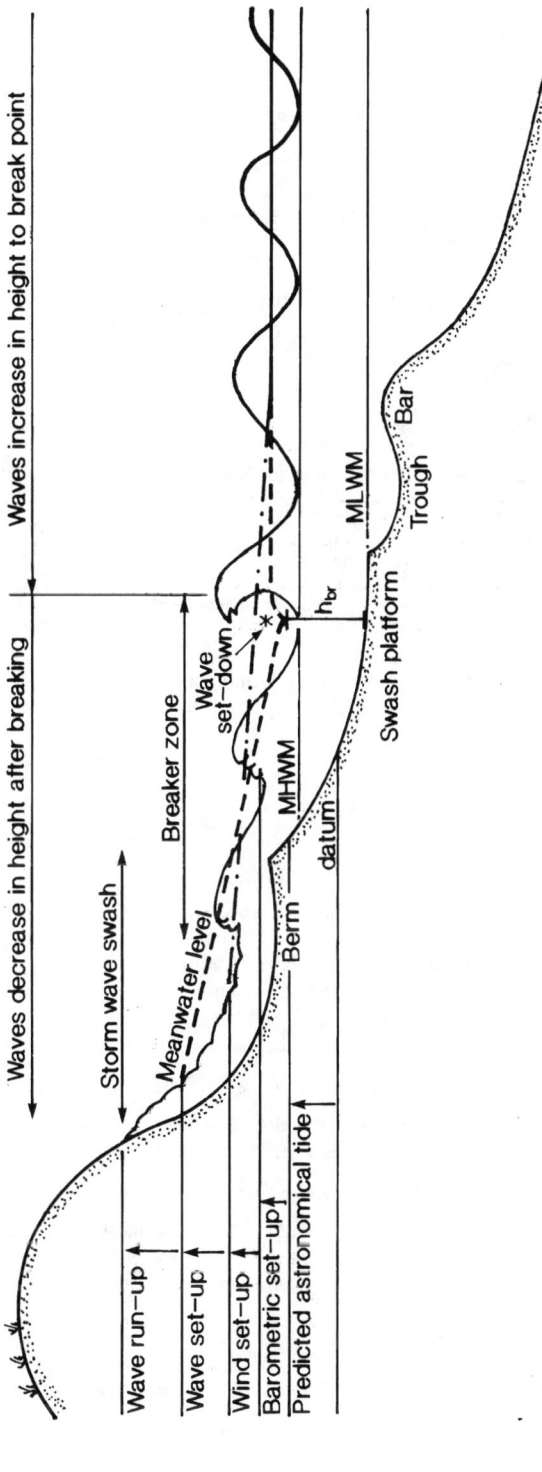

FIGURE 19.5. Components of storm surge and run-up adapted from Frisby and Goldberg (1981).

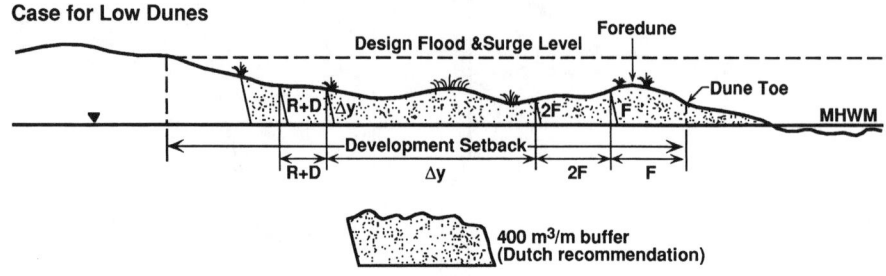

FIGURE 19.6. Application of the CHZ model detailing the various components for different shoreline topographies.

$$h_w = \frac{Kh_1 u^2 W}{gh_1^2 \left(1 - \frac{h_2}{h_1}\right)} \ln\left(\frac{h_1}{h_2}\right) \tag{17}$$

where K is a constant (3×10^{-6}); u is the wind speed (m^{-1}); W is the width of the continental shelf; h_1 is the depth at shelf break and h_2 is the depth at the shore or point of interest. It is necessary to ensure that the depth at the shoreline is nonzero ($h_2 \neq 0$) by using a standard depth of 1 m or the depth of water at breaking (h_b).

(iv) *Wave set-up* (h_s). The Sorenson (1978) formula for wave set-up is given below and assumes that wave set-up is solely a function of wave height and steepness.

$$h_s = 0.19 \left[1 - 2.82 \, (H_b/gT^2)^{0.5}\right] H_b \tag{18}$$

where H_b is calculated using the Komar and Gaughan (Komar, 1976) relationship for breaking waves, assuming that there is no refraction or diffraction:

$$H_b = 0.563 \, H_0/(H_0/L_0)^{0.2} \tag{19}$$

where H_0 is the deepwater wave height, usually $H_{1\%}$ (in m), and L_0 is the deepwater wave length (m) = $1.56 \, T^2$ (where T = the deep water wave period, s).

(v) *Wave run-up* (h_r). The *Shore Protection Manual* (1984) method for wave run-up is given as:

$$h_r = (H_b L_0)^{1/2} \tan \beta \tag{20}$$

where H_b is the breaking wave height (m), L_0 the deep water wavelength (m), and β is the gradient of the offshore bar.

(vi) The total predicted storm wave run-up (h_{total}) in meters is the sum of all the above factors, i.e.,

$$h_{total} = h_T + h_{\Delta p} + h_w + h_s + h_r \tag{21}$$

To apply this test for the initially delineated CHZ setback,

$$E_f > h_{total} \tag{22}$$

where E_f is the elevation (height) of the frontal dune and h_{total} is the storm surge and wave run-up elevation. If $E_f < h_{total}$, then the coastal hazard/setback line would need to be extended inland at the storm surge elevation until it intersects with land.

Design storm waves and surge for New Zealand's Bay of Plenty are based on assumptions of deepwater waves of significant wave height, $H_s = 9.9$ m and period $T = 9.5$ sec (Frisby and Goldberg, 1981). Assuming a Rayleigh distribution, for the 1% highest waves, $H_s = 15$ m. However, on the basis that the very largest few waves are unlikely to do extensive damage to structures on the land, but that the highest 10% of waves likely will, the design deepwater significant wave height is taken as $H_s = 12.6$ m. From the Komar and Gaughan relationship, this results in a design breaking wave of $H_b = 11.5$ m, assuming no refraction. The total wave run-up level above datum is composed of the following components. From the *New Zealand Nautical Almanac,* for a spring tide along this coast we obtain 0.80 m; the barometric set-up for a severe storm is about 0.35 m; the wind set-up from Silvester's formula for a 40 knot wind blowing over an inner shelf width of 25 km is calculated as 0.23 m; the wave set-up for waves of $H_b = 11.5$ m from the Sorenson formula is 1.39 m for the 10% highest waves, and the wave run-up from Hunt's formula depends upon the beach slope. Thus total design wave run-up varies from about 4–5 m for this coastal sector (Healy, 1993). This accords closely to the elevation of the flotsam line resulting from cyclone Bola of March 1988, observed on the frontal dunes (Healy, 1988).

For the Florida coastlines, equivalent design considerations are presented in Chiu and Dean (1984) and Dean and Grant (1989) based upon numerical modeling. Hurricane surge levels are close to 4 m based on probabilities of return periods of storm surge and wave run-up levels for a 100 year storm event, and this likely results in an order of 12 m of dune erosion.

Possible Tsunami Hazard

Historical impact and experience of distantly generated tsunamis and understanding of likely mechanisms of locally generated tsunamis is an additional coastal hazard in certain tsunami prone coasts of the world (see de Lange and Healy, 1986a,b), particularly around the Pacific Ocean (Bernard, 1991). The details of tsunami run-ups are not well known except for Hawaii and Japan, and the likelihood of having a sufficient data base to utilize potential run-up levels is rare. However where such data are available, the initially delineated CHZ should be tested as for storm surge and flood levels, such that the elevation of the frontal dune (E_f) should be higher than the known tsunami run-up level:

$$E_f > h_{ts} \tag{23}$$

where h_{ts} is the tsunami run-up inundation level. Synolakis (1991) has shown that the approximations of linear theory predict tsunami run-up at near-identical values to nonlinear theory. Accordingly, h_{ts} can be obtained from

$$h_{ts} = 2.831 \sqrt{\cot\beta} \, H_{ts}^{5/4} \tag{24}$$

where H_{ts} is the tsunami nearshore wave height, and β is the beach slope. In a practical sense, β would be taken from the top of the dunes to beyond the break point bar.

APPLICATION TO NEW ZEALAND'S BAY OF PLENTY COAST

The first setback established for the Bay of Plenty arose from the Town and Country Planning Appeal Board (now the Environment Court) case at Piripai (Healy, 1980a,b, 1981). At that case, the Appeal Board applied a 60 m setback. Subsequently for the case at Matarangi Beach, the Appeal Board imposed a 100 m setback, with dwellings between 100–200 m back from the beach to be capable of being relocated should excessive erosion occur. At the 1982 Planning Tribunal hearing for the Papamoa dunes (Healy, 1982) argument was presented for a setback of 70 m minimum from the toe of the frontal or foredune, which was accepted by the Planning Tribunal.

The determination of such setback distances was based upon a set of general premises relating to coastal foreshore planning and management:

- The seaward toe of the frontal dune is the baseline against which to measure development setback. Although a variable morphological feature, it is less variable than the mean high water mark (MHWM), and where it is difficult to identify, the vegetation line may be utilized.

- The need to retain the foredune in its natural state to provide a reservoir of sand to satisfy normal "cut and fill" cyclical erosion, and to ensure that the dune remains well vegetated to prevent loss of sand from the reservoir by wind erosion (Healy 1974, 1980b, 1981).

- The preservation of the natural character of the coastal environment and the protection of it from inappropriate subdivision, use, and development, following the requirements originally listed within Section 3 of the Town and Country Planning Act of 1977 (now encompassed within S5 and S6 of the Resource Management Act 1991).

- Generally providing for other potential coastal hazards that have a reasonable probability of impacting on the coast. This concept is now enshrined within the Resource Management Act of 1991 (S3), and follows from legal precedence in which the court ruled that hazards with an impact that had a significant probability of occurrence must be allowed for in planning of coastal developments.

- The planning should take cognizance of a development life span of at least 100 years (National Research Council, 1987).

WHAT IS THE APPROPRIATE PROJECTION FOR SEA LEVEL RISE?

Experience has shown that delineation of a CHZ by the method explicated above is very sensitive to the application of the Bruun rule. Accordingly the value selected for the 100 year sea level rise projection is an issue.

Much has been aired during the late 1980s and early 1990s about potential rises in sea level associated with the "greenhouse effect," whereby increased atmospheric CO_2 and chlorofluorocarbon concentrations induce a substantial atmospheric warming followed by presumed consequent polar ice melting and sea level rise. Early predictions of sea level change by the U.S. Environmental Protection Agency (Hoffman et al., 1983; Barth and Titus, 1984; Titus, 1986a,b) and the National Research Council (Thomas, 1986) were for increases of between 0.6 and 2.3 m by the year 2100, with a most likely value of about 1 m. Such prognostications were modified as they became subjected to more critical assessment and now seem rather extreme in mainstream scientific thinking, although there always were the doubters (Idso, 1987).

Until 1994, the accepted "assessment" of expected sea level change to the year 2100 was that provided by the Intergovernmental Panel on Climatic Change (IPCC '90) (Houghton et al., 1990; Warrick and Oerlemans, 1990). The IPCC '90 assessment suggested a "business as usual" scenario prediction of a 66 cm sea level rise by the year 2100 (Figure 19.7). This prediction had wide scientific peer acceptance, with some 175 active scientists working in the field from 25 different countries participating in preparation of the main assessment; a further 200 scientists were involved in peer review of the draft report, which has led to a high degree of scientific consensus. Thus the IPCC '90 Assessment was regarded as the authoritative statement of the views of the international scientific community at that time.

In the interim, advances have been made on general circulation models (GCMs), but as concluded by Vellinga and Klein (1993, p. 28), "because of the relatively broad range of the results, the IPCC '90 report can still be considered valid, and one can still apply the projections on average global sea level rise that were formulated."

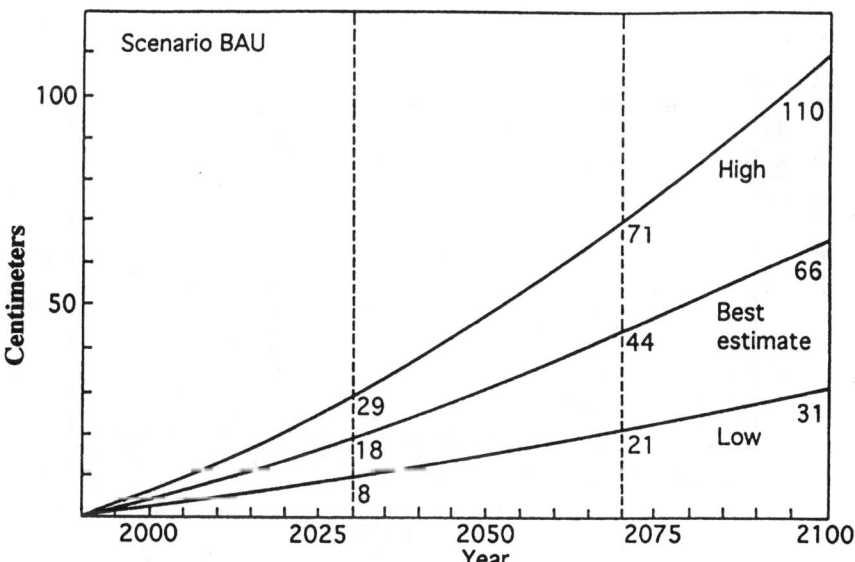

FIGURE 19.7. ICPP '90 global sea level rise scenario indicating a "best guess" rise of 66 cm by 2100.

Thus at the time of writing, the IPCC '90 retains that authoritative international acceptance because (McLean and Mimura, 1993; Warrick, 1993a,b; Warrick et al., 1993):

- No more recent assessments have been through wide scientific review and consensus.
- All subsequent best estimates are well within the range of uncertainty of IPCC '90.
- Even if the most recent predictions (Wigley and Raper, 1992, 1993a) are lower, it does not necessarily mean that future estimates based on more recent knowledge and theory may not produce higher sea level estimates.
- The IPCC '90 estimate is a best estimate for existing "business as usual," but future global CO_2 emissions are likely to be higher than existing "business as usual."

In 1993, preparations started for a Second Assessment Report, to be published in 1995. At time of writing (1995), although a full review assessment has not been completed, the subsequent modeling with additional refinements of Wigley and Raper (1993a,b) resulted in a lower "business as usual" sea level change scenario (Figure 19.8). Wigley and Raper (1993a) perspicaciously point out the wide uncertainties in the GCMs relating to future greenhouse gas concentrations, future natural climate perturbations, climate sensitivity, relaxation time of the system, downwelling and circulation fluxes in the ocean, and ice melt processes. Nevertheless, the general trend in sea level projections does seem to be reducing (Warrick, 1993a,b) and a projection akin to that in Figure 19.8 is likely to

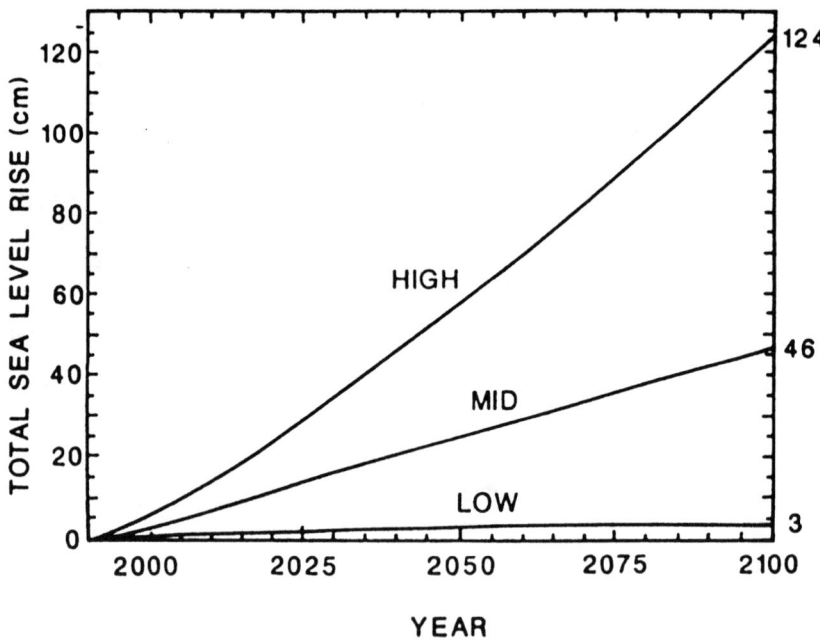

FIGURE 19.8. ICPP '92 modified sea level rise projection, after Wigley and Raper (1993a,b). Note that the U.S. EPA (1995) median estimate of global sea level rise is closely similar at 45 cm by 2100.

become the new IPCC '95 best guess scenario. This is consistent with the recently released U.S. Environmental Protection Agency report "Probability of Sea Level Rise" (Titus and Narayanan, 1995), which gives a median best guess estimate for global sea level rise of 45 cm by 2100.

Warrick (1993a) points out that recent GCM numerical modelling results, including dynamic ocean models, show clearly that global sea level topography changes will not be globally uniform because of the dynamic effects of atmosphere and oceanic circulation changes. So for practical purposes, "global mean" sea level change values should be recognized to have limited utility in local and regional contexts. Nevertheless, in coastal sectors where there is a lack of detailed tectonic information, for purposes of establishing coastal setback it is believed appropriate to apply the "best guess" IPCC global mean sea level rise projection as a first approximation (Warrick, 1993b).

OTHER FUNCTIONS OF A CHZ AND DEVELOPMENT SETBACK

While the CHZ is, as the name suggests, determined on the basis of identifiable hazard, the development setback is able to incorporate both the concept of coastal hazard as well as preservation of the natural character of the coastal environment, including outstanding natural features and the protection of specialized habitats of high ecological value, and other issues.

To date, although the concept of the "preservation of the natural character of the coastline" has been a supporting argument for determining development setback (see Healy, 1980b, 1981, 1993), there are no specific guidelines for the quantitative delineation of setback relating to protection of outstanding features or preservation of natural character.

Where the coastal landscape of open duned coast beaches is accepted as an "outstanding natural feature," then additional restrictions to development for presently undeveloped coasts may be needed based on landscape architecture criteria. In such cases, it may be appropriate that:

> for new subdivisional developments the development should not impose on the landscape, as defined by being visible when viewed by a person of average height standing on the beach berm, or otherwise viewed from the middle of the beach at low tide if no berm exists. (Healy, 1993)

There are two implications arising from such a recommendation:

1. People enjoying the beach should not be imposed upon by dwellings overlooking the beach. For western-style subdivisions typical to date—i.e., single-story houses—there would be incentive to ensure that the frontal dunes were built up to enhance their role as a reservoir of sand and defense against coastal hazard, as well as enhance their landscape values.

2. High-rise buildings and condominiums should be restricted to already developed areas.

Irrespective of implementation of landscape values in coastal management, it cannot be denied that the land involved in the CHZ, which may be several tens of meters wide, possesses additional potential uses. These may include:

- Passive recreation activities
- Protection of coastal ecosystems
- Protection of the natural character of the coastline
- Protection of cultural and historical sites

CONCLUDING ISSUES

It is important to emphasize that having established a setback line, there is still no guarantee that within the next 100 years the coast in question could not be pounded by a series of destructive storms capable of removing the frontal dune and sand reservoir, or some other coastal hazard, placing the development at peril. Neither can it be guaranteed that having imposed a setback line, substantial progradation creating additional new land might not occur.

Moreover, it is axiomatic that a coastal hazard zone setback does not of itself constitute a "magical" safety zone immediately on one side of it, and a zone of "total hazard" or "impending destruction" on the other. Rather, it is a line on the ground beyond which, on the balance of evidence, and in the light of scientific knowledge of the moment, it would be prudent to restrict development.

This raises the issue of the "accuracy" of the setback calculations and the ultimate delineation of the line on the ground. The methodology used to produce the setback distance is based on a two-dimensional profile and is then averaged alongshore so as to be representative of the prevailing geomorphology, as is current practice for implementing setbacks by many coastal states (National Research Council, 1990). It must be emphasized that the overall CHZ is a "robust" calculation and each parameter is considered orthogonal or independent of the other. But to expend great effort to try to refine a 50 m setback by, say, 3 m is to assign greater accuracy to the calculation than it merits; there is significant "uncertainty" in components of the calculation already, so that overemphasis on attempts to refine it is counterproductive. It is preferable to be conservative in philosophy and designate an appropriately wide zone. In delineating the CHZ it is, therefore, better to err in the direction of too much sand in the coastal reserve than too little. It is not in the interests either of the people who in good faith buy into such developments, or of ethical developers and concerned local authorities, for the developers or local authorities to take risks.

Concerning the issue of extracting the historical local relative sea level rise (ΔS_h) from the projected local relative sea level rise (Δs) for application of Equation 1, such logic may be appropriate to natural shorelines where the coastal sedimentary processes follow equilibrium beach profile theory, and are not influenced substantially by human activities. But in other situations, the human impacts would likely not be distinguishable from any sea level effect. For example, dune erosion downdrift from an inlet with large jetties that trap sand updrift, or a dredged inlet channel that inhibits bypassing, is more a function of longshore rather than diabathic effects. Application of the Bruun rule in such locations is questionable (Bruun, 1989).

Because a coastal hazard zone is initially established in the light of scientific knowledge of the moment, it may be reasonable to periodically reassess the CHZ. In South Carolina and Florida, for example, the CCCL setback is reassessed every 10 years. Scientific knowledge advances through time, note, for example, the efforts of the past decade to refine the projected sea level rise for the next 100 years, and for coastal setback and applying the Bruun rule, the projected sea level rise is a major influence.

Where public funds may be called upon to undertake remedial works within a CHZ, it is appropriate that urban development be strictly controlled (Gibb, 1981; Pilkey et al., 1982). But where such development exists prior to establishment of a CHZ, discouragement of development may be difficult to justify if public funds are not at risk. The National Research Council (1990) recommendation is clear that where it is possible to make an appropriately conservative coastal hazard zonation, this should be done prior to the development and at the planning stage; it is too costly to achieve after the subdivision or development has occurred.

A further important issue arises as to what constitutes a credible coastal hazard risk, i.e., what is the probability of occurrence of the variously identified coastal hazards and their effect on the coast? In response, reference is made to the New Zealand Resource Management Act (1991), which defines the meaning of 'effect' in relation to the use, development, or protection of natural and physical resources or in relation to the environment, as "any positive or adverse effect, any temporary or permanent or any past present or future effect, or any cumulative effect regardless of the scale intensity duration or frequency. It thus includes any potential effect of high probability and any potential effect of low probability which has high potential impact." In this respect, the incredible impact of rare natural hazard events of low probability have recently been demonstrated by the Mississippi River floods, estimated by Dr M. F. Meyer (Manager of the Natural Hazards Research and Applications Center of the University of Colorado, personal communication), as demonstrating flood heights of between a 1 in 100 and 1 in 500 year event. The impact of those floods was, of course, colossal.

It is emphasized that in the light of the above definitions, the concepts of "coastal hazard zone" and "development setback" may be applied as a coastal management strategy to protect the development, and avoid the costs to which the local government authority will be put should the hazards eventuate and the local authority be required to undertake remedial action, in particular construction of expensive solid wall protection.

ACKNOWLEDGMENTS

This paper was completed while Terry Healy was a Fulbright Research Fellow at the Department of Coastal and Oceanographic Engineering at the University of Florida. Cynthia Vey is thanked for her careful assistance in preparing the manuscript.

REFERENCES

Aubrey, D. G. and Emery, K. O. 1993. Global Recent Sea Levels and Land Levels. In *Climate and Sea Level Change: Observations, Projections and Implications,* ed. R. A. Warrick, E. M. Barrow, and T. M. L. Wigley, pp. 45–56. Cambridge, UK: Cambridge University Press.

Barnett, T. P. 1983. The Estimates of "Global" Sea Level Change: A Problem of Uniqueness. *Journal of Geophysical Research 89*: 7980–7988.

Barth, M. C. and Titus, J. G. 1984. *Greenhouse Effect and Sea Level Rise.* New York: Van Nostrand Reinhold.

Beatley, T. 1994. *Coastal Hazard Mitigation in the U.S.: State-of-the-Art and Future Directions.* Report of the Proceedings of the U.S.–Spain Workshop on Natural Hazards, Universitat Polytecnica de Catalunya and National Science Foundation, Barcelona, Spain, pp. 304–309.

Bernard, E. N. (Ed.). 1991. *Tsunami Hazard. A Practical Guide for Tsunami Hazard Reduction.* Dordrecht, The Netherlands: Kluwer Academic Publishers.

Boesch, D. F., Josselyn, M. N., Mehta, A. J., Morris, J. T., Nuttle, W. K., Simenstad, C. A., and Swift, D. J. 1994. Scientific Assessment of Coastal Wetland Loss, Restoration and Management in Louisiana. *Journal of Coastal Research,* Special Issue 20: 103.
Bradshaw, B. 1991. Nearshore and Inner Shelf Sedimentation on the East Coromandel Coast, New Zealand. Dr. Phil. Thesis, University of Waikato, Hamilton, New Zealand, 565 pp.
Bradshaw, B., Healy, T. R., Dell, P. M., and Bolstad, W. M. 1991. Inner Shelf Dynamics on a Storm Dominated Coast, East Coromandel, New Zealand. *Journal of Coastal Research* 7(1): 11–30.
Brindell, J. R. 1990. Florida Coastal Management Moves to Local Government. *Journal of Coastal Research* 6(3): 727–733.
Bruun, P. 1962. Sea Level Rise as a Cause of Shore Erosion. *Journal of the Waterways and Harbors Division,* American Society of Civil Engineers, *88*: 117–130.
Bruun, P. 1964. Setback Lines. Engineering Progress at the University of Florida XVIII. p. 5. Gainesville, Florida.
Bruun, P. 1983. Review of Conditions for Uses of the Bruun Rule of Erosion. *Coastal Engineering* 7: 77–89.
Bruun, P. 1986. Worldwide Impact of Sea Level Rise on Shorelines. In *Effects of Changes in Stratospheric Ozone and Global Climate, Vol. 4, Sea Level Rise,* ed. J. G. Titus, pp. 99–128. Washington: U.S. Environmental Protection Agency.
Bruun, P. 1987. Sea Level Rise Effect on Shoreline Position. *Journal of Waterways, Ports, Coastal and Ocean Engineering 113*(5): 550–553.
Bruun, P. 1989. The Bruun Rule of Erosion by Sea Level Rise: A Discussion on Large-Scale Two and Three Dimensional Uses. *Journal of Coastal Research,* 4: 627–648.
Bruun, P. 1990. Worldwide Impact of Sea-Level Rise on Shorelines. In *Handbook of Coastal and Ocean Engineering, Vol. 1, Wave Phenomena and Coastal Structures,* ed. J. B. Herbich, pp. 647–762. Houston, Texas: Gulf Publishing Company.
Camfield, F. E. 1994. Tsunami Effects on Coastal Structures. In *Coastal Hazards Perception Susceptibility and Mitigation,* ed. C. W. Finkl, pp. 177–187. Coastal Education and Research Foundation,
Chiu, T. Y. and Dean, R. G. 1984. Methodology on Coastal Construction Control Line Study. Beaches and Shores Resource Center, Florida State University, Tallahassee, Florida, 166 pp.
CUR, 1987. Manual on Artificial Beach Renourishment. Report 130, Center for Civil Engineering Research, Codes and Specifications, Rijkswaterstaat, and Delft Hydraulics, The Netherlands, 195 pp.
CUR, 1989. Guide to the Assessment of the Safety of Dunes as a Sea Defence. Report 140, Center for Civil Engineering Research, Codes and Specifications, Rijkswaterstaat, and Delft Hydraulics, The Netherlands, 30 pp.
Dahm, J. and Healy, T. R. 1980. A Study of Dredge Spoil Dispersion Off the Entrance to Tauranga Harbour. A Report to the Bay of Plenty Harbour Board, Tauranga, New Zealand.
Dahm, J. and Healy, T. R. 1985. Dredge Spoil Dispersal and Implications for Nearshore Sedimentation in Northern Bay of Plenty, New Zealand. In *Proceedings of the 1985 Australasian Conference on Coastal and Ocean Engineering,* Christchurch, New Zealand, pp. 339–346.
Dean, R. G. 1983. Principles of Beach Nourishment," in *Handbook of Coastal Processes and Erosion.* ed. P. D. Komar, pp. 217–232. Boca Raton, FL: CRC Press.
Dean, R. G. 1988. Sediment Interaction at Modified Coastal Inlets: Processes and Policies. In *Hydrodynamics and Sediment Dynamics of Tidal Inlets, Lecture Notes on Coastal Estuarine Studies, 29,* ed. D. Aubrey and L. Weishar, pp. 412–439. New York: Springer-Verlag.
Dean, R. G. 1990. Beach Response to Sea Level Change. In *The Sea, Volume 9, Ocean Engineering Science,* ed. B. LeMehaute and D. Hanes, pp. 869–887. New York: Wiley.
Dean, R. G. 1991. Equilibrium Beach Profiles and Applications. *Journal of Coastal Research* 7: 53–84.
Dean, R. G. 1994. Shoreline and Sea Level Changes in Florida: Evaluation of a Long Term Coastal Hazard. Report of the Proceedings of the U.S.–Spain Workshop on Natural Hazards, Universitat Polytecnica de Catalunya and National Science Foundation, Barcelona, Spain, pp. 269–275.
Dean, R. G. and Grant, J. 1989. Development of Methodology for Thirty-Year Shoreline Projections in the Vicinity of Beach Nourishment Projects. Rept. No. UFL/COEL-89/026, Coastal and Oceanographic Engineering Department, University of Florida, Gainesville, 153 pp.

Dean, R. G., Healy, T. R., and Dommerholt, A. 1993. A "Blind-Folded" Test of Equilibrium Beach Profile Concepts with New Zealand Data. *Marine Geology 109*: 253–266.
Dean, R. G. and Maurmeyer, E. M. 1983. Models for Beach Profile Response. In *Handbook of Coastal Processes and Erosion,* ed. P. D. Komar, pp. 151–166. Boca Raton, FL: CRC Press.
Dean, R. G. and Yoo, C. H. 1992. Beach Nourishment Performance Predictions. *Journal of the Waterways, Ports, Coastal and Ocean Engineering 118*(6): 567–586.
de Lange, W. and Healy, T. R. 1986a. Tsunami Hazards in the Bay of Plenty, New Zealand: An Example of Hazard Analysis Using Numerical Models. *Journal of Shoreline Management 2*: 177–197.
de Lange, W. P. and Healy, T. R. 1986b. New Zealand Tsunamis, 1840–1982. *New Zealand Journal of Geology and Geophysics 29*: 115–134.
de Lange, W. P. and Healy, T. R. 1994. Assessing the Stability of Inner Shelf Dredge Spoil Mounds Using Spreadsheet Applications on Personal Computers. *Journal of Coastal Research 10*(4): 946–958.
Dubois, R. N. 1977. Predicting Beach Erosion as a Function of Rising Water Level. *Journal of Geology 85*: 470–475.
Everts, C. H. 1985. Sea Level Rise Effects on Shoreline Position. *Journal of Waterways, Ports, Coastal and Ocean Engineering 111*(6): 985–999.
Finkl, C. W. 1994a. Tidal Inlets in Florida: Their Morphodynamics and Role in Coastal Sand Management. In *Proceedings of the Hornafjördur International Coastal Symposium,* ed. G. Viggosson, Höfn, Iceland, pp. 67–85.
Finkl, C. W. (Ed.), 1994b. *Coastal Hazards Perception, Susceptibility and Mitigation,* Florida: Coastal Education and Research Foundation, 372 pp.
Foster, G. A. . 1991. Beach Nourishment from a Nearshore Dredge Spoil Dump at Mt. Maunganui Beach. M. Sc. Thesis, Department of Earth Sciences, University of Waikato, New Zealand, 156 pp.
Foster, D., Warren, S., and Healy, T. 1991. Environmental Aspects of a Recent Dredge Spoil Dumping Operation off Tauranga, New Zealand. In *Coastal Engineering—Climate for Change, Proceedings of the 10th Australasian Conference on Coastal and Ocean Engineering,* Auckland, pp. 141–146.
Foster, G., Healy, T., and de Lange, W. 1994. Sediment Budget and Equilibrium Beach Profiles Applied to Renourishment of an Ebb-Tidal Delta Adjacent Beach, Mt. Maunganui, New Zealand. *Journal of Coastal Research 10*: 564–575.
Frisby, R. B. and Goldberg, E. 1981. Storm Wave Run-Up Levels at Onepoto Bay, East Coast, North Island, New Zealand. In *Coastal Hazard Mapping as a Planning Technique for Waiapu County,* ed. J. G. Gibb, pp. 59–63. Wellington, New Zealand: National Water and Soil Conservation Organisation.
Gibb, J. G. 1981. Coastal Hazard Mapping as a Planning Technique for Waiapu County. Water and Soil Technical Publication 21, National Water and Soil Conservation Organisation, Wellington, 63 pp.
Gibb, J. G. and Aburn, J. H. 1986. Shoreline Fluctuations and an Assessment of a Coastal Hazard Zone Along Pauanui Beach, Eastern Coromandel Peninsula, New Zealand. Water and Soil Technical Publication 27, National Water and Soil Conservation Organisation, Wellington, New Zealand, 48 pp.
Gornitz, V. 1993. Mean Sea Level Changes in the Recent Past. In *Climate and Sea Level Change: Observations Projections and Implications,* ed. R. A. Warrick, E. M. Barrow, and T. M. L. Wigley, pp. 25–44. Cambridge, UK: Cambridge University Press.
Graaff, J., van de, 1986. Probabalistic Design of Dunes: An Example from the Netherlands. *Coastal Engineering 9*(5): 479–500.
Graaff, J., van de, 1994. Coastal and Dune Erosion Under Extreme Conditions. In *Coastal Hazards Perception, Susceptibility and Mitigation,* ed. C. W. Finkl, pp. 253–262. Coastal Education and Research Foundation.
Griggs, G. B. 1994. California's Coastal Hazards. In *Coastal Hazards Perception, Susceptibility and Mitigation,* ed. C. W. Finkl, pp. 1–15. Coastal Education and Research Foundation.
Griggs, G. B. and Fulton-Bennett, K. 1988. Rip Rap Revetments and Seawalls and Their Effectiveness Along the Central California Coast. *Shore and Beach 56*(2): 3–11.

Hallermeier, R. J. 1981. A Profile Zonation for Seasonal Sand Beaches from Wave Climate. *Coastal Engineering 4*: 253–277.
Harray, K. G. and Healy, T. R. 1978. Beach Erosion at Waihi Beach, Bay of Plenty, New Zealand. *New Zealand Journal of Marine and Freshwater Research 12*(2): 99–107.
Hattersley, R. T. and Foster, D. N. 1968. Problems of Beach Erosion and Some Solutions. *Australian Civil Engineering 9*: 1–8.
Healy, T. R. 1974. The Equilibrium Beach: A Model for Real Estate Development and Management of the Coastal Zone in the Northeast of New Zealand. *Proceedings of the International Geographical Union Regional Conference and 8th New Zealand Geography Conference*, Palmerston North, New Zealand, pp. 319–324.
Healy, T. R. 1978a. Beach Surveys 1977–78, Bay of Plenty Coastal Survey Report 78/2. Bay of Plenty Catchment Commission, Wakatane, New Zealand, 37 pp.
Healy, T. R. 1978b. Nearshore Hydrographic Survey of Beach Bars Bay of Plenty Coastal Survey Report 78/3. Bay of Plenty Catchment Commission, New Zealand, 38 pp.
Healy, T. R. 1980a. Erosion and Sediment Drift on the Bay of Plenty Coast. *Soil and Water 16*: 12–14.
Healy, T. R. 1980b. Conservation and Management of Coastal Resources—The Earth Science Basis. In *The Land Our Future,* ed. A. G. Anderson, pp. 239–260. Auckland: Longman Paul.
Healy, T. R. 1981. Coastal Erosion and the Siting of Subdivisions. *Proceedings of the 11th New Zealand Geography Conference,* Wellington, August, 1981, pp. 135–139.
Healy, T. R. 1982. Statement of Evidence to the Planning Tribunal in the Case Tauranga Electric Power Board et al. vs Mt Maunganui Borough Council (unpublished).
Healy, T. R. 1988. Coastal Hazard Setback and Management Recommendations for Development of the Papamoa Dunes [Part section 53 Block I]. A Report to the Planning Committee, Tauranga County Council, Tauranga, 15 pp.
Healy, T. R. 1991. Coastal Erosion and Sea Level Rise. *Zeitschrift für Geomorphologie,* Supplementband 81: 15–29.
Healy, T. R. 1993. Coastal Erosion, Setback Determination, and Recommendations for Management of the Waihi-Bowentown and Pukehina Beach and Dunes. A Report to the Western Bay of Plenty District Council, 55 pp.
Healy, T., Harms, C., and de Lange, W. 1991. Dredge Spoil and Inner Shelf Investigations Off Tauranga Harbour, Bay of Plenty, New Zealand. *Coastal Sediments '91,* Seattle: American Society of Civil Engineers, pp. 2037–2051.
Healy, T. R., Harray, K. G., and Richmond, B. 1977. The Bay of Plenty Coastal Erosion Survey. Occasional Report 3, Department of Earth Sciences, University of Waikato, Hamilton, 64 pp.
Hoffman, J. S., Keyes, J., and Titus, J. G. 1983. Projecting Future Sea Level Rise. Washington: U.S. Environmental Protection Agency, 121 pp.
Houghton, J. T., Jenkins, G. J., and Ephraums, J. J. (Eds.), 1990. *Climatic Change: The IPCC Scientific Assessment,* Cambridge, UK: Cambridge University Press, 362 pp.
Idso, S. B. 1987. CO_2 and Sea Level. *Journal of Coastal Research 3*(4): ii–iii [editorial].
Intergovernmental Panel on Climatic Change, 1990. Strategies for Adaption to Sea Level Rise. Report of the Coastal Zone Management Subgroup, 122 pp.
Kaufman, W. and Pilkey, O. H. Jr. 1983. *The Beaches are Moving,* Durham, NC: Duke University Press, 336 pp.
Kay, R. C., Healy, T. R., Foster, G. A., and Sheffield, A. T. 1993. Tauranga Coastal Hazards Project. Report to the Tauranga District Council and Bay of Plenty Regional Council, Center for Environmental and Resource Studies, University of Waikato, Hamilton, New Zealand.
Komar, P. D. 1976. *Beach Processes and Sedimentation.* Englewood Cliffs: Prentice Hall, 429 pp.
Komar, P. D., Lanfredi, N., Baba, M., Dean, R. G., Dyer, K., Healy, T. R., Ibe, I. C., Terwindt, J., and Thom, B. 1991. The Response of Beaches to Sea-Level Changes: A Review of Predictive Models. *Journal of Coastal Research 7*: 895–921.
Kriebel, D. L. and Dean, R. G. 1985. Numerical Simulation of Time-Dependent Beach and Dune Erosion. *Coastal Engineering 9*: 221–245.
Leatherman, S. P. 1988. Barrier Island Handbook, University of Maryland, College Park, Maryland, 93 pp.
McLean, R. and Mimura, N. (Eds.), 1993. Vulnerability Assessment to Sea-Level Rise and Coastal Zone Management. *Proceedings of the IPCC Eastern Hemisphere Workshop,* Tsukuba, Japan, 427 pp.

Mörner, N. A. 1976. Eustacy and Geoid Changes. *Journal of Geology 84*: 123–151.
National Research Council, 1987. *Responding to Changes in Sea Level—Engineering Implications.* Washington: National Academy Press, 140 pp.
National Research Council, 1990. *Managing Coastal Erosion,* Washington: National Academy Press, 182 pp.
Pilkey, O. H. Jr., Neal, W. J., Pilkey, O. H., Sr., and Riggs, S. R. 1982. *From Currituck to Calabash,* Durham, NC: Duke University Press, 245 pp.
Pilkey, O. H. Sr., Pilkey, W. D., Pilkey, O. H, Jr., and Neal, W. J. 1983. *Coastal Design. A Guide for Builders, Planners and Home Owners,* New York: Van Nostrand Reinhold, 224 pp.
Pirazzoli, P. A. 1986. Secular Trends of Relative Sea-Level (RSL) Changes Indicated by Tide-Gauge Records. *Journal of Coastal Research,* Special Issue No. 1, 1–26.
Schwartz, M. L. 1967. The Bruun Theory of Sea Level Rise as a Cause of Shore Erosion. *Journal of Geology 75*: 76–92.
Schwartz, M. L. 1968. The Scale of Shore Erosion. *Journal of Geology 76*: 508–517.
Schwartz, M. L. 1987. The Bruun Rule—Twenty Years Later. *Journal of Coastal Research* 3(3): ii–iii [editorial].
Sherman, D. J. and Nordstrom, K. F. 1994. Hazards of Wind-Blown Sand and Coastal Sand Drifts. In *Coastal Hazards Perception, Susceptibility and Mitigation,* ed. C. W. Finkl, Coastal Education and Research Foundation, pp. 263–275.
Shore Protection Manual, 1984. Washington, U.S. Army Corps of Engineers, Coastal Engineering Research Center.
Silvester, R. A. 1974. *Coastal Engineering.* Vol. 2. Amsterdam: Elsevier, 338 pp.
Sorensen, R. M., 1978, *Basic Coastal Engineering,* New York: Wiley, 227 pp.
Stewart, R. W., Kjerfve, B., Milliman, J., and Dwivedi, S. N. 1990. Relative Sea-Level Change: A Critical Evaluation. *UNESCO Reports in Marine Science 54,* 22 pp.
Synolakis, C. E. 1991. Tsunami Runup on Steep Slopes: How Good Linear Theory Really Is. In *Tsunami Hazard. A Practical Guide for Tsunami Hazard Reduction,* ed. E. N. Bernard, pp. 221–234. Dordrecht, The Netherlands: Kluwer Academic Publishers.
Thomas, R. 1986. Future Sea Level Rise and Its Early Detection by Satellite Remote Sensing. In *Effects of Changes in Stratospheric Ozone and Global Climate. Vol. 4, Sea Level Rise,* ed. J. G. Titus, pp. 19–36. Washington: U.S. Environmental Protection Agency.
Titus, J. G. (Ed.), 1986a. *Effects of Changes in Stratospheric Ozone and Global Climate. Vol. 4, Sea Level Rise,* Washington: U.S. Environmental Protection Agency, 192 pp.
Titus, J. G. 1986b. Greenhouse Effect, Sea Level Rise, and Coastal Zone Management. *Coastal Zone Management Journal 14*: 147–172.
Titus, J. G. and Narayanan, V. K. 1995. *The Probability of Sea Level Rise.* Washington: U.S. Environmental Protection Agency, 186 pp.
Tooley, M. J. 1993. Long Term Changes in Eustatic Sea Level. In *Climate and Sea Level Change: Observations, Projections and Implications,* ed. R. A. Warrick, E. M. Barrow, and T. M. L. Wigley, Cambridge, UK: Cambridge University Press, pp. 81–107.
Vellinga, P. and Klein, R. J. T. 1993. Climate Change, Sea Level Rise and Integrated Coastal Management: An IPCC Approach. In *Vulnerability Assessment to Sea-Level Rise and Coastal Zone Management.* ed. R. McLean and N. Mimura, Proceedings of the IPCC Eastern Hemisphere Workshop, Tsukuba, Japan, pp. 27–47.
Warrick, R. A. 1993a. Climate and Sea Level Change: A Synthesis. In *Climate and Sea Level Change: Observations, Projections and Implications,* ed. R. A. Warrick, E. M. Barrow, and T. M. L. Wigley, Cambridge, UK: Cambridge University Press, pp. 3–24.
Warrick, R. A. 1993b. Projections of Future Sea-Level Rise: An Update. In *Vulnerability Assessment to Sea-Level Rise and Coastal Zone Management,* ed. R. McLean and N. Mimura, Proceedings of the IPCC Eastern Hemisphere Workshop, Tsukuba, Japan, pp. 51–67.
Warrick, R. A. and Oerlemans, J. 1990. Sea Level Rise. In *Climatic Change, The IPCC Scientific Assessment,* ed. J. T. Houghten, G. J. Jenkins, and J. J. Ephraums, Cambridge, UK: Cambridge University Press, pp. 257–281.
Warrick, R. A., Barrow, E. M., and Wigley, T. M. L. (Eds.), 1993. *Climate and Sea Level Change: Observations, Projections and Implications.* Cambridge, UK: Cambridge University Press, 424 pp.
Weggel, R. J. 1979. A Method for Estimating Long-Term Erosion Rates from a Long Term Rise in

Water Level. Coastal Engineering Technical Aid 79-2, Coastal Engineering Research Center, U.S. Army Corps of Engineers, Vicksburg, Mississippi.

Wigley, T. M. L. and Raper, S. C. B. 1992. Implications for Climate and Sea Level of Revised IPCC Emissions Scenarios. *Nature 357*: 293–300.

Wigley, T. M. L. and Raper, S. C. B. 1993a. Future Changes in Global Mean Temperature and Sea Level. In *Climate and Sea Level Change: Observations, Projections and Implications,* ed. R. A. Warrick, E. M. Barrow, and T. M. L. Wigley, Cambridge, UK: Cambridge University Press, pp. 111–133.

Wigley, T. M. L. and Raper, S. C. B. 1993b. Global Mean Temperature and Sea Level Projections Under the 1992 IPCC Emissions Scenario. In *Climate and Sea Level Change: Observations, Projections and Implications,* ed. R. A. Warrick, E. M. Barrow, and T. M. L. Wigley, Cambridge, UK: Cambridge University Press, pp. 401–404.

Yeh, H. H. 1991. Tsunami Bore Runup. In *Tsunami Hazard. A Practical Guide for Tsunami Hazard Reduction,* ed. E. N. Bernard, Dordrecht, The Netherlands: Kluwer Academic Publishers, pp. 209–220.

APPENDIX A
AUTOMATED COASTAL ENGINEERING SYSTEM (ACES)*

Introduction	A.4
Contents of ACES	A.5
Windspeed Adjustment and Wave Growth	A.6
Description	A.6
Introduction	A.6
General Assumptions and Limitations	A.6
Wind Adjustment	A.7
Wave Growth	A.11
References and Bibliography	A.15
Beta-Rayleigh Distribution	A.16
Description	A.16
Introduction	A.16
General Assumptions and Limitations	A.17
Rayleigh Distribution	A.17
Shallow-Water Distributions	A.18
Beta-Rayleigh Distribution	A.19
Parameterization	A.20
Application	A.23
References and Bibliography	A.23
Extremal Significant Wave Height Analysis	A.24
Description	A.24
General Assumptions and Limitations	A.24
Data for Extremal Analysis	A.24
Extremal Distributions	A.25
Probability of Wave Occurrence	A.27
Selection of a Distribution Function	A.28
References and Bibliography	A.28
Constituent Tide Record Generation	A.28
Description	A.28
Introduction	A.29
General Assumptions and Limitations	A.29
The Tide Prediction Equation	A.30
References and Bibliography	A.30
Linear Wave Theory	A.30
Description	A.30
Introduction	A.31
General Assumptions and Limitations	A.31
Governing Equation	A.32

*Developed by the Coastal Engineering Research Center, U.S. Army Corps of Engineers, Vicksburg, Mississippi.

Boundary Conditions	A.32
Additional Simplifying Assumptions	A.33
Solution of the Governing Equation	A.33
Common Variables of Interest	A.34
References and Bibliography	A.35
Cnoidal Wave Theory	A.35
Description	A.35
Introduction	A.35
General Assumptions and Limitations	A.36
Governing Equation	A.37
Boundary Conditions	A.37
Cnoidal Wave Theory Considerations	A.38
Solution of the Governing Equations	A.39
Results from the Theory	A.41
References and Bibliography	A.44
Fourier Series Wave Theory	A.45
Description	A.45
Introduction	A.45
General Assumptions and Terminology	A.46
Governing Equation and Boundary Conditions	A.47
Additional Considerations Relative to the Dispersion Relation	A.48
Solution and Method	A.49
Derived Results	A.53
Integral Properties	A.54
References and Bibliography	A.55
Linear Wave Theory with Snell's Law	A.56
Description	A.56
Introduction	A.56
General Assumptions and Limitations	A.56
Shoaling and Refraction Coefficients	A.57
References and Bibliography	A.58
Irregular Wave Transformation (Goda's Method)	A.58
Description	A.58
General Assumptions and Limitations	A.59
Theory	A.59
References and Bibliography	A.62
Combined Diffraction and Reflection by a Vertical Wedge	A.62
Description	A.62
Introduction	A.62
General Assumptions and Limitations	A.63
Theoretical Background	A.63
References	A.66
Breakwater Design Using Hudson and Related Equations	A.66
Description	A.66
Introduction	A.67
General Assumptions and Limitations	A.67
Stability of Rubble Structures	A.67
References and Bibliography	A.69
Toe Protection Design	A.69
Description	A.69
Introduction	A.69
General Assumptions and Limitations	A.70
Width of Toe Apron	A.70
Toe Stone Weight	A.72

References and Bibliography	A.73
Nonbreaking Wave Forces at Vertical Walls	A.73
Description	A.73
Introduction	A.73
General Assumptions and Limitations	A.74
Sainflou Method	A.74
Miche–Rundgren Method	A.76
Implementation	A.79
References and Bibliography	A.80
Rubble-Mound Revetment Design	A.81
Description	A.81
Introduction	A.81
General Assumptions and Limitations	A.81
Riprap Armor Stability Formula	A.81
Revetment Design	A.85
Irregular Wave Runup on Riprap	A.87
References and Bibliography	A.88
Irregular Wave Runup on Beaches	A.89
Description	A.89
General Assumptions and Limitations	A.89
Wave Runup Equation	A.90
References and Bibliography	A.90
Wave Runup And Overtopping on Impermeable Structures	A.90
Description	A.90
Introduction	A.91
General Assumptions and Limitations	A.91
Wave Runup	A.92
Rough Slope Runup	A.92
Overtopping Rate	A.94
References and Bibliography	A.95
Wave Transmission on Impermeable Structures	A.96
Description	A.96
Introduction	A.96
General Assumptions and Limitations	A.97
Wave Transmission by Overtopping	A.97
References and Bibliography	A.100
Wave Transmission Through Permeable Structures	A.101
Description	A.101
Introduction	A.102
General Assumptions and Limitations	A.102
Total Wave Transmission	A.102
References and Bibliography	A.114
Longshore Sediment Transport	A.114
Description	A.114
Introduction	A.114
General Assumptions and Limitations	A.115
General Energy Flux Equation	A.115
References and Bibliography	A.119
Numerical Simulation of Time-Dependent Beach and Dune Erosion	A.120
Description	A.120
Introduction	A.120
General Assumptions and Limitations	A.120
Theoretical Development of the Model	A.120
Numerical Solution	A.122

Initial Boundary Conditions	A.124
Model Capabilities	A.125
References and Bibliography	A.126
Calculation of Composite Grain-Size Distributions	A.126
Description	A.126
Introduction	A.126
Sediment Analysis	A.127
Soil Classification	A.128
Statistical Analysis of Sediment Data	A.129
References and Bibliography	A.130
Beach Nourishment Overfill Ratio and Volume	A.131
Description	A.131
Introduction	A.131
General Assumptions and Limitations	A.132
Soil Classification	A.132
Grain-Size Distributions	A.133
Beach-Fill Models	A.134
References and Bibliography	A.136
A Spatially Integrated Numerical Model for Inlet Hydraulics	A.136
Description	A.136
Introduction	A.136
General Assumptions and Limitations	A.137
Numerical Model	A.137
References and Bibliography	A.142
Miscellaneous Routines	A.142
Introduction	A.142
Wave Steepness	A.142
Monochromatic Wave Breaking	A.143
References and Bibliography	A.144

INTRODUCTION

A summary of the various computer programs from the ACES User's Guide is presented here. The ACES Program is also available on the world wide web. The procedure is as follows:

    First get to the Coastal and Hydraulics homepage by keying in the following address:
        http://chl.wes.army.mil/ (there is no www involved).
    On this page, scroll down and click on "software" to produce the page
        http://chl.wes.army.mil/software/
    On this page, scroll down and click on "ACES" to produce the page
        http://chl.wes.army.mil/software/aces/
    On this page, scroll down and click on "ACES" again to produce the page
        http://chl.wes.army.mil/software/aces/aces/
    On this page, click on "download ACES" to produce
        http://chl.wes.army.mil/software/aces/aces/dl.htp
and then follow the directions on this page to unzip all the ACES programs. Of course, one could go directly to this last address and avoid all the intermediate steps from the CHL homepage if desired.

The system is an interactive computer-based design and analysis system in the field of coastal engineering.

CONTENTS OF ACES

Reflecting the nature of coastal engineering, methodologies contained in this release of the ACES are richly diverse in sophistication and origin. The contents range from simple algebraic expressions, both theoretical and empirical in origin, to numerically intense algorithms spawned by the increasing power and affordability of computers. Historically, the methods range from classical theory describing wave motion, to expressions resulting from nests of structures in wave flumes, and to recent numerical models describing the exchange of energy from the atmosphere to the sea surface. In a general procedural sense, much has been taken from previous individual programs on both mainframes and microcompters.

The various methodologies included in ACES are called applications and are organized into categories called functional areas, differentiated according to general relevant physical processes and design or analysis activities. A list of the applications currently resident in the ACES is given in Table A1.

TABLE A1. Current ACES Applications

Functional Area	Application Name
Wave Prediction	Windspeed Adjustment and Wave Growth
	Beta-Rayleigh Distribution
	Extremal Significant Wave Height Analysis
	Constituent Tide Record
Wave Theory	Linear Wave Theory
	Cnoidal Wave Theory
	Fourier Series Wave Theory
Wave Transformation	Linear Wave Theory with Snell's Law
	Irregular Wave Transformation (Goda's method)
	Combined Diffraction and Reflection by a Vertical Wedge
Structural Design	Breakwater Design Using Hudson and Related Equations
	Toe Protection Design
	Nonbreaking Wave Forces on Vertical Walls
	Rubble-Mound Revetment Design
Wave Runup, Transmission, and Overtopping	Irregular Wave Runup on Beaches
	Wave Runup and Overtopping on Impermeable Structures
	Wave Transmission on Impermeable Structures
	Wave Transmission Through Permeable Structures
Littoral Processes	Longshore Sediment Transport
	Numerical Simulation of Time-Dependent Beach and Dune Erosion
	Calculation of Composite Grain-Size Distribution
	Beach Nourishment Overfill Ratio and Volume
Inlet Processes	A Spatially Integrated Numerical Model for Inlet Hydraulics

The ACES Program is menu-driven. Menus are displayed on the screen, and in general, single keystrokes are required to select activities or options in the system. The programs are updated as needs arise.

Either U.S. customary (ft, lbs) or metric (m, kg) systems of units are displayed.

Fresh or seawater is used in numerical examples.

WINDSPEED ADJUSTMENT AND WAVE GROWTH

Description

The methodologies represented in this ACES application provide quick and simple estimates for wave growth over open-water and restricted fetches in deep and shallow water. Also, improved methods (over those given in the *Shore Protection Manual* (SPM), 1984) are included for adjusting the observed winds to those required by wave growth formulas.

Introduction

Wind-generated wave growth is a complex process of considerable practical interest. Although the process is only partially understood, substantial demand remains for quick estimates required for design and analysis procedures. The most accurate estimates available are those provided by sophisticated numerical models such as those presented in Cardone et al. (1976), Hasselmann et al. (1976), Resio (1981), and Resio (1987). Yet many studies, especially at the preliminary level, attempt to describe wind-generated wave growth without the benefit of intensive large-scale modeling efforts. The prediction methods that follow present a first-order estimate for the process, but their simplification of the more complex physics should always be considered.

Methods are included for adjusting observed winds of varying character and location to the conditions required by wave growth formulas. A model depicting an idealized atmospheric boundary layer over the water surface is employed to estimate the low-level winds above the water surface. Stability effects (air-sea temperature gradient) are included, but barotropic effects (horizontal temperature gradient) are ignored. The numerical descriptions of the planetary boundary layer model are based upon similitude theory. Additional corrections are provided for the observed bias of ship-based wind observations as well as short fetches. Formulas for estimating winds of alternate durations are also included. The methodology for this portion of the application is largely taken from Resio, Vincent, and Corson (1982).

The simplified wave growth formulas predict deepwater wave growth according to fetch- and duration-limited criteria and are bounded (at the upper limit) by the estimates for a fully developed spectrum. The shallow-water formulations are based partly upon the fetch-limited deepwater forms and do not encompass duration effects. The methods described are essentially those in Vincent (1984), the SPM (1984), and Smith (1991).

Unless otherwise annotated, metric units are assumed for the following discussion.

General Assumptions and Limitations

The deep- and shallow-water wave growth curves are based on limited field data that have been generalized and extended on the basis of dimensionless analysts. The wind estimation procedures are based on a combination of boundary layer theory and limited field data largely from the Great Lakes. Wind transformation from land to water tends to be highly site and condition specific. The derivation of an individual site from these generalized conditions can create significant errors. Collection of site-specific field data to calibrate the techniques is suggested.

Wind Adjustment

The methodology for preparing wind observations for use in the wave growth formulas is based upon an idealized model of the planetary boundary layer depicted in Figure 1-1-1. For typical mid-latitude conditions, this planetary boundary layer exists in the lowest kilometer of the atmosphere and contains about 10 percent of the atmospheric mass (Holton, 1979).

Low-level winds directly over the water surface are considered to exist in a region characterized as having relatively constant stress at the air-sea interface. This surface layer will be designated the constant stress region for the remainder of this discussion.

Above the constant stress region is the Ekman layer, where the additional forces of Coriolis force, pressure gradient force, viscous stress, and convectively driven mixing are considered important.

Finally, above the Ekman region, geostrophic winds are considered to exist which result from considering the balance between pressure gradient forces and Coriolis force for synoptic scale systems.

Observed winds for use in the wave growth equations are considered to be characterized by six categories summarized in Table 1-1-1.

Although the above six wind observation categories are presented for user convenience, only two separate cases are ultimately considered by the methodology: low-level winds observed within the constant stress region and known or estimated geostrophic winds. In the ACES application, adjustments for ship-based observations are made before proceeding with a solution in the constant stress region, and geostrophic winds are estimated for cases where low-level observed winds are predominantly over land masses. The case of observed winds blowing onshore and measured at the shoreline is considered to be effectively identical to the case of winds observed over water. Similarly, winds observed at the shoreline but blowing from the land mass in an offshore direction are considered effectively equivalent to winds observed at a more inland location. Complex wind

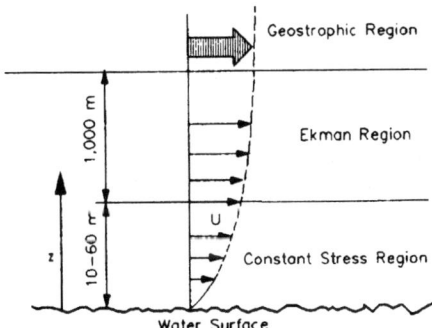

FIGURE 1-1-1. Idealised atmospheric boundary layer over water.

TABLE 1-1-1. Character and action for wind observations

Observation type	Initial action	Solution domain
Over water (non-ship obs)	—	Constant stress layer
Over water (ship obs)	Adjusted	Constant stress layer
At shoreline (onshore winds)	—	Constant stress layer
At shoreline (offshore winds)	Geostrophic wind estimated	Full PBL* model
Over land	Geostrophic wind estimated	Full PBL model
Geostrophic wind	—	Full PBL model

*PBL = Planetary Boundary Layer

patterns caused by local frictional characteristics or topography are obviously not considered by these simplifications.

Initial Adjustments and Estimates

Wind observations over water are typically the most desirable choice of available data sources for wave prediction. Observers on ships at sea frequently record such data and make qualitative estimates. Cardone (1969) reviewed the bias of ship-based observations and suggested the following adjustment:

$$U = 1.864 U_{obs}^{7/9} \text{ (mps)} \tag{1}$$

where

U = adjusted ship-based wind speed
U_{obs} = ship-based observations

For cases where the observed winds are predominantly over land surfaces, similar models of the boundary layer are sometimes employed for other prediction purposes. However, in this application, the following simple estimate for geostrophic winds is made from low-level wind observations (cgs units):

$$V_g = \frac{U^*}{\sqrt{C_{D\,land}}} \tag{2}$$

where

U^* = friction velocity

$$= \frac{kU_{obs}}{\ln\left(\frac{z_{obs}}{z_0}\right)} \tag{3}$$

k = von Karman constant ($k \sim 0.4$)
z_{obs} = elevation of wind observation
z_0 = surface roughness length (assumed = 30 cm)
$C_{D land}$ = drag coefficient over land

$$C_{D land} \sim 0.00255 z_0^{0.1639} \tag{4}$$

Constant Stress Region

The major features of the constant stress region can be summarized as follows:

- The constant stress region is confined to the lowest few meters of the boundary layer.
- Wind flow is assumed parallel to the water surface.

- The wind velocity is adjusted so that the horizontal frictional stress is nearly independent of height.
- The stress remains constant within the layer and is characterized by the friction velocity U^*.

Stability (air-sea temperature gradient) has an important effect on wave growth. The wind profile within this region is described by the following modified logarithmic form:

$$U_z = \frac{U^*}{k}\left[\ln\left(\frac{z}{z_0}\right) - \Psi\left(\frac{z}{L'}\right)\right] \qquad (5)$$

where

U_z = wind velocity at elevation z
z_0 = surface roughness length

$$= \frac{C_1}{U^*} + C_2 U^{*2} + C_3 \qquad (6)$$

$$\left(C_1 = 0.1525,\ C_2 = \frac{0.019}{980},\ C_3 = -0.00371\right) \qquad (7)$$

Ψ = universal similarity function KEYPS formula (Lumley and Panofsky, 1964)
L' = Obukov stability length

$$= 1.79\frac{U^{*2}}{\Delta T}\left[\ln\left(\frac{z}{z_0}\right) - \Psi\left(\frac{z}{L'}\right)\right] \qquad (8)$$

ΔT = air-sea temperature gradient

$$\left.\begin{array}{ll} \Psi = 0 & \Delta T = 0 \\[4pt] \Psi = C\dfrac{z}{L'} & \dfrac{z}{L'} > 0 \\[6pt] \Psi = 1 - \phi_u - 3\ln\phi_u + 2\ln\left(\dfrac{1+\phi_u}{2}\right) + 2\tan^{-1}\phi_u - \dfrac{\pi}{2} + \ln\left(\dfrac{1+\phi_u^2}{2}\right) & \dfrac{z}{L'} \le 0 \end{array}\right\} \qquad (9)$$

$$\phi_u = \frac{1}{1 - 18R_i^{1/4}} \qquad (10)$$

$$R_i = \frac{z}{L'}(1 - 18R_i)^{1/4} \qquad (11)$$

The solution of the above equations is an iterative process that converges very rapidly. The convergence criterion (ϵ) for U^* and L' are given below:

$$\epsilon_{U^*} \to 0.1(\text{cm/sec}) \quad \text{and} \quad \epsilon_L \to 1(\text{cm}) \qquad (12)$$

The wave growth equations discussed later require the equivalent wind speed at a 10-m elevation under conditions of neutral stability ($\Delta T = 0$). Having solved the equations in the constant stress region for U^*, the required equivalent neutral wind speed U^* may be easily obtained from Equation 5 using ($U^*, z = 10$ m, $\Delta T = 0$):

$$U_{e_{1000}} = \frac{U^*}{k}\left[\ln\left(\frac{1000}{z_0}\right) - 0\right] \qquad (13)$$

Full Boundary Layer

For cases where the geostrophic winds are known or have been estimated, the similitude equations describing the entire planetary boundary layer are solved. In addition to the relations described above for the constant stress region, the following relationships describe the model from water surface level to the geostrophic level:

$$\ln \frac{|\vec{V}_g|}{f z_0} = A - \ln \frac{U^*}{|\vec{V}_g|} + \sqrt{\frac{k^2 |\vec{V}_g|^2}{U^{*2}} - B^2} \tag{14}$$

$$\sin \theta = \frac{B U^*}{k |\vec{V}_g|} \tag{15}$$

where

$\vec{V}_g$ = geostrophic wind
f = Coriolis acceleration
A, B = nondimensional functions of stability

$$\begin{aligned} A &= A_0[1 - e^{(0.015\mu)}] \\ B &= B_0 - B_1[1 - e^{(0.03\mu)}] \end{aligned} \qquad \mu \leq 0 \tag{16}$$

$$\begin{aligned} A &= A_0 - 0.96\sqrt{\mu} + \ln(\mu + 1) \\ B &= B_0 + 0.7\sqrt{\mu} \end{aligned} \qquad \mu > 0 \tag{17}$$

μ = dimensionless stability parameter

$$= \frac{kU^*}{fL'} \tag{18}$$

A_0, B_0, B_1 = constants
θ = angle between $\vec{V}_g$ and the surface stress

Equations 14–18 are solved simultaneously together with Equations 5–11 until the convergence of U^*, L', and A is obtained. A slightly different value of ($C_2 = 0.0144/980$) in Equation 7 is used (Dr. C. Linwood Vincent, CERC, personal communication, September 1989). The convergence criteria for the iterative solution to the equations are as follows:

$$\epsilon_{U^*} \to 0.1 \text{(cm/sec)} \quad \text{and} \quad \epsilon_{L'} \to \text{(cm)} \quad \text{and} \quad \epsilon_A \to 0.1 \tag{19}$$

The solution procedure converges very rapidly. As before, Equation 13 is then used to determine the equivalent neutral wind speed at the 10-in elevation using (U^*, $z = 10$ m, $\Delta t = 0$).

Final Adjustments

An additional adjustment is made for situations having relatively short fetch lengths before application of the wave growth equations. For fetch lengths shorter than 16 km, the following reduction is applied:

$$U_e = 0.9 U_e \tag{20}$$

Finally, it is necessary to evaluate the effects of winds of varying duration, t_i, on the wave growth equations. The following expressions are used to adjust the wind speed to a duration of interest:

$$\frac{U_i}{U_{3600}} = 1.277 + 0.296 \tanh\left(0.9 \log \frac{45}{t_i}\right) \quad (1 < t_i < 3600 \text{ sec}) \tag{21}$$

$$\frac{U_i}{U_{3600}} = -0.15 \log t_i + 1.5334 \qquad (3600 < t_i < 36000 \text{ sec}) \qquad (22)$$

The 1-hr wind speed U_{3600} is first determined (using $t_i = t_{obs}$). The wind speed U_i at the desired duration of interest is then determined by selecting the desired t_i and using the appropriate equation.

Wave Growth

Having estimated the winds above the water surface at a duration of interest, the objective is to provide an estimate of the wave growth caused by the winds. The simple wave growth formulas that follow provide quick estimates for wind-wave growth in deep and shallow water. The open-water expressions correspond to those listed in the SPM (1984) and Vincent (1984). The restricted fetch deepwater expressions can be found in Smith (1991). It should be noted that the drag law (Garratt, 1977) employed differs from that in the SPM. The major assumptions regarding the use of the simplified wave growth expressions include:

- Energy from the presence of other existing wave trains is neglected.
- Relatively short fetch geometries ($F \leq 75$mi).
- Relatively constant wind speed ($\Delta U \leq 5$ kts) and direction ($\Delta \alpha \leq 15°$).
- Winds prescribed at the *10-m* elevation ($z = 10$ m).
- Neutral stability conditions.
- Fixed value of drag coefficient ($C_D = 0.001$).

The wind adjustment methodology described earlier in this report adjusts the observed wind, U_{obs}, to the 10-m elevation under neutrally stable conditions U_e. Vincent (1984) maintains the wind speed should be adjusted to consider the nonlinear effect on the wind stress creating the waves. The drag law reported by Garratt (1977) is used:

$$\tau = \rho_a C_D U^2 \qquad (23)$$

where

ρ = air density

$$C_D = 0.001(0.75 + 0.067U) \qquad (24)$$

The equivalent neutral wind speed, then, is adjusted (or linearized) to a constant drag coefficient ($C_D = 0.001$) before application in the wave growth formulas:

$$U_a = U_e \sqrt{\frac{C_D}{0.001}} \qquad (25)$$

Fetch Considerations
The wave growth formulations which follow are segregated into four categories: deep and shallow-water forms for both simple open-water fetches and for more complex, limiting geometries (designated "restricted fetch"). A brief discussion of fetch delineation is useful.

Open-Water Fetches
In open water, wave generation is limited by the dimensions of the subject meteorological event, and fetch widths are of the same order of magnitude as the fetch length. The sim-

plified estimates for wave growth in open water attribute significance to the fetch length (but not width or shape). The wave growth is assumed to occur along the fetch in the direction of the wind.

Restricted Fetches
The more limiting or complex geometries of water bodies such as lakes, rivers, bays, and reservoirs have an impact on wind-wave generation. This restricted fetch methodology applies the concept of wave development in off-wind directions and considers the shape of the basin. The details of the method are reported by Smith (1991), and are based upon a concept reported by Donelan (1980) whereby the wave period (as a function of fetch lengths at off-wind directions) is maximized. For this approach, the radial fetch lengths (as measured from various points along the shoreline of the basin to the point of interest) are used to describe the geometry of the basin. In addition, the wind direction must be specified. Figure 1-1-2 illustrates the relevant geometric data required for the restricted fetch approach.

The conventions used for specifying wind direction and fetch geometry are illustrated in Figure 1-1-3. The approach wind direction (α) as well as the radial fetch angles (β_1), and ($\Delta\beta$) should be specified in a clockwise direction from north from the point of interest where wave growth prediction is required.

From the specified radial fetch data, intermediate values are interpolated at 1-deg increments around the entire 360-deg compass. These interpolated fetches are subsequently averaged over 15-deg arcs centered at each whole 1-deg value.

The direction of wave development (θ) is solved by maximizing the product

$$F_\phi^{0.28} \cdot (\cos \phi)^{0.44} \tag{26}$$

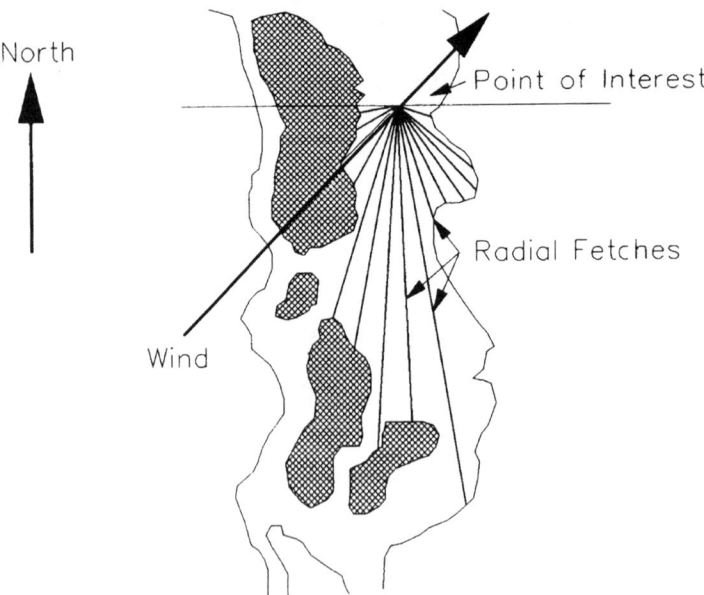

FIGURE 1-1-2. Restricted fetch geometry data.

AUTOMATED COASTAL ENGINEERING SYSTEM (ACES)   A.13

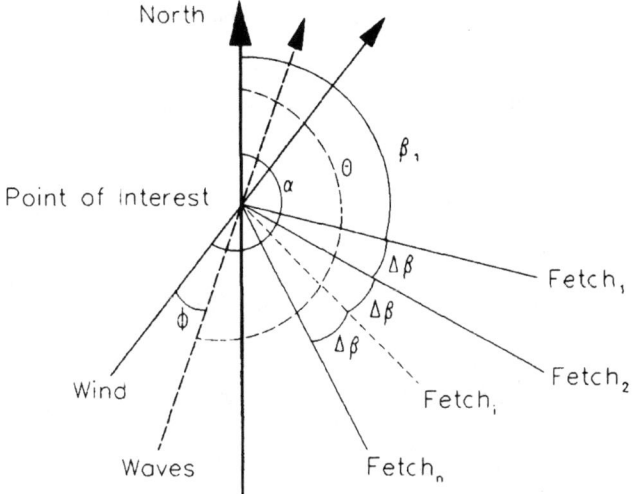

FIGURE 1-1-3. Restricted fetch conventions.

This procedure maximizes the relevant terms in the expression for wave period (T_p) (Equation 36). The angle (ϕ) is defined as the off-wind direction angle associated with the interpolated averaged fetch length value (F_ϕ). Product results (Equation 26) are evaluated from ($\phi = 0 \pm 90°$) at 1-deg increments. When the product (Equation 26) is maximized, (ϕ) represents the angle between the wind and waves, and (θ) represents the compass direction from which wave development occurs along (F_ϕ). For a specified wind direction, there will be a corresponding wave development direction where (T_p) is maximized by Equation 26.

Deepwater Wave Growth

The formulas for wave growth in deep water encompass the effects of fetch and duration. The open-water formulas for fetch- and duration-limited wave growth are taken from Vincent (1984) and are based upon the spectrally based results given in Hasselmann et al. (1973, 1976). The fetch-limited and fully developed forms are also tabulated in the SPM (1984). The expressions for restricted fetch wave growth in deep water are from Smith (1991). In all cases, the wave growth estimates are bounded by the expressions for a fully developed equilibrium spectrum. The procedure is outlined as follows:

- Determine the minimum duration, t_{fetch}, required for a wave field to become fetch-limited:

  Open Water

  $$t_{fetch} = 68.8 \frac{\bar{F}^{2/3}}{g^{1/3} U_a^{1/3}} \tag{27}$$

  Restricted Fetch

  $$t_{fetch} = 51.09 \frac{F^{0.72}}{g^{0.28} \hat{U}_a^{0.44}} \tag{28}$$

- Determine the character of the wave growth (duration-limited or fetch-limited):

Open Water   **Restricted Fetch**

$$H = 0.0000851\left(\frac{U_a^2}{g}\right)\left(\frac{gt_i}{U_a}\right)^{5/7} \quad (29)$$

Duration Limited

$$H = 0.000103\left(\frac{\hat{U}_a^2}{g}\right)\left(\frac{gt_i}{\hat{U}_a}\right)^{0.69} \quad (30)$$

$$T = 0.0702\left(\frac{U_a}{g}\right)\left(\frac{gt_i}{U_a}\right)^{0.411} \quad (31)$$

$(t_i < t_{fetch})$

$$T = 0.082\left(\frac{\hat{U}_a}{g}\right)\left(\frac{gt_i}{\hat{U}_a}\right)^{0.39} \quad (32)$$

or

$$H = 0.0016\left(\frac{U_a^2}{g}\right)\left(\frac{gF}{U_a^2}\right)^{1/2} \quad (33)$$

Fetch Limited

$$H = 0.0015\left(\frac{\hat{U}_a^2}{g}\right)\left(\frac{gF}{\hat{U}_a^2}\right)^{1/2} \quad (34)$$

$$T = 0.2857\left(\frac{U_a}{g}\right)\left(\frac{gF}{U_a^2}\right)^{1/3} \quad (35)$$

$(t_i \geq t_{fetch})$

$$T = 0.3704\left(\frac{\hat{U}_a}{g}\right)\left(\frac{gF}{\hat{U}_a^2}\right)^{0.28} \quad (36)$$

- Determine the "fully developed" condition:

Open Water   **Restricted Fetch**

$$H_{fd} = 0.2433\left(\frac{U_e^2}{g}\right) \quad (37)$$

Fully Developed

$$H_{fd} = 0.2433\left(\frac{\hat{U}_e^2}{g}\right) \quad (38)$$

$$T_{fd} = 8.134\left(\frac{U_e}{g}\right) \quad (39)$$

$$T_{fd} = 8.134\left(\frac{\hat{U}_e}{g}\right) \quad (40)$$

- Ensure that the "fully developed" condition is not exceeded:

$$H_{m_0} = \min(H, H_{fd}) \quad (41)$$

$$T_p = \min(T, T_{fd}) \quad (42)$$

where

- g = acceleration due to gravity
- t_i = wind duration used in duration-limited expressions
- F = fetch length used in fetch-limited expressions
- $\hat{U}_a = U_a \cos(\phi)$ = fetch-parallel component of U_a for restricted fetch approach
- $\hat{U}_e = U_e \cos(\phi)$ = fetch-parallel component of U_e for restricted fetch approach
- H = wave height determined by duration-limited or fetch-limited expressions
- T = wave period determined by duration-limited or fetch-limited expressions
- H_{fd} = wave height limited by fully developed spectrum criteria
- T_{fd} = wave period limited by fully developed spectrum criteria
- H_{m_0} = final wave height determined from spectrally based methods
- T_p = final wave period determined from spectrally based methods

Shallow-Water Wave Growth

Estimates for wave growth in shallow water are based upon the fetch-limited deepwater formulas, but modified to include the effects of bottom friction and percolation (Bretschneider and Reid, 1954). Water depth is assumed to be constant over the fetch. Duration-limited effects are not embodied by these formulas. The relationships have not been verified and may (or may not) be appropriate for the conditions and assumptions of the original Bretschneider-Reid work. **The expressions represent an interim method pending results of further research.** The open-water forms are also presented in the SPM (1984).

Open-Water Forms:

$$H_{m_0} = \frac{U_a^2}{g} 0.283 \tanh\left[0.530\left(\frac{gd}{\hat{U}_a^2}\right)^{0.75}\right] \tanh\left\{\frac{\frac{0.0016}{0.283}\left(\frac{gF}{U_a^2}\right)^{0.5}}{\tanh\left[0.530\left(\frac{gd}{U_a^2}\right)^{0.75}\right]}\right\} \quad (43)$$

$$T_p = \frac{U_a}{g} 7.54 \tanh\left[0.833\left(\frac{gd}{\hat{U}_a^2}\right)^{0.375}\right] \tanh\left\{\frac{\frac{0.2857}{7.54}\left(\frac{gF}{\hat{U}_a^2}\right)^{0.333}}{\tanh\left[0.833\left(\frac{gd}{\hat{U}_a^2}\right)^{0.375}\right]}\right\} \quad (44)$$

Restricted Fetch Forms:

$$H_{m_0} = \frac{U_a^2}{g} 0.283 \tanh\left[0.530\left(\frac{gd}{\hat{U}_a^2}\right)^{0.75}\right] \tanh\left\{\frac{\frac{0.0015}{0.283}\left(\frac{gF}{\hat{U}_a^2}\right)^{0.5}}{\tanh\left[0.530\left(\frac{gd}{\hat{U}_a^2}\right)^{0.75}\right]}\right\} \quad (45)$$

$$T_p = \frac{U_a}{g} 7.54 \tanh\left[0.833\left(\frac{gd}{\hat{U}_a^2}\right)^{0.375}\right] \tanh\left\{\frac{\frac{0.3704}{7.54}\left(\frac{gF}{\hat{U}_a^2}\right)^{0.28}}{\tanh\left[0.833\left(\frac{gd}{\hat{U}_a^2}\right)^{0.375}\right]}\right\} \quad (46)$$

References and Bibliography

Bretschneider, C. L., and Reid, R. 0. 1954. "Modification of Wave Height Due to Bottom Friction, Perlocation and Refraction," Technical Report 50-I, The Agricultural and Mechanical College of Texas, College Station, TX.

Cardone, V. J. 1969. "Specification of the Wind Distribution in the Marine Boundary Layer for Wave Forecasting," TR-69-1, Geophysical Sciences Laboratory, Department of Meteorology and Oceanography, School of Engineering and Science, New York University, New York.

Cardone, V. J., et al. 1976 "Hindcasting the Directional Spectra of Hurricane-Generated Waves," *Journal of Petroleum Technology*, American Institute of Mining and Metallurgical Engineers, No. 261, pp. 385–394.

Donelan, M.A. 1980. "Similarity Theory Applied to the Forecasting of Wave Heights, Periods, and Directions," *Proceedings of the Canadian Coastal Conference*, National Research Council, Canada, pp. 46–61.

Garratt, J. R., Jr. 1977. "Review of Drag Coefficients over Oceans and Continents," *Monthly Weather Review*, Vol. 105, pp. 915–929.

Hasselmann, K., Barnett, T. P., Bonws, E., Carlson H., Cartwright, D. C., Enke, K., Ewing, J., Gienapp, H., Hasselmann, D. E., Kruseman, P., Meerburg, A., Muller, P., Olbers, D. J., Richter, K., Sell, W., and Walden, H. 1973. "Measurements of Wind-Wave Growth and Swell Decay During the Joint North Sea Wave Project (JONSWAP)," Deutches Hydrographisches Institut, Hamburg, 95 pp.

Hasselmann, K., Ross, D. B., Muller, P., and Sell, W. 1976. "A Parametric Prediction Model," *Journal of Physical Oceanography*, Vol. 6, pp. 200–228.

Holton, J. R. 1979. *An Introduction to Dynamic Meteorology*, Academic Press, Inc., New York, pp. 102–118.
Lumley, J. L., and Panofsky, H. A. 1964. *The Structure of Atmospheric Turbulence*, Wiley, New York.
Mitsuyasu, H. 1968. "On the Growth of the Spectrum of Wind-Generated Waves (I)," Reports of the Research Institute of Applied Mechanics, Kyushu University, Fukuoka, Japan, Vol. 16, No. 55, pp. 459–482.
Resio, D. T. 1981. "The Estimation of Wind Wave Generation in a Discrete Model," *Journal of Physical Oceanography*, Vol. II, pp. 510–525.
Resio, D. T. 1987. "Shallow Water Waves. I: Theory," *Journal of Waterway, Port, Coastal and Ocean Engineering*, American Society of Civil Engineers, Vol. 113, No. 3, pp. 264–281.
Resio, D. T., Vincent, C. L., and Corson, W. D. 1982. "Objective Specification of Atlantic Ocean Wind Fields from Historical Data," Wave Information Study Report No. 4, US Army Engineer Waterways Experiment Station, Vicksburg, MS.
Shore Protection Manual. 1984. 4th ed., 2 Vols., US Army Engineer Waterways Experiment Station, Coastal Engineering Research Center, US Government Printing Office, Washington, DC, Chapter 3, pp. 24–66.
Smith, J. M. 1991. Wind-Wave Generation on Restricted Fetches," Miscellaneous Paper CERC-91-2, US Army Engineer Waterways Experiment Station, Vicksburg, MS.
Vincent, C. L. 1984. "Deepwater Wind Wave Growth with Fetch and Duration," Miscellaneous Paper CERC-84-13, US Army Engineer Waterways Experiment Station, Vicksburg, MS.

BETA-RAYLEIGH DISTRIBUTION

Description

This application provides a statistical representation for a shallow-water wave height distribution. The Beta-Rayleigh distribution is expressed in familiar wave parameters: H_{mo} (energy-based wave height), T_p (peak spectral wave period), and d (water depth). After constructing the distribution, other statistically based wave height estimates such as H_{rms}, H_{mean}, $H_{1/10}$ can be easily computed. The Beta-Rayleigh distribution features a finite upper bound corresponding to the breaking wave height, and the expression collapses to the Rayleigh distribution in the deepwater limit. The methodology for this portion of the application is taken exclusively from Hughes and Borgman (1987).

Introduction

Economic coastal engineering design requires accurate specification of the characteristics of the irregular wave field in nearshore waters. In the absence of a fully deterministic understanding of irregular water waves, coastal engineers often describe the sea surface in terms of meaningful wave field parameters which appear to vary in a consistent and predictable manner.

An important and useful statistical descriptor of irregular waves is the wave height distribution. It provides knowledge of the range of wave heights under a given sea condition, as well as the probability of occurrence of a particular wave height within the range. This knowledge can be used to better represent the effects of irregular waves in coastal engineering calculations.

The Rayleigh distribution has proven to be a reliable measure of the wave height distribution for waves in deep water. It is theoretically valid for a wave field composed of a very large number of superimposed deepwater, small-amplitude sinusoidal waves with random phasing and with frequencies very narrowly spread about a single value. For deepwater waves, only the tail in the region of the highest waves exhibits any noticeable deviation from the theory (SPM, 1984, Figure 3-4), and this tendency occurs primarily in the most energetic conditions such as hurricanes (Earle, 1975; Forristall, 1978). Nevertheless, application of the Rayleigh distribution to field wave data has shown repeatedly that the distribution does a remarkable job of predicting the wave height distribution in cases well outside its strict theoretical limits.

General Assumptions and Limitations

All wave parameters derived from the Beta-Rayleigh distribution are limited by the data used to derive this parametric formulation. This derivation was based on preliminary examination of surf zone wave height distributions which indicated that the Beta-Rayleigh distribution can provide a reasonable fit to the data. If the input information H_{mo}, T_p, and d do not fall within the range of data used to derive the parametric formulation, inaccuracies may result. Also, the assumption that the maximum individual wave condition can equal the water depth (upper limit of the Beta-Rayleigh distribution) was suggested by Hughes and Borgman (1987), but they concluded it was only a recommendation until further research is conducted. Because the Beta-Rayleigh distribution reverts to the Rayleigh distribution as depth increases, parameters given by this application when $d/gT^2 \geq 0.01$ are calculated from the Rayleigh distribution.

Rayleigh Distribution

One of the most significant contributions to the parameterization of ocean waves was the demonstration that the wave height distribution of a narrowly banded Gaussian sea state is described by the Rayleigh distribution (Longuet-Higgins, 1952) as

$$p(H) = \frac{2H}{H_{rms}^2} \exp\left[-\left(\frac{H}{H_{rms}}\right)^2\right] \tag{1}$$

where

$p(H)$ = probability density function of wave heights
H = wave height (vertical distance between wave trough and peak)
H_{rms} = root-mean-squared wave height
$= \left[\frac{1}{N}\sum_{i=1}^{N} H_i^2\right]^{1/2}$

The Rayleigh probability distribution also can be written as

$$p(H) = \frac{2H}{8\sigma^2} \exp\left[\frac{-H^2}{8\sigma^2}\right] \tag{2}$$

where

σ^2 = variance of the sea surface elevations

Equating Equations 1 and 2 gives

$$H_{rms} = 2\sqrt{2}\sigma \tag{3}$$

The frequently used definition of significant wave height is given by

$$H_{m_0} = 4\sigma \tag{4}$$

The Rayleigh distribution also has been used to describe wave heights in finite depth water with reasonable success if the assumption of a Gaussian sea state is not violated to a great extent. However, as waves approach depth-limited breaking, they become highly nonsinusoidal in shape, they interact with other nonlinear waves (Scheffner, 1986), and the larger waves in the sea state start breaking. Hence, the process becomes very non-Gaussian, and the wave height distributions deviate from the Rayleigh theory. This deviation can become significant during high-energy wave conditions as evidenced in the Atlantic Ocean Remote Sensing Land-Ocean Experiment (ARSLOE) data set (Ochi, Malakar, and Wang, 1982), in data collected during the monsoonal season in India (Dattatri, 1973), and in data from the US coastline (Thompson, 1974). The Rayleigh distribution of wave heights shows a definite weighting toward the higher end of the distribution. For example, Figure 1-2-1 presents two wave height distribution functions measured in shallow water during the DUCK 85 experiment (Ebersole and Hughes, 1987).

The solid curve in Figure 1-2-1 is the Rayleigh prediction using the measured H_{rms} (root mean square) and the dashed curve is the Rayleigh distribution using the variance as the governing parameter. Theoretically, the two curves should coincide, but in shallow water the nonlinear shape of the waves causes the statistical measure of the wave heights H_{rms} to be larger than that predicted by Equation 3. This difference becomes pronounced as the waves approach breaking. From Figure 1-2-1, it is seen that the shallow-water wave height distribution shows a tendency to skew toward the high-wave side of the Rayleigh distribution; that is, there are more of the higher waves than predicted by the Rayleigh theory. Also, using the statistical parameter, H_{rms}, in the Rayleigh distribution provides a better (though still not good) estimation of the observed wave height distributions.

Shallow-Water Distributions

Previous investigators have suggested improved wave height distributions for shallow water and for the surf zone that are either modified Rayleigh distributions or mathematical forms that have the Rayleigh distribution as a deepwater asymptote. Collins (1970)

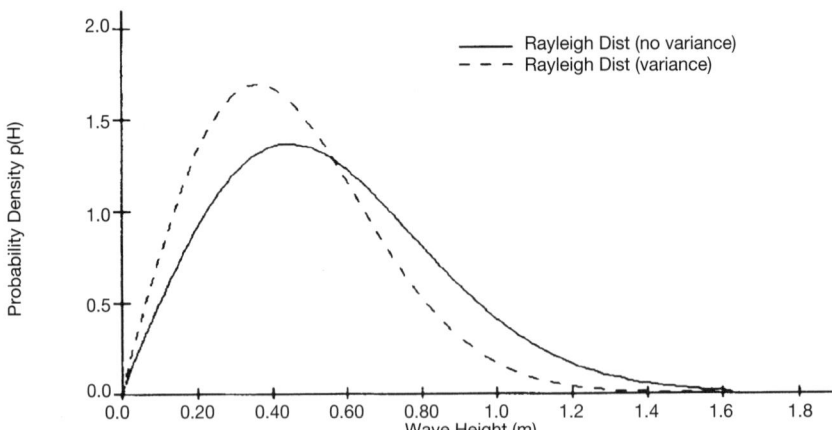

FIGURE 1-2-1. Rayleigh distribution comparison (H_{m_0} = 0.72 m; T_p = 10.9 sec; d = 1.63 m).

and Batties (1972) assumed that the Rayleigh distribution was valid up until the point that depth-limited breaking begins to occur. They then truncated the Rayleigh distribution at the breaking wave height and assumed that all broken waves have a height equal to the local breaking wave height. Hence, the truncated probability density is represented by a delta function located at the breaking wave height. This modification becomes pronounced within the surf zone. Kuo and Kuo (1974) truncated the Rayleigh distribution at the breaking wave height and redistributed the truncated probability density over the remaining wave heights. Goda (1975) assumed wave breaking occurs over a small range of wave heights together with a redistribution of the truncated probability density over the smaller wave heights.

The three examples of modified Rayleigh distributions discussed above result in probability density functions (*pdf*) which have a maximum in about the same location as the original Rayleigh distribution; thus, these distributions do not adequately model the characteristic skewing of the shallow-water wave height distribution toward the higher waves as observed in energetic sea conditions.

Beta-Rayleigh Distribution

To better characterize the wave height distribution for shallow-water waves, a *pdf* with the following attributes is desirable:

- The *pdf* should be bounded by a maximum wave height.
- The *pdf* should have the capability to skew toward higher wave heights than predicted by the Rayleigh distribution, as the sea becomes increasingly non-Gaussian.
- The *pdf* should transform into the Rayleigh distribution as the maximum wave height approaches infinity (i.e., deep water).
- The *pdf* should be mathematically tractable and not too complicated.
- The *pdf* should have some physical justification, either from a deterministic or stochastic viewpoint.

In Hughes and Borgman (1987), a plausible candidate for a shallow-water wave height *pdf* that satisfies these five criteria is derived. The resulting shallow-water *pdf* is referred to as the *Beta-Rayleigh distribution*, and it is expressed as a variation of the beta distribution as

$$p_{BR}(H) = \frac{2\Gamma(\alpha+\beta)}{\Gamma(\alpha)\Gamma(\beta)} \frac{H^{2\alpha-1}}{H_b^{2\alpha}} \left(1 - \frac{H^2}{H_b^2}\right)^{\beta-1} \tag{5}$$

valid in the range $0 < H < H_b$. In Equation 5, H_b denotes the maximum (or breaking) wave height, and α and β are related to the rms wave height by the expression

$$\overline{H^2} = H_{rms}^2 = \int_0^{H_b} H^2 p_{BR}(H) dH = \frac{\alpha H_b^2}{\alpha + \beta} \tag{6}$$

or

$$\beta = \alpha \left(\frac{H_b^2}{H_{rms}^2} - 1\right) \tag{7}$$

Other moments of the *pdf* given in Equation 5 also yield relationships between α and β. The first and third moments are complicated by gamma functions, but the fourth moment provides a simple relationship given by

$$\overline{H^4} = H_{rmq}^2 = \int_0^{H_b} H^4 p_{BR}(H)\,dh = \frac{\alpha(\alpha+1)H_b^4}{(\alpha+\beta)(\alpha+\beta+1)} \tag{8}$$

where

H_{rmq} = *root-mean-quad* wave height

$$= \left[\frac{1}{N}\sum_{i=1}^{N} H_i^4\right]^{1/2} \tag{9}$$

Solving Equations 6 and 8 provides expressions for α and β in terms of physical parameters of the wave height distribution.

$$\alpha = \frac{K_1(K_2 - K_1)}{K_1^2 - K_2} \tag{10}$$

$$\beta = \frac{(1 - K_1)(K_2 - K_1)}{K_1^2 - K_2} \tag{11}$$

where

$$K_1 = \frac{H_{rms}^2}{H_b^2}$$

$$K_2 = \frac{H_{rmq}^2}{H_b^4}$$

Thus, the *pdf* of Equation 5 can be expressed in terms of three parameters, H_b, H_{rmq}, and H_{rms}. Since this is a three-parameter distribution, Equation 5 will provide a better realization than the Rayleigh *pdf* for shallow-water wave heights, provided the three parameters can be correlated to the sea state.

The transition of Equation 5 to the Rayleigh distribution can be seen by taking the limit, as H_b approaches infinity, of the *pdf* given by Equations 5, 10, and 11. This procedure yields (Hughes and Borgman, 1987) a *generalized Rayleigh distribution* given by

$$p_{GR}(H) = \frac{2H^{2\alpha_0 - 1}}{b_0^{\alpha_0}\Gamma(\alpha_0)}\exp\left(\frac{-H^2}{b_0}\right) \tag{12}$$

where

$$\alpha_0 = \frac{H_{rms}^4}{H_{rmq}^2 - H_{rms}^4} \tag{13}$$

$$b_0 = \frac{H_{rmq}^2 - H_{rms}^4}{H_{rms}^2} \tag{14}$$

By noting that $H_{rmq} = \sqrt{2}H_{rms}^2$ for the Rayleigh distribution, Equation 12 reverts to the Rayleigh form given by Equation 1.

Parameterization

The most often used parameters for describing sea states in shallow water are the energy-based significant wave height H_{m_0}, the peak spectral period T_p, and the water depth d.

Less often used parameters include the width of the spectral peak, some indicators of wave groupiness, and the skewness and kurtosis of sea surface elevations.

Root-Mean-Square Wave Height
Following Thompson and Vincent (1985), the deviation of the ratio H_{rms}/H_{m_0} as a function of relative depth, d/gT_p^2, was examined. For the deepwater Rayleigh case, Equations 3 and 4 can be combined to give

$$\frac{H_{rms}}{H_{m_0}} = \frac{1}{\sqrt{2}} = 0.707 \tag{15}$$

However, data obtained from the field (Hughes and Ebersole, 1987) and from laboratory tests indicate that this ratio increases as relative depth decreases. A fit to the data is suggested by the following equation:

$$\frac{H_{rms}}{H_{m_0}} = \frac{1}{\sqrt{2}} \exp\left[a\left(\frac{d}{gT_p^2}\right)^{-b}\right] \tag{16}$$

where
a, b = fitting parameters
g = gravitational acceleration

Linear regression of Equation 16 to laboratory and field data yields values of

$$a = 0.00089 \quad \text{and} \quad b = 0.834$$

with a correlation coefficient of $r = 0.848$. Applying linear regression to Equation 16 weights the lower values of H_{rms}/H_{m_0} more than the higher values. This procedure tends to force the fit through the data at the larger values of relative depth, which serves as a good control in view of the observed scatter at smaller values of relative depth. A reasonable upper envelope to the data is given by Equation 16 when $a = 0.00136$ and b remains as before.

Root-Mean-Quad Wave Height
As previously mentioned, the Rayleigh distribution has a theoretical value of

$$H_{rmq} = \sqrt{2} H_{rms}^2 \tag{17}$$

Combining Equations 15 and 17 provides a deepwater Rayleigh limit for the ratio of H_{rmq} to H_{rms}^2,

$$\frac{H_{rmq}}{H_{m_0}^2} = \frac{1}{\sqrt{2}} = 0.707 \tag{18}$$

Plotting this ratio based on laboratory and field data showed a strong deviation from its theoretical value of 0.707 as relative depth decreased. This deviation was most likely caused by the increasing nonlinearity of the waves as they move into shallow water. As the wave crests peak up and the troughs flatten out, individual wave heights increase without a corresponding increase in sea surface variance, as represented by H_{m_0}.

The quantity H_{rmq} is sensitive to the accurate determination of wave height because it is calculated from wave heights raised to the fourth power. This sensitivity might account for some of the scatter found in the data; however, it is entirely possible that H_{rmq} depends on more than the three parameters investigated here. Surprisingly, the scatter was not reduced by including skewness of sea surface elevations in the parameterization.

Linear regression of the curve

$$\frac{H_{rmq}}{H_{m_0}^2} = \frac{1}{\sqrt{2}} \exp\left[a\left(\frac{d}{gT_p^2}\right)^{-b}\right] \qquad (19)$$

resulted in a fit to the data with

$$a = 0.000098 \quad \text{and} \quad b = 1.208$$

and a correlation coefficient of $r = 0.7863$ Here, the weighting has resulted in a fit that will tend to underpredict H_{rmq} at lower values of relative depth. A reasonable upper envelope to the data is given by Equation 19 when $a = 0.00023$ and b remains as before.

Breaking Wave Height
There are numerous methods for determining a breaking wave height for use in the Beta-Rayleigh distribution. Because the standard predictive expression of

$$H_b = 0.78d \qquad (20)$$

appears to underestimate the breaking wave height, as observed in the photopole data (Ebersole and Hughes, 1987), a representative breaking wave height is suggested of simply

$$H_b = 0.9d \qquad (21)$$

To investigate the effect of the choice for breaking wave height on the Beta-Rayleigh distribution, comparisons were made graphically (Hughes and Borgman, 1987), using Equations 20 and 21. The differences between Equations 20 and 21 were minimal at the lower values of H_{m_0}/d, but a distinct deviation occurred at the higher values of H_{m_0}/d. Until further research is conducted, it is recommended that H_b be taken as equal to the depth, based on the field data obtained from Ebersole and Hughes (1987) and as expressed by Equation 21.

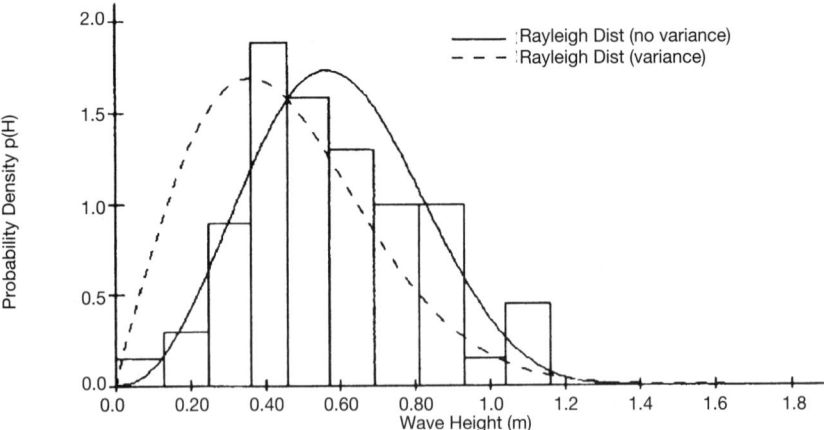

FIGURE 1-2-2. Beta-Rayleigh predictions ($H_{m_0} = 0.72$ m; $T_p = 10.9$ sec; $d = 1.63$ m).

Application

Application of the Beta-Rayleigh distribution requires specification of the water depth, d; the energy-based significant wave height, H_{m_0}; and the peak spectral period, T_p. These parameters are used in Equations 16 and 19 to determine H_{rms} and H_{rmq}, respectively. Next, α and β can be computed by Equations 10 and 11 using a value of H_b equal to the depth (Equation 21). Finally, the Beta-Rayleigh pdf is determined from Equation 5.

The relative improvement of this method over the Rayleigh distribution is illustrated in Figure 1-2-2, where a prebreaking wave height distribution histogram from the photopole experiment is plotted together with the Rayleigh prediction based on H_{m_0} (dashed line) and the Beta-Rayleigh prediction (solid line). In this example (typical of the other cases studied), the Beta-Rayleigh pdf provides a better estimate, as somewhat expected since these data are part of the set used in the parameterization. The Rayleigh prediction to the data based on H_{rms} has already been presented in Figure 1-2-1.

References and Bibliography

Battjes, J. A. 1972. "Set-Up Due to Irregular Waves," *Proceedings of the 13th International Conference on Coastal Engineering*, American Society of Civil Engineers, pp. 1993-2004.

Collins, J. I. 1970. "Probabilities of Breaking Wave Characteristics," *Proceedings of the 12th International Conference on Coastal Engineering*, American Society of Civil Engineers, pp. 399-412.

Dattatri, J. 1973. "Waves off Mangalore Harbor - West Coast of India," *Journal of the Waterway, Port, Coastal, and Ocean Engineering Division*, American Society of Civil Engineers, Vol. 99, No. 1, pp. 39-58.

Earle, M. D. 1975. "Extreme Wave Conditions During Hurricane Camille," *Journal of Geophysical Research*, Vol. 80, No. 3, pp. 377-379.

Ebersole, B. A., and Hughes, S. A. 1987. "DUCK85 Photopole Experiment," Miscellaneous Paper CERC-87-18, US Army Engineer Waterways Experiment Station, Vicksburg, MS.

Forristall, G. Z. 1978. "On the Statistical Distribution of Wave Heights in a Storm," *Journal of Geophysical Research*, Vol. 83, No. C5, pp. 2353-2358.

Goda, Y. 1975. "Irregular Wave Deformation in the Surf Zone," *Coastal Engineering in Japan*, Vol. 18, pp. 13-26.

Hughes, S. A., and Borgman, L. E. 1987. "Beta-Rayleigh Distribution for Shallow Water Wave Heights," *Proceedings of the American Society of Civil Engineers Specialty Conference on Coastal Hydrodynamics*, American Society of Civil Engineers, pp. 17-31.

Kuo, C. T., and Kuo, S. T. 1974. "Effect of Wave Breaking on Statistical Distribution of Wave Heights," *Proceedings of Civil Engineering in the Oceans. III*, American Society of Civil Engineers, pp. 1211-1231.

Longuet-Higgins, M. S. 1952. "On the Statistical Distribution of the Heights of Sea Waves," *Journal of Marine Research*, Vol. II, No. 3, pp. 245-266.

Ochi, M. K., Malakar, S. B., and Wang, W. C. 1982. "Statistical Analysis of Coastal Waves Observed During the ARSLOE Project," UFL/COEL/TR-045, Coastal and Oceanographic Engineering Department, University of Florida, Gainesville, FL.

Scheffner, N. W. 1986. "Biperiodic Waves in Shallow Water," *Proceedings of the 20th International Conference on Coastal Engineering*, American Society of Civil Engineers, pp. 724-736.

Shore Protection Manual. 1984. 4th ed., 2 Vols., US Army Engineer Waterways Experiment Station, Coastal Engineering Research Center, US Government Printing Office, Washington, DC.

Thompson, E. F. 1974. "Results from the CERC Wave Measurement Program," *Proceedings of the International Symposium on Ocean Wave Measurement and Analysis*, American Society of Civil Engineers, pp. 836-855.

Thompson, E. F., and Vincent, C. L. 1985. "Significant Wave Height for Shallow Water Design," *Journal of the Waterway, Port, Coastal, and Ocean Engineering Division*, American Society of Civil Engineers, Vol. 111, No. 5, pp. 828-842.

EXTREMAL SIGNIFICANT WAVE HEIGHT ANALYSIS

Description

This application provides significant wave height estimates for various return periods. Confidence intervals are also provided. The approach developed by Goda (1988) is used to fit five candidate probability distributions to an input array of extreme significant wave heights. Candidate distribution functions are Fisher-Tippett Type I and Weibull with exponents ranging from 0.75 to 2.0. Goodness-of-fit information is provided for identifying the distributions which best match the input data.

General Assumptions and Limitations

General assumptions used in this approach are:

- All extreme wave heights come from a single statistical population of storm events. For example, the heights might represent only extratropical storms in the Northern Hemisphere at a site.
- Wave height properties for an event are reasonably represented by the significant height.
- Extreme wave heights are not limited by any physical factors such as shallow-water depth.

Data for Extremal Analysis

The input array of significant wave heights for extremal analysis is extracted from a long-term data source of measurements, hindcasts, or observations. The reliability of predicted extremes is directly related to the accuracy of available data and the number of years of record. Therefore, the longest high-quality data source available should be used. As a general rule-of-thumb, heights can be extrapolated to return periods up to 3 times the length of record, K. Thus, 20 years of data can be used to estimate significant heights up to a 60-year return period. If return periods longer than $3K$ are requested, a warning message is generated.

Each significant height typically represents the maximum from a storm event. The array of significant heights is referred to as a partial duration series. Often only the more severe storms, with significant height above some minimum value, are considered. The total number, N_T, of events from the population during the length of record must be estimated. The parameter N_T can only be approximated, but results are fairly insensitive to the precise value. It is advisable to consider an average of at least one event per year in the partial duration series. If the average number of events per year, N_T/K, is less than one, a warning message is generated.

Another approach is to use only the highest significant height from each year to form an annual maximum series. The partial duration series is usually preferable for waves. Most sources cover relatively short time periods. The additional information available from a partial duration series can be helpful in increasing confidence in the extremal estimates and extrapolating to rare events.

Extremal Distributions

Probability Distribution Functions

There is no strong physical, theoretical, or empirical evidence for selecting a particular *pdf* for extreme wave heights. The approach commonly used is to try several candidate distributions with each data set and select the one that fits best. The candidate distributions suggested by Goda (1988) and used in this application are as follows:

- Fisher-Tippett Type I (FT-I) Distribution:

$$F(H_s \leq \hat{H}_s) = e^{-e^{-\left(\frac{\hat{H}_s - B}{A}\right)}} \tag{1}$$

- Weibull Distribution:

$$F(H_s \leq \hat{H}_s) = 1 - e^{-\left(\frac{\hat{H}_s - B}{A}\right)^k} \tag{2}$$

where

$F(H_s \leq \hat{H}_s)$ = probability of $\hat{H}_s$ not being exceeded
H_s = significant wave height
$\hat{H}_s$ = particular value of significant wave height
B = location parameter
A = scale parameter
k = shape parameter

The input data are ranked in descending order of significant height. A probability, or *plotting position*, is assigned to each height as follows (Goda 1988 based on previous work by Gringorten, 1963; Muir and El-Shaarawi, 1986; and Petrauskas and Aagaard, 1970):

$$F(H_s \leq H_{sm}) = 1 - \frac{m - 0.44}{N_T + 0.12}; FT-I \tag{3}$$

$$F(H_s \leq H_{sm}) = 1 - \frac{m - 0.20 - \frac{0.27}{\sqrt{k}}}{N_T + 0.20 + \frac{0.23}{\sqrt{k}}}; Weibull$$

where

$F(H_s \leq H_{sm})$ = probability of the mth significant height not being exceeded
H_{sm} = m^{th} value in the ranked significant heights
m = rank of a significant height value
    = 1, 2, ..., N
N_T = total number of events during the length of record (which may exceed the number of input significant heights)

The parameters A and B in Equations 1 and 2 are estimated by computing a least squares fit of the five candidate distribution functions to the data. Computations are based on linear regression analysis of the relationship

$$H_{sm} = \hat{A} y_m + \hat{B}, m = 1, 2, \ldots, N \tag{4}$$

where

$$y_m = -\ln[-\ln F(H_s \le H_{sm})], FT-I$$
$$y_m = \{-\ln[1 - F(H_s \le H_{sm})]\}^{1/k}, Weibull \tag{5}$$

$\hat{A}$ and $\hat{B}$ = estimates of the scale and location parameters from linear regression analysis

Return Period

Return period is defined as the average time interval between successive events of an extreme significant wave height being equalled or exceeded. For example, the 25-year significant height can be expected to be equaled or exceeded an average of once every 25 years. Significant heights for various return periods are calculated from the probability distribution functions by the following equations:

$$H_{sr} = \hat{A}y_r + \hat{B} \tag{6}$$

where

H_{sr} = significant wave height with return period T_r

$$y_r = -\ln\left[-\ln\left(1 - \frac{1}{\lambda T_r}\right)\right], FT-I$$
$$y_r = [\ln(\lambda T_r)]^{1/k}, Weibull \tag{7}$$

λ = average number of events per year

$$= \frac{N_T}{K}$$

T_r = return period (years)
K = length of record (years)

Confidence Intervals

Estimation of confidence intervals is an essential part of extremal wave analysis. Typically the period of record is short, and the level of uncertainty in extremal estimates with long return periods is high. Confidence intervals give a quantitative indicator of the level of uncertainty in estimated extremal wave heights. The approach of Gumbel (1958) and Goda (1988) for estimating standard deviation of return value when the true distribution is known is used. The normalized standard deviation is calculated by

$$\sigma_{nr} = \frac{1}{\sqrt{N}}[1.0 + a(y_r - c + \epsilon \ln v)^2]^{1/2} \tag{8}$$

where

σ_{nr} = normalized standard deviation of significant wave height with return period r
N = number of input significant heights

$$a = a_1 e^{a_2 N^{-1.3} + \kappa\sqrt{-\ln v}} \tag{9}$$

$a_1, a_2, c, \epsilon, \kappa$ = empirical coefficients (Table 1-3-1)
v = censoring parameter

$$= \frac{N}{N_T}$$

TABLE 1-3-1. Coefficients of empirical standard deviation formula for extreme significant height (Goda, 1988)

Distribution	a_1	a_2	κ	c	ϵ
$FT-I$	0.64	9.0	0.93	0.0	1.33
Weibull ($k = 0.75$)	1.65	11.4	−0.63	0.0	1.15
Weibull ($k = 1.0$)	1.92	11.4	0.00	0.3	0.90
Weibull ($k = 1.4$)	2.05	11.4	0.69	0.4	0.72
Weibull ($k = 2.0$)	2.24	11.4	1.34	0.5	0.54

The absolute magnitude of the standard deviation of significant wave height is calculated by

$$\sigma_r = \sigma_{nr}\sigma_{H_s} \tag{10}$$

where

σ_r = standard error of significant wave height with return period r
σ_{H_s} = standard deviation of input significant heights

Confidence intervals are calculated by assuming that significant height estimates at any particular return period are normally distributed about the assumed distribution function. Factors by which to multiply the standard error (Equation 10) to get bounds with various levels of confidence are given in Table 1-3-2. It is important to note that the width of confidence intervals depends on the distribution function, N, and ν; but it is not related to how well the data fit the distribution function.

Probability of Wave Occurrence

The probability of wave occurrence is defined as the probability that a height with given return period will be equaled or exceeded during some time period. An example is the probability that the 25-year wave height will occur during a 10-year period. The probability of wave occurrence, expressed as a percent chance of occurrence, is calculated as (Headquarters, Department of the Army, 1989)

$$P_e = 100\left[1 - \left(1 - \frac{1}{T_r}\right)^L\right] \tag{11}$$

TABLE 1-3-2. Confidence interval bounds for extreme significant height

Confidence level (%)	Confidence interval bounds around H_{sr}	Probability of exceeding upper bound (%)
80	$\pm 1.28\sigma_r$	10.0
85	$\pm 1.44\sigma_r$	7.5
90	$\pm 1.65\sigma_r$	5.0
95	$\pm 1.96\sigma_r$	2.5
99	$\pm 2.58\sigma_r$	0.5

where

P_e = percent chance of occurrence
L = time period of concern (years)

Selection of a Distribution Function

The five distribution functions considered in this analysis are sufficiently different that only one or two can be expected to provide a wood fit to any particular data set. Two statistics are provided to assist in selecting the best fit dlstribution function. The correlation between variables in the linear Equation 4 is the primary selection criterion. The distribution function that gives the highest correlation should be selected. The sum of the squares of residuals,

$$\sum_{m=1}^{N} [H_{sm} - (\hat{A}y_m + \hat{B})]^2 \qquad (12)$$

is also provided. The sum is usually smallest for the distribution function with the highest correlation. Plots are also available to help judge the fit between data and distribution functions and the width of confidence intervals. If a second distribution function fits nearly as well as the best fit (i.e., the correlation is nearly as high and the sum of the squares of residuals is comparable), then it would be appropriate to consider extremal heights from both distributions. Extreme heights from the two distribution functions could be averaged together. Alternatively, the higher of the two could be used if a conservative estimate is desired.

References and Bibliography

Goda, Y. 1988. "On the Methodology of Selecting Design Wave Height," *Proceedings, Twenty-first Coastal Engineering Conference*, American Society of Civil Engineers, Costa del Sol-Malaga, Spain, pp. 899–913.
Gringorten, I. I. 1963. "A Plotting Rule for Extreme Probability Paper," *Journal of Geophysical Research*, Vol. 68, No. 3, pp. 813–814.
Gumbel, E. J. 1958. *Statistics of Extremes*, Columbia University Press, New York.
Muir, L. R., and El-Shaarawi, A. H. 1986. "On the Calculation of Extreme Wave Heights: A Review," *Ocean Engineering*, Vol. 13, No. 1, pp. 93–118.
Petrauskas, C., and Aagaard, P. M. 1970. "Extrapolation of Historical Storm Data for Estimating Design Wave Heights," *Proceedings, 2nd Offshore Technology Conference*, OTC1190.
Headquarters, Department of the Army. 1989. "Water Levels and Wave Heights for Coastal Engineering Design," Engineer Manual 1110-2-1414, Washington, DC, Chapter 5, pp. 72–80.

CONSTITUENT TIDE RECORD GENERATION

Description

This ACES application predicts a tide elevation record at a specific time and locale using known amplitudes and epochs for individual harmonic constituents.

Introduction

Tides are periodic in nature and are caused by the gravitational attraction of the sun and moon acting upon the rotating earth. The irregular distribution of land and water upon the planet, as well as the effects of friction, inertia, and the interactions of superimposed standing waves and progressive waves tend to complicate actual tidal motions induced by the gravitational forces. Around the year 1867, Lord Kelvin devised the method of reducing tides into constituents using harmonic analysis (Schureman, 1971). Harmonic analysis (as applied to tides) is the process by which observed tidal data at a location are reduced into a number of harmonic constituents. The quantities produced by such an analysis are known as harmonic constants and consist of amplitudes and phase relationships for each constituent. Harmonic prediction (the subject of this application) is accomplished by re-assembling the individual constituents in accordance with the appropriate astronomical relationships during the prediction period.

Harmonic tide prediction is a convenient tool for scheduling construction or maintenance work that may be affected by spring or neap tide conditions. It is also a useful tool in the application and verification of certain types of physical and hydrodynamic models where tide elevations are required either as boundary conditions or as stations for calibrating model parameters and verifying model performance.

General Assumptions and Limitations

The method is based upon the assumption that the tide elevation at a location can be expressed as a series of harmonic terms which are in turn dependent upon the gage location as well as the astronomical conditions during the prediction period. An individual harmonic term is represented as a simple cosine function consisting of an amplitude and phase relationship:

$$h_n = A_n \cos(a_n t + \alpha_n) \tag{1}$$

Equation 1 expresses the contribution of an individual constituent (n) to the complete tide elevation. The expression contains an amplitude (A_n) and a phase relationship (the argument of the cosine function). The phase relationship is a function of the speed of the constituent (a_n), the time (t) measured from some initial epoch, and (α_n), which represents the initial phase of the constituent at the initial epoch ($t = 0$).

An important assumption in applying this methodology is that the observed record (from which the harmonic constants were derived) was of sufficient length, resolution, and quality to resolve the harmonic data adequately. Data from site-specific studies are usually short term in duration. Permanent recording stations provide longer and often more reliable records. Typical records chosen for analysis consist of hourly data analyzed in series lengths of 29 and 369 days. The latter series length is considered highly desirable for resolving most of the short period constituents and eliminating seasonal effects. Harmonic constants are often available from site-specific studies and from the National Ocean Survey.

The accuracy of the tide prediction is dependent upon the number of constituents included in the equation. In the United States, a maximum of 37 constituents (listed in Table A-5, Appendix A) are typically considered. However, in some areas, more constituents are included because of regional complexities. In many cases, a small number of constituents will represent the significant portion of the tide. For example, tides along the eastern coast of the United States are dominated by the M_2 constituent. Care should be taken to examine the amplitudes of the individual constituents in order to select those having the most significant influence on the amplitude and phase of the tide.

The Tide Prediction Equation

The general equation for the height of the tide at any time t can be determined from the following formula:

$$h = H_0 + \sum_{n=1}^{N} f_n H_n \cos[a_n t + (V_0 + u)_n - \kappa_n] \qquad (2)$$

where

h = height of tide at time t
H_0 = mean height of water level above datum used for prediction
N = number of constituents to be considered
f_n = node factor of constituent n
H_n = amplitude of constituent n
a_n = speed of constituent n
t = time reckoned from some initial epoch
$(V_0 + u)_n$ = value of local equilibrium argument of constituent n at the initial epoch ($t = 0$)
κ_n = value of local phase lag or epoch of constituent n

In Equation 2, all quantities except h and t may be considered constant for a particular time period and location. Values of H_n and κ_n are the harmonic constants derived from analysis of past records. They must be specified (as input) for each of the constituents to be considered in the summation. The methodology implemented in ACES has algorithms for determining f_n and $(V_0 + u)_n$ for each constituent as functions of the date and time of the initial epoch ($t = 0$), the total record length, and the longitude of the gage record. Essentially, they are determined by various combinations of astronomical functions (which are functions of time). Additionally, the equilibrium terms are also a function of longitude. In this application, their values are determined at Greenwich and translated to the gage longitude specified. The details of the computations are well described in Schureman (1971).

It should be noted that the results of some harmonic analysis are reported yielding a value of (κ'_n) which is a modified epoch (relative to Greenwich). In these cases the value of longitude at Greenwich (= 0) should be specified when using (κ'_n).

References and Bibliography

Harris, D. L. 1981. "Tides and Tidal Datums in the United States," Special Report SR-7, US Army Engineer Waterways Experiment Station, Vicksburg, MS.
Headquarters, Department of the Army. 1989. "Water Levels and Wave Heights for Coastal Engineering Design," Engineer Manual 1110-2-1414, Washington, DC, Chapter 2, pp. 5–10.
Schureman, P. 1971 (reprinted). "Manual of Harmonic Analysis and Prediction of Tides," Coast and Geodetic Survey Special Publication No. 98, Revised (1940) Edition, US Government Printing Office, Washington, DC.

LINEAR WAVE THEORY

Description

This application yields first-order approximations for various parameters of wave motion as predicted by the wave theory bearing the same name (also known as small amplitude, sinusoidal, or Airy theory). It provides estimates for engineering quantities such as water surface elevation, general wave properties, particle kinematics, and pressure as functions of wave height and period, water depth, and position in the wave form.

Introduction

The effects of water waves are of major importance in the field of coastal engineering. Waves are a major factor in determining geometry and composition of beaches and significantly influence planning and design of harbors, waterways, shore protection measures, coastal structures, and other coastal works.

In general, actual water-wave phenomena are complex and difficult to describe mathematically because of nonlinearities, three-dimensional characteristics, and apparent random behavior. The most elementary wave theory, referred to as small-amplitude or linear wave theory, was developed by Airy (1845). This nomenclature derives from the simplifying assumptions of its derivation. Additionally, it represents a first approximation resulting from a formal perturbation procedure for waves of finite amplitude.

General Assumptions and Limitations

A typical representation of a wave is depicted in Figure 2-1-1.
Common terminology for wave discussions includes the following:

d = still-water depth
η = free surface elevation relative to still water ($z = 0$)
a = wave amplitude
H = wave height = $2a$ for small-amplitude waves
L = wavelength
T = wave period
c = velocity of wave propagation (celerity) = L/T
k = wave number = $2\pi/L$

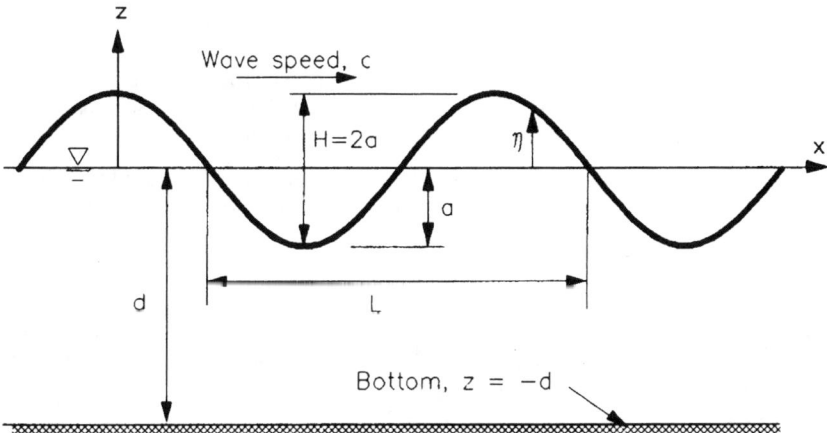

FIGURE 2-1-1. Progressive wave.

ω = wave angular frequency = $2\pi/T$
u, w = horizontal and vertical components of velocity vector $\vec{u}$
ϕ = velocity potential $\vec{u} = \vec{\nabla}\phi$
g = acceleration of gravity
ρ = density of water

General assumptions and limitations used to derive wave theories are:

- Waves are two-dimensional (2-D) in the x-z plane.
- Waves propagate in a permanent form over a smooth horizontal bed of constant depth in the positive x-direction.
- There is no underlying current.
- Fluid is inviscid and incompressible, having no surface tension.
- Flow is irrotational.
- Coriolis effect is neglected.

Governing Equation

The assumption of irrotational flow leads to the existence of a velocity potential. An ideal fluid must satisfy the mass continuity equation and can be expressed in primitive variables as

$$\frac{\partial u}{\partial x} + \frac{\partial w}{\partial z} = 0 \qquad (1)$$

or, in terms of the velocity potential

$$\nabla^2 \phi = \frac{\partial^2 \phi}{\partial x^2} + \frac{\partial^2 \phi}{\partial z^2} = 0 \qquad (2)$$

Equation 2 is the 2-D Laplace equation.

Boundary Conditions

The governing equation describes a boundary value problem. The various boundary conditions of the problem domain affect the form and complexity of the solution of the Laplace equation. The boundary conditions can be summarized as:

Bottom Boundary Condition (BBC). Fluid must not pass through the seafloor. Therefore at the sea bottom, the vertical component of the water particle velocity must vanish.

$$\frac{\partial \phi}{\partial z} = 0 \qquad \text{at } z = -d \qquad (3)$$

Kinematic Free Surface Boundary Condition (KFSBC). The fluid particle velocity normal to the free surface is equal to the velocity of the free surface itself. This condition implies that a water particle on the free surface remains there.

$$\frac{\partial \eta}{\partial t} + \frac{\partial \phi}{\partial x}\frac{\partial \eta}{\partial x} - \frac{\partial \phi}{\partial z} = 0 \qquad \text{at } z = \eta \qquad (4)$$

Dynamic Free Surface Boundary Condition (DFSBC). Expressed as the Bernoulli equation, the pressure at the free surface is constant. This requirement follows from an assumption that atmospheric pressure above the fluid is itself constant and no surface tension is present at the air-water interface.

$$\frac{\partial \phi}{\partial t} + \frac{1}{2}\left[\left(\frac{\partial \phi}{\partial x}\right)^2 + \left(\frac{\partial \phi}{\partial z}\right)^2\right] + g\eta = f(t) \qquad \text{at } z = \eta \qquad (5)$$

Periodicity. The wave is periodic in time and space.

$$\phi(x, t) = \phi(x + L, t)$$
$$\phi(x, t) = \phi(x, t + T) \qquad (6)$$

Additional Simplifying Assumptions

The partial differential equations are difficult to solve because of problems associated with the free surface boundary conditions. They are nonlinear and occur at location $z = \eta$, which is initially unknown. Linear wave theory derives from applying simplifying assumptions to the free surface boundary conditions. The still-water wave height H is assumed to be very small relative to both the wavelength L and the still-water depth d:

$$H \ll L \quad \text{and} \quad H \ll d$$

The result of these assumptions is that the nonlinear terms which contain products of terms of order of H are then negligible in comparison with remaining linear terms of order H. Also, the free surface boundary conditions may now be applied at the still-water level $z = 0$. The simplified or linearized free surface boundary conditions then reduce to:

Kinematic Free Surface Boundary Condition (KFSBC).

$$\frac{\partial \phi}{\partial z} - \frac{\partial \eta}{\partial t} = 0 \qquad \text{at } z = 0 \qquad (7)$$

Dynamic Free Surface Boundary Condition (DFSBC).

$$\frac{\partial \phi}{\partial t} + g\eta = 0 \qquad \text{at } z = 0 \qquad (8)$$

Solution of the Governing Equation

The solution for the velocity potential is obtained by applying the method of separation of variables and the given boundary conditions. The resulting solution is stated below as well as the linear dispersion relation:

$$\phi = \frac{\pi H}{kT}\frac{\cosh(ks)}{\sinh(kd)}\sin\theta \quad \text{or} \quad \phi = \frac{gH}{2\omega}\frac{\cosh(ks)}{\cosh(kd)}\sin\theta \qquad (9)$$

$$\omega^2 = gk\tanh(kd) \quad \text{or} \quad c^2 = \frac{g}{k}\tanh(kd) \qquad (10)$$

where

$\theta = k(x - ct) = kx - wt$ = wave phase angle
$s = z + d$ measured upwards from the seabed

The dispersion relation describes the manner in which a field of propagating waves consisting of many frequencies would separate or "disperse" due to the different celerities of the various frequency components (Dean and Dalrymple, 1984). The linear dispersion relation is a transcendental function but may be readily evaluated using a Pade approximation (Hunt, 1979):

$$c^2 = gd\left[y + \left(1 + \sum_{n=1}^{9} d_n y^n\right)^{-1}\right]^{-1} \tag{11}$$

where

$$y = \frac{\omega^2 d}{g}$$

d_n = constants
$d_1 = 0.66667$ $d_4 = 0.06320$ $d_7 = 0.00171$
$d_2 = 0.35550$ $d_5 = 0.02174$ $d_8 = 0.00039$
$d_3 = 0.16084$ $d_6 = 0.00654$ $d_9 = 0.00011$

This approximation has an accuracy better than 0.01 percent over the range $0 \leq y \leq \infty$.

Common Variables of Interest

The solution for ϕ and the linear dispersion relation provide a foundation for deriving expressions for other common variables of interest. These equations are listed below and can be found in many texts including Sarpkaya and Isaacson (1981) and the SPM (1984).

Wavelength: $\qquad L = cT \qquad$ (12)

Group velocity: $\qquad C_g = \frac{1}{2}\left[1 + \frac{2kd}{\sinh(2kd)}\right]c \qquad$ (13)

Water surface elevation: $\qquad \eta = \frac{H}{2}\cos\theta \qquad$ (14)

Average energy density: $\qquad E = \frac{1}{8}\rho g H^2 \qquad$ (15)

Energy flux: $\qquad P = EC_g \qquad$ (16)

Pressure: $\qquad p = -\rho g z + \frac{1}{2}\rho g H \frac{\cosh(ks)}{\cosh(kd)}\cos\theta \qquad$ (17)

Horizontal particle displacement: $\qquad \xi = -\frac{H}{2}\frac{\cosh(ks)}{\sinh(kd)}\sin\theta \qquad$ (18)

Vertical particle displacement: $\qquad \zeta = \frac{H}{2}\frac{\sinh(ks)}{\sinh(kd)}\cos\theta \qquad$ (19)

Horizontal particle velocity: $\qquad u = \frac{\pi H}{T}\frac{\cosh(ks)}{\sinh(kd)}\cos\theta \qquad$ (20)

Vertical particle velocity: $$w = \frac{\pi H}{T} \frac{\sinh(ks)}{\sinh(kd)} \sin\theta \qquad (21)$$

Horizontal particle acceleration: $$\frac{\partial u}{\partial t} = \frac{2\pi^2 H}{T^2} \frac{\cosh(ks)}{\sinh(kd)} \sin\theta \qquad (22)$$

Vertical particle acceleration: $$\frac{\partial w}{\partial t} = -\frac{2\pi^2 H}{T^2} \frac{\sinh(ks)}{\sinh(kd)} \cos\theta \qquad (23)$$

A common parameter for the applicability of various wave theories is the Stokes (1847) or Ursell (1953) parameter, which is defined as

$$U_r = \frac{HL^2}{d^3} \qquad (24)$$

This parameter is reported for convenience.

References and Bibliography

Airy, G. B. 1845. "Tides and Waves," *Encyclopaedia Metropolitana*, Vol. 192, pp. 241–396.
Dean, R. G., and Dalrymple, R. A. 1984. *Water Wave Mechanics for Engineers and Scientists*, Prentice-Hall, Englewood Cliffs, NJ, pp. 41–86.
Hunt, J. N. 1979. "Direct Solution of Wave Dispersion Equation," *Journal of Waterway, Port, Coastal and Ocean Division*, American Society of Civil Engineers, Vol. 105, No. WW4, pp. 457–459.
Sarpkaya, T., and Isaacson, M. 1981. *Mechanics of Wave Forces on Offshore Structures*, Van Nostrand Reinhold, New York, pp. 150–168.
Shore Protection Manual. 1984. 4th ed., 2 Vols., US Army Engineer Waterways Experiment Station, Coastal Engineering Research Center, US Government Printing Office, Washington, DC, Chapter 2, pp. 6–33.
Stokes, G. G. 1847. "On the Theory of Oscillatory Waves," *Transactions of the Cambridge Philosophical Society*. Vol. 8, pp. 441–455.
Ursell, F. 1953. "The Long-Wave Paradox in the Theory of Gravity Waves," *Proceedings of the Cambridge Philosophical Society*, Vol. 49, pp. 685–694.

CNOIDAL WAVE THEORY

Description

This application yields various parameters of wave motion as predicted by first-order (Isobe, 1985) and second-order (Hardy and Kraus, 1987) approximations for cnoidal wave theory. It provides estimates for common items of interest such as water surface elevation, general wave properties, kinematics, and pressure as functions of wave height and period, water depth, and position in the wave form.

Introduction

The accurate description of waves in the nearshore region is an important element in the design and analysis process. The complexity of nonlinear wave theories has apparently discouraged their common application, especially for reconnaissance level investigations; yet they often provide significant improvements over linear wave theory descriptions. Today's common desktop microcomputers offer adequate computational power to implement many nonlinear wave theories.

Solutions for wave attributes described by this theory are expressed in terms of the Jacobian elliptic function cn, thus providing an explanation for the theory's name. The original source for the theory was a paper by the Dutch mathematicians Korteweg and de Vries (1895). More recent derivations for cnoidal wave theory include those of Keulegan and Patterson (1940), Keller (1948), Laitone (1960), Chappelear (1962), Fenton (1979), and Isobe and Kraus (1983) among others. Differences between the derivations consist primarily of choices in perturbation parameter, definition of celerity, and order of the solution. The following discussion is taken principally from Hardy and Kraus (1987), and Isobe and Kraus (1983).

General Assumptions and Limitations

A representative cnoidal wave profile is illustrated in Figure 2-2-1.

The assumptions and terminology for general wave theories will be reviewed first, followed by specific assumptions and treatments associated with cnoidal wave theory. Terms common to wave discussions include:

d = still-water depth
η = free surface elevation relative to still water ($z = 0$)
a = wave amplitude
H = wave height = $2a$ for small-amplitude waves
L = wavelength
T = wave period
c = velocity of wave propagation (celerity) = L/T
$\vec{u}$ = water particle velocity vector
u, w = horizontal and vertical components of velocity vector $\vec{u}$

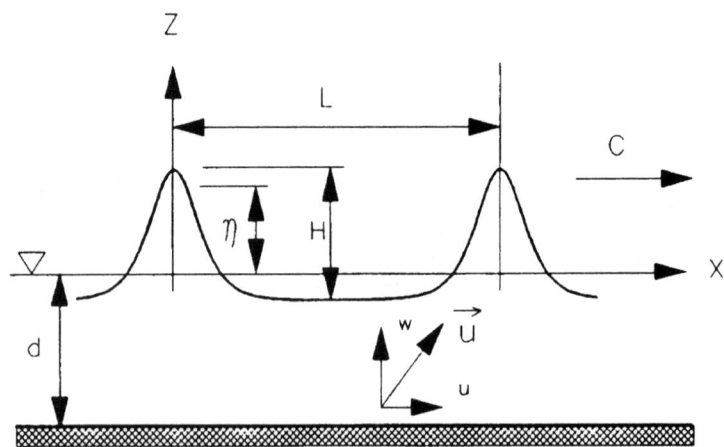

FIGURE 2-2-1. Cnoidal wave profile.

ϕ = velocity potential $\vec{u} = \vec{\nabla}\phi$
ψ = stream function
g = acceleration of gravity
ρ = density of water
p_B = Bernoulli constant

General assumptions and limitations used to derive wave theories include:

- Waves are two-dimensional in the x-z plane.
- Waves propagate in a permanent form over a smooth horizontal bed of constant depth in the positive x-direction.
- There is no underlying current.
- Fluid is inviscid and incompressible, having no surface tension.
- Flow is irrotational.

Governing Equation

The assumption of irrotational flow leads to the existence of a velocity potential. An ideal fluid must satisfy the mass continuity equation and can be expressed in terms of the velocity potential as:

$$\nabla^2 \phi = 0 \tag{1}$$

This is the 2-D Laplace equation.

Boundary Conditions

The governing equation describes a boundary value problem. The various boundary conditions of the problem domain affect the form and complexity of the solution of the Laplace equation. The boundary conditions can be summarized as:

Bottom Boundary Condition (BBC). Fluid must not pass through the seafloor. Therefore, at the sea bottom, the vertical component of the water particle velocity must vanish.

$$\phi_z = 0 \qquad \text{at } z = -d \tag{2}$$

Kinematic Free Surface Boundary Condition (KFSBC). The fluid particle velocity normal to the free surface is equal to the velocity of the free surface itself. This condition implies that a water particle on the free surface remains there. It also implies that the wave does not break.

$$\eta_t + \phi_x \eta_x - \phi_z = 0 \qquad \text{at } z = \eta \tag{3}$$

Dynamic Free Surface Boundary Condition (DFSBC). Expressed as the Bernoulli equation, the pressure at the free surface is constant. This requirement follows from an assumption that atmospheric pressure above the fluid is itself constant and no surface tension is present at the air-water interface

$$-\phi_t + \frac{1}{2}(\phi_x^2 + \phi_z^2) + gz = \frac{p_b}{\rho} \qquad \text{at } z = \eta \tag{4}$$

Periodicity. The wave is periodic in time and space.

$$\phi(x, t) = \phi(x + L, t) \qquad \phi(x, t) = \phi(x, t + T) \tag{5}$$

Cnoidal Wave Theory Considerations

Cnoidal wave theory is valid in relatively shallow water. For deriving waves of finite amplitude, the simplifying assumptions which characterize small-amplitude (linear) wave theory ($H \ll L$) and ($H \ll d$) are not appropriate. For shallow water, the second assumption ($H \ll d$) is invalid. Consequently, the nonlinear terms of the free surface boundary conditions are retained in the solution of the boundary value problem.

The free surface boundary conditions can be simplified by adopting a moving coordinate system having the same velocity as the wave celerity. The unsteady terms (time derivatives) are then eliminated. However, this procedure requires an additional assumption for an initial definition of celerity. The two most common assumptions were originally proposed by Stokes (1847) and are stated below:

Stokes Definition 1 (avg horizontal velocity = 0)

$$\frac{1}{L}\int_0^L u\, dx = 0 \qquad (6)$$

Stokes Definition 2 (avg mass flux = 0)

$$\frac{1}{L}\int_0^L \int_{-d}^{\eta} u\, dz\, dx = 0$$

The second definition of wave celerity was selected for this derivation. The tabulated approximations resulting from the derivation will be expressed relative to a fixed coordinate system.

Wave theories are often derived in any of three sets of variables: primitive variables (u, w), velocity potential ϕ, or stream function ψ. The assumption of 2-D, irrotational flow of an incompressible fluid leads to the following relationships between the three variables:

$$\psi_z = \phi_x = u \qquad -\psi_x = \phi_z = w \qquad (7)$$

Use of the stream function automatically satisfies the KFSBC (Equation 3) and consequently simplifies its form. The governing equation and boundary equations can then be restated in terms of stream function as

Laplace Equation: $\qquad \nabla^2 \psi = 0 \qquad\qquad\qquad\qquad (8)$

BBC: $\qquad\qquad\qquad \psi = 0 \qquad\qquad \text{at } z = -d \qquad (9)$

KFSBC: $\qquad\qquad\quad \psi = q \qquad\qquad \text{at } z = \eta \qquad (10)$

DFSBC: $\qquad\quad \dfrac{1}{2}(\psi_x^2 + \psi_z^2) + g\eta = \dfrac{p_b}{\rho} \qquad \text{at } z = \eta \qquad (11)$

Other Conditions: $\qquad \overline{\eta} = 0 \qquad\qquad\qquad\qquad (12)$

$$\eta(0) - \eta\left(\frac{L}{2}\right) = H \qquad (13)$$

Celerity Definition: $\qquad c = -\dfrac{q}{d} \qquad\qquad\qquad (14)$

The variable q in the above expressions represents the unit flow rate. Expressions for the periodic nature of the wave have been recast into alternate forms. Equation 12 requires the average value of η taken over a wavelength to be zero, while Equation 13 defines the wave height.

In shallow water, the vertical and horizontal length scales are of different orders of magnitude. For convenience, the variables are nondimensionalized using the horizontal length scale (L), vertical length scale (H), and velocity scale $\sqrt{gd}$:

$$X = \frac{x}{L} \quad Z = \frac{z}{d} \quad N = \frac{\eta}{d} \quad \Psi = \frac{\psi}{d\sqrt{gd}}$$

$$Q = \frac{q}{d\sqrt{gd}} \quad P = \frac{p}{\rho g d} \quad P_B = \frac{p_B}{\rho g d}$$

(15)

The complete boundary value problem restated in terms of these nondimensional variables is:

Governing Equation: $\quad \Psi_{ZZ} + \left(\dfrac{d}{L}\right)^2 \Psi_{XX} = 0 \quad\quad$ (16)

BBC: $\quad \Psi = 0 \quad$ at $z = -1 \quad$ (17)

KFSBC: $\quad \Psi = q \quad$ at $z + N \quad$ (18)

DFSBC: $\quad \dfrac{1}{2}[\Psi_Z^2 + \left(\dfrac{d}{L}\right)^2 \Psi_X^2] + N = P_B \quad$ at $z + N \quad$ (19)

Other Conditions: $\quad \overline{N} = 0 \quad\quad$ (20)

$$N(0) - N\left(\frac{1}{2}\right) = \frac{H}{d}$$

(21)

Solution of the Governing Equations

The boundary value problem consists of a set of nonlinear partial differential equations. The principal approach used to solve finite-amplitude wave behavior is the perturbation method. Dependent variables in the problem are assumed to be functions of an auxiliary parameter, δ, and are expanded in a power series of a small perturbation parameter, ϵ.

Waves of permanent form can be described by three independent variables, H, L, and d, from which two independent nondimensional quantities can be formed: (H/L) and (H/d). One of these ratios is often selected as the perturbation parameter in finite-amplitude wave theories. In this derivation of cnoidal wave theory, the second quantity is selected as the perturbation parameter:

$$\epsilon = \frac{H}{d}$$

(22)

The modulus of the elliptic integral, κ, is selected as the auxiliary parameter.

The nondimensional variables of the boundary value problem are expanded as power series about ϵ:

$$\Psi(X, Z, \kappa, \epsilon) = \sum_{n=0}^{\infty} \Psi_n(X, Z, \kappa)\epsilon^n = \Psi_0 + \epsilon\Psi_1 + \epsilon^2\Psi_2 + \epsilon^3\Psi_3 + \cdots$$

(23)

$$N(X, Z, \kappa, \epsilon) = \sum_{n=1}^{\infty} N_n(X, \kappa)\epsilon^n = \epsilon N_1 + \epsilon^2 N_2 + \epsilon^3 N_3 + \cdots \qquad (24)$$

$$Q(\kappa, \epsilon) = \sum_{n=0}^{\infty} Q_n(\kappa)\epsilon^n = Q_0 + \epsilon Q_1 + \epsilon^2 Q_2 + \epsilon^3 Q_3 + \cdots \qquad (25)$$

$$P_B(\kappa, \epsilon) = \sum_{n=0}^{\infty} P_n(\kappa)\epsilon^n = P_0 + \epsilon P_1 + \epsilon^2 P_2 + \epsilon^3 P_3 + \cdots \qquad (26)$$

$$\left(\frac{d}{L}\right)^2(\kappa, \epsilon) = \sum_{n=1}^{\infty} \delta_n(\kappa)\epsilon^n = \epsilon\delta_1 + \epsilon^2\delta_2 + \epsilon^3\delta_3 + \cdots \qquad (27)$$

The series expansions for N and $(d/L)^2$ (Equations 24 and 27) do not have *zero* terms because they are always smaller than ϵ.

The expanded forms for variables (Equations 23-27) are inserted into the Equations 16-21. For example, substituting the expanded form for Ψ in the governing Equation 16 yields

$$\begin{aligned}(\Psi_{0_{ZZ}} + \epsilon\Psi_{1_{ZZ}} + \epsilon^2\Psi_{2_{ZZ}} + \epsilon^3\Psi_{3_{ZZ}} + \cdots) \\ + (\epsilon\delta_1 + \epsilon^2\delta_2 + \epsilon^3\delta_3 + \cdots) \times (\Psi_{0_{XX}} + \epsilon\Psi_{1_{XX}} + \epsilon^2\Psi_{2_{XX}} + \epsilon^3\Psi_{3_{XX}} + \cdots) = 0\end{aligned} \qquad (28)$$

and at the bottom boundary (Equation 17):

$$\Psi_0 + \epsilon\Psi_1 + \epsilon^2\Psi_2 + \epsilon^3\Psi_3 + \cdots = 0 \qquad (29)$$

Equation 28 also provides some useful initial information. With the wave height equal to zero ($\epsilon = 0$), uniform flow conditions exist as a result of the moving coordinate system. Further, it can be seen that

$$\Psi_0 = b_{00}Z \quad \Rightarrow \quad \Psi_{0_{XX}} = \Psi_{0_{ZZ}} = 0 \qquad (30)$$

The term b_{00} is a constant to be determined later.

The free surface boundary conditions require special treatment because they occur at a location which is initially unknown. Instead of expansion with a power series, variables requiring evaluation at the free surface are expanded using a Taylor series about the known still-water level ($Z = 0$). For example, the Taylor series expansion for Ψ at the free surface is:

$$\Psi(X, N) = \Psi(X, 0) + N\Psi_Z(X, 0) + \frac{N^2}{2}\Psi_{ZZ}(X, 0) + \cdots \qquad (31)$$

Continuing with the perturbation technique, for a given equation, terms are grouped by equal powers of ϵ, and terms of an individual order on the left side must equal terms of like order on the right side. For example, again consider the governing Equation 28:

$$\epsilon \text{ terms:} \qquad \Psi_{1_{ZZ}} = 0 \qquad (32)$$

$$\epsilon^2 \text{ terms:} \qquad \Psi_{2_{ZZ}} + \delta_1\Psi_{1_{XX}} = 0 \qquad (33)$$

$$\epsilon^3 \text{ terms:} \qquad \Psi_{3_{ZZ}} + \delta_1\Psi_{2_{XX}} + \delta_2\Psi_{1_{XX}} = 0 \qquad (34)$$

This procedure is also applied to the remaining five boundary condition equations (17–21). The result is a set of equations for each order of ϵ. For example, the set of equations of second-order (ϵ^2) terms is

$$\Psi_{2ZZ} + \delta_1 \Psi_{1XX} = 0$$

$$\Psi_2 = 0 \quad \text{at } Z = -1$$

$$\Psi_2 + b_{00}N_2 + N_1\Psi_{1Z} = Q_2 \quad \text{at } Z = N$$

$$N_2 + b_{00}(\Psi_{2Z} + N_1\Psi_{1ZZ}) + \frac{1}{2}(\Psi_{1Z})^2 = P^2 \quad \text{at } Z = N \tag{35}$$

$$\overline{N}_2 = 0$$

$$N_2(0) - N_2\left(\frac{1}{2}\right) = 0$$

The solution of nth order equations requires information from the equations of $(n + 1)$th order. In total, the algebraic manipulations are quite lengthy and are beyond the intent of this report. A short summary of the procedure follows. The equations of 0th order determine P_0 and Q_0 once b_{00} is determined. The equations of 1st order determine b_{00} such that a nontrivial solution of N_1 and Ψ_1 exists. In similar fashion, the 2nd order equations determine N_1 and Ψ_1 if a nontrivial solution of N_2 and Ψ_2 exists. The process continues upward to the desired order of solution. This particular ACES application will provide results for first and second order solutions.

The solution for the cnoidal theory contains elliptic integrals and Jacobian elliptic functions which arise from the choice of κ as the auxiliary parameter and from the solution of certain nonlinear differential equations. Standard numerical methods described in Abramowitz and Stegun (1972) are employed for approximating these quantities.

Results from the Theory

The resulting approximations for the critical elements of the wave description are summarized below. Integral properties, water particle kinematics, and other traditional quantities of interest are included in the tabulations. All quantities are relative to a fixed frame of reference.

First-Order Solutions

Dispersion relation:
$$\frac{16\kappa^2 K^2}{3} = \frac{gHT^2}{d^2} \tag{36}$$

Celerity:
$$c = \sqrt{gd}(C_0 + \epsilon C_1) \tag{37}$$

$$C_0 = 1 \tag{37.1}$$

$$C_1 = \frac{1 + 2\lambda - 3\mu}{2} \tag{37.2}$$

Wavelength:
$$L = cT \tag{38}$$

Water surface elevation:
$$\eta = d(A_0 + A_1 \text{cn}^2\theta) \tag{39}$$

$$A_0 = \epsilon(\lambda - \mu) \tag{39.1}$$

$$A_1 = \epsilon \tag{39.2}$$

Average energy density:
$$E = \rho g H^2 E_0 \tag{40}$$
$$E_0 = \frac{-\lambda + 2\mu + 4\lambda\mu - \lambda^2 - 3\mu^2}{3} \tag{40.1}$$

Energy flux:
$$F = \rho g H^2 \sqrt{gd} F_0 \tag{41}$$
$$F_0 = E_0 \tag{41.1}$$

Pressure:
$$p = p_b - \frac{\rho}{2}[(u-c)^2 + w^2] - g\rho(z+d) \tag{42}$$
$$p_b = \rho g d(P_0 + \epsilon P_1) \tag{42.1}$$
$$P_0 = \frac{3}{2} \tag{42.2}$$
$$P_1 = \frac{1 + 2\lambda - 3\mu}{2} \tag{42.3}$$

Horizontal velocity:
$$u = \sqrt{gd}(B_{00} + B_{10}\operatorname{cn}^2\theta) \tag{43}$$
$$B_{00} = \epsilon(\lambda - \mu) \tag{43.1}$$
$$B_{10} = \epsilon \tag{43.2}$$

Vertical velocity:
$$w = \sqrt{gd}\,\frac{4Kd\,\operatorname{csd}}{L}\left(\frac{z+d}{d}\right)B_{10} \tag{44}$$

Horizontal acceleration:
$$\frac{\partial u}{\partial t} = \sqrt{gd}\,B_{10}\frac{4K}{T}\operatorname{csd} \tag{45}$$

Vertical acceleration:
$$\frac{\partial w}{\partial t} = \sqrt{gd}\frac{4Kd}{L}\left(\frac{z+d}{d}\right)B_{10}\frac{2K}{T}(\operatorname{sn}\theta\operatorname{dn}\theta - \operatorname{cn}\theta\operatorname{dn}\theta + \kappa^2\operatorname{sn}\theta\operatorname{cn}\theta) \tag{46}$$

The following general symbols also require definition:
$$\kappa' = \sqrt{1 - \kappa^2} \tag{47}$$
$$\lambda = \frac{\kappa'^2}{\kappa^2} \tag{48}$$
$$\mu = \frac{E}{\kappa^2 K} \tag{49}$$
$$\theta = 2K\left[\left(\frac{x}{L}\right) - \left(\frac{t}{T}\right)\right] \tag{50}$$

K and E_e are the complete elliptic integrals of the first and second kind, respectively; $\operatorname{sn}\theta$, $\operatorname{cn}\theta$, $\operatorname{dn}\theta$ are the Jacobian elliptic functions, and
$$\operatorname{csd} = \operatorname{cn}\theta\operatorname{sn}\theta\operatorname{dn}\theta \tag{51}$$

Second-Order Approximations

Dispersion relation:
$$\frac{16\kappa^2 K^2}{3} = \frac{gHT^2}{d^2}\left[1 - \epsilon\left(\frac{1+2\lambda}{4}\right)\right] \quad (52)$$

Celerity:
$$c = \sqrt{gd}(C_0 + \epsilon C_1 + \epsilon^2 C_2) \quad (53)$$

$$C_0 = 1 \quad (53.1)$$

$$C_1 = \frac{1 + 2\lambda - 3\mu}{2} \quad (53.2)$$

$$C_2 = \frac{-6 - 16\lambda + 5\mu - 16\lambda^2 + 10\lambda\mu + 15\mu^2}{40} \quad (53.3)$$

Water surface elevation:
$$\eta = d(A_0 + A_1 \text{cn}^2\theta + A_2 \text{cn}^4\theta) \quad (54)$$

$$A_0 = \epsilon(\lambda - \mu) + \epsilon^2\left(\frac{-2\lambda + \mu - 2\lambda^2 + 2\lambda\mu}{4}\right) \quad (54.1)$$

$$A_1 = \epsilon - \frac{3}{4}\epsilon^2 \quad (54.2)$$

$$A_2 = \frac{3}{4}\epsilon^2 \quad (54.3)$$

Average energy density:
$$E = \rho g H^2 (E_0 + \epsilon E_1) \quad (55)$$

$$E_0 = \frac{-\lambda + 2\mu + 4\lambda\mu - \lambda^2 - 3\mu^2}{3} \quad (55.1)$$

$$E_1 = \frac{1}{30}(\lambda - 2\mu - 17\lambda\mu + 3\lambda^2 - 17\lambda^2\mu + 2\lambda^3 + 15\mu^3) \quad (55.2)$$

Energy flux:
$$F = \rho g H^2 \sqrt{gd}(F_0 + \epsilon F_1) \quad (56)$$

$$F_0 = E_0 \quad (56.1)$$

$$F_1 = \frac{1}{30}(-4\lambda + 8\mu + 53\lambda\mu - 12\lambda^2 - 60\mu^2 + 53\lambda^2\mu - 120\lambda\mu^2 - 8\lambda^3 + 75\mu^3) \quad (56.2)$$

Pressure:
$$p = p_b - \frac{\rho}{2}[(u-c)^2 + w^2] - g\rho(z+d) \quad (57)$$

$$p_b = \rho g d(P_0 + \epsilon P_1 + \epsilon^2 P_2) \quad (57.1)$$

$$P_0 = \frac{3}{2} \quad (57.2)$$

$$P_1 = \frac{1 + 2\lambda - 3\mu}{2} \quad (57.3)$$

$$P_2 = \frac{-1 - 16\lambda + 15\mu - 16\lambda^2 + 30\lambda\mu}{40} \quad (57.4)$$

Horizontal velocity:

$$u = \sqrt{gd}[(B_{00} + B_{10}\text{cn}^2\theta + B_{20}\text{cn}^4\theta) - \frac{1}{2}\left(\frac{z+d}{d}\right)^2(B_{01} + B_{11}\text{cn}^2\theta + B_{21}\text{cn}^4\theta)] \tag{58}$$

$$B_{00} = \epsilon(\lambda - \mu) + \epsilon^2\left(\frac{\lambda - \mu - 2\lambda^2 + 2\mu^2}{4}\right) \tag{58.1}$$

$$B_{10} = \epsilon + \epsilon^2\left(\frac{1 - 6\lambda + 2\mu}{4}\right) \tag{58.2}$$

$$B_{20} = -\epsilon^2 \tag{58.3}$$

$$B_{01} = \frac{3\lambda}{2}\epsilon^2 \tag{58.4}$$

$$B_{11} = 3\epsilon^2(1 - \lambda) \tag{58.5}$$

$$B_{21} = -\frac{9}{2}\epsilon^2 \tag{58.6}$$

Vertical velocity:

$$w = \sqrt{gd}\,\frac{4Kd\,\text{csd}}{L} \times \left[\left(\frac{z+d}{d}\right)(B_{10} + 2B_{20}\text{cn}^2\theta) - \frac{1}{6}\left(\frac{z+d}{d}\right)^3(B_{11} + 2B_{21}\text{cn}^2\theta)\right] \tag{59}$$

Horizontal acceleration:
$$\frac{\partial u}{\partial t} = \sqrt{gd}\left\{\left[B_{10} - \frac{1}{2}\left(\frac{z+d}{d}\right)^2 B_{11}\right]\frac{4K}{T}\text{csd} \right.$$
$$\left. + \left[B_{20} - \frac{1}{2}\left(\frac{z+d}{d}\right)^2 B_{21}\right]\frac{8K}{T}\text{cn}^2\theta\,\text{csd}\right\} \tag{60}$$

Vertical acceleration:
$$\frac{\partial w}{\partial t} = \sqrt{gd}\,\frac{4Kd}{L}\left\{\frac{8K}{T}\text{csd}^2\left[\left(\frac{z+d}{d}\right)B_{20} - \frac{1}{6}\left(\frac{z+d}{d}\right)^3 B_{21}\right] \right.$$
$$+ \left[\left(\frac{z+d}{d}\right)(B_{10} + 2B_{20}\text{cn}^2\theta) - \frac{1}{6}\left(\frac{z+d}{d}\right)^3(B_{11} + 2B_{21}\text{cn}^2\theta)\right] \tag{61}$$
$$\left. \times [\text{sn}\theta\,\text{dn}\theta - \text{cn}\theta\,\text{dn}\theta + \kappa^2\text{sn}\theta\,\text{cn}\theta]\right\}$$

References and Bibliography

Abramowitz, M., and Stegun, I. A. 1972. *Handbook of Mathematical Functions*, Dover Publications, New York, 1046 pp.

Chappelear, J. E. 1962. "Shallow Water Waves," *Journal of Geophysical Research*, Vol. 67, No. 12, pp. 4693–4704.

Davis, H. T. 1962. *Introduction to Nonlinear Differential and Integral Equations*, Dover Publications, New York, 596 pp.

Fenton, J. D. 1979. "A High Order Cnoidal Wave Theory," *Journal of Fluid Mechanics*, Vol. 94, pp. 129–161.

Hardy, T. A., and Kraus, N. C. 1987. "A Numerical Model for Shoaling and Refraction of Second-Order Cnoidal Waves over an Irregular Bottom," Miscellaneous Paper CERC-87-9, US Army Engineer Waterways Experiment Station, Vicksburg, MS.

Isobe, M. 1985. "Calculation and Application of First-Order Cnoidal Wave Theory," *Coastal Engineering*, Vol. 9, pp. 309–325.

Isobe, M., and Kraus, N. C. 1983. "Derivation of a Second-Order Cnoidal Wave Theory," Hydraulics Laboratory Report No. YNU-HY-83-2, Department of Civil Engineering, Yokohama National University, 43 pp.

Keller, J. B. 1948. "The Solitary Wave and Periodic Waves in Shallow Water," *Communication of Pure and Applied Mathematics*, Vol. 1, pp. 323–339.

Keulegan, G. H., and Patterson, G. W. 1940. "Mathematical Theory of Irrotational Translation Waves," *Journal of Research of the National Bureau of Standards*, Vol. 24, pp. 47–101.

Korteweg, D. J., and de Vries, G. 1895. "On the Change of Form of Long Waves Advancing in a Rectangular Canal, and on a New Type of Long Stationary Waves," *Philosophy Magazine*, Series 5, Vol. 39, pp. 422–443.

Laitone, E. V. 1960. "The Second Approximation to Cnoidal and Solitary Waves," *Journal of Fluid Mechanics*, Vol. 9, pp. 430–444.

Shore Protection Manual. 1984. 4th ed., 2 Vols., US Army Engineer Waterways Experiment Station, Coastal Engineering Research Center, US Government Printing Office, Washington, DC, Chapter 2, pp. 44–55.

Stokes, G. G. 1847. "On the Theory of Oscillatory Waves," *Transactions of the Cambridge Philosophical Society*, Vol. 8, pp. 441–455.

FOURIER SERIES WAVE THEORY

Description

This application yields various parameters for progressive waves of permanent form, as predicted by Fourier series approximation. It provides estimates for common engineering parameters such as water surface elevation, integral wave properties, and kinematics as functions of wave height, period, water depth, and position in the wave form which is assumed to exist on a uniform co-flowing current. Stokes first and second approximations for celerity (i.e., values of the mean Eulerian current or mean mass transport rate) may be specified. Fourier series of up to 25 terms may be selected to approximate the wave. In addition to providing kinematics at a given point in the wave, this application provides graphical presentations of kinematics over two wavelengths (at a given z coordinate), and the vertical profile of selected kinematics under the wave crest. The methodology is based upon a series of papers by J. D. Fenton (Reinecker and Fenton, 1981; Fenton, 1988a; Fenton, 1988b; Fenton, 1990) and R. J. Sobey (Sobey, Goodwin, Thieke, and Westberg, 1987). LINPACK routines (Dongarra et al., 1979) are used to solve the set of up to 60 simultaneous equations to determine the Fourier coefficients for the series.

Introduction

The accurate description of the steady wave is an important element in the design and analysis process. While linear wave theory has traditionally (and often correctly) been used for a large group of wave applications, nonlinear wave theories often provide significant improvements over linear wave theory descriptions, particularly where wavelengths are long or short relative to water depth. Recent developments and technology enable engineers to easily employ higher order wave theories to increase the accuracy of their estimates of wave properties. Today's common desktop microcomputers offer adequate computational power to implement many nonlinear wave theories.

Common practice includes the use of Stokes' theory in deep water, cnoidal theory in shallow water, and Fourier series wave theory in deep, transitional, and shallow water. Contemporary versions of Stokes' and cnoidal theory are readily accessible at fifth order, and Fourier series theory at any order, though 15th to 20th order appears adequate to resolve even highly nonlinear waves. All are capable of incorporating a uniform underlying current. Stokes and cnoidal theory have limited regions of physical validity, and at higher orders, can be quite accurate. The domain of applicability for Fourier approximation wave theory includes that of both Stokes and cnoidal theory. It is based upon a numerical approach, and substitutes computational effort for the limited domain of the above perturbation methods. It is easily tractable at any desired order (subject to machine precision), and has become a well-established, robust, and accurate engineering tool for the steady wave problem.

Several variations of Fourier series wave theory have been presented in recent years: (Chappelear, 1961; Dean, 1965, and 1974; Schwartz, 1974; Dalrymple, 1974; Cokelet, 1977; Chaplin, 1980; Reinecker and Fenton, 1981; Le Mehaute, et al., 1984; Fenton, 1988a, and 1988b). Differences between the various approaches are well summarized by Sobey (Sobey, et al., 1987), and can be generally categorized by choice of field variable (stream function versus velocity potential), form of the terms in the cosine series and behavior in deep water, inclusion of a current (none, uniform, or shear), and formulation, extensibility, treatment, and efficiency of numerical approaches for the problem. The methodology of Fenton (1988a and 1988b) is utilized here to formulate the problem and determine the Fourier series coefficients, while revised derivations for wave properties have been taken from Sobey (1988) and Klopman (1990).

General Assumptions and Terminology

A representative wave profile is illustrated in Figure 2-3-1.

Assumptions applicable to the above progressive wave include:

- The wave is two-dimensional in the x-z plane.
- It propagates in a permanent form over a smooth horizontal bed of constant depth in the positive x-direction.
- There is a uniform underlying current.
- The fluid is inviscid and incompressible, having no surface tension.
- The flow is irrotational.

Terminology relevant to the following discussion includes:

(X, Z) = coordinates in Fixed (non-translating) Reference Frame
$(x, z) = (X - ct, Z)$ = coordinates in Steady Reference Frame (which moves at wave speed c)
d = still-water depth ($z = 0$)
$\eta(x)$ = water surface elevation
H = wave height
L = wavelength
T = wave period
$k = \dfrac{2\pi}{L}$ = wave number

AUTOMATED COASTAL ENGINEERING SYSTEM (ACES)   **A.47**

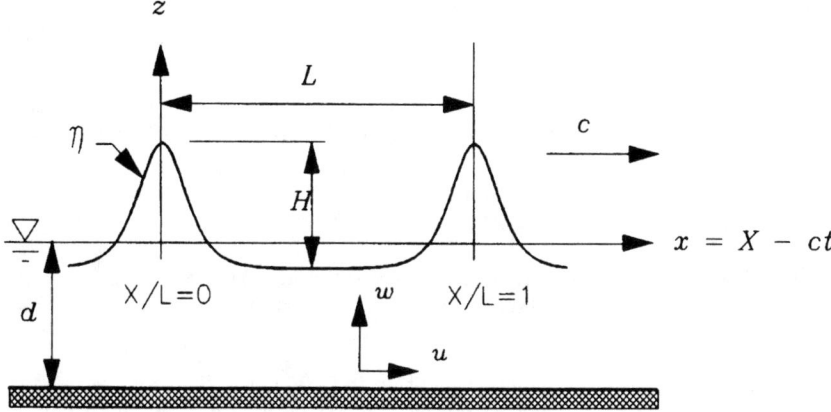

FIGURE 2-3-1. Wave in steady reference frame (which moves at wave speed c).

c = wave speed
$\bar{u}$ = mean value of horizontal fluid velocity for a constant value of z, over one wavelength
$\bar{u}_1$ = time-mean Eulerian current (corresponding to Stokes First approximation for c)
$\bar{u}_2$ = depth-averaged mass transport velocity (corresponding to Stokes Second approximation for c)
$\psi(x, z)$ = stream function
Q = constant volume flow rate per unit width under the steady wave
$q = \bar{u}d - Q$ = constant volume flow rate per unit width due to wave
R = Bernoulli constant
$r = R - gd$ = Bernoulli constant (separated for convenience in treatment of deep water)
(U, W) = velocity components in Fixed Reference Frame
$(u, w) = (U - c, W)$ = velocity components in Steady Reference Frame
g = acceleration of gravity
ρ = density of water

Governing Equation and Boundary Conditions

The steady wave problem is a boundary value problem consisting of the two-dimensional Laplace equation as the governing field equation, boundary conditions at the free surface and bottom, and the assertion that the wave is periodic in space and time.

The wave problem has traditionally been formulated in three sets of variables: primitive variables (u, w), velocity potential ϕ, or stream function ψ. The assumption of two-dimensional, irrotational flow of an incompressible fluid leads to the following relationships between the three variables:

$$\psi_z = \phi_x = u \qquad -\psi_x = \phi_z = w \tag{1}$$

The formulation in stream function, and the adoption of a steady (i.e., moving at wave speed c) coordinate system (as depicted in Figure 2-3-1) simplifies the form of some of the boundary conditions for the problem. The governing equation and boundary conditions formulated in stream function for the selected coordinate system can be summarized as follows:

Governing (Laplace) Equation. The governing field equation is the two-dimensional Laplace equation. It represents the conservation of mass and momentum within the fluid field.

$$\frac{\partial^2 \psi(x, z)}{\partial x^2} + \frac{\partial^2 \psi(x, z)}{\partial z^2} = 0 \tag{2}$$

Bottom Boundary Condition (BBC). The seabed is considered impermeable. Therefore, at the sea bottom, the flow rate through the bottom boundary must be 0.

$$\psi(x, -d) = 0 \qquad \text{at } z = -d \tag{3}$$

Kinematic Free Surface Boundary Condition (KFSBC). The fluid particle velocity normal to the free surface is equal to the velocity of the free surface itself. This condition implies that all water particles on the free surface remain there. Therefore, at the free surface, the flow rate through the boundary must also be 0. It also implies that the wave does not break.

$$\psi(x, \eta) = -Q \qquad \text{at } z = \eta \tag{4}$$

Dynamic Free Surface Boundary Condition (DFSBC). Expressed as the Bernoulli equation, this boundary condition asserts that the pressure at the free surface is constant (equal to atmospheric pressure). This requirement follows from the preliminary assumptions that atmospheric pressure above the free surface is itself constant and that surface tension may be neglected at the air-sea interface.

$$\frac{1}{2}\left\{\left(\frac{\partial \psi(x, \eta)}{\partial x}\right)^2 + \left(\frac{\partial \psi(x, \eta)}{\partial z}\right)^2\right\} + g\eta(x) = R \qquad \text{at } z = \eta \tag{5}$$

Periodicity. The wave is periodic in time and space.

$$\psi(x, t) = \psi(x + L, t) \qquad \psi(x, t) = \psi(x, t + T) \tag{6}$$

Additional Considerations Relative to the Dispersion Relation

The adoption of a moving coordinate system (having the same velocity as the wave phase speed) simplifies the form of the KFSBC by eliminating a time derivative. However, this simplification yields some additional complexities which have been emphasized by Fenton (1988a, 1988b, 1990) and Sobey et al. (1987). The exact wave speed is correctly predicted by the wave theory only for the singular case of zero underlying current. It is more rational to assert that the wave theory predicts the wave phase speed relative to the underlying current, which is seldom zero in a realistic environment. A principal consequence of an underlying current is a Doppler shift of the apparent wave period (relative to a stationary observer or gage). Hence, the underlying current velocity must also be known in order to resolve the wave problem.

Some confusion results from consideration of the above concerns relative to the traditional definitions of Stokes (1847) for wave speed. Concerning the first definition, consider a given time-mean Eulerian current ($\overline{u_1}$) which is a time-mean horizontal fluid velocity at any point (X, Z) wholly in the fluid (within the fixed, non-moving frame of reference). This corresponds to the current recorded by a stationary meter. By considering

$-\bar{u}$ to be the mean fluid velocity (i.e., at a fixed z coordinate, but averaged over one wavelength) in the steady (moving) frame of reference, the relationship between the wave speed and the time-mean Eulerian current applied to Stokes' first definition is:

$$c = \bar{u} + \overline{u_1} \tag{7}$$

The special case of $\overline{u_1} = 0$ corresponds exactly to Stokes first definition of wave speed where $c = \bar{u}$.

An alternate estimate of the wave phase speed is a function of the depth-averaged mass transport velocity (Stokes' drift velocity) ($\overline{u_2}$). By considering Q to be the volume flow rate (per unit width) under the wave in the steady (moving) frame of reference, the depth-averaged fluid velocity is then $-Q/d$. The relationship between wave speed and the mass transport velocity is suggested by Stokes' second definition as:

$$c = \frac{Q}{d} + \overline{u_2} \tag{8}$$

The special case of $\overline{u_2} = 0$ corresponds to Stokes' second definition of wave speed where $c = Q/d$. It is a condition that is commonly imposed in wave flumes or any situation where mass transport is actually or assumed to be restricted. It is not, however, generally true in open water sea states.

The above two equations include the principle of an underlying current in the dispersion relation and can represent the shift in wave period due to the current. A final consideration in formulating the dispersion relationship involves the boundary condition requiring the wave to be periodic in space and time. This condition implicitly requires the wave speed to satisfy the familiar relation $c = L/T$. The dispersion relation is declared in the wave problem by selecting the definition of wave speed, and providing the appropriate value for either the mean Eulerian velocity, or the Stokes drift velocity:

Stokes' First Approximation of Wave Speed

$$c = \frac{L}{T} = \frac{2\pi}{kT} = \bar{u} + \overline{u_1}$$

(specify mean Eulerian velocity $\overline{u_1}$)

(9)

Stokes' Second Approximation of Wave Speed

$$c = \frac{L}{T} = \frac{2\pi}{kT} = \frac{Q}{d} + \overline{u_2}$$

(specify Stokes' drift velocity $\overline{u_2}$)

Solution and Method

Fourier series wave theory derives its name from an approximate (but potentially very accurate) solution to the governing wave equation using a Fourier cosine series in kx, as follows:

$$\psi(x, z) = -\bar{u}(d + z) + \left(\frac{g}{k^3}\right)^{1/2} \sum_{j=1}^{N} B_j \frac{\sinh jk(d + z)}{\cosh jkd} \cos jkx \tag{10}$$

where

B_j = dimensionless Fourier coefficients

The solution is obtained by numerically computing the N Fourier coefficients to satisfy a system of simultaneous equations that consist of the two free-surface boundary conditions (evaluated at $N + 1$ evenly spaced points on the free surface between a crest and following trough) as well as the dispersion relation when given the wave height, wave period, water depth, and either the mean Eulerian velocity, or the Stokes drift velocity. The set of dimensionless variables and equations as described by Fenton (1988a and 1988b) are listed in Tables 2-3-1 through 2-3-3.

Table 2-3-1 lists the name and order of dimensionless variables for use in the system of equations. Table 2-3-2 lists the relationships being formalized and their order as a system of equations to be solved simultaneously. Table 2-3-3 identifies some special forms of the equations used for numerical stability or speed in deepwater environments. Deepwater simplifications are automatically employed when $(d/L > 3/2)$ with L initially estimated by a linear wave, theory approximation.

Tables 2-3-1 and 2-3-2 describe a system of $2N + 10$ equations:

$$\mathbf{f(z)} = \{f_i(\mathbf{z}), i = 1 \ldots 2N + 10\} = 0 \tag{11}$$

The system of equations is solved iteratively using Newton's method:

$$\left[\frac{\partial f_i}{\partial z_i}\right]^{(n)} \cdot (\mathbf{z}^{(n+1)} - \mathbf{z}^{(n)}) = -\mathbf{f}(\mathbf{z}^{(n)}) \tag{12}$$

where

n = Iteration index
$\mathbf{z}^{(n)}$ = Values of $(z_i, i = 1 \ldots 2N + 10)$ at iteration n
$\mathbf{z}^{(n+1)}$ = Values $(z_i, i = 1 \ldots 2N + 10)$ at iteration $n + 1$

TABLE 2-3-1. Vector of dimensionless problem variables (Fenton, 1988)

Variable	Dimensionless Form	Variable	Dimensionless Form
depth	$z_1 = kd$	free surface elevation at crest $(x = 0)$	$z_{10} = k\eta_0$
wave height	$z_2 = kH$	free surface elevation at $\left(x = \dfrac{L}{2N}\right)$	$z_{11} = k\eta_1$
wave period	$z_3 = T(gk)^{1/2}$	free surface elevation at $\left(x = \dfrac{L}{N}\right)$	$z_{12} = k\eta_2$
wave speed	$z_4 = c(k/g)^{1/2}$	...	...
mean Eulerian velocity	$z_5 = \overline{u_1}(k/g)^{1/2}$	free surface elevation at trough $\left(x = \dfrac{L}{2}\right)$	$z_{N+10} = k\eta_N$
Stokes' drift velocity	$z_6 = \overline{u_2}(k/g)^{1/2}$	Fourier series coefficient	$z_{N+11} = \beta_1$
mean fluid velocity	$z_7 = \overline{u}(k/g)^{1/2}$	...	$z_{N+12} = \beta_2$
volume flow rate per unit width	$z_8 = q(k^3/g)^{1/2}$	...	...
Bernoulli constant	$z_9 = r(k/g)$	...	$z_{2N+10} = \beta_N$

TABLE 2-3-2. System of equations (Fenton, 1988)

Relation	Equation
H to d	$f_1(z_{1,2}) = kH - (H/d)kd = 0$
H to T	$f_2(z_{2,3}) = kH - \left(\dfrac{H}{gT^2}\right)(T(gk)^{1/2})^2 = 0$
$c = \dfrac{L}{T} = \dfrac{2\pi}{kT}$	$f_3(z_{3,4}) = c(k/g)^{1/2}T(gk)^{1/2} - 2\pi = 0$
$c = \bar{u} + \overline{u_1}$	$f_4(z_{4,5,7}) = \overline{u_1}(k/g)^{1/2} + \bar{u}(k/g)^{1/2} - c(k/g)^{1/2} = 0$
$c = \dfrac{Q}{d} + \overline{u_2}$	$f_5(z_{1,4,6,7,8}) = \overline{u_2}(k/g)^{1/2} + \bar{u}(k/g)^{1/2} - c(k/g)^{1/2} - \dfrac{q(k^3/g)^{1/2}}{kd} = 0$
apply selected approx for c $u_c = \overline{u_1}$ or $\overline{u_2}$	$f_6(z_{2,5 \text{ or } 6}) = u_c(k/g)^{1/2} - \dfrac{u_c}{(gH)^{1/2}}(gH)^{1/2} = 0$
$\overline{\eta(x)} = 0$	$f_7(z_{10,N+10,m+10}) = k\eta_0 + k\eta_N + 2\sum_{m=1}^{N-1} k\eta_m = 0$
$H = \eta_0 - \eta_N$	$f_8(z_{2,10,N+10}) = k\eta_0 - k\eta_N - kH = 0$
KFSBC: solved at $m = 0 \ldots N$ points on free surface	$f_{m+9}(z_{1,7,8,m+10,N+j+10}) = -q\left(\dfrac{k^3}{g}\right)^{1/2} - k\eta_m \bar{u}\left(\dfrac{k}{g}\right)^{1/2}$ $+ \sum_{j=1}^{N} B_j\left[\dfrac{\sinh j(kd + k\eta_m)}{\cosh jkd}\right]\cos\dfrac{jm\pi}{N} = 0$
DFSBC: solved at $m = 0 \ldots N$ points on free surface	$f_{N+10+m}(z_{1,7,9,10,m+10,N+j+10}) = k\eta_m - \dfrac{rk}{g}$ $+ \dfrac{1}{2}\left(-\bar{u}\left(\dfrac{k}{g}\right)^{1/2} + \sum_{j=1}^{N} jB_j\left[\dfrac{\cosh j(kd + k\eta_m)}{\cosh jkd}\right]\cos\dfrac{jm\pi}{N}\right)^2$ $+ \dfrac{1}{2}\left(\sum_{j=1}^{N} jB_j\left[\dfrac{\sinh j(kd + k\eta_m)}{\cosh jkd}\right]\sin\dfrac{jm\pi}{N}\right)^2 = 0$

$\mathbf{f}(\mathbf{z}^{(n)})$ = Values of $\{f_i(\mathbf{z}), i = 1 \ldots 2N + 10\}$ evaluated at iteration n

$\left[\dfrac{\partial f_i}{\partial z_j}\right]^{(n)}$ = Jacobian matrix evaluated at iteration n as

$$\dfrac{\partial f_i}{\partial z_i} \approx \dfrac{f_i(z_1, \cdots, z_j + \Delta_j, \cdots, z_{2N+10}) - f_i(z_1, \cdots, z_j, \cdots, z_{2N+10})}{\Delta_j} \qquad (13)$$

Note: The Jacobian matrix may be evaluated analytically or numerically. As indicated by Equation (13), in this implementation the Jacobian matrix is evaluated numerically. Each of $2N + 10$ equations were evaluated a total of $2N + 10$ times with the value of Δ_j as follows:

TABLE 2-3-3. System of equations (special variations for deep water) (Fenton, 1988)

Relation	Equation
proxy	$f_1(z_1) = kd + 1 = 0$
$c = \dfrac{Q}{d} + \overline{u_2}$	$f_5(z_{4,6,7}) = \overline{u_2}(k/g)^{1/2} + \overline{u}(k/g)^{1/2} - c(k/g)^{1/2} = 0$
KFSBC: solved at $m = 0 \ldots N$ points on free surface	$f_{m+9}(z_{7,8,m+10,N+j+10}) = -q\left(\dfrac{k^3}{g}\right)^{1/2} - k\eta_m \overline{u}\left(\dfrac{k}{g}\right)^{1/2}$ $\quad + \sum\limits_{j=1}^{N} B_j e^{jk\eta_m} \cos\dfrac{jm\pi}{N} = 0$
DFSBC: solved at $m = 0 \ldots N$ points on free surface	$f_{N+10+m}(z_{7,9,m+10,N+j+10}) = k\eta_m - \dfrac{rk}{g}$ $\quad + \dfrac{1}{2}\left(-\overline{u}\left(\dfrac{k}{g}\right)^{1/2} + \sum\limits_{j=1}^{N} jB_j e^{jk\eta_m} \cos\dfrac{jm\pi}{N}\right)^2$ $\quad + \dfrac{1}{2}\left(\sum\limits_{j=1}^{N} jB_j e^{jk\eta_m} \sin\dfrac{jm\pi}{N}\right)^2 = 0$

$$\Delta_j = z_j/100 \quad \text{for } z_j > 10^{-4}$$
$$\Delta_j = 10^{-5} \quad \text{for } z_j \leq 10^{-4}$$

Iterations continue until the solution converges to the following criteria:

$$\sum_{j=1}^{2N+10} |z_j^{(n+1)} - z_j^{(n)}| < 10^{-6} \tag{14}$$

within ($n_{max} \leq 9$) iterations. Double precision LINPACK (Dongarra et al; 1979) routines are employed to solve the matrix Equation (12) at each iteration. As currently implemented in ACES, the square matrix is dimensioned to a maximum rank of 60, and the solved matrix rank is $2N = 10$, which allows a maximum of ($N_{max} = 25$) terms in the Fourier series approximation.

Solution Aids and Insights
Several researchers (Dalrymple and Solana, 1986), (Fenton, 1988) and (Sobey, 1988) have noted that Fourier theory formulations can potentially admit the odd harmonics (L, $L/3$, $L/5$, ... L/N) singularly, or in combinations, particularly in shallow water. Sobey (1988) notes that two features of the formulation (symmetry about crest, and periodicity at lateral boundary conditions) permit the manifestation of these apparent multicrested solutions. Several strategies for achieving the fundamental solution (rather than higher odd harmonics) have been discussed in the cited literature, and the method employed by Fenton (1988) is implemented in this version. A "ramping" mechanism is provided to solve the problem by initially (internally) solving for a lower wave height, and approaching the specified wave height in a specified (as input) number of evenly spaced steps. The approach eliminates the problem at the expense of additional iterations to achieve the final fundamental solution for the specified wave height.

Maximum Wave Checks

Most wave theories (both analytic and numerically based) are capable of yielding valid mathematical solutions to physically implausible data; particularly with regards to wave steepness and depth-related breaking. In part, this is a consequence of some of the assumptions imposed upon the boundary problem formulation. As an aid in restricting solutions to an observed physically valid domain, empirical data and formulations are often employed to estimate the validity of the given wave. The following expression (Fenton, 1990) is used for estimating the greatest wave as a function of both wavelength and depth:

$$H_{max} = d \left\{ \frac{.141063\frac{L}{d} + .0095721\left(\frac{L}{d}\right)^2 + .0077829\left(\frac{L}{d}\right)^3}{1 + .078834\frac{L}{d} + .0317567\left(\frac{L}{d}\right)^2 + .0093407\left(\frac{L}{d}\right)^3} \right\} \quad (15)$$

In the limits, the leading term in the numerator of the above expression provides the familiar steepness limit for short waves ($H_{max}/L \rightarrow .141063$), and as ($\lim L/d \rightarrow \infty$) the ratio of coefficients of the cubic terms provides the familiar ratio ($.0077829(L/d)^3/.0093407(L/d)^3 \rightarrow .83322$). This simple empirical test is applied using the given water depth, and solved wavelength as a rough filter for implausible wave specifications.

Derived Results

Traditional engineering quantities of interest about the wave are derived from the solution of the governing equation. Since the solution is expressed as a Fourier series, many of the derived quantities will also be functions of the series. Formulas for kinematics, integral properties, and other relevant items are included in the following tabulations. All quantities are relative to the stationary (non-moving) frame of reference.

Kinematics and Other Derived Variables

Velocities:

Horizontal: $\quad u(x, z) = \frac{\partial \psi}{\partial z} = -\bar{u} + \left(\frac{g}{k}\right)^{1/2} \sum_{j=1}^{N} jB_j \frac{\cosh jk(d + z)}{\cosh jkd} \cos jkx \quad (16)$

Vertical: $\quad w(x, z) = -\frac{\partial \psi}{\partial x} = \left(\frac{g}{k}\right)^{1/2} \sum_{j=1}^{N} jB_j \frac{\sinh jk(d + z)}{\cosh jkd} \sin jkx \quad (17)$

Accelerations:

Horizontal: $\quad a_x(x, y) = \frac{du}{dt} = u\frac{\partial u}{\partial x} + w\frac{\partial u}{\partial z} \quad (18)$

Vertical: $\quad a_z(x, y) = \frac{dw}{dt} = u\frac{\partial w}{\partial x} + w\frac{\partial w}{\partial z} = u\frac{\partial u}{\partial z} - w\frac{\partial u}{\partial x} \quad (19)$

where

$$\frac{\partial u}{\partial x} = -(gk)^{1/2} \sum_{j=1}^{N} j^2 B_j \frac{\cosh jk(d + z)}{\cosh jkd} \sin jkx$$

$$\frac{\partial u}{\partial z} = (gk)^{1/2} \sum_{j=1}^{N} j^2 B_j \frac{\sinh jk(d + z)}{\cosh jkd} \cos jkx$$

Pressure:
$$p(x, y) = pr - \rho g z - \frac{1}{2}\rho(u^2 + w^2) \tag{20}$$

Water Surface:
$$\eta(x) = \sum_{j=1}^{N-1} f_j \cos jkx + \frac{1}{2} f_N \cos Nkx \tag{21}$$

where
$$f_j = \frac{2}{N}\left\{\frac{1}{2}\eta_0 + \sum_{m=1}^{N-1} \eta_m \cos \frac{jm\pi}{N} + \frac{1}{2}\eta_N \cos j\pi \right\}$$

Notes: $X = x + ct$, $Z = z$ $\quad r = R - gd$
$U(X, Z, t) = u(x, z) + c$
$W(X, Z, t) = w(x, z)$
$P(X, Z, t) = p(x, z)$

Integral Properties

Potential Energy: (per unit horizontal area)
$$E_P = \overline{\int_0^\eta (\rho g Z)dZ}$$
$$E_P = \frac{1}{2}\rho g \overline{\eta^2} \tag{22}$$

Momentum: (per unit horizontal area) (Impulse)
$$I = \overline{\int_{-d}^\eta (\rho U)dZ}$$
$$I = \rho(cd - Q) \tag{23}$$

Kinetic Energy: (per unit horizontal area)
$$E_K = \overline{\int_{-d}^\eta \left(\frac{1}{2}\rho(U^2 + W^2)\right)dZ}$$
$$E_K = \frac{1}{2}(cI - \rho \overline{u_1} Q) \tag{24}$$

Mean Square of Bed Velocity:
$$\overline{U_b^2} = \frac{1}{L}\int_0^L U^2(X, -d, t)dX$$
$$\overline{U_b^2} = 2(R - gd) - c^2 + 2\overline{u_1}c \tag{25}$$

Energy Flux: (per unit length of crest) (Wave Power)
$$F = \overline{\int_{-d}^\eta \left(P + \frac{1}{2}\rho(U^2 + W^2) + \rho g Z\right)UdZ}$$
$$F = (3E_K - 2E_P - 2\overline{u_1}I)c + \frac{1}{2}\overline{U_b^2}(I + \rho c d) \tag{26}$$

Radiation Stress:

$$S_{xx} = \int_{-d}^{\eta}(P + \rho U^2)dZ - \frac{1}{2}\rho g d^2$$

$$S_{xx} = 4E_K - 3E_P + \rho d\,\overline{U_b^2} - 2\overline{u_1}I$$

(27)

Notes: $X = x + ct$, $Z = z$ $P(X, Z, t) = p(x, z)$
$U(X, Z, t) = u(x, z) + c$ $\overline{f(\ )} \Rightarrow$ averaged over one wavelength
$W(X, Z, t) = w(x, z)$

References and Bibliography

Chaplin, J. R., 1980. "Developments of Stream-Function Wave Theory," *Coastal Engineering*, Vol. 3, pp. 179–205.
Chappelear, J. E., 1961. "Direct Numerical Calculation of Wave Properties," *Journal of Geophysical Research*, Vol. 66, No. 2, pp. 501–508.
Cokelet, E. D., 1977. "Steep Gravity Waves in Water of Arbitrary Uniform Depth," *Proceedings of the Royal Society of London*, Series A, Vol. 286, pp. 183–230.
Dalrymple, R. A., 1974. "A Finite Amplitude Wave on a Linear Shear Current," *Journal of Geophysical Research*, Vol. 79, No. 30, pp. 4498–4504.
Dalrymple, R. A., and Solana, P., 1986. "Nonuniqueness in Stream Function Wave Theory," *Journal of Waterway, Port, Coastal and Ocean Division*, American Society of Civil Engineers, Vol. 112, No. 2, pp. 333–337.
Dean, R. G., 1965. "Stream Function Representation of Nonlinear Ocean Waves," *Journal of Geophysical Research*, Vol. 70, No. 18, pp. 4561–4572.
Dean, R. G., 1974. "Evaluation and Development of Water Wave Theories for Engineering Application," Special Report No. 1, Coastal Engineering Research Center, 2 Vols.
Dongarra, J. J., Moler, C. B., Bunch, J. R., and Stewart, G. W., 1979. *LINPACK User's Guide*, S. I. A. M., Philadelphia.
Fenton, J. D., 1988a. "The Numerical Solution of Steady Water Wave Problems," *Computers and Geoscience*, Vol. 14, No. 3, pp. 357–368.
Fenton, J. D., 1988b. Discussion of "Nonuniqueness in Stream Function Wave Theory," by R. A. Dalrymple and P. Solana, *Journal of Waterway, Port, Coastal and Ocean Division*, American Society of Civil Engineers, Vol. 114, No. 1, pp. 110–112.
Fenton, J. D., 1990. "Nonlinear Wave Theories," *Ocean Engineering Science, The Sea*, Vol. 9, Part A, Edited by Le Mehaute, B., and Hanes, D., John Wiley and Sons, New York, pp. 3–25.
Klopman, G., 1990. "A Note on Integral Properties of Periodic Gravity Waves in the Case of a Nonzero Mean Eulerian Velocity," *Journal of Fluid Mechanics*, Vol. 211, pp. 609–615.
Le Mehaute, B., Lu, C. C., and Ulmer, E. W., 1984. "Parametized Solution to Nonlinear Wave Problem," *Journal of Waterway, Port, Coastal and Ocean Division*, American Society of Civil Engineers, Vol. 110, No. 3, pp. 309–320.
Reinecker, M. M., and Fenton, J. D., 1981. "A Fourier Approximation Method for Steady Water Waves," *Journal of Fluid Mechanics*, Vol. 104, pp. 119–137.
Schwartz, L. W., 1974. "Computer Extension and Analytical Continuation of Stokes' Expansion for Gravity Waves," *Journal of Fluid Mechanics*, Vol. 62, Part 3, pp. 553–578.
Sobey, R. J., 1988. Discussion of "Nonuniqueness in Stream Function Wave Theory," by R. A. Dalrymple and P. Solana, *Journal of Waterway, Port, Coastal and Ocean Division*, American Society of Civil Engineers, Vol. 114, No. 1, pp. 112–114.
Sobey, R. J., Goodwin, P., Thieke, R. J., Westberg, R. J. Jr., 1987. "Application of Stokes, Cnoidal, and Fourier Wave Theories," *Journal of Waterway, Port, Coastal and Ocean Division*, American Society of Civil Engineers, Vol. 113, No. 6, pp. 565–587.
Stokes, G. G., 1847. "On the Theory of Oscillatory Waves," *Transactions of the Cambridge Philosophical Society*, Vol. 8, pp. 441–455.

LINEAR WAVE THEORY WITH SNELL'S LAW

Description

This application provides a simple estimate for the transformation of monochromatic waves. It considers two common processes of wave transformation: refraction (using Snell's law) and shoaling using wave properties predicted by linear wave theory (Airy, 1845). Given wave properties and a crest angle at a known depth, this application predicts the values in deep water and at a subject location specified by a new water depth. An important assumption is that all depth contours are assumed to be straight and parallel. In addition to the wave transformation results, this application reports common bulk wave properties from linear wave theory. More detailed discussion of these methods can be found in the SPM (1984), Dean and Dalrymple (1984), Sarpkaya and Isaacson (1981), as well the section of this reference manual entitled **Linear Wave Theory.**

Introduction

In deep water, waves often referred to as ocean swell have a profile that is very nearly sinusoidal, with long, low crests. As the waves propagate into shallow water, they undergo a transformation, starting where the waves are affected by the seabed at a depth approximately one-half of the deepwater wavelength. The wave velocity, height, and length alter. This process is called wave shoaling.

When waves travel at an angle to underwater contours, the portion of the wave in deeper water is moving faster than the part in shallower water. This variation causes the wave crest to bend toward alignment with the contours. This process is called wave refraction.

General Assumptions and Limitations

Important assumptions and limitations made in this shoaling and refraction discussion are:

- Wave energy between wave orthogonals remains constant. A wave orthogonal is a locus of points that define the minimum time of travel for wave propagation between two points (Le Mehaute, 1976). The wave orthogonals (Figure 3-1-1) are drawn perpendicular to the wave crest.

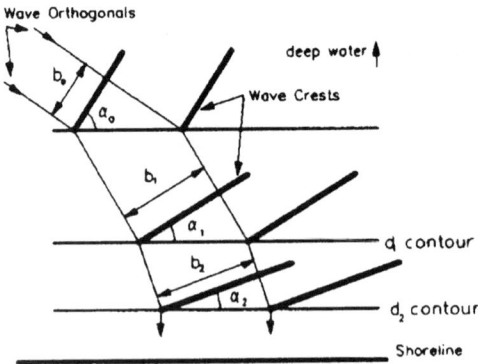

FIGURE 3-1-1. Snell's law and wave refraction.

- Direction of wave advance is perpendicular to the wave crest.
- Speed of a wave of a given period at a particular location depends only on the depth at that location.
- Changes in bottom topography are gradual.
- Waves are based on small-amplitude wave theory.
- Effects of current, winds, and reflections from beaches and underwater topographic variations are ignored.
- Offshore contours are straight and parallel to the shoreline.

Shoaling and Refraction Coefficients

It has been observed that the decrease in wave celerity with decreasing water depth is analogous to the decrease in the speed of light with an increase in the refractive index of the transmitting medium. Using this analogy, O'Brien (1942) suggested the use of Snell's law of geometrical optics for addressing the problem of water-wave refraction by changes in depth.

Using deep water as a reference, the general form of Snell's law is:

$$\frac{c}{c_0} = \frac{\sin \alpha}{\sin \alpha_0} \tag{1}$$

where

c, c_0 = wave velocity at the depth contour
α, α_0 = angle between a wave crest and the depth contour

As stated earlier, the rate of energy transfer between wave orthogonals is assumed to remain constant and is given by average energy flux:

$$\overline{P}_0 = \overline{P} \tag{2}$$

where

Deep Water	Shallow Water	Item
$\overline{P}_0 = \overline{E}_0 C_{g0}$	$\overline{P} = \overline{E} C_g$	Energy flux
b_0	b	Distance between orthogonals
$\overline{E}_0 = \dfrac{\rho g H_0^2}{8}$	$\overline{E} = \dfrac{\rho g H^2}{8}$	Average energy density
C_{g0}	C_g	Group velocity

Rearranging terms and solving for H/H_0 yields:

$$\frac{H}{H_0} = \sqrt{\frac{C_{g0}}{C_g}} \sqrt{\left(\frac{b_0}{b}\right)} \tag{3}$$

The term $\sqrt{C_{g0}/C_g}$ is known as the *shoaling coefficient* K_S, and the term $\sqrt{b_0/b}$ is known as the *refraction coefficient* K_R.

The assumption of straight and parallel depth contours leads to a simple geometrical relationship between b and α and a resulting expression for K_R:

$$K_R = \sqrt{\frac{b_0}{b}} = \sqrt{\frac{\cos \alpha_0}{\cos \alpha}} \tag{4}$$

A final expression for this simplified wave transformation approach is then:

$$\frac{H}{H_0} = K_R K_S \tag{5}$$

References and Bibliography

Airy, G. B. 1845. "Tides and Waves," *Encyclopaedia Metropolitana*, Vol. 192, pp. 241–396.
Dean, R. G., and Dalrymple, R. A. 1984. *Water Wave Mechanics for Engineers and Scientists*, Prentice-Hall, Englewood Cliffs, NJ, pp. 41–86, 104–105.
Hunt, J. N. 1979. "Direct Solution of Dispersion Equation," *Journal of Waterway, Port, Coastal and Ocean Division*, American Society of Civil Engineers, Vol. 107, No. WW4, pp. 457–459.
Le Mehaute, B. 1976. *An introduction to Hydrodynamics and Water Waves*, Springer-Verlag, New York, pp. 228–232.
O'Brien, M. P. 1942. "A Summary of the Theory of Oscillatory Waves," Technical Report No. 2, US Army Corps of Engineers, Beach Erosion Board, Washington, DC.
Sarpkaya, T., and Isaacson, M. 1981. *Mechanics of Wave Forces on Offshore Structures.* Van Nostrand Reinhold, New York, pp. 150–168, 237–242.
Shore Protection Manual. 1984. 4th ed., 2 Vols., US Army Engineer Waterways Experiment Station, Coastal Engineering Research Center, US Government Printing Office, Washington, DC, Chapter 2, pp. 6–33, 60–66.
Singamsetti, S. R., and Wind, H. G. 1980. "Characteristics of Shoaling and Breaking Periodic Waves Normally Incident to Plane Beaches of Constant Slope," Breaking Waves Publication No. M1371, Waterstaat, The Netherlands, pp. 23–27.
Weggel, J. R. 1972. "Maximum Breaker Height," *Journal of Waterways, Harbors and Coastal Engineering Division*, American Society of Civil Engineers, Vol. 98, No. WW4, pp. 529–548.

IRREGULAR WAVE TRANSFORMATION (GODA'S METHOD)

Description

This application yields cumulative probability distributions of wave heights as a field of irregular waves propagate from deep water through the surf zone. The application is based on two random-wave theories by Yoshimi Goda (1975 and 1984). The 1975 paper concerns transformation of random waves shoaling over a plane bottom with straight parallel contours. This analysis treated breaking and broken waves and resulted in cumulative probability distributions for wave heights given a water depth. It did not include refraction, however. The 1984 article details a refraction procedure for random waves propagating over a plane bottom with straight parallel contours assuming a particular incident spectrum. This ACES application combines the two approaches by treating directional random waves propagating over a plane bottom with straight parallel contours. This application also uses the theory of Shuto (1974) for the shoaling calculation. The theories assume a Rayleigh distribution of wave heights in the nearshore zone and a Bretschneider-Mitsuyasu incident directional spectrum. The processes modeled include:

- Wave refraction
- Wave shoaling
- Wave breaking
- Wave setup
- Surf beat

AUTOMATED COASTAL ENGINEERING SYSTEM (ACES)   **A.59**

General Assumptions and Limitations

General assumptions and limitations in this irregular wave transformation discussion are:

- The incident wave spectrum is of the Bretschnieder–Mitsuyasu type.
- The incident wave height distribution is of the Rayleigh type.
- Waves propagate over a smooth, absorbent beach (no reflection) with straight, parallel contours.
- Irregular wave shoaling may be approximated by shoaling of monochromatic waves.
- The probability distribution of broken wave heights is proportional to that of unbroken wave heights.
- An empirical relation determines the surf beat level.

These assumptions dictate the following theoretical limitations:

- The Bretschnieder–Mitsuyasu wave spectrum is a narrow-banded spectrum. Thus, this application should not be used where broad-banded spectra or multiple-peaked spectra are present.
- This application should not be used in areas with complex bathymetry; contours should be roughly parallel to shore.

In addition, the following limitations were implemented:

- Peak period must not exceed 16 sec.
- The smallest depth of interest that can be represented is 10 ft or 3.04 m.
- Principal direction of incidence must not exceed 75 deg from shore normal.

Theory

The development of the theory begins by considering the incident Bretschnieder–Mitsuyasu spectrum:

$$S(f) = 0.257(H_{1/3})^2 T_{1/3}(T_{1/3}f)^{-5} e^{[-1.03(T_{1/3}f)^{-4}]} \qquad (1)$$

where

$S(f)$ = spectral density (m² sec)
$H_{1/3}$ = significant wave height (in)
$T_{1/3}$ = significant wave period (sec)
f = wave frequency (rad/sec)

This incident wave spectrum is assumed to propagate over straight parallel bathymetric contours at a principal direction to shore normal. The spectrum is also assumed to have a directional spread of 135 deg, 67.5 deg to either side of the principal direction. The spreading function $G(f, \theta)$ used in this application is that of Mitsuyasu (1975):

$$G(f, \theta) = G_0 \cos^{2s}\left(\frac{\theta}{2}\right) \qquad (2)$$

where

θ = angular deviation from principal direction

$$G_0 = \frac{2^{2s-1}}{\pi} \frac{\Gamma^2(s+1)}{\Gamma(2s+1)} \qquad (3)$$

s = parameter representing directional energy concentration around a peak
s_{max} = peak value of s
   = 10 for wind waves
   = 25 for steep swell
   = 75 for flat swell
Γ = Gamma function

The effective refraction coefficient for the spectrum is defined as the *average* refraction coefficient for the entire spectrum. The *directional* Bretschnieder–Mitsuyasu spectrum is defined as (Goda, 1984):

$$S(f, \theta) = S(f)G(f, \theta) \tag{4}$$

From this an expression for the effective refraction coefficient $(K_r)_{eff}$ is given as:

$$(K_r)_{eff} = \left[\frac{1}{m} \int_0^\infty \int_{\theta_{min}}^{\theta_{max}} S(f, \theta) K_s^2(f) K_r^2(f, \theta) d\theta \, df \right]^{1/2} \tag{5}$$

where

$$m = \int_0^\infty \int_{\theta_{min}}^{\theta_{max}} S(f, \theta) K_s^2(f) d\theta \, df \tag{6}$$

$K_s(f)$ = individual shoaling coefficient
$K_r(f, \theta)$ = individual refraction coefficient

As the incident wave spectrum propagates into shallow water, individual waves within the spectrum refract, shoal, and break at rates dependent upon their individual heights, periods, and directions. Since these processes would occur at different rates (due to the irregularity of the wave field), any wave height prediction model would also have to be irregular. One way to describe the wave field during transformation and still retain irregularity is by using a probability distribution of wave heights. In this application, Goda's irregular wave height distribution model (1975) is used.

In the absence of wave breaking and assuming a Rayleigh distribution of wave heights, a probability density function of wave height normalized by H_0' is given by:

$$P_0(x) = 2a^2 x \, e^{\left(-\frac{a^2}{x^2}\right)} \tag{7}$$

where

$P_0(x)$ = probability density function

$x = \dfrac{H}{H_0'}$ = normalized wave height

$a = \dfrac{1.416}{K_s}$

K_s = shoaling coefficient

The Rayleigh distribution does not have an upper limit on the normalized wave height; it approaches zero asymptotically. However, normalized wave height in nature is limited by wave breaking. This is the primary difference between Goda's model and the ideal Rayleigh model.

If wave breaking is assumed to occur between x_2 and x_1, the unbroken wave heights are restricted in Equation 7 by the following equations:

AUTOMATED COASTAL ENGINEERING SYSTEM (ACES) A.61

$$P_r(x) = \begin{vmatrix} P_0(x) & x \le x_2 \\ P_0(x) - \left(\dfrac{x-x_2}{x_1-x_2}\right)P_0(x_1) & x_2 < x \le x_1 \\ 0 & x_1 \le x \end{vmatrix} \quad (8)$$

It is assumed that broken waves carry a small amount of energy. This assumption is accounted for by redistributing the energy along the top of the distribution. This redistribution leads to Goda's form of the probability density function for irregular waves:

$$P(x) = \alpha\, P_r(x)$$

where

$$\frac{1}{\alpha} = \int_0^{x_1} P_r(x)dx = 1 - [1 + a^2 x_1(x_1 - x_2)]e^{(-a^2 x_1^2)} \quad (10)$$

Note: $P_r(x)$ is restricted by Equation 8 and a is defined in Equation 7.

At this point, essentially two different theories are represented in the method: Goda's refraction algorithm (Equations 1–6) and Goda's irregular wave height distribution model (Equations 7–10). The wave parameter in common with the two theories is the wave height, more specifically, the breaking wave height. Goda (1975) gave the following expression for normalized breaking wave height based on wave shoaling only:

$$X_b = \frac{H_b}{H_0'} = 0.17 \frac{L_0}{H_0'}\left\{1 - e^{\left[\frac{-1.5\pi h}{H_0'}\frac{H_0'}{L_0}(1+15\tan^{4/3} n)\right]}\right\} \quad (11)$$

where

H_b = breaking wave height
H_0' = equivalent deepwater significant wave height
L_0 = deepwater wavelength
h = water depth of interest
n = bottom slope

By multiplying Equation 11 by $(K_r)_{eff}$, Goda's irregular wave distribution model is extended to include refraction effects. Shoaling, surf beat, and wave setup are also included in the application and are used to determine water depth and wave height for input into Equation 11. The equations for surf beat, wave setup, and wave shoaling follow:

Surf beat

$$\xi_{rms} = \frac{0.01 H_0'}{\sqrt{\dfrac{H_0'}{L_0}\left(1 + \dfrac{h}{H_0'}\right)}} \quad (12)$$

Wave setup

$$\frac{d\overline{\eta}}{dx} = \frac{-1}{(\overline{\eta}+h)}\frac{d}{dx}\left[\frac{1}{8}H_{rms}^2\left(\frac{1}{2} + \frac{2kh}{\sinh 2kh}\right)\right] \quad (13)$$

where

$\dfrac{d\overline{\eta}}{dx}$ = set-up gradient
x = offshore-onshore coordinate
$\overline{\eta}$ = wave setup
k = wave number

Wave shoaling (Shuto, 1974)

$$0 < \frac{gH(T_{1/3})^2}{h^2} \leq 30: \text{use linear wave theory}$$

$$30 < \frac{gH(T_{1/3})^2}{h^2} \leq 50: Hh^{2/7} = \text{constant} \qquad (14)$$

$$50 < \frac{gH(T_{1/3})^2}{h^2} < \infty: Hh^{5/2}\left(\sqrt{\frac{gH(T_{1/3})^2}{h^2}} - 2\sqrt{3}\right) = \text{constant}$$

where

g = gravitational acceleration

References and Bibliography

Goda, Y. 1975. "Irregular Wave Deformation in the Surf Zone," *Coastal Engineering in Japan*, Vol. 18, pp. 13–26.
Goda, Y. 1984. *Random Seas and Design of Maritime Structures*, University of Tokyo Press, Tokyo, Japan, pp. 41–46.
Mitsuyasu, H. 1975. "Observation of the Directional Spectrum of Ocean Waves Using a Cloverleaf Buoy," *Journal of Physical Oceanography*, Vol. 5, No. 4, pp. 750–760.
Shuto, N. 1974. "Nonlinear Long Waves in a Channel of Variable Section," *Coastal Engineering in Japan*, Vol. 17, pp. 1–12.

COMBINED DIFFRACTION AND REFLECTION BY A VERTICAL WEDGE

Description

This application estimates wave height modification due to combined diffraction and reflection near jettied harbor entrances, quay walls, and other such structures. Jetties and breakwaters are approximated as a single straight, semi-infinite breakwater by setting the wedge angle to zero. Corners of docks and quay walls may be represented by setting the wedge angle equal to 90 deg. Additionally, such natural diffracting and reflecting obstacles as rocky headlands can be approximated by setting a particular value for the wedge angle.

Introduction

In coastal and ocean engineering practice, it is often important to be able to determine wave height near such coastal structures as jetties, breakwaters, platforms, and docks. Such information aids engineers in evaluating coastal structure designs, especially in the areas of energy transmission, sediment transport, and structural strength. Wave heights in the vicinity of a structure have traditionally been presented in the form of dimensionless wave diffraction and reflection coefficients defined as the ratio of diffracted/reflected wave height to incident wave height. Early studies presented a dimensionless graphical solution based on theory by Penny and Price (1952) for diffraction by a semi-infinite breakwater (Wiegel, 1962). Subsequently these diagrams have been incorporated into every edition of the SPM (1984). They remain useful tools for preliminary engineering design.

Limitations of the traditional diagrams include monochromatic and unidirectional wave assumption, constant water depth assumption, no reflection, and simple structure shape with vertical walls. Recently Chen (1987) presented an analytical solution for wave height modification in the vicinity of a structure. The solution is more general than the traditional approach in that it includes reflection as well as diffraction and it allows a wedge-shaped structure with vertical walls. Other limitations of the traditional approach still apply. Chen's (1987) solution was implemented in a computer code, PCDFRAC, by Kaihatu and Chen (1988). Output consists of a wave height modification coefficient (ratio of combined diffracted and reflected height to incident height) and a wave phase for any selected point in the vicinity of the structure. The code PCDFRAC has been modified to reside in ACES.

General Assumptions and Limitations

Assumptions inherent in the approach include linear, monochromatic, unidirectional waves, and constant water depth. The structure is assumed to be straight and semi-infinite in length with vertical walls. The walls are treated as fully reflecting.

Modified wave heights in areas strongly affected by reflection, such as the area immediately in front of the wedge, are variable in space because of interference between the incident and reflected waves. Such variability would not be expected in natural wave conditions.

Solutions are not available beyond about 10 wavelengths from the wedge tip. The approach is based on an analytical solution which includes summation of an infinite number of terms. The summation is computed with as many terms as needed to satisfy a convergence criterion. There is a limit on the maximum number of terms allowed. More terms are required for convergence as distance from the wedge tip increases. The limit is usually reached at distances of 10 to 15 wavelengths.

Theoretical Background

The general boundary value problem of linear wave reflection and diffraction by a vertical wedge of arbitrary angle has been well formulated and presented by Stoker (1957) among many other investigators. The technique to obtain an analytical solution for the problem is also given by Stoker (1957). However, analytical solutions were available only for the special case of wave diffraction by a thin semi-infinite breakwater, that is, a wedge with angle equal to zero. The more general solution by Chen (1987) is the basis for the model PCDFRAC.

A cylindrical coordinate system (r, θ, z) is adopted, where $z = 0$ represents the undisturbed free surface and the positive z-axis is positioned vertically upward. The tip of the wedge is chosen to be the origin of the coordinate system, and the two rigid walls of the wedge are at $\theta = 0$ and $\theta = \theta_0$, respectively (Figure 3-3-1). Cartesian coordinates (x,y,z), also shown in Figure 3-3-1, are used for specifying input to the routine. The wedge angle is defined as $2\pi - \theta_0$, while the water domain is defined as $0 \leq \theta \leq \theta_0$ and $0 \geq z \geq -h$.

For the problem at hand, the velocity field for wave motion in an ideal fluid may be represented by the velocity potential $\Phi(r, \theta, z, t)$. This may be expressed as follows:

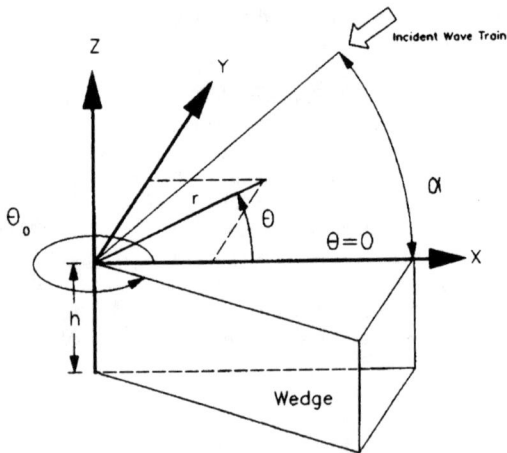

FIGURE 3-3-1. A vertical wedge of arbitrary angle.

$$\Phi(r, \theta, z, t) = A_0 \frac{\cosh k(z + h)}{\cosh(kh)} \phi(r, \theta) e^{iwt} \tag{1}$$

where

$\phi(r, \theta)$ = velocity potential function in the horizontal plane
t = time variable

$A_0 = \dfrac{-iga_0}{\omega}$

$i = \sqrt{-1}$
g = gravitational acceleration
a_0 = incident wave amplitude
ω = radian frequency
k = wave number

The wave number k must be real and satisfy the following linear dispersion equation:

$$\omega^2 = gk \tanh(kh) \tag{2}$$

Several properties of wave mechanics are dependent on the velocity potential component $\phi(r, \theta)$. For example, the free surface elevation η may be expressed in terms of $\phi(r, \theta)$ as follows:

$$\eta = a_0 \phi(r, \theta) e^{iwt} \tag{3}$$

The flow velocity u may be expressed in polar coordinates as follows:

$$u_r = \frac{\partial \phi}{\partial r} = A_0 \frac{\cosh k(z + h)}{\cosh(kh)} \frac{\partial \phi}{\partial r} e^{iwt} \tag{4}$$

$$u_\theta = \frac{1}{r} \frac{\partial \phi}{\partial \theta} = A_0 \frac{\cosh k(z + h)}{\cosh(kh)} \frac{1}{r} \frac{\partial \phi}{\partial \theta} e^{iwt} \tag{5}$$

where the subscripts, r and θ, refer to the $r-$ and $\theta-$ directions. The absolute values of u_r and u_θ are the maximum flow velocities for each component direction, while the phases of u_r and u_θ contain the wave phase information. It is clear the solution to Equation 1 and Equations 3–5 lies in finding $\phi(r, \theta)$.

For an incident plane wave train coming from the a direction (Figure 3-3-1), the free surface elevation of the incident wave may be described by

$$\eta_i = a_0 e^{ikr_i \cos(\alpha + \omega t)} \tag{6}$$

where r_i defines a point on the incident wave.

The analytical solution for a wave field by a vertical wedge of arbitrary wedge angle, based on the linearized wave theory, may be written as follows (Chen, 1987):

$$\phi(r, \theta) = \frac{2}{v}\left[J_0(kr) + 2\sum_{n=1}^{\infty} e^{\frac{in\pi}{2v}} J_{n/v}(kr) \cos\frac{n\alpha}{v} \cos\frac{n\theta}{v}\right] \tag{7}$$

where

$$v = \frac{\theta_0}{\pi}$$

J_0 = zero order Bessel function of the first kind
$J_{n/v}$ = n/v order Bessel function of the first kind

The semi-infinite breakwater is a special case of the diffraction/reflection problem, where the wedge angle is equal to zero ($\theta_0 = 2\pi$) and v = 2. The solution of Equation 1 for this case is

$$\phi(r, \theta) = J_0(kr) + 2\sum_{n=1}^{\infty} e^{\frac{in\pi}{4}} J_{n/2}(kr) \cos\frac{n\alpha}{2} \cos\frac{n\theta}{2} \tag{9}$$

Note: Equations 7 and 9 show a summation of an infinite number of terms. This has been accommodated in the program by carrying the summation out to a term followed by eight successive terms in which the absolute value of the Bessel function is 10^{-8} or less. If the value of the solution is of the order one, this corresponds to a truncation error of 10^{-8} or less.

The velocity potential function $\phi(r, \theta)$ in Equations 7 and 9 is a complex function and may be expressed as

$$\phi = |\phi|e^{i\beta} \tag{10}$$

where

$$|\phi|^2 = [Im(\phi)]^2 + [Re(\phi)]^2 \qquad \text{(amplitude of } \phi\text{)} \tag{11}$$

$Im\ \phi$ = imaginary part of $\phi(r, \theta)$
$Re\ \phi$ = real part of $\phi(r, \theta)$

$$\beta = \tan^{-1}\left[\frac{Im(\phi)}{Re(\phi)}\right] \qquad \text{(phase of } \phi\text{)} \tag{12}$$

Substituting Equation 10 into Equation 3, the following is obtained:

$$\eta = a_0|\phi|e^{i(\beta + \omega t)} \tag{13}$$

This expression represents the actual water surface elevation at a point in the water domain bounded by a vertical wedge of arbitrary wedge angle. Since the incident wave train is expressed in Equation 6, the normalized water surface elevation in the near field may be expressed as

$$\frac{\eta}{\eta_i} = |\phi|e^{i(\beta - kr_i \cos\alpha)} \qquad (14)$$

where

η_i = incident free surface elevation

It is clear that the term $|\phi|e^{i(\beta - kr_i \cos\alpha)}$ is a factor that modifies the incident wave elevation to account for reflection and diffraction effects. The amplitude of the normalized surface elevations, which is comparable to diffraction and reflection coefficients defined in the SPM (1984), may be expressed as the following wave diffraction/reflection coefficient:

$$\left|\frac{\eta}{\eta_i}\right| = |\phi| \qquad (15)$$

The phase of Equation 14 is the phase difference between incident and modified waves.

$$\text{Phase difference} \quad \frac{\eta}{\eta_i} = \beta - kr_i \cos\alpha \qquad (16)$$

Output of PCDFRAC is composed of $|\phi|$ and β. The diffraction/reflection coefficient $|\phi|$ (modification factor) is then multiplied by the incident wave height to obtain the modified wave height. The phase difference β related to the modified wave is a quantity not usually required in engineering practice; however, it may be useful on some occasions.

References

Chen, H. 5. 1987. "Combined Reflection and Diffraction by a Vertical Wedge," Technical Report CERC-87-16, US Army Engineer Waterways Experiment Station, Vicksburg, MS.

Chen, H. S., and Thompson, E. F. 1985. "Iterative and Pade Solutions for the Water–Wave Dispersion Relation," Miscellaneous Paper CERC-85-4, US Army Engineer Waterways Experiment Station, Vicksburg, MS.

Kaihatu, J. A., and Chen, H. 5. 1988. "Combined Diffraction and Reflection by a Vertical Wedge: PCDFRAC User's Manual," Technical Report CERC-88-9, US Army Engineer Waterways Experiment Station, Vicksburg, MS.

Morris, A. H., Jr. 1984. "NSWC Library of Mathematics Subroutines," NSWC TR 84 -143, Strategic Systems Department, Naval Surface Weapons Center, Dahlgren, Va.

Penny, W. G., and Price, A. T. 1952 "The Diffraction Theory of Sea Waves by Breakwaters, and the Shelter Afforded by Breakwaters", *Philosophical Transactions, Royal Society* (London), Series A, Vol. 244, pp. 236–253

Stoker, J. J. 1957. *Water Waves*, Interscience Publishers Inc., New York, pp. 109–133.

Shore Protection Manual. 1984. 4th ed., 2 Vols., US Army Engineer Waterways Experiment Station, Coastal Engineering Research Center, US Government Printing Office, Washington, DC, Chapter 2, pp. 75–109.

Wiegel, R. L. 1962. "Diffraction of Waves by a Semi-Infinite Breakwater," *Journal of the Hydraulics Division*, Vol. 88, No. HYI, pp. 27–44.

BREAKWATER DESIGN USING HUDSON AND RELATED EQUATIONS

Description

A rubble structure is often composed of several layers of random-shaped or random-placed stones, protected with a cover layer of selected armor units of either quarrystone or specially shaped concrete units. This ACES application provides estimates for the armor weight, minimum crest width, armor thickness, and the number of armor units per unit area of a breakwater using Hudson's and related equations. The material presented herein can be found in Chapter 7 of the SPM (1984).

Introduction

Until about 1930, design of rubble structures was based only on experience and general knowledge of site conditions. Empirical methods have been developed that, if used with care, will give satisfactory determination of the stability characteristics of these structures when under attack by storm waves.

General Assumptions and Limitations

Empirical formulas that were developed for the design of rubble-mound structures are generally expressed in terms of the stone weight required to withstand design wave conditions. These formulas have been largely derived from physical model studies. They are guides and must be used with experience and engineering judgment. Physical modeling is often a cost-effective measure to determine the final cross-section design for most rubble-mound structures.

Stability of Rubble Structures

A proposed breakwater may necessarily be designed for either nonbreaking or breaking waves depending upon positioning of the breakwater and severity of anticipated wave action during its economic life. Some local wave conditions may be of such severity that the protective cover layer must consist of specially shaped concrete armor units in order to provide economic construction of a stable breakwater. The following four sections describe empirical formulas used in the ACES package for design of rubble structures.

Weight of Primary Armor Unit
Comprehensive investigations were made by Hudson (1953, 1959, 1961a, 1961b) to develop a formula to determine the stability of armor units on rubble structures. The stability formula, based on the results of extensive small-scale model testing and some preliminary verification by large-scale model test, is

$$W = \frac{w_r H_i^3}{K_D (S_r - 1)^3 \cot \theta} \tag{1}$$

where
W = weight of individual armor unit in the primary cover layer
w_r = unit weight of armor unit material
H_i = design wave height
K_D = armor unit stability coefficient (see Table A-I of Appendix A)
$S_r = w_r/w_w$ = specific gravity of armor material
w_w = unit weight of water
θ = angle between seaward structure slope and horizontal

The dimensionless stability coefficient, K_D, accounts for factors other than structure slope, wave height, and the specific gravity of water at the site. The most important of these variables include

- Shape of armor units.
- Number of units comprising the thickness of the armor layer.
- Manner of placing armor units.
- Surface roughness and sharpness of edges of the armor units (degree of interlocking of the armor units).
- Type of wave attacking structure (breaking or nonbreaking).
- Part of the structure being attacked (trunk or head).
- Angle of incident wave attack.

These stability coefficients (Table A-1 of Appendix A) were derived from large- and small-scale tests that used many various shapes and sizes of both natural and artificial armor units. Values are reasonably definitive and are recommended for use in the design of rubble-mound structures, supplemented by physical model test results when economically warranted.

Crest Width
The width of the crest depends greatly on the degree of allowable overtopping; where there will be no overtopping, crest width is not critical. Little study has been made of crest width of a rubble structure subject to overtopping. As a general guide for overtopping conditions, the minimum crest width should equal the combined widths of three armor units ($n = 3$). The crest should be wide enough to accommodate any construction and maintenance equipment that may be operated from the structure. Crest width is obtained from the following equation:

$$B = nk_\Delta \left(\frac{W}{w_r}\right)^{1/3} \quad (2)$$

where

B = crest width
n = number of armor units (ACES application sets $n = 3$)
k_Δ = layer coefficient (see Table A-2 of Appendix A)
W = weight of individual armor unit in the primary cover layer
w_r = unit weight of armor unit material

Thickness of the Armor Layer
The thickness of the cover layer is determined from the following formula:

$$r = nk_\Delta \left(\frac{W}{w_r}\right)^{1/3} \quad (3)$$

where
r = average layer thickness
n = number of layers of armor units

Armor Unit Placement Density
The placing density is given by the following formula:

$$N_r = Ank_\Delta \left(1 - \frac{p}{100}\right)\left(\frac{w_r}{W}\right)^{2/3} \quad (4)$$

where

N_r = number of armor units for a given surface area
A = surface area (assumed as 1000)
p = average porosity of cover layer (see Table A-2 of Appendix A)

References and Bibliography

Headquarters, Department of the Army. 1986. "Design of Breakwaters and Jetties," Engineer Manual 1110-2-2904, Washington, DC, p. 4–10.
Hudson, R. Y. 1953. "Wave Forces on Breakwaters," *Transactions of the American Society of Civil Engineers*, American Society of Civil Engineers, Vol. 118, p. 653.
Hudson, R. Y. 1959. "Laboratory Investigations of Rubble-Mound Breakwaters," *Proceedings of the American Society of Civil Engineers*, American Society of Civil Engineers, Waterways and Harbors Division, Vol. 85, NO. WW3, Paper No. 2171.
Hudson, R. Y. 1961a. "Laboratory Investigation of Rubble-Mound Breakwaters," *Transactions of the American Society of Civil Engineers*, American Society of Civil Engineers, Vol. 126, Pt IV.
Hudson, R. Y. 1961b. "Wave Forces on Rubble-Mound Breakwaters and Jetties," Miscellaneous Paper 2-453, US Army Engineer Waterways Experiment Station, Vicksburg, MS.
Markel, D. G., and Davidson, D. D. 1979. "Placed-Stone Stability Tests, Tillamook, Oregon; Hydraulic Model Investigation," Technical Report HL-79- 16, US Army Engineer Waterways Experiment Station, Vicksburg, MS.
Shore Protection Manual. 1984. 4th ed., 2 Vols., US Army Engineer Waterways Experiment Station, Coastal Engineering Research Center, US Government Printing Office, Washington, DC, Chapter 7, pp. 202–242.
Smith, 0. P. 1986. "Cost-Effective Optimization of Rubble-Mound Breakwater Cross Sections," Technical Report CERC-86-2, US Army Engineer Waterways Experiment Station, Vicksburg, MS. p. 48.
Zwamborn, J. A., and Van Niekerk, M. 1982. *Additional Model Tests—Dolos Packing Density and Effect of Relative Block Density*, CSIR Research Report 554, Council for Scientific and Industrial Research, National Research Institute for Oceanology, Coastal Engineering and Hydraulics Division, Stellenbosch, South Africa.

TOE PROTECTION DESIGN

Description

Toe protection consists of armor for the beach or bottom material fronting a structure to prevent wave scour. This application determines armor stone size and width of a toe protection apron for *vertical* faced structures such as sea walls, bulkheads, quay walls, breakwaters, and groins. Apron width is determined by the geotechnical and hydraulic guidelines specified in Engineer Manual 1110-2-1614 (Headquarters, Department of the Army, 1985). Stone size is determined by a method (Tanimoto, Yagyu, and Goda, 1982) whereby a stability equation is applied to a single rubble unit placed at a position equal to the width of the toe apron and subjected to standing waves.

Introduction

Coastal structures rely upon the foundation material for vertical support. Some types of retaining walls also rely upon the bottom material for lateral support. Wave action resulting in loss of bottom material can cause damage and ultimate collapse of a protective structure. While a variety of methods for wave scour protection are employed in practice, this application addresses a simple toe protection design using an apron of armor stones fronting a structure with a vertical seaward face. Unbroken waves are assumed to be normally incident to the structure and are assumed to produce standing waves above the toe protection apron. Stone size is determined by consideration of the stability of a single stone subjected to the standing waves and situated at the seaward edge of the apron.

General Assumptions and Limitations

The methodology represented in this application is a composite of largely empirical guidance for the width of the toe apron, and a semi-empirical formulation for the toe stone weight. General assumptions include the following:

- Waves are normally incident to the structure.
- Standing waves form as a result of wave interaction with a vertical (seaward) face of the structure and remain unbroken in the water depth above the toe apron.
- Linear wave theory approximations are adequate to predict the standing wave properties.
- Rankine theory is adequate for evaluating the stability of the soil wedge beneath the toe apron.
- Rubble-mound material is used as the toe protection material.

In general, this application considers the stability of an armor stone at the seaward crest of the toe protection apron. It does not offer any guidance for preparation or detailed protection of foundation material at the dredge line.

The method for stone weight (Tanimoto, Yagyu, and Goda, 1982) is based upon irregular waves (characterized by significant wave height, H_s) acting on composite breakwaters. To be formally consistent with Tanimoto's approach, the normally incident wave train should use H_s values and structure configurations limited to those tested in the original research. Sheet-pile designs were not considered in the development of the method. In general, Tanimoto's method has not been verified by model tests within the Corps of Engineers. The empirical guidance used for estimating the toe protection width (Engineer Manual 1110-2-1614) has no physical coupling with Tanimoto's method for determining the toe stone weight, yet has significant impact on the formulas for stone weight.

In using Rankine theory as part of the guidance for determining toe protection width for pile structures, a common practice is to use the effective penetration depth of the pile which is less than the full driven depth. The actual computations will apply the user-specified value, and engineering judgment should be applied to adjust d_e as an input parameter.

Official guidance regarding toe protection is being revised to a less conservative design, and this interim methodology will be revised at a later date.

Width of Toe Apron

The width of the toe apron is estimated using hydraulic and geotechnical factors. Sketches depicting geotechnical and hydraulic factors considered in the estimation of the apron width for a typical sheet-pile wall and a gravity type breakwater are shown in Figures 4-2-1 and 4-2-2 respectively.

The minimum width of the toe apron, B, from a geotechnical perspective is associated

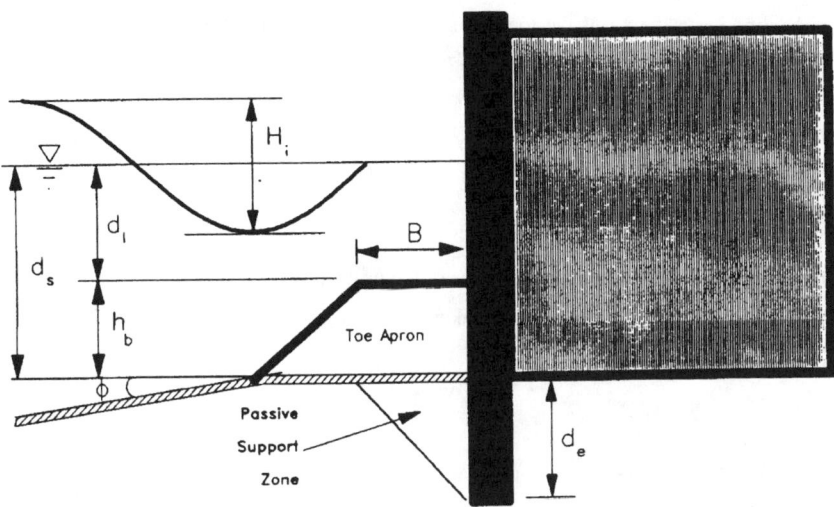

FIGURE 4-2-1. Typical toe apron for sheet-pile walls.

with structure–soil equilibrium considerations and is estimated using Rankine theory (Eckert, 1983 and Eckert and Callendar, 1987):

$$B_1 = K_p d_e \tag{1}$$

where

K_p = coefficient of passive earth pressure
d_e = sheet-pile penetration depth (0 if no pile)

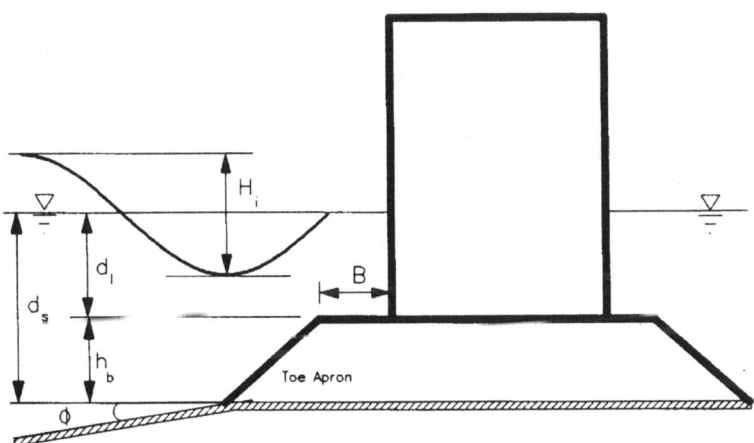

FIGURE 4-2-2. Typical toe apron for breakwaters

A.72   HANDBOOK OF COASTAL ENGINEERING

The minimum width of the toe apron, B, from a hydraulics perspective is estimated from simple criteria stated as (EM 1110-2-1614):

$$B_2 = 2H_i \tag{2}$$

or

$$B_3 = 0.4d_s \tag{3}$$

where

H_i = incident wave height
d_s = water depth at structure (in .absence of toe protection)

The design width for the apron is taken as the largest value predicted by the various criteria considered in Equations 1–3.

$$B = \max(B1, B2, B3) \tag{4}$$

Toe Stone Weight

The stability equation (SPM, 1984) used to determine the armor stone weight in the toe protection apron is cast in the form of the equation derived by Hudson (1959) for armor units for rubble-mound breakwaters.

$$W = \frac{w_r H_i^3}{N_s^3 (S_r - 1)^3} \tag{5}$$

where

W = weight of individual armor stone
w_r = unit weight of armor stone
N_s = stability number
$S_r = w_r/w_w$ = specific gravity of armor stone
w_w = unit weight of water

Stability Number

The stability number, N_s, is semi-empirically formulated on the basis of irregular wave tests conducted by Tanimoto, Yagyu, and Goda (1982). It is determined as:

$$N_s = \max\left\{1.3\left(\frac{1-K}{K^{1/3}}\right)\left(\frac{d_l}{H_i}\right) + 1.8e^{\left[-1.5\frac{(1-K)^2}{K^{1/3}}\frac{d_l}{H_i}\right]}; 1.8\right\} \tag{6}$$

where
K = parameter associated with the maximum horizontal velocity at the edge of the apron determined from standing waves as described by linear wave theory

$$= \frac{\dfrac{4\pi d_l}{L}}{\sinh\left(\dfrac{4\pi d_l}{L}\right)}\left(\sin\frac{2\pi B}{L}\right)^2$$

d_l = water depth (at top of toe protection)
L = wavelength at depth d_l predicted by linear wave theory

References and Bibliography

Eckert, J. W. 1983. "Design of Toe Protection for Coastal Structures," *Proceedings of the Coastal Structures '83 Conference*, American Society of Civil Engineers, Arlington, VA, pp. 331–341.

Eckert, J. W., and Callendar, G. 1987. "Geotechnical Engineering in the Coastal Zone," Instructional Report CERC-87- 1, US Army Engineer Waterways Experiment Station, Vicksburg, MS. Chapter 8, pp. 36–38.

Headquarters, Department of the Army. 1985. "Design of Coastal Revetments, Seawalls, and Bulkheads," Engineer Manual 1110-2-1614, Washington, DC, Chapter 2, pp. 15–19.

Hudson, R. Y. 1959. "Laboratory Investigations of Rubble-Mound Breakwaters," *Proceedings of the American Society of Civil Engineers*, American Society of Civil Engineers, Waterways and Harbors Division, Vol. 85, NO. WW3, Paper No. 2171.

Shore Protection Manual. 1984. 4th ed., 2 Vols., US Army Engineer Waterways Experiment Station, Coastal Engineering Research Center, US Government Printing Office, Washington, DC, Chapter 7, pp. 242–249.

Tanimoto, K., Yagyu, T., and Goda, Y. 1982. "Irregular Wave Tests for Composite Breakwater Foundations," *Proceedings of the 18th Coastal Engineering Conference*, American Society of Civil Engineers, Cape Town, Republic of South Africa, Vol. III, pp. 2144–2161.

NONBREAKING WAVE FORCES AT VERTICAL WALLS

Description

This application provides the pressure distribution and resultant force and moment loading on a vertical wall caused by normally incident, nonbreaking regular waves. The results can be used to design vertical structures in protected or fetch-limited regions when the water depth at the structure is greater than about 1.5 times the maximum expected wave height. The application provides the same results as found using the design curves given in Chapter 7 of the SPM (1984).

Introduction

The pressure distribution on the seaward side of a vertical wall exposed to wave action is composed of two components, the hydrostatic pressure due to the depth of water at the wall and the wave-induced dynamic pressure caused by acceleration of the fluid particles. Estimates of wave-induced pressure are required to design vertical walls that will resist the applied loads without loss of functionality.

Design curves presented in the SPM (1984) provide a means of determining nonbreaking wave forces and moments as a function of water depth, water specific weight, incident wave height, and wave period. One set of curves applies to the case of complete reflection ($\chi = 1.0$) whereas the other set is for a slightly less ($\chi = 0.9$) reflective case.

The curves in the SPM represent a composite method using two solution methods for the wave induced pressure distribution on a vertical wall. The Miche–Rundgren method provides better fit to laboratory data for steep, nonbreaking waves, but the theory begins to overpredict as the wavelength is increased. On the other hand, the Sainflou method provides better estimates for long, low-steepness waves, but it overpredicts as the waves become steeper. Therefore, the curves in the SPM use the Sainflou method for low-steepness waves and the Miche–Rundgren method for steeper waves. Transition from one method to the other is determined simply by whichever method provides the minimum force or moment for given values of wave steepness and wave height-to-depth ratio.

The major disadvantage of using the design curves in the SPM (other than inconvenience and potential for error) is determination of resultant forces and moments for values of wave height-to-depth ratio that fall between the curves. Additionally, the designer is restricted to using a reflection coefficient of either $\chi = 1.0$ or $\chi = 0.9$.

General Assumptions and Limitations

Hydrodynamic assumptions invoked in derivation of both the Sainflou and the Miche–Rundgren equations are typical for theoretical wave motion theories. These assumptions include:

- Nonviscous flow, i.e., ideal fluid in which there are no tangential stresses between adjacent water particles.
- Incompressible fluid of constant density.
- Irrotational flow in two dimensions.
- The absolute pressure at the surface is equal to atmospheric pressure everywhere (so surface pressure is defined as zero).

The above assumptions allow the wave motion problem at a reflecting vertical wall to be expressed in terms of potential flow theory.

Both methods are derived in the Lagrangian system, which follows individual water particles rather than remaining stationary. The primary advantage of Lagrangian coordinates is better representation of the hydrodynamic pressure above the still-water level when the crest of the standing wave is at the wall. Hydrodynamic pressure exerted on the vertical wall is hydrostatic pressure due to the depth of water, and wave pressure due to transformation of wave kinetic energy into pressure energy when wave motion acts on the wall. Most applications of the theories usually omit the hydrostatic pressure component of the total pressure, and this custom has been followed in this ACES application.

Both methods assume normally incident, monochromatic waves being reflected by a vertical wall. The waves are nonbreaking and of constant form, and it is assumed that no overtopping of the wall occurs. The bottom fronting the vertical wall is assumed horizontal. A definition sketch for a nonbreaking, normally incident, monochromatic wave being reflected from a vertical wall is shown on Figure 4-3-1.

Primary drawbacks to using the Sainflou or Miche–Rundgren method of calculating wave-induced pressure distributions and resulting forces and moments are the facts that they are monochromatic theories and they do not give breaking wave forces. There may be waves in an *equivalent* irregular wave train that produce greater force loading than predicted using monochromatic waves. Also waves breaking directly on the wall can produce significantly higher forces than nonbreaking waves.

The only limitation in applying the Miche–Rundgren method is the restriction that the reflection coefficient can vary only over the range $0.9 < \chi < 1.0$. This range is based on recommendations given in the SPM. The Sainflou method accommodates only a reflection coefficient equal to unity $\chi = 1.0$. Consequently, predictions from this method may sometimes be conservative.

Sainflou Method

Sainflou (1928) published a theoretical solution for the pressure distribution at a vertical wall for a perfectly reflected monochromatic wave of incident height, H_i. His derivation was based on the classical hydrodynamic equations of continuity and momentum. Sain-

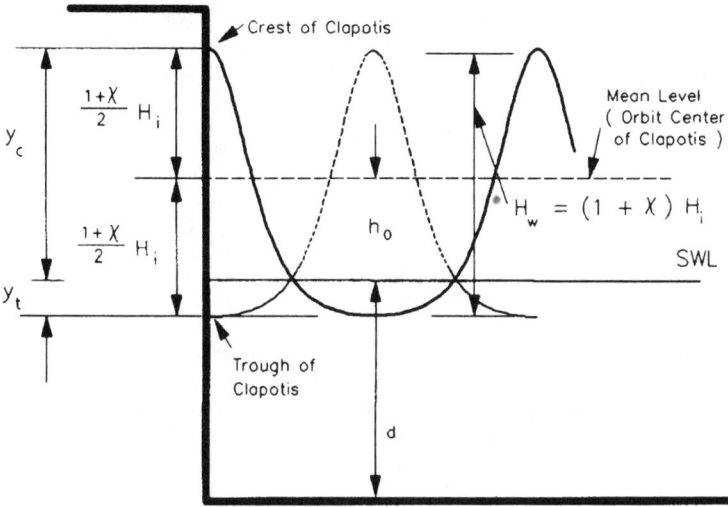

FIGURE 4-3-1. Definition sketch.

flou developed his equations as a simplification of second-order wave theory in which he omitted some of the second-order terms in the expressions. Consequently, his solution is partly of second-order in the parameter H/L. He assumed that ambient atmospheric pressure is constant everywhere on the free surface, and his expression for pressure due to the wave motion does not include any atmospheric pressure contribution.

Free Surface Elevation
In the Lagrangian coordinate system, Sainflou derived the following expressions for the vertical elevations of water particles of a standing wave at a vertical wall under the crest and trough:

Crest

$$y_{cr} = y_0 + H_i \frac{\sinh\left[\frac{2\pi}{L}(d+y_0)\right]}{\sinh\left(\frac{2\pi d}{L}\right)} + \pi H_i \left(\frac{H_i}{L}\right) \frac{\sinh\left[\frac{2\pi}{L}(d+y_0)\right] \cosh\left[\frac{2\pi}{L}(d+y_0)\right]}{\sinh\left(\frac{2\pi d}{L}\right) \sinh\left(\frac{2\pi d}{L}\right)} \quad (1)$$

Trough

$$y_{tr} = y_0 - H_i \frac{\sinh\left[\frac{2\pi}{L}(d+y_0)\right]}{\sinh\left(\frac{2\pi d}{L}\right)} + \pi H_i \left(\frac{H_i}{L}\right) \frac{\sinh\left[\frac{2\pi}{L}(d+y_0)\right] \cosh\left[\frac{2\pi}{L}(d+y_0)\right]}{\sinh\left(\frac{2\pi d}{L}\right) \sinh\left(\frac{2\pi d}{L}\right)} \quad (2)$$

where

H_i = incident wave height
L = wavelength
d = water depth
y_0 = initial vertical elevation of a water particle at rest

The free surface at rest is given by $y_0 = 0$. Therefore, the elevation of the free surface relative to still-water level (SWL) when the crest or trough of the standing wave is at the wall is found by substituting $y_0 = 0$ into Equations 1 and 2, yielding the following:

Crest

$$\eta_{cr} = H_i + \pi H_i \left(\frac{H_i}{L}\right) \coth\left(\frac{2\pi d}{L}\right) \tag{3}$$

Trough

$$\eta_{tr} = -H_i + \pi H_i \left(\frac{H_i}{L}\right) \coth\left(\frac{2\pi d}{L}\right) \tag{4}$$

The first term is the first-order contribution, whereas the second term is a second-order component representing the elevation of the particle orbit center about SWL. The second-order term arises from the continuity requirement of constant water volume. Thus, the elevation of the orbit center (see Figure 4-3-1) is given as:

$$\frac{h_0}{H_i} = \pi \left(\frac{H_i}{L}\right) \coth\left(\frac{2\pi d}{L}\right) \tag{5}$$

Sainflou Pressure

The pressure (expressed as force per unit area) at any vertical elevation at the wall under a standing wave when the crest or trough is at the vertical wall was given by Sainflou as:

Crest

$$\frac{p_{cr}}{\gamma} = -y_0 - H_i \frac{\sinh\left(\frac{2\pi y_0}{L}\right)}{\sinh\left(\frac{2\pi d}{L}\right) \cosh\left(\frac{2\pi d}{L}\right)} \tag{6}$$

Trough

$$\frac{p_{tr}}{\gamma} = -y_0 + H_i \frac{\sinh\left(\frac{2\pi y_0}{L}\right)}{\sinh\left(\frac{2\pi d}{L}\right) \cosh\left(\frac{2\pi d}{L}\right)} \tag{7}$$

where

γ = specific weight of water

Miche–Rundgren Method

Miche (1944) used the classical hydrodynamic equations to derive a theoretical description for the pressure distribution at a vertical wall for perfectly reflected monochromatic waves of incident height, H_i. Miche's theory is expressed in the Lagrangian coordinate system (he also transformed the equations to Eulerian coordinates), and the equations are complete to second order in the parameter H/L. The assumption of constant atmospheric pressure on the free surface was also adopted by Miche, and the pressure equation does not include the contribution due to atmospheric pressure.

Rundgren (1958) further developed the equations of Miche (1944) by including provision for a reflection coefficient (χ) less than unity. Rundgren's equations reduce to those of Miche when the reflection coefficient of $\chi = 1.0$ is substituted.

Rundgren performed laboratory experiments and obtained pressure measurements that were compared with predictions from a number of theoretical methods expressed in both Lagrangian and Eulerian coordinates. He concluded that equations in the Lagrangian system offered the most complete description of the pressure distribution because they include the pressure contribution above SWL. Generally, he observed that the fully second-order expressions gave better comparisons than the Sainflou method, which in turn performed better than first-order theory.

Free Surface Elevation

Rundgren's equations in Lagrangian coordinates for the vertical elevations of standing wave water particles when the crest (Equation 8) or the trough (Equation 9) is at the vertical wall are given as

Crest

$$y_{cr} = y_0 + \frac{H_i}{2}(1+\chi)\frac{\sinh\left[\frac{2\pi}{L}(d+y_0)\right]}{\sinh\left(\frac{2\pi d}{L}\right)}$$

$$+ \frac{\pi H_i}{4}\left(\frac{H_i}{L}\right)\frac{\sinh\left[\frac{2\pi}{L}(d+y_0)\right]}{\sinh\left(\frac{2\pi d}{L}\right)}\frac{\cosh\left[\frac{2\pi}{L}(d+y_0)\right]}{\sinh\left(\frac{2\pi d}{L}\right)}[(1+\chi)^2\Theta_1 + (1-\chi)^2\Theta_2]$$

(8)

Trough

$$y_{tr} = y_0 - \frac{H_i}{2}(1+\chi)\frac{\sinh\left[\frac{2\pi}{L}(d+y_0)\right]}{\sinh\left(\frac{2\pi d}{L}\right)}$$

$$+ \frac{\pi H_i}{4}\left(\frac{H_i}{L}\right)\frac{\sinh\left[\frac{2\pi}{L}(d+y_0)\right]}{\sinh\left(\frac{2\pi d}{L}\right)}\frac{\cosh\left[\frac{2\pi}{L}(d+y_0)\right]}{\sinh\left(\frac{2\pi d}{L}\right)}[(1+\chi)^2\Theta_1 + (1-\chi)^2\Theta_2]$$

(9)

where

$$\Theta_1 = 1 + \frac{3}{4\sinh^2\left(\frac{2\pi d}{L}\right)} - \frac{1}{4\cosh^2\left(\frac{2\pi d}{L}\right)}$$

(10)

$$\Theta_2 = \frac{3}{4\sinh^2\left(\frac{2\pi d}{L}\right)} + \frac{1}{4\cosh^2\left(\frac{2\pi d}{L}\right)}$$

(11)

The free surface elevation (relative to SWL) when the crest or trough is at the wall is found by substituting $y_0 = 0$ into Equations 8 and 9, resulting in the following expressions:

Crest

$$\eta_{cr} = \frac{H_i}{2}(1 + \chi) + \frac{\pi H_i}{4}\left(\frac{H_i}{L}\right)\coth\left(\frac{2\pi d}{L}\right)[(1 + \chi)^2 \Theta_1 + (1 - \chi)^2 \Theta_2] \quad (12)$$

Trough

$$\eta_{tr} = -\frac{H_i}{2}(1 + \chi) + \frac{\pi H_i}{4}\left(\frac{H_i}{L}\right)\coth\left(\frac{2\pi d}{L}\right)[(1 + \chi)^2 \Theta_1 + (1 - \chi)^2 \Theta_2] \quad (13)$$

In the Miche–Rundgren formulation, the height of the orbit center above the SWL is represented by the second term in Equations 12 and 13, i.e.,

$$\frac{h_0}{H_i} = \frac{\pi}{4}\left(\frac{H_i}{L}\right)\coth\left(\frac{2\pi d}{L}\right)[(1 + \chi)^2 \Theta_1 + (1 - \chi)^2 \Theta_2] \quad (14)$$

The similarity between the Miche–Rundgren equations and the Sainflou equations is easily recognized. In the case of complete reflection when $\chi = 1.0$, the only difference between the corresponding equations is the Θ_1-term present in the Miche–Rundgren equations.

Miche–Rundgren Pressure

The pressure at any vertical elevation under a standing wave when the crest or trough is at the vertical wall was given by Rundgren (1958) as:

Crest

$$\frac{p_{cr}}{\gamma} = -y_0 - \frac{H_i}{2}(1 + \chi)\frac{\sinh\left(\frac{2\pi y_0}{L}\right)}{\sinh\left(\frac{2\pi d}{L}\right)\cosh\left(\frac{2\pi d}{L}\right)}$$

$$-\frac{\pi H_i}{4}\left(\frac{H_i}{L}\right)\frac{\sinh\left(\frac{2\pi y_0}{L}\right)}{\sinh^2\left(\frac{2\pi d}{L}\right)}[(1 + \chi)^2 \Theta_3 + (1 - \chi)^2 \Theta_4] \quad (15)$$

Trough

$$\frac{p_{tr}}{\gamma} = -y_0 + \frac{H_i}{2}(1 + \chi)\frac{\sinh\left(\frac{2\pi y_0}{L}\right)}{\sinh\left(\frac{2\pi d}{L}\right)\cosh\left(\frac{2\pi d}{L}\right)}$$

$$-\frac{\pi H_i}{4}\left(\frac{H_i}{L}\right)\frac{\sinh\left(\frac{2\pi y_0}{L}\right)}{\sinh^2\left(\frac{2\pi d}{L}\right)}[(1 + \chi)^2 \Theta_3 + (1 - \chi)^2 \Theta_4] \quad (16)$$

where

$$\Theta_3 = \left[1 - \frac{1}{4\cosh^2\left(\frac{2\pi d}{L}\right)}\right]\cosh\left[\frac{2\pi}{L}(2d+y_0)\right] - 2\tanh\left(\frac{2\pi d}{L}\right)\sinh\left[\frac{2\pi}{L}(2d+y_0)\right]$$

$$+ \frac{3}{4}\left[\frac{\cosh\left(\frac{2\pi y_0}{L}\right)}{\sinh^2\left(\frac{2\pi d}{L}\right)} - \frac{2\cosh\left[\frac{2\pi}{L}(d+y_0)\right]}{\cosh\left(\frac{2\pi d}{L}\right)}\right] \qquad (17)$$

$$\Theta_4 = \frac{\cosh\left[\frac{2\pi}{L}(2d+y_0)\right]}{4\cosh^2\left(\frac{2\pi d}{L}\right)} - 2\tanh\left(\frac{2\pi d}{L}\right)\sinh\left[\frac{2\pi}{L}(2d+y_0)\right]$$

$$+ \frac{3}{4}\left[\frac{\cosh\left(\frac{2\pi y_0}{L}\right)}{\sinh^2\left(\frac{2\pi d}{L}\right)} - \frac{2\cosh\left[\frac{2\pi}{L}(d+y_0)\right]}{\cosh\left(\frac{2\pi d}{L}\right)}\right] \qquad (18)$$

Implementation

The determination of wave forces and moments acting on a vertical wall (when the crest or trough of a standing wave is at the wall) is accomplished in this ACES application by numerically integrating the equations given in the preceding sections.

Numerical Implementation

After performing initial bookkeeping chores, the numerical code first determines the sea surface elevation for when the crest and when the trough are at the wall (corresponds to $y_0 = 0$). This calculation is performed using Equations 12 and 13 from the Miche–Rundgren method. The resulting values for crest and trough elevation are taken as valid for both the Miche–Rundgren solution and for the Sainflou solution. The corresponding Sainflou equations (Equations 3 and 4) would give similar results, but the increased accuracy produced by the fully second-order Miche–Rundgren equations, and the fact that reduced reflection could be accommodated, led to the decision of using only the Miche–Rundgren result. The nomogram method in the SPM also follows this convention.

Next, the water depth is divided into 90 equal increments for calculation of wave pressure. Each incremental depth represents an at-rest value of y_0 for use in the Lagrangian equations for vertical elevation. During each program loop at an incremental depth, the following values are calculated and stored:

Sainflou Method

- Vertical elevation of particle when crest is at wall (Equation 1).
- Vertical elevation of particle when trough is at wall (Equation 2).
- Pressure when crest is at wall (Equation 6).

- Pressure when trough is at wall (Equation 7).
- Incremental component to overturning moment about the bottom of the wall, given as $p_{cr}(y_{cr} + d)$ and $p_{tr}(y_{tr} + d)$ for the crest and trough, respectively.

Miche–Rundgren Method

- Vertical elevation of particle when crest is at wall (Equation 8).
- Vertical elevation of particle when trough is at wall (Equation 9).
- Pressure when crest is at wall (Equation 15).
- Pressure when trough is at wall (Equation 16).
- Incremental component to overturning moment about the bottom of the wall, given as $p_{cr}(y_{cr} + d)$ and $p_{tr}(y_{tr} + d)$ for the crest and trough, respectively.

After all loops have been completed, the total force per unit horizontal width of the wall is found for both methods by integrating the pressure over the entire water column (summing the incremental pressure values calculated at each depth). Corresponding overturning moments are found by integrating the incremental moment components.

Implementation Notes

Forces and moments per unit length of wall calculated by this program are those caused by the wave action and hydrostatic water pressure. The designer must add any other external force loads due to soil pressure from backfill, etc.

As mentioned previously, the Miche–Rundgren method overpredicts when the waves are very long (low steepness), whereas the Sainflou method provides a more realistic answer for this condition. Conversely, the Miche–Rundgren method is better for steeper wave conditions. This application reports answers from both methods, with the recommendation of using the smaller values for force and moment.

The Sainflou method *does not* include provision for reflection coefficients less than one. Therefore, it gives conservative results if reflection is less than complete. The SPM (1984) gives the following guidance on choosing the reflection coefficient:

It should be assumed that smooth vertical walls completely reflect incident waves and $\chi = 1.0$. Where wales, tiebacks, or other structural elements increase the surface roughness of the wall by retarding vertical motion of the water, a lower value of χ may be used. A lower value of χ also may be assumed when the wall is built on a rubble base or when rubble has been placed seaward of the structure toe. Any value of χ less than 0.9 should not be used for design purposes.

This application should provide results that closely match the design curves given in the SPM. An exception may occur in determination of crest and trough elevation at very low values of H_i/gT^2 using SPM Figures 7-90 or 7-93. The SPM method first determines h_0 and then calculates η_{cr} and η_{tr} as a distance $H_i(1 + \chi)/2$ above or below h_0. As H_i/gT^2 approaches zero, the curves for h_0 in the SPM were forced to a value of $h_0/H_i = 1.0$, representing the limit of a solitary wave. On the other hand, this ACES application calculates η_{cr} and η_{tr} directly, and the values of h_0 obtained from Equations S and 14 may be different from the SPM at very small values of wave steepness.

References and Bibliography

Miche, R. 1944. "Mouvements ondulatoires de la mer en profondeur constante ou decroissante," *Annales des Ponts et Chaussees*, Paris, Vol. 114.

Rundgren, L. 1958. "Water Wave Forces," Bulletin No. 54, Royal Institute of Technology, Division of Hydraulics, Stockholm, Sweden.

Sainflou, M. 1928. "Essay on Vertical Breakwaters," *Annals des Ponts et Chaussees*, Paris (Translated by Clarence R. Hatch, Western Reserve University, Cleveland, OH).

Shore Protection Manual. 1984. 4th ed., 2 Vols., US Army Engineer Waterways Experiment Station, Coastal Engineering Research Center, US Government Printing Office, Washington, DC, Chapter 7, pp. 161-173.

RUBBLE-MOUND REVETMENT DESIGN

Description

Quarrystone is the most commonly used material for protecting earth embankments from wave attack because high-quality stone, where available, provides a stable and unusually durable revetment armor material at relatively low cost. This ACES application provides estimates for revetment armor and bedding layer stone sizes, thicknesses, and gradation characteristics. Also calculated are two values of runup on the revetment, an expected extreme and a conservative runup value.

Introduction

Structures are often needed along either bluff or beach shorelines to provide protection from wave action or to retain in situ soil or fill. Vertical structures are classified as either sea walls or bulkheads, according to their function, and protective materials laid on slopes are called revetments.

Revetments are generally constructed of durable stone or other material that will provide sufficient armoring for protected slopes. They consist of an armor layer, filter layer, and toe protection. The filter layer assures drainage and retention of the underlaying soil. Toe protection is needed to provide stability against undermining at the bottom of the structure.

General Assumptions and Limitations

Empirical formulas that were developed for the design of rubble-mound structures are generally expressed in terms of the stone weight required to withstand design wave conditions. These formulas have been largely derived from physical model studies. They are guides and must be used with experience and engineering judgment. Physical modeling is often a cost-effective measure to determine the final cross-section design for most rubble-mound structures. A definition sketch for some of the terms used in this section is shown in Figure 4-4-1.

Riprap Armor Stability Formula

For irregular wave conditions, a stability formula is defined that is similar to the one developed by Hudson (1958) for monochromatic waves:

$$W_{50} = w_r \left[\frac{H_s}{N_s \left(\frac{w_r}{w_w} - 1 \right)} \right]^3 \quad (1)$$

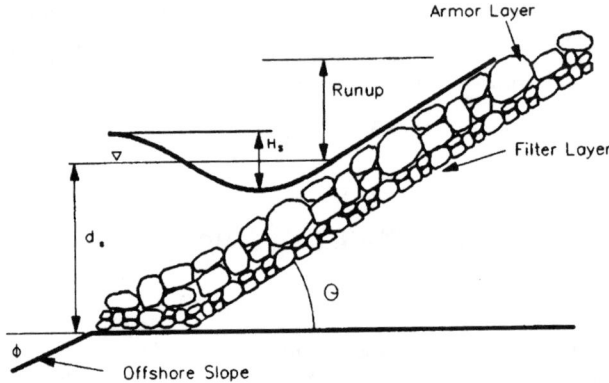

FIGURE 4-4-1. Definition sketch for riprap revetments.

where

W_{50} = median weight of the armor stone
w_r = unit weight of the armor stone
H_s = significant wave height
N_s = stability number
w_w = unit weight of water

The following sections present two stability numbers for use in Equation 1. These stability numbers are based on riprap stability studies conducted at CERC and by the Dutch. This ACES application uses the larger of the two in Equation 1 to compute the required stone weight for the revetment.

CERC Stability Number
Based on findings of Broderick (1983), the *zero-damage* riprap stability number given in Ahrens (1981) as:

$$N_{s-zero} = 1.45(\cot \theta)^{1/6} \qquad (2)$$

should be changed to

$$N_{s-zero} = \frac{1.45}{1.27}(\cot \theta)^{1/6} \qquad (3)$$

where

$\cot \theta$ = cotangent of structure slope

The factor, 1.27, is the ratio between the average of the highest 10 percent of the waves and the significant wave height in a Rayleigh distribution. By changing the coefficient in the stability number equation, the significant wave height can continue to be used in the stability formula, and at the same time the essential findings of Broderick can be addressed. Broderick found that a wave height greater than the significant gave better correlation to damage observed in laboratory tests of riprap stability. Broderick suggested that the average of the highest 10 percent of the waves was the appropriate wave height to use in a riprap stability formula.

AUTOMATED COASTAL ENGINEERING SYSTEM (ACES)   A.83

Dutch Stability Number

Based on findings of Van der Meer and Pilarczyk (1987) and Van der Meer (1988a, 1988b) two formulas for riprap stability numbers were derived, one for plunging waves and one for surging (nonbreaking) waves. Basic assumptions for the formulas are:

- A rubble-mound structure with an armor layer consisting of rock.
- Little or no overtopping (less than 10 to 15 percent of the waves).
- The slope of the structure should be generally uniform.

The formulas are:

Plunging Waves

$$N_s = 6.2 P^{0.18} \left(\frac{S}{\sqrt{N}} \right)^{0.2} (\zeta_z)^{-0.5} \qquad (4)$$

Surging (Nonbreaking) Waves

$$N_s = 0.1 P^{-0.13} \left(\frac{S}{\sqrt{N}} \right)^{0.2} \sqrt{\cot\theta}\, \zeta_z^P \qquad (5)$$

where

P = permeability coefficient (Figure 4-4-2)
S = damage level (Table 4-4-1)
N = number of waves
Note: The equations are valid in the range $1{,}000 < N < 7{,}000$, so $N = 7{,}000$ represents a logical limiting value that is used in this ACES application and should be conservative.
ζ_z = surf similarity parameter (see Equation 6)

The permeability coefficient P was introduced to describe the influence of the permeability of the structure on its stability. Van der Meer investigated three structures. The lower boundary value of P is that given by an impermeable core (clay or sand) (see Figure 4-4-2(a)). The ratio of armor/filter stone diameter was 4.5. With this impermeable core, a value of $P = 0.1$ was assumed. The upper boundary value of P is that given by a homogeneous structure, consisting only of armor stones (see Figure 4-4-2(d)). For this structure, a value of $P = 0.6$ was assumed. The third structure consisted of a two-diameter-thick armor layer on a permeable core. The ratio of armor/core stone diameter was 3.2 (see Figure 4-4-2(c)). For this structure, a value of $P = 0.5$ was assumed. The value of P for other structures with, for example, more than one layer of stones (Figure 4-4-2(b) or a thicker armor layer must be estimated from the values established for t'e three specific structures. The design engineer's experience is obviously important when selecting a value of P.

Equations 4 and 5 were developed for deepwater wave conditions, and Van der Meer (1988a) recommends a correction for shallow-water conditions. This ACES application uses a factor of 1.2 for the correction

The next section describes the surf similarity parameter and how it is used in this ACES application.

Surf Similarity Parameter

The surf similarity parameter has been found to be a very useful variable to characterize the breaker conditions on coastal structures or beaches. For irregular waves the surf similarity parameter can be defined as (Battjes, 1974):

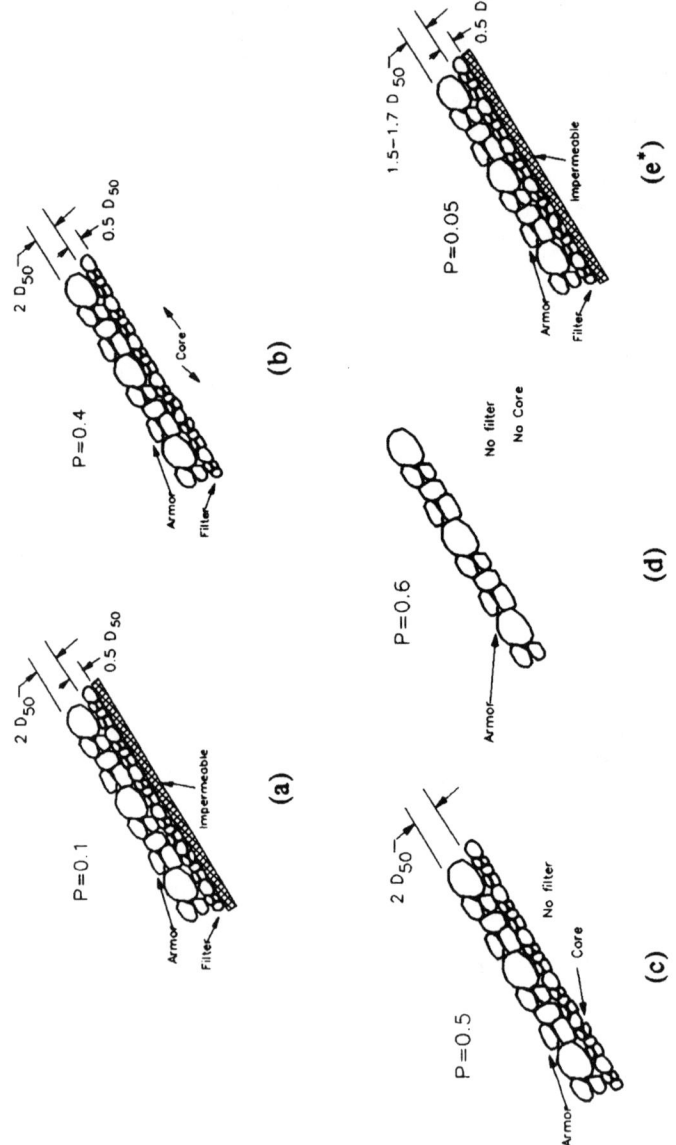

FIGURE 4-4-2. Permeability coefficient (Van der Meer, 1988a; Bradbury, Allsop, and Latham, 1990*).

TABLE 4-4-1. Damage levels for two diameter thick rock slopes (Van der Meer, 1988a)

	Damage Level S	
cot θ	Start of Damage	Failure (Filter Layer Visible)
2.0	2	8
3.0	2	12
4.0	3	17
6.0	3	17

$$\zeta_z = \frac{\tan\theta}{\left(\dfrac{2\pi H_s}{gT_z^2}\right)^{1/2}} \tag{6}$$

where

T_z = average wave period

$= T_s\left(\dfrac{0.67}{0.80}\right)$

Note: This ratio, (0.67/0.80), is based on laboratory data collected by Ahrens (1987).

The surf similarity parameter is used to determine which of the Van der Meer/Pilarczyk formulas for stability number (Equations 4 or 5) should be compared with the CERC stability number and finally used in the stability formula (Equation 1). Van der Meer (1988a) derives the following formula for the surf similarity parameter, ζ_{ztp} at the transition point from plunging to surging waves:

$$\zeta_{ztp} = (6.2P^{0.31}\sqrt{\tan\theta})^{1/(P-0.5)} \tag{7}$$

The following criteria are used to determine which of the Van der Meer/Pilarczyk stability numbers (Equations 4 or 5) is to be compared with the CERC stability number (Equation 3):

$\zeta_z \leq \zeta_{ztp} \rightarrow$ use Equation 4

$\zeta_z > \zeta_{ztp} \rightarrow$ use Equation 5

Revetment Design

The following sections describe the formulas for median weight and other percentiles of stone, armor and filter layer thicknesses, and stone dimensions used in this ACES application for design of rubble-mound revetments.

Weight of Armor Unit
The median weight of the armor unit is computed using Equation 1.

$$W_{50} = w_r\left[\dfrac{H_s}{N_s\left(\dfrac{w_r}{w_w}-1\right)}\right]^3$$

The stability number, N_s, used in Equation 1 is the larger of the CERC stability number (Equation 3) or the Dutch stability number (Equation 4 or 5).

Armor Layer Thickness
The minimum armor layer thickness is given as

$$r_{armor} = 2\left(\frac{W_{50}}{w_r}\right)^{1/3} \tag{8}$$

Filter Layer Thickness
The filter layer thickness is given as the maximum of

$$r_{filter} = \frac{r_{armor}}{4} \text{ or } 1 \text{ foot} \tag{9}$$

The total horizontal thickness of the armor layer and first underlayer, l, must satisfy the following relation:

$$l \geq 2H_s \tag{10}$$

where

$$l = r_t\sqrt{1 + \cot^2\theta} \tag{11}$$

$$r_t = r_{armor} + r_{filter} \tag{12}$$

The purpose of Equation 10 is to ensure there is sufficient stone between the violent wave attack on the surface of the riprap and the geotextile filter cloth to dissipate a considerable portion of the wave energy. A geotextile filter cloth is relatively permeable to ground-water seepage but not to the short duration loads and high velocity impacts of breaking waves.

Stone Sizes (Gradation)
Armor Layer. Gradation is based on guidance given in EM 1110-2-2300 (1971), which specifies that the maximum and minimum weight of the riprap stone is given by:

$$W_{max} = 4W_{50} \tag{13}$$

$$W_{min} = \frac{1}{8}W_{50} \tag{14}$$

where W_{max} and W_{min} are the weight of the largest and smallest stone respectively in the gradation.

In addition, laboratory tests (Ahrens, 1975) provide the following approximate relations:

$$W_{85} = 1.96W_{50} \tag{15}$$

$$W_{15} = 0.4W_{50} \tag{16}$$

where the subscript indicates the percentage of the total weight of gradation contributed by stones of lesser weight.

Stone dimensions, D, are computed by the following relationship:

$$D_x = \left(\frac{W_x}{w_r}\right)^{1/3} \tag{17}$$

where the subscript x indicates the percentage of the weight of the total gradation contributed by stones of lesser weight.

Filter Layer. The ratio of the filter layer stone size to armor stone size is given by Ahrens (1981) as

$$\frac{D_{15(\text{armor})}}{D_{85(\text{filter})}} = 4.0 \qquad (18)$$

Knowing $D_{85(\text{filter})}$ the following relationship is used to calculate the median stone dimension, $D_{50(\text{filter})}$, of the filter layer:

$$\frac{D_x}{D_{50}} = e^{(0.01157x - 0.5785)} \qquad (19)$$

where

$$x = 85$$

Knowing $D_{50(\text{filter})}$ Equation 19 is used to determine stone dimensions for the 0 (minimum), 15, and 100 (maximum) percentile size of the filter layer. Equation 17 is then used to determine corresponding stone weights for the filter layer.

Irregular Wave Runup on Riprap

Recent research by Ahrens and Heimbaugh (1988) provides an improved method to estimate the maximum runup caused by irregular waves on riprap revetments. An unusual advantage of this method is that it works well for both shallow and deep water at the toe of the revetment. The approach is based on the surf parameter discussed earlier. In this instance the surf parameter is calculated using the energy-based variables H_{m_0} and T_p. The energy-based surf parameter, ζ, is defined as

$$\zeta = \frac{\tan\theta}{\left(\frac{2\pi H_{m_0}}{gT_p^2}\right)^{1/2}} \qquad (20)$$

where

T_p = period of peak energy density of the wave spectrum

$$= \frac{T_s}{0.80} \qquad (21)$$

H_{m_0} = energy-based zero-moment wave height
T_s = average period of the highest one-third of the waves

In this ACES application, the energy-based zero-moment wave height, H_{m_0}, is computed by two methods, and the smaller value is then selected for use in the runup equation. The two methods for calculating H_{m_0} are

$$H_{m_0} = 0.10 L_p \tanh\left(\frac{2\pi d_s}{L_p}\right) \qquad (22)$$

or

$$H_{m_0} = \frac{H_s}{\exp\left[C_0 \left(\frac{d}{gT_p^2}\right)^{-C1}\right]} \qquad (23)$$

where

$C_0 = 0.00089$
$C_1 = 0.834$

The expected maximum runup is calculated using the following equation:

$$R_{\max} = H_{m_0} \frac{a\zeta}{1 + b\zeta} \qquad (24)$$

where

a and b = dimensionless runup coefficients

In research for improved estimates of runup, Ahrens and Heimbaugh (1988) conducted investigations of the systematic error in predicting the maximum runup. They found approximately 25 percent of their tests had a percent error greater than ±10 percent. Because of this, they suggest that it may be useful in some critical situations to use a conservative value of runup. Therefore, two sets of runup coefficients are provided in this ACES application, one for the maximum runup and the other for a conservative runup:

Expected Maximum Runup

$a = 1.022; b = 0.247$

Conservative Runup

$a = 1.286; b = 0.247$

References and Bibliography

Ahrens, J. P. 1975. "Large Wave Tank Tests of Riprap Stability," CERC Technical Memorandum 51, US Army Engineer Waterways Experiment Station, Vicksburg, MS.
Ahrens, J. P. 1977. "Prediction of Irregular Wave Overtopping," CERC CETA 77-7, US Army Engineer Waterways Experiment Station, Vicksburg, MS.
Ahrens, J. P. 1981. "Design of Riprap Revetments for Protection Against Wave Attack," CERC TP 81-5, US Army Engineer Waterways Experiment Station, Vicksburg, MS.
Ahrens, J. P. 1987. "Characteristics of Reef Breakwaters," Technical Report CERC-87-17, US Army Engineer Waterways Experiment Station, Vicksburg, MS.
Ahrens, J. P., and Heimbaugh, M. S. 1988. "Approximate Upperlimit of Irregular Wave Runup on Riprap," Technical Report CERC-88-5, US Army Engineer Waterways Experiment Station, Vicksburg, MS.
Ahrens, J. P., and McCartney B. L. 1975. "Wave Period Effect on the Stability of Riprap," *Proceedings of Civil Engineering in the Oceans/III*, American Society of Civil Engineers. pp. 1019–1034.
Batties, J. A. 1974. "Surf Similarity," *Proceedings of the 14th Coastal Engineering Conference*, Copenhagen, Denmark.
Bradbury, A. P., Allsop, N. W. H., and Latham, L-P. 1990. "Rock Armor Stability Formulae-Influence of Stone Shape and Layer Thickness," *Proceedings of the 22nd International Conference on Coastal Engineering*, Delft, The Netherlands.
Broderick, L. L. 1983. "Riprap Stability, A Progress Report," *Proceedings of the Coastal Structures '83 Conference*, American Society of Civil Engineers, Arlington, VA, pp. 320–330.
Broderick, L. L., and Ahrens, J. P. 1982. "Riprap Stability Scale Effects," CERC TP 82-3, US Army Engineer Waterways Experiment Station, Vicksburg, MS.
Headquarters, Department of the Army. 1971. "Earth and Rock-Fill Dams, General Design and Constructions Operations," Engineer Manual 1110-2-2300, Washington, DC.
Hudson, R. Y. 1958. "Design of Quarry Stone Cover Layers for Rubble Mound Breakwaters," Research Report 2-2, US Army Engineer Waterways Experiment Station, Vicksburg, MS.

Shore Protection Manual. 1984. 4th ed., 2 Vols., US Army Engineer Waterways Experiment Station, Coastal Engineering Research Center, US Government Printing Office, Washington, DC, Chapter 7.

Van der Meer, J. W., and Pilarczyk, K. W. 1987. "Stability of Breakwater Armor Layers Deterministic and Probabilistic Design," Delft Hydraulics Communication No. 378, Delft, The Netherlands.

Van der Meer, J. W. 1988a. "Deterministic and Probabilistic Design of Breakwater Armor Layers, *Journal of Waterways, Port, Coastal, and Ocean Engineering,* American Society of Civil Engineers, Vol. 114, No. 1, pp. 66–80.

Van der Meer, J. W. 1988b. "Rock Slopes and Gravel Beaches Under Wave Attack," Ph.d. Thesis, Department of Civil Engineering, Delft Technical University; also Delft Hydraulics Communication No. 396, Delft, The Netherlands.

IRREGULAR WAVE RUNUP ON BEACHES

Description

This application provides an approach to calculate runup statistical parameters for wave runup on smooth slope linear beaches. To account for permeable and rough slope natural beaches, the present approach needs to be modified by multiplying the results for the smooth slope linear beaches by a reduction factor. However, there is no guidance for such a reduction due to the sparsity of good field data on wave runup. The approach used in this ACES application is based on existing laboratory data on irregular wave runup (Mase and Iwagaki, 1984; Mase, 1989).

General Assumptions and Limitations

At present there are no theoretical approaches to calculate either monochromatic or irregular wave runup on beaches. The lack of a theoretical approach to solve the problem is due to the numerous difficulties inherent in the runup prediction problem such as:

- Nonlinear transformation of wave energy in the breaking wave zone.
- Wave reflection effects.
- Three-dimensional effects such as standing or progressive edge waves.
- Permeability.
- Porosity.
- Roughness.
- Ground-water table level.

Present approaches to calculating monochromatic wave runup on smooth steep slope coastal structures have been limited to empirical expressions of a Hunt (1959) equation form with limiting runup as determined via analytical breaking wave steepness limiting expressions (Walton and Ahrens, 1989; Walton, et al., 1989). Additionally, empirical nonlinear power law expressions exist for predicting irregular wave runup on smooth linear slopes (Mase, 1989) for the following runup statistics in a stationary wave train:

R_{max} = maximum wave runup
R_2 = runup value exceeded by 2 percent of the runups
$R_{1/10}$ = average of the highest one-tenth of the wave runups
$R_{1/3}$ = significant or average of the highest third of the runups
$\overline{R}$ = average wave runup

This ACES application calculates these runup statistics and quantiles based on the coefficients provided by Mase (1989).

Wave Runup Equation

The methodology is based on a fitting of the relative wave runup, R, via an equation of the form

$$R_p = H_{S0} a_p I^{(b_p)} \qquad (1)$$

where

R_p = quantile or statistic value desired (max, 2%, $\frac{1}{3}$, $\frac{1}{10}$, $average$)
a_p, b_p = constants based on the statistic or quantile value of desired runup
I = Iribarren number

$$= \frac{\tan\theta}{\left(\dfrac{H_{S0}}{L_0}\right)^{1/2}} \qquad (2)$$

$\tan\theta$ = tangent of the beach slope
H_{S0} = deepwater significant wave height
L_0 = deepwater wavelength

Until further research is conducted, it is suggested that the beach foreshore slope be used as the required beach slope in Equation (1).

References and Bibliography

Hunt, I. A. 1959. "Design of Seawalls and Breakwaters," *Journal of the Waterway, Port, Coastal, and Ocean Engineering Division*, American Society Civil Engineers, Vol. 85, No. 3, pp. 123–152.

Mase, H. 1989. "Random Wave Runup Height on Gentle Slopes," *Journal of the Waterway, Port, Coastal, and Ocean Engineering Division*, American Society Civil Engineers, Vol. 115, No. 5, pp. 649–661.

Mase, H., and Iwagaki, Y. 1984. "Runup of Random Waves on Gentle Slopes," *Proceedings of the 19th International Conference on Coastal Engineering*, Houston, TX, American Society Civil Engineers, pp. 593–609.

Walton, T. L., Jr., and Ahrens, J. P. 1989. "Maximum Periodic Wave Run-Up on Smooth Slopes," *Journal of the Waterway, Port, Coastal, and Ocean Engineering Division*, American Society Civil Engineers, Vol. 115, No. 5, pp. 703–708.

Walton, T. L., Jr., Ahrens, J. P., Truitt, C. L., and Dean, R. G. 1989. "Criteria for Evaluating Coastal Flood-Protection Structures," Technical Report CERC-89-15, US Army Engineer Waterways Experiment Station, Vicksburg, MS.

WAVE RUNUP AND OVERTOPPING ON IMPERMEABLE STRUCTURES

Description

This application provides estimates of wave runup and overtopping on rough and smooth slope structures that are assumed to be impermeable. Run-up heights and overtopping rates are estimated independently or jointly for monochromatic or irregular waves specified at the toe of the structure. The empirical equations suggested by Ahrens and McCartney (1975), Ahrens and Titus (1985), and Ahrens and Burke (unpublished report) are used to predict runup, and Weggel (1976) to predict overtopping. Irregular waves are represented by a significant wave height and are assumed to conform to a Rayleigh distribution (Ahrens, 1977). The overtopping rate is estimated by summing the overtopping contributions from individual runups in the distribution. Portions of the material presented herein can also be found in Chapter 7 of the SPM (1984).

Introduction

As waves encounter certain types of coastal structures, the water rushes up and sometimes over the structure. These closely related phenomena, wave runup and wave overtopping, often strongly influence the design and the cost of coastal projects. Wave runup is defined as the vertical height above still-water level to which a wave will rise on the structure (of assumed infinite height). Overtopping is the flow rate of water over the top of the finite height structure as a result of wave runup. Waves are assumed to be normally incident to the structure.

General Assumptions and Limitations

The various relationships for runup and overtopping employed in this application are empirically derived from physical model studies originally conducted for specific structures and wave climates. General assumptions applicable to the various expressions can be summarized as:

- Waves are normally incident to the structure and are unbroken in the vicinity of the structure toe.
- Waves are considered to be monochromatic. Irregular wave conditions are characterized by significant wave height H_s.
- Waves are specified at the structure location. Linear wave theory is applied to determine unrefracted deepwater wave height where necessary.
- The crest of the structure must be above still-water level.
- For run-up estimates, structures are considered to be impermeable and to have infinite height and simple plane slopes.
- For overtopping estimates, the actual finite structure height is employed. Results for additional structure configurations (such as curved and recurved walls) can be obtained if runup is known.

As reported in the references, the expressions for runup were primarily determined by empirical curve fitting procedures and consequently do not formally represent "best fit" curves derived by statistical procedures. The exception is the expression for smooth slope runup for nonbreaking wave conditions that was developed using regression analysis.

Similarly, the expression for overtopping rate was originally derived by a graphical curve fitting procedure (Weggel, 1976). In practice, the empirical coefficients required for the overtopping rate equation are often difficult to obtain. While a representative value of a is easy to estimate as a function of the structure slope, no satisfactory functional approximations for Q_0^* are available. An estimate is usually made by interpolation or extrapolation of the values presented in the SPM (1984), which are tabulations derived from the original data set and curve fitting procedure.

While expressions empirically derived from model data represent a useful and valid technology, engineering judgment should always be applied to the results, particularly when applying the formulas in situations much different from the bounds and character of the original data from which they were derived.

Wave Runup

Numerous laboratory tests have been conducted over the years resulting in data for wave runup. Figure 5-2-1 shows parameters involved in discussing wave runup, and the next two sections present equations used in ACES for rough and smooth slopes.

Rough Slope Runup

Ahrens and McCartney (1975) present an empirical method for estimating the runup on structures protected by various types of primary armor faces. In their method, the runup is predicted as a nonlinear function of the surf parameter, ξ.

$$R = H_i \frac{a\xi}{1 + b\xi} \quad (1)$$

where
  R = runup
  H_i = incident wave height
  a, b = empirical coefficients associated with corresponding types of armor unit (see Table A-3 of Appendix A)

$$\xi = \frac{\tan\theta}{\sqrt{\frac{H_i}{L_0}}} \quad (2)$$

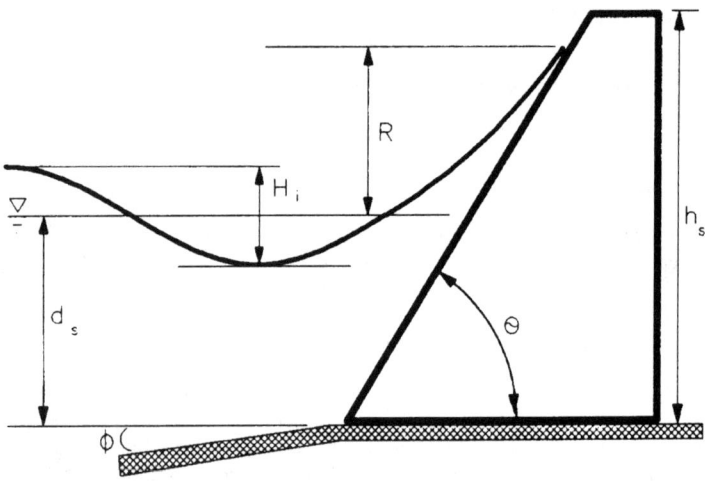

FIGURE 5-2-1. Wave runup and overtopping.

θ = angle between structure seaward face and horizontal (a measure of structure slope)
L_0 = deepwater wavelength

Smooth Slope Runup
Ahrens and Titus (1985) recommend the following general equation for runup on smooth slopes:

$$R = CH_i \tag{3}$$

The coefficient C is characterized by the surf parameter ξ according to the following three wave-structure regimes:

- ($\xi \leq 2$) waves plunging directly on the run-up slope.
- ($\xi \geq 3.5$) wave conditions that are nonbreaking and are regarded as standing or surging waves.
- ($2 < \xi < 3.5$) transition conditions where breaking characteristics are difficult to define.

Recommended expressions for coefficient C corresponding to these regimes are then:

- *Plunging wave conditions* ($\xi \leq 2$)

$$C_p = 1.002\xi \tag{4}$$

- *Nonbreaking wave conditions* ($\xi \geq 3.5$)

$$C_{nb} = 1.181 \left(\frac{\pi}{2\theta}\right)^{0.375} \exp\left[3.187\left(\frac{\eta_c}{H_i} - 0.5\right)^2\right] \tag{5}$$

where

η_c = crest height of the wave above the still-water level calculated using the Stream Function Wave Theory (Dean, 1974)

- *Transitional wave conditions* ($2 < \xi < 3.5$)

$$C_t = \left(\frac{3.5 - \xi}{1.5}\right)C_p + \left(\frac{\xi - 2}{1.5}\right)C_{nb} \tag{6}$$

This ACES application uses a more convenient but less accurate expression for the coefficient under nonbreaking conditions derived by Ahrens and Burke (unpublished report):

$$C_{nb} = 1.087\sqrt{\frac{\pi}{2\theta}} + 0.775\Pi \tag{7}$$

where

Π = Goda's (1983) nonlinearity parameter

$$= \frac{\frac{H_i}{L}}{\tanh^3\left(\frac{2\pi d_s}{L}\right)} \tag{8}$$

L = incident wavelength

Overtopping Rate

Several consequences of overtopping are important to engineers designing coastal structures. For structures along the shoreline (sea walls, bulkheads, or revetments), the volume of water that flows over the structure significantly impacts backside flooding. For breakwaters, wave transmission to the leeward side is an important criterion in harbor design. Also for breakwaters, the stability of armor material on the backslope of the structure is an important consideration. This ACES methodology estimates the overtopping flow rate for simple structures.

Monochromatic Wave Overtopping

The method implemented within this ACES application was developed by Weggel (1976) using data reported by Saville (1955) and by Saville and Caldwell (1953). It consists of an empirical expression for the monochromatic-wave overtopping rate:

$$Q = C_w \sqrt{gQ_0^* H_0^3} \left(\frac{R+F}{R-F} \right)^{-0.1085/\alpha} \quad (9)$$

where

Q = overtopping rate/unit length of structure
C_w = wind correction factor
g = gravitational acceleration
Q_0^*, α = empirical coefficients (see SPM (1984) figures)

Note: An average value for $\bar{\alpha}$ as a function of structure slope may be approximated by:

$$\bar{\alpha} = 0.06 - 0.01431 \ln(\sin\theta)$$

This option is available in the application.

H_0 = unrefracted deepwater wave height
R = runup
$F = h_s - d_s$ = freeboard
h_s = height of structure
d_s = water depth at structure

Wind Effects

Onshore winds can increase the overtopping rate at a barrier. The effect is dependent upon wind velocity, direction with respect to the axis of the structure, and structure characteristics. This increased overtopping rate is approximated by adjusting the above value for Q with a wind correction factor C_w (SPM, 1984):

$$C_w = 1 + W_f \left(\frac{F}{R} + 0.1 \right) \sin\theta \quad (10)$$

where

$$W_f = \frac{U^2}{1800} \quad (11)$$

U = onshore wind speed (mph)

Irregular Wave Overtopping

Douglass (1986) presents a summary of methods available for estimating overtopping rates from irregular waves. The method summarized below is that of Ahrens (1977) and embodies the following assumptions:

- Run-up values caused by an irregular sea will follow a Rayleigh distribution.
- Significant deepwater wave, $H_{1/3}$, causes the significant runup, $R_{1/3}$
- α, Q_0^*, and H_0 in Weggel's overtopping equation remain constant for all members of the distribution.

Ahrens estimates the overtopping rate by summing the overtopping contributions from the individual members of the run-up distribution:

$$Q = \frac{1}{199}\sum_{i=1}^{199} Q_i \qquad (12)$$

where

Q = volume rate of overtopping caused by irregular waves
Q_i = volume rate of overtopping caused by one runup on the run-up distribution

$$= C_w\sqrt{gQ_0^*(H_{so})^3}\left(\frac{R_i + F}{R_i - F}\right)^{-0.1085/\alpha} \qquad (13)$$

H_{so} = deepwater significant wave height
R_i = run-up value having exceedance probability p

$$= \sqrt{\frac{\ln\frac{1}{p}}{2}}\, R_s \qquad (14)$$

$p = 0.005*i, i = 1, 2, 3, \ldots, 199$
R_s = runup with a given deepwater significant wave height and period

These equations *modify* Weggel's monochromatic expressions to account for the effect of irregular waves when the freeboard, F, is less than the runup of the significant wave, R_s. When the freeboard is greater than the runup, Weggel's equations yield no overtopping while larger runups in the distribution may still overtop the structure. For these relatively high freeboards, the run-up distribution is broken into 999 elements, instead of 199, to better account for the effect of the higher runups. The overtopping equation for this larger distribution becomes

$$Q = \frac{1}{999}\sum_{i=1}^{999} Q_i \qquad (15)$$

where

$p = 0.001*i, i = 1, 2, 3, \ldots, 999$

References and Bibliography

Ahrens, J. P. 1977, "Prediction of Irregular Wave Overtopping," CERC CETA 77-7, US Army Engineer Waterways Experiment Station, Vicksburg, MS.

Ahrens, J. P., and Burke, C. E. 1987. Unpublished report of modifications to method cited in above reference.

Ahrens, J. P., and Titus, M. F. 1985. "Wave Runup Formulas for Smooth Slopes," *Journal of Waterways, Port, Coastal and Ocean Engineering*, American Society of Civil Engineers, Vol. 111, No. 1, pp. 128–133.

Battjes, J. A. 1974. "Surf Similarity," *Proceedings of the 14th Coastal Engineering Conference*, Copenhagen, Denmark.

Cross, R., and Sollitt, C. 1971. "Wave Transmission by Overtopping," Technical Note No. 15, Massachusetts Institute of Technology, Ralph M. Parsons Laboratory, Boston.
Douglass, S. L. 1986. "Review and Comparison of Methods for Estimating Irregular Wave Overtopping Rates," Technical Report CERC-86-12, US Army Engineer Waterways Experiment Station, Vicksburg, MS, pp. 6–14.
Goda, Y. 1969. "Reanalysis of Laboratory Data on Wave Transmission over Breakwaters," *Report of the Port and Harbour Research Institute*, Vol. 8, No. 3.
Goda, Y. 1983. "A Unified Nonlinearity Parameter of Water Waves," *Report of the Port and Harbour Research Institute*, Vol. 22, No. 3, pp. 3–30.
Goda, Y., Takeda, H., and Moriya, Y. 1967. "Laboratory Investigation of Wave Transmission over Breakwaters," *Report of the Port and Harbour Research Institute*, No. 13.
Saville, T., Jr. 1955. "Laboratory Data on Wave Run-Up and Overtopping on Shore Structures," TM No. 64, US Army Corps of Engineers, Beach Erosion Board, Washington, DC.
Seelig, W. N. 1976. "A Simplified Method for Determining Vertical Breakwater Crest Elevation Considering Wave Height Transmitted by Overtopping," CERC CDM 76-1, US Army Engineer Waterways Experiment Station, Vicksburg, MS.
Seelig, W. N. 1980. "Two-Dimensional Tests of Wave Transmission and Reflection Characteristics of Laboratory Breakwaters," CERC TR 80-1, US Army Engineer Waterways Experiment Station, Vicksburg, MS.
Shore Protection Manual. 1984. 4th ed., 2 Vols., US Army Engineer Waterways Experiment Station, Coastal Engineering Research Center, US Government Printing Office, Washington, DC, Chapter 7, pp. 43–58.
Smith, O. P. 1986. "Cost-Effective Optimization of Rubble-Mound Breakwater Cross Sections," Technical Report CERC-86-2, US Army Engineer Waterways Experiment Station, Vicksburg, MS, pp. 45–53.
Weggel, J. R. 1972. "Maximum Breaker Height," *Journal of Waterways, Harbors and Coastal Engineering Division*, American Society of Civil Engineers, Vol. 98, No. WW4, pp. 529–548.

WAVE TRANSMISSION ON IMPERMEABLE STRUCTURES

Description

This application provides estimates of wave runup and transmission on rough and smooth slope structures. It also addresses wave transmission over impermeable vertical walls and composite structures. In all cases, monochromatic waves are specified at the toe of a structure that is assumed to be impermeable. For sloped structures, a method suggested by Ahrens and Titus (1985) and Ahrens and Burke (1987) is used to predict runup, while the method of Cross and Sollitt (1971) as modified by Seelig (1980) is used to predict overtopping. For vertical wall and composite structures, a method proposed by Goda, Takeda, and Moriya (1967) and Goda (1969) is used to predict wave transmission.

Introduction

The transmission of wave energy beyond protective structures involves a number of complex processes. Some incident wave energy may be reflected by the structure, some wave energy may be dissipated by turbulent interaction with primary armor units (if present), some may be dissipated internally by the finer materials beneath the armor layers of an impermeable structure, and some may be transmitted through or over the structure with resultant wave regeneration. Important factors identifiable in the process include the shape and material composition of the structure, the incident wave environment, as well as the degree of immergence or submergence of the structure.

General Assumptions and Limitations

The various relationships for runup and transmission employed in this application are empirically derived from physical model studies originally conducted for specific structures and wave climates. For sloped structures, the run-up methodology is described in the section entitled "Wave Runup and Overtopping on Impermeable Structures" of this *ACES Technical Reference*. For convenience, the pertinent assumptions and limitations are restated below. General assumptions applicable to the various methods can be summarized as:

- Waves are monochromatic, normally incident to the structure, and unbroken in the vicinity of the structure toe.
- Waves are specified at the structure location.
- All structure types are considered to be impermeable.
- For sloped structures the crest of the structure must be above still-water level. For vertical and composite structures, partial and complete submersion of the structure is considered.
- Run-up estimates on sloped structures require the assumption of infinite structure height and a simple plane slope.
- The expressions for transmission by overtopping use the actual finite structure height.

As reported in the references, the expressions for runup were primarily determined by empirical curve fitting procedures and consequently do not formally represent "best fit" curves derived by statistical procedures. The exception is the expression for smooth slope runup for nonbreaking wave conditions that was developed using regression analysis.

The methodology for wave transmission was also empirically derived. The transmission predicted by the expression for sloped structures with freeboard was tested over the range of ($0 \leq B/h_s \leq 0.86$) for smooth impermeable structures, and ($0.88 \leq B/h_s \leq 3.2$) for rough impermeable breakwaters (Seelig, 1980). Seelig also recommended that the expression be applied in the range ($0.006 \leq d_s/gT^2 \leq 0.03$). For transmission over vertical or composite structures, the empirical coefficients a and 13 were determined from laboratory experiments for three breakwater types and wave conditions in the range ($0.14 \leq d_s/L \leq 0.5$) (Goda, 1969, and Seelig, 1976).

While expressions empirically derived from model data represent a useful and valid technology, engineering judgment should always be applied to the results, particularly when applying the formulas in situations much different from the bounds and character of the original data from which they were derived. Familiarity with the history, techniques, and data bounds of original experimental results should complement the use of sound engineering judgment when applying such procedures.

Wave Transmission by Overtopping

In general, wave transmission at structures is characterized by the following expression:

$$H_T = K_{TO} H_i \tag{1}$$

where

H_T = transmitted wave height
K_{TO} = wave transmission coefficient (overtopping)
H_i = incident wave height

The next two sections discuss the wave transmission coefficient for the simple, idealized impermeable structures.

Transmission Coefficient for Sloped Structures with Freeboard

Wave transmission over a sloped breakwater occurs when runup exceeds the freeboard. Some of the pertinent parameters for the following discussion are shown in Figure 5-3-1 below.

The transmission coefficient for sloped structures subject to wave overtopping is estimated by an empirical equation based on the work of Cross and Sollitt (1971) and on 2-D laboratory tests conducted by Seelig (1980).

$$K_{TO} = C\left(1 - \frac{F}{R}\right) \quad (2)$$

where

C = empirical coefficient
$= 0.51 - 0.11\dfrac{B}{h_s}$ (3)
B = crest width of structure
h_s = structure height
$F = h_s - d_s$ = freeboard
d_s = water depth at structure
R = runup

As stated previously, runup is calculated as outlined in the section entitled "Wave Runup and Overtopping on Impermeable Structures" of this *ACES Technical Reference*.

Transmission Coefficient for Vertical or Composite Structures

The transmission coefficient for impermeable vertical-faced structures is estimated by an empirical equation based on the work of Goda, Takeda, and Moriya (1967) and Goda (1969). The equation is presented in Seelig (1976) as:

$$K_{TO} = 0.5\left\{1 - \sin\left[\frac{\pi}{2\alpha}\left(\frac{F}{H_i} + \beta\right)\right]\right\} \quad (4)$$

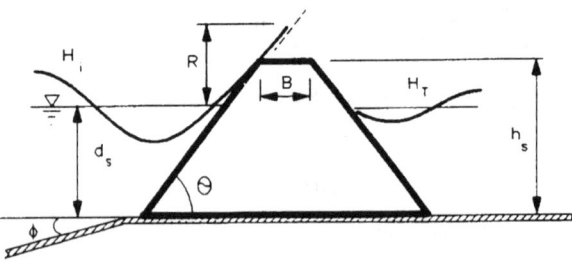

FIGURE 5-3-1. Transmission over sloped structures.

The empirical coefficients, α and β, were determined from laboratory experiments for three breakwater types for water depth to wavelength ratios of

$$0.14 \leq \frac{d_s}{L} \leq 0.5 \tag{5}$$

The breakwater types and definition of terms and symbols are shown in Figure 5-3-2, which is taken from Seelig (1976).
The range of applicability of K_{TO} is given in Table 5-3-I.

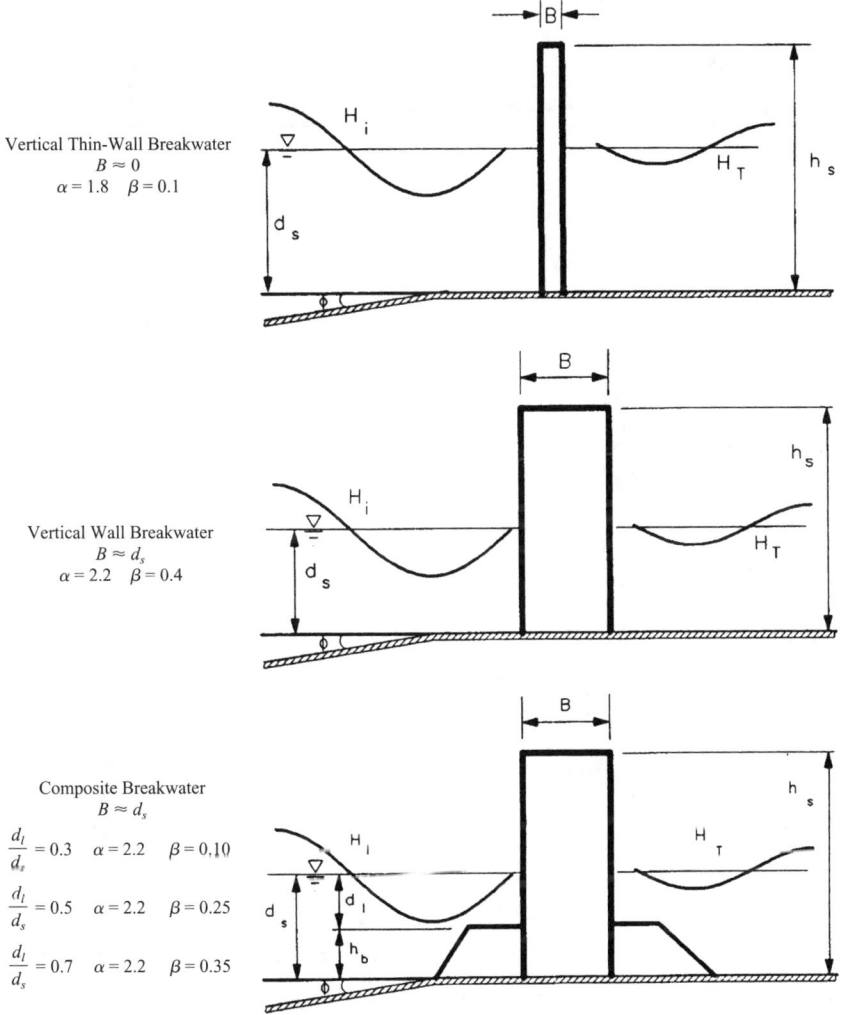

FIGURE 5-3-2. Transmission over vertical/composite breakwaters (Seelig, 1976).

TABLE 5-3-1. K_{TO} for vertical and composite breakwaters

K_{TO}	Domain
$K_{TO} = 1.0$	$\dfrac{F}{H_i} \leq -(\alpha + \beta)$
$K_{TO} = 0.5\left\{1 - \sin\left[\dfrac{\pi}{2\alpha}\left(\dfrac{F}{H_i} + \beta\right)\right]\right\}$	$-(\alpha + \beta) \dfrac{F}{H_i} < (\alpha - \beta)$
$K_{TO} = 0.0$	$\dfrac{F}{H_i} \geq (\alpha - \beta)$

TABLE 5-3-2. α and β_1, β_2 for vertical and composite breakwaters

Coefficient	Domain—Vertical Breakwaters		Domain—Composite Breakwaters	
	$0 \leq \dfrac{B}{d_s} < 1.0$	$\dfrac{B}{d_s} \geq 1.0$	$\dfrac{d_l}{d_s} \leq 0.3$	$0.3 < \dfrac{d_l}{d_s} \leq 1.0$
α	$1.8 + 0.4\left(\dfrac{B}{d_s}\right)$	2.2	2.2	2.2
β_1	$0.1 + 0.3\left(\dfrac{B}{d_s}\right)$	0.4	N/A	N/A
β_2	N/A	N/A	0.1	$0.527 - \dfrac{0.130}{\left(\dfrac{d_l}{d_s}\right)}$

d_l = water depth above berm or toe (see Figure 5-3-2).

β is defined as

$$\beta = C_1\beta_1 + C_2\beta_2 \qquad (6)$$

where

$$C_1 = \max\left(0, 1 - \dfrac{B}{d_s}\right) \quad \text{and} \quad C_2 = \min\left(1, \dfrac{B}{d_s}\right) \qquad (7)$$

Table 5-3-2 presents equations for α and β_1, and β_2 used to calculate the transmission coefficient for the various breakwater types shown in Figure 5-3-2. These equations were established from analysis of the data in Figure 5-3-2.

References and Bibliography

Ahrens, J. P. 1977. "Prediction of Irregular Wave Overtopping," CERC CETA 77-7, US Army Engineer Waterways Experiment Station, Vicksburg, MS.

Ahrens, J. P., and Burke, C. E. 1987. Unpublished report of modifications to method cited in above reference.

Ahrens, J. P., and Titus, M. F. 1985. "Wave Runup Formulas for Smooth Slopes," *Journal of Waterway, Port, Coastal and Ocean Engineering*, American Society of Civil Engineers, Vol. 111, No. 1, pp. 128-133.
Battjes, J. A. 1974. "Surf Similarity," *Proceedings of the 14th Coastal Engineering Conference*, Copenhagen, Denmark.
Cross, R., and Sollitt, C. 1971. "Wave Transmission by Overtopping," Technical Note No. 15, Massachusetts Institute of Technology, Ralph M. Parsons Laboratory, Boston.
Douglass, S. L. 1986. "Review and Comparison of Methods for Estimating Irregular Wave Overtopping Rates," Technical Report CERC-86-12, US Army Engineer Waterways Experiment Station, Vicksburg, MS. pp. 6-14.
Goda, Y. 1969. "Reanalysis of Laboratory Data on Wave Transmission over Breakwaters," *Report of the Port and Harbour Research Institute*, Vol. 8, No. 3.
Goda, Y. 1983. "A Unified Nonlinearity Parameter of Water Waves," *Report of the Port and Harbour Research Institute*, Vol. 22, No. 3, pp. 3-30.
Goda, Y., Takeda, H., and Moriya, Y. 1967. "Laboratory Investigation of Wave Transmission over Breakwaters," *Report of the Port and Harbour Research Institute*, No. 13.
Saville, T., Jr. 1955. "Laboratory Data on Wave Run-Up and Overtopping on Shore Structures," TM No. 64, US Army Corps of Engineers, Beach Erosion Board, Washington, DC.
Seelig, W. N. 1976. "A Simplified Method for Determining Vertical Breakwater Crest Elevation Considering Wave Height Transmitted by Overtopping," CERC CDM 76-1, US Army Engineer Waterways Experiment Station, Vicksburg, MS.
Seelig, W. N. 1980. "Two-Dimensional Tests of Wave Transmission and Reflection Characteristics of Laboratory Breakwaters," CERC TR 80-1, US Army Engineer Waterways Experiment Station, Vicksburg, MS.
Shore Protection Manual. 1984. 4th ed., 2 Vols., US Army Engineer Waterways Experiment Station, Coastal Engineering Research Center, US Government Printing Office, Washington, DC, Chapter 7, pp. 61-80.
Smith, O. P. 1986. "Cost-Effective Optimization of Rubble-Mound Breakwater Cross Sections," Technical Report CERC-86-2, US Army Engineer Waterways Experiment Station, Vicksburg, MS, pp. 45-53.
Weggel, J. R. 1972. "Maximum Breaker Height," *Journal of Waterways, Harbors and Coastal Engineering Division*, American Society of Civil Engineers, Vol. 98, No. WW4, pp. 529-548.

WAVE TRANSMISSION THROUGH PERMEABLE STRUCTURES

Description

Porous rubble-mound structures consisting of quarry stones of various sizes often offer an attractive solution to the problem of protecting a harbor against wave action. It is important to assess the effectiveness of a given breakwater design by predicting the amount of wave energy transmitted by the structure. This application determines wave transmission coefficients and transmitted wave heights for permeable breakwaters with crest elevations at or above the still-water level. This application can be used with breakwaters armored with stone or artificial armor units. The application uses a method developed for predicting wave transmission by overtopping coefficients using the ratio of breakwater freeboard to wave runup (suggested by Cross and Sollitt, 1971). The wave transmission by overtopping prediction method is then combined with the model of wave reflection and wave transmission through permeable structures of Madsen and White (1976). Seelig (1979,1980) had developed a similar version for mainframe processors.

The material presented here is intended as a brief summary of the methodology. More detailed discussion is presented in Madsen and White (1976) and Seelig (1980).

Introduction

The transmission of wave energy beyond protective structures involves a number of complex processes. Some incident wave energy may be reflected by the structure, some wave energy may be dissipated by turbulent interaction with primary armor units (if present), some may be dissipated internally by the finer materials beneath the armor layers of a permeable structure, and some may be transmitted through or over the structure with resultant wave regeneration. Important factors identifiable in the process include the shape and material composition of the structure, the incident wave environments, and the degree of immergence or submergence of the structure.

The methodology summarized in the following sections provides an attempt to account for the various processes of wave transmission at an unsubmerged rubble-mound structure subjected to relatively long-period waves. Figure 5-4-1 presents general symbology for the following discussions.

General Assumptions and Limitations

General assumptions and limitations for the model are:

- Incident waves are periodic, relatively long, and normally incident.
- Fluid motion is adequately described by the linearized governing equations.
- The model can be used only for breakwaters with crests above the still-water line.
- The model can be used for unbroken waves.
- Tests of breakwaters armored with dolos units suggest that the model can be used for artificial armor units.
- Laboratory data showed that the model gives best predictions for shallow-water waves.
- Predictions of transmission coefficients tend to be conservative for transitional or deepwater waves.

Total Wave Transmission

A common measure of breakwater performance is that of wave transmission, expressed as a transmission coefficient generally defined as the ratio of the transmitted wave height to the incident wave height.

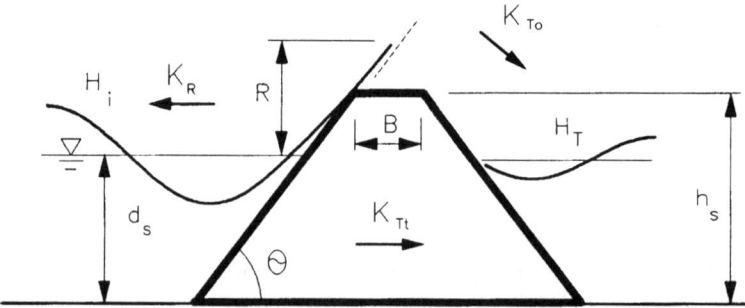

FIGURE 5-4-1. Wave transmission over and through permeable breakwaters.

$$K_T = \frac{H_T}{H_i} \quad (1)$$

where

H_T = transmitted wave height
H_i = incident wave height

There are two basic types of wave transmission considered in this ACES application:

- Wave regeneration caused by overtopping of the structure crest.
- Wave energy transmitted through the permeable materials of the structure.

A total wave transmission coefficient, K_T, is given by

$$K_T = \sqrt{(K_{To})^2 + (K_{Tt})^2} \quad (2)$$

where

K_{To} = overtopping coefficient
K_{Tt} = transmission coefficient

The following sections discuss the formulations for both the overtopping and transmission coefficients.

Overtopping Coefficient
The overtopping transmission coefficient is estimated by an empirical equation based on 2-D laboratory tests (Seelig, 1980).

$$K_{To} = C\left(1 - \frac{F}{R}\right) \quad (3)$$

where

C = empirical coefficient = $0.51 - 0.11(B/h_s)$
$F = h_s - d_s$ = freeboard
h_s = structure height
d_s = water depth
R = wave runup

Wave runup is estimated using the following formula (Ahrens and McCartney, 1975).

$$R = H_I\left(\frac{a\zeta}{1 + b\zeta}\right) \quad (4)$$

where

$a = 0.692$
ζ = surf parameter
$$= \frac{\tan\theta}{\sqrt{\frac{H_I}{L_o}}} \quad (5)$$
θ = angle of seaward face of breakwater
L_0 = deepwater wavelength (linear wave theory)
$b = 0.504$

Transmission Coefficient

The coefficient of wave transmission through permeable breakwaters, K_{Tl}, is estimated using the analytical model of Madsen and White (1976). In this model the transmission coefficient is related to a complex function of the following parameters:

- Size, porosity, and placement of materials in the breakwater
- Breakwater geometry
- Seaward slope of the structure
- Water depth
- Wave height and period
- Kinematic viscosity of the water

The Madsen and White model combines an analytical treatment with empirical relationships for the hydraulic characteristics of the porous material and for the friction factor representing energy dissipation on the seaward face of the breakwater. Important assumptions of the model include the following:

- Incident waves are periodic, relatively long, unbroken, and normally incident.
- Fluid motion is adequately described by the linear long-wave equations.

Madsen and White base their analytical solution on the fundamental argument that the problem of reflection from and transmission through a structure may be regarded as one of determining the partition of incident wave energy among reflected, transmitted, and dissipated energy. The analytical model is divided into four analyses.

- Internal energy dissipation—Idealized by considering the problem of the interaction of waves with a homogeneous porous structure of rectangular cross section that is "hydraulically" equivalent to the trapezoidal, multilayered breakwater.
- External energy dissipation—Based upon the associated problem of energy dissipation on a rough impermeable slope.
- Synthesis of the two analyses—Combines the two analyses into a rational procedure for the estimation of reflection and transmission coefficients of trapezoidal, multilayered breakwaters.
- Equivalent breakwater analysis—A simple method to determine characteristics of an idealized homogeneous rectangular breakwater that is hydraulically equivalent to a trapezoidal, multilayered breakwater.

Each of these analyses is briefly discussed in the next four sections.

Internal Energy Dissipation

This section presents a discussion of the treatment of internal energy dissipation within the porous media of the structure. The actual geometry of this portion of the breakwater is replaced by an idealized rectangular crib-style structure of homogeneous material of known properties. The theory also embodies the following additional assumptions:

- The idealized structure is subject to relatively long, normally incident unbroken waves described by linear wave theory.
- The flow resistance within the porous structure is a linear function of velocity.

The theoretical consideration for internal energy dissipation is based upon an analytic solution to simplified long wave equations. The problem domain is depicted in Figure 5-4-2.

Internal energy dissipation is represented by a friction term in the momentum equation

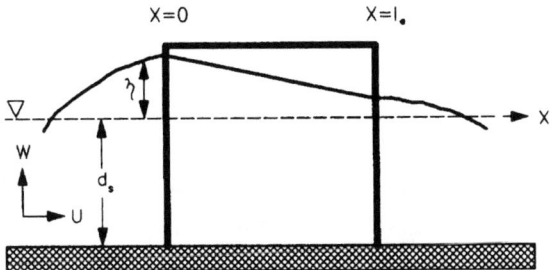

FIGURE 5-4-2. Wave transmission through a rectangular porous breakwater.

only for the subdomain that involves the porous rectangular structure. Within the subdomain representing the structure, a flow resistance of the Dupuit-Forchheimer type (Bear, et al. 1968) is assumed, and an empirical relationship relating flow resistance to stone size, porosity, and fluid viscosity is used to provide a representation of experimentally observed hydraulic properties of porous media. Adopting this empirical formulation of the flow resistance for a porous medium in conjunction with Lorentz' principle of equivalent work leads to a determination of a linearized flow resistance factor in terms of the characteristics of the porous material and the incident wave characteristics.

The resulting analytic solutions to the long-wave equations are manipulated to provide reflection and transmission coefficients for the idealized crib-style breakwater. These coefficients will be used in the synthesis of the separate analyses for energy dissipation and transmission.

The equations governing the motion outside and within the structure are

Outside of structure Within porous structure

$$\frac{\partial \eta}{\partial t} + d_s\left(\frac{\partial U}{\partial x}\right) = 0 \quad \text{continuity} \quad n\frac{\partial \eta}{\partial t} + d_s\left(\frac{\partial U}{\partial x}\right) = 0 \tag{6}$$

$$\frac{\partial U}{\partial t} + g\left(\frac{\partial \eta}{\partial x}\right) = 0 \quad \begin{array}{c}\text{conservation of}\\\text{momentum}\end{array} \quad \frac{S}{n}\left(\frac{\partial U}{\partial t}\right) + g\left(\frac{\partial \eta}{\partial x}\right) + f\frac{\omega}{n}U = 0 \tag{7}$$

where

η = free surface elevation
d_s = water depth
U = horizontal water particle velocity
g = acceleration due to gravity
S = factor for the effect of unsteady motion (taken as 1)
f = nondimensional friction factor
$\omega = 2\pi/T$ = angular frequency
n = porosity of the porous medium

Since the equations are linear, complex variables may be used. Requiring a periodic solution in terms of radian frequency ω, the following are used:

$$\eta = \zeta(x)e^{i\omega t} \tag{8}$$

$$U = u(x)e^{i\omega t} \tag{9}$$

In these equations $i = \sqrt{-1}$, and the amplitude functions ζ and u are complex functions of x only. Only the real part of the complex solutions for ζ and u constitutes the physical solutions.

General solutions for the governing Partial Differential Equations by region are

$$\zeta = a_i e^{-ik_0 x} + a_r e^{ik_0 x} \tag{10}$$

$$(x \leq 0)$$

$$u = \sqrt{\frac{g}{d_s}} (a_i e^{-ik_0 x} - a_r e^{ik_0 x}) \tag{11}$$

$$\zeta = a_t e^{-ik_0(x - l_e)} \tag{12}$$

$$(x \geq l_e)$$

$$u = \sqrt{\frac{g}{d_s}} [a_t e^{-ik_0(x - l_e)}] \tag{13}$$

$$\zeta = a_+ e^{-ikx} + a_- e^{ik(x - l_e)} \tag{14}$$

$$(0 \leq x \leq l_e)$$

$$u = \sqrt{\frac{g}{d_s}} \frac{n}{\sqrt{s - if}} [a_+ e^{-ikx} - a_- e^{ik(x - l_e)}] \tag{15}$$

where

a_i = complex incident wave amplitude
k_0 = wave number

$$= \frac{\omega}{\sqrt{g d_s}} \tag{16}$$

a_r = complex reflected wave amplitude
a_t = complex transmitted wave amplitude
a_+ = complex amplitude of wave propagating in the positive x-direction within the structure

$$k = \text{complex wave number} = nk_0 \sqrt{S - if} \tag{17}$$

a_- = complex amplitude of wave propagating in the negative x-direction within the structure

The general solutions for the motions in the three regions, given Equations 10 through 15, show the problem to involve four unknown quantities. These unknowns are the complex wave amplitudes a_r, a_t, a_+, and a_-. They may be determined by matching solutions at the common boundaries of the regions and further manipulated to eliminate a_+ and a_- and provide expressions for the complex amplitudes of the transmitted and reflected waves:

$$\frac{a_t}{a_i} = \frac{4\epsilon}{(1 + \epsilon)^2 e^{ikl_e} - (1 - \epsilon)^2 e^{-ikl_e}} \tag{18}$$

$$\frac{a_r}{a_i} = \frac{(1 - \epsilon^2)(e^{ikl_e} - e^{-ikl_e})}{(1 + \epsilon)^2 e^{ikl_e} - (1 - \epsilon)^2 e^{-ikl_e}} \tag{19}$$

where

$$\epsilon = \frac{\frac{n}{\sqrt{s}}}{\sqrt{\left(1 - \frac{if}{s}\right)}} \qquad (20)$$

$$f = \frac{n_r}{k_0 l_e}\left\{\left[1 + \left(1 + \frac{170}{R_d}\right)\frac{16\beta_r}{3\pi}a_i\frac{l_e}{d_s}\right]^{1/2} - 1\right\} \qquad (21)$$

n_r = porosity of reference material = 0.435
l_e = width of idealized breakwater
R_d = particle Reynolds number

$$= \frac{|u_s|d_r}{v} \qquad (22)$$

$|u_s|$ = horizontal velocity within structure

$$= a_i\sqrt{\frac{g}{d_s}}\left(\frac{1}{1+\lambda}\right) \qquad (23)$$

$$\lambda = \frac{k_0 l_e f}{2 n_r} \qquad (24)$$

d_r = 1/2 mean diameter of reference material
v = kinematic viscosity = 0.0000141
β_r = hydrodynamic characteristic of reference material

$$= 2.7\left(\frac{1-n_r}{n_r^3}\right)\frac{1}{d_r} \qquad (25)$$

General solutions for the transmission and reflection coefficients for this idealized structure follow directly:

$$T_I = \frac{|a_t|}{a_i} \qquad (26)$$

$$R_I = \frac{|a_r|}{a_i} \qquad (27)$$

Numerically, T_I, R_I, are solved iteratively by first assuming a value for λ and solving for u_s, R_d, f, then solving for a new value of λ, and repeating the procedure until convergence is achieved.

External Energy Dissipation

In the previous section, an analytical solution for the idealized problem of wave transmission through and reflection from rectangular breakwaters was obtained. Since most breakwaters are of trapezoidal, rather than rectangular cross section, a considerable amount of energy may be dissipated on the seaward slope of the breakwater. This external dissipation of energy is not accounted for in the analysis of porous crib-style breakwaters. To account for the external dissipation of energy on the seaward slope of a trapezoidal breakwater, theoretical and empirical relationships are presented for the problem of energy dissipation on a rough, impermeable slope. A theoretical analysis of this problem is based on the following assumptions:

- Relatively long, normally incident unbroken waves described by linear wave theory.
- Energy dissipation on the rough impermeable slope may be represented as bottom friction.

The problem to be considered is illustrated in Figure 5-4-3.

The linear long-wave equations for the problem subdomain bounded by the structure slope ($x < l_s$) are given below. These equations also describe the subdomain approaching the structure $x \geq l_s$ with the omission of the friction term in the momentum equation. The orientation of the x-axis has reversed from the previous notation.

$$\text{continuity} \qquad\qquad \text{conservation of momentum}$$

$$\frac{\partial \eta}{\partial t} + \frac{\partial}{\partial x}(dU) = 0 \qquad \frac{\partial U}{\partial t} + g\left(\frac{\partial \eta}{\partial x}\right) + f_b w U = 0 \qquad (28)$$

where

η = surface elevation relative to still water
d = water depth along sloping breakwater face
U = horizontal velocity component
g = acceleration due to gravity
f_b = linearized bottom friction factor

$$= \frac{\frac{1}{2}f_w|U|}{\omega d} \qquad (29)$$

f_w = wave friction factor relating bottom shear stress, τ_b, fluid density, ρ, and velocity

$$\tau_b = \frac{1}{2}\rho f_w|U|U \qquad (30)$$

$$w = \frac{2\pi}{T} = \text{radian frequency}$$

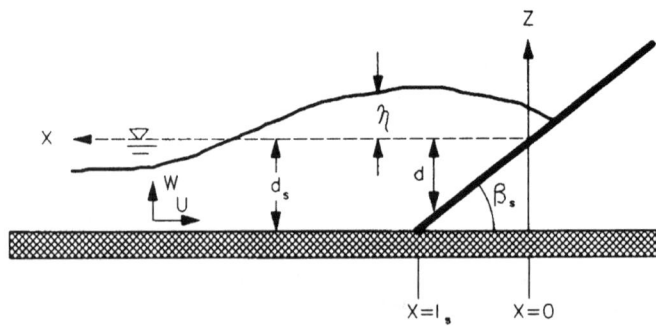

FIGURE 5-4-3. Wave runup and energy dissipation on impermeable slope.

Like the previous procedure, the equations are solved by assuming a periodic solution of radian frequency, w, and introducing complex variables:

$$\eta = \zeta(x)e^{iwt} \quad (31)$$

$$U = u(x)e^{iwt} \quad (32)$$

General solutions for the governing Partial Differential Equations for the two subdomains are

$$\zeta = a_i e^{ik_0 x} + a_r e^{-ik_0 x} \quad (33)$$

$$(x \geq l_s)$$

$$u = -\sqrt{\frac{g}{d_s}} (a_i e^{ik_0 x} - a_r e^{-ik_0 x}) \quad (34)$$

$$\zeta = A J_0 2 \left[\frac{\omega^2 (1 - if_b) x}{g \tan \beta_s} \right]^{1/2} \quad (35)$$

$$(0 \leq x \geq l_s)$$

$$u = -iA \left[\frac{g}{(1 - if_b) x \tan \beta_s} \right]^{1/2} J_1 2 \left[\frac{w^2(1 - if_b) x}{g \tan \beta_s} \right]^{1/2} \quad (36)$$

where

a_i = amplitude of incident wave

$$k_0 = \frac{\omega}{\sqrt{gd_s}} \quad (37)$$

d_s = constant water depth seaward of breakwater
a_r = complex amplitude of reflected wave
l_s = submerged horizontal length of the impermeable slope
A = arbitrary constant that is the complex vertical amplitude of the wave motion at the intersection of the still-water level and the slope. The modulus (or magnitude) $|A|$ is treated as an approximate value of the runup on the slope.
J_0 = Bessel function of the first kind of order zero
β_s = angle of impermeable slope
J_1 = Bessel function of the first kind of order one

Simultaneous solution at their common boundary ($x = l_s$) yields:

$$a_i e^{ik_0 l_s} + a_r e^{-ik_0 l_s} = A J_0 2 k_0 l_s \sqrt{1 - if_b} \quad (38)$$

$$a_i e^{ik_0 l_s} - a_r e^{-ik_0 l_s} = A \frac{i}{\sqrt{1 - if_b}} J_1 2 k_0 l_s \sqrt{1 - if_b} \quad (39)$$

The above complex solutions may be manipulated to produce expressions for reflection and runup on rough impermeable slopes:

$$\frac{a_r}{a_i} = \left(\frac{J_0 2 k_0 l_s \sqrt{1 - if_b} - \dfrac{i}{\sqrt{1 - if_b}} 2 J_1 k_0 l_s \sqrt{1 - if_b}}{J_0 2 k_0 l_s \sqrt{1 - if_b} + \dfrac{i}{\sqrt{1 - if_b}} 2 J_1 k_0 l_s \sqrt{1 - if_b}} \right) e^{i 2 k_0 l_s} \quad (40)$$

$$\frac{A}{2a_i} = \frac{e^{ik_0 l_s}}{J_0 2k_0 l_s \sqrt{1-if_b} + \dfrac{i}{\sqrt{1-if_b}} 2J_1 k_0 l_s \sqrt{1-if_b}} \qquad (41)$$

Expressions for a reflection coefficient and nondimensional run-up amplitude follow directly:

$$R_{II} = \frac{|a_r|}{a_i} \qquad (42)$$

$$R_u = \frac{|A|}{2a_i} \qquad (43)$$

The important parameters in determining the reflected wave amplitude and the run-up amplitude are the submerged horizontal length of the slope relative to incident wavelength, l_s/L and the linearized friction factor, f_b. Since the linearized friction factor appears in the form $\sqrt{1-if_b}$, it is expedient to introduce the friction angle ϕ defined by

$$\tan 2\phi = f_b \qquad (44)$$

since

$$\sqrt{1-if_b} = (1 + \tan^2 2\phi)^{1/4} e^{-i\phi} \qquad (45)$$

The expression for the friction angle, ϕ, is

$$\tan 2\phi = f_w \frac{|A|}{d_s} \frac{1}{\tan \beta_s} F_s \qquad (46)$$

where

f_w = wave friction factor (empirically determined)

$$= 0.29 \left(\frac{d}{d_s}\right)^{-0.5} \left(\frac{d \tan \beta_s}{|A|}\right)^{0.7} \qquad (47)$$

d = average stone diameter
F_s = slope friction constant

$$= \frac{4}{3\pi} \frac{\int_0^1 \left(\dfrac{J_1 2\psi\sqrt{y}}{\psi\sqrt{y}}\right)^3 dy}{\int_0^1 y \left(\dfrac{J_1 2\psi\sqrt{y}}{\psi\sqrt{y}}\right)^2 dy} \qquad (48)$$

$$\psi = k_0 l_s \sqrt{1 - i \tan 2\phi} \qquad (49)$$

$$y = \frac{x}{l_s} \qquad (50)$$

These equations are solved iteratively by first assuming a value of ϕ, next evaluating R_u, and F_s, then calculating a new ϕ, and repeating the procedure until convergence is achieved.

Synthesis of the Two Preceding Analyses

This section discusses the synthesis of the results of the two preceding analyses. The procedure provides approximate values for wave reflection and transmission for trapezoidal multilayered rubble-mound breakwaters.

For most trapezoidal, multilayered breakwaters, the stone size in the layer under the armor layer of the seaward slope is small relative to the material of the armor layer. As a first approximation, the structure may therefore be treated as having an impermeable rough slope. Thus, with incident wave characteristics, rubble-mound armor, and seaward structure slope, the procedure developed in the section entitled **"External Energy Dissipation"** may be used to approximately account for the energy dissipation on the seaward slope. The remaining wave energy may be expressed as the energy associated with a progressive wave of the following amplitude:

$$a_I = R_{II} a_i \tag{51}$$

where

a_I = wave amplitude representing remaining energy after dissipation on the seaward slope of the breakwater

R_{II} = reflection coefficient of a rough impermeable sloped structure (from external energy dissipation analysis)

a_i = amplitude of incident wave

This remaining energy (represented by wave amplitude a_I) is partitioned among the reflected, transmitted, and internally dissipated energy of a hydraulically equivalent homogeneous rectangular breakwater for which a reflection coefficient R_I and transmission coefficient T_I have been determined as shown in the section entitled "Internal Energy Dissipation." A rational method for obtaining a homogeneous rectangular breakwater that is hydraulically equivalent to a trapezoidal, multilayered breakwater is developed in the next section entitled "Hydraulically Equivalent Rectangular Breakwater."

Having now accounted for the external as well as the internal energy dissipation, the amplitude of the reflected wave is found to be

$$|a_r| = R_I a_I = R_I R_{II} a_i \tag{52}$$

The transmitted wave amplitude is

$$|a_t| = T_I a_I = T_I R_{II} a_i \tag{53}$$

Therefore, the approximate values of the reflection R and transmission coefficients K_{Tt} of a trapezoidal, multilayered breakwater are

$$R = \frac{|a_r|}{a_i} = R_I R_{II} \tag{54}$$

$$K_{Tt} = \frac{|a_t|}{a_i} = T_I R_{II} \tag{55}$$

Hydraulically Equivalent Rectangular Breakwater

This section will present a method for determining an idealized homogeneous rectangular breakwater that is hydraulically equivalent to a trapezoidal, multilayered breakwater.

A hydraulically equivalent breakwater is taken as a homogeneous rectangular breakwater that yields the same discharge as would an actual trapezoidal, multilayered breakwater *with its top layer of stones on the seaward slope removed*. This definition of the equivalent breakwater is illustrated schematically in Figure 5-4-4. Typical realistic

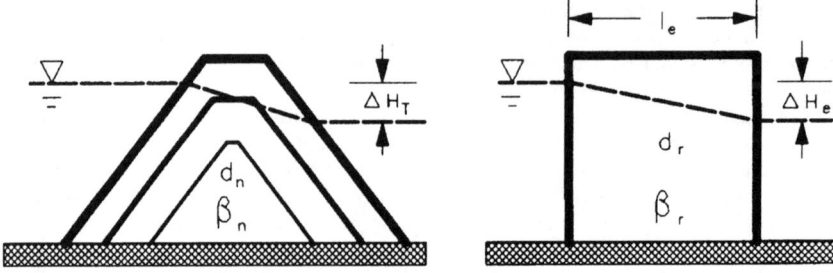

Tropezoidal Breakwater Equivalent Rectangular Breakwater

FIGURE 5-4-4. A Multilayered trapezoidal breakwater and its idealized homogeneous rectangular equivalent.

breakwaters consist of several different porous materials that are identified by their stone size, d_n, and their hydraulic characteristics, β_n. The idealized homogeneous rectangular breakwater consists of a reference material of stone size, d_n, and hydraulic characteristics, β_r. The reference material is considered representative of the porous materials of the multilayered breakwater.

The flow through the structure is assumed to be one-dimensional (1-D), and the discharge per unit length of the equivalent rectangular breakwater is

$$Q_{\substack{\text{equivalent} \\ \text{stucture}}} = \sqrt{\left(\frac{g\Delta H_e}{\beta_r}\right)\frac{d_s}{\sqrt{l_e}}} \quad (56)$$

where

g = acceleration due to gravity
ΔH_e = head difference
β_r = hydrodynamic characteristic of reference breakwater
d_s = water depth
l_e = width of the equivalent breakwater

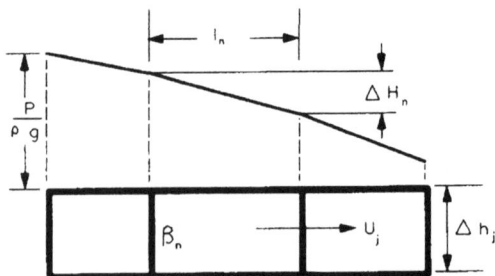

FIGURE 5-4-5. Illustrative slice of a multilayered trapezoidal breakwater section.

AUTOMATED COASTAL ENGINEERING SYSTEM (ACES) **A.113**

To evaluate the discharge per unit length of the multilayered trapezoidal breakwater, the structure is segmented into horizontal slices. The slices may be selected arbitrarily; however, it is expeditious to place them at material boundaries or at elevations where slopes change. A typical horizontal slice of height, Δh_j, is shown in Figure 5-4-5. Each slice consists of segments of different porous materials with individual hydraulic characteristics, β_n, and lengths, l_n.

By summing contributions from all horizontal slices of the trapezoidal breakwater, the total discharge is

$$Q_{\substack{\text{trapezoidal} \\ \text{breakwater}}} = \sqrt{\left(\frac{g\Delta H_T}{\beta_r}\right)} d_s \Sigma_j \left[\frac{1}{\sqrt{\Sigma_n\left(\frac{\beta_n}{\beta_r}l_n\right)}} \left(\frac{\Delta h_j}{d_s}\right) \right] \tag{57}$$

It is required that the discharges per unit length for the rectangular and trapezoidal breakwaters be identical:

$$Q_{\substack{\text{equivalent} \\ \text{stucture}}} = Q_{\substack{\text{trapezoidal} \\ \text{breakwater}}} \tag{58}$$

The above relationship is reordered to solve for l_e, the equivalent rectangular breakwater.

$$l_e = \left\{ \Sigma_j \left[\frac{1}{\sqrt{\Sigma_n\left(\frac{\beta_n}{\beta_r}l_n\right)}} \left(\frac{\Delta h_j}{d_s}\right) \right] \right\}^{-2} \left(\frac{\Delta H_e}{\Delta H_T}\right) \tag{59}$$

where

Σ_j = summation over number of layers in the breakwater
Σ_n = summation over number of materials in the breakwater

$$\beta_n = \beta_0 \left(\frac{1-n_n}{n_n^3}\right) \frac{1}{d_n} \tag{60}$$

$\beta_o = 2.7$
d_n = mean diameter of individual material

$$\beta_r = \beta_0 \left(\frac{1-n_r}{n_r^3}\right) \frac{1}{d_r} \tag{61}$$

$n_r = 0.435$
d_r = mean diameter of reference material

The equation for l_e shows that the width of the equivalent breakwater may be determined from knowledge of the configuration of the trapezoidal, multilayered breakwater and the corresponding head differences, ΔH_e and ΔH_T.

Runup on the seaward face of both the equivalent and trapezoidal breakwater is taken as a representative value of the head difference.

$$\Delta H_e = (1 + R_I)a_I = (1 + R_I)R_{II}a_i \tag{62}$$

$$\Delta H_T = R_u H_i = 2R_u a_i \tag{63}$$

where

$H_i = 2a_i$ = incident wave height

$$\frac{\Delta H_e}{\Delta H_T} = \frac{(1 + R_I)R_{II}}{2R_u} \tag{64}$$

The head difference ratio is a function of the reflection coefficient R_I of the equivalent breakwater, which cannot be determined till the width of the equivalent breakwater l_e is known (i.e., an iterative procedure).

References and Bibliography

Ahrens, J. P., and McCartney B. L. 1975. "Wave Period Effect on the Stability of Riprap," *Proceedings of Civil Engineering in the Oceans/III*, American Society of Civil Engineers, pp. 1019–1034.

Bear, J., et al. 1968. *Physical Principles of Water Percolation and Seepage*, United Nations Educational, Scientific and Cultural Organization.

Cross, R., and Sollitt, C. 1971. "Wave Transmission by Overtopping," Technical Note No. 15, Ralph M. Parsons Laboratory, Massachusetts Institute of Technology, Boston.

Madsen, O. S., and White, S. M. 1976. "Reflection and Transmission Characteristics of Porous Rubble-Mound Breakwaters," CERC MR 76-5, US Army Engineer Waterways Experiment Station, Vicksburg, MS.

Morris, A. H. 1981. "NSWC/DL Library of Mathematics Subroutines," NSWC-TR-81-410, Naval Surface Weapons Center, Dahlgren, VA.

Seelig, W. N. 1979. "Estimation of Wave Transmission Coefficients for Permeable Breakwaters," CERC CETA 79-6, US Army Engineer Waterways Experiment Station, Vicksburg, MS.

Seelig, W. N. 1980. "Two-Dimensional Tests of Wave Transmission and Reflection Characteristics of Laboratory Breakwaters," CERC TR 80-1, US Army Engineer Waterways Experiment Station, Vicksburg, MS.

LONGSHORE SEDIMENT TRANSPORT

Description

This application provides estimates of the *potential* longshore transport rate under the action of waves. The method used is based on the empirical relationship between the longshore component of wave energy flux entering the surf zone and the immersed weight of sand moved (Galvin, 1979). Three methods are available to the user depending on whether available input data are breaker wave height and direction, deepwater wave height and direction, or using a Wave Information Study (WIS) hindcast data file created by the Coastal Engineering Data Retrieval System (CEDRS). The material presented herein can be found in Chapter 4 of the *Shore Protection Manual* (1984) and in Gravens (1988).

Introduction

The longshore transport rate, Q, is the volumetric rate of movement of sand parallel to the shoreline. Much longshore transport occurs in or near the surf zone and is caused by the approach of waves at an angle to the shoreline. Q is expressed in terms of sand volume per unit time (such as cubic yards per year or cubic meters per year). The method used to calculate the longshore transport in this ACES application is based on the assumption that longshore transport rate Q is dependent upon the longshore component of wave energy flux P_{ls} entering the surf zone (Equation 4-49 of the SPM (1984)).

$$Q = \frac{K}{(\rho_s - \rho)ga'} P_{ls} \quad \text{(unit volume per sec)} \tag{1}$$

where

K = dimensionless empirical coefficient (based on field measurements)
   = 0.39
ρ_s = density of sand
ρ = density of water
g = acceleration due to gravity
a' = ratio of the volume of solids to total volume, accounting for sand porosity
   = 0.6

General Assumptions and Limitations

General assumptions and limitations used to derive the longshore energy flux in the surf zone are

- Conservation of energy flux in shoaling waves.
- Linear wave theory.
- Evaluating the energy flux relation at the breaker position.
- Breaker characteristics described by solitary wave theory.
- Straight bathymetric contours parallel to the shoreline.

Judgment is required in using the empirical relationship between longshore transport rate and the energy flux factor (Equation 1). The accuracy of Q found using the energy flux factor can be estimated to be ±50 percent (SPM, 1984).

General Energy Flux Equation

The energy flux per unit length of wave crest or, equivalently, the rate at which wave energy is transmitted across a plane of unit width perpendicular to the direction of wave advance is

$$P = EC_g \tag{2}$$

where

E = wave energy density
C_g = wave group speed

The wave energy density is calculated by

$$E = \frac{\rho g H^2}{8} \tag{3}$$

where

ρ = mass density of water
g = acceleration of gravity
H = wave height

If the wave crests make an angle a with the shoreline, the energy flux in the direction of wave advance *per unit length of beach* is

$$P \cos \alpha = \frac{\rho g H^2}{8} C_g \cos \alpha \qquad (4)$$

The longshore component of wave energy flux is

$$P_l = P \cos \alpha \sin \alpha = \frac{\rho g H^2}{8} C_g \cos \alpha \sin \alpha \qquad (5)$$

By use of the identity

$$\cos \alpha \sin \alpha = \frac{1}{2} \sin 2\alpha \qquad (6)$$

the general equation for P_l becomes

$$P_l = \frac{\rho g}{16} H^2 C_g \sin 2\alpha \qquad (7)$$

Up to this point, all result have been for small-amplitude linear wave theory. However, the assumed relation between longshore transport and energy flux in the surf zone requires that P_l be evaluated at the breaker line, where small-amplitude theory is less valid. To indicate approximations for waves entering the surf zone, the symbol P_{ls} will be used in place of P_l. this approximation is called the energy flux factor P_{ls} in the SPM (1984), and like P_l it is measured in units of energy per second per unit length of shoreline. The next two sections derive approximate fomiulas for computing the longshore energy flux factor P_{ls} entering the surf zone based on known breaking or deepwater significant wave conditions.

Energy Flux Equation (Breaking Wave Conditions)
If the breaker values of the wave characteristics (H_{sb} significant breaker height and α_b angle of approach) are applied to Equation 7, the energy flux factor results.

$$P_{ls} = \frac{\rho g}{16} H_{sb}^2 C_{gb} \sin 2\alpha_b \qquad (8)$$

From linear wave theory

$$C_{gb} = nC_b \qquad (9)$$

where

$n \approx 1.0$ (in shallow water)
C_b = wave phase speed at breaking

Group velocity equals wave speed at breaking, and breaking speed is given by solitary wave theory according to the approximation (Galvin and Schweppe, 1980):

$$C_b = 8.02 \sqrt{H_b} \qquad (10)$$

Substituting Equation 10 into Equation 8 and simplifying yields the longshore energy flux factor for breaking wave conditions (Equation 4-44 of the SPM (1984)).

$$P_{ls} = 0.0884 \, \rho g^{3/2} H_{sb}^{5/2} \sin 2\alpha_b \qquad (11)$$

Energy Flux Equation (Deepwater Wave Conditions)

Using deepwater values of wave characteristics (H_{s0} significant wave height and α_0 angle of approach), Equation 7 becomes

$$P_{ls} = \frac{\rho g}{16} H^2_{s0} C_{gb} \sin 2\alpha_0 \qquad (12)$$

Again, from linear wave theory:

$$C_{gb} = n C_b \qquad (13)$$

where

$$n \approx \frac{1}{2} \text{ (in deep water)}$$

Again, using solitary wave theory for breaker characteristics (Galvin and Schweppe, 1980):

$$C_b = 8.02\sqrt{H_b} \qquad (14)$$

Local wave height H_b can be related to deepwater height H_0 by refraction and shoaling coefficients, where the coefficients are evaluated at the breaker position.

$$\sqrt{H_b} = \sqrt{K_r K_s H_0} \qquad (15)$$

where

K_r = refraction coefficient

$$\sqrt{K_r} = \left(\frac{\cos \alpha_0}{\cos \alpha_b}\right)^{1/4} \qquad (16)$$

K_s = shoaling coefficient (approximated by the breaker height index) = 1.3 (Galvin and Schweppe, 1980)

In Equation 16, cos α_b equals 1.0 to a good approximation. For example, if α_b has a high value of 20°, then $(\cos 20°)^{1/4} = 0.98$.

Substituting the above variables into Equation 11 for P_{ls} and simplifying yields the longshore energy flux factor entering the surf zone using deepwater wave conditions (Equation 4-45 of the SPM (1984)).

$$P_{ls} = 0.05 \rho g^{3/2} H^{5/2}_{s0} (\cos \alpha_0)^{1/4} \sin 2\alpha_0 \qquad (17)$$

Estimating Potential Sand Transport Rates Using WIS-CEDRS Data

The following paragraphs describe the procedure that is used in this ACES application for calculating potential longshore transport rates using WIS hindcast wave estimates as provided in a specially formatted data file by the CEDRS (see section titled "Coastal Engineering Data Retrieval System"),

The potential longshore sand transport rate using WIS-CEDRS data is calculated using Equations 1 and 11 which require the breaking wave height and incident angle with respect to the shoreline. WIS-CEDRS hindcast estimates, however, are given for water depths greater than or equal to 10 meters (Jensen, 1983). Therefore, a transformation of the WIS-CEDRS hindcast wave estimates to breaking conditions is necessary. Refraction and shoaling of incident waves provided by WIS-CEDRS is accomplished using linear wave theory and numerically solving Snell's law for wave direction and the equation of conservation of wave energy flux for wave height. The governing equations are given below (Coastal Engineering Technical Note-II-19, 1989). The subscripts 0 and b denote values in deep water and at breaking, respectively.

Wave direction is obtained through Snell's law:

$$\frac{\sin(\alpha_0)}{L_0} = \frac{\sin(\alpha_b)}{L_b} \tag{18}$$

where

α_0 = deepwater wave approach angle (extracted from the CEDRS data file)
α_b = wave breaker angle
L = wavelength

$$= \frac{gT^2}{2\pi} \tanh\left(\frac{2\pi d}{L}\right) \tag{19}$$

T = wave period (extracted from the CEDRS data file)

The wave height is obtained by invoking the conservation of wave energy flux directed onshore

$$E_0 C_{g0} \cos(\alpha_0) = E_b C_{gb} \cos(\alpha_b) \tag{20}$$

where

$E_{0/b}$ = wave energy

$$= \frac{1}{8} \rho g H_{0/b}^2)$$

H_0 = deepwater wave height (extracted from the CEDRS data file)
H_b = breaking wave height
C_g = wave group speed

$$= \frac{L}{2T}\left[1 + \frac{\frac{4\pi d}{L}}{\sinh\frac{4\pi d}{L}}\right]$$

The breaking wave height is linearly related to the depth at breaking as

$$H_b = \gamma d_b \tag{21}$$

where

γ = wave breaking index
= 0.78

Wave Transformation Procedure

The first step in the wave transformation procedure (Gravens, 1988) is to calculate the wavelength (Equation 19) at the location (denoted by subscript 0) where the wave height, incident angle, period, and water depth are known. These parameters are read by this ACES application directly from the *formatted* WIS-CEDRS data file. Equation 19 is solved by using a Newton-Raphson iteration.

The second step is to determine wave height, water depth, and incident angle at breaking (denoted by subscript b in Equations 18 and 20). Equation 18 is first solved for cos α_b and substituted together with Equation 21 into the right-hand side of Equation 20. This yields an expression for the conservation of wave energy flux in terms of the known wave characteristics (left-hand side of Equation 20) and the unknown wave characteristics at breaking (right-hand side of Equation 20). Equation 19 evaluated at breaking gives another expression in terms of the wavelength and water depth at breaking. A Newton-Raphson type solution can be used to iterate for the two unknown variables of wavelength and water depth at breaking.

Having determined the wavelength and depth at breaking, breaking wave height is calculated using Equation 21, and the breaking wave angle is calculated using Equation 18. The potential longshore sand transport rate is then estimated using Equations 1 and 11.

References and Bibliography

Coastal Engineering Technical Note II-19. 1989. "Estimating Potential Longshore Sand Transport Rates Using WIS Data," US Army Engineer Waterways Experiment Station, Vicksburg, MS.

Galvin, C. J. 1979. "Relation Between Immersed Weight and Volume Rates of Longshore Transport," CERC TP 79-1, US Army Engineer Waterways Experiment Station, Vicksburg, MS.

Galvin, C. J., and Schweppe, C. R. 1980. "The SPM Energy Flux Method for Predicting Longshore Transport Rate," CERC TP 80-4, US Army Engineer Waterways Experiment Station, Vicksburg, MS.

Gravens, M. B. 1988. "Use of Hindcast Wave Data for Estimation of Longshore Sediment Transport," *Proceedings of the Symposium on Coastal Water Resources*, American Water Resources Association, Wilmington, NC, pp. 63–72.

Shore Protection Manual. 1984. 4th ed., 2 Vols., US Army Engineer Waterways Experiment Station, Coastal Engineering Research Center, US Government Printing Office, Washington, DC, Chapter 4, pp. 89–107.

Vitale, P. 1980. "A Guide for Estimating Longshore Transport Rate Using Four SPM Methods," CERC CETA 80-6, US Army Engineer Waterways Experiment Station, Vicksburg, MS.

Wave Information Studies of US Coastlines

Corson, W. D., Able, C. E., Brooks, R. M., Farrar, P. D., Groves, B. J., Payne, J. B., McAneny, D. S., and Tracy, B. A. 1987. "Pacific Coast Hindcast Phase II Wave Information," Wave Information Studies of US Coastlines, WIS Report 16, US Army Engineer Waterways Experiment Station, Vicksburg, MS.

Driver, D. B., Reinhard, R. D., and Hubertz, J. M. 1991. "Hindcast Wave Information for the Great Lakes: Lake Erie," Wave Information Studies of US Coastlines, WIS Report 22, US Army Engineer Waterways Experiment Station, Vicksburg, MS.

Driver, D. B., Reinhard, R. D., and Hubertz, J. M. 1992. "Hindcast Wave Information for the Great Lakes: Lake Superior," Wave Information Studies of US Coastlines, WIS Report 23, US Army Engineer Waterways Experiment Station, Vicksburg, MS.

Hubertz, J. M., and Brooks, R. M. 1989. "Gulf of Mexico Hindcast Wave Information," Wave Information Studies of US Coastlines, WIS Report 18, US Army Engineer Waterways Experiment Station, Vicksburg, MS.

Hubertz, J. M., Driver, D. B., and Reinhard, R. D. 1991. "Hindcast Wave Information for the Great Lakes: Lake Michigan," Wave Information Studies of US Coastlines, WIS Report 24, US Army Engineer Waterways Experiment Station, Vicksburg, MS.

Jensen, R. E. 1983. "Atlantic Coast Hindcast, Shallow-Water Significant Wave Information," Wave Information Studies of US Coastlines, WIS Report 9, US Army Engineer Waterways Experiment Station, Vicksburg, MS.

Reinhard, R. D., Driver, D. B., and Hubertz, J. M. 1991. "Hindcast Wave Information for the Great Lakes: Lake Huron," Wave Information Studies of US Coastlines, WIS Report 26, US Army Engineer Waterways Experiment Station, Vicksburg, MS.

NUMERICAL SIMULATION OF TIME-DEPENDENT BEACH AND DUNE EROSION

Description

This application is a numerical beach and dune erosion model that predicts the evolution of an equilibrium beach profile from variations in water level and breaking wave height as occur during a storm. The model is one-dimensional (only onshore-offshore sediment transport is represented). It is based on the theory that an equilibrium profile results from uniform wave energy dissipation per unit volume of water in the surf zone. The general characteristics of the model are based on a model described by Kriebel (1982, 1984a, 1984b, 1986). Because of the complexity of this methodology and the input requirements, familiarization with the above references is strongly recommended.

Introduction

The model described herein, called XSHORE, is a 1-D model in that only cross-shore sediment transport is represented. This implies that the effect of longshore transport is negligible as, for example, if the gradient of longshore transport were near zero. The model is based on the theory that the equilibrium shape of a beach profile is related to uniform wave energy dissipation per unit volume of water in the surf zone under breaking waves (Dean, 1977).

General Assumptions and Limitations

General assumptions and limitations adopted in developing the beach and dune erosion numerical model are:

- An equilibrium beach profile can be attained if a beach is exposed to constant wave and water level conditions for a sufficiently long time.
- Longshore sediment transport is neglected, and beach and dune erosion results solely from cross-shore transport.
- The model is restricted to calculating beach and dune erosion; i.e., recovery (accretive) processes are not well represented.
- Cross-shore sediment transport is caused by breaking of short-period waves.
- Wave transformation in the nearshore is approximated using the assumptions of straight and parallel bathymetric contours and linear wave theory.
- Surf zone wave heights are assumed to be proportional to the local water depth.
- Wave runup and setup can be neglected
- Waves are considered normally incident inside the surf zone.
- Median grain size (sand-size material) is constant along the profile.

Theoretical Development of the Model

The development of the model begins by assuming that the nearshore profile can be described by a monotonic function of depth, h, which increases in the seaward direction, x, across the actively modified portion of the profile according to the relationship (Bruun, 1954; Dean, 1977)

$$h = Ax^{2/3} \tag{1}$$

where A is a profile shape coefficient that has been shown to be correlated with median grain size (Moore, 1982). The coefficient A governs the steepness of the subaqueous beach profile and has been related to a unique value of the wave energy dissipation per unit volume, D_{eq}, which exists everywhere in the surf zone if the profile is in equilibrium (Dean, 1977).

Based on these concepts and using shallow-water linear wave theory, the cross-shore sediment transport rate, Q, is expressed in terms of the excess energy dissipation per unit volume of water across the surf zone as

$$Q = K(D - D_{eq}) \tag{2}$$

where

K = empirically determined sand transport rate coefficient
D, D_{eq} = actual and equilibrium energy dissipation per unit volume of water under given local wave and water level conditions

Moore (1982) determined a design curve for K by simulating the large-scale laboratory tests of Saville (1957) (see Kraus and Larson, 1988) using a numerical model based on Equation 2. Profile evolution was best reproduced for the value of

$$K = 1.144 \times 10^{-3} \text{ ft}^4/\text{lb} \quad \text{or} \quad 2.2 \times 10^{-6} \text{ m}^4/\text{N}$$

In differential form, wave energy dissipation per unit volume of water in the surf zone may be written as

$$D = \frac{1}{h} \frac{\partial P}{\partial x} \tag{3}$$

where

P = wave energy flux

This wave energy flux is given by

$$P = EC_g \tag{4}$$

where

E = total wave energy in one wavelength per unit crest width
C_g = wave group speed

The total wave energy is given by

$$E = \frac{1}{8} \rho g H^2 \tag{5}$$

where

ρ = fluid density
g = acceleration of gravity
H = wave height

The quantity C_g is the wave group speed and in shallow water is given by

$$C_g = (gh)^{1/2} \tag{6}$$

Substituting the wave energy E and group speed C_g into Equation 4 produces the following relationship:

$$P = \frac{1}{8}\rho g H^2 (gh)^{1/2} \qquad (7)$$

Breaking wave heights are assumed to be proportional to local water depth and given by

$$H = \gamma h \qquad (8)$$

where γ is a dimensionless breaking wave constant, equal to 0.78 in the model. Wave energy flux in the surf zone can now be written as a function of total water depth.

$$P = \frac{1}{8}\rho g^{3/2} \gamma^2 h^{5/2} \qquad (9)$$

By differentiating Equation 9 according to Equation 3, an expression for energy dissipation per unit volume of water can be derived.

$$d = \frac{5}{16}\rho g^{3/2} \gamma^2 h^{1/2} \frac{\partial h}{\partial x} \qquad (10)$$

Assuming a profile shape in equilibrium (which implies $D = D_{eq}$) with uniform energy dissipation per unit volume of water, integrating Equation 10 provides a relationship between water depth, h, and distance offshore, x.

$$h = \left[\frac{24}{5}\left(\frac{D_{eq}}{\rho \gamma^2 g^{3/2}}\right)\right]^{2/3} x^{2/3} \qquad (11)$$

Using the equilibrium profile shape given by Equation 1, Equation 11 can be solved for equilibrium energy dissipation, D_{eq}, within the surf zone.

$$D_{eq} = \frac{5}{24}\rho \gamma^2 g^{3/2} A^{3/2} \qquad (12)$$

Numerical Solution

In the present model, time-dependent profile response is determined by an explicit finite difference solution of the equation for continuity of sand in the onshore-offshore direction.

$$\frac{\partial h}{\partial t} = \frac{\partial Q}{\partial x} \qquad (13)$$

The surf zone is represented by a series of cells in which the incremental change in cross-shore distance from a baseline, Δx, is uniform, and h is the total water depth. Figure 6-2-1 is a schematic representation of the surf zone showing sediment transport over a horizontal grid of uniform width Δx.

Since water level can fluctuate with respect to storm surge and tide, then

$$h_i = \eta_i - d_i \qquad (14)$$

η_i = water surface elevation due to storm surge and tide
d_i = bottom elevation

Both η_i and d_i are defined at h_i and are relative to mean water level (MWL). Wave setup in the surf zone and wave runup on the beach face are not represented.

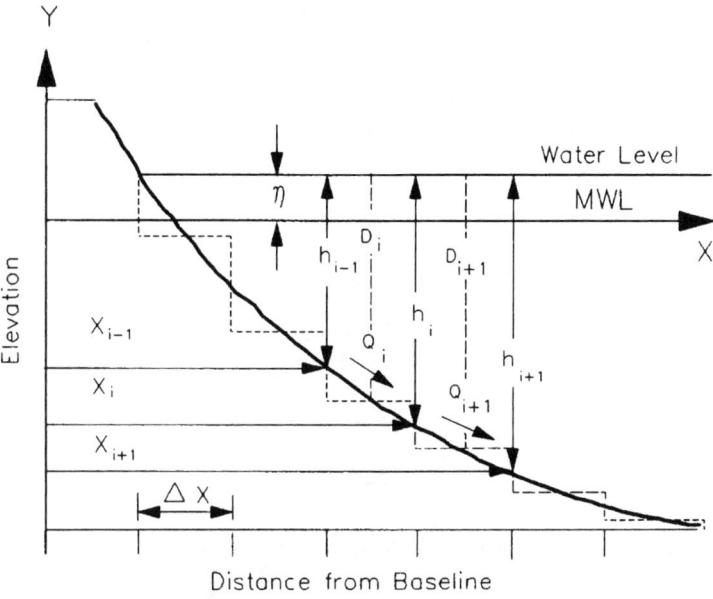

FIGURE 6-2-1. Numerical representation of the beach profile.

Since D and Q vary with water depth and local bottom slope, the rate of change at each discrete cell differs from adjacent cells. Therefore, Equation 13 can be written as

$$\frac{\Delta h_i}{\Delta t} = \frac{Q_{i+1} - Q_i}{\Delta x} \qquad (15)$$

From Equation 2

$$Q_i = K(D_i - D_{eq}) \qquad (16)$$

$$Q_{i+1} = K(D_{i+1} - D_{eq}) \qquad (17)$$

where

D_i = energy dissipation per unit volume of water at i
D_{i+1} = energy dissipation per unit volume of water at $i + 1$

Substituting Equations 16 and 17 into Equation 15

$$\frac{\Delta h_i}{\Delta t} = \frac{K(D_{i+1} - D_{eq}) - K(D_i - D_{eq})}{\Delta x} \qquad (18)$$

Since $\Delta h_i = h_i' - h_i$ at an internal grid point and D_{eq} is constant across the surf zone

$$h_i' = h_i + \frac{K\Delta t}{\Delta x}(D_{i+1} - D_i) \qquad (19)$$

where

h_i' = water depth at the succeeding time step

From this expression, if D_{i+1} is greater than D_i over time Δt, there will be a net increase in water depth at the succeeding time step.

As a wave breaks over a sloping bottom, the energy dissipated in a distance Δx may be expressed as the difference between the energy flux P entering and exiting a control volume defined by Δx and the water depth, where P_i and P_{i+1} are defined at h_i and h_{i+1}. Therefore,

$$D_{i+1} = \frac{P_{i+1} - P_i}{h_{i+1/2} \Delta x} \tag{20}$$

Combining Equations 9 and 20, the energy dissipation per unit volume of water in the cell between grid points i and $i + 1$ is

$$D_{i+1} = \alpha \left[\frac{h_{i+1}^{5/2} - h_i^{5/2}}{0.5(h_{i+1} + h_i)\Delta x} \right] \tag{21}$$

where

$$\alpha = \frac{\rho g^{3/2} \gamma^2}{8} \tag{22}$$

Substituting Equation 21 into the finite difference form of the continuity Equation 19 yields

$$h_i' = h_i + \frac{2\alpha K \Delta t}{(\Delta x)^2} \left[\frac{(h_{i+1}^{2.5} - h_i^{2.5})}{(h_{i+1} + h_i)} - \frac{(h_i^{2.5} - h_{i-1}^{2.5})}{(h_i + h_{i-1})} \right] \tag{23}$$

where

h_i' = calculated water depth at position x_i for the succeeding time step

Initial Boundary Conditions

Figure 6-2-2 is a schematic representation of the macro-morphologic features of a beach-dune system, illustrating initial parameters required for constructing an idealized beach profile.

A linearly sloping beach face is assumed to intersect an equilibrium profile at a depth, h_t, such that there is a continuous decrease in depth in the offshore direction. This parameter is calculated in the model knowing the beach face slope and the equilibrium profile shape factor, A, which has been shown to be a function of median grain diameter (D_{50}) (Moore, 1982). Once this depth has been defined, erosion of the subaerial beach is assumed to respond geometrically, as determined solely from continuity, from h_t to the top of the berm or top of the dune, depending on profile geometry and water level elevation. When actual profile data are supplied, the user defines the initial landward boundary condition, and all other geometrical parameters are calculated internally.

The offshore boundary condition for the profile response portion of the model is defined at the depth of wave breaking, shoreward of which refraction is assumed to be negligible. The region between the point of wave breaking and li 1is governed by energy dissipation and may be thought of as the dynamic zone. Beyond the breaking depth, energy dissipation is assumed negligible, and sediment transport is defined as zero. Seaward of wave breaking, wave refraction and shoaling are calculated from data input supplied by the user (significant wave height, peak period, and wave angle relative to the shoreline).

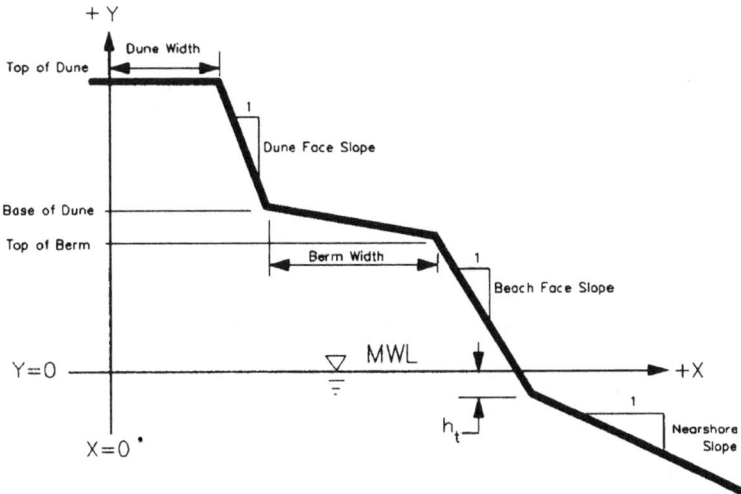

FIGURE 6-2-2. Idealized berm, dune, and offshore system.

Model Capabilities

The preceding discussion summarizes the basic equations and conditions used for simulating beach and dune erosion caused by varying water levels and wave heights, as occur during a storm. The present model differs from that of Kriebel (1982) in two significant ways:

- A horizontal grid is used (Δx constant) rather than a vertical grid (Δy constant).
- An explicit finite difference calculation scheme is used rather than an implicit scheme.

Model characteristics that have been implemented in XSHORE, include:

- Capability for generating an idealized beach profile (see schematization shown in Figure 6-2-2) according to criteria provided by the user (e.g. dune width, dune height, dune slope, height at base of dune, berm width, berm height, beach slope, nearshore slope) or use of actual profile data to be entered as x,y pairs.
- Use of time-dependent water level data (tide and/or storm surge) as recorded by tide gages or from predicted tidal variations.
- Use of a time series of wave height, peak period, and wave angle data representing the primary forcing parameters used to operate the model.
- A limitation on the maximum run time of 5 days, since most storm events (periods of assumed dominant erosion through cross-shore sand transport) have a duration less than this period of time.

The smallest time step for data entry and output is 1 hr. Summary statistics include change in shoreline position at the 0, +5, +10, +15 ft contours and associated adjustments in sand volume (yd^3/ft) above mean water level for integer multiples of 1 hr. Simulated profile change data are also written to an ASCII file at the chosen time interval for the user to view with an available graphics software package.

References and Bibliography

Birkemeier, W. 1984. "A User's Guide to ISRP: The Interactive Survey Reduction Program, Instruction Report CERC-84-1, US Army Engineer Waterways Experiment Station, Vicksburg, MS.
Bruun, P. 1954. "Coast Erosion and the Development of Beach Profiles," Technical Memorandum No. 44, Beach Erosion Board, US Army Engineer Waterways Experiment Station, Vicksburg, MS.
Dean, R. G. 1977. "Equilibrium Beach Profiles: U.S. Atlantic and Gulf Coasts," Ocean Engineering Report No. 12, Department of Civil Engineering, University of Delaware, Newark, DE.
Kraus, N. C., and Larson, M. 1988. "Beach Profile Change Measured in the Tank for Large Waves, 1956–1957 and 1962," Technical Report CERC-88-6, US Army Engineer Waterways Experiment Station, Vicksburg, MS.
Kriebel, D. L. 1982. "Beach and Dune Response to Hurricanes," M. S. Thesis, Department of Civil Engineering, University of Delaware, Newark, NJ.
Kriebel, D. L. 1984a. "Beach Erosion Model (EBEACH) Users Manual, Volume I: Description of Computer Model," Beach and Shores Technical and Design Memorandum No. 84-5-I, Division of Beaches and Shores, Florida Department of Natural Resources, Tallahassee, FL.
Kriebel, D. L. 1984b. "Beach Erosion Model (EBEACH) Users Manual, Volume II: Theory and Background," Beach and Shores Technical and Design Memorandum No. 84-5-Il, Division of Beaches and Shores, Florida Department of Natural Resources, Tallahassee, FL.
Kriebel, D. L. 1986. "Verification Study of a Dune Erosion Model," *Shore and Beach*, Vol. 54, No. 3, pp. 13–21.
Moore, B. 1982. "Beach Profile Evolution in Response to Changes in Water Level and Wave Height," M.S. Thesis, Department of Civil Engineering, University of Delaware, Newark, DE.
Saville, T. 1957. "Scale Effects in Two-Dimensional Beach Studies," *Transactions 7th Meeting of International Association of Hydraulic Research, Lisbon, Portugal*, Vol. 10, pp. A3-1 through A3-l0.

CALCULATION OF COMPOSITE GRAIN-SIZE DISTRIBUTIONS

Description

The major concern in the design of a sediment sampling plan for beach-fill purposes is determining the composite grain-size characteristics of both the native beach and the potential borrow site. This application calculates a composite grain-size distribution that reflects textural variability of the samples collected at the native beach or the potential borrow area.

Introduction

Sediment of a similar grain size (particle diameter) to that of the project, or native beach, is used to construct a beach fill. Particle diameters smaller than that of the native beach will eventually be transported (winnowed) and lost from the project site. Sediment containing finer particle diameters can still be used for construction by placing an overage amount with the assumption that, after the winnowing process is complete, the volume of material remaining is equal to the design volume. This overage is known as the overfill ratio or fill factor.

The fill factor is computed by comparing the grain-size distributions (mean and standard deviation) of the native and borrow sediments (see the ACES application, "Beach Nourishment Overfill Ratio and Volume"). Since only one grain-size distribution (GSD) is used to characterize each population, several (5 to 100) individual sediment samples are mathematically combined into one to form a composite sample. Obviously, the number of samples and the spatial and temporal distribution are critical in determining the composite so that each population is properly represented.

Sediment Analysis

Both native beach and borrow sediments should be analyzed using standard sieving techniques. The sieving process mechanically separates a sediment sample containing an infinite number of particle diameters into finite particle size classes. Table A-4 in Appendix A is a list of standard size classes and corresponding sieve mesh sizes. Statistical parameters such as the median, mean, standard deviation, kurtosis, and skewness can be mathematically computed using data from the GSD.

In this application the individual sediment weight retained on each sieve has been normalized so that the *sum* of the individual sediment weights retained on all sieves equals 100. Stated mathematically, normalized sediment weight for each ϕ size is

$$w_{\phi\,\text{normalized}} = w_\phi \frac{100}{\Sigma w} \qquad (1)$$

where

$w_{\phi\,\text{normalized}}$ = normalized sediment weight for a specific ϕ
w_ϕ = sediment weight on each sieve
Σw = total weight of sample

Composite Surface Samples

The sediment characteristics of the native beach can vary considerably.

- Across the beach profile through varied energy zones.
- Along the beach.
- At depths within the active profile.
- Between seasons.

The native beach composite must reflect these components. A typical plan would call for surficial sediment samples to be taken at regular vertical elevation intervals across the beach profile such as +12, +9, +6, +3, 0, –3, etc., offshore to the point of profile closure. Depending upon the longshore variability, this plan would be repeated along 1 to 4 transects for every mile of the project. To capture the seasonal variation, the beach would be sampled 2 to 3 times a year during the winter storm erosion and summer accretion periods. As an alternative, shallow cores could be taken along profile transects that would represent the spatial and temporal variations captured by the surface samples.

Surface samples are combined into one composite *average* GSD by summing the weights retained on each sieve interval and then dividing by the number of samples. Stated mathematically, the composite sediment weight for a given size is

$$w_{\phi\,\text{composite}} = \frac{w_{\phi s_1} + w_{\phi s_2} + w_{\phi s_3} + \cdots + w_{\phi s_n}}{n} \qquad (2)$$

where

$w_{\phi\,\text{composite}}$ = composite weight for a specific sieve
$w_{\phi s_n}$ = sediment weight retained on a specific sieve for each sample
n = number of samples

Composite Core Samples

Cores are usually used to sample potential borrow sites and in some instances the native beach. Cores are brought back to the laboratory to be split, photographed, subsampled, and sieved using standard sieving techniques. Cores are typically subsampled based upon unique lithologic units. A 20-ft core may contain five or more unique units. Some units may be only inches in length, whereas others may be several feet in length.

The composite GSD of each lithologic unit is mathematically weighted proportionally based on its length relative to the core length. For example, if a unit is 1-ft in length from a 10-ft core, each sieve weight for the entire GSD is multiplied by 0.1 (1/10). This proportional weighting is done for all samples in the core resulting in a composite GSD for each core. This scheme is repeated for all cores that are to be included in the core composite. After all individual core composites are created, the entire GSD (consisting of individual sieve weights) for each core is summed and divided by the number of cores. This final step is identical to the procedure in which surface samples are combined into one composite.

A slight variation of the previously described procedure occurs when the user specifies the core composite by an upper and lower elevation limit. For example, a core is a total of 10 ft in length and contains two lithologic units, one 2 ft in length and the other 8 ft in length. The 2-ft unit represents a lithologic unit from an elevation of +11 to +9 ft, and the 8-ft unit represents a lithologic unit from an elevation of +9 to +1 ft. If for example, the user specifies a composite from an elevation of +10 to 0 ft, 1 ft of the 2-ft unit is above the requested upper elevation limit (+10 ft) and 1 ft below the limit. Proportionally, this units represents 10 percent (1/10) of the core within the requested elevation limits and is thereby weighted accordingly; i.e. the GSD is multiplied by 0.1. The 8-ft unit represents 80 percent (8/10) of the core within the elevation limits; thereby its GSD is multiplied by 0.8. Since no sample was present for the +1 to 0 elevation, the weighting scheme accounts for only 90 (10 + 80) percent of the entire core. This procedure is done for all cores located within the elevation limits.

Soil Classification

A grade scale commonly used for classifying sediments is the phi (ϕ) scale devised by Krumbein (1934, 1938). The phi scale is a geometric scale to the power of 2 with individual size classes defined as "twice as large or half as large" as some other class (Table B-4, Appendix B). The phi transformation is given by

$$\phi = -\log_2 d \quad \text{or} \quad 2^{-(\phi)} = d \qquad (3)$$

where

d = grain-size particle diameter in -millimeters

The advantages of the phi scale are

- The distribution of particle sizes can be plotted on arithmetic graph paper, obviating the necessity of logarithmic graph paper; thus the graphical method of computing sediment statistics is simplified.
- Particle diameters for each size class become whole numbers instead of fractions of millimeters.

The disadvantages of the phi scale are

- Frequent unfamiliarity on the part of users.
- No intuitive relation to numerical results of sieve analysis.

- Progression of numerical scales is counter-intuitive (larger value means smaller grain size).

Sediments are classified as GRAVEL, COARSE SAND, MEDIUM SAND, FINE SAND, SILT, and CLAY to indicate the dominant particle size diameter of the sample. The two size schemes used today by coastal engineers are the Unified and Wentworth Soils classifications. These classifications assign similar but different size ranges to each category (Table A-4, Appendix A).

Statistical Analysis of Sediment Data

There are two basic methods of obtaining statistical parameters of a GSD. The first method is known as the "Folk Graphk Method," which was developed before the advent of advanced computers. The Folk method involves plotting a cumulative curve of the sample and determining the particle diameter that corresponds with certain cumulative percentages, i.e. 5, 16, 25, 50, 75, 84, and 95. The second method of obtaining statistical parameters is called the "Method of Moments." It is a computational (not graphic) method using every size class, and it provides a more accurate measure than the graphic method, which relies on only a few selected percentages.

Folk's Graphical Method

Median (M_d). The graphic median is the size class in which half of the particles by weight are coarser than the median and half are finer.

Mean (μ). The mean is probably the best statistical parameter for determining the overall particle size and is given by

$$\mu = \frac{(\phi_{16} + \phi_{50} + \phi_{84})}{3} \tag{4}$$

Standard Deviation (σ). The inclusive graphic standard deviation is a measure of sorting and is given by

$$\sigma = \frac{\phi_{84} - \phi_{16}}{4} + \frac{\phi_{95} - \phi_5}{6.6} \tag{5}$$

Most beach sands have a standard deviation ranging from 0.5 to 2.0.

Skewness (S_k). The inclusive graphic skewness is a measure of asymmetry and is given by

$$S_k = \frac{\phi_{16} + \phi_{84} - 2(\phi_{50})}{2(\phi_{84} - \phi_{16})} + \frac{\phi_5 + \phi_{95} - 2(\phi_{50})}{2(\phi_{95} - \phi_5)} \tag{6}$$

A positive value indicates the sediment has an excess amount of fines, whereas a negative value indicates an excess amount of coarse material. The mathematical limits of skewness are -1.0 to $+1.0$.

Kurtosis (K). The graphic kurtosis is a measure of the ratio between the sorting of the *tails* (ends) of the curve and the sorting of the central portion. When the central portion is better sorted (values greater than 1.5), the curve is said to be peaked or leptokurtic. If the tails are better sorted than the central portion, the curve is said to be flat-peaked or platykurtic (values between 0.0 and 1.1). Kurtosis is given by

$$K = \frac{\phi_{95} - \phi_5}{2.44(\phi_{75} - \phi_{25})} \tag{7}$$

Method of Moments

First Moment. By definition, the first moment equals the mean $\overline{X}$ and is expressed as

$$\overline{X} = \frac{\Sigma f m_\phi}{100} \qquad (8)$$

where

f = frequency in percentage for each size class
m_ϕ = midpoint of each ϕ size class

Second Moment. The second moment is a measure of the dispersion about the mean and is expressed as

$$\frac{\Sigma f(m_\phi - \overline{X})^2}{100} \qquad (9)$$

The second moment represents the numerical value of the standard deviation; therefore, the standard deviation is expressed as

$$\sigma = \sqrt{\frac{\Sigma f(m_\phi - \overline{X})^2}{100}} \qquad (10)$$

Third Moment. The third moment (known as the *mean cubed deviation*) is the measure of symmetry about the mean and is expressed as

$$\frac{\Sigma f(m_\phi - \overline{X})^3}{100} \qquad (11)$$

Skewness is computed by dividing the *mean cubed deviation* by the cube of the standard deviation and is expressed as:

$$S_k = \frac{\Sigma f(m_\phi - \overline{X})^3}{100\sigma^3} \qquad (12)$$

Fourth Moment. The fourth moment is the distribution about the mean and is expressed as

$$\frac{\Sigma f(m_\phi - \overline{X})^4}{100} \qquad (13)$$

Kurtosis is derived by dividing the fourth moment by the standard deviation raised to the fourth power:

$$S_k = \frac{\Sigma f(m_\phi - \overline{X})^4}{100\sigma^4} \qquad (14)$$

References and Bibliography

Folk, R. L. 1974. *Petrology of Sedimentary Rocks*, Hemphill Publishing Company, Austin, TX, pp. 183.

Friedman, G. M., and Sanders, J. E. 1978. *Principles of Sedimentology*, John Wiley & Sons, New York, NY, Chapter 3.

Hobson, R. D. 1977. "Review of Design Elements for Beach Fill Evaluation," Technical Paper 77-6, US Army Engineer Waterways Experiment Station, Vicksburg, MS.

James, W. R. 1974. "Beach Fill Stability and Borrow Material Texture," *Proceedings of the j4th International Conference on Coastal Engineering*, American Society of Civil Engineers, pp. 1334–1349.

James, W. R. 1975. "Techniques in Evaluating Suitability of Borrow Material for Beach Nourishment," Technical Memorandum No. 60, US Army Engineer Waterways Experiment Station, Vicksburg, MS.

Krumbein, W. C. 1934. "Size Frequency Distribution of Sediments," *Journal of Sedimentary Petrology*, Vol. 4, pp. 65–77.

Krumbein, W. C. 1938. "Size Frequency Distributions of Sediments and the Normal Phi Curve," *Journal of Sedimentary Petrology*, Vol. 18, pp. 84–90.

Krumbein, W. C. 1957. "A Method for Specification of Sand for Beach Fills," Technical Memorandum No. 102, Beach Erosion Board, US Army Engineer Waterways Experiment Station, Vicksburg, MS.

Moussa T. M. 1977. "Phi Mean and Phi Standard Deviation of Grain-Size Distribution in Sediments: Method of Moments," *Journal of Sedimentary Petrology*, Vol. 47, No. 3, pp. 1295–1298.

Shore Protection Manual. 1984. 4th ed., 2 Vols., US Army Engineer Waterways Experiment Station, Coastal Engneering Research Center, US Government Printing Office, Washington, DC, Chapter 5, pp. 6–24.

BEACH NOURISHMENT OVERFILL RATIO AND VOLUME

Description

The methodologies represented in this ACES application provide two approaches to the planning and design of nourishment projects. The first approach is the calculation of the *overfill ratio*, which is defined as the volume of actual borrow material required to produce a unit volume of usable fill. The second approach is the calculation of a *renourishment factor*, which is germane to the long-term maintenance of a project, and addresses the basic question of how often renourishment will be required if a particular borrow source is selected that is texturally different from the native beach sand.

Introduction

Beaches can effectively dissipate wave energy and are classified as shore protection structures of adjacent uplands when maintained at proper dimensions. Existing beaches are part of the natural coastal system, and their wave dissipation usually occurs without creating adverse environmental effects. Since most beach erosion problems occur when there is a deficiency in the natural supply of sand, the placement of borrow material on the shore should be considered as one shore stabilization measure.

An important question arises in these situations with respect to the volume of material taken from the borrow zone that will be effectively lost from the project during and following placement due to sorting processes. The design engineer is required to estimate the proportion of a proposed borrow material that will serve a useful function in a specific project requiring beach fill.

Quantitative methods for evaluating the suitability of borrow material as beach fill are those that give overfill ratios or factors or renourishment factors. Overfill ratios apply when the beach in the project area is expected to be stable if composed of sand of certain characteristics, or will be stabilized with engineering structures. Renourishment factors apply where the beach is undergoing erosion and the project requires periodic nourishment for beach stabilization. The methods described can be found in James (1975) and the SPM (1984).

General Assumptions and Limitations

The overfill ratio and renourishment factor described herein are not physically related. Each value results from unique models of predicted beach-fill behavior that are dissimilar, although both use the comparison of native and borrow sand texture as input. The overfill ratio is calculated on the assumption that some portion of the borrow material is absolutely stable and hence a finite proportion of the original material will remain on the beach indefinitely. The renourishment factor is calculated on the opposing assumption that no material is absolutely stable, but that a finer material is less stable than coarse material, and hence a coarse beach fill will require renourishment less frequently than a fine one. The models address the different problems in determining nourishment requirements when fill that is dissimilar to native sediments is to be used (overfill ratio) and in predicting how quickly a particular fill will erode (renourishment factor). For design purposes, the overfill ratio, R_A, should be applied to adjust both initial and renourishment volumes. The renourishment factor, R_J, should be considered an independent evaluation of when renourishment will be required.

It is recommended that selection of borrow material be based on all available historical information on the project area. Computations such as those performed by this ACES application provide useful supplemental inputs for planning and designing, but should not be regarded as accurate predictors. Engineering judgment and experience must accompany design application.

The procedures described herein require enough core borings and samples in the borrow zone, on the beach, and in the nearshore zones to adequately describe the size distribution of borrow and beach material. Size analyses of the borings and samples are used to compute composite size distributions for the two types of materials. These composite distributions are compared to determine the suitability of the borrow material. The concept of composite material properties is discussed in Krumbein (1957) and Hobson (1977).

Soil Classification

A grade scale most commonly used for classifying sediments is the phi (ϕ) scale devised by Krumbein (1934, 1938). The phi scale is a geometric scale to the power of 2 with individual size classes defined as "twice as large or half as large" as some other class (Table A-4, Appendix A). The phi transformation is given by:

$$\phi = -\log_2 d \quad \text{or} \quad 2^{-(\phi)} = d \tag{1}$$

where

d = grain-size particle diameter in millimeters

The advantages of the phi scale are

- The distribution of particle sizes can be plotted on arithmetic graph paper, obviating the necessity of logarithmic graph paper; thus the graphical method of computing sediment statistics is simplified.

- Particle diameters for each size class become whole numbers instead of fractions of millimeters.

The disadvantages of the phi scale are

- Frequent unfamiliarity on the part of users.
- No intuitive relation to numerical results of sieve analysis.
- Progression of numerical scales is counter-intuitive (larger value means smaller grain size).

Sediments are classified as GRAVEL, COARSE SAND, MEDIUM SAND, FINE SAND, SILT, and CLAY to indicate the dominant particle size diameter of the sample. The two size schemes used today by coastal engineers are the Unified and Wentworth Soils classifications. These classifications assign similar but different size ranges to each category (Table B-4, Appendix B).

Grain-Size Distributions

The first stage in specifying beach-fill material is analysis of the sampling material from the existing beach to determine grain-size characteristics (*grain-size distribution*) that will be used as a basis for specification of the fill.

There are two basic methods of obtaining statistical parameters of a *grain-size distribution*. The first is known as the Folk Graphic Method, which was developed before the advent of advanced computers. The Folk method involves plotting a cumulative curve of the sample and determining the particle diameter that corresponds with certain cumulative percentages, i.e. 5, 16, 25, 50, 75, 84, and 95. The second method of obtaining statistical parameters is called the Method of Moments. It is a computational (not graphic) method using every size class, and it provides a more accurate measure than the graphic method, which relies on only a few selected percentages. These two methods are described in more detail in the ACES application titled "**Calculation of Composite Grain-Size Distributions.**"

Phi Mean and Phi Sorting

Grain-size distributions are usually characterized by two parameters.

- The phi mean, M_ϕ, which is a measure of the location of the central tendency of the *grain-size distribution*.

- The phi sorting or phi standard deviation, σ_ϕ, which is a measure of the gradation or scale of the spread of the grain size about the phi mean. A low value (< 0.5 ϕ) indicates that the *grain-size distribution* contains only a narrow range of sizes; it is *well sorted* or *poorly graded*. Conversely, a high value of phi sorting (> 1.0 ϕ) indicates the presence of a wide range of grain sizes. Material of this type is *poorly sorted* or *well graded*.

Comparison Parameters

It is convenient to define two statistical parameters used to compare the *grain-size distributions* of native and borrow materials. The first parameter, δ, is the phi mean difference and is defined as a scaled difference between borrow and native phi means:

$$\delta = \frac{M_{\phi b} - M_{\phi n}}{\sigma_{\phi n}} \tag{2}$$

where

$-_b$ = subscript *b* refers to borrow material
$-_n$ = subscript *n* refers to natural sand on beach

M_ϕ = the phi mean

$$= \frac{(\phi_{84} + \phi_{16})}{2} \tag{3}$$

ϕ_{84} = 84th percentile in phi units
ϕ_{16} = 16th percentile in phi units
σ_ϕ = standard deviation

$$= \frac{(\phi_{84} - \phi_{16})}{2} \tag{4}$$

The phi mean difference has positive values where borrow materials are, on the average, finer than native materials, and negative values where borrow materials have a phi mean coarser than that of the native materials.

The second comparison parameter, σ, is the phi sorting ratio (phi standard deviation) and is defined as phi sorting of the borrow material over phi sorting of the native material:

$$\sigma = \frac{\sigma_{\phi b}}{\sigma_{\phi n}} \tag{5}$$

Borrow materials more poorly sorted than native materials have values of σ greater than unity; where borrow materials are well sorted in comparison with native materials, σ is less than unity.

Beach-Fill Models

Two basic types of mathematical models have been proposed to handle beach-fill problems. The first model (Overfill Ratio) enables calculation of a fill factor that is an estimate of the volume of a specific fill material needed to create a unit volume of *stable* native beach material. In most cases, fill factors exceed one, indicating that the particular borrow material is less than ideal and that winnowing processes will selectively remove unsuitable parts from the fill until it becomes compatible with existing beach sediments. The second model (Renourishment Factor) enables calculation of a factor that is used to estimate how often placement of a particular fill will be required to maintain specific beach dimensions. A more detailed and formalized discussion of the two methods can be found in James (1975) and the SPM (1984).

Overfill Ratio

There are four possible combinations that result from a comparison of the composite grain-size distribution statistical parameters (mean and standard deviation) of native material and borrow material. These are listed in Table 6-4-1.

The overfill ratio is given by

$$\frac{1}{R_A} = 1 - F\left(\frac{\theta_2 - \delta}{\sigma}\right) + F\left(\frac{\theta_1 - \delta}{\sigma}\right) + \left[\frac{F(\theta_2) - F(\theta_1)}{\sigma}\right] \exp\left\{\frac{1}{2}\left[\theta_1^2 - \left(\frac{\theta_1 - \delta}{\sigma}\right)^2\right]\right\} \tag{6}$$

where

F = integral of the standard normal curve

TABLE 6-4-1. Relationships of phi means and phi standard deviations

Category	Phi Means	Phi Standard Deviations
I	$M_{\phi b} > M_{\phi n}$ Borrow is finer than native material	$\sigma_{\phi b} > \sigma_{\phi n}$ Borrow material is more poorly sorted than native material
I	$M_{\phi b} < M_{\phi n}$ Borrow is coarser than native material	
III	$M_{\phi b} < M_{\phi n}$ Borrow is coarser than native material	$\sigma_{\phi b} < \sigma_{\phi n}$ Borrow material is better sorted than native material
IV	$M_{\phi b} > M_{\phi n}$ Borrow is finer than native material	

Case	θ_1	θ_2
I and II	$\text{Max}\left\{-1, \dfrac{-\delta}{\sigma^2 - 1}\right\}$	∞
III and IV	-1	$\text{Max}\left\{-1, 1 + \dfrac{2\delta}{1-\sigma^2}\right\}$

Renourishment Factor

The renourishment factor (R_J) is a dynamic approach to answering how beach processes can be expected to modify specific fill sediments. It is an estimate of how often renourishment will be needed and helps evaluate the long-term performance of different fill materials with regard to suitability, maintenance, and expense. The conceptual approach is that the active beach system can be viewed as a compartment which receives sediments through longshore transport and from gradual erosion of the inactive *reservoir* of sediments that form the backshore. The compartment loses sediments by longshore and offshore transport beyond its boundaries. In this scheme, a fill is viewed essentially as an increase to the backshore reservoir. Sediment particle residence time in the compartment is longer for coarse-grained sediments than for fine; thus, a comparison between *composite size distributions* of native and borrow sediments can be used to predict the lifetime of a fill. The scheme thus becomes a "bookkeeping" problem of monitoring material going in and out of the system by using mass-balance equations that are similar to the more familiar sediment budget calculations.

To determine periodic renourishment requirements, James (1975) defines a renourishment factor, R_J, which is the ratio of the rate at which borrow material will erode to the rate at which natural beach material is eroding. The renourishment factor is given as

$$R_J = \exp\left[\Delta\left(\frac{M_{\phi b} - M_{\phi n}}{\sigma_{\phi n}}\right) - \frac{\Delta^2}{2}\left(\frac{\sigma_{\phi b}^2}{(\sigma_{\phi n}^2)} - 1\right)\right] \tag{7}$$

where

Δ = winnowing function = 1.0 (recommended value)

A renourishment factor of one-third implies that the borrow material is three times as stable as the native, or that renourishment with this borrow material would be required one-third as often as renourishment with nativelike sediments. However, an R_J of 3 indicates the borrow is one-third as stable, and if used as beach fill, will require renourishment three times as often as the nativelike sediments.

References and Bibliography

Hobson, R. D. 1977. "Review of Design Elements for Beach Fill Evaluation," Technical Paper 77-6, US Army Engineer Waterways Experiment Station, Vicksburg, MS.

James, W. R. 1974. "Beach Fill Stability and Borrow Material Texture," *Proceedings of the 14th International Conference on Coastal Engineering*, American Society of Civil Engineers, pp.1334–1349.

James, W. R. 1975. "Techniques in Evaluating Suitability of Borrow Material for Beach Nourishment," Technical Memorandum No. 60, US Army Engineer Waterways Experiment Station, Vicksburg, MS.

Krumbein, W. C. 1934. "Size Frequency Distribution of Sediments," *Journal of Sedimentary Petrology*, Vol. 4, pp. 65–77.

Krumbein, W. C. 1938. "Size Frequency Distributions of Sediments and the Normal Phi Curve," *Journal of Sedimentary Petrology*, Vol. 18, pp. 84–90.

Krumbein, W. C. 1957. "A Method for Specification of Sand for Beach Fills," Technical Memorandum No. 102, Beach Erosion Board, US Army Engineer Waterways Experiment Station, Vicksburg, MS.

Shore Protection Manual. 1984. 4th ed., 2 Vols., US Army Engineer Waterways Experiment Station, Coastal Engineering Research Center, US Government Printing Office, Washington, DC, Chapter 5, pp. 6–24.

A SPATIALLY INTEGRATED NUMERICAL MODEL FOR INLET HYDRAULICS

Description

This application is a numerical model that estimates coastal inlet velocities, discharges, and bay levels as functions of time for a given time-dependent sea level fluctuation. Inlet hydraulics are predicted in this model by simultaneously solving the time-dependent momentum equation for flow in the inlet and the continuity equation relating the bay and sea levels to inlet discharge. The model is designed for cases where the bay water level fluctuates uniformly throughout the bay and the volume of water stored in the inlet between high and low water is negligible compared with the tidal prism of water that moves through the inlet and is stored in the bay. The model has been previously described by Seelig (1977) and Seelig, Harris, and Herchenroder (1977).

Introduction

Quick, inexpensive estimates of inlet velocities and bay water surface levels associated with tidal or storm event sea level fluctuations are needed in planning the design, construction, and maintenance of coastal inlets. Field data are often unavailable or incomplete. In addition, hydraulic characteristics for proposed inlets are unavailable and must be predicted. The numerical model described herein will predict bay levels, inlet velocities, and discharge as a function of time given the geometry of the system and the water level fluctuations in the sea.

An inlet-bay system typically consists of a *sea* (ocean or lake) connected to a *bay* by one or more *inlets*. Possible system configurations considered in this version of the model include:

- 1-Sea–1-Inlet–1-Bay System
- 1-Sea–2-Inlet–1-Bay System
- 2-Sea Boundary Conditions–2-Inlet–1-Bay System

AUTOMATED COASTAL ENGINEERING SYSTEM (ACES)   A.137

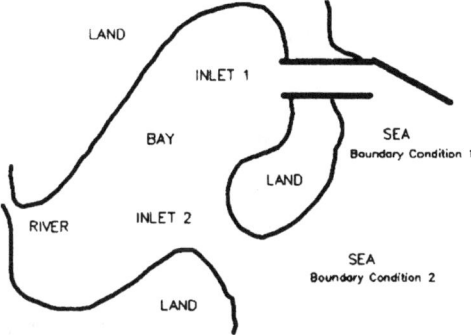

FIGURE 7-1-1. Conceptual 2-sea, 2-inlet, and 1-bay system.

Figure 7-1-1 depicts a conceptual two-sea, two-inlet, and one-bay system.
Seaward boundary conditions for the model are specified as water level fluctuations associated with astronomical tides, storm surges, seiches, and tsunamis.

General Assumptions and Limitations

General assumptions and limitations for the model include:

- Sea level is specified as a function of time.
- Bay water level remains horizontal.
- The bay is connected to the sea by one or two inlets.
- Bay water surface area is a function of bay water level.
- The water level slope in the inlet is assumed to be linearly related to the friction loss along the inlet between the sea and the bay levels.
- There is a water level drop along the inlet that is proportional to the unrecovered velocity head lost through turbulent eddy diffusion in the bay (flood flow) or sea (ebb flow).
- Storage of water in the inlet is negligible.
- Wind stress on the inlet and bay surface is negligible.
- Water has constant properties throughout the inlet and bay.
- Radiation stress and Coriolis effects are neglected.
- The bay and inlet must contain water throughout the water level cycle.
- The simulation begins with a sea level of zero and zero current.

Numerical Model

This model is based upon an area-averaged, one-dimensional momentum equation for the inlet and a continuity equation for the bay.

Momentum Equation

The model is primarily based upon the 1-D momentum equation through the inlet (Harris and Bodine, 1977):

$$\frac{\partial \bar{u}}{\partial t} + \frac{1}{2}\frac{\partial}{\partial x}\bar{u}^2 + g\frac{\partial h}{\partial x} + \frac{1}{A_c}\int_{y_1}^{y_2}(\tau_{zx})_z\,dy = 0 \qquad (1)$$

where

$\bar{u}$ = cross-sectional mean water velocity in the inlet (positive for flood flow, negative for ebb flow)
t = time
x = distance along the main axis of inlet
g = gravitational acceleration
h = water level above datum
A_c = inlet cross-section flow area at x
$(\tau_{zx})_z$ = bottom stress tensor component in the direction of the main axis of the inlet

The first term in the above equation is the temporal acceleration, the second term is the advective acceleration, the third term is the slope of the water surface along the inlet, and the fourth term represents bottom friction. The equation is integrated over the problem domain (length of the inlet between the sea and the bay), with x_s, and x_b, as the respective limits:

$$\int_{x_s}^{x_b}\frac{\partial \bar{u}}{\partial t}dx + \int_{x_s}^{x_b}\frac{1}{2}\frac{\partial \bar{u}^2}{\partial x}dx + \int_{x_s}^{x_b}g\frac{\partial h}{\partial x}dx + \int_{x_s}^{x_b}\frac{1}{A_c}\int_{y_1}^{y_2}(\tau_{zx})_z\,dy\,dx = 0 \qquad (2)$$

Carrying out some of the integrations and rearranging, this equation becomes

$$\frac{\partial}{\partial t}\int_{x_s}^{x_b}\bar{u}\,dx + \frac{1}{2}(\bar{u}_b^2 - \bar{u}_s^2) + g(h_b - h_s) + \int_{x_s}^{x_b}\frac{1}{A_c}\int_{y_1}^{y_2}(\tau_{zx})_z\,dy\,dx = 0 \qquad (3)$$

where

h_b, h_s = water levels at the bay and seaward boundaries

From continuity the cross-sectional mean inlet water velocity is equal to the inlet discharge, Q, divided by the inlet cross-sectional area, A_c.

$$\bar{u} = \frac{Q}{A_c} \qquad (4)$$

Substituting Equation 4 into 3 and using the product rule for integration, the temporal acceleration of Equation 3 can be expanded to

$$\int_{x_s}^{x_b}\frac{\partial\left(\frac{Q}{A_c}\right)}{\partial t}dx = \frac{\partial Q}{\partial t}\int_{x_s}^{x_b}\frac{dx}{A_c} + Q\frac{\partial}{\partial t}\left(\int_{x_s}^{x_b}\frac{dx}{A_c}\right) \qquad (5)$$

where the second term

$$Q\frac{\partial}{\partial t}\left(\int_{x_s}^{x_b}\frac{dx}{A_c}\right) = 0$$

since channel storage is neglected.

Substitution of Equation 4 into 3 yields the following expression for the sum of the advective and slope surface terms:

$$\frac{1}{2}\left(\frac{1}{A_b^2} - \frac{1}{A_s^2}\right)Q^2 + g(h_b - h_s) \qquad (6)$$

where

A_b, A_s = cross-sectional areas of the inlet at the bay and seaward boundaries

The bottom friction term of Equation 3 is evaluated by using Manning's equation:

$$(\tau_{zx})_z = \frac{gn^2}{kd^{1/3}}|u|u \qquad (7)$$

where

n = Manning's coefficient of friction
  = $C1 - C2\, d$

	4ft < d < 30ft		d < 4ft
C1	0.037770		0.055
C2	0.000667		0.005

d = water depth
k = conversion factor to adapt Manning's equation to the applicable system of units
u = water velocity in the inlet

The bottom stress is approximated by determining the water velocity, u, at a number of locations throughout the inlet. The inlet is discretized into a flow net or grid consisting of channels (j) and cross sections (i). The major axis of each channel is drawn approximately perpendicular to the flow (Figure 7-1-2). Inlet cross sections are indexed from $i = 1$ (seaward boundary) to $i = IS$ (last cross section at the bay end of the inlet). Channels are indexed $j = 1 \ldots IC$ from left to right across the inlet from a seaward perspective. A typical grid cell, denoted as cell (i, j), consists of that part of the channel j situated between

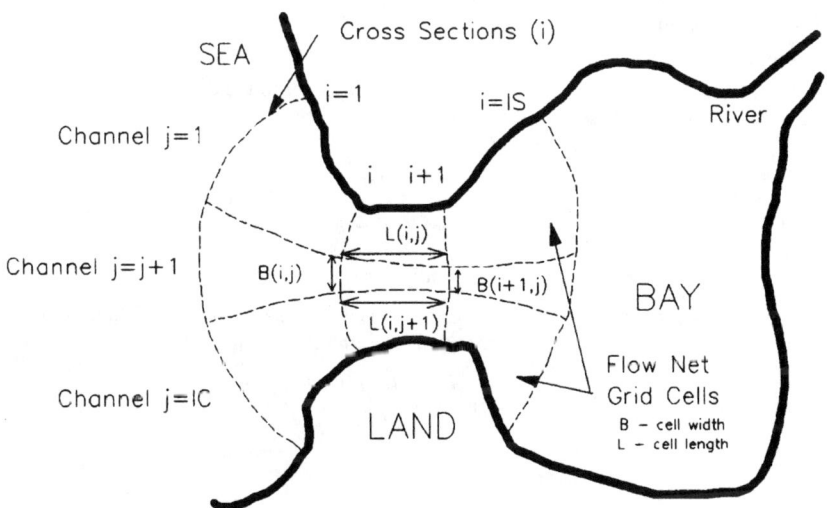

FIGURE 7-1-2. Inlet grid system.

cross sections i and $i+1$. The water velocity in the cell, u_{ij}, is assumed to be at the center of cell (i,j) and to act parallel to the axis of channel j.

A weighting function W_j is used to determine the fraction of the total inlet flow Q which passes through a channel at an instant in time.

$$Q_{ij} = W_j Q \tag{8}$$

where

Q_{ij} = discharge in cell (i,j)

The weighting function provides a systematic method of distributing flow throughout the inlet for use in evaluating the bottom stress. Major assumptions in developing the weighting function include:

- Flow is parallel to the streamlines of the flow net.
- Discharge is distributed in each channel of the inlet to minimize total friction within the solution domain.

The weighting function per channel is defined as

$$W_j = \frac{C_j}{\sum_{i=j}^{IC} C_j} \tag{9}$$

where

$$C_j = \frac{A_j^2 d_j^{1/3}}{n_j^2 Q^2 B_j L_j} \tag{10}$$

A_j = channel cross-sectional area
d_j = water depth of the channel
n_j = Manning's coefficient of friction
Q = inlet discharge
B_j = channel width
L_j = channel length

The mean water velocity in a cell u_{ij} is assumed equal to the discharge in the cell divided by the mean cross-sectional area of the cell perpendicular to flow:

$$u_{ij} = W_j \frac{Q}{A_{ij}} \tag{11}$$

where

A_{ij} = cross-sectional area of the cell perpendicular to flow

The total bottom friction term for Equation 3 is evaluated by using Equations 7 and 11 and integrating across the cross sections and through the inlet to yield

$$F = \sum_{i=1}^{IS-1} \frac{1}{\sum_{j=1}^{IC} A_{ij}} \sum_{j=1}^{IC} \frac{g n_{ij}^2 |W_j Q| W_j Q B_{ij} L_{ij}}{k d_{ij}^{1/3} A_{ij}^2} \tag{12}$$

where

n_{ij} = Manning's coefficient for a cell
B_{ij} = mean cell width

L_{ij} = mean cell length
d_{ij} = mean instantaneous water depth in a cell

All of the parameters are defined at the center of the cell. For convenience, let I_g be a geometry integral defined as

$$I_g = \frac{1}{\int_{x_s}^{x_b} \frac{dx}{A_c}} = \frac{1}{\sum_{i=1}^{IS-1}\left(\frac{\sum_{j=1}^{IC} \frac{L_{ij}}{IC}}{\sum_{j=1}^{IC} A_{ij}}\right)} \quad (13)$$

Substituting Equations 5, 6, and 12 into Equation 3 and multiplying by I_g results in a convenient form for the spatially integrated momentum equation:

$$\frac{dQ}{dt} = -\frac{I_g}{2}\left(\frac{1}{A_b^2} - \frac{1}{A_s^2}\right)Q^2 - gI_g(h_b - h_s) - I_g F \quad (14)$$

In this version of the model, the convective acceleration (first term to the right of the equals sign in Equation 14) is expressed in terms of the empirical loss coefficients. A simple approach (Seelig, Harris, and Herchenroder, 1977) is taken to approximately consider this loss and further simplify the momentum equation. The assumptions are:

- A controlling cross section (throat) exists for the inlet where the cross-sectional area is minimum and losses are effectively encountered.
- Water velocity far from the controlling cross section of the inlet is small.
- The loss can be approximated by multiplying the convective term by a combined empirical ebb- and flood-loss coefficient, C_D (Keulegan, 1967) evaluated at the controlling cross section.

The spatially integrated 1-D momentum equation for flow in the inlet then is

$$\frac{dQ}{dt} = \frac{I_g}{2} C_D \frac{Q|Q|}{(A_{\min})^2} - gI_g(h_b - h_s) - I_g F \quad (15)$$

where

$A_{\min}$ = minimum inlet cross-sectional area (at the throat)

Continuity Equation

The rate of change of water level in the bay is related to inlet discharge plus discharge into the bay from other sources by a simple mass continuity equation:

$$\frac{dh_b}{dt} = \frac{Q_T}{A_{\text{bay}}} + \frac{Q_{\text{inflow}}}{A_{\text{bay}}} \quad (16)$$

where

Q_T = total inlet discharge 2
$= \sum_{m=1}^{2} Q_m$

Q_m = discharge of the mth inlet (limited to two inlets)

A_{bay} = instantaneous surface area of the bay
$= A_0(1 + \beta h_b)$
Q_{inflow} = discharge into bay from sources other than the inlet(s) (such as rivers, bayous, pumped inflows, etc.)
A_0 = initial surface area of the bay
β = bay variation parameter

Solution Method

Equations 15 and 16 comprise a set of simultaneous differential equations. There are several methods for solving these differential equations. The numerical technique used here is a fourth order Runge-Kutta-Gill method. Advantages of this method are that it is self-starting, extremely stable, can handle a long time step, has wide application, and converges quickly. The details of the method are not presented here. The original scheme adopted by Seelig, Harris, and Herchenroder (1977) was a modification of routines published by International Business Machines (1970).

References and Bibliography

Harris, D. L., and Bodine, B. R. 1977. "Comparison of Numerical and Physical Models, Masonboro Inlet, North Carolina," CERC GITI Report 6, US Army Engineer Waterways Experiment Station, Vicksburg, MS.

International Business Machines. 1970. "System/360 Scientific Subroutine Package, Version II Programmer's Manual," White Plains, NY.

Keulegan, G. H. 1967. "Tidal Flow in Entrances, Water-Level Fluctuations of Basins in Communication with Seas," Technical Bulletin No. 14, Committee on Tidal Hydraulics, US Army Corps of Engineers, Vicksburg, MS.

Masch, F. D., Brandes, R. J., and Reagan, J. D. 1977. "Numerical Simulation of Hydrodynamics (WRE)," Appendix 2, CERC GITI Report 6, US Army Engineer Waterways Experiment Station, Vicksburg, MS.

Seelig, W. N. 1977. "A Simple Computer Model for Evaluating Coastal Inlet Hydraulics," CERC CETA 77-1, US Army Engineer Waterways Experiment Station, Vicksburg, MS.

Seelig, W. N., Harris, D. L., and Herchenroder, B. E. 1977. "A Spatially Integrated Numerical Model of Inlet Hydraulics," CERC GITI Report 14, US Army Engineer Waterways Experiment Station, Vicksburg, MS.

MISCELLANEOUS ROUTINES

Introduction

This section of the reference manual contains descriptions of miscellaneous routines in the ACES package that support several methodologies throughout the software. The nature of these routines is to provide checks associated with wave stability to assure compliance with limiting assumptions shared by several methodologies.

Wave Steepness

The maximum height of a wave is limited by a maximum wave steepness for which the wave form can remain stable. Waves reaching the limiting steepness will begin to break and dissipate a portion of their energy. A simple formula for approximating a limiting wave steepness in uniform finite depths was proposed by Miche (1944):

$$\frac{H}{L} = 0.142 \tanh(kd) \quad (1)$$

where

H = wave height
L = wavelength
k = wave number = $2\pi/L$
d = still-water depth

Monochromatic Wave Breaking

A number of methodologies assume unbroken monochromatic wave conditions. Wave breaking is a significant yet complex phenomenon difficult to measure or accurately predict. No one theory or empirical expression adequately addresses the variability of observed data, yet some estimate of the process is required for design purposes. The following sections discuss the breaker criteria adopted for nearshore regions as well as at structure locations within ACES.

Nearshore Region

For situations where the nearshore slope is considered flat ($m = 0$), breaker height H_b is estimated using the expression (McCowan, 1894)

$$H_b = 0.78d \quad (2)$$

where

d = water depth

The above expression is also used when nearshore slope is unknown.

More recent estimates for breaker parameters in the nearshore region include the effects of wave steepness (H_0/L_0) and beach slope (m). Expressions for either breaker height or breaker index ($\gamma_b = H_b/d_b$) as functions of (H_0/L_0, m) have been given by Iversen (1952), Weggel (1972), Singamsetti and Wind (1980), and Sunamura (1981). Smith (1986) summarized additional data available from seven other independent investigations. Testing of the expressions of the above investigators against these additional data resulted in a selection for the expressions used in ACES. It is reemphasized that these expressions provide only rough estimates for design purposes.

For cases where a finite nearshore slope is known ($m > 0$), the breaker height is estimated as (Singamsetti and Wind, 1980):

$$H_b = H_0 \left[0.575 m^{0.031} \left(\frac{H_0}{L_0} \right)^{-0.254} \right] \quad (3)$$

where

H_0 = deepwater wave height
$m = \tan\phi$ = nearshore slope
L_0 = deepwater wavelength

and the breaker depth, d_b, is estimated as (Weggel, 1972):

$$d_b = \frac{H_b}{b - a\frac{H_b}{T^2}} \quad (4)$$

where

$$b = \frac{1}{0.64(1 + e^{-19.5m})}$$
$$a = 1.36(1 - e^{-19m})$$
T = wave period

Structure Vicinity
The maximum breaker height to which a structure might be subjected is estimated as (Weggel, 1972)

$$H_b = \frac{d_s}{ma(18.5m - 8)}\left[P - \sqrt{P^2 \frac{4mba}{(d_s/T^2)}(9.25m - 4)}\right] \quad (5)$$

where
d_s = water depth at the structure
$$P = a + \frac{1 + 9.25m^2b - 4mb}{\left(\dfrac{d_s}{T^2}\right)}$$
$$a = 1.36(1 - e^{-19m})$$
$$b = \frac{1}{0.64(1 + e^{-19.5m})}$$

References and Bibliography

Iversen, H. W. 1952. "Laboratory Study of Breakers," Circular 521, National Bureau of Standards, Washington, DC, pp. 9–32.
McCowan, J. 1894. "On the Highest Wave of Permanent Type," *Philosophical Magazine*, 5th Series, Vol. 38, pp. 351–357.
Miche, R. 1944. "Mouvements ondulatoires de la mer en profondeur constante ou decroissante," *Annales des Ponts et Chaussees*, Vol. 114.
Munk, W. H. 1949. "The Solitary Wave Theory and Its Application to Surf Problems," *Annals of the New York Academy of Sciences*, Vol. 51, pp. 376–462.
Singamsetti, S. R., and Wind, H. G. 1980. "Characteristics of Shoaling and Breaking Periodic Waves Normally Incident to Plane Beaches of Constant Slope," Breaking Waves Publication No. M1371, Waterstaat, The Netherlands, pp. 23–27.
Smith, J. M. 1986 (Dec). "An Analytical Model of Wave-Induced Longshore Current Based on Power Law Wave Height Decay," M.S. Thesis, Mississippi State University, Mississippi State, MS.
Sunamura, T. 1981. "A Laboratory Study of Offshore Transport of Sediment Transport and a Model for Eroding Beaches," *Proceedings of the 17th Coastal Engineering Conference*, American Society of Civil Engineers, pp. 1051–1070.
Weggel, J. R. 1972. "Maximum Breaker Height," *Journal of Waterways, Harbors and Coastal Engineering Division*. American Society of Civil Engineers, Vol. 98, No. WW4, pp. 529–548.

APPENDIX B
TABLES

TABLE B-1. K_D values for use in determining armor unit weight (source: EM 1110-2-2904)

| Armor Units | $n^{(2)}$ | Placement | Structure Trunk[7] | | Structure Head | | Slope cot θ |
			Breaking Wave	Nonbreaking Wave	Breaking Wave	Nonbreaking Wave	
Quarrystone							
Smooth rounded	2	Random	$1.2^{(1)}$	2.4	$1.1^{(1)}$	1.9	$1.5-3.0^{(8)}$
Smooth rounded	>3	Random	$1.6^{(1)}$	$3.2^{(1)}$	$1.4^{(1)}$	$2.3^{(1)}$	$1.5-3.0^{(8)}$
Rough angular	1	Random[3]	—[3]	$2.9^{(1)}$	—[3]	$2.3^{(1)}$	$1.5-3.0^{(8)}$
					$1.9^{(1)}$	3.2	1.5
Rough angular	2	Random	2.0	4.0	$1.6^{(1)}$	2.8	2.0
					1.3	2.3	3.0
Rough angular	>3	Random	$2.2^{(1)}$	$4.5^{(1)}$	$2.1^{(1)}$	$4.2^{(1)}$	$1.5-3.0^{(8)}$
Rough angular	2	Special[4]	5.8	7.0	$5.3^{(1)}$	6.4	$1.5-3.0^{(8)}$
Parallelepiped[9]	2	Special	7.0–20.0	$8.5-24.0^{(1)}$	—	—	1.0–3.0
Tetrapod					$5.0^{(1)}$	6.0	1.5
and	2	Random	7.0	8.0	$4.5^{(1)}$	5.5	2.0
Quadripod					$3.5^{(1)}$	4.0	3.0
					$8.3^{(1)}$	9.0	1.5
Tribar	2	Random	$9.0^{(1)}$	10.0	$7.8^{(1)}$	8.5	2.0
					6.0	6.5	3.0
Dolos	2	Random	$15.0^{(6)}$	$31.0^{(6)}$	$8.0^{(1)}$	$16.0^{(1)}$	$2.0^{(5)}$
					7.0	$14.0^{(1)}$	3.0
Modified cube	2	Random	$6.5^{(1)}$	7.5	—	$5.0^{(1)}$	$1.5-3.0^{(8)}$
Hexapod	2	Random	$8.0^{(1)}$	9.5	$5.0^{(1)}$	$7.0^{(1)}$	$1.5-3.0^{(8)}$
Toskane	2	Random	$11.0^{(1)}$	22.0	—	—	$1.5-3.0^{(8)}$
Tribar	1	Uniform	12.0	15.0	$7.5^{(1)}$	$9.5^{(1)}$	$1.5-3.0^{(8)}$
Quarrystone-graded angular riprap	—	Random	2.2	2.5	—	—	—

(1) *CAUTION:* These K_D values are unsupported and are provided only for preliminary design.
(2) n is the number of units comprising the thickness of the armor layer.
(3) The use of single layer of quarrystone armor units is not recommended for structures subject to breaking waves, and only under special conditions for structures subject to nonbreaking waves. When it is used, the stone should be carefully placed.
(4) Special placement with long axis of stone placed perpendicular to structure face.
(5) Stability of dolosse on slopes steeper than 1 on 2 should be substantiated by site-specific tests.
(6) Refers to no-damage criteria (<5 percent displacement, rocking, etc.); if no rocking (<2 percent) is desired, reduce K_D 50 percent (Zwamborn and Van Niekerk, 1982).
(7) Applicable to slopes ranging from 1 on 1.5 to 1 on 5.
(8) Until more information is available, the use of K_D should be limited to slopes ranging from 1 on 1.5 to 1 on 3. Some armor units tested on a structure head indicate a K_D-slope dependence.
(9) Parallelepiped-shaped stone: long slab-like stone with long dimension approximately three times the shortest dimension (Markle and Davidson, 1979).

TABLE B-2. Layer coefficient and porosity for various armor units (source: SPM)

Armor Unit	n	Placement	Layer Coefficient	Porosity %
Quarrystone (smooth)	2	Random	1.02	38
Quarrystone (rough)	2	Random	1.00	37
Quarrystone (rough)	>3	Random	1.00	40
Quarrystone (parallelepiped)	2	Special	—	27
Cube (modified)	2	Random	1.10	47
Tetrapod	2	Random	1.04	50
Quadripod	2	Random	0.95	49
Hexipod	2	Random	1.15	47
Tribar	2	Random	1.02	54
Dolos	2	Random	0.94	56
Toskane	2	Random	1.03	52
Tribar	1	Uniform	1.13	47
Quarrystone	Graded	Random	—	37

TABLE B-3. Rough slope run-up coefficients (source: Smith, 1986)

Armor Material	a	b
Riprap	0.956	0.398
Rubble (Permeable—No Core)	0.692	0.504
Rubble (2 Layers—Impermeable Core)	0.775	0.361
Modified Cubes	0.950	0.690
Tetrapods	1.010	0.910
Quadripods	0.590	0.350
Hexapods	0.820	0.630
Tribars	1.810	1.570
Dolosse	0.988	0.703

TABLE B-4. Grain-size scales (soil classification)

	Unified Soils Classification	ASTM Mesh	PHI	MM	Wentworth Classification	
			–8.00	256.00		
	Cobble		–7.00	128.00		
			–6.75	107.60	Cobble	
			–6.50	90.51		
			–6.25	76.11		
			–6.00	64.00		
			–5.75	53.82		
	Coarse Gravel		–5.50	45.26		G
			–5.25	38.06		
			–5.00	32.00		R
			–4.75	26.91		
			–4.50	22.63		A
			–4.25	19.00		
			–4.00	16.00	Pebble	V
			–3.75	13.45		
	Fine Gravel		–3.50	11.31		E
			–3.25	9.51		
		2.5	–3.00	8.00		L
		3	–2.75	6.73		
		3.5	–2.50	5.66		
		4	–2.25	4.76		
		5	–2.00	4.00		
	Coarse	6	–1.75	3.36		
		7	–1.50	2.83	Granule	
		8	–1.25	2.38		
		10	–1.00	2.00		
		12	–0.75	1.68		
		14	–0.50	1.41	Very Coarse	
		16	–0.25	1.19		
S	Medium	15	0.00	1.00		
		20	0.25	0.84		
A		25	0.50	0.71	Coarse	
		30	0.75	0.59		5
N		35	1.00	0.50		
		40	1.25	0.42		A
D		45	1.50	0.35	Medium	
		50	1.75	0.30		N
		60	2.00	0.25		
		70	2.25	0.21		D
	Fine	80	2.50	0.177	Fine	
		100	2.75	0.149		
		120	3.00	0.125		
		140	3.25	0.105		
		170	3.50	0.088	Very Fine	
		200	3.75	0.074		
		230	4.00	0.0625		
		270	4.25	0.0526		
		325	4.50	0.0442		
	Silt	400	4.75	0.0372	Silt	M
			5.00	0.0313		
			6.00	0.0156		U
			7.00	0.0078		
			8.00	0.0039		D
			9.00	0.0020		
	Clay		10.00	0.0009	Clay	
			12.00	0.0002		

TABLE B-5. Major tidal constituents

Symbol	Constituent Name	Frequency (degrees/hour)
M_2	Lunar semidiurnal	28.984
S_2	Principal solar semidiurnal	30.000
N_2	Larger lunar elliptic semidiurnal	28.439
K_1	Lunisolar diurnal	15.041
M_4	Shallow-water overtide of principal lunar	57.968
O_1	Principal lunar diurnal	13.943
M_6	Shallow-water overtide of principal lunar	86.952
MK_3	Shallow-water compound	44.025
S_4	Shallow-water overtide of principal solar	60.000
MN_4	Shallow-water compound	57.423
ν_2	Larger lunar evectional	28.512
S_6	Shallow-water overtide of principal solar	90.000
$\mu 2$	Variational	27.968
$2N_2$	Lunar elliptic aemidiurnal (second order)	27.895
00_1	Lunar diurnal (second order)	16.139
$\lambda 2$	Smaller lunar evectional	29.455
S_1	Solar diurnal	15.000
M_1	Smaller lunar elliptic diurnal	14.496
J_1	Smaller lunar elliptic diurnal	15.585
M_m	Lunar monthly	0.544
S_{sa}	Solar semidiurnal	0.082
S_a	Solar annual	0.041
M_{sf}	Luninsolar synodic fortnightly	1.015
M_f	Lunar fortnightly	1.098
$\rho 1$	Larger lunar evectional diurnal	13.471
Q_1	Larger lunar elliptic diurnal	13.398
T_2	Larger solar elliptic	29.958
R_1	Smaller solar elliptic	30.041
$2Q_1$	Lunar elliptic diurnal (second order)	12.854
P_1	Solar diurnal	14.958
$2SM_2$	Shallow-water compound	31.015
M_3	Lunar terdiurnal	43.476
L_2	Smaller lunar elliptic semidiurnal	29.528
$2MK_3$	Shallow-water compound	42.927
K_2	Lunisolar semidiurnal	30.082
M_8	Shallow-water overtide of principal lunar	115.936
MS_4	Shallow-water compound	58.984

AUTHOR INDEX

Aagaard, T., 7.19
Abbott, M. B., 1.5, 1.50, 1.59
Abdala, S., 4.25
Abdelrahman, S. M., 7.17
Aberg, B., 14.15, 14.16
Abramson, L. W., 14.10
Aburn, J. H., 19.6, 19.11, 19.14
Ackers, P., 16.12
Adams, C., 5.14, 5.65
Adams, W. J., 17.105
Ahrens, J. P., 5.67, 5.75, 8.39, 8.40
Alfageme, S., 8.29
Allan, R. J., 17.110
Allen, J. R. L., 16.12
Allsop, N. W. H., 2.1, 2.12, 2.19, 4.4, 4.5, 4.8, 4.11, 4.14, 4.15, 4.22-4.24, 4.26, 4.28–4.30, 4.32-4.34, 4.40, 4.41, 6.8, 6.9, 6.13, 6.15, 6.16
Al-Salem, A. A., 7.5
Anastasiou, K., 1.50
Aoki, H., 9.26, 9.27
Arad, M., 1.28
Asakawa, T., 9.14, 9.20, 9.21, 9.24, 9.26
Atkinson, J. A., 11.13
Aubrey, D. G., 19.10
Ault, J. S., 15.51
Averett, D. E., 13.21, 14.5

Bagnold, R. A., 4.33
Bailand, J. A., 7.19-7.22
Bakker, K. J., 3. 80
Bank, R. B., 14.12
Banyard, L., 6.7, 6.9, 6.13, 6.15
Barnett, T. P., 19.10
Barras, C. B., 11.32
Barret, M. G., 3. 92
Barth, M. C., 19.10, 19.21
Battjes, S. A., 4.18, 4.20, 4.21, 4.29-4.31
Beach, R. A., 7.16
Beatly, T., 19.5
Bedient, P. B., 14.12
Behrendet, L., 1.39
Bélorgg, M., 4.9
Benque, J. P., 1.59
Berenguer, J., 5.66, 5.67, 5.75, 8.44, 8.45
Berkeley-Thorn, R., 6.10
Berkhoff, J. C. W., 1.8, 1.9, 1.13, 1.38, 1.39
Besley, P. B., 2.16, 4.23

Bettess, P., 1.35, 1.36, 1.39
Bezuijen, A., 2.10, 2.12, 2.14-2.16, 3.61-3.65, 3.74, 3.76
Bijker, E. W., 12.3, 16.12
Bird, E. C. F., 7.1
Birkemeier, W. A., 10.14
Bixby, A., 17.115
Blaauw, H. G., 14.14
Blackmore, P. A., 4.25
Blaschak, J. G., 1.22
Bodge, K. R., 8.44, 8.45
Boesch, D. F., 19.11
Bokuniewicz, H. J., 13.9, 13.11, 16.3, 16.5, 18.3, 18.5
Borgman, L. E., 16.5, 16.11, 16.16
Bosworth, W. S., 13.26
Bowles, J. E., 8.18
Boyer, L. F., 13.17
Bradbury, A. P., 2.12, 6.13
Bradshaw, B., 19.14
Brandsma, M. G., 16.6, 16.9
Brannon, J., 13.26
Bray, R. N., 4.4, 4.5
Bridges, T. S., 15.55
Brindel, J. R., 19.7
Brunt, D. H., III, 10.9, 10.17
Bruun, P., 8.12, 9.7, 9.20, 10.12, 19.5, 19.10, 19.11, 19.13, 19.24
Buchberger, C., 17.22
Bullock, G. N., 4.9
Burcharth, H. F., 4.4
Burger, A. M., 3.62
Burton, G. A., 17.113, 17.114

Calabrese, M., 4.12, 4.22, 4.26, 4.29, 4.34, 4.40
Calderoni, A., 17.110
Camfield, F. E., 19.8
Campbell, N. P., 11.18, 11.26, 11.35, 11.47, 11.56
Campbell, T. J., 10.12
Cara, T., 15.55
Cargill, K. W., 14.9, 16.13
Chamberlain, P. G., 1.8
Chan, C. T., 1.50
Chang, Y. C., 16.6-16.8
Channell, A. R., 2.19
Chapman, P. M., 17.105
Chen, C. J., 1.24

AUTHOR INDEX

Chen, H. C., 1.24, 1.29
Chen, H. S., 1.21, 1.35, 1.38, 1.39
Chin, I., 9.32, 9.34
Chiu, T. Y., 19.7, 19.8, 19.14, 19.19
Choot, G. E. B., 18.8
Chou, L., 16.5
Churchill, J. H., 18.7
Chwang, A. T., 1.29
Clark, D., 5.65
Clarke, D. G., 15.19, 15.51
Clausner, J. E., 14.5, 14.6, 15.49, 15.53, 17.18
Clayton, T. D., 10.6, 10.29
Clifford, J. E., 4.4
Collis, L., 18.1, 18.3, 18.4, 18.8
Copeland, G. J. M., 1.9, 1.45, 1.49
Cords, D. A., 5.11, 5.31, 5.33, 5.34, 5.79
Cox, J., 5.65, 8.39, 8.40
Cox, P. J., 11.47
Crane, L. C., 11.20, 11.22
Cross, R. H., 1.21
Cruickshank, M. J., 18.8
Curtis, W. R., 9.29
Cushing, B. S., 17.102

d'Angremond, K., 13.15
Dahm, J., 19.14
Daily, J., 8.39, 8.40
Dally, W. R., 5.29, 5.62, 7.19
Dalrymple, R. A., 1.18, 5.25, 5.69, 5.72, 5.73, 8.6, 8.7, 8.12, 8.20, 8.25, 8.33, 8.39, 8.40, 10.26
Das, B. M., 14.9, 14.10
Dattatri, J., 5.12
Davidenkoff, R., 14.15
De Abreu-Lima, C., 14.15
de Groot, S. J., 18.1
de Lange, W. P., 19.14, 19.19
de Quelerij, L., 3. 55
de Waal, J. P., 3.34, 3.35, 3.40, 6.6-6.8, 6.15
Dean, J. L., 5.66
Dean, R. G., 3. 23, 7.2, 7.3, 8.6, 8.7, 8.18, 8.20, 8.25, 8.33, 8.40, 8.41, 10.6, 10.12, 10.14, 10.17-10.20, 10.22, 10.29, 19.3, 19.7-19.9, 19.14, 19.15, 19.19
Dean-Rosati, J., 8.25, 8.26
Dette, H. H., 10.25
Diaz, R. J., 17.114
Dibajma, M., 7.19, 7.20, 7.22
Dickerson, D. D., 15.51
Dingemans, M. W., 1.15
Dixon, K., 10.29
Dornhelm, R. B., 10.6
Driscoll, S. K., 15.55
Drogosz-Wawrzyniak, L., 3.33
Drucker, B. R., 18.3, 18.7
Dubois, R. N., 19.13
Durand, N., 4.15, 4.40

Earhart, G., 13.15
Ebersole, B. A., 10.23
Edge, B., 10.9
Elliot, D. O., 13.13
Elliott, G. M., 17.19
Emery, K. O., 19.10
Emmott, B., 9.2
Engquist, B., 1.22
Enriquez, F. J., 8.44, 8.45
Enriquez, J., 5.66, 5.67, 5.75
Ergin, A., 4.24
Evans, C., 9.56, 9.57
Everts, C. H., 19.8
Eysink, W. D., 12.3-12.5, 12.7-12.14

Fairchild, J. F., 17.114
Fairweather, G., 1.46, 1.48
Fetter, C. W., 14.12
Finkl, C. W., 19.8
Fischer, D. W., 3. 92
Flach, B., 14.5
Fong, M., 13.19, 16.6, 16.8
Foster, D. N., 19.8, 19.9
Foster, G. A., 19.14
Fowler, J., 17.18
Francingues, N. R., 13.2, 14.5
Franco, L., 4.4, 4.5, 4.22, 4.30, 4.40, 6.9, 6.15–6.17, 6.19
Frautschy, J. D., 5.93, 8.40
Fredette, T. J., 13.26, 14.5, 14.6, 15.49
Fried, I., 5.10
Frisby, R. B., 19.17, 19.19
Fuchs, R.A., 5.12
Führböter, A. U., 3.37
Führböter, I. A., 4.4
Fukuchi, T., 9.16, 9.53
Fukuda, N. T., 3. 41
Fukuya, M., 9.2, 9.5, 9.51
Fulton-Bennett, K., 19.8
Funke, E. R., 5.15

Gailani, J. Z., 15.53
Galvin, C. J., 1.18
Gaudet, C. L., 17.103
Germano, J. D., 13.17
Gibb, J. G., 19.5, 19.6, 19.11, 19.14, 19.25
Gibson, R. E., 16.13
Gilmur, R. C., 17.44
Givoli, D., 1. 22
Goda, Y., 1.18, 4.5, 4.10, 4.11, 4.15, 4.16, 4.18, 4.19, 4.24, 4.29, 4.33, 4.34, 4.36, 4.37, 4.40, 5.86, 6.14, 8.7–8.9, 8.12, 8.29, 8.53, 9.5
Goldberg, E., 19.17, 19.19
Gomez-Pina, G., 10.6
Gormitz, V., 19.10
Gourlay, M. R., 5.60, 8.39

Graham, J. E., 14.8
Grant, J., 19.14, 19.19
Gravens, M. B., 10.19
Greenwood, B., 7.11, 7.19
Grieve, G., 11.15
Griggs, G. B., 19.8, 19.11
Groothuizen, B., 18.2
Guilford, C. M., 18.5
Gunn, J. R., 8.44, 8.45
Gunnison, D., 13.23, 14.5, 14.12, 17.17

Hakanson, L., 17.110
Halber, D., 11.10, 11.28, 11.30
Hales, L. Z., 3.51
Hall, B., 4.5
Hallermeier, R. J., 5.68, 5.76, 8.40, 10.14, 19.13, 19.14
Halpern, L., 1. 22
Hamilton, R. P., 10.26
Hanich, L., 1. 58
Hanson, H., 5.19, 5.21, 8.14, 10.19-10.21, 10.29
Hardaway, C. S., 8.44, 8.45
Harray, K. G., 19.14
Harris, M. M., 5.11, 5.17, 5.23, 5.29, 5.34, 5.69, 5.70-5.72, 5.74, 5.79
Harris, R. L., 9.61
Harrison, D. J., 18.8
Haruta, T., 9.35, 9.36
Hattersley, R. T., 19.8, 19.9
Hawkes, P. J., 2. 16
Hayes, D. F., 17.21
He, S., 1.57, 1.59
Headland, J. R., 8.21, 8.29, 8.58, 8.60, 12.1, 12.13
Healy, T. R., 19.6, 19.8-19.10, 19.14, 19.19, 19.20, 19.23
Hench, J. L., 16.5, 16.16
Herbert, D. M., 6.7, 6.9, 6.13, 6.15, 6.19
Herbich, J. B., 5.12, 5.17, 5.18, 5.23, 5.29, 5.69-5.72, 5.74, 5.79, 11.24, 12.1, 13.21, 14.7, 14.14, 17.21, 17.29
Herron, W. J., 9.61
Hess, H. D., 18.8
Hewson, P., 4.35
Hiroi, I., 4.15, 9.62
Hitachi, S., 4.5
Hitchcock, D. R., 18.3, 18.7
Iljorthaes-Pedersen, A. G. I., 3. 61
Ho, S. K., 9.58
Ho, S., 5.70, 5.72
Hoffman, J. S., 19.21
Hoffmans, G. I. C. M., 3. 50
Hogben, N., 11.8
Hom-ma, M., 9.2, 9.6, 9.11, 9.35, 9.37, 9.41, 9.42, 9.45, 9.47, 9.48
Horikawa, K., 1. 18, 7.2, 7.3, 7.5, 7.6, 7.11,
7.19-7.22, 9.2, 9.5, 9.6, 9.8, 9.10-9.13, 9.35, 9.37, 9.45, 9.47, 9.48, 9.55, 9.70, 9.72
Houghton, J. T., 19.21
Houston, J. R., 1.39, 3.51, 5.6, 10.14, 10.19, 10.28
Howarth, M. W., 4.33
Hsu, J. R. C., 8.42, 8.43, 9.7, 9.14, 9.16, 9.18, 9.21, 9.36, 9.37, 9.56, 9.57, 9.59, 9.60, 9.62-9.67, 9.71
Hudson, R. Y., 2.10, 2.11, 2.14
Hughes, T. J. R., 1.33
Hull, J., 14.5
Hummer, C. W., 17.20
Hwang, L. S., 1.21, 1.29

Idso, S. B., 19.21
Igarashi, T., 9.34, 9.37, 9.40, 9.44
Ingersoll, C., 17.114
Inman, L. O., 5.58, 8.40
Inouye, L. S., 15.49
Iribarren, J. R., 11.30
Irie, I., 9.13, 9.30-9.32, 9.45, 9.46, 9.70
Ishihara, T., 9.16, 9.17, 9.70
Ito, Y., 1.6, 1.30, 1.45, 1.49, 4.15
Iwagaki, Y., 7.2, 9.2, 9.11, 9.70

James, W. R., 10.6, 10.7
Janssen, F. M., 3.27, 3.35, 3.41
Jetter, K., 14.15
Johnson, B. H., 13.19, 15.51, 16.3, 16.5, 16.6, 16.8, 16.9
Johnson, J. W., 5.12, 7.2
Jonsson, I. G., 1.16, 1.39
Juhl, J., 6.9

Kabelac, O., 5.12
Kabiling, M. B., 1.19
Kamphuis, J. W., 1.16
Kanazawa, H., 9.29, 9.30
Karpoff, K. P., 14.15
Katayama, T., 9.29, 9.30
Katoh, K., 7.18, 7.19, 9.5, 9.28-9.30
Katopodes, N., 1.54, 1.55, 1.58
Kaufman, W., 19.3, 19.8
Kawahara, M., 1. 50
Kawata, Y., 9.2, 9.5, 9.8, 9.13, 9.14, 9,16, 9 18, 9.28-9.30, 9.69
Kay, R. C., 19.9
Kenney, T. C., 14.15, 14.16
Kieslich, J. M., 10.9, 10.17
Kikegawa, K., 13.4, 13.15, 14.5
Kilmer, F. A., 1. 39
Kim, C., 14.5, 14.16
Kimon, P. M., 11.32
Kirby, J. T., 1.61, 1.65, 8.12
Kirkgoz, M. S., 4.8

AUTHOR INDEX

Klein Breteler, M., 2.10, 2.12-2.14, 2.16, 3.62, 3.64, 3.65, 3.76
Klein, R. J. T., 19.21
Kleinbloesem, W. C. H., 13.3
Knauss, J., 3.52, 3.53
Koh, R. C. Y., 9.35, 16.6-16.8
Koike, K., 9.2, 9.5
Komar, P. D., 7.2, 9.61, 19.11, 19.18
Kortch, M. S., 13.8, 13.29
Kortenhaus, A., 4.9, 4.23, 4.25, 4.41
Kotulak, P., 8.29, 8.58, 8.60, 12.1
Koutsourais, M. M., 2.10, 2.14
Krafft, K. M., 5.12, 5.23
Kraus, N. C., 3.9, 3.23, 5.19, 5.21, 7.3, 7.4, 8.14, 9.48, 9.49, 9.66, 10.6, 10.9, 10,15, 10.16, 10.19-10.21, 16.9
Kressner, B., 8.37
Kriebel, D. L., 7.2, 8.18, 19.7, 19.8, 19.10
Kriegsmann, G. A., 1.22
Krishnappan, V., 14.7, 14.8
Krone, R. B., 16.12
Kroon, A., 10.6
Krumbein, W. C., 8.41, 10.6

La Fleur, J., 14.15
La Point, T. W., 17.114
Laboyrie, H., 14.5
Lai, C., 1.55, 1.58
Lai, P., 1.50, 1.52, 1.57, 1.59
Lamberson, J. O., 17.105
Lambre, T. W., 3. 55
Lapidus, L., 1. 28
Larsen, J., 1.9, 1.49
Larson, J., 8.18, 8.35
Larson, M., 7.20, 10.15, 10.16, 16.12
Lau, D., 14.15, 14.16
Laustrup, C., 10.6
Lean, G. H., 12.2
Leatherman, S. P., 19.9
Lee, C. R., 15.49
Lee, H., 17.114
Lee, J. J., 1.21
Lekic, S., 10.1
Lendertse, J. J., 1.60
Leonard, L. A., 10.29
Leschinsky, D., 14.10
Lesnick, J. R., 5.9
Li, P., 1.24
Li, W., 14.5
Lillycrop, L., 14.5, 14.6
Lillycrop, W. J., 10.29
Lin, P., 1.50, 1.52, 1.57, 1.59
Lindeman, K. C., 15.51
Lindenberg, J., 2. 14, 3.76
Ling, H. E., 14.5, 14.6, 14.10
Ling, H. I., 13.10
Liu, S. K., 1.60

Long, E. R., 17.105, 17.108, 17.110, 17.111
Loring, D. H., 17.112
Louaked, M., 1.58
Luettich, R. A., 16.11
Luke, J. C., 1.11
Lumb, F. E. 11.8
Lyzlov, A. I., 5.12

MacDonald, D. D., 17.105, 17.106, 17.111, 17.112
Madsen, O. S., 16.9
Madsen, P. A., 1.9, 1.11, 1.18, 1.49
Mah, F. T. S., 17.110
Mahmood, K., 1. 0, 1.58
Majda, A., 1. 22
Manampevi, H. D. S., 14.15
Mansard, E. P. D., 5.15
Maruyama, K., 1.6-1.8, 1.19, 1.44, 1.45, 1.49
Maurmeyer, E. M., 19.8
Maynord, S. T., 14.6, 14.16
Mayor-Mora, D., 12.2
McCant, D. D., 15.49
McConnell, K. J., 2.1, 2.3, 2.9-2.10, 4.4, 4.26, 4.35
McFarland, V. A., 15.49
McKenna, J. E., 4.8, 4.11, 4.23
McKnight, A. L., 13.11, 13.15, 13.16
McLean, R., 19.22
McLellan, T. N., 10.9, 13.15
McNair, C., 15.55
Mei, C. C., 1.4, 1.13, 1.21, 1.37, 1.39
Meijer, D. G., 3.49
Meijers, P., 3.80
Miles, J., 1.39
Miller, J., 17.17
Mimura, N., 5.10, 7.3, 7.4, 19.22
Mingham, C. G., 1.50
Minikin, R. R., 4.23, 4.24
Mitchell, A. R., 1.46, 1.48
Mitsunobu, N., 7.17, 7.19, 7.21
Mizumura, K., 9.48, 9.66
Mohan, R. K., 14.5-14.7, 14.15
Moller, J. P., 5.11
Montgomery, R. L., 13.5
Moore, B. D., 10.14
Moore, D. W., 15.55
Morgan, K., 1.34
Morgan, L. G., 17.110
Morgan, S., 14.15
Morison, J. R., 5.12
Moritz, H. R., 13.19, 13.20, 16.5, 16.13
Mörner, N. A., 19.11
Morton, R. W., 13.11
Mosello, R., 17.110
Moutzouris, C., 9.62
Müller, G., 4.24
Munk, W., 1.39

Muraki, Y., 4.33
Myers, T. E., 13.26, 15.49, 17.16

Nadaoka, K., 1.22
Nagao, Y., 9.69, 9.72
Naguchi, K., 9.35-9.37, 9.67
Narayaran, V. K., 19.23
Narumi, Y., 18.1, 18.8
Neal, R. W., 13.32
Nelson, E. E., 14.5
Nicholson, J., 11.10
Nir, Y., 5.9, 5.10, 5.29, 5.60, 5.72, 5.74, 8.40
Nishimura, H., 7.8, 9.73
Noble, R. M., 5.68, 8.40
Noda, E. K., 5.11, 5.68
Noda, H., 7.2
Nordstrom, K. F., 19.8
Nwogu, O., 1.11, 1.61

O'Connor, J. M., 13.11
O'Connor, S. G., 13.11
Oerlemans, J., 19.21
Ogata, A., 14.12
Okajima, Y., 9.41, 9.43, 9.70
Orlanski, I., 1.22
Osborne, P. D., 7.11
Ostendorf, D. W., 1.18
Oswalt, R., 14.6
Oumeraci, H., 4.5, 4.8-4.11, 4.23, 4.24
Owen, M. W., 2.16, 2.17, 2.19, 3.37, 4.15, 4.40, 6.3, 6.6, 6.8-6.10, 6.16
Oyama, T., 1.22
Ozaki, A., 9.51, 9.52

Packer, T., 18.3
Palermo, M. R., 13.2, 13.3, 13.5-13.9, 13.14, 13.19, 13.29, 14.5-14.7, 14.14, 15.5, 15.49, 17.5, 17.17, 18.6
Panchang, V. G., 1.30
Parchure, T. M., 15.51
Partenscky, H. W., 4.24, 4.25
Partheniades, E., 16.12
Peck, K. B., 14.8-14.10
Peddicord, D., 15.55
Peerbolte, E. B., 10.25
Pelnard-Considere, R., 8.32, 8.36
Peregrine, D. H., 1.10, 4.32
Perlin, M., 5.10
Persaud, D. R., 17.105, 17.113, 17.114
Petit, H. A. H., 3.59
Philipse, L., 3.48
Phipps, G. L., 17.113
Pilarczyk, K. W., 2.10, 2.14, 3.4, 3.12, 3.24, 3.44-3.45, 3.48-3.50, 3.52, 4.5, 8.9, 8.10, 8.38, 14.14, 17.18
Pilkey, O. H., 3.4, 10.6, 10.29, 19.3, 19.8, 19.25
Pinder, G. F., 1.28

Pirazzoli, P. A., 19.10
Poindexter-Rollings, M. E., 13.31, 16.3
Pope, J., 5.29, 5.62, 5.65, 5.66, 8.39, 8.40, 8.41
Porter, D., 1.8
Pos, J. D., 1.39
Powell, K. A., 2.19, 2.21
Prickett, T. L., 15.53
Puckett, P. T., 15.51

Quelerij, L., 3.91, 3.97

Radder, A. C., 1.8
Ramirez, J. L., 10.6
Randall, R. E., 13.12, 13.19, 13.20, 14.5, 14.6, 15.49, 16.5, 16.13
Rantala, R. T. T., 17.112
Raper, S. C. B., 19.22
Reid, R. O., 1.65
Reinalda, R., 12.8
Reine, K. J., 15.51
Reynoldson, T. B., 17.114
Rhoads, D. C., 13.17
Ribberink, J. A., 7.5
Roberts, A. C., 6.10
Roelfzema, A., 12.7
Roelvink, J. A., 7.17, 7.19
Rogers, R., 17.42
Rollings, M. P., 15.49
Romagnoli, R., 17.101, 17.102
Rosati, J. D., 5.57, 5.70, 5.75, 5.76, 8.38, 15.24
Rosen, D. S., 5.17, 5.31, 5.61
Rosser, J. B., 1.28-1.29
Rouville, M. A., 4.22, 4.23
Ruiz, C., 15.55
Russell, K. S., 11.10

Saathoff, D. D., 17.44
Saito, K., 9.22, 9.23, 9.50
Sakai, F., 1.39
Sanderson, W. H., 13.11, 13.15, 13.16
Sato, E., 17.24
Sato, M., 9.29, 9.50, 9.62
Sato, N., 1.19, 1.22
Sato, S., 7.5, 7.6, 7.11, 7.17, 7.19, 7.21
Sauer, T. C., 16.9
Savenÿe, P. Ph. A. C., 11.52, 11.55
Sawaragi, T., 1.18, 9.16, 9.17
Schaffer, H. A., 1.13, 1.19
Scheffner, N. W., 13.8, 13.20, 16.5, 16.11, 16.16
Schmidt, R., 4.11, 4.22
Schoones, J. S., 5.11
Schroeder, P. R., 13.19, 14.12
Schropp, S. G., 17.112
Schwartz, M. L., 19.11
Scott, K. J., 13.17
Seabergh, W. C., 5.32

Sedji, M. T., 8.40
Seelig, W. N., 2.19
Seijffert, J. W. W., 3.48
Seiji, M., 5.51, 5.67, 5.73, 5.74, 9.19
Seo, G., 9.16, 9.53
Shapiro, H. A., 9.5, 9.69
Shaw, J., 14.5
Sherand, J. L., 14.15
Sherman, D. J., 19.8
Sherman, W. C., 14.15
Shi, N. C., 7.10
Shibayama, T., 7.19-7.22, 9.2, 9.8, 9.13, 9.14, 9.16, 9.18, 9.69
Shields, F. D., 13.5
Shinohara, K., 5.11, 5.17, 5.31, 5.34, 9.19
Short, A. D., 7.19
Silva, A. J., 14.6
Silvester, R., 5.11, 5.51, 5.70, 5.72, 8.41-8.45, 9.5-9.7, 9.14-9.16, 9.20, 9.21, 9.28, 9.50, 9.56–9.60, 9.62-9.68, 9.70, 19.16, 19.19
Simm, J. D., 2.9, 2.12, 2.16, 4.4, 4.10, 4.15, 6.8, 6.18
Sivaram, B., 14.9
Sleath, J. F. A., 1.16, 1.18
Sloth, P., 6.9
Smith, A. W., 8.58
Smith, D. S., 14.10
Smith, G. M., 3.49, 6.18, 6.19
Smith, M. R., 18.1, 18.3, 18.4, 18.8
Smith, O. P., 8.54
Smith, R., 1.13
Smits, J., 13.15
Sollitt, C. K., 1.21
Sonu, C. J., 5.14, 5.69, 9.2
Sorensen, O. R., 1.11
Sorenson, R. M., 19.19
Sprague, C. J., 2.10, 2.14
Sprinks, T., 1.13
Stauble, D. K., 10.26
Steele, A. A., 6.10
Steinberg, R. W., 7.16
Stemmer, B. L., 17.113
Stewart, R. W., 19.11
Stive, M. J. F., 7.17, 7.19
Stoker, J. J., 1.9, 1.50, 1.59
Strating, J., 11.52, 11.55
Strelkoff, T., 1.54-1.55
Stronge, W. B., 10.28
Sturgis, T., 13.23, 14.5, 17.17
Sudar, R. A., 10.6
Suh, K., 5.25, 5.69, 5.72, 5.73, 8.39, 8.40
Sullivan, J., 17.115
Sumeri, A., 13.7, 13.11-13.13, 13.26, 14.5
Sunamura, T., 7.3, 7.19, 7.20, 7.22
Suyama, H. T., 9.13
Swamee, A., 14.9
Swart, D. H., 16.12

Swartz, R. C., 17.105, 17.114
Synder, G. W., 17.31
Synolakis, C. E., 19.8, 19.19

Takahashi, S., 4.5
Tallent, J. R., 16.11
Tanaka, N., 5.14, 5.67, 5.73, 5.74, 9.50, 9.62
Tanimoto, K., 1.6, 1.30, 1.45, 1.49, 4.5, 4.10
Tatham, P. F., 4.5
Tavolaro, J. F., 16.2
Taylor, A., 17.33
Taylor, R. L., 1. 50
Tchbangolous, G., 14.12
Teeter, A. M., 13.8, 13.29
Terzaghi, K., 14.8-14.10, 14.15
Thevenot, M. M., 11.26
Thibodeaux, L. J., 13.26, 14.5, 14.6
Thoma, G. J., 13.26, 14.6
Thomas, R. S., 4.5
Thomas, R., 19.21
Thornton, E. B., 7.17, 7.19
Titus, J. G., 19.20, 19.21, 19.23
Togashi, H., 1.34, 1.39, 13.4, 13.16
Tondello, M., 10.25
Tooley, M. J., 19.11
Torrini, L., 4.8, 4.23
Toshinaga, K., 17.96
Townson, J. M., 1.54, 1.58
Toyoshima, O., 5.10, 5.13, 5.23, 5.63, 5.64, 9.4, 9.7, 9.8, 9.14-9.16, 9.18, 9.20, 9.32, 9.33, 9.37, 9.39, 9.40
Trefethen, L. N., 1. 22
Trembanis, A. C., 10.6
Truitt, C. L., 5.70, 5.75, 5.76, 13.3, 13.7, 13.9, 13.15, 13.18, 13.24, 16.2, 17.17
Tsubaki, T., 5.11, 5.17, 5.31, 5.34, 9.19
Tsuchiya, Y., 9.20, 9.21, 9.29, 9.56, 9.62, 9.70
Tsukioka, K., 1.39
Tuck, E. O., 1.21

Uda, T., 5.67, 5.73, 5.74, 9.7, 9.18, 9.19, 9.22, 9.23, 9.33–9.35, 9.37, 9.44, 9.47-9.50, 9.53, 9.61
Uren, M., 18.3

Vajda, M., 5.17, 5.31, 5.61
van de Graaf, J., 12.8, 12.9, 19.8, 19.16
van de Kaa, E. J., 14.14
van der Meer, J. W., 2.10, 2.11, 2.15, 3.27, 3.34, 3.35, 3.40, 3.41, 4.4, 5.16, 6.3, 6.6-6.8, 8.48, 8.51-8.53, 8.56, 8.60, 8.61
van Gent, M. R. A., 3.59
van Hijum, E., 3.91, 3.92
Van Hoff, J., 9.68
Van Noorwijk, J. M., 10.25
Van Os, A. G., 12.7
Van Rijn, L. C., 8.10

Van Wijck, J., 13.15
Van Wyk, A. C., 11.26, 11.34, 11.35
Vanoni, V. A., 14.13, 14.15
Vastano, A. C., 1. 65
Velinga, P., 19.21
Verheij, H., 3.49, 3.50
Vermaas, H., 12.3, 12.4, 12.7, 12.9
Verruijt, A., 3.75
Vicinanza, D., 4.26, 4.30, 4.34
Vickerman, M. J., 11.11
von Stackelberg, K., 15.55
Voortman, H. G., 4.8
Vorhees, D., 15.55
Vrijling, J. K., 4.8

Walkden, J. A., 4.32
Walker, J., 5.9, 5.65
Waller, M. N. H., 4.42
Wang, X. Q., 14.5 14.6
Warrick, R. A., 19.10, 19.21-19.23
Warwar, J. F., 5.69
Watanabe, A., 1.6-1.8, 1.18, 1.19, 1.44, 1.45, 7.19, 7.20, 7.22
Weggel, J. R., 4.15, 4.40, 10.6, 19.13
Wei, G., 1.11, 1.61, 1.65
Weijde, R. W., 13.3
Wemelsfelder, P. J., 3. 9
White, R. W., 16.12
Whitman, R. W., 3. 55
Wiegel, R. L., 10.6
Wigley, T. M. L., 19.22
Wikramanayake, P. N., 16.9

Wiley, M. B., 13.34
Williams, G. L., 15.49, 15.53
Williams, R. G., 1. 30
Williamson, G. A., 11.11, 11.12
Winfield, L. E., 15.49
Wise, R. A., 10.6, 10.25
Witting, J. M., 1.11
Worman, A., 14.16
Wright, S. J., 14.5, 14.16

Yagima, M., 9.30, 9.31
Yanagishima, S., 7.18, 7.19, 9.29, 9.30
Yarde, A. J., 2.9, 2.10, 2.12-2.16
Yasso, W. E., 8.41, 9.56, 9.58
Yeh, H. H., 19.8
Yevjevich, V., 1.50, 1.58
Yoo, C. H., 10.6, 10.20, 19.3
Yoshioka, K., 9.26
Young, W., 11.44
Yu, X., 1.18, 1.22, 1.30, 1.39, 1.49, 1.57

Zappi, M. E., 17.16
Zarull, M. A., 17.114
Zeeman, A. J., 14.8
Zeidler, R. B., 8.9, 8.10, 8.38
Zienkiewicz, O. C., 1. 34-1. 36, 1. 39
Zwamborn, J. A., 5.13, 11.7, 11.10, 11.15, 11.26, 11.34, 11.35, 11.47, 11.56, 11.61, 11.64
Zweck, H., 14.15
Zwemmer, D., 9.65

SUBJECT INDEX

ADCIRC, 16.5, 16.16
Aerobic bioremediation, 15.8
Aids to navigation, 11.43
Analysis, 4.39
Armor
   damage, 8.58
   rock, 2.10; *see also* revetments
   size, 8.53, 8.82
ASCII, 16.14, 16.15
Auxiliary craft, 11.13
Avalanching, 16.12

Bathymetry, 8.9, 13.7
Bay shape, 9.56, 9.60, 9.63, 9.64
   equation, 9.56
Bayed beaches, 9.58
Beach nourishment, 8.19
Beach, 7.1
   causes, 9.6
   erosion, 5.2, 9.4, 19.8
   state of, 9.7
   nourishment, 19.3
   profile, 7.2, 7.18, 7.19
      classification, 7.2–7.4
      equilibrium, 10.71
      factor, 10.17
      response, 10.18
   replenishment, 18.4
      experience, 10.6
      process, 10.11
      projects, 10.3, 10.28
Beneficial uses of dredged material, 15.34, 15.35
Benefits of beach nourishment, 10.27–10.29
Berm, 6.5
   erosion, 7.18
Biodegradation, 17.13
Bioturbation, 13.23
Bituminous systems, 3.47
Blocks
   concrete, 2.10
Block mats, 3.45, 3.88
Block revetment, 3.45
Blockwork, 2.14–2.16
Boundaries
   nonreflective, 1.21
Boundary
   conditions, 3.14–3.17

   geotechnical, 3.14
   hydraulic, 3.14
   value
      problems, 1.23, 1.43
Breaking
   criteria, 1.18
Breakwater
   emergent, 5.5, 5.9, 5.16
   length, 5.27, 5.30, 5.32, 5.33
   multiple, 5.9, 5.10, 5.62–5.64
   offshore, 5.2, 5.3, 5.7, 8.38, 8.39
   porous, 1.21
   reef-type, 5.15
   single, 5.7, 5.62–5.64
   submerged, 5.12–5.14
Breakwaters
   composite, 4.13
   crown walls, 4.13
   detached, 9.8, 9.18, 9.19, 9.29, 9.30, 9.32, 9.34–9.36, 9.38, 9.40, 9.67
   L-shaped, 9.48
   offshore, 8.38, 8.39
   submerged, 9.45
   vertical, 4.13

Caisson, 4.31, 6.14
   effect of length, 4.31
Canadian Sediment Quality Standards, 17.103
Cap
   areal coverage, 13.24
   monitoring, 13.29
   stability, 13.27
   thickness, 13.24
Cap components, 14.5
   armor layer, 14.5
   base isolation layer, 14.5
   base stabilizing layer, 14.5
   filter layer, 14.5
Capping, 14.4, 14.5, 17.50, 17.59, 17.83
   cap components, 14.5
   design criteria, 14.6
Case studies, 9.30–9.55
Channel design, 11.13
   alignment, 11.15
   bends, 11.15, 11.25
   depth, 11.30, 11.42
   entry maneuver, 11.23, 11.50
   length, 11.16

Channel design *(continued)*
    ship handling, 11.18
    stopping distance, 11.20
    width, 11.23–11.25
Chemical
    isolation, 13.22
    stability analysis, 14.11
    contaminant release, 14.11
Chromium, 17.33
CHZ model, 19.18
Classification methodologies, 17.7–17.9
Cleanup cost, 17.12, 17.75, 17.77, 17.102
Coastal, 19.4
    hazard zone, 19.4, 19.5, 19.8, 19.9
    planning, 19.9
Coefficient
    roughness, 2.18
    run-up, 2.18
    stability, 2.12
    values for Owen's formula, 2.18
Coefficients
    empirical, 6.5
    roughness, 6.9
Collapse
    dynamic, 16.3, 16.7, 16.8
Confined Aquatic Disposal (CAD), 13.3
Consolidation, 13.24, 13.28, 16.13
Construction
    materials, 8.62
    methods, 8.60
    practice, 3.10, 10.9
        deterministic, 3.11
        probabilistic, 3.11
        quasiprobabilistic, 3.11
Contaminant flux, 13.24, 13.26, 13.27
Contaminants
    PAH, 15.8
    PCB, 15.8
    PCDD, 15.8
    PCDF, 15.8
Contamination, 14.4
    CERCLA, 17.2
    CWA, 17.2
    DDT, 17.5
    extent, 17.5
    lead, 17.33
    legislative authority, 17.2
    mercury, 17.95, 1797
    MPRSA, 17.2
    PAH, 17.5
    PCB, 14.22, 14.23, 17.5, 17.64, 17.68, 17.70, 17.73, 17.83, 17.86, 17.88, 17.91, 17.101
Copper, 17.33
COSED IV, 16.5
Co-spectrum, 7.14, 7.16
Crest walls, 6.10
    on permeable sea walls, 6.13
Currents, 8.8, 13.7
    cross shore, 8.8, 8.9
    longshore, 8.8
    tidal, 8.8

Data base, 16.10–16.12
    input, 16.16
Density current exchange, 12.7, 12.9–12.11
Descent
    convective, 16.3, 16.6, 16.7
Design, 2.6, 3.12, 3.81–3.86
    armor, 8.50
    berm fill, 10.10
    berm plus dune, 10.9, 10.10
    checklist, 3.24
    conditions, 2.6
    considerations, 5.57–5.72, 5.79
    depth, 5.76
    empirical relationships, 5.57–5.72, 5.77
    example, 5.79, 5.83
    functional, 8.26
    life, 2.6
    methodology, 3.12
        construction method, 3.12
        function of the structure, 3.12
        operation and maintenance, 3.12
        physical environment, 3.12
    methods, 2.10, 2.14
        hydraulic performance, 2.16
    nearshore berm, 10.10
    of dikes, 3.23, 3.25
    of sloping seawalls, 3.23, 3.25
    optimization, 8.45–8.59
    probabilistic, 8.45, 8.59, 8.60
    profile fill, 10.10
    project progress, 10.13
    requirements, 3.14
    rubble mound, 8.50
    shore protection, 8.5
    steps in design, 10.12
    storm tide, 8.5
    structural, 8.45
    template, 10.9
Design issues
    capping, 13.4, 13.21
        material placement, 13.10
        sequence, 13.5, 13.6
Design procedures, 4.39, 4.40
Design ships, 11.10
    choice, 11.10
    dimensions, 11.12
    loading conditions, 11.12
    propulsion, 11.11
    type, 11.11
Development setback, 19.6
    acts, 19.6, 19.7

Florida, 19.7, 19.8
  New Zealand, 19.6
  need for, 19.6
Diffuser
  submerged, 13.15
Diffusion, 1.15
Dike, 3.5, 3.8, 3.10
  failure mechanisms, 3.11
  instability, 3.10
  overtopping, 3.10
  protection, 3.4, 3.6
Dioxin, 15.6
Dispersion, 13.18
Disposal, 17.15, 17.33, 17.37
  aquatic, 17.15, 17.17, 17.18, 17.92
  confined, 15.8, 15.10–15.16, 17.15, 17.16, 17.18, 17.56, 17.81
  land, 17.15
  management, 15.8, 15.14
Dissipation, 1.15
Dredged material
  decision support system (DMDSS), 15.37
  disposal, 16.2, 16.9, 16.14, 16.19, 16.21–16.23
  long-term fate, 16.3
  numerical modeling, 16.4
  placement, 16.2
  short-term fate, 16.3
Dredging, 9.8, 9.30, 9.31, 9.67, 15.3, 15.17, 17.20, 17.33, 17.35, 17.37, 17.44, 17.51, 17.59, 17.61, 17.65, 17.77, 17.89, 17.96, 18.2
  equipment, 17.20, 17.21, 18.2
    hydraulic, 17.21, 18.2
    mechanical, 18.2, 18.3
  innovative technology, 15.38, 15.39
  marine minerals, 18.2
  mechanical, 17.21
  mechanical-hydraulic, 17.21, 17.44
  methodology, 18.3
  pneumatic, 17.22
  quantities, 18.4
Dump, 16.14, 16.15
Dune, 3.5
Dune line trends, 19.10
  long-term, 19.10
  short-term, 19.10
Dune stability factor, 19.15

EBERM, 16.5
Effect of
  dikes on the beach, 3.21
  seawalls on the beach, 3.21
Environmental concerns, 17.38, 17.48
  disturbance of benthic organisms, 18.7
  large borrow pits, 18.6
  redeposition of unwanted sediments, 18.6
  removal of substrate, 18.6
  turbidity, 18.6, 18.7
Equation
  Boussinesq, 1.10, 1.59
  Euler, 1.3
Erosion, 13.12, 13.28

Fabric containers, 3.48
Finite element
  methods, 1.30, 1.34
Flooding, 19.16
Foundation
  impermeable, 2.16
Frontal dunes,
  blowouts, 19.8
  role, 19.9

Gabion
  baskets, 3.47
  mattresses, 3.47
Gap, 5.7
  effect of, 5.17, 5.18, 5.33
  erosion, 5.67, 5.74, 5.75
GCM model, 19.21
Geomorphology
  coastal, 9.55
Geotechnical
  consideration, 13.10
  stability analysis, 14.8
    bearing capacity, 14.9
    settlement, 14.10
    slope stability, 14.10
Grass mats, 3.48
Groins, 9.15, 9.18, 9.30, 9.35, 9.37, 9.50, 9.54
Groundwater table, 9.29

Harbor, 11.4
  currents, 11.10
  design ships, 11.5
  environmental data, 11.5, 11.7
  geotechnical data, 11.5, 11.7
  nautical requirements, 11.5, 11.6
  site, 11.5
  sedimentology, 11.5, 11.8
    channel/basin dredging, 11.5
    wave protection, 11.5, 11.8
    water salinity
    tide, 11.9
    wind, 11.9
  wave induced ship motions, 11.34, 11.35, 11.42
Headland
  control, 9.20, 9.23, 9.52
Headlands
  artificial, 8.41
Horizontal eddy exchange, 12.7

HPDPRE, 16.5
HPDSIM, 16.5, 16.14
Hydraulic exchange, 12.7, 12.8
 total, 12.13
Hydraulic filtering analysis, 14.15
 filter design, 14.16

Impoldering, 3.7
 of marshlands, 3.7
Infinite elements, 1.35

Leachate, 15.9
Leading lights, 11.44
Level bottom capping (LBC), 13.3
Littoral
 drift, 9.8
Loading
 strength concept, 3.20
 zone, 3.17
Location
 distance offshore, 5.31
LTFATE, 16.4, 16.5, 16.10, 16.14

Maintenance, 3.91, 3.92, 10.13, 10.25, 15.3
 navigation channels, 15.3
Material, 2.4
 asphaltic, 2.14
 types, 2.4
Mattress, 2.14
 concrete, 2.14
Mathematical
 models, 1.2
MDFATE, 13.19, 13.20, 16.4, 16.5,
  16.13–16.16, 16.18, 16.24
Mining
 marine aggregates, 18.1, 18.2
 projects,
  Hong Kong airport, 18.5
  Honolulu airport, 18.5
  Kansai airport, 18.5
  Newark airport, 18.5
  The Netherlands, 18.3
  United Kingdom, 18.3
  United States, 18.6
Model, 4.41
 ADDAMS, 13.19
 basin sedimentation, 12.15
 CAPSTABL, 14.17, 14.21
 channel sedimentation, 12.6
 dissipation, 1.15
 EDUNE, 8.9
 FAST-TABS, 12.16, 12.18
 GENESIS, 5.6, 5.18, 5.19, 5.22, 8.14
 Goda random wave, 8.7, 8.9
 hydraulic, 4.16, 5.9–5.11, 5.16
 MIKE21MT, 12.17, 12.18
 numerical, 5.6, 5.10

 numerical erosion, 8.18
 PROVERBS, 4.9
 RELEASE, 14.8
 SBEACH, 8.9
 STFATE, 13.19
 SUBTRENCH, 12.5
 Monitoring, 10.13, 10.25
 Morphological response, 5.72
Mound
 apron, 13.20
 armored rubble, 6.7
 crest, 13.20
 development, 13.18
 geometry, 13.20
 inner flank, 13.20
 MDFATE, 13.19, 13.20
 outer flank, 13.20
 underwater, 5.13

Navigation buoys, 11.44
NEPA, 17.5, 17.6, 17.48, 17.72, 17.79, 17.82,
 17.84–17.86
NOAA, 17.3
Numerical modeling, 1.1, 1.2, 7.19, 7.20,
 15.21
 ADDAMS, 15.31
 bottom friction, 1.16
 breaking, 1.18, 1.20
 dispersion from pipeline disposal, 15.31
 DMSMART, 15.30, 15.36
 EDUNE, 10.16
 GENESIS, 10.16, 10.19–10.24
 LTFATE, 15.30
 SBEACH, 10.15, 10.16, 10.25
 STFATE, 15.30
 surf zone, 1.19
 suspended sediments, 15.21–15.24

ODMDS, 16.2, 16.5, 16.14–16.16,
 16.18–16.20, 16.24
Overfill factor, 10.7
Overtopping, 2.4, 2.8, 2.16, 3.37–3.42, 3.51,
 4.9, 4.10, 6.2, 6.3, 6.7, 6.13, 6.15, 8.47,
 8.49, 8.58
 breaking waves, 3.39
 discharge, 2.7, 3.38, 6.17–6.19
  peak, 6.17, 6.18
  single, 6.19
  tolerable, 6.17, 6.18
 nonbreaking waves, 3.40
 reduction of, 2.4, 2.6
 rough armored slopes, 6.8

Performance measures, 10.3
 post-storm damage, 10.3
 protection capacity, 10.3
 volume of dry sand, 10.3

width of dry beach, 10.3
Period, 2.6
　return, 2.6
Permeability, 2.15
Physical model, 11.26
　VESDYN, 11.35
Physical stability analysis, 14.3
　erosion of sediments, 14.15
　ice loading, 14.14
　propeller wash, 14.14
　river flow, 14.13
　wave action, 14.13
Placement, 15.27
　accurate, 13.7
　analysis, 14.7
　　horizontal displacement of the slug, 14.8
　　impact velocity, 14.7
　　requirement numer of lifts, 14.8
　　thickness of the slug, 14.7
　hopper dredge pump-down, 13.16
　nearshore/aquatic, 15.27–15.34
　pipeline with baffle plate, 13.12
　pipeline with "sand box", 13.14
　spreading by barge, 13.11, 13.15
　spreading by hopper dredges, 13.12
　submerged discharge, 13.13
　surface discharge, 13.11
　trémie (downpipe), 13.16
Polders, 3.7
Port
　Haifa, 11.61–11.66
　in natural bays, 11.4
　lagoon, 11.4
　open coastal, 11.4
　operation, 11.45–11.47
　Richards Bay, 11.48, 11.51–11.54, 11.56–11.61
　river, 11.4
Prediction methods, 4.23, 4.33
　Hannover/Braunschweig, 4.24
　Minikin, 4.23
　prediction methods for impact pressures, 4.26
Pressure distribution, 4.17
　gradients, 4.28
Probability, 2.6
　encounter, 2.6
Profile
　construction, 10.11
　design fill, 10.11
Program, 15.3
　DMRP, 15.44
　DOER, 15.3, 15.4
　　contaminated sediments, 15.4, 15.7
　　ecological risk management, 15.4
　　environmental windows, 15.4, 15.18, 15.21

　　instrumentation for dredge, 15.4, 15.24–15.27
　　nearshore/aquatic placement, 15.4
　DOTS, 15.45, 15.46
　FVP, 15.46
　LEDO, 15.47
　ORP, 15.44
　Project, 17.33
　analysis, 5.50
　　statistical, 5.51, 5.54, 5.56
　Antwerp, Belgium, 17.89
　Bayou Bonfouca, Louisiana, 17.33
　Cold Spring, New York, 17.35
　Commencement Bay, Washington, 17.40
　Eagle Harbor, Washington, 17.50
　foreign, 5.47–5.49
　Grand Calumet River, Indiana, 17.50
　Ketelmeer, the Netherlands, 17.91
　Manistique River, Michigan, 17.55
　Massena, New York (General Motors Site), 17.62
　Massena, New York (Reynolds Metals Facility), 17.71
　Minamata Bay, Japan, 17.95
　New Bedford Harbor, Massachusetts, 17.76
　Penetanguishene, Canada, 17.97
　Scarborough Bluffs, Canada, 17.98
　Southampton, United Kingdom, 17.99
　Waukegan, Illinois, 17.86
　United States, 5.34–5.47, 5.52, 5.53
　Zeebrugge, Belgium, 17.99
Protection, 8.19
　coastal, 8.1
　shore, 8.1
Protective systems, 3.46

Recommendations for beach replenishment, 10.5
Reefs
　artificial, 9.24, 9.26
Reflection coefficient, 4.14
　composite walls, 4.14
　low-crested walls, 4.14
　simple vertical walls, 4.14
Regime
　littoral, 8.10
Remediation of contaminated sediments technology, 17.12, 17.14, 17.19, 17.52
Renourishment factor, 10.8, 10.13
Return period, 8.5, 8.9
Revetment, 2.3
　components, 2.4
　design, 2.9
　failure modes, 2.6, 2.7
　flexible, 2.6
　performance, 2.4
　system, 2.3, 2.4

Revetments
  modes of failure, 3.43
  placed-block, 3.55
  slope protection, 3.42, 3.43
  with granular filter, 3.61
Risk
  assessment, 17.19
  management, 15.40
    environmental, 15.42
Role of fixed structure, 10.4
Rubble mound, 8.50
Runup, 19.17
  effect of
    angle of attack, 3.34
    roughness, 3.32
    shallow foreshore, 3.31
    stepped slope, 3.29

Salient, 5.5, 5.7, 5.26, 5.27, 5.62, 5.66, 5.73,
    5.78, 8.39–8.41, 9.36, 9.39, 9.60, 9.68
Sand bars, 7.19
Scale effect corrections, 4.32, 4.41
  impact pressure, 4.33
Scour, 2.7, 3.50
  depth, 2.21
  local, 2.22
  toe, 2.19
Sea dikes, 9.14, 9.53
Sea level rise, 19.10, 19.20–19.22
Sea walls, 9.14, 9.20, 9.35, 9.39, 9.48
  armored, 6.7, 6.9
  bermed, 6.7
  impermeable, 6.10
  vertical, 6.14, 6.16
Seaward forces, 4.35
Sediment
  budget, 8.11
  category, 17.7
  characteristic, 8.14
  characterization, 13.5
  contaminated 17.10
  deposition, 5.28, 5.29, 12.3
  evaluation, 17.10
  flux, 7.9, 7.11, 12.3
  inventory, 17.10
  movement, 9.8
  settling velocity, 12.2
  supply, 9.8
  suspended, 7.11, 7.16, 12.6
  transport, 7.9, 7.10, 12.2
Sedimentation
  channels, 12.2, 12.4
  harbors, 12.2, 12.7
  local, 12.5
  rate, 12.6, 12.14
Seiche, 8.5
Setback, 19.3, 19.15

development, 19.3, 19.6, 19.9, 19.23
reassessment, 19.24
Ship motions, 11.35, 11.37–11.41
Shoaling, 8.7
Shoreline
  adjustment, 19.12
  changes, 5.9, 5.13, 8.10, 8.11
  crenulate changes, 8.43
  long-term, 8.10
  protection methods, 9.13, 9.28, 9.46
  response, 5.25
  storm-induced, 8.18
Silt curtains, 17.63
Simulation, 11.26, 11.28, 11.29, 11.31
  Ashdod Port, 11.28
Site selection, 13.7
Slab, 2.12, 2.15
  stability, 2.12, 2.14–2.15
  thickness, 2.12
Slope
  impermeable, 6.3
  simply-sloping, 6.3, 6.4
  smooth, 6.3, 6.4, 6.16
"Solid structure" response, 19.3
Splash area, 3.52
Squat, 11.32
Stability criteria for
  cover layer, 3.59
  current attack, 3.49
  geotechnical, 3.72
  wave attack, 3.44
STFATE, 16.4, 16.5, 16.8–16.10, 16.14
Stone
  grouted, 3.46
Storm
  surge, 3.9, 9.2, 9.6, 19.8, 19.16
Structures
  dikes, 8.23–8.28
  dunes, 8.23–8.28
  groins, 8.26, 8.30, 8.32
  offshore breakwater, 8.38
  perched beach, 8.23, 8.25, 8.28
Studies
  field, 5.23
  model, 5.23
Subaqueous caps, 15.13
Superfund, 17.2
"Sustainable management", 19.3, 19.6

T-groins, 8.41, 8.44
Tidal prism exchange, 12.7
Tide
  astronomical, 8.3
  storm, 8.3
TIDEPRED, 16.14
Toe
  geometry, 8.55

protection, 3.50
Tombolo, 5.5, 5.10, 5.22, 5.54, 5.55, 5.58, 5.63, 5.68, 5.78, 8.39–8.41, 9.33, 9.39
Transport
  diffusion, 16.8
  longshore sediment, 8.12
  passive, 16.3
  sediment, 16.12
Treatment technologies for CDF, 15.13
Tsunami, 8.3, 8.5, 9.2, 19.8
  hazard, 19.19
Turbidity, 17.100
Typhoon, 9.2, 9.6, 9.42

Underkeel, 11.35, 11.40
  allowance, 11.47
  clearance, 11.49
Uplift, 3.63

Velocities
  asymmetry in velocity variation, 7.5, 7.6
  near bottom, 7.5
  undertow, 7.5, 7.7, 7.8

Wall
  structures, 4.5
    composite, 4.5
    vertical, 4.5
  recurved, 6.10
  wave-return, 6.11, 6.12
Water
  depth, 13.8
Water levels, 8.3
  long-term, 8.5
  MHWS, 8.3
  MHHW, 8.3
  MHW, 8.3
  MSL, 8.3
  MTL, 8.3
Waterborne transportation, 15.3
Wave, 8.6
  action, 3.8
  breaker, 8.7
  classical, 1.5
  equations, 1.5
    by variational approach, 1.11, 1.32

  mild slope, 1.6, 1.12
    time dependent, 1.8
  quasilinear hyperbolic, 1.50
  shallow water, 1.9, 1.13
  height, 2.10, 8.7
  impact, 2.8
  length, 5.32
  oblique, 4.18, 4.19, 4.29
  overtopping, 4.9, 4.10
  performance, 2.20
  period, 2.10, 2.15
  plunging, 2.11, 2.16, 8.51
  reflection, 2.19
  run-up, 3.8, 3.26–3.28, 8.47–8.49
  short-crested, 4.17, 4.20, 4.21, 4.29
  spectrum, 8.6
  steepness, 5.32
  surging, 2.11, 8.51
  transmissivity, 5.21
  uplift, 2.8
  wave makers
    numerical, 1.22
  wave-structure
    interaction, 3.19
    overtopping, 3.35
  weighted residual
    methods, 1.33
Wave attack
  angled, 6.13
Wave forces on walls, 4.5, 4.8, 4.37
  broken waves, 4.7, 4.10, 4.35
  horizontal, 4.15
  impulsive breaking impact, 4.7, 4.10, 4.12, 4.22, 4.27, 4.28
  pulsating, 4.7, 4.15, 4.30
Wave run-up, 6.2, 6.7, 6.8
  long-crested, 6.7
  short-crested, 6.7
Waves
  infragravity, 7.18
  long, 7.7, 7.9, 7.15, 7.17, 7.18
  short, 7.6, 7.17
WIS, 16.5, 16.10

Zuyder Zee (Zuyder Sea), 3.9

About the editor

John B. Herbich, Ph.D., P.E., is the W. H. Bauer Professor Emeritus, Civil and Ocean Engineering, Texas A&M University, College Station, Texas. Dr. Herbich is a Fellow and Life Member of the American Society of Civil Engineers and many other engineering societies. He received a B.Sc. degree in Civil Engineering from the University of Edinburgh, Scotland, a M.S.C.E. degree in hydrodynamics from the University of Minnesota, and a Ph.D. in Civil Engineering from the Pennsylvania State University.

Prior to joining Texas A&M University, Dr. Herbich was on the faculty of Lehigh University, Bethlehem, Pennsylvania (1957–1967), and was a research engineer at the University of Delft, The Netherlands (1949–1950). He served as project manager of a United Nations Development Program in Poona, India (1972–1973), and was a visiting professor at the U.S. Army Corps of Engineers Waterways Experiment Station in Vicksburg, Mississippi (1987–1988). He is a consultant for many U.S. and international governments and industries. Dr. Herbich served on several committees of the National Research Council and is the recipient of the "International Coastal Engineering Award," American Society of Civil Engineers (1993), and the "1995 Dredger of the Year Award," Western Dredging Association. He is the author and editor of numerous engineering books and is a registered professional engineer in Texas.